Handbook of Affinity Chromatography

Second Edition

CHROMATOGRAPHIC SCIENCE SERIES

A Series of Textbooks and Reference Books

Editor: JACK CAZES

Handbook of Affinity Chromatography

Second Edition

edited by
David S. Hage
University of Nebraska
Lincoln, Nebraska, U.S.A.

Taylor & Francis
Taylor & Francis Group

Boca Raton London New York Singapore

A CRC title, part of the Taylor & Francis imprint, a member of the
Taylor & Francis Group, the academic division of T&F Informa plc.

Published in 2006 by
CRC Press
Taylor & Francis Group
6000 Broken Sound Parkway NW, Suite 300
Boca Raton, FL 33487-2742

Library of Congress Cataloging-in-Publication Data

Catalog record is available from the Library of Congress

Taylor & Francis Group
is the Academic Division of T&F Informa plc.

Visit the Taylor & Francis Web site at
http://www.taylorandfrancis.com

and the CRC Press Web site at
http://www.crcpress.com

Preface

Welcome to the *Handbook of Affinity Chromatography, Second Edition*. The purpose of this book is to guide scientists, students, and laboratory workers in the theory, use, and applications of affinity chromatography. Since its original use almost 100 years ago, affinity chromatography has become an important tool in a large variety of areas. This is illustrated in this book through examples and topics that range from the fields of biology, biotechnology, biochemistry, and molecular biology to analytical chemistry, pharmaceutical science, environmental science, and clinical chemistry.

This handbook is divided into six sections. Section I (Chapters 1 to 4) provides an overview of affinity chromatography and discusses important factors to consider in the development of affinity methods, including the choice of support material, immobilization method, and application or elution conditions. Section II (Chapters 5 to 10) provides detailed reviews of common affinity methods, such as bioaffinity chromatography, immunoaffinity chromatography, DNA affinity chromatography, boronate affinity chromatography, dye-ligand and biomimetic affinity chromatography, and immobilized metal-ion affinity chromatography.

Sections III, IV, and V of this book (Chapters 11 to 16, 17 to 21, and 22 to 25, respectively) explore the preparative, analytical, and biophysical applications of affinity methods in areas such as biochemistry, molecular biology, biotechnology, pharmaceutical analysis, proteomics, clinical testing, and environmental analysis. Section VI (Chapters 26 to 30) considers recent developments in this field, including the use of affinity ligands in capillary electrophoresis, mass spectrometry, microanalytical systems, chromatographic immunoassays, and molecularly imprinted polymers.

This book is the result of a collaborative effort involving 48 scientists and students from 23 laboratories and organizations located throughout the world. I would like to thank each of these individuals for their contributions to this project. I would also like to thank my wife Jill and my sons Ben and Brian for their assistance in the preparation of this text. I hope that all who use this handbook will find it a valuable addition to their work in affinity chromatography or related areas.

David S. Hage

Contents

Contents

Editor

David S. Hage is a professor of analytical and bioanalytical chemistry at the University of Nebraska. He received his B.S. in chemistry and biology from the University of Wisconsin–La Crosse in 1983 and a Ph.D. in analytical chemistry from Iowa State University in 1987. After finishing a postdoctoral position at the Mayo Clinic in 1989, he joined the faculty at the University of Nebraska. His interests include the study of biological interactions by affinity methods and analytical applications of affinity chromatography for biological and environmental agents. He is the author of over 120 research articles and reviews on the topic of affinity chromatography and received the 1995 Young Investigator Award from the American Association for Clinical Chemistry for his work. He has presented numerous seminars and workshops on affinity methods and is on the editorial boards of several journals in the field of chemical separation and analysis.

Contributors

Pascal Bailon
Hoffmann-LaRoche Inc.
Nutley, New Jersey

Min Bian
Department of Chemistry
University of Nebraska
Lincoln, Nebraska

Chad J. Briscoe
MDS Pharma Services
Lincoln, Nebraska

Raychelle Burks
Department of Chemistry
University of Nebraska
Lincoln, Nebraska

Jianzhong Chen
Department of Chemistry
University of Nebraska
Lincoln, Nebraska

William Clarke
Department of Pathology
Johns Hopkins School of Medicine
Baltimore, Maryland

Y. D. Clonis
Laboratory of Enzyme Technology
Department of Agricultural Biotechnology
Agricultural University of Athens
Athens, Greece

George K. Ehrlich
Hoffmann-LaRoche Inc.
Nutley, New Jersey

Felix Friedberg
Department of Biochemistry and
    Molecular Biology
College of Medicine
Howard University
Washington, DC

Himanshu Gadgil
Department of Biochemistry
University of Tennessee Health Science
    Center
Memphis, Tennessee

Per-Erik Gustavsson
Department of Pure and Applied
    Biochemistry
Center for Chemistry and Chemical
    Engineering
University of Lund
Lund, Sweden

David S. Hage
Department of Chemistry
University of Nebraska
Lincoln, Nebraska

Karsten Haupt
Compiègne University of Technology
Centre de Recherches de Royallieu
Compiègne, France

Niels Heegaard
Department of Autoimmunology
Statens Serum Institute
Copenhagen, Denmark

Harry W. Jarrett
Department of Biochemistry
University of Tennessee Health Science
  Center
Memphis, Tennessee

Neil Jordan
Department of Chemistry and Chemical
  Biology
Northeastern University
Boston, Massachusetts

Luis A. Jurado
Department of Biochemistry
University of Tennessee Health Science
  Center
Memphis, Tennessee

Elizabeth Karle
Department of Chemistry
University of Nebraska
Lincoln, Nebraska

Hee Seung Kim
Department of Chemistry
University of Nebraska
Lincoln, Nebraska

Ira S. Krull
Department of Chemistry and Chemical
  Biology
Northeastern University
Boston, Massachusetts

N. E. Labrou
Laboratory of Enzyme Technology
Department of Agricultural Biotechnology
Agricultural University of Athens
Athens, Greece

Per-Olof Larsson
Department of Pure and Applied
  Biochemistry
Center for Chemistry and Chemical
  Engineering
University of Lund
Lund, Sweden

Xiao-Chuan Liu
Department of Chemistry
California State Polytechnic University
Pomona, California

Sheree D. Long
Biacore Inc.
Piscataway, New Jersey

W. John Lough
Institute of Pharmacy and Chemistry
University of Sunderland
Sunderland, United Kingdom

K. Mazitsos
Laboratory of Enzyme Technology
Department of Agricultural Biotechnology
Agricultural University of Athens
Athens, Greece

Annette C. Moser
Department of Chemistry
University of Nebraska
Lincoln, Nebraska

Robert A. Moxley
Department of Biochemistry
University of Tennessee Health Science
  Center
Memphis, Tennessee

David G. Myszka
Center for Biomolecular Interaction
  Analysis
University of Utah
Salt Lake City, Utah

Michele Nachman-Clewner
Hoffmann-LaRoche Inc.
Nutley, New Jersey

Mary Anne Nelson
Department of Chemistry
University of Nebraska
Lincoln, Nebraska

Shilpa Oak
Department of Biochemistry
University of Tennessee Health Science
    Center
Memphis, Tennessee

Corey Ohnmacht
Department of Chemistry
University of Nebraska
Lincoln, Nebraska

Sharvil Patel
Bioanalytical and Drug Discovery Unit
National Institute on Aging
National Institutes of Health
Baltimore, Maryland

Terry M. Phillips
Ultramicro Analytical Immunochemistry
    Resource
Division of Bioengineering and Physical
    Sciences
Office of Research Services
National Institutes of Health
Bethesda, Maryland

Allen R. Rhoads
Department of Biochemistry and
    Molecular Biology
College of Medicine
Howard University
Washington, DC

Peggy F. Ruhn
MDS Pharma Services
Lincoln, Nebraska

William H. Scouten
College of Sciences
University of Texas at San Antonio
San Antonio, Texas

Christian Schou
Department of Autoimmunology
Statens Serum Institute
Copenhagen, Denmark

Cheryl L. Spence
Hoffmann-LaRoche Inc.
Nutley, New Jersey

Anuradha Subramanian
Department of Chemical Engineering
University of Nebraska
Lincoln, Nebraska

Daniela Todorova
Molecular Interaction & Separation
    Technology Labs
Université de Technologie de Compiègne
Compiègne, France

Mookambeswaran A. Vijayalakshmi
Molecular Interaction & Separation
    Technology Labs
Université de Technologie de Compiègne
Compiègne, France

Chunling Wa
Department of Chemistry
University of Nebraska
Lincoln, Nebraska

Irving W. Wainer
Bioanalytical and Drug Discovery Unit
National Institute on Aging
National Institutes of Health
Baltimore, Maryland

Donald J. Winzor
Department of Biochemistry
School of Molecular and Microbial
    Sciences
University of Queensland
Brisbane, Queensland, Australia

Carrie A. C. Wolfe
Division of Science and Mathematics
Union College
Lincoln, Nebraska

Hai Xuan
Department of Chemistry
University of Nebraska
Lincoln, Nebraska

Section I

Introduction and Basic Concepts

1

An Introduction to Affinity Chromatography

David S. Hage

Department of Chemistry, University of Nebraska, Lincoln, NE

Peggy F. Ruhn

MDS Pharma Services, Lincoln, NE

CONTENTS

1.1   INTRODUCTION

Chemical separation is an essential component of modern research and is widely used to process complex samples. Examples range from the trace analysis of a drug or hormone in blood to the large-scale isolation of a recombinant protein. The method of liquid chromatography has become particularly popular for these separations because of its ability to work with a wide range of substances. When combined with appropriate support materials, this technique can be used in either high-performance separations for chemical detection and measurement or in systems designed to purify a desired product. The wide range of stationary phases and mobile phases that can be employed in liquid chromatography also makes this method quite flexible in terms of the types of chemical or physical properties that can be used as the basis for these separations.

One of the most versatile forms of liquid chromatography is the technique known as *affinity chromatography*, which can generally be defined as a liquid chromatographic technique that uses a specific binding agent for the purification or analysis of sample components [1–6]. This technique makes use of the selective and reversible interactions that occur in many biological systems, such as the binding of an enzyme with a substrate or an antibody with an antigen. These interactions are used in affinity chromatography by immobilizing one of a pair of interacting molecules onto a solid support and placing it into a column. The immobilized molecule is referred to as the *affinity ligand*. This makes up the stationary phase of the affinity column.

Figure 1.1 shows a typical scheme used to perform affinity chromatography. In this approach, a sample containing the compound of interest is injected onto the affinity

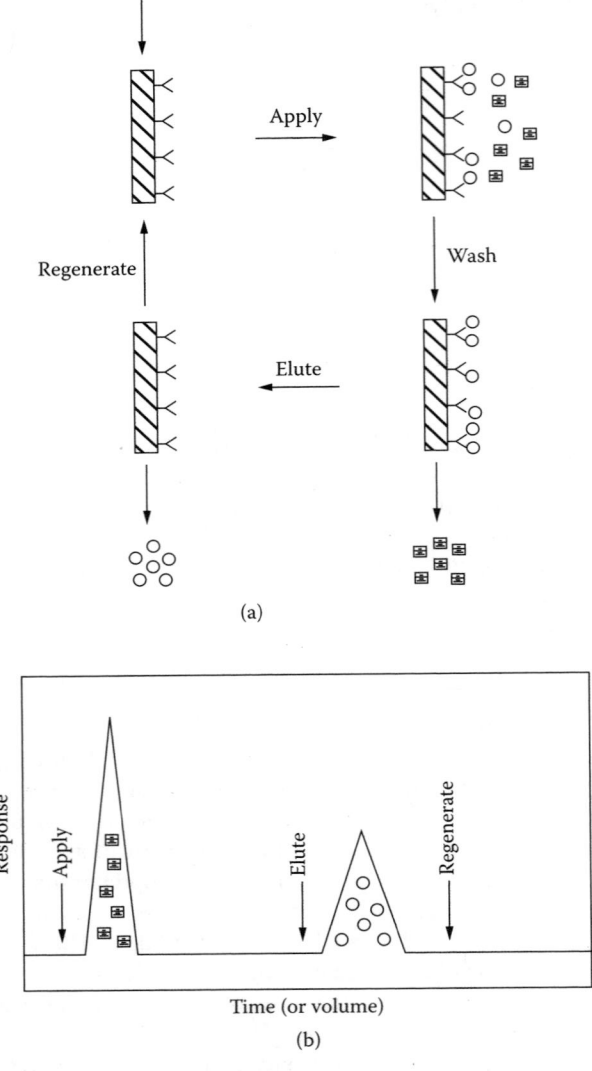

(a)

(b)

Figure 1.1 Typical separation scheme for affinity chromatography. (Reproduced with permission from Hage, D.S., in *Handbook of HPLC*, Katz, E., Eksteen, R., Shoenmakers, P., and Miller, N., Eds., Marcel Dekker, New York, 1998, pp. 483–498.)

column in the presence of a mobile phase that has the right pH, ionic strength, and solvent composition for solute-ligand binding. This solvent, which represents the weak mobile phase of an affinity column, is referred to as the *application buffer.* As the sample passes through the column under these conditions, compounds that are complementary to the affinity ligand will bind. However, due to the high selectivity of this interaction, other solutes in the sample will tend to wash or elute from the column as a nonretained peak.

After all nonretained components have been washed from the column, the retained solutes are then eluted by applying a solvent that displaces them from the column or that promotes dissociation of the solute-ligand complex. This solvent, which represents the strong mobile phase for the column, is known as the *elution buffer.* As the solutes of interest elute from the column, they are either quantitated directly or collected for later use. The application buffer is then reapplied to the system and the column is allowed to regenerate prior to the next sample injection.

Due to the strong and selective binding that characterizes many affinity ligands, solutes that are quantitated or purified by these ligands can often be separated with little or no interference from other sample components. In many cases the solute of interest can be isolated in only one or two steps, with purification yields of 100-fold to several thousand-fold being common [2–6]. In work with hormone receptors, purification yields approaching one million-fold have even been reported with affinity-based separations [5].

The wide range of ligands available for affinity chromatography makes this method a valuable tool for the purification and analysis of compounds present in complex samples. Areas in which affinity chromatography has been used include biochemistry, pharmaceutical science, clinical chemistry, and environmental testing. This book will examine the variety and types of affinity ligands used in these areas and the main factors to consider in the development of such a method. Several specific applications will also be considered, including the use of affinity chromatography in small- or large-scale purification, analyte detection, and the characterization of biological interactions. Finally, a number of new developments in this field and in related areas of work will be discussed.

1.2 HISTORY OF AFFINITY CHROMATOGRAPHY

1.2.1 Origins of Affinity Chromatography

Although some consider affinity chromatography to be a relatively new method, it is actually one of the oldest forms of liquid chromatography. For instance, the earliest use of this method was just seven years after Michael Tswett reported the first known use of column liquid chromatography [7]. This occurred in 1910 when Emil Starkenstein examined the binding of insoluble starch to the enzyme α-amylase [8]. This is also the first known case in which liquid chromatography was used for a separation involving a protein.

The original studies with affinity chromatography all made use of insoluble materials that acted as both the stationary phase and support material. This is not surprising, since this is the simplest form of affinity separation. For instance, the insoluble starch used by Starkenstein acted as both a support material and as a substrate for amylase, thus leading to this enzyme's binding and retention. Similar work with starch and amylase was conducted in the 1920s through 1940s by other investigators [9–12], with a 300-fold purification being obtained in one of these studies [12]. Other examples include the use of polygalacturonase as a support and ligand for the adsorption of alginic acid [13], the purification of pepsin through the use of edestin, a crystalline protein [14], and the isolation of porcine elastase with powdered elastin [15].

As this suggests, much of the earliest work with affinity supports involved its use in the purification of enzymes. But research was also being conducted at this time in the selective purification of antibodies with biological ligands. This arose from the work by Landsteiner, who showed in 1920 that antibodies can recognize and bind substances with a specific structure, referred to as "antigens" [16]. This type of binding plus the ability of polyclonal antibodies to form insoluble complexes with antigens led to the growth in the 1930s of immunoprecipitation as an important technique for antibody purification [17–19]. For instance, this approach was used by Kirk and Sumner to isolate antibodies against urease and to demonstrate that these antibodies were proteins [17]. This approach was similar to the work being performed in the purification of enzymes in that a ligand (i.e., the antigen) was used to create a specific biological interaction with the target of interest (the antibodies). However, rather than having the ligand be the same as the solid used for this isolation, this solid was formed as a result of the binding process.

1.2.2   Early Immobilization Methods

Although the use of insoluble ligands and immunoprecipitation proved the potential for biological interactions as a means for chemical isolation, this approach was still limited to solutes for which an appropriate ligand and/or support was available. However, this began to change in the 1940s and 1950s as synthetic techniques became available for placing a broader range of ligands onto insoluble materials.

These efforts began by employing solids that contained a noncovalently adsorbed layer of ligand. For instance, in 1935 D'Alessandro and Sofia used antigens coated on kaolin and charcoal for the isolation of antibodies associated with syphilis and tuberculosis [20]. A similar type of support was used by Meyer and Pic in 1936 [21].

However, it was soon realized that a more stable system would be obtained by chemically bonding the ligand to the support. This was first used by Landsteiner and van der Scheer in 1936, when they adapted a diazo-coupling technique used to prepare hapten conjugates [22]. In this case, they attached a number of haptens to a solid material based on chicken erythrocyte stroma, with this material then being used for the isolation of antibodies for these haptens (see Figure 1.2).

Work with other, more durable support materials also began to appear. A key development in this area occurred in 1951, when Campbell and coworkers used an activated form

1 cc of suberanilic acid immune serum absorbed with azostromata made from	Azoproteins made from casein and		
	p-Aminoadipanilic acid	*p*-Aminosuberanilic acid	*p*-Aminosebacanilic acid
p-Aminoadipanilic acid	0	++±	+±
	0	+++±	++
p-Aminosebacanilic acid	+	+++	0
	+±	++++	0
Unabsorbed immune serum	+++	++++	+++
	+++±	++++	+++±

Figure 1.2   An early example of the use of immobilized antigens for antibody purification. These were used by Landsteiner and van der Scheer in 1936 for performing cross-reactivity studies of immune sera. (Reproduced with permission from Landsteiner, K. and van der Scheer, J., *J. Exp. Med.*, 63, 325–339, 1936.)

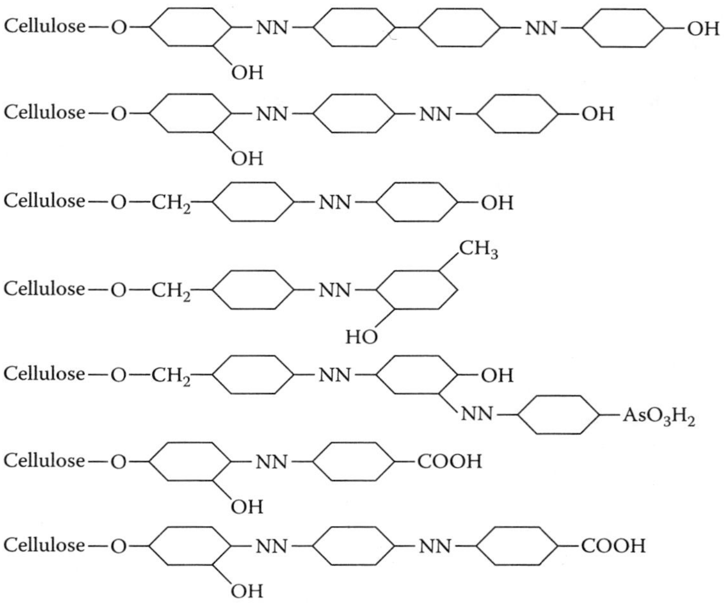

Cellulose—O—〈 〉—NN—〈 〉—〈 〉—NN—〈 〉—OH
                    OH

Cellulose—O—〈 〉—NN—〈 〉—NN—〈 〉—OH
                    OH

Cellulose—O—CH$_2$—〈 〉—NN—〈 〉—OH

Cellulose—O—CH$_2$—〈 〉—NN—〈 〉 CH$_3$
                                  HO

Cellulose—O—CH$_2$—〈 〉—NN—〈 〉—OH
                                  NN—〈 〉—AsO$_3$H$_2$

Cellulose—O—〈 〉—NN—〈 〉—COOH
                    OH

Cellulose—O—〈 〉—NN—〈 〉—NN—〈 〉—COOH
                    OH

Figure 1.3 Early examples of immobilized inhibitors that were used by Lerman in 1953 for the isolation of tyrosinase. (Reproduced with permission from Lerman, L.S., *Proc. Natl. Acad. Sci. U.S.A.*, 39, 232–236, 1953.)

of cellulose (*p*-aminobenzylcellulose) for immobilizing the protein serum albumin. This material was then used to isolate antialbumin antibodies from rabbit serum [23]. Similar work appeared by Lerman in the preparation of immobilized hapten supports [24] and in 1953 through the use of a covalently immobilized ligand for purifying the enzyme mushroom tyrosinase (see Figure 1.3) [25].

Other studies in immunopurification later began to appear with ligands attached to such supports as polyaminostyrene [26] and glass beads [27]. By 1966, several reviews on such work had appeared [28–34]. After the report by Lerman [25], the use of immobilized ligands for enzyme isolation received only limited attention for some time, probably due to the rise of ion exchange with cellulose in the 1950s and 1960s as a separation tool in enzymology. However, this began to change in the mid-1960s, when Arsenis and McCormick began to use cellulose-based affinity supports for the isolation of flavokinase and flavin mononucleotide-dependent enzymes [35, 36].

1.2.3 Modern Era of Affinity Chromatography
The next major development came about in the late 1960s. There were three things that led to this event. The first was the creation of beaded agarose supports by Hjerten in the mid-1960s [37]. This provided a more efficient and flexible support than cellulose for use with biopolymers in liquid chromatography. The second development was the discovery of the cyanogen bromide immobilization method. This was reported in 1967 by Axen, Porath, and Ernback, providing a more convenient and general approach that could be used to attach proteins and peptides to polysaccharides [38]. The third development was a 1968 report by Cuatrecasas, Wilchek, and Anfinsen in which agarose and the cyanogen bromide method were used together to create immobilized nuclease inhibitor columns [39]. These columns were then used for purifying the enzymes staphylococcal nuclease, α-chymotrypsin,

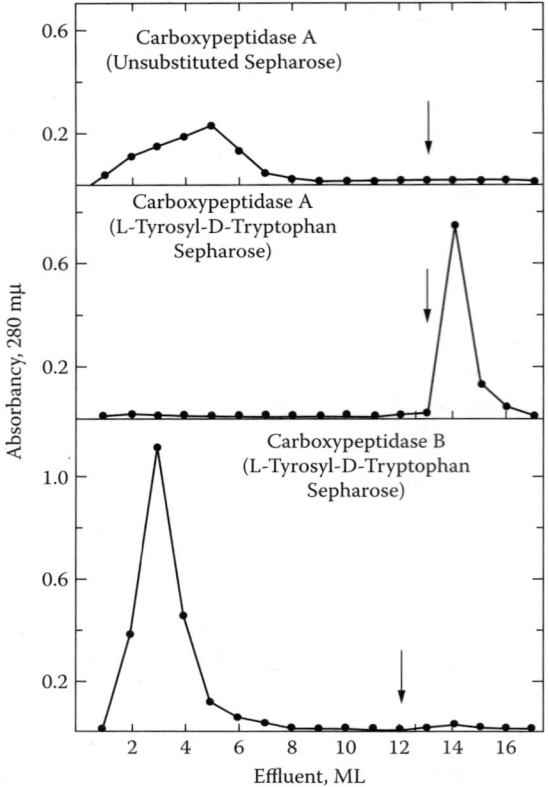

Figure 1.4 One of the first examples of modern affinity chromatography (Reproduced with permission from Cuatrecasas, P., Wilchek, M., and Anfinsen, C.B., *Proc. Natl. Acad. Sci. U.S.A.*, 68, 636–643, 1968.)

and carboxypeptidase A. An example of such a separation is shown in Figure 1.4. This was the same report in which the term "affinity chromatography" was first used to describe this separation technique [39].

The techniques and applications that appeared in this last report led to a rapid increase in interest in the use of affinity ligands in liquid chromatography. This is illustrated by Figure 1.5, which shows the number of publications that have appeared between 1968 and 2001 that have contained the phrase "affinity chromatography." The number of such reports has grown from four in 1968 to approximately 1000 per year over most of the last decade. An even greater number of reports (i.e., roughly three times the number shown in Figure 1.5) have included the use of related phrases like "immunoaffinity chromatography." These results clearly demonstrate the widespread use of affinity chromatography and the impact it has had on modern separations.

Following the reintroduction of affinity chromatography in 1968, a variety of ligands and applications for this method began to appear. For instance, the next six years saw the introduction of such methods as DNA-cellulose affinity chromatography [40], boronate affinity chromatography [41], dye-ligand affinity chromatography [42], and immobilized metal-ion affinity chromatography [43]. The ligands that are used in these techniques and the types of targets they retain are summarized in Table 1.1. This same table also shows other ligands and approaches that are now commonly used in affinity chromatography.

Figure 1.5 Number of articles per year that appeared from 1968 to 2001 and contained the phrase "affinity chromatography." A total of 26,132 articles were identified in this search, which was performed using CAPLUS and MEDLINE. The inclusion of other phrases, such as "affinity chromatographic," "immunoaffinity chromatography," and "immunoaffinity chromatographic" led to the identification of 75,742 articles from this same period of time.

Table 1.1   Common Ligands Used in Affinity Chromatography

Method	Ligand	Retained Solutes
Immunoaffinity chromatography	Antibodies	Antigens (e.g., drugs, hormones, peptides, proteins, viruses, cell components)
Affinity chromatography of enzymes	Antigens	Antibodies against the antigens
	Inhibitors, substrates, cofactors, coenzymes, and miscellaneous ligands	Enzymes
Lectin affinity chromatography	Lectins	Sugars, glycoproteins, and glycolipids
Protein A/protein G affinity chromatography	Protein A, protein G, and related ligands	Antibodies and antibody fragments
DNA affinity chromatography	DNA or RNA	DNA/RNA-binding proteins, complementary nucleotides
Boronate affinity chromatography	Boronates	Carbohydrates, nucleosides, nucleotides, nucleic acids, glycoproteins, and catechols
Dye-ligand affinity chromatography	Synthetic dyes	Proteins and enzymes
Biomimetic affinity chromatography	Ligands generated using combinatorial chemistry or from peptide, phage display, ribosome display, or aptamer libraries	Various targets, including proteins and enzymes
Immobilized metal-ion affinity chromatography	Metal-ion chelates	Metal-binding amino acids, peptides, proteins, and nucleotides

1.3 OVERVIEW OF HANDBOOK

This book is designed to introduce the reader to the topic of affinity chromatography and to the most common methods that are employed in this technique. The first section of this handbook deals with the basic components of an affinity chromatographic system. This begins with a discussion of support materials that are used in this method (Chapter 2), followed by a review of immobilization methods for attaching ligands to such supports (Chapter 3). The selection of application and elution conditions are then examined in Chapter 4.

Common ligands employed in affinity chromatography are examined in Section II (Chapters 5 to 10). A variety of biological ligands are first considered in Chapters 5 to 7 under the topics of bioaffinity, immunoaffinity, and DNA affinity chromatography. Examples of synthetic or nonbiological ligands are then given in Chapters 8 to 10, including boronates, dyes and biomimetic compounds, and immobilized metal-ion chelates, respectively.

Section III (Chapters 11 to 16) discusses the use of affinity chromatography for the isolation of targets. This was the original application of affinity chromatography and remains one of the largest uses for this technique. In this section, a review of general items to consider in the design of preparative chromatographic systems is provided in Chapter 11. More-detailed presentations are then made in Chapters 12 to 16 regarding the use of affinity ligands in the isolation of enzymes, recombinant proteins, antibodies and antigens, regulatory or signal-transducing proteins, and targets for receptors.

Along with its traditional use in preparative work, affinity chromatography has also been playing an increasing role in the analysis of chemicals. Section IV (Chapters 17 to 21) looks at some specific areas in which this method continues to play a significant role in research and analysis, including clinical testing and pharmaceutical analysis, biotechnology, environmental analysis, molecular biology, and chiral separations, respectively.

Another growing field has been the use of affinity columns to characterize and study biological interactions. This approach is sometimes referred to as quantitative affinity chromatography. This area is examined in Section V (Chapters 22 to 25). First, various practical aspects of the experiments involved in such studies are considered (Chapter 22). An in-depth look at the theory behind this method is then given (Chapter 23), and its use in the study of enzymes and plasma proteins is reviewed (Chapter 24). Finally, the use and development of affinity-based biosensors for such work is considered (Chapter 25).

Section VI (Chapters 26 to 30) presents several new methods that are closely related to affinity chromatography or are special applications of this technique. The discussion in this final section includes the use of affinity ligands in such areas as capillary electrophoresis (Chapter 26) and mass spectrometry (Chapter 27). The creation of microanalytical systems based on affinity chromatography and the use of affinity columns in chromatographic-based immunoassays are also described in this section (Chapters 28 and 29). The last chapter (Chapter 30) explores a new approach for creating affinity columns using molecularly imprinted polymers.

REFERENCES

1. Ettre, L.S., Nomenclature for chromatography, *Pure Appl. Chem.*, 65, 819–872, 1993.
2. Hage, D.S., Affinity chromatography, in *Handbook of HPLC*, Katz, E., Eksteen, R., Shoenmakers, P., and Miller, N., Eds., Marcel Dekker, New York, 1998, pp. 483–498.
3. Turkova, J., *Affinity Chromatography*, Elsevier, Amsterdam, 1978.
4. Scouten, W.H., *Affinity Chromatography: Bioselective Adsorption on Inert Matrices*, Wiley, New York, 1981.

5. Parikh, I. and Cuatrecasas, P., Affinity chromatography, *Chem. Eng. News,* 63, 17–32, 1985.
6. Walters, R.R., Affinity chromatography, *Anal. Chem.,* 57, 1099A–1114A, 1985.
7. Tswett, M., The chemistry of chlorophyll, phylloxanthin, phyllocyanin, and chlorophyllane, *Biochem. Zeit.,* 5, 6–32, 1907.
8. Starkenstein, E., Ferment action and the influence upon it of neutral salts, *Biochem. Z.,* 24, 210–218, 1910.
9. Ambard, L., Amylase: Its estimation and the mechanism of its action, *Bull. Soc. Chim. Biol.,* 3, 51–65, 1921.
10. Holmbergh, O., Adsorption of α-amylase from malt by starch, *Biochem. Z.,* 258, 134–140, 1933.
11. Tokuoka, Y., Koji amylase, IX: Existence of β-amylase, *J. Agric. Chem. Soc. Japan,* 13, 586–594, 1937.
12. Hockenhull, D.J.D. and Herbert, D., The amylase and maltase of *Clostridium acetobutylcium, Biochem. J.,* 39, 102–106, 1945.
13. Lineweaver, H., Jang, R., and Jansen, E.F., Specificity and purification of polygalacturonase, *Arch. Biochem.,* 20, 137–152, 1949.
14. Northrup, J.H., Crystalline pepsin, VI: Inactivation by β- and γ-rays from radium and by ultraviolet light, *J. Gen. Physiol.,* 17, 359–363, 1934.
15. Grant, N.H. and Robbins, K.C., Porcine elastase and proelastase, *Arch. Biochem. Biophys.,* 66, 396–403, 1957.
16. Landsteiner, K., Specific serum reactions induced by the addition of substances of known constitution (organic acids), XVI: Antigens and serological specificity, *Biochem. Z.,* 104, 280–299, 1920.
17. Kirk, J.S. and Sumner, J.B., The reaction between crystalline urease and antiurease, *J. Immunol.,* 26, 495–504, 1934.
18. Marrack, J.R. and Smith, F.C., Quantitative aspects of immunity reactions: The combination of antibodies with simple haptenes, *Brit. J. Exp. Pathol.,* 13, 394–402, 1932.
19. Heidelberger, M. and Kabat, E.A., Quantitative studies on antibody purification, II: The dissociation of antibody from pneumococcus-specific precipitates and specifically agglutinated pneumococci, *J. Exp. Med.,* 67, 181–199, 1938.
20. d'Alessandro, G. and Sofia, F., The adsorption of antibodies from the sera of syphilitics and tuberculosis patients, *Z. lmmunitats.,* 84, 237–250, 1935.
21. Meyer, K. and Pic, A., Isolation of antibodies by fixation on an adsorbent-antigen system with subsequent regeneration, *Ann. Inst. Pasteur,* 56, 401–412, 1936.
22. Landsteiner, K. and van der Scheer, J., Cross reactions of immune sera to azoproteins, *J. Exp. Med.,* 63, 325–339, 1936.
23. Campbell, D.H., Luescher, E., and Lerman, L.S., Immunologic adsorbents, I: Isolation of antibody by means of a cellulose-protein antigen, *Proc. Natl. Acad. Sci. U.S.A.,* 37, 575–578, 1951.
24. Lerman, L.S., Antibody chromatography on an immunologically specific adsorbent, *Nature,* 172, 635–636, 1953.
25. Lerman, L.S., A biochemically specific method for enzyme isolation, *Proc. Natl. Acad. Sci. U.S.A.,* 39, 232–236, 1953.
26. Manecke, G. and Gillert, K.E., Serologically specific adsorbents, *Naturwissenschaften,* 42, 212–213, 1955.
27. Sutherland, G.B. and Campbell, D.H., The use of antigen-coated glass as a specific adsorbent for antibody, *J. Immunol.,* 80, 294–298, 1958.
28. Isliker, H.C., Chemical nature of antibodies, *Adv. Prot. Chem.,* 12, 387–463, 1957.
29. Kabat, E.A. and Mayer, M.M., *Experimental Immunochemistry,* 2nd ed., Charles C Thomas, Springfield, IL, 1961, pp. 781–797.
30. Manecke, G., Reactive polymers and their use for the preparation of antibody and enzyme resins, *Pure Appl. Chem.,* 4, 507–520, 1962.
31. Sehon, A.H., Physicochemical and immunochemical methods for the isolation and characterization of antibodies, *Brit. Med. Bull.,* 19, 183–191, 1963.

32. Weliky, N., Weetall, H.H., Gilden, R.V., and Campbell, D.H., Synthesis and use of some insoluble immunologically specific adsorbents, *Immunochemistry,* 1, 219–229, 1964.
33. Weliky, N. and Weetall, H.H., Chemistry and use of cellulose derivatives for the study of biological systems, *Immunochemistry,* 2, 293–322, 1965.
34. Silman, I.H. and Katchalski, E., Water-insoluble derivatives of enzymes, antigens, and antibodies, *Annu. Rev. Biochem.,* 35, 873–908, 1966.
35. Arsenis, C. and McCormick, D.B., Purification of liver flavokinase by column chromatography on flavine-cellulose compounds, *J. Biol. Chem.,* 239, 3093–3097, 1964.
36. Arsenis, C. and McCormick, D.B., Purification of flavin mononucleotide-dependent enzymes by column chromatography on flavin phosphate cellulose compounds, *J. Biol. Chem.,* 241, 330–334, 1966.
37. Hjerten, S., The preparation of agarose spheres for chromatography of molecules and particles, *Biochem. Biophys. Acta,* 79, 393–398, 1964.
38. Axen, R., Porath, J., and Ernback, S., Chemical coupling of peptides and proteins to polysaccharides by means of cyanogen halides, *Nature,* 214, 1302–1304, 1967.
39. Cuatrecasas, P., Wilchek, M., and Anfinsen, C.B., Selective enzyme purification by affinity chromatography, *Proc. Natl. Acad. Sci. U.S.A.,* 68, 636–643, 1968.
40. Alberts, B.M., Amodio, F.J., Jenkins, M., Gutmann, E.D., and Ferris, F.L., Studies with DNA-cellulose chromatography, I: DNA-binding proteins from *Escherichia coli, Cold Spring Harbor Symp. Quant. Biol.,* 33, 289–305, 1968.
41. Weith, H.L., Wiebers, J.L., and Gilham, P.T., Synthesis of cellulose derivatives containing the dihydroxyboryl group and a study of their capacity to form specific complexes with sugars and nucleic acid components, *Biochemistry,* 9, 4396–4401, 1970.
42. Staal, G., Koster, J., Kamp, H., Van Milligen-Boersma, L., and Veeger, C., Human erythrocyte pyruvate kinase, its purification and some properties, *Biochem. Biophys. Acta,* 227, 86–92, 1971.
43. Porath, J., Carlsson, J., Olsson, I., and Belfrage, B., Metal chelate affinity chromatography, a new approach to protein fraction, *Nature,* 258, 598–599, 1975.

2

Support Materials for Affinity Chromatography

Per-Erik Gustavsson and Per-Olof Larsson

Department of Pure and Applied Biochemistry,
Lund University, Lund, Sweden

CONTENTS

2.1 INTRODUCTION

A key item in the selection and design of an affinity chromatographic method is the choice of support material. The supports used in affinity chromatography must meet several requirements. Ideally, such a support should be inexpensive and allow solutes to have rapid, unhindered access to the immobilized affinity ligand. In addition, the support should play a completely passive role during the separation while also being able to couple the desired affinity ligand.

These are not trivial requirements. For example, quick access of solutes to the ligand will require small support particles. But if these solutes are large biomolecules, the support must also have large pores. For the support to play a passive role, it must be chemically inert toward the chromatographic solvents and solutes, have no ionic or hydrophobic groups, and be resistant toward the mechanical strains exerted during the chromatographic process. And yet, for the support to be used for ligand attachment, it must be able to undergo chemical modification or somehow adsorb the ligand.

In reality, no true ideal support exists for affinity chromatography. This is because many of these requirements are in direct conflict with each other. As a result, all current affinity supports involve some compromise in these properties and are generally geared toward a particular application. Some extreme examples of this are micrometer-sized nonporous particles, which are optimized for rapid analytical or micropreparative separations, and 0.2-mm expanded-bed particles or 0.4-mm Big Beads, which are optimized for separating crude extracts in the early stages of a purification process. This chapter examines the various properties that should be considered when selecting a support for affinity chromatography. By being familiar with these properties, one can make an informed choice in selecting a support for a given application.

2.2 PROPERTIES OF SUPPORT MATERIALS

2.2.1 Chemical Inertness

The first important property for a successful affinity column is that it should firmly and specifically bind the desired solute while leaving all other molecules in the sample or process stream untouched. This requires that the support within the column contain an affinity ligand that is capable of forming a suitably strong complex with the solute of interest. Second, the support material must be inert to other solutes to avoid the simultaneous binding of nondesired sample components. This requires that the support have a chemical character that is very similar to that of the medium in which it is operating. Since almost all affinity separations occur in aqueous solutions, the support should thus be as hydrophilic as possible. As a rule, the mobile phase used in affinity separations has a low ionic strength. The support should therefore contain as few charges as possible to prevent ionic interactions.

Many supports that are available today fulfill these requirements. This is because either their basic structure has the desired properties or because they are provided with a hydrophilic coating that gives them such properties. A well-known example is the polysaccharide agarose, which was used in the first modern application of affinity chromatography [1]. This is sold under several trade names, like Sepharose Fast Flow from Amersham Biosciences or Affi-Gel from Bio-Rad Laboratories (see Table 2.1). Agarose consists of polymeric chains of the disaccharide agarobiose, which in turn is made up of D-galactose and 3,6-anhydro-L-galactose, as shown in Figure 2.1a. The individual polymeric chains in agarose are clustered together in bundles. These form a porous and hydrophilic network,

Table 2.1 Examples of Traditional Affinity Supports

Trade Name	Material	Average Particle Diameter (µm)	Examples of Available Ligands	Preactivated Form Available?	Manufacturer/Supplier
Sepharose HP	Agarose	34	Protein A, heparin, Cibacron Blue	Yes	Amersham Biosciences
Sepharose FF	Agarose	90	Protein A, heparin, Cibacron Blue	Yes	Amersham Biosciences
Mimetic series	Agarose	105	Synthetic ligands	No	Prometic Biosciences/ACL
Affi-Gel	Agarose	150–300, 75–150	Protein A, heparin, Cibacron Blue	Yes	Bio-Rad Laboratories
Sepharose Big Beads	Agarose	200	IMAC[a]	No	Amersham Biosciences
Cellthru Big Beads	Agarose	400	Cibacron Blue, heparin, IMAC[a]	Yes	Sterogene
TSK-Gel	Polymethacrylate	10	Boronate, IMAC[a], heparin	Yes	TosoHaas/Supelco
Affi-Prep	Polymethacrylate	10, 50	Protein A, polymyxin	No	Bio-Rad Laboratories
SigmaChrom AF	Polymethacrylate	20	Protein A, IMAC[a], Cibacron Blue	No	Supelco
Fractogel	Polymethacrylate	30, 65	IMAC[a], heparin	Yes	Merck
ProteinPak	Silica	40	No	Yes	Waters
Bakerbond	Silica	40	No	Yes	J.T. Baker
Trisacryl	Polyacrylamide derivative	60, 200	Cibacron Blue, Basiline Blue	No	BioSepra/Pall
Affi-Gel 601	Polyacrylamide	Not specified	Boronate	No	Bio-Rad Laboratories
Cellufine	Cellulose	85, 90, 170	Heparin, IMAC[a], gelatin	Yes	Chisso/Amicon/Millipore

[a] IMAC, immobilized metal-ion affinity chromatography.

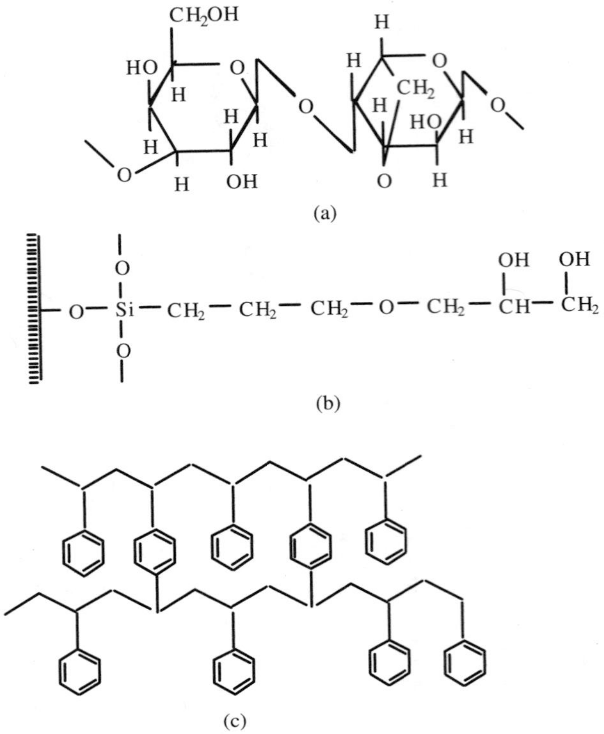

Figure 2.1 Common materials used in the preparation of affinity supports. The structure in (a) shows the repeating unit of agarose: D-galactose and 3,6-anhydro-L-galactose. The structure in (b) shows diol-bonded silica, and (c) is the structure of polystyrene/divinylbenzene.

where the groups facing the solvent have a minimum tendency to attract sample components. Cellulose is another example of a polysaccharide support that is used in affinity chromatography.

Another example of a support used in affinity chromatography is silica. This is used in the method of high-performance liquid affinity chromatography (HPLAC) or high-performance affinity chromatography (HPAC), which was first reported in 1978 [2]. Although silica-based materials are certainly hydrophilic, they are unsuitable for affinity chromatography unless they have first been modified at their surface. This is the case because the native surface of silica is primarily covered with silanol groups. These groups are weak acids that give silica's surface a strong negative charge at neutral pH. These charges, in combination with other binding forces, often result in the irreversible adsorption of solutes like proteins to native silica. However, several schemes can be used to render this surface inert toward such solutes, including polymer coating techniques and reactions between silica and alcohols or trialkoxysilanes [3, 4]. An example of such a scheme is the reaction of silica with γ-glycidoxypropyltrimethoxysilane [5], followed by acid hydrolysis of the resulting epoxy groups; this leaves the modified silica with a hydrophilic, noncharged surface (shown in Figure 2.1b) that has little or no binding to many biological compounds.

The polymeric support polystyrene (illustrated in Figure 2.1c) is also unsuitable in its original form for affinity separations due to the highly hydrophobic character of this material. Native polystyrene, which is often used as a reversed-phase material, must first be rendered hydrophilic by one of various surface-coating techniques before it can be used in other chromatographic methods [6, 7]. Polymeric supports based on polymethacrylate [8, 9] are more hydrophilic than polystyrene supports and can be used directly in affinity chromatography. Examples of these and other polymeric affinity supports, such as those based on polyacrylamide, are listed in Table 2.1.

A support material should be inert toward solutes, but to make it suitable for affinity chromatography, it should also be easy to couple to a ligand. Since most support materials are rich in hydroxyl groups, the chemistries developed for the attachment of ligands have focused mainly on using these regions as anchoring points. Many such methods are available [10, 11], with these being described in greater detail in Chapter 3.

A support material with a tendency to adsorb solutes in a nonspecific manner may still be useful, provided that the surrounding medium is modified accordingly. For instance, if one changes the buffer in an affinity column so that the buffer's chemical character (e.g., hydrophilicity and ion strength) matches that of the support, there will be a minimum tendency toward nonspecific adsorption. As an example, when a support has charged groups present that can create undesired ionic interactions, a mobile phase buffer with an ionic strength of about 0.15 M can be used to suppress such binding [12, 13].

Even if the support is essentially free of nonspecific interactions with sample components, the final affinity adsorbent obtained with this support could have a dramatically different character. This occurs because the procedure used to couple the ligand to the support may introduce undesired groups that might then act as a new source of nonspecific binding. A well-known example occurs during the activation of agarose and other polysaccharide supports with cyanogen bromide. In one standard protocol for this method, a large excess of cyanogen bromide is used. This excess is necessary because the alkaline conditions used for the reaction lead to a substantial breakdown of the active cyanate ester groups (see Figure 2.2a). Several breakdown products (including charged ones) may form when this occurs, which later leads to nonspecific interactions between the support and injected solutes. To avoid this problem, a modified protocol can be utilized that involves activating the support at a low temperature in the presence of a cyanogen transfer agent (e.g., triethylamine), which results in a less-altered support [14]. The cyanogen transfer agent in this alternative technique increases the electrophilicity of the cyanogen bromide by complex formation (as shown in Figure 2.2b). This allows the activation to be carried out at a neutral pH, which, in turn, minimizes hydrolysis of the cyanate ester groups and increases the overall reaction yield.

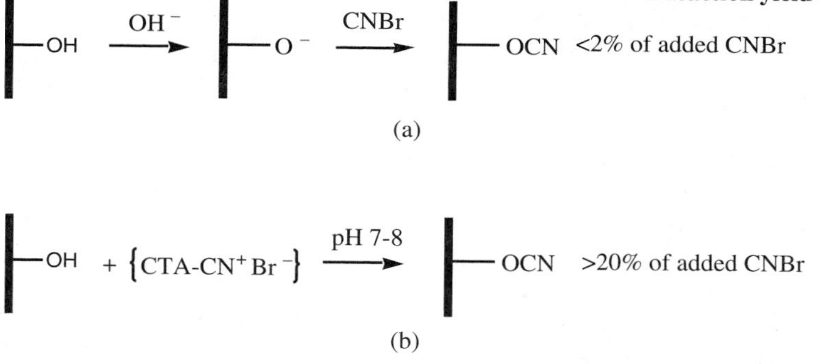

Figure 2.2 Reaction schemes showing (a) the classical cyanogen bromide (CNBr) activation technique and (b) the modified cyanogen bromide activation technique using a cyano transfer agent (CTA).

The ligand itself may be responsible for unwanted interactions on the final support. For instance, suppose that some agarose beads contained an immobilized derivative of adenosine triphosphate (ATP) and that the final support had a ligand concentration of 10 to 15 µmol ATP per milliliter of packed-bed volume. This is a normal degree of substitution and would be suitable for retaining ATP-dependent proteins. However, this modification has also made the agarose into a cation-exchange resin through the presence of the phosphate groups on the ATP. At pH 7, the charge concentration on this gel would be about 50 µmol per milliliter of packed bed. Taking into account the multivalency of the charged triphosphate groups, the mobile phase used with this support would need to contain a 0.1 to 0.2 M buffer to cancel the ion-exchange properties imposed by the ATP affinity ligand. Such effects should be considered whenever a new affinity adsorbent is being designed. To avoid these nonspecific effects and thereby improve selectivity, it may also be wise to avoid working at unnecessarily high ligand concentrations.

A related source of unwanted binding can occur via the ligand spacer. The spacer molecule serves to make an immobilized ligand more accessible to an injected macromolecule. Hexamethylenediamine is often used as spacer for this purpose [10, 11], but this introduces both charged groups and a patch of hydrophobicity to the support's surface. Sometimes these extra functionalities may act with the ligand to give even stronger binding for the desired sample components. However, in most cases they will lower the selectivity of the adsorbent, due to its retention of undesired sample components. This phenomenon was observed early in the development of affinity chromatography and eventually led to the development of a separation mode known as *hydrophobic interaction chromatography* (HIC) [15].

2.2.2 Chemical Stability

Obviously, the affinity adsorbent should be chemically stable under the operating conditions that will be used in the column. This includes stability toward enzymes and microbes that might be present in the process stream. This also means that the support must be stable in the presence of the elution buffers, regenerating solvents, and cleaning agents that will be used with the column. Agarose-based supports are almost ideal in this respect, especially when they are in their cross-linked form (e.g., Sepharose Fast Flow from Amersham Biosciences). Such supports are not attacked by enzymes, can be used between pH 3 and 12, and can withstand all commonly used water-based eluants without shrinking or swelling [11].

An additional attractive feature of cross-linked agarose is that it easily withstands sanitation with 0.5 M sodium hydroxide. This last feature is especially important in industry, where regular sanitation is used to achieve reliable performance and good product quality. A preferred way of doing this is cleaning in place (CIP) with a strong sodium hydroxide solution. This strong alkaline solution is an active bactericide and removes otherwise irreversibly deposited material from the column, such as particles, denatured proteins, lipids, and other compounds. If not removed, such deposits can contaminate the desired product and lead to column clogging. The fact that cross-linked agarose is stable at high temperatures is also important, since this allows it to be sterilized by autoclaving at 121°C.

Inorganic materials like porous glass and silica are vulnerable to hydrolytic damage. These supports should not be used above pH 8, and preferably not above pH 7, for any prolonged period of time, since these conditions can lead to breakdown of their silica structure [11]. However, silica materials are rarely used without the presence of a coating to make them more inert toward solutes, as discussed in the previous section. To a certain extent, these coatings also shield and protect the silica from hydrolysis. Furthermore, certain brands of silica supports are treated to incorporate zirconium or aluminum into their surfaces, which can considerably improve the stability of these supports in an alkaline environment [16, 17].

Up to this point, it may appear that affinity supports based on agarose will be satisfactorily stable under all situations encountered. But this is not the case, since often the weak point is not the matrix itself but rather the ligand or the attachment between a ligand and the support. For example, derivatives of adenosine monophosphate (AMP) and nicotinamide adenine dinucleotide (NAD$^+$) are both efficient ligands for purifying NAD$^+$-dependent dehydrogenases. But these are also delicate molecules prone to breakdown, either spontaneously or in the presence of hydrolytic enzymes. In this case, less efficient but more stable ligands, such as the dye Cibacron Blue, may be preferred, especially in large-scale applications where the cost of frequently replacing the separation material may otherwise be prohibitive.

An example where the anchoring between a ligand and matrix is the weak point on a support can be observed in the cyanogen bromide method. Although this method is convenient to use, it does lead to an isourea bridge between the support and ligand, which will be slowly hydrolyzed at an alkaline pH. This problem can be overcome by alternative immobilization techniques that give a more stable product. Examples include methods that couple through ether linkages using bisepoxides [10, 11] or various other approaches, as outlined in Chapter 3.

2.2.3 Mechanical Stability

The mechanical stability of an affinity chromatographic support should be sufficient to withstand the pressure drop across a column when the column is run at an optimum speed and flow rate for a separation. Most packing materials meet this requirement in well-behaved systems. However, in preparative-scale work, the sample or feed stream can contain many substances that may foul the separation bed. Deposits of lipids, denatured proteins, and particulate contaminants may restrict flow and quickly raise the column backpressure to unacceptable levels. In the case of soft gels like standard agarose beads, high backpressures will compress the column bed, which will increase the pressure even further and ultimately cause a collapse of this bed.

Stronger supports such as silica or heavily cross-linked polymers will not collapse at such pressures, but even here very high backpressures are undesirable. Analytical-scale chromatography is often performed using short- to medium-sized columns with small-sized beads. In this situation, supports like silica and polystyrene are preferred, since the pressure drops in these columns can be high, often extending up to several hundred bars [18, 19]. The higher pressures in these applications are due, in part, to the desire for good mass-transfer properties and fast analysis times, which are generally obtained through the use of fast flow rates and small-diameter supports.

2.2.4 Pore Size

It was stated earlier that the ideal affinity support should allow unhindered access of a solute to the immobilized ligand. For a macromolecular solute, this requires a support that has large pores. But just how large must these pores be? An answer to this question is given by the Renkin equation [20], which allows one to estimate the effective diffusion coefficient (D_{eff}) of a solute in a porous material.

$$D_{eff} = DK_D \varepsilon_p \left[1 - 2.10(R_s/R_p) + 2.09(R_s/R_p)^3 - 0.95(R_s/R_p)^5 \right]/\tau \qquad (2.1)$$

In this equation, R_s/R_p is the ratio of the solute's radius (R_s) to the pore radius (R_p), ε_p is the particle porosity, τ is the tortuosity factor, K_D is the distribution coefficient for the solute, and D is the diffusion coefficient for the solute in free solution. By inserting different values for the ratio R_s/R_p, one finds that the pore diameter should be at least five times the diameter of the solute to avoid severely restricted rates of diffusion.

For a protein of normal size (i.e., a diameter around 60 Å), a ratio of five for R_p/R_s means that the support pores should be in the range of 300 Å. Several common supports are available with such pore sizes. For example, supports based on 4 and 6% agarose have pore sizes of about 700 and 300 Å, respectively. Silica particles are available with pores in the range of 40 to 4000 Å. Polymethacrylate particles can be obtained with pores that are 100, 200, 500, or 1000 Å.

Support materials with very large pores give essentially unhindered diffusion for most solutes, but they also have a smaller surface area per milliliter of bed volume than supports with smaller pores. This reduced surface area leads to a diminished binding capacity. As a rule, a pore size of 300 to 700 Å is usually a good compromise in most situations encountered in affinity chromatography, since this gives fairly unrestricted diffusion for most biomolecules while also providing a relatively large surface area for retention.

2.2.5 Particle Size

Affinity supports are available in a wide variety of particle diameters. These range from HPLC-type materials with diameters of 10 μm or less [18, 19] to large particles for preparative work that have diameters of 400 μm. But which particle size is preferred for a given application? The answer will depend on the purpose of the separation, the mechanical properties of the support, and the characteristics of the sample.

From a theoretical viewpoint, it is always advantageous to have a small particle size, since this will promote fast mass transfer of a solute between the outer flow stream and interior of a support particle. Figure 2.3 gives a simplified view of the various steps that are involved in this process. In this model, sample molecules are transported down through the column by the flow of the mobile phase in the spaces between the support particles. To reach the affinity ligands, these molecules must diffuse through the stagnant mobile-phase layer surrounding the particles (i.e., the film model) and proceed to the inside pore network. It is here that the sample molecules will finally bind to the affinity ligand. When the retained molecules are eluted, the same steps occur but in a reversed order.

In this model, smaller support particles mean shorter diffusion distances, since they have shorter pores and a thinner stagnant mobile phase layer around and in the support. This, in turn, results in shorter times being needed for diffusion [21]. According to the Einstein equation [21], the diffusion time (t_d) that is required for a molecule to travel a given mean

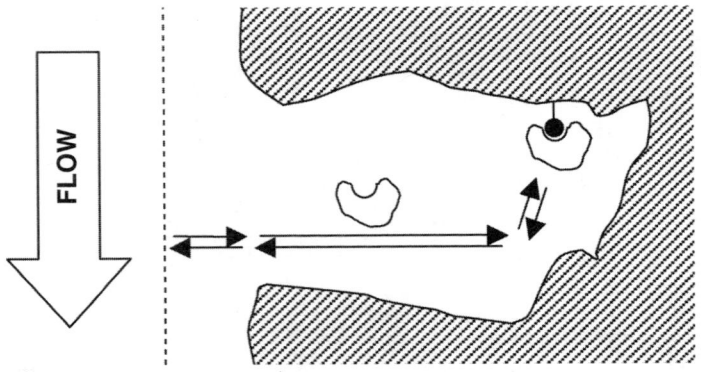

Figure 2.3 Transport processes that occur in a chromatographic column.

distance (d) will be proportional to the square of the molecule's diffusion distance, or

$$t_d = d^2/2D_{eff} \qquad (2.2)$$

where D_{eff} is the effective diffusion coefficient of the molecule in its surrounding medium.

In preparative affinity chromatography, relatively large support particles are often used, making intraparticle diffusion the main factor limiting efficiency. In this case, diminishing the particle size will increase the rate of movement of solutes between the support and surrounding flow stream, giving an improved column performance. It is this effect that was the original driving force behind the use of smaller supports in affinity columns, thus giving rise to the technique of HPLAC [2, 5]. Under such conditions, a decrease in particle size by a factor of five can make it possible to increase the flow rate by up to 25-fold and still retain good chromatographic performance. This results in a dramatic improvement in the productivity of the system.

However, a point is eventually reached when a decrease in particle diameter no longer gives a proportional improvement in an affinity column's performance. This has been observed in many analytical-scale systems that use HPLC-type supports with particle sizes less than 10 μm in diameter. Under these conditions, diffusion in the particle is now relatively fast, and it is the adsorption/desorption of sample molecules to and from the affinity ligand that becomes the limiting factor in speed and efficiency [19].

Although better efficiency is always obtained with small support particles, using a small particle size is not without difficulties. One problem is the much higher flow resistance of these smaller particles. According to the Kozeny-Carman relationship, the pressure drop over a column (Δp) is inversely proportional to the square of the support's particle size (d_p),

$$\Delta p/L = C\eta u/(d_p)^2 \qquad (2.3)$$

where L is the bed height, η is the mobile phase viscosity, u is the linear flow velocity, and C is a constant that depends on the bed porosity. This equation indicates that decreasing a support's particle size by a factor of five will increase the pressure drop across this support by a factor of 25. This increased flow resistance may lead to bed collapse when using soft gels such as agarose. And, although supports like silica can tolerate the higher pressures that result, these will require the use of more expensive pumps to work at such pressure, as is generally done in HPLC. Another route that could be taken with small affinity supports is to use a short and wide column instead of a long and narrow one. The advantages of this are that the shorter, wider column can be run at higher flow rates without creating high-pressure drops.

Another drawback with small particle sizes, especially in preparative work, is the increased danger of fouling that exists when particulate contaminants are in the feed stream or sample. This occurs because the interstitial spaces in a bed of small particles can be too narrow for such agents to pass through. Such fouling will increase the flow resistance and may lead to bed collapse if the support material does not have sufficient mechanical strength, as discussed earlier in Section 2.2.3.

As a result of these various requirements, the particle size to pick when designing a new affinity adsorbent will be a compromise between the desired chromatographic performance, properties of the feed stream, and the mechanical strength of the support. Some common selections made in specific cases will be described in the next few sections.

2.2.5.1 Zonal Elution Chromatography

Zonal elution chromatography (also known as *weak* or *dynamic affinity chromatography*) is mainly used in analytical work and in basic studies of affinity phenomena. It is seldom used for preparative purposes, due to its relatively low capacity compared with other methods. The sample molecules in this case are injected as a narrow plug. These are then monitored as they travel through the bed at speeds that are dependent on the strengths of their respective interactions with the ligand.

To obtain a good separation in this type of system, it is important to have a column with a good efficiency, as is true in HPLC-based affinity methods. Ideally, this requires that the support be based on particles that are as small as reasonably possible. The use of small particles will reduce diffusion times, as discussed earlier, which also promotes the formation of narrow solute peaks. This can be performed by using HPLC-grade silica that has been modified for use in affinity chromatography, as discussed in Section 2.2.1. Another alternative is to use ultrasmall, nonporous particles based on micron-sized silica or polystyrene [18]; however, these materials are not yet commercially available in the affinity mode. More information on zonal elution chromatography and its applications in affinity separations can be found in Chapter 4 and in Chapters 18 to 24.

2.2.5.2 Adsorption-Desorption Affinity Purification

The most common way of performing affinity chromatography for preparative-scale work is to use this as an adsorption-desorption process. In this strategy, the target molecule alone is bound tightly by the ligand, and the adsorbent has a high binding capacity. After performing a suitable wash to release weakly bound substances from the sample, the conditions on the column are drastically changed (e.g., by using a substantial pH shift) so that the target molecule loses its affinity for the ligand and is eluted.

Such a protocol works well with large support particles, since there is no absolute need for high chromatographic efficiency. Still, small support particles can certainly help speed up the adsorption step, make the washing step more efficient, and permit the target molecule to be eluted in a smaller volume (i.e., as a narrower peak). This is especially valuable in analytical applications, where sharper peaks allow for more convenient quantitation, better limits of detection, and faster analysis times. In preparative-scale work, the greater ease of handling larger particles is often preferred in practice, with particle sizes in the range 30 to 100 μm being common in such work.

2.2.5.3 Partial Loading in Adsorption-Desorption Affinity Purification

Partial loading is a method for improving the performance of large particles in preparative-scale affinity chromatography. The trick in this approach is to utilize only the outer shell of the support particle and thus diminish the distances required for solute diffusion. Binding to a thin outer layer of the support takes only a fraction of the time needed to saturate the entire particle. However, this outer shell still represents a substantial fraction of the support's total binding capacity.

Figure 2.4 illustrates results for the partial loading technique. These results were determined based on the following equation [21],

$$t_{shell}/t_{total} = 1 - 3(1 - a/r)^2 + 2(1 - a/r)^3 \qquad (2.4)$$

where t_{total} is the time needed to saturate the whole particle, t_{shell} is the time needed to saturate a layer with thickness of "a" in the particle, and r is the particle radius. According to Figure 2.4, an outer shell with a depth equal to one fifth the total support-particle radius ($a/r = 0.2$) will contain half the particle volume ($V_{shell}/V_{total} = 0.5$). But this figure also shows that such a shell will take only one tenth the time needed to fill the whole particle ($t_{shell}/t_{total} = 0.1$).

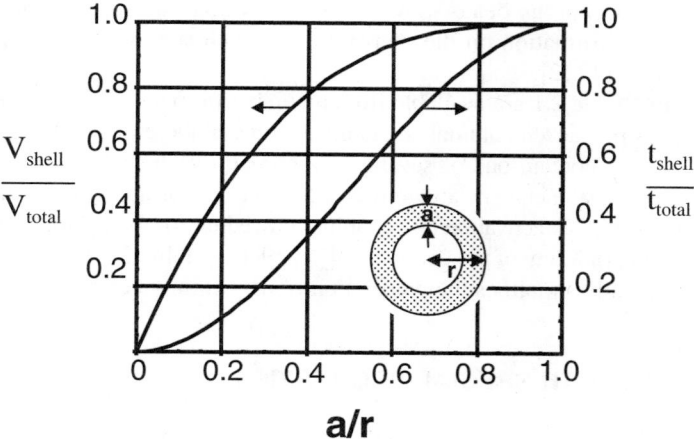

Figure 2.4 The shrinking-core model for protein adsorption. It is assumed in this model that the protein binds strongly to the adsorbent (step adsorption) and that the binding kinetics are fast compared with diffusion (i.e., the normal situation in preparative affinity chromatography operated according to the adsorption-desorption principle). The radius of the entire support particle is given by r, and the thickness of the outer absorbed layer or shell is given by a. The relative volume of the shell is given by the ratio V_{shell}/V_{total}, where V_{total} is the total volume of the support particle and V_{shell} is the volume occupied by the outer layer occupied by the adsorbed molecule. The relative time needed to fill the exterior shell is t_{shell}/t_{total}, where t_{shell} is the time need to fill the outer layer and t_{total} is the time needed to completely saturate the support particle.

Admittedly, special arrangements have to be made to make full use of this principle. For instance, fast recirculation through the bed is needed to ensure that all the support particles experience the same loading concentration. Also, the washing and elution steps are not substantially improved by this approach, since the released substances (impurities or the desired compound) will still have access to the entire particle.

2.2.5.4 *Particulate Contaminants: Large Beads/Expanded Beds*

In large-scale purification schemes, particulate contaminants such as cells and cell debris are a problem, since they tend to clog affinity columns. This problem becomes worse as the size of the packing materials decreases, as noted earlier. One solution to this problem is to use extra-large particles. This strategy has been adopted in products like Sepharose Big Beads from Amersham Biosciences and Cellthru Big Beads from Sterogene, which have particle sizes of 100 to 300 μm and 300 to 500 μm, respectively. These very large particles should be run in the partial-loading mode described in the previous section. If this is not done, their performance will suffer from their long diffusion distances.

Another solution to particulate contaminants is to use the expanded-bed principle. In this technique, the distance between the adsorbent particles is increased to allow contaminants to pass through freely. The adsorbents necessary for this mode of operation will be described later in Section 2.3.

2.2.6 Standard Commercially Available Affinity Supports

The previous sections of this chapter described the properties that an affinity support should have, such as a hydrophilic character, good chemical and physical stability, and an appropriate pore size for the target solute. Most of these criteria are presently met by commercial adsorbents based on porous beads. Several examples of these supports are shown in Table 2.1. For more information on the preparation of such support materials, see the reviews in the literature [4, 22].

The adsorbents shown in Table 2.1 are available from a number of manufacturers and are supplied either in bulk or as prepacked columns. In addition, some of these materials can be obtained in forms that already contain one of several common affinity ligands, such as protein A, Cibacron Blue, heparin, or metal chelating groups. If other ligands are required, it may be necessary to first find a suitable activation and coupling procedure for the ligand (see Chapter 3 or the literature [10, 11]). Many of the commercial suppliers in Table 2.1 also offer preactivated supports, which greatly simplifies the preparation of new affinity columns.

2.3 AFFINITY SUPPORTS WITH SPECIAL PROPERTIES

Porous supports like agarose, polymethacrylate, or silica beads are the main workhorses in most current applications of affinity chromatography. However, in the past several years other types of supports have also become available commercially (Table 2.2) or have been described for use in affinity chromatography. Many of these newer materials have properties that give them superior performance in certain applications. Materials that fall in this category include nonporous supports, membranes, flow-through beads, continuous beds, and expanded-bed particles.

2.3.1 Nonporous Supports

Nonporous beads with diameters of 1 to 3 µm can be an optimum choice for fast analytical or micropreparative separations, since the limiting factor of pore diffusion is virtually eliminated in these materials. Such beads may also be the best choice for fundamental or quantitative studies of affinity interactions [18], since the binding and dissociation behavior seen with these materials should be more directly linked with the interactions occurring between solutes and the affinity ligand.

However, there is a substantial loss of surface area and binding capacity that occurs through the elimination of internal pores. For instance, 1.0-µm nonporous particles have a surface area of about 5 m^2 per milliliter of packed bed, but the corresponding value for porous silica with 300-Å pores is about ten times higher. This difference becomes even more accentuated when comparing larger beads. In addition, as was discussed earlier, a smaller particle size leads to larger column backpressures. Thus, micron-sized particles are usually used in shallow beds to avoid high system pressures. Difficulties with backpressure can also be minimized by using monodisperse particles, which will create fewer problems than polydisperse supports [23].

Nonporous fibers are another category of materials that have been used as supports for affinity chromatography. These can have very high dynamic capacities [24]. Another advantage of these supports is that, in spite of the fact that the fibers are submicron in diameter, their backpressures are low due to the low packing density of the overall fiber bed.

2.3.2 Membranes

Membranes have been used for affinity chromatography in various formats, such as stacked sheets, in rolled geometries, or as hollow fibers [25, 26]. Materials that are commonly used for these membranes are cellulose, polysulfone, and polyamide [25, 26]. Because of their lack of diffusion pores, the surface area in these materials is as low as it is in nonporous beads. However, the flat geometry and shallow bed depth of membranes keep the pressure drop across them to a minimum. This means that high flow rates can be used, which makes

Table 2.2 Examples of Commercial Affinity Supports with Special Properties

Trade Name	Format	Material	Examples of Available Ligands	Preactivated Form Available?	Manufacturer/Supplier
Sartobind	Stacked membranes	Regenerated cellulose	Protein A, IMAC	Yes	Sartorius
Poros	Flow-through beads (20- or 50-μm diameter)	Polystyrene/Divinylbenzene	Protein A, protein G, heparin, IMAC[a]	Yes	Applied Biosystems
CIM	Monolithic disk or tubes	Polymethacrylate	Protein A, protein G	Yes	BiaSeparations
Streamline	Composite beads (120- or 200-μm diameter)	Agarose plus quartz or stainless steel	IMAC[a], heparin, protein A	No	Amersham Biosciences
UFC	Composite beads (30- or 200-μm diameter)	Agarose plus quartz or stainless steel	IMAC[a], lysine, biotin	Yes	UpFront Chromatography

[a] IMAC, immobilized metal-ion affinity chromatography.

these membranes especially well-suited for capturing proteins from dilute feed streams. One commercial membrane, Sartobind from Sartorius, is even available in a preactivated form for use in affinity methods.

2.3.3 Flow-Through Beads

As stated earlier, porous supports with a larger diameter facilitate low column backpressures and allow easy passage of contaminants through the column. But it is also necessary to keep the diameter of these supports as small as possible to diminish diffusion distances and thereby improve their chromatographic performance. One solution to these contradictory requirements is to use particles that allow the flow of mobile phase directly through some of the pores. This is done in materials known as *perfusion media* or *through-pore particles*. Flow-through particles were initially developed in the early 1990s for ion-exchange chromatography [27] and were later adapted for use in affinity chromatography [28–30].

Flow-through particles have a bimodal pore configuration, in which both small diffusion pores and large flow-through pores are present (see Figure 2.5). Substances applied to a bed of this support are transported by mobile phase flow to the interior of each particle, leaving only short distances to be covered by diffusion to the support's surface. This combination leads to a dramatic improvement in performance compared with standard porous particles of the same size. This improvement is most pronounced in situations where slow diffusion is a limiting factor, such as in the chromatography of large molecules (e.g., proteins) at high flow rates.

The presence of the flow-through pores can diminish the available surface area and, therefore, can decrease the static binding capacity (expressed as mg protein/g absorbent) of these materials versus traditional porous supports. However, the improved performance of these alternative particles also leads to a much higher dynamic capacity (expressed as mg protein/g adsorbent/h).

In a few instances it has been found that the static binding capacity of flow-through particles can be equal to [29] or even higher [31] than the capacity of standard particles of a similar size. This may be due to differences in pore blockage [32]. For instance, when

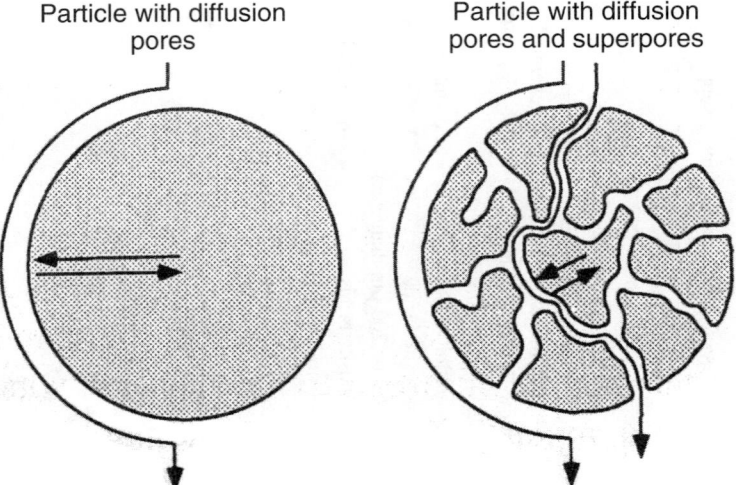

Figure 2.5 Comparison of a particle with normal porosity versus a particle that contains flow-through pores. The normal particle has long diffusion distances, whereas the flow-through particle has short diffusion distances.

large molecules are binding to large ligands (e.g. immunoglobulins), the resulting complexes may be large enough to block small pores, especially if the complex is situated at a narrow passage. Ligands that lie farther down the pore will then be difficult to reach, leading to a low static-binding capacity. This problem should be much less pronounced with particles that contain flow-through pores, since the length of the small diffusion pores is now much shorter and the statistical chance that ligands will become unreachable by a large solute is therefore lower.

2.3.4 Continuous Beds

Another format developed in the 1990s was the *continuous bed* or *monolithic support*. Continuous bed supports consist of a single piece of material intersected by pores large enough to support chromatographic flow through the bed. Continuous beds have been developed using many well-known chromatographic materials, such as polyacrylamide [33], silica [34], polystyrene/polymethacrylates [35, 36], cellulose [37], and agarose [38]. Most of these continuous beds have two types of pores: large flow-carrying pores and smaller diffusion pores. Figure 2.6 shows some examples of continuous bed structures made from agarose [38].

The preparation of a continuous bed is usually straightforward. These beds can often be prepared directly in a chromatographic column, thereby avoiding the time-consuming steps of size classification and column packing that are normally needed with particle-based supports. Reports using continuous beds in affinity chromatography [33, 35, 36, 38] have shown that the efficiency of these materials is as good as that for particle-based supports.

Commercially available continuous beds with affinity ligands are manufactured by BiaSeparations (see Table 2.2). These materials, known as convective interaction media (CIM), are based on polymethacrylate. They have flow-carrying pores with diameters of 1.5 μm and diffusion pores with sizes below 100 nm. The CIM disks are intended for analytical or micropreparative purposes and are operated axially, whereas the CIM tubes are designed for preparative applications and are operated in the radial mode.

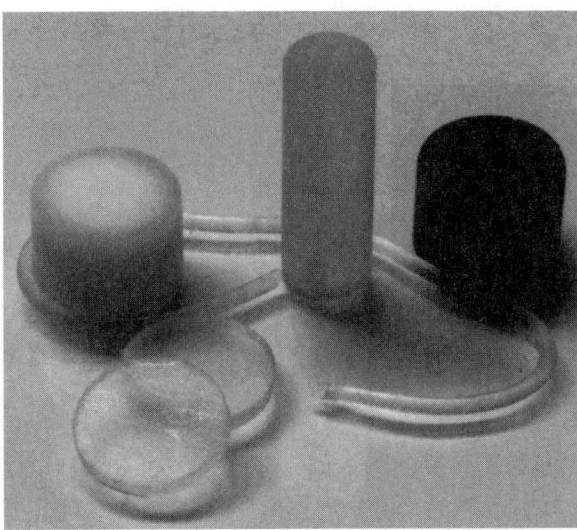

Figure 2.6 Examples of continuous-bed structures made of agarose. The supports shown in this figure (from left to right) are a continuous bed, membranes, a continuous bed, a continuous bed derivatized with Cibacron blue, and a fiber. (Reprinted with permission from Gustavsson, P.-E. and Larsson, P.-O., *J. Chromatogr. A*, 832, 29–39, 1999.)

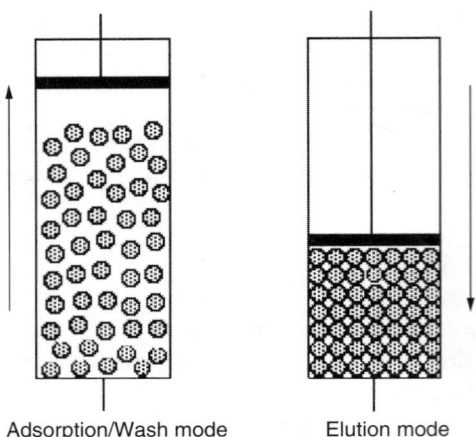

Adsorption/Wash mode   Elution mode

Figure 2.7   Expanded-bed chromatography. During the adsorption-wash step, the flow is directed upward and the bed is expanded. The elution step is carried out in a normal packed-bed mode with downward flow.

2.3.5   Expanded-Bed Adsorbents

To avoid column clogging, various pretreatment methods like filtration and centrifugation are often necessary to remove particulate matter from samples. To cut down on the need for such methods, a new class of adsorbents has recently been developed to handle viscous and particle-containing feed streams [39]. These materials are known as *expanded-bed adsorbents*. In expanded-bed chromatography the direction of mobile phase flow is upward through the column and is fast enough to fluidize the support particles in the column. This causes the column bed to expand, as depicted in Figure 2.7. This expansion makes the interstitial spaces in the column bed larger so that solid contaminants like cells and cell debris can pass through, thereby avoiding column clogging.

To more easily control the expansion of the column bed and to hinder support particles from leaving the top of the column, the support used in expanded-bed chromatography must have a relatively high density. This can be accomplished by adding a dense material like quartz to a low-density base material like agarose, as is done in the Streamline adsorbents from Amersham Biosciences and UFC-agarose from UpFront Chromatography. Another way of obtaining an appropriate support for expanded-beds is to employ a base material of intrinsically high density. An example of this is the use of porous mineral oxides [40].

Another new type of expanded-bed adsorbent uses a thin layer of active material (i.e., derivatized agarose) that surrounds a heavy core [41, 42]. These adsorbents have small diffusion distances for biomolecules along with a higher density than other expanded-bed particles. The advantage of this combination is that it allows better chromatographic efficiencies to be obtained at higher flow rates.

2.4   SUMMARY AND CONCLUSIONS

The discussion in this chapter highlights the various aspects that must be considered when selecting a support material for affinity chromatography. The selection process should be governed by the properties of the affinity support and its intended use. Properties to consider for the support are its inertness toward solutes; mechanical, chemical, and micro-biological stability; pore size and particle size; and the availability of suitable chemistries for placing the affinity ligand on the support. In large-scale preparative applications, the cost of the support is another important item to consider. Since some of these factors have optimum conditions that are contradictory, the final choice of a support for any given application will represent a balanced compromise of these requirements.

A considerable number of affinity support materials are now available commercially. Some of these can be purchased with ligands already attached to the support, but the number of ligands available in this fashion is not large. Another alternative is to use a preactivated support to carry out the necessary coupling chemistry. And finally, the supports can be activated directly in the lab, but this is usually a more laborious procedure, often requiring special techniques and equipment to achieve consistent results.

During the last few years, a number of new and innovative support materials have appeared for affinity columns. Examples are expanded-bed supports for particle-contaminated process streams, flow-through pore materials and nonporous supports with good mass-transfer properties, and nonstandard formats such as membranes and continuous (or monolithic) beds. Further developments are expected for many of these materials and formats, and these should produce even better supports for affinity chromatography.

SYMBOLS AND ABBREVIATIONS

a	Thickness of an adsorbed protein (or adsorbed molecule) layer
AMP	Adenosine monophosphate
ATP	Adenosine triphosphate
C	Constant for bed porosity in the Kozeny-Carman equation
CIM	Convective interaction media
CIP	Cleaning in place
CNBr	Cyanogen bromide
CTA	Cyanogen transfer agent
d	Diffusion distance
d_p	Particle diameter
D_{eff}	Effective diffusion coefficient of a solute in a porous material
D	Diffusion coefficient of a solute in free solution
HIC	Hydrophobic interaction chromatography
HPAC	High-performance affinity chromatography
HPLAC	High-performance liquid affinity chromatography
HPLC	High-performance liquid chromatography
IMAC	Immobilized metal-ion affinity chromatography
K_D	Distribution coefficient of a solute
L	Column length or bed height
NAD$^+$	Nicotinamide adenine dinucleotide
r	Particle radius
R_p	Pore radius
R_s	Solute radius
t_d	Diffusion time for a solute
t_{shell}	Time to produce a given shell layer thickness for an adsorbed molecule in a support particle
t_{total}	Time to saturate an entire support particle
u	Linear velocity
V_{shell}	Volume of a given shell about a support particle
V_{total}	Total volume of a particle

Δp Pressure drop across a column
ε_p Particle porosity
η Mobile-phase viscosity
t Tortuosity factor

REFERENCES

1. Cuatrecasas, P., Wilchek, M., and Anfinsen, C.B., Selective enzyme purification by affinity chromatography, *Proc. Natl. Acad. Sci. U.S.A.*, 61, 636–643, 1968.
2. Ohlson, S., Hansson, L., Larsson, P.-O., and Mosbach, K., High-performance liquid affinity chromatography (HPLAC) and its application to the separation of enzymes and antigens, *FEBS Lett.*, 93, 5–9, 1978.
3. Arshady, R., Beaded polymer supports and gels, II: Physicochemical criteria and functionalization, *J. Chromatogr.*, 586, 199–219, 1991.
4. Buchmeiser, M.R., New synthetic ways for the preparation of high-performance liquid chromatography supports, *J. Chromatogr. A*, 918, 233–266, 2001.
5. Larsson, P.-O., Glad, M., Hansson, L., Månsson, M.-O., Ohlson, S., and Mosbach, K., High-performance liquid affinity chromatography, in *Advances in Chromatography*, Vol. 21, Giddings, J.C., Gruschka, E., Cazes, J., and Brown, P.R., Eds., Marcel Dekker, New York, 1983, pp. 41–85.
6. Varady, L., Mu, N., Yang, Y.-B., Cook, S.E., Afeyan, N., and Regnier, F.E., Fimbriated stationary phases for proteins, *J. Chromatogr.*, 631, 107–114, 1993.
7. Nash, D.C. and Chase, H.A., Modification of polystyrenic matrices for the purification of proteins, III: Effects of poly(vinyl alcohol) modification on the characteristics of protein adsorption on conventional and perfusion polystyrenic matrices, *J. Chromatogr. A*, 776, 65–73, 1997.
8. Boschetti, E., Advanced sorbents for preparative protein separation purposes, *J. Chromatogr. A*, 658, 207–236, 1994.
9. Leonard, M., New packing materials for protein chromatography, *J. Chromatogr. B*, 699, 3–27, 1997.
10. Carlsson, J., Jansson, J.-C., and Sparrman, M., Affinity chromatography, in *Protein Purification, Principles, High Resolution Methods, and Applications*, 2nd ed., Jansson, J.-C. and Rydén, L., Eds., John Wiley, New York, 1998, pp. 375–442.
11. Hermanson, G.T., Mallia, A.K., and Smith, P.K., *Immobilized Affinity Ligand Techniques*, Academic Press, San Diego, 1992.
12. Hagel, L., Gel filtration, in *Protein Purification, Principles, High Resolution Methods, and Applications*, 2nd ed., Jansson, J.-C. and Rydén, L., Eds., John Wiley, New York, 1998, p. 90.
13. Hermanson, G.T., Mallia, A.K., and Smith, P.K., *Immobilized Affinity Ligand Techniques*, Academic Press, San Diego, 1992, p. 305.
14. Kohn, J. and Wilchek, M., A new approach (cyano-transfer) for cyanogen bromide activation of Sepharose at neutral pH, which yields activated resins, free of interfering nitrogen derivatives, *Biochem. Biophys. Res. Commun.*, 107, 878–884, 1982.
15. Er-el, Z., Zaidenzaig, Y., and Shaltiel, S., Hydrocarbon-coated Sepharoses: Use in the purification of glycogen phosphorylase, *Biochem. Biophys. Res. Commun.*, 49, 383–390, 1972.
16. Stout, R.W., Sivakoff, S.I., Ricker, R.D., Palmer, H.C., Jackson, M.A., and Odiorne, T.J., New ion-exchange packings based on zirconium oxide surface-stabilized, diol-bonded, silica substrates, *J. Chromatogr.*, 352, 381–397, 1986.
17. Pfannkoch, E.A., Switzer, B.S., and Kopaciewicz, W., Aluminum ion-mediated stabilization of silica-based anion-exchange packings to caustic regenerants, *J. Chromatogr.*, 503, 385–401, 1990.
18. Lee, W.-C., Protein separation using non-porous sorbents, *J. Chromatogr. B*, 699, 29–45, 1997.

19. Clarke, W.S. and Hage, D.S., Development of sandwich HPLC microcolumns for analyte adsorption on the millisecond time scale, *Anal. Chem.*, 73, 1366–1373, 2001.

20. Renkin, E.M., Filtration, diffusion and molecular sieving through porous cellulose membranes, *J. Gen. Physiol.*, 38, 225–243, 1954.

21. Jansson, J.-C. and Pettersson, T., Large-scale chromatography of proteins, in *Preparative and Production Scale Chromatography*, Ganetsos, G. and Barker, P.E., Eds., Marcel Dekker, New York, 1992, p. 567.

22. Arshady, R., Beaded polymer supports and gels, I: Manufacturing techniques, *J. Chromatogr.*, 586, 181–197, 1991.

23. Anspach, B., Unger, K.K., Davies, J., and Hearn, M.T.W., Affinity chromatography with triazine dyes immobilized onto activated non-porous monodisperse silicas, *J. Chromatogr.*, 457, 195–204, 1988.

24. Wikström, P. and Larsson, P.-O., Affinity fibre: A new support for rapid enzyme purification by high-performance liquid affinity chromatography, *J. Chromatogr.*, 388, 123–134, 1987.

25. Charcosset, C., Purification of proteins by membrane chromatography, *J. Chem. Technol. Biotechnol.*, 71, 95–110, 1998.

26. Roper, D.K. and Lightfoot, E.N., Separation of biomolecules using adsorptive membranes, *J. Chromatogr. A*, 702, 3–26, 1995.

27. Afeyan, N.B., Gordon, N.F., Mazsaroff, I., Varady, L., Fulton, S.P., Yang, Y.B., and Regnier, F.E., Flow-through particles for the high-performance liquid chromatographic separation of biomolecules: perfusion chromatography, *J. Chromatogr.*, 519, 1–29, 1990.

28. Fulton, S.P., Meys, M., Varady, L., Jansen, R., and Afeyan, N.B., Antibody quantitation in seconds using affinity perfusion chromatography, *BioTechniques*, 11, 226–231, 1991.

29. Gustavsson, P.-E., Mosbach, K., Nilsson, K., and Larsson, P.-O., Superporous agarose as an affinity chromatography support, *J. Chromatogr. A*, 776, 197–203, 1997.

30. Pålsson, E., Smeds, A.-L., Petersson, A., and Larsson, P.-O., Faster isolation of recombinant factor VIII SQ, with a superporous agarose matrix, *J. Chromatogr. A*, 840, 39–50, 1999.

31. Gustavsson, P.-E. and Larsson, P.-O., Superporous Agarose Beads Give Higher Binding Capacity (Static and Dynamic) Than Homogeneous Agarose Beads, report, Pure and Applied Biochemistry Department, Lund University, November 24, 1997.

32. Horstmann, B.J., Kenney, C.N., and Chase, H.A., Adsorption of proteins on Sepharose affinity adsorbents of varying particle size, *J. Chromatogr.*, 361, 179–190, 1986.

33. Mohammad, J., Zeerak, A., and Hjertén, S., Dye-ligand affinity chromatography on continuous beds, *Biomed. Chromatogr.*, 9, 80–84, 1995.

34. Minakuchi, H., Nakanishi, K., Soga, N., Ishizuka, N., and Tanaka, N., Octadecylsilylated porous silica rods as separation media for reversed-phase liquid chromatography, *Anal. Chem.*, 68, 3498–3501, 1996.

35. Petro, M., Svec, F., and Frechet, J.M.J., Immobilization of trypsin onto "molded" macroporous poly(glycidyl methacrylate-co-ethylene dimethacrylate) rods and use of the conjugates as bioreactors and for affinity chromatography, *Biotechnol. Bioeng.*, 49, 355–363, 1996.

36. Josic, D., Schwinn, H., Strancar, A., Podgornik, A., Barut, M., Lim, Y.-P., and Vodopivec, M., Use of compact, porous units with immobilized ligands with high molecular masses in affinity chromatography and enzymatic conversion of substrates with high and low molecular masses, *J. Chromatogr. A*, 803, 61–71, 1998.

37. Sepragen, SepraSorb; available on-line at www.sepragen.com; accessed March 26, 2005.

38. Gustavsson, P.-E. and Larsson, P.-O., Continuous superporous agarose beds for chromatography and electrophoresis, *J. Chromatogr. A*, 832, 29–39, 1999.

39. Mattiasson, B., Ed., Expanded bed chromatography (special issue), *Bioseparation*, 8, 1–271, 1999.

40. Voute, N. and Boschetti, E., Highly dense beaded sorbents suitable for fluidized bed applications, *Bioseparation*, 8, 115–120, 1999.

41. Pålsson, E., Nandakumar, M.P., Mattiasson, B., and Larsson, P.-O., Miniaturised expanded-bed column with low dispersion suitable for fast-flow ELISA analyses, *Biotechnol. Lett.*, 22, 245–250, 2000.

42. Theodossiou, I., Olander, M.A., Sondergaard, M., and Thomas, O.R.T., New expanded bed adsorbents for the recovery of DNA, *Biotechnol. Lett.*, 22, 1929–1933, 2000.

3

Immobilization Methods for Affinity Chromatography

Hee Seung Kim and David S. Hage

Department of Chemistry, University of Nebraska, Lincoln, NE

CONTENTS

3.1   INTRODUCTION

A key factor in the success of any type of affinity chromatography is the way in which the ligand is attached to its support [1–3]. Ideally, the chosen ligand should mimic its interactions within its natural environment. Thus, it is important to consider the way in which the ligand is placed within the column. The success of this process depends on the selection of an appropriate immobilization method. This chapter examines several immobilization methods for affinity chromatography and discusses the relative advantages and disadvantages of each.

The term *immobilization* as it is used in this chapter refers to the means by which a ligand is attached to or incorporated within a chromatographic support. It is important to select this method carefully, since the ligand could be altered or denatured if the wrong technique is employed. For example, the optimum pH for an enzyme's activity may change by as much as two pH units when the enzyme is immobilized [4–7]. In addition, the use of cross-linking agents during immobilization can cause bond formation between ligands, altering their activity and even affecting the support's porosity [8–10].

Figure 3.1 summarizes the general approaches used for ligand immobilization. This includes noncovalent methods like nonspecific and biospecific adsorption, covalent coupling techniques, entrapment, and molecular imprinting. In this chapter, the basic principles of immobilization will be discussed, and various approaches for placing ligands on supports will be examined. General factors that affect the selection of an immobilization method will be considered, along with approaches for measuring and characterizing immobilized ligands.

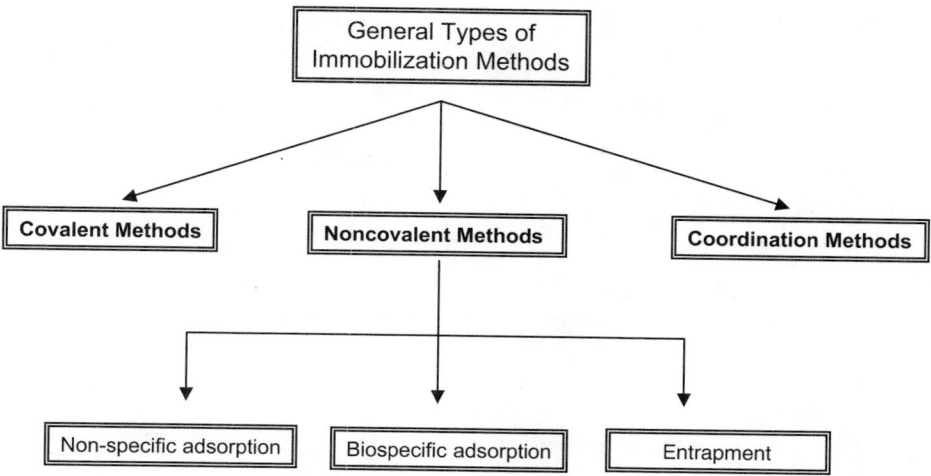

Figure 3.1 General strategies for the immobilization of ligands in affinity chromatography.

3.2 GENERAL CONSIDERATIONS

There are many factors to consider when selecting and using an immobilization method. This section first examines the nonideal effects that occur during immobilization and then discusses how these effects can be avoided. The choice of a support material for immobilization is then considered, along with factors that determine the speed, selectivity, and stability of an immobilization method. In addition, several techniques are presented for characterizing an immobilized ligand.

3.2.1 Immobilization Effects

The proper choice of an immobilization method is important, since it can affect the actual or apparent activity of a ligand. If the correct coupling procedure is not used, several nondesirable effects can result. These include multisite attachment, improper orientation, and steric hindrance. Figure 3.2 illustrates each of these effects.

3.2.1.1 Multisite Attachment

One immobilization effect that often occurs is *multisite attachment*. This refers to the coupling of a ligand to the support through more than one functional group on the same ligand molecule. This has been demonstrated in a study that examined the immobilization of carboxypeptidase A to aldehyde-activated agarose [11]. When amino acid analysis was performed on the hydrolyzed products of soluble and immobilized carboxypeptidase A, it was found that four lysine residues per carboxypeptidase were linked to the agarose, indicating that multisite attachment had occurred. Multisite attachment can be a positive feature if it creates a more stable linkage than single-site immobilization. However, it can also lead to distortion and denaturation of a ligand at its active site [12].

Multisite binding can be minimized by coupling a ligand through functional groups that occur in only a few places in its structure. One example is the use of carbohydrate chains in antibodies for their site-selective attachment to hydrazide-containing supports. Another example is the use of site-direct mutagenesis to introduce cysteine residues into otherwise cysteine-free proteins. This last technique has been used on several occasions to immobilize enzymes at locations distant from their active sites [13]. Another approach

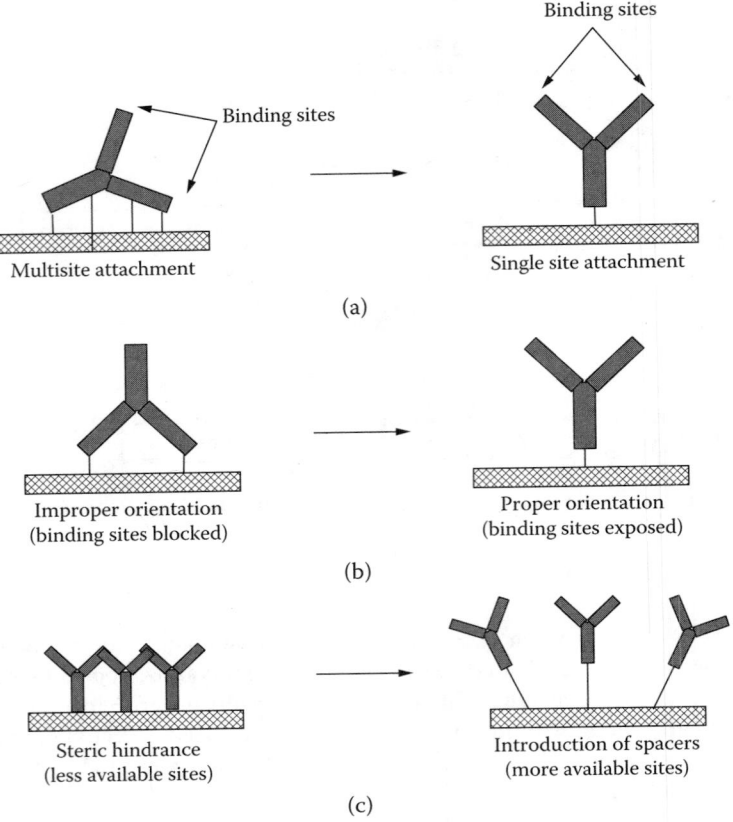

Figure 3.2 The effects of several nonideal processes during ligand immobilization: (a) multisite attachment, (b) random or improper orientation, and (c) steric hindrance.

for minimizing multisite binding is to use a support with a low density of reactive sites. However, for large ligands this will often require the use of a very small amount of such sites, since a single macromolecule can cover a large number of these groups.

3.2.1.2 Improper Orientation

Another immobilization effect is *improper orientation*. This occurs when the site of attachment for a ligand is at or near its active region. Like multisite attachment, this can lead to a decrease or complete loss in the ligand's activity. This is commonly seen in immobilization methods that involve the use of primary amine groups on a protein, which can occur at many locations on such a ligand.

The most common way of avoiding improper orientation is to use a coupling method that makes use of sites that are distant from a ligand's active region. Again, a good example of this is the use of carbohydrate chains on antibodies for their coupling to hydrazide-containing supports [14–16]. Similar results can be obtained for other proteins by using site-directed mutagenesis [17]. In one example, a single cysteine was placed into the structure of subtilisin at a region known to be distant from its active site. The modified subtilisin was then immobilized through this residue by reacting it with a cysteine-specific reagent, N-γ-maleimidobutyryl-oxysuccimide ester. The final product gave a catalytic activity almost three times higher than seen when using an amine-based coupling method.

If an appropriate group is not available on a ligand for site-selective attachment, it is still possible in many cases to adjust the coupling site by varying the immobilization conditions. For proteins, this is often accomplished by varying the pH of the reaction buffer. This is based on the fact that only amine groups with low pK_a values will be available at a lower pH for immobilization. The same approach can be used to promote the reaction of sulfhydryl groups over amines, as might be used for the attachment of antibody F_{ab} fragments through their free cysteine residues.

3.2.1.3 Steric Hindrance

The third type of immobilization effect is *steric hindrance*. This occurs when the active site of a ligand is blocked by the support or by neighboring ligands. Again, this can lead to an apparent loss of ligand activity. Steric hindrance due to neighboring ligands can be minimized by employing a support with a low coverage of ligands. Such an effect was indicated in the purification of guanidinobenzoatase on an immobilized agmatine column. It was found here that the adsorption of pure guanidinobenzoatase depended strongly on the concentration of agmatine coupled to the support, with the amount of adsorbed target decreasing by 75% as the effective concentration of ligand increased from 2 to 15 μmol/ml [18].

Steric hindrance due to the support can be reduced by adding a *spacer arm*, or tether, between the ligand and supporting material. The presence of a spacer arm is particularly important when using a small ligand for the retention of a large target. Common spacer arms used in affinity chromatography are 6-aminocaproic acid, diaminodipropylamine, 1,6-diaminohexane, ethylenediamine, and succinic acid anhydride [9]. One study investigated the effect of spacer arm length on the enzymatic hydrolysis of 1-fatty acyl-2(12-aminododecyl) phosphatidylcholine (APC) when APC was immobilized on *N*-hydroxysuccinimide-activated agarose [19]. It was found that APC attached through an eight-atom spacer arm (i.e., *N*-hydroxysuccinimidylester-6-aminohexanoic acid) gave a ten-fold faster rate of hydrolysis by *Crotalus adamanteus* phospholipase A$_2$ than APC immobilized through a one-atom spacer arm (i.e., *N*-hydroxysuccinimidyl ester).

Spacer arms can also have an effect on the binding of solutes to large ligands. This has been noted in the use of dextrans as long, hydrophilic spacers for improving the activity of immobilized rennin and protein A [20]. It was found that the caseinolytic activity of rennin on dextran-coated agarose was 15-fold higher than the activity for rennin immobilized to agarose through a two-atom spacer arm. The amount of immunoglobulin G (IgG) that could bind to immobilized protein A was also affected by the presence of a dextran spacer. Protein A immobilized on dextran agarose was able to bind two moles of IgG per mole of protein A, but only one mole of IgG per mole of protein A was noted when a two-atom spacer was used [20].

One problem that may be encountered when using spacer arms is that these can nonspecifically retain some solutes. In fact, it was this effect that led to the original discovery of *hydrophobic interaction chromatography* (HIC). A recent example is work in which the retention of bovine serum albumin was examined on protein A immobilized to a monolithic column both with and without the use of the six-carbon spacer hexanediamine, which has been used as a ligand for HIC [21].

3.2.2 Other Factors to Consider During Immobilization

3.2.2.1 Choice of Support

Another factor to consider in affinity chromatography is the material used to hold the ligand within the column. As discussed in Chapter 2, this support should have minimal nonspecific binding but be easy to modify for ligand attachment. This material should also be stable

Table 3.1 Relationship of Support Pore Size and Surface Area to Amount
of Immobilized Papain

Support (Corning Glass)	Pore Size of Support (Å)	Surface Area of Support (m²/g)	Amount of Immobilized Papain (mg eq/g)
7740 (test tube)	Nonporous	Low	0
7740 (F-frit)	50,000	0.2	0.01
X740BXY	900	20	0.33
7930	92	60	0.08
7930	68	118	0.07

Note: These results were obtained for papain, which has a molecular mass of 21 kDa. A similar
trend but different optimum pore size would be noted for other ligands.

Source: Messing, R.A., *Enzymologia*, 39, 12–14, 1970.

under the flow rate, pressure, and solvent conditions needed for the final application. In
addition, this support should be readily available and simple to use in method development.

One reason the choice of support is important is that it will affect the amount of ligand
that can be immobilized. For instance, the surface area of a support will decrease as its pore
size increases; this, in turn, gives a smaller area for ligand attachment. Ligand size also needs
to be considered during the selection of a support because the ligand will not be able to
enter pores smaller than its cross-sectional diameter. This means there will be some optimum
pore size for obtaining maximum ligand coverage. Such an effect is illustrated in Table 3.1
for the immobilization of papain to supports with various pore sizes [22]. In this case, the
amount of adsorbed papain increased as the support's pore size decreased to 900 Å and its
surface area increased to 20 m²/g. However, the amount of adsorbed ligand then decreased
as the pore size decreased to 60 Å, due to restricted access of papain into the pores. Similar
results have been observed in other studies [13, 16]. This includes work with the covalent
immobilization of antibodies on silica, in which maximum coverage was observed with pores
that were roughly three to five times the diameter of the ligand [23].

Another way the support will affect immobilization is in the types of coupling methods
that can be successfully used with this material. Most supports for affinity chromatography
can be used with amine-based coupling techniques like the Schiff base or carbonyldiimidazole methods, as discussed later in this chapter. Agarose supports were made popular when
the amine-based cyanogen bromide method was introduced for this material [24], but it was
not until much later that the same method was adapted for use with silica [25].

Some supports can be used directly for immobilization, while others require a coating
or derivatization to give them suitable surface properties. Carbohydrate-based materials
used for low- or medium-performance affinity chromatography (e.g., agarose and cellulose)
can be activated directly, since they are already in a hydrophilic form that has low nonspecific binding for biomolecules. However, this is not the case with most supports for
high-performance affinity chromatography (HPAC), including silica, glass, and polystyrene-
based perfusion media. In these cases, the untreated supports may either be hydrophobic
(e.g., polystyrene) or have strong polar interactions with biomolecules (e.g., silica and
glass). These difficulties can be overcome by placing a more appropriate covering on the
support's surface. For glass and silica, this is generally accomplished by converting them
into a diol-bonded form. For polystyrene, a glycol coating is similarly used to produce a
hydroxylated surface.

3.2.2.2   Specificity and Stability of the Immobilization Method

As noted earlier, the specificity of an immobilization method can play a large role in determining the final activity of a ligand. Most immobilization methods involve reactions with amine, sulfhydryl, or hydroxyl groups [26]. The specificity of these methods will be determined by the relative reactivity of these groups and the coupling conditions. For example, the reaction of maleimides with sulfhydryl groups near a neutral pH (6.5 to 7.5) gives a rate of reaction 1000 times faster than it is for amines. However, as the pH is increased, the reaction of maleimide with amines becomes more significant, and this reagent becomes less specific for sulfhydryl groups.

When both the relative abundance and reactivity of the functional groups on a protein are considered, it is generally the amine groups that are chosen for covalent modification. Coupling through these groups is usually performed under slightly basic conditions (pH 9.0 to 10.0), but lower pHs can also be employed. Although many amines can be present on a single protein, immobilization based on these residues can be made specific if the pH, temperature, and other conditions are carefully optimized. For instance, it has been found in the conjugation of fluorescein isothiocyanate to insulin that one to four tags can be placed on insulin by adjusting the reaction conditions [27].

The stability of the bonds or forces that hold the ligand onto a support is another item to consider when selecting an immobilization method. As would be expected, a covalent bond between the ligand and support is usually more stable than a noncovalent bond or coordination complex. However, there are also differences in the stability of bonds produced by covalent immobilization. As an example, thioester linkages are much less stable than esters, which are less stable than substituted amines [28].

3.2.2.3   Rate of Immobilization

A factor often overlooked in immobilization is the rate of this process. This is particularly important to consider when dealing with nucleophilic reactions in water, since the rate of ligand immobilization may compete with the loss of activated sites due to hydrolysis. Some immobilization methods, like those based on N-hydroxysuccinimide (NHS) ester or maleimide, have fast rates of ligand attachment along with relatively fast rates of hydrolysis. Other methods, like reductive amination, may take longer to couple ligands but also have better long-term stability for the activated groups on the support.

The rate of immobilization will be affected by the specificity of this process and by the number of sites on the ligand that can take part in this reaction. This is illustrated in Figure 3.3, which shows how the rate of immobilization changes for antibodies as the average number of reactive sites on these ligands is varied [16]. In this case, the reactive sites were aldehyde groups in the antibody carbohydrate chains that were produced by periodate oxidation. When there was an average of less than one reactive site per antibody, immobilization was relatively slow, and a significant fraction of ligand remained in solution even after 14 days. When more aldehyde groups were placed on the antibodies, complete immobilization was observed in 1 to 2 days, with the rate increasing with the number of available coupling sites.

3.2.2.4   Other Considerations

Several other factors can play a role in the selection of an immobilization method. For instance, it is desirable to have an immobilization technique that is straightforward and reproducible. The easiest such approaches are noncovalent methods, but these are also the least specific for ligand attachment. Methods like covalent immobilization are generally more selective but involve more steps, more reagents, and require a better understanding of the reaction.

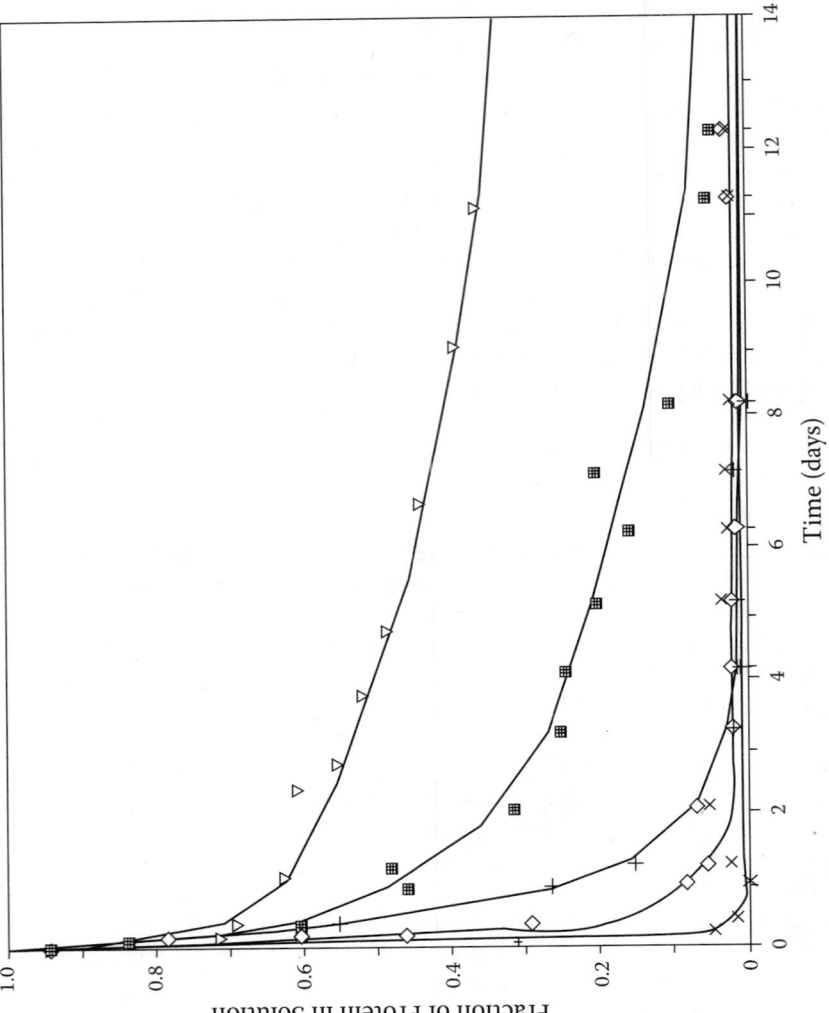

Figure 3.3 Rate of immobilization seen for the reaction of a dihydrazide-activated support with oxidized rabbit IgG antibodies that contained various amounts of reactive aldehyde groups. The average number of aldehyde groups on these oxidized antibodies were (from bottom to top) 4.6, 3.6, 2.0, 0.9, and 0.5.

The cost of the immobilization method is an additional item to consider. Ideally, the reagents and supports required should be inexpensive and readily available. The cost and amount of ligand that will be needed should be considered as well. Noncovalent techniques tend to require periodic treatment of the support with new amounts of ligand, which can slowly desorb over time. This makes covalent immobilization more appealing for rare or expensive ligands because they are attached in a more stable manner to the support.

Safety is another factor to address when selecting an immobilization technique. Many covalent procedures require the use of volatile or flammable chemicals, since their activation step is usually performed under nonaqueous conditions. The reagents or products of the immobilization process may also be volatile and toxic, as occurs in the cyanogen bromide and tosyl chloride procedures. Other chemicals used during immobilization (e.g., epichlorohydrin, acrylamide, and sodium borohydride) can be carcinogenic, mutagenic, or strong reducing agents. Thus, proper facilities for the use and storage of these chemicals should always be employed.

3.2.3 Characterization of Immobilized Ligands

Once immobilization has been completed, it is helpful to perform tests to ensure that the ligand has been attached to the support and has the desired activity. This information, in turn, is valuable in optimizing use of the ligand in an affinity separation. This section examines some general approaches utilized in this characterization, including methods for determining the amount of an immobilized ligand, the binding capacity of an affinity support, and the thermodynamic and kinetic properties of the ligand.

3.2.3.1 Amount of Immobilized Ligand

There are a variety of methods for measuring the amount of a ligand on an affinity support. For instance, the bicinchoninic acid (BCA) assay is a well-known method for examining immobilized proteins [29]. In this method, Cu^{2+} is reduced to Cu^+ in the presence of proteins or peptides, with the Cu^+ forming a colored complex that can be detected at 562 nm.

Ligands with amine groups can also be measured using ninhydrin [30]. This is based on the reduction of ninhydrin and the subsequent formation of a colored product that is detected at 570 nm. Besides being used to measure ligands, both the BCA method and ninhydrin can be employed in studying supports that contain amines or related groups (e.g., hydrazide residues).

If an adequate amount of protein is present on a support, it can be examined by placing the support into a refractive-index-matched medium and using absorbance spectroscopy. This is often accomplished by placing the support in a 50% solution of glycerol in water [9] or in a saturated sucrose solution. This approach can also be used to measure the active sites on some supports, such as dihydrazide-activated silica [15]. Another approach for measuring immobilized proteins is to look at their binding to a suitable dye. Coomassie Brilliant Blue is one dye used for this purpose [31].

DTNB (i.e., 5,5-dithio-*bis*-(2-nitrobenzoic acid), or Elman's reagent) can be used to measure a ligand with free sulfhydryl groups [32]. In this method, DTNB reacts with a sulfhydryl to produce a mixed disulfide and 2-nitro-5-thiobenzoic acid (TNB) under slightly alkaline conditions. The TNB is then measured at 412 nm. This method can be used for ligands with disulfide bonds if they are first reduced with sodium borohydride. In addition, DTNB can be used to examine thiol groups on activated supports.

Trinitrobenzenesulfonic acid (TNBS) is another reagent used to examine ligands and supports. TNBS can measure amine, hydrazide, or sulfhydryl groups [33]. For some affinity supports, a specific reagent may be available for ligand detection. A good example is the use of biotin-*p*-nitrophenyl ester for examining immobilized avidin [34]. In this case, the reagent is first allowed to bind avidin, with the bound reagent then being treated under basic conditions to release *p*-nitrophenol for detection.

3.2.3.2 Binding Capacity

A second item to characterize for an affinity support is its *binding capacity*. This refers to the amount of target solute that can bind to the support under the desired application conditions. Although the binding capacity is related to the amount of immobilized ligand, it is generally smaller than this value, since not all the ligands may be active or have proper orientation and spacing for binding to occur.

One way the binding capacity can be determined is by incubating the affinity support with a solution of the target. The amount of target remaining in solution is then measured by a technique like absorbance or fluorescence spectroscopy [35, 36]. Next, this value is compared with the total known quantity of the added target to give the amount that has been bound by the support. Alternatively, some targets (e.g., those with radiolabels) can be measured directly on the support after they have been removed from any nonadsorbed chemicals in solution. Either of these approaches provides a value known as the *static binding capacity*. This represents the maximum amount of solute that can be adsorbed by the ligand at equilibrium.

The binding capacity of an affinity column can also be determined by *frontal analysis* [1, 3]. In this method, a solution containing a known concentration of solute is continuously applied to an affinity column. As the solute binds to the immobilized ligand, the column becomes saturated, and the amount of solute eluting from the column gradually increases, forming a breakthrough curve. If fast association and dissociation kinetics are present, the mean position of this curve will be related to both the amount of active ligand in the column and the association equilibrium constant for the solute. Further information on this method and its use can be found in Chapters 22 and 23.

If frontal analysis is performed at a relatively slow flow rate, it should provide the static binding capacity of the system. However, as the flow rate is increased, kinetic effects will prevent some of the applied solute from having sufficient time to interact with the immobilized ligand. This gives an apparent capacity that is less than that expected at equilibrium. This flow-dependent value is known as the *dynamic binding capacity* and is important to consider when affinity columns are used at high flow rates or in systems for large-scale processing. A further discussion of binding capacity and its role in determining the performance of an affinity column is given in Chapter 11.

3.2.3.3 Ligand Affinity and Adsorption Rate

An item related to the binding capacity is the equilibrium constant for the interaction of a solute with the immobilized ligand. This is commonly represented by using either the *association equilibrium constant* (K_a) or *dissociation equilibrium constant* ($K_d = 1/K_a$) for the solute-ligand interaction. This is one factor that determines the static binding capacity of an affinity support [9]. As a result, the measurement of binding capacity, whether it involves static measurements or frontal analysis, can often be used to give both the amount of immobilized ligand in an affinity column and the equilibrium constant for this reaction, as described in Chapters 11 and 22.

When dealing with a dynamic binding capacity, the binding rate for the target also needs to be considered. This depends on the rates of mass transfer for the target within the column and the association rate for this compound to the ligand. Some methods that can be used for examining these rates include (1) flow-based biosensors, (2) frontal analysis, and (3) band-broadening, split-peak, or peak-decay analysis. More details on these approaches can be found in Chapters 22, 23, and 25.

3.3 NONCOVALENT IMMOBILIZATION TECHNIQUES

Some of the earliest immobilization methods were based on noncovalent attachment. This can involve the simple adsorption of a ligand to a surface, binding to a secondary ligand, or ligand immobilization through a coordination complex. Each of these approaches are examined in this section.

3.3.1 Nonspecific Adsorption

Nonspecific adsorption is based on the attachment of a ligand to a support that has not been specifically functionalized for covalent attachment [37, 38]. This is accomplished as shown in Figure 3.4. Adsorption of the ligand to a support depends on the chemical characteristics of both the ligand and support. Forces involved in this process can include coulombic interactions, hydrogen bonding, and hydrophobic interactions.

Nonspecific immobilization has been used for a long time [39–51]. Early work in this field was conducted by Stone, who patented procedures for preparing immobilized enzymes [52]. Since that time, there have been numerous supports using nonspecific adsorption for affinity ligands. Several workers have reported using natural polymers based on polysaccharides (e.g., dextran) and proteins (e.g., collagen and gelatin), as well as inorganic materials like alumina and silica [53–63]. These supports can be used directly or after treatment to enhance their adsorption of a ligand. For example, the adsorption of some enzymes on cross-linked dextran is too weak for use in noncovalent immobilization; however, if hydrophobic groups are first placed in these gels, strong adsorption for the same enzymes can be observed [64].

Methods for nonspecific adsorption can be placed into four categories: (1) the static procedure, (2) electrodeposition, (3) in-column loading, and (4) the mixing or shaking method [36]. The *static procedure* is the simplest but least efficient of these procedures. This involves placing the support in a solution of the ligand and allowing these to react

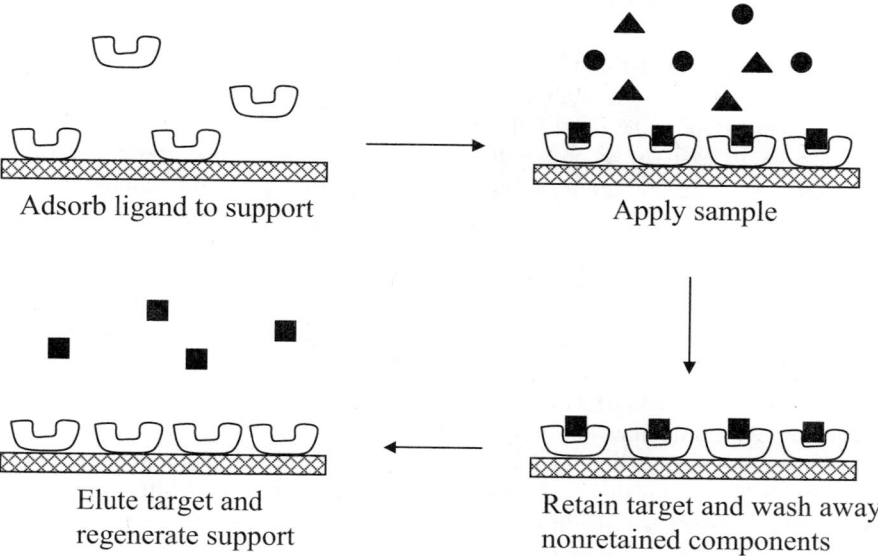

Adsorb ligand to support   Apply sample

Elute target and   Retain target and wash away
regenerate support   nonretained components

Figure 3.4 General process involved in the immobilization of a ligand through nonspecific adsorption.

under the control of diffusion without agitation or shaking. In *electrodeposition*, the support is placed proximal to an electrode in contact with a solution of the ligand. The ligand is then deposited on the support as a result of electromigration when a current is applied to this system. *In-column loading* is often used in industrial applications and involves the continuous introduction of a ligand to a support by means of a mechanical pump. The *mixing* or *shaking method* is similar to the static procedure, except the ligand and support are continuously agitated, resulting in a more rapid rate of immobilization and a more uniform distribution of the ligand.

The physical and chemical properties of the ligand are important to consider during nonspecific adsorption. For instance, it has been found in the adsorption of proteins to porous glass that the linear inclusion rate of a protein by this support is inversely proportional to the protein's mass [36, 65]. The same report suggested that the inclusion rate was dependent on the isoelectric point (pI) of a protein, where proteins with a high pI value give more rapid adsorption due to the formation of ionic bonds between positive groups on the protein and negative silanol groups on the support's surface [63].

Other factors that can affect the adsorption of ligands include the pH, ionic strength, and temperature. The pH and ionic strength are particularly important if coulombic interactions play a large role in ligand adsorption. For instance, a rapid decline in the apparent activity of a support containing chymotrypsin was noted above pH 9.6, which was attributed to the reduced adsorption of this enzyme as the pH went above its pI [66].

One problem with nonspecific adsorption is that it often causes a loss of ligand activity due to random orientation. This has been noted for trypsin adsorbed onto porous glass [67]. Some control of this effect can be achieved by adjusting the pH and ionic strength. Another means for minimizing loss of activity is to protect the ligand's active site by mixing this with a target prior to adsorption.

When using nonspecific adsorption, the activity of the affinity support should be carefully monitored during its use. If this activity falls below an acceptable level, the support can be treated to remove the old ligands, and a fresh batch of ligands can be applied. The relative simplicity of this process makes nonspecific adsorption attractive for the immobilization of inexpensive and abundant ligands.

3.3.2 Biospecific Adsorption

Another type of noncovalent immobilization method is *biospecific adsorption*. This makes use of the binding between the ligand of interest and a secondary ligand attached to the support. Although a variety of secondary ligands can be used for this purpose, two of the most common are avidin and streptavidin for the adsorption of biotin-containing compounds and protein A or protein G for the adsorption of antibodies [34, 68–72].

3.3.2.1 Avidin and Streptavidin

Avidin and streptavidin have become popular in many applications due to their ability to specifically bind biotin-containing molecules with a high affinity [68]. Figure 3.5 shows the general way in which these can be used to adsorb ligands to create affinity supports. This first involves the covalent immobilization of avidin or streptavidin to a support. A solution containing the ligand of interest with a biotin tag is then applied. As the biotin tag binds to the immobilized avidin or streptavidin, the ligand is captured and immobilized within the column.

The various properties of the biotin-avidin system are discussed in Chapter 5. Biotin is a 244-Da molecule found in tissue and blood. Due to this relatively small size, a biotin tag tends to have only a minimal effect on activity when it is attached to a large ligand. Avidin is a glycoprotein from egg whites that has a molecular weight of 68 kDa. It has

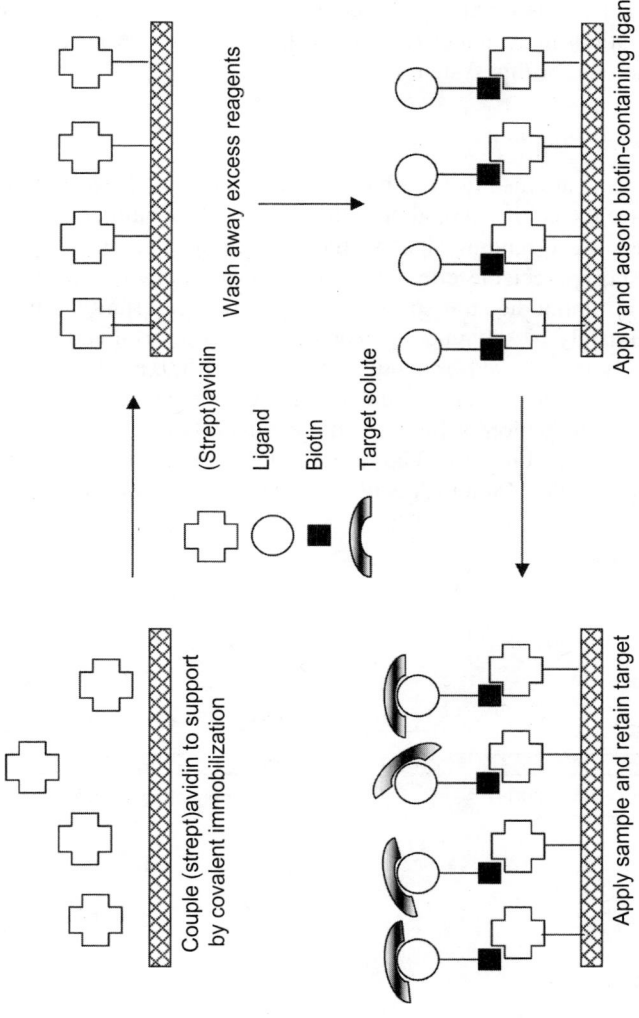

Figure 3.5 Use of avidin or streptavidin as a secondary ligand for the biospecific adsorption of biotin-containing ligands.

four identical subunits, each of which can bind one biotin with an affinity of 10^{15} M^{-1}. Streptavidin has similar binding properties but is obtained from *Streptomyces avidinii* and has a mass of 60 kDa. Like avidin, it has four identical subunits that can each bind one biotin; however, it has a slightly lower affinity for this interaction (i.e., 10^{13} M^{-1}). Advantages of avidin include its higher solubility than streptavidin [73] and its stability over a wide range of pHs and temperatures. In addition, avidin is less expensive than streptavidin, and extensive modification has little effect on its activity [74].

Both avidin and streptavidin can be immobilized on supports through amine-reactive methods. The fact that avidin is glycosylated means that it can also be immobilized through its carbohydrate residues. The binding of (strept)avidin with biotin was first exploited for histochemical applications in the mid-1970s [75, 76]. Since that time, several reviews have appeared on applications for this system [77–80].

3.3.2.2 *Protein A and Protein G*

Protein A and protein G are also used in biospecific adsorption. These are bacterial cell wall proteins that have the ability to bind the constant regions of antibodies. These regions are distant from the antigen binding sites, so the adsorption of an antibody to protein A or G tends to have no appreciable effect on the antibody's activity. The interactions of protein A and G with antibodies are specific and have a high affinity, with association constants of approximately 10^8 M^{-1} under physiological conditions. In addition, the bound antibodies can quickly be released by eluting at pH 2.5 to 3.0, as shown in Figure 3.6. These properties have made protein A and protein G popular for the purification of antibodies and for the production of immunoaffinity supports [69, 81–91].

Although protein A and protein G are used in the same manner, they do have some differences in their properties. Protein A is obtained from *Staphylococcus aureus* and has

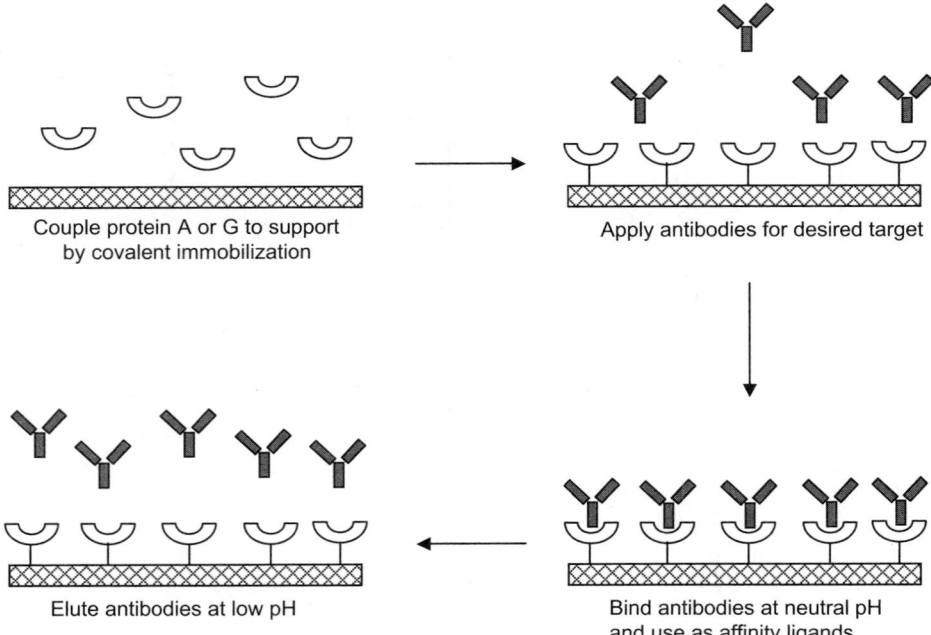

Figure 3.6 Use of protein A or protein G as ligands for the reversible capture of antibodies.

a molecular weight of 42 kDa. It is capable of binding many types of IgG-class antibodies, as well as some antibodies from other classes (see Chapter 5) [69, 92]. One molecule of protein A contains four identical sites and can bind at least two antibodies [93]. It is relatively stable to heat and retains its native conformation even after exposure to denaturing reagents like 4 M urea, 4 M thiocyanate, and 6 M guanidine hydrochloride [73].

Protein G is a surface receptor from *Streptococcus aureus*. It has a similar size to protein A. Protein G can bind IgG-class antibodies from numerous species, including all human and mouse IgG subclasses [70]. Unlike protein A, it does not recognize IgM, IgA, or IgE. Each protein G molecule can bind up to two antibodies. Although native protein G also has some affinity for albumin, recombinant forms of protein G are available in which the domain responsible for this binding has been removed.

The immobilization of protein A is usually performed through its amine groups by either reductive amination, the cyanogen bromide technique, or the glutaraldehyde method [94–98]. The same approaches are used for coupling protein G to supports. A further discussion of these immobilization methods can be found in Section 3.4 of this chapter.

3.3.3 Coordination Complexes

A coordination complex can be used on some occasions to prepare an immobilized ligand. This is used to place metal ions in columns for immobilized metal-ion affinity chromatography (IMAC). A detailed discussion of IMAC is given in Chapter 10.

IMAC is based on the formation of a complex between a metal ion and electron donor groups. The most common metal ions used for this purpose are Zn^{2+}, Cu^{2+}, Ni^{2+}, and Co^{2+}, which are all electron-pair acceptors. Such ions can be placed within a column by using an immobilized chelating group like iminodiacetic acid (IDA) or nitrilotriacetic acid (NTA). The remaining coordination sites on the metal ion are normally occupied by water, but this can be exchanged with electron donor groups on a protein or peptide [99]. Although several amino acids have electron donor capabilities (e.g., tryptophan, histidine, tyrosine, and cysteine), histidine is the main amino acid that interacts with chelated metal ions at a neutral pH.

Figure 3.7 shows how the chelating agent IDA can be attached to a support for use in IMAC. This chelating group is then allowed to complex with the desired metal ion by passing a solution of this ion through the IMAC column. Commercial IMAC supports are generally prepared by reacting an epoxide group on the support with the amine of IDA. Other chelating compounds that can be coupled to supports by this approach include 1,4,7-triazocyclononane (TACN), 2,6-diaminomethylpyridine, and *tris*(2-aminoethyl)amine (TREN) [100–102].

3.4 COVALENT IMMOBILIZATION METHODS

Covalent immobilization is the most popular means for placing a ligand in an affinity column. To do this, it is necessary to first activate the ligand and/or the support. Activation of the ligand can be employed when it is desired to couple this ligand through a specific region. An example is the creation of aldehydes in the carbohydrate regions of an antibody for its attachment to a support that contains amines or hydrazide groups. The use of an activated support is more common for ligand immobilization but tends to be less specific in nature. Examples include the immobilization of proteins through their amine groups to supports activated with *N*-hydroxysuccinimide or carbonyldiimidazole.

The support used for covalent immobilization must meet several requirements. First, it should have a sufficient number of groups for activation and ligand attachment. Hydroxyl

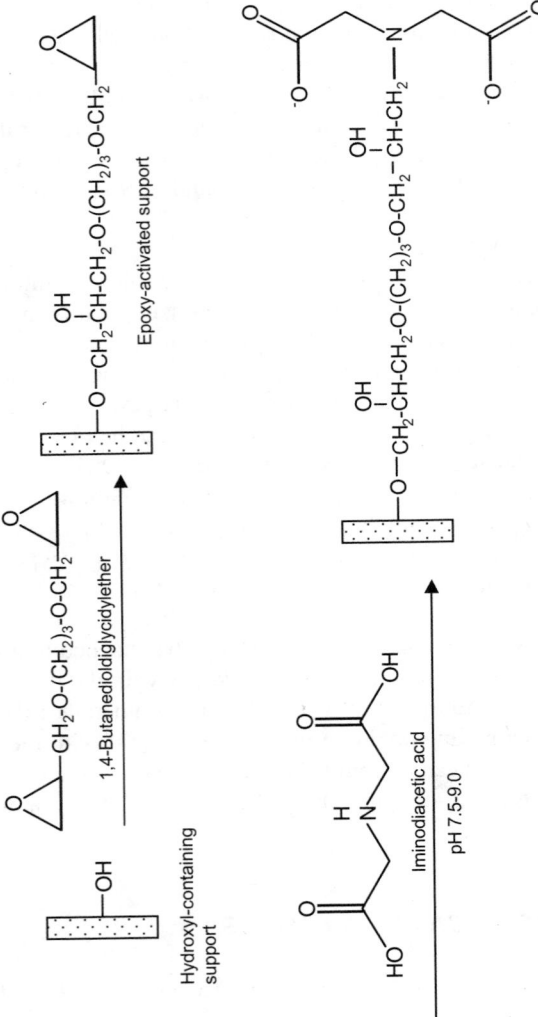

Figure 3.7 Method for the attachment of iminodiacetic acid (IDA) to a support for use in immobilized metal-ion affinity chromatography.

groups on the support are employed in most covalent coupling methods. Depending on how its surface is activated, a support can be used to immobilize ligands through their amine, sulfhydryl, hydroxyl, or carbonyl groups, among others. The methods used for each of these groups will now be examined in more detail.

3.4.1 Amine-Reactive Methods

The use of amine groups is often employed for the immobilization of proteins and peptides [9, 28]. Table 3.2 lists several techniques for this approach. Specific methods discussed will include the cyanogen bromide method, reductive amination, the N-hydroxysuccinimide technique, and the carbonyldiimidazole method. A more detailed description of the experimental protocols for these techniques and other amine-reactive methods can be found in the references given in Table 3.2.

Table 3.2 Covalent Immobilization Methods for Affinity Chromatography

Group/Compound	Immobilization Technique	References
Amine groups	Cyanogen bromide (CNBr) method	[103–109]
	Schiff base (reductive amination) method	[110–116]
	N-hydroxysuccinimide (NHS) method	[118–126]
	Carbonyldiimidazole (CDI) method	[129–134]
	Cyanuric chloride method	[135–139]
	Azalactone method (for Emphaze supports)	[23]
	Divinylsulfone (DVS)	[140]
	Epoxy (bisoxirane) method	[141]
	Ethyl dimethylaminopropyl carbodiimide (EDC) method	[142]
	Tresyl chloride/tosyl chloride method	[143]
Sulfhydryl groups	Azalactone method (for Emphaze supports)	[9]
	Divinylsulfone method	[9, 139, 165]
	Epoxy (bisoxirane) method	[9]
	Iodoacetyl/bromoacetyl method	[146–151]
	Maleimide method	[152–157]
	Pyridyl disulfide method	[158–164]
	TNB-thiol method	[165, 166]
	Tresyl chloride/tosyl chloride method	[9]
Hydroxyl groups	Cyanuric chloride method	[165]
	Divinylsulfone method	[9, 28]
	Epoxy (bisoxirane) method	[9, 28]
Aldehyde groups	Hydrazide method	[169–174]
Carboxyl groups	Ethyl dimethylaminopropyl carbodiimide (EDC) method	[175–177]

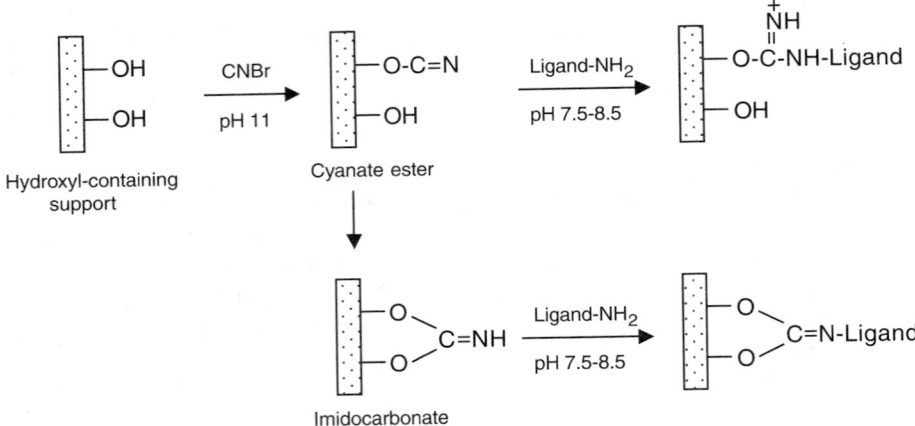

Figure 3.8 Possible pathways in the cyanogen bromide (CNBr) immobilization method when used for an amine-containing ligand (ligand-NH$_2$).

3.4.1.1 Cyanogen Bromide Method

The *cyanogen bromide (CNBr) method* was the first technique used on a large scale for immobilizing amine-containing ligands [103–109]. This approach is illustrated in Figure 3.8. It involves the derivatization of hydroxyl groups on the surface of a support to form an active cyanate ester or an imidocarbonate group [9]. Both of these active groups can couple ligands through primary amines, but the cyanate ester is more reactive than the imidocarbonate.

The CNBr method uses relatively mild conditions for ligand attachment, making it suitable for many sensitive biomolecules. But one problem with this approach is that the isourea linkages obtained by the reaction of CNBr with the support are positively charged at a neutral pH. This means that these groups can act as anion exchangers and give rise to nonspecific binding. Other problems with this method include the toxicity of CNBr, requiring the use of adequate safety precautions during the activation process, and the leakage of ligands that can result from CNBr-activated supports.

3.4.1.2 Reductive Amination

Reductive amination (also known as the *Schiff base method*) couples ligands with amine groups to an activated support that contains aldehyde residues [110–116]. The reactions involved in this process are shown in Figure 3.9. During the activation step, periodic acid (H$_5$IO$_6$, or HIO$_4$·2H$_2$O) or sodium periodate is used to oxidize diol groups on the support's surface to give aldehydes. This can be performed directly on carbohydrate-based supports like dextran or cellulose. However, materials like silica or glass must first be treated to place diols on their surface. This can be accomplished by reacting the silica or glass with γ-glycidoxypropyltrimethoxysilane, followed by acid hydrolysis [15, 117].

When an amine-containing ligand reacts with the aldehyde groups, the resulting product is known as a Schiff base. Since this is a reversible reaction, the Schiff base must be converted into a more stable form. This is accomplished by including sodium cyanoborohydride in the reaction mixture. Cyanoborohydride is a weak reducing agent that converts the Schiff base into a secondary amine without affecting the aldehydes on the support. After the coupling reaction has been completed, the remaining aldehyde groups can be removed by treating the support for a short period of time with a stronger reducing agent (i.e., sodium

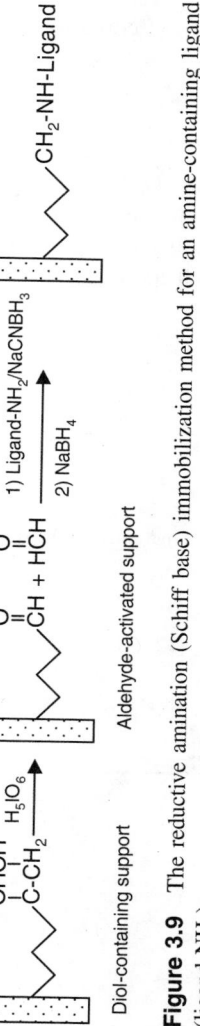

Figure 3.9 The reductive amination (Schiff base) immobilization method for an amine-containing ligand (ligand-NH$_2$).

borohydride) or by reacting these groups with an excess of a small amine-containing agent (e.g., ethanolamine).

The Schiff base method is relatively easy to perform and often gives a higher ligand activity than other amine-based coupling methods. This also results in ligands that have stable linkages to the support and that can be used for long periods of time. However, there are some disadvantages to this method. One is the need to work with relatively hazardous agents such as sodium cyanoborohydride and sodium borohydride. Thus, care must be taken to perform this technique with proper ventilation and safety precautions. The use of sodium borohydride for the removal of excess aldehyde groups must also be performed with caution, since the use of conditions that are too harsh may result in the loss of ligand activity.

3.4.1.3 N-Hydroxysuccinimide Method

The *N-hydroxysuccinimide (NHS) method* is another technique often employed when immobilizing biomolecules through amine groups. This gives rise to the formation of a stable amide bond. There are a number of ways a support can be activated with NHS [118–126]. Figure 3.10 shows one such approach, starting with an amine-containing support

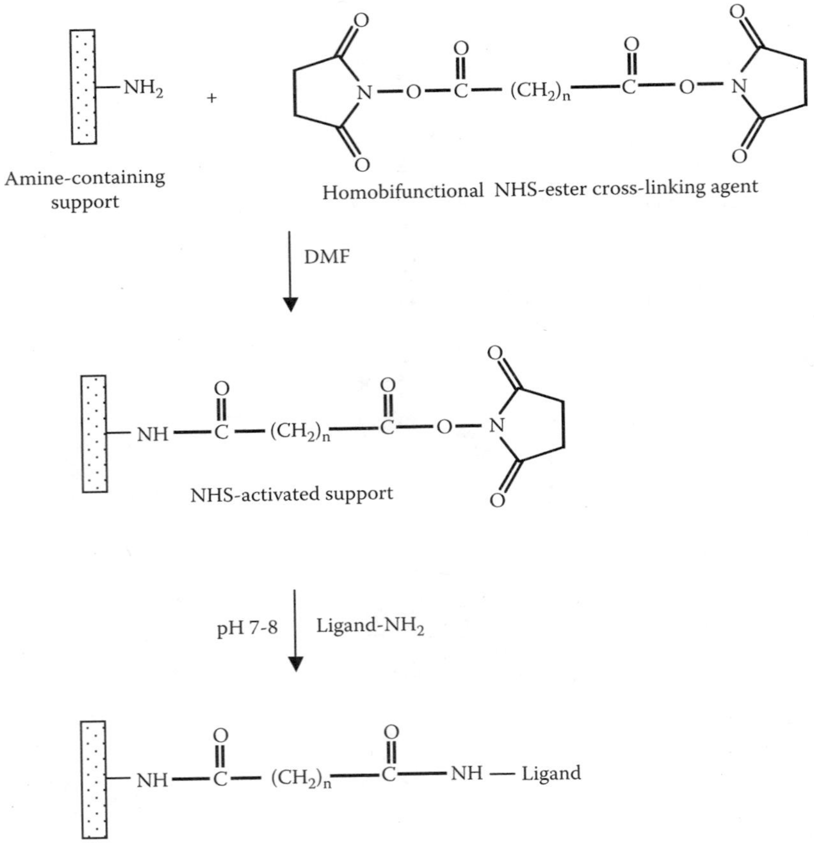

Figure 3.10 The *N*-hydroxysuccinimide (NHS) immobilization method for an amine-containing ligand (ligand-NH$_2$).

Table 3.3 Homo- and Heterobifunctional Cross-Linking Agents Containing NHS Esters

Cross-Linking Agent	Reacts with	Comments
Disuccinimidyl glutarate (DSG)	Amine	Noncleavable with five-carbon spacer
Disuccinimidyl suberate (DSS)	Amine	Noncleavable with eight-carbon spacer
Bis-(sulfosuccinimidyl) suberate (BS)	Amine	Water-soluble analog of DSS
Dithio-*bis*-(succinimidyl propionate) (DSP)	Amine	Thiol cleavable
Ethylene glycobis (succinimidyl succinate) (EGS)	Amine	Cleavable with 1 *M* hydroxylamine
Succinimidyl 4-(*N*-maleimidomethyl) cyclohexane-carboxylate (SMCC)	Sulfhydryl	Noncleavable, water insoluble
Sulfosuccinimidyl 4-(*N*-maleimidomethyl) cyclohexane-carboxylate (sulfo-SMCC)	Sulfhydryl	Noncleavable, water-soluble analog of SMCC
N-succinimidyl (4-iodoacetyl) aminobenzoate (SIAB)	Sulfhydryl	Involves an active halogen (iodoacetate)
N-succinimidyl-3-(2-pyridyldithio) propionate (SPDP)	Sulfhydryl	Makes use of disulfide exchange

and using a bifunctional cross-linking agent that contains NHS on at least one of its two ends. Examples of other bifunctional agents that can be used for this purpose are given in Table 3.3.

The relative ease with which activated supports can be prepared is one advantage of the NHS method. But the fast hydrolysis of NHS esters tends to compete with the immobilization of ligands. This rate of hydrolysis increases with pH and is particularly important when dealing with dilute protein solutions. The half-life of these NHS groups at pH 7 and 0°C is approximately 4 to 5 h [127] and decreases to as little as 10 min at pH 8.6 and 4°C [128].

3.4.1.4 Carbonyldiimidazole Method

Carbonyldiimidazole (CDI) can also be used to activate supports for the immobilization of amine-containing ligands. This reagent can react with materials that contain hydroxyl or carboxyl groups on their surface (see Figure 3.11) [129–134]. Supports with carboxyl groups will react with CDI to produce an acylimidazole, which can react with primary amines on a ligand to give an amide linkage. Supports with hydroxyl groups will react with CDI to produce an imidazolylcarbamate, which reacts with primary amines at pH 8.5 to 10.0.

The CDI method is relatively simple and easy to perform. In addition, supports that have been activated to produce imidazolylcarbamate groups are more stable to hydrolysis than those activated by the NHS method. A CDI-activated support is stable when stored in dry dioxane, with a half-life of greater than 14 weeks [28]. Another advantage of this method is that the amide linkages formed by this technique (as well as those created by the NHS method) are more stable than the isourea linkages obtained by CNBr immobilization. One disadvantage of the CDI method is that it tends to produce ligands with a lower activity than alternative techniques (e.g., reductive amination).

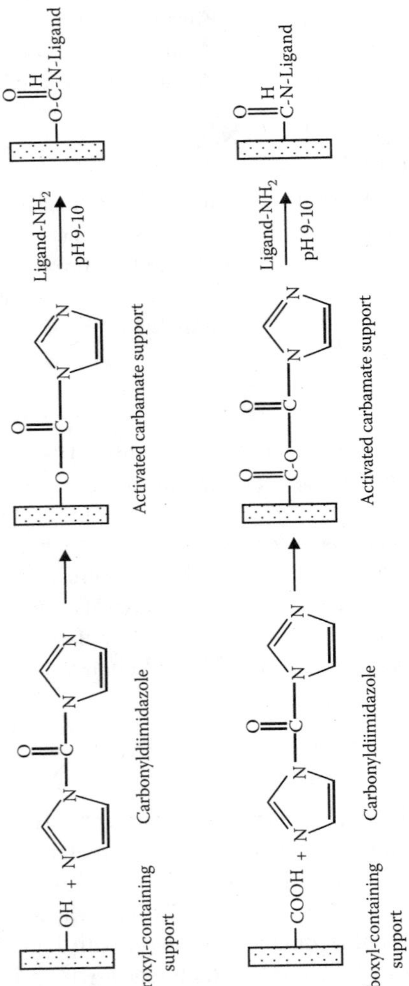

Figure 3.11 The carbonyldiimidazole (CDI) immobilization method for an amine-containing ligand (ligand-NH_2).

3.4.1.5 Other Methods

Table 3.2 lists many other techniques that can be used for immobilizing amine-containing ligands. One example is the use of cyanuric chloride (or 2,4,6-trichlorotriazine) to activate hydroxyl- or amine-containing supports for ligand attachment [135–139]. Cyanuric chloride has been widely employed as a cross-linking agent and as a reagent for protein modification. It has three reactive acyl-like chlorines, each of which has a different chemical reactivity. The first chlorine is reactive toward hydroxyls and amines at 4°C and pH 9. After the first chlorine has reacted, the second requires a slightly higher temperature for its reaction (20°C), and the third chlorine needs an even higher temperature (80°C).

Other techniques for amine-containing ligands include the azalactone [23], divinyl-sulfone [140], bisoxylane [141], ethyldimethylaminopropyl carbodiimide [142], and tresyl chloride-tosyl chloride methods [143]. Some of these methods are specific for certain supports (e.g., the azalactone method), while others can be used with a variety of materials. In choosing between these approaches, the final selection will often depend on the type of ligand being immobilized, the support desired for this ligand, and the conditions that can be tolerated by both the ligand and support during the immobilization process.

3.4.2 Sulfhydryl-Reactive Methods

The use of sulfhydryl groups on ligands is another approach for preparing affinity supports. Examples of such methods are listed in Table 3.2. If a ligand has a free sulfhydryl group on its surface, using this group is advantageous, since it often gives site-specific immobilization and a cleavable product. If the ligand is a protein or peptide that has no free sulfhydryl groups but that does have a disulfide bond, this bond can be reduced to allow ligand attachment. It is also possible to introduce sulfhydryl groups on a ligand by thiolating amines or carboxyl groups [144]. Table 3.4 gives several reagents that can be used for this purpose.

Table 3.4 Reagents for Introducing Specific Functional Groups on a Ligand or Support

Reagent	Reactive Site on Ligand	Final Added Group
2-Iminothiolane	Amine	Sulfhydryl
Succinimidyl acetylthiopropionate	Amine	Sulfhydryl
Cystamine	Carboxyl or phosphate	Disulfide
Dithiothreitol (DTT)	Disulfide	Sulfhydryl
Anhydride	Amine	Carboxyl
Succinic anhydride	Amine	Carboxyl
Maleic anhydride	Amine	Carboxyl
Diamines with EDC	Carboxyl or carbohydrate	Amine
Diamine	Aldehyde	Amine
N-(iodoethyl)trifluoroacetamide	Sulfhydryl	Amine
Sodium periodate	Glycol or carbohydrate	Aldehyde
Succinimidyl p-formylbenzoate (SFB)	Amine	Aldehyde
Glutaraldehyde	Amine	Aldehyde
Adipic dihydrazide (*bis*-hydrazide)	Aldehyde	Hydrazide
Bis-hydrazide with EDC	Carboxylate	Hydrazide

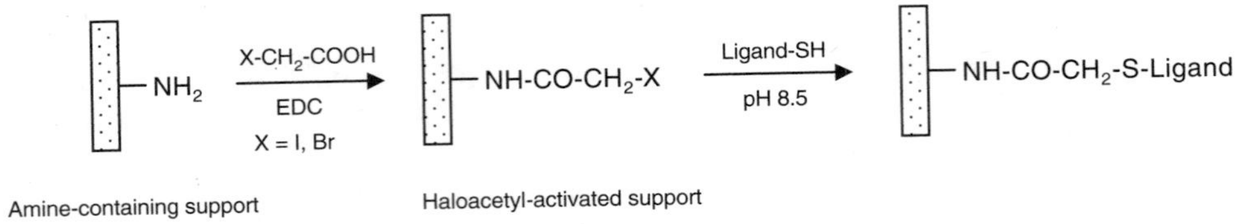

Amine-containing support   Haloacetyl-activated support

Figure 3.12 General scheme for the haloacetyl immobilization method, in which the activated support is shown reacting with a sulfhydryl-containing ligand (ligand-SH).

A support can be activated in several ways for the immobilization of ligands through sulfhydryl groups. Unlike amine-reactive methods, where hydroxyl groups on the support are generally used, most sulfhydryl-reactive methods require the introduction of an amine, carboxyl group, or some other intermediate site onto the support. For example, silica can not be used directly with sulfhydryl-reactive methods but must be reacted with aminopropyltriethoxysilane or mercaptopropyltrimethoxysilane to convert it into a suitable form [145].

There are various approaches that can be used to immobilize ligands with sulfhydryl groups. The following subsections examine some of these techniques, including the haloacetyl, maleimide, and pyridyl disulfide methods. More details on these and related techniques can be found in the references given in Table 3.2.

3.4.2.1   Haloacetyl Method

The haloacetyl method uses supports that contain iodoacetyl or bromoacetyl groups for the immobilization of ligands through sulfhydryl residues [146–151]. These supports are usually prepared by reacting an amine-containing material with iodoacetic or bromoacetic acid in the presence of ethyldimethylaminopropyl carbodiimide (EDC) at pH 4 to 5, as shown in Figure 3.12. In this reaction, EDC reacts with the carboxylic acid in iodo- or bromoacetic acid to form a reactive ester, which can react with primary amine groups on the support.

The second part of this process involves combination of the haloacetyl-activated support with a ligand containing a sulfhydryl group. This reaction proceeds by nucleophilic substitution and produces a thioether. The resulting bond is comparable in stability to an amide linkage. Although the reactivity of haloacetyl-activated supports toward sulfhydryls is relatively selective, these can react with methionine, histidine, or tyrosine under appropriate conditions. If the immobilization is carried out above pH 8, amines can also react with these supports.

3.4.2.2   Maleimide Method

Maleimides are another group of reagents employed for the selective coupling of a ligand through sulfhydryl groups [152–157]. These tend to be more selective than a haloacetyl for such a reaction. The activation of a support with a maleimide is accomplished by using a homobifunctional or heterobifunctional cross-linking agent. One agent employed for this purpose is *bis*-maleimidohexane (BMH), which is a homobifunctional cross-linker with a maleimide group on both ends. The first of these groups can react with a support that has a sulfhydryl group. After the excess BMH has been washed away, the maleimide at the other end can react with a sulfhydryl group on a ligand. This occurs through the addition of a thiol group across the double bond of maleimide to produce a thioether linkage.

Heterobifunctional cross-linkers can be used in place of BMH if the support does not contain any sulfhydryl groups on its surface. If the support contains amine groups,

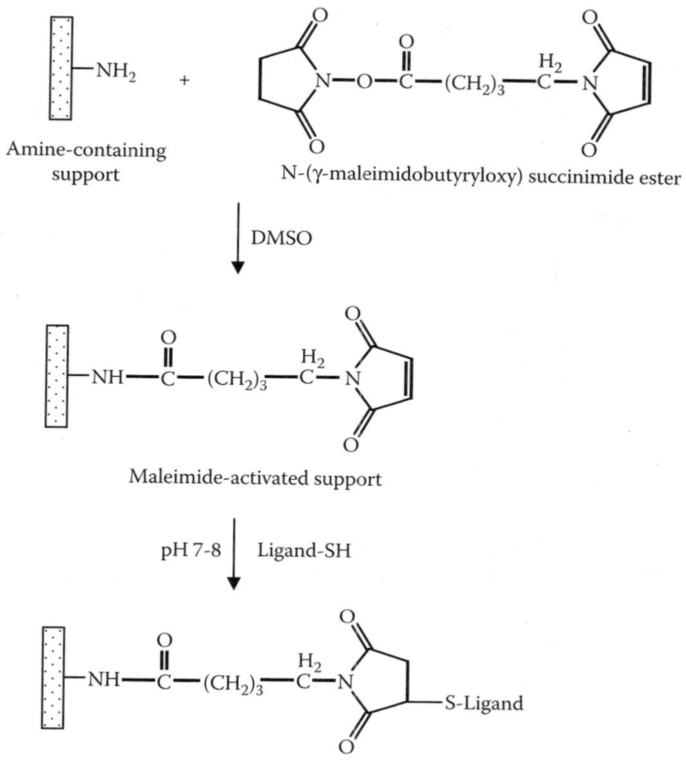

Figure 3.13 An example of a maleimide immobilization method for a sulfhydryl-containing ligand (ligand-SH). This particular reaction uses the cross-linking agent *N*-(γ-maleimidobutyryloxy) succinimide ester (GMBS) to attach the ligand to a support that originally has amine groups on its surface.

cross-linking agents like *N*-(γ-maleimidobutyryloxy) succinimide ester (GMBS) can be employed. This takes place through the reaction illustrated in Figure 3.13, where the NHS group of GMBS reacts with amine groups on the support to activate this material for ligand attachment. A sulfhydryl group on the ligand can then react with the double bond on the cross-linking agent to form a thioether bond.

The low solubility of BMH and GMBS limits the use of these reagents under aqueous solutions. However, more water-soluble reagents are available [73]. One of these is sulfosuccinimidyl-4-(*N*-maleimidomethyl) cyclohexane-1-carboxylate, or sulfo-SMCC. Sulfo-SMCC has a sulfo-NHS group on one end that can react with an amine-containing support. The other end contains a maleimide group that can react with ligands that have sulfhydryl residues. One problem with this and other water-soluble cross-linking agents is that they are often susceptible to hydrolysis, giving them a low coupling efficiency. To minimize this hydrolysis, the ligand should be added immediately after the support has been activated and excess cross-linking reagent has been removed.

One precaution that must be followed when using maleimide-activated supports is that these can react with amine groups if the pH is above 8. Hydrolysis of the maleimide groups can also be significant and compete with ligand immobilization under these pH conditions. However, maleimide-activated materials are essentially specific for sulfhydryl groups if the pH is between 6.5 and 7.5.

3.4.2.3 Pyridyl Disulfide Method

Pyridyl disulfide (or 2,2′-dipyridyldisulfide) is a homobifunctional cross-linking agent used for immobilizing ligands with sulfhydryl groups to supports that contain sulfhydryls on their surface [144]. Activation of the support is accomplished by disulfide exchange between the sulfhydryl groups on the support and pyridyl disulfide, giving rise to the release of pyridyl-2-thione, as shown in Figure 3.14. Further disulfide exchange between the activated support and ligand gives a disulfide linkage [158–164].

This immobilization reaction is usually carried out at pH 4 to 5. It is possible to use a neutral or slightly alkaline pH, but this also gives a slower reaction. The by-product of this reaction, pyridyl-2-thione, can be measured by using its absorbance at 343 nm, thus giving a means for determining how many bonds have formed between the ligand and support. Another useful feature of this method is that the immobilized ligand can later be cleaved and removed from the support by using a reducing agent like mercaptoethanol or $NaBH_4$.

3.4.2.4 Other Methods

Table 3.4 shows a number of other techniques that can be used for immobilizing sulfhydryl-containing ligands. For instance, divinylsulfone (DVS) can be used to activate a hydroxyl-containing support by introducing a reactive vinylsulfonyl group on its surface at pH 10 to 11 [139, 165]. This support can then be reacted with ligands that contain sulfhydryl, amine, or hydroxyl groups [9], with the rate of this reaction following the order –SH > –NH > –OH. Although the resulting bond for a sulfhydryl group is labile, the linkage for amine-containing ligands is more stable [164].

3.4.3 Hydroxyl-Reactive Methods

A number of methods have been used to couple ligands through hydroxyl groups [9, 166–168]. Examples are given in Table 3.4. However, unlike many amine- and sulfhydryl-reactive methods, techniques for hydroxyl-containing ligands are not as selective. For example, the divinylsulfone method can be used for coupling an amine-, sulfhydryl-, or hydroxyl-containing ligand.

Many supports used in affinity chromatography already contain hydroxyl groups on their surface. One way these groups can be activated is by introducing bisoxirane (epoxy) groups. The most frequently used oxirane for this purpose is 1,4-butanediol diglycidyl ether, which contains two epoxy groups. One epoxy can react with the hydroxyl groups on a support while the other is used for coupling ligands containing sulfhydryl, amine, or hydroxyl groups. The reactions involved in this process are shown in Figure 3.15. The reactivity of the terminal epoxide to other groups follows the order –SH > –NH > –OH. Strong alkaline conditions (pH 11) allow for coupling by this method through hydroxyl groups, while amines and sulfhydryl groups can react at a lower pH (pH 7 to 8) [9].

Cyanuric chloride is another agent used for attaching a hydroxyl-containing ligand to a support. This can only be used effectively in the absence of amine groups due to the higher reactivity of these groups in this method [165]. As mentioned earlier, divinylsulfone can be used for coupling hydroxyl-containing ligands. This, however, is not usually performed if the immobilized ligand is present at a pH higher than 9 to 10 [9].

3.4.4 Carbonyl-Reactive Methods

Although most immobilization techniques involve coupling ligands through amine or sulfhydryl groups, the large number of such groups can create a problem with improper orientation or multipoint attachment. This can be avoided by using alternative groups that

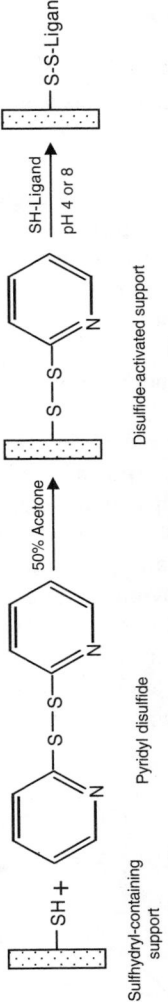

Figure 3.14 The pyridyl disulfide method for the immobilization of a ligand with sulfhydryl groups (ligand-SH) to a support that also contains sulfhydryl groups.

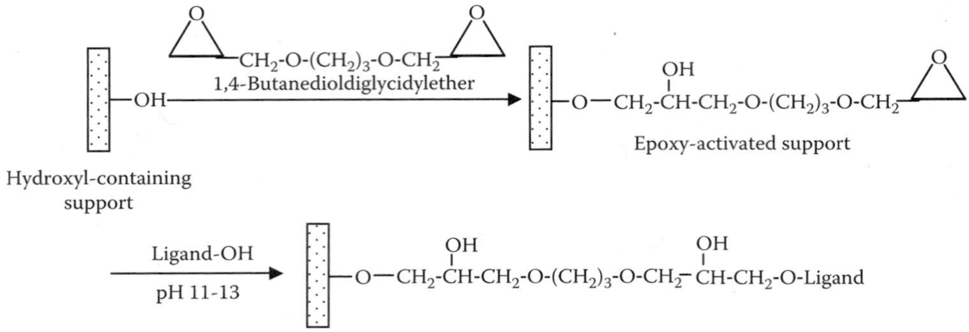

Figure 3.15 The use of oxirane activation for the immobilization of a hydroxyl-containing ligand (ligand-OH).

occur only in specific locations on the ligand. One example is the immobilization of antibodies through their carbohydrate residues. To use the carbohydrate groups of an antibody (or any other glycoprotein) for immobilization, these groups must first be oxidized to form reactive aldehyde groups. This can be accomplished by enzymatic treatment; however, it is usually performed through mild treatment with periodate. These aldehyde groups are then reacted with a support containing amine or hydrazide groups for ligand immobilization [169–174]. This approach has been used not only for antibodies but also for glycoenzymes, RNA, and sugars [15].

Supports with amine groups can be used for coupling aldehyde-containing ligands by reductive amination. This is the same process as shown in Figure 3.9 but with the positions of the amine and aldehyde groups on the ligand and support now being reversed. Hydrazide-activated supports can also be employed for immobilizing ligands with aldehyde groups. Such supports can be prepared by forming aldehyde groups on the support and reacting these with an excess of a dihydrazide (e.g., oxalic or adipic dihydrazide). This process is shown in Figure 3.16. Once activated, the terminal hydrazide on the surface can be reacted with the aldehyde groups on an oxidized antibody, glycoenzyme, or RNA [15]. An advantage of using a hydrazide-activated support is that no reducing agent is needed to stabilize the linkage between the ligand and support, as is required in reductive amination.

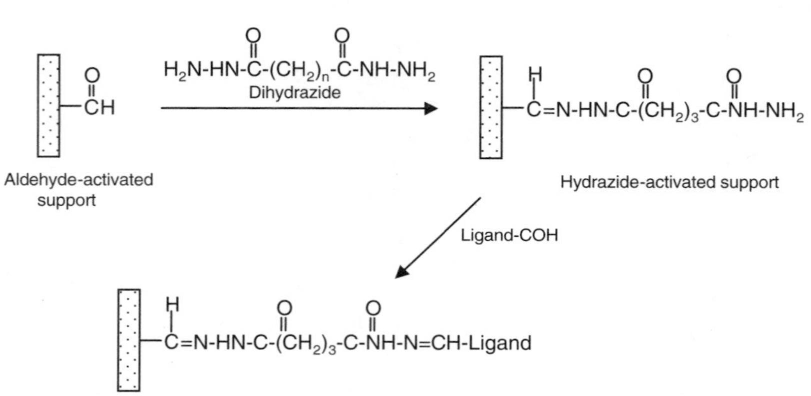

Figure 3.16 The activation of an aldehyde-containing support with a dihydrazide and the subsequent use of this support for immobilizing a ligand with oxidized carbohydrate residues (ligand-COH).

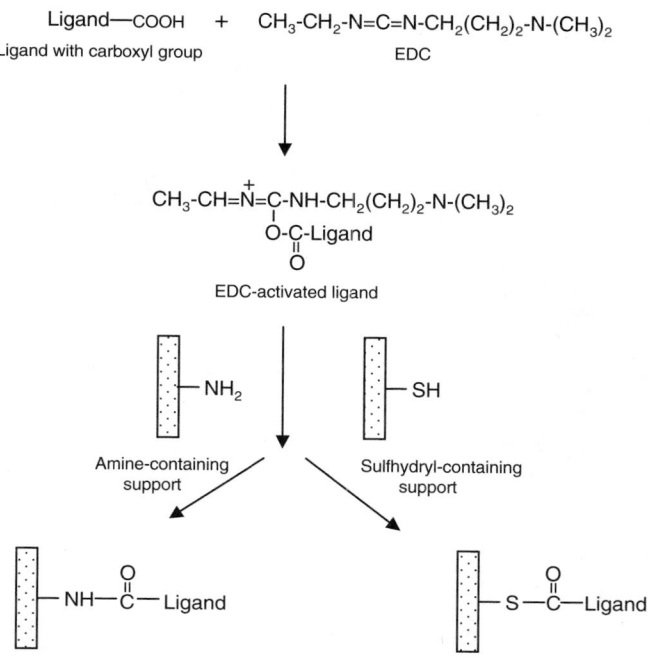

Figure 3.17 The activation of immobilization of a ligand containing a carboxyl group through the use of 1-ethyl-3-(dimethylaminopropyl) carbodiimide (EDC) as an activation agent. Reactions are shown for this activated ligand with supports containing either amines or sulfhydryl groups on their surfaces.

3.4.5 Carboxyl-Reactive Methods

There are currently no activated supports that react specifically with a ligand containing carboxyl groups. This is a result of the low nucleophilicity of carboxyl groups in an aqueous solution. However, there are reagents that will react with carboxylic acids and allow them to be activated for ligand attachment. 1-Ethyl-3-(dimethylaminopropyl) carbodiimide is an example of such a reagent [9, 175–177]. Figure 3.17 shows the reactions involved in the activation of a ligand with EDC and in the coupling of this ligand to a support that contains amine or sulfhydryl groups. One problem with this process is that severe cross-linking is possible, since amine groups as well as carboxyl groups can react if excess EDC is present. In addition, the activated derivative formed, *O*-acylisourea, is not stable in an aqueous environment. This means that the activated ligand must be used immediately for immobilization without further purification.

3.5 OTHER IMMOBILIZATION TECHNIQUES

Along with noncovalent and covalent immobilization methods, other techniques have been developed for the creation of affinity supports. Such methods include entrapment, molecular imprinting, and the use of the ligands as both the support and stationary phase. Although these methods are not as common as the approaches already examined, they have important advantages in some applications.

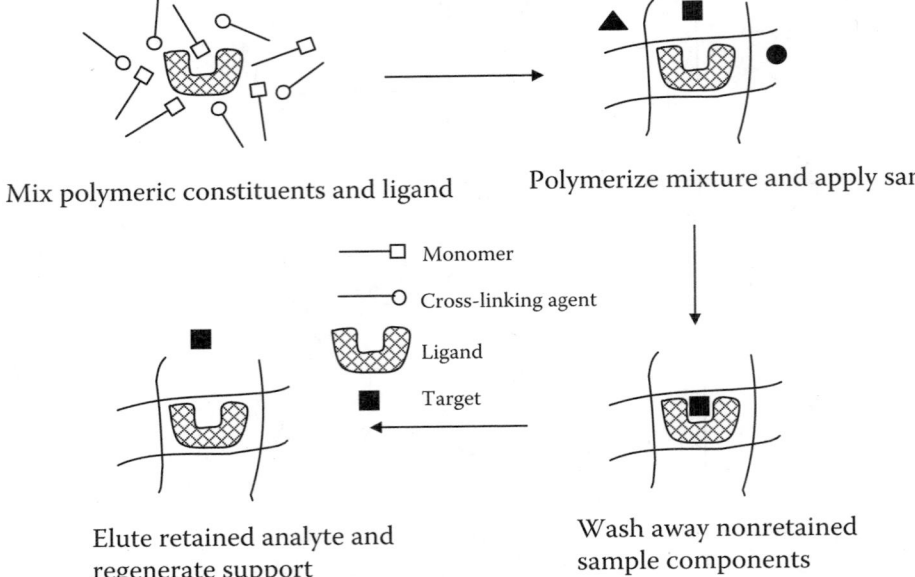

Figure 3.18 shows the labels:
Mix polymeric constituents and ligand

Polymerize mixture and apply sample

———□ Monomer

———○ Cross-linking agent

Ligand

■ Target

Elute retained analyte and
regenerate support

Wash away nonretained
sample components

Figure 3.18 General approach for the entrapment of a ligand in a polymerized support and the use of this support for binding to a target solute.

3.5.1 Entrapment

The first alternative method that will be considered is *entrapment*, or *encapsulation*. This involves the physical containment of a ligand in a support that contains small pores or a highly cross-linked polymer network. Figure 3.18 shows the approach usually employed in this work. Bernfeld and Wan first proposed this method as a means for the immobilization of enzymes in polyacrylamide gels in 1963 [178]. It has since been used with a variety of substances, including proteins, liposomes, and cells.

3.5.1.1 General Techniques for Protein Entrapment

Three basic approaches can be used for the immobilization of a protein through entrapment [179]. The first is to form a cross-linked polymer around the ligand. This is often accomplished by combining a monomer, cross-linking agent, and polymerization initiator with the ligand. Entrapment performed in this manner is often carried out in the absence of oxygen. Proper temperature control is also important to prevent thermal denaturation of the ligand, since some polymerization processes can release significant amounts of heat [178]. In addition, proper selection of the monomer and cross-linking agent is important, since these will determine the physical and chemical properties of the final support.

Protein A and IgG have been entrapped in gels using acrylamide as the monomer and *N,N*-methylene-*bis*-acrylamide as the cross-linking agent [180]. It was found that the amount of immobilized ligand increased as the ratio of monomer to cross-linking agent was increased. However, the apparent binding ability of immobilized IgG decreased as the monomer concentration was increased. This was probably due to the smaller pores present in the polymer, leading to steric hindrance and restricted movement of the target.

The second method for entrapment is to place a ligand in the support and enclose it by cross-linking or altering the ability of the ligand to leave the polymer [178]. As an example, a chiral stationary phase for the separation of (±)-benzoin was prepared using bovine serum albumin (BSA) entrapped on silica after cross-linking it with glutaraldehyde [180, 181].

The most widely used support for ligand entrapment is polyacrylamide. One benefit of using polyacrylamide is its neutral charge. This avoids nonspecific binding by charged solutes and gives a similar pH profile for the entrapped and nonimmobilized forms of the ligand [182]. A disadvantage of polyacrylamide is that its mechanical stability is not sufficient for use in high-performance liquid chromatography. These mechanical properties can be improved by using 2-hydroxylmethacrylate as the monomer and ethyleneglycol dimethacrylate as the cross-linking agent [183]. The pore size of these materials can be altered by changing the amount of monomer versus cross-linking agent. This can also be controlled by dehydration of the gel [184]. In this latter case, the effect of dehydration on ligand stability should be considered [185].

3.5.1.2 Entrapment with Liposomes

Liposomes can be used for the entrapment of some ligands [186–188]. Liposomes are assemblies of phospholipids and other lipids that can sustain a particular configuration [189]. Liposomes form when water-insoluble polar lipids (e.g., phosphatidylethanolamine, -choline, and -serine, plus cardiolipins and phosphatidic acids) are placed in contact with water. Under these conditions, the lipids conglomerate to give a highly ordered structure that persists in the presence of water [190, 191].

Various solutes and biomolecules can be entrapped within liposomes. The extent to which a solute can be entrapped is proportional to the volume of the aqueous phase within the liposome. This, in turn, is related to the charge of the inner or outer surface of the lipid bilayer and to the ionic strength of the medium used for liposome formation [192]. As a result, an increase in the internal volume of water or the size of an entrapped solute will require an increase in the ratio of the lipid's charge to the ionic strength of entrapping media.

The following procedure can be used to entrap ligands in liposomes. First, lipids such as phospholipids are dissolved in an organic solvent. After subsequent removal of the organic solvent under inert conditions, the dried lipid film is dispersed in a solution containing the ligand. This results in formation of a liposome containing the entrapped ligand. If desired, the size of the liposome obtained during this process can be reduced by sonication.

During the period of liposome entrapment, it is important to use a low ratio for the amount of lipid versus ligand. This prevents penetration of the ligand's hydrophobic regions into the lipid phase. Minimal adsorption of the entrapped ligand onto the liposome's surface is also necessary to avoid hindering liposome formation. This can be accomplished by adjusting the pH to make the net charges on the lipid and ligand equal [193].

3.5.1.3 Cell Entrapment

The use of entrapment for the immobilization of cells has been a topic of great interest over the past few decades [194–203]. This involves the localization of intact cells within a defined region of a support that preserves the cells' activity [204]. Advantages of entrapped versus nonimmobilized cells include an increase in stability, greater protection from contamination, and the ability to use these as part of an automated system.

Many of the immobilization methods discussed earlier can be used for cells (e.g., adsorption and covalent attachment), but entrapment has been the most widely used approach. This is due to the low leakage of cells in this technique and the minimal effects it has on the intact cell. Polyacrylamide is the material most commonly used for cell entrapment. The procedure for this is described in the literature [205]. The purification of lectins using

Candida lipolylica and the isolation of immunoglobulins using *Staphylococcus aureus* are two applications in which immobilized cells have been employed [204]. The use of immobilized cells for chemical analysis has also been described in the literature [206–208].

3.5.1.4   Sol Gels

Another way in which ligands can be entrapped is through the use of a sol gel [209–211]. This technique makes use of ceramic or glass materials in which there is a transition from a liquid solution or sol to a solid or gel phase. Such materials have been used with a variety of biological agents [212–214]. An example in affinity chromatography is a sol gel that was used for immunoaffinity purification based on entrapped anti-dinitrophenyl antibodies [215].

3.5.2   Molecular Imprinting

Molecular imprinting is a newer approach for creating immobilized ligands. This topic is discussed in detail in Chapter 30. This involves a template polymerization process that produces highly specific cavities that act as artificial ligand sites (see Figure 3.19). During this process, functional monomers are allowed to interact with a template molecule. The functional monomers are then fixed in place by polymerizing them with a cross-linking agent. After removing the template, a polymer is obtained with cavities complementary in size and shape to the template. In this way, a ligand site is introduced into the polymer that can now selectively bind the target [216–218]. The resulting material is known as a *molecularly imprinted polymer* (*MIP*).

Molecularly imprinted polymers can be obtained by using either a covalent or a noncovalent approach. The covalent approach requires that the template be a polymerizable derivative of the target of interest. This derivatized template reacts with a functional monomer to form a degradable covalent bond, which can later be cleaved after copolymerization with a cross-linking agent. The most common linkages used for this purpose are esters of carboxylic

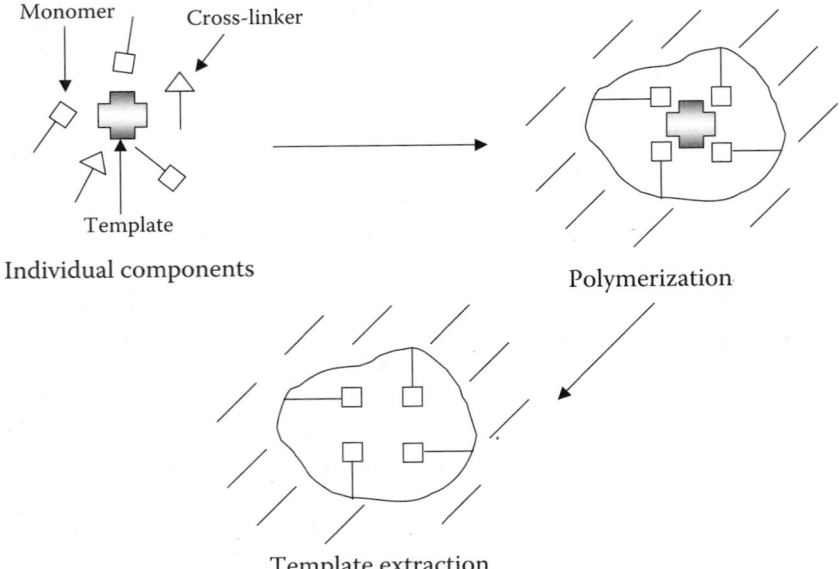

Figure 3.19   An approach for creating a molecularly imprinted polymer.

or boronic acids, ketals, and imines [219]. The noncovalent approach utilizes interactions such as hydrogen bonding, π-π interactions, hydrophobic interactions, and van der Waals forces to obtain a molecular recognition memory. More details on both of these methods can be found in Chapter 30.

Molecularly imprinted polymers have been used in a number of applications. Many of these have involved the creation of supports for small targets like pesticides and amino acids [220–223]. However, work with macromolecules (e.g., proteins) is also feasible [224]. The first application of these materials was in the area of affinity chromatography, and these have been extensively studied as chiral stationary phases [225, 226]. For instance, a molecularly imprinted polymer has been used to separate the enantiomers of *N*-acetyl-Trp-Phe-OMe, giving rise to a separation factor of 17.8 [227].

3.5.3 Ligand-Based Supports

There are some occasions in which the ligand itself can be used as the support in affinity chromatography. In fact, this was employed in the earliest use of affinity chromatography when Starkenstein used starch for the purification of amylase [228]. Similar work has been performed by others with starch or modified starch columns [229, 230].

Cross-linked proteins have been used in some studies as both ligands and supports. For instance, Margolin and Vilenchik created cross-linked bovine serum albumin crystals for use as affinity supports [231]. These crystals were found to have good mechanical strength, a well-developed porous structure, and significant affinity and chiral selectivity. These materials were also extremely stable in both aqueous and organic solvents.

Affinity capillary electrochromatography is one field that has made use of cross-linked proteins as supports. As an example, D- and L-tryptophan were separated using cross-linked BSA by this approach [232]. In other work, BSA and cellobiohydrolase I were used in a mixed cross-linked gel for the separation of β-adrenergic blockers [233]. One reason that packed capillaries might be chosen over traditional columns for this work is the improved efficiency that has been reported when using an electric field instead of pressure as the driving force behind solvent flow [234]. However, it is not yet clear whether this advantage pertains to immobilized or cross-linked protein supports.

3.6 SUMMARY AND CONCLUSIONS

This chapter examined several immobilization methods for affinity ligands and presented many factors to consider in the selection of such an approach. This included a discussion of noncovalent and covalent attachment methods, as well as entrapment, molecular imprinting, and the use of ligands as affinity supports. The characterization of ligands was examined along with the role played by multisite attachment, improper orientation, and steric hindrance during the immobilization process. Based on this information, general guidelines were given for the preparation of affinity columns. These guidelines should be of some assistance in the generation of new affinity supports and in the optimization of current ligands for affinity separations.

SYMBOLS AND ABBREVIATIONS

APC	1-Fatty acyl-2(12-aminododecyl) phosphatidylcholine
BCA	Bicinchoninic acid
BMH	*Bis*-Maleimidohexane

BS	*Bis*-(Sulfosuccinimidyl) suberate
BSA	Bovine serum albumin
CDI	Carbonyldiimidazole
CNBr	Cyanogen bromide
DSG	Disuccinimidyl glutarate
DSP	Dithio-*bis*-(succinimidyl propionate)
DSS	Disuccinimidyl suberate
DTNB	5,5-Dithio-*bis*-(2-nitrobenzoic acid), or Elman's reagent
DTT	Dithiothreitol
DVS	Divinylsulfone
EDC	Ethyl dimethylaminopropyl carbodiimide
EGS	Ethylene glycobis (succinimidyl succinate)
GMBS	*N*-(γ-Maleimidobutyryloxy) succinimide ester
HIC	Hydrophobic interaction chromatography
HPAC	High-performance affinity chromatography
IDA	Iminodiacetic acid
IgA	Immunoglobulin A
IgE	Immunoglobulin E
IgG	Immunoglobulin G
IgM	Immunoglobulin M
IMAC	Immobilized metal-ion affinity chromatography
K_a	Association-equilibrium constant
K_d	Dissociation-equilibrium constant
MIP	Molecularly imprinted polymer
NHS	*N*-Hydroxysuccinimide
NTA	Nitrilotriacetic acid
SFB	Succinimidyl *p*-formylbenzoate
SIAB	*N*-Succinimidyl (4-iodoacetyl) aminobenzoate
SMCC	Succinimidyl-4-(*N*-maleimidomethyl) cyclohexane-1-carboxylate
SPDP	*N*-Succinimidyl-3-(2-pyridyldithio) propionate
Sulfo-SMCC	Sulfosuccinimidyl-4-(*N*-maleimidomethyl) cyclohexane-1-carboxylate
TACN	1,4,7-Triazocyclononane
TNB	2-Nitro-5-thiobenzoic acid
TREN	*Tris*-(2-Aminoethyl)amine

REFERENCES

1. Turkova, J., *Affinity Chromatography,* Elsevier, Amsterdam, 1978.
2. Scouten, W.H., *Affinity Chromatography: Bioselective Adsorption on Inert Matrices,* Wiley, New York, 1981.
3. Chaiken, I.M., *Analytical Affinity Chromatography,* CRC Press, Boca Raton, FL, 1987.
4. Quiquampoix, H., Servagent-Noinville, S., and Baron, M.H., Enzyme adsorption on soil mineral surfaces and consequences for the catalytic activity, In *Enzymes in the Environment,* Burns, R.G. and Dick, R.P., Eds., Marcel Dekker, New York, 2002, 285–306.
5. Quiquampoix, H., Mechanisms of protein adsorption on surfaces and consequences for extracellular enzyme activity in soil, *Soil Biochem.,* 10, 171–206, 2000.
6. Chibata, I., *Immobilized Enzymes,* Wiley, New York, 1978.
7. McLaren, A.D., pH dependence of enzyme reactions on cells, particulates and in solution, *Science,* 125, 697, 1957.

8. Kabra, B.G. and Gehrke, S.H., Microporous crosslinked polymer gels showing fast reversible volume change and their uses, *PCT Int. Appl.,* p. 77, 1995.

9. Hermanson, G.T., Mallia, A.K., and Smith, P.K., *Immobilized Affinity Ligand Techniques,* Academic Press, New York, 1992.

10. Dunlap, R.B., *Immobilized Biochemicals and Affinity Chromatography,* Plenum Press, New York, 1974.

11. Tardioli, P.W., Fernandez-Lafuente, R., Guisan, J.M., and Giordano, R.L.C., Design of new immobilized-stabilized carboxypeptidase A derivative for production of aromatic free hydrolysates of proteins, *Biotechnol. Prog.,* 19, 565–574, 2003.

12. Wimalasena, R.L. and Wilson, G.S., Factors affecting the specific activity of immobilized antibodies and their biologically active fragments, *J. Chromatogr.,* 572, 85–102, 1991.

13. Mansfeld, J. and Ulbrich-Hofmann, R., Site-specific and random immobilization of thermolysin-like proteases reflected in the thermal inactivation kinetics, *Biotechnol. Appl. Biochem.,* 32, 189–195, 2000.

14. Matson, R.S. and Little, M.C., Strategy for the immobilization of monoclonal antibodies on solid-phase supports, *J. Chromatogr.,* 458, 67–77, 1988.

15. Ruhn, P.F., Garver, S., and Hage, D.S., Development of dihydrazide-activated silica supports for high-performance affinity chromatography, *J. Chromatogr. A,* 669, 9–19, 1994.

16. Hage, D.S., Periodate oxidation of antibodies for site-selective immobilization in immunoaffinity chromatography, *Methods Mol. Biol.,* 147, 69–82, 2000.

17. Huang, W., Wang, J., Bhattacharyya, D., and Bachas, L.G., Improving the activity of immobilized subtilisin by site-specific attachment to surfaces, *Anal. Chem.,* 69, 4601–4607, 1997.

18. Murza, A., Fernandez-Lafuente, R., and Guisan, J.M., Essential role of the concentration of immobilized ligands in affinity chromatography: Purification of guanidinobenzoatase on an ionized ligand, *J. Chromatogr. B,* 740, 211–218, 2000.

19. Delfino, J.M., Florin-Christensen, J., Florin-Christensen, M., and Richards, F.M., Differential hydrolysis of immobilized phosphatidylcholines by phospholipases A2 and C, *Biochem. Biophys. Res. Comm.,* 205, 113–119, 1994.

20. Penzol, G., Armisen, P., Fernandez-Lafuente, R., Rodes, L., and Guisan, J.M., Use of dextrans as long and hydrophilic spacer arms to improve the performance of immobilized proteins acting on macromolecules, *Biotech. Bioeng.,* 60, 518–523, 1998.

21. Luo, Q., Zou, H., Zhang, Q., Xiao, X., and Ni, J., High-performance affinity chromatography with immobilization of protein A and L-histidine on molded monolith, *Biotech. Bioeng.,* 80, 481–489, 2002.

22. Messing, R.A., Relation of pore size and surface area to quantity of stabilized enzyme bound to glass, *Enzymologia,* 39, 12–14, 1970.

23. Clarke, W., Beckwith, J.D., Jackson, A., Reynolds, B., Karle, E.M., and Hage, D.S., Antibody immobilization to high-performance liquid chromatography supports: Characterization of maximum loading capacity for intact immunoglobulin G and Fab fragments, *J. Chromatogr. A,* 888, 13–22, 2000.

24. Cuatrecasas, P., Protein purification by affinity chromatography: Derivatization of agarose and polyacrylamide beads, *J. Biol. Chem.,* 245, 3059–3065, 1970.

25. Chockalingam, P.S., Jurado, L.A., and Jarrett, H.W., DNA affinity chromatography, *Mol. Biotechnol.,* 19, 189–199, 2001.

26. Means, G. and Feeney, R.E., *Chemical Modification of Proteins,* Holden-Day, San Francisco, 1971.

27. Hentz, N.G., Richardson, J.M., Sportsman, J.R., Daijo, J., and Sittampalam, G.S., Synthesis and characterization of insulin-fluorescein derivatives for bioanalytical applications, *Anal. Chem.,* 69, 4994–5000, 1997.

28. Taylor, R.F., *Protein Immobilization: Fundamentals and Applications,* Marcel Dekker, New York, 1991.

29. Smith, P.K., Krohn, R.I., Hermanson, G.T., Mallia, A.K., Gartner, F.H., Provenzano, M.D., Fujimoto, E.K., Goeke, N.M., Olson, B.J., and Klenk, D.C., Measurement of protein using bicinchoninic acid, *Anal. Biochem.,* 150, 76–85, 1985.

30. Starcher, B., A ninhydrin-based assay to quantitate the total protein content of tissue samples, *Anal. Biochem.,* 292, 125–129, 2001.

31. Hische, E.A.H., Van der Helm, H.J., Van Meegen, M.T., and Blanken, H.I.G., Protein estimation in cerebrospinal fluid with Coomassie brilliant blue, *Clin. Chem.,* 28, 1236–1237, 1982.

32. Smith, I.K., Vierheller, T.L., and Thorne, C.A., Assay of glutathione reductase in crude tissue homogenates using 5,5'-dithiobis(2-nitrobenzoic acid), *Anal. Biochem.,* 175, 408–413, 1988.

33. Inman, J.K. and Dintzis, H.M., The derivatization of cross-linked polyacrylamide beads: Controlled introduction of functional groups for the preparation of special-purpose, biochemical adsorbents, *Biochemistry,* 8, 4074–4082, 1969.

34. Wilchek, M. and Bayer, E.A., Introduction to avidin-biotin technology, *Methods Enzymol.,* 184, 5–13, 1990.

35. Langone, J.J., Applications of immobilized protein A in immunochemical techniques, *J. Immunol. Methods,* 55, 277–296, 1982.

36. Guilbault, G.G. and Mascini, M., *Analytical Uses of Immobilized Biological Compounds for Detection, Medical, and Industrial Uses,* Dordrecht, Boston, 1988.

37. Chibata, I., *Immobilized Enzymes,* Wiley, New York, 1978.

38. Messing, R.A., Adsorption and inorganic bridge formations, *Methods Enzymol.,* 44, 148–169, 1976.

39. Liu, Y.C., Wang, C.M., and Hsiung, K.P., Comparison of different protein immobilization methods on quartz crystal microbalance surface in flow injection immunoassay, *Anal. Biochem.,* 299, 130–135, 2001.

40. Topoglidis, E., Campbell, C.J., Cass, A.E.G., and Durrant, J.R., Factors that affect protein adsorption on nanostructured titania films: A novel spectroelectrochemical application to sensing, *Langmuir,* 17, 7899–7906, 2001.

41. Mateo, C., Fernandez-Lorente, G., Abian, O., Fernandez-Lafuente, R., and Guisan, J.M., Multifunctional epoxy supports: A new tool to improve the covalent immobilization of proteins: The promotion of physical adsorptions of proteins on the supports before their covalent linkage, *Biomacromolecule,* 1, 739–745, 2000.

42. Shimizu, M., Komatsu, T., Shimizu, M., and Kase, M., Immobilization of a Lipolytic Enzyme by Adsorption onto a Porous, Anion-Exchange Resin and Its Use for Lipolysis and Esterification of Fats and Oils, European patent application, 2000, p. 8.

43. Fernandez-Lafuente, R., Armisen, P., Sabuquillo, P., Fernandez-Lorente, G., and Guisan, J.M., Immobilization of lipases by selective adsorption on hydrophobic supports, *Chem. Phys. Lipids,* 93, 185–197, 1998.

44. Yingfan, W. and Dubin, P.L., Protein binding on polyelectrolyte-treated glass: Effect of structure of adsorbed polyelectrolyte, *J. Chromatogr. A,* 808, 61–70, 1998.

45. Wheatley, J.B. and Schmidt, D.E., Jr., Salt-induced immobilization of proteins on a HPLC epoxide affinity support, *J. Chromatogr.,* 644, 11–16, 1993.

46. Orberger, G., Gessner, R., Fuchs, H., Volz, B., Kottgen, E., and Tauber, R., Enzymatic modeling of the oligosaccharide chains of glycoproteins immobilized onto polystyrene surface, *Anal. Biochem.,* 214, 195–204, 1993.

47. Tyagi, R. and Gupta, M.N., Noncovalent and reversible immobilization of chemically modified amyloglucosidase and beta-glucosidase on DEAE-cellulose, *Proc. Biochem.,* 29, 443–448, 1994.

48. Kovalenko, G.A. and Sokolovskii, V.D., Immobilization of oxyreductases on inorganic supports based on alumina-immobilization of alcohol dehydrogenase on nonmodified and modified alumina, *Biotech. Bioeng.,* 25, 3177–3184, 1983.

49. Angelino, S.A.G.F., Muller, F., and Van der Plas, H.C., The use of enzymes in organic synthesis, Part 14: Purification and immobilization of rabbit liver aldehyde oxidase, *Biotech. Bioeng.,* 27, 447–455, 1985.

50. Vandamme, E.J., Peptide antibiotic production through immobilized biocatalyst technology, *Enzyme Microb. Technol.,* 5, 403–416, 1983.

51. Gekas, V.C., Artificial membranes as carriers for the immobilization of biocatalysts, *Enzyme Microb. Technol.,* 8, 450–460, 1986.

52. Stone, I., Dextrose Preparation by Use of Starch-Glucogenase Enzymes, U.S. patent 2,717,852, 1955.

53. Tosa, T., Mori, T., and Chibata, I., Studies on continuous enzyme reactions, VII: Activation of water-insoluble aminoacylase by protein denaturing agents, *Enzymologia,* 40, 49–63, 1971.

54. Tosa, T., Mori, T., and Chibata, I., Continuous enzyme reactions, VIII: Kinetics and pressure drop of an aminoacylase column, *Hakko Kogaku Zasshi,* 49, 522–528, 1971.

55. Tosi, T., Sato, T., Mori T., Matuo, Y., and Chibata, I., Immobilized enzymes, XI: Continuous production of L-aspartic acid by immobilized aspartase, *Biotech. Bioeng.,* 15, 69–84, 1973.

56. Hofstee, B.H.J. and Otillio, N.F., Immobilization of enzymes through noncovalent binding to substituted agaroses, *Biochim. Biophys. Res. Comm.,* 53, 1137–1144, 1973.

57. Venkatsubramanian, K., Vieth, W.R., and Wang, S.S., Lysozyme immobilized on collagen, *Hakko Kagaku Zasshi,* 50, 600–614, 1972.

58. Solomon, B. and Levin, Y., Adsorption of amyloglucosidase on inorganic carriers, *Biotech. Bioeng.,* 17, 1323–1333, 1975.

59. Serralha, F.N., Lopes, J.M., Aires-Barros, M.R., Prazeres, D.M.F., Cabral, J.M.S., Lemos, F., and Ramoa Ribeiro, F., Stability of a recombinant cutinase immobilized on zeolites, *Enzyme Microb. Tech.,* 31, 29–34, 2002.

60. Abdel-Naby, M.A., Hashem, A.M., Esawy, M.A., and Abdel-Fattah, A.F., Immobilization of *Bacillus subtilis* α-amylase and characterization of its enzymic properties, *Microb. Res.,* 153, 319–325, 1999.

61. Hyndman, D., Lever, G., Burrell, R., and Flynn, T.G., Protein immobilization to alumina supports, I: Characterization of alumina-organophosphate ligand interactions and use in the attachment of papain, *Biotech. Bioeng.,* 40, 1319–1327, 1992.

62. Torchilin, V.P., Galka, M., and Ostrowski, W., Comparative studies on immobilization of human prostatic acid phosphatase, *Biochim. Biophys. Acta,* 483, 331–336, 1977.

63. Thust, M., Schoning, M.J., Schroth, P., Malkoc, U., Dicker, C.I., Steffen, A., Kordos, P., and Luth, H., Enzyme immobilization on planar and porous silicon substrates for biosensor applications, *J. Mol. Catal. B,* 7, 77–83, 1999.

64. Porath, J. and Axen, R., Immobilization of enzymes to agar, agarose and Sephadex supports, *Methods Enzymol.,* 44, 19–45, 1976.

65. Messing, R.A., Molecular inclusions: Adsorption of macromolecules on porous glass membranes, *J. Am. Chem. Soc.,* 91, 2370–2371, 1969.

66. McLaren, A.D. and Estermann, E.F., Influence of pH on the activity of chymotrypsin at a solid-liquid interface, *Arch. Biochem. Biophys.,* 68, 157–160, 1957.

67. Messing, R.A., Immobilized RNase by adsorption on porous glass, *Enzymologia,* 38, 370–372, 1970.

68. Wilchek, M. and Bayer, E.A., Avidin-biotin immobilization systems, in *Immobilized Biomolecules in Analysis,* Cass, T. and Ligler, F.S., Eds., Springer, Oxford, U.K., 1998.

69. Bayer, E.A. and Wilchek, M., The avidin-biotin system, in *Immunoassay,* Diamandis, E.P. and Chrisopoulos, T.K., Eds., Academic Press, San Diego, 1996.

70. Page, M. and Thorpe, R., Purification of IgG using protein A or protein G, in *Protein Protocols Handbook,* 2nd ed., Walker, J.M., Ed., Humana Press, Totowa, NJ, 2002.

71. Akerstrom, B. and Bjorck, L., A physicochemical study of protein G, a molecule with unique immunoglobulin G-binding properties, *J. Biol. Chem.,* 261, 10240–10247, 1986.

72. Sjoquist, J., Meloun, B., and Hjelm, H., Protein A isolated from *Staphylococcus aureus* after digestion with lysostaphin, *Eur. J. Biochem.,* 29, 572–578, 1972.

73. Masseyeff, R.F., Albert, W.H., and Staines, N.A., Eds., *Method of Immunological Analysis,* Vol. 1, VCH, New York, 1993.

74. Anon., *Pierce Catalogue and Handbook,* Pierce, Rockford, IL, 2003.

75. Heitzmann, H. and Richards, F.M., Use of the avidin-biotin complex for specific staining of biological membranes in electron microscopy, *Proc. Natl. Acad. Sci. U.S.A.,* 71, 3537–3541, 1974.

76. Becker, J.M. and Wilchek, M., Inactivation by avidin of biotin-modified bacteriophage, *Biochim. Biophys. Acta,* 264, 165–170, 1972.

77. Schetters, H., Avidin and streptavidin in clinical diagnostics, *Biomol. Eng.,* 16, 73–78, 1999.

78. Hnatowich, D.J., Virzi, F., and Rusckowski, M., Investigations of avidin and biotin for imaging applications, *J. Nucl. Med.,* 28, 1294–1302, 1987.

79. La Rochelle, W.J. and Froehner, S.C., Immunochemical detection of proteins biotinylated on nitrocellulose replicas, *J. Immunol. Methods,* 92, 65–71, 1986.

80. Cole, S.R., Ashman, L.K., and Ey, P.L., Biotinylation: An alternative to radioiodination for the identification of cell surface antigens in immunoprecipitates, *Mol. Immunol.,* 24, 699–705, 1987.

81. Stoebel, K., Schoenberg, A., and Staak, C., A new non-species dependent ELISA for detection of antibodies to *Borrelia burgdorferis,* I, in zoo animals, *Int. J. Med. Microb.,* 291, 88–99, 2002.

82. Heelan, B., Purification of monoclonal antibodies using protein A/G, *Methods Mol. Med.,* 40, 281–288, 2000.

83. Sugiura, T., Imagawa, H., and Kondo, T., Purification of horse immunoglobulin isotypes based on differential elution properties of isotypes from protein A and protein G columns, *J. Chromatogr. B,* 742, 327–334, 2000.

84. Nedonchelle, E., Pitiot, O., and Vijayalakshmi, M.A., A preliminary study for isolation of catalytic antibodies by histidine ligand affinity chromatography as an alternative to conventional protein A/G methods, *Appl. Biochem. Biotech.,* 83, 287–295, 2000.

85. Hage, D.S., Affinity chromatography: A review of clinical applications, *Clin. Chem.,* 45, 593–615, 1999.

86. Fratamico, P.M., Strobaugh, T.P., Medina, M.B., and Gehring, A.G., Real-time detection of *Escherichia coli* O157:H7 using a surface plasmon resonance biosensor, in *Book of Abstracts,* 214th ACS National Meeting, ACS, Washington, DC, 1997.

87. Lu, B., Malcolm, M.R., and O'Kennedy, R., Immunological activities of IgG antibody on pre-coated Fc receptor surfaces, *Anal. Chim. Acta,* 331, 97–102, 1996.

88. Jones, R.H.V., Rademacher, T.W., and Williams, P.J., Bias in murine IgG isotype immobilization: Implications for IgG glycoform analysis ELISA procedures, *J. Immunol. Methods,* 197, 109–120, 1996.

89. Peng, Z., Becker, A.B., and Simons, F.E.R., Binding properties of protein A and protein G for human IgE, *Int. Arch. Allergy Immunol.,* 104, 204–206, 1994.

90. Widjojoatmodjo, M.N., Fluit, A.C., Torensma, R., and Verhoef, J., Comparison of immunomagnetic beads coated with protein A, protein G, or goat anti-mouse immunoglobulins: Applications in enzyme immunoassays and immunomagnetic separations, *J. Immunol. Methods,* 165, 11–19, 1993.

91. Peng, Z., Simons, F.E.R., and Becker, A.B., Differential binding properties of protein A and protein G for dog immunoglobulins, *J. Immunol. Methods,* 145, 255–258, 1991.

92. Boyle, M.D.P. and Reis, K.J., Bacterial Fc receptors, *Biotechnology,* 5, 697–703, 1987.

93. Sjöquist, J., Meloun, B., and Hjelm, H., Protein A isolated from *Staphylococcus aureus* after digestion with lysostaphin, *Eur. J. Biochem.,* 29, 572–578, 1972.

94. Ayhan, H., Kesenci, K., and Piskin, E., Protein A immobilization and hIgG adsorption onto porous/nonporous and swellable HEMA-incorporated polyEGDMA microspheres, *J. Biomater. Sci.,* 11, 13–25, 2000.

95. Denizli, A., Rad, A.Y., and Piskin, E., Protein A immobilized on polyhydroxyethylmethacrylate beads for affinity sorption of human immunoglobulin G, *J. Chromatogr. B,* 668, 13–19, 1995.

96. Denizli, A. and Arica, Y., Protein A-immobilized microporous polyhydroxyethyl methacrylate affinity membranes for selective sorption of human-immunoglobulin-G from human plasma, *J. Biomater. Sci.,* 11, 367–382, 2000.

97. Weiner, C., Sara, M., Dasgupta, G., and Sleytr, U.B., Affinity cross-flow filtration: Purification of IgG with a novel protein A affinity matrix prepared from two-dimensional protein crystals, *Biotech. Bioeng.,* 44, 55–65, 1994.

98. Terman, D.S., Preparation of protein A immobilized on collodion-coated charcoal and plasma perfusion system for treatment of cancer, *Methods Enzymol.,* 137, 496–515, 1988.

99. Gaberc-Porekar, V. and Menart, V., Perspective of immobilized-metal affinity chromatography, *J. Biochem. Biophys. Methods,* 49, 335–360, 2001.

100. Jiang, W., Graham, B., Spiccia, L., and Hearn, M.T.W., Protein selectivity with immobilized metal ion-tacn sorbents: Chromatographic studies with human serum proteins and several other globular proteins, *Anal. Biochem.*, 255, 47–58, 1998.

101. Chaouk, H. and Hearn, M.T.W., Examination of the protein binding behavior of immobilized copper (II)-2,6-diaminomethylpyridine and its application in the immobilized metal ion affinity chromatographic separation of several human serum proteins, *J. Biochem. Biophys. Methods*, 39, 161–177, 1999.

102. Winzerling, J.J., Berna, P., and Porath, J., How to use immobilized metal ion affinity chromatography, *Methods*, 4, 4–13, 1992.

103. Fiddler, M.B. and Gray, G.R., Immobilization of proteins on aldehyde-activated polyacrylamide supports, *Anal. Biochem.*, 86, 716–724, 1978.

104. Denizli, A. and Arica, Y., Protein A-immobilized microporous polyhydroxyethylmethacrylate affinity membranes for selective sorption of human-immunoglobulin-G from human plasma, *J. Biomater. Sci.*, 11, 367–382, 2000.

105. Murphy, R.F., Conlon, J.M., Imam, A., and Kelly, G.J.C., Comparison of non-biospecific effects in immunoaffinity chromatography using cyanogen bromide and bifunctional oxirane as immobilizing agents, *J. Chromatogr.*, 135, 427–433, 1977.

106. Kuemel, G., Daus, H., and Mauch, H., Improved method for the cyanogen bromide activation of agarose beads, *J. Chromatogr.*, 172, 221–226, 1979.

107. Yager, T.D. and Barrett, D., Coupling dilute protein to cyanogen bromide-agarose, *Biochim. Biophys. Acta*, 802, 215–220, 1984.

108. Asther, M. and Meunier, J.C., Immobilization as a tool for the stabilization of lignin peroxidase produced by *Phanerochaete chrysosporium* INA-12, *Appl. Biochem. Biotech.*, 38, 57–67, 1993.

109. Potempa, L.A., Motie, M., Anderson, B., Klein, E., and Baurmeister, U., Conjugation of a modified form of human C-reactive protein to affinity membranes for extracorporeal adsorption, *Clin. Mater.*, 11, 105–117, 1992.

110. Suda, Y., Nakamura, M., Koshida, S., Kusumoto, S., and Sobel, M., Novel photoaffinity crosslinking resin for the isolation of heparin binding proteins, *J. Bioact. Compat. Polym.*, 15, 468–477, 2000.

111. Suzuki, N., Quesenberry, M.S., Wang, J.K., Lee, R.T., Kobayashi, K., and Lee, Y.C., Efficient immobilization of proteins by modification of plate surface with polystyrene derivatives, *Anal. Biochem.*, 247, 412–416, 1997.

112. Bjoerklund, M. and Hearn, M.T.W., High-performance liquid chromatography of amino acids, peptides, and proteins, 149: Synthesis of silica-based heparin-affinity adsorbents, *J. Chromatogr. A*, 728, 149–169, 1996.

113. Frey, T., Cosio, E.G., and Ebel, J., Affinity purification and characterization of a binding protein for a hepta-β-glucoside phytoalexin elicitor in soybean, *Phytochemistry*, 32, 543–550, 1993.

114. Thomas, D.H., Beck-Westermeyer, M., and Hage, D.S., Determination of atrazine in water using tandem high-performance immunoaffinity chromatography and reversed-phase liquid chromatography, *Anal. Chem.*, 66, 3823–3829, 1994.

115. Stults, N.L., Asta, L.M., and Lee, Y.C., Immobilization of proteins on oxidized crosslinked Sepharose preparations by reductive amination, *Anal. Biochem.*, 180, 114–119, 1989.

116. Hornsey, V.S., Prowse, C.V., and Pepper, D.S., Reductive amination for solid-phase coupling of protein: A practical alternative to cyanogen bromide, *J. Immunol. Methods*, 93, 83–88, 1986.

117. Cardiano, P., Sergi, S., Lazzari, M., and Piraino, P., Epoxy-silica polymers as restoration materials, *Polymer*, 43, 6635–6640, 2002.

118. Jarrett, H.W., Development of *N*-hydroxysuccinimide ester silica: A novel support for high-performance affinity chromatography, *J. Chromatogr.*, 405, 179–189, 1987.

119. Chockalingam, P.S., Gadgil, H., and Jarrett, H.W., DNA-support coupling for transcription factor purification: Comparison of aldehyde, cyanogen bromide, and *N*-hydroxysuccinimide chemistries, *J. Chromatogr. A*, 942, 167–175, 2002.

120. Murza, A., Fernandez-Lafuente, R., and Guisan, J.M., Essential role of the concentration of immobilized ligands in affinity chromatography: Purification of guanidinobenzoatase on an ionized ligand, *J. Chromatogr. B*, 740, 211–218, 2000.

121. Patel, N., Davies, M.C., Hartshorne, M., Heaton, R.J., Roberts, C.J., Tendler, S.J.B., and Williams, P.M., Immobilization of protein molecules onto homogeneous and mixed carboxylate-terminated self-assembled monolayers, *Langmuir,* 13, 6485–6490, 1997.

122. Wilchek, M. and Miron, T., Limitations of *N*-hydroxysuccinimide esters in affinity chromatography and protein immobilization, *Biochemistry,* 26, 2155–2161, 1987.

123. Peng, L., Calton, G.J., and Burnett, J.W., Stability of antibody attachment in immunosorbent chromatography, *Enzyme Microb. Tech.,* 8, 681–685, 1986.

124. Frost, R.G., Monthony, J.F., Engelhorn, S.C., and Siebert, C.J., Covalent immobilization of proteins to *N*-hydroxysuccinimide ester derivatives of agarose: Effect of protein charge on immobilization, *Biochim. Biophys. Acta,* 670, 163–169, 1981.

125. Besselink, G. and de Korte, D., Sephadex-based cell-affinity adsorbents: Preparation and performance, *Biotech. Appl. Biochem.,* 35, 55–60, 2002.

126. Murza, A., Fernandez-Lafuente, R., and Guisan, J.M., Essential role of the concentration of immobilized ligands in affinity chromatography: Purification of guanidinobenzoatase on an ionized ligand, *J. Chromatogr. B,* 740, 211–218, 2000.

127. Cuatrecasas, P. and Parikh, I., Adsorbents for affinity chromatography: Use of *N*-hydroxysuccinimide esters of agarose, *Biochemistry,* 11, 2291–2299, 1972.

128. Lomant, A.J. and Fairbanks, G., Chemical probes of extended biological structure: Synthesis and properties of the cleavable protein cross-linking reagent (35S) dithiobis(succinimidyl propionate), *J. Mol. Biol.,* 104, 243–261, 1976.

129. Koyama, T. and Terauchi, K., Synthesis and application of boronic acid-immobilized porous polymer particles: A novel packing for high-performance liquid affinity chromatography, *J. Chromatogr. B,* 679, 31–40, 1996.

130. Burton, S.J., Stead, C.V., and Lowe, C.R., Design and applications of biomimetic anthraquinone dyes, III: Anthraquinone-immobilized C.I. Reactive Blue 2 analogs and their interaction with horse liver alcohol dehydrogenase and other adenine nucleotide-binding proteins, *J. Chromatogr.,* 508, 109–125, 1990.

131. Taylor, R.F., A comparison of various commercially available liquid chromatographic supports for immobilization of enzymes and immunoglobulins, *Anal. Chim. Acta,* 172, 241–248, 1985.

132. Hearn, M.T.W., Smith, P.K., Mallia, A.K., and Hermanson, G.T., Preparative and analytical applications of CDI-mediated affinity chromatography, in *Affinity Chromatography of Biological Recognition,* Chaiken, I.M., Wilchek, M., and Parikh, I., Eds., Academic Press, Orlando, FL, 1983.

133. Sudi, P., Dala, E., and Szajani, B., Preparation, characterization, and application of a novel immobilized carboxypeptidase B, *Appl. Biochem. Biotech.,* 22, 31–43, 1989.

134. Potempa, L.A., Motie, M., Anderson, B., Klein, E., and Baurmeister, U., Conjugation of a modified form of human C-reactive protein to affinity membranes for extracorporeal adsorption, *Clin. Mater.,* 11, 105–117, 1992.

135. Tiller, J.C., Rieseler, R., Berlin, P., and Klemm, D., Stabilization of activity of oxidoreductases by their immobilization onto special functionalized glass and novel aminocellulose film using different coupling reagents, *Biomacromolecule,* 3, 1021–1029, 2002.

136. Mirsky, V.M., Riepl, M., and Wolfbeis, O.S., Capacitive monitoring of protein immobilization and antigen-antibody reactions on monomolecular alkylthiol films on gold electrodes, *Biosens. Bioelectron.,* 12, 977–989, 1997.

137. Chellapandian, M. and Sastry, C.A., Covalent linking of alkaline protease on trichlorotriazine-activated nylon, *Bioproc. Eng.,* 15, 95–98, 1996.

138. Bisse, E. and Wieland, H., Coupling of *m*-aminophenylboronic acid to *s*-triazine-activated Sephacryl: Use in the affinity chromatography of glycated hemoglobins, *J. Chromatogr.,* 575, 223–228, 1992.

139. Smith, N.L. and Lenhoff, H.M., Covalent binding of proteins and glucose 6-phosphate dehydrogenase to cellulosic carriers activated with *s*-triazine trichloride, *Anal. Biochem.,* 61, 392–415, 1974.

140. Lihme, A., Schafer-Nielsen, C., Larsen, K.P., Mueller, K.G., and Boeg-Hansen, T.C., Divinylsulfone-activated agarose: Formation of stable and non-leaking affinity matrixes by immobilization of immunoglobulins and other proteins, *J. Chromatogr.,* 376, 299–305, 1986.

141. Leckband, D. and Langer, R., An approach for the stable immobilization of proteins, *Biotechnol. Bioeng.,* 37, 227–237, 1991.
142. Kobayashi, M., Yanagihara, S., Kitae, T., and Ichishima, E., Use of water-soluble carbodi-imide (EDC) for immobilization of EDC-sensitive dextranase, *Agric. Biol. Chem.,* 53, 2211–2216, 1989.
143. Nilsson, K. and Mosbach, K., p-Toluenesulfonyl chloride as an activating agent of agarose for the preparation of immobilized affinity ligands and proteins, *Eur. J. Biochem.,* 112, 397–402, 1980.
144. Hermanson, G.T., *Bioconjugate Techniques,* Academic Press, New York, 1996.
145. Leyden, D.E. and Collins, W.T., *Silylated Surfaces,* Gordon and Breach, New York, 1980.
146. Rehbock, B. and Berger, R.G., Covalent immobilization of a hydroperoxide lyase from mung beans (*Phaseolus radiatus L.*), *Biotech. Tech.,* 12, 539–544, 1998.
147. Lee, Y.W., Reed-Mundell, J., Zull, J.E., and Sukenik, C.N., Electrophilic siloxane-based self-assembled monolayers for thiol-mediated anchoring of peptides and proteins, *Langmuir,* 9, 3009–3014, 1993.
148. Domen, P.L., Nevens, J.R., Mallia, A.K., Hermanson, G.T., and Klenk, D.C., Site-directed immobilization of proteins, *J. Chromatogr.,* 510, 293–302, 1990.
149. Chu, V.P. and Tarcha, P.J., Protein-reactive, molded polystyrene surfaces having applications to immunoassay formats, *J. Appl. Polym. Sci.,* 34, 1917–1924, 1987.
150. Leckband, D. and Langer, R., An approach for the stable immobilization of proteins, *Biotech. Bioeng.,* 37, 227–237, 1991.
151. Kato, S., Aizawa, M., and Suzuki, S., Effects of chemical modifications of membranes on transmembrane potential, *J. Membrane Sci.,* 3, 29–38, 1978.
152. Tada, T., Mano, K., Yoshida, E., Tanaka, N., and Kunugi, S., SH-group introduction to the *N*-terminal of subtilisin and preparation of immobilized and dimeric enzymes, *Bull. Chem. Soc. Japan,* 75, 2247–2251, 2002.
153. Hughes, K.A., Lucas, D.D., Stolowitz, M.L., and Wiley, J.P., Novel affinity tools for protein immobilization: Implications for proteomics, *Am. Biotech. Lab.,* 19, 36–38, 2001.
154. Rezania, A., Johnson, R., Lefkow, A.R., and Healy, K.E., Bioactivation of metal oxide surfaces, 1: Surface characterization and cell response, *Langmuir,* 15, 6931–6939, 1999.
155. Tournier, E.J.M., Wallach, J., and Blond, P., Sulfosuccinimidyl 4-(*N*-maleimidomethyl)-1-cyclohexane carboxylate as a bifunctional immobilization agent: Optimization of the coupling conditions, *Anal. Chim. Acta,* 361, 33–44, 1998.
156. Yeung, C. and Leckband, C., Molecular level characterization of microenvironmental influences on the properties of immobilized proteins, *Langmuir,* 13, 6746–6754, 1997.
157. Prisyazhnoi, V.S., Fusek, M., and Alakhov, Y.B., Synthesis of high-capacity immunoaffinity sorbents with oriented immobilized immunoglobulins or their Fab' fragments for isolation of proteins, *J. Chromatogr.,* 424, 243–253, 1998.
158. May, L.M. and Russell, D.A., The characterization of biomolecular secondary structures by surface plasmon resonance, *Analyst,* 127, 1589–1595, 2002.
159. Basinska, T. and Caldwell, K.D., Colloidal particles as immunodiagnostics: Preparation and FFF characterization, *ACS Symp. Ser.,* 731, 162–177, 1999.
160. Thompson, M. and McGovern, M.E., High-surface-density covalent immobilization of oligonucleotide monolayers, *PCT Int. Appl.,* 1999.
161. Ringler, P., Kessler, P., Menez, A., and Brisson, A., Purification of the nicotinic acetylcholine receptor protein by affinity chromatography using a regioselectively modified and reversibly immobilized α-toxin from *Naja nigricollis, Biochim. Biophys. Acta,* 1324, 37–46, 1997.
162. Li, J.T., Carlsson, J., Lin, J.N., and Caldwell, K.D., Chemical modification of surface-active poly(ethylene oxide)-poly(propylene oxide) triblock copolymers, *Bioconjugate Chem.,* 7, 592–599, 1996.
163. Brena, B.M., Ovsejevi, K., Luna, B., and Batista-Viera, F., Thiolation and reversible immobilization of sweet potato β-amylase on thiolsulfonate-agarose, *J. Mol. Catal.,* 84, 381–390, 1993.
164. Egorov, T.A., Svenson, A., Ryden, L., and Carlsson, J., Rapid and specific method for isolation of thiol-containing peptides from large proteins by thiol-disulfide exchange on a solid support, *Proc. Natl. Acad. Sci. U.S.A.,* 72, 3029–3033, 1975.

165. Porath, J. and Axen, R., Immobilization of enzymes to agar, agarose, and Sephadex supports, *Methods Enzymol.*, 44, 19–45, 1976.

166. Finlay, T.H., Troll, V., Levy, M., Johnson, A.J., and Hodgins, L.T., New methods for the preparation of biospecific adsorbents and immobilized enzymes utilizing trichloro-*s*-triazine, *Anal. Biochem.*, 87, 77–90, 1978.

167. Porath, J., General methods and coupling procedures, *Methods Enzymol.*, 34, 13–30, 1974.

168. Larsson, P.O., High-performance liquid affinity chromatography, *Methods Enzymol.*, 104, 212–223, 1984.

169. Bilkova, Z., Slovakova, M., Horak, D., Lenfeld, J., and Churacek, J., Enzymes immobilized on magnetic carriers: Efficient and selective system for protein modification, *J. Chromatogr. B*, 770, 177–181, 2002.

170. Vankova, H., Kucerova, Z., and Turkova, J., Reversed-phase high-performance liquid chromatography of peptides of porcine pepsin prepared by the use of various forms of immobilized α-chymotrypsin, *J. Chromatogr. B*, 753, 37–43, 2001.

171. van Bommel, M.R., de Jong, A.P.J.M., Tjaden, U.R., Irth, H., and van der Greef, J., Enzyme amplification as detection tool in continuous-flow systems, II: Development of an enzyme-amplified biochemical detection system coupled on-line to flow-injection analysis, *J. Chromatogr. A*, 855, 383–396, 1999.

172. Pande, C.S. and Gupta, N., Gamma-radiation-induced graft copolymerization of acrylamide onto crosslinked poly(*N*-vinylpyrrolidone), *J. Appl. Polym. Sci.*, 71, 2163–2168, 1999.

173. Turkova, J., Kucerova, Z., Vankova, H., and Benes, M.J., Stabilization and oriented immobilization of glycoproteins, *Int. J. BioChromatogr.*, 3, 45–55, 1997.

174. O'Shannessy, D.J. and Hoffman, W.L., Site-directed immobilization of glycoproteins on hydrazide-containing solid supports, *Biotech. Appl. Biochem.*, 9, 488–496, 1987.

175. Su, C.C., Wu, T.Z., Chen, L.K., Yang, H.H., and Tai, D.F., Development of immunochips for the detection of dengue viral antigens, *Anal. Chim. Acta*, 479, 117–123, 2003.

176. Wade, J.D., Domagala, T., Rothacker, J., Catimel, B., and Nice, E., Use of thiazolidine-mediated ligation for site specific biotinylation of mouse EGF for biosensor immobilization, *Lett. Peptide Sci.*, 8, 211–220, 2001.

177. Salnikow, J., Solid-phase sequencing of peptides and proteins, in *Protein Structure Analysis*, Kamp, R.M., Choli-Papadopoulou, T., and Wittmann-Liebold, B., Eds., Springer, New York, 1997, 153–165.

178. Bernfeld, P. and Wan, J., Antigens and enzymes made insoluble by entrapping them into lattices of synthetic polymers, *Science*, 142, 678–679, 1963.

179. O'Driscoll, K.F., Techniques of enzyme entrapment in gels, *Methods Enzymol.*, 44, 169–183, 1976.

180. Lindmark, R., Larsson, E., Nilsson, K., and Sjoquist, J., Immobilization of proteins by entrapment in polyacrylamide microbeads, *J. Immunol. Methods*, 49, 159–177, 1982.

181. Thomson, R.A., Andersson, S., and Allenmark, S., Direct liquid chromatographic separation of enantiomers on immobilized protein stationary phases, VII: Sorbent obtained by entrapment of cross-linked bovine serum albumin in silica, *J. Chromatogr.*, 465, 263–270, 1989.

182. Trevan, M.D. and Grover, S., Relationship of the enzymic activity of polyacrylamide-gel-entrapped trypsin to pH, *Biochem. Soc. Trans.*, 71, 28–30, 1979.

183. Yasuda, H., Lamaze, C.E., and Peterlin, A., Diffusive and hydraulic permeabilities of water in water-swollen polymer membranes, *J. Polym. Sci. Polym. Phys. Ed.*, 9, 1117–1131, 1971.

184. Westermann, R., Simple drying of polyacrylamide gels for fluorography and storage, *Electrophoresis*, 6, 136–137, 1985.

185. Nunez-Olea, J. and Sanchez-Ruiz, J.M., The effect of gradual dehydration on the thermal stability of a protein entrapped in a polymeric network, *J. Chem. Soc. Perkin Trans. 2*, 4, 643–644, 1995.

186. Cardile, V., Renis, M., Gentile, B., and Panico, A.M., Activity of liposome-entrapped immunomodulator oligopeptides on human epithelial thymic cells, *Pharm. Pharmacol. Com.*, 6, 381–386, 2000.

187. Gregoriadis, G., McCormack, B., Obrenovic, M., Perrie, Y., and Saffie, R., Liposomes as immunological adjuvants and vaccine carriers, *Methods Mol. Med.*, 42, 137–150, 2000.

188. Karau, C., Pongpaibul, Y., and Schmidt, P.C., Quantitative evaluation of human leukocyte interferon-α entrapped in liposomes, *Drug Delivery,* 3, 59–62, 1996.
189. Bangham, A.D., Hill, M.W., and Miller, N.G.A., Preparation and use of liposomes as models of biological membranes, *Methods Membrane Biol.,* 1, 1–68, 1974.
190. Gregoriadis, G., Enzyme entrapment in liposomes, *Methods Enzymol.,* 44, 218–227, 1976.
191. Kierstan, M.P.J. and Coughlan, M.P., Immobilization of proteins by noncovalent procedures: Principles and application, in *Protein Immobilization,* Taylor, R.F., Ed., Marcel Dekker, New York, 1991.
192. Papahadjopoulos, D. and Miller, N., Phospholipid model membranes, I: Structural characteristics of hydrated liquid crystals, *Biochim. Biophys. Acta,* 135, 624–638, 1967.
193. Gregoriadis, G. and Allison, A.C., Entrapment of proteins in liposomes prevents allergic reactions in preimmunized mice, *FEBS Lett.,* 45, 71–74, 1974.
194. Taylor, R.F., Commercially available supports for protein immobilization, in *Protein Immobilization,* Taylor, R.F., Ed., Marcel Dekker, New York, 1991.
195. Lu, Z., Lu, F., Bie, X., and Fujimurai, T., Immobilization of yeast cells with polymeric carrier cross-linked using radiation technique, *J. Agric. Food Chem.,* 50, 2798–2801, 2002.
196. Norouzian, D., Javadpour, S., Moazami, N., and Akbarzadeh, A., Immobilization of whole cell penicillin G acylase in open pore gelatin matrix, *Enzyme Microb. Tech.,* 30, 26–29, 2002.
197. Perrot, F., Hebraud, M., Charlionet, R., Junter, G.A., and Jouenne, T., Cell immobilization induces changes in the protein response of *Escherichia coli* K-12 to a cold shock, *Electrophoresis,* 22, 2110–2119, 2001.
198. Wang, A.A., Mulchandani, A., and Chen, W., Whole-cell immobilization using cell surface-exposed cellulose-binding domain, *Biotech. Prog.,* 17, 407–411, 2001.
199. Bickerstaff, G.F., Immobilization of enzymes and cells: Some practical considerations, *Methods Biotech.,* 1, 1–11, 1997.
200. Champagne, C.P., Immobilized cell technology in food processing, *Prog. Biotech.,* 11, 633–640, 1996.
201. Li, R.H., Altreuter, D.H., and Gentile, F.T., Transport characterization of hydrogel matrixes for cell encapsulation, *Biotech. Bioeng.,* 50, 365–373, 1996.
202. Jen, A.C., Wake, M.C., and Mikos, A.G., Review: Hydrogels for cell immobilization, *Biotech. Bioeng.,* 50, 357–364, 1996.
203. Krisch, J., Buzas, Z., Dallman, K., Toth, M., Gimesi, I., and Szajani, B., Application of preformed cellulose beads as a support in cell immobilization, *Biotech. Tech.,* 9, 221–224, 1995.
204. Karel, S.F., Libicki, S.B., and Robertson, C.R., The immobilization of whole cells: Engineering principles, *Chem. Eng. Sci.,* 40, 1321–1354, 1985.
205. Birnbaum, S., Larsson, P.O., and Mosbach, K., Immobilized cells, in *Solid Phase Biochemistry,* Scouten, W.H., Ed., Wiley, New York, 1983.
206. Hikuma, M., Kubo, T., Yasuda, T., Karube, I., and Suzuki, S., Amperometric determination of acetic acid with immobilized *Trichosporon brassicae, Anal. Chim. Acta,* 109, 33–38, 1979.
207. Walters, R.R., Moriarty, B.E., and Buck, R.P., Pseudomonas bacterial electrode for determination of L-histidine, *Anal. Chem.,* 52, 1680–1684, 1980.
208. Jensen, M.A. and Rechnitz, G.A., Bacterial membrane electrode for L-cysteine, *Anal. Chim. Acta,* 101, 125–130, 1978.
209. Lin, J. and Brown, C.W., Sol-gel glass as a matrix for chemical and biochemical sensing, *TrAC,* 16, 200–211, 1997.
210. Kauffmann, C. and Mandelbaum, R.T., Entrapment of atrazine chlorohydrolase in sol-gel glass matrix, *J. Biotechnol.,* 62, 169–176, 1998.
211. Zusman, R., Beckman, D.A., Zusman, I., and Brent, R.L., Purification of sheep immunoglobulin G using protein A trapped in sol-gel glass, *Anal. Biochem.,* 201, 103–106, 1992.
212. Narang, U., Rahman, M.H., Wang, J.H., Prasad, P.N., and Bright, F.V., Removal of ribonucleases from solution using an inhibitor-based sol-gel-derived biogel, *Anal. Chem.,* 67, 1935–1939, 1995.
213. Xu, J., Dong, H., Feng, Q., and Wei, Y., Direct immobilization of horseradish peroxidase in hybrid mesoporous sol-gel materials, *Polym. Prepr.,* 41, 1044–1045, 2000.

214. Flora, K. and Brennan, J.D., Comparison of formats for the development of fiber-optic biosensors utilizing sol-gel derived materials entrapping fluorescently labelled protein, *Analyst*, 124, 1455–1462, 1999.

215. Bronshtein, A., Aharonson, N., Turniansky, A., and Altstein, M., Sol-gel based immunoaffinity chromatography: Application to nitroaromatic compounds, *Chem. Mater.*, 12, 2050–2058, 2000.

216. Kriz, D., Ramström, O., and Mosbach, K., Molecular imprinting: New possibilities for sensor technology, *Anal. Chem.*, 69, 345A–349A, 1997.

217. Sellergren, B., Imprinted polymers with memory for small molecules, proteins, or crystals, *Angew. Chem. Int. Ed.*, 39, 1031–1037, 2000.

218. Bruggemann, O., Haupt, K., Ye, L., Yilmaz, E., and Mosbach, K., New configurations and applications of molecularly imprinted polymers, *J. Chromatogr. A*, 889, 15–24, 2000.

219. Mosbach, K. and Ramström, O., The emerging technique of molecular imprinting and its future impact on biotechnology, *BioTechnology*, 14, 163–170, 1996.

220. Matsui, J., Miyoshi, Y., Doblhoff-Dief, O., and Takeuchi, T., A molecularly imprinted synthetic polymer receptor selective for atrazine, *Anal. Chem.*, 67, 4404–4408, 1995.

221. Takeuchi, T. and Haginaka, J., Separation and sensing based on molecular recognition using molecularly imprinted polymers, *J. Chromatogr. B*, 728, 1–20, 1999.

222. Kempe, M., Antibody-mimicking polymers as chiral stationary phases in HPLC, *Anal. Chem.*, 68, 1948–1953, 1996.

223. Hjertén, S., Liao, J.L., Nakazato, K., Yang, Y., Zamaratskaia, G., and Zhang, H.X., Gels mimicking antibodies in their selective recognition of proteins, *Chromatographia*, 44, 227–234, 1997.

224. Shi, H.Q., Tsai, W.B., Garrison, M.D., Ferrari, S., and Ratner, B.D., Template-imprinted nanostructured surfaces for protein recognition, *Nature*, 398, 593–597, 1999.

225. Schweitz, L., Andersson, L.I., and Nilsson, S., Molecular imprinting for chiral separations and drug screening purposes using monolithic stationary phases in CEC, *Chromatographia*, 49, S93–S94, 1999.

226. Nilsson, S., Schweitz, L., and Petersson, M., Three approaches to enantiomer separation of beta-adrenergic antagonists by capillary electrochromatography, *Electrophoresis*, 18, 884–890, 1997.

227. Ramströem, O., Nicholls, I.A., and Mosbach, K., Synthetic peptide receptor mimics: Highly stereoselective recognition in non-covalent molecularly imprinted polymers, *Tetrahedron: Asymmetry*, 5, 649–656, 1994.

228. Starkenstein, E., Uber fermentwirkung und deren beeinflussung durch neutralsalze, *Biochim. Zeitschrift*, 24, 210–218, 1910.

229. Lee, S.K., Park, S.Y., and Yang, C.H., An improved method for the purification of α-amylase by affinity chromatography, *Korean Biochem. J.*, 27, 576–578, 1994.

230. Weber, M., Foglietti, M.J., and Percheron, F., Purification of α-amylases by affinity chromatography on crosslinked starch, *Biochimie*, 58, 1299–1302, 1976.

231. Margolin, A.L. and Vilenchik, L.Z., Crosslinked protein crystals as universal separation media, *PCT Int. Appl.*, p. 115, 1988.

232. Birnbaum, S. and Nilsson, S., Protein-based capillary affinity gel-electrophoresis for the separation of optical isomers, *Anal. Chem.*, 64, 2872–2874, 1992.

233. Ljungberg, H. and Nilsson, S., Protein-based capillary affinity gel-electrophoresis for chiral separation of beta-adrenergic blockers, *J. Liq. Chromatogr.*, 18, 3685–3698, 1995.

234. Knox, J.H. and Grant, I.H., Electrochromatography in packed tubes using 1.5- to 50-μm silica gels and ODS-bonded silica gels, *Chromatographia*, 32, 317–328, 1991.

4

Application and Elution in Affinity Chromatography

David S. Hage, Hai Xuan, and Mary Anne Nelson
Department of Chemistry, University of Nebraska, Lincoln, NE

CONTENTS

4.1 INTRODUCTION

The previous chapters have examined the support materials and immobilization methods that are often used in affinity chromatography, and the ligands used in these columns will be discussed in Chapters 5 to 10. However, another set of factors in the use of affinity chromatography must also be considered: the application and elution conditions. If these conditions are not selected properly, the target of interest may not bind to the affinity column or may not be recovered in an appropriate form.

As stated in Chapter 1, the sample in affinity chromatography is usually injected or applied in the presence of a mobile phase that has the appropriate pH, ionic strength, and solvent composition for solute-ligand binding. This solvent, which represents the weak mobile phase for the column, is known as the *application buffer*. As the sample passes through the column under these conditions, compounds that are complementary to the affinity ligand will bind. However, due to the high selectivity of this interaction, other solutes in the sample will tend to wash through the column as a nonretained peak.

After all nonretained components have been removed from the column, the retained solutes are eluted by applying a solvent that dissociates them from the ligand. This solvent, which represents the strong mobile phase, is known as the *elution buffer*. As the solutes of interest elute from the column, they are either quantitated directly or collected for later use. The application buffer is then reapplied to the system and the column is allowed to regenerate prior to the next sample application.

A typical scheme used for this process is the *step-elution mode*, as shown in Figure 4.1a. This is also known as the "on/off" mode of affinity chromatography. This

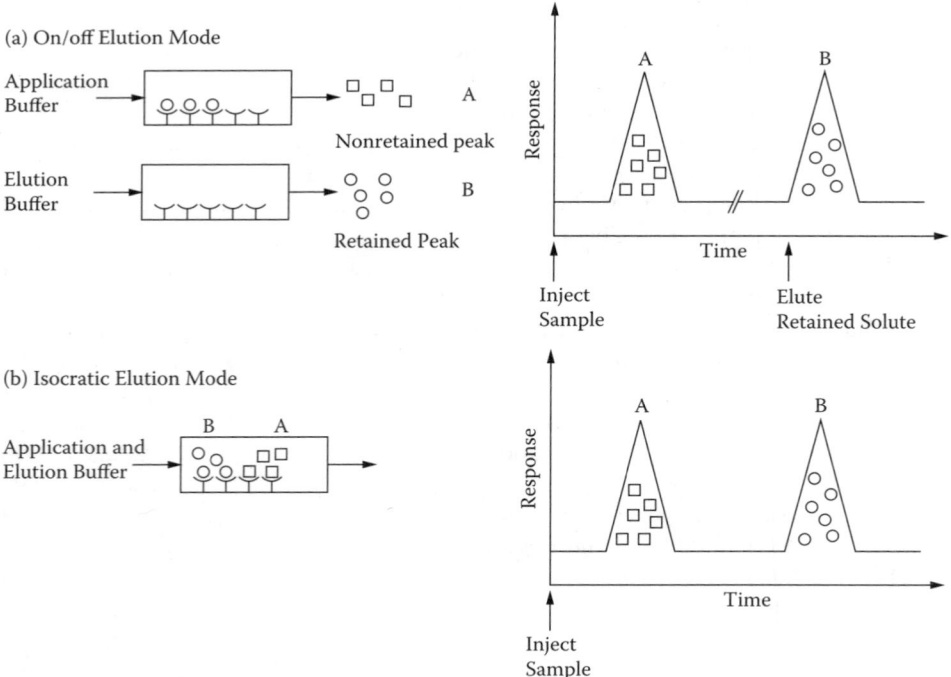

Figure 4.1 Separation schemes for affinity chromatography involving (a) the traditional on/off or step-elution mode of sample application and elution, or (b) isocratic elution (i.e., weak-affinity chromatography).

approach is employed when working with ligands that have a high affinity for the target under the application conditions [1–4]. However, there are some occasions in which isocratic elution can also be used in affinity chromatography. This is possible if the target and ligand have reasonably weak interactions, allowing the same solvent to be utilized for both sample application and elution. This approach is illustrated in Figure 4.1b and is known as *weak affinity chromatography* (*WAC*) or *dynamic affinity chromatography* [5–10]. This chapter examines the theory of sample application and elution in affinity chromatography and discusses several practical considerations in the selection of these items.

4.2 THEORY OF SAMPLE APPLICATION AND ELUTION

In order to properly use an affinity column, it is necessary to understand the basic processes that occur during sample application and elution. This section discusses the general factors that affect the retention of a solute on an affinity column and considers how these factors are related to the affinity ligand. The similarities and differences between affinity chromatography and other types of liquid chromatography are also examined.

4.2.1 General Factors Affecting Retention

A number of factors are important in determining the retention and elution of a compound on an affinity column. These factors include the strength of the solute-ligand interaction, the amount of immobilized ligand present, and the kinetics of solute-ligand association and dissociation. In the case of a solute (*A*) that has single-site binding to a ligand (*L*), the following equations can be used to describe the interactions between the solute and ligand in an affinity column [4].

$$A + L \underset{k_d}{\overset{k_a}{\rightleftharpoons}} A - L \tag{4.1}$$

$$K_a = k_a/k_d = \{A - L\}/[A]\{L\} \tag{4.2}$$

In these equations, K_a is the association equilibrium constant for the binding of A with L, and $A - L$ is the resulting solute-ligand complex. The term $[A]$ is the mobile-phase concentration of A at equilibrium, while $\{L\}$ and $\{A - L\}$ represent the surface concentrations of the ligand and solute-ligand complex at equilibrium. The term k_a is the second-order association rate constant for solute-ligand binding, and k_d is the first-order dissociation rate constant for the solute-ligand complex.

At equilibrium, the retention of a solute in this column will be described by Equations 4.3 and 4.4,

$$k = K_a m_L/V_M \tag{4.3}$$

$$= (t_R/t_M) - 1 \tag{4.4}$$

where k is the retention factor for the injected solute, t_R is the solute's retention time, and t_M is the void time of the column. The factor m_L in Equation 4.3 represents the moles of active ligand in the column, and V_M is the column void volume. These equations indicate that the retention factor (and retention time) for a solute injected onto an affinity column will depend on both the strength of its binding to the ligand (as described by K_a) and the amount of ligand in the column (as given by m_L/V_M).

The strong retention seen for many solutes with affinity columns can be explained by using these equations along with the association constants and binding capacities that

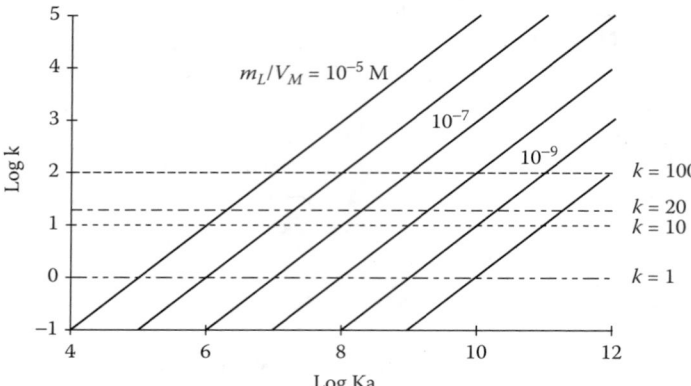

Figure 4.2   Effect of the association constant between a solute and ligand (K_a) and the effective concentration of ligand (m_L/V_M) on the predicted retention factor for the solute on an affinity column containing the ligand. These results were calculated using Equation 4.3. This assumes that a 1:1 interaction is present between the solute and ligand, with no other source of solute retention being present in the system.

are typically present for such separations. This is illustrated in Figure 4.2. For example, in the case of a column containing immobilized polyclonal antibodies, a typical association constant would be 10^8 to 10^{10} M^{-1}, and the concentration of active ligand is often around 10 μM [4]. Using these values with Equation 4.3 gives a calculated retention factor of 1,000 to 100,000. For a 10 cm $\times$ 4.1 mm I.D. column operated at 1.0 ml/min, this would give a mean retention time in Equation 4.4 of 1 to 90 days. In this situation, the only way of eluting the retained solute in a reasonable time is to change the mobile phase or column conditions to lower the association constant for solute-ligand binding. This is accomplished by using the on/off elution scheme in Figure 4.1a.

If the ligand has weaker binding for its target, it may be possible to use isocratic elution. This becomes possible when the target's retention time is on the order of minutes instead of hours or days. Fast association and dissociation kinetics are also required to allow a large number of interactions to occur between the target and ligand as the sample passes through the column. This helps increase the number of theoretical plates in the system and makes it easier to resolve solutes with similar binding properties. Such an elution scheme generally becomes possible when the association constant is less than 10^6 M^{-1} for low-capacity columns or 10^4 M^{-1} for high-capacity columns [4–9].

4.2.2   Relationship of Elution Method to the Affinity Ligand

As Figure 4.2 suggests, the association constant between the solute and ligand is a key factor in determining which elution method can be used with an affinity column. All affinity ligands can be placed into one of two categories: (1) high-specificity ligands and (2) general, or group-specific, ligands [3, 4]. The high-specificity ligands include agents like antibodies and enzyme inhibitors, which tend to bind only one or a few closely related targets. This specific binding usually requires that the ligand have a large association constant for the target and a low binding affinity for other sample components. As a result, these ligands generally require the use of step elution.

General ligands are less specific in nature and are used to bind a family of related molecules. These ligands can be of either biological or nonbiological origin and range from agents like protein A and protein G to boronates, synthetic dyes, and immobilized metal-ion chelates. Since these ligands interact with many substances, they tend to have a lower selectivity and weaker binding than high-specificity ligands. This often allows the use of isocratic conditions with such ligands. However, there are exceptions to this rule. For example, protein A has an association constant of over 10^8 M^{-1} for some antibodies, making it necessary to use step elution [4, 11]. Also, some ligands usually considered high-specificity ligands, such as antibodies, can be selected to have weak binding [5, 8] or be used to retain a broad class of solutes if they recognize a feature common to all of the desired targets [12, 13].

4.2.3 Affinity Chromatography versus Other Chromatographic Methods

The strong retention and selectivity of many affinity ligands gives affinity chromatography some unique properties when compared with other chromatographic methods. This can be illustrated by the general resolution equation for chromatography, as shown below [14].

$$R_s = (N^{1/2}/4)[(\alpha - 1)/\alpha][k_2/(1 + k_2)] \tag{4.5}$$

In this equation, R_s is the resolution between two neighboring peaks with approximately equal widths, N is the number of theoretical plates for the chromatographic system, and k_2 is the retention factor for the second eluting compound. The term α is the separation factor for the two peaks, where $\alpha = k_2/k_1$, and k_1 is the retention factor for the first eluting substance.

According to Equation 4.5, there are three approaches that can be used to improve the resolution of a chromatographic separation. The first involves increasing the efficiency of the system, as represented by N. This approach is often employed in gas chromatography (GC) or reversed-phase liquid chromatography (RPLC), where longer columns or more efficient supports are used to increase the resolution or speed of a separation. The second item that can be varied is the retention, as represented by k_2, where an increase in retention leads to improved resolution. In GC, this is often achieved by using a lower column temperature, while in RPLC this involves the use of a weaker mobile phase or a stationary phase with stronger solute retention. The final item that can be used to improve resolution is the selectivity. This is represented by the term α and is adjusted by altering the nature of the mobile phase (in liquid chromatography) or stationary phase (in liquid or gas chromatography) to alter the retention of one solute versus another.

Although any of these methods could be used to improve resolution in affinity chromatography, the high selectivity of affinity ligands and strong retention for their targets means that α and k_2 are usually the dominant factors in controlling resolution. As a result, the efficiency of the system takes a secondary role, with even relatively low-performance supports giving rise to good separations. This is in contrast to GC and RPLC, where the presence of a much lower selectivity and retention makes high efficiency essential to achieving many separations [3].

An important difference between affinity chromatography and other liquid chromatographic methods is that the former technique generally involves many types of interactions between the ligand and target. These interactions can include steric effects, hydrogen bonding, ionic interactions, van der Waals forces, dipole-dipole interactions, and even covalent bonds. In comparison, methods like RPLC, ion-exchange chromatography (IEC), and size-exclusion chromatography (SEC) emphasize just one or a few of these interactions. It is this combination of multiple interactions that gives affinity chromatography its high retention and selectivity, but this feature also makes it more difficult to determine the most effective conditions for eluting solutes from affinity columns.

Another difference between an affinity ligand and other stationary phases for liquid chromatography concerns the amount of stationary phase in the column. In affinity chromatography, this is determined by the amount of active ligand present. However, since many affinity ligands are biomacromolecules, these ligands occupy a much larger area than the alkane chains or ion-exchange sites used in RPLC or IEC. This gives affinity columns a much lower sample capacity, making it important to characterize this item whenever such a column is used for preparative or analytical work [3].

A related item that should be considered is that the binding capacity for an affinity column can vary even for two closely related compounds binding to the same ligand. This has been demonstrated in work examining the retention of (R)- and (S)-warfarin on an immobilized human serum albumin (HSA) column. Although these compounds have the same general binding region on HSA, they are believed to have slightly different points of interactions with this site. As a result, HSA has a measurable difference in binding capacity for these solutes, which helps determine their overall retention and degree of separation on an HSA column [15].

4.3   SELECTION OF APPLICATION CONDITIONS

The successful use of affinity chromatography requires that the target first be able to bind effectively to the immobilized ligand. This means that the solvent conditions must be adequate for such binding to occur. In addition, the column must be able to retain the amount of target that is being passed through the column with minimal binding by other sample components. This section examines these factors and discusses how they can be optimized for a given affinity separation.

4.3.1   Solvent Considerations

The first item to consider in choosing application conditions is the mobile phase that will be used to apply the sample to the column. Most application buffers in affinity chromatography are solvents that mimic the pH, ionic strength, and polarity experienced by the solute and ligand in their natural environment. Any cofactors or metal ions required for solute-ligand binding should also be present in this solvent. Under these conditions, the solute will probably have its highest association constant for the ligand and its highest degree of retention on the affinity column.

Although the application buffer for many biological agents is one that possesses a neutral pH, this does vary from one ligand to the next. For instance, the immunoglobulin binding agent protein A has optimum binding at pH 8.5 but also has sufficient binding at pH 7 to 7.5 for use in antibody isolation. However, another agent used for the same purpose, protein G, has optimum binding at a pH as low as 5.0 (see Chapter 5).

Many nonbiological ligands also have optimum binding under conditions that differ from physiological conditions. One common example is a boronate-based ligand, which generally has its best binding to solutes at a pH of 8.0 or greater. In some situations, the ligand may even require a nonaqueous application buffer. For instance, many molecularly imprinted polymers (MIPs) are formed and used in the presence of an organic solvent to maximize their ability to form hydrogen bonds with solutes. More details on these various ligands can be found in Chapters 8 and 30.

The proper choice of an application buffer can help minimize nonspecific binding due to undesired sample components. For example, coulombic interactions between solutes and the support can often be decreased by altering the ionic strength and pH of the application buffer. In addition, surfactants and blocking agents (e.g., Triton X-100, Tween-20, bovine serum albumin, gelatin, etc.) may be added to the buffer to prevent nonspecific retention of solutes on the support or affinity ligand [4].

4.3.2 Factors Related to the Column and Sample

The activity of the immobilized ligand should be considered in determining how much sample can be applied to the affinity column. A rough indication of the maximum column binding capacity can be made by assaying the total amount of ligand present. However, a better approach is to actually measure the ligand's activity. This can be accomplished by continuously applying a known concentration of solute to the affinity column (i.e., performing frontal analysis), as described in Chapters 11 and 22. A third approach that can be used for this purpose is to combine the immobilized ligand with a known excess of solute and measure the amount of free solute that remains after binding has occurred.

With some affinity systems, it is possible to see a large amount of nonretained solute during the application step, even when the amount of injected target is significantly less than the column binding capacity. This phenomenon, known as the *split-peak effect*, is caused by the presence of slow adsorption or mass-transfer kinetics within the column [16–18]. An example of such an effect is shown in Figure 4.3. This effect has been reported with many types of affinity ligands, but tends to occur with high-performance supports because of the more rapid flow rates often used with these materials [1, 4, 16–18]. This effect can be minimized by reducing the flow rate used for sample injection, increasing the column size, or placing a more efficient support within the column. In some cases, changing to a different immobilization method may help provide a ligand with more rapid binding kinetics [17].

The amount of applied sample is another item to consider in the design of an affinity method. For preparative applications, it is generally desirable to pass the maximum amount of target possible through the column while also providing good separation times and minimizing the amount of target that passes nonretained through the system. Some guidelines for the selection of conditions for preparative-scale affinity columns are given in

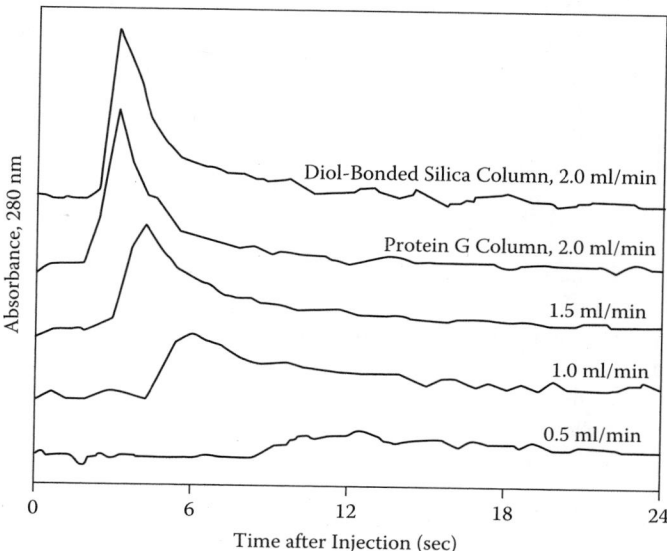

Figure 4.3 The split-peak effect, in which the amount of nonretained solute increases as a function of flow rate, even when applied to an affinity column with an excess of ligand. These chromatograms were obtained for the injection of rabbit immunoglobulin G onto a protein G column and an inert diol-bonded silica control column. (Reproduced with permission from Rollag, J.G. and Hage, D.S., *J. Chromatogr. A*, 795, 185–198, 1998.)

Chapter 11. In analysis, the goal is usually to inject a small quantity of sample that allows for good analyte detection but avoids overloading the chromatographic system. This requires that the amount of injected target be low enough so that there is no measurable effect on the observed retention and efficiency for this target on the affinity column [19]. Approaches for identifying such conditions are discussed in Chapter 22.

The nonspecific binding of other sample components should also be considered during the selection of application conditions. Such binding would be expected to increase as more sample is applied to the column and as more time is allowed for this interaction. This can be minimized by using columns with low surface areas, employing immobilization methods that provide inert surfaces after ligand attachment, using mobile-phase additives that block nonspecific binding sites, and adding a wash step between sample application and elution. This issue is particularly important when the isolation of a high-purity product is desired or when the affinity column is to be used for the detection of a trace component in a complex mixture.

4.3.3   Description of Sample Retention

In the design of a new affinity method, it is valuable to use chromatographic theory as an aid in the selection and optimization of sample application conditions. For instance, in the case of isocratic elution, where relatively fast association and dissociation occur between the target and ligand, Equation 4.3 is useful for estimating the retention factor for the solute as it passes through the column. In fact, such an approach has been shown to be an effective aid in understanding the mechanisms of systems with weak affinity interactions by using this and related expressions to study how changes in the temperature or mobile-phase composition affect solute-ligand binding [19]. An example of such work is given in Figure 4.4 [20].

When step elution is employed, the target generally has tight binding to the ligand during the application step and usually has only slow dissociation during this process. In this situation, the binding of the target to the ligand can often be approximated by an irreversible adsorption reaction, in which the dissociation of A from L is considered negligible on the time scale of the experiment. If the actual binding of A to L is the rate-limiting step in this process, as opposed to mass transfer of A from the mobile phase to the stationary phase, the following expression can be used to estimate the relative fraction of $A(f)$ that will not bind to L for a given load of sample (Load A) and flow rate [21–23].

$$f = (S_0/\text{Load A})\ln[1 + (e^{\text{Load A}/S_0} - 1)\, e^{-1/S_0}] \qquad (4.6)$$

In this equation, Load A is the ratio of the moles of A applied versus the moles of active ligand in the column. The term S_0 is a factor that represents the effects of flow rate (F) and solute-ligand association kinetics, as given by the relationship $S_0 = F/(k_a m_L)$. This equation shows that the degree of target binding will be highest at low flow rates (small S_0 values) and low sample loads (small Load A). The presence of fast association kinetics, which will be aided by large values for k_a and m_L, also increases the efficiency of target capture. An illustration of how these various items alter solute binding to an affinity column is given in Chapter 29.

Equation 4.6 is especially useful when dealing with HPLC supports, which have relatively fast mass-transfer properties. For low- or medium-performance supports, alternative equations must be used, since mass transfer now becomes the rate-limiting step in target binding [16]. One way this can be determined is by estimating the average time it takes the target to diffuse a given distance within the column. As discussed in Chapter 16, this can be determined by using the following equation, where t_D (in sec) is the time allowed for diffusion, D is the effective diffusion coefficient (in cm^2/sec) of the target in

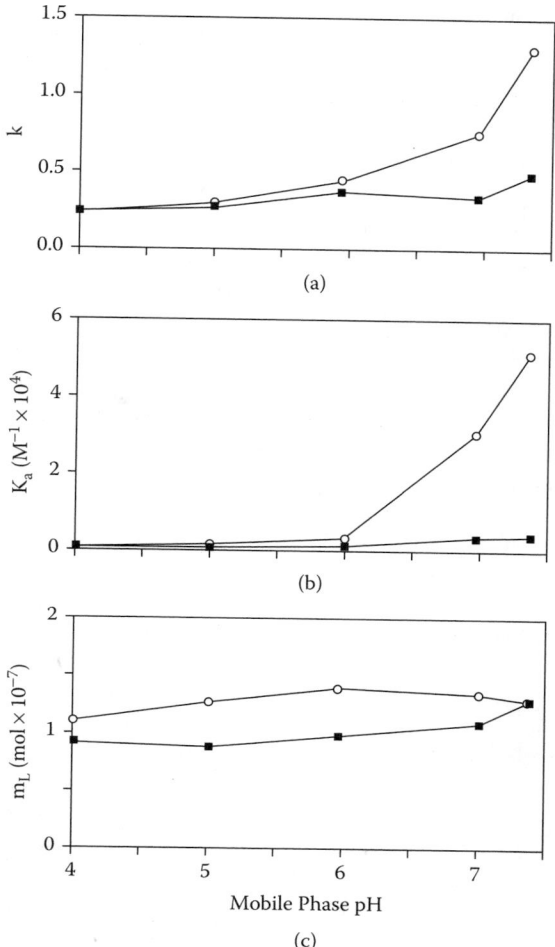

Figure 4.4 Illustration of how the change in retention factor (k), association constant (K_a), and moles of ligand sites (m_L) for D-tryptophan (■) and L-tryptophan (○) on a human serum albumin column change with the mobile phase pH. (Reproduced with permission from Yang, J. and Hage, D.S., *J. Chromatogr. A*, 725, 273–285, 1996.)

the chromatographic support, and L_D is the average distance (in cm) traveled by the target in time t_D.

$$t_D = L_D^2/D \tag{4.7}$$

For a column with slow mass-transfer kinetics, the distance that must be traveled by a molecule to get from the mobile phase to the stationary phase (L_D in Equation 4.7) will be large, giving rise to a large value for t_D. Under these conditions, the diameter of the support plays a large role in determining the effective rate of target capture. However, as L_D and t_D decrease, a point is eventually reached where the rate of capture becomes independent of mass transfer and now relies on the rate of association between the target and ligand. It is at this point that Equation 4.6 would then be used.

4.4 SELECTION OF ELUTION CONDITIONS

The elution step is another critical component of a successful affinity separation. This section examines the methods of biospecific elution and nonspecific elution and considers the relative advantages and disadvantages of each. In addition, several factors to consider in the selection of elution conditions are presented, and the theory of target elution is discussed.

4.4.1 General Approaches for Elution

The conditions used for removal of retained solutes is another item that should be considered in affinity chromatography. Just as the application conditions are selected to maximize specific solute-ligand interactions, the elution conditions are chosen to promote fast or gentle removal of solute from the column. The elution buffer used in affinity chromatography can be either a solvent that produces weak solute-ligand binding (i.e., a small association constant) or a solvent that decreases the extent of this binding by using a competing agent that displaces solute from the column. These approaches are known as *nonspecific elution* and *biospecific elution*, respectively [3].

Biospecific elution is the gentler of these two methods, since it is carried out under essentially the same solvent conditions as used for sample application. This makes this approach attractive for purification work, where a high recovery of active solute is desired. As shown in Figure 4.5, biospecific elution can be performed either by adding an agent that competes with ligand for the solute (i.e., normal-role elution) or by adding an agent that competes with solute for the ligand (i.e., reversed-role elution). In both cases, retained solutes are eventually eluted from the column by displacement and mass action [3, 4].

The main advantage of biospecific elution is its ability to gently remove a target from the column. The main disadvantages include its slow elution times and broad solute peaks. Another limitation is the need to remove competing agent from the eluted solute. A further difficulty that may be encountered in analytical applications is the need to use a competing agent that does not produce a large background signal under the conditions used for analyte detection.

Many of the problems of biospecific elution can be overcome by using nonspecific elution. This approach involves changing the column conditions to weaken interactions between retained solutes and the immobilized ligand. This can be done by changing the pH, ionic strength, or polarity of the mobile phase. The addition of denaturing or chaotropic agents can also be used. This results in an alteration in the structure of the solute or ligand, leading to a lower association constant and lower solute retention.

Nonspecific elution tends to be much faster than biospecific elution in removing analytes from affinity columns. This results in sharper peaks, which in turn produces lower limits of detection and shorter analysis times. For these reasons, nonspecific elution is commonly used in analytical applications of affinity chromatography. This elution method can also be used in purifying solutes, but there is a greater risk of solute denaturation than there is with biospecific elution. Also, care must be taken in nonspecific elution to avoid using conditions that are too harsh for the column. If this is not considered, it may result in long column regeneration times or an irreversible loss of ligand activity [3, 4].

4.4.2 Selection of Elution Buffers

The selection of a solvent for biospecific elution will depend on the type of target and ligand being used. This solvent usually has a pH and ionic composition similar to the application buffer but now contains a competing agent to displace the target from the column through mass action. For reversed-role elution, an additive that is similar but distinct from the target in its ligand binding is desired. This approach is employed when

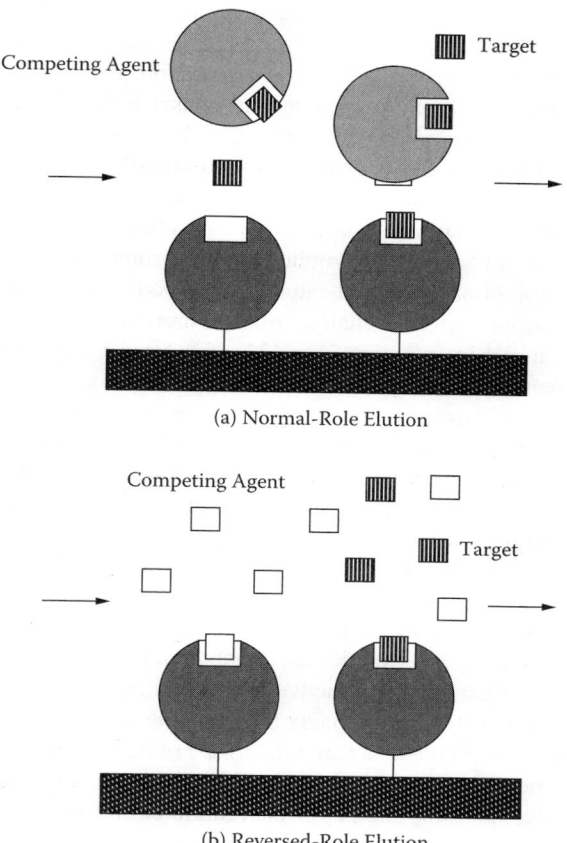

(a) Normal-Role Elution

(b) Reversed-Role Elution

Figure 4.5 (a) Normal-role and (b) reversed-role biospecific elution.

the target is a small compound. Ideally, the competing agent for this process should be readily available in an inexpensive form and be soluble in the elution buffer. When used for target purification, it must also be possible to separate the target from the additive after these have passed through the column. This last item can be particularly difficult to achieve for solutes that have few available competing agents.

Normal-role elution is often used to isolate macromolecules by affinity chromatography. This is particularly common for work with lectins and enzymes, as discussed in Chapters 5 and 12. As was true for reversed-role elution, it must be possible to remove the competing agent from the target when the affinity column is used for purification. However, this task is generally much simpler in normal-role elution, since there is often a significant difference in size between the target and eluting agent.

Nonspecific elution can employ a broad range of mobile phase additives. The goal here is to change the nature of the target-ligand interaction rather than to simply move the target off the column through mass action. This can be accomplished in several ways. For instance, altering the pH of the mobile phase will affect ionic interactions in targets and ligands that interact through weakly acidic or basic groups. A change in pH may also alter the conformation of the target or ligand. Either an increase or decrease in pH can, in theory, be used for this purpose, but a decrease in pH is more common [3, 4, 24, 25]. Although a change in pH can lead to fast elution, care must be taken to avoid using conditions that irreversibly denature the target, ligand, or support. This effect can be minimized by collecting the eluted target in a neutral pH buffer and by regenerating the column as soon as possible after the elution step.

Altering the ionic strength of the mobile phase is a second means for achieving nonspecific elution. In this method, an increase in ionic strength and salt concentration is used to disrupt ionic interactions [24]. However, this approach also promotes hydrophobic interactions and can actually lead to an increase in retention for some solutes [15]. Chaotropic salts are also useful for altering the retention of targets on affinity columns. These salts disrupt the stability of water and decrease hydrophobic interactions. Such salts are often used when working with protein-based ligands, such as antibodies [25]. The main advantage of using either a chaotropic salt or a change in ionic strength is that this usually leads to gentle elution of the target in an active form [24, 25]. After elution, the excess salt must then be removed prior to further use of the isolated compound.

In some situations, denaturing agents have been used for elution in affinity chromatography. Examples of these agents include urea, guanidine hydrochloride, and sodium dodecyl sulfate [24]. However, these should only be used (a) in analytical applications where the ligand is quite stable or (b) in preparative work where both the ligand and target are relatively stable and can recover their activity after such elution.

For work with affinity columns in chiral separations, it is common to have a small amount of a water-miscible organic solvent in the mobile phase, such as 1-propanol [19]. This is used to adjust solute retention and produce narrow peaks for good resolution. Organic solvents are also sometimes used for the isolation of targets by affinity chromatography. Polyols like ethylene glycol are especially attractive for this purpose [24].

The choice of a nonspecific elution buffer can be made through several approaches. Usually this is based on information in the literature on a particular ligand or target or on past experience with these substances. For new targets and ligands, a more systematic approach can also be employed. This may involve adjusting the pH or concentration of a particular buffer until effective elution is obtained [17, 26]. Sometimes, a wide range of buffers and eluants can be tested and compared [27–30]. To decrease the time needed for this process, a few reports have used an enzyme-linked immunosorbent assay (ELISA) in an *ELISA-elution assay* to examine the binding of a given ligand and target in the presence of an array of buffers [31–33]. However, the results of this last approach do not always directly correlate with those obtained when using the same ligand and target in an affinity column [34].

4.4.3   Description of Sample Elution

A theoretical description of the elution process can be helpful in determining the amount of time needed for target recovery or for the removal of retained substances from an affinity column. In the case of isocratic elution, Equation 4.3 would again be used for this process, as discussed earlier in this chapter. However, a modified form of this equation must be used when biospecific elution is employed. In this situation, an additional term must be added to adjust the observed retention factor for the presence of the competing agent (I). In the case of either normal- or reversed-role elution, the change in retention factor (k) for the target as the competing agent's concentration is varied can be described by Equations 4.8 or 4.9 [35–37],

$$k = K_a \, m_L/(V_M \, (1 + K_I \, [I])) \tag{4.8}$$

$$1/k = V_M/(K_a \, m_L) + (V_M \, K_I \, [I])/(K_a \, m_L) \tag{4.9}$$

where K_I is the association constant for the competing agent with the target (in normal-role elution) or the ligand (in reversed-role elution) and [I] is the concentration of this

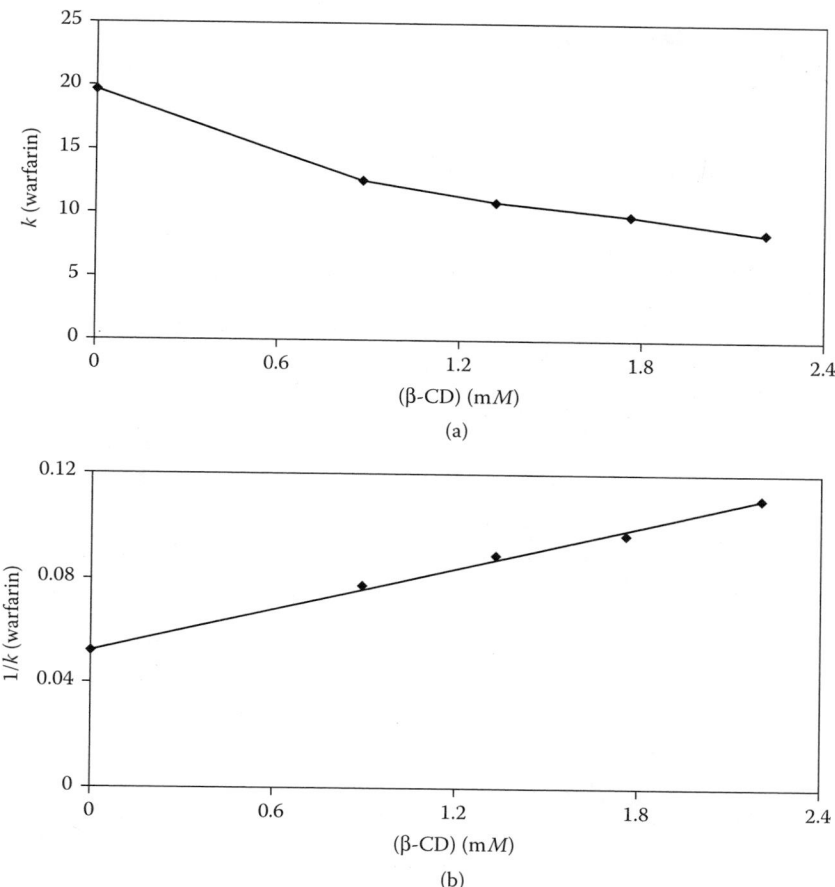

Figure 4.6 Change in the retention factor (k) for racemic warfarin when eluted from a human serum albumin column using normal-role biospecific elution with β-cyclodextrin being employed as a competing ligand in the mobile phase. (Reproduced with permission from Chen, J., Ohnmacht, C.M., and Hage, D.S., *J. Chromatogr A*, 1033, 115–126, 2004.)

agent in the mobile phase. These equations assume that 1:1 interactions are present and that the target is present in a small amount versus I or L. An example in which the above equations were used to analyze and describe the elution of a solute on an affinity column is given in Figure 4.6.

For nonspecific elution, it has been shown that the dissociation of a target from an affinity column can often be described as a first-order decay process [38, 39]. This is represented by the reaction shown in Equation 4.10,

$$A - L^* \xrightarrow{k_d^*} A^* + L^* \tag{4.10}$$

where L^*, A^*, and $A - L^*$ are the ligand, target-ligand complex, and free target in the elution buffer, with k_d^* being the dissociation rate constant for $A - L^*$ in this buffer. This model assumes that the rate of rebinding between the target and ligand is negligible under the elution conditions, or that the association constant approaches zero. The corresponding

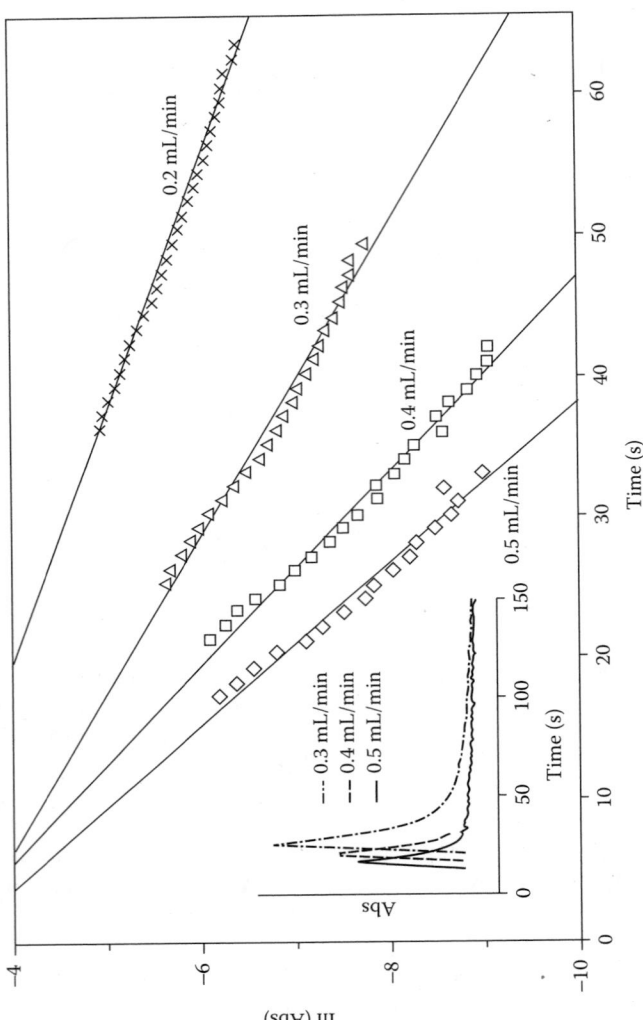

Figure 4.7 Determination of the net dissociation rate for the herbicide 2,4-D from anti-2,4-D immobilized antibodies at various flow rates and in the presence of a pH 2.5, 0.10 M phosphate elution buffer. The absorbance of the eluting solute was assumed to be directly proportional to [A^*] in Equation 4.12. (Reproduced with permission from Nelson, M.A., dissertation, University of Nebraska–Lincoln, 2003.)

first-order rate law and integrated rate equation are given in Equations 4.11 and 4.12 [38], where $[A^*]_0$ is the concentration of A^* at the beginning of the elution step.

$$-d\{A - L^*\}/dt = d[A^*]/dt = k_d^* [A^*] \tag{4.11}$$

$$\ln[A^*] = -k_d^* t + \ln[A^*]_0 \tag{4.12}$$

According to these equations, a plot of $\ln[A^*]$ versus time (t) should be linear with a slope equal to $-k_d^*$, giving the dissociation rate constant. An example of such a study is given in Figure 4.7, where the release of a chlorophenoxyacetic acid herbicide (2,4-D) from an anti-2,4-D antibody column was examined in the presence of various flow rates and a pH 2.5 elution buffer [38].

4.5 COLUMN REGENERATION

A third item associated with the application and elution steps is the procedure used for column regeneration. Although this step is seldom discussed in the literature, it is essential to consider when using an affinity column for multiple sample applications [24, 38].

4.5.1 Choice of Regeneration Conditions

The solvent used for regenerating an affinity column is usually the application buffer. However, there are some occasions in which additional solvents may be used between the elution and application steps. For instance, if there are highly retained contaminants that remain on the column after target elution, an additional solvent may be needed to remove these before the application buffer is reapplied to the system. In addition, it is possible in some cases that an intermediate solvent may be needed between the application and elution buffers to avoid problems with precipitation of mobile-phase additives within the affinity column.

The time allowed for column regeneration and the flow rate of the regeneration buffer are other factors to consider. In the case of nonspecific elution, the regeneration time and volume of regeneration solvent should be sufficiently large to allow most of the ligand to convert back to its initial form for binding or to allow any mobile-phase additives to be washed from the system. In the case of nonspecific elution, these items should be large enough to allow any competing agents to be completely removed from the system. Although a long regeneration time or large volume of applied solvent is best for these purposes, this increases the total cycle time of the method and its consumption of mobile phase. Thus, some compromise generally has to be reached between regeneration and the desired time or expense of the separation.

4.5.2 Description of Column Regeneration

A method for the quantitative description of column regeneration in the case of nonspecific elution has recently been reported based on a first-order process [38]. This is represented by the following reaction,

$$L^* \xrightarrow{k_r} L \tag{4.13}$$

where L^* and L represent the ligand in its elution and application configurations, and k_r is the rate constant for the conversion of L^* to L. This process can be described by the following first-order rate law and integrated rate equation [38], where $\{L^*\}$ is the surface concentration of L^* remaining at time t, and $\{L^*\}_0$ is the surface concentration of this

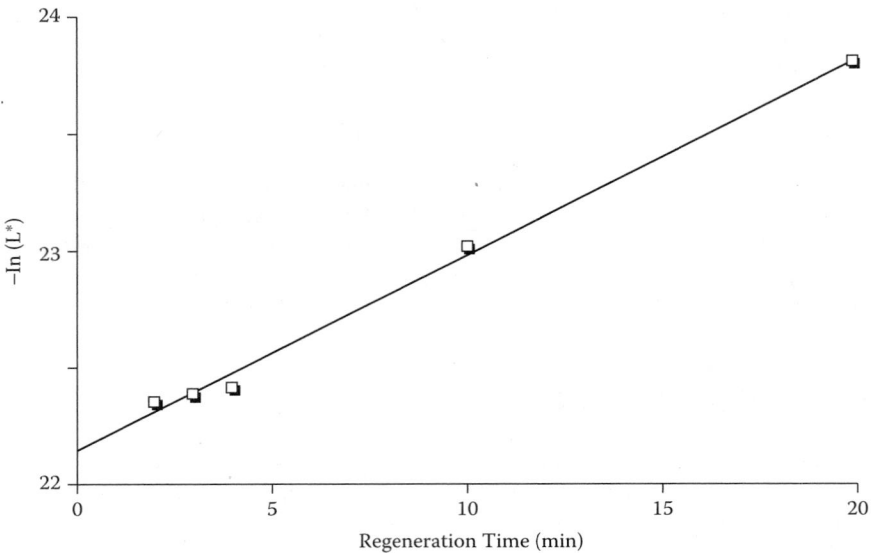

Figure 4.8 Calculation of the regeneration rate constant for an anti-2,4-D antibody column using a pH 7.0, 0.1 M potassium phosphate buffer as the regeneration solvent applied at 0.5 ml/min. (Reproduced with permission from Nelson, M.A., dissertation, University of Nebraska–Lincoln, 2003.)

ligand prior to column regeneration.

$$d\{L\}/dt = -d\{L^*\}/dt = k_r\,\{L^*\} \tag{4.14}$$

$$\ln\{L^*\} = -k_r\,t + \ln\{L^*\}_0 \tag{4.15}$$

This second equation has been used to examine the regeneration of a monoclonal antibody in going from an acidic to neutral pH phosphate buffer. The results, as indicated by Figure 4.8, gave good agreement with this model. This makes this approach potentially useful in predicting column regeneration and in optimizing the ability to reuse affinity columns [38].

4.6 SUMMARY AND CONCLUSIONS

This chapter has examined the topic of application and elution in affinity chromatography and has discussed the conditions commonly used for these processes. Both practical aspects and theoretical considerations were discussed for these items. In addition, the process of column regeneration was examined. A comparison was also made between the behavior of affinity chromatography and other types of chromatography. Although the choice of application and elution conditions in affinity methods can be a challenge in some situations, the high specificity and retention of affinity ligands often makes this worth the effort. As will be seen throughout the remainder of this text, this has led to the use of affinity chromatography in numerous applications involving target isolation, analysis, and characterization.

SYMBOLS AND ABBREVIATIONS

D Diffusion coefficient
f Free fraction of a solute passing through a column

F	Flow rate of sample application
GC	Gas chromatography
HSA	Human serum albumin
IEC	Ion-exchange chromatography
k	Retention factor
K_a	Association equilibrium constant
k_a	Association rate constant
k_d	Dissociation rate constant
k_d^*	Dissociation rate constant under elution conditions
k_r	Regeneration rate constant
L_D	Diffusion distance
Load A	Relative moles of solute applied to a column versus the moles of ligand present
m_L	Moles of active ligand in an affinity column
N	Number of theoretical plates for a chromatographic column
RPLC	Reversed-phase liquid chromatography
R_s	Resolution of a chromatographic separation
S_0	Term describing the effects of flow rate and adsorption kinetics on solute retention
SEC	Size-exclusion chromatography
t	Time
t_D	Time allowed/required for diffusion
t_M	Column void time
t_R	Retention time
V_M	Column void volume
WAC	Weak affinity chromatography
α	Separation factor, where $\alpha = k_2/k_1$

ACKNOWLEDGMENTS

This work was funded in part by the National Institutes of Health under grant R01 GM44931 and by an exploratory research grant from the U.S. Environmental Protection Agency.

REFERENCES

1. Turkova, J., *Affinity Chromatography,* Elsevier, Amsterdam, 1978.
2. Scouten, W.H., *Affinity Chromatography: Bioselective Adsorption on Inert Matrices,* Wiley, New York, 1981.
3. Walters, R.R., Affinity chromatography, *Anal. Chem.,* 57, 1099A–1114A, 1985.
4. Hage, D.S., Affinity chromatography, in *Handbook of HPLC,* Katz, E., Eksteen, R., Schoenmakers, P., and Miller, N., Eds., Marcel Dekker, New York, 1998, chap. 13.
5. Ohlson, S., Lundblad, A., and Zopf, D., Novel approach to affinity chromatography using "weak" monoclonal antibodies, *Anal. Biochem.,* 169, 204–208, 1988.
6. Wikstroem, M. and Ohlson, S., Computer simulation of weak affinity chromatography, *J. Chromatogr.,* 597, 83–92, 1992.
7. Ohlson, S., Bergstroem, M., Leickt, L., and Zopf, D., Weak affinity chromatography of small saccharides with immobilized wheat germ agglutinin and its application to monitoring of carbohydrate transferase activity, *Bioseparation,* 7, 101–105, 1998.
8. Ohlson, S., Bergstroem, M., Pahlsson, P., and Lundblad, A., Use of monoclonal antibodies for weak affinity chromatography, *J. Chromatogr. A,* 758, 199–208, 1997.
9. Strandh, M., Andersson, H.S., and Ohlson, S., Weak affinity chromatography, *Methods Mol. Biol.,* 147, 7–23, 2000.

10. Bergstrom, M. and Ohlson, S., Use of perfusive supports in weak affinity chromatography (WAC), *Int. J. Bio-Chromatogr.,* 6, 163–172, 2001.

11. Lindmark, R., Biriell, C., and Sjoequist, J., Quantitation of specific IgG antibodies in rabbits by a solid-phase radioimmunoassay with ^{125}I-protein A from *Staphylococcus aureus, Scand. J. Immunol.,* 14, 409–420, 1981.

12. de Frutos, M. and Regnier, F.E., Tandem chromatographic-immunological analyses, *Anal. Chem.,* 65, 17A–25A, 1993.

13. Hage, D.S., Survey of recent advances in analytical applications of immunoaffinity chromatography, *J. Chromatogr. B,* 715, 3–28, 1998.

14. Poole, C.F. and Poole, S.K., *Chromatography Today,* Elsevier, New York, 1991.

15. Loun, B. and Hage, D.S., Chiral separation mechanisms in protein-based HPLC columns, 1: Thermodynamic studies of (*R*)- and (*S*)-warfarin binding to immobilized human serum albumin, *Anal. Chem.,* 66, 3814–3822, 1994.

16. Hage, D.S., Walters, R.R., and Hethcote, H.W., Split-peak affinity chromatographic studies of the immobilization-dependent adsorption kinetics of protein A, *Anal. Chem.,* 58, 274–279, 1986.

17. Hage, D.S. and Walters, R.R., Dual-column determination of albumin and immunoglobulin G in serum by high-performance affinity chromatography, *J. Chromatogr.,* 386, 37–49, 1987.

18. Rollag, J.G. and Hage, D.S., Non-linear elution effects in split-peak chromatography, II: Role of ligand heterogeneity in solute binding to columns with adsorption-limited kinetics, *J. Chromatogr. A,* 795, 185–198, 1998.

19. Hage, D.S., High-performance affinity chromatography: A powerful tool for studying serum protein binding, *J. Chromatogr. B,* 768, 3–30, 2002.

20. Yang, J. and Hage, D.S., Role of binding capacity versus binding strength in the separation of chiral compounds on protein-based high-performance liquid chromatographic columns: Interactions of *D*- and *L*-tryptophan with human serum albumin, *J. Chromatogr. A,* 725, 273–285, 1996.

21. Hage, D.S., Thomas, D.H., and Beck, M.S., Theory of a sequential addition competitive binding immunoassay based on high-performance immunoaffinity chromatography, *Anal. Chem.,* 65, 1622–1630, 1993.

22. Jaulmes, A. and Vidal-Madjar, C., Split-peak phenomenon in nonlinear chromatography, 1: A theoretical model for irreversible adsorption, *Anal. Chem.,* 63, 1165–1174, 1991.

23. Hage, D.S., Thomas, D.H., Chowdhuri, A.R., and Clarke, W., Development of a theoretical model for chromatographic-based competitive binding immunoassays with simultaneous injection of sample and label, *Anal. Chem.,* 71, 2965–2975, 1999.

24. Firer, M.A., Efficient elution of functional proteins in affinity chromatography, *J. Biochem. Biophys. Methods,* 49, 433–442, 2001.

25. Phillips, T.M., High-performance immunoaffinity chromatography, *Adv. Chromatogr.,* 29, 133–173, 1989.

26. Santos, E., Tahara, S.M., and Kaback, H.R., Monoclonal antibodies against the membrane-bound, flavin-linked *D*-lactate dehydrogenase of *Escherichia coli:* Preparation, characterization and the use of immunoaffinity chromatography, *Biochemistry,* 24, 3006–3011, 1985.

27. Tsang, V.C.W. and Wilkins, P.P., Optimum dissociating condition for immunoaffinity and preferential isolation of antibodies with high specific activity, *J. Immunol. Methods,* 138, 291–299, 1991.

28. Downham, M., Busby, S., Jefferis, R., and Lyddiatt, A., Immunoaffinity chromatography in biorecovery: An application of recombinant DNA technology to generic adsorption processes, *J. Chromatogr.,* 584, 59–67, 1992.

29. Ben-David, A. and Firer, M.A., Immunoaffinity purification of monoclonal antibodies: In search of an elution buffer of general applicability, *Biotechnol. Tech.,* 10, 799–802, 1996.

30. Narhi, L.O., Caughey, D.J., Horan, T.P., Kita, Y., Chang, D., and Arakawa, T., Effect of three elution buffers on the recovery and structure of monoclonal antibodies, *Anal. Biochem.,* 253, 236–245, 1997.

31. Kummer, A. and Li-Chan, E.C., Application of an ELISA elution assay as a screening tool for dissociation of yolk antibody-antigen complexes, *J. Immunol. Methods,* 211, 125–137, 1998.

32. Boyd, S. and Yamakazi, H., Efficient and gentle elution of antibodies from an immobilized polypeptide antigen by saturated magnesium chloride, *Biotechnol. Tech.,* 7, 827–832, 1993.

33. Agraz, A., Duarte, C.A., Costa, L., Perez, L., Paez, R., Pujol, V., and Fontirrochi, G., Immunoaffinity purification of recombinant hepatitis B surface antigen from yeast using a monoclonal antibody, *J. Chromatogr. A,* 672, 25–33, 1994.

34. Kellogg, D.R. and Alberts, B.M., Purification of a multiprotein complex containing centrosomal proteins from the *Drosophila* embryo by chromatography with low-affinity polyclonal antibodies, *Mol. Biol. Cell,* 3, 1–11, 1992.

35. Chaiken, I.M., Ed., *Analytical Affinity Chromatography,* CRC Press, Boca Raton, FL, 1987.

36. Loun, B. and Hage, D.S., Characterization of thyroxine-albumin binding using high-performance affinity chromatography, *J. Chromatogr.,* 579, 225–235, 1992.

37. Chen, J., Ohnmacht, C.M., and Hage, D.S., Characterization of drug interactions with soluble β-cyclodextrin by high-performance affinity chromatography, *J. Chromatogr. A,* 1033, 115–126, 2004.

38. Nelson, M.A., Studies of Portable Immunochromatographic Methods for Analysis of Pesticide Residues, dissertation, University of Nebraska–Lincoln, 2003.

39. Lowe, C.R. and Dean, P.D.G., *Affinity Chromatography,* Wiley, New York, 1974.

Section II

General Affinity Ligands and Methods

5

Bioaffinity Chromatography

**David S. Hage, Min Bian, Raychelle Burks, Elizabeth Karle,
Corey Ohnmacht, and Chunling Wa**
Department of Chemistry, University of Nebraska, Lincoln, NE

CONTENTS

5.1 INTRODUCTION

Although affinity chromatography can employ a variety of ligands, most of these tend to be of biological origin. The use of a biological ligand within a column is sometimes known as *bioaffinity chromatography* or *biospecific adsorption*. This was the first type of affinity chromatography developed, as used by Starkenstein in 1910 when he purified α-amylase using insoluble starch [1]. The specificity of most biological ligands makes bioaffinity chromatography an ideal technique for the purification of numerous compounds. This method has also become important as a tool in the analysis and characterization of biological samples.

The variety of biological agents that exist in nature means that bioaffinity chromatography is an inherently diverse method. But there is a smaller set of ligands that are commonly used as stationary phases in this approach. This chapter examines several of these ligands, including immunoglobulin-binding proteins, enzymes, lectins, carbohydrates, serum proteins, and the avidin/biotin system. Other examples of such agents are given in Chapters 6 and 7, which explore, respectively, the use of antibodies as ligands in immunoaffinity chromatography and nucleic acids as ligands in DNA affinity chromatography.

5.2 IMMUNOGLOBULIN BINDING PROTEINS

One important set of ligands in bioaffinity chromatography includes the immunoglobulin-binding proteins. Two common examples are protein A and protein G. These ligands have the ability to bind to the constant region of many types of immunoglobulins, or antibodies, which makes these ligands useful in the purification of antibodies and in the detection or use of antibodies in analytical methods.

5.2.1 General Properties

Protein A is produced by the bacterium *Staphylococcus aureus*; protein G is a similar protein found in groups C and G *streptococci* [2–4]. Both proteins are found on the bacterial cell surface, where they form binding sites for the F_c regions of immunoglobulins in the blood of an infected host. It is believed that this binding protects a bacterial cell from being identified as a foreign agent and destroyed by the host's immune system.

There are five classes of antibodies in humans: immunoglobulins A, D, E, G, and M (i.e., IgA, IgD, IgE, IgG, and IgM). Protein A is able to bind human IgG as well as IgM, IgD, and IgA; however, protein G binds to only IgG-class antibodies in humans. As shown in Table 5.1, these two ligands also have differences in their binding to antibodies from other species. For instance, protein G is able to bind all of the subclasses of IgG in humans and mice, while protein A has strong binding to only some types of IgG in these species [1].

Table 5.1 Binding of Immunoglobulins from Various Species to Proteins A, G, A/G, and L

Type of Immunoglobulin		Interactions with Immunoglobulin Binding Proteins			
		Protein A	Protein G	Protein A/G	Protein L
Bovine	IgG$_1$	Weak	Strong	Strong	No binding
	IgG$_2$	Strong	Strong	Strong	No binding
Cat	IgG	Strong	Weak	Strong	Unknown
Dog	IgG	Strong	Weak	Strong	Unknown
Donkey	IgG	Mild	Strong	Strong	Unknown
Goat	IgG$_1$	Weak	Strong	Strong	No binding
	IgG$_2$	Strong	Strong	Strong	No binding
Guinea Pig	IgG	Strong	Weak	Strong	Unknown
Hamster	IgG	Mild	Unknown	Unknown	Unknown
Horse	IgG$_{(ab)}$	Weak	No binding	Weak	Unknown
	IgG$_{(c)}$	Weak	No binding	Weak	Unknown
	IgG$_{(T)}$	No binding	Strong	Strong	Unknown
Human	IgG$_1$	Strong	Strong	Strong	Strong
	IgG$_2$	Strong	Strong	Strong	Strong
	IgG$_3$	Weak	Strong	Strong	Strong
	IgG$_4$	Strong	Strong	Strong	Strong
	IgM	Weak	No Binding	Weak	Strong
	IgA$_1$	Weak	No Binding	Weak	Strong
	IgA$_2$	Weak	No Binding	Weak	Strong
	IgE	Mild	No Binding	Mild	Strong
	IgD	No Binding	No Binding	Weak	Strong
Monkey (Rh)	IgG	Strong	Strong	Strong	Unknown
Mouse	IgG$_1$	Weak	Mild	Mild	Strong
	IgG$_{2a}$	Strong	Strong	Strong	Strong
	IgG$_{2b}$	Strong	Strong	Strong	Strong
	IgG$_3$	Strong	Strong	Strong	Strong
Pig	IgG	Strong	Weak	Strong	Strong
Rabbit	IgG	Strong	Strong	Strong	Weak
Rat	IgG$_1$	Weak	Mild	Mild	Unknown
	IgG$_{2a}$	No Binding	Strong	Strong	Unknown
	IgG$_{2b}$	No Binding	Weak	Weak	Unknown
	IgG$_{2c}$	Strong	Strong	Strong	Unknown
Sheep	IgG$_1$	Weak	Strong	Strong	No binding
	IgG$_2$	Strong	Strong	Strong	No binding

Source: Pierce Chemical Co.

A recombinant form of these two proteins, known as protein A/G (mass, 45 to 47 kDa), has also been produced. This results in a ligand with a broader range of interactions and applications than either protein A or protein G alone [5, 6].

Native protein A has a mass of 42 kDa and is composed of five domains, four of which can bind to the F_c region of IgG. When protein A is immobilized, only two of these domains are available for binding at any given time [7]. The interaction of protein A with an immunoglobulin occurs mainly between the CH_2 and CH_3 domains of an antibody's F_c region [8–10]. However, some interactions have also been reported between protein A and the F_{ab} region of an immunoglobulin [11].

Protein G is produced in two forms that are distinguished by their number of internal repeats. In its native form, protein G binds both immunoglobulins and albumin. However, in most affinity applications, a recombinant form of protein G is used that has been altered to eliminate its albumin-binding region. This recombinant form has a mass of 23 kDa. Native protein G contains three F_c-binding regions (C_1, C_2, and C_3), while the recombinant protein G contains only two F_c-binding regions (B_1 and B_2) [12]. In the recombinant form, the interaction between the B_1 domain of protein G and the F_c region of an antibody is dominant [13]. The association constant of this interaction for human IgG is approximately 3.3 to $8.4 \times 10^8 \ M^{-1}$ [7]. Some researchers have suggested that protein G and protein A bind to the same location on an antibody's F_c region [14], while others believe that these interact at different sites [12].

Other examples of antibody-binding proteins include protein B and mannan-binding protein. Protein B is a β antigen protein found on the surface of group A *streptococci* bacteria. It is useful in binding to several types of IgA-class antibodies, such as human IgA_1, IgA_2, and secretory IgA. However, it does not bind to mouse IgA or IgA from many other mammals [5, 15]. Mannan-binding protein is found in the serum of various mammals [16] and binds to IgM-class antibodies in mouse ascites fluid. This makes this ligand of potential use for the isolation of IgM monoclonal antibodies produced in mice [17].

Another useful ligand for immunoglobulins is protein L [18–20]. This is a cell wall protein produced by *Peptostreptococcus magnus*. Like protein A and protein G, this ligand is available in a recombinant form (mass, 35.8 kDa). Protein L has four binding sites for antibodies and has the unique ability to bind to the kappa light chains of antibodies without affecting their antigen-binding sites. This makes protein L valuable in binding antibody fragments that contain kappa light chains but not the F_c region. Examples of such agents include F_{ab} fragments and single-chain variable-region fragments (scF_v).

5.2.2   Applications in Affinity Chromatography

An important application of protein A, protein G, and related ligands is their use in the purification of immunoglobulins. This topic is discussed in more detail in Chapter 14. Table 5.1 is useful in selecting a ligand for this purpose by identifying the strength of retention for the immunoglobulin type and class of interest. For instance, protein G or protein A/G could be employed for the isolation of goat IgG-class antibodies, while proteins A, G, or A/G could be used for rabbit IgG. However, the isolation of mouse IgG_1-class antibodies would be best performed with protein L.

The ability of these proteins to bind antibodies also makes these ligands valuable for the analysis of immunoglobulins. As an example, proteins A and G have been used in high-performance liquid chromatography for the measurement of IgG in human serum [21–23]. There has also been work involving a combination of multiple affinity columns, one containing immobilized protein A and the other containing antihuman serum albumin antibodies, for the simultaneous analysis of IgG and albumin in humans for the determination of albumin/IgG ratios [24].

Yet another application of proteins A and G has been as secondary ligands for the adsorption of antibodies onto supports used in immunoaffinity chromatography or chromatographic immunoassays (see Chapters 6 and 29). One reason for using this approach is that it helps provide antibodies with a specific site of attachment to the support through a region that is distant from their antigen-binding sites (see Figure 5.1). This results in a high activity for the immobilized antibodies and a known orientation for these ligands on the surface [7]. The fact that protein A and protein G will release these antibodies under

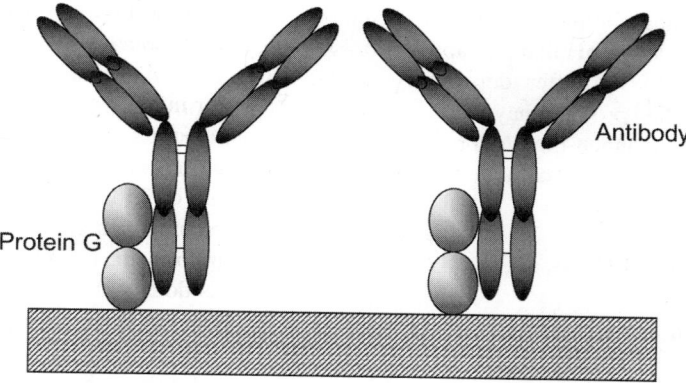

Figure 5.1 A simple model for the site-selective adsorption of antibodies to a solid support using recombinant protein G as a secondary binding agent.

acidic conditions also makes this approach appealing when it is desirable to frequently replace the antibodies in the immunoaffinity column [25–27].

These ligands have also been used for the removal of specific interferences from samples. For instance, protein A and antimouse immunoglobulin supports have been used to remove human antimouse antibodies from samples prior to the use of an immunoassay [28]. Also, antihuman immunoglobulin immunoaffinity chromatography and protein A supports have been utilized to selectively adsorb enzyme-immune complexes (i.e., macro-enzymes) from patient samples [29].

Besides being used as part of liquid chromatographic systems, protein A and protein G can also be used in other analytical methods. One example of such a technique is capillary electrophoresis (CE). For example, a capillary packed with a protein G chromatographic support has been used to adsorb antibodies for the extraction and concentration of insulin from serum prior to the quantitation of the insulin by CE [30].

5.2.3 Immobilization and Column Preparation

There are a number of commercial supports available that contain immobilized protein A and protein G. Examples include materials that can be obtained from BioRad, Pharmacia, Pierce, and Sigma. Protein L and protein A/G supports are also available commercially. In addition, procedures can be found in the literature for immobilizing these ligands onto various materials. Detailed descriptions of such methods are provided by Hermanson et al. [5]. These methods are typically amine-based coupling methods, as described in Chapter 3. The Schiff base method is commonly used for protein A, since this tends to give a higher recovery of activity for this protein than other amine-based coupling approaches [31]. The same method is often used for protein G [5].

5.2.4 Application and Elution Conditions

Most immunoglobulin binding proteins have their strongest binding to antibodies at or near a neutral pH. However, these ligands do differ somewhat in their optimum pH for sample application. Protein L has its strongest binding at pH 7.5, while protein A has optimum binding at pH 8.2 but gives good retention at pH 7 to 7.5. Recombinant protein G has its strongest binding at approximately pH 5, but can also be used at pH 7 to 7.5. Because it combines the features of protein A and protein G, protein A/G has an even broader range for its pH optimum, with values of 5 to 8.2 being adequate for the binding of antibodies to this ligand.

The elution of immunoglobulins from proteins A, G, A/G, and L is usually performed by using a buffer that has a lower pH than the application solvent. This is typically accomplished by using a pH 2.5 to 3.0 buffer, depending on which ligand is being employed. Other immunoglobulin-binding proteins may require other elution conditions. As an example, the binding of antibodies to mannan-binding protein is calcium-dependent, so these can be eluted by adding EDTA as a calcium-complexing agent to the mobile phase [5].

5.3 IMMOBILIZED ENZYMES

Immobilized enzymes and enzyme inhibitors have been used for many years in affinity chromatography. The topic of enzyme purification by employing inhibitors, cofactors, or other agents that can bind to an enzyme is discussed in Chapter 12. However, the use of immobilized enzymes has also been an area of great interest. In this case, the enzymes are physically confined or localized on a support in a fashion that retains their catalytic activity and specificity [32, 33]. This has many advantages over methods utilizing soluble enzymes. For instance, immobilized enzymes can be reused many times and placed into an automated system. They are also not subject to autolysis. In addition, solid-phase enzymes allow easy removal of enzymatic products and can often be used at higher concentrations than soluble enzyme preparations.

5.3.1 General Properties

The term *enzymes* is used here to refer to protein-based catalysts. This catalytic ability makes enzymes useful in generating a number of biological products. Their ability to selectively recognize a particular substrate is also valuable for the purification of compounds or in the isolation of enzymes by employing an immobilized substrate, inhibitor, or cofactor.

The first reported use of an immobilized enzyme was almost 90 years ago [34]. This occurred when Nelson and Griffin found that the activity of invertase was not affected when this enzyme was absorbed to charcoal or to a colloid like saponin, serum, or egg albumin. Since then, many enzymes have been immobilized to a wide range of carriers. For example, amino acylase adsorbed onto diethyl aminoethyl cellulose or Sephadex has been used to separate L amino acids from their biologically inactive D forms [35]. Another example is glucose isomerase, which has been adsorbed to DEAE-cellulose or a porous ceramic carrier and used in the production of high-fructose syrups [36].

5.3.2 Applications in Affinity Chromatography

Immobilized enzymes have been used in industry for work with pharmaceuticals and foods, as well as in agricultural and environmental applications. The use of immobilized enzymes in these fields relies either on the enzyme's catalytic activity or its binding specificity, as illustrated in Figure 5.2. For instance, in an enzyme reactor, the immobilized enzyme is used as an efficient catalyst. Alternatively, the enzyme might be used as an affinity ligand for the purification of an enzyme inhibitor. A third possible application for an immobilized enzyme is its use as a scavenger to remove impurities. Examples of such applications are given in Table 5.2.

Because of high specificity of enzyme binding, immobilized enzyme inhibitors are widely used for enzyme purification. Inhibitors used for this purpose include p-aminophenyl-β-D-thiogalactopyranoside [51], N-5-carboxypentyl-1-deoxymannojirimycin (CP-dMM) [52], novobiocin [53], 2-[3-(2-ammonioethoxy)benzoyl]ethyltrimethylammonium bromide [54], guanosine triphosphate (GTP) [55], hadacidin [56], deoxyadenosine triphosphate (dATP) [57], arabinofuranosyluracil-5′-triphosphate (araUTP) [58], amidines [59], cilastatin [60],

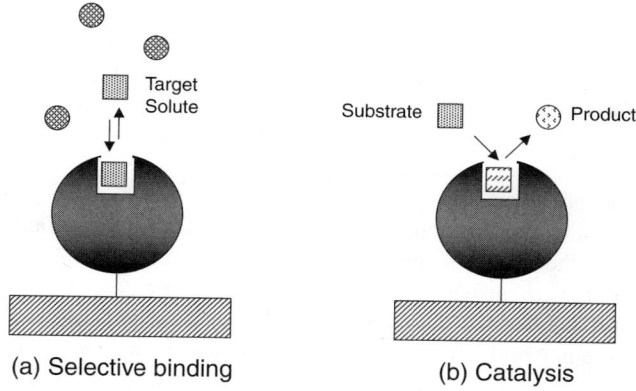

Figure 5.2 Use of an immobilized enzyme for (a) selective binding or (b) the catalytic conversion of a substrate into a product.

pepstatins [61], peptide aldehyde inhibitors [62], *t*-butyloxycarbonyl-gly-phe-NHCH$_2$CN [63], cycloheptaamylose [64], ovomucoid [65], and D-tryptophan methylester [66]. Further details on many of these inhibitors and their applications can be found in Chapter 12.

Although they have a variety of possible applications, the use of immobilized enzymes in purification and scavenging has been limited in part by the cost of the enzymes. For instance, these are usually synthesized in only small quantities by cells, so their isolation involves a separation from large amounts of inactive material. In addition, an immobilized enzyme may slowly leach from its support over time [67]. This is not a major problem when the enzyme is used in analytical or semipreparative work. However, this is an issue when the enzyme is used to isolate or treat a pharmaceutical protein on a large scale, where the presence of even a trace amount of such a contaminant will lead to lack of approval for the product by regulatory agencies like the U.S. Food and Drug Administration [68].

Table 5.2 Applications of Immobilized Enzymes in Purification and Scavenging

Enzyme	Application	References
Catalase	Removal of H$_2$O$_2$	[37]
Glucose oxidase + catalase	Removal of glucose	[38]
Sulfhydryl oxidase	Removal of sulfhydryl groups in skim milk	[39]
Rhodanase	Removal of cyanides in effluents and air	[40]
Urease	Removal of urea	[40]
Catalase	Removal of perborate	[41]
Horseradish peroxidase	Removal of chlorophenol from wastewater	[42]
Trypsin	Purification of trypsin inhibitors	[43]
α-Chymotrypsin	Purification of chicken ovoinhibitor	[44]
RNase	Purification of PUDP peptide	[45]
Staphylococcal nuclease	Purification of its antibody	[46]
Isoleucyl-tRNA synthetase	Purification of isoleucyl-tRNA	[47]
Peroxidase	Purification of Jack bean agglutinin	[48]
Glutamic-aspartic transaminase	Purification of an antienzyme	[49]
DD-peptidase	Purification of penicillin	[50]

5.3.3 Immobilization and Column Preparation

Common methods for enzyme immobilization include the attachment of these ligands to prefabricated carriers, their cross-linking by bi- or multifunctional reagents, and their encapsulation in a solid matrix [69]. Binding to a prefabricated carrier is currently the preferred method. The coupling of an enzyme to such a carrier can be carried out in various ways, such as through covalent bond formation, ionic interactions, adsorption, complexation with metal chelates, or binding to secondary ligands [70].

The enzymes used in analytical applications are mainly immobilized through covalent bonds [71]. There are two stages for the immobilization of enzymes in this approach. First, the support is activated with a bridge or carrier group. Any of the supports described in Chapter 2 can be used for this purpose, but the most common ones used for enzymes are cellulose and polysaccharides. These materials can be activated with many agents, including cyanuric chloride, glutaraldehyde, sodium periodate, carbodiimide, diisocyanates, difluorodinitrophenyl sulfone, diazonium salts, p-benzoquinone, and N,N-o-phenylenedimaleimide.

In the second step of this process, the enzyme is covalently linked to the carrier group. This can be performed by using various reactive side chains on the enzyme, such as the amine groups of lysine and arginine, the carboxyl groups of aspartic and glutamic acid, the hydroxyl groups of serine and threonine, the phenolic group of tyrosine, and the imidazole group of histidine [72].

The activity of the final immobilized enzyme will depend on the nature of the support, the coupling method, and the conditions used for coupling. In choosing these items, the side chain used for linking the enzyme to the support should not be one that is essential for the enzyme's binding to a substrate or its catalytic function. This can be a challenge, since side chains in these regions tend to be more reactive than similar groups in other parts of the enzyme. As a result, blocking these groups prior to immobilization may be required. More information on immobilization effects and approaches for dealing with these can be found in Chapter 3.

5.3.4 Application and Elution Conditions

The typical application conditions used for an immobilized enzyme mimic the pH and composition of its native environment. When the enzyme is used as a catalyst, this includes the presence of any cofactors, metal ions, or other items needed for the enzyme's function. Many of the same substances will also be required when using the enzyme as a selective binding agent.

The optimal conditions for desorbing a compound from an immobilized enzyme must often be determined experimentally, but they can often be achieved by making changes in the pH, ionic strength, or temperature of the mobile phase. Biospecific elution, a method described in Chapter 4, is particularly common with immobilized enzymes. This can be carried out by using a soluble substrate, competitive inhibitor, coenzyme, or allosteric effector for the enzyme. Although biospecific elution is slower than nonspecific elution methods, this approach is more specific and yields a final product that generally has a higher purity and activity.

5.4 LECTINS

Lectins are another class of ligands that have been widely used in affinity chromatography. These are nonimmune system proteins that have the ability to recognize and bind certain carbohydrate residues [73, 74]. Two lectins often placed into affinity columns are concanavalin A and wheat germ agglutinin. Other lectins employed are jacalin and the lectins

Table 5.3 Examples of Carbohydrate Residues That Bind to Lectins

Carbohydrate Residue	Interacting Lectins
D-*N*-Acetyl galactosamine	Lectin 60 from *Ricinus communis*
	Soybean lectin
	Phytohemagglutinin E_4
	Phytohemagglutinin L_4
α-D-Galactose	Jacalin
	Peanut lectin
	Soybean lectin
β-D-Galactose	*Ricinus communis* lectins 60 and 120
α-D-Glucose	Concanavalin A
	Lentil lectin
	Pea lectin
α-D-Mannose	Concanavalin A
	Lentil lectin
	Pea lectin

Source: Data obtained from Hermanson, G.T., Mallia, A.K., and Smith, P.K., *Immobilized Affinity Ligand Techniques*, Academic Press, San Diego, 1992.

found in peas, peanuts, or soybeans. These ligands are commonly used in the isolation of carbohydrate-containing compounds such as polysaccharides, glycoproteins, and glycolipids [5, 75]. When these agents are used as ligands in affinity columns, the resulting method is often known as *lectin-affinity chromatography.*

5.4.1 General Properties

The binding properties of several lectins are summarized in Table 5.3. Concanavalin A (Con A) is one of the most commonly used lectins. This is a hemagglutinin produced by the jack bean (*Canavalia ensiformis*). It exists as a tetramer of four identical subunits (mass, 26 kDa per subunit) at a neutral or basic pH. However, at a pH of 5.6 or less it is present as an active dimer with a total mass of 52 kDa. As shown in Table 5.3, Con A has the ability to bind to various carbohydrate residues, including those that contain α-D-glucopyranosyl or α-D-mannopyranosyl residues [5]. It also requires Ca^{2+} or Mn^{2+} for optimal binding.

Two other lectins often used in affinity columns are wheat germ agglutinin (WGA) and jacalin. As its name implies, WGA is obtained from wheat germ (*Triticum vulgaris*). It has a mass of 36 kDa and consists of two identical subunits. It is useful in binding to substances that possess an *N*-acetylglucosamine residue. Jacalin is found in the seeds of the jack fruit (*Artocarpus integrifolia*). It contains four subunits, two with masses of 10 kDa each and two with masses of 16 kDa each. It is useful in binding to compounds that contain α-D-galactosyl groups. This has made it useful in the isolation of IgA- and IgD-class antibodies.

5.4.2 Applications in Affinity Chromatography

Lectins have frequently been used in affinity chromatography to isolate biological compounds that contain carbohydrate residues. This has included their use with polysaccharides, glycoproteins, glycopeptides, glycolipids, and enzyme-antibody conjugates. In addition, lectins have been employed as tools in identifying the types of carbohydrate groups that are associated with a given biological agent [5, 75].

Lectins have further been used in several analytical applications. One specific example has been their use in affinity chromatography for the separation and analysis of isoenzymes. This has been accomplished by using an HPLC column that contained immobilized wheat germ agglutinin for the separation of liver- and bone-derived isoenzymes of alkaline phosphatase in human serum [76]. This method gave an improved resolution of these isoenzymes versus a low-performance affinity column [77] and good correlation for patient samples when compared with a solid-phase immunoassay for alkaline phosphatase [78].

A variety of other glycoproteins have been studied and quantitated by lectin-affinity chromatography. For instance, low-performance columns based on concanavalin A have been used to separate ApoA- and Apo B-containing lipoproteins in human plasma [79] and to study the microheterogeneity of serum transferrin during alcoholic liver disease [80]. Such columns have also been used to characterize the carbohydrate structure of follicle-stimulating hormone and luteinizing hormone under various conditions [81]. Other work involving a combination of concanavalin A and wheat germ agglutinin columns has been used to identify changes that occur during prostate cancer in asparagine-linked sugars on human prostatic acid phosphatase [82].

Another application for lectins has been as ligands for the affinity extraction of samples. In one study, a method based on a wheat germ agglutinin extraction column and a high-performance anion-exchange column was employed to purify and analyze angiotensinase A and aminopeptidase M in human urine and kidney samples [83].

5.4.3 Immobilization and Column Preparation

There are several commercial suppliers for immobilized lectin supports and columns (e.g., Sigma and Pharmacia). Furthermore, procedures for preparing several types of immobilized lectins are given in the literature [5]. For Con A and jacalin, this can be accomplished by using the Schiff base method or other amine-based coupling techniques. A coupling pH of 7 to 8 is often employed for this purpose. To protect the active sites of these ligands, the sugar they interact with is generally included in a soluble form in the immobilization buffer [5].

5.4.4 Application and Elution Conditions

The application of solutes to the lectin columns is often carried out in the presence of a neutral pH buffer. Some lectins, such as Con A, require the presence of divalent metal ions like Ca^{2+} or Mn^{2+} for binding. For elution, the same buffer is often employed, but with an excess of a competing sugar now being added to the mobile phase. This allows for gentle biospecific elution of retained targets from the column. However, it is also necessary with this approach to later remove the competing sugar from the recovered target. When the target is a large biomolecule, this can be accomplished by size-exclusion chromatography, dialysis, or some alternative size-separation method. With some lectins (e.g., WGA), nonspecific elution based on a low pH buffer can also be utilized.

5.5 CARBOHYDRATES

Another group of biological compounds that have been used as affinity ligands are the carbohydrates. This includes simple as well as complex carbohydrates and glycosaminoglycans. Such agents have been used for both the preparation and analysis of chemicals. Some carbohydrates, particularly cyclodextrins, have also been used in a variety of separations for small chiral solutes.

5.5.1 GENERAL PROPERTIES

Carbohydrates have the basic formula $(CH_2O)_n$ and can be classified into three groups. The first of these are the monosaccharides, which are simple sugars with multiple hydroxyl groups. The second group is made up of the oligosaccharides, which contain two to ten monosaccharides covalently bonded to one another. The third group consists of the polysaccharides, which are larger polymers consisting of chains of monosaccharides or disaccharides and their derivatives. All of these categories have been used in affinity chromatography as ligands.

A special subset of carbohydrates are the glycosaminoglycans (GAGs). These are polymers that contain repeating units of special disaccharides. Within these chains, the disaccharides are often modified to give them acidic or amino groups. Several GAGs of physiological significance include hyaluronic acid, dermatan sulfate, chondroitin sulfate, heparin, heparan sulfate, and keratin sulfate. When covalently linked to specific proteins, GAGs form large complexes called proteoglycans. The modification of glycosaminoglycans gives them a large negative charge and an extended conformation that results in a high viscosity in solution. Along with their high viscosity, GAGs have low compressibility, which makes them ideal for use as a lubricating fluid in the joints. In addition, their rigidity provides structural integrity to cells and provides passageways between cells.

Another subset of carbohydrates is the cyclodextrins. As shown in Figure 5.3, these are cyclic oligosaccharides that are generally composed of six, seven, or eight α-D-glucose units. This gives rise to α-, β-, or γ-cyclodextrin, respectively. These have an overall shape that resembles a truncated cone with a hydrophobic interior and polar regions at the upper and lower edges. Due to their relatively hydrophobic interiors, cyclodextrins can form inclusion complexes with a wide range of substrates in aqueous solution. This property makes

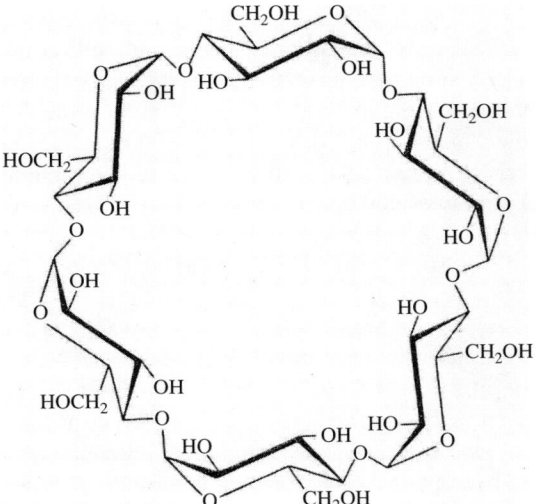

Figure 5.3 The structure of α-cyclodextrin.

cyclodextrins useful as solubilizing agents and as mimics for enzyme-binding sites. The fact that they are chiral also makes cyclodextrins useful as ligands for the separation of small chiral solutes.

5.5.2 Applications in Affinity Chromatography

There are several applications for immobilized carbohydrates in affinity chromatography. One of these is their use for isolating carbohydrate-binding proteins like lectins. However, many other proteins have also been isolated or examined with immobilized carbohydrate columns. For example, the protein ficolin has been isolated from human serum by using a Sepharose column that contained N-acetyl-D-glucosamine (GlcNAc) as a ligand [84]. Insect-cell culture-expressed influenza HA protein has also been purified by an affinity column containing immobilized carbohydrates [85]. A proteoglycan was isolated from a human placenta by using fibronectin immobilized on agarose [86], and GAG affinity chromatography has been used to purify perlecan for use in various assays [87].

In the field of capillary electrophoresis, an immobilized glycosaminoglycan was developed for the resolution of synthetic heparin-binding peptides. This made use of heparin and heparan sulfate coupled to fused silica capillaries through biotin-neutravidin conjugation. These capillaries exhibited markedly reduced electroosmotic flow and were able to distinguish peptides from the heparin-binding domain of acidic fibroblast growth factor that differed only in the stereochemistry of their proline residues [88].

Carbohydrate columns can also be used as a means for investigating the ability of proteins to bind specific sugars, such as the Charcot-Leyden crystal (CLC) protein [89]. In another study, a 130-residue peptide segment near the C terminus of proteoglycan core protein was tested for its ability to interact with carbohydrates using affinity chromatography [90]. Other examples can be found in the literature [91–95].

These interactions can be investigated in a more quantitative manner by using an approach such as frontal analysis (see Chapter 22 for further details on this technique). This has been conducted using lectin columns to which labeled glycans are applied or by using glycan columns to which lectins have been applied. Of these two methods, the first is more frequently employed, since it allows a panel of oligosaccharides to be analyzed simultaneously [96–99].

There have been other reports in which carbohydrate columns have been used in clinical studies. For instance, mononuclear cell factor was purified by an immobilized D-galactose column using a supernatant from LPS-stimulated macrophages as the sample [100]. Another report chronicled the use of sugars such as alginic acid, inulin, and gum arabic that selectively adsorb B lymphocytes and show a high affinity for monocytes and granulocytes in the separation of leukocytes [101]. In cancer research, the carbohydrate-binding proteins in rat mammary tumors, a fibroadenoma, and two tubulopapillary adenocarcinomas were compared in their binding to immobilized α- and β-galactosyl, α-mannosyl, and α-fucosyl residues [102]. Other examples of immobilized-sugar columns in clinical applications can be found in the literature [103–108].

Another application for immobilized sugars has been in chiral chromatography [109]. This has been particularly true for the cyclodextrins, which have been used in more than 26,000 reported separations. The chiral recognition of solutes by cyclodextrins is based on several factors [110, 111]. First, these ligands have numerous chiral centers with various orientations that can interact with a chiral solute. As an example, β-cyclodextrin has 35 different chiral recognition sites. Cyclodextrins also have a hydrophobic cavity that can inform an inclusion complex with many small organic compounds, giving yet another way in which this ligand can bind to solutes. In addition, cyclodextrins can change their shape to interact more intimately with a complexed compound, giving a further increase in binding strength and chiral selectivity.

Table 5.4 Examples of Carbohydrates Available on Agarose Supports

N-Acetyl-D-galactosamine	N-Acetyl-D-glucosamine
p-Aminobenzyl 1-thio-β-D-galactopyranoside	Cellobiose
p-Aminophenyl β-D-thiogalactopyranoside-	Fetuin
L-Fucose	β-D-Glucose
Isomaltose	α-Lactose
Maltoheptose	Maltose
Maltotriose	D-Mannose
Mannose-6-phosphate	D-Melibiose
α-Methyl D-mannoside	Saccharolactone

Source: Information in this table is based on products available from Sigma.

5.5.3 Immobilization and Column Preparation

Several types of sugars attached to agarose supports can be obtained commercially (see Table 5.4). The same is true for cyclodextrin columns. In addition, various approaches can be used to attach sugars to chromatographic supports. Examples of techniques for this purpose are given in Chapter 3 and in the literature [112, 113]. For instance, a monosaccharide can be immobilized by reacting cyanogen bromide-activated agarose with a derivative of a monosaccharide that contains an amino group [112]. The major problem with this approach is the need to synthesize or obtain an amino derivative of the monosaccharide.

5.5.4 Application and Elution Conditions

The literature [5, 114] provides several examples of separations involving carbohydrates and lectins on carbohydrate-immobilized resins. The application and elution conditions used for lectins on these columns are the same as described in the previous section on lectin-affinity chromatography. One difference between immobilized carbohydrates and other affinity ligands is that an organic solvent is often used with carbohydrate ligands to maximize their ability to hydrogen bond with other compounds. For instance, this approach is often used with immobilized cyclodextrins for the separation of chiral compounds. However, aqueous solvents are also used with such ligands, especially when they are used for the isolation of biomolecules. Many immobilized carbohydrates have moderate-to-weak binding for their targets, allowing the use of isocratic elution for the passage of solutes through columns that contain these ligands.

5.6 SERUM PROTEINS

Immobilized serum proteins have also been used in affinity chromatography. Proteins such as human serum albumin, bovine serum albumin, and α_1-acid glycoprotein are a few examples that have been successfully utilized as stationary phases for this purpose. This section examines some practical aspects of performing affinity chromatography with such proteins. Further details on these ligands and their applications can be found in Chapters 21 and 22.

5.6.1 General Properties

The most abundant protein in the plasma of vertebrates is albumin. Human serum albumin (HSA) has a typical concentration in plasma of 40 g/l [115] and has the structure shown

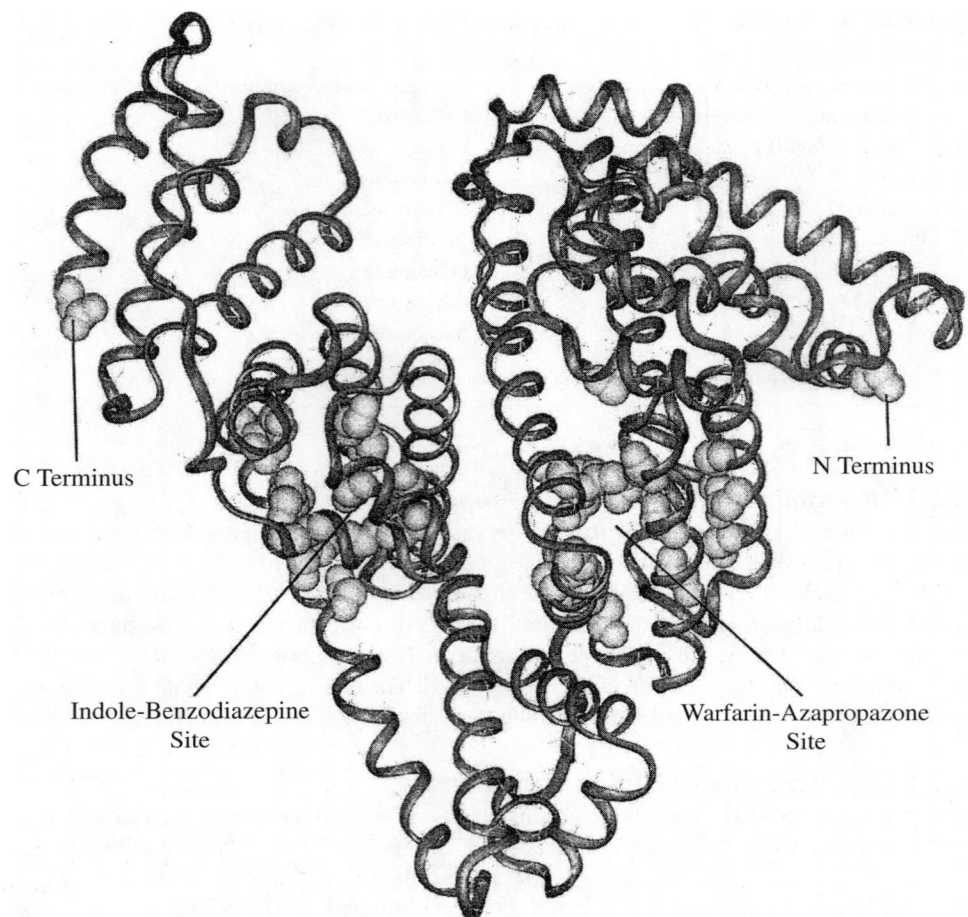

C Terminus

N Terminus

Indole-Benzodiazepine
Site

Warfarin-Azapropazone
Site

Figure 5.4 The crystal structure of human serum albumin. This image was generated using file PDB ID 1A06 in the protein data bank of the Research Collaboratory for Structure Bioinformatics. (Reproduced with permission from Hage, D.S. and Austin, J., *J. Chromatogr. B*, 739, 39–54, 2000.)

in Figure 5.4 [116]. This is a 66.5-kDa single-chain protein with 585 amino acids that is produced by the liver. Bovine serum albumin (BSA) has a similar mass but only 583 amino acids. The structure of HSA is stabilized by 17 disulfide bridges. The close coupling of adjacent half-cysteine residues that is associated with these disulfide bridges gives albumin both flexibility and resistance to extreme conditions. These bridges also form the pseudo-nine (eight and one-half) double loops that are usually grouped together as three homologous domains (I, II, and III) [115].

Albumin serves many functions in the body. For example, it is involved in the regulation of osmotic pressure and plays a large role in providing almost all of the anions in plasma. In addition, it acts as the primary buffering system for extravascular fluids and protects low-density lipoproteins from peroxidative effects [115]. Another function of albumin is to bind and deliver a number of substances within the body. Besides organic anions and long-chain fatty acids, albumin also binds to vitamins and acidic drugs. Table 5.5 gives a list of some substances that bind to this protein.

Table 5.5 Binding of Various Ligands to Specific Sites on Serum Albumin

Site I	Site II	Cationic Drugs
Warfarin	(S)-Diazepam	Chlorpromazine
Digitoxin	Ibuprofen	Imipramine
Phenytoin	Naproxen	Quinidine
Phenol Red	Octanoate	—

Source: Information in this table was obtained from Peters, T., Jr., *All about Albumin: Biochemistry, Genetics and Medical Applications*, Academic Press, New York, 1996.

Another important plasma protein is α_1-acid glycoprotein (AGP) [115, 117]. Like albumin, AGP is produced by the liver. It is a fairly heterogeneous protein with an approximate molecular weight of 41 kDa [118]. AGP contains a single polypeptide chain with 181 amino acids and five carbohydrate groups. There are many ways in which these carbohydrate groups can potentially be attached to AGP, but only 12 to 20 combinations have been detected. The extent of this glycosylation and the arrangement of these carbohydrate groups is dependent on the disease state of the body. This, in turn, affects the structure of AGP and its binding properties. For instance, it has been proposed for some solutes that binding occurs through sialic acid residues in the carbohydrate chains rather than through interactions with the protein core of AGP [119]. While it is generally accepted that the binding of most solutes occurs at a single hydrophobic core of this protein, the presence of more than one site on this ligand has also been postulated.

AGP has many biological functions in serum. Many of these functions are shared with albumin, including its involvement in transportation and protection. Normal plasma levels of AGP are around 0.50 to 1.0 g/l, but its levels increase during disease. This has led investigators to predict that there is a direct involvement of AGP in the immune response [118]. Although serum albumin is present in plasma at a greater concentration and has greater binding capacity than AGP, this latter protein is more important as a binding agent for basic and neutral drugs.

5.6.2 Applications in Affinity Chromatography

One application of HSA and BSA in affinity chromatography has been their use in the study of solute-protein interactions. For instance, an immobilized HSA support has been used to characterize how L-thyroxine interacts with the warfarin and indole sites of albumin [120], as well as the binding of digitoxin and acetyldigitoxin to HSA [121]. Other examples can be found in Chapters 21 and 22 and in the literature [122]. Similar work has been reported using AGP immobilized under mild conditions that allow this protein to mimic the behavior of AGP in solution [123].

Another use for HSA, BSA, and AGP columns has been as chiral stationary phases for the separation of drugs [124]. Examples of such work can be found in Chapter 21 and in the literature [124–134]. HSA and BSA work well for neutral and basic drugs, while AGP can be used for neutral and acidic compounds. AGP columns have been particularly common in such studies. The main advantage of using AGP over albumin or other proteins is AGP's higher affinity for its target solutes, giving it better retention and resolution.

5.6.3   Immobilization and Column Preparation

BSA and HSA can be purchased in many forms from suppliers like Sigma, Serva, and Pierce. These can then be immobilized for use in affinity chromatography in a variety of ways [127–129]. For instance, albumin can be covalently attached to diol-bonded silica by a two-step or three-step Schiff base method or by using modified silica that has been activated with *N*-hydroxysuccinimide ester [122, 130]. Further details on these immobilization procedures can be found in Chapter 3. Other techniques that can be used to immobilize albumin include adsorption and entrapment [122, 124].

AGP columns for chiral separations can be purchased from ChromTech, Astec, or Regis Technologies. Supports containing AGP can also be prepared by several immobilization techniques [130]. A commercially available AGP chiral stationary phase has been made by ionically binding a monolayer of AGP to diethylaminoethyl silica. This step is followed by oxidizing the protein with periodate to form aldehyde groups that can cross-link AGP through Schiff base formation. Alternatively, AGP can be immobilized directly through its carbohydrate residues by oxidizing these groups with periodate and reacting the resulting aldehyde groups with a support containing free amine or hydrazide groups [123].

5.6.4   Application and Elution Conditions

When using immobilized serum proteins for drug-binding studies, isocratic conditions are generally employed in which the mobile phase is a pH 7.4 phosphate buffer. However, alternative conditions can be used when these proteins are employed as chiral binding agents. For instance, displacing agents can be added to the mobile phase to control the retention of solutes on such columns or to examine the binding of solutes at specific sites on the immobilized proteins [122].

If necessary, a change in pH can be used to alter solute retention on serum protein columns. For albumin, a change in conformation occurs as it is taken to either a more acidic or basic pH. Below pH 4 to 5, albumin undergoes a conformation change known as the N-F transition. An additional change occurs above pH 9, but this is rarely used to alter retention on HPLC columns containing serum proteins, since the support materials within such columns are not stable under these conditions.

A small amount of organic solvent can be used with serum protein columns. This can alter hydrophobic interactions between a solute and the protein or act as a protein denaturant. In addition, temperature can be varied to alter retention. For instance, column temperatures of 4 to 45°C have been successfully used to study and alter the retention of (*R/S*)-warfarin [131] and D/L-tryptophan [132] on immobilized HSA columns.

5.7   AVIDIN AND BIOTIN

The strong binding that occurs between the proteins avidin and streptavidin with biotin has made the avidin/biotin system useful in a variety of applications. This section focuses on the use of avidin and biotin in affinity chromatography. A more in-depth discussion of this system and its applications in other fields can be found in the literature [135].

5.7.1   General Properties

The structure of biotin is given in Figure 5.5. This is also known as vitamin H or vitamin B_7. It is a cofactor present in every living cell, usually in amounts less than 0.001% [5]. Normally, biotin is bound to proteins through either covalent or noncovalent interactions. There are a large number of proteins that bind biotin covalently, such as pyruvate carboxylase.

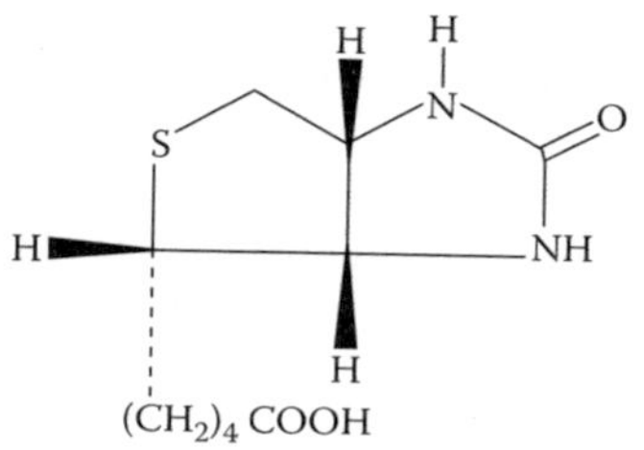

H H

S   N

O

H

NH

H

(CH$_2$)$_4$ COOH

Figure 5.5 The structure of biotin.

However, most applications involving biotin in affinity chromatography are based on its noncovalent binding with the proteins avidin and streptavidin [135].

Avidin has a mass of 66 kDa and is a protein found in egg whites. Streptavidin, which has a mass of 54 kDa, is the bacterial counterpart of avidin and is produced by *Streptomyces avidinii* [136]. These two proteins have several common features. Both are tetrameric proteins with two-fold symmetry and share a similar monomer construction of eight anti-parallel β-strands that form a β-barrel [135, 136]. These proteins also have high affinities for biotin, with association constants of approximately 10^{13} to 10^{15} M^{-1} and with each monomer having a biotin-binding site positioned near one end of the β-barrel [136, 137].

One difference between these proteins is the fact that avidin is a glycoprotein, while streptavidin is not. In addition, avidin is more basic, having a pI of approximately 10.5 [135]. The abundance of positively charged residues in avidin and the presence of its carbohydrate moiety are both thought to be responsible for avidin's nonspecific binding to nucleic acids and negatively charged cell membranes. Carbohydrate-free avidin can be produced by using deglycosylating enzymes to remove such moieties, followed by the use of a lectin column to capture these groups. This deglycosylation apparently does not affect the function of avidin [135].

Although streptavidin has less nonspecific binding than avidin [135], it has been shown to bind at low levels and with high affinity to cell surfaces. This occurs due to the presence of the amino acid sequence Arg-Tyr-Asp, which mimics the universal recognition domain present in fibronectin and other adhesion-related molecules [138]. However, this interaction does not affect streptavidin's binding to biotin and is more of a concern in histochemical and cytochemical applications rather than in affinity chromatography.

While avidin and streptavidin share similar biotin sites and binding activities, there are some other notable differences. For instance, both proteins have noncontact aromatic residues in their binding sites, giving rise to a hydrophobic "box," but avidin contains an additional aromatic residue in its binding site. Another difference is the degree of hydrogen bonding between the valeryl carboxylate group of biotin and these binding sites. Avidin is able to form five hydrogen bonds with biotin carboxylate, while streptavidin only forms two. These differences are thought to be the reason for streptavidin's larger association constant for biotin (1.6×10^{13} M^{-1} for avidin versus 2.5×10^{15} M^{-1} for streptavidin).

5.7.2 Applications in Affinity Chromatography

The extraordinary affinity for biotin exhibited by avidin and streptavidin has spurred the use of this system in numerous applications. For instance, biotin can be placed as a tag on a desired target and used to capture and enrich this target on an immobilized avidin or streptavidin column. A common example is the use of biotin in *isotopically coded affinity tags* (*ICATs*) for quantitative proteomic studies, as discussed in Chapters 18 and 27.

Immobilized-biotin columns can also be used to recover biotin-binding proteins from solutions (e.g., avidin from egg whites) or biotinylated protein-avidin complexes from solution [5]. Another use of the avidin/biotin system in affinity chromatography is as a scaffold for small-affinity ligands. For example, this approach has been used to attach aptamers to solid supports by first biotinylating them and then adsorbing them to an immobilized avidin or streptavidin support [139].

5.7.3 Immobilization and Column Preparation

Either biotin, avidin, or streptavidin can be immobilized for use in affinity chromatography. For instance, Hermanson et al. [5] provide an immobilization protocol for coupling either avidin or streptavidin to Sepharose through the cyanogen bromide method. Similar methods for generating an immobilized avidin column are described by Wilchek and Bayer [135].

The immobilization of avidin or streptavidin to a support does have some effect on the binding of these proteins. For this reason, an assay should be performed to examine their biotin sites after immobilization, as described by Hermanson et al. [5]. The N-hydroxysuccinimide ester of biotin (also known as BNHS) is often used to tag proteins with biotin for their capture on avidin or streptavidin columns. A more detailed description of biotin-containing compounds available for such work can be found in the literature [135].

To prepare an immobilized biotin column, either native biotin or 2-iminobiotin (i.e., a biotin analog) can be coupled to a support through its valeric acid side chain. This is generally accomplished by using an amine-containing spacer arm. A detailed description of how such ligands can be placed onto Sepharose is given in the literature [135].

5.7.4 Application and Elution Conditions

Although the strong interaction between biotin and avidin or streptavidin is a major advantage of this system, this same property has hindered the use of these ligands in affinity chromatography. This occurs because harsh elution conditions (e.g., 6 M guanidine hydrochloride at pH 1.5) are generally necessary to disrupt this binding [5, 135]. However, these conditions can also denature the immobilized avidin when it is being used to retrieve a biotinylated target [140]. In addition, such conditions are far too extreme for the recovery of most proteins in an active form [5].

This difficulty has led to modified versions of the avidin/biotin system that allow the use of milder elution conditions. This has been accomplished in two ways: (1) modifying avidin or streptavidin and (2) using a biotin analog. One common way to modify avidin or streptavidin is to use their monomeric forms [5]. This gives a binding constant for biotin that is greatly reduced compared with the tetrameric species [5, 135, 136, 141]. For example, one study reported association constants of 10^8 and $10^7 \, M^{-1}$ for the monomeric and dimeric forms of streptavidin with biotin [140].

5.8 SUMMARY AND CONCLUSIONS

This chapter examined several types of ligands that are commonly used in the method of bioaffinity chromatography. First, immunoglobulin-binding proteins were considered, such as protein A, protein G, and protein L. Other ligands examined included enzymes, lectins, carbohydrates, and serum proteins such as HSA and AGP. Finally, the avidin/biotin system was discussed. In each case, various applications of these ligands were provided. The conditions required for immobilization and use of these agents in affinity columns were also presented. Similar information is provided in other chapters of this handbook for other biological ligands, such as antibodies (Chapter 6) and nucleic acids (Chapter 7).

SYMBOLS AND ABBREVIATIONS

AGP	α_1-Acid glycoprotein
araUTP	Arabinofuranosyluracil-5′-triphosphate
BNHS	Biotin N-hydroxysuccinimide ester
BSA	Bovine serum albumin
CE	Capillary electrophoresis
CLC	Charcot-Leyden crystal protein
Con A	Concanavalin A
CP-dMM	N-5-Carboxypentyl-l-deoxymannojirimycin
dATP	Deoxyadenosine triphosphate
DEAE-cellulose	Diethylaminoethyl-cellulose
EDTA	Ethylenediaminetetraacetic acid
GAG	Glycosaminoglycan
GlcNAc	N-Acetyl-D-glucosamine
GTP	Guanosine triphosphate
F_{ab}	Antigen-binding fragment of an antibody
F_c	Constant (or crystallizable) fragment of an antibody
HPLC	High-performance liquid chromatography
HSA	Human serum albumin
ICAT	Isotopically coded affinity tag
IgA	Immunoglobulin A
IgD	Immunoglobulin D
IgE	Immunoglobulin E
IgG	Immunoglobulin G
IgM	Immunoglobulin M
kDa	Kilodaltons
pI	Isoelectric point
scF_v	Single-chain variable-region fragment of an antibody
WGA	Wheat germ agglutinin

REFERENCES

1. Starkenstein, E., Ferment action and the influence upon it of neutral salts, *Biochem Z.*, 24, 210–218, 1910.
2. Lindmark, R., Biriell, C., and Sjoequist, J., Quantitation of specific IgG antibodies in rabbits by a solid-phase radioimmunoassay with ^{125}I-protein A from Staphylococcus aureus, *Scand. J. Immunol.*, 14, 409–420, 1981.
3. Ey, P.L., Prowse, S.J., and Jenkin, C.R., Isolation of pure IgG, IgG$_{2a}$ and IgG$_{2b}$ immunoglobulins from mouse serum using protein A-Sepharose, *Immunochemistry*, 15, 429–436, 1978.
4. Bjorck, L. and Kronvall, G., Purification and some properties of streptococcal protein G, a novel IgG-binding reagent, *J. Immunol.*, 133, 969–974, 1984.
5. Hermanson, G.T., Mallia, A.K., and Smith, P.K., *Immobilized Affinity Ligand Techniques*, Academic Press, San Diego, 1992.
6. Eliasson, M., Olsson, A., Palmcrantz, E., Wibers, K., Inganas, M., Guss, B., Lindberg, M., and Uhlen, M., Chimeric IgG-binding receptors engineered from staphylococcal protein A and streptococcal protein G, *J. Biol. Chem.*, 263, 4323–4327, 1988.
7. Turkova, J., Oriented immobilization of biologically active proteins as a tool for revealing protein interactions and function, *J. Chromatogr. B*, 722, 11–31, 1999.

8. Lian, L.Y., Derrick, J.P., Sutcliffe, M.J., Yang, J.C., and Roberts, G.C.K., Determination of the solution structures of domains II and III of protein G from *Streptococcus* by ^{1}H nuclear magnetic resonance, *Mol. Biol.,* 228, 1219–1234, 1992.

9. Jendeberg, L., Tashiro, M., Tejero, R., Lyons, B.A., Uhlen, M., Montelione, G.T., and Nilsson, B., The mechanism of binding staphylococcal protein A to immunoglobulin G does not involve helix unwinding, *Biochemistry,* 35, 22–31, 1996.

10. Derrick, J.P. and Wigley, D.B., The third IgG-binding domain from streptococcal protein G: An analysis by X-ray crystallography of the structure alone and in a complex with Fab, *J. Mol. Biol.,* 243, 906–918, 1994.

11. Graille, M., Stura, E.A., Corper, A.L., Sutton, B.J., Taussig, M.J., Charbonnier, J.B., and Silverman, G.J., Crystal structure of a *Staphylococcus aureus* protein A domain complexed with the Fab fragment of a human IgM antibody: Structural basis for the recognition of B-cell receptors and superantigen activity, *Proc. Natl. Acad. Sci. U.S.A.,* 97, 5399–5404, 2000.

12. Sauer-Eriksson, A.E., Kleywegt, G.J., Uhlen, M., and Jones, T.A., Crystal structure of the C2 fragment of streptococcal protein G in complex with the Fc domain of human IgG, *Structure,* 3, 265–278, 1995.

13. Viera, C., Yang, H., and Etzel, M.R., Affinity membranes: Competitive binding of the human IgG subclasses to immobilized protein G, *Ind. Eng. Chem. Res.,* 39, 3356–3363, 2000.

14. Gronenborn, A.M. and Clore, G.M., Identification of the contact surface of a streptococcal protein G domain complexed with a human Fc fragment, *J. Mol. Biol.,* 233, 331–335, 1993.

15. Faulmann, E.L., Duval, J.L., and Boyle, M.D.P., Protein B: A versatile bacterial Fc-binding protein selective for human IgA, *BioTechniques,* 10, 748–750, 1991.

16 Kawasaki, T., Kawasaki, N., and Yamashina, I., Mannose/*N*-acetylglucosamine-binding proteins from mammalian sera, *Methods Enzymol.,* 179, 310–321, 1989.

17. Nevens, J.R., Mallia, A.K., Wendt, M.W., and Smith, P.K., Affinity chromatographic purification of immunoglobulin M antibodies utilizing immobilized mannan-binding protein, *J. Chromatogr.,* 597, 247–256, 1992.

18. Bjorck, L., Protein L: A novel bacterial cell wall protein with affinity for IgL chains, *J. Immunol.,* 140, 1194–1197, 1988.

19. Akerstrom, B. and Bjorck, L., Protein L: An immunoglobulin light-chain-binding bacterial protein: Characterization of binding and physicochemical properties, *J. Biol. Chem.,* 264, 19740–19746, 1989.

20. Nilson, B.H., Logdberg, L., Kastern, W., Bjorck, L., and Akerstrom, B., Purification of antibodies using protein *L*-binding framework structures in the light chain variable domain, *J. Immunol. Methods,* 164, 33–40, 1993.

21. Ohlson, S., High-performance liquid affinity chromatography (HPLAC) with protein A-silica, in *Affinity Chromatography and Biological Recognition,* Chaiken, I.M., Wilchek, M., and Parikh, M., Eds., Academic Press, New York, 1983, pp. 255–256.

22. Crowley, S.C. and Walters, R.R., Determination of immunoglobulins in blood serum by high-performance affinity chromatography, *J. Chromatogr.,* 266, 157–162, 1983.

23. Cassulis, P., Magasic, M.V., and DeBari, V.A., Ligand affinity chromatographic separation of serum IgG on recombinant protein G-silica, *Clin. Chem.,* 37, 882–886, 1991.

24. Hage, D.S. and Walters, R.R., Dual-column determination of albumin and immunoglobulin G in serum by high-performance affinity chromatography, *J. Chromatogr.,* 386, 37–49, 1987.

25. Hage, D.S., Survey of recent advances in analytical applications of immunoaffinity chromatography, *J. Chromatogr.,* 715, 3–28, 1998.

26. de Frutos, M. and Regnier, F.E., Tandem chromatographic-immunological analyses, *Anal. Chem.,* 65, 17A–25A, 1993.

27. Phillips, T.M., High-performance immunoaffinity chromatography, *LC Mag.,* 3, 962–972, 1985.

28. Madry, N., Auerbach, B., and Schelp, C., Measures to overcome HAMA interferences in immunoassays, *Anticancer Res.,* 17, 2883, 1997.

29. Remaley, A.T. and Wilding, P., Macroenzymes: Biochemical characterization, clinical significance, and laboratory detection, *Clin. Chem.,* 35, 2261–2270, 1989.

30. Cole, L.J. and Kennedy, R.T., Selective preconcentration for capillary-zone electrophoresis using protein G immunoaffinity capillary chromatography, *Electrophoresis,* 16, 549–556, 1995.

31. Hage, D.S., Walters, R.R., and Hethcote, H.W., Split-peak affinity chromatographic studies of the immobilization-dependent adsorption kinetics of protein A, *Anal. Chem.,* 58, 274–279, 1986.

32. Zaborsky, O.R., *Immobilized Enzymes,* CRC Press, Cleveland, 1973.

33. Chibata, I., *Immobilized Enzymes: Research and Development,* Wiley, New York, 1978.

34. Nelson, J.M. and Griffin, E.G., Adsorption of invertase, *J. Am. Chem. Soc.,* 38, 1109–1115, 1916.

35. Tosa, T., Mori, T., Fuse, N., and Chibata, I., Continuous enzyme reactions, V: Kinetics and industrial application of aminoacylase columns for continuous optical resolution of acyl-DL-amino acids, *Agric. Biol. Chem.,* 33, 1047–1052, 1969.

36. Weetall, H.H., Immobilized enzyme technology, *Cereal Foods World,* 21, 581–587, 1976.

37. Wang, S.S., Gallili, G.E., Gilbert, S.G., and Leeder, J.G., Inactivation and regeneration of immobilized catalase, *J. Food Sci.,* 39, 338–341, 1974.

38. Brodelius, P., Industrial applications of immobilized biocatalysts, *Adv. Biochem. Eng.,* 10, 75–129, 1978.

39. Godfrey, T. and Reichelt, J., *Industrial Enzymology: The Application of Enzymes in Industry,* Nature Press, New York, 1983.

40. Wiseman, A., Potential for immobilized enzymes in water and air pollution, *Topics Enzyme Ferment. Biotechnol.,* 7, 264–270, 1983.

41. Chang, T.M.S. and Poznonsky, M.J., Semipermeable microcapsules containing catalase for enzyme replacement in acatalasaemic mice, *Nature,* 218, 243–245, 1968.

42. Tatsumi, K., Wada, S., and Ichikawa, H., Removal of chlorophenols from wastewater by immobilized horseradish peroxidase, *Biotechnol. Bioeng.,* 51, 126–130, 1996.

43. Fritz, H., Gebhardt, M., Meister, R., and Schult, H., Preparation of modified protease inhibitors using water-insoluble trypsin resins, *Hoppe-Seyler's Z. Physiol. Chem.,* 351, 1119–1122, 1970.

44. Feinstein, G., Isolation of chicken ovoinhibitor by affinity chromatography on chymotrypsin-Sepharose, *Biochim. Biophys. Acta,* 236, 73–77, 1971.

45. Givol, D., Weinstein, Y., Gorecki, M., and Wilchek, M., A general method for the isolation of labelled peptides from affinity-labelled proteins, *Biochem. Biophys. Res. Comm.,* 38, 825–830, 1970.

46. Omenn, G.S., Ontjes, D.A., and Anfinsen, C.B., Fractionation of antibodies against staphylococcal nuclease on "Sepharose" immunoadsorbents, *Nature,* 225, 189–190, 1970.

47. Denburg, J. and DeLuca, M., Purification of a specific tRNA by Sepharose-bound enzyme, *Proc. Natl. Acad. Sci. U.S.A.,* 67, 1057–1062, 1970.

48. Avrameas, S. and Guilbert, B., Biologically active water-insoluble protein polymers: Their use for the isolation of specifically interacting proteins, *Biochimie,* 53, 603–614, 1971.

49. Goldman, R. and Lenhoff, H.M., Glucose 6-phosphate dehydrogenase adsorbed on collodion membranes, *Biochim. Biophys. Acta,* 242, 514–518, 1971.

50. Eng, G.Y., Jones, L., and Medina, M., Characterization of an immobilized protein matrix for use in an affinity method for β-lactam antibiotics, *Biotechnol. Appl. Biochem.,* 22, 129–144, 1995.

51. Steers, E., Jr. and Cuatrecasas, P., Isolation of beta-galactosidase by chromatography, *Methods Enzymol.,* 34, 350–358, 1974.

52. Schweden, J. and Bause, E., Characterization of trimming Man9-Mannosidase from pig liver: Purification of a catalytically active fragment and evidence for the transmembrane nature of the intact 65kDa enzyme, *Biochem. J.,* 264, 347–355, 1989.

53. Straudenbauer, W.L. and Orr, E., DNA gyrase: Affinity chromatography on novobiocin-Sepharose and catalytic properties, *Nucleic Acids Res.,* 9, 3589–3603, 1981.

54. Raeber, A.J., Riggio, G., and Waser, P.G., Purification and isolation of choline acetyltransferase from the electric organ of *Torpedo marmorata* by affinity chromatography, *Eur. J. Biochem.,* 186, 487–492, 1989.

55. Dadabay, C.Y. and Pike, L.J., Purification and characterization of a cytosolic transglutaminase from a cultured human tumour-cell line, *Biochem. J.,* 264, 679–685, 1989.

56. Matsuda, Y., Shimura, K., Shiraki, H., and Nakagawa, H., Purification and properties of adenylosuccinate synthetase from Yoshida sarcoma ascites tumor cells, *Biochim. Biophys. Acta,* 616, 340–350, 1980.

57. Eliasson, R., Fontecave, M., Joernvall, H., Krook, M., Pontis, E., and Reichard, P., The anaerobic ribonucleoside triphosphate reductase from *Escherichia coli* requires S-adenosylmethionine as a cofactor, *Proc. Natl. Acad. Sci. U.S.A.,* 87, 3314–3318, 1990.

58. Izuta, S. and Saneyoshi, M., AraUTP-Affi-Gel 10: A novel affinity absorbent for the specific purification of DNA polymerase alpha-primase, *Anal. Biochem.,* 174, 318–324, 1988.

59. Holmberg, L., Bladh, B., and Astedt, B., Purification of urokinase by affinity chromatography, *Biochim. Biophys. Acta,* 445, 215–222, 1976.

60. Campbell, B.J., Forrester, L.J., Zahler, W.L., and Burks, M., Beta-lactamase activity of purified and partially characterized human renal dipeptidase, *J. Biol. Chem.,* 259, 14586–14590, 1984.

61. Huang, J.S., Huang, S.S., and Tang, J., Cathepsin D isozymes from porcine spleens: Large-scale purification and polypeptide chain arrangements, *J. Biol. Chem.,* 254, 11405–11417, 1979.

62. Rich, D.H., Brown, M.A., and Barrett, A.J., Purification of cathepsin B by a new form of affinity chromatography, *Biochem. J.,* 235, 731–734, 1986.

63. Buttle, D.J., Kembhavi, A.A., Sharp, S.L., Shute, R.E., Rich, D.H., and Barrett, A.J., Affinity purification of the novel cysteine proteinase papaya proteinase IV and papain from papaya latex, *Biochem. J.,* 261, 469–476, 1989.

64. Tibbot, B.K., Wong, D.W.S., and Roberston, G.H., A functional raw starch-binding domain of barley alpha-amylase expressed in *Escherichia coli, J. Protein Chem.,* 19, 663–669, 2000.

65. Feinstein, G., Purification of trypsin by affinity chromatography on ovomucoid-Sepharose resin, *FEBS Lett.,* 7, 353–355, 1970.

66. Cuatrecasas, P., Wilchek, M., and Anfinsen, C.B., Selective enzyme purification by affinity chromatography, *Proc. Natl. Acad. Sci. U.S.A.,* 61, 636–643, 1968.

67. Tyagi, R., Roy, I., Agarwal, R., and Gupta, M.N., Carbodiimide coupling of enzymes to the reversibly soluble insoluble polymer Eudragit S-100, *Biotechnol. Appl. Biochem.,* 28, 201–206, 1998.

68. Roy, I. and Gupta, M.N., Current trends in affinity-based separations of proteins/enzymes, *Curr. Sci.,* 78, 587–591, 2000.

69. Tischer, W. and Wedekind, F., Immobilized enzymes: Methods and applications, *Topics Curr. Chem.,* 200, 95–126, 1999.

70. Saleemuddin, M., Bioaffinity-based immobilization of enzymes, *Adv. Biochem. Eng. Biotechnol.,* 64, 203–226, 1999.

71. Carr, P.W. and Bowers, L.D., *Immobilized Enzymes in Analytical and Clinical Chemistry,* Wiley-Interscience, New York, 1980.

72. Ngo, T.T., *Molecular Interactions in Bioseparations,* Plenum Press, New York, 1993, pp. 49–68.

73. Liener, I.E., Sharon, N., and Goldstein, I.J., *The Lectins: Properties, Functions and Applications in Biology and Medicine,* Academic Press, London, 1986.

74. Nilsson, C.L., Lectins: Proteins that interpret the sugar code, *Anal. Chem.,* 75, 348A–353A, 2003.

75. Hage, D.S., Affinity chromatography, in *Handbook of HPLC,* Katz, E., Eksteen, R., Shoenmakers, P., and Miller, N., Eds., Marcel Dekker, New York, 1998, chap. 13.

76. Anderson, D.J., Branum, E.L., and O'Brien, J.F., Liver- and bone-derived isoenzymes of alkaline phosphatase in serum as determined by high-performance affinity chromatography, *Clin. Chem.,* 36, 240–246, 1990.

77. Gonchoroff, D.G., Branum, E.L., and O'Brien, J.F., Alkaline phosphatase isoenzymes of liver and bone origin are incompletely resolved by wheat-germ-lectin affinity chromatography, *Clin. Chem.,* 35, 29–32, 1989.

78. Gonchoroff, D.G., Branum, E.L., Cedel, S.L., Riggs, B.L., and O'Brien, J.F., Clinical evaluation of high-performance affinity chromatography for the separation of bone and liver alkaline phosphatase isoenzymes, *Clin. Chim. Acta,* 199, 43–50, 1991.
79. Tavella, M., Alaupovic, P., Knight-Gibson, C., Tournier, H., Schinella, G., and Mercuri, O., Separation of ApoA- and ApoB-containing lipoproteins of human plasma by affinity chromatography on concanavalin A, *Prog. Lipid Res.,* 30, 181–187, 1991.
80. Inoue, T., Yamauchi, M., Toda, G., and Ohkawa, K., Microheterogeneity with concanavalin A affinity of serum transferrin in patients with alcoholic liver disease, *Alcohol Clin. Exp. Res.,* 20, 363A–365A, 1996.
81. Papandreou, M.J., Asteria, C., Pettersson, K., Ronin, C., and Beck-Peccoz, P., Concanavalin A affinity chromatography of human serum gonadotropins: Evidence for changes of carbohydrate structure in different clinical conditions, *J. Clin. Endocrinol. Metab.,* 76, 1008–1013, 1993.
82. Yoshida, K.I., Honda, M., Arai, K., Hosoya, Y., Moriguchi, H., Sumi, S., Ueda, Y., and Kitahara, S., Serial lectin affinity chromatography with concanavalin A and wheat germ agglutinin demonstrates altered asparagine-linked sugar-chain structures of prostatic acid phosphatase in human prostate carcinoma, *J. Chromatogr. B,* 695, 439–443, 1997.
83. Scherberich, J.E., Wiemer, J., Herzig, C., Fischer, P., and Schoeppe, W., Isolation and partial characterization of angiotensinase A and aminopeptidase M from urine and human kidney by lectin affinity chromatography and high-performance liquid chromatography, *J. Chromatogr.,* 521, 279–289, 1990.
84. Le, Y., Tan, S.M., Lee, S.H., Kon, O.L., and Lu, J., Purification and binding properties of a human ficolin-like protein, *J. Immunol. Methods,* 204, 43–49, 1997.
85. Ishama, S., Iketa, K., Shimamura, T., Nomura, T., Saeki, Y., and Makino, H., Recombinant influenza HA protein purified from insect cell culture with chromatography column containing immobilized carbohydrate for use as vaccine, *Jpn. Kokai Tokkyo Koho,* 1995, 8 pp.
86. Isemura, M., Sato, N., Yamaguchi, Y., Aikawa, J., Munakata, H., Hayashi, N., Yosizawa, Z., Nakamura, T., Kubota, A., and Arakawa, M., Isolation and characterization of fibronectin-binding proteoglycan carrying both heparan sulfate and dermatan sulfate chains from human placenta, *J. Biol. Chem.,* 262, 8926–8933, 1987.
87. Castillo, G. and Snow, A.D., Composition and methods for treating Alzheimer's disease and other amyloidoses, *PCT Int. Appl.* 1998, 58 pp.
88. VanderNoot, V.A., Hileman, R.E., Dordick, J.S., and Linhardt, R.J., Affinity capillary electrophoresis employing immobilized glycosaminoglycan to resolve heparin-binding peptides, *Electrophoresis,* 19, 437–441, 1998.
89. Hirabayashi, J., Charcot-Leyden crystal protein, a new member of the galectin family, is a lysophospholipase having weak carbohydrate-binding activity, *Trends Glycosci. Glycotechnol.,* 8, 135–136, 1996.
90. Halberg, D.F., Proulx, G., Doege, K., Yamada, Y., and Drickamer, K., A segment of the cartilage proteoglycan core protein has lectin-like activity, *J. Biol. Chem.,* 263, 9486–9490, 1988.
91. Griegel, S., Rajewsky, M.F., Ciesiolka, T., and Gabius, H.J., Endogenous sugar receptor (lectin) profiles of human retinoblastoma and retinoblast cell lines analyzed by cytological markers, affinity chromatography and neoglycoprotein-targeted photolysis, *Anticancer Res.,* 9, 723–730, 1989.
92. Gao-Uozumi, C.X., Uozumi, N., Miyoshi, E., Nagai, K., Ikeda, Y., Teshima, T., Noda, K., Shiba, T., Honke, K., and Taniguchi, N., A novel carbohydrate binding activity of annexin V toward a bisecting *N*-acetylglucosamine, *Glycobiology,* 10, 1209–1216, 2000.
93. Redmond, J.W., Packer, N.H., Gooley, A.A., Williams, K.L., Batley, M., Kett, W.C., Pisano, A., Tweeddale, H.J., and Cooper, C.A., Characterization of individual *N*- and *O*-linked glycosylation sites using Edman degradation, *PCT Int. Appl.* 1995, 112 pp.
94. Gabius, H.J., Engelhardt, R., Hellmann, T., Midoux, P., Monsigny, M., Nagel, G.A., and Vehmeyer, K., Characterization of membrane lectins in human colon carcinoma cells by flow cytofluorometry, drug targeting and affinity chromatography, *Anticancer Res.,* 7, 109–112, 1987.

95. Caron, M., Joubert-Caron, R., Cartier, J.R., Chadli, A., and Bladier, D., Study of lectin-ganglioside interactions by high-performance liquid affinity chromatography, *J. Chromatogr.*, 646, 327–333, 1993.

96. Ohyama, Y., Kasai, K., Nomoto, H., and Inoue, Y., Frontal affinity chromatography of ovalbumin glycoasparagines on a concanavalin A-sepharose column. A quantitative study of the binding specificity of the lectin, *J. Biol. Chem.*, 260, 6882–6887, 1985.

97. Hirabayashi, J., Arata, Y., and Kasai, Y., Reinforcement of frontal affinity chromatography for effective analysis of lectin-oligosaccharide interactions, *J. Chromatogr. A*, 890, 261–271, 2000.

98. Arata, Y., Hirabayashi, J., and Kasai, K., Sugar binding properties of the two lectin domains of the tandem repeat-type galectin LEC-1 (N32) of *Caenorhabditis elegans*, *J. Biol. Chem.*, 276, 3068–3077, 2001.

99. Schriemer, D.C., Bundle, D.R., Li, L., and Hindsgaul, O., Micro-scale frontal affinity chromatography with mass spectrometric detection: A new method for the screening of compound libraries, *Angew. Chem. Int. Ed.*, 37, 3383–3387, 1998.

100. Dias-Baruffi, M., Roque-Barreira, M.C., Cunha, F.Q., and Ferreira, S.H., Isolation and partial chemical characterization of macrophage-derived neutrophil chemotactic factor, *Mediators Inflamm.*, 4, 257–262, 1995.

101. Yura, H., Samejima, T., Nagoya, M., Yamamoto, Y., and Akaike, T., Method and absorbents for separation of leukocytes, *Jpn. Kokai Tokkyo Koho*, 1990. 12 pp.

102. Gabius, H.J., Engelhardt, R., Rehm, S., Deerberg, F., and Cramer, F., Differences in the pattern of endogenous lectins from spontaneous rat mammary tumours, *Tumor Biol.*, 7, 71–81, 1986.

103. Gabius, H.J., Engelhardt, R., Sartoris, D.J., and Cramer, F., Pattern of endogenous lectins of a human sarcoma (Ewing's sarcoma) reveals differences to human normal tissues and tumors of epithelial and germ cell origin, *Cancer Lett.*, 31, 139–145, 1986.

104. Gabius, H.J., Engelhardt, R., Rehm, S., and Cramer, F., Biochemical characterization of endogenous carbohydrate-binding proteins from spontaneous murine rhabdomyosarcoma, mammary adenocarcinoma, and ovarian teratoma, *J. Natl. Cancer Inst.*, 73, 1349–1357, 1984.

105. Feldman, S.A., Audet, S., and Beeler, J.A., The fusion glycoprotein of human respiratory syncytial virus facilitates virus attachment and infectivity via an interaction with cellular heparan sulfate, *J. Virol.*, 74, 6442–6447, 2000.

106. Alberdi, E., Hyde, C.C., and Becerra, S.P., Pigment epithelium-derived factor (PEDF) binds to glycosaminoglycans: Analysis of the binding site, *Biochemistry*, 37, 10643–10652, 1998.

107. Krusat, T. and Streckert, H.J., Heparin-dependent attachment of respiratory syncytial virus (RSV) to host cells, *Arch. Virol.*, 142, 1247–1254, 1997.

108. Vanderplasschen, A., Bublot, M., Dubuisson, J., Pastoret, P.P., and Thiry, E., Attachment of the gammaherpesvirus bovine herpesvirus 4 is mediated by the interaction of gp8 glycoprotein with heparinlike moieties on the cell surface, *Virology*, 196, 232–240, 1993.

109. Juvancz, Z. and Szejtli, J., The role of cyclodextrins in chiral selective chromatography, *Trends Anal. Chem.*, 21, 379–388, 2002.

110. Szejtli, J., Introduction and general overview of cyclodextrin chemistry, *Chem. Rev.*, 98, 1743–1753, 1998.

111. Juvancz, Z. and Peterson, P., Enantioselective gas chromatography, *J. Microcol. Sep.*, 8, 99, 1996.

112. Barker, R., Chiang, C.K., Trayer, I.P., and Hill, R.L., Monosaccharides attached to agarose, *Methods Enzymol.*, 34, 317–328, 1974.

113. Agarwal, B.B. and Goldstein, I.J., Concanavalin A, the jack bean (*Canavalia ensiformis*) phytohemaglutinin, *Methods Enzymol.*, 28, 313–318, 1972.

114. Honda, S., Suzuki, S., and Kakehi, K., Separation of carbohydrates and lectins on carbohydrate-immobilized resins, *J. Chromatogr.*, 396, 93–100, 1987.

115. Peters, T., Jr., *All about Albumin: Biochemistry, Genetics and Medical Applications*, Academic Press, New York, 1996.

116. Hage, D.S. and Austin, J., High-performance affinity chromatography and immobilized serum albumin as probes for drug- and hormone-protein binding, *J. Chromatogr. B*, 739, 39–54, 2000.

117. Gottschalk, A., Ed., *Glycoproteins: Their Composition, Structure and Function,* Part A, 2nd ed., Elsevier, New York, 1972, Chaps. 1 and 3.

118. Kremer, J.M.H., Wilting, J., and Janssen, L.H.M., Drug binding to human alpha-1-acid glycoprotein in health and disease, *Pharmacol. Rev.,* 40, 1–47, 1988.

119. Yehl, P.M., O'Brien, T.P., Moeder, C.W., Grinberg, N., Bicker, G., and Wyvratt, J., Mechanisms of retention of pyrrolidinyl norephedrine on immobilized α_1-acid glycoprotein, *Chirality,* 12, 107–113, 2000.

120. Loun, B. and Hage, D.S., Characterization of thyroxine-albumin binding using high-performance affinity chromatography, I: Interaction at the warfarin and indole sites of albumin, *J. Chromatogr.,* 579, 225–235, 1992.

121. Sengupta, A. and Hage, D.S., Characterization of the binding of digitoxin and acetyldigitoxin to human serum albumin by high-performance affinity chromatography, *J. Chromatogr. B,* 724, 91–100, 1999.

122. Hage, D.S., High-performance affinity chromatography: A powerful tool for studying serum protein binding, *J. Chromatogr. B,* 768, 3–30, 2002.

123. Xuan, H., Drug Binding Studies of Alpha 1-Acid Glycoprotein Immobilized to High-Performance Liquid Chromatographic Supports, M.S. thesis, University of Nebraska-Lincoln, 2003.

124. Allenmark, S., *Chromatographic Enantioseparation: Methods and Applications,* 2nd ed. Ellis Horwood, New York, 1991, chap 7.

125. Aubry, A.F., Gimenez, F., Farinotti, R., and Wainer, I.W., Enantioselective chromatography of the antimalarial agents chloroquine, mefloquine, and enpiroline on a α_1-acid glycoprotein chiral stationary phase, *Chirality,* 4, 30–35, 1992.

126. Mills, M.H., Mather, L.E., Gu, X.S., and Huang, J.L., Determination of ketorolac enantiomers in plasma using enantioselective liquid chromatography on an α_1-acid glycoprotein chiral stationary phase and ultraviolet detection, *J. Chromatogr. B,* 658, 117–182, 1994.

127. Andersson, S., Thompson, R.A., and Allenmark, S.G., Direct liquid chromatographic separation of enantiomers on immobilized protein stationary phases, IX: Influence of the cross-linking reagent on the retentive and enantioselective properties of chiral sorbents based on bovine serum albumin, *J. Chromatogr.,* 591, 65–73, 1992.

128. Thompson, R.A., Andersson, S., and Allenmark, S., Direct liquid chromatographic separation of enantiomers on immobilized protein stationary phases, VII: Sorbents obtained by entrapment of cross-linked bovine serum albumin in silica, *J. Chromatogr.,* 465, 263–270, 1989.

129. Ladaviere, C., Delair, T., Domard, A., Pichot, C., and Mandrand, B., Covalent immobilization of bovine serum albumin onto (maleic anhydride-alt-methyl vinyl ether) copolymers, *J. Appl. Polym. Sci.,* 72, 1565–1572, 1999.

130. Hermanson, G.T., Mallia, A.K., and Smith, P.K., *Immobilized Affinity Ligand Techniques,* Academic Press, San Diego, 1992.

131. Loun, B. and Hage, D.S., Chiral separation mechanisms in protein-based HPLC columns, 2: Kinetic studies of (*R*)- and (*S*)-warfarin binding to immobilized human serum albumin, *Anal. Chem.,* 68, 1218–1225, 1996.

132. Yang, J. and Hage, D.S., Characterization of the binding and chiral separation of *D*- and *L*-tryptophan on a high-performance immobilized human serum albumin column, *J. Chromatogr.,* 645, 241–250, 1993.

133. Gyimesi-Forras, K., Szasz, G., Bathory, G., Meszaros, G., and Gergely, A., Study on the sorption properties of α_1-acid glycoprotein (AGP)-based stationary phase modified by organic solvents, *Chirality,* 15, 377–381, 2003.

134. Gotmar, G., Albareda, N.R., and Fornstedt, T., Investigation of the heterogeneous adsorption behavior of selected enantiomers on immobilized α_1-acid glycoprotein, *Anal. Chem.,* 74, 2950–2959, 2002.

135. Wilchek, M. and Bayer, E., Introduction to avidin-biotin technology, *Methods Enzymol.,* 184, 5–13, 1990.

136. Livnah, O., Bayer, E.A., Wilchek, M., and Sussman, J.L., Three-dimensional structures of avidin and the avidin-biotin complex, *Proc. Natl. Acad. Sci. U.S.A.,* 90, 5076–5080, 1993.

137. Weber, P.C., Ohlendorf, D.H., Wendoloski, J.J., and Salemme, F.R., Structural origins of high-affinity biotin binding to streptavidin, *Science,* 243, 85–88, 1989.

138. Alon, R., Bayer, E.A., and Wilchek, M., Streptavidin contains an RYD sequence which mimics the RGD receptor domain of fibronectin, *Biochem. Biophys. Res. Commun.,* 170, 1236–1241, 1990.

139. Liss, M., Petersen, B., Wolf, H., and Prohaska, E., An aptamer-based quartz crystal protein, *Anal. Chem.,* 74, 4488–4495, 2002.

140. Sano, T., Vajda, S., and Cantor, C., Genetic engineering of streptavidin, a versatile affinity tag, *J. Chromatogr. B,* 715, 85–91, 1998.

141. Sano, T. and Cantor, C.R., Intersubunit contacts made by tryptophan 120 with biotin are essential for both strong biotin-binding and biotin-induced tighter subunit association of streptavidin, *Proc. Natl. Acad. Sci. U.S.A.,* 92, 3180–3184, 1995.

6

Immunoaffinity Chromatography

David S. Hage

Department of Chemistry, University of Nebraska, Lincoln, NE

Terry M. Phillips

Ultramicro Analytical Immunochemistry Resource, National Institutes
of Health, Bethesda, MD

CONTENTS

6.1   INTRODUCTION

Immunoaffinity chromatography (IAC) refers to any chromatographic method in which the stationary phase consists of antibodies or substances that interact with antibodies. The high selectivity of antibodies in their interactions with other molecules, and the ability to produce antibodies against a wide range of solutes, has made IAC a popular tool for the purification of biological compounds.

Work with the use of immobilized substances for antibody purification appeared as early as 1935, when d'Allesandro and Sofia used antigens adsorbed on kaolin and charcoal for the isolation of antibodies associated with syphilis and tuberculosis [1]. In 1936, Landsteiner and coworkers began to immobilize haptens for antibodies by using a diazo coupling method and chicken erythrocyte stroma as the support material [2]. However, the first modern use of immunoaffinity chromatography is generally credited to Campbell et al. in 1951, who employed immobilized serum albumin on *p*-aminobenzylcellulose for antibody purification [3]. There are thousands of IAC methods that have been developed for the isolation of antibodies, hormones, peptides, enzymes, recombinant proteins, receptors, viruses, and subcellular components [4–11]. In recent years, the high selectivity of IAC has also made it appealing as a means for the development of a variety of semipreparative and analytical methods [12]. This chapter discusses the basic principles of IAC and examines its various applications.

6.1.1   Antibody Structure

The key component of any IAC method is the antibody preparation used as the stationary phase. An *antibody* is simply a type of glycoprotein produced by the body in response to a foreign agent, or *antigen*. It has been estimated that the body has the capability of producing between 10^6 and 10^8 types of antibodies, each with the ability to bind to a different antigen. The basic structure of a typical antibody (i.e., immunoglobulin G, or IgG) consists of four polypeptides (two identical heavy chains and two identical light chains) linked by disulfide bonds. These polypeptide chains form a Y- or T-shaped structure. The amino acid composition in the lower stem region is highly conserved within the same general group of antibodies (e.g., IgG-class antibodies). The stem region also contains a number of carbohydrate residues that are covalently linked to the antibody's polypeptide chains. Two equivalent antigen-binding sites are present at the two upper ends of the antibody. Located within or near these binding sites are areas in which the amino acid composition is highly variable from one type of antibody to the next; it is this difference in composition that allows the body to produce antibodies with a variety of different binding affinities and specificities to foreign agents that enter the body [13].

As shown in Figure 6.1, an antibody can be chemically modified to produce a number of reactive fragments. For instance, cleavage of an antibody through the use of the enzyme papain produces two identical antigen-binding fragments (i.e., F_{ab} fragments) plus a single

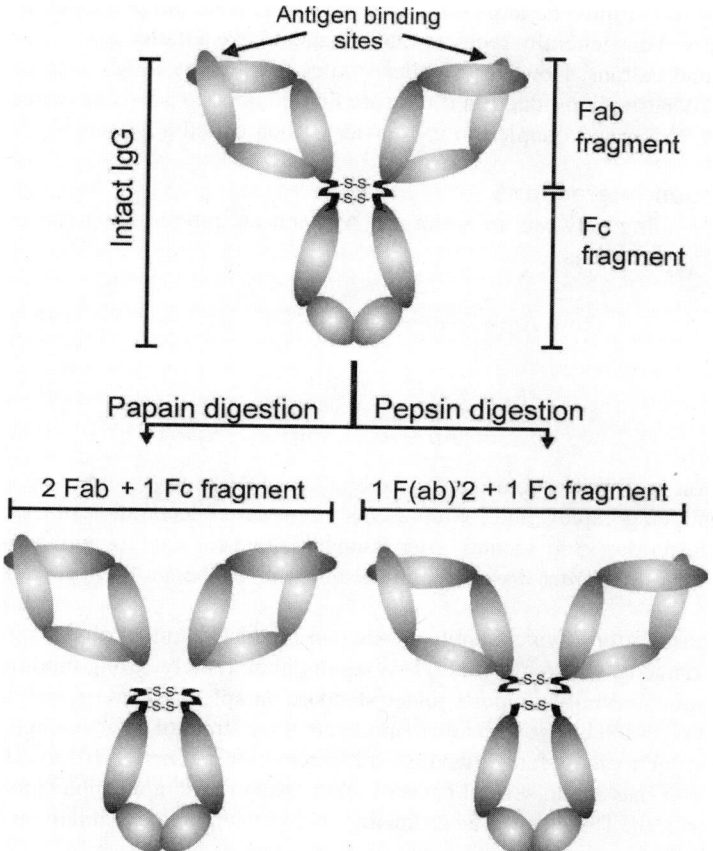

Figure 6.1 Fragments of immunoglobulin G (IgG) produced by papain or pepsin digestion.

stem (crystallizable) fragment, or F_c fragment [14]. This essentially produces monovalent antibodies, which are useful for performing quantitative binding studies. Digestion of an intact antibody with another enzyme, pepsin, produces an $F_{(ab')2}$ fragment and F_c fragment [15]. The $F_{(ab')2}$ fragment still possesses the activity of the original antibody, with an antigen-binding capacity of two moles per mole of antibody. Pepsin cleavage takes place in such a way that the two arms of the antibody plus a single disulfide bond remain intact. Reducing this disulfide bond with agents such as dithiothreitol (DTT) or diethanolamine (DEA) gives rise to two F_{ab} fragments, each complete with a reactive thiol group. This single reactive thiol group is extremely useful for immobilizing the reduced F_{ab} fragments to free thiol groups on chromatographic supports, as discussed later in this chapter. Like enzyme-generated F_{ab} fragments, those produced by reduction of $F_{(ab')2}$ fragments are useful for quantitative work where single valence reactions are required.

Any foreign agent that is capable on its own of giving rise to antibody production once it has entered the body is referred to as an antigen. Examples of common antigens include bacteria, viral particles, and foreign proteins (e.g., proteins from animals, plants, or food that produce an allergic response). Since all of these agents are rather large when compared with the size of a typical antibody binding site, each is capable of binding many different types of antibodies at a variety of locations along its structure. The individual locations on an antigen that can potentially bind to antibodies are called *epitopes*. For a foreign agent to be antigenic, it must be large enough to be recognized and processed by the body's immune system. This generally requires that the agent have a molecular weight of at least several thousand daltons. However, smaller solutes (e.g., many drugs or pesticides) can also give rise to antibody production if they are first coupled to a larger species, such as a carrier protein. This agent coupled to the carrier is then called a *hapten*.

6.1.2 Antibody-Antigen Interactions

The general process of binding between an antibody (Ab) and an antigen (Ag) can be described by the following equations:

$$Ab + Ag \overset{k_a}{\underset{k_d}{\leftrightarrow}} Ab - Ag \qquad (6.1)$$

$$K_a = k_a/k_d \qquad (6.2)$$

$$= [Ab - Ag]/[Ab][Ag] \qquad (6.3)$$

where K_a is the association equilibrium constant for the binding of Ab with Ag, Ab − Ag is the resulting antibody-antigen complex, and [ ] represents the molar concentration of each species at equilibrium. The term k_a is the second-order association rate constant for antibody-antigen binding and k_d is the first-order dissociation rate constant for the antibody-antigen complex.

The association constant for a typical antibody-antigen interaction under physiological conditions is in the range of 10^8 to 10^{12} M^{-1}. This results in extremely strong binding between analytes and immunoaffinity supports under standard sample application conditions (i.e., a neutral pH application buffer with low to moderate ionic strength). For example, in a typical IAC column, such values for K_a would lead to retention factors of 10^3 to 10^7 and mean retention times of one day to several decades when using the sample application buffer for isocratic elution [16]. The result is a chromatographic system that essentially has irreversible binding to the analyte under common sample-injection conditions.

The basis for the formation of an antibody-antigen complex lies in the ability of the antibody's binding sites to form pockets that conform to the electron cloud of the antigenic epitope [17]. The importance of this conformational matching lies in the need for "closeness of fit" between the antigen electron cloud shape and the "space fill" of the area within the antibody's antigen-binding site.

Obtaining a close fit for these intermolecular processes is essential because the bonds that hold antibody-antigen interactions together are the collective forces of four relatively weak forces: (1) ionic interactions, (2) hydrogen bonding, (3) hydrophobic interactions, and (4) van der Waals forces. These are illustrated in Figure 6.2. Individually, each of these forces is weak, but when the antibody and antigen come into close contact, the sum of these factors can exert an overall strong binding force.

Once an epitope has come into close contact with the specific antibody-antigen receptor, these binding forces come into play. Attraction between two ionic groups of opposite charge pulls the molecules into closer contact, which in turn elicits the formation of hydrogen bonds. Once this occurs, water molecules are excluded from the intramolecular space, causing hydrophobic attractive forces that help keep the molecules in close contact. Finally, as the molecules become closer together, van der Waals forces increase. This occurs when the electron cloud of the binding site on the antibody interacts with the electron cloud of the epitope and forms temporary dipoles. This creates an attractive force that holds the molecules in close proximity to each other. Since this force is inversely proportional to the seventh power of the distance between the two interacting molecules, the "closeness of fit" between the epitope and antibody dictates the strength of this force.

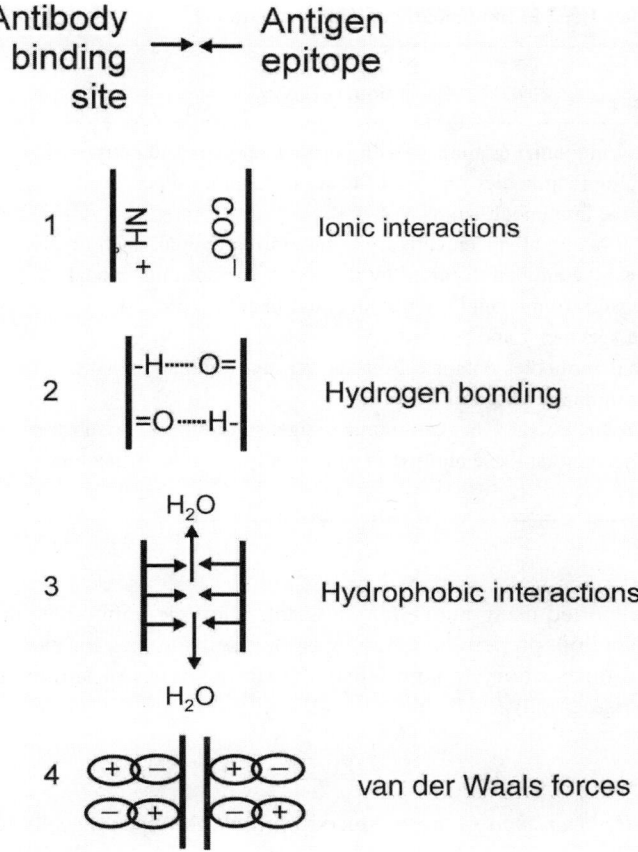

| | Antibody binding site | Antigen epitope | |

Figure 6.2 Types of interactions involved in antibody-antigen binding.

Other forces also contribute to the overall strength of this interaction. For instance, steric repulsion arises from the interpenetration of electron clouds in nonbound regions of the epitope and antibody [18]. Therefore, when the shapes of an antibody binding site and epitope are complementary, their repulsive forces will be weak but their attractive forces will be strong. This situation results in a strong affinity between these two reactants. In fact, at equilibrium, the affinity of an antibody can be thought of as representing the sum of both the attractive and repulsive forces arising from the interaction between the binding site on the antibody and epitope on an antigen.

6.2 DEVELOPMENT OF AN IMMUNOAFFINITY METHOD

There are several items to consider in the creation of an immunoaffinity column and a method that makes use of this column. These items include the production of the antibodies, the selection of a support and immobilization method for the antibodies, and the determination of appropriate application and elution conditions for the immunoaffinity column.

6.2.1 Antibody Production

There are several types of ligands that can be used in immunoaffinity chromatography. The most common examples are polyclonal antibodies and monoclonal antibodies. However,

Table 6.1 Types of Antibodies Used in Immunoaffinity Chromatography

Type of Antibody	Source
Polyclonal antibody	Raised by immunizing animals with purified antigens and cells; the resulting antibodies represent the animal's entire repertoire of reactive immunoglobulins
Monoclonal antibody	Raised by fusion of spleen cells from immunized animals with myeloma cell lines; the antibodies raised by this technique are monoclonal (i.e., arising from a single cell clone), and therefore possess a single specificity and strength
Autoantibodies	Polyclonal antibodies obtained from the serum of human patients with autoimmune diseases
Anti-idiotypic antibodies	Specialized antibodies that can mimic antigens, hormones, or substrates for cell receptors; these antibodies can be polyclonal or monoclonal in nature

applications have also been reported using autoantibodies, anti-idiotypic antibodies, and antibody fragments. The production, properties, advantages, or disadvantages for each of these ligands are examined in this section. Table 6.1 lists the various types of antibodies used in immunoaffinity chromatography.

6.2.1.1   Polyclonal Antibodies

As their name implies, *polyclonal antibodies* are produced by multiple types of cells (or clonal lines) within the body. These are generated against a specific antigen by immunization of an animal with this antigen. This is accomplished by injecting into a laboratory animal a solution of the antigen either alone, in combination with a carrier protein, or mixed with an enhancing agent called an *adjuvant*. Mice, rabbits, goats, and sheep are often used for this purpose. The most common approach for this procedure involves intramuscular injection, although intradermal, subcutaneous, and intraperitoneal routes can also be used.

After this injection, samples of the animal's blood are collected at 3 to 4 weeks and tested for the presence of antibodies directed against the desired antigen (see Figure 6.3). The animal is then given another injection of antigen (i.e., a booster shot) and retested 10 days later. If the antibody levels are satisfactory, the animal can be rebled at 3 weeks, rested for 2 weeks, and given another booster injection. This booster/bleed routine can be repeated several times until antibody concentrations reach a peak. At this point, the animal is usually bled out, with the entire antibody-containing serum being collected and stored for later use.

Highly specific antibodies are usually produced by short-term immunization programs, while long-term programs with repeated antigen injections usually produce more cross-reactive antibodies or antibodies that are reactive to more common antigenic determinants. This method generally results in a heterogeneous mixture of antibodies that bind with a variety of strengths and to various epitopes on the antigen or hapten/carrier complex, where each type of antibody is produced by a different cell line.

When an animal is exposed (or immunized) to a foreign agent, the first antibodies produced are immunoglobulin M (IgM)-class antibodies. However, if this exposure is prolonged, the IgM response will eventually be replaced by the production of IgG-class antibodies.

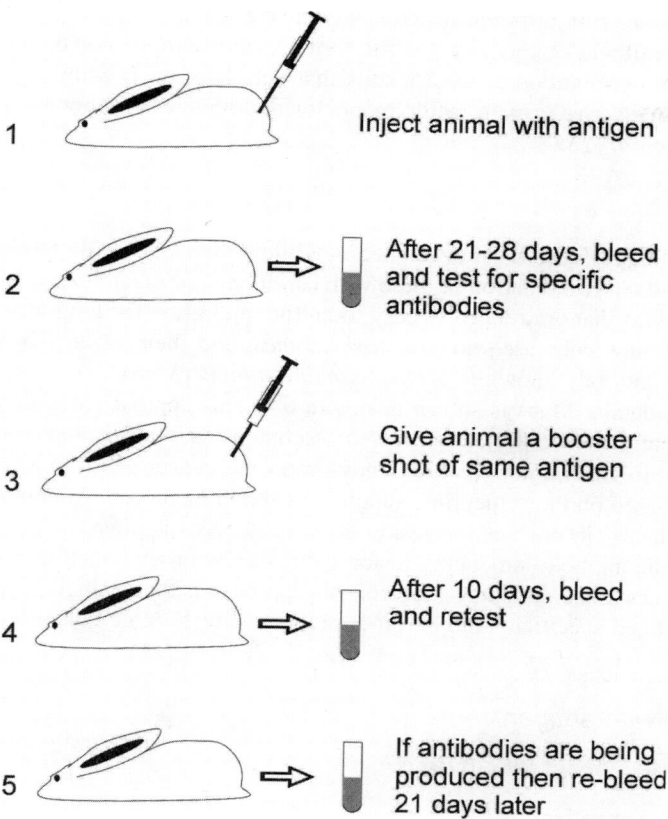

1   Inject animal with antigen

2   After 21-28 days, bleed and test for specific antibodies

3   Give animal a booster shot of same antigen

4   After 10 days, bleed and retest

5   If antibodies are being produced then re-bleed 21 days later

Figure 6.3   Basic procedure for producing polyclonal antibodies.

It is this secondary response with IgG that produces antibodies with affinities and properties that are the most suitable for use in IAC.

Following the collection of polyclonal antibodies, some cleanup is usually needed before these can be used in IAC. Examples of such methods can be found in Chapter 14. These cleanup techniques usually consist of isolating the IgG fraction from the serum by using ion-exchange chromatography [19], salt precipitation with ammonium or dextran sulfate [20], or isolation with a protein A/G column [21]. The IgG-rich fraction obtained by any of these techniques will also contain IgG antibodies that are nonreactive with the antigen. These can be removed by using an immobilized antigen affinity column, thus producing antigen-specific antibodies.

One advantage of polyclonal antibodies is the relative ease with which they can be produced, requiring access to only a few animals and some purified antigen. However, perhaps the most important advantage of polyclonal preparations is that they contain a collection of antibodies directed against more antigenic determinants than monoclonal antibodies, as described in the following section. This characteristic can be helpful when one is using the antibodies for screening purposes. For instance, it was shown that polyclonal antibodies raised against gp185HER2 peptide products of the HER2 breast tumor-associated oncogene could identify a subset of malignant tumors that were not detected when using anti-gp185HER2 monoclonal antibodies [22].

Polyclonals can also react with different forms (e.g., active and inactive) of a given protein, while monoclonal antibodies target one specific form. As an example, polyclonal antiphospholipid transfer protein antibodies gave results that correlated well with other techniques for determining the active protein, while monoclonal antibodies preferentially reacted with the inactive protein [23].

6.2.1.2 Monoclonal Antibodies

Over the last 25 years, techniques have been developed that allow the isolation of single antibody-producing cells and the combination of these with carcinoma, or myeloma, cells. This produces hybrid cell lines that are relatively easy to culture and grow for long-term antibody production. These new cells are known as *hybridomas*, and their product is a single type of well-defined antibody known as a *monoclonal antibody (Mab)*.

The technique for producing Mabs is shown in Figure 6.4. This approach was first reported by Kohler and Milstein in 1975 [24]. Basically, this technique uses genetic material from a primed immune cell to teach a nonsense-antibody-producing cancer cell to manufacture a specific antibody based on the "blueprint" supplied by the immune cell. Although this principle is simple in theory, in practice the fusion of two cell types is often difficult, and the selection of a specific antibody-producing fusion clone can be an extremely time-consuming and difficult process. The main advantage of this approach is that a single type of antibody with a well-defined specificity can be produced in relatively large quantities.

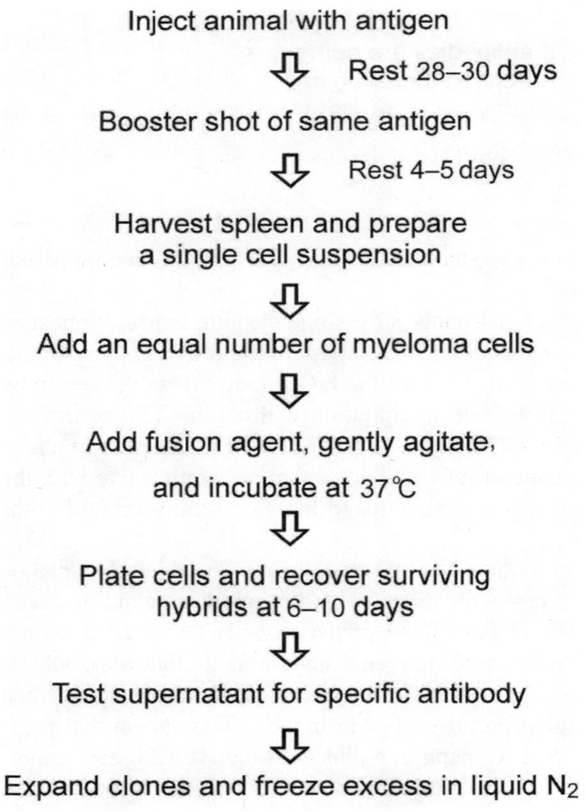

Inject animal with antigen

⇩ Rest 28–30 days

Booster shot of same antigen

⇩ Rest 4–5 days

Harvest spleen and prepare a single cell suspension

⇩

Add an equal number of myeloma cells

⇩

Add fusion agent, gently agitate, and incubate at 37°C

⇩

Plate cells and recover surviving hybrids at 6–10 days

⇩

Test supernatant for specific antibody

⇩

Expand clones and freeze excess in liquid N₂

Figure 6.4 Basic procedure for producing monoclonal antibodies (Mabs).

The list of myeloma cell lines suitable for use in this method has increased greatly in recent years, and active hybrids for many different species can now be produced.

As shown in Figure 6.4, monoclonal production begins by mixing a soluble antigen with an adjuvant and injecting this subcutaneously into an animal. The animal is then rested for four weeks, with a booster shot of antigen being given 4 to 5 days before the animal is killed and its spleen harvested. A single-cell suspension is next prepared from the isolated spleen by gently teasing cells out of the tissue and mixing them with an equal number of myeloma cells. Cell fusion is promoted by adding polyethylene glycol to these cells and incubating this mixture at 37°C. The cells are then plated, and the surviving hybrids are examined after 6 to 10 days. A portion of the supernatant from each surviving hybrid is then tested for the presence of antigen-specific antibodies. A recent variation on this procedure has been described for the production of polyol-responsive Mabs, which are claimed to show improved recovery of analyte when used in immunoaffinity chromatography [25].

The commercial availability of Mabs has made these reagents important in biochemistry. For instance, Mabs have been used extensively to probe and isolate membrane receptors. Although most studies have used Mabs for immunocytochemical localization or for immunological isolation of proteins by precipitation techniques, some investigators have used Mabs for immunoaffinity isolation of specific membrane proteins, especially adrenergic receptors [26], viral receptors [27], and complement receptors on immunocompetent cells [28].

The monospecificity of Mabs makes these extremely useful when preparing IAC columns. The restricted specificity and clonal properties of these antibody preparations also obviate the need for additional cleanup procedures, as are required for polyclonal preparations. However, as stated earlier, the restricted reactivity of Mabs may not be ideal for situations where these antibodies are to be used in a screening assay. Another disadvantage is that the production of Mabs requires cell culture facilities and special equipment for expanding and maintaining functional clones [29–31]. This often includes the need for liquid-nitrogen storage facilities for archiving and maintaining clonal libraries.

6.2.1.3 Autoantibodies

Autoantibodies are naturally occurring antibodies produced against components of the host's body that have become altered during a disease process and are no longer considered "self." In most cases, these antibodies are used as biomarkers for clinical disease. However, when these unique antibodies are made against specific cellular structures, they can also be used as probes for investigating antigenic structure and function. In such studies, they have an advantage over both polyclonal antibodies and Mabs because their reactive antigens are selected by the host and reflect the true antigens involved in an autoimmune disease, rather than an artificially isolated antigen.

One example of the use of autoantibodies in IAC is work in which affinity-purified parietal cell autoantibodies were used in an IAC column for the purification of a 60- to 90-kDa gastric parietal cell antigen targeted in autoimmune gastritis. The antigen was purified from a biochemically precleared Triton X-100 membrane extract of porcine gastric parietal cells by IAC on an IgG autoantibody-Sepharose column, followed by preparative sodium dodecyl sulfate-polyacrylamide gel electrophoresis. It was claimed in this report that this approach was a general one that could be used to isolate specific autoantigens for further studies of human and animal models of autoimmune diseases [32].

Perhaps the most important autoantibodies are those produced against cell membrane receptors, which can be used to isolate the receptor from human cells. Blecher [33] reviewed this area and stated that autoantibodies against acetylcholine, insulin, thyrotropin, β-adrenergic, and prolactin receptors are all naturally produced and can be used for biochemical investigations. Such examples indicate that autoantibodies can be useful reagents for binding studies and in the IAC isolation of receptors and clinically important proteins. However, a disadvantage of this type of antibody is its availability and the difficulty of its isolation and purification. For instance, to obtain a specific autoantibody, its presence in a certain disease state must be established and, once this has been accomplished, access to the serum or plasma for patients with this disease will be required. This last aspect tends to restrict the usefulness of autoantibodies to researchers working in the clinical arena, and especially those investigating autoimmune diseases.

6.2.1.4 Anti-Idiotypic Antibodies

Another exciting area in immunology is the discovery of the idiotypic network and the isolation of *anti-idiotypic antibodies* (*anti-ids*) that can mimic receptor substrates and bind to membrane receptors. Anti-ids have been used to isolate receptors from tissues by IAC. For instance, two studies have employed IAC columns with monoclonal anti-ids that were specific for the combining site on an antinicotine Mab. These columns were then used to isolate nicotinic receptors from rat brain tissue [34, 35].

In autoimmune diseases, cellular hyperactivity can arise from the stimulation of cell membrane receptors by anti-id antibodies. The increasing sophistication of Mab technology makes it possible to make anti-id antibodies that are reactive with the original idiotype and that can be used to isolate specific antibodies in the absence of the original antigen. Likewise, immobilized idiotypic antibodies can be used to isolate anti-ids by IAC [36].

Anti-id antibodies can be obtained by four different routes: (1) immunization with the original idiotype, (2) immunization with fixed cells that produce the original idiotype, (3) immunization with immune complexes containing the idiotype and rheumatoid factor, or (4) screening the serum of hosts that have been repeatedly immunized over a long period of time. Both idiotypes and anti-ids have been used as ligands for the isolation of receptors from activated lymphocytes [37] and specific anti-id antibodies from cancer patient samples [38].

The advantages of anti-id antibodies include their unique specificity for defined epitopes on their target immunoglobulin and their ability to manipulate immune responses. However, this is not too useful in affinity chromatography unless one is interested in examining epitopes reactive with idiotypic antibodies or in mapping epitopes within the antigen-binding site of antibodies. Thus, these antibodies currently hold only limited practical value for IAC except in specialized cases, such as isolation of membrane receptors and labeled antibodies. For instance, anti-id Mabs have been found to be useful for the isolation of isotope-labeled antitumor antibodies by IAC. The anti-ids in this case were immobilized for the purification of radiolabeled antibodies for later use in imaging and immunotherapy [38].

6.2.1.5 Related Ligands

In addition to traditional polyclonal and monoclonal antibodies, there has also been recent work aimed at using genetic engineering and bacteria or bacteriophages for the generation of antibodies or antibody-like molecules [39–42]. Although these types of ligands have currently been used in only a few IAC applications, they do represent an attractive alternative for the large-scale production of immunoaffinity supports [43]. Engineered reagents for IAC can come in several different forms. Examples include bifunctional (or bispecific) antibodies, antibodies incorporating binding proteins or tags, or a single-chain variable-region fragment (scF_v).

Preparing F_{ab} fragments from two different antibodies and allowing them to recombine to form a bispecific molecule is one way to construct a bifunctional antibody. Although this reasonably easy immunochemical procedure is adequate, the yield of truly bifunctional antibodies is often low. One study used a solid-phase synthesis to chemically cross-link two F_{ab} fragments via free thiol groups [44], while another used genetic engineering to produce bifunctional antibodies as well as scF_vs [45]. An example of recombinant antibodies incorporating a tag was reported by Weiss et al., in which biotinylated antibody F_{ab} fragments were prepared in *Escherichia coli* and immobilized onto streptavidin-coated agarose. This support was then used in the one-step purification of recombinant TNFα by IAC [46].

The advent of genetic engineering has also allowed the development of recombinant single-chain antibodies expressed in *E. coli* [47], which can be developed with either single or bivalent binding [48]. In one report, Kleymann et al. developed an IAC method based on engineered F_v fragments expressed in *E. coli*. These fragments were engineered to incorporate a strep-tag affinity peptide, allowing the scF_v to be immobilized onto a streptavidin matrix for the isolation of cytochrome c oxidase, ubiquinol:cytochrome c oxidoreductase, and subcomplexes from *Paracoccus denitrificans* [49]. Similarly, heat-shock protein gp96-specific recombinant scF_v antibodies were created from a semisynthetic phage display library and immobilized on Sepharose beads for the purification of native gp96 molecules from mouse and human tumor cell lysates [50].

Recently, recombinant antibody engineering has produced a novel approach for antibody immobilization. This has involved the construction of engineered scF_v fragments that contain chitin-binding domains capable of being directly linked to chitin beads. It has been claimed that this approach will enable these fragments to be used for all types of batch purifications, ranging from small to production-scale [51].

6.2.2 Supports for Immunoaffinity Chromatography

The support is another item to consider in the development of a successful IAC method. Traditionally, most IAC applications have been based on low-performance supports. This began with the work by Campbell et al., who used a protein antigen immobilized to diazotized aminobenzylcellulose [3]. Others have explored the use of immunoaffinity supports based on bromoacetylcellulose, ester-derivatized cellulose, or related materials [52–54]. Since these early studies, IAC has been performed with dextran [55], polyacrylamide [56], rigid plastic beads [57], and derivatized silica or glass beads [58–60]. Other techniques have involved cross-linking of the ligand with glutaraldehyde [61] or carbodiimides [62]. Reviews and descriptions of this early work can be found in the literature [58–60, 63–65].

Low-performance supports for IAC are generally carbohydrate-related materials (i.e., agarose or cellulose) or synthetic organic supports (e.g., acrylamide polymers, copolymers, or derivatives; polymethacrylate derivatives; and polyethersulfone matrixes). There are many commercial supports that fit into this category, as shown in Table 6.2 [16, 66]. The low back pressure of these supports means that they can be operated under gravity flow, a slight applied vacuum, or peristaltic flow. This makes these gels relatively simple and inexpensive to use for IAC, particularly in off-line immunoextraction methods or techniques that involve the use of IAC with flow-injection analysis systems. The disadvantages of these materials are their slow mass-transfer properties and their limited stability at high flow rates and pressures. These factors limit the usefulness of such supports when performing IAC in standard systems for high-performance liquid chromatography (HPLC).

IAC can be used as an HPLC method, but the materials that are employed must be more rigid and have higher efficiency. Supports that have been developed for this purpose

Table 6.2 Commercially Available Supports for Immunoaffinity Chromatography

Low- or Medium-Performance Chromatographic Supports (Supplier)
Affi-Gel (BioRad)
Affinica Agarose/Polymeric Supports (Schleicher & Schuell)
AvidGel (BioProbe)
Bio-Gel (BioRad)
Fractogel (EM Separations)
HEMA-AFC (Alltech)
Reacti-Gel (Pierce)
Sephacryl (Pharmacia)
Sepharose (Pharmacia)
Superose (Pharmacia)
Trisacryl (IBF)
TSK Gel Toyopearl (TosoHaas)
Ultrogel (IBF)

High-Performance Chromatographic Supports (Supplier)
AvidGel CPG (BioProbe)
HiPAC (ChromatoChem)
Protein-Pak Affinity Packing (Waters)
Ultraaffinity-EP (Bodman)
Emphaze (3M Corp./Pierce)
POROS (ABI/PerSeptive Biosystems)

include derivatized silica, glass, and certain organic matrixes such as azalactone beads or polystyrene-based perfusion media. The use of these supports along with an antibody or related ligand is referred to as *high-performance immunoaffinity chromatography* (*HPIAC*) [6, 16]. The mechanical stability and efficiency of these materials makes them attractive for use with standard HPLC equipment, which in turn helps to provide improved speed and precision for analytical applications of IAC.

Besides efficiency and mechanical stability, other items to consider when choosing a support for IAC include its degree of nonspecific binding and the ease with which it can be modified for ligand attachment. Both of these items are discussed in more detail in Chapter 2. Another item to consider for a porous material is the size of its pores and the ability of an antibody to gain access to its surface for immobilization. This is illustrated in Figure 6.5. In this example, supports with small pores have the largest surface area, but much of this is unavailable for immobilization, since it can not be reached by the antibodies. On the other hand, supports with larger pores do not have an accessibility problem, but they do have a lower total surface area. Thus, the maximum coverage for antibodies in this case occurs with supports that have an intermediate pore size (i.e., roughly three to five times the diameter of an antibody, or 300 to 500 Å) [67].

Along with traditional carbohydrate supports and HPLC materials like silica, there has been recent interest in immobilizing antibodies to other types of media. Examples include disks, fibers, and monolithic rods [68]. Although the use of these supports is currently an area of active research and development, such materials do offer some unique advantages in terms of their mass-transfer and flow properties. A further discussion of these and alternative types of chromatographic media for affinity chromatography can be found in Chapter 2.

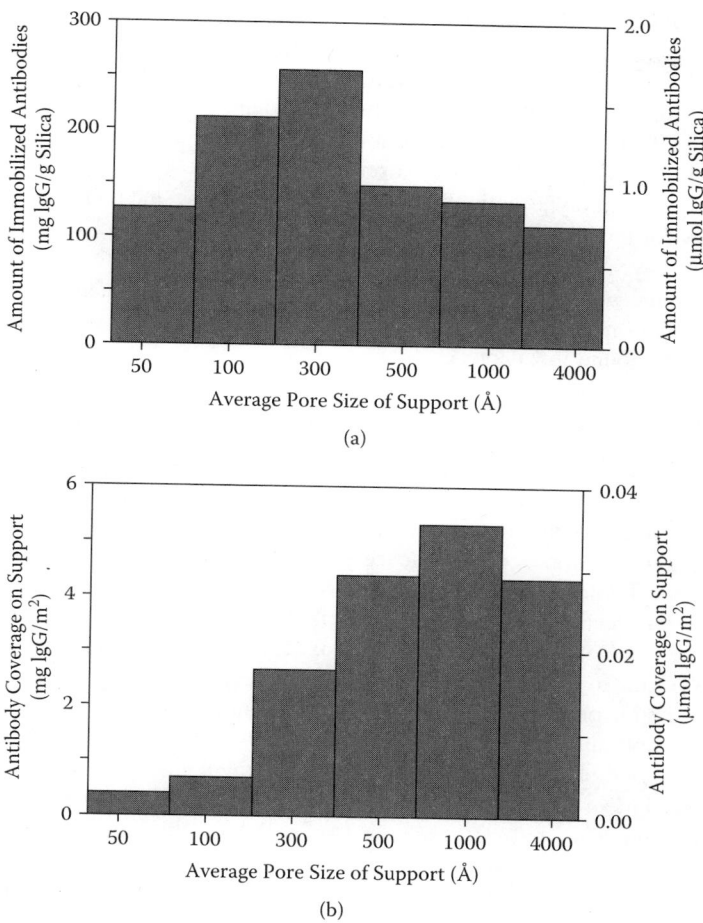

Figure 6.5   Total amount of (a) immobilized antibodies and (b) immobilized antibodies per unit surface area for rabbit immunoglobulin G attached to HPLC-grade silica with various pore sizes. (Adapted with permission from Clarke, W., Beckwith, J.D., Jackson, A., Reynolds, B., Karle, E.M., and Hage, D.S., *J. Chromatogr. A*, 888, 13–22, 2000.)

6.2.3   Antibody Immobilization

There are many different methods that can be used for antibody immobilization to both low- and high-performance supports. These techniques range from random attachment via amino or carboxyl groups to site-directed attachment using carbohydrate moieties and thiol groups. In addition, many investigators have employed secondary ligands such as protein A, protein G, or streptavidin to produce immobilized antibodies with a specific orientation on the support.

6.2.3.1   *Random Immobilization Methods*

One common approach involves direct, covalent attachment of the antibodies to the chromatographic support. This is often achieved by reacting free amine groups on the antibodies with supports that are activated with agents such as N, N'-carbonyldiimidazole, cyanogen bromide, N-hydroxysuccinimide, or tresyl chloride/tosyl chloride, or with supports that have been treated to produce reactive epoxide or aldehyde groups on their surfaces [16, 66, 69].

The use of antibody amine groups is probably the easiest route to immobilization, but this can give rise to less-than-optimum activity due to random orientation and blocking of the antigen-binding sites on the immobilized antibodies. This effect was noted with an anti-idiotypic antibody immobilized to the surface of a biosensor [70]. In this study, the immobilization of a monomeric anti-idiotype antibody by an amine coupling method gave a reduced binding efficiency for the target antibody versus thiol-based site-directed immobilization. Other investigators have noted that the immobilization of capture ligands (especially antibodies) via primary amine groups can lead to steric hindrance in their interactions, causing a decrease or inhibition of binding [71, 72]. More details on these and other immobilization effects can be found in Chapter 3.

6.2.3.2 Site-Selective Immobilization

Antibodies, or antibody fragments (F_{ab}), can also be covalently immobilized through more site-selective methods. For example, free sulfhydryl groups that are generated during the production of F_{ab} can be used to couple these fragments to supports by using a variety of techniques, including the divinylsulfone, epoxy, iodoacetyl/bromoacetyl, maleimide, TNB-thiol, and tresyl chloride/tosyl chloride methods [16, 66, 69].

Another route for site-selective immobilization involves the coupling of antibodies through their carbohydrate residues. This is done by mild oxidation of these residues with periodate or enzymatic systems to produce aldehyde residues. These aldehyde groups can then be reacted with a hydrazide or amine-containing support for antibody immobilization [16, 66]. The advantage of both the sulfhydryl and carbohydrate-based approaches is that they are believed to produce immobilized antibodies or antibody fragments that have fairly well-defined points of attachment and thus greater accessibility of the antibody binding regions to antigens in solution. This, in turn, results in IAC columns with higher relative binding activities than comparable columns made by amine-coupling methods. An illustration of the carbohydrate-based immobilization method is given in Figure 6.6, with further details being given in Chapter 3 and the literature [73–76].

6.2.3.3 Adsorption to Secondary Ligands

Noncovalent immobilization can also be used for the site-selective coupling of antibodies to IAC supports. For example, antibodies that have been oxidized to produce aldehyde groups in their carbohydrate residues can be reacted with biotin-hydrazide, which in turn can be used to bind the antibodies noncovalently to an immobilized avidin or streptavidin support [66].

As stated in Chapter 5, avidin is a protein isolated from chicken egg whites that has a natural affinity for the vitamin biotin. Avidin has four subunits, each with a functional receptor for a single biotin molecule. In ideal situations, avidin has an association constant for biotin in excess of 10^{15} M^{-1} [77], suggesting that this binding can withstand many types of harsh elution conditions without dissociation. This makes avidin an ideal secondary ligand for IAC applications. Streptavidin is a related ligand produced by the bacterium *Streptomyces avidinii* [78]. It exhibits all of the properties of avidin but has been shown to possess fewer nonspecific interactions with tissue extracts. Likewise, a recombinant avidin called Neutravidin (from Pierce) is available commercially. This reagent, like streptavidin, has been engineered to reduce nonspecific binding and can be a useful agent when IAC techniques are employed for the isolation of analytes from tissue extracts or cell cytosol samples.

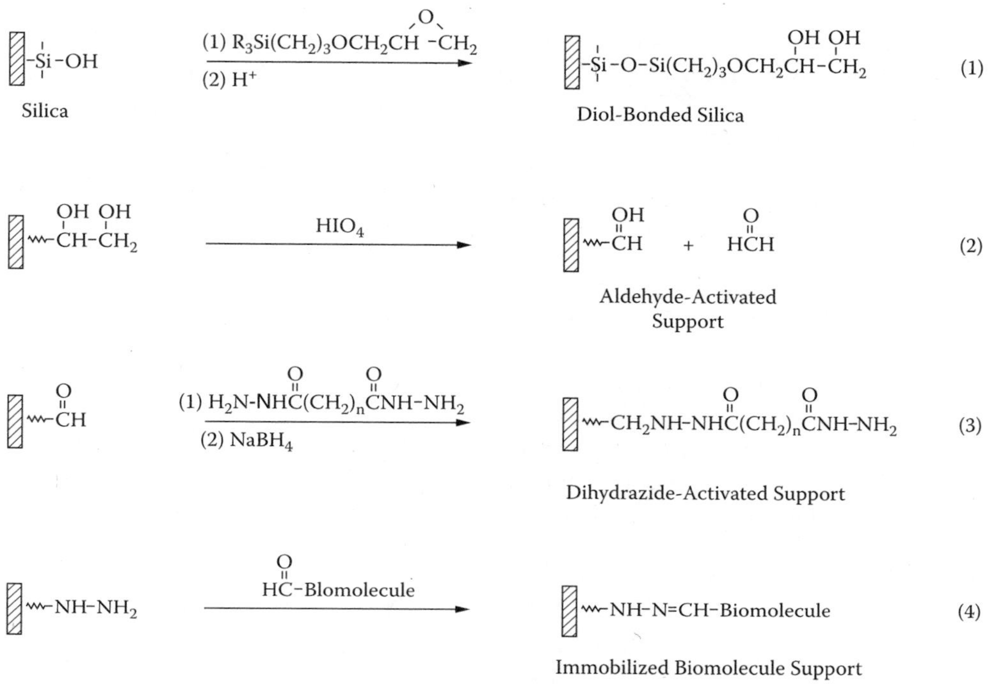

$$\text{Silica} \quad \boxed{}\text{-Si-OH} \quad \xrightarrow[\text{(2) } H^+]{\text{(1) } R_3Si(CH_2)_3OCH_2CH-CH_2 \; (O)} \quad \boxed{}\text{-Si-O-Si(CH_2)_3OCH_2CH-CH_2} \quad (1)$$

Silica

Diol-Bonded Silica

$$\boxed{}\text{-CH-CH_2} \; \overset{OH \; OH}{} \quad \xrightarrow{HIO_4} \quad \boxed{}\text{-CH} \; \overset{OH}{} \; + \; HCH \overset{O}{} \quad (2)$$

Aldehyde-Activated Support

$$\boxed{}\text{-CH} \overset{O}{} \quad \xrightarrow[\text{(2) } NaBH_4]{\text{(1) } H_2N-NHC(CH_2)_nCNH-NH_2} \quad \boxed{}\text{-CH_2NH-NHC(CH_2)_nCNH-NH_2} \quad (3)$$

Dihydrazide-Activated Support

$$\boxed{}\text{-NH-NH_2} \quad \xrightarrow{HC-Biomolecule \; (O)} \quad \boxed{}\text{-NH-N=CH-Biomolecule} \quad (4)$$

Immobilized Biomolecule Support

Figure 6.6 Method for the preparation of dihydrazide-activated silica and its use in the immobilization of an antibody or other biomolecule after the mild oxidation of carbohydrate groups within this ligand with periodate. (Reproduced with permission from Ruhn, P.F., Garver, S., and Hage, D.S., *J. Chromatogr. A*, 669, 9–19, 1994.)

Before antibodies can be attached to immobilized avidin or streptavidin, they have to be labeled with biotin, or biotinylated. This can be performed using one of several commercial biotin derivatives. The most popular technique is to incubate antibodies with *N*-hydroxysuccinimide-*D*-biotin in carbonate buffer at pH 9. Once labeled, the antibodies will then bind to immobilized avidin, forming a stable complex on the support's surface.

One disadvantage of this technique is that little control is available over the distribution of the attached biotin on the antibody, which can affect the adsorbed antibody's binding activity. This effect can arise from two sources. One of these occurs when biotin is attached in the vicinity of the antibody binding sites and interferes with their interactions with the antigen. The other source is due to the random attachment of the biotinylated antibody to the avidin support, causing some of the antibodies to be incorrectly oriented for optimum binding.

One biotin derivative that minimizes this problem is hydrazide-biotin [78]. This can be selectively attached to the carbohydrate portion of an antibody after mild oxidation of this region with periodate, as described in the previous section. This approach has been used to biotinylate Mabs for use in the HPIAC isolation of specific receptors of primed lymphocytes [36, 37]. The advantage of this type of biotinylation is that the modified carbohydrate moieties mainly reside in the F_c region and away from the antibody's binding sites. One disadvantage of this approach is that the antibody must first be oxidized with periodate. Also, some engineered antibodies (e.g., Mabs and scF$_v$s) can be poorly glycosylated or have no carbohydrate residues, making it impossible to use this method with these ligands.

Another approach for indirect immobilization involves adsorbing the antibody to a secondary ligand such as protein A or protein G. As discussed in Chapters 5, 14, and 21, protein A and protein G are bacterial cell wall proteins that have the ability to bind to the F_c region of many types of antibodies. This binding is quite strong under physiological conditions but can be easily disrupted by decreasing the pH of the surrounding solution [6, 16, 66].

The use of protein A or protein G is appealing in cases where high antibody activity is needed and where it is desirable to have frequent replacement of the antibodies in the IAC column. This provides good long-term reproducibility for the IAC column binding capacity, but does require the use of much larger amounts of antibody than direct immobilization methods. If desired, supports that have a permanent coating of antibodies can be prepared by cross-linking the antibodies to the immobilized protein A or G through the use of carbodiimide [79] or dimethyl pimelimidate [80, 81].

6.2.4 Selection of Application and Elution Conditions

Although it is possible in IAC to perform isocratic elution on a reasonable time scale by using a competing agent in the mobile phase and low-affinity antibodies (see Chapter 4), this does not work for the high- or moderate-affinity antibodies that are used in the vast majority of IAC columns. The only way solutes can usually be eluted from these antibodies is to change the column conditions to lower the effective value of K_a (i.e., increase the relative value of k_d versus k_a) for antibody-analyte binding.

The proper choice of an elution buffer and elution scheme is important in analytical applications of IAC, since it is usually desirable to elute the analyte as quickly as possible while avoiding any irreversible damage to the immobilized antibody support. This currently needs to be addressed on a case-by-case basis and is essential to consider if the same IAC column is to be reused for a large number of samples. In addition, the choice of elution conditions has to take into consideration the nature of the analyte and whether it must be isolated in an active form. For instance, delicate compounds like membrane receptors can be damaged by elution at a low pH and often require a more gentle eluting solvent, such as one containing a chaotropic agent [82].

To help in this process, several investigators have used enzyme-linked immunosorbent assays (ELISAs) for finding elution conditions that can recover antigens from antibodies immobilized within plastic microtiter plates [83–85]. In some cases, knowledge of the properties of the antigen has been used for selecting the elution conditions. An example of this is work performed with human serum albumin, in which a pH elution step was created based on a reversible change in conformation that occurs for this protein under acidic conditions [86]. In other cases, it may be possible to select a ligand with given elution properties from a larger population of binding agents. An example of this is shown in Figure 6.7, in which a series of pH step changes and an immobilized antigen column were used to select and purify a group of antibodies against the antigen that dissociated from this compound over a specified range of pH [87].

There are several common approaches used for elution in IAC. These include the use of a change in mobile-phase pH, the addition of chaotropic ions to the mobile phase, and the use of substances that change the polarity of the mobile phase. Each of these approaches is examined in greater detail in the following subsections.

6.2.4.1 pH Elution

One of the most popular techniques in IAC for dissociating antibody-antigen complexes is altering the pH of the mobile phase. This is often accomplished by going from a neutral

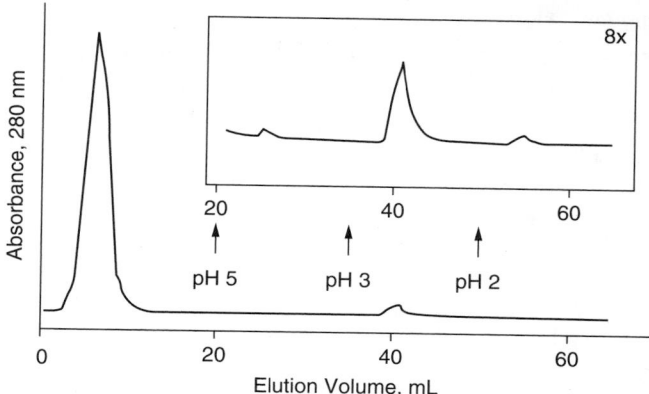

Figure 6.7 Purification of polyclonal antibodies against parathyroid hormone using a Sepharose column containing an immobilized 44–68 PTH fragment. A series of pH elution steps were used to elute the retained antibodies and obtained a subpopulation that desorbed between pH 3 and 5. (Reproduced with permission from Hage, D.S. and Kao, P.C., *Anal. Chem.*, 63, 586–595, 1991.)

to acidic pH, with an elution pH of 1 to 2.5 being common [12, 88–90]. This is generally accomplished by using a gradient of phosphate, citric, formic, or acetic acid [12, 65]. A combination of Tris and HCl has also been shown to act as an efficient elution agent [91].

Alkaline elution conditions have been used with low-performance IAC with a high degree of success [92, 93]. However, in some cases a high pH can have a denaturing effect on proteins [94, 95]. In addition, common supports used in HPIAC, such as silica and glass beads, can not be used at a pH above 8.0. This means that alkaline elution can generally not be used with such materials [12].

6.2.4.2   Elution with Chaotropic Agents

Elution with chaotropic agents often gives the most effective results for IAC while avoiding the denaturing effects often observed when pH changes are used for elution. Chaotropic salts can be used for this purpose, since they disrupt the structure of water around the analyte or antibody and reduce hydrophobic interactions. Several salts exhibit chaotropic properties, including thiocyanate (SCN^-), trifluoroacetate (CF_3COO^-), perchlorate (ClO_4^-), iodide (I^-), and chloride (Cl^-). The effectiveness of these ions as dissociation agents for antibody-antigen complexes follows the approximate order $SCN^- > CF_3COO^- > ClO_4^- > I^- > Cl^-$.

When used at concentrations of 1.5 to 8 M, these salts have been shown to be effective in the dissociation of high-affinity antibody-antigen complexes [96] and in the recovery of receptors from affinity columns [97]. The most widely used chaotropic salts are potassium and sodium thiocyanate, which are used at concentrations up to 3 M [98, 99]. Other reagents used for this purpose are sodium iodide [100] and a polyvinylpyrolidone-iodine mixture [91, 101], which are used at concentrations of 2.5 to 3.0 M.

Sodium chloride is a weak chaotropic ion that can be used at concentrations of 2 M or higher for elution in IAC. For instance, the dissociation of antibody-antigen complexes in solution has been reported in the presence of 4 to 8 M sodium chloride [102–104]. Although such conditions often work well in IAC, it must be remembered that use of a high-concentration chaotropic solution can still be a source of some denaturation in both the immobilized antibody and recovered antigen.

6.2.4.3   Other Elution Techniques

Other agents used for the recovery of solutes from IAC columns include polarity-reducing agents such as methanol, dioxane, and ethylene glycol [105, 106]. These reduce the hydrophobic forces responsible for holding an antigen in close proximity to its antibody binding sites. However, one precaution that must be observed with such agents is that they can denature the antibody and antigen as their concentration is increased. In addition, these substances will alter the solubility of the antigen in the mobile phase.

Antigen or epitope elution has also been used for target recovery in IAC. This is a subset of biospecific elution, as described in Chapter 4, and allows the use of gentle elution conditions and optimum activity for the recovered target. However, this type of elution can also be costly and take a long period of time to perform. These factors hinder the use of this approach in the quantitative recovery or analysis of an eluting substance [107, 108].

6.3   IMMUNOAFFINITY PURIFICATION AND SAMPLE PRETREATMENT

The ability to selectively isolate a given substance from a complex biological sample has made IAC a powerful tool in many fields of science. For instance, IAC is often used by biochemists for the target isolation on a preparative scale. However, this method can also be employed as a semipreparative method for sample pretreatment prior to the use of an analytical technique like HPLC, gas chromatography (GC), mass spectrometry, or capillary electrophoresis (CE).

6.3.1   Preparative Applications

One of the most common applications of IAC has been in the isolation and purification of biological compounds. This is illustrated by the broad array of compounds that have been isolated by this approach, as shown in Table 6.3 [109–181]. This list ranges from proteins and glycoproteins to carbohydrates, lipids, infectious agents, biomarkers, drugs, and environmental agents. An example of such an application is given in Figure 6.8.

An area in which IAC has been particularly useful has been in the isolation of immunological reagents. This is the optimum approach for purifying either specific antibodies or antigens, wherein an appropriate ligand (i.e., either an antibody or purified antigen) is used to isolate its counterpart from a sample. This results in the isolation of an immunologically specific agent, regardless of its chemical or biochemical nature. Although IAC often provides a fast and simple approach to the isolation of immunological reagents, it does have the disadvantage of often being expensive, making it prohibitive when production-scale purifications are required. A further discussion of this area and related applications can be found in Chapters 13 and 14.

6.3.2   Off-Line Immunoextraction

The use of IAC for sample pretreatment prior to analysis is another field that has seen rapid growth in recent years [12]. When IAC is used in this fashion, it is also sometimes referred to as *immunoextraction (IE)*. In this approach, an immunoaffinity column is used for the removal of a specific solute or group of solutes from a sample prior to determination by a second analytical method. This employs the same general operating scheme as other types of IAC, but it involves combining the immunoaffinity column either off-line or on-line with some other method for the actual quantitation of analytes.

Table 6.3 Examples of Substances That Have Been Purified by Immunoaffinity Chromatography

Class of Substances	Specific Examples	References
Peptides and proteins	Antipeptide antibodies, enzymes, antigens, recombinant human erythropoietin isoforms, alpha-tubulin isoforms, multiprotein complexes	[109–114]
Glycoproteins	Herpesvirus hydrophobic proteins or glycoproteins, protein S, tumor necrosis factor receptor immunoadhesin, CEACAM1, human dipeptidyl peptidase IV/CD26	[115–119]
Lipoproteins	Myelin proteolipid protein, lipoproteins in cerebrospinal fluid	[120–122]
Toxins	Mycotoxins, aflatoxin M1, ochratoxin A, microcystins	[123–127]
Hormones	Prolactins, hypophyseal growth hormone, bovine growth hormone releasing factor, human erythropoietin isoforms, melatonin	[128–132]
Cytokines	Interferon, interleukin-2, interleukin-18, bovine stem cell factor	[133–138]
Enzymes	Human bilirubin/phenol UDP-glucuronosyltransferases, against ascorbate peroxidase isoenzymes	[139, 140]
Vitamins	Plasma 1α,25-dihydroxyvitamin D3	[141, 142]
Arachidonic acids	Arachidonate 5-lipoxygenase, prostaglandin E2 and leukotriene C4, leukotriene E4	[143–145]
Environmental agents	DNA adducts, PAH metabolites, chlorimuron-ethyl, taxanes, bisphenol A, s-triazines	[146–155]
Pharmaceutical agents	Penicillins, clenbuterol, sulfamethazine, avermectins, ractopamine	[156–161]
Cells	Lymphocyte subpopulations, CD34+ cells	[162–164]
Receptors	Membrane proteins, human thrombin receptor, membrane glucocorticoid receptor, G-protein-coupled receptors, human alpha(2)-adrenergic receptor subtype C2	[165–169]
Viral components	Papillomavirus structural polypeptides, avian sarcoma and leukemia virus gag-containing proteins, rabies virus glycoprotein, poliovirus type 1	[170–174]
Bacterial components	Staphylococcal enterotoxins A and E, *Vibrio cholerae* non-O1 hemagglutinin/protease, hemolysin of *Actinobacillus pleuropneumoniae*, enterotoxin produced by *Bacillus cereus*, sodium-coupled branched-chain amino acid carrier of *Pseudomonas aeruginosa*, listeriolysin O from *Listeria monocytogenes*, immunogenic outer membrane proteins from *Mannheimia haemolytica*	[175–181]

Off-line methods are generally the easiest way to combine immunoaffinity columns with other analytical techniques. This approach typically involves the use of antibodies that are immobilized onto a low-performance support packed into a small disposable syringe or solid-phase extraction cartridge. After conditioning the immunoaffinity column with the necessary application buffer or conditioning solvents, the sample is applied and

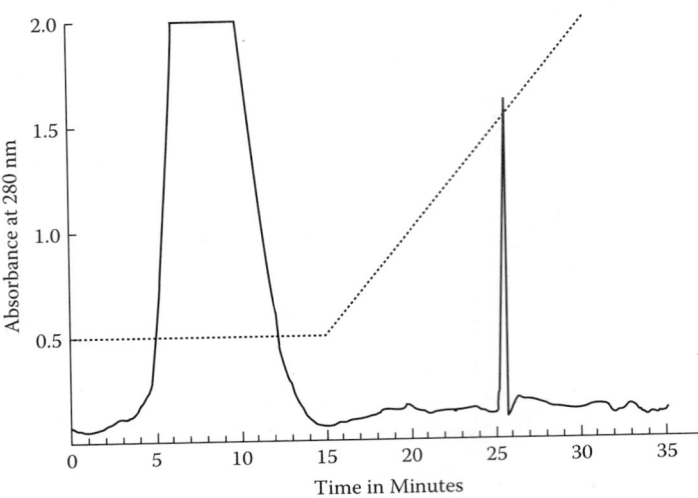

Figure 6.8 Use of HPIAC for the isolation of complement component C3 from human serum. The first peak is due to nonretained sample components, and the second peak is due to the C3 eluted in the presence of a linear gradient of mobile phases containing 0 to 2.5 M sodium thiocyanate.

undesired sample components removed. An elution buffer is then applied, and the test analyte is collected as it is removed from the column. In some cases, this eluted fraction is analyzed directly by a second technique, but in most situations the collected fraction is first dried down and reconstituted in a solvent that is compatible with the method to be used for later quantitation. If needed, the collected solute fraction can also be derivatized before it is examined by other techniques in order to obtain improved detection or more appropriate physical properties (e.g., an increase in solute volatility prior to separation and analysis by GC).

The use of off-line IE for sample preparation has been the subject of several recent reviews [12, 182–184]. Several examples involving its use with HPLC and GC are given in Table 6.4. Like any IAC method, off-line IE requires the availability of an antibody preparation that is selective for the desired analyte or group of analytes. If such antibodies are available, then IE offers the potential of much greater specificity than traditional liquid-liquid or solid-phase extraction methods. However, it should always be kept in mind that, when using IE, most antibodies will show some binding, or cross-reactivity, with solutes that are close to the desired analyte in structure. Ideally, this cross-reactivity should be evaluated for each IE support by performing binding and interference studies with any solutes or metabolites that are related to the analyte and that may be present in the samples of interest. However, even if several solutes do bind to the same IAC column, this will not present a problem as long as the analyte can be resolved or discriminated from these other compounds by the method that is to be used for quantitation.

The ability of an IAC column to bind to several types of solutes has actually been used as an advantage in a number of procedures involving IE. For instance, the ability of antibodies to cross-react with a parent compound and related agents or metabolites has been used for the development of IE methods for 17β- and 17α-trenbolone [185]; 17β- and 17α-nortestosterone [186]; and diethylstilbestrol, dienestrol, and hexestrol [187]. This idea can be taken one step further by placing multiple types of antibodies into the same IAC column and using this for multiresidue analysis. Such an approach is sometimes referred

Table 6.4 Examples of Off-Line Immunoextraction

Analysis Method	Analytes (samples)
Off-Line Immunoextraction in HPLC	
RPLC and absorbance detection	Cytokinins (plant extracts); ivermectin and avermectin (serum, plasma, meat, fruit); nortestosterone (urine, bile); phenylurea herbicides (food extracts); Sendai virus protein (viral and tissue extracts); trenbolone (urine); triazine herbicides (plant extracts)
RPLC and fluorescence detection	Aflatoxin (food and animal feed, urine, dust, milk, cheese); albuterol (plasma); fumonisin (corn); ochratoxin A (serum, plasma, milk)
RPLC and ESI-MS detection	Benzodiazepines (synthetic libraries); human chorionic gonadotropin (urine)
RPLC and coulometric detection	Oxytocin (culture media)
Off-Line Immunoextraction in Gas Chromatography	
GC/MS	Alkylated DNA adducts (DNA extracts, urine); dexamethasone (urine); estrogens (plasma, urine); flumethasone (urine); nortestosterone (meat samples); prostaglandins and thromboxanes (urine)
GC/ECD	Chloramphenicol (urine, tissue samples)

Note: ECD = electron capture detector; ESI-MS = electrospray ionization mass spectrometry; GC = gas chromatography; MS = mass spectrometry; RPLC = reversed-phase liquid chromatography.

Source: The information in this table was obtained from Hage, D.S., *J. Chromatogr. B*, 715, 3–28, 1998.

to as *multi-immunoaffinity chromatography* (*MIAC*) and has been used in situations where several classes of compounds are to be analyzed simultaneously [12].

6.3.3 On-Line Immunoextraction

The direct coupling of IAC with other methods is also possible. In this case, the IAC column is often used for sample pretreatment prior to an analyte's detection, quantitation, or characterization. This can be performed with a variety of methods, but the direct coupling of IE with HPLC has been of particular interest. The relative ease with which an IE column can be incorporated into an HPLC system makes this appealing as a means of automating sample pretreatment and reducing the time required for this process. In addition, the high reproducibility seen with modern HPLC systems gives on-line IE a better precision than off-line immunoextraction methods.

The topic of on-line IE/HPLC has been discussed in previous reviews [12, 188]. Specific examples of such techniques are provided in Table 6.5. The majority of these applications have involved the use of IAC along with reversed-phase liquid chromatography (RPLC), but there have also been techniques involving protein analytes that have used IAC coupled to size-exclusion chromatography (SEC) or ion-exchange chromatography (IEC). Most of the examples in Table 6.5 have employed antibodies that are coupled to high-performance supports; however, low-performance immunoaffinity media have also been used in some of these applications. These supports can be based either on covalently immobilized antibodies or antibodies that are adsorbed to protein A or on protein G as secondary ligands.

Table 6.5   Examples of On-Line Immunoextraction in HPLC

Analysis Method	Analytes (samples)
On-Line Immunoextraction in RPLC	
RPLC and absorbance detection	α_1-Antitrypsin (plasma); atrazine and its degradation products (water); benzylpenicilloyl-peptides (tryptic digests); bovine serum albumin (standards); carbendazim (water, food); chloramphenicol (milk, muscle tissue); clenbuterol (urine); cortisol (plasma, urine, milk, saliva); dexamethasone (urine); diethylstilbestrol (urine); estrogens (plasma, urine); hemoglobin (standards); interferon α-2 (cell extracts); LSD and metabolites (urine); lysozyme variants (standards); 17β- and 17α-19-nortestosterone (urine, bile, tissue samples); phenytoin (plasma); propranolol (urine); Δ^9-tetrahydrocannibinol (saliva); transferrin (serum, plasma); 17β- and 17α-trenbolone (standards)
RPLC and fluorescence detection	Aflatoxin M1 (milk); digoxin (serum); human epidermal growth factor fluorescence (bacterial extracts, urine)
RPLC and mass spectrometry	β-Agonists (urine); carbendazim (water); carbofuran (water, potato extract); clenbuterol (urine); dexamethasone (urine); diethylstilbestrol (urine); LSD and metabolites (urine); propranolol (urine); tolubuterol (urine)
On-line Immunoextraction in Other HPLC Methods	
SEC and absorbance detection	Human growth hormone variants (standards)
IEC and absorbance detection	Lysozyme variants (standards)

Note: IEC = ion-exchange chromatography; LSD = lysergic acid diethylamide; RPLC = reversed-phase liquid chromatography; SEC = size-exclusion chromatography.

Source: The information in this table was obtained from Hage, D.S., *J. Chromatogr. B*, 715, 3–28, 1998.

One reason for the large number of reports involving the combination of on-line IE with RPLC undoubtedly has to do with the popularity of RPLC in routine chemical separations, but there is also a more fundamental reason that involves the underlying nature of both IAC and RPLC. For instance, the fact that the elution buffer for an IAC column is an aqueous solvent that generally contains little or no organic modifier is convenient, since this same elution buffer will act as a weak mobile phase for RPLC. In other words, as a solute elutes from an IAC column, it will tend to have strong retention on any on-line reversed-phase support, leading to analyte reconcentration. This reconcentration phenomenon is valuable in dealing with analytes that have slow desorption from their immobilized antibodies, a factor that makes it impractical to analyze these solutes by the more traditional on/off mode of IAC.

A general scheme for performing on-line IE in RPLC is shown in Figure 6.9 [189]. Similar systems are used when coupling IAC with size-exclusion or ion-exchange supports

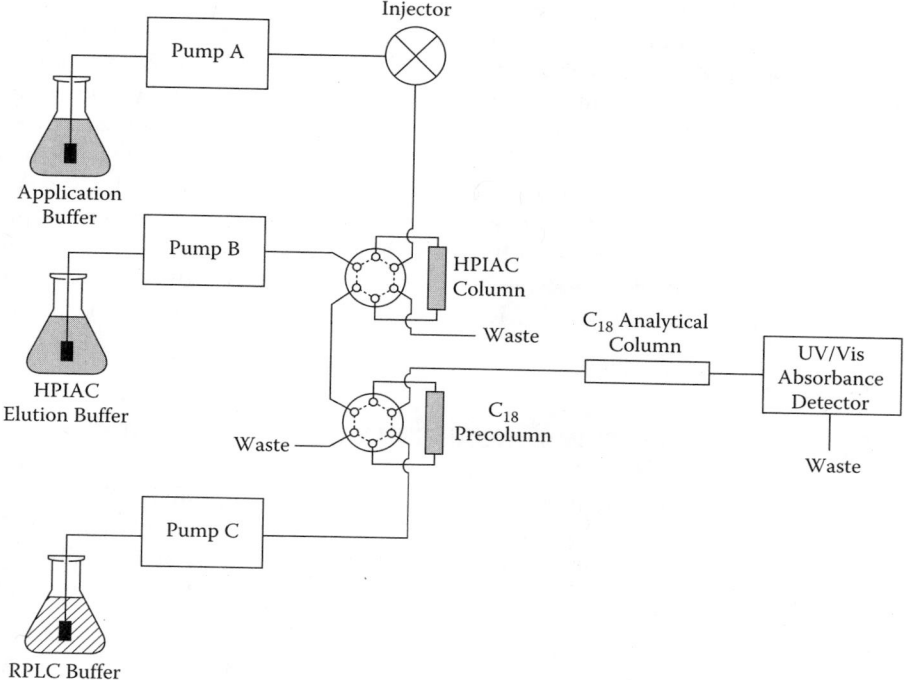

Figure 6.9 General system for the on-line immunoextraction of small solutes and their separation by RPLC. In this example, an HPIAC column is used for the immunoextraction process. (Reproduced with permission from Thomas, D.H., Beck-Westermeyer, M., and Hage, D.S., *Anal. Chem.*, 66, 3823–3829, 1994.)

[12, 188]. The general format for using such a system involves injecting the sample onto the IE column, with the nonretained components being allowed to go to a waste container. The IE column is then switched on-line with an RPLC column, and an elution buffer is applied to dissociate any retained analyte. As these analytes elute, they are captured and reconcentrated at the head of an RPLC precolumn. After all solutes have left the IE column, this column is then switched back off-line and regenerated by passing through the initial application buffer. Meanwhile, the RPLC precolumn is placed in parallel with a longer analytical column and developed with either an isocratic or gradient elution scheme that uses a mobile phase with an increased organic modifier content. This causes analytes on the RPLC columns to be separated based on their differences in polarity. As these solutes elute, they are monitored by an on-line detector.

Like off-line IE, on-line techniques can be used to monitor either a single analyte or a group of analytes. Such techniques are attractive, since they allow both simple sample pretreatment and good resolution between closely related compounds. Along with these types of applications, there have also been several alternative uses reported for on-line IE/HPLC. For example, on-line IE has been used to remove major proteins from plasma prior to the analysis of minor proteins by RPLC [190]. It has also been used to examine the covalent binding sites of benzylpenicillin on human serum albumin (HSA) [191]. Other applications have included the use of IE/RPLC with an on-line dialysis system [192] and the use of IE/HPLC to analyze protein variants and aggregates [193–195].

Along with HPLC, IE has also been used on-line with CE and mass spectrometry, as discussed in Chapters 28 and 29, respectively. There has also been a report in which on-line IE has been used in GC [196]. This was performed by using an IE column that was first coupled to an RPLC column. After the sample was applied to and washed from the IE column, the retained solutes were passed onto the RPLC column, as described earlier for IE/HPLC. The analytes on the RPLC column were then eluted with a volatile organic solvent (i.e., ethyl acetate) and passed into a GC injection gap. As this solvent plug entered the GC system, a temperature program was then started for solute separation. One advantage of this approach, as with IE/HPLC, is that large volumes of sample can be applied to the IE column, thus allowing low detection limits to be obtained. The main disadvantage of using on-line IE with GC is the greater complexity of this method versus off-line IE.

6.4 SAMPLE ANALYSIS BY IMMUNOAFFINITY CHROMATOGRAPHY

In addition to its use in purification or sample pretreatment, immunoaffinity chromatography has emerged as an important analytical tool for trace analysis. This can be performed either by using an antibody column for the direct detection of analytes or by combining this with a scheme for indirect analyte detection [12]. Both approaches are examined in the following subsections.

6.4.1 Direct Analyte Detection

The simplest format for IAC in analytical applications involves the adsorption of test solutes by an immobilized antibody column, followed by the later release and detection of these solutes. This makes use of the traditional on/off elution mode of affinity chromatography. In this scheme, the sample of interest is first injected onto the IAC column under mobile-phase conditions in which the analyte has a strong binding to the immobilized antibodies in the column. After nonretained sample components have been removed, an elution buffer is passed through the column, and the analyte is detected as it dissociates from the column. The advantages of this approach include its speed and simplicity, especially when it is performed as part of an HPLC system.

Examples of compounds that have been determined by this approach are given in Table 6.6. These methods have been shown in many studies to have good correlation versus reference techniques, such as immunoassays or electrophoresis, but typically take much less time to perform. In addition, the precision of these methods is generally in the range of 1 to 5% when performed by HPLC, and the immunoaffinity columns have often been shown to be stable for up to several hundred sample applications [12, 197–200].

Figure 6.10 shows an example of a typical chromatogram generated when using HPIAC in the direct-detection mode. This approach ideally uses antibodies in the IAC column that are selective for only the analyte(s) of interest and that have little or no binding to other sample components. Because of this, the resulting chromatograms generally contain only two peaks, with the first peak representing all nonretained sample components that elute during the application/wash step and the second peak representing solutes that were retained by the IAC column and later dissociated when the elution buffer was passed through the column. It is usually quite easy to obtain baseline resolution between the nonretained and retained peaks by changing the time at which the elution buffer is applied to the column.

Many of the theoretical and practical aspects of direct detection by IAC are discussed in Chapter 29. One requirement of this mode is that there must be some way to monitor

Table 6.6 Examples of Direct Analyte Detection by Immunoaffinity Chromatography

Analysis Method	Analyte (sample)
Absorbance detection	Anti-idiotypic antibodies (serum); bovine growth hormone (aqueous standards); fibrinogen (plasma); glutamine synthetase with enzyme assay (bacterial extracts); human serum albumin (serum, urine); immunoglobulin G antibodies (CSF); immunoglobulin E antibodies with ELISA (plasma, serum); interferon (plasma, serum, urine, saliva, CSF, fermentation broth, bacterial extract); lymphocyte receptors (cell extracts); β_2-microglobulin with EIA (plasma); transferrin (serum)
Fluorescence detection	Antithrombin III (cell culture); granulocyte colony stimulating factor (plasma, CSF, BMAF); interleukin-2 with receptor assay (tissue samples); tissue-type plasminogen activator (cell cultures)
Pulsed amperometric detection	Fungal carbohydrate antigens (fungal isolates); glucose tetrasaccharide (serum, urine)
Radiolabel detection	Group A-active oligosaccharides (standards)

Note: BMAF = bone marrow aspirate fluid; CSF = cerebrospinal fluid; EIA = enzyme immunoassay; ELISA = enzyme-linked immunosorbent assay.

Source: The information in this table was obtained from Hage, D.S., *J. Chromatogr. B*, 715, 3–28, 1998.

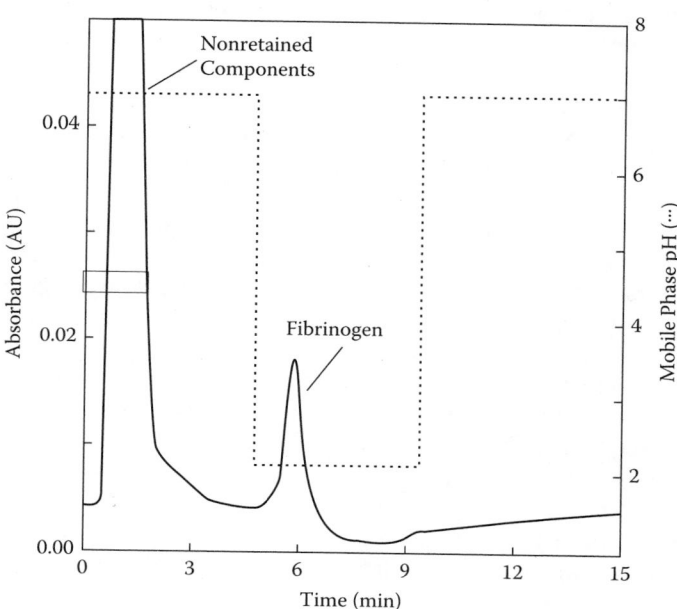

Figure 6.10 Determination of fibrinogen in human plasma using direct detection by HPIAC. The dotted line shows the times at which the application buffer and elution buffers were passed through the column. (Adapted with permission from McConnell, J.P. and Anderson, D.J., *J. Chromatogr.*, 615, 67–75, 1993.)

the analyte as it leaves the column. The various detection methods that have been reported for use in HPIAC are shown in Table 6.5. For carbohydrates, pulsed-amperometric detection has been employed in several cases for direct detection; for protein solutes, UV absorbance measurements at 210 to 215 nm or 280 nm are most commonly used. All of these approaches need enough analyte to give a measurable signal. In some cases, minor sample components have been monitored by using precolumn derivatization to place labels on these substances prior to injection [12]. This has included the use of fluorescent tags and radiolabels. As an alternative, fractions of the column eluant can be collected and later analyzed by a separate technique, such as an immunoassay or an assay for biological activity.

Because of the strong binding and slow dissociation of most antibody-analyte interactions, IAC columns tend to produce a response that is related to the moles of applied solute rather than to the solute's initial sample concentration [189, 201, 202]. This means that only small sample volumes are needed to provide a detectable signal if the analyte is present at intermediate or high concentrations. For more-dilute solutions, the IAC column can be used to concentrate the desired test substance from large sample volumes prior to detection [202]. However, caution must be used in this approach since there is also an increased chance of other sample components adsorbing nonspecifically to the column, thus giving rise to an increase in the background signal.

The choice of elution conditions is important in determining both the detectability of the analyte and the overall speed of the IAC method. Step elution is commonly used in the on/off mode of HPIAC because of its speed and sharp analyte peaks, but gradient elution is also employed for cases where there is a need for gentler elution conditions, a more gradual change in background signal, or better resolution between the eluting solutes. In general, direct detection works best for analytes that dissociate from the column within a few minutes, which will give sharp peaks and allow for easy detection. For analytes with longer elution times and broader peaks, it is generally recommended that an alternative approach be used, such as IE coupled on-line with RPLC, as described earlier in this chapter.

Although step gradients or linear elutions are often needed to disrupt the strong analyte-antibody interactions that occur under typical application conditions, there are some cases in which isocratic elution is possible with an IAC column. This has been reported for systems that use antibodies with weak affinities (i.e., $K_a < 10^4 M^{-1}$) and fast solute association/dissociation kinetics. This approach, which is a type of weak-affinity chromatography (see Chapter 4), has been used for the analysis of various carbohydrates [203–206] and is similar to traditional types of HPLC in its operation. One advantage of using isocratic elution and weak-affinity antibodies is that several related solutes can be separated in a single run by HPIAC. Isocratic methods are also attractive, since they potentially allow less harsh conditions to be used for analyte elution, thereby increasing IAC column lifetimes. The main disadvantage of this approach is the increased time required for solute analysis versus step- or gradient-elution methods.

Along with the basic on/off format of IAC, there have been a number of reports in which the on/off mode of IAC has been combined with other analytical methods. Some examples are the off-line immunoextraction methods described earlier in Section 6.3. Another example is work in which HPIAC was used along with flow-injection analysis (FIA) for the determination of urinary albumin [198]. In this method, a system for performing FIA was attached to the outlet of an HPIAC column for measuring creatinine in the nonretained peak. This creatinine level was then used to correct for fluctuations in the urine concentration of HSA due to normal variations in urine output and volume. A third example is a report in which an HPIAC column followed by an immobilized receptor cartridge was used for the measurement of both total and bioactive interleukin-2 in tissue samples [199]. An example of such a scheme is shown in Figure 6.11.

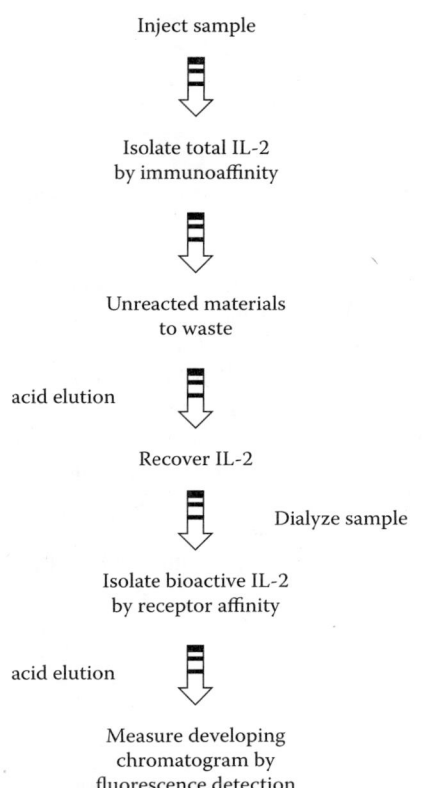

Inject sample

Isolate total IL-2
by immunoaffinity

Unreacted materials
to waste

acid elution

Recover IL-2

Dialyze sample

Isolate bioactive IL-2
by receptor affinity

acid elution

Measure developing
chromatogram by
fluorescence detection

Figure 6.11 Flowchart for the analysis of total and bioactive interleukin-2 (IL-2) by HPIAC and a receptor assay. (Reproduced with permission from Phillips, T.M., *Biomed. Chromatogr.*, 11, 200–204, 1997.)

6.4.2 Indirect Analyte Detection

Another area that has received increasing attention in recent years has been the use of immobilized-antibody (or immobilized-antigen) columns to perform various types of immunoassays. Such an approach is known as a *chromatographic* (or *flow-injection*) *immunoassay*. A detailed discussion of this area is given in Chapter 29, with other reviews of this method being given in the literature [207–210]. Examples of some specific applications are provided in Table 6.7, with other examples being provided in Chapter 29.

The use of IAC to perform immunoassays is particularly valuable in determining trace analytes that may not produce a readily detectable signal. This problem is overcome in chromatographic immunoassays by using a labeled antibody or labeled analyte analog that can be used for indirect analyte detection. As discussed in Chapter 29, there are several types of formats, as well as labels, that can be used in such methods. Examples of these formats are competitive-binding immunoassays, sandwich immunoassays, and one-site immunometric assays.

A competitive-binding assay is the common method used in performing immunoassays by IAC [207, 211–216]. The basic principle behind a competitive-binding immunoassay involves the incubation of analyte in the sample with a fixed amount of a labeled analyte analog in the presence of a limited amount of antibodies that bind to both the native analyte and labeled species. Because there is only a limited amount of antibodies

Table 6.7 Examples of Chromatographic Immunoassays

Analysis Method	Analyte/Label (sample)
Competitive-Binding Immunoassays	
Absorbance detection	α-Amylase with HRP label (standards); human chorionic gonadotropin with HRP label (serum); human serum albumin with and without HRP label (standards, serum); isoproturon with HRP label (water); thyroid-stimulating hormone with HRP label (serum); transferrin (serum)
Fluorescence detection	Adrenocorticotropic hormone (standards); atrazine and triazines with HRP label (water); 2,4-dinitrophenyl lysine with fluorescein label (standards); immunoglobulin G with fluorescein label (serum); immunoglobulin G with fluorescein label (standards); testosterone with Texas red label (serum); theophylline with liposome labels (serum); transferrin with Lucifer yellow label (serum); trinitrotoluene with fluorescein label (water)
Radiolabel detection	2,4-Dinitrophenyl lysine with I^{125} label (standards)
Electrochemical detection	Immunoglobulin G with GOD label (serum); theophylline with ALP label (serum)
Thermometric detection	Insulin with ALP label (standards)
Chemiluminescence detection	Thyroxine with HRP label (standards)
Sandwich Immunoassays	
Fluorescence detection	Human serum albumin with fluorescein label (fermentation broth); mouse immunoglobulin G with HRP label (standards)
Electrochemical detection	Antibovine IgG antibodies with GOD label (serum); interferon with GOD label (cell culture)
Chemiluminescence detection	Bovine immunoglobulin G with HRP label (serum); human immunoglobulin G with acridinium ester label (serum); mouse immunoglobulin G with acridinium ester label (standards); parathyroid hormone with acridinium ester label (plasma)
One-Site Immunometric Assays	
Fluorescence detection	α-(Difluoromethyl)ornithine with HRP label (plasma); 17-β-estradiol with liposome label (standards)

Note: ALP = alkaline phosphatase; GOD = glucose oxidase; HRP = horseradish peroxidase.

Source: The information in this table was obtained from Hage, D.S., *J. Chromatogr. B*, 715, 3–28, 1998.

present, the sample and labeled analyte molecules must compete for binding sites on these antibodies. After this competition has been allowed to take place, the analyte and labeled analog bound to the antibodies are separated from the analyte as well as the labeled analog molecules that remain free in solution. The amount of the labeled analog that is present in either the bound or free fraction is then measured. In the absence of any sample analyte, the largest amount of labeled analyte in the bound fraction should be observed. However, as the amount of sample analyte increases, the level of bound labeled analyte will decrease, giving rise to an indirect measure of the amount of native analyte in the sample.

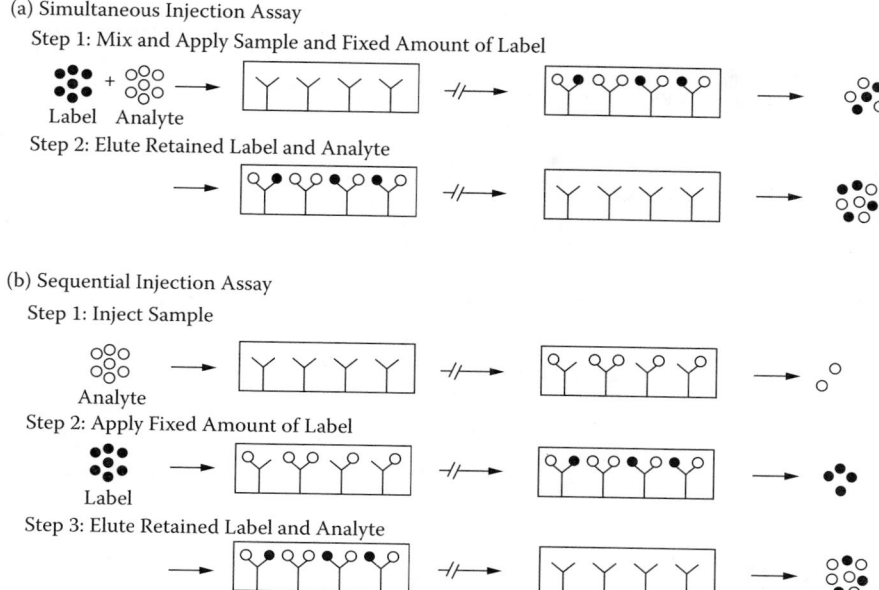

Figure 6.12 Two approaches for performing a competitive-binding immunoassay by IAC, where (○) represents the analyte and (●) represents its labeled analog. (Reproduced with permission from Nelson, M.A., Reiter, W.S., and Hage, D.S., *Biomed. Chromatogr.*, 17, 188–200, 2003.)

There are several different ways in which competitive-binding immunoassays can be performed by IAC (see Figure 6.12). The simplest approach is to mix the sample and labeled analyte analog (i.e., the "label") and simultaneously apply these to the IAC column. This is known as a *simultaneous-injection competitive-binding immunoassay*. Detection can be performed by either examining the amount of label that elutes nonretained from the column or by measuring the labeled species that dissociate from the IAC column during the elution step. An alternative format involves the application of only the sample to the IAC column, followed later by a separate injection of the label. This method is known as a *sequential-injection competitive-binding immunoassay*. The advantage of this approach is that even an unlabeled preparation of analyte can potentially be used as the label, provided that this species produces a sufficient signal for detection. In addition, there are no matrix interferences present during detection of the label in the nonretained fraction [207].

A third format for a competitive-binding method is the *displacement competitive-binding immunoassay*. In this technique, the IAC column is first saturated with the labeled analog, followed by application of the sample to the column. As the sample passes through the column, the unlabeled analyte can bind to any antibody-binding regions that are momentarily unoccupied by the label as this label undergoes local dissociation/reassociation. The net result is displacement of the label from the column by mass action, with the degree of displacement increasing with the amount of applied sample analyte [207, 217–219].

A *sandwich immunoassay*, or *two-site immunometric assay*, involves the use of two different types of antibodies that each bind to the analyte of interest. The first of these two antibodies is attached to a solid-phase support and is used for extraction of the analyte from samples (see Figure 6.13). The second antibody contains an easily measured tag (e.g., an enzyme or fluorescent label) and is added in solution to the analyte either before or after this extraction. This second antibody then serves to place a label onto the analyte, thus allowing the amount of analyte on the immunoaffinity support to be quantitated. An important

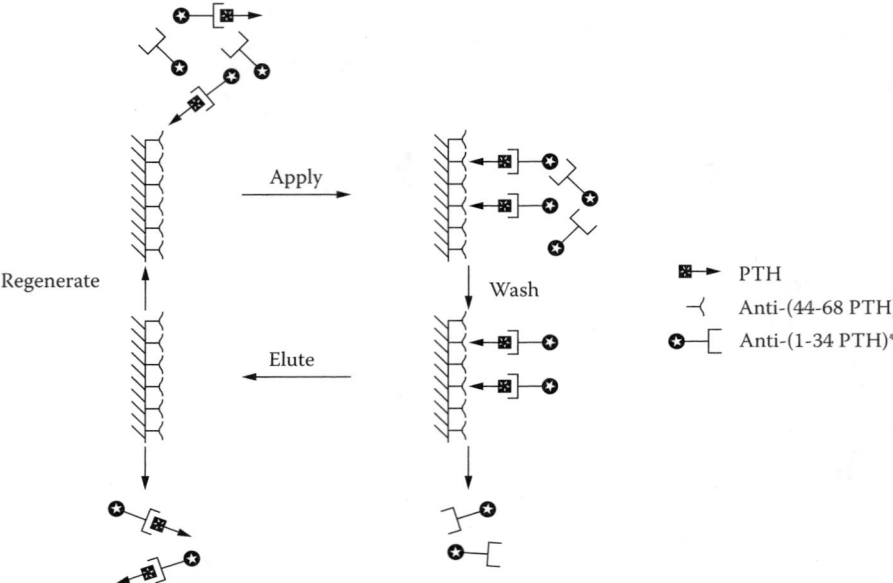

Figure 6.13 Format for a chromatographic sandwich immunoassay in which the analyte (▩) is preincubated with a labeled antibody (circles with stars), with this mixture later being injected onto an IAC with a second immobilized antibody against the analyte. (Reproduced with permission from Hage, D.S. and Kao, P.C., *Anal. Chem.*, 63, 586–595, 1991.)

advantage of sandwich immunoassays is that they produce a signal for the bound label directly proportional to the amount of sample analyte. The fact that two types of antibodies are used in this assay also gives this technique higher selectivity than competitive-binding immunoassays [87, 207, 220–223].

Another format for an IAC-based immunoassay is the *one-site immunometric assay* [207, 224–226]. In this technique, the sample is first incubated with a known excess of labeled antibodies or F_{ab} fragments specific to the analyte of interest. After binding between the sample analyte and antibodies has occurred, this mixture is applied to a column containing an immobilized analog of the analyte. This column serves to extract any antibodies or F_{ab} fragments not bound to sample analyte. Meanwhile, those antibodies or F_{ab} fragments that are bound to the analyte will pass through the column in the nonretained peak. Detection is then performed either by looking at the nonretained labeled antibody/F_{ab} fragments, which will give a signal directly proportional to the sample analyte concentration, or by monitoring the amount of excess antibody/F_{ab} fragments that later dissociate from the column during the elution step. Like a competitive-binding immunoassay, this method is able to detect both small and large solutes. Also, like a sandwich immunoassay, it gives a signal for the bound-label fraction that is directly proportional to the amount of analyte in the original sample.

6.4.3 Postcolumn Immunodetection

Direct detection and chromatographic immunoassays can be used either as stand-alone systems or as a means for monitoring the presence of a specific solute eluting from other chromatographic columns. The latter approach is often referred to as *immunodetection*. This technique typically involves the use of a postcolumn reactor and an immobilized antibody or antigen column attached to the exit of an analytical HPLC column. Reviews of this topic can be found in the literature [12, 227].

The direct-detection mode of IAC is the simplest approach for postcolumn immuno-detection if the analyte of interest is capable of generating a sufficient signal for detection. One example of where direct postcolumn immunodetection has been used is in the monitoring of acetylcholinesterase (AChE) in amniotic fluid by size-exclusion chromatography [228]. This method used an IAC column containing anti-AChE antibodies to capture AChE as it eluted from the analytical column. After the AChE was adsorbed to the IAC column, a substrate solution for AChE was passed through the column, and an on-line absorbance detector was used to detect the resulting colored product. Another example is work reported with bovine growth-hormone-releasing factor (GHRF) [229].

One item that must be considered in such work is the need to adjust the eluant of the HPLC analytical column to a particular pH, ionic strength, and polarity appropriate for an IAC application buffer [229]. This item is especially important when using immunodetection for RPLC, where an appreciable amount of organic modifier may be present in the mobile phase leaving the HPLC analytical column. One solution to this problem is to combine the analytical column eluant with a dilution buffer prior to sample application onto the IAC detection column. Immunodetection by the on/off mode also requires that the eluting analyte be present in a conformation that is recognized by antibodies in the IAC column. This latter item is of particular concern when detecting the elution of proteins from RPLC columns, since many proteins can undergo denaturation in the presence of the organic solvent levels used in RPLC mobile phases.

Chromatographic sandwich immunoassays have also been used for postcolumn immunodetection. For example, this mode was used with RPLC for the analysis of GHRF [229]. In this case, an IAC column containing antibodies against GHRF was placed after the RPLC analytical column and used to capture any eluting GHRF. This captured GHRF was then quantitated by applying excess fluorescein-labeled antibodies capable of binding GHRF. After washing the excess labeled antibodies from the IAC column, the GHRF and associated labeled antibodies were eluted and monitored by an on-line fluorescence detector.

The one-site immunometric assay is the most common approach for performing postcolumn immunodetection. As shown in Figure 6.14, this involves taking the analytical HPLC column eluant and combining it with a solution of labeled antibodies or F_{ab} fragments that will bind the analyte of interest. The column eluant and antibody or F_{ab} mixture is then allowed to react in a mixing coil and passed through an immunodetection column that contains an immobilized analog of the analyte. The antibodies or F_{ab} fragments bound to the analyte will pass through this column and to the detector, where they will provide a signal that is proportional to the amount of bound analyte. If desired, the immunodetection column can later be washed with an eluting solvent to dissociate the retained antibodies or F_{ab} fragments, but a sufficiently high binding capacity is generally used so that a reasonably large amount of analytical column eluant can be analyzed before the immunodetection column must be regenerated.

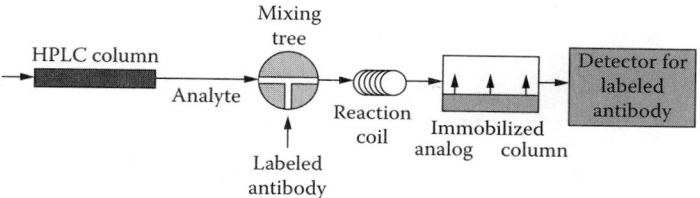

Figure 6.14 Typical system for performing postcolumn immunodetection with HPLC using a one-site immunometric format. (Reproduced with permission from Hage, D.S. and Nelson, M.A., *Anal. Chem.*, 73, 198A–205A, 2001.)

The one-site immunometric approach was originally used to quantitate digoxin and digoxigenin as they eluted from a standard RPLC column [230]. This was performed using fluorescein-labeled F_{ab} fragments (raised against digoxigenin) and an immobilized digoxin support in the postcolumn detection system. This method was then successfully used to monitor both digoxin and its metabolites in plasma and urine samples [230]. The same general system was later used along with a restricted-access RPLC column to monitor digoxin, digoxigenin, and related metabolites in serum samples [231]. A similar approach with a traditional RPLC column has been used for the detection of human methionyl granulocyte colony-stimulating factor and its derivatives [232].

6.5   NEW DEVELOPMENTS IN IMMUNOAFFINITY CHROMATOGRAPHY

New developments in IAC technology are constantly being reported in the literature. Some examples include the coupling of IAC with capillary electrophoresis and mass spectrometry, as discussed in Chapters 27 to 29. The use of IAC as a tool in such fields as biotechnology, molecular biology, and clinical testing has also seen recent growth. A further discussion of these areas can be found in other chapters within this text.

One particular development that has recently been reported is the use of IAC for ultrafast sample extraction [233–235]. This has been illustrated in work involving the isolation of drugs and other small solutes from biological mixtures. Such studies use small IAC columns that can be operated at moderate to high flow rates and that have a high-volume density of active antibodies. For particulate supports, microsandwich columns are used for this purpose, allowing sample residence times as low as a few milliseconds [233]. Monolithic supports can also be employed [68]. With such columns, it has been shown that compounds like fluorescein, warfarin, and thyroxine can be quantitatively extracted in as little as 60 to 120 msec (see Figure 6.15). This has made it possible to use such columns for the direct detection of free drug and hormone fractions in protein/drug mixtures and in clinical samples [234, 235].

Other areas of interest in IAC include the use of this method for multianalyte detection and in miniaturized systems. The idea of using a series of chromatographic ligands to

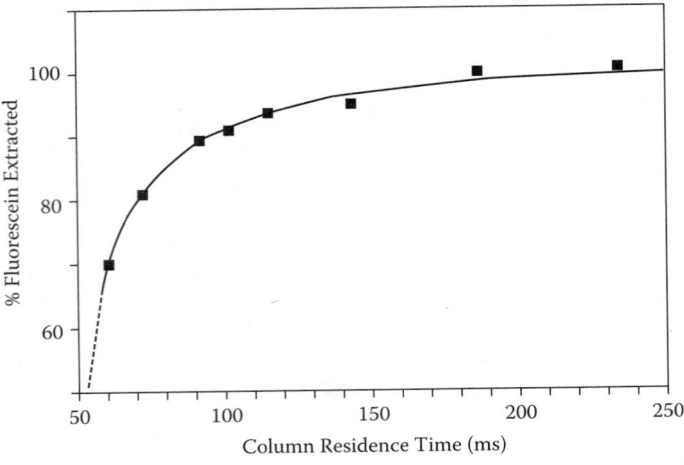

Figure 6.15   Adsorption of fluorescein using antifluorescein antibodies within 124- to 620-μm-long microsandwich IAC columns operated at various flow rates. Quantitative extraction of fluorescein was obtained in as little as 100 to 120 msec with this system. (Reproduced with permission from Clarke, W. and Hage, D.S., *Anal. Chem.*, 73, 2157–2164, 2001.)

separate different analytes has been in existence for some years, but the last few years have seen a growing need for these techniques in multianalyte detection in medical research and biochemistry [236, 237]. This is particularly true when analyzing valuable archival materials, low-density cell cultures, or samples from small experimental animals.

As an example, in one study, samples were passed through microtiter plates containing different immobilized anticytokine antibodies to analyze several cytokines within the same sample [238]. The same approach was employed to isolate and measure ten different cytokines from 25-μl samples of whole dried blood, plasma, and urine. This latter work was performed through the use of a series of immunoaffinity columns linked in series by microswitching valves, as shown in Figure 6.16 [239]. This technique was further

(a) System during sample injection

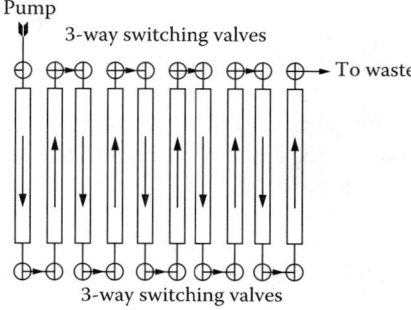

(b) System during stepwise column elution

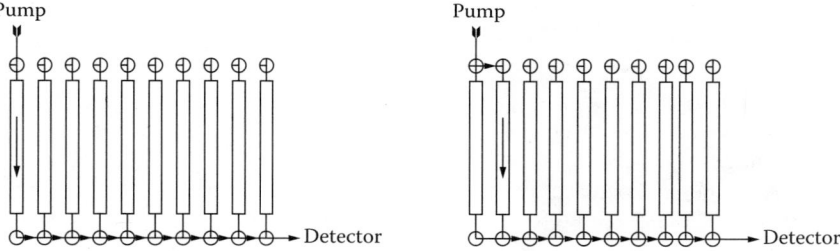

(c) Typical chromatogram

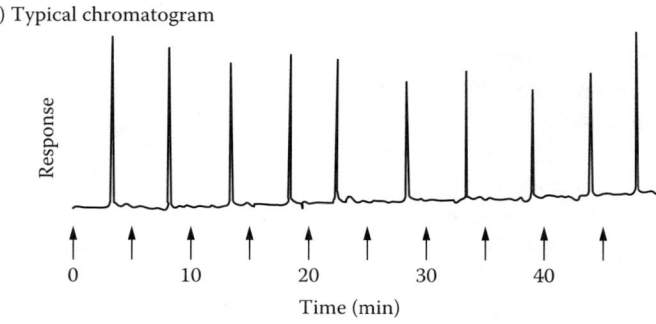

Figure 6.16 A recycling immunoaffinity system for the determination of multiple analytes. The peaks shown in the bottom chromatogram (from left to right) represent interleukin-1 (IL-1), IL-2, IL-4, IL-5, IL-6, IL-10, IL-12, IL-13, tumor necrosis factor-α, and γ-interferon, respectively. (Reproduced with permission from Phillips, T.M. and Krum, J.M., *J. Chromatogr. B*, 715, 55–63, 1998.)

applied to the analysis of a series of dried human blood spots, in which a total of 53 analytes were measured [240, 241]. Other micro-HPLC systems have ranged from laboratory-built IAC systems using commercially available components [242] to chip-based chromatography systems [243] and the concepts of a lab-on-a-chip. For a more detailed discussion of these recent advances in this technology as applied to IAC, the reader is directed to Chapter 28 of this book.

6.6 SUMMARY AND CONCLUSIONS

This chapter has shown that IAC is an important tool for the purification, pretreatment, and analysis of a wide range of samples. The basic principles of IAC and the various factors to consider in the development of an IAC method were examined. Examples of preparative applications included the use of IAC in isolating various target compounds and its use in the off-line or on-line extraction of samples. Quantitative applications involving direct analyte detection, chromatographic immunoassays, or postcolumn immunodetection were also discussed. Such techniques have already been reported for a large variety of biological and nonbiological agents in fields ranging from clinical chemistry and biochemical research to food science and environmental analysis. The selectivity of immunoaffinity methods makes them particularly appealing for the development of fast and simple methods for complex samples. The ability to combine IAC with other methods, such as HPLC, GC, or CE, is also attractive as a means for obtaining even greater selectivity or information on a group of structurally related solutes. Because of these advantages, it is expected that IAC will continue to see even greater use in the future for the purification and quantitation of analytes in real-world samples.

SYMBOLS AND ABBREVIATIONS

Ab	Antibody
AChE	Acetylcholinesterase
Ag	Antigen
ALP	Alkaline phosphatase
Anti-id	Anti-idiotypic antibody
BMAF	Bone marrow aspirate fluid
CE	Capillary electrophoresis
CSF	Cerebrospinal fluid
DEA	Diethanolamine
DTT	Dithiothreitol
ECD	Electron capture detector
EIA	Enzyme immunoassay
ELISA	Enzyme-linked immunosorbent assay
ESI-MS	Electrospray-ionization mass spectrometry
F_{ab}	Antigen-binding fragment of an antibody
$F_{(ab')2}$	Antibody fragment with two antigen-binding regions but no F_c region
F_c	Single-stem (crystallizable) fragment of an antibody
FIA	Flow-injection analysis
GC	Gas chromatography
GHRF	Growth-hormone-releasing factor
GOD	Glucose oxidase
HPIAC	High-performance immunoaffinity chromatography

HPLC	High-performance liquid chromatography
HRP	Horseradish peroxidase
IAC	Immunoaffinity chromatography
IE	Immunoextraction
IEC	Ion-exchange chromatography
IgG	Immunoglobulin G
IgM	Immunoglobulin M
K_a	Association equilibrium constant
k_a	Association rate constant
k_d	Dissociation rate constant
LSD	Lysergic acid diethylamide
Mab	Monoclonal antibody
MIAC	Multi-immunoaffinity chromatography
MS	Mass spectrometry
RPLC	Reversed-phase liquid chromatography
scF_v	Single-chain variable-region fragment
SEC	Size-exclusion chromatography

REFERENCES

1. d'Allesandro, G. and Sofia, F., The adsorption of antibodies from the sera of syphilitics and tuberculosis patients, *Z. Immunitats*, 84, 237–250, 1935.
2. Landsteiner, K. and van der Scheer, J., Cross reactions of immune sera to azoproteins, *J. Exp. Med.*, 63, 325–339, 1936.
3. Campbell, D.H., Luescher, E., and Lerman, L.S., Immunologic adsorbents, I: Isolation of antibody by means of a cellulose-protein antigen, *Proc. Natl. Acad. Sci. U.S.A.*, 37, 575–578, 1951.
4. Wilchek, M., Miron, T., and Kohn, J., Affinity chromatography, *Methods Enzymol.*, 104, 3–55, 1984.
5. Calton, G.J., Immunosorbent separations, *Methods Enzymol.*, 104, 381–387, 1984.
6. Phillips, T.M., High performance immunoaffinity chromatography: An introduction, *LC Mag.*, 3, 962–972, 1985.
7. Ehle, H. and Horn, A., Immunoaffinity chromatography of enzymes, *Bioseparation*, 1, 97–110, 1990.
8. Phillips, T.M., Isolation and recovery of biologically active proteins by high performance immunoaffinity chromatography, *Recept. Biochem. Methodol.*, 14, 129–154, 1989.
9. Bailon, P., Roy, S.K., and Swapan, K., Recovery of recombinant proteins by immunoaffinity chromatography, *ACS Symp. Ser.*, 427, 150–167, 1990.
10. Howell, K.E., Gruenberg, J., Ito, A., and Palade, G.E., Immuno-isolation of subcellular components, *Prog. Clin. Biol. Res.*, 270, 77–90, 1988.
11. Nakajima, M. and Yamaguchi, I., Purification of plant hormones by immunoaffinity chromatography, *Kagaku Seibutsu*, 29, 270–275, 1991.
12. Hage, D.S., Survey of recent advances in analytical applications of immunoaffinity chromatography, *J. Chromatogr. B*, 715, 3–28, 1998.
13. Janeway, C.A. and Travers, P., *Immunobiology: The Immune System in Health and Disease*, Current Biology Ltd., London, 1996, pp. 3.1–3.42.
14. Porter, R.R., The hydrolysis of rabbit-globulin and antibodies with crystalline papain, *Biochem. J.*, 73, 119–126, 1959.
15. Stein, S.R., Palmer, J.L., and Nisonoff, A., Re-formation of interchain bonds linking half-molecules of rabbit gamma-globulin, *J. Biol. Chem.*, 239, 2872–2877, 1964.
16. Hage, D.S., Affinity chromatography, in *Handbook of HPLC*, Katz, E., Eksteen, R., Schoenmakers, P., and Miller, N., Eds., Marcel Dekker, New York, 1998, chap. 13.

17. Roitt, I.M., *Essential Immunology,* Blackwell, Oxford, U.K., 1985.
18. DeLisi, C., A theory of precipitation and agglutination reactions in immunological systems, *J. Theoret. Biol.,* 45, 555–575, 1974.
19. Boyle, M.D.P. and Langone, J.J., A simple procedure to use whole serum as a source of either IgG- or IgM-specific antibody, *J. Immunol. Methods,* 32, 51–58, 1980.
20. Kabat, E.A. and Mayer, M., *Experimental Immunochemistry,* Charles C Thomas, Springfield, IL, 1961.
21. Eliasson, M., Andersson, R., Olsson, A., Wigzell, H., and Uhlen, M., Differential IgG-binding characteristics of staphylococcal protein A, streptococcal protein G, and a chimeric protein AG, *J. Immunol.,* 142, 575–581, 1989.
22. Di Modugno, F., Buglioni, S., Mottolese, M., Bello, D.D., Cascioli, S., Chersi, A., Santoni, A., and Nistico, P., Polyclonal antibodies against: Their putative role in the identification of a particular HER2 status in patients with breast cancer, *J. Immunother.,* 24, 221–231, 2001.
23. Murdoch, S.J., Wolfbauer, G., Kennedy, H., Marcovina, S.M., Carr, M.C., and Albers, J.J., Differences in reactivity of antibodies to active versus inactive PLTP significantly impacts PLTP measurement, *J. Lipid Res.,* 43, 281–289, 2002.
24. Kohler, G. and Milstein, C., Continuous cultures of fused cells secreting antibody of predefined specificity, *Nature,* 256, 495–497, 1975.
25. Burgess, R.R. and Thompson, N.E., Advances in gentle immunoaffinity chromatography, *Curr. Opinions Biotechnol.,* 13, 304–308, 2002.
26. Venter, J.C., in *Monoclonal and Anti-Idiotypic Antibodies: Probes for Receptor Structure and Function,* Venter, J.C., Fraser, C.M., and Lindstrom, J., Eds., Alan R. Liss, New York, 1984, p. 117.
27. Greene, M.I., Weiner, H.L., Dichter, M., and Fields, B., in *Monoclonal and Anti-Idiotypic Antibodies: Probes for Receptor Structure and Function,* Venter, J.C., Fraser, C.M., and Lindstrom, J., Eds., Alan R. Liss, New York, 1984, p. 177.
28. Weis, J.J. and Fearon, D.T., in *Hybridoma Technology in the Biosciences and Medicine,* Spring, T.A., Ed., Plenum, New York, 1985, p. 207.
29. Lee, G.M. and Palsson, B.O., Monoclonal antibody production using free-suspended and entrapped hybridoma cells, *Biotechnol. Genet. Eng. Rev.,* 12, 509–533, 1994.
30. Hendriksen, C.F. and de Leeuw, W., Production of monoclonal antibodies by the ascites method in laboratory animals, *Res. Immunol.,* 149, 535–542, 1998.
31. Nelson, P.N., Reynolds, G.M., Waldron, E.E., Ward, E., Giannopoulos, K., and Murray, P.G., Monoclonal antibodies, *Mol. Pathol.,* 53, 111–117, 2000.
32. Goldkorn, I., Gleeson, P.A., and Toh, B.H., Reverse immunoaffinity chromatography: Application to the purification of the 60- to 90-kDa gastric parietal cell autoantigen associated with autoimmune gastritis, *Anal. Biochem.,* 194, 433–438, 1991.
33. Blecher, M., in *Receptors, Antibodies and Disease,* Evered, D. and Whelan, J., Eds., Pitman, London, 1982, p. 279.
34. Abood, L.G., Langone, J.J., Bjercke, R., Lu, X., and Banerjee, S., Characterization of a purified nicotinic receptor from rat brain by using idiotypic and anti-idiotypic antibodies, *Proc. Natl. Acad. Sci. U.S.A.,* 84, 6587–6590, 1987.
35. Bjercke, R.J. and Langone, J.J., Anti-idiotypic antibody probes of neuronal nicotinic receptors, *Biochem. Biophys. Res. Commun.,* 162, 1085–1092, 1989.
36. Phillips, T.M., High-performance immunoaffinity chromatographic detection of immunoregulatory anti-idiotypic antibodies in cancer patients receiving immunotherapy, *Clin. Chem.,* 34, 1689–1692, 1988.
37. Phillips, T.M. and Frantz, S.C., Isolation of specific lymphocyte receptors by high-performance immunoaffinity chromatography, *J. Chromatogr.,* 444, 13–20, 1988.
38. Hamby, C.V., Chinol, M., Manzo, C., and Ferrone, S., Purification by affinity chromatography with anti-idiotypic monoclonal antibodies of immunoreactive monoclonal antibodies following labeling with 188Re, *Hybridoma,* 16, 27–31, 1997.
39. Hage, D.S., Immunoassays, *Anal. Chem.,* 65, 420R–424R, 1993.
40. Hage, D.S., Immunoassays, *Anal. Chem.,* 67, 455R–462R, 1995.
41. Hage, D.S., Immunoassays, *Anal. Chem.,* 69, 165R–171R, 1997.

42. Hage, D.S., Immunoassays, *Anal. Chem.,* 71, 294R–304R, 1999.

43. Berry, M.J. and Davies, J., Use of antibody fragments in immunoaffinity chromatography: Comparison of Fv fragments, Vh fragments and paralog peptides, *J. Chromatogr.,* 597, 239–245, 1992.

44. DeSilva, B.S. and Wilson, G.S., Synthesis of bifunctional antibodies for immunoassays, *Methods,* 22, 33–43, 2000.

45. Kriangkum, J., Xu, B., Nagata, L.P., Fulton, R.E., and Suresh, M.R., Bispecific and bifunctional single chain recombinant antibodies, *Biomol. Eng.,* 18, 31–40, 2001.

46. Weiss, E., Chatellier, J., and Orfanoudakis, G., *In vivo* biotinylated recombinant antibodies: Construction, characterization, and application of a bifunctional Fab-BCCP fusion protein produced in *Escherichia coli, Protein Exp. Purif.,* 5, 509–517, 1994.

47. Shibui, T. and Nagahari, K., Antibody produced by using *Escherichia coli* expression systems, *Bioprocess. Technol.,* 19, 253–268, 1994.

48. McGregor, D.P., Molloy, P.E., Cunningham, C., and Harris, W.J., Spontaneous assembly of bivalent single chain antibody fragments in *Escherichia coli, Mol. Immunol.,* 31, 219–226, 1994.

49. Kleymann, G., Ostermeier, C., Ludwig, B., Skerra, A., and Michel, H., Engineered Fv fragments as a tool for the one-step purification of integral multisubunit membrane protein complexes, *Biotechnology,* 13, 155–160, 1995.

50. Arnold-Schild, D., Kleist, C., Welschof, M., Opelz, G., Rammensee, H.G., Schild, H., and Terness, P., One-step single-chain Fv recombinant antibody-based purification of gp96 for vaccine development, *Cancer Res.,* 60, 4175–4178, 2000.

51. Blank, K., Lindner, P., Diefenbach, B., and Pluckthun, A., Self-immobilizing recombinant antibody fragments for immunoaffinity chromatography: Generic, parallel, and scalable protein purification, *Protein Exp. Purif.,* 24, 313–322, 2002.

52. Robbins, J.B., Haimovich, J., and Sela, M., Purification of antibodies with immunoadsorbents prepared using bromoacetyl cellulose, *Immunochemistry,* 4, 11–22, 1967.

53. Behrens, M.M., Inman, J.K., and Vannier, W.E., Protein-cellulose derivatives for use as immunoadsorbents: Preparation employing an active ester intermediate, *Arch. Biochem. Biophys.,* 119, 411–419, 1967.

54. Kisil, F.T., Centeno, E.R., Attallah, N.A., and Sehon, A.H., Demonstration of regain-allergen complexes formed on elution of regains from immunosorbent with allergens, *Int. Arch. Allergy Appl. Immunol.,* 42, 40–49, 1972.

55. Axen, R., Porath, J., and Ernback, S., Chemical coupling of peptides and proteins to polysaccharides by means of cyanogen halides, *Nature,* 214, 1302–1304, 1967.

56. Inman, J.K. and Dintzis, H.M., Derivatization of cross-linked polyacrylamide beads: Controlled introduction of functional groups for the preparation of special-purpose, biochemical adsorbents, *Biochemistry,* 8, 4074–4082, 1969.

57. Inman, J.K., Covalent linkage of functional groups, ligands, and proteins to polyacrylamide beads, *Methods Enzymol.,* 34, 30–58, 1974.

58. Weetall, H.H. and Hersh, L.S., Urease covalently coupled to porous glass, *Biochim. Biophys. Acta,* 185, 464–465, 1969.

59. Weetall, H.H., Preparation and characterization of antigen and antibody adsorbents covalently coupled to an inorganic carrier, *Biochem. J.,* 117, 257–261, 1970.

60. Weetall, H.H. and Filbert, A.M., Porous glass for affinity chromatography applications, *Methods Enzymol.,* 34, 59–72, 1974.

61. Avrameas, S. and Ternynck, T., Cross-linking of proteins with glutaraldehyde and its use for the preparation of immunoadsorbents, *Immunochemistry,* 6, 53–66, 1969.

62. Cuatrecasas, P., Protein purification by affinity chromatography: Derivatizations of agarose and polyacrylamide beads, *J. Biol. Chem.,* 245, 3059–3065, 1970.

63. Silman, I.H. and Katchalski, E., Water-insoluble derivatives of enzymes, antigens, and antibodies, *Ann. Rev. Biochem.,* 35, 873–908, 1966.

64. Scouten, W.H., *Affinity Chromatography,* Wiley, New York, 1981.

65. Mohr, P. and Pommering, K., *Affinity Chromatography,* Marcel Dekker, New York, 1985.

66. Hermanson, G.T., Mallia, A.K., and Smith, P.K., *Immobilized Affinity Ligand Techniques,* Academic Press, New York, 1992.

67. Clarke, W., Beckwith, J.D., Jackson, A., Reynolds, B., Karle, E.M., and Hage, D.S., Antibody immobilization to high-performance liquid chromatography supports: Characterization of maximum loading capacity for intact immunoglobulin G and Fab fragments, *J. Chromatogr. A,* 888, 13–22, 2000.

68. Jiang, T., Mallik, R., and Hage, D.S., Affinity monoliths for ultrafast immunoextraction, *Anal. Chem.,* in press.

69. Larsson, P.O., High-performance liquid affinity chromatography, *Methods Enzymol.,* 104, 212–223, 1984.

70. Kortt, A.A., Oddie, G.W., Iliades, P., Gruen, L.C., and Hudson, P.J., Nonspecific amine immobilization of ligand can be a potential source of error in BIAcore binding experiments and may reduce binding affinities, *Anal. Biochem.,* 253, 103–111, 1997.

71. Wilcheck, M. and Miron, T., Limitations of N-hydroxy-succinimide esters in affinity chromatography and protein immobilization, *Biochemistry,* 26, 2155–2161, 1987.

72. Phillips, T.M., Isolation and recovery of biologically active proteins by high-performance immunoaffinity chromatography, in *The Use of HPLC in Receptor Biochemistry,* Venter, C.J. and Harrison, L.C., Eds., Alan R. Liss, New York, 1998, pp. 129–154.

73. Wolfe, C.A.C. and Hage, D.S., Studies on the rate and control of antibody oxidation by periodate, *Anal. Biochem.,* 231, 123–130, 1995.

74. Ruhn, P.F., Garver, S., and Hage, D.S., Development of dihydrazide-activated silica supports for high-performance affinity chromatography, *J. Chromatogr. A,* 669, 9–19, 1994.

75. Hage, D.S., Wolfe, C.A.C., and Oates, M.R., Development of a kinetic model to describe the effective rate of antibody oxidation by periodate, *Bioconj. Chem.,* 8, 914–920, 1997.

76. Hage, D.S., Periodate oxidation of antibodies for site-selective immobilization in immunoaffinity chromatography, in *Affinity Chromatography: Methods and Protocols,* Bailon, P., Ehrlich, G.W., Fung, W.J., and Berthold, W., Eds., Humana Press, Totowa, NJ, 2000, chap. 7.

77. Bayer, E.A. and Wilchek, M., The use of the avidin-biotin complex as a tool in molecular biology, *Methods Biochem. Anal.,* 26, 1–45, 1980.

78. O'Shannessy, D.J. and Quarles, R.H., Labeling of the oligosaccharide moieties of immunoglobulins, *J. Immunol. Methods,* 99, 153–161, 1987.

79. Phillips, T.M., Queen, W.D., More, N.S., and Thompson, A.M., Protein A-coated glass beads: Universal support medium for high-performance immunoaffinity chromatography, *J. Chromatogr.,* 327, 213–219, 1985.

80. Schneider, C., Newman, R.A., Sutherland, D.R., Asser, U., and Greaves, M.F., A one-step purification of membrane proteins using a high efficiency immunomatrix, *J. Biol. Chem.,* 257, 10766–10769, 1982.

81. Sisson, T.H. and Castor, C.W., An improved method for immobilizing IgG antibodies on protein A-agarose, *Immunol. Methods,* 127, 215–220, 1990.

82. Litti, S., Matikainen, M.T., Scheinin, M., Glumoff, T., and Goldman, A., Immunoaffinity purification and reconstitution of human alpha(2)-adrenergic receptor subtype C2 into phospholipid vesicles, *Protein Exp. Purif.,* 22, 1–10, 2001.

83. Yarmush, M.L., Antonsen, K.P., Sundaram, S., and Yarmush, D.M., Immunoadsorption: Strategies for antigen elution and production of reusable adsorbents, *Biotechnol. Prog.,* 8, 168–178, 1992.

84. Ferreira, M.U. and Katzin, A.M., The assessment of antibody affinity distribution by thiocyanate elution: A simple dose-response approach, *J. Immunol. Methods,* 187, 297–305, 1995.

85. Yang, J., Moyana, T., and Xiang, J., Enzyme-linked immunosorbent assay-based selection and optimization of elution buffer for TAG72-affinity chromatography, *J. Chromatogr. B,* 731, 299–308, 1999.

86. Hage, D.S. and Walters, R.R., Dual-column determination of albumin and immunoglobulin G in serum by high-performance affinity chromatography, *J. Chromatogr.,* 386, 37–49, 1987.

87. Hage, D.S. and Kao, P.C., High-performance immunoaffinity chromatography and chemiluminescent detection in the automation of a parathyroid hormone sandwich immunoassay, *Anal. Chem.,* 63, 586–595, 1991.

88. Kim, H.O., Durance, T.D., and Li-Chan, E.C., Reusability of avidin-biotinylated immunoglobulin Y columns in immunoaffinity chromatography, *Anal. Biochem.,* 268, 383–97, 1999.

89. Guzman, N.A., Determination of immunoreactive gonadotropin-releasing hormone in serum and urine by on-line immunoaffinity capillary electrophoresis coupled to mass spectrometry, *J. Chromatogr. B,* 749, 197–213, 2000.

90. Shelver, W.L. and Smith, D.J., Immunoaffinity column as sample cleanup method for determination of the beta-adrenergic agonist ractopamine and its metabolites, *J. AOAC Int.,* 85, 1302–1307, 2002.

91. Phillips, T.M., Queen, W.D., More, N.S., and Thompson, A.M., *J. Chromatogr.,* 327, 213, 1985.

92. Ibarra, N., Caballero, A., Gonzalez, E., and Valdes, R., Comparison of different elution conditions for the immunopurification of recombinant hepatitis B surface antigen, *J. Chromatogr. B,* 735, 271–277, 1999.

93. Cong, J., Thompson, V.F., and Goll, D.E., Immunoaffinity purification of the calpains, *Protein Exp. Purif.,* 25, 283–290, 2002.

94. Bai, J.H., Wang, H.J., and Zhou, H.M., Alkaline-induced unfolding and salt-induced folding of pig heart lactate dehydrogenase under high pH conditions, *Int. J. Biol. Macromol.,* 23, 127–133, 1998.

95. Prajapati, S., Bhakuni, V., Babu, K.R., and Jain, S.K., Alkaline unfolding and salt-induced folding of bovine liver catalase at high pH, *Eur. J. Biochem.,* 255, 178–184, 1998.

96. Kristiansen, T., in *Affinity Chromatography,* Hoffman-Ostenhof, O., Breitbach, M., Koller, F., Kraft, D., and Scheiner, O., Eds., Pergamon, New York, 1978, p. 191.

97. Sica, V., Puca, G.A., Molinari, M., Buonaguro, F.M., and Bresciani, F., Effect of chemical perturbation with sodium thiocyanate on receptor-estradiol interaction: A new exchange assay at low temperature, *Biochemistry,* 19, 83–88, 1980.

98. Kannan, K., Lalitha, P., Rao, K.V., Narayanan, R.B., and Kaliraj, P., Optimisation of immunoaffinity purification of *Wuchereria bancrofti* specific antibodies from human sera, *Indian J. Exp. Biol.,* 35, 1076–1079, 1997.

99. Liitti, S., Matikainen, M.T., Scheinin, M., Glumoff, T., and Goldman, A., Immunoaffinity purification and reconstitution of human alpha(2)-adrenergic receptor subtype C2 into phospholipid vesicles, *Protein Exp. Purif.,* 22, 1–10, 2001.

100. Avrameas, S. and Ternynck, T., Biologically active water-insoluble protein polymers, I: Their use for the isolation of antigens and antibodies, *J. Biol. Chem.,* 242, 1651–1659, 1967.

101. Phillips, T.M., More, N.S., Queen, W.D., and Thompson, A.M., Isolation and quantitation of serum IgE levels by high-performance immunoaffinity chromatography, *J. Chromatogr.,* 327, 205–211, 1985.

102. Phillips, T.M., MacDonalds, J.S., and Lewis, M.G., in *Immune Complexes and Plasma Exchanges in Cancer Patients,* Serrou, B. and Rosenfeld, C., Eds., Elsevier/North-Holland, New York, 1982, p. 3.

103. Dandliker, W.B., Alonso, R., de Sausserre, V.A., Kierszenbaum, F., Leviaon, S.A., and Shapiro, H.C., The effect of chaotropic ions on the dissociation of antibody-antigen complexes, *Biochemistry,* 6, 1460–1467, 1967.

104. Nishi, S. and Hirai, H., Purification of human, dog and rabbit α-fetoprotein by immunoadsorbents of Sepharose coupled with anti-human α-fetoprotein, *Biochim. Biophys. Acta,* 278, 293–298, 1972.

105. Ikegawa, S., Itoh, M., Murao, N., Kijima, H., Suzuki, M., Fujiyama, T., and Goto, J., Immunoaffinity extraction for liquid chromatographic determination of equilin and its metabolites in plasma, *Biomed. Chromatogr.,* 10, 73–77, 1996.

106. Crabbe, P., Haasnoot, W., Kohen, F., Salden, M., and Van Peteghem, C., Production and characterization of polyclonal antibodies to sulfamethazine and their potential use in immunoaffinity chromatography for urine sample pre-treatment, *Analyst,* 124, 1569–1575, 1999.

107. Katoh, S., Terashima, M., and Miyaoku, K., Purification of alpha-amylase by specific elution from anti-peptide antibodies, *Appl. Microbiol. Biotechnol.,* 47, 521–524, 1997.

108. Powell, C.E., Watson, C.S., and Gametchu, B., Immunoaffinity isolation of native membrane glucocorticoid receptor from S-49++ lymphoma cells: Biochemical characterization and interaction with Hsp 70 and Hsp 90, *Endocrine,* 10, 271–280, 1999.

109. Parekh, B.S., Schwimmbeck, P.W., and Buchmeier, M.J., High efficiency immunoaffinity purification of anti-peptide antibodies on thiopropyl Sepharose immunoadsorbents, *Pept. Res.*, 2, 249–252, 1989.

110. Ehle, H. and Horn, A., Immunoaffinity chromatography of enzymes, *Bioseparation*, 1, 97–110, 1990.

111. Brassfield, A.L., Antigen purification by monoclonal antibody immunoaffinity chromatography, *Methods Mol. Biol.*, 45, 195–203, 1995.

112. Gokana, A., Winchenne, J.J., Ben-Ghanem, A., Ahaded, A., Cartron, J.P., and Lambin, P., Chromatographic separation of recombinant human erythropoietin isoforms, *J. Chromatogr. A*, 791, 109–118, 1997.

113. Banerjee, A., A monoclonal antibody to alpha-tubulin: Purification of functionally active alpha-tubulin isoforms, *Biochemistry*, 38, 5438–5446, 1999.

114. Kellogg, D.R. and Moazed, D., Protein- and immunoaffinity purification of multiprotein complexes, *Methods Enzymol.*, 351, 172–183, 2002.

115. Gretch, D.R., Suter, M., and Stinski, M.F., The use of biotinylated monoclonal antibodies and streptavidin affinity chromatography to isolate herpesvirus hydrophobic proteins or glycoproteins, *Anal Biochem.*, 163, 270–277, 1987.

116. Bovill, E.G., Landesman, M.M., Busch, S.A., Fregeau, G.R., Mann, K.G., and Tracy, R.P., Studies on the measurement of protein S in plasma, *Clin. Chem.*, 37, 1708–1714, 1991.

117. Battersby, J.E., Vanderlaan, M., and Jones, A.J., Purification and quantitation of tumor necrosis factor receptor immunoadhesin using a combination of immunoaffinity and reversed-phase chromatography, *J. Chromatogr. B*, 728, 21–33, 1999.

118. Muchova, L., Jirsa, M., Kuroki, M., Dudkova, L., Benes, M.J., Marecek, Z., and Smid, F., Immunoaffinity isolation of CEACAM1 on hydrazide-derivatized cellulose with immobilized monoclonal anti-CEA antibody, *Biomed. Chromatogr.*, 15, 418–422, 2001.

119. Dobers, J., Zimmermann-Kordmann, M., Leddermann, M., Schewe, T., Reutter, W., and Fan, H., Expression, purification, and characterization of human dipeptidyl peptidase IV/CD26 in Sf9 insect cells, *Protein Exp. Purif.*, 25, 527–532, 2002.

120. Tadey, T. and Purdy, W.C., Chromatographic techniques for the isolation and purification of lipoproteins, *J. Chromatogr. B*, 671, 237–253, 1995.

121. Fukuzono, S., Takeshita, T., Sakamoto, T., Hisada, A., Shimizu, N., and Mikoshiba, K., Overproduction and immuno-affinity purification of myelin proteolipid protein (PLP), an inositol hexakisphosphate-binding protein, in a baculovirus expression system, *Biochem. Biophys. Res. Commun.*, 249, 66–72, 1998.

122. Koch, S., Donarski, N., Goetze, K., Kreckel, M., Stuerenburg, H.J., Buhmann, C., and Beisiegel, U., Characterization of four lipoprotein classes in human cerebrospinal fluid, *J. Lipid Res.*, 42, 1143–1151, 2001.

123. Gilbert, J., Recent advances in analytical methods for mycotoxins, *Food Addit. Contam.*, 10, 37–48, 1993.

124. Scott, P.M. and Trucksess, M.W., Application of immunoaffinity columns to mycotoxin analysis, *J. AOAC Int.*, 80, 941–949, 1997.

125. Cathey, C.G., Huang, Z.G., Sarr, A.B., Clement, B.A., and Phillips, T.D., Development and evaluation of a minicolumn assay for the detection of aflatoxin M1 in milk, *J. Dairy Sci.*, 77, 1223–1231, 1994.

126. Visconti, A., Pascale, M., and Centonze, G., Determination of ochratoxin A in wine by means of immunoaffinity column clean-up and high-performance liquid chromatography, *J. Chromatogr. A*, 864, 89–101, 1999.

127. Tsutsumi, T., Nagata, S., Hasegawa, A., and Ueno, Y., Immunoaffinity column as clean-up tool for determination of trace amounts of microcystins in tap water, *Food Chem. Toxicol.*, 38, 593–597, 2000.

128. Brooks, C.L., Isaacs, L.A., and Wicks, J.R., Preparative purification of phosphorylated and nonphosphorylated bovine prolactins, *Mol. Cell Endocrinol.*, 99, 301–305, 1994.

129. Berghman, L.R., Lescroart, O., Roelants, I., Ollevier, F., Kuhn, E.R., Verhaert, P.D., De Loof, A., van Leuven, F., and Vandesande, F., One-step immunoaffinity purification and

partial characterization of hypophyseal growth hormone from the African catfish, *Clarias gariepinus* (Burchell), *Comp. Biochem. Physiol. B Biochem. Mol. Biol.,* 113, 773–780, 1996.

130. Cho, B.Y., Zou, H., Strong, R., Fisher, D.H., Nappier, J., and Krull, I.S., Immunochromatographic analysis of bovine growth hormone releasing factor involving reversed-phase high-performance liquid chromatography-immunodetection, *J. Chromatogr. A,* 743, 181–194, 1996.

131. Gokana, A., Winchenne, J.J., Ben-Ghanem, A., Ahaded, A., Cartron, J.P., and Lambin, P., Chromatographic separation of recombinant human erythropoietin isoforms, *J. Chromatogr. A,* 791, 109–118, 1997.

132. Rolcik, J., Lenobel, R., Siglerova, V., and Strnad, M., Isolation of melatonin by immunoaffinity chromatography, *J. Chromatogr. B,* 775, 9–15, 2002.

133. Novick, D., Eshhar, Z., Fischer, D.G., Friedlander, J., and Rubinstein, M., Monoclonal antibodies to human interferon-gamma: Production, affinity purification and radioimmunoassay, *EMBO J.,* 2, 1527–1530, 1983.

134. Phillips, T.M., Measurement of recombinant interferon levels by high-performance immunoaffinity chromatography in body fluids of cancer patients on interferon therapy, *Biomed. Chromatogr.,* 6, 287–290, 1992.

135. Grether, J.K., Nelson, K.B., Dambrosia, J.M., and Phillips, T.M., Interferons and cerebral palsy, *J. Pediatr.,* 134, 324–332, 1999.

136. Phillips, T.M., Measurement of total and bioactive interleukin-2 in tissue samples by immunoaffinity-receptor affinity chromatography, *Biomed. Chromatogr.,* 11, 200–204, 1997.

137. Muneta, Y., Shimoji, Y., Yokomizo, Y., and Mori, Y., Production of monoclonal antibodies to porcine interleukin-18 and their use for immunoaffinity purification of recombinant porcine interleukin-18, *J. Immunol. Methods,* 236, 99–104, 2000.

138. Hikono, H., Zhou, J.H., Ohta, M., Inumaru, S., Momotani, E., and Sakurai, M., Production of a monoclonal antibody that recognizes bovine stem cell factor (SCF) and its use in the detection and quantitation of native soluble bovine SCF in fetal bovine serum, *J. Interferon Cytokine Res.,* 22, 231–235, 2002.

139. Seppen, J., Jansen, P.L., and Oude Elferink, R.P., Immunoaffinity purification and reconstitution of the human bilirubin/phenol UDP-glucuronosyltransferase family, *Protein Exp. Purif.,* 6, 149–154, 1995.

140. Yoshimura, K., Ishikawa, T., Wada, K., Takeda, T., Kamata, Y., Tada, T., Nishimura, K., Nakano, Y., and Shigeoka, S., Characterization of monoclonal antibodies against ascorbate peroxidase isoenzymes: Purification and epitope-mapping using immunoaffinity column chromatography, *Biochim. Biophys. Acta,* 1526, 168–174, 2001.

141. Kobayashi, N., Mano, H., Imazu, T., and Shimada, K., Tandem immunoaffinity chromatography for plasma 1α,25-dihydroxyvitamin D3 utilizing two antibodies having different specificities: A novel and powerful pretreatment tool for 1α,25-dihydroxyvitamin D3 radioreceptor assays, *J. Steroid Biochem. Mol. Biol.,* 54, 217–226, 1995.

142. Kobayashi, N., Imazu, T., Kitahori, J., Mano, H., and Shimada, K., A selective immunoaffinity chromatography for determination of plasma 1α,25-dihydroxyvitamin D3: Application of specific antibodies raised against a 1α,25-dihydroxyvitamin D3-bovine serum albumin conjugate linked through the 11α-position, *Anal. Biochem.,* 244, 374–383, 1997.

143. Ueda, N. and Yamamoto, S., Immunoaffinity purification of arachidonate 5-lipoxygenase from porcine leukocytes, *Methods Enzymol.,* 187, 338–343, 1990.

144. Matsumoto, S., Hayashi, Y., Kinoshita, I., Ikata, T., and Yamamoto, S., Immunoaffinity purification of prostaglandin E2 and leukotriene C4 prior to radioimmunoassay: Application to human synovial fluid, *Ann. Clin. Biochem.,* 30, 60–68, 1993.

145. Westcott, J.Y., Maxey, K.M., MacDonald, J., and Wenzel, S.E., Immunoaffinity resin for purification of urinary leukotriene E4, *Prostaglandins Other Lipid Mediat.,* 55, 301–321, 1998.

146. King, M.M., Cuzick, J., Jenkins, D., Routledge, M.N., and Garner, R.C., Immunoaffinity concentration of human lung DNA adducts using an anti-benzo[a]pyrene-diol-epoxide-DNA antibody: Analysis by 32P-postlabelling or ELISA, *Mutation Res.,* 292, 113–122, 1993.

147. Weston, A., Bowman, E.D., Carr, P., Rothman, N., and Strickland, P.T., Detection of metabolites of polycyclic aromatic hydrocarbons in human urine, *Carcinogenesis,* 14, 1053–1055,
1993.
148. Dalluge, J., Hankemeier, T., Vreuls, R.J., and Brinkman, U.A., On-line coupling of
immunoaffinity-based solid-phase extraction and gas chromatography for the determination
of s-triazines in aqueous samples, *J. Chromatogr. A,* 830, 377–386, 1999.
149. Cichna, M., Markl, P., Knopp, D., and Niessner, R., On-line coupling of sol-gel-generated
immunoaffinity columns with high-performance liquid chromatography, *J. Chromatogr. A,*
919, 51–58, 2001.
150. Sheedy, C. and Hall, J.C., Immunoaffinity purification of chlorimuron-ethyl from soil
extracts prior to quantitation by enzyme-linked immunosorbent assay, *J. Agric. Food Chem.,*
49, 1151–1157, 2001.
151. Theodoridis, G., Haasnoot, W., Cazemier, G., Schilt, R., Jaziri, M., Diallo, B., Papadoyannis,
I.N., and de Jong, G.J., Immunoaffinity chromatography for the sample pretreatment of
Taxus plant and cell extracts prior to analysis of taxanes by high-performance liquid
chromatography, *J. Chromatogr. A,* 948, 177–185, 2002.
152. Zhao, M., Liu, Y., Li, Y., Zhang, X., and Chang, W., Development and characterization of
an immunoaffinity column for the selective extraction of bisphenol A from serum samples,
J. Chromatogr. B, 783, 401–410, 2003.
153. Van Emon, J.M., Gerlach, C.L., and Bowman, K., Bioseparation and bioanalytical techniques
in environmental monitoring, *J. Chromatogr. B,* 715, 211–228, 1998.
154. Van Emon, J.M., Immunochemical applications in environmental science, *J. AOAC Int.,* 84,
125–133, 2001.
155. Stalikas, C., Knopp, D., and Niessner, R., Sol-gel glass immunosorbent-based determination
of s-triazines in water and soil samples using gas chromatography with a nitrogen phosphorus
detection system, *Environ. Sci. Technol.,* 36, 3372–3377, 2002.
156. Katz, S.E. and Siewierski, M., Drug residue analysis using immunoaffinity chromatography,
J. Chromatogr., 624, 403–409, 1992.
157. Dietrich, R., Usleber, E., and Martlbauer, E., The potential of monoclonal antibodies against
ampicillin for the preparation of a multi-immunoaffinity chromatography for penicillins,
Analyst, 123, 2749–2754, 1998.
158. Rashid, B.A., Kwasowski, P., and Stevenson, D., Solid-phase extraction of clenbuterol from
plasma using immunoaffinity followed by HPLC, *J. Pharm. Biomed. Anal.,* 21, 635–639,
1999.
159. Crabbe, P., Haasnoot, W., Kohen, F., Salden, M., and Van Peteghem, C., Production and
characterization of polyclonal antibodies to sulfamethazine and their potential use in immunoaffinity chromatography for urine sample pre-treatment, *Analyst,* 124, 1569–1575, 1999.
160. Wu, Z., Li, J., Zhu, L., Luo, H., and Xu, X., Multi-residue analysis of avermectins in swine
liver by immunoaffinity extraction and liquid chromatography-mass spectrometry, *J. Chromatogr. B,* 755, 361–366, 2001.
161. Shelver, W.L. and Smith, D.J., Immunoaffinity column as sample cleanup method for
determination of the beta-adrenergic agonist ractopamine and its metabolites, *J. AOAC Int.,*
85, 1302–1307, 2002.
162. Au, A.M. and Varon, S., Neural cell sequestration on immunoaffinity columns, *Exp. Cell
Res.,* 120, 269–276, 1979.
163. Kataoka, K., Sakurai, Y., Hanai, T., Maruyama, A., and Tsuruta, T., Immunoaffinity chromatography of lymphocyte subpopulations using tert-amine derived matrices with adsorbed
antibodies, *Biomaterials,* 9, 218–224, 1988.
164. Watts, M.J., Sullivan, A.M., Ings, S.J., Leverett, D., Peniket, A.J., Perry, A.R., Williams, C.D.,
Devereux, S., Goldstone, A.H., and Linch, D.C., Evaluation of clinical scale CD34+ cell
purification: Experience of 71 immunoaffinity column procedures, *Bone Marrow Transplant,*
20, 157–162, 1997.
165. Phillips, T.M., Isolation and measurement of membrane proteins by high-performance
immunoaffinity chromatography, in *High-Performance Liquid Chromatography of Peptides*

and Proteins: Separation, Analysis, and Conformation, Mant, C.T. and Hodges, R.S., Eds., CRC Press, Boca Raton, FL, 1991, pp. 517–527.

166. Chinni, C., Bottomley, S.P., Duffy, E.J., Hemmings, B.A., and Stone, S.R., Expression and purification of the human thrombin receptor, *Protein Exp. Purif.,* 13, 9–15, 1998.

167. Powell, C.E., Watson, C.S., and Gametchu, B., Immunoaffinity isolation of native membrane glucocorticoid receptor from S-49++ lymphoma cells: Biochemical characterization and interaction with Hsp 70 and Hsp 90, *Endocrine,* 10, 271–280, 1999.

168. Eckard, C.P. and Beck-Sickinger, A.G., Characterisation of G-protein-coupled receptors by antibodies, *Curr. Med. Chem.,* 7, 897–910, 2000.

169. Liitti, S., Matikainen, M.T., Scheinin, M., Glumoff, T., and Goldman, A., Immunoaffinity purification and reconstitution of human alpha(2)-adrenergic receptor subtype C2 into phospholipid vesicles, *Protein Exp. Purif.,* 22, 1–10, 2001.

170. Nakai, Y., Lancaster, W.D., and Jenson, A.B., Purification of papillomavirus structural polypeptides from papillomas by immunoaffinity chromatography, *J. Gen. Virol.,* 68, 1891–1896, 1987.

171. Potts, W.M. and Vogt, V.M., A simple method for immunoaffinity purification of nondenatured avian sarcoma and leukemia virus gag-containing proteins, *Virology,* 160, 494–497, 1987.

172. Santucci, A., Rustici, M., Bracci, L., Lozzi, L., Soldani, P., and Neri, P., HPLC immunoaffinity purification of rabies virus glycoprotein using immobilized antipeptide antibodies, *J. Immunol. Methods,* 127, 131–138, 1990.

173. Cameron-Smith, R. and Harbour, C., Removal of poliovirus type 1 from a protein mixture using an immunoaffinity chromatography column, *Biomed. Chromatogr.,* 15, 471–483, 2001.

174. White, A.R., Dutton, N.S., and Cook, R.D., Isolation and growth of a cytopathic agent from multiple sclerosis brain tissue, *J. Neurovirol.,* 8, 111–121, 2002.

175. Shinagawa, K., Mitsumori, M., Matsusaka, N., and Sugii, S., Purification of staphylococcal enterotoxins A and E by immunoaffinity chromatography using a murine monoclonal antibody with dual specificity for both of these toxins, *J. Immunol. Methods,* 139, 49–53, 1991.

176. Naka, A., Honda, T., and Miwatani, T., A simple purification method of *Vibrio cholerae* non-O1 hemagglutinin/protease by immunoaffinity column chromatography using a monoclonal antibody, *Microbiol. Immunol.,* 36, 419–423, 1992.

177. Ma, J. and Inzana, T.J., Rapid purification of a 110-kilodalton hemolysin of *Actinobacillus pleuropneumoniae* by monoclonal antibody-affinity chromatography, *Am. J. Vet. Res.,* 53, 59–62, 1992.

178. Shinagawa, K., Takechi, T., Matsusaka, N., and Sugii, S., Purification of an enterotoxin produced by *Bacillus cereus* by immunoaffinity chromatography using a monoclonal antibody, *Can. J. Microbiol.,* 38, 153–156, 1992.

179. Uratani, Y., Immunoaffinity purification and reconstitution of sodium-coupled branched-chain amino acid carrier of *Pseudomonas aeruginosa, J. Biol. Chem.,* 267, 5177–5183, 1992.

180. Matar, G.M., Bibb, W.F., Helsel, L., Dewitt, W., and Swaminathan, B., Immunoaffinity purification, stabilization and comparative characterization of listeriolysin O from *Listeria monocytogenes* serotypes 1/2a and 4b, *Res. Microbiol.,* 143, 489–498, 1992.

181. McVicker, J.K. and Tabatabai, L.B., Isolation of immunogenic outer membrane proteins from *Mannheimia haemolytica* serotype 1 by use of selective extraction and immunoaffinity chromatography, *Am. J. Vet. Res.,* 63, 1634–1640, 2002.

182. van Ginkel, L.A., Immunoaffinity chromatography, its applicability and limitations in multiresidue analysis of anabolizing and doping agents, *J. Chromatogr.,* 564, 363–384, 1991.

183. Haagsma, N. and van de Water, C., in *Analysis of Antibiotic Drug Residues in Food Products of Animal Origin,* Agarwal, V.K., Ed., Plenum Press, New York, 1992, p. 81.

184. Katz, S.E. and Siewierski, M., Drug residue analysis using immunoaffinity chromatography, *J. Chromatogr.,* 624, 403–409, 1992.

185. van Ginkel, L.A., van Blitterswijk, H., Zoontjes, P.W., van den Bosch, D., and Stephany, R.W., Assay for trenbolone and its metabolite 17a-trenbolone in bovine urine based on immunoaffinity chromatographic clean-up and off-line high-performance liquid chromatography-thin-layer chromatography, *J. Chromatogr.,* 445, 385–392, 1988.

186. van Ginkel, L.A., Stephany, R.W., van Rossum, H.J., van Blitterswijk, H., Zoontjes, P.W., Hooijshuur, R.C.M., and Zuydendorp, J., Effective monitoring of residues of nortestosterone and its major metabolite in bovine urine and bile, *J. Chromatogr.,* 489, 95–104, 1989.

187. Bagnati, R., Castelli, M.G., Airoldi, L., Oriundi, M.P., Ubaldi, A., and Fanelli, R., Analysis of diethylstilbestrol, dienestrol and hexestrol in biological samples by immunoaffinity extraction and gas chromatography negative-ion chemical ionization mass spectrometry, *J. Chromatogr.,* 527, 267–278, 1990.

188. de Frutos, M. and Regnier, F.E., Tandem chromatographic-immunological analyses, *Anal. Chem.,* 65, 17A–25A, 1993.

189. Thomas, D.H., Beck-Westermeyer, M., and Hage, D.S., Determination of atrazine in water using tandem high-performance immunoaffinity chromatography and reversed-phase liquid chromatography, *Anal. Chem.,* 66, 3823–3829, 1994.

190. Flurer, C.L. and Novotny, M., Dual microcolumn immunoaffinity liquid chromatography: An analytical application to human plasma proteins, *Anal. Chem.,* 65, 817–821, 1993.

191. Yvon, M. and Wal, J.M., Tandem immunoaffinity and reversed-phase high-performance liquid chromatography for the identification of the specific binding sites of a hapten on a proteic carrier, *J. Chromatogr.,* 539, 363–371, 1991.

192. Farjam, A., van de Merbel, N.C., Nieman, A.A., Lingeman, H., and Brinkman, U.A.T., Determination of aflatoxin M1 using a dialysis-based immunoaffinity sample pretreatment system coupled online to liquid chromatography: Reusable immunoaffinity columns, *J. Chromatogr.,* 589, 141–149, 1992.

193. Janis, L.J. and Regnier, F.E., Immunological-chromatographic analysis, *J. Chromatogr.,* 444, 1–11, 1988.

194. Janis, L.J., Grott, A., Regnier, F.E., and Smith-Gill, S.J., Immunological-chromatographic analysis of lysozyme variants, *J. Chromatogr.,* 476, 235–244, 1989.

195. Riggin, A., Sportsman, J.R., and Regnier, F.E., Immunochromatographic analysis of proteins: Identification, characterization and purity determination, *J. Chromatogr.,* 632, 37–44, 1993.

196. Farjam, A., Vreuls, J.J., Cuppen, W.J.G.M., Brinkman, U.A.T., and de Jong, G.J., Direct introduction of large-volume urine samples into an on-line immunoaffinity sample pretreatment-capillary gas chromatography system, *Anal. Chem.,* 63, 2481–2487, 1991.

197. McConnell, J.P. and Anderson, D.J., Determination of fibrinogen in plasma by high-performance immunoaffinity chromatography, *J. Chromatogr.,* 615, 67–75, 1993.

198. Ruhn, P.F., Taylor, J.D., and Hage, D.S., Determination of urinary albumin using high-performance immunoaffinity chromatography and flow injection analysis, *Anal. Chem.,* 66, 4265–4271, 1994.

199. Phillips, T.M., Measurement of total and bioactive interleukin-2 in tissue samples by immunoaffinity-receptor affinity chromatography, *Biomed. Chromatogr.,* 11, 200–204, 1997.

200. Ohlson, S., Gudmundsson, B.M., Wikstrom, P., and Larsson, P.O., High-performance liquid affinity chromatography: Rapid immunoanalysis of transferrin in serum, *Clin. Chem.,* 34, 2039–2043, 1988.

201. Hage, D.S., Rollag, J.G., and Thomas, D.H., in *Immunochemical Technology for Environmental Applications,* Aga, D.S. and Thurman, E.M., Eds., ACS Press, Washington, DC, 1997, chap. 10.

202. Rollag, J.G., Beck-Westermeyer, M., and Hage, D.S., Analysis of pesticide degradation products by tandem high-performance immunoaffinity chromatography and reversed-phase liquid chromatography, *Anal. Chem.,* 68, 3631–3637, 1996.

203. De Ruiter, G.A., Smid, P., Schols, H.A., Van Boom, J.H., and Rombouts, F.M., Detection of fungal carbohydrate antigens by high-performance immunoaffinity chromatography using a protein A column with covalently linked immunoglobulin G, *J. Chromatogr.,* 584, 69–75, 1992.

204. Wang, W.T., Kumlien, J., Ohlson, S., Lundblad, A., and Zopf, D., Analysis of a glucose-containing tetrasaccharide by high-performance liquid affinity chromatography, *Anal. Biochem.,* 182, 48–53, 1989.

205. Zopf, D., Ohlson, S., Dakour, J., Wang, W., and Lundblad, A., Analysis and purification of oligosaccharides by high-performance liquid affinity chromatography, *Methods Enzymol.,* 179, 55–64, 1989.

206. Dakour, J., Lundblad, A., and Zopf, D., Separation of blood group A-active oligosaccharides by high-pressure liquid affinity chromatography using a monoclonal antibody bound to concanavalin A silica, *Anal. Biochem.,* 161, 140–143, 1987.

207. Hage, D.S. and Nelson, M.A., Chromatographic immunoassays, *Anal. Chem.,* 73, 198A–205A, 2001.

208. Mattiasson, B., Nilsson, M., Berden, P., and Hakanson, H., Flow-ELISA: binding assays for process control, *Trends Anal. Chem.,* 9, 317–321, 1990.

209. Gubitz, G. and Shellum, C., Flow-injection immunoassays, *Anal. Chim. Acta,* 283, 421–428, 1993.

210. Pollema, C.H., Ruzicka, J., Lernmark, A., and Christian, G.D., Flow-injection immunoassays: Present and future, *Microchem. J.,* 45, 121–128, 1992.

211. Nelson, M.A., Reiter, W.S., and Hage, D.S., Chromatographic competitive binding immunoassays: A comparison of the sequential and simultaneous injection methods, *Biomed. Chromatogr.,* 17, 188–200, 2003.

212. Locascio-Brown, L., Plant, A.L., Chesler, R., Kroll, M., Ruddel, M., and Durst, R.A., Liposome-based flow-injection immunoassay for determining theophylline in serum, *Clin. Chem.,* 39, 386–391, 1993.

213. De Alwis, U. and Wilson, G.S., Rapid heterogeneous competitive electrochemical immunoassay for IgG in the picomole range, *Anal. Chem.,* 59, 2786–2789, 1987.

214. Palmer, D.A., Xuezhen, R., Fernandez-Hernando, P., and Miller, J.N., A model on-line flow injection fluorescence immunoassay using a protein A immunoreactor and Lucifer yellow, *Anal. Lett.,* 26, 2543–2553, 1993.

215. Hage, D.S., Thomas, D.H., and Beck, M.S., Theory of a sequential-addition competitive binding immunoassay based on high-performance immunoaffinity chromatography, *Anal. Chem.,* 65, 1622–1630, 1993.

216. Cassidy, S.A., Janis, L.J., and Regnier, F.E., Kinetic chromatographic sequential-addition immunoassays using protein A affinity chromatography, *Anal. Chem.,* 64, 1973–1977, 1992.

217. Kusterbeck, A.W., Wemhoff, G.A., Charles, P.T., Yeager, D.A., Bredehorst, R., Vogel, C.W., and Ligler, F.S., A continuous flow immunoassay for rapid and sensitive detection of small molecules, *J. Immunol. Methods,* 135, 191–197, 1990.

218. Whelan, J.P., Kusterbeck, A.W., Wemhoff, G.A., Bredehorst, R., and Ligler, F.S., Continuous-flow immunosensor for detection of explosives, *Anal. Chem.,* 65, 3561–3565, 1993.

219. Kaptein, W.A., Zwaagstra, J.J., Venema, K., Ruiters, M.H.J., and Korf, J., Analysis of cortisol with a flow displacement immunoassay, *Sens. Actuators B,* B45, 63–69, 1997.

220. De Alwis, W.U. and Wilson, G.S., Rapid sub-picomole electrochemical enzyme immunoassay for immunoglobulin G, *Anal. Chem.,* 57, 2754–2756, 1985.

221. Hacker, A., Hinterleitner, M., Shellum, C., and Gubitz, G., Development of an automated flow injection chemiluminescence immunoassay for human immunoglobulin G, *Frenz. J. Anal. Chem.,* 352, 793–796, 1995.

222. Hage, D.S., Taylor, B., and Kao, P.C., Intact parathyroid hormone: Performance and clinical utility of an automated assay based on high-performance immunoaffinity chromatography and chemiluminescence detection, *Clin. Chem.,* 38, 1494–1500, 1992.

223. Shellum, C. and Gubitz, G., Flow-injection immunoassays with acridinium ester-based chemiluminescence detection, *Anal. Chim. Acta,* 227, 97–107, 1989.

224. Gunaratna, P.C. and Wilson, G.S., Noncompetitive flow injection immunoassay for a hapten, α-(difluoromethyl)ornithine, *Anal. Chem.,* 65, 1152–1157, 1993.

225. Locascio-Brown, L. and Choquette, S.J., Measuring estrogens using flow injection immunoanalysis with liposome amplification, *Talanta,* 40, 1899–1904, 1993.

226. Oates, M.R., Clarke, W., Zimlich, A., II, and Hage, D.S., Optimization and development of a high-performance liquid chromatography-based one-site immunometric assay with chemiluminescence detection, *Anal. Chim. Acta,* 470, 37–50, 2002.

227. Irth, H., Oosterkamp, A.J., Tjaden, U.R., and van der Greef, J., Strategies for online coupling of immunoassays to HPLC, *Trends Anal. Chem.*, 14, 355–361, 1995.

228. Vanderlaan, M., Lotti, R., Siek, G., King, D., and Goldstein, M., Perfusion immunoassay for acetylcholinesterase: Analyte detection based on intrinsic activity, *J. Chromatogr. A*, 711, 23–31, 1995.

229. Cho, B.Y., Zou, H., Strong, R., Fisher, D.H., Nappier, J., and Krull, I.S., Immunochromatographic analysis of bovine growth hormone releasing factor involving reversed-phase high-performance liquid chromatography-immunodetection, *J. Chromatogr. A*, 743, 181–194, 1996.

230. Irth, H., Oosterkamp, A.J., van der Welle, W., Tjaden, U.R., and van der Greef, J., Online immunochemical detection in liquid chromatography using fluorescein-labeled antibodies, *J. Chromatogr.*, 633, 65–72, 1993.

231. Oosterkamp, A.J., Irth, H., Beth, M., Unger, K.K., Tjaden, U.R., and van der Greef, J., Bioanalysis of digoxin and its metabolites using direct serum injection combined with liquid chromatography and online immunochemical detection, *J. Chromatogr. B*, 653, 55–61, 1994.

232. Miller, K.J. and Herman, A.C., Affinity chromatography with immunochemical detection applied to the analysis of human methionyl granulocyte colony-stimulating factor in serum, *Anal. Chem.*, 68, 3077–3082, 1996.

233. Clarke, W. and Hage, D.S., Development of sandwich HPLC microcolumns for analyte adsorption on the millisecond time scale, *Anal. Chem.*, 73, 2157–2164, 2001.

234. Clarke, W., Chowdhuri, A.R., and Hage, D.S., Analysis of free drug fractions by ultrafast immunoaffinity chromatography, *Anal. Chem.*, 73, 2157–2164, 2001.

235. Clarke, W., Schiel, J.E., Moser, A., and Hage, D.S., Analysis of free hormone fractions by an ultrafast immunoextraction/displacement immunoassay: studies using free thyroxine as a model system, *Anal. Chem.*, 77, 1859–1866, 2005.

236. Chen, G. and Ewing, A.G., Chemical analysis of single cells and exocytosis, *Crit. Rev. Neurobiol.*, 11, 59–90, 1997.

237. Kricka, L.J., Multianalyte testing, *Clin. Chem.*, 38, 327–328, 1992.

238. Steffen, M.J. and Ebersole, J.L., Sequential ELISA for cytokine levels in limited volumes of biological fluids, *BioTechniques*, 21, 504–509, 1996.

239. Phillips, T.M. and Krum, J.M., Recycling immunoaffinity chromatography for multiple analyte analysis in biological samples, *J. Chromatogr. B*, 715, 55–63, 1998.

240. Grether, J.K., Nelson, K.B., Dambrosia, J.M., and Phillips, T.M., Interferons and cerebral palsy, *J. Pediatr.*, 134, 324–332, 1999.

241. Nelson, K.B., Dambrosia, J.M., Grether, J.K., and Phillips, T.M., Neonatal cytokines and coagulation factors in children with cerebral palsy, *Ann. Neurology*, 44, 665–675, 1998.

242. Phillips, T.M. and Smith, P.D., Immunoaffinity analysis of substance P in complex biological fluids: Analysis of sub-microliter samples, *J. Liq. Chromatogr. Rel. Technol.*, 25, 2889–2900, 2002.

243. Regnier, F.E., He, B., Lin, S., and Busse, J., Chromatography and electrophoresis on chips: Critical elements of future integrated, microfluidic analytical systems for life science, *Trends Biotechnol.*, 17, 101–106, 1999.

7

DNA Affinity Chromatography

Robert A. Moxley, Shilpa Oak, Himanshu Gadgil, and Harry W. Jarrett
Department of Biochemistry, University of Tennessee, Memphis, TN

CONTENTS

7.1   INTRODUCTION

DNA affinity chromatography is a technique used to purify DNA-binding proteins [1–10]. Usually, DNA coupled to a suitable solid matrix is used for this purpose. Elution of the retained proteins can later be achieved through one of many methods, including the use of a salt or heparin, oligonucleotide competition, ligand-specific competition, or a change in temperature. This chapter discusses the various types of DNA affinity chromatography and examines some of its applications.

7.1.1   Types of DNA Affinity Chromatography

All types of DNA affinity chromatography can be categorized into one of two classes: nonspecific DNA affinity chromatography or sequence-specific DNA affinity chromatography. Columns for nonspecific DNA affinity chromatography are made by coupling fragmented nuclear DNA (e.g., salmon sperm DNA or calf thymus DNA) to a solid support [3]. The ability to couple DNA to cellulose, as first reported by Alberts et al. [4, 5], was an important advance in this field. The use of such a support is simple: a DNA-cellulose matrix interacts with DNA-binding proteins, while the binding of non-DNA-related proteins is avoided under appropriate application conditions. DNA-binding proteins are then later eluted in an active form from this material. DNA-cellulose can be made with single-stranded or double-stranded DNA by adsorbing either native (double-stranded) DNA or thermally denatured (single-stranded) DNA to cellulose.

Sequence-specific DNA affinity columns are made by coupling synthetic oligonucleotides [9], plasmids, or concatemers (i.e., tandem repeats of a sequence) [10] to a solid support. In this case, the DNA sequence used in the column is chosen to contain an element consensus sequence, restriction site, or other DNA motif that is recognized by the protein of interest. The support that is used most often for this purpose is Sepharose [10]. However, silica, Teflon, latex, and other materials have also been employed.

7.1.2   General Applications of DNA Affinity Chromatography

One common application of DNA affinity chromatography is its use in the purification of DNA-binding proteins. This application is becoming less empirical as past work in this field is used to select rational approaches for the purification and characterization of newly discovered proteins. In addition, new and more powerful approaches are being developed at a rapid pace in this area.

Dissecting the DNA repair apparatus is another area where DNA affinity chromatography has been important. An understanding of DNA repair is important in work with many genetic diseases and cancer, and in understanding the roles of nutrition and environment in pathology. As shown by the literature in this field, there are many similarities in the types of chromatography that are used when purifying DNA-binding proteins and the proteins involved in DNA repair. For instance, sequence-specific DNA columns and adjunct methods like heparin-Sepharose or phosphocellulose columns, are commonly used for both DNA-binding proteins such as transcription factors and for repair proteins. Thus, the use of DNA affinity chromatography for the purification of DNA-repair proteins will be another topic covered in this chapter.

DNA affinity chromatography has also played an important role in purifying and characterizing the components involved in the cell machinery for duplicating and transcribing DNA. This chapter discusses the role that this method has played in purifying primases, helicases, polymerases, and other proteins. Furthermore, there will be a discussion of which procedures work best for each of these various classes of proteins.

Several recent advances in DNA affinity chromatography will also be discussed. One such advance is the method of catalytic chromatography, which is an approach that has been developed for purifying enzymes that catalyze reactions involving DNA [2]. This is a powerful technique for improving both the purity and yield of such enzymes and provides a valuable alternative to the salt elution schemes that are often used with such substances. Another area that will be discussed is the use of aptamers as affinity ligands. These are RNA or DNA sequences that have been selected to bind specific proteins, nucleotides, or small molecules. This is of interest, since it allows the extension of DNA- and RNA-based affinity chromatography to non-DNA-binding proteins and other target solutes.

7.2 HETEROGENEOUS DNA AFFINITY CHROMATOGRAPHY

The first technique to be discussed in this chapter is heterogeneous DNA affinity chromatography. This section examines the applications of this technique, along with the stationary phases, supports, mobile phases, and elution methods that are used in this approach.

7.2.1 Stationary Phases and Supports

Heterogeneous DNA affinity columns are made by coupling fragmented nuclear DNA, such as salmon sperm DNA or calf thymus DNA, to a solid support [3]. Cellulose is the material most often used for this purpose. As mentioned earlier, the use of this support with DNA ligands was originally developed by Alberts and Herrick [4] and further characterized by Alberts et al. [5]. After it has been prepared, the DNA-cellulose matrix is then used to isolate DNA-binding proteins and later elute them in an active form.

This is a popular method for preparing DNA-binding proteins due to the ease of column preparation. In this technique, the DNA to be used as a ligand is extracted from cells, fractionated, and fragmented. It is then mixed with cellulose and thoroughly dried. After reconstitution in a buffer and washing of the support, about 50% of the added DNA will have adsorbed to the cellulose. Since this DNA is adsorbed rather than covalently immobilized, it does slowly release over time. However, the typical lifetime of this adsorbed ligand is fairly long, with a half-life of about 400 h. DNA-cellulose can be prepared using either double-stranded (native) DNA or single-stranded (thermally denatured) DNA. It is also possible to covalently attach DNA to agarose [6], which gives a material that has a more stable ligand.

Due to the diverse population of DNA sequences used in these columns, they are selective for solutes that can bind to DNA but that are not specific for a given DNA sequence. It is for this reason that these columns are referred to as either *heterogeneous* or *nonspecific DNA affinity columns.* This lack of sequence specificity can be an advantage in the purification of proteins like histones, DNA polymerase, and helicase, which show little preference for a specific DNA sequence. For instance, such columns have been used to identify nuclear proteins that bind to genomic DNA [7]. Although nonspecific DNA affinity columns are limited in their selectivity for a particular DNA-binding protein, they can still be used to give a severalfold increase in purity for the protein of interest [4].

A problem with nonspecific DNA affinity chromatography is that this method alone rarely yields a homogeneous protein preparation. Instead, this approach is often used as the last step in combination with other methods, such as phosphocellulose and heparin-agarose chromatography [8]. These additional purification steps not only enrich the protein of interest, but also minimize the presence of endogenous DNA, which can interfere with the DNA affinity chromatography. Endogenous DNA can also be removed by treatment with DNase or through precipitation using streptomycin or polyethylene glycol [4] if DNA affinity chromatography is used as the first step in the purification process.

7.2.2 Mobile Phases and Elution Methods

In heterogeneous DNA affinity chromatography, a crude or partially purified protein extract is loaded onto a DNA affinity column that has been previously equilibrated with a low-salt buffer (i.e., typically containing about 0.1 M sodium chloride or potassium chloride). This column is then washed with several column volumes of buffer to wash off any nonbinding proteins. Nondenaturing detergents like Tween or Triton can be included in the mobile phase to further reduce the retention of non-DNA-binding proteins. Preclearing of the samples is also effective in removing proteins that show nonspecific binding and is performed by passing the samples over the same support as used in the affinity column but with no DNA being present. However, preclearing may not be necessary with some supports (e.g., Sepharose).

Binding proteins that are retained by a DNA affinity column can be eluted with salt by either a linear or step gradient. This makes use of the fact that when a protein binds to DNA, it will displace counterions from the DNA. Thus, adding more salt to the mobile phase can be used to reverse this process and dissociate the protein. Gradient salt elution can separate weak- from strong-binding proteins; however, the resolution of this separation is low due to the heterogeneity of the support. Some proteins do not elute from DNA columns even at salt concentrations as high as 2 M, but these can instead be eluted by using DNase to digest the immobilized DNA in the column.

7.3 SEQUENCE-SPECIFIC DNA COLUMNS

Sequence-specific DNA can also be used in affinity columns. This section examines the stationary phases and supports used in this method, as well as common mobile phases and elution methods. Some applications of this approach are also discussed.

7.3.1 Stationary Phases and Supports

As mentioned earlier, sequence-specific DNA affinity columns are made by coupling synthetic oligonucleotides [9], plasmids, or concatemers [10] to a solid support. This support can be Sepharose [10], silica, or a variety of other materials. The ligand in this case is a DNA sequence that contains an element consensus sequence, restriction site, or DNA motif that is bound by the protein of interest. It has been found that DNA affinity columns made from a discrete consensus sequence alone fail to bind to one transcription factor (Sp1). It has also been proposed that longer DNA sequences, as obtained by either the catenation of short consensus sequences into concatemers or by introducing the consensus sequence into a plasmid, are required for proper binding [10]. This has led to the common use of concatemers in DNA affinity chromatography.

The major disadvantage of using longer, more complex DNA in affinity columns is that nonspecific DNA-binding proteins may bind to the additional sequences that have been introduced during catenation. This can result in a lower specificity for the stationary phase and a decreased purity for the protein of interest. However, discrete sequence columns are routinely and successfully used. Furthermore, recent studies have shown that columns containing discrete consensus sequence oligonucleotides function well for binding transcription factors such as Sp1. For example, columns made from discrete sequences extended with homopolymeric extensions like T_{18} give a higher purity and yield than concatemeric columns that have been used for isolating the lac repressor, C/EBP, and Sp1 proteins [11]. In addition, discrete sequences with homopolymeric extensions coupled to Sepharose or silica have been used in simple purification schemes that have resulted in high-purity products [11].

7.3.2 Mobile Phases and Elution Methods

The mobile phase used with sequence-specific DNA affinity columns usually consists of a buffer with a low salt content and a near-neutral pH. The temperature is usually maintained at 4°C to prevent protein denaturation or degradation during the separation. Depending on the length of the coupled DNA, this low temperature may also prevent melting of double-stranded DNA in the column. Detergents, protein stabilizers, and nonspecific or specific competitors are sometimes included in the mobile phase to improve the purity and stability of the isolated proteins.

Nondenaturing detergents like Nonidet P-40 (NP40) [12–14], CHAPS (3-[(3-cholamidopropyl)dimethylamino]-2-hydroxy-1-propane sulfonate) [15], DOC (deoxycholate) [16], Tween 20, and Triton X are often included in the mobile phase. These detergents reduce the nonspecific binding of proteins to the DNA, support, and chromatographic system. They also help prevent the aggregation of proteins, which can interfere with the binding of a protein to DNA. In addition, detergents can wet the surfaces within the system, leading to improved binding and a better yield for the purified protein.

During purification, a protein solution can become more dilute as contaminants are removed. Under such conditions, a carrier protein can help stabilize proteins that may denature at low concentrations while also minimizing the retention of nonspecific proteins. Bovine serum albumin (BSA), casein, and insulin are sometimes used as carrier proteins for this purpose. However, it should be kept in mind that the carrier protein itself is a contaminant, albeit an intentional one, and will eventually have to be separated from the protein of interest. This requires the use of an additional step in the purification. To simplify this process, the carrier protein should have properties that make it easy to remove from the protein of interest.

The use of nonspecific and specific competitor DNA can greatly improve the purity of proteins obtained from DNA affinity columns. Fragmented genomic DNA such as salmon sperm DNA or calf thymus DNA [10], dI:dC [10], a specific mutant oligonucleotide [17], and single-stranded DNA have all been used as mobile-phase additives for such work. Alternatively, crude protein extracts can be incubated with these competitors and then loaded onto a DNA affinity column. Most of these competitors have been successfully used to reduce or even completely eliminate nonspecific interactions in an electrophoretic mobility

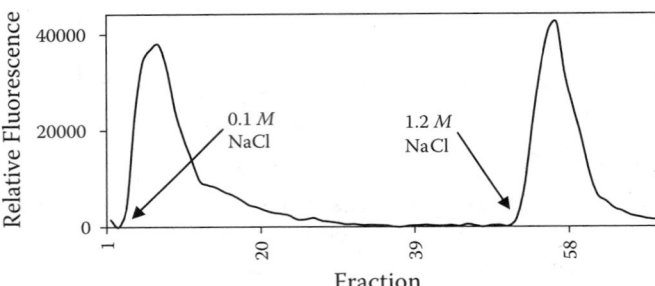

Figure 7.1 Elution of GFP-C/EBP with salt from DNA-Sepharose. This illustrates the salt elution of a recombinant protein from a sequence-specific DNA-Sepharose column. The column was washed with 0.1 M sodium chloride, followed by the application of a bacterial extract containing GFP-C/EBP. The retained protein was later eluted with 1.2 M sodium chloride. The second peak in this chromatogram represents the retained GFP-C/EBP, while the first peak represents nonretained proteins that did not bind to the DNA-Sepharose column.

shift assay (also known as an EMSA or "gel shift" assay). This means that their proper use in chromatography could be quite important. However, despite the use of such competitors, the single-step purification of proteins has only rarely been achieved with DNA affinity chromatography.

DNA-binding proteins can be eluted from DNA affinity columns by using high salt concentrations or by ligand-specific elution. Salt elution is the most widely used method for the elution of DNA-binding proteins [1, 10, 11, 18, 19]. As mentioned earlier, transcription factors and other DNA-binding proteins displace sodium ions and other counterions when these proteins bind to DNA. High salt concentrations counteract this effect and result in the elution of proteins from a DNA column, as shown in Figure 7.1. Since the mechanism of DNA binding is unique to each DNA-protein combination, elution with several steps of increasing salt concentrations or with a continuous gradient can help resolve a protein from its contaminants. However, salt elution alone rarely results in a homogeneous protein preparation [11, 19].

Oligonucleotide elution is another possibility. In this case, a specific-sequence DNA affinity column is eluted by including DNA in the mobile phase. However, this has been used only rarely because the additional DNA can create an interference in techniques that are later used to detect a DNA-binding protein (e.g., EMSA and DNase I footprinting). Another problem with this approach is the cost associated with using DNA for elution.

Ligands other than DNA can also allow specific elution. This makes use of the observation that the DNA-binding affinity of certain transcription factors can be changed in the presence of certain non-DNA ligands. For instance, the lac repressor protein has a very low affinity for its operator DNA sequence (Op1) in the presence of isopropyl-thiogalactopyranoside (IPTG) or lactose. As a result, IPTG can be used to elute the lac repressor protein from Op1-Sepharose columns. The fact that these ligands affect only the specific proteins to which they bind makes this type of elution highly specific [20]. However, only a small number of proteins respond to a specific ligand, which means that this method is not widely applicable.

The ability of heparin to mimic some of the properties of DNA makes this useful for another type of ligand-based elution. This elution is based on the presence of a polyanionic sugar structure in heparin, which mimics the negatively charged DNA ribose-phosphate

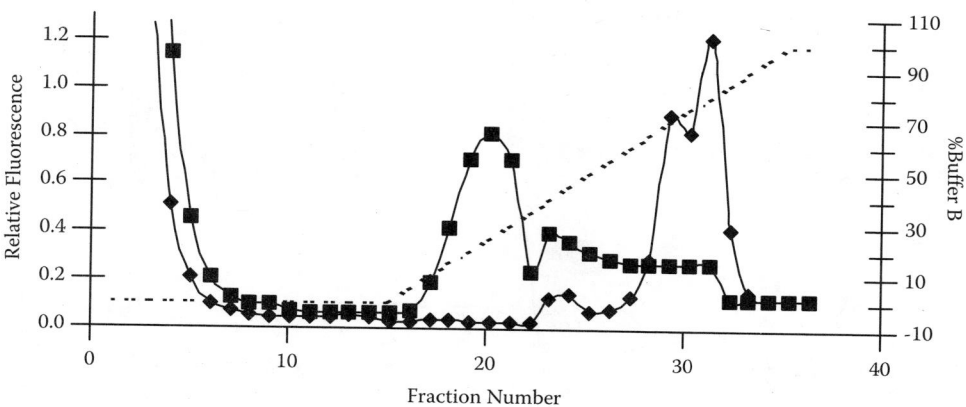

Figure 7.2 Elution of GFP-C/EBP with heparin from DNA-Sepharose. A crude bacterial extract containing GFP-C/EBP was loaded onto a DNA-Sepharose column in the presence of 0.1 M sodium chloride. This protein was eluted either with a 20-ml gradient (■) from zero to 40 mg/ml heparin in 0.1 M sodium chloride, or by using a 20-ml salt gradient (♦) from 0.1 M to 1.2 M sodium chloride. (Reproduced with permission from Gadgil, H. and Jarrett, H.W., *J. Chromatogr. A*, 848, 131–138, 1999.)

backbone [20]. As a result of this structure, many DNA-binding proteins also bind to heparin. This has made heparin-agarose chromatography an integral part of most DNA-binding protein purifications [1, 21–23], and has made it possible to elute such proteins from DNA affinity columns with a moderate to high concentration of heparin in the mobile phase [20]. An example of this latter approach is given in Figure 7.2, where C/EBP is eluted from a DNA affinity column in the presence of 40 mg/ml heparin. The effectiveness of this elution is directly proportional to the heparin concentration in the mobile phase and is inversely proportional to the concentration of DNA on the column. This suggests that the mechanism for the heparin elution is competitive, where heparin in the mobile phase competes with the immobilized DNA for its binding to retained proteins such as transcription factors. As a result, elution with heparin operates in a different way than salt elution, making it possible to use these two methods in conjunction with one another.

DNA-protein interactions are governed by hydrophobic interactions, salt displacement, and changes in heat capacity, making these interactions highly temperature dependent [24–28]. To illustrate this, it has been shown that the concentration of salt needed to elute three different transcription factors from specific DNA affinity columns will vary with temperature [29]. This effect has been used to develop a "temperature jump" elution method, as demonstrated in Figure 7.3. In this method, proteins are loaded onto a column at a salt concentration and temperature that favor binding. After washing the column under these conditions, the temperature is abruptly changed to elute the retained proteins. Since the same salt concentration is used throughout this procedure, the ionic elution of contaminants is minimal. With this method, higher purities can be obtained for proteins than when using a simple salt elution scheme [29]. Furthermore, it has been found that moderate temperatures (i.e., 0 to 40°C) uniquely affect each of the proteins that have been studied, suggesting that a single salt–temperature combination might give high resolution for DNA-binding proteins.

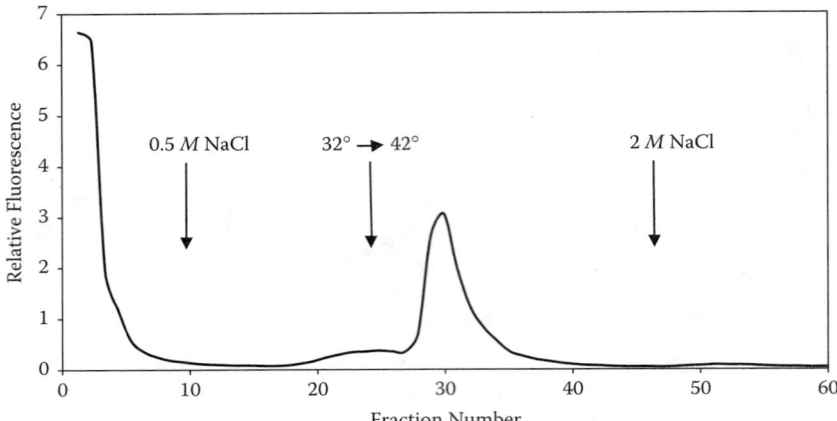

Figure 7.3 Temperature-jump elution of GFP-C/EBP from DNA-Sepharose. In this example, two water baths were set at 32° and 42°C. An EP18-Sepharose column was placed into the 32°C water bath, loaded with a sample containing a GFP-C/EBP fusion protein, and washed with 0.1 M sodium chloride, followed by 0.5 M sodium chloride. The flow rate was 0.3 ml/min, and 0.5-ml fractions were collected. After 12.5 ml (25 fractions), the column was rapidly transferred to the second water bath at 42°C. The column was later washed with 2 M sodium chloride. This chromatogram shows that the protein stayed on the column in the presence of 0.5 M sodium chloride at 32°C but eluted when the temperature was raised to 42°C. (Reproduced with permission from Jarrett, H.W., *Anal. Biochem.*, 279, 209–217, 2000.)

7.4 PURIFICATION OF DNA REPAIR PROTEINS

Cells are constantly under threat from the cytotoxic and mutagenic effects of DNA-damaging agents. These agents can either be from an exogenous source or formed within the cells. Environmental agents that can damage DNA include ultraviolet light, x-rays, and other ionizing radiation that can produce photoadducts. This, in turn, can lead to breaks in single-stranded or double-stranded DNA (ssDNA and dsDNA), abasic sites, and intrastrand crosslinks. A variety of chemicals encountered in food or as airborne and waterborne agents can also change the chemical composition of DNA through alkylation, deamination, or the insertion of bulky chemical moieties. Endogenous damaging agents include methylating species and the reactive oxygen species that arise during respiration and metabolism [30]. The biological consequences of damaged DNA include cell death, inhibition of reproduction, mutagenesis, carcinogenesis, inhibition of enzyme induction, and an alteration or inactivation of protein synthesis.

As shown in Figure 7.4, cells have several modes of defense to prevent DNA damage from causing deleterious effects. DNA repair is found in all life and is required for the protection of an organism from agents that cause DNA damage. As shown in Figure 7.4, four modes of DNA repair are known: base excision repair (BER), mismatch repair, nucleotide excision repair (NER), and recombination repair. Although an in-depth discussion of DNA repair is beyond the scope of this chapter, more information is available from other sources [31–35]. Recent research in the field of DNA repair has focused on determining the carcinogenic effects of damaged DNA due to a failed repair pathway and in understanding how DNA repair pathways function and cross-communicate.

Several types of chromatography are used in the purification of DNA repair enzymes. These include ion-exchange, gel-filtration, and heparin-agarose chromatography. A few

DNA Damage Repair Systems

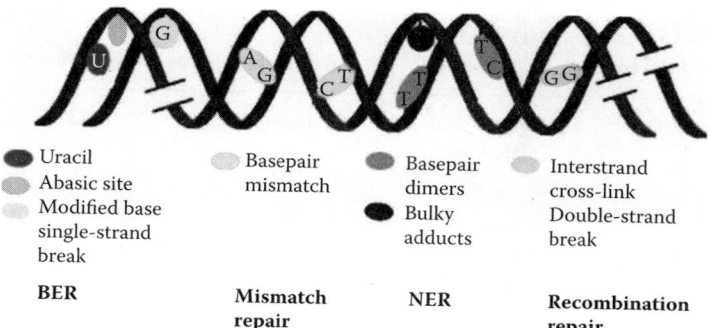

Uracil
Abasic site
Modified base
single-strand
break

Basepair
mismatch

Basepair
dimers
Bulky
adducts

Interstrand
cross-link
Double-strand
break

BER

**Mismatch
repair**

NER

**Recombination
repair**

Figure 7.4 DNA damage and repair mechanism. The top portion of this figure shows examples of DNA lesions, and the bottom shows the repair pathways most often used for such DNA damage. It is important to note that different repair pathways can be induced by the same damage, and that the damage spectrum of different pathways may overlap.

methods involving DNA affinity chromatography have also been employed in this area (see Table 7.1). Sequence-specific DNA columns based on agarose [3] and DNA-cellulose [5, 6, 36] have been used for this purpose, as well as heterogeneous DNA on cellulose. The following subsections discuss these and other DNA affinity methods as they relate to DNA repair proteins.

7.4.1 Proteins in Base Excision Repair

The base excision repair pathway protects cells from the deleterious effects of DNA damage. Some of the major forms of DNA damage arising from endogenous agents are hydrolytic depurination (apurinic site formation), hydrolytic deamination and alkylation, and the formation of covalent adducts with DNA, as well as oxidative damage to bases and the phosphodiester backbone of DNA.

Base excision repair relies on DNA glycosylases to recognize aberrant base residues in DNA and to excise these through cleavage of the *N*-glycosidic bond linking the base to its deoxyribose sugar. Many types of glycosylases exist that recognize specific aberrations in base structure. Major classes of DNA glycosylases include those that recognize uracil, alkylated bases, oxidized bases, apurinic sites, and pyrimidine dimers. Once a damaged base is excised, the apurinic/apyrimidinic (AP) site is removed by an AP-endonuclease and an AP-lyase, which cleave the DNA strand 5′ and 3′ to the AP site, respectively. The resulting gap is then filled by DNA polymerase and joined by DNA ligase.

Several base excision repair proteins have been purified by DNA affinity chromatography to apparent homogeneity. One of these is endonuclease III from *Escherichia coli*. This is a glycosylase that cleaves pyrimidine hydrates from the DNA backbone with its AP endonuclease and AP lyase activities [37]. Endonuclease III, like most other enzymes involved in DNA repair, has been purified using single-stranded (i.e., denatured, heterogeneous) DNA-cellulose, as shown in Table 7.1.

Another base excision repair protein that has been purified by DNA affinity chromatography is 8-oxoguanine glycosylase. This enzyme removes 8-oxoguanine from damaged DNA [38]. 8-Oxoguanine, which is induced by reactive oxygen species, is mutagenic because it mispairs with adenine. One step in the purification of 8-oxoguanine glycosylase has involved the covalent trapping of this enzyme with biotinylated DNA. This trapping

Table 7.1 DNA Affinity Purifications of DNA Repair Enzymes

DNA Repair Enzyme	Source	DNA Affinity Method	Additional Steps	References
Endonuclease III	Calf thymus tissue	Single-stranded DNA-cellulose	SP fast-flow octylsepharose Mono S Gel filtration	[37]
5-Hydroxymethyl-uracil DNA N-glycosylase	Calf thymus tissue	Single-stranded DNA-cellulose	Ammonium sulfate Superdex Mono S Heparin Sepharose	[150]
8-Oxoguanine glycosylase	Escherichia coli	Biotinylated oligomers	Polymin P Ammonium sulfate DEAE-cellulose Superdex-75	[38]
Uracil DNA glycosylase	Rat liver nuclear extract	Single-stranded DNA-agarose	Ammonium sulfate DEAE-Sephadex Sephadex G-75 Phosphocellulose	[151]
Uracil DNA glycosylase	Dictyostelium discoideum	Double-stranded DNA-cellulose	DEAE-cellulose Hydroxyapatite	[152]
hUracil DNA glycosylase	Human fibroblasts	Double-stranded DNA-cellulose	DEAE-cellulose Phosphocellulose Sephadex G-100 Immunoaffinity	[153]
hUracil DNA glycosylase	Human placenta tissue	Double-stranded DNA-cellulose	DE-52 cellulose Phosphocellulose Immunoaffinity Sephadex G-100	[154]
UvrA	Escherichia coli	Single-stranded DNA-cellulose	Q-Sepharose Mono-Q FPLC Gel filtration Phosphocellulose	[155–157]
UvrC	Escherichia coli	Single-stranded DNA-cellulose	Polymin P Ammonium sulfate Phosphocellulose AcA34 gel filtration AcA44 gel filtration Q-Sepharose	[157–160]
Ercc1	HeLa cell nuclear extract	Single-stranded DNA-cellulose	Phosphocellulose Hydroxyapatite Sepharose Mono-Q Mono-S	[56]

Table 7.1 DNA Affinity Purifications of DNA Repair Enzymes (Continued)

DNA Repair Enzyme	Source	DNA Affinity Method	Additional Steps	References
HHR23B	HeLa cell nuclear extract	Single-stranded DNA-cellulose	Phosphocellulose CM cosmogel Mono-Q	[55]
Replication protein A (RPA)	Sf9 cells	Single-stranded DNA-cellulose	Affi-Gel blue	[52]
Replication protein A (yRPA)	*Saccharomyces cerevisiae*	Single-stranded DNA-cellulose	Affi-Gel blue	[53]
Replication protein A (RPA)	Tobacco BY-2	Single-stranded DNA-cellulose	None	[161]
XPC	Fibroblast cell extract	Single-stranded DNA-cellulose	Phosphocellulose DEAE	[54]
UvrD	*Escherichia coli*	Single-stranded DNA-agarose	Ammonium sulfate Hydroxyapatite Heparin-agarose	[58]
MSH2-MSH6	*Saccharomyces cerevisiae*	Single-stranded DNA-cellulose	PBE-94 Superose	[59, 60]
hMutSα	Raji cell extract	Single-stranded DNA-cellulose	Ammonium sulfate	[61]
MtMYH	Calf liver mitochondrial extract	Single-stranded DNA-cellulose	Ammonium sulfate Phosphocellulose Q-Sepharose Superose	[62]
Thymine DNA glycosylase	HeLa cell whole-cell extract	Biotinylated oligonucleotide	DEAE-Sepharose Phosphocellulose Sepharose Mono-Q	[63]
RecBCD	*Escherichia coli*	Single-stranded DNA-cellulose	Ammonium sulfate Q-Sepharose Heparin-Sepharose	[73]
RecG	*Escherichia coli*	Single-stranded DNA-cellulose	Ammonium sulfate Heparin-agarose Phenyl-Sepharose Q-Sepharose Sephacryl	[70]
RecGK302A	*Escherichia coli*	Double-stranded DNA-cellulose Single-stranded DNA-cellulose	Ammonium sulfate Heparin-agarose Phenyl-Sepharose Q-Sepharose Sephacryl	[162]
RecQ	*Escherichia coli*	Single-stranded DNA-cellulose	Sephacryl Hydroxyapatite	[72]

(Continued)

Table 7.1 DNA Affinity Purifications of DNA Repair Enzymes (Continued)

DNA Repair Enzyme	Source	DNA Affinity Method	Additional Steps	References
RusA	*Escherichia coli*	Double-stranded DNA-cellulose	Phosphocellulose DEAE Bio Gel Heparin agarose	[71]
DNA ligase IV	HeLa cell extract	Double-stranded DNA-cellulose	Phosphocellulose Ultrogel AcA34 Mono-S Mono-Q	[80]
Ku86	Sf9 cell extract	Double-stranded DNA-cellulose	Mono-Q	[79]
Rad51	*Escherichia coli*	Single-stranded DNA-cellulose	Q-Sepharose Bio-Gel HTP Mono-Q	[77]
Rad52	*Escherichia coli*	Single stranded DNA-cellulose	Ni-NTA agarose	[78]

reaction was carried out at room temperature for 4 h in a buffer containing sodium cyanoborohydride and 5′-GACGAATTCGCGATCG*TCGACTCGAGCTCAG-3′ as the oligonucleotide substrate (where G* represents the oxoguanine base). A 5′-^{32}P-labeled oligonucleotide substrate was included in this reaction mixture as a tracer. The result was a covalent adduct of DNA-bound protein. In this method, the 3′ terminus of the complementary strand in the duplex oligonucleotide was biotinylated. This allowed the substrate DNA to be separated later by using streptavidin on magnetic beads. After mixing these beads with the sample, they were washed several times with 1 M sodium chloride and 1% sodium dodecyl sulfate. The material retained on the beads was then recovered by boiling these beads in a sample buffer for sodium dodecyl sulfate polyacrylamide gel electrophoresis (SDS-PAGE), which was then used to analyze the recovered material and identify the captured enzyme.

Several other DNA repair enzymes have also been purified by using DNA affinity chromatography in combination with other separation methods. For instance, 5-hydroxymethyluracil DNA N-glycosylase is an enzyme that removes the 5-hydroxymethyluracil formed in DNA upon exposure to ionizing radiation. This has been purified using a single-stranded DNA-cellulose along with ammonium sulfate precipitation and chromatography based on Superdex, Mono S, and Heparin Sepharose columns [38]. Similar work has been performed with uracil DNA glycosylase, which is an enzyme that recognizes and removes uracil from DNA [39]. This uracil might appear in DNA during synthesis by the enzyme DNA polymerase or by chemical modification (deamination) of an existing cytosine residue.

7.4.2 Proteins in Nucleotide Excision Repair

Nucleotide excision repair is the process where DNA damage is removed as a part of an oligonucleotide fragment. This is followed by its replacement with new DNA using the intact strand as a template. The nucleotide excision repair machinery has very broad specificity and is able to recognize a variety of chemical alterations to DNA that result in large local distortions of the DNA structure.

Prokaryotic nucleotide excision repair in *E. coli* involves six proteins: UvrA, UvrB, UvrC, UvrD, polymerase A, and ligase. These perform the functions of DNA damage recognition, damaged DNA excision, excised DNA displacement, polymerization, and ligation. A summary of this overall process is as follows. First, a dimer of UvrA binds to UvrB and then to the DNA [40, 41]. The 5′-3′ helicase activity of this UvrA$_2$B-DNA complex allows limited ATP-dependent one-dimensional movement along the DNA duplex [42]. Upon encountering damage, the UvrA$_2$B heterotrimer kinks the DNA and causes local denaturation [43]. UvrA$_2$ then dissociates, leaving a stable complex consisting of UvrB bound to DNA. This complex is then bound by UvrC. In the UvrBC-DNA complex, incisions are made by UvrB at points four to five phosphodiester bonds past the 3′ end of the lesion. Other incisions are then made eight bonds past the 5′ end of the lesion by UvrC, releasing a 12-14 base pair oligonucleotide that contains the lesion [44–46]. ATP is required for most of the steps in this damage recognition and incision. UvrD then binds at the free 3′ end of the damaged oligonucleotide and displaces it through its intrinsic 3′-5′ helicase activity. Next, this gap is filled by polymerase A and sealed by DNA ligase. As indicated by Table 7.1, two of the proteins involved in this process, UvrA and UvrC, have been purified to apparent homogeneity by DNA affinity chromatography [155–160].

Eukaryotic nucleotide excision repair is more complex, requiring 20 to 30 proteins instead of the 6 needed in prokaryotes. Some of the most well-characterized proteins in this repair pathway are XPC, XPE, XPA, RPA, XPB, XPD, XPF, and XPG. XPC, XPA, and XPE are thought to be the initial proteins to recognize DNA damage, but they differ in the type of damage they recognize. Replication protein A (RPA) associates with XPA in the recognition of DNA damage [47]. XPB is an ATPase-driven 3′-5′ helicase, while XPD is a 5′-3′ helicase [48, 49]. Both XPB and XPD are subunits of transcription factor IIH (TFIIH). XPF is a 5′ endonuclease and XPG is a 3′ endonuclease. Together, these proteins remove a 24-32 base-pair oligonucleotide that contains the damaged DNA [50, 51]. Nucleotide excision repair is an important means of maintaining the integrity of the genome. Deficiencies in this repair system can lead to several devastating and often fatal diseases in humans, such as xeroderma pigmentosum (i.e., the source of the "XP" in many of these protein names), trichothiodystrophy, and Cockayne syndrome.

Many of the eukaryotic nucleotide excision repair proteins have been purified by DNA affinity chromatography to apparent homogeneity. One of these is RPA, which is also known as human single-stranded DNA-binding protein. This is a three-subunit protein complex made up of subunits p70, p11, and p34 [52]. Another nucleotide excision repair protein that has been purified by DNA affinity chromatography is yRPA, which is the homolog of human RPA found in *Saccharomyces cerevisiae* [53]. Additional proteins that have been isolated using DNA affinity methods include XPC, HHR23B, and Ercc1. The XPC (or xeroderma pigmentosum complementation group C) gene encodes a 125-kDa protein that is essential to nucleotide excision repair and that associates closely with HHR23B [54]. HHR23B is a 58-kDa human homolog of the *Saccharomyces cerevisiae* protein RAD23, a damage-binding factor from the Rad4-Rad23 complex [33]; HHR23B is required for XPC function [55]. Ercc1 (which stands for excision repair cross-complementation) is a protein that associates with and is required for the endonuclease activity of XPF [56].

7.4.3 Proteins in Mismatch Repair

DNA mismatch repair systems exist that fix any mispaired bases formed during replication or genetic recombination and as a result of damage to DNA. Many components of these systems are conserved in prokaryotes and eukaryotes. Furthermore, genetic defects in mismatch-repair genes play a role in common cancer-susceptible syndromes and sporadic cancer.

The prokaryotic mismatch repair system consists of MutH, MutL, MutS, MutM, UvrD, DNA polymerase III, and ligase I. Together, these are called the MutHLS system. No enzymatic activity has been assigned to MutL, although it interacts with MutS and is required for activation of MutH. MutH is an endonuclease that nicks hemimethylated DNA on the unmethylated strand and is activated by MutS and MutL in the presence of a mismatch. MutS recognizes the mismatch bases. UvrD, as mentioned previously, is an ATP-dependent DNA helicase [57] and has been purified by DNA affinity chromatography to homogeneity [58].

Eukaryotes have a broad-spectrum mismatch repair system related to the bacterial MutHLS system. Homologs of MutS and MutL have been identified in many eukaryotic organisms, including *Saccharomyces cerevisiae* and humans [59]. *Escherichia coli* MutS homologs, MSH1-MSH6, and two *E. coli* MutL homologs, MLH1 and PMS1, have been identified. However, no MutH homologs have yet been described [60]. Human MutSα is a protein complex consisting of at least four subunits, two of which are MSH2 and MSH6 [61]. The eukaryotic mismatch repair proteins MSH2 and MSH6 have been purified by DNA affinity chromatography to apparent homogeneity. MtMYH is a mitochondrial protein that is homologous to *E. coli* MutY. MutY is a glycosylase that specifically binds mispairs of all bases with 8-oxoguanine (G*). MtMYH has binding activity with A/G* and A/G base pairs [62]. Thymine DNA glycosylase is a 55-kDa eukaryotic enzyme capable of hydrolyzing mismatched thymines (G/T, C/T, and T/T). This enzyme initiates the repair process by excising the mispaired thymine from damaged DNA to generate an apyrimidinic site [63].

For the purification of thymine DNA glycosylase, a G/U affinity support was prepared by annealing two 14-mer oligonucleotides (5′-GATCCGTCGACCTG-3′ and 5′-GATCCAGGTUGACG-3′). The annealed duplexes were ligated end-to-end to form concatemers. The remaining 5′ overhangs (i.e., GATC) were filled in the presence of biotin-16-dUTP, creating a biotin-tagged end. Thymine DNA glycosylase was then fractionated on a Mono Q ion-exchange column, followed by its incubation with the biotinylated DNA concatemers on streptavidin-Dynabeads. After incubation, the beads were washed and eluted with a sodium chloride step gradient. The thymine DNA glycosylase purified by this method was found to be suitable for use in band shift, glycosylase activity, and AP endonuclease assays.

7.4.4   Proteins in Recombination Repair

Homologous recombination is one of the most important processes involved in repairing DNA damage and in ensuring that correct genetic information is passed to offspring. Double-stranded breaks and single-stranded gaps are efficiently repaired by mechanisms associated with recombination [64].

Prokaryotic recombination repair involves at least 13 proteins (i.e., RecA, RecB, RecC, RecD, RecF, RecG, RecO, RecQ, RecR, RuvA, RuvB, RuvC, and Rus), which are highly homologous to some eukaryotic proteins for recombination repair. A thorough review of prokaryotic recombination repair is beyond the scope of this chapter, but more information can be found in the literature [64–69].

RecA forms a nucleoprotein on DNA and carries out an ATP-dependent homology search. This is followed by strand-exchange reactions involving RecF, RecO, and RecR. RecBCD forms a multifunctional protein that has helicase and exonuclease activity, and binds to the ends of single-stranded DNA tails during recombination. Holliday junctions are processed by RuvA, RuvB, RuvC, Rus, and RecG. RecG is a DNA helicase that may target stalled replication forks to help generate four-stranded Holliday junctions [70]. RusA is an endonuclease that resolves Holliday junction intermediates in recombination and DNA repair [71]. RecQ has a DNA-dependent ATPase and a DNA helicase activity [72]. As shown in Table 7.1, RecBCD [73], RecG [70], RecQ [72], and RusA [71] have all been purified using DNA affinity chromatography.

Eukaryotic recombination repair differs from prokaryotic recombination repair in two important aspects. First, eukaryotes possess three pathways for repairing double-stranded breaks instead of the one pathway found in prokaryotes. Second, many more proteins are associated with repair in eukaryotes, such as Xrs2, Mre11, Rad52, Rad51, DNA PK, Ku70, Ku80, XRCC4-7, and DNA ligase [30]. The three pathways found in eukaryotes will now be outlined. More-detailed reviews can be found in the literature [74–76].

Two of these pathways, homologous repair and single-stranded annealing, are classified as recombination repair. Homologous repair and single-stranded annealing are similar in mechanism and require many of the same proteins, including Xrs2, Mre11, Rad52, Rad51, and Rad50. The third pathway is called nonhomologous end joining. This involves very different proteins and uses a separate mechanism of repair from homologous repair and single-stranded annealing.

Several eukaryotic recombination repair proteins have been purified by DNA affinity chromatography. The first was Rad51, which is an ATP-dependent 38-kDa protein; this is a yeast homolog of the bacterial homologous repair protein, RecA [77]. A second repair protein in this group that has been isolated by a DNA affinity method is yeast Rad52, which is also an ATP-dependent protein [78]. All types of homologous recombination in yeast are virtually eliminated in Rad52 mutants. Ku86 (a Ku80-like protein) binds to DNA ends and is a factor for the DNA-dependent protein kinase. Ku86 has been purified by a DNA affinity column from a *Baculovirus*-infected Sf9 cell extract [79]. Another protein that has been purified by DNA affinity chromatography is DNA ligase IV. This enzyme is involved in one of the last steps of homologous DNA repair, the ligation of the two strands to form a continuous duplex [80].

7.4.5 Alternative Methods for Separating DNA Repair Proteins

As can be seen from the previous sections, many proteins associated with DNA repair have been prepared by DNA affinity chromatography. This is true for proteins from each of the four known pathways for DNA repair, with these proteins being used in a variety of repair assays. Up to this point, however, DNA affinity chromatography has always been used as only one of several steps for purifying the repair proteins of interest. Furthermore, these methods have made use of nonspecific DNA interactions by the DNA-binding proteins.

However, an alternative technique for DNA affinity chromatography can also be used. Many of the DNA repair proteins recognize specific DNA aberrations. This might involve the recognition of base modification and AP sites in base excision repair. Alternatively, it might involve the recognition of bulky adducts in nucleotide excision repair, mismatched bases in mismatch repair, and single- or double-stranded breaks in recombination repair. Several reports in the area of DNA repair enzyme purification have employed assays that rely upon or characterize these specific DNA-protein interactions.

Such work suggests that a purification scheme could be developed that takes advantage of these specific DNA interactions for repair proteins. In this approach, DNA that has been altered in a specific manner could be used to obtain specific DNA repair proteins that can bind the altered DNA. Two routes could then be followed at this point. In the first, specifically bound DNA-binding proteins could be eluted by disrupting the protein-DNA interaction, as might be accomplished by using a higher salt concentration or an elevated temperature. In the second possible route, the DNA-bound protein could be eluted by making use of its catalytic mechanism after DNA binding occurs. This latter technique can be performed using a method known as *catalytic chromatography.*

As described in a recent paper [2], catalytic chromatography was used with EcoRI restriction endonuclease and Klenow large fragment of DNA polymerase. Both of these proteins were shown to elute from DNA columns by completing their catalytic mechanism upon the addition of the required factors Mg^{2+} and dATP, respectively. Thus, upon the binding of a repair protein to a specific aberrant DNA site, such proteins can be eluted by allowing their catalytic function to take place, followed by the subsequent release of their catalyzed products. This approach allows the DNA affinity support to be used as a tool for obtaining a high-purity protein as well as a method for the study of specific repair mechanisms. While this approach is clearly feasible, it has not yet been tested.

7.5 PURIFICATION OF TRANSCRIPTION FACTORS

Unlike many DNA-binding proteins, transcription factors have an affinity that is often several orders of magnitude higher for a specific DNA sequence than for a random DNA sequence. In this situation, the specific sequence to which transcription factors bind is called the "element" (or sometimes the "consensus sequence" or "footprint"). This has made sequence-specific DNA affinity chromatography a valuable tool for this large group of DNA-binding proteins. This topic has recently been reviewed in the literature [1, 81], with recent examples being given in Table 7.2. More extensive lists of agents that have been purified by this approach can be found in the literature [1, 10].

Table 7.2 DNA Affinity Purification of Transcription Factors

Transcription Factor	Source	Sequence Obtained	Purification Steps	References
TF2	*Dictyostelium discoideum* strain AX3K	Yes	Ion exchange Sequence-specific DNA affinity chromatography Gel filtration	[163]
MEF1	HL60/VCR nuclear extract	No	Sequence-specific DNA affinity chromatography	[164]
CBF-A	Rat mammary cells	Yes	Sequence-specific DNA affinity chromatography	[165]
LuxT	*Vibrio harveyi*	Yes	Sequence-specific DNA affinity chromatography Reversed-phase HPLC	[166, 167]
CERC	K562 nuclear extract	—	Biotinylated dsDNA or mutant oligonucleotide on streptavidin beads	[168]

7.6 PURIFICATION OF OTHER PROTEINS

A variety of other proteins have been purified by DNA affinity chromatography. Examples that will be considered in this section are histones, helicases, topoisomerases, primases, and DNA polymerases.

7.6.1 Histones

Histones are the key proteins responsible for packaging genomic DNA in eukaryotes to form chromatin, a highly organized form of DNA. Without histones, chromosomal segregation during cellular reproduction would not be possible. Nucleosomes, formed by wrapping DNA around histones, are the basic repeating units of chromatin, which are organized into polynucleosomal arrays. Polynucleosomal arrays are further assembled into a higher-order chromatin structure with the assistance of other proteins. Nucleosomes consist of five proteins that fall into two categories: nucleosomal histones and H1 histones.

Nucleosomal histones (i.e., H2A, H2B, H3, and H4) are highly conserved, small, basic proteins with 102 to 135 amino acids. These are responsible for wrapping DNA into nucleosomes and forming the 10-nm DNA fiber. Nucleosomal histones form an octamer, (H2A-H2B-H3-H4)$_2$, which is wrapped by 146 base pairs of DNA. While this nucleosome functions to package and organize DNA, it also serves to regulate transcription. Chromatin can be either transcriptionally active or inactive, depending, in part, on the local structure of the nucleosomes within a region of chromatin. In addition, nucleosomal DNA can be either tightly or loosely associated with the core histones. Protein acetylation dictates the state of DNA association. Acetylation of lysine residues in the core histones results in a more open structure of chromatin, making the nucleosomal DNA more accessible to transcription factors.

H1 histones are found in the "linker" region of polynucleosomal DNA between nucleosomes. These are larger than nucleosomal histones and contain about 220 amino acids. The primary function of H1 histones is to contact other nucleosomes and pull together polynucleosomal DNA into a regular repeating array. Upon the binding of H1 histones to the linker DNA, the polynucleosomal fiber folds into a 30-nm-diameter chromatin fiber. H1 histones are regulated by a cycle of phosphorylation that coincides with the M phase. This phosphorylation reduces H1 DNA binding and results in a less tightly wound polynucleosomal fiber.

Two histones have been purified by DNA affinity chromatography, histone H1 and histone H2B (see Table 7.3). This has involved the use of two distinct methods: DNA-agarose

Table 7.3 DNA Affinity Purification of Histones

Histone Protein	Source	DNA Affinity Method	Additional Steps	References
H1	Rat liver cell nuclear extract	Nonspecific DNA-cellulose	Heparin-agarose	[82]
H1	Rat liver cell nuclear extract	Specific DNA-cellulose Biotinylated oligomers	—	[84]
MDBP-2	Hen/rooster liver nuclear extract	Biotinylated oligomers	Heparin-Sepharose	[85]
H2B	*Drosophila* embryo nuclear extract	Specific DNA-cellulose	Heparin-Sepharose	[86]

columns and oligonucleotide biotinylation. More details on each of these purification procedures will be given later in this section. Current work in the histone field has involved determining the role of histone modifications, specifically in transcriptional regulation. DNA affinity chromatography should prove useful for these studies. Another potentially valuable tool for this type of research is the ChIP assay, which is discussed in Chapter 20.

Histone H1 was shown to bind to the nuclear factor I (NF-I) recognition sequence in the mouse $\alpha_2(I)$ collagen promoter and was purified using both sequence-specific and nonspecific DNA-cellulose affinity chromatography [82, 83]. To do this, rat liver extracts were applied to a heparin-agarose column, with the eluants then being loaded onto a nonspecific DNA-cellulose affinity matrix prepared with high-molecular-weight E. coli DNA. The next step in this purification used a specific DNA-cellulose affinity matrix. This column was prepared with a linearized plasmid containing 88 tandem copies of the NF-I recognition sequence (TGGATTGAAGCCAA). The final purified H1 was then used in DNase footprinting assays.

To demonstrate that histone H1 is capable of hydrolyzing NTPs, a specific sequence of DNA was used for its purification [84]. In this experiment, an affinity matrix was generated using a 25 base-pair oligonucleotide that contained the NF-I recognition sequence. This oligomer was biotinylated and attached to a streptavidin-agarose matrix. This gave rise to a single-step purification method that could purify histone H1 to apparent homogeneity from rat liver cell nuclear extracts. This purified protein was then used in NTP hydrolysis, nucleotide binding, and immunoprecipitation assays.

MDBP-2 is a repressor that binds preferentially to methylated DNA. This has been determined to be a member of the histone H1 family [85]. To isolate this protein, hen and rooster liver nuclear extracts were loaded onto a heparin-Sepharose column. The eluants were later mixed with a biotinylated oligonucleotide duplex (5'-CTAGACTATTCACCTT-^{m}CGCTATGAGGGGGATCATACTGGCT-3' and 5'-TGCCAGTATGATCCCCCCTCAT-A^{m}GCGAAGGTGAATAGTCTAG-3') and added to streptavidin-Dynabeads. In this duplex, methylated cytosine (^{m}C) was on one strand and methylated guanosine (^{m}G) was on the other. It was found that MDBP-2 did not bind to this oligonucleotide through sequence-specific recognition, but rather through recognition of the methylated C-G pair. The purified protein was later used in ultraviolet DNA-protein cross-linking and Southern blot assays.

Histone H2B was purified inadvertently by means of its binding to the Drosophila minimum enhancer region of the gene knirps [86]. In this work, Drosophila embryo nuclear extracts were loaded onto a heparin-Sepharose column. The eluants were next loaded onto a specific-sequence DNA-agarose column [87] containing one of two portions of the knirps minimal enhancer region, (5'-GATCTCGTTACCTAATCGCGG-3') and (5'-GATCCCCG-CATTAGGTAACGA-3'). The purified protein was later used in DNase footprinting and in vitro transcription assays.

7.6.2 Helicases

DNA helicase enzymes unwind duplex DNA, which is essential for the progression of the replication fork during DNA replication [88, 89]. Helicases break the hydrogen bonds that hold two complementary DNA strands together but do not affect the covalent backbone of the two DNA strands. The activity of helicase is ATP-dependent, with the energy for its reaction being provided by the hydrolysis of ATP. Helicases can bind to single-stranded DNA, double-stranded DNA, and ATP. This means that dsDNA-cellulose, ssDNA-cellulose, and ATP-agarose affinity columns can all be used for their isolation. Heparin-agarose and phosphocellulose columns will also bind to helicases and can be used for their purification.

Several protocols have been developed for the purification of helicase without DNA affinity chromatography [90]. However, the use of heterogeneous DNA affinity columns for such work gives rise to a severalfold increase in purity. As one example, Tuteja et al. used a five-step protocol for the purification of human DNA helicase VI [91]. Their procedure involved (1) ammonium sulfate precipitation, (2) Bio-Rex 70 chromatography, (3) MonoQ column chromatography, and (4) DNA affinity chromatography using dsDNA-Sepharose and ssDNA-Sepharose. Helicase with a specific activity of 20,000 units/mg was obtained after the first three purification steps. But this specific activity increased over fourfold after using dsDNA-Sepharose and increased another 13-fold after using ssDNA-Sepharose, giving a final activity of 1,200,000 units/mg.

7.6.3 Topoisomerases

Topoisomerases are enzymes that catalyze the interconversion of various topological iso-forms of DNA [92]. Topoisomerases play an important role in the replication, transcription, and recombination of DNA [92]. The types of reactions catalyzed by topoisomerases include DNA relaxation, supercoiling, knotting, and catenation [92].

Type I topoisomerases operate by making a single knick in one of the DNA strands and passing the other strand through the first. Type II topoisomerases, on the other hand, make transient double-stranded breaks and pass one DNA helix through. Heterogeneous DNA affinity columns have been used with other techniques for the purification of topoisomerases.

Bergerat et al. used a six-step protocol for the purification of DNA topoisomerase II from *Sulfolobus shibatae* [93]. This protocol included: (1) ammonium sulfate precipitation, (2) phenyl-Sepharose chromatography, (3) phosphocellulose chromatography, (4) heparin-Sepharose chromatography, (5) DNA-cellulose chromatography, and (6) hydroxyapatite chromatography. Topoisomerase with a specific activity of 53 units/mg was obtained after the first four purification steps. The next step based on DNA affinity chromatography improved the purity by almost threefold and yielded topoisomerase with a specific activity of 184 units/mg. However, later purification with a hydroxyapatite column was required to obtain topo-isomerase with a specific activity of 300 units/mg.

Markauskas et al. used triazine dyes for the purification of topoisomerase [94]. These authors characterized 12 triazine dyes for their binding to this enzyme, with Bordeaux 4ST-Sepharose CL-6B being selected for the purification of topoisomerase. Melendy and Ray used novobiocin affinity chromatography for the purification of type II mitochondrial topoisomerase from *Crithidia fasciculate* [95]. Novobiocin is a competitive inhibitor of the ATP-binding moiety of type II topoisomerase and, thus, can act as an affinity ligand for this enzyme. Novobiocin-Sepharose columns have been used for this purpose, with the retained protein being eluted with a buffer containing 20 mM ATP and 20 mM magnesium chloride. This resulted in a 300-fold purification of topoisomerase, with further steps based on S-Sepharose chromatography and glycerol density centrifugation being required to obtain a homogeneous protein.

7.6.4 Primases

During DNA replication, the double-stranded DNA is separated into single-stranded DNA, giving rise to the leading and lagging strands. Replication occurs at the fork of DNA strand separation and requires many proteins in addition to DNA polymerase. DNA is synthesized at the fork in a continuous fashion on the leading strand and in a discontinuous fashion on the lagging strand, generating short Okazaki fragments. All known DNA polymerases require a primer for the initiation of DNA synthesis. The requirement for a separate activity to initiate nascent DNA arises from the incapability of DNA polymerases to initiate DNA chains *de novo*. More information on the replication of DNA and complementary primers is available in the literature [96–99].

In many cases, an RNA polymerase called DNA primase carries out the priming of DNA chains. Primase is the single-stranded DNA-dependent RNA polymerase that synthesizes primers during DNA replication. As with all DNA and RNA polymerases, primase has structural and functional features involved in polymer formation. Some distinguishing characteristics of primases are their synthesis of a specific-size oligonucleotide primer (approximately ten base pairs), their ability to incorporate deoxyribonucleotides into a primer, and their ability to synthesize a primer in the absence of complementary ribonucleotides.

The prokaryotic primase appears to function as a single protein unassociated with any other polymerase. In eukaryotes, however, primase is tightly associated with DNA polymerase α. Eukaryotic and prokaryotic primases share little homology between kingdoms. However, primases within a kingdom are highly homologous. Though the primase sequence is not conserved between kingdoms, the reaction mechanism is conserved and is similar to other polymerases and transferases. Primase initiates synthesis at many template sequences, but the factors that regulate sequence recognition of the template are unknown, and sequence recognition appears nonspecific. The only sequence requirement is that the first nucleotide of the template must be a pyrimidine.

As is true for RNA polymerases in transcription, primase must initiate primer synthesis by polymerizing two NTPs. Primase binds single-stranded DNA and then two NTPs to form a primase-DNA-NTP-NTP quaternary complex. After synthesis of the dinucleotide, additional NTPs (complementary to the DNA template strand) are polymerized to generate a 7-10 base-pair primer. A remarkable feature of primase is its ability to regulate primer length during synthesis. The mechanism that regulates primer length is not clear. Presently, research in the primase field includes determination of protein-protein interactions of the replisome, primase stimulatory and inhibitory factors, and the DNA primer length regulatory mechanism.

Several methods have been used in the purification of primase, including ion exchange, gel filtration, and precipitation. However, only a few methods employing DNA affinity chromatography have been used for purifying primase. As is shown in Table 7.4, these procedures usually involve the use of DNA-agarose [3], DNA-Sepharose [100], or DNA-cellulose [4–6, 35]. Because eukaryotic primase is closely associated with DNA polymerase α, both DNA polymerase α and primase usually copurify. However, in many cases primase and DNA polymerase α can be separated by using a high concentration of salt or polyethyleneglycol followed by gradient centrifugation.

7.6.5 DNA Polymerases

DNA polymerases can synthesize a new strand of DNA on a template strand. Multiple DNA polymerase enzymes are present in prokaryotes and eukaryotes. A search of the *E. coli* gene database reveals the presence of five different DNA polymerases (I to V) in this organism. Out of these five polymerases, three have been characterized. DNA polymerase I (pol I) is involved in DNA repair and has a subsidiary role in replication. DNA polymerase II is also involved in repair, while DNA polymerase III (a multisubunit protein) is responsible for the *de novo* synthesis of DNA. The first DNA polymerase (pol I) was discovered and purified from *E. coli* by Bessman and coworkers in 1958 [101]. The original procedure for its purification included several steps, such as streptomycin precipitation, DNase digestion, adsorption to alumina gel, and ammonium sulfate precipitation followed by chromatography on DEAE-cellulose.

Table 7.4 DNA Affinity Purification of Primases

Source	DNA Affinity Method	Additional Steps	References
Nicotiana tabacum leaf extract	Single-stranded DNA-cellulose	Ammonium sulfate Phosphocellulose Q-Sepharose Heparin-Sepharose	[169]
293 Cell nuclear extract	Single-stranded DNA-cellulose	Hydroxyapatite Phosphocellulose Mono-Q	[170, 171]
HeLa S$_3$ cell nuclear extract	Single-stranded DNA-cellulose Double-stranded DNA-cellulose	Poly(ethylene glycol) DEAE	[172]
Saccharomyces cerevisiae	Single-stranded DNA-agarose	Ammonium sulfate Phosphocellulose ATP-agarose	[173]
Bovine thymus	Single-stranded DNA-cellulose	Ammonium sulfate Q-Sepharose	[174]
Simian CV-1 cells	DNA-Sepharose	Ammonium sulfate DEAE Sepharose Hydroxyapatite Butyl-agarose	[100]
Mouse hybridoma cells	Single-stranded DNA-cellulose	Sepharose Phosphocellulose Hydroxyapatite DEAE Blue-agarose	[175]
Drosophila melanogaster	Single-stranded DNA-cellulose	Phosphocellulose Hydroxyapatite Mono-Q	[104]
Drosophila melanogaster	Single-stranded DNA-cellulose	Ammonium sulfate Phosphocellulose Hydroxyapatite Sepharose	[176]

Immobilized poly(U), poly(A), GTP, single-stranded DNA, and double-stranded DNA columns can retain DNA-binding proteins like polymerases under controlled ionic strength conditions. Hepatitis C polymerase has been purified using poly(U)-Sepharose columns [102], while poliovirus polymerase has been purified with a GTP-agarose column [103]. Since DNA polymerases do not show specificity for a particular DNA sequence, sequence-specific DNA affinity columns are not generally used for their purification. Instead, single-stranded heterogeneous DNA affinity chromatography is often used for the purification of these enzymes [104, 105].

However, DNA affinity chromatography alone does not lead to the complete purification of polymerases. This means that other tools like hydroxyapatite or phosphocellulose columns, heparin-Sepharose, ion exchange, and immunoaffinity chromatography are often used in conjunction with this method. The most widely used steps in the purification of polymerases are phosphocellulose and heparin-Sepharose chromatography. Immunoaffinity chromatography also gives a high degree of purification for polymerases and can be used if an antibody against one of the DNA polymerase subunits is available [106].

Mono Q and phenyl-Sepharose columns are useful in removing nucleic acids and cellular nucleases from DNA polymerases [107]. FPLC columns containing Mono S or Mono Q (from Pharmacia) are used in the final stages of this purification because of their high resolution. These columns have been employed in the purification of HSV polymerase [108] and human DNA polymerases [108, 109]. Even though the availability of these other separation methods means that DNA affinity chromatography is not always necessary for the purification of DNA polymerases [110, 111], it has been shown that this approach does greatly increase the specific activity that is obtained for polymerases [104, 105].

7.7   MECHANISM-BASED AFFINITY CHROMATOGRAPHY

Another tool that has been used with DNA-binding proteins is mechanism-based affinity chromatography. This has been successfully applied to the purification of *Herpes simplex* virus (HSV) type I polymerase [112] by using acyclovir triphosphate as a ligand. This ligand is a specific inhibitor of HSV-1 polymerase. It forms a dead-end complex by incorporating acyclovir monophosphate into the 3' end of newly synthesized DNA in the presence of the appropriate deoxynucleotide 5'-triphosphates encoded by the template.

Reardon used DNA affinity columns that contained acyclovir monophosphate coupled at the 3' end [112]. In the presence of Mg^{2+} and the next required nucleotide encoded by the template (in this case dGTP), the affinity of polymerase increased toward the immobilized acyclovir monophosphate template, leading to formation of the dead-end complex. The polymerase was retained on this column even in the presence of 1 M sodium chloride. However, in the absence of dGTP and Mg^{2+}, HSV type I polymerase could be eluted at lower salt concentrations, such as 0.4 M sodium chloride. This allowed the column to be washed at high salt concentrations in the presence of dGTP and Mg^{2+} to remove contaminant proteins and then eluted at a lower salt concentration in the absence of dGTP and Mg^{2+} to recover the polymerase.

Another example of a mechanism-based separation is the purification of DNA polymerase by using 5'-AMP as a ligand. Nucleoside 5'-monophosphates are selective and competitive inhibitors of the 3'-5' exonuclease activity of *E. coli* DNA polymerase I, which has a binding site for 5'-AMP. However, these agents do not affect this enzyme's polymerase activity. Lee and Whyte [113] have shown that columns containing immobilized 5'-AMP can be used to retain *E. coli* polymerase I. Several 5'-AMP supports based on agarose and cellulose were studied. This includes supports that immobilized the AMP through its N-6 or C-8 atoms, phosphodiester linkage, or ribose ring (after oxidative treatment with periodate). DNA polymerases possessing 3'-5' exonuclease activities were retained by the agarose supports in which the 5'-phosphoryl group and 3'-hydroxyl group of AMP were unsubstituted. The retained enzymes were eluted by increasing the ionic strength of the mobile phase or by using 5'-dGMP as a competing agent. Magnesium was found to reinforce binding of the DNA polymerases to these affinity columns. This method was also useful in separating mammalian DNA polymerase δ from DNA polymerase α, since DNA polymerase δ has 3′-5′ exonuclease activity and was retained by 5′-AMP agarose columns, while DNA polymerase α lacks this activity and was not retained.

7.8 PURIFICATION OF RNA POLYMERASES

RNA polymerases synthesize cellular RNA molecules on DNA and RNA templates. The RNA polymerase of bacteria consists of several subunits. The core enzyme in this case is composed of $\alpha_2\beta\beta'$ subunits. Different subunits associate with this core enzyme and create specificity for promoter recognition [114–116].

In eukaryotes, there are three RNA polymerases: pol I, pol II, and pol III [117]. Each of these polymerases is responsible for the transcription of a subclass of nuclear genes. RNA pol I transcribes only genes encoding large ribosomal RNAs. RNA pol II transcribes all of the cell's protein-coding messenger RNAs and most small nuclear RNAs (snRNAs). The only exception to this is U6 RNA, which is transcribed by RNA pol III. RNA pol III also synthesizes tRNAs and the 5S RNA component of ribosomes. All RNA polymerases function through the formation of a preinitiation complex. In the case of RNA pol II, the preinitiation complex consists of several general transcription factors (such as TFIIB, TFIID, TFIIE, and TFIIH) and promoter-specific transcription factors (such as Sp1, AP1, and others).

RNA polymerase holoenzyme does not show a high specificity toward a particular DNA sequence. Therefore, specific DNA affinity columns are not used for its purification. Instead, heterogeneous DNA-cellulose chromatography is often used. By themselves, these types of columns allow for only the partial purification of RNA polymerase holoenzyme and require the use of other methods for complete purification.

Since both prokaryotic and eukaryotic RNA polymerases form high-molecular-weight preinitiation complexes, size-exclusion supports like Sepharose 6B and Sephacryl S-300 can be used to separate these from low-molecular-weight proteins. For example, in one study, supercoiled DNA containing a promoter sequence was incubated with extracts that allowed formation of the RNA polymerase preinitiation complex. This extract was then subjected to size-exclusion chromatography using a Sepharose C6-2B column. The DNA and large preinitiation complex proteins were eluted in the column void volume, separating them from smaller contaminants in the sample [118]. Heparin-agarose is another common approach used to purify RNA polymerase. This has been used to purify several forms of RNA polymerases from *Bacillus subtilis*, with the different forms of RNA polymerase being resolved upon elution from heparin-agarose or a DNA-cellulose column in the presence of a salt gradient [119].

In another report, a biotinylated single-stranded oligonucleotide containing a sequence from a specific promoter region was bound to streptavidin-coated magnetic beads and streptavidin-agarose. This was used to purify the preinitiation complex in an immobilized-DNA template method [118]. To do this, a crude extract containing the preinitiation complex was passed over the oligonucleotide support in the absence of rNTPs to avoid initiation. After washing the support with a solution that had a low salt concentration to remove nonspecific proteins, the preinitiation complex was eluted with a sample buffer for SDS-PAGE. The eluted proteins were then subjected to immuno-blotting to detect the various components of the preinitiation complex. However, the preinitiation complex obtained by this method was in a denatured state that could not be used for functional studies. In addition, the streptavidin in the affinity column was found to bind to several other proteins in the sample. This binding occurred both to proteins that contained a biocytin prosthetic group and to proteins that bound nonspecifically, such as cellular actin. Since such proteins could also have been retained on the oligonucleotide column, they may have interfered with the results obtained with this support.

Eukaryotic RNA pol II can be separated from its associated transcription factors by using DEAE-cellulose, phosphocellulose, CM-Sepharose, and Mono S columns. Although this is a useful technique, the entire preinitiation complex is required for the functional studies. The reconstituted preinitiation complex can be used for some studies, but there are many unique transcription factors and coactivators for a particular promoter. Thus, a preinitiation complex that has been isolated in the presence of the promoter is better suited for functional studies.

7.9   PURIFICATION OF TELOMERASE

The complete replication of the ends of eukaryotic chromosomes (or telomeres) poses special problems to the conventional DNA replication machinery. Conventional DNA polymerases cannot prime DNA synthesis *de novo*, which results in a loss of telomeric sequence with each cell division. Telomerase is a specialized ribonucleoprotein (RNP) that helps complete the replication of linear chromosomes through the *de novo* synthesis of telomeric repeats. It does this by using a specific sequence in its RNA component as a template [120, 121]. This RNA component is composed of a conserved sequence, which is 3'-AACCCCAAC-5' in the case of *Tetrahymena* [122].

Human telomerase has been purified in a single step using DNA affinity chromatography. This was accomplished using an immobilized oligonucleotide that was complementary to the conserved region of the RNA component of the telomerase. Elution of the bound telomerase was carried out by its displacement in the presence of a 2.5-fold excess of soluble oligonucleotide that was complementary to the sequence on the column support [123]. This gave rise to a 1320-fold purification of telomerase.

Telomerase from nuclear extracts of *Euplotes aediculatus* was enriched on an Affi-Gel-heparin column and further purified on an affinity column that contained an antisense oligonucleotide complementary to the template region for telomerase [122, 124]. Telomerase has also been purified using hydroxyapatite, DEAE-Sepharose, heparin-agarose, phenyl-agarose, and spermine-agarose [125–128]. In addition, glycerol gradient sedimentation along with hydroxyapatite, spermine-agarose, Sepharose CL-6B, phenyl-Sepharose, and DEAE-agarose, or Q-Sepharose have been used to purify *Tetrahymena* telomerase [129].

The affinity purification of telomerase is very effective compared with other methods. However, the affinity columns needed for this application consist of single-stranded DNA. As a result, several single-stranded DNA-binding proteins can bind to the column. To avoid this problem, nonspecific competitor tRNA [123], yeast tRNA, and specific competitor RNAs [122] have been employed. Contamination by single-stranded DNA-binding proteins can also be reduced by adding a large excess of single-stranded DNA to the mobile phase.

7.10   PURIFICATION OF RESTRICTION ENZYMES

Restriction enzymes such as EcoRI exhibit sequence-specific DNA binding, with each restriction endonuclease having a particular target in duplex DNA. This target is generally a specific sequence of four to eight base pairs. The DNA sequence to which these enzymes bind is referred to as a "restriction site" and is often smaller than the consensus sequence that is recognized by transcription factors.

Affinity techniques other than DNA affinity chromatography can be used to purify some restriction enzymes. For instance, Bouriotis et al. [130] used a rapid, two-step affinity procedure for purifying three restriction enzymes (BanII, SacI, and SphI). In this method, inexpensive adsorbents such as agarose-immobilized Procion blue H-B, Procion red H-8BN, or phosphocellulose were used in the first step, followed by high-performance anion-exchange chromatography on a Mono Q column. The SphI and SacI were purified in the first step using Procion red H-8BN and Procion blue H-B, respectively, while phospho-cellulose was used for the purification of BanII. In each case, a gradient with an increasing sodium chloride concentration was used for elution.

7.10.1 DNA Affinity Purification

Since restriction enzymes bind to specific DNA sequences, these sequences can be used as ligands for isolating restriction enzymes. For example, Vlatakis and Bouriotis [131] used sequence-specific DNA affinity columns to purify restriction endonucleases to near-homogeneity. This was demonstrated with the enzymes EcoRI and SphI, which were purified by a two-step method that involved heterogeneous DNA-cellulose chromatography and sequence-specific duplex-oligonucleotide affinity chromatography. The oligonucleotides in this case were ligated to form concatemers and were designed in such a way to give the resulting DNA recognition sequences for one or two restriction enzymes. Since catalysis by restriction enzymes requires the presence of divalent metal ions, all the buffers used with these columns contained EDTA as a chelating agent to prevent column DNA digestion (note: these enzymes still show that DNA binding occurred in the presence of EDTA). A 68.7-fold purification and 20% yield were obtained for EcoRI by this method, and a 157-fold purification and 75% yield were obtained for SphI. Furthermore, the affinity column employed in this study could be used more than 30 times without affecting the degree of enzyme purity that was obtained.

The enzymes HpaI, SphI, and EcoRI were eluted at a low potassium chloride concentration ($0.1\ M$) from columns that did not contain their recognition sequence. But these same enzymes required a higher potassium chloride concentration ($0.3\ M$) to elute them from columns that contained their restriction site. In theory, a group of specific ligands could be constructed that contain restriction sites for a number of different restriction enzymes. This would reduce the cost of such an adsorbent by allowing it to be used to recover multiple enzymes [131].

7.10.2 Catalytic Chromatography

Catalytic chromatography has also been used for the isolation of restriction enzymes. Jurado et al. [2] purified EcoRI restriction endonuclease to apparent homogeneity in a single step using catalytic chromatography, as illustrated in Figure 7.5. This chromatographic technique is similar to that described by Vlatakis and Bouriotis [131], except that elution was instead based only on the catalytic function of the retained enzyme. Since catalysis is the most specific characteristic of an enzyme, it should also provide the greatest chromatographic selectivity for the enzyme. For instance, this method was found to lead to a significantly greater yield (85%) and purification (344-fold purification) compared with standard affinity chromatography (35% yield and 99-fold purification) [2].

One disadvantage of catalytic chromatography is that the purified enzyme will be contaminated by its hydrolyzed DNA product. In the case of EcoRI, the coeluting product did not appear to affect the enzyme's activity; however, the thermodynamic and kinetic properties of other enzymes might be affected if such products are present. If possible, these products should be removed by dialysis after the enzyme has been eluted from the affinity column. Another disadvantage of catalytic chromatography is that the support is

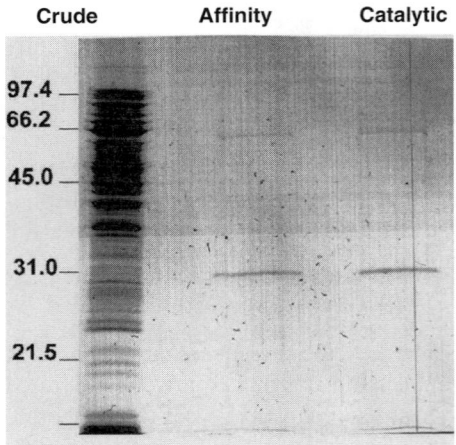

Figure 7.5 Purification of EcoRI from a crude RY13 extract (left lane) by traditional affinity chromatography (central lane) and catalytic chromatography (right lane). In this work, the active fractions eluted by affinity or catalytic chromatography were pooled, dialyzed, lyophilized, separated by SDS-PAGE, and stained with Coomassie Blue. (Reproduced with permission from Jurado, L.A., Drummond, J.T., and Jarrett, H.W., *Anal. Biochem.*, 282, 39–45, 2000.)

destroyed during the catalytic process. This means that the column can be used only once, but this loss is balanced by the value of the purified enzyme and the simplicity and additional purity that can be obtained through catalytic elution.

7.11   RECENT ADVANCES IN DNA AFFINITY CHROMATOGRAPHYS

Despite its high specificity, the use of DNA affinity chromatography alone rarely yields homogeneous proteins. Examples of this have been provided throughout this chapter. However, recent advances have made single-step purifications by DNA affinity chromatography more feasible. Several of these advances are discussed in this section, including the bi-column method, oligonucleotide trapping, and tandem DNA affinity chromatography. The use of aptamers as affinity ligands is also discussed, and further consideration is given to the method of catalytic chromatography.

7.11.1   The Bi-Column Method

The bi-column method is one of the newer approaches that has been developed for DNA affinity chromatography. This method allows the repeated use of DNA affinity columns, utilizing two columns with different specificities and two different elution methods (i.e., heparin elution and salt elution) [132]. This approach, illustrated in Figure 7.6, is based on the observation that different heparin concentrations are needed to elute a given protein from different DNA sequences. Given this fact, the first module of the bi-column method consists of a low-complexity (i.e., short length) DNA or mutant DNA that has a low affinity for the protein to be purified. The second column contains high-complexity (longer sequence) DNA or nonmutant DNA that has a higher affinity for the protein.

In this technique, a cell extract is loaded onto the first column, and nonspecific proteins are washed off using a low-salt buffer. The two columns are then connected in

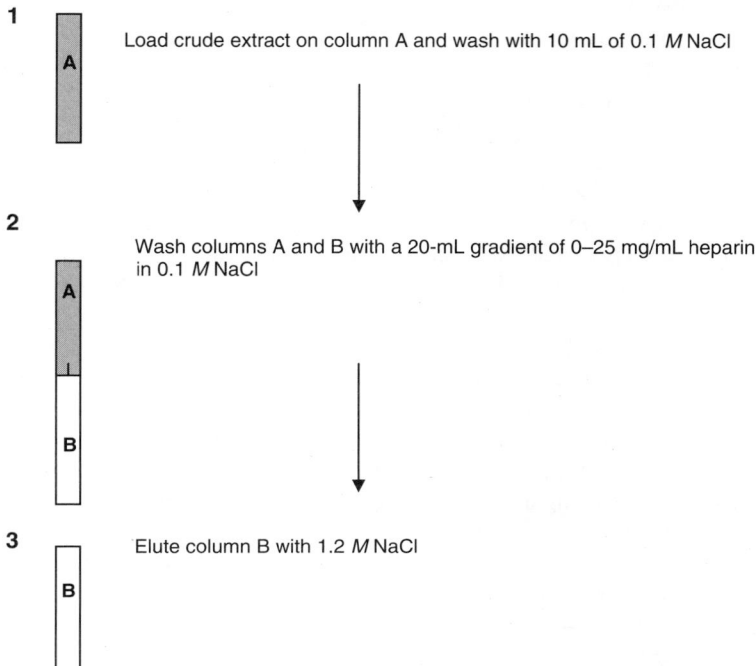

1 Load crude extract on column A and wash with 10 mL of 0.1 *M* NaCl

2 Wash columns A and B with a 20-mL gradient of 0–25 mg/mL heparin in 0.1 *M* NaCl

3 Elute column B with 1.2 *M* NaCl

Figure 7.6 The bi-column method. In this example, column A contains a low-affinity DNA sequence and is washed with 0.1 *M* sodium chloride. Column A is then coupled to column B, which contains a higher affinity DNA sequence. Nonspecific binding proteins will wash out of both columns A and B, while specific binding proteins will remain bound to column B and elute only at higher salt concentrations. (Reproduced with permission from Gadgil, H., Taylor, W.L., and Jarrett, H.W., *J. Chromatogr. A*, 905, 133–139, 2001.)

tandem so that the low-affinity column is on top. These columns are then washed with a buffer containing a heparin concentration that is just high enough to elute the protein of interest from the first column but not from the second. The two columns are then separated, and the protein is eluted from the second column using a buffer with a high salt concentration. Based on this approach, there have been reports in which three different transcription factors have been isolated with a high degree of purity [132].

7.11.2 Oligonucleotide Trapping Method

It is possible that the frequent use of a high concentration of immobilized DNA in a DNA affinity column is problematic. While the amount of coupled DNA in such columns varies widely from one report to the next, the final DNA concentration in these columns is usually higher than the micromolar level. However, a large number of proteins can bind to virtually any DNA sequence with at least a micromolar dissociation constant. Thus, these proteins can bind to the DNA column, decreasing its capacity and making it necessary to resolve these later from the actual protein of interest. This effect has remained a challenge in this type of protein purification.

While lowering the amount of DNA in these affinity columns may be desirable, this also lowers the column-binding capacity and decreases the yield for a protein that is to be isolated by such a column. DNA-protein interactions can occur in solution at low DNA concentrations by using a large volume of solution and a proportionally large amount of DNA.

This has been used with the streptavidin-biotin system, in which biotinylated DNA was incubated with a protein extract in solution [133], with the resulting protein-DNA complex then being trapped by an immobilized streptavidin column. In this technique, appreciable DNA concentrations in the column occur only when most of the protein-DNA complex is trapped. The main drawback of this approach is that streptavidin can also bind nonspecifically to many of the proteins in a crude extract (as discussed previously in this chapter), which will decrease the purity of the desired protein.

Another oligonucleotide-trapping method has been reported in which a double-stranded DNA element has been prepared with a single-stranded tail containing the sequence TGTGTGTGTG (i.e., TG_5). This method is illustrated in Figure 7.7. This DNA element was incubated with a crude extract containing the corresponding transcription factor. The DNA concentrations in this approach were maintained at a low level (50 nM or lower) to discourage binding by nonspecific proteins but still allow binding by the protein of interest. The DNA/extract mixture was next incubated on ice to allow formation of the DNA-protein complex. This mixture was then passed over a column containing the "trap," a single-stranded sequence (CA_5) that was complementary to the single-stranded tail on the DNA. This trap rapidly annealed with the tail and retained the DNA-protein complex. The bound protein was later eluted with a high concentration of salt or heparin. Alternatively, the entire DNA-protein complex in this method could be eluted using a reduced level of salt and a temperature that is high enough to disrupt interactions between the tail and trap. Homogeneous *Xenopus* transcription factor B3 was purified by this trapping method when using heparin and T_{18} as mobile-phase competitors [134]. An advantage of this approach is that competitor concentrations that are compatible with an electrophoretic-mobility-shift assay can be used for trapping.

7.11.3 Tandem DNA Affinity Chromatography

Tandem DNA affinity chromatography is yet another method that has been recently developed. This was used by Ostrowski and Bomsztyk for purifying χ-binding phosphoprotein [135]. This method involved the use of three to six DNA affinity columns connected in series. The top columns contained mutant DNA oligonucleotides, which are unable to bind the transcription factor of interest. The bottom one to two columns contained a specific wild-type sequence that could be bound by the transcription factor.

To use these columns, crude extracts were loaded when all the columns were connected in tandem. After thorough washing, the bottom (wild-type) columns were taken off-line and eluted with a buffer containing a high salt concentration to obtain the purified protein. It has been suggested that this same method could be used to purify multiple transcription factors from a single extract by using different types of wild-type columns.

7.11.4 Catalytic Chromatography

Another new method that has already been mentioned is catalytic chromatography. This makes use of the catalytic activity of an enzyme to allow its specific elution from an affinity column. One application that has already been discussed for this method has been its use in the purification of DNA polymerase [2]. This made use of a DNA affinity column that contained a self-priming template (5′-NH$_2$-ethyl-TTTTTTTTTTGGGGGCCCCC-3′) coupled to Sepharose, as shown in Figure 7.8. It was shown by Jurado et al. [2] that Klenow large fragment could bind to this column in the presence of Mg^{2+} and dCTP, which also prevented degradation of this column by the 3′-5′ exonuclease activity of DNA polymerase. This was the case, since the penultimate 3′ nucleotide is C, so if the exonuclease activity removes it, the polymerase activity will use dCTP to replace it. This allows the enzyme to bind to the column in a futile catalytic cycle.

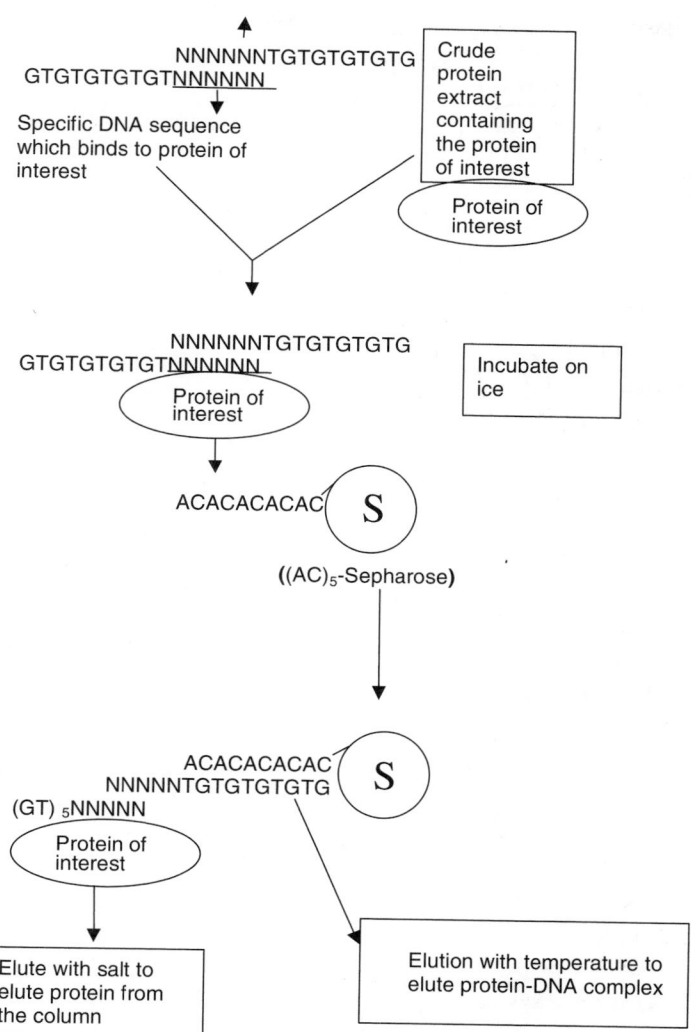

Figure 7.7 Scheme for oligonucleotide trapping. The circled S represents the chromatographic support (in this case, Sepharose). In this method, a 5′-aminoethyl (AC$_5$) oligonucleotide is first chemically coupled to the support. A footprint region having a TG$_5$ extension on both strands is next incubated with an extract containing the protein of interest. This mixture is then passed over the AC$_5$ column. The retained protein can then be eluted by using a buffer containing a high salt concentration, or the DNA protein complex can be eluted at a moderate temperature. The sequence NNNNNN in this diagram represents the bound element (i.e., the footprint). (Reproduced with permission from Gadgil, H. and Jarrett, H.W., *J. Chromatogr. A, 966, 99–110, 2002.)

In this report, the bound polymerase was eluted by including dCTP and dATP in the mobile phase. This elution resulted from the 5′-3′ polymerase activity of the enzyme, which used dATP for the synthesis of a complementary poly(A) region to the column DNA template, followed by enzyme dissociation from the completed product. Using this approach, five well-resolved peaks with DNA polymerase activities were obtained from a crude extract of *E. coli* strain RY13.

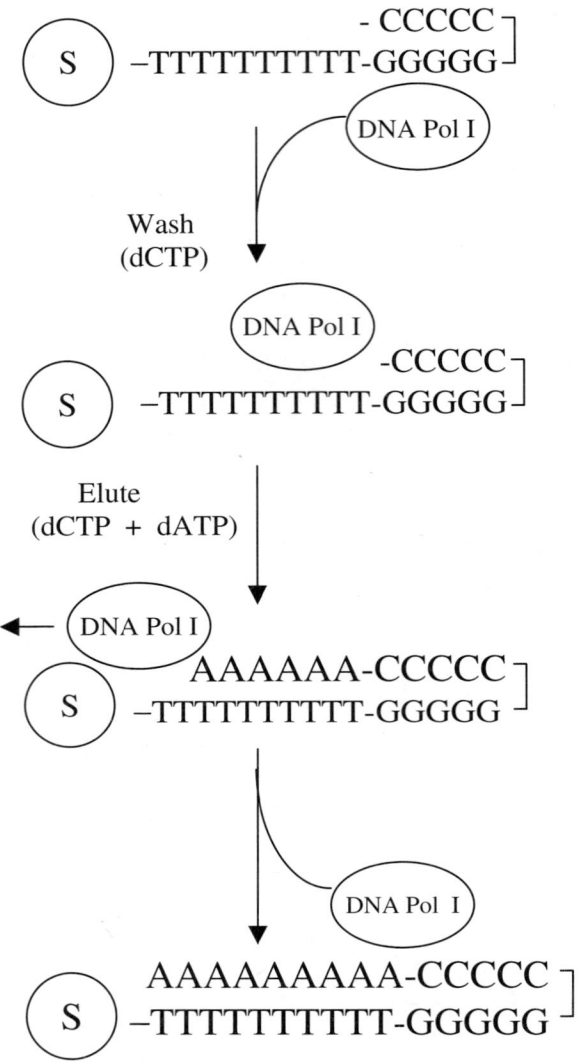

Figure 7.8 Strategy used for catalytic chromatography of DNA polymerase I (DNA Pol I). The DNA sequence shown in this figure was synthesized using the AminoLink reagent (Applied Biosystems) on the last cycle to introduce a 5′-aminoethyl moiety. This sequence was then coupled to cyanogen bromide-activated Sepharose 4B, represented by the support (S). (Reproduced with permission from Jurado, L.A., Drummond, J.T., and Jarrett, H.W., *Anal. Biochem.*, 282, 39–45, 2000.)

As illustrated in Figure 7.9 and Table 7.5, the catalytic elution of EcoR1 [2] can also give a better yield and purity than conventional DNA affinity chromatography. Catalytic chromatography could also find applications in the purification of enzymes. The main requirement is that it must be possible to immobilize the substrate and delay the catalytic process long enough to wash away sample impurities. However, this method does have one major disadvantage in that the affinity column cannot be used for more than one sample, since it is broken down as part of the elution process.

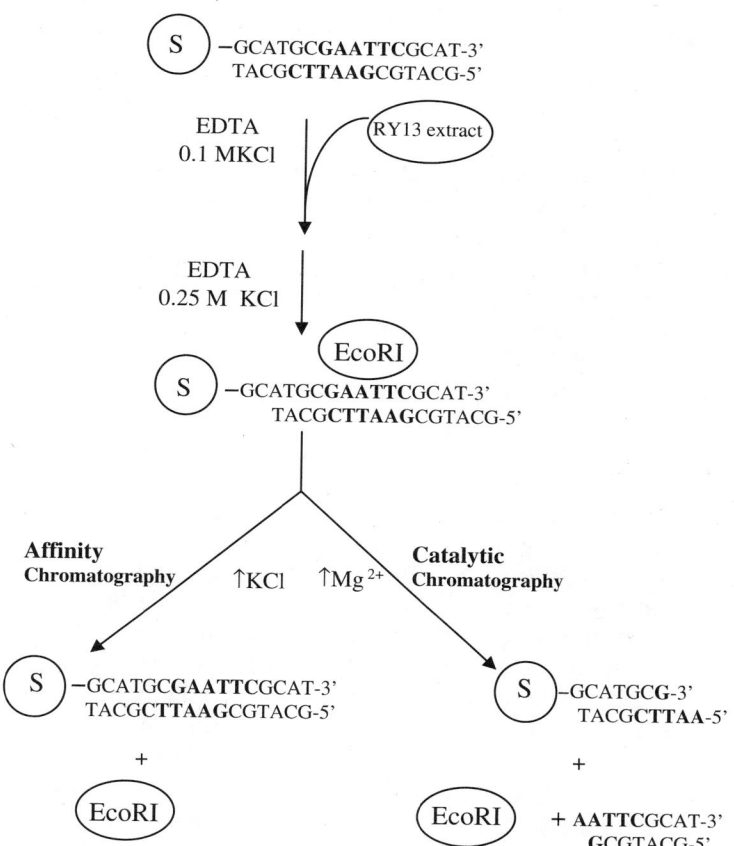

Figure 7.9 Strategy in performing affinity and catalytic chromatography for EcoRI restriction endonuclease. The symbol S denotes the support (Sepharose in this case). In this method, the column was 3 ml in volume and contained about 25 nmol of the DNA oligonucleotide sequence per milliliter of resin. (Reproduced with permission from Jurado, L.A., Drummond, J.T., and Jarrett, H.W., *Anal. Biochem.,* 282, 39–45, 2000.)

Table 7.5 Catalytic Purification of EcoRI

Sample	Total Protein (mg)	Total Units[a]	Specific Activity (Units/mg)	Yield (%)	Fold-Purification
Crude	2.975	7,500	2,521	100	1
Affinity chromatography	0.011	2,660	241,818	35.5 ± 13.7	98.5 ± 12.4
Catalytic chromatography	0.0075	6,378	850,400	85.0 ± 9.5	343.5 ± 48.8

Source: Reproduced with permission from Jurado, L.A., Drummond, J.T., and Jarrett, H.W., *Anal. Biochem.,* 282, 39–45, 2000.

7.11.5 Aptamers

The use of aptamers as affinity ligands is another recent development in affinity chroma-tography. Aptamers are oligonucleotides that have been engineered to have a high affinity for a particular target. Because of this property, aptamers can be used as ligands for the purification of target proteins [136]. To obtain aptamers, oligonucleotides that bind to a given target are separated from a random single-stranded DNA library and enriched by using the SELEX process (systematic evolution of ligands by exponential enrichment). Since the SELEX process can be used to generate aptamers against non-DNA-binding proteins, this enables the use of aptamers as binding agents for a variety of target proteins. A further discussion of aptamers and aptamer development can be found in Chapter 9.

While the details of how aptamer selection is conducted may vary, the overall process is conceptually simple and is an *in vitro* process that mimics molecular evolution. This selection process is performed as follows. First, a protein (or other desired target) is attached to a column. A random library of short (typically less than 60-mer) RNA or DNA oligo-nucleotides is passed through this column. After washing away nonretained RNA or DNA, the bound oligonucleotides are eluted and amplified. This population is then applied to a fresh affinity column for additional rounds of selection and enrichment. After this has been done many times, the final enriched sequences are examined, and a consensus sequence that binds with high affinity to the target is deduced from the population of isolated ligands.

Aptamers have many potential uses in pharmacology [137], chemical analysis [138, 139], and cell biology. For instance, aptamers can be attached to a chromatographic support and used for DNA (or RNA) affinity chromatography. This would allow proteins that do not normally bind DNA (or RNA) to be purified using DNA or RNA affinity chromatog-raphy. Due to their precise three-dimensional structures, aptamers can also be used to separate nontarget compounds. For example, the G-quartet structure of the thrombin aptamer [140] has been used for the stereochemical separation of amino acids [141].

Bahler et al. [142] used an aptamer to isolate a specific RNA-binding protein in a simple and fast one-step purification method. This involved the incubation of a crude sample extract with an *in vitro*-transcribed hybrid RNA containing both an aptamer sequence with a high binding specificity to the antibiotic streptomycin and a putative protein-binding RNA sequence. This mixture was then passed over an affinity column containing streptomycin, which retained the aptamer and any associated RNA-protein complex. The bound proteins were then eluted by adding streptomycin to the mobile phase. In this work, the appropriate RNA motifs containing the hybrid RNA from a crude yeast extract were isolated and used as affinity matrix to purify RNA-binding proteins such as spliceosomal U1A protein and bacteriophage MS2 coat. This gave high yields and purities in only one round of purification. This method was also found to be independent of the exact source of the sample, and it was subsequently used in an *in vitro*-selected antibiotic binding RNA as a tag (i.e., a "Strepto Tag") that could be used for purifying RNA-binding proteins from other organisms.

Romig et al. [136] used a DNA aptamer specific for human L-selectin to create an affinity column for purifying a recombinant human L-selectin-Ig fusion protein. In this study, L-selectin was eluted from the aptamer column by a sodium chloride gradient, a sodium chloride gradient in the presence of magnesium chloride, and an EDTA gradient. Elution with EDTA was found to be the most effective of these methods. When this aptamer column was used as an initial purification step, the L-selectin was obtained with a 1500-fold purification and 83% recovery.

Even though aptamers have been used only since the early 1990s, their applications are rapidly expanding. It is expected in the future that this will become an important component of DNA affinity chromatography. Since DNA is much more stable than RNA, DNA aptamers will most likely find the greatest use for preparative and analytical work. The ability to select an aptamer for any pure biochemical should allow these ligands to extend DNA affinity chromatography to many types of separation schemes.

7.12 SUMMARY AND CONCLUSIONS

It has been shown throughout this chapter that DNA affinity chromatography is an evolving technique. This field began with the use of heterogeneous DNA absorbed on cellulose for the retention of proteins, with elution being conducted at high salt concentrations [4, 5, 143]. Although somewhat crude, this approach has worked well with DNA polymerases, histones, and other nonsequence-specific proteins.

In later work, specific DNA sequences were covalently attached to agarose [144] and used for the purification of transcription factors and restriction enzymes. Another major advance was the use of competitor DNA in the mobile phase [10, 145]. Other improvements have involved the development of more-selective ways for eluting proteins from these columns. Examples include the use of heparin, temperature jumps, and catalytic chromatography.

The stationary phases for DNA affinity columns have also been improving. For example, it is now possible to use HPLC-grade silica for DNA affinity chromatography [146–148]. In the past, only two coupling chemistries [149] were available for such work (i.e., the N-hydroxysuccinimide and aldehyde methods), but neither was well-characterized. However, the cyanogen bromide method has also recently been adapted to silica [177]. These developments are of interest, since the high resolutions that are obtainable with HPLC should greatly improve the separations obtained by DNA affinity chromatography.

The concepts of stationary and mobile phases are also undergoing some changes in this field. For instance, static and covalently coupled DNA may not be the most desirable mode for such separations. Instead, it has been shown that there are clear advantages to performing protein-DNA binding in solution and using affinity chromatography as a means for "trapping" or recovering the resulting complex. This approach has been proved to be especially powerful when combined with the use of competitor DNA and heparin for elution [134]. Another new development in the area of stationary phases has been the rising popularity of aptamers as ligands. With these agents, it is now possible to prepare DNA affinity supports for virtually any biochemical substance.

Perhaps one of the more surprising things about this field is how little sequence-specific DNA affinity chromatography has been used to purify and study DNA repair proteins. This technique has been used to great advantage in a few cases, such as the purification of thymine DNA glycosylase. However, most repair enzymes have instead been purified using denatured, or single-stranded, heterogeneous DNA-cellulose. This approach has been successfully used for many such proteins, but often only in combination with additional purification steps. Synthetic apurinic sites containing oligonucleotides, oligonucleotides containing mismatches, or those containing thymine dimers can be used to facilitate the purification of repair complexes and might also be valuable tools for investigating repair mechanisms.

The combination of catalytic chromatography with a sequence-specific approach might be especially interesting. For example, a double-stranded oligonucleotide containing a mismatch could be coupled to a chromatographic support, packed in a column, and used to isolate the mismatch-repair complex. The components of this complex would conceivably change as ATP, divalent cations, and deoxynucleoside triphosphates are added and these components elute. Such a column could help resolve the complex from other cellular DNA-binding proteins. This could also be used to resolve the added components as they complete their role in catalyzing repair and provide details about the sequence of events that lead to DNA repair.

In summary, DNA affinity chromatography is a widely used and valuable tool. Without this tool, it is hard to imagine further progress being made in understanding genetic regulation (e.g., transcription factors), repair from environmental damage (DNA repair), DNA synthesis (DNA polymerases), and translation (RNA polymerases). This method has also been a key tool in providing highly purified enzymes (e.g., restriction enzymes and polymerases) for use as reagents in molecular and cellular biology. New improvements are constantly being made, and additional applications should continue to appear as work continues with this method.

SYMBOLS AND ABBREVIATIONS

AP	Apurinic/apyrimidinic
AMP	Adenosine monophosphate
ATP	Adenosine triphosphate
BER	Base excision repair
BSA	Bovine serum albumin
^{m}C	Methyl C, or methylated cytosine
CA$_5$	Cytosine-adenine oligonucleotide (CACACACACA)
C/EBP	Creb-enhancer binding protein
CHAPS	3-[(3-Cholamidopropyl) dimethylamino]-2-hydroxy-1-propane sulfonate
ChIP	Chromatin immunoprecipitation
CM-Sepharose	Carboxymethyl-Sepharose
dATP	Deoxyadenosine triphosphate
dCTP	Deoxycytidine triphosphate
dGMP	Deoxyguanosine monophosphate
dGTP	Deoxyguanosine triphosphate
DEAE	Diethylaminoethyl
DNA	Deoxyribonucleic acid
DOC	Deoxycholate
dsDNA	Double-stranded DNA
EDTA	Ethylenediaminetetraacetic acid
EMSA	Electrophoretic mobility shift assay
Ercc1	Excision repair cross-complementation
FPLC	Fast-performance liquid chromatography
G*	Oxoguanine
^{m}G	MethylG, or methylated guanosine
GFP-C/EBP	Green fluorescent protein fused to first 100 amino acids of rat C/EBP
GTP	Guanosine triphosphate
HSV	Herpes simplex virus
IPTG	Isopropyl-thiogalactopyranoside
mRNA	Messenger RNA
NER	Nucleotide excision repair
NF-I	Nuclear factor I
NP40	Nonidet P-40
NTP	Designation for any of the four mononucleotides
Op1	Operator one of the *lac* operon
Pol	Polymerase

Poly(A)	Polyadenylic acid
Poly(U)	Polyuridylic acid
RNA	Ribonucleic acid
RNP	Ribonucleoprotein
RPA	Replication protein A
SDS-PAGE	Sodium dodecyl sulfate polyacrylamide gel electrophoresis
SELEX	Systematic evolution of ligands by exponential enrichment
snRNA	Small nuclear RNA
ssDNA	Single-stranded DNA
T_{18}	Thymidine oligonucleotide 18-mer (TTTTTTTTTTTTTTTTTT)
TE	10 mM Tris-HCl buffer (pH 7.5) with 0.1 mM EDTA
TFIIH	Transcription factor IIH
TG_5	Thymidine-guanidine oligonucleotide (TGTGTGTGTG)
tRNA	Transfer RNA
XPC	Xeroderma pigmentosum complementation group C

REFERENCES

1. Gadgil, H., Jurado, L.A., and Jarrett, H.W., DNA affinity chromatography of transcription factors, *Anal. Biochem.,* 290, 147–178, 2001.
2. Jurado, L.A., Drummond, J.T., and Jarrett, H.W., Catalytic chromatography, *Anal. Biochem.,* 282, 39–45, 2000.
3. Weissbach, A. and Poonian, A., Nucleic acids attached to solid matrices, *Methods Enzymol.,* 34, 463–475, 1974.
4. Alberts, B.M. and Herrick, G., DNA-cellulose chromatography, *Methods Enzymol.,* 21, 198–217, 1971.
5. Alberts, B.M., Amodio, F.J., Jenkins, M., Gutmann, E.D., and Ferris, F.L., Studies with DNA-cellulose chromatography, I: DNA-binding proteins from *Escherichia coli, Cold Spring Harb. Symp. Quant. Biol.,* 33, 289–305, 1968.
6. Gilham, P.T., The synthesis of celluloses containing covalently bound nucleotides, polynucleotides, and nucleic acids, *Biochemistry,* 7, 2809–2813, 1968.
7. Ghislain, J.J. and Fish, E.N., Application of genomic DNA affinity chromatography identifies multiple interferon-alpha-regulated Stat2 complexes, *J. Biol. Chem.,* 271, 12408–12413, 1996.
8. Govindan, M.V. and Gronemeyer, H., Characterization of the rat liver glucocorticoid receptor purified by DNA-cellulose and ligand affinity chromatography, *J. Biol. Chem.,* 259, 12915–12924, 1984.
9. Jarrett, H.W. and Taylor, W.L., Transcription factor-green fluorescent protein chimeric fusion proteins and their use in studies of DNA affinity chromatography, *J. Chromatogr. A,* 803, 131–139, 1998.
10. Kadonaga, J.T., Purification of sequence-specific binding proteins by DNA affinity chromatography, *Methods Enzymol.,* 208, 10–23, 1991.
11. Gadgil, H., Taylor, W.L., and Jarrett, H.W., Comparative studies on discrete and concatemeric DNA-Sepharose columns for the purification of transcription factors, *J. Chromatogr. A,* 917, 43–53, 2001.
12. Benyajati, C., Ewel, A., McKeon, J., Chovav, M., and Juan, E., Characterization and purification of Adh distal promoter factor 2, Adf-2, a cell specific and promoter-specific repressor in *Drosophila, Nucleic Acids Res.,* 20, 4481–4489, 1992.
13. West, M., Mikovits, J., Princler, G., Liu, Y., Ruscetti, F.W., Kung, H., and Raziuddin, Characterization and purification of a novel transcriptional repressor from HeLa cell nuclear extracts recognizing the negative regulatory element region of human immunodeficiency virus-1 long terminal repeat, *J. Biol. Chem.,* 267, 24948–24952, 1992.

14. Sze, J. and Kohlaw, G.B., Purification and structural characterization of transcriptional regulator Leu3 of yeast, *J. Biol. Chem.,* 268, 2505–2512, 1993.

15. Hansen, S.K., Baeuerle, P.A., and Blasi, F., Purification, reconstitution, and IkB association of the c-Rel-p65(RelA) complex, a strong activator of transcription, *Mol. Cell Biol.,* 14, 2593–2603, 1994.

16. Huber, H.E., Edwards, G., Goodhart, P.J., Patrick, D.R., Huang, P.S., Ivey-Hoyle, M., Barnett, S.F., Oliff, A., and Heimbrook, D.C., Transcription factor E2F binds DNA as a heterodimer, *Proc. Natl. Acad. Sci. U.S.A.,* 90, 3525–3529, 1993.

17. Kroeger, K.M. and Abraham, L.J., Magnetic bead purification of specific transcription factors using mutant competitor oligonucleotide, *Anal. Biochem.,* 250, 127–129, 1997.

18. Briggs, M.R., Kadonaga, J.T., Bell, S.P., and Tjian, R., Purification and biochemical characterization of the promoter-specific transcription factor, Sp1, *Science,* 234, 47–52, 1986.

19. Robinson, F.D., Gadgil, H., and Jarrett, H.W., Comparative studies on chemically and enzymatically coupled DNA-Sepharose for purification of a lac repressor chimeric fusion protein, *J. Chromatogr. A,* 849, 403–412, 1999.

20. Gadgil, H. and Jarrett, H.W., Heparin elution of transcription factors from DNA-Sepharose columns, *J. Chromatogr. A,* 848, 131–138, 1999.

21. He, F., Narayan, S., and Wilson, S.H., Purification and characterization of a DNA polymerase beta promoter initiator element-binding transcription factor from bovine testis, *Biochemistry,* 35, 1775–1782, 1996.

22. Gao, B., Jiang, L., and Kunos, G., Transcriptional regulation of alpha1b adrenergic receptors (alpha1bAR) by nuclear factor 1 (NF1): A decline in the concentration of NF1 correlates with the downregulation of alpha1bAR gene expression in regenerating liver, *Mol. Cell Biol.,* 16, 5997–6008, 1996.

23. Fu, G.K. and Markovitz, D.M., Purification of the pets factor: A nuclear protein that binds to the inducible TG-rich element of the human immunodeficiency virus type 2 enhancer, *J. Biol. Chem.,* 271, 19599–19605, 1996.

24. Takeda, Y., Ross, P.D., and Mudd, C.P., Thermodynamics of Cro protein-DNA interactions, *Proc. Natl. Acad. Sci. U.S.A.,* 89, 8180–8184, 1992.

25. Ha, J.-H., Spolar, R.S., and Record, M.T., Jr., Role of the hydrophobic effect in stability of site-specific protein-DNA complexes, *J. Mol. Biol.,* 209, 801–816, 1989.

26. Bracken, C., Carr, P.A., Cavanagh, J., and Palmer, A.G., Temperature dependence of intramolecular dynamics of the basic leucine zipper of GCN4: Implications for the entropy of association with DNA, *J. Mol. Biol.,* 285, 2133–2146, 1999.

27. Foguel, D. and Silva, J.L., Cold denaturation of a repressor-operator complex: The role of entropy in protein-DNA recognition, *Proc. Natl. Acad. Sci. U.S.A.,* 91, 8244–8247, 1994.

28. Spolar, R.S. and Record, M.T., Coupling of local folding to site-specific binding of proteins to DNA, *Science,* 263, 777–784, 1994.

29. Jarrett, H.W., Temperature-dependence of the DNA affinity chromatography of transcription factors, *Anal. Biochem.,* 279, 209–217, 2000.

30. Norbury, C.J. and Hickson, I.D., Cellular responses to DNA damage, *Ann. Rev. Pharmacol. Toxicol.,* 41, 367–401, 2001.

31. Wood, R.D., DNA repair in eukaryotes, *Ann. Rev. Biochem.,* 65, 135–167, 1996.

32. de Laat, W.L., Jaspers, N.G., and Hoeijmakers, J.H., Molecular mechanism of nucleotide excision repair, *Genes Dev.,* 13, 768–785, 1999.

33. Batty, D.P. and Wood, R.D., Damage recognition in nucleotide excision repair of DNA, *Gene,* 241, 193–204, 2000.

34. Fishel, R., Signaling mismatch repair in cancer, *Nat. Med.,* 5, 1239–1241, 1999.

35. Buermeyer, A.B., Deschenes, S.M., Baker, S.M., and Liskay, R.M., Mammalian DNA mismatch repair, *Ann. Rev. Genet.,* 33, 533–564, 1999.

36. Gilham, P.T., Covalent binding of nucleotides, polynucleotides, and nucleic acids to cellulose, *Methods Enzymol.,* 21, 191–197, 1971.

37. Hilbert, T.P., Boorstein, R.J., Kung, H.C., Bolton, P.H., Xing, D., Cunningham, R.P., and Teebor, G.W., Purification of a mammalian homologue of *Escherichia coli* endonuclease III:

Identification of a bovine pyrimidine hydrate-thymine glycol DNAse/AP lyase by irreversible cross linking to a thymine glycol-containing oligoxynucleotide, *Biochemistry*, 35, 2505–2511, 1996.

38. Hazra, T.K., Izumi, T., Venkataraman, R., Kow, Y.W., Dizdaroglu, M., and Mitra, S., Characterization of a novel 8-oxoguanine-DNA glycosylase activity in *Escherichia coli* and identification of the enzyme as endonuclease VIII, *J. Biol. Chem.*, 275, 27762–27767, 2000.

39. Caradonna, S.J. and Cheng, Y.C., DNA glycosylases, *Mol. Cell Biochem.*, 46, 49–63, 1982.

40. Orren, D.K. and Sancar, A., The (A)BC excinuclease of *Escherichia coli* has only the UvrB and UvrC subunits in the incision complex, *Proc. Natl. Acad. Sci. U.S.A.*, 86, 5237–5241, 1989.

41. Orren, D.K. and Sancar, A., Formation and enzymatic properties of the UvrB.DNA complex, *J. Biol. Chem.*, 265, 15796–15803, 1990.

42. Thiagalingam, S. and Grossman, L., Both ATPase sites of *Escherichia coli* UvrA have functional roles in nucleotide excision repair, *J. Biol. Chem.*, 266, 11395–11403, 1991.

43. Oh, E.Y. and Grossman, L., The effect of *Escherichia coli* Uvr protein binding on the topology of supercoiled DNA, *Nucleic Acids Res.*, 14, 8557–8571, 1986.

44. Lin, J.J., Phillips, A.M., Hearst, J.E., and Sancar, A., Active site of (A)BC excinuclease, II: Binding, bending, and catalysis mutants of UvrB reveal a direct role in 3′ and an indirect role in 5′ incision, *J. Biol. Chem.*, 267, 17693–17700, 1992.

45. Zou, Y., Liu, T.M., Geacintov, N.E., and Van Houten, B., Interaction of the UvrABC nuclease system with a DNA duplex containing a single stereoisomer of dG-(+)- or dG-(−)-anti-BPDE, *Biochemistry*, 34, 13582–13593, 1995.

46. Moolenaar, G.F., Franken, K.L., Dijkstra, D.M., Thomas-Oates, J.E., Visse, R., van de Putte, P., and Goosen, N., The C-terminal region of the UvrB protein of *Escherichia coli* contains an important determinant for UvrC binding to the preincision complex but not the catalytic site for 3′-incision, *J. Biol. Chem.*, 270, 30508–30515, 1995.

47. He, Z., Henricksen, L.A., Wold, M.S., and Ingles, C.J., RPA involvement in the damage-recognition and incision steps of nucleotide excision repair, *Nature*, 374, 566–569, 1995.

48. Schaeffer, L., Roy, R., Humbert, S., Moncollin, V., Vermeulen, W., Hoeijmakers, J.H., Chambon, P., and Egly, J.M., DNA repair helicase: A component of BTF2 (TFIIH) basic transcription factor, *Science*, 260, 58–63, 1993.

49. Schaeffer, L., Moncollin, V., Roy, R., Staub, A., Mezzina, M., Sarasin, A., Weeda, G., Hoeijmakers, J.H., and Egly, J.M., The ERCC2/DNA repair protein is associated with the class II BTF2/TFIIH transcription factor, *EMBO J.*, 13, 2388–2392, 1994.

50. Sijbers, A.M., de Laat, W.L., Ariza, R.R., Biggerstaff, M., Wei, Y.F., Moggs, J.G., Carter, K.C., Shell, B.K., Evans, E., de Jong, M.C., Rademakers, S., de Rooij, J., Jaspers, N.G., Hoeijmakers, J.H., and Wood, R.D., Xeroderma pigmentosum group F caused by a defect in a structure-specific DNA repair endonuclease, *Cell*, 86, 811–822, 1996.

51. Donovan, O., Davies, A.A., Moggs, J.G., West, S.C., and Wood, R.D., XPG endonuclease makes the 3′ incision in human DNA nucleotide excision repair, *Nature*, 371, 432–435, 1994.

52. Kim, D.K., Stigger, E., and Lee, S.H., Role of the 70-kDa subunit of human replication protein A (I): Single-stranded DNA binding activity, but not polymerase stimulatory activity, is required for DNA replication, *J. Biol. Chem.*, 271, 15124–15129, 1996.

53. Alani, E., Thresher, R., Griffith, J.D., and Kolodner, R.D., Characterization of DNA-binding and strand-exchange stimulation properties of y-RPA, a yeast single-strand-DNA-binding protein, *J. Mol. Biol.*, 227, 54–71, 1992.

54. Reardon, J.T., Ma, D., and Sancar, A., Overproduction, purification, and characterization of the XPC subunit of the human DNA repair excision nuclease, *J. Biol. Chem.*, 271, 19451–19456, 1996.

55. Masutani, C., Sugasawa, K., Yanagisawa, J., Sonoyama, T., Ui, M., Enomoto, T., Takio, K., Tanaka, K., van der Spek, P.J., and Bootsma, D., Purification and cloning of a nucleotide excision repair complex involving the xeroderma pigmentosum group C protein and a human homologue of yeast RAD23, *EMBO J.*, 13, 1831–1843, 1994.

56. Aboussekhra, A., Biggerstaff, M., Shivji, M.K., Vilpo, J.A., Moncollin, V., Podust, V.N., Protic, M., Hubscher, U., Egly, J.M., and Wood, R.D., Mammalian DNA nucleotide excision repair reconstituted with purified protein components, *Cell*, 80, 859–868, 1995.

57. Kolodner, R.D., Mismatch repair: Mechanisms and relationship to cancer susceptibility, *Trends Biochem. Sci.,* 20, 397–401, 1995.

58. Hickson, I.D., Arthur, H.M., Bramhill, D., and Emmerson, P.T., The *E. coli* uvrD gene product is DNA helicase II, *Mol. Gen. Genet.,* 190, 265–270, 1983.

59. Alani, E., The *Saccharomyces cerevisiae* Msh2 and Msh6 proteins form a complex that specifically binds to duplex oligonucleotides containing mismatched DNA base pairs, *Mol. Cell Biol.,* 16, 5604–5615, 1996.

60. Habraken, Y., Sung, P., Prakash, L., and Prakash, S., ATP-dependent assembly of a ternary complex consisting of a DNA mismatch and the yeast MSH2-MSH6 and MLH1-PMS1 protein complexes, *J. Biol. Chem.,* 273, 9837–9841, 1998.

61. Macpherson, P., Humbert, O., and Karran, P., Frameshift mismatch recognition by the human MutS alpha complex, *Mutat. Res.,* 408, 55–66, 1998.

62. Parker, A., Gu, Y., and Lu, A.L., Purification and characterization of a mammalian homolog of *Escherichia coli* MutY mismatch repair protein from calf liver mitochondria, *Nucleic Acids Res.,* 28, 3206–3215, 2000.

63. Neddermann, P. and Jiricny, J., The purification of a mismatch-specific thymine-DNA glycosylase from HeLa cells, *J. Biol. Chem.,* 268, 21218–21224, 1993.

64. Shinohara, A. and Ogawa, T., Homologous recombination and the roles of double-strand breaks, *Trends Biochem. Sci.,* 20, 387–391, 1995.

65. Kowalczykowski, S.C., Biochemical and biological function of *Escherichia coli* RecA protein: Behavior of mutant RecA proteins, *Biochimie,* 73, 289–304, 1991.

66. West, S.C., Enzymes and molecular mechanisms of genetic recombination, *Ann. Rev. Biochem.,* 61, 603–640, 1992.

67. Cox, M.M., Relating biochemistry to biology: How the recombinational repair function of RecA protein is manifested in its molecular properties, *BioEssays,* 15, 617–623, 1993.

68. Kowalczykowski, S.C., Dixon, D.A., Eggleston, A.K., Lauder, S.D., and Rehrauer, W.M., Biochemistry of homologous recombination in *Escherichia coli, Microbiol. Rev.,* 58, 401–465, 1994.

69. Friedberg, E.C., *DNA Repair and Mutagenesis,* ASM Press, Washington, DC, 1995, pp. 407–593.

70. McGlynn, P., Mahdi, A.A., and Lloyd, R.G., Characterisation of the catalytically active form of RecG helicase, *Nucleic Acids Res.,* 28, 2324–2332, 2000.

71. Chan, S.N., Harris, L., Bolt, E.L., Whitby, M.C., and Lloyd, R.G., Sequence specificity and biochemical characterization of the RusA Holliday junction resolvase of *Escherichia coli, J. Biol. Chem.,* 272, 14873–14882, 1997.

72. Umezu, K., Nakayama, K., and Nakayama, H., *Escherichia coli* RecQ protein is a DNA helicase, *Proc. Natl. Acad. Sci. U.S.A.,* 87, 5363–5367, 1990.

73. Masterson, C., Boehmer, P.E., McDonald, F., Chaudhuri, S., Hickson, I.D., and Emmerson, P.T., Reconstitution of the activities of the RecBCD holoenzyme of *Escherichia coli* from the purified subunits, *J. Biol. Chem.,* 267, 13564–13572, 1992.

74. Haber, J.E., Partners and pathways repairing a double-strand break, *Trends Genet.,* 16, 259–264, 2000.

75. Karran, P., DNA double strand break repair in mammalian cells, *Curr. Opin. Genet. Dev.,* 10, 144–150, 2000.

76. Thompson, L.H. and Jeggo, P.A., Nomenclature of human genes involved in ionizing radiation sensitivity, *Mutat. Res.,* 337, 131–134, 1995.

77. Gupta, R.C., Bazemore, L.R., Golub, E.T., and Radding, C.M., Activities of human recombination protein Rad51, *Proc. Natl. Acad. Sci. U.S.A.,* 94, 463–468, 1997.

78. Mortensen, U.H., Bendixen, C., Sunjevaric, I., and Rothstein, R., DNA strand annealing is promoted by the yeast Rad52 protein, *Proc. Natl. Acad. Sci. U.S.A.,* 93, 10729–10734, 1996.

79. Ramsden, D.A. and Gellert, M., Ku protein stimulates DNA end joining by mammalian DNA ligases: A direct role for Ku in repair of DNA double-strand breaks, *EMBO J.,* 17, 609–614, 1998.

80. Robins, P. and Lindahl, T., DNA ligase IV from HeLa cell nuclei, *J. Biol. Chem.,* 271, 24257–24261, 1996.

81. Gadgil, H., Oak, S.A., and Jarrett, H.W., Affinity purification of DNA binding proteins, *J. Biochem. Biophys. Meth.,* 49, 607–624, 2001.

82. Ristiniemi, J. and Oikarinen, J., Histone H1 binds to the putative nuclear factor I recognition sequence in the mouse alpha 2(I) collagen promoter, *J. Biol. Chem.,* 264, 2164–2174, 1989.

83. Rosenfeld, P.J. and Kelly, T.J., Purification of nuclear factor I by DNA recognition site affinity chromatography, *J. Biol. Chem.,* 261, 1398–1408, 1986.

84. Mannermaa, R.M. and Oikarinen, J., Nucleoside triphosphate binding and hydrolysis by histone H1, *Biochem. Biophys. Res. Commun.,* 182, 309–317, 1992.

85. Jost, J.P. and Hofsteenge, J., The repressor MDBP-2 is a member of the histone H1 family that binds preferentially *in vitro* and *in vivo* to methylated nonspecific DNA sequences, *Proc. Natl. Acad. Sci. U.S.A.,* 89, 9499–9503, 1992.

86. Kerrigan, L.A. and Kadonaga, J.T., Periodic binding of individual core histones to DNA: Inadvertent purification of the core histone H2B as a putative enhancer-binding factor, *Nucleic Acids Res.,* 20, 6673–6680, 1992.

87. Kadonaga, J.T., Purification of sequence-specific binding proteins by DNA affinity chromatography, *Methods Enzymol.,* 208, 10–23, 1991.

88. Soultanas, P. and Wigley, D.B., Unwinding the "Gordian knot" of helicase action, *Trends Biochem. Sci.,* 26, 47–54, 2001.

89. Hall, M.C. and Matson, S.W., Helicase motifs: The engine that powers DNA unwinding, *Mol. Microbiol.,* 34, 867–877, 1999.

90. Thommes, P. and Hubscher, U., DNA helicase from calf thymus: Purification to apparent homogeneity and biochemical characterization of the enzyme, *J. Biol. Chem.,* 265, 14347–14354, 1990.

91. Tuteja, N., Ochem, A., Taneja, P., Tuteja, R., Skopac, D., and Falaschi, A., Purification and properties of human DNA helicase VI, *Nucleic Acids Res.,* 23, 2457–2463, 1995.

92. Wang, J.C., DNA topoisomerases, *Ann. Rev. Biochem.,* 54, 665–697, 1985.

93. Bergerat, A., Gadelle, D., and Forterre, P., Purification of a DNA topoisomerase II from the hyperthermophilic archaeon *Sulfolobus shibatae:* A thermostable enzyme with both bacterial and eucaryal features, *J. Biol. Chem.,* 269, 27663–27669, 1994.

94. Markauskas, A., Tiknius, V., and Marcisauskas, R., Use of immobilized triazine dyes in the purification of DNA topoisomerase I (Topo I) and terminal deoxynucleotidyl transferase (TdT) from calf thymus, *J. Chromatogr.,* 539, 525–529, 1991.

95. Melendy, T. and Ray, D.S., Novobiocin affinity purification of a mitochondrial type II topoisomerase from the trypanosomatid *Crithidia fasciculata, J. Biol. Chem.,* 264, 1870–1876, 1989.

96. Kornberg, A., DNA replication, *J. Biol. Chem.,* 263, 1–4, 1988.

97. Bryant, J.A., Moore, K., and Aves, S.J., Origins and complexes: The initiation of DNA replication, *J. Exp. Bot.,* 52, 193–202, 2001.

98. Arezi, B. and Kuchta, R.D., Eukaryotic DNA primase, *Trends Biochem. Sci.,* 25, 572–576, 2000.

99. Zavitz, K.H. and Marians, K.J., Dissecting the functional role of PriA protein-catalysed primosome assembly in *Escherichia coli* DNA replication, *Mol. Microbiol.,* 5, 2869–2873, 1991.

100. Yamaguchi, M., Hendrickson, E.A., and DePamphilis, M.L., DNA primase-DNA polymerase alpha from simian cells: Modulation of RNA primer synthesis by ribonucleoside triphosphates, *J. Biol. Chem.,* 260, 6254–6263, 1985.

101. Bessman, M.J., Lehman, I.R., Simms, E.S., and Kornberg, A., Enzymatic synthesis of deoxyribonucleic acid, II: General properties of the reaction, *J. Biol. Chem.,* 233, 171–177, 1958.

102. De Francesco, R., Behrens, S.E., Tomei, L., Altamura, S., and Jiricny, J., RNA-dependent RNA polymerase of hepatitis C virus, *Methods Enzymol.,* 275, 58–67, 1996.

103. Barton, D.J., Morasco, B.J., and Flanegan, J.B., Assays for poliovirus polymerase, 3D(Pol), and authentic RNA replication in HeLa S10 extracts, *Methods Enzymol.,* 275, 35–57, 1996.

104. Mitsis, P.G., Chiang, C.S., and Lehman, I.R., Purification of DNA polymerase-primase (DNA polymerase alpha) and DNA polymerase delta from embryos of *Drosophila melanogaster, Methods Enzymol.,* 262, 62–77, 1995.

105. Wang, T.S., Copeland, W.C., Rogge, L., and Dong, Q., Purification of mammalian DNA polymerases: DNA polymerase alpha, *Methods Enzymol.,* 262, 77–84, 1995.

106. Jiang, Y., Zhang, S.J., Wu, S.M., and Lee, M.Y., Immunoaffinity purification of DNA polymerase delta, *Arch. Biochem. Biophys.,* 320, 297–304, 1995.

107. Burgers, P.M., DNA polymerase from *Saccharomyces cerevisiae*, *Methods Enzymol.*, 262, 49–62, 1995.
108. Malkas, L.H. and Hickey, R.J., Expression, purification, and characterization of DNA polymerases involved in papovavirus replication, *Methods Enzymol.*, 275, 133–167, 1996.
109. Lee, M.Y., Jiang, Y.Q., Zhang, S.J., and Toomey, N.L., Characterization of human DNA polymerase delta and its immunochemical relationships with DNA polymerase alpha and epsilon, *J. Biol. Chem.*, 266, 2423–2429, 1991.
110. Joyce, C.M. and Derbyshire, V., Purification of *Escherichia coli* DNA polymerase I and Klenow fragment, *Methods Enzymol.*, 262, 3–13, 1995.
111. Cai, H., Yu, H., McEntee, K., and Goodman, M.F., Purification and properties of DNA polymerase II from *Escherichia coli*, *Methods Enzymol.*, 262, 13–21, 1995.
112. Reardon, J.E., Herpes simplex virus type 1 DNA polymerase: Mechanism-based affinity chromatography, *J. Biol. Chem.*, 265, 7112–7115, 1990.
113. Lee, M.Y. and Whyte, W.A., Selective affinity chromatography of DNA polymerases with associated 3′ to 5′ exonuclease activities, *Anal. Biochem.*, 138, 291–297, 1984.
114. Haldenwang, W.G., The sigma factors of *Bacillus subtilis*, *Microbiol. Rev.*, 59, 1–30, 1995.
115. Zhou, Y.N. and Gross, C.A., How a mutation in the gene encoding sigma 70 suppresses the defective heat shock response caused by a mutation in the gene encoding sigma 32, *J. Bacteriol.*, 174, 7128–7137, 1992.
116. Lonetto, M., Gribskov, M., and Gross, C.A., The sigma 70 family: Sequence conservation and evolutionary relationships, *J. Bacteriol.*, 174, 3843–3849, 1992.
117. Zawel, L. and Reinberg, D., Common themes in assembly and function of eukaryotic transcription complexes, *Ann. Rev. Biochem.*, 64, 533–561, 1995.
118. Roberts, S.G. and Green, M.R., Purification and analysis of functional preinitiation complexes, *Methods Enzymol.*, 273, 110–118, 1996.
119. Wiggs, J.L., Gilman, M.Z., and Chamberlin, M.J., Heterogeneity of RNA polymerase in *Bacillus subtilis*: Evidence for an additional sigma factor in vegetative cells, *Proc. Natl. Acad. Sci. U.S.A.*, 78, 2762–2766, 1981.
120. Greider, C.W. and Blackburn, E.H., A telomeric sequence in the RNA of Tetrahymena telomerase required for telomere repeat synthesis, *Nature*, 337, 331–337, 1989.
121. Shippen-Lentz, D. and Blackburn, E.H., Functional evidence for an RNA template in telomerase, *Science*, 247, 546–552, 1990.
122. Lingner, J. and Cech, T.R., Purification of telomerase from *Euplotes aediculatus*: Requirement of a primer 3′ overhang, *Proc. Natl. Acad. Sci. U.S.A.*, 93, 10712–10717, 1996.
123. Schnapp, G., Rodi, H.P., Rettig, W.J., Schnapp, A., and Damm, K., One-step affinity purification protocol for human telomerase, *Nucleic Acids Res.*, 26, 3311–3313, 1998.
124. Cech, T.R. and Lingner, J., Telomerase and the chromosome end replication problem, *CIBA Found. Symp.*, 211, 20–34, 1997.
125. Tsao, D.A., Wu, C.W., and Lin, Y.S., Molecular cloning of bovine telomerase RNA, *Gene*, 221, 51–58, 1998.
126. Niu, H., Xia, J., and Lue, N.F., Characterization of the interaction between the nuclease and reverse-transcriptase activity of the yeast telomerase complex, *Mol. Cell Biol.*, 20, 6806–6815, 2000.
127. Cohn, M. and Blackburn, E.H., Telomerase in yeast, *Science*, 269, 396–400, 1995.
128. Lue, N.F. and Xia, J., Species-specific and sequence-specific recognition of the dG-rich strand of telomeres by yeast telomerase, *Nucleic Acids Res.*, 26, 1495–1502, 1998.
129. Collins, K., Kobayashi, R., and Greider, C.W., Purification of Tetrahymena telomerase and cloning of genes encoding the two protein components of the enzyme, *Cell*, 81, 677–686, 1995.
130. Bouriotis, V., Zafeiropoulos, A., and Clonis, Y.D., High-performance liquid chromatography for the purification of restriction endonucleases: Application to BanII, SacI, and SphI, *Anal. Biochem.*, 160, 127–134, 1987.
131. Vlatakis, G. and Bouriotis, V., Sequence-specific DNA affinity chromatography: Application to the purification of EcoRI and SphI, *Anal. Biochem.*, 195, 352–357, 1991.
132. Gadgil, H., Taylor, W.L., and Jarrett, H.W., Bi-column method for the purification of transcription factors, *J. Chromatogr. A*, 905, 133–139, 2001.

133. Leblond-Francillard, M., Dreyfus, M., and Rougeon, F., Isolation of DNA-protein complexes based on streptavidin and biotin interaction, *Eur. J. Biochem.*, 166, 351–355, 1987.

134. Gadgil, H. and Jarrett, H.W., An oligonucleotide trapping method for purification of transcription factors, *J. Chromatogr. A*, 966, 99–110, 2002.

135. Ostrowski, J. and Bomsztyk, K., Purification of DNA-binding proteins using tandem DNA-affinity column, *Nucleic Acids Res.*, 21, 1045–1046, 1993.

136. Romig, T.S., Bell, C., and Drolet, D.W., Aptamer affinity chromatography: Combinatorial chemistry applied to protein purification, *J. Chromatogr. B*, 731, 275–284, 1999.

137. Bock, L.C., Griffin, L.C., Latham, J.A., Vermaas, E.H., and Toole, J.J., Selection of single-stranded DNA molecules that bind and inhibit human thrombin, *Nature*, 355, 564–566, 1992.

138. German, I., Buchanan, D.D., and Kennedy, R.T., Aptamers as ligands in affinity probe capillary electrophoresis, *Anal. Chem.*, 70, 4540–4545, 1998.

139. Potyrailo, R.A., Conrad, R.C., Ellington, A.D., and Hieftje, G.M., Adapting selected nucleic acid ligands (aptamers) to biosensors, *Anal. Chem.*, 70, 3419–3425, 1998.

140. Schultze, P., Macaya, R.F., and Feigon, J., Three-dimensional solution structure of the thrombin-binding DNA aptamer d(GGTTGGTGTGGTTGG), *J. Mol. Biol.*, 235, 1532–1547, 1994.

141. Kotia, R.B., Li, L., and McGown, L.B., Separation of nontarget compounds by DNA aptamers, *Anal. Chem.*, 72, 827–831, 2000.

142. Bahler, M., Kroschewski, R., Stoffler, H.E., and Behrmann, T., Rat myr 4 defines a novel subclass of Myosin: Identification, distribution, localization, and mapping of calmodulin-binding sites with differential calcium sensitivity, *J. Cell. Biol.*, 126, 375–389, 1994.

143. Alberts, B.M., Amodio, F.J., Jenkins, M., Gutman, E.D., and Ferris, F.J., Studies with DNA-cellulose chromatography, I: DNA-binding proteins from *Escherichia coli*, *Cold Spring Harbor Symp. Quant. Biol.*, 33, 289–305, 1968.

144. Arndt-Jovin, D.J., Jovin, T.M., Bahr, W., Frischauf, A.-M., and Marquardt, M., Covalent attachment of DNA to agarose: Improved synthesis and use in affinity chromatography, *Eur. J. Biochem.*, 54, 411–418, 1975.

145. Kadonaga, J.T. and Tjian, R., Affinity purification of sequence-specific DNA binding proteins, *Proc. Natl. Acad. Sci. U.S.A.*, 83, 5889–5893, 1986.

146. Goss, T.A., Bard, M., and Jarrett, H.W., High-performance affinity chromatography of DNA, *J. Chromatogr. A*, 508, 279–287, 1990.

147. Goss, T.A., Bard, M., and Jarrett, H.W., High-performance affinity chromatography of messenger RNA, *J. Chromatogr. A*, 588, 157–164, 1991.

148. Solomon, L.R., Massom, L.R., and Jarrett, H.W., Enzymatic syntheses of DNA-silicas using DNA polymerase, *Anal. Biochem.*, 203, 58–69, 1992.

149. Chockalingam, P.S., Gadgil, H., and Jarrett, H.W., DNA-support coupling for transcription factor purification: Comparison of aldehyde, cyanogen bromide, and *N*-hydroxysuccinimide chemistries, *J. Chromatogr. A*, 942, 167–175, 2002.

150. Boorstein, R.J., Cummings, A., Jr., Marenstein, D.R., Chan, M.K., Ma, Y., Neubert, T.A., Brown, S.M., and Teebor, G.W., Definitive identification of mammalian 5-hydroxymethyluracil DNA *N*-glycosylase activity as SMUG1, *J. Biol. Chem.*, 276, 41991–41997, 2001.

151. Hazra, T.K., Izumi, T., Venkataraman, R., Kow, Y.W., Dizdaroglu, M., and Mitra, S., Characterization of a novel 8-oxoguanine-DNA glycosylase activity in *Escherichia coli* and identification of the enzyme as endonuclease VIII, *J. Biol. Chem.*, 275, 27762–27767, 2000.

152. Guyer, R.B., Nonnemaker, J.M., and Deering, R.A., Uracil-DNA glycosylase activity from *Dictyostelium discoideum*, *Biochim. Biophys. Acta*, 868, 262–264, 1986.

153. Seal, G., Tallarida, R.J., and Sirover, M.A., Purification and properties of the uracil DNA glycosylase from Bloom's syndrome, *Biochim. Biophys. Acta*, 1097, 299–308, 1991.

154. Seal, G., Arenaz, P., and Sirover, M.A., Purification and properties of the human placental uracil DNA glycosylase, *Biochim. Biophys. Acta*, 925, 226–233, 1987.

155. Tang, M.S., Nazimiec, M.E., Doisy, R.P., Pierce, J.R., Hurley, L.H., and Alderete, B.E., Repair of helix-stabilizing anthramycin-N2 guanine DNA adducts by UVRA and UVRB proteins, *J. Mol. Biol.*, 220, 855–866, 1991.

156. Myles, G.M. and Sancar, A., Isolation and characterization of functional domains of UvrA, *Biochemistry*, 30, 3834–3840, 1991.

157. Zou, Y., Walker, R., Bassett, H., Geacintov, N.E., and Van Houten, B., Formation of DNA repair intermediates and incision by the ATP-dependent UvrB-UvrC endonuclease, *J. Biol. Chem.*, 272, 4820–4827, 1997.

158. Thomas, D.C., Levy, M., and Sancar, A., Amplification and purification of UvrA, UvrB, and UvrC proteins of *Escherichia coli, J. Biol. Chem.*, 260, 9875–9883, 1985.

159. Nazimiec, M., Lee, C.S., Tang, Y.L., Ye, X., Case, R., and Tang, M., Sequence-dependent interactions of two forms of UvrC with DNA helix-stabilizing CC-1065-N3-adenine adducts, *Biochemistry*, 40, 11073–11081, 2001.

160. Lin, J.J. and Sancar, A., Reconstitution of nucleotide excision nuclease with UvrA and UvrB proteins from *Escherichia coli* and UvrC protein from *Bacillus subtilis, J. Biol. Chem.*, 265, 21337–21341, 1990.

161. Garcia-Maya, M.M. and Buck, K.W., Isolation and characterization of replication protein A (RP-A) from tobacco cells, *FEBS Lett.*, 413, 181–184, 1997.

162. McGlynn, P. and Lloyd, R.G., RecG helicase activity at three- and four-strand DNA structures, *Nucleic Acids Res.*, 27, 3049–3056, 1999.

163. Warner, N. and Rutherford, C.L., Purification and cloning of TF2: A novel protein that binds a regulatory site of the gp2 promoter in dictyostelium, *Arch. Biochem. Biophys.*, 373, 462–470, 2000.

164. Ogretmen, B. and Safa, A.R., Identification and characterization of the MDR1 promoter-enhancing factor 1 (MEF1) in the multidrug resistant HL60/VCR human acute myeloid leukemia cell line, *Biochemistry*, 39, 194–204, 2000.

165. Mikheev, A.M., Mikheev, S.A., Zhang, Y., Aebersold, R., and Zarbl, H., CArG binding factor A (CBF-A) is involved in transcriptional regulation of the rat Ha-ras promoter, *Nucleic Acids Res.*, 28, 3762–3770, 2000.

166. Lin, Y.H., Miyamoto, C., and Meighen, E.A., Purification and characterization of a luxO promoter binding protein LuxT from *Vibrio harveyi, Protein Expr. Purif.*, 20, 87–94, 2000.

167. Lin, Y.H., Miyamoto, C., and Meighen, E.A., Cloning and functional studies of a luxO regulator LuxT from *Vibrio harveyi, Biochim. Biophys. Acta*, 1494, 226–235, 2000.

168. Polanowska, J., Fabbrizio, E., Le Cam, L., Trouche, D., Emiliani, S., Herrera, R., and Sardet, C., The periodic down regulation of Cyclin E gene expression from exit of mitosis to end of G(1) is controlled by a deacetylase- and E2F-associated bipartite repressor element, *Oncogene*, 20, 4115–4127, 2001.

169. Garcia-Maya, M.M. and Buck, K.W., Purification and properties of a DNA primase from *Nicotiana tabacum, Planta*, 204, 93–99, 1998.

170. Waga, S., Bauer, G., and Stillman, B., Reconstitution of complete SV40 DNA replication with purified replication factors, *J. Biol. Chem.*, 269, 10923–10934, 1994.

171. Tsurimoto, T. and Stillman, B., Purification of a cellular replication factor, RF-C, that is required for coordinated synthesis of leading and lagging strands during simian virus 40 DNA replication *in vitro, Mol. Cell Biol.*, 9, 609–619, 1989.

172. Cao, Q.P., McGrath, C.A., Baril, E.F., Quesenberry, P.J., and Reddy, G.P., The 68 kDa calmodulin-binding protein is tightly associated with the multiprotein DNA polymerase alpha-primase complex in HeLa cells, *Biochemistry*, 34, 3878–3883, 1995.

173. Jazwinski, S.M. and Edelman, G.M., A DNA primase from yeast: Purification and partial characterization, *J. Biol. Chem.*, 260, 4995–5002, 1985.

174. Itaya, A., Hironaka, T., Tanaka, Y., Yoshihara, K., and Kamiya, T., Purification and properties of a specific primase-stimulating factor of bovine thymus, *Eur. J. Biochem.*, 174, 261–266, 1988.

175. Tseng, B.Y. and Ahlem, C.N., A DNA primase from mouse cells: Purification and partial characterization, *J. Biol. Chem.*, 258, 9845–9849, 1983.

176. Kaguni, L.S., Rossignol, J.M., Conaway, R.C., and Lehman, I., Isolation of an intact DNA polymerase-primase from embryos of *Drosophila melanogaster, Proc. Natl. Acad. Sci. U.S.A.*, 80, 2221–2225, 1983.

177. Jurado, L.A., Mosely, J., and Jarrett, H.W., Cyanogen bromide activation and coupling of ligands to diol-containing silica for high-performance affinity chromatography: Optimization of conditions, *J. Chromatogr. A*, 971, 95–104, 2002.

8

Boronate Affinity Chromatography

Xiao-Chuan Liu

Department of Chemistry, California State Polytechnic University,
Pomona, CA

William H. Scouten

College of Sciences, University of Texas at San Antonio,
San Antonio, TX

CONTENTS

8.1   INTRODUCTION

Boronate affinity chromatography can be defined as a chromatographic method that uses a boronic acid as an affinity ligand. Retention in this technique is based mainly on the interaction between boronic acids and *cis*-diol compounds. The binding between borate and *cis*-diols was discovered more than 100 years ago. By the 1940s, the interaction between borate and *cis*-diols had been employed as a tool in the analysis of carbohydrates [1]. In the 1950s, borate/*cis*-diol interactions were used for separations in zone electrophoresis by employing borate buffers [2], and in the 1960s these same buffers were used with borate/*cis*-diol systems in ion-exchange chromatography [3]. However, it was not until the 1970s that researchers developed immobilized boronate columns [4].

Boronate affinity columns were first employed for the separation of sugars and nucleic acid components by Weith et al. in 1970 [4]. Since then, this technique has been exploited for the separation of a wide variety of *cis*-diol compounds, including nucleosides, nucleotides, nucleic acids, carbohydrates, glycoproteins, and enzymes. Earlier reviews of this field are given in the literature [5, 6], while specific experimental protocols used in boronate affinity chromatography are described by Liu and Scouten [7]. A few examples of recent applications include the separation of mistletoe lectins [8] and neoglycoproteins [9], the analysis of glycohemoglobin [10] and the determination of hemoglobin HbA1c in diabetic patients [11], the purification of catechol siderophores from bacterial culture supernatants [12], the purification of thymine glycol DNA and nucleotides [13], assays for glycated lipoproteins [14], and affinity chromatography of serine proteases using peptide boronic acids as ligands [15].

Other methods besides affinity chromatography have made use of the interactions between boronates and *cis*-diols. Examples include an enzyme-linked boronate immunoassay (ELBIA) for glycated albumin [16], fluorescent sensors for carbohydrates [17], and photometric sensors for boronic acids [18]. Borate/*cis*-diol interactions have also been used in molecular imprinting [19, 20], a technique for synthesizing tailor-made affinity matrices and enzyme mimics [21, 22].

8.2   BORONATE/ANALYTE INTERACTIONS

8.2.1   Primary Interactions

The key interaction in boronate chromatography is the esterification that occurs between a boronate ligand and a *cis*-diol compound. Ideally, this esterification requires that the two hydroxyl groups of the diol be on adjacent carbon atoms and in an approximately coplanar configuration (i.e., they should occur as a 1,2-*cis*-diol). Interactions between boronates and 1,3-*cis*-diols or trident interactions between boronates and *cis*-inositol or triethanolamine can occur as well, but 1,2-*cis*-diols give the strongest ester bonds [23, 24]. Since catechol-containing compounds have two adjacent and coplanar hydroxyl groups, these chemicals can also interact readily with boronate ligands. In addition, trigonal boronates can react directly with *cis*-diols in organic solvents, but the rate of esterification is several orders of magnitude lower than it is in an aqueous solution [20].

The mechanism of interaction between boronic acids and *cis*-diols is not fully understood. In aqueous solution under basic conditions, the boronate is hydroxylated and goes from a trigonal coplanar form to a tetrahedral boronate anion, which can then form esters with *cis*-diols. This process is illustrated in Figure 8.1. The resulting cyclic diester can be hydrolyzed under acidic conditions, thus reversing the reaction. This reaction is unusual in that, although two covalent bonds form, it is quite reversible in aqueous solution.

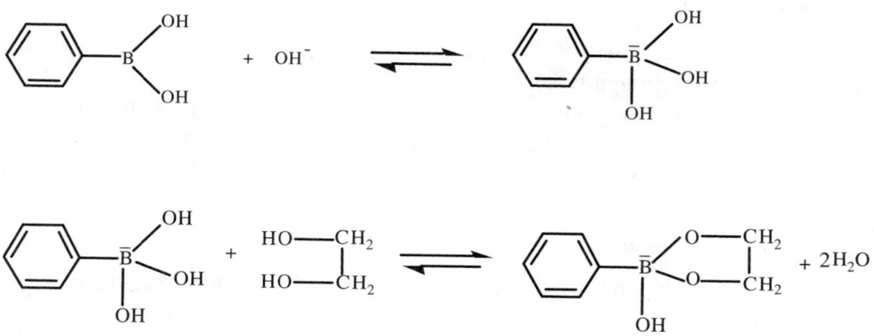

Figure 8.1 The interaction between a boronic acid and a *cis*-diol in aqueous solution.

The boronate diester bond strength is not well-characterized, and only a few dissociation constants have been reported for phenylboronic acid diesters. Those that have been reported include dissociation constants of 2.3×10^{-3} M for adonitol, 1.1×10^{-3} M for dulcitol [25], 3.3×10^{-3} M for mannitol [26], and 5.9×10^{-3} M for NADH [27]. In addition, it is known that the dissociation constant for 4-(*N*-methyl)carboxamido-benzeneboronic acid with D-fructose diester is 1.2×10^{-4} M [28]. This type of binding is relatively weak compared with that observed for many of the biological ligands used in affinity chromatography and is more typical of the interaction strengths that are used in weak-affinity chromatography, as discussed in Chapter 4.

8.2.2 Secondary Interactions

Although boronate/*cis*-diol ester formation is the main basis for boronate affinity chromatography, there are several secondary interactions that sometimes play an important role in retention. Four types of secondary processes that can occur on boronate columns are hydrophobic interactions, ionic interactions, hydrogen bonding, and coordination interactions.

8.2.2.1 Hydrophobic Interactions

Most boronate ligands used in affinity columns are aromatic boronates. These contain a phenyl ring, which can give rise to hydrophobic interactions or aromatic π-π interactions. Hydrophobic interactions can, in certain cases, provide additional selectivity to the column. But they can also lead to the nonspecific adsorption of analytes such as proteins. As a result, there have been many efforts to synthesize aliphatic boronate ligands to overcome such adsorption [29]. Another way to reduce hydrophobic effects in boronate columns is to keep the ionic strength of the mobile phase low. For reasons given in the next section, ionic strengths down to 50 mM are generally recommended for such columns [7].

8.2.2.2 Ionic Interactions

The negative charge of the active tetrahedral boronate can produce coulombic attraction or repulsion for ionic analytes. In general, this effect is weaker than boronate/*cis*-diol ester formation. As was true for hydrophobic effects, there are some cases in which ionic interactions can provide additional selectivity for boronate columns. But in other cases, these secondary interactions are undesirable. To decrease ionic effects, the ionic strength of the mobile phase should be kept high; however, this will also lead to stronger hydrophobic interactions. A good compromise for minimizing both hydrophobic and ionic interactions is to use mobile phases with ionic strengths between 50 and 500 mM [7].

8.2.2.3   Hydrogen Bonding

Since a boronate acid has two hydroxyls (or three in the active tetrahedral form), it has several possible sites for hydrogen bonding. Although the effect of this added interaction is usually small, in special circumstances hydrogen bonding can be strong enough to be the main mechanism for retention on a boronate column. An example of this occurs during the isolation of serine proteases by boronate affinity chromatography [30].

8.2.2.4   Coordination Interactions

Because a trigonal uncharged boronate contains a boron atom with an empty orbital, this can serve as an electron receptor for a coordination interaction. Unprotonated amines are good electron donors, and when an amine donates a pair of electrons to boron, the boron atom becomes tetrahedral (i.e., the more active form); this explains why amines can serve to promote boronate/*cis*-diol esterification. However, if there is a hydroxyl group adjacent to the amine, this hydroxyl group can also interact with the boronate, thus blocking esterification between the boronate ligand and analyte. It is for this reason that Tris and ethanolamine derivatives should be avoided in buffers used for boronate affinity chromatography. Furthermore, carboxyl groups can serve as electron donors for these coordination interactions. Together with α-hydroxyl groups, carboxyls can form stable complexes with boronates, as demonstrated by the esterification of lactic acid or salicylic acid with boronate affinity columns [31].

8.3   BORONATE LIGANDS AND SUPPORTS

8.3.1   Boronate Ligands

The most common ligand used in boronate affinity chromatography is 3-aminophenylboronic acid, also known as 3aPBA (see Figure 8.2a). The anilino group on this ligand is used to couple it to solid supports. The use of a *meta*-amino substitution also lowers this ligand's pK_a. The pK_a of 3aPBA is 8.8 [32], so for optimum binding to analytes, the pH of the mobile phase should be as high as reasonably possible. All current applications using 3aPBA have a mobile-phase pH greater than 8.0. However, in many cases analytes can lose their biological activities at such high pH values. The pH stability of the support that contains the boronate ligand must also be considered, as discussed in Chapter 2. Thus, the current need for relatively high pH buffers is a major limitation to the use of 3aPBA as an affinity ligand.

A number of efforts have been made to produce better boronate ligands. For instance, Yurkevich et al. prepared dextran and cellulose containing bipolar 2-(((4-boronophenyl)-methyl)-ethylammonio)ethyl and 2-(((4-boronophenyl)-methyl)-diethylammonio)ethyl [33]. These supports were used for separating ribonucleosides such as adenosine, cytidine, guanosine, and uridine at pH 8.0. These matrices were found to bind adenosine in a nearly pH-independent manner from pH 2.5 to 8.2. However, the most likely mechanism for this adsorption was ionic interactions and not boronate/*cis*-diol esterification.

Akparov and Stepanov coupled *p*-(ω-aminoethyl)phenyl-boronate to CH-Sepharose for performing chromatography with serine proteases at pH 7.5 [30]. However, retention in this system was due to coordinated hydrogen bonding in the active site where the boronate served as a transition-state intermediate [34]. A similar phenomenon was observed with porcine pancreatic lipase [35]. In another study, Elliger et al. prepared poly(*p*-vinylbenzeneboronic acid)-coated porous polystyrene beads to separate vicinal diols near pH 8 [36]. All analytes considered were small organic compounds. As expected, the observed interaction between *cis*-diols and these boronate matrices was highly pH-dependent.

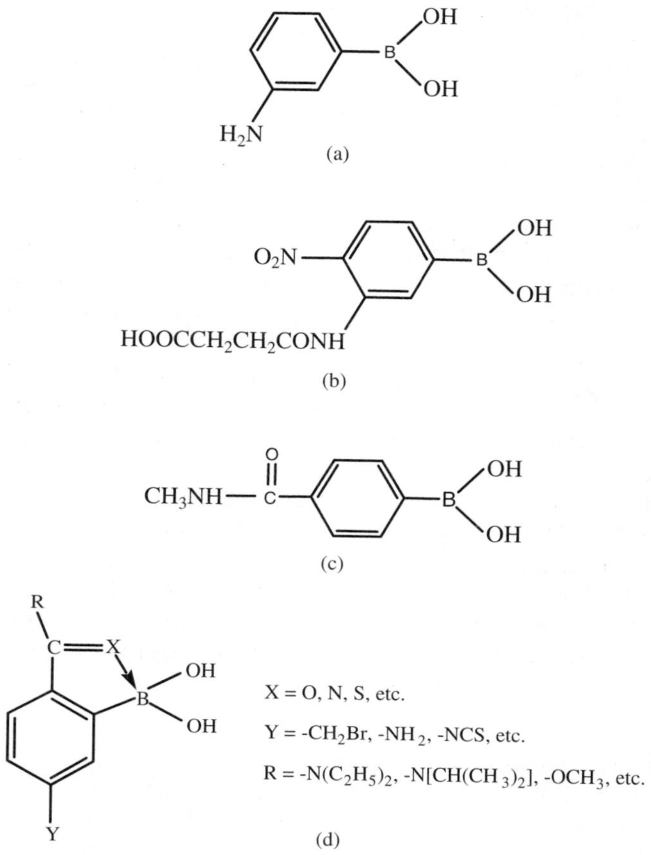

Figure 8.2 Structures of various boronate ligands: (a) 3-aminophenylboronic acid, (b) *N*-(4-nitro-3-dihydroxyborylphenyl)succinamic acid, (c) 4-(*N*-methyl)carboxamido-benzeneboronic acid, (d) a new type of boronate ligand with an internal coordination bond.

Wulff synthesized 2-dimethylaminomethylphenyl-boronic acid, in which the intramolecular B-N bond was established by ^{11}B NMR [20]. This ligand was shown to be highly specific for aliphatic *cis*-diols, but bound poorly to aromatic *cis*-diols, amines, and monoalcohols. No data have been reported on the interactions of this ligand with large molecules such as glycoproteins or carbohydrates.

A number of researchers have attempted to introduce a strong electron withdrawing group into the phenyl ring of aromatic boronates, thus lowering the pK_as of these ligands. Myohanen et al. coupled 3-nitro-4-carboxamidobenzeneboronic acid to Sepharose-CL6B and found that the glycoprotein α-glucosidase was retained by this support at pH 7.4 [37]. However, at pH 8.5 and 6.5, the chromatographic behavior of this material was totally different. This was thought to have been caused by the unstable amide linkage between the ligand and support, which would have been rapidly hydrolyzed at an alkaline pH [28]. Johnson coupled a mixture of 2-nitro-3-succinamido-benzeneboronic acid and 3-succinamido-4-nitro-benzeneboronic acid to aminoethyl polyacrylamide beads and used these to separate tRNAs at pH 7 [38]. Unfortunately, it appears that 2-nitro-3-succinamido-benzeneboronic acid gradually hydrolyzes, and the freed amine groups can create strong ionic interactions. A report by Singhal et al. described a similar approach in which N-(4-nitro-3-dihydroxyborylphenyl) succinamic acid was coupled to a porous, semirigid spherical gel of vinyl polymer (see Figure 8.2b) [39]. They found that this matrix offers enhanced binding of many cis-diol compounds.

Hageman and coworkers synthesized 4-(N-methyl) carboxamido-benzeneboronic acid (see Figure 8.2c) and determined that it had a lower pK_a (7.86) than 3aPBA [28]. This gave strong cis-diol ester formation with D-fructose, with a binding constant of 8600 M^{-1}. They suggested that this ligand should provide sufficient acidity, binding capacity, and hydrolytic stability to make it an excellent boronate ligand for affinity supports. However, this ligand has not yet been coupled to such a support, and no further publications have appeared describing its use.

Liu and Scouten reported another new type of boronate ligand with the general structure shown in Figure 8.2d [40]. X-ray crystallography showed that an internal coordination bond was formed between atom X and the boron atom in this ligand [41]. This means that the boron is in a tetrahedral conformation, which is favorable for esterification between a boronate and cis-diol. In solution, this type of ligand can esterify catechol at neutral pH, as demonstrated by ^{11}B NMR studies. Subsequently, when this ligand is immobilized onto a solid matrix, chromatographic analysis of catechol and horseradish peroxidase can be performed at neutral pH. Aliphatic boronate ligands were studied by this research group [29]. Their results indicate that although these ligands can be used in affinity chromatography for a short period of time, their lack of stability is an obstacle to prolonged use.

8.3.2 Solid Supports

As in other types of affinity chromatography [42], the support material can play an important role in boronate-based separations. Materials that have been used in boronate affinity chromatography include dextran [33], cellulose [33, 40], agarose [30, 37, 40], polyacrylamide [38], silica [43], polystyrene [36], and polymethacrylate [44]. Boronate affinity matrices are commercially available from several major chemical and biochemical suppliers. The list given in Table 8.1 is representative of the current market. The ligand for all these products is 3-aminophenylboronic acid. Users can choose their own products based upon properties such as ligand capacity, mechanical stability, hydrophilicity/hydrophobicity, porosity, and cost. A further discussion of these general types of supports and their properties can be found in Chapter 2.

Table 8.1 Commercially Available Boronate Matrices

Supplier	Name of Boronate Affinity Gel	Support Material
Aldrich	Boric acid gel	Polymethacrylate
BioRad	Affi-Gel 601 gel	Polyacrylamide
Pierce	Immobilized boronic acid gel	Polyacrylamide
Sigma	m-Aminophenylboronic acid-acrylate	Acrylic beads
Tosoh	m-Aminophenylboronic acid-agarose	Agarose
Tosoh	TSKgel Boronate-5PW Column	Polymethacrylate

Note: Data collected March 2005.

8.4   APPLICATIONS OF BORONATE AFFINITY CHROMATOGRAPHY

Boronate affinity chromatography has been employed in the separation of several types of molecules. Specific categories of analytes for which this method has been used include (1) carbohydrates; (2) nucleosides, nucleotides, and nucleic acids; (3) glycoproteins and enzymes; and (4) miscellaneous small molecules. Further details on each of these applications are given in the following sections.

8.4.1   Carbohydrates

Carbohydrates containing *cis*-diols can bind to boronate ligands with a strength that is proportional to the number of *cis*-diol groups on the analyte. In the 1960s, Bourne et al., found that adding phenylboronic acid to developing solvents in paper chromatography enhanced the mobility of sugars that had *cis*-diol or *cis*-triol groups [45]. This technique is also suitable for separating free sugars (that have *cis*-diols) from their reduced polyol counterparts, since the reduced sugars may not have the *cis*-diols needed to esterify with boronate.

Boronate chromatography has been used for separating monosaccharides and oligosaccharides, with compounds containing a 1,2-*cis*-diol binding the most strongly to boronate gels [4, 43, 46]. Because of their strong binding to boronates, carbohydrates such as sorbitol or mannitol are often used as competing agents for the elution of other analytes from boronate columns. In addition, isomeric pentose phosphates have been separated by boronate affinity chromatography [4, 43, 46, 47]. Resolution of D- and L-mannopyranoside racemates using boronate affinity chromatography was reported by Wulff and Vesper using a matrix that was synthesized by molecular imprinting (or templated polymerization) technology [48]. Figure 8.3 shows a chromatogram obtained for a racemic mixture of D/L-mannopyranoside in such an application [49].

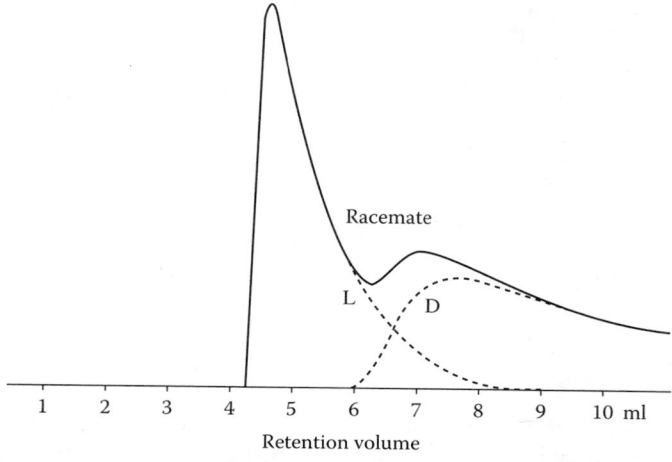

Figure 8.3   Chromatographic separation of D- and L-4-nitrophenyl mannopyranoside on a boronate affinity column. Conditions: solvent, methanol/piperidine (98:2); temperature, 65°C. The separation factor between these peaks was 1.85 and the optical yield was 87%. (Reproduced with permission from Wulff, G., Dederichs, W., Grotstollen, R., and Jupe, C., in *Affinity Chromatography and Related Techniques*, Gribnau, T.C.J., Visser, J., and Nivard, R.J.F., Eds., Elsevier, Amsterdam, 1982, pp. 207–216.)

There have been few applications involving polysaccharides on boronate columns. This is because the internal glycosidic linkages of polysaccharides minimizes the number of *cis*-diols they contain and makes only the terminal residuals available for boronate esterification. This might be the reason why many common polysaccharides such as starch, agarose, methylcellulose, and inulin do not interact with boronate. However, other compounds like gums and mucilages do react with borate [26].

8.4.2 Nucleosides, Nucleotides, and Nucleic Acids

Ribose has a 1,2-*cis*-diol group at the 2′,3′ position that can give rise to strong interactions with boronate. As a result, boronate affinity chromatography has been used successfully in separating various compounds that contain this sugar, such as ribonucleosides, ribonucleotides, and RNA. Figure 8.4 shows a chromatogram for the purification of tRNA from *Escherichia coli* [50].

Since there is no 2′-hydroxyl in DNA, it does not esterify with boronate matrices. Thus, boronate affinity chromatography can easily separate RNA from DNA [50, 51]. Since 3′-phosphorylated ribonucleotides do not bind to boronate gels for the same reason, these also can be readily separated from RNA [52]. Oligonucleotides of RNA and large RNA molecules only have a *cis*-diol at their 3′-end, so their binding to boronate gels is relatively weak. In this case, longer RNA chain lengths correlate with weaker binding.

Messenger RNA (mRNA) can be isolated using boronate affinity chromatography [53]. Furthermore, boronate affinity chromatography can be used for separating aminoacylated

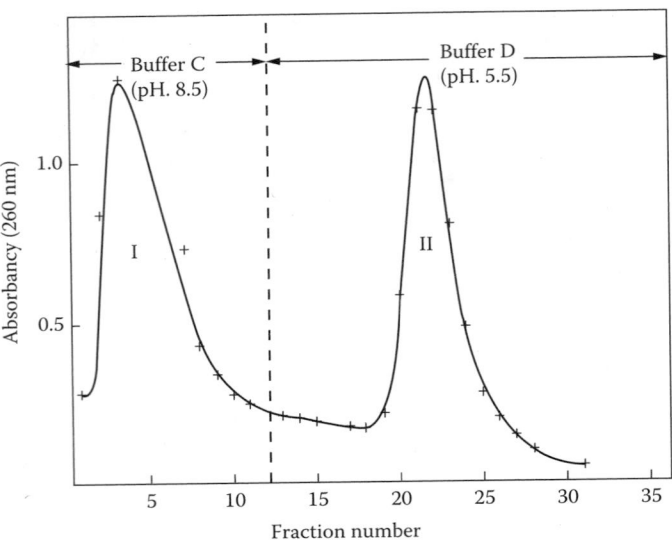

Figure 8.4 Purification of crude *E. coli* tRNA from unchargeable polynucleotides by a borate gel column. Peak II was produced by the tRNA, while peak I was due to other polynucleotides. Buffer C was pH 8.5, 0.05 *M* morpholine HCl containing 0.1 *M* MgCl and 1 m*M* 2-mercaptoethanol; buffer D was pH 5.5, 0.05 *M* sodium morpholinoethanesulfonate containing 0.1 *M* MgCl and 1 m*M* 2-mercaptoethanol. In this separation, 90 A_{260} units of tRNA dissolved in about 1 ml of buffer C was applied to a 25 cm × 1 cm column. Elution was carried out with buffers C and D, as indicated by the arrows. The flow rate was 50 to 60 ml/h, and 4.6 ml was collected in each fraction. (Reproduced with permission from Schott, H., Rudloff, E., Schmidt, P., Roychoudhury, R., and Kossel, H., *Biochemistry*, 12, 932–938, 1973.)

tRNA from non-aminoacylated tRNA [54]. Unfortunately, the high pH required in boronate chromatography may hydrolyze the amino acid-tRNA bond [50]. Since dinucleotide cofactors, such as NAD(H) and FAD, have more than one accessible *cis*-diol, they bind more strongly to boronate chromatography gels than do mononucleotides or oligonucleotides [55].

Secondary interactions are important in the binding between nucleic acid components and boronates. The negatively charged phosphate groups on these analytes cause strong ionic repulsion with the boronate ligand, which may weaken or prevent binding of *cis*-diols on other parts of the analyte. To minimize this effect, divalent cations such as Mg^{2+} are commonly used in buffers for these analytes in boronate affinity chromatography. The nature of the base in the nucleic acid is also important, since purine bases (adenine and guanine) have stronger binding than pyrimidine bases (cytosine and thymine). This is probably due to differences in the hydrophobic interactions and hydrogen bonding between aromatic boronates and these bases [54].

8.4.3 Glycoproteins and Enzymes

One of the most important uses of boronate affinity chromatography is in the separation of glycosylated hemoglobin from nonglycosylated hemoglobin [56–61]. In humans, glucose and other aldoses can glycosylate hemoglobin at its amino terminal end and certain lysine residues. The concentration of glycosylated hemoglobin that is present in blood is known to reflect the blood glucose concentration over the past one or two months. Thus, measurement of glycosylated hemoglobin levels is an important tool in the clinical management of diabetes [6]. Boronate affinity chromatography is a rapid, simple, and accurate means for assaying glycosylated hemoglobin for this purpose. In addition, glycosylated albumin, which can also be used for evaluation of diabetic states, can be determined by boronate chromatography [62]. Clinical applications of affinity chromatography, including boronate chromatography, were recently reviewed by Hage [63]. An example of the clinical use of boronate affinity chromatography in glycated hemoglobin analysis is provided in Chapter 17.

Other glycosylated proteins have also been isolated by boronate affinity chromatography. These include human immunoglobulins [64, 65], γ-glutamyltransferase [66], human platelet glycocalicin [67], α-glucosidase from yeast [37], 3,4-dihydroxyphenylalanine-containing proteins [68], membrane glycoproteins from human lymphocytes [69], horseradish peroxidase, and glucose oxidase [27]. Figure 8.5 shows a chromatogram for the separation of mistletoe lectins I and III using boronate affinity chromatography [8].

Boronic acid derivatives are potent transition-state inhibitors of serine proteases [34, 35]. This means that these ligands can also be used for isolation and purification of serine proteases. Examples of such applications include separations that have been reported for α-chymotrypsin, trypsin, and subtilisin [30], as well as for human neutrophil elastase, human cathepsin G, and porcine pancreatic elastase [15].

When using boronate affinity chromatography with glycoproteins, secondary interactions (especially hydrophobic and ionic interactions) must be considered. Since hydrophobic interactions can cause the nonspecific binding of undesired proteins, detergents are sometimes added to the mobile-phase buffer to decrease such binding [69]. Ionic interactions involving the negatively charged boronate ligand can prevent binding by anionic proteins or cause nonspecific adsorption by cationic proteins. Divalent cations such as Mg^{2+} are often added in low concentrations to the mobile phase to reduce these ionic effects without also enhancing hydrophobic interactions.

Proteins that do not normally interact with boronate can be isolated by "ligand-mediated" chromatography. In this process, the boronate column is first saturated with a second affinity ligand that contains a *cis*-diol. Note in this scheme that the second affinity

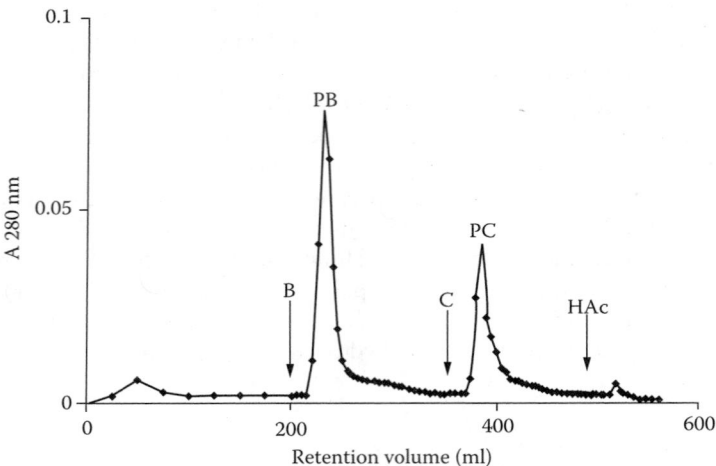

Figure 8.5 Boronate affinity chromatography for a mixture of mistletoe lectin I (MLI, represented by peak PC) and mistletoe lectin III (MLIII, represented by peak PB), using stepwise elution with Tris buffer. Conditions: column size, 20 × 1.0 cm I.D.; matrix, *m*-aminophenyl boronic acid agarose (15 ml); loading buffer, 0.02 MEPPS-NaOH + 0.5 *M* NaCl, pH 8.5; elution buffer B, 0.02 *M* EPPS-NaOH + 0.5 *M* NaCl + 0.07 *M* Tris-HCl, pH 8.5; elution buffer C, 0.02 *M* EPPS-NaOH + 0.5 *M* NaCl + 1.0 *M* Tris-HCl, pH 8.5. The column was rinsed with 0.05 *M* acetic acid (HAc), pH 4.5. The flow rate was 30 ml/h, and the sample contained 0.8 mg of MLI and 1.9 mg MLIII. MLI and MLIII had been separately dialyzed against the loading buffer before mixing. The protein absorption was measured at 280 nm. (Reproduced with permission from Li, Y., Pfuller, U., Larsson, E.L., Jungvid, H., Galaev, I.Y., and Mattiasson, B., *J. Chromatogr. A*, 925, 115–121, 2001.)

ligand is not boronate but rather a binding agent that has been selected to have specific interactions with the target proteins. The function of the boronate in this situation is merely to provide a means for coupling the secondary ligand to the matrix through the formation of a boronate/*cis*-diol ester. After the secondary ligand has been adsorbed to the column, a sample containing the target protein is applied. This allows the target protein to be retained while other proteins are washed through the column. Later, the bound target can be eluted by washing with an appropriate buffer or by applying a mobile phase that contains a soluble affinity ligand or competing *cis*-diol compound.

There are several examples of protein separations based on the boronate columns and the ligand-mediated approach. For example, concanavalin A has been isolated using methyl α-D-glucopyranoside as the secondary ligand [27], and glucose-6-phosphate dehydrogenase from yeast has been isolated using NADP+ adsorbed to boronate (see Figure 8.6) [70]. Other examples include the isolation of yeast hexokinase using ATP as the secondary ligand [70], lactate dehydrogenase using NAD+ as the secondary ligand [71], and UDP-glucose pyrophosphorylase using adsorbed UTP [71]. Affinity ligands that do not normally have diols can also be used in this format if they are first derivatized to contain such a group within their structure [72].

8.4.4 Miscellaneous Small Molecules

Other groups of small biological molecules can also be isolated by boronate affinity chromatography. Examples include catechol-containing compounds, such as D/L-dopa, 5-*S*-cysteinyldopa, dopamine, epinephrine, and norepinephrine [31, 36, 73, 74]. Additional small molecules that bind to boronate columns are α-hydroxycarboxylic acids, such as

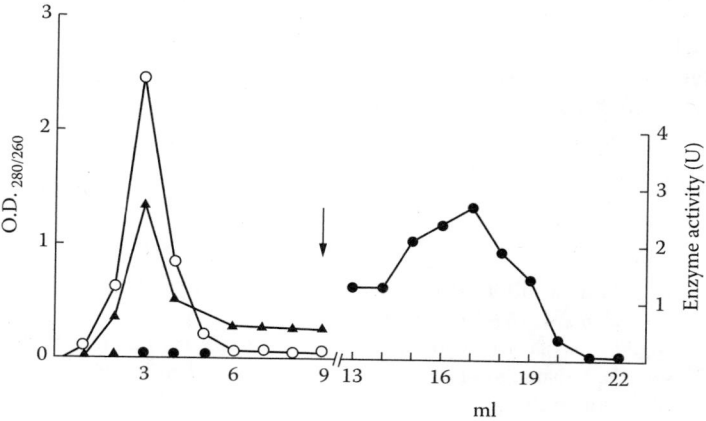

Figure 8.6 Boronate affinity chromatography of yeast glucose-6-phosphate dehydrogenase (G-6-PDH) by NADP⁺ on a presaturated matrix gel containing phenylboronic acid. Yeast enzyme concentrate (7.5 mg in 0.5 ml) in buffer (ligand concentration, 41 μmoles/ml) was applied to the boronate column (1 ml) that had been pre-equilibrated with NADP⁺. The concentrate contained 19.4 units of G-6-PDH. Fractions of 1 ml were collected and assayed for enzyme activity (•) and their optical density (O.D.) at 260 nm (▲) and 280 nm (○). Elution was achieved by applying 2 m*M* NADP⁺ at the point indicated by the arrow. (Reproduced with permission from Bouriotis, V., Galpin, I.J., and Dean, P.D.G., *J. Chromatogr.*, 210, 267–278, 1981.)

lactic acid and salicylic acid [75]. Such columns have also been used for pyridoxal [71], quercetin [76], and ecdysteroids [77].

8.5 SUMMARY AND CONCLUSIONS

In summary, boronic acid and its derivatives are versatile ligands that allow for selective binding of a variety of biological molecules. This chapter has examined the way in which these ligands bind to other agents and the various interactions that lead to retention on boronate affinity columns. The types of ligands and supports that are used in boronate affinity chromatography were also considered, as well as the various applications of this method. It was shown that the unique and reversible interactions of boronates with *cis*-diol compounds offer many advantages for affinity chromatography. Further work with boronates in both this and related fields is expected to continue, including work with these ligands in the oriented, reversible immobilization of enzymes [78], the creation of molecular imprints [21, 22], and the development of synthetic receptors and probes [79, 80].

SYMBOLS AND ABBREVIATIONS

3aPBA	3-Aminophenylboronic acid
ATP	Adenosine triphosphate
DNA	Deoxyribonucleic acid
ELBIA	Enzyme-linked boronate immunoassay
FAD	Flavin adenine dinucleotide
G-6-PDH	Glucose-6-phosphate dehydrogenase
NADH	Nicotinamide adenine dinucleotide

NADP	Nicotinamide adenine dinucleotide phosphate
RNA	Ribonucleic acid
tRNA	Transfer ribonucleic acid
UDP	Uridine diphosphate
UTP	Uridine triphosphate

REFERENCES

1. Boeseken, J., The use of boric acid for the determination of the configuration of carbohydrates, *Adv. Carbohydr. Chem.,* 4, 189–210, 1949.
2. Foster, A., Zone electrophoresis of carbohydrates, *Adv. Carbohydr. Chem.,* 12, 81–115, 1957.
3. Mattok, G. and Wilson, D., Separation of catecholamines and metanephrine and normetanephrine using a weak cation-exchange resin, *Anal. Biochem.,* 11, 575–579, 1965.
4. Weith, H.L., Wiebers, J.L., and Gilham, P.T., Synthesis of cellulose derivatives containing the dihydroxyboryl group and a study of their capacity to form specific complexes with sugars and nucleic acid components, *Biochemistry,* 9, 4396–4401, 1970.
5. Bergold, A. and Scouten, W.H., Borate chromatography, in *Solid Phase Biochemistry,* Scouten, W.H., Ed., Wiley, New York, 1983, pp. 149–187.
6. Benes, M.J., Stambergova, A., and Scouten, W.H., Affinity chromatography with immobilized benzeneboronates, in *Molecular Interactions in Bioseparations,* Ngo, T.T., Ed., Plenum Press, New York, 1993, pp. 313–322.
7. Liu, X.-C. and Scouten, W.H., Boronate affinity chromatography, in *Affinity Chromatography,* Bailon, P., Ehrlich, G.K., Fung, W.-J., and Berthold, W., Eds., Humana Press, Totowa, NJ, 2000, pp. 119–128.
8. Li, Y., Pfuller, U., Larsson, E.L., Jungvid, H., Galaev, I.Y., and Mattiasson, B., Separation of mistletoe lectins based on the degree of glycosylation using boronate affinity chromatography, *J. Chromatogr. A,* 925, 115–121, 2001.
9. Li, Y.C., Larsson, E.L., Jungvid, H., Galaev, I.Y., and Mattiasson, B., Separation of neoglycoproteins with different degrees of glycosylation by boronate chromatography, *Chromatographia,* 54, 213–217, 2001.
10. Roberts, W.L., Frank, E.L., Moulton, L., Papadea, C., Noffsinger, J.K., and Ou, C.-N., Effects of nine hemoglobin variants on five glycohemoglobin methods, *Clin. Chem.,* 46, 569–572, 2000.
11. Schnedl, W.J., Trinker, M., Krause, R., Lipp, R.W., Halwachs-Baumann, G., and Krejs, G.J., Evaluation of HbA1c determination methods in patients with hemoglobinopathies, *Diabetes Care,* 23, 339–344, 2000.
12. Barnes, H.H. and Ishimaru, C.A., Purification of catechol siderophores by boronate affinity chromatography: Identification of chrysobactin from *Erwinia carotovora* subspecies *carotovora, BioMetals,* 12, 83–87, 1999.
13. Jerkovic, B., Kung, H.C., and Bolton, P.H., Purification of thymine glycol DNA and nucleosides by use of boronate chromatography, *Anal. Biochem.,* 255, 90–94, 1998.
14. Tanaka, A., Yui, K., Tomie, N., Baba, T., Tamura, M., Makita, T., Numano, F., Nakatani, S., and Kato, Y., New assay for glycated lipoproteins by high-performance liquid chromatography, *Ann. N.Y. Acad. Sci.,* 811, 385–394, 1997.
15. Zembower, D.E., Neudauer, C.L., Wick, M.J., and Ames, M.M., Versatile synthetic ligands for affinity chromatography of serine proteinases, *Int. J. Pept. Protein Res.,* 47, 405–413, 1996.
16. Ikeda, K., Sakamoto, Y., Kawasaki, Y., Miyake, T., Tanaka, K., Urata, T., Katayama, Y., Ueda, S., and Horiuchi, S., Determination of glycated albumin by enzyme-linked boronate immunoassay (ELBIA), *Clin. Chem.,* 44, 256–263, 1998.
17. Gao, S., Wang, W., and Wang, B., Building fluorescent sensors for carbohydrates using template-directed polymerizations, *Bioorg. Chem.,* 29, 308–320, 2001.
18. Springsteen, G., Ballard, C.E., Gao, S., and Wang, B., The development of photometric sensors for boronic acids, *Bioorg. Chem.,* 29, 259–270, 2001.

19. Liu, X.-C. and Mosbach, K., Surface imprinting of RNase B using boronate-carbohydrate interaction, *Am. Biotech. Lab.*, 16 (10), 90, 1998.

20. Wulff, G., Selective binding to polymers via covalent bonds: The construction of chiral cavities as specific receptor sites, *Pure Appl. Chem.*, 54, 2093–2102, 1982.

21. Mosbach, K., Haupt, K., Liu, X.-C., Cormack, P., and Ramstrom, O., Molecular imprinting: *Status artis et quo vadere?* in *Molecular and Ionic Recognition with Imprinted Polymers*, Barsch, R.A. and Maeda, M., Eds., ACS, Washington, DC, 1998, pp. 29–48.

22. Wulff, G., Molecular imprinting in cross-linked materials with the aid of molecular templates: A way towards artificial antibodies, *Angew. Chem. Int. Ed. Engl.*, 34, 1812–1832, 1995.

23. Angyal, S., Greeves, D., and Pickles, V., The stereochemistry of complex formation of polyols with borate and periodate anions, and with metal cations, *Carbohydr. Res.*, 35, 165–173, 1974.

24. Ferrier, R.J., Carbohydrate boronates, *Adv. Carb. Chem. Biochem.*, 35, 31–80, 1978.

25. Evans, W., McCourtney, E., and Carney, W., A comparative analysis of the interaction of borate ion with various polyols, *Anal. Biochem.*, 95, 383–386, 1979.

26. Zittle, C., Reaction of borate with substances of biological interest, *Adv. Enzymol.*, 12, 493–527, 1951.

27. Fulton, S., *Boronate Ligands in Biochemical Separations*, Amicon Corp., Danvers, MA, 1981.

28. Soundararajan, S., Badawi, M., Kohlrust, C.M., Montano, C., and Hageman, J., Boronic acids for affinity chromatography: Spectral methods for determinations of ionization and diol-binding constants, *Anal. Biochem.*, 178, 125–134, 1989.

29. Adamek, V., Liu, X.-C., Zhang, Y.-A., Adamkova, K., and Scouten, W.H., New aliphatic boronate ligands for affinity chromatography, *J. Chromatogr.*, 625, 91–99, 1992.

30. Akparov, V. and Stepanov, V., Phenylboronic acid as a ligand for biospecific chromatography of serine proteases, *J. Chromatogr.*, 155, 329–336, 1978.

31. Higa, S., Suzuki, T., Hayashi, A., Tsuge, I., and Yamamura, Y., Isolation of catecholamines in biological fluids by boric acid gel, *Anal. Biochem.*, 77, 18–24, 1977.

32. Boit, H.-G., *Belstein 16*, Springer, Berlin, 1974, p. 1284.

33. Yurkevich, A., Kolodkina, I., Ivanova, E., and Pichuzhkina, E., Study of the interaction of polyols with polymers containing *N*-substituted [(4-boronophenyl)methyl-ammonio] groups, *Carbohydr. Res.*, 43, 215–224, 1975.

34. Matthews, D., Alden, R., Birktoft, J., and Freer, J.K.S., X-ray crystallographic study of boronic acid adducts with subtilisin BPN′ (Novo), *J. Biol. Chem*, 250, 7120–7126, 1975.

35. Garner, C.W., Boronic acid inhibitors of porcine pancreatic lipase, *J. Biol. Chem*, 255, 5064–5068, 1980.

36. Elliger, C., Chan, B., and Stanley, W., The *p*-vinylbenzeneboronic acid polymers for separation of vicinal diols, *J. Chromatogr.*, 104, 57–61, 1975.

37. Myoehanen, T.A., Bouriotis, V., and Dean, P.D.G., Affinity chromatography of yeast α-glucosidase using ligand-mediated chromatography on immobilized phenylboronic acids, *Biochem. J.*, 197, 683–688, 1981.

38. Johnson, B.J.B., Synthesis of a nitrobenzeneboronic acid substituted polyacrylamide and its use in purifying isoaccepting transfer ribonucleic acids, *Biochemistry*, 20, 6103–6108, 1981.

39. Singhal, R.P., Ramamurthy, B., Govindraj, N., and Sarwar, Y., New ligands for boronate affinity chromatography: Synthesis and properties, *J. Chromatogr.*, 543, 17–38, 1991.

40. Liu, X.-C. and Scouten, W.H., New ligands for boronate affinity chromatography, *J. Chromatogr. A*, 687, 61–69, 1994.

41. Liu, X.-C., Hubbard, J.L., and Scouten, W.H., Synthesis and structural investigation of two potential boronate affinity chromatography ligands: Catechol [2-(diisopropylamino)carbonyl]phenyl-boronate and catechol [2-(diethylamino)carbonyl,4-methyl]phenylboronate, *J. Organomet. Chem.*, 493, 91–94, 1995.

42. Larsson, P.O., Support materials for affinity chromatography, in *Handbook of Affinity Chromatography*, Marcel Dekker, New York, 1993, pp. 61–75.

43. Glad, M., Ohlson, S., Hansson, L., Mansson, M., and Mosbach, K., High-performance liquid affinity chromatography of nucleosides, nucleotides and carbohydrates with boronic acid-substituted microparticulate silica, *J. Chromatogr.*, 200, 254–260, 1980.

44. Koyama, T. and Terauchi, K.-I., Synthesis and application of boronic acid-immobilized porous polymer particles: A novel packing for high-performance liquid affinity chromatography, *J. Chromatogr. B*, 679, 31–40, 1996.

45. Bourne, E.J., Lees, E.M., and Weigel, H., Paper chromatography of carbohydrates and related compounds in the presence of benzeneboronic acid, *J. Chromatogr.*, 11, 253–257, 1963.

46. Schott, H., New dihydroxyboryl-substituted polymers for column-chromatographic separation of ribonucleoside-deoxyribonucleoside mixtures, *Angew. Chem. Int. Ed. Engl.*, 11, 824–825, 1972.

47. Gascon, A., Wood, T., and Chitemerese, L., The separation of isomeric pentose phosphates from each other and the preparation of D-xylulose 5-phosphate and D-ribulose 5-phosphate by column chromatography, *Anal. Biochem.*, 118, 4–9, 1981.

48. Wulff, G. and Vesper, W., Preparation of chromatographic sorbents with chiral cavities for racemic resolution, *J. Chromatogr.*, 167, 171–186, 1978.

49. Wulff, G., Dederichs, W., Grotstollen, R., and Jupe, C., On the chemistry of binding sites, Part II: Specific binding of substances to polymers by fast and reversible covalent interactions, in *Affinity Chromatography and Related Techniques*, Gribnau, T.C.J., Visser, J., and Nivard, R.J.F., Eds., Elsevier, Amsterdam, 1982, pp. 207–216.

50. Schott, H., Rudloff, E., Schmidt, P., Roychoudhury, R., and Kossel, H., A dihydroxyboryl-substituted methacrylic polymer for the column chromatographic separation of mononucleotides, oligonucleotides, and transfer ribonucleic acid, *Biochemistry*, 12, 932–938, 1973.

51. Ackerman, S., Cool, B., and Furth, J., Removal of DNA from RNA by chromatography on acetylated N-[N'-(m-dihydroxylborylphenyl)succinamyl]aminoethyl cellulose, *Anal. Biochem.*, 100, 174–178, 1979.

52. Rosenberg, M. and Gilham, P.T., Isolation of 3'-terminal polynucleotides from RNA molecules, *Biochem. Biophys. Acta*, 246, 337–340, 1971.

53. Wilk, H.E., Kecskemethy, N., and Schaefer, K.P., m-Aminophenylboronate agarose specifically binds capped snRNA and mRNA, *Nucleic Acids Res.*, 10, 7621–7633, 1982.

54. Rosenberg, M., Wiebers, J., and Gilham, P.T., Studies on the interactions of nucleotides, polynucleotides, and nucleic acids with dihydroxylboryl-substituted cellulose, *Biochemistry*, 11, 3623–3628, 1972.

55. Van Ness, B., Howard, J., and Bodley, J., ADP-ribosylation of elongation factor 2 by diphtheria toxin: NMR spectra and proposed structures of ribosyl-diphthamide and its hydrolysis products, *J. Biol. Chem*, 255, 10717–10720, 1980.

56. Gould, B.J., Hall, P.M., and Cook, G.H., Measurement of glycosylated hemoglobins using an affinity chromatography method, *Clin. Chim. Acta*, 125, 41–48, 1982.

57. Fluckiger, R., Woodtli, T., and Berger, W., Quantitation of glycosylated hemoglobin by boronate affinity chromatography, *Diabetes*, 33, 73–76, 1984.

58. Klenk, D.C., Hermanson, G.T., Krohn, R.I., Fujimoto, E.K., Malia, A.K., Smith, P.K., England, J.D., Wiedmeyer, H.M., Little, R.R., and Goldstein, D.E., Determination of glycosylated hemoglobin by affinity chromatography: Comparison with colorimetric and ion-exchange methods, and effects of common interferences, *Clin. Chem.*, 28, 2088–2094, 1982.

59. Hjerten, S. and Li, J., High-performance liquid chromatography of proteins on deformed non-porous agarose beads, *J. Chromatogr.*, 500, 543–553, 1990.

60. Bisse, E. and Wieland, H., Coupling of m-aminophenylboronic acid to s-triazine-activated Sephacryl: Use in the affinity chromatography of glycated hemoglobins, *J. Chromatogr.*, 575, 223–228, 1992.

61. Herold, C.D., Andree, K., Herold, D.A., and Felder, R.A., Robotic chromatography: Development and evaluation of automated instrumentation for assay of glycohemoglobin, *Clin. Chem.*, 39, 143–147, 1993.

62. Yasukawa, K., Abe, F., Shida, N., Koizumi, Y., Uchida, T., Noguchi, K., and Shima, K., High-performance affinity chromatography system for the rapid, efficient assay of glycated albumin, *J. Chromatogr.*, 597, 271–275, 1992.

63. Hage, D.S., Affinity chromatography: A review of clinical applications, *Clin. Chem.*, 45, 593–615, 1999.

64. Ugelstad, J., Stenstad, P., Kilaas, L., Prestvik, W.S., Rian, A., Nustad, K., Herje, R., and Berge, A., Biochemical and biomedical application of monodisperse polymer particles, *Macromol. Symp.*, 101, 491–500, 1996.

65. Brena, B.M., Batista-Viera, F., Ryden, L., and Porath, J., Selective adsorption of immuno-globulins and glycosylated proteins on phenylboronate-agarose, *J. Chromatogr.*, 604, 109–115, 1992.

66. Yamamoto, T., Amuro, Y., Matsuda, Y., Nakaoka, H., Shimomura, S., Hada, T., and Higashino, K., Boronate affinity chromatography of γ-glutamyltransferase in patients with hepatocel-lular carcinoma, *Am. J. Gastroenterol.*, 86, 495–499, 1991.

67. DeCristofaro, R., Landolfi, R., Bizzi, B., and Castagnola, M., Human platelet glycocalicin purification by phenyl boronate affinity chromatography coupled to anion-exchange high-performance liquid chromatography, *J. Chromatogr.*, 426, 376–380, 1988.

68. Hawkins, C.J., Lavin, M.F., Parry, D.L., and Ross, I.L., Isolation of 3,4-dihydroxyphenyla-lanine-containing proteins using boronate affinity chromatography, *Anal. Biochem.*, 159, 187–190, 1986.

69. Williams, G.T., Johnstone, A.P., and Dean, P.D.G., Fractionation of membrane proteins on phenylboronic acid-agarose, *Biochem. J.*, 205, 167–171, 1982.

70. Bouriotis, V., Galpin, I.J., and Dean, P.D.G., Applications of immobilized phenylboronic acids as supports for group-specific ligands in the affinity chromatography of enzymes, *J. Chromatogr.*, 210, 267–278, 1981.

71. Maestas, R., Prieto, J., Duehn, G., and Hageman, J., Polyacrylamide-boronate beads satu-rated with biomolecules: A new general support for affinity chromatography of enzymes, *J. Chromatogr.*, 189, 225–231, 1980.

72. Ho, N., Duncan, R., and Gilham, P., Esterification of terminal phosphate groups in nucleic acids with sorbitol and its application to the isolation of terminal polynucleotide fragments, *Biochemistry*, 20, 64–67, 1981.

73. Hansson, C., Kaagedal, B., and Kaellberg, M., Determination of 5-S-cysteinyldopa in human urine by direct injection in coupled-column high-performance liquid chromatography, *J. Chromatogr.*, 420, 146–151, 1987.

74. Hansson, C., Agrup, G., Rorsman, H., Rosengren, A., and Rosengren, E., Chromatographic separation of catecholic amino acids and catecholamines on immobilized phenylboronic acid, *J. Chromatogr.*, 161, 352–355, 1978.

75. Higa, S. and Kishimoto, S., Isolation of 2-hydroxy carboxylic acids with a boronate affinity gel, *Anal. Biochem.*, 154, 71–74, 1986.

76. Bongartz, D. and Hesse, A., Selective extraction of quercetin in vegetable drugs and urine by off-line coupling of boronic acid affinity chromatography and high-performance liquid chromatography, *J. Chromatogr. B*, 673, 223–230, 1995.

77. Pis, J. and Harmatha, J., Phenylboronic acid as a versatile derivatization agent for chroma-tography of ecdysteroids, *J. Chromatogr.*, 596, 271–275, 1992.

78. Liu, X.-C. and Scouten, W.H., Studies on oriented and reversible immobilization of glyco-protein using novel boronate affinity gel, *J. Mol. Recogn.*, 9, 462–467, 1996.

79. Hartley, J.H., James, T.D., and Ward, C.J., Synthetic receptors, *J. Chem. Soc. Perkin Trans.*, 1, 3155–3184, 2000.

80. DiCesare, N. and Lakowicz, J., Fluorescent probe for monosaccharides based on a func-tionalized boron-dipyrromethene with a boronic acid group, *Tetrahedron Lett.*, 42, 9105–9108, 2001.

9

Dye-Ligand and Biomimetic Affinity Chromatography

N.E. Labrou, K. Mazitsos, and Y.D. Clonis
Department of Agricultural Biotechnology, Agricultural
University of Athens, Athens, Greece

CONTENTS

9.1   INTRODUCTION

In the last 20 years, protein-based drugs have become significant weapons in the war against disease [1]. Approximately 1049 million grams of protein-based drugs were produced worldwide in 1999, and their production is expected to rise to 1150 million grams by 2004 [2]. Within a decade, it is estimated that approximately 25% of all drugs in developed countries will fall in this category, especially as the number of candidate therapeutic targets is increased [2].

Given the worldwide demand for pure therapeutic proteins, the biotechnology industry will soon suffer from a severe shortage of manufacturing capacity for such agents [3]. To meet this demand, technical improvements will be needed in protein purification methods to give both good purification efficiency and high protein yields [3, 4]. In addition, now that the human genome has been determined, scientists are faced with the challenge of studying large protein libraries as they examine the molecular workings of the cell [5]. This will also require improved methods and materials for the purification of proteins [6–8].

One approach for dealing with these needs is to develop better affinity ligands for protein isolation. Synthetic compounds such as dyes and chlorotriazine-linked biomimetic agents are particularly appealing for use as ligands in large-scale affinity chromatography and high-throughput protein screening. Reasons for this include the low cost of these ligands, their resistance to chemical or biological degradation, and their ability to be sterilized and cleaned *in situ*. In addition, they can be readily immobilized to generate affinity adsorbents that have high binding capacities for proteins [4, 9–11].

Dye-ligand affinity chromatography and *biomimetic affinity chromatography* are two chromatographic methods that are being explored for this purpose. These use either synthetic dyes or biomimetic molecules as immobilized ligands. The discovery of dye-ligands occurred accidentally in the late 1960s. At that time, scientists purifying the enzyme pyruvate kinase were puzzled when they found that a small dye (Blue Dextran) coeluted with this enzyme from a gel filtration column [12]. It was later learned that this was caused by binding between the enzyme and dye, thus causing the dye to elute with the enzyme. This soon led to the first reported application of dyes in affinity chromatography when Staal et al. used a Blue Dextran column in 1971 to purify pyruvate kinase from human erythrocytes [13].

The dyes originally employed as ligands in dye-affinity chromatography were obtained from the textile industry. In particular, chlorotriazine polysulfonated aromatic molecules, often referred to as triazine dyes, were used for this purpose. Over the past few decades, these dyes have found wide application as general ligands for the purification of albumin and other blood proteins, as well as for oxidoreductases, decarboxylases, glycolytic enzymes, nucleases, hydrolases, lyases, synthetases, and transferases [9, 10]. Many dye-ligand adsorbents are now available from suppliers like Sigma-Aldrich and Amersham Pharmacia Biotech. Examples of these adsorbents include CB3GA-Agarose, Blue-Trisacryl, Reactive Brown 10-Sepharose, Reactive Green 19-Sepharose, Reactive Red 120-Sepharose, and Reactive Yellow 3-Sepharose.

To increase the specificity of these supports, the concept of a biomimetic dye-ligand was later introduced [11, 14, 15]. In this approach, new dyes are designed that mimic the natural ligand of a targeted protein, such as by substituting the terminal 2-aminobenzene sulfonate moiety of the dye Cibacron Blue 3GA (CB3GA). It has been found that these biomimetic dyes exhibit increased purification ability and specificity versus traditional dye-ligands, thus making them useful tools for simple and effective purification protocols.

This chapter examines the design and use of dye-ligands and biomimetic ligands for protein purification. In addition, a number of specific applications are given for these methods, along with a discussion of some practical issues to consider in their use.

9.2 DYE-LIGAND AFFINITY CHROMATOGRAPHY

Triazine dyes are the most common binding agents used in dye-ligand chromatography. The structures of these dyes consist of two distinct units joined through an amino-bridge, as illustrated in Figure 9.1. The first unit in this structure, the chromophore, contributes

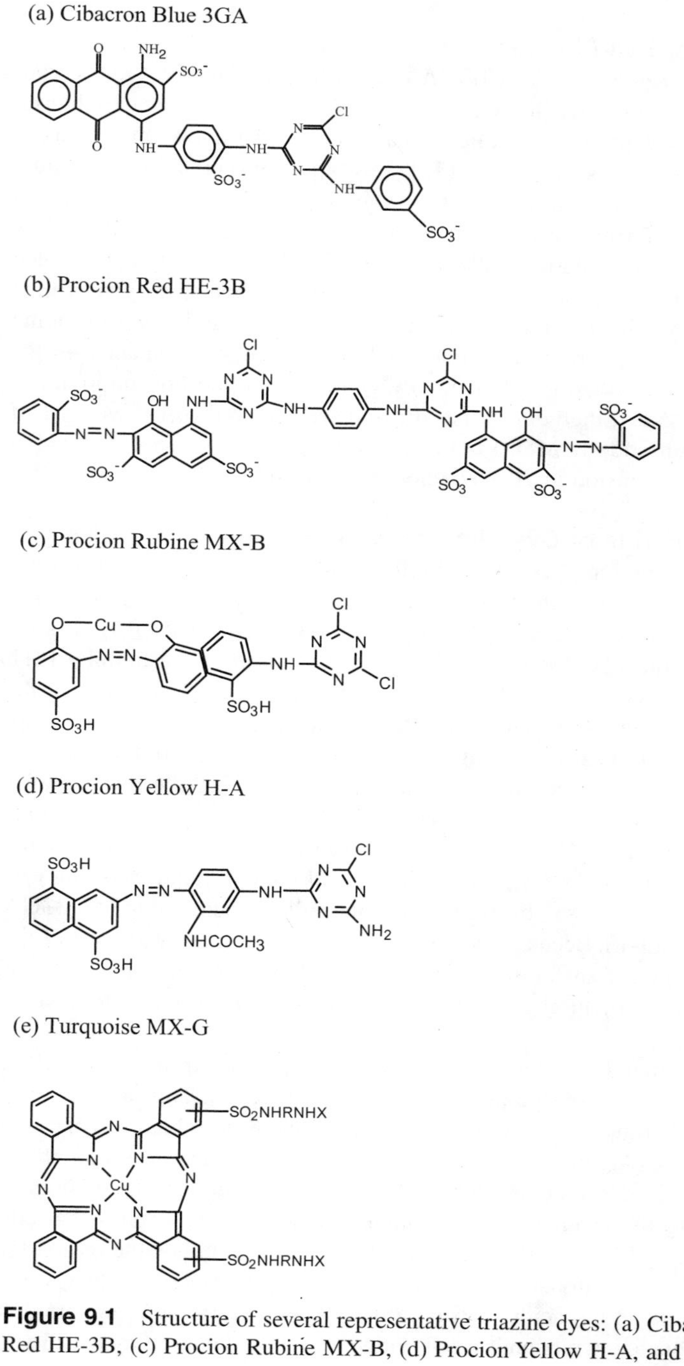

(a) Cibacron Blue 3GA

(b) Procion Red HE-3B

(c) Procion Rubine MX-B

(d) Procion Yellow H-A

(e) Turquoise MX-G

Figure 9.1 Structure of several representative triazine dyes: (a) Cibacron Blue 3GA, (b) Procion Red HE-3B, (c) Procion Rubine MX-B, (d) Procion Yellow H-A, and (e) Turquoise MX-G.

the color (usually based on an anthraquinone, azo, or phthalocyanine group). The other unit (usually 1,3,5-*sym*-trichlorotriazine) provides the site for covalent attachment to an insoluble support [10, 16]. Anthraquinone triazine dyes are the most widely used dye-ligands in enzyme and protein purification. Of this group, the triazine dye CB3GA has received the most attention. Therefore, the following discussion on dye-ligand affinity chromatography focuses primarily on this dye and its analogs.

9.2.1 Immobilization of Dye-Ligands

Covalent attachment of a triazine dye like CB3GA onto an affinity support (e.g., carbohydrates such as agarose, dextran, and cellulose, as discussed in Chapter 2) can be achieved under alkaline conditions by using nucleophilic displacement of the dye's chlorine atom by hydroxyl groups on the support's surface [9, 17, 18]. A typical protocol for performing this with an agarose support is as follows. First, one gram of a washed agarose gel is combined with a solution of purified dye in water (5 to 20 mg dye in 1 ml), followed by the addition of a sodium chloride solution (22% w/v, 0.2 ml). This suspension is tumbled for 30 min at room temperature prior to adding solid sodium carbonate to give a final concentration for this salt of 1%. The reaction is then allowed to continue with shaking at 25 to 60°C for 1 to 8 h [18]. The actual time used in this last step depends on the reactivity of the dye molecule, with 4 to 8 h generally giving 2 to 4 μmol of immobilized dye per gram of moist gel. Next, the gel is washed extensively with distilled water, 1 *M* KCl, and 20% dimethyl sulfoxide (DMSO) until the washings are colorless. The gel is then stored until use as a suspension in 20% ethanol at 4°C [18].

9.2.2 Selective Interactions of Dye-Ligands with Proteins

The first x-ray crystal structure that gave detailed information on dye-protein interactions was for the complex of CB3GA with horse liver alcohol dehydrogenase [19]. More recently, two other structures have been solved: one for the CB3GA-NAD(P)H:quinone reductase complex [20] and the other for a CB3GA-glutathione *S*-transferase complex [21].

The results for the CB3GA-NAD(P)H:quinone reductase complex indicate that the dye CB3GA and the AMP moiety of NADP⁺ interact very similarly with the enzyme, with three of the ring systems of CB3GA mimicking AMP binding. In the case of the CB3GA-glutathione *S*-transferase complex, the anthraquinone chromophore of the dye occupies the H-site of each enzyme subunit and makes van der Waals contacts with amino acids Phe-8, Val-10, Ile-104, Tyr-108, and Gly-205 (see Figure 9.2). In addition, there is possible hydrogen bonding between the hydroxyl group of Tyr-7 and the anthraquinone ring carbonyl group of CB3GA, as well as a salt bridge between the sulfonic acid group of CB3GA and the guanidyl group of Arg-13. Because there is no observed density for the rest of the dye-ligand in the glutathione *S*-transferase complex, it is believed that the remainder of the dye molecule exhibits a structural mobility that is too high to be characterized by crystallography [21].

Recently, matrix-assisted laser desorption/ionization mass spectrometry (MALDI MS) was also applied to the study of dye-protein interactions [22, 23]. This approach has shown that the sulfonic acid groups of CB3GA bind only to arginine side chains, as opposed to binding to all basic sites (Arg, Lys, and His). This specific interaction is thought to represent the driving force in dye-protein binding. To explain this finding, the authors of this study noted that among all the amino acids, arginine has the highest gas-phase basicity and highest pK_a value ($pK_a = 12.48$ for the guanidino group). This could be the reason for the preferential selectivity of a sulfonate group (with a typical pK_a much less than zero) for arginine over other basic amino acids. It was further suggested by these authors that the principal arginine-sulfonate interaction is a salt bridge, which is enhanced by ion-dipole

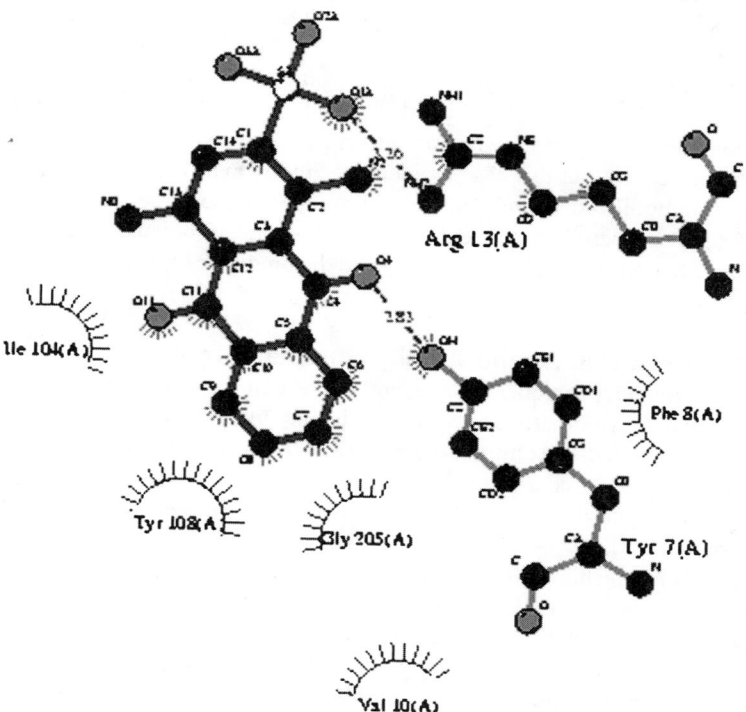

Figure 9.2 A LigPlot diagram of the Cibacron Blue 3GA/glutathione *S*-transferase complex (PDB code 20gs), illustrating some of the residues involved in van der Waals interactions.

interactions and hydrogen bonds that are especially favorable because of the near-perfect fit between these two bidentate binding groups [22].

9.2.3 Development of a Dye-Ligand Purification Method

The selection of a dye adsorbent for a particular macromolecule is currently an empirical process. The suitability of a range of dyes and adsorbents can be evaluated chromatographically by using small affinity columns (i.e., 0.5 to 1 ml in size). Detailed screening procedures that can be used for this purpose have been described elsewhere [17].

During the development and optimization of a dye-ligand purification, attention should be paid to variables such as the immobilized ligand concentration; the mobile phase pH, ionic strength, and composition; temperature; and sample size [9, 17]. The pH, ionic strength, and temperature of the initial equilibration buffer are all important factors that will influence the adsorption and elution of target proteins on a dye-ligand column. The influence of pH and ionic strength arises from the cation-exchange character of common dye-ligands, which results from the presence of negatively charged sulfonic groups in these dyes [9].

The elution conditions used for a bound biomolecule should be compatible with the dye-ligand adsorbent, as well as effective in desorbing the retained molecule in a good yield and in its native state. The elution of bound proteins can be performed in either a nonspecific or biospecific manner, as discussed in Chapter 4 [9, 24]. Nonspecific elution usually involves changing the pH or ionic strength of the mobile phase. An increase in pH generally weakens adsorption and promotes the elution of retained molecules. The ionic strength of the mobile phase is usually increased by adding a salt like KCl or NaCl. If the hydrophobic contribution to dye-protein binding is large, it is also possible to promote elution by altering the polarity of the mobile phase, such as by adding ethylene glycol or other organic solvents.

Nonspecific elution methods, although economical, generally give a lower degree of purification due to the coelution of nonspecifically bound proteins. Thus, biospecific elution is instead recommended for group-specific adsorbents like dye-ligand supports. Biospecific elution can be achieved in this case by including in the mobile phase an agent that competes with the immobilized ligand for its binding site on the target enzyme or protein. Examples of such competing agents can include substrates, products, cofactors, inhibitors, or compounds that produce allosteric changes in the binding of the immobilized dye to its target [24].

9.2.4 Recent Progress in Dye-Ligand Affinity Chromatography

New developments in dye-ligand techniques have recently appeared for both the traditional chromatographic mode and for alternative separation schemes like affinity partitioning, affinity precipitation, and expanded-bed chromatography [25–28]. As one example, a PEG-CB3GA conjugate was used in an aqueous two-phase system for the isolation of IgG, where the partitioning of IgG into the top phase was obtained in the presence of a 10 mM potassium phosphate buffer [25].

Affinity precipitation is a technique that integrates specific affinity interactions into a conventional precipitation method. This technique involves the use of a soluble ligand that binds the target molecule to form a complex that can be selectively precipitated, thus allowing the target molecule to be subsequently eluted in higher purity. This can be accomplished by using heterobifunctional ligands and reversibly soluble polymers (e.g., carboxymethyl cellulose) onto which the affinity ligand has been conjugated. Precipitation of a polymer-ligand-target protein complex in this method can be induced by a change in such conditions as temperature, pH, or ionic strength [26]. For instance, CB3GA conjugated to carboxymethyl cellulose was used for the affinity precipitation of lactate dehydrogenase. This gave a 23-fold purification from a crude porcine extract in a single precipitation step [26].

The efficiency of a chromatographic process can be significantly improved by pretreatment of an affinity matrix with certain water-soluble polymers [27, 28]. This technique is known as *polymer-shielded dye-affinity chromatography*. In this method, the dye matrix complexes with a nonionic water-soluble polymer, such as poly(vinylpyrrolidone) or poly(vinyl alcohol), prior to the application of a sample. The purpose of these polymers is to shield the dye and prevent nonspecific interactions between the target protein and dye, as illustrated in Figure 9.3. This provides a faster purification than can be achieved on an unshielded column. Polymer-shielded dye-affinity chromatography has successfully been used in purifying a number of enzymes [28]. For example, polymer-shielded dye-affinity chromatography was used in an expanded-bed system with the support Streamline-CB3GA for the isolation of lactate dehydrogenase from a crude porcine muscle extract, resulting in 4.1-fold purification [28].

Dye-ligand affinity chromatography has also found wide use in the purification of pharmaceutical proteins [29–34]. Examples include work with this method in purifying human component factor B, factor C2, factor II, factor IX, trypsin, chymotrypsin, and proteinase 3 [28]. For instance, human recombinant alpha-interferon was purified in a single step on a mimetic dye-ligand matrix, yielding monomeric alpha-interferon with a specific activity of 2.8×10^8 IU/mg [30]. Other examples have involved the purification of follicle-stimulating hormone [31], pituitary gonadotropins [32], ricin A chain [33], and human serum albumin [34]. In the case of the human serum albumin, a CB3GA-Sepharose adsorbent was used on a pilot scale to produce 250 g of albumin on a 50-l column, giving a product with a purity of 98 to 100% at a yield of 82%.

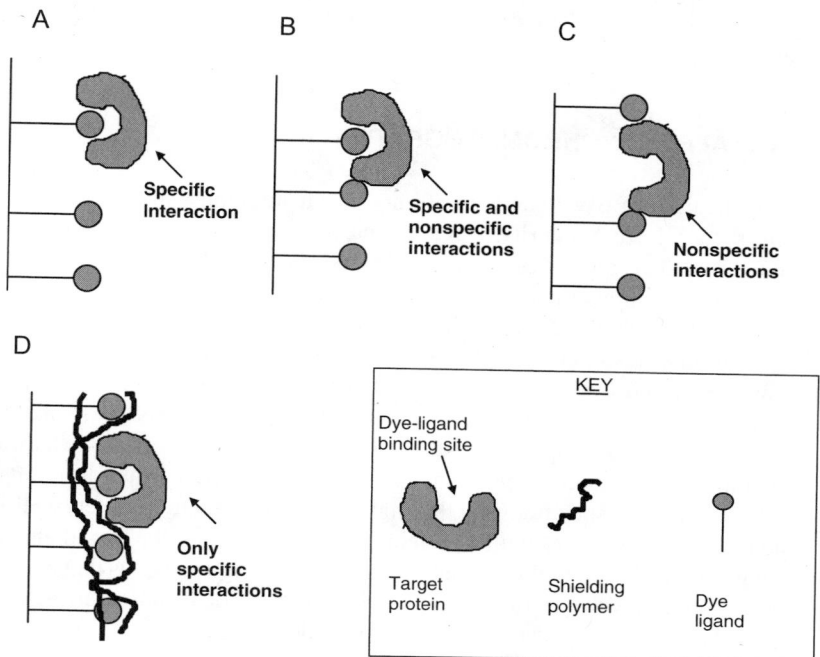

Figure 9.3 Polymer-shielded dye-affinity chromatography.

The long lifetime of dye-ligand adsorbents makes these attractive and economical for such purifications. Unfortunately, regulations dealing with the safety and toxicity issues that are associated with CB3GA have prevented albumin prepared by CB3GA adsorbents from being used in clinical applications. However, this type of albumin can be used as a component of cell culture media or as stabilizer for recombinant products [35].

Another application of dye-ligands has been in the removal of small hydrophobic molecules and ions from biological fluids using conventional affinity chromatography. For example, CB3GA coupled to poly(EGDMA-2-hydroxyethyl methacrylate) (pHEMA) micro-beads was used to remove bilirubin from human plasma in a packed-bed system [36]. The same adsorbent was also used for the removal of Al^{3+} and Fe^{3+} from human plasma [37–39]. In addition, a dye-ligand column has been employed in the separation of G-DNA structures that are formed by guanosine-rich oligodeoxyribonucleotide. This particular separation was performed on a Reactive Green 19-agarose resin in the presence of Li^+ ions [40].

Recently, an affinity extraction with a reversed-micelle system was used in the extraction of lysozyme. The reversed micelle in this method was generated by using a CB3GA-lecithin conjugate in n-hexane and gave an overall recovery for lysozyme of 87% [41]. This approach gave high selectivity, resulting from the presence of both biospecific and steric-hindrance effects, and it was found to be well suited for the purification of low-molecular-weight proteins [41].

A future application for dye-ligands may be their use in the removal of toxic macromolecules from biological fluids, such as prion proteins, human immunodeficiency virus-1, or hepatitis B particles. Prion proteins have been shown to bind specifically to the azo dye Congo Red [42]. It is also known that immobilized CB3GA is able to adsorb envelope glycoproteins, such as the 140-kDa precursor and mature 120-kDa envelope glycoproteins from human immunodeficiency virus-1 [43]. In addition, it has been shown that this same dye can retain more than 99.5% of the hepatitis B particles from a human plasma sample [44].

9.3 BIOMIMETIC AFFINITY CHROMATOGRAPHY

Despite the wide application of dye-ligands in affinity chromatography, there have been concerns regarding their selectivity, purity, leakage, and toxicity when used in biopharmaceutical manufacturing. This has led to an interest in designing improved "biomimetic" ligands through the use of computational and combinatorial chemistry.

Choosing an immobilized ligand for affinity chromatography is the most challenging part of preparing an affinity adsorbent [8]. But the selection of an appropriate target site and the design, synthesis, and evaluation of a complementary affinity ligand is only a semirational process. This is because numerous factors affect the final behavior of the ligand, including the nature of the support and the activation/coupling chemistry used to attach the ligand to this matrix. Some items that need to be considered when designing biomimetic ligands include the specificity of this ligand for the protein of interest, the reversibility of the ligand's interactions with a protein, and the stability of the ligand under the desired operating conditions (including its stability towards biological agents that might appear in the sample). The following section describes various theoretical and experimental approaches that have been used as part of this ligand-development process.

9.3.1 Methods Based on Protein Structure

In the past, ligand discovery usually consisted of the systematic modification of a promising chemical structure, as based on standard chemical methods and several cycles of synthesis and evaluation. However, the rapid growth of computational tools and the ability to exploit protein structural data (as derived from x-ray crystallography and NMR) has made the rational design of affinity ligands more feasible [8, 42, 43, 45, 46]. This direct approach makes use of a full knowledge of a target's three-dimensional structure.

In contrast to this, an indirect approach would involve the creation of a ligand that incorporates those structural features thought to be important for binding a target without directly resorting to the target's three-dimensional structure. Features considered in this approach might include a certain molecular shape, general structures derived from other known ligands for the target (e.g., substrates or inhibitors), or the use of specific functional groups known to react with a broad class of targets. Examples of this would include the use of hydroxamic acids as inhibitors of metalloproteases or phosphonic acid as an inhibitor of phosphatases [18, 47–49].

The design of biomimetic ligands now makes extensive use of biocomputing. This has become a powerful tool for designing synthetic ligands that interact with the protein in a similar manner to that of its natural substrates. But there are two key requirements needed for all computational algorithms used to model ligand-target docking. First, these algorithms must have the ability to correctly predict the conformation of the docked ligand. Second, they must have the ability to accurately predict the binding affinity of a given ligand [45]. The use of advanced molecular modeling programs for three-dimensional visualization (such as MOLSCRIPT or LIGPLOT), for investigating and calculating molecular parameters of protein-ligand interactions (such as GROMOS, CHARMM, MM3, or AMBER), and for automated docking (DOCK, LUDI, FlexX, and GOLD) have now made this approach feasible. However, only a few docking tools (FlexX and GOLD) have been thoroughly tested on a large number of different ligand-target complexes [50].

9.3.1.1   Biomimetic Dye-Ligands

One area in which molecular modeling has been applied to the design of affinity ligands has been its use with biomimetic dye-ligands for enzymes. Examples are dye-ligands that have been reported for enzymes that recognize (keto)carboxyl groups [51], glutathione [49], and galactose [52–54]. Each of these examples will be examined in more detail in this section.

9.3.1.1.1   Biomimetic Dye-Ligands for (Keto)Carboxyl-Group-Recognizing Enzymes.

Biomimetic dyes for (keto)carboxyl-recognizing enzymes [18, 48, 51] were designed to contain two enzyme-recognition moieties: an anthraquinone chromophore that functions as a nucleotide pseudoanalog and a triazine ring-linked substrate-mimetic moiety. The mimetic moieties have (keto)carboxyl (KC) structures similar to those of the natural substrates or inhibitors of carboxyl-recognizing enzymes. Examples of these biomimetic dyes are shown in Figure 9.4. In their immobilized form, these ligands have been used to purify several (keto)carboxyl-group-recognizing enzymes. For instance, KC-BM6 has been used to purify formate dehydrogenase from *Candida boidinii* in a single chromatographic step, leading to a specific activity greater than 7 U/mg and a recovery of more than 60% [18]. KC-BM7 has been used to isolate oxalate oxidase from *Hordeum vulgare* in two chromatographic steps, with a specific activity greater than 30 U/mg and a recovery of more than 40% [55]. And KC-BM1 has been used with oxaloacetate decarboxylase from *Pseudomonas stutzeri* to give a specific activity exceeding 300 U/mg and a recovery of over 30% in three steps [56].

An even wider group of applications has been reported for the biomimetic adsorbent KC-BM5. This was used for the purification of L-malate dehydrogenase (MDH) and L-lactate dehydrogenase (LDH) from a bovine heart extract [54]. This adsorbent was also incorporated into a two-step purification protocol for MDH, leading to a specific activity for the purified enzyme of 1300 U/mg and a 35.7-fold purification [57]. Furthermore, this ligand was used in a two-step chromatographic procedure for the purification of a new MDH from the bacterium *Pseudomonas stutzeri*, giving an 82-fold purification and 48% recovery [58].

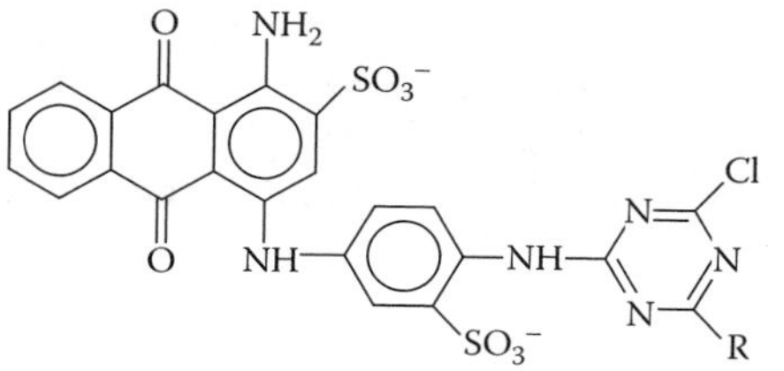

Dye-ligand	(-R)
KC-BM1	$-m$-HN-C$_6$H$_4$-COO$^-$
KC-BM2	$-m$-HN-C$_6$H$_4$-CH$_2$COO$^-$
KC-BM3	-HNCH$_2$COO$^-$
KC-BM4	-HN(CH$_2$)$_2$COO$^-$
KC-BM5	$-p$-HN-C$_6$H$_4$-NHCOCOO$^-$
KC-BM6	-SCH$_2$COCOO$^-$
KC-BM7	-HN(CH$_2$)$_2$NHCOCOO$^-$

Figure 9.4   Structures of the anthraquinone (keto)carboxyl-biomimetic dyes KC-BM (1-7) for (keto)carboxyl-recognizing enzymes.

In one study, this adsorbent was used for the isolation of MDH and LDH from the same bovine heart extract, following their separation by an ion-exchange resin [59]. In this procedure, the purified MDH had a specific activity greater than 1330 U/mg and a yield of 45%, while the purified LDH had a specific activity of 500 U/mg and a yield of 40%.

More recently, a new biomimetic dye-ligand was designed for LDH [60]. This was accomplished by exploiting the known three-dimensional structure of the porcine heart LDH-(*S*)-lactate-NAD$^+$ complex for the modeling procedure. The new ligand had 2-(4-aminophenyl) ethyloxamic acid as the terminal biomimetic moiety. The design of this ligand was based on a match between the polar and hydrophobic regions of the binding site on LDH with the biomimetic moiety of the ligand [60]. The length of the ketoacid biomimetic moiety was determined to approach the enzyme catalytic site and to form charge-charge interactions with Arg-171. When immobilized, this biomimetic dye gave an improved purification for LDH from bovine heart when compared with a traditional dye-ligand adsorbent containing CB3GA. When this new adsorbent was used in a two-step purification of LDH from bovine heart tissue, this led to a final specific activity of 600 U/mg and a recovery of 56%. The same adsorbent was found to be effective in the purification of LDH from other sources, such as chicken liver and muscle, pig muscle, bovine pancreas, and pea seeds [60].

9.3.1.1.2 Biomimetic Dye-Ligands for Glutathione-Recognizing Enzymes.

Three glutathionyl-biomimetic (GSH-BM) dyes have been designed and synthesized for purifying glutathione-recognizing enzymes. The structures of these dyes are given in Figure 9.5 [49]. This again was accomplished through the use of molecular modeling and ligand-docking studies. The dyes that were created in this work had glutathione analogs (such as glutathione-sulfonic acid, *S*-methyl-glutathione, and glutathione) as terminal biomimetic moieties. The immobilized forms of these ligands were then examined for their ability to recognize and purify (1) glutathione-dependent formaldehyde dehydrogenase (FaDH) from *Candida boidinii*, (2) NAD(P)$^+$-dependent glutathione reductase (GSHR) from *Saccharomyces cerevisiae*, and (3) recombinant maize glutathione *S*-transferase I (GSTI).

All of these GSH-BM adsorbents exhibited higher purifying ability than seen for a control adsorbent containing CB3GA. The GSH-BM1 adsorbent, bearing glutathione-sulfonic acid as the terminal biomimetic moiety, gave the best purification for FaDH (15.0-fold) and GSTI (66.3-fold purification, with an essentially homogeneous preparation when examined by SDS-PAGE). But the GSH-BM2 adsorbent, bearing methyl-glutathione as the terminal biomimetic moiety, gave the best separation for GSHR (15.0-fold purification).

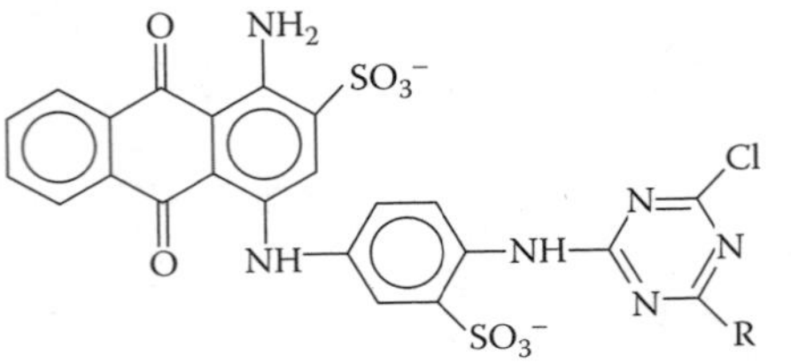

Dye-ligand	(-R)
GSH-BM1	$-HNC_{10}H_{15}N_2O_6SO_3^-$
GSH-BM2	$-HNC_{10}H_{14}N_2O_6SCH_3$
GSH-BM3	$-SC_{10}H_{14}N_2O_6NH_2$

Figure 9.5 Structures of the anthraquinone glutathionyl-biomimetic dyes GSH-BM (1-3) for the glutathione-recognizing enzymes.

The GSH-BM1 adsorbent has also been used as part of a two-step purification for FaDH, leading to a final specific activity of 79 U/mg and an overall yield of 51%.

9.3.1.1.3 Biomimetic Dye-Ligands for Galactose-Recognizing Enzymes.

Four anthraquinone galactosyl-biomimetic dye-ligands have been designed and synthesized with the aid of biocomputing. These have exhibited high selectivity for two galactose-recognizing enzymes: galactose oxidase (GaO) and galactose dehydrogenase (GalDh). Two of these new ligands were designed for GaO and bore the galactose analogs 1-amino-1-deoxy-β-D-galactose and D(+)-galactosamine as terminal biomimetic moieties (Figure 9.6) [52]. A 1.7-Å resolution structure of GaO (PDB code 1GOG) was used for this modeling procedure.

Previous hypotheses about the binding of galactose to GaO were used to guide the positioning of the biomimetic portion of these ligands. Modeling was first carried out with four biomimetic structures that contained either α-D-galactose or β-D-galactose. One possible point of interaction between galactose and GaO is between the alcohol group on carbon 6 of galactose and the copper atom at the active site of GaO. Additional hydrogen bonds can be made between Arg-330 on GaO and the alcohol groups on carbons 3 and 4 of galactose, along with hydrophobic contacts between the galactose ring and phenylalanine residues 194 and 464 on GaO.

Molecular modeling showed in this work that, unlike previous cases, the synthesized biomimetic dyes were composed of only one enzyme-recognition moiety. The galactosyl terminal biomimetic moiety mimicked the enzyme substrate. However, the anthraquinone chromophore moiety, which usually acts as a nucleotide coenzyme pseudoanalog, did not seem to play such a role with this particular enzyme. Instead, this chromophore probably acted as a mixed-function spacer that protruded the galactosyl moiety into the solvent, thus facilitating this group's interaction with GaO.

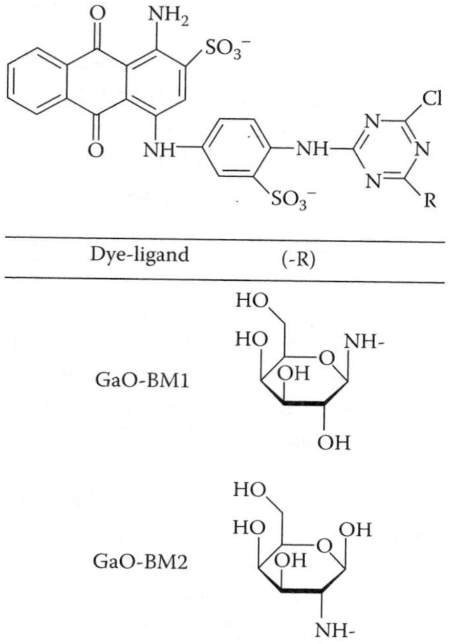

Figure 9.6 Structures of the anthraquinone galactosyl-biomimetic dyes GaO-BM (1-2) for galactose oxidase.

The immobilized forms of these ligands were examined for their ability to recognize and purify GaO from *Dactylium dendroides*. These adsorbents exhibited an increased purifying ability compared with a control adsorbent carrying CB3GA [52]. The affinity adsorbent that contained the biomimetic moiety 1-amino-1-deoxy-β-D-galactose (GaO-BM1) gave the highest purifying ability for GaO. This was then used in a two-step purification procedure for GaO from *Dactylium dendroides*. The enzyme prepared by this approach had a specific activity of 2038 U/mg, a 74% overall recovery, and gave a single band in SDS-PAGE analysis.

Two biomimetic dye-ligands (GalDh-BM) have also been designed and synthesized for another galactose-recognizing enzyme, β-galactose dehydrogenase (GalDh) from *Pseudomonas fluorescens* (Figure 9.7) [53]. These dye-ligands contained a hydrophilic spacer molecule and a terminal biomimetic moiety made up of D(+)-galactosamine or shikimic acid, with these moieties being attached to the triazine ring through a linker molecule. The design and docking of these ligands were based on the structure of *Zymomonas mobilis* glucose-fructose oxidoreductase (the best available template for model construction). The three-ring systems of the Cibacron Blue-derived portion of these ligands were first docked into the GalDh model, followed by (1) the galactose or the shikimic acid and (2) the link between the triazine ring and the galactose or shikimate. The docked positions of the dye and galactose or shikimic acid were then used to predict suitable lengths for the linker to be used in joining the ligand portions in the cofactor- and substrate-binding sites. The connection between the triazine ring and chromatographic matrix was also considered to predict an appropriate spacer between these regions. Since the path from the catalytic site to the protein surface is highly hydrophilic, a hydrophilic connection was predicted to be favored over a hydrophobic one. Based on the above analyses, entire models of GalDh bound to GalDh-BM1 (Figure 9.8) and GalDh-BM2 were constructed and energy-minimized [53].

Once the designed ligands had been made, they were immobilized and examined for their ability to purify GalDh from *Pseudomonas fluorescens* [53, 54]. Both biomimetic adsorbents exhibited an increased purifying ability compared with control nonbiomimetic adsorbents. The affinity adsorbent containing the biomimetic moiety D(+)-galactosamine (GalDh-BM1 in Figure 9.7) exhibited the highest purifying ability for GalDh (41.9-fold), but the adsorbent containing shikimic acid (GalDh-BM2) also exhibited a remarkable purifying ability for this enzyme (37.5-fold). It was also demonstrated that this second adsorbent was effective in purifying GalDh from other sources, such as baker's yeast (9.1-fold) and green peas (17.4-fold) [54].

9.3.1.2 Chlorotriazine-Linked Synthetic Biomimetic Ligands

A combination of structural data and computational tools were again used to design chlorotriazine-linked biomimetic ligands for tissue kallikrein [61] and elastase [62]. In the first case, a ligand was designed that was based on a Phe-Arg dipeptide substrate, which contained phenethylamine and *p*-aminobenzamidine on a triazine scaffold (see Figure 9.9). This was used to create hydrophobic and π-π interactions with Trp-215 and Tyr-99 on kallikrein. This also provided electrostatic interactions between the cationic side chain of the biomimetic ligand and Asp-189 of kallikrein. The ligand that was created was successfully used to purify kallikrein from acetone-dried pancreatic powder, giving a purification of tenfold [61]. A similar approach was used to mimic the interactions between turkey ovomucoid inhibitor and elastase [62]. A ligand was designed in this situation by mimicking the heptapeptide PACTLEY from the third domain of ovomucoid inhibitor (a region known to inhibit elastase) by placing various aromatic amines on the ligand's triazine ring [62].

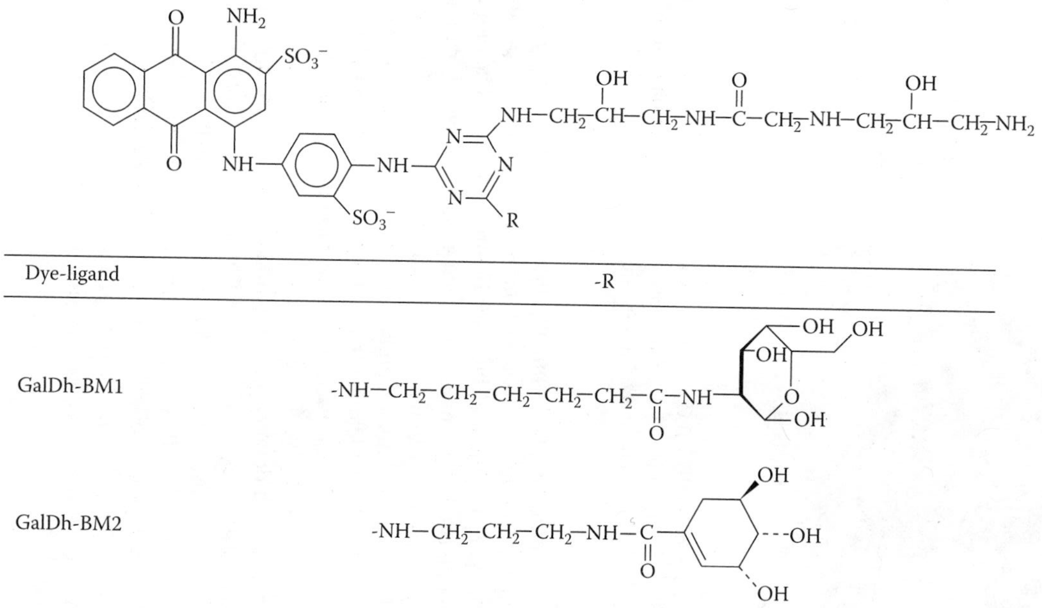

Figure 9.7 Structures of the anthraquinone galactosyl-biomimetic dyes GalDh-BM (1-2) for β-galactose dehydrogenase.

cofactor binding domain

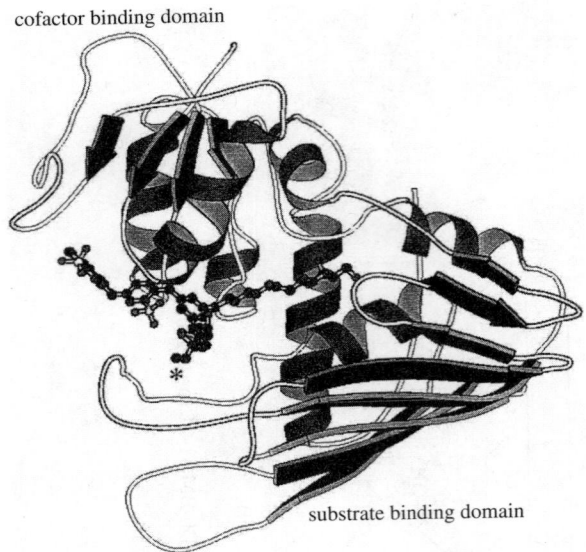

substrate binding domain

Figure 9.8   Fit of the ligand GalDh-BM1, shown as a ball-and-stick model, to the surface of GalDh. The numbers 1, 2, and 3 refer to the anthraquinone, 2,4-diaminosulfonic and triazine moieties of the dye, respectively. The "G" refers to the biomimetic moiety of the dye.

More recently, an immunoglobulin-binding ligand was designed that mimics the key dipeptide Phe-132 and Tyr-133 on fragment B of *Staphylococcus aureus* protein A, a region known to interact with the F_c fragment of IgG (see Figure 9.9b) [63]. The biomimetic ligand used in this case was composed of 3-aminophenol and 4-amino-1-naphthol moieties oriented on a triazine ring. This ligand was found to bind human IgG with a theoretical maximum capacity of 152 mg IgG per g of moist-weight gel. The IgG that was bound to this ligand could be eluted with a 67 to 69% yield and had a final purity of 97.3 to 99.2% [63].

Similar approaches have been used to design biomimetic ligands for recombinant insulin precursor MI3 [64] and glycoproteins [65]. In the case of the glycoproteins, the rational design was based on a detailed investigation of protein-carbohydrate interactions in known x-ray crystal structures. This was used to identify the key features needed to obtain specific binding to monosaccharides. A triazine ligand that was *bis*-substituted with 5-aminoindan was then identified that had a high affinity for the mannoside residues on glycoproteins (see Figure 9.9c). The interaction of this ligand with a mannoside moiety was studied using ^{1}H-NMR, molecular modeling, and partition-coefficient studies. It was found that this ligand displayed specificity similar to that of the lectin concanavalin A toward neutral hexoses [65].

9.3.2   Combinatorial Approaches

The rapid growth of combinatorial techniques for generating large numbers of novel compounds (e.g., synthetic peptides) and the availability of more powerful computers have made it more feasible to use computer-aided screening for ligand libraries [46]. This screening involves an evaluation of electrostatic interactions and steric effects, along with more complex energy terms, in addition to molecular-modeling and docking techniques. With this approach, it is possible to examine about 10 to 100 ligands per minute. This enables the examination of a database of 100,000 compounds in less than a week [66].

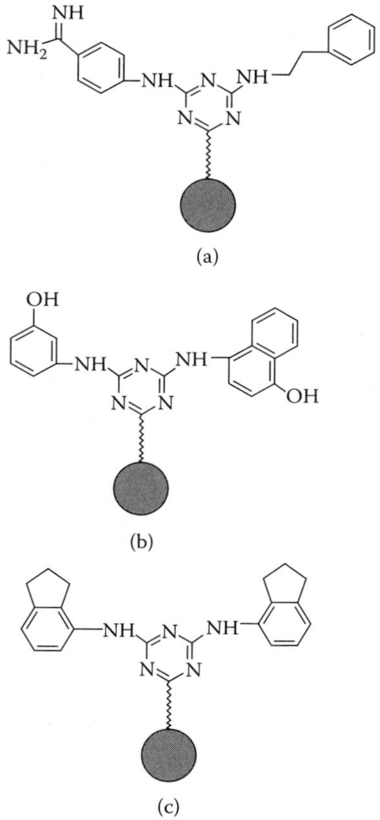

(a)

(b)

(c)

Figure 9.9 Structures of some biomimetic ligands: (a) kallikrein ligand; (b) IgG-binding ligand; (c) glycoprotein-binding ligand.

The combined use of structure-based design and combinatorial chemistry, known as *combinatorial docking*, has emerged as a new and promising approach to ligand discovery [46]. The goal of this approach is to use three-dimensional structural information to focus synthetic efforts on those molecules that are most likely to bind to a target. Although such methods have been successful, the need to simplify the energy function used in the screening process is a severe limitation, since this requires a complementary experimental method for *in vitro* verification of molecular recognition. This is needed because studies in free solution with soluble ligands do not adequately reflect the chemical, geometrical, and steric constraints that are imposed by the three-dimensional matrix used in ligand screening [45].

9.3.2.1 Affinity Chromatographic Screening of Combinatorial Peptide Libraries

Libraries of random synthetic peptides can be generated chemically using solid-phase synthesis methods. The synthesized peptides are then either cleaved from the support to form a soluble library or are allowed to remain on the solid support to give a one-bead–one-structure peptide library [67–69].

A basic *in vitro* screen for a combinatorial library of possible ligands is to pass the library mixture over a surface (such as in a chromatographic column) onto which the protein or target of interest has been immobilized [67]. The ligands that bind to the target will be retarded in their passage over this surface. Thus, the elution profile of the ligands will provide direct information on structure-affinity relationships. For instance, the longer the retention time is for a library member, the higher this member's affinity must be for the immobilized target. This means that late-eluting compounds on the support will form a pool of good candidates for further development as affinity ligands (see Figure 9.10).

Elution in this screening approach can be achieved either specifically or nonspecifically. Specific elution is usually accomplished by adding an agent known to bind the desired target. Examples of such agents for an enzyme target would be a substrate, inhibitor,

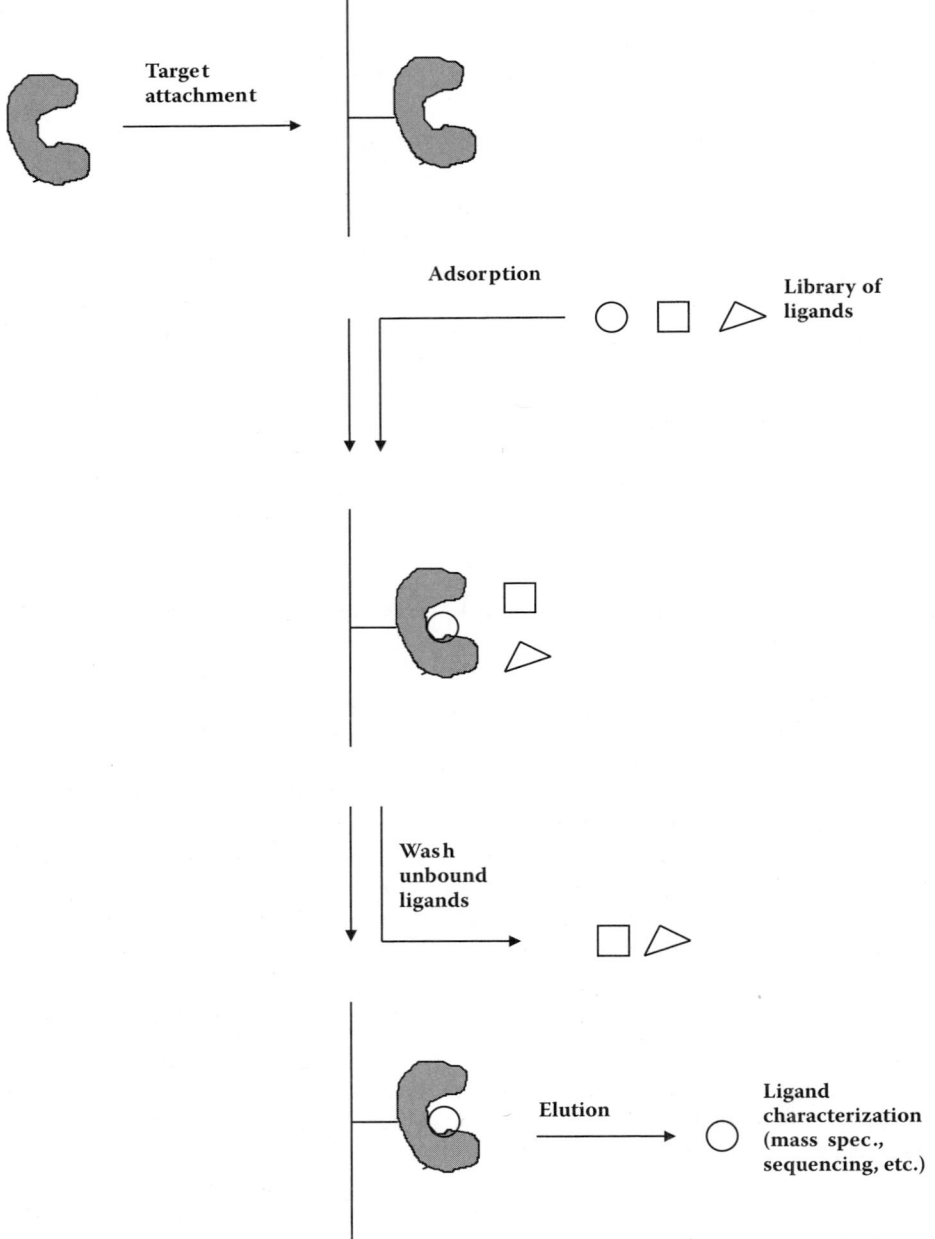

Figure 9.10   Affinity chromatographic screening of combinatorial peptide libraries.

activator, or some combination of these compounds. For example, Huang and Carbonell [67] demonstrated an affinity chromatographic process for screening peptide ligands from two soluble libraries that contained 46,656 and 104,976 peptides. These libraries were screened using an immobilized S-protein as the target. The retained peptides were then separated by reversed-phase HPLC, and their sequences were determined by using fast-atom-bombardment mass spectrometry [67].

Another recent approach for screening combinatorial peptide libraries has been to use peptide arrays [70]. In this method, peptides are attached to a positionally defined support like a cellulose acetate sheet [71]. For instance, this technique has been used with synthetic peptide tetramers that mimic protein A in its ability to recognize the F_c fragment of IgG as well as its affinity purification of human IgG, mouse ascetic fluid IgE, chicken yolk IgY, and IgA from cell culture supernatants [72]. In another example, the screening of peptide libraries on spots has been used to yield peptides that are selective for the light and heavy chains of human blood coagulation factor VII [71].

9.3.2.2 Phage-Display Technology and Biopanning

Phage-display technology is another method that is becoming a versatile tool for the discovery of new affinity ligands [73]. This technology relies on the use of phage-display libraries in a screening process known as *biopanning*, as illustrated in Figure 9.11. In biopanning, a phage-display library is first incubated with a target molecule. This library can be incubated directly with an immobilized target or preincubated with a target prior to capture on a solid support. (e.g., by using streptavidin to bind to biotin). The noninteracting peptides are then washed away, and any retained peptides or proteins are eluted. The interacting and captured phage-display peptides can then be amplified by bacterial infection to increase their copy number. This screening and amplification process can be repeated as necessary to obtain higher-affinity phage-display peptides. Once a ligand with the desired affinity has been obtained, its primary structure can be determined by performing DNA sequencing on the phage.

Peptide libraries that are displayed on phages could represent a general source of affinity ligands for protein purification. For example, the immunoglobulin-binding domain of protein A has been displayed on the surface of a phage. This raises the potential for screening mutant forms of protein A that might have improved specificity or that are able to work under milder elution properties for use in the purification of monoclonal antibodies [73]. In one study, the target for phage display was the constant region of an IgG_1 humanized monoclonal antibody that had been raised against the low-affinity p55 subunit of the interleukin-2 receptor. Two peptides with the sequences SPAPSDS and EPIHRST-LTALL were selected by this approach and were used in affinity chromatography. When immobilized, the first peptide had a binding capacity of 120 μg IgG per gram of gel, and the other had a binding capacity of 320 μg IgG per gram of gel.

9.3.2.3 Ribosome Display Technology

Another combinatorial approach that has been used in ligand design is that of *ribosome display technology*. The principles underlying this method are depicted in Figure 9.12 [74]. In this method, a DNA library (encoding a peptide in a ribosome display cassette) is used either directly for *in vitro* transcription-translation or is first transcribed *in vitro* to give mRNA, which is then purified and used for *in vitro* translation [74].

In this technique, the affinity selection of ribosomal complexes can be performed in two different ways. For instance, a ligand could be immobilized on a surface, such as in panning tubes or in microtiter wells. Alternatively, biotinylated ligands could be used

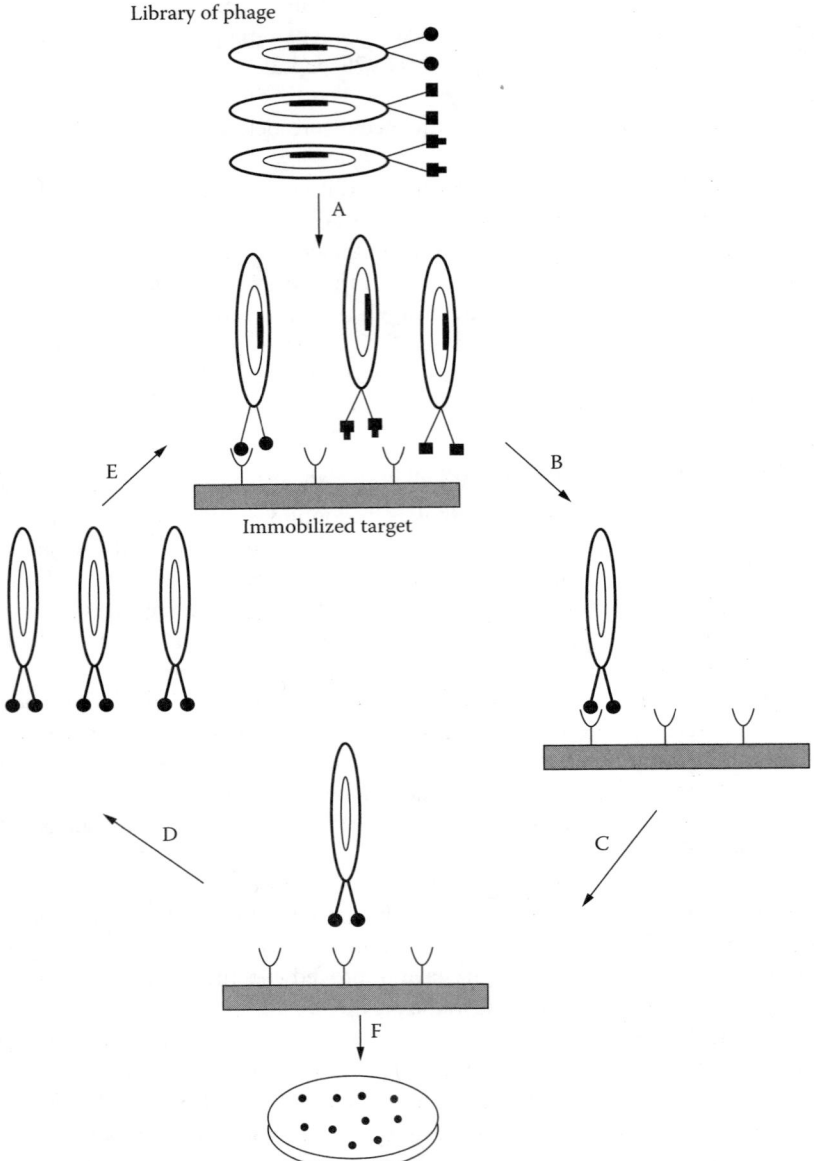

Figure 9.11 General scheme of selecting phages from a phage-displayed combinatorial peptide library. In step A, the displaying peptides are introduced into microtiter plate wells containing an immobilized target. After incubation (step B), the nonbinding phage is washed away. The bound phage is then recovered (step C). The phage particles are then transferred to another tube, and the bacteria are infected and more phage particles produced (step D). The amplified phage is then rescreened (step E) to complete one cycle of affinity selection. After three or more rounds of screening, the phage is plated out (step F) such that there are individual plaques (clones) for additional analysis.

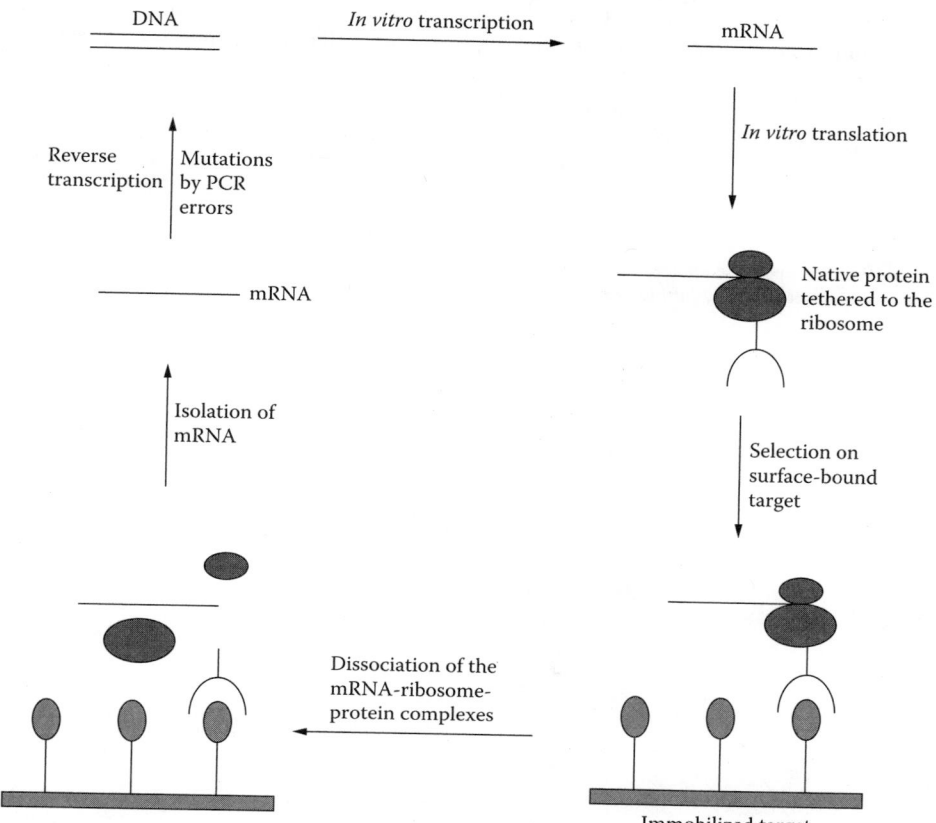

Figure 9.12 Schematic diagram of ribosome display. A library of peptides is transcribed and translated *in vitro*. The resulting mRNA lacks a stop codon, giving rise to linked mRNA-ribosome-protein complexes that are used for selection on the immobilized target. The bound mRNA is eluted, purified, and characterized. (Adapted from Pluckthun, A., Schaffitzel, C., Hanes, J., and Jermutus, L., *Adv. Prot. Chem.*, 55, 367–403, 2001. With permission.)

to bind to the proteins displayed on the ribosomal complexes, thus allowing their capture by streptavidin-coated magnetic beads. Because the peptide sequences used in this approach are very short, few errors are introduced by the error-prone PCR techniques. Therefore, a cassette-mutagenesis strategy should be adopted.

Ribosome display is particularly appropriate for the screening and selection of folded proteins [74]. Protein domains from fibronectin, V domains, α-helical bacterial receptor domains, and lipocalin domains have all been shown to produce specific binding agents by this technique. Dissociation constants obtained for these ligands have been in the millimolar range [75].

9.3.2.4   SELEX

A fourth combinatorial technique that can be used to generate affinity ligands is the *SELEX method*. Introduced in 1990 by Tuerk and Gold, SELEX is an acronym for the Systematic Evolution of Ligands by Exponential Enrichment [76]. With this technique, it is possible to exponentially enrich and obtain nucleic acid-based ligands from a random oligonucleotide pool. Currently, this technology is used to screen for nucleic acid ligands (called *aptamers*) that can bind protein targets with potential applications in diagnostics and biotechnology [74, 77, 78].

The basis of the SELEX method is shown in Figure 9.13. In this approach, a DNA library is first synthesized. This library contains a random or mutagenized region that is

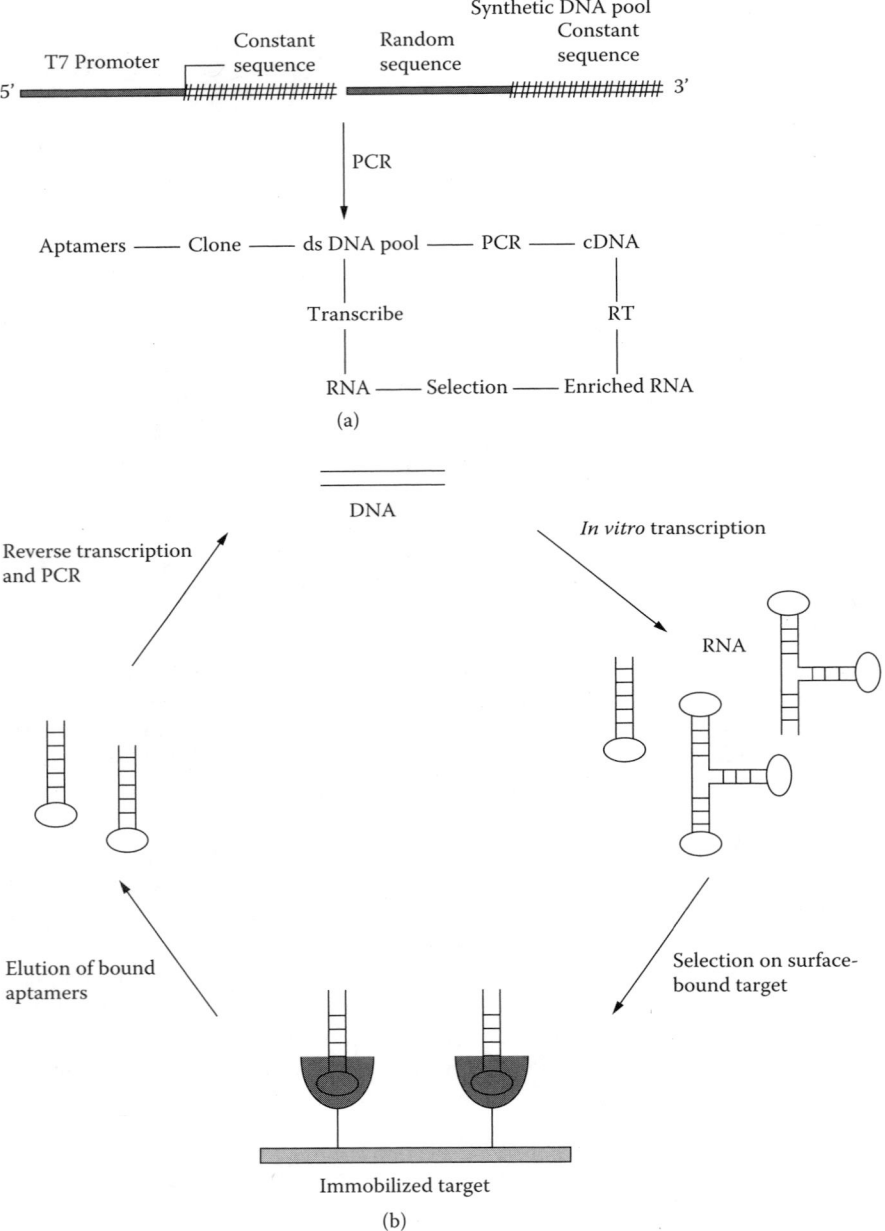

Figure 9.13 (a) Steps in a typical SELEX process and (b) schematic diagram of the SELEX process. (Adapted from Pluckthun, A., Schaffitzel, C., Hanes, J., and Jermutus, L., *Adv. Prot. Chem.*, 55, 367–403, 2001. With permission.)

flanked on each end by a constant sequence. A T7 RNA polymerase promoter region is also present at the 5′ end. This DNA is amplified by a few cycles of PCR and is then transcribed *in vitro* to make an RNA pool. The individual RNA molecules are then screened for their binding to a given target, such as by passing them through an affinity column that contains this target in an immobilized form. The retained RNAs are then eluted, reverse-transcribed, amplified by PCR, and transcribed back to RNA. These are then used again in the selection and screening cycle until a ligand preparation with the final desired properties is obtained.

In vitro selection has been used to obtain aptamers to targets that cover a wide range of sizes. These targets range from simple ions, peptides, proteins, organelles, and viruses to even entire cells (see the literature [77] for a review). The successful use of this technology for affinity chromatography has been recently reported [79]. In this case, a DNA aptamer was developed for the purification of a recombinant human L-selectin-immunoglobulin fusion protein from Chinese hamster ovary cells. An impressive purification of 1500-fold was achieved with 83% recovery using this ligand.

Other than the example given in the literature [79], SELEX has not yet found wide applications in affinity separations. However, the versatility of this method and its ability to rapidly generate high-affinity ligands from very large libraries (10^{15} to 10^{16} sequences) should help increase the role of the technique in future uses of affinity chromatography.

9.4 SUMMARY AND CONCLUSIONS

This chapter examined the use of both dye-ligands and biomimetic ligands in affinity chromatography. Dye-ligands that are used for such work are generally based on triazine dyes, which are widely used in protein purification. The advantages of using these dyes for such work include their stability, ease of immobilization, low cost, availability, and high binding capacities.

However, a desire to improve the selectivity of these dyes has led to the use of rational molecular design techniques for the generation of biomimetic ligands. These are designed to resemble the structure of biological ligands or to mimic the interactions of these ligands with the binding sites of target proteins. The use of combinatorial techniques and molecular modeling for the discovery of better biomimetic ligands has opened new avenues for the development of more efficient, less expensive, and safer procedures for protein purification at the industrial level. Many of these same approaches should also lead to the creation of better affinity columns for analytical-scale work.

SYMBOLS AND ABBREVIATIONS

CB3GA	Cibacron Blue 3GA
FaDH	Formaldehyde dehydrogenase
GaO	Galactose oxidase
GaO-BM	Biomimetic dyes for galactose oxidase
GalDh	Galactose dehydrogenase
GalDh-BM	Biomimetic dyes for β-galactose dehydrogenase
GSH-BM	Glutathionyl-biomimetic dyes
GSHR	Glutathione reductase
GSTI	Glutathione *S*-transferase I
KC-BM	(Keto)carboxyl-biomimetic dyes

LDH	L-Lactate dehydrogenase
MALDI MS	Matrix-assisted laser desorption/ionization mass spectrometry
MDH	L-Malate dehydrogenase
PHEMA	Poly(EGDMA-2-hydroxyethyl methacrylate)
SELEX	Systematic evolution of ligands by exponential enrichment

REFERENCES

1. Thomas, R., Drug discovery of the future: the implications of the human genome project, *Trends Biotechnol.,* 19, 496–499, 2001.
2. Business Communications Company, Protein Drugs: Manufacturing Technologies, RB-114R, Norwalk, CT, 2000.
3. Garber, K., Biotech industry faces new bottleneck, *Nat. Biotechnol.,* 19, 184–185, 2001.
4. Lowe, C.R., Combinatorial approaches to affinity chromatography, *Curr. Opin. Chem. Biol.,* 5, 248–256, 2001.
5. Geoffrey, S. and McCarty, J., Personalised medicine: Revolutionizing drug discovery and patient care, *Trends Biotechnol.,* 19, 491–496, 2001.
6. Wehr, T., Separation technology in proteomics, *LC-GC,* 19, 702–711, 2001.
7. Johnston, K.A., Chemical attraction: Affinity techniques aid in new drug discovery, *PharmaGenomics,* Oct/Nov, 28–39, 2001.
8. Kodadek, T., Protein microarrays: Prospects and problems, *Chem. Biol.,* 8, 105–115, 2001.
9. Labrou, N.E. and Clonis, Y.D., Immobilised synthetic dyes in affinity chromatography, in *Theory and Practice of Biochromatography,* Vijayalakshmi, M.A., Ed., Taylor & Francis, London, 2002, pp. 335–351.
10. Labrou, N.E. and Clonis, Y.D., The affinity technology in downstream processing, *J. Biotechnol.,* 36, 95–119, 1994.
11. Clonis, Y.D., Labrou, N.E., Kotsira, V., Mazitsos, K., Melissis, S., and Gogolas, G., Biomimetic dyes as affinity chromatography tools in enzyme purification, *J. Chromatogr. A,* 891, 33–44, 2000.
12. Haeckel, R., Hess, B., Lauterborn, W., and Wurster, K.-H., Purification and allosteric properties of yeast pyruvate kinase, *Hoppe-Seyler's Z. Physiol. Chem.,* 349, 699–714, 1968.
13. Staal, G., Koster, J., Kamp, H., Van Milligen-Boersma, L., and Veeger, C., Human erythrocyte pyruvate kinase, its purification and some properties, *Biochem. Biophys. Acta,* 227, 86–92, 1971.
14. Lowe, C.R., Burton, S.J., Burton, N.P., Alderton, W.K., Pitts, J.M., and Thomas, J.A., Designer dyes as affinity chromatography tools in enzyme purification, *Trends Biotechnol.,* 10, 442–448, 1992.
15. Lowe, C.R., Burton, S.J., Pearson, J., Clonis, Y.D., and Stead, C.V., Design and application of biomimetic dyes in biotechnology, *J. Chromatogr.,* 376, 121–130, 1986.
16. Denizli, A. and Pikin, E., Dye-ligand affinity systems, *J. Biochem. Biophys. Methods,* 49, 391–416, 2001.
17. Labrou, N.E., Dye-ligand affinity chromatography for protein separation and purification, in *Methods in Molecular Biology, Affinity Chromatography: Methods and Protocols,* Walker, J.M., Ed., Humana Press, Totowa, NJ, 2000, pp. 129–139.
18. Labrou, N.E., Karagouni, A., and Clonis, Y.D., Biomimetic-dye affinity adsorbents for enzyme purification: Application to the one-step purification of *Candida boidinii* formate dehydrogenase, *Biotech. Bioeng.,* 48, 278–288, 1995.
19. Biellmann, J.F., Samama, J.P., Braden, C., and Eklund, H., X-Ray studies of the binding of Cibacron Blue F3GA to liver alcohol dehydrogenase, *Eur. J. Biochem.,* 102, 107–110, 1979.
20. Li, R., Bianchet, M.A., Talalay, P., and Amzel, L.M., The three-dimensional structure of NAD(P)H:quinone reductase, a flavoprotein involved in cancer chemoprotection and chemotherapy: Mechanism of the two electron reduction, *Proc. Natl. Acad. Sci.,* 92, 8846–8850, 1995.

21. Oakley, A.J., Lo Bello, M., Nuccetelli, M., Mazzetti, A.P., and Parker, M.W., The ligandin (non-substrate) binding site of human Pi class glutathione transferase is located in the electrophile binding site (H-site), *J. Mol. Biol.*, 291, 913–926, 1999.

22. Friess, S.D. and Zenobi, R., Protein structure information from mass spectrometry? Selective titration of arginine residues by sulfonates, *J. Am. Soc. Mass Spectrom.*, 12, 810–818, 2001.

23. Salih, B. and Zenobi, R., MALDI mass spectrometry of dye-peptide and dye-protein complexes, *Anal. Chem.*, 70, 1536–1543, 1998.

24. Firer, M.A., Efficient elution of functional proteins in affinity chromatography, *J. Biochem. Biophys. Methods*, 49, 433–442, 2001.

25. Zijlstra, G.M., Michielsen, M.J., de Gooijer, C.D., van der Pol, L.A., and Tramper, J., IgG and hybridoma partitioning in aqueous two-phase systems containing a dye-ligand, *Bioseparation*, 7, 117–126, 1998.

26. Lali, A., Balan, S., John, R., and D'Souza, F., Carboxymethyl cellulose as a new heterobifunctional ligand carrier for affinity precipitation of proteins, *Bioseparation*, 7, 195–205, 1999.

27. Mattiasson, B., Galaev, Y.I., and Garg, N., Polymer-shielded dye affinity chromatography, *J. Mol. Recognit.*, 9, 509–514, 1996.

28. Garg, N., Galaev, I.Y., and Mattiasson, B., Polymer-shielded dye affinity chromatography of lactate dehydrogenase from porcine muscle in an expanded bed system, *Bioseparation*, 6, 193–199, 1996.

29. Koch, C., Borg, L., Skjodt, K., and Houen, G., Affinity chromatography of serine proteases on the triazine dye ligand Cibacron Blue F3G-A, *J. Chromatogr. B*, 718, 41–46, 1998.

30. Swaminathan S. and Khanna, N., Affinity purification of recombinant interferon-alpha on a mimetic ligand adsorbent, *Protein Exp. Purif.*, 15, 236–242, 1999.

31. Moore, L.G., Ng-Chie, W., Lun, S., Lawrence, S.B., Young, W., and McNatty, K.P., Follicle-stimulating hormone in the brushtail (*Trichosurus vulpecula*): Purification, characterization, and radioimmunoassay, *Gen. Comp. Endocrinol.*, 106, 30–38, 1997.

32. Govoroum, M.S., Huet, J.C., Pernollet, J.C., and Breton, B., Use of immobilized metal ion affinity chromatography and dye-ligand chromatography for the separation and purification of rainbow trout pituitary gonadotropins, GTH I and GTH II, *J. Chromatogr. B*, 698, 35–46, 1997.

33. Alderton, W.K., Thatcher, D., and Lowe, C.R., Affinity labeling of recombinant ricin A chain with Procion Blue MX-R, *Eur. J. Biochem.*, 233, 880–885, 1995.

34. Allary, M., Saint-Blancard, J., Boschetti, E., and Girot, P., Large scale production of human albumin: Three years experience of an affinity chromatography process, *Bioseparation*, 2, 167–175, 1991.

35. Burnouf, T. and Radosevich, M., Affinity chromatography in the industrial purification of plasma proteins for therapeutic use, *J. Biochem. Biophys. Methods*, 49, 575–586, 2001.

36. Denizli, A., Kocakulak, M., and Piskin, E., Bilirubin removal from human plasma in a packed-bed column system with dye-affinity microbeads, *J. Chromatogr. B*, 707, 25–31, 1998.

37. Denizli, A., Salih, B., Kavakli, C., and Piskin, E., Dye-incorporated poly(EGDMA-HEMA) microspheres as specific sorbents for aluminum removal, *J. Chromatogr. B*, 698, 89–96, 1997.

38. Denizli, A., Salih, B., and Piskin, E., New chelate-forming polymer microspheres carrying dyes as chelators for iron overload, *J. Biomater. Sci. Polym. Ed.*, 9, 175–187, 1998.

39. Arica, M.Y., Testereci, H.N., and Denizli, A., Dye-ligand and metal chelate poly(2-hydroxyethylmethacrylate) membranes for affinity separation of proteins, *J. Chromatogr. A*, 799, 83–91, 1998.

40. Alberghina, G., Fisichella, S., and Renda, E., Separation of G structures formed by a 27-mer guanosine-rich oligodeoxyribonucleotide by dye-ligand affinity chromatography, *J. Chromatogr. A*, 840, 51–58, 1999.

41. Sun, Y., Ichikawa, S., Sugiura, S., and Furusaki, S., Affinity extraction of proteins with a reversed micellar system composed of Cibacron Blue-modified lecithin, *Biotechnol. Bioeng.*, 58, 58–64, 1998.

42. Caspi, S., Halimi, M., Yanai, A., Sasson, S.B., Taraboulos, A., and Gabizon, R., The anti-prion activity of Congo Red, *J. Biol. Chem.*, 273, 3484–3489, 1998.

43. Hattori, T., Zhang, X., Weiss, C., Xu, Y., Kubo, T., Sato, Y., Nishikawa, S., Sakaida, H., and Uchiyama, T., Triazine dyes inhibit HIV-1 entry by binding to envelope glycoproteins, *Microbiol. Immunol.*, 41, 717–724, 1997.

44. Brown, R.A. and Combridge, B.S., Binding of hepatitis virus particles to immobilized Procion Blue and Cibacron Blue 3GA, *J. Virol. Methods*, 14, 267–274, 1986.

45. Lowe, C.R., Lowe, A.R., and Gupta, G., New developments in affinity chromatography with potential application in the production of biopharmaceuticals, *J. Biochem. Biophys. Meth.*, 49, 561–574, 2001.

46. Bohm, H.-J. and Stahl, M., Structure-based library design: Molecular modelling merges with combinatorial chemistry, *Curr. Opin. Chem. Biol.*, 4, 283–286, 2000.

47. Lindner, N.M., Jeffcoat, R., and Lowe, C.R., Design and application of biomimetic dyes: Purification of calf intestinal alkaline phosphatase with immobilized terminal ring analogues of C.I. Reactive Blue 2, *J. Chromatogr.*, 473, 227–240, 1989.

48. Labrou, N.E. and Clonis, Y.D., Biomimetic-dye affinity chromatography for the purification of bovine heart L-lactate dehydrogenase, *J. Chromatogr. A*, 718, 35–44, 1995.

49. Melissis, S., Rigden, D.J., and Clonis, Y.D., A new family of glutathionyl-biomimetic ligands for affinity chromatography of glutathione-recognizing enzymes, *J. Chromatogr. A*, 917, 29–43, 2001.

50. Jones, G., Willet, P., Glen, R.C., and Taylor, R., Development and validation of a genetic algorithm for flexible docking, *J. Mol. Biol.*, 267, 727–748, 1997.

51. Labrou, N.E., Eliopoulos, E., and Clonis, Y.D., Molecular modeling for the design of dye-ligands and their interaction with mitochondrial malate dehydrogenase, *Biochem. J.*, 315, 695–703, 1996.

52. Mazitsos, C.F., Rigden, D.J., Tsoungas, P., and Clonis, Y.D., Galactosyl-biomimetic dye ligands for the purification of Dactylium dendroides galactose oxidase, *J. Chromatogr. A*, 954, 137–150, 2002.

53. Mazitsos, C.F., Rigden, D.J., Tsoungas, P., and Clonis, Y.D., Galactosyl-mimodye ligands for *Pseudomonas fluorescens* beta-galactose dehydrogenase, *Eur. J. Biochem.*, 269 (22), 5391–5405, 2002.

54. Mazitsos, C.F., Rigden, D.J., and Clonis, Y.D., Galactosyl-biomimetic dye ligands for the purification of *Dactylium dendroides* galactose dehydrogenase, submitted.

55. Kotsira, V.P. and Clonis, Y.D., Oxalate oxidase from barley roots: Purification to homogeneity and study of some molecular, catalytic, and binding properties, *Arch. Biochem. Biophys.*, 340, 239–249, 1997.

56. Labrou, N.E. and Clonis, Y.D., Oxaloacetate decarboxylase from *Pseudomonas stutzeri*: Purification and characterization, *Arch. Biochem. Biophys.*, 365, 17–24, 1999.

57. Labrou, N.E. and Clonis, Y.D., Biomimetic-dye affinity chromatography for the purification of mitochondrial L-malate dehydrogenase from bovine heart, *J. Biotechnol.*, 45, 185–194, 1996.

58. Labrou, N.E. and Clonis, Y.D., L-malate dehydrogenase from *Pseudomonas stutzeri*: Purification and characterization, *Arch. Biochem. Biophys.*, 337, 103–114, 1997.

59. Labrou, N.E. and Clonis, Y.D., Biomimetic-dye affinity chromatography for simultaneous separation and purification of lactate dehydrogenase and malate dehydrogenase from bovine heart, *Bioprocess Eng.*, 16, 157–161, 1997.

60. Labrou, N.E., Eliopoulos, E., and Clonis, Y.D., Molecular modelling for the design of a biomimetic chimeric ligand: Application to the purification of bovine heart L-lactate dehydrogenase, *Biotechnol. Bioeng.*, 63, 322–332, 1999.

61. Burton, N.P. and Lowe, C.R., Design of novel affinity adsorbents for the purification of trypsin-like proteases, *J. Mol. Recogn.*, 5, 55–68, 1992.

62. Filippusson, H., Erlendsson, L.S., and Lowe, C.R., Synthesis of biomimetic affinity ligands for elastases and their use for the purification of porcine and cod elastases, *J. Mol. Recogn.*, 13, 370–381, 2000.

63. Teng, S.-F., Sproule, K., Hussain, A., and Lowe, C.R., Affinity chromatography on immobilized "biomimetic" ligands: Synthesis, immobilization and chromatographic assessment of an immunoglobulin G-binding ligand, *J. Chromatogr. B*, 740, 1–15, 2000.

64. Sproule, K., Morrill, P., Pearson, J.C., Burton, S.J., Hejnas, K.R., Valore, H., Ludvigsen, S., and Lowe, C.R., New strategy for the design of ligands for the purification of pharmaceutical proteins by affinity chromatography, *J. Chromatogr. B,* 740, 17–33, 2000.

65. Palanisamy, U.D., Winzor, D.J., and Lowe, C.R., Synthesis and evaluation of affinity adsorbents for glycoproteins, *J. Chromatogr. B,* 746, 265–281, 2000.

66. Whittle, P.J. and Blundell, T.L., Protein structure-based drug design, *Ann. Rev. Biophys. Biomol. Struct.,* 23, 349–375, 1994.

67. Huang, P.Y. and Carbonell, R.G., Affinity chromatographic screening of soluble combinatorial peptide libraries, *Biotechnol. Bioeng.,* 63, 633–641, 2000.

68. Lebl, M., Krchnak, V., Sepetov, N.F., Seligmann, B., Strop, P., Felder, S., and Lam, K.S., One-bead-one-structure combinatorial libraries, *Biopolymers,* 37, 177–198, 1995.

69. Kaufman, D.B., Hentsch, M.E., Baumbach, G.A., Buettner, J.A., Dadd, C.A., Huang, P.Y., Hammond, D.J., and Carbonell, R.G., Affinity purification of fibrinogen using a ligand from a peptide library, *Biotechnol. Bioeng.,* 77, 278–289, 2002.

70. Thorpe, D.S. and Walle, S., Combinatorial chemistry defines general properties of linkers for the optimal display of peptide ligands for binding soluble protein targets to TentaGel microscopic beads, *Biochem. Biophys. Res. Commun.,* 269, 591–595, 2000.

71. Amatschek, K., Necina, R., Hahn, R., Schallaun, E., Schwinn, H., Josic, D., and Jungbauer, A., Affinity chromatography of human blood coagulation factor VII on monoliths with peptides from a combinatorial library, *HRC-J. High Res. Chromatogr.,* 23, 47–58, 2000.

72. Fassina, G., Verdoliva, A., Palombo, G., Ruvo, M., and Cassini, G., Immunoglobulin specificity of TG19318: A novel synthetic ligand for antibody affinity purification, *J. Mol. Recogn.,* 11, 107–109, 1998.

73. Ehrlich, G.K. and Bailon, P., Identification of peptides that bind to the constant region of a humanized IgG$_1$ monoclonal antibody using phage display, *J. Mol. Recogn.,* 11, 121–125, 1998.

74. Pluckthun, A., Schaffitzel, C., Hanes, J., and Jermutus, L., *In vitro* selection and evolution of proteins, *Adv. Prot. Chem.,* 55, 367–403, 2001.

75. Nygren, P.A. and Uhlen, M., Scaffolds for engineering novel binding sites in proteins, *Curr. Opin. Struct. Biol.,* 7, 463–469, 1997.

76. Tuerk, C. and Gold, L., Systematic evolution of ligands by exponential enrichment: RNA ligands to bacteriophage T4 DNA polymerase, *Science,* 249, 505–510, 1990.

77. Wilson, D.S. and Szostak, J.W., *In vitro* selection of functional nucleic acids, *Annu. Rev. Biochem.,* 68, 611–647, 1999.

78. Wen, J.D., Gray, C.W., and Gray, D.M., SELEX selection of high-affinity oligonucleotides for bacteriophage Ff gene 5 protein, *Biochemistry,* 40, 9300–9310, 2001.

79. Roming, T.S., Bell, C., and Drolet, D.W., Aptamer affinity chromatography: Combinatorial chemistry applied to protein purification, *J. Chromatogr. B,* 731, 275–284, 1999.

10

Immobilized Metal-Ion Affinity Chromatography

Daniela Todorova and Mookambeswaran A. Vijayalakshmi

Molecular Interaction & Separation Technology Labs,
Université de Technologie de Compiègne, Compiègne, France

CONTENTS

10.1   INTRODUCTION

The idea of using immobilized metal-chelate complexes in ligand-exchange chromatography for the separation of amino acids, nucleic acids, peptides, and proteins was developed in the early 1960s [1–3]. In 1975, Porath and coworkers [4] introduced immobilized metal-ion affinity chromatography (IMAC) as a purification technique for biomolecules. IMAC is based on the solution-phase interactions of chelated metal ions with proteins and other biological compounds. Other terms used to describe this method are *metal-chelate chromatography* and *metal-ion interaction chromatography.*

Since the work by Porath and his colleagues, many investigators have examined and documented the high selectivity of IMAC adsorbents, helping to develop this mode of chromatography into a popular approach for protein purification [5–7]. This chapter discusses the basic concepts behind this method and looks at various practical issues that should be considered in its use. Numerous applications of IMAC are also examined, with both traditional and newer uses being presented.

10.1.1   Basic Principles of IMAC

The primary basis for separations in IMAC is an effect known as *immobilized-metal affinity* (IMA). This makes use of the ability of certain amino acids on proteins or purine and pyrimidine bases on nucleic acids to bind chelated transition metals, as shown in Figure 10.1. To make use of this interaction in IMAC, metal ions are immobilized through coordination bonds to a metal chelator, which in turn is covalently bound to an insoluble matrix. In principle, there are many metal ions that can be used in IMAC, but the most popular ones for studying and making use of protein-metal interactions are the borderline transition metals, including copper (Cu(II)), zinc (Zn(II)), nickel (Ni(II)), and cobalt (Co(II)) [4, 8, 9].

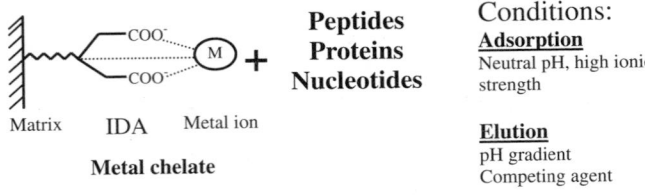

Figure 10.1   The basic principles behind immobilized metal-ion affinity chromatography (IMAC). The abbreviation IDA refers to iminodiacetic acid, which is a common chelating agent used in IMAC. Common metal ions (M) that are used in IMAC include Cu(II), Ni(II), Zn(II), or Co(II). Examples of typical biomolecules and adsorption or elution conditions that are used in IMAC are listed to the right of this figure.

Table 10.1 Influence of Histidine Topography on Protein Retention in IMAC

Ligand	Type of Metal Ion			
	Cu(II)	Ni(II)	Zn(II)	Co(II)
No histidines (His) or nonaccessible histidines	–	–	–	–
One histidine	+	–	–	–
His-(X)$_n$-His	++	+	–	–
His-(X)$_n$-His With n = 2, 3 in an α-Helix	+++	+	+	+
--His…His-- In a vicinal position in a protein's 3-D structure	+++	+	+	+

Note: The values shown in this table represent a semiquantitative estimation of the strength of metal-affinity binding, where "+++" indicates strong binding, "++" or "+" refer to intermediate binding strengths, and "–" represents no interactions.

Source: From Sulkowski, E., *BioEssays*, 10, 170–175, 1989. With permission.

One recent definition proposed for metal affinity chromatography posits that it is "a special form of ligand-exchange chromatography used to separate biopolymers with a particular affinity for a specific metal cation, typically copper(II), zinc(II), and iron(II)" [10]. However, this definition does not completely cover the full potential of IMAC and does not include its use with nickel(II), the metal ion most often employed in this technique.

When a protein is applied to an IMAC column, it will be retained if it has surface-accessible amino acid residues with electron-donating properties. Examples of such residues are histidine, tryptophan, and cysteine, which are sometimes called "Porath's triad." Terminal α-amine groups can also be involved in this process [11]. Of these residues, studies have shown that histidine is probably the main amino acid that leads to the binding of proteins to chelated transition metals at or near a neutral pH and at high ionic strengths (the latter condition often being used in IMAC to suppress electrostatic interactions) [12].

The number of histidine residues on a protein, their surface accessibility, and the nature of their microenvironment are all important in determining the retention of a protein by a chelated metal. In work with several model proteins, Sulkowski [13] established the ground rules for protein recognition by transition metal ions that are chelated to iminodiacetic acid (IDA), a common agent used in IMAC columns. The results obtained in his studies are summarized in Table 10.1.

Nucleotide–transition metal interactions have also been investigated for use in IMAC. For instance, the group fractionation of pyrimidine and purine mono- and dinucleotides has been achieved by using agarose or silica-based copper chelate adsorbents [14–18]. The adsorption process involved in such separations appears to be complex, since this may involve electron donor-acceptor interactions as well as hydrophobic and electrostatic interactions. But it has been noted that the copper ions have an affinity for purine that is much greater than that for pyrimidine (with adenine > guanine, and cytosine being roughly the same as thymidine and uracil). In addition, bases generally show much greater affinity to these chelates than nucleosides or nucleotides [14–16].

10.1.2 Recent Developments in the Use of Immobilized-Metal Affinity

Immobilized-metal affinity has now been developed for the separation of amino acids, peptides [19, 20], a variety of native proteins [12, 21, 22], genetically modified proteins

[23], and even whole cells [24, 25]. Today, IMAC is the most common method used for the purification of engineered proteins that contain histidine affinity tags [26], with it often being possible to use IMAC for the single-step isolation of His-tagged proteins [27, 28]. After this purification, the chelating His-tag handle can then be removed either chemically or enzymatically to generate a protein with the final desired amino acid sequence.

As will be shown later, another application of IMAC has been its use for mapping the surface topography of proteins [29, 30]. Moreover, immobilized-metal affinity recognition has been extended to various microanalytical methods, such as *immobilized metal-ion affinity gel electrophoresis* (IMAGE) and *immobilized metal-ion affinity capillary electrophoresis* (IMACE) [31, 32]. Many studies have shown that IMAGE and IMACE can be utilized to quantify ligand-protein interactions [33], to analyze conformational changes after the chemical modification or oligomerization of proteins [34–36], and to study the histidine topography of genetically engineered proteins or the mechanism of protein-protein interactions [37]. Further discussion on each of these applications is provided in later sections of this chapter.

10.2   EXPERIMENTAL CONSIDERATIONS

As is true in any type of affinity chromatography, there are several factors that should be considered during the design or optimization of an IMAC separation. This includes items such as the support used in the IMAC column, the type of chelating agent and metal ion placed onto this support, and the methods used to apply and elute solutes on this column. Each of these factors will be considered in greater detail in the following subsections.

10.2.1   Support Selection

Examples of support materials that can be employed in IMAC columns are listed in Table 10.2. These supports can be divided into two groups, depending on the pressure

Table 10.2   Commercially Available IMAC Supports

Name of Support	Manufacturer	Support Material [a]
Chelating Sepharose fast flow	Pharmacia (Sweden)	Cross-linked agarose
Chelating Superose	Pharmacia (Sweden)	Cross-linked agarose
Hi-Trap chelating	Pharmacia (Sweden)	High-density Sepharose
Chelating Sepharose 6B	Sigma (USA)	Sepharose 6B
Chelex	Sigma (USA)	Polystyrene beads
NTA-resin [b]	Qiagen (USA)	Sepharose-CL-6
Novarose Act [high]	Inovata (Sweden)	Cross-linked agarose
Novarose Act [low]	Inovata (Sweden)	Cross-linked agarose
Immobilized iminodiacetic acid	Pierce (USA)	Cross-linked 4% beaded agarose
TALON Superflow [c]	Clontech Laboratories (USA)	Rigid, highly porous agarose
TSK-GEL (HPLC) [d]	Tosoh (Japan)	Small, spherical particles
Ni-NTA silica [b]	Qiagen (USA)	Chelated Ni-silica

[a] Unless stated otherwise, the chelating group is iminodiacetic acid (IDA).

[b] The chelating group for the NTA-resin and Ni-NTA silica is nitrilotriacetic acid (NTA).

[c] The chelating group for the TALON Superflow is a Co(II)-chelating tetradentate agent [38].

[d] This support offers high resolution (even at high flow rates), excellent reproducibility, and long column lifetimes.

that will be used during the IMAC separation. Low-pressure separations based on IMAC often use cross-linked agarose matrices, while hydrophilized silica [16] or a TSK gel can be used in IMAC systems that are used in high-performance liquid chromatography (HPLC).

IMAC is usually performed with gel matrices for both analytical and preparative procedures. The rapid screening of conditions for such separations (e.g., the choice of affinity matrix, metal ion, buffer system, etc.) can be accomplished by using nonclassical IMAC systems that make use of centrifugation or batch adsorption [39]. In fact, the miniaturization of IMAC equipment through the use of pipette tips, centrifuge tubes, or related items is a promising area for the creation of rapid immobilized-metal affinity procedures [40–42].

HPLC supports are used in IMAC when it is desirable to work at elevated pressures or flow rates that allow a more efficient fractionation of biomolecules on an analytical or preparative scale [43]. For instance, greater speed is often of interest in preparative work, since protein yields can be improved by shortening the length of a purification procedure. The combination of HPLC with IMAC gives a method known as *high-performance immobilized metal-ion affinity chromatography* (*HP-IMAC*). Applications of HP-IMAC include its use in the rapid and high-resolution fractionation of peptides, His-tagged proteins, recombinant proteins, and protein mixtures [23, 44].

One new trend in IMAC has been to use metal-chelate columns in association with other columns as part of a tandem procedure. For instance, two metal-chelate columns containing IDA-Zn(II) and IDA-Cu(II) have been used in series for the adsorption of human serum proteins. This approach has also been used with success for the fractionation of proteins with different affinities for chelated metals [45, 46].

Porath [47] has recently introduced an approach that combines size-exclusion chromatography (SEC) with adsorption, giving rise to a single operation called *adsorptive SEC* (AdSEC). This has been used by Vijayalakshmi et al. [48] with a gel that possesses both exclusion properties and metal-ion affinity for the removal of β_2-microglobulin from the ultrafiltrate of long-term dialyzed patients. This approach, which combines IMAC with SEC, is referred to as *immobilized metal-ion adsorptive size-exclusion chromatography*, or ImadSEC.

Other commercial supports that have been developed for IMAC include membranes [49] and magnetic beads [50]. The combination of polysaccharides with inorganic beads has resulted in several additional support materials from Streamline Chelating and Pharmacia Biotech that are applicable to *expanded-bed adsorption* (EBA), which enables the recovery of proteins from unclarified cell suspensions and homogenates (see Chapter 2 for more details) [51].

One alternative to gels for the fast separation of proteins is an affinity membrane that can be functionalized with an appropriate ligand [49]. This has allowed the creation of *immobilized-metal affinity membrane adsorbents* (IMAMA). As discussed in Chapter 2, the use of membranes is a promising alternative to conventional chromatography in terms of its speed and simple scale-up [52]. Moreover, affinity membranes operate in a convective mode, which can significantly reduce the diffusion limitations that are commonly encountered in column chromatography. Membranes are also capable of handling unclarified solutions and thus can be applied in the earlier stages of downstream processing [53].

10.2.2 Choice of Chelating Agents and Adsorption Center

The properties of the affinity center in an IMAC column depend on the type of chelating agent that is present. The number, type, and exposure of donor atoms (N, S, and O) in the chelating group will determine the force and stability of the metal complex. In general, when the chelating agent is polydentate (i.e., tetradentate or higher), it will strongly retain metal ions and prevent leakage of these ions or metal scavenging by metalloproteins. However, this also weakens the interaction of the metal ions with biomolecules in solution.

Examples of some chelating agents that are used in IMAC are given in Table 10.3. The most widely used of these agents is iminodiacetic acid (IDA), which is a tridentate agent that provides three coordination sites for a metal ion. A chelating gel for IMAC can be prepared with this agent by coupling IDA to a support such as agarose through a spacer arm, such as epichlorohydrin or bisoxirane (see Figure 10.2) [8, 56]. Adsorbents containing IDA have a negative charge and are strong cation exchangers [57]. Thus, a high-ionic-strength mobile phase (>0.5 M NaCl) must be used to quench electrostatic interactions when proteins are applied to such a support.

Other chelating groups, such as nitrilotriacetic acid (NTA, a tetradentate agent) or *tris*(carboxymethyl) ethylene-diamine (TED, a pentadentate agent), result in metal-chelating gels that have stronger retention than IDA for metal ions and weaker retention for proteins. NTA chelated with Ni(II) has been used for the purification of His-tagged proteins. Higher dentate chelating ligands like *tris*(2-aminoethyl)amine (TREN) or diethylenetriamine pentaacetate (DTPA) can be used for entrapping cations in the presence of a strong metal chelator like EDTA [58, 59]. In addition, peptides with metal-binding domains have been used as chelating agents by immobilizing these to agarose [60].

10.2.3   Selection of Metal Ions

All metal ions that interact specifically with proteins can be employed in immobilized-metal affinity methods. As shown in Table 10.4, metal ions can be grouped by Pearson's polarizability-based classification as being soft, hard, or borderline Lewis acids [61]. Typical metal ions that fall under the "soft" category are Hg^{2+}, Cd^{2+}, Cu^+, Ag^+, Pb^{2+}, and the platinum group of metal ions. These metal ions prefer sulfur as a coordination partner, and the coordinate bonds that involve these metals are essentially covalent. Up to the present, immobilized soft metal ions have received only a little attention as adsorbents for proteins, but they are now seeing more interest. For instance, the groups of Porath and Vijayalakshmi are working with Au and Pd for the specific recognition of sulfur-containing residues and aromatic amino acids in proteins and polypeptides. Moreover, chelated Hg(II) has been examined for use in the separation of nucleotides; however, at a neutral pH, no discrimination between mononucleotides was found to occur with this agent (Vijayalakshmi, unpublished results).

Ions of group II and III elements are hard Lewis acids. Among these, Ca^{2+} and Mg^{2+} are of particular biological interest [62]. The data on the coordination chemistry of these ions as related to IMAC is not extensive. It is known that oxygen is the preferred bonding partner, with the metal-oxygen bond of these agents being essentially ionic. The heavier group III metal ions tend to have partial covalency intermixed in the complexation. This is also reflected by the behavior of these ions in IMAC columns. For example, among the group III B3 immobilized metal ions, Tl(III) is similar to Zn(II) and Cu(II) in that it is a borderline Lewis acid, whereas Al(III) is a hard Lewis acid [63].

The specificity of terbium/rare earth calcium-mimicking gels toward calcium-binding proteins was demonstrated when these gels were used to separate a mixture of human serum albumin (which has no site for calcium) and trypsin (which can take part in calcium binding) [64]. The selectivity of tandem columns containing lanthanide metals (Sm(III) and La(III)) for plasma proteins has also been demonstrated [64]. In this latter case, it was shown that the chelated Sm(III) column specifically adsorbed immunoglobulin G fragments, while the chelated La(III) specifically adsorbed fibrinogen.

Table 10.3 Chelating Agents Used in IMAC

Chelating Agent	Reference
 Iminodiacetic acid (IDA)	Porath et al. [4]
 Tris (carboxymethyl) ethylendiamine (TED)	Porath and Olin [8]
 Nitrilotriacetate (NTA)	Hochuli et al. [26]
 Carboxymethylated aspartic acid (CM-Asp)	Mantovaara et al. [54]
 Tris (2-aminoethyl) amine (TREN)	Winzerling et al. [6]
GHHPHG GHHPHGHHPHG GHHPHGHHPHGHHPHG (GHHPH)nG Metal chelating peptides	Yip and Hutchens [60]
 Diethylene triamine pentaacetate (DTPA)	Boden et al. [59]

(Continued)

Table 10.3   Chelating Agents Used in IMAC (Continued)

Chelating Agent	Reference
	Jiang et al. [55]

Triazacyclononane (TCAN)

Other studies have been conducted with immobilized iron ions, with Fe(III) being a harder Lewis acid than Fe(II). In this work, it was found that an immobilized Fe(III) support can be used for the selective adsorption of phosphoproteins and organophosphates [57, 65, 66], as discussed later in Section 10.3.4 of this chapter.

Zachariou and Hearn [67] have investigated the interactions of cytochrome c and myoglobin variants, as well as hen egg-white lysozyme, with the hard Lewis acids Al^{3+}, Ca^{2+}, Fe^{3+}, and Yb^{3+}, and with the borderline Lewis acid Cu^{2+}. Chaga and coworkers [68] have reported the fractionation of calcium-binding proteins using IMAC with hard metal ions (Mn^{2+}, La^{3+}, Nd^{3+}, Eu^{3+}). In addition, the application of hard Lewis metal ions such as Al^{3+}, Ca^{2+}, Fe^{3+}, and Yb^{3+} has been recently extended to the selective fractionation of human serum proteins [69].

Among the borderline metal ions, those belonging to the 3d-block elements, such as Zn(II), Cu(II), Ni(II), and Co(II), can coordinate nitrogen, oxygen, and sulfur. These have been extensively studied for use in IMAC. Zinc and nickel ions (Zn^{2+} and Ni^{2+}) are electrochemically stable under the conditions used in this type of chromatography. But copper and cobalt ions (Cu^{2+} and Co^{2+}) can be readily reduced or oxidized by redox-active solutes. Although these ions can act as catalysts in decarboxylation and hydrolytic or redox reactions, no adverse reactions have yet been reported with these ions when they are present as an immobilized chelate. Nevertheless, when the retention of a thiol protein (e.g., thioredoxin from *Escherichia coli*) was studied on a Sepharose-IDA-Cu(II) column, there was an observed loss of the protein's functional properties that might have been due to the oxidation of sulfhydryl groups at its active site (Andersson and Vijayalakshmi, unpublished data).

IDA-Agarose

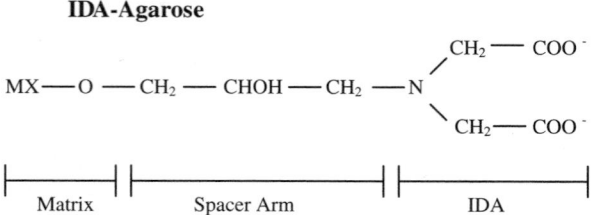

Figure 10.2   The structure of IDA-agarose, a common material used in IMAC columns. (From Winzerling, J.J., Berna, P., and Porath, J., *Methods: Companion Methods Enzymol.*, 4, 4–13, 1992. With permission.)

Table 10.4 Classification of Metal Ions and Their Ligands

Classification	Metal Ions (Lewis Acids)	Principal Ligands (Lewis Bases)
Hard	H^+, Li^+, Na^+, Mg^{2+}, Ca^{2+}, Mn^{2+}, Cr^{3+}, Co^{3+}, Al^{3+}, Ga^{3+}, La^{3+}, Nd^{3+}, Eu^{3+}	H_2O, OH, OH^-, RO^-, NH_3, RNH_2, CO_3^{2-}, $RCOO^-$ (oxygen ligands)
Soft	Cu^+, Ag^+, Au^+, Ti^+, Pd^{2+}, Pt^{2+}, Cd^{2+}, Hg^+, Hg^{2+}	RSH, RS, R_2S, CN^-, H^-, I^- (sulfur ligands)
Borderline	Zn^{2+}, Cu^{2+}, Fe^{2+}, Fe^{3+}, Ni^{2+}, Co^{3+}, Sn^{2+}, Pb^{2+}, Rh^{3+}, Ir^{3+}, In^{3+}, Ru^{3+}	Imidazole, pyridine, N_2, N_2^-, N_3^-, Br^- (nitrogen ligands)

Source: Vijayalakshmi, M.A., Ed., *Biochromatography: Theory and Practice*, CRC Press, Boca Raton, FL, 2002. With permission.

10.2.4 Mobile-Phase Selection

The choice of solvent for the mobile phase in IMAC will depend on the protein or molecule to be separated. One requirement for the target protein is that it should have an electron-donor group on its surface that is at least partially unprotonated. This generally means that the sample containing this protein should be applied to the IMAC column at or near a neutral pH, although some proteins can also be retained at lower pH values [70]. The solvated side chain of histidine has a pK_a of about 6. Thus, at a pH near 6, half of the surface histidines on a protein will be unprotonated and able to form coordination bonds to chelated metals. In contrast to this, other amino acids that contain electron-donor groups have much higher pK_a values. For instance, the thiol group of cysteine has a pK_a of about 9 and a pK_a of 15 has been estimated for the side-chain nitrogen of tryptophan [71]. This means that at or near a neutral pH, histidine is the predominant amino acid that will lead to a protein's retention in IMAC. Thus, if surface histidines are present on a protein, this should bind to a typical IMAC column that contains IDA and a chelated "+2" metal ion (a combination that will be referred to as an "IDA-M(II)" column).

Another requirement for the mobile phase in IMAC is that it should include a salt (usually >0.5 *M* NaCl). This not only helps quench electrostatic interactions between the sample components and the column, but it also enhances selectivity and increases the stability of the electron donor-acceptor complex that is formed between the chelated metal and the amino acids on the protein's surface [21].

The mobile phases used for sample adsorption in IMAC are typically aqueous buffers that contain MOPS, MES, HEPES, acetate, or phosphate salts. Compounds that have an affinity for metal ions, like Tricine or citrate, should be avoided in the application buffer. To a lesser extent, Tris buffer can also act as a weak competitor for metal coordination sites in an IMAC column [29].

As will be discussed later, the elution of proteins from an IMAC column can be achieved by reducing the pH of the mobile phase. Since histidine becomes protonated with a decreasing pH, the binding of this group to a chelated metal will weaken until the protein is released from the stationary phase. This suggests that special care must be taken in the selection and preparation of solvents for good pH control. The preparation of buffer mixtures containing equimolar amounts (20 m*M*) of MOPS, MES, and acetic acid (a mixture referred to as "20 m*M* MMA") can be used to cover the pH range of 4 to 8 [23, 70]. Furthermore, reproducible gradients over the pH range of 4 to 7 can be obtained by mixing predetermined amounts of pH 7.0, 20 m*M* MMA and pH 4, 20 m*M* MMA.

The elution of retained substances in IMAC can also involve the use of imidazole as a competing agent for the immobilized metal chelate. The active species of imidazole (pK_a = 6.95 at 25°C) in this case is the unprotonated form. Consequently, elution buffers that make use of imidazole as a competing agent should have a neutral to slightly alkaline pH.

The elution of some proteins may require mobile-phase additives other than imidazole. For example, competing agents like glycine or ammonia and additives like urea, ethylene glycol, detergents, and alcohols can be included in the column equilibration and elution buffers. IMAC is also compatible with substances like nonionic surfactants, urea, ethylene glycol, and dimethyl sulfoxide that are often used to solubilize membrane proteins and other hydrophobic proteins or peptides [43, 72].

10.2.5 Packing and Preparing IMAC Columns

IMAC supports for both low-pressure and high-pressure liquid chromatography are usually packed in columns with adjustable plungers or packing pumps. After being washed with water, the desired amount of support (e.g., an IDA-agarose gel) is resuspended in a solution at high ionic strength ($\geq$0.2 M NaCl) before packing. This step is performed because an IDA gel will be negatively charged and will require electrostatic quenching for adequate packing. When this gel has settled in the column, packing is continued with at least five column volumes of the high-ionic-strength solution.

After the column has been packed, it is then washed with another five column volumes of water and brought to a final pH of 5 with an appropriate buffer. This acidification of the support is to prevent the formation of weak bonds with metal ions and provides a more homogeneous distribution of metal ligation states. In addition, divalent metal ions are known to undergo hydrolysis at an alkaline pH, leading to precipitation of insoluble metal hydroxide, which is avoided by working at a lower pH value. After acidification, the column is then saturated with a solution of the desired metal ion. Excess and weakly bound metal ions are removed by washing the support with at least five column volumes of pH 5 buffer, and the column is then placed into the final desired buffer for storage or use.

10.2.6 Elution Strategies

10.2.6.1   pH Elution Schemes

As mentioned earlier, a decrease in mobile-phase pH is a common means of elution used in IMAC. Since histidine becomes protonated with a decrease in pH, the binding of this group to a chelated metal will weaken at lower pHs and cause a retained protein to be released from an IMAC column. The conditions needed for a stepwise change in pH or isocratic elution in IMAC must be determined experimentally. There are three major items to consider in selecting this protocol: (1) the stability of the chelated metal at various pHs, (2) the stability of the protein or retained molecule over the pH range to be used for elution, and (3) the isoelectric point of the retained protein. This last factor is important, since the precipitation of a protein at or near its isoelectric point can lead to column clogging and poor recovery.

IDA-Cu(II) and IDA-Ni(II) gels are stable when the pH is lowered to 4. However, the use of pH elution is more problematic with columns that contain IDA-Zn(II) or IDA-Co(II) supports [21, 23]. For example, bleaching of an IDA-Co(II) column can be observed at pH 5. Thus, for Zn(II)- or Co(II)-containing columns, it is recommended that proteins and other retained targets be adsorbed at a neutral pH and then eluted with pHs in the range of 5 to 7, which helps prevent the leakage of metal ions. This also avoids changes in the metal-ion content of the IMAC column and can prevent the loss of biological activity when IMAC is used to purify enzymes that are inhibited by metal ions.

10.2.6.2   Addition of Competing Agents

The elution of proteins from an IMAC column at or near a neutral pH can be accomplished by using competing agents that act as electron donors. Imidazole (Im) is often used for this purpose. However, because imidazole forms a relatively stable metal complex with an IDA-M(II) support, it is advisable to first saturate such a column with an imidazole solution (such as 100 mM Im) and to equilibrate the column with a buffer containing 1 mM Im [21].

Recently Sulkowski [73, 74] showed that the complexation of imidazole with an IDA-M(II) metal-chelate column will result in acidification of the mobile phase. This occurs because, at a pH less than 8 to 8.5, the protonated form of imidazole (ImH$^+$) can combine with the chelated metal (M) to give ImM and a proton, as shown in Equation 10.1.

$$ImH^+ + M \leftrightarrow ImM + H^+ \tag{10.1}$$

This IDA-M(II) imidazole "proton pump" can induce errors in the interpretation of elution profiles in IMAC, where the elution of an adsorbed protein may now reflect changes in protonation rather than competitive displacement by imidazole. Because of this, it is recommended that 1 mM Im (or more, if the protein of interest allows) be added to the column equilibration buffer to minimize this effect.

The use of ammonia salts as competing agents to elute human serum proteins was reported by Porath and Olin [8]. Recently, Fitton and Santarelli [75] employed a mobile phase containing NH$_4$Cl to obtain good recovery and activity of penicillin acylase on an IMAC column. Glycine [76], histidine [77], and histamine [78] have also been used as competing agents in IMAC. However, it should be noted that the use of histidine as a competing agent might result in depletion of metal ions from such columns (Vijayalakshmi, unpublished data).

Chelate annihilation is an approach in which both a retained protein and the metal ions on an IMAC column are eluted with a chelating agent solution, such as pH 7, 0.05 M NaEDTA. However, this is not a selective method and is generally used when the protein of interest is the only retained species on the column [77, 79]. Washing with EDTA is also used in IMAC as a final step for cleaning columns of all metal ions and any strongly bound contaminants. This is done prior to changing metal ions or reloading the column with more of the same metal ions.

10.3   APPLICATIONS

There are a variety of applications for IMAC. Both theoretical and practical approaches concerning metal affinity separations have been described in many reviews. For example, Vijayalakshmi [5] examined the mechanism and practical implications of the interactions between proteins and immobilized metals, dyes, and histidine in "pseudobioaffinity" chromatography. Winzerling et al. [6] and Porath [80] discussed the basic principles of IMAC with regard to protein separations. Yip and Hutchens [60] summarized protocols for IMAC separations. On the 25th anniversary of IMAC, Chaga [7] published a review of its past, present, and future in which he described the historical developments in IMAC and some recent applications in genomics and postgenomic studies. And recently, Mrabet and Vijayalakshmi [81] described the basic principles and structural aspects of molecular recognition in IMAC. Applications that will be considered in this section include the use of IMAC in protein purification, studies of protein surface topography or modifications, microanalytical methods, and genomic or postgenomic research.

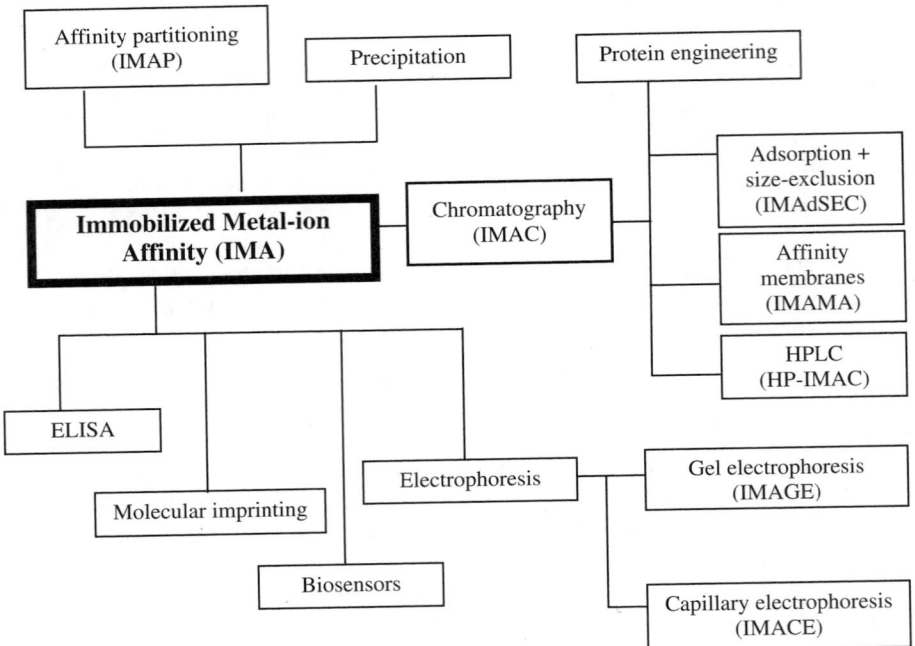

Figure 10.3 Uses of immobilized metal-ion affinity in biotechnology. The abbreviations used in this figure are defined within the text and are summarized at the end of the chapter.

10.3.1 Protein Purification by IMAC

Since the introduction of IMAC in 1975, it has become a widely used method for the purification of native and recombinant proteins. This has been discussed by many authors and has been reviewed by Mrabet and Vijayalakshmi [81]. A summary of these and other applications is given in Figure 10.3. The specific applications that are discussed in this section include the use of IMAC with proteins that contain surface histidines, the isolation of His-tagged proteins by IMAC, and the separation of plant proteins by IMAC columns.

10.3.1.1 Proteins with Surface Histidines

Numerous proteins contain histidine residues in their amino acid sequences. Even though histidine is not a prevalent amino acid (making up less than 2% of all amino acids in proteins), it does often play an important role in affecting biological activity. Data regarding the number and arrangement of histidine residues in proteins with known three-dimensional structures can be obtained from protein data banks. This can serve as a basis for forecasting protein behavior in IMAC. In 1985 Sulkowski [21] worked with model proteins of known structure (like bovine pancreatic RNase A, cytochrome c, lysozymes, albumins, and interferons) and determined the major rules for protein–metal-chelate recognition. Since then, IMAC has become one of the most popular methods for purifying native and recombinant proteins. Some examples of proteins that have been purified by IMAC are presented in Table 10.5. When working with an unknown protein, the conditions to be used for purification by IMAC can be determined by selecting the adsorbent, adsorption center, and elution conditions, as described previously in this chapter. Moreover, a protein's chromatographic behavior with different metal chelates gives structural information on

Table 10.5 Examples of Native Proteins That Have Been Purified by IMAC

Type of Protein [a]	Metal Chelate	Reference
Microbial proteins:		
Carboxypeptidases (*A. niger*)	IDA-Cu(II)	Krishnan and Vijayalakshmi [70]
Glucose isomerase (*A. missouriensis*)	IDA-Cu(II)	Mrabet [23]
Human proteins:		
α-Fetoproteins, albumin	IDA-Ni(II)	Andersson et al. [45]
Serum amyloid P component	CM-Asp-Ca(II)[b]	Mantovaara et al. [62]
Animal proteins:		
Goat immunoglobulins	TREN-Cu(II)	Boden et al. [59]
Bovine calmodulin	TED-Eu(III)	Chaga et al. [68]
Plant proteins:		
Sporamin from sweet potato	IDA-Cu(II)	Boulis et al. [42]
α-Amylases from potato	IDA-Cu(II)	Todorova and Vijayalakshmi [90]

[a] All of these examples were purified from crude extracts.
[b] CM-Asp-Ca(II) refers to a carboxymethylated aspartic acid agarose that is loaded with Ca(II).

the number and arrangement of histidine residues that are involved in metal-chelate recognition, according to IMAC ground rules established by Sulkowski [13], or gives information on the role of such residues in the biological activity of proteins [46, 70].

10.3.1.2   His-Tagged Proteins

IMAC has also gained broad popularity for the purification of recombinant proteins that have been engineered to contain histidine or that contain histidine clusters attached to their amino or carboxy terminal ends. Such proteins are commonly referred to as "His-tagged" proteins. In making such a protein, the accessibility of the histidine tag on the protein's surface is important for its efficient binding to a metal chelate. Whether the His-tag should be placed on the N- or C-terminal end for optimum binding must be determined experimentally. However, N-terminal tagging is most frequently employed because several endoproteases are available for the precise cleavage of such a tag after the protein's purification.

A NTA-Ni(II) support is commonly used as the affinity matrix for His-tagged proteins. Since the early 1990s, this has been exploited for such applications as the purification of His-tagged proteins, the immobilization of enzymes, and the identification of hexahistidine fusion proteins [82]. Many companies like Qiagen, Invitrogen, Novagen, and Clontech are developing new matrixes for the recovery of His-tagged proteins by IMAC (see Table 10.2). Examples of His-tagged proteins that have been purified by this method are listed in Table 10.6. This technique permits protein purification in the presence of 6 M guanidinium hydrochloride, making it useful for His-tagged proteins that are to be extracted from inclusion bodies. For instance, $(His)_6$-dihydrofolate reductase was recovered with at least 90% purity from inclusion bodies by means of IMAC [83].

10.3.1.3   Plant Proteins

Transgenic plants are now being targeted for the production of therapeutic proteins. This makes the use of IMAC for the purification of plant proteins of great interest. IMAC has

Table 10.6   Examples of His-Tagged Proteins That Have Been Purified by IMAC

Protein	Affinity Tag	Metal Ion	References
Hormones, recombinant proinsulin trpLE' [a]	His-Trp	Ni	Smith et al. [83]
Dihydrofolate reductase	(His)$_6$	Ni	Hochuli et al. [26]
Protein A, ZZ [b]	Ala-His-Gly-His-Arg-Pro	Zn	Ljungquist et al. [84]
Single chain Fv	Glu-Leu-Lys-(His)$_5$	Zn	Skerra et al. [85]
Human oncostatin M	(His)$_6$	Ni	Sporeno et al. [86]
Human prolactin	Ala-Ser–(His)$_6$-Ile-Glu-Gly-Arg	Ni	Morganti et al. [87]
DNA polymerase from *Thermus thermophilus*	(His)$_6$	Ni	Dabrowski and Kur [88]
Myoglobin-like aerotaxis transducer	(His)$_6$	Co	Piatibratov et al. [27]
Human tryptophanyl-tRNA synthetase	(His)$_6$	Ni	Xu et al. [28]

[a] Bacterial TrpLE' protein — leader peptide.
[b] IgG-binding protein ZZ derived from staphylococcal protein A.

been used for the recovery of both native and recombinant proteins from plants such as tobacco and potato. As an example, Boulis et al. [42] used IMAC to purify native sporamin, the storage protein present in sweet potatoes. This method was then applied to the isolation of several different sporamin constructs expressed in transgenic tobacco plants [89]. By combining IMAC with electrospray-ionization mass spectrometry (ESI-MS), differences in the conformation and processing (proteolytic cleavage) could be observed between the variant forms of the targeted sporamins.

Native amylolytic enzymes from potato (*Solanum tuberosum, cv. Bintje* and *cv. Desirée*), have also been studied by IMAC using an IDA-Ni(II) column [90, 91]. Three groups of proteins with starch-degrading activity were separated by a decreasing protonation gradient. This indicated that these proteins had different histidine surface topographies and, thus, different conformations in their enzymatically active forms. Moreover, molecular heterogeneity and appearance of new amylolytic species in aged germinated potato tubers was found by IMAC through the use of the chelated Ni(II) matrix. Different IMAC behavior was found for amylolytic enzymes from different origins (thermostable microbial, human saliva, porcine pancreas, barley, and potato) and provided clues concerning the difference in the three-dimensional structure of these functionally related enzymes.

10.3.2   Studies of Protein Surface Topography and Posttranslational Modifications

As already mentioned, another important application of IMAC is its use in relating protein surface structure to the protein's binding on immobilized metal ligands. Moreover, this procedure can be used along with other analytical methods like crystallography and molecular modeling to give complementary information about protein surface topography.

Using model proteins, Sulkowski documented the correlation between the availability of electron-donating amino acids on a protein's surface and the protein's chromatographic behavior in IMAC, which led to the ground rules for recognition given earlier in Table 10.1 [12, 13, 21, 92]. These observations have since been used to predict the IMAC behavior of proteins that have known structures by using columns containing iminodiacetate and borderline transition metals such as Cu(II), Ni(II), Zn(II), and Co(II). Moreover, by using different metal chelates, the rules given in Table 10.1 can be used to obtain information on histidine topography for a protein with an unknown structure. For instance, Mrabet [23] used an IDA-Cu(II) column in a one-step purification of xylose isomerase from *Actinoplanes missouriensis* and proved that only His-41 (out of four histidines in this protein's sequence) was accessible for binding to the affinity ligand, as determined by using molecular modeling, an IDA-Cu(II) spherical probe, and site-directed mutagenesis. A review of such work, in which molecular modeling and IMAC are used to study protein-metal interactions, has been written by Mrabet and Vijayalakshmi [81].

IMAC can also be used to investigate conformational changes of proteins. Histidine residues are usually important amino acids in determining the functional properties of enzymes. This makes IMAC an attractive method for studying structure-function relationships in enzymes. Areas in which IMAC has been applied for this purpose include its use in studying the histidine accessibility of isoforms or mutants [23, 29], posttranslational modifications in proteins (such as glycosylation or phosphorylation) [35], subunit oligomerization [36], conformation changes in proteins (e.g., allosteric effects, denaturation, and renaturation) [34], as well as studies of the active sites on proteins and protein-ligand interactions [30].

One report that combined the use of IMAC with molecular modeling for studying structure-function relationships in proteins was that of Berna et al. [29]. These authors studied the IMAC recognition of a model protein, bovine chymotrypsin, by using a high-capacity gel consisting of Novarose-IDA-Cu(II). With this matrix, they achieved the separation of several chymotrypsin subspecies (chymotrypsinogen, π, κ, δ, and α/γ) by using a gradient with an increasing Tris concentration. Molecular modeling was then used to determine if His-40 or the catalytic His-57 residue was involved in the metal recognition process. Although His-57 was predicted to have sufficient accessibility (as calculated by ASA computations) to metal chelates, His-40 was identified as the actual interaction partner. This contradiction can be explained by the existence of strong hydrogen bonding that precluded the interaction of His-57 with the metal-chelate complex. These authors also showed through molecular modeling that histidine accessibility alters during the maturation of chymotrypsinogen, where His-40 is engaged in the hydrogen bonds in chymotrypsinogen and is gradually liberated during maturation. The microanalytical method IMACE, as discussed in the next section, has also been used to investigate chymotrypsin subspecies and has detected microheterogeneity in this histidine-containing protein [32].

Similar studies by IMAC have been carried out using other serine proteases, including bovine and porcine trypsins, subtilisins, and thrombin. Contrary to the case of chymotrypsin, all these other serine proteases have showed at least a partial involvement of the catalytic triad histidine in protein retention on a Novarose-IDA-Cu(II) column [30]. This can be correlated to the weaker hydrogen bonding of the catalytic histidine residue in these proteases compared with His-57 of the catalytic triad in chymotrypsin.

10.3.3 Microanalytical Methods

Another application of metal-ion affinity interactions has been their use in various microanalytical methods. Examples of these methods are immobilized metal-ion gel electrophoresis and immobilized metal-ion capillary electrophoresis. These methods are discussed in more detail in the following subsections.

10.3.3.1   Immobilized Metal-Ion Affinity Gel Electrophoresis

Goubran-Botros and Vijayalakshmi [31] extended the use of metal-ion affinity recognition to gel electrophoresis in 1991. As mentioned earlier, this method is referred to as immobilized metal-ion affinity gel electrophoresis, or IMAGE. Many studies have shown that IMAGE is a useful analytical tool for the quantitative determination of binding constants between a protein and a chelated metal ion. This can also be used for the design of preparative-scale methods for the purification of proteins, for the detection of protein unfolding, and for the study of structure-function relationships in native and modified proteins [33, 34, 37, 93].

Several methods to incorporate metal chelate ligands into agarose or polyacrylamide gels have been reported [31, 33]. In addition, preliminary studies with unmodified model proteins such as ribonucleases A and B (RNases A and B), cytochrome c, and chymotrypsin have been used to determine the usefulness of IMAGE for determining the affinity constants for ligand-protein interactions [33]. In such work, it has been found that separations in IMAGE are based on the same general principles as established by Sulkowski [13] for IMAC (see Table 10.1).

It has been shown that IMAGE is a sensitive method for studying conformational changes that occur after the chemical modifications of proteins. For instance, subtle differences in the microenvironment and accessibility of histidine residues on RNase A or RNase B and chemically glycosylated RNase A have been detected using IMAGE and IDA-Cu(II), as shown in Figure 10.4 [34]. In addition, the dissociation constant (K_D) for chemically glycosylated RNase A has been found by IMAGE to decrease slightly when compared with the native RNase A (see Table 10.7); this is indicated by an increase in the affinity of glycosylated RNase A to IDA-Cu(II). It was further noted that glycosylated RNase A had a dramatically enhanced thermostability at 90°C when compared with its native counterpart. This increase in thermal stability was correlated with an increased affinity to the metal chelate ligand, reflecting minor protonation modifications of the active-site histidine residue on this protein. Such work indicates that affinity constants determined by IMAGE can be used to examine subtle conformational modifications and help determine favorable conformations for the tertiary structure of proteins.

Recently, IMAGE was employed to study structure-function relationships of genetically engineered proteins and to examine the mechanisms of protein-protein recognition [37]. In this case, the affinities of native Binase, Barnase, and modified Binase were measured on a copper chelate. This was done to give information on the changes in surface histidine topography for Binase that had been altered by site-directed mutagenesis and photooxidation. Protein-protein interactions in the enzyme-inhibitor (Barstar) complexes were also related to changes in histidine accessibility and enzymatic activity.

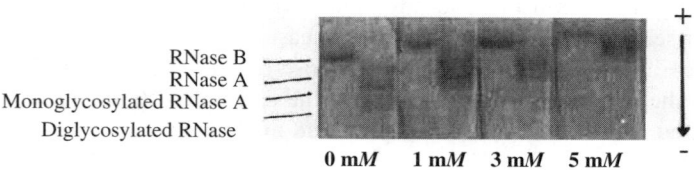

Figure 10.4   Migration patterns for native RNase A, native RNase B, and mono- or diglycosylated RNase A in an IMAGE experiment. The ligand concentrations used for PEG-IDA-Cu(II) in these experiments are given at the bottom of each lane. Other conditions were as follows: 7.5%T–3%C polyacrylamide gel containing pH 7.2, 0.05 M Tris-acetate; electrode buffer, 0.1 M Tris-acetate; electrical field strength, 8 V/cm; migration time, 6 h at 20°C. (Reproduced with permission from Baek, W.O. and Vijayalakshmi, M.A., *Biochim. Biophys. Acta*, 1336, 394–402, 1997.)

Table 10.7 Dissociation Constants (K_D)
Determined by IMAGE for RNase A, RNase B,
and Chemically Glycosylated RNase A

Type of RNase	K_D (mM)
RNase A	2.85
RNase B	2.99
Monoglycosylated RNase A	2.37
Diglycosylated RNase A	2.21

Source: Baek, W.O. and Vijayalakshmi, M.A., *Biochim.*
Biophys. Acta, 1336, 394–402, 1997. With permission.

10.3.3.2 *Immobilized Metal-Ion Affinity Capillary Electrophoresis*

It was noted earlier that the use of metal-ion chelates in capillary electrophoresis gives rise to a method known as immobilized metal-ion affinity capillary electrophoresis (IMACE). As was true for the use of metal-ion chelates in gel electrophoresis, the same rules that were given in Table 10.1 for protein interactions in IMAC also apply to IMACE. But one difference in these methods is that such interactions are created in IMACE by using a ligand such as IDA-Cu(II) that has been covalently coupled to polyethylene glycol as a soluble and replaceable polymer.

Like IMAGE, the technique of IMACE can be used for determining protein–metal-ligand affinities and dissociation constants. This can be accomplished by using a modified Langmuir adsorption isotherm equation that is applicable to interactions with fast on/off kinetics. The affinity of a protein for a metal chelate is related in IMACE to the change in the protein's migration in the presence of the chelating ligand. This change in migration, in turn, is related to the amount of time that the protein spends in its complexed and free forms during electrophoresis [34]. Any observed change in protein migration as it encounters the ligand can be linked to the numerous association and dissociation steps that occur during the electrophoretic run. At the center of the migration peak, this can be regarded as a steady-state situation that is comparable with that used in measuring binding constants by frontal analysis or zonal-elution affinity chromatography (see Chapters 22 and 23).

Mathematically, the differential migration times for the protein can be substituted for the dependent variable in the Langmuir isotherm, and the ligand concentration in the electrophoretic buffer can be used to represent the free ligand concentration at equilibrium. Using a differential migration time of $\Delta t_N = t_N - t_{No}$ (where $t_N = t/t_{EOF}$, t is the observed migration time, t_{EOF} is the migration time for an electroosmotic flow marker, and t_{No} is the normalized migration time in the absence of the affinity ligand), this gives the following form for the Langmuir isotherm,

$$\Delta t_N = \frac{\Delta t_{N\max} C}{K_D + C} \tag{10.2}$$

where C is the ligand concentration, and K_D is the dissociation constant for the interaction.

Figure 10.5 shows the adsorption isotherms obtained in IMACE for three model proteins: cytochrome C from tuna heart, horse heart, and *Candida krusei*, which have zero, one, and two accessible histidines on their surfaces, respectively [21]. The dissociation

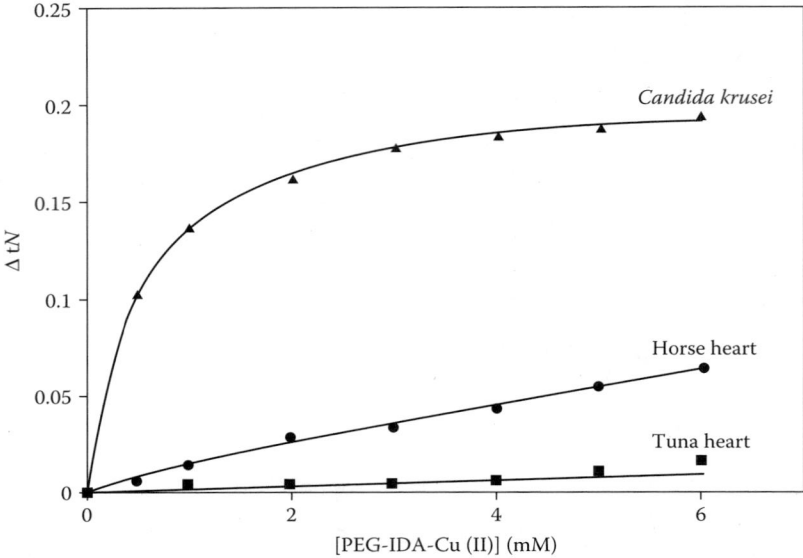

Figure 10.5 Adsorption isotherms for cytochrome C from horse heart, tuna heart, and *Candida crusei* as measured by their normalized migration times in IMACE at different concentrations of m-PEG 5000-IDA-Cu(II) in the running buffer. Other conditions were as follows: fused silica capillary, 27 cm × 20 μm I.D.; electric field strength, 185 V/cm; voltage separation with the run being continued after 8 min with low-pressure assistance; electrophoretic buffer, pH 7.2, 0.1 *M* HEPES containing 0.3 *M* NaCl, with a total PEG concentration of 10 m*M*. (Reproduced with permission from Haupt, K., Roy, F., and Vijayalakshmi, M.A., *Anal. Biochem.*, 234, 149–154, 1996.)

constants that were measured from these proteins are given in Table 10.8. These data show that the highest affinity (lowest K_D) was obtained for cytochrome C from *C. krusei*, which has two accessible histidines, followed by horse heart cytochrome C and tuna heart cytochrome c, the latter of which showed no detectable binding. These data correlate well with those generated chromatographically by a classical IMAC approach. This indicates that IMACE can be used as an analytical tool for studying protein interactions with such metal chelates.

At a supramolecular level, the oligomeric forms of the enzyme dimeric RNase have also been distinguished by IMAGE and IMACE, using oligomeric preparations that were

Table 10.8 Dissociation Constants (K_D) Determined by IMACE for Cytochrome C Binding to m-PEG 5000-IDA-Cu(II)

Protein	K_D (m*M*)
Cytochrome C (*Candida krusei*)	0.55 ± 0.03
Cytochrome C (horse heart)	29.0 ± 10
Cytochrome C (tuna heart)	Below detection threshold

Source: Haupt, K., Roy, F., and Vijayalakshmi, M.A., *Anal. Biochem.*, 234, 149–154, 1996. With permission.

produced naturally and purified by IMAC or that were produced by physicochemical means. This suggests that an interesting correlation may exist between enzyme oligomerization, the induction of new biological properties (e.g., an increased activity toward double-stranded RNA, cytotoxicity, etc.), and surface histidine topography [36].

10.3.3.3   Immobilized Metal-Ion Affinity Partitioning

Another impressive application of immobilized metal-ion affinity is its use to study conformational changes in proteins on cell surfaces. This can be performed *in situ* by means of a method known as *immobilized metal-ion affinity partitioning* (IMAP). Using IMAP, Nanak et al. [95] were able to differentiate healthy cells from infected malaria red blood cells, cancerous fibroblasts, and lymphoma cells. This was accomplished by using PEG-IDA-Ni(II) as an affinity ligand. In the case of red blood cells, the mechanism of immobilized metal-ion recognition has been attributed to the affinity of membrane glycoproteins for metal chelates [25, 96, 97]. In addition, the differences in surface histidine residues and their microenvironments on cell surface proteins seems to be the main factor responsible for cell discrimination by the PEG-IDA-Cu(II) ligand. This makes IMAP attractive for *in situ* studies of protein conformational changes in pathological states like cancer and infectious diseases.

10.3.4   Applications of IMAC in Genomics and Postgenomic Studies

With the completion of the Human Genome Project, there will be an increasing need for tools that can carry out improved nucleic acid separations and large-scale, parallel protein-expression experiments. With regard to nucleic acids, IMAC has already been employed for the separation of mono-, di-, and oligonucleotides [14–17]. As one example, Fanou-Ayi and Vijayalakshmi [16] compared the separation of mono- and dinucleotides using IDA-Cu(II) in low-pressure and high-performance liquid chromatographic systems (see Figure 10.6). In both cases, a clear separation of purine and pyrimidine mononucleotides was seen, with stronger retention being noted for purine-containing compounds. A clear cooperative effect was also noted in such separations, since purine dinucleotides were more strongly retained than pyrimidine dinucleotides. Moreover, the position of the purine base played an important role, making these columns useful for the separation of dinucleotides, as indicated in Figure 10.6b.

Dobrowolska et al. [17] have examined the chromatographic behavior of nucleotides and related compounds on an immobilized Fe(III) chelate column. Their results indicate that the interaction of nucleotides with immobilized Fe(III) ions is due to free phosphate groups, where the oxygen atoms in the phosphate must be accessible and a phospho monoester must be present for retention.

IMAC has also been employed in the purification of DNA and oligonucleotide derivatives that are tagged with six successive 6-histaminylpurine residues [98, 99]. This gives rise to the selective adsorption of such derivatives to NTA-Ni(II) resins. Phosphoramidites modified with one or two histidyl residues can be directly incorporated into primers or oligonucleotides and utilized for the purification of PCR products or base-pairing oligonucleotide chains. Willson et al. [18] found that plasmid and genomic DNA have a very limited IMAC binding affinity, while RNA and oligonucleotides bind strongly to metal chelate matrices. This indicates that IMAC could play a role in separating plasmid DNA.

One promising future application of IMAC is in transcriptomics and proteomics (see Chapters 18 and 29). In the identification of proteins, the selective enrichment of proteins that contain specific functional groups and that occur at low concentrations is invaluable for the analysis of posttranslational modifications. As a result, the major application of IMAC in proteomics so far has been its use in isolating phosphorylated proteins and peptides. This is accomplished by using immobilized Fe(III) and Ga(III), but intermediate metal ions may also

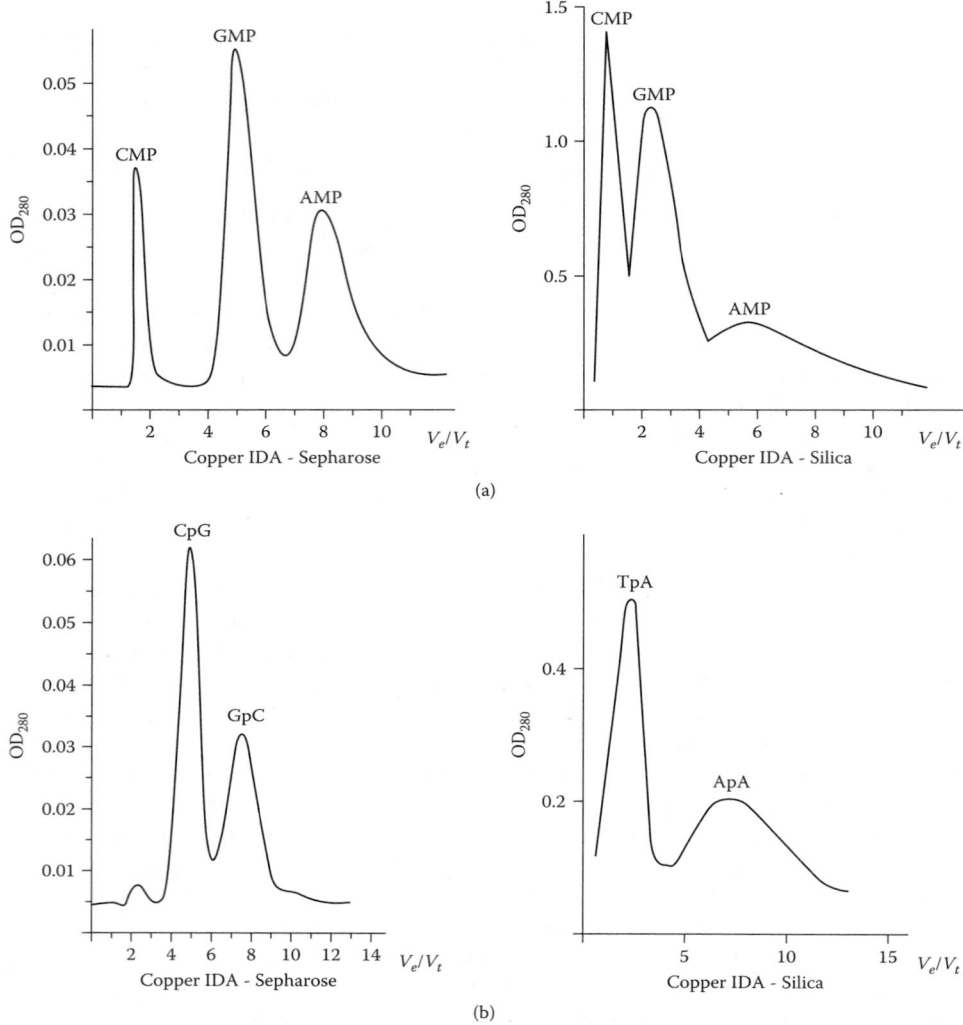

Figure 10.6 (a) Separation of mononucleotides on IDA-Cu(II) Sepharose (elution with 0.05 M Tris-HCl buffer at pH 7.0 and 20°C) and IDA-Cu(II) silica (elution with 0.05 M ethylene morpholine acetate buffer at pH 6.0 and 20°C); and (b) dinucleotide separations on these supports under the same elution conditions as in (a). Abbreviations: CMP, cytidine 5′-monophosphate; GMP, guanidine 5′-monophosphate; AMP, adenosine 5′-monophosphate; CpG, cytinylyl(3′-5′) guanidine; GpC, guanylyl(3′-5′) cytidine; TpA, thymidinylyl(3′-5′) adenosine; ApA, adenylyl(3′-5′) adenosine; OD_{280}, optical density at 280 nm; V_e, elution volume; V_t, total volume. (Reproduced with permission from Fanou-Ayi, A. and Vijayalakshmi, M.A., *Biochem. Eng. Ann. N.Y. Acad. Sci.*, 413, 300–306, 1983.)

play a future role in this approach [100, 101]. Such work is of interest given the important role that phosphorylation plays in the regulation of life processes (e.g., the control of signal transduction, gene expression, cytoskeletal regulation, and apoptosis). In this application, IMAC has been used as a prefractionation technique for complex protein mixtures prior to analysis by mass spectrometry [102, 103]. In addition, IMAC ligands on protein chips have been used for the specific enrichment of proteins prior to their examination by surface-enhanced laser desorption ionization time-of-flight mass spectrometry (SELDI TOF MS) [104].

10.4 SUMMARY AND CONCLUSIONS

Although IMAC is a relatively new purification method, it is already a popular technique for proteins. For instance, the use of genetically engineered histidine tags has made IMAC a common tool for the recovery of recombinant proteins. The incorporation of IMAC in large-scale purification procedures should also continue as companies that produce proteins pay more attention to the possible incorporation of purification tags in their downstream processes.

Another use of immobilized metal-ion affinity has been in microanalytical studies of protein–metal-ligand complexes and protein tertiary or quaternary structures. Subtle changes in the histidine topography of native versus recombinant proteins can be detected by this approach, thus providing data that can be used for structural studies. In the future, it is expected that immobilized metal-ion affinity methods will also find a place in diagnostics, since these can be used to examine *in situ* the conformational changes of a protein in different pathologic states.

Another area in which IMAC will surely play a large role is in genomic and postgenomic research. This should include a broader use of IMAC for RNA and oligonucleotide separations. Continued use of IMAC is also expected in the prefractionation of protein mixtures prior to their analysis by mass spectrometry, as might be performed to analyze the phosphorylation of a protein or to examine the effects of a drug on protein expression.

IMAC should continue to develop as its number of applications grows. New types of matrixes and better combinations of chelating agents and metal ions should deliver improved specificity and selectivity to these separations. This, in turn, should make IMAC useful in an even greater range of areas in biology, biotechnology, and pharmaceutical science.

ACKNOWLEDGMENTS

The authors thank Prof. J. Porath from the Biochemical Separation Center in Uppsala, Sweden, for his fruitful discussions and helpful comments on the manuscript.

SYMBOLS AND ABBREVIATIONS

AdSEC	Adsorptive size-exclusion chromatography
C	Concentration of an affinity ligand in an electrophoretic buffer
CM-Asp	Carboxymethylated aspartic acid
DTPA	Diethylenetriamine pentaacetic acid
EBA	Expanded-bed adsorption
EDTA	Ethylenediamine tetraacetic acid
ESI-MS	Electrospray-ionization mass spectrometry
HEPES	*N*-[2-Hydroxyethyl]piperazine-*N′*-[2-ethanesulfonic acid])
His	Histidine
HP-IMAC	High-performance immobilized metal-ion affinity chromatography
HPLC	High-performance liquid chromatography
IDA	Iminodiacetic acid
IDA-M(II)	Iminodiacetic metal chelate
Im	Imidazole

IMA	Immobilized metal-ion affinity
IMAC	Immobilized metal-ion affinity chromatography
IMACE	Immobilized metal-ion affinity capillary electrophoresis
ImadSEC	Immobilized metal-ion adsorptive size-exclusion chromatography
IMAGE	Immobilized metal-ion affinity gel electrophoresis
IMAMA	Immobilized metal-ion affinity membrane adsorbents
IMAP	Immobilized metal-ion affinity partitioning
K_D	Dissociation equilibrium constant for a protein-ligand complex
MES	2-[N-Morpholino]ethanesulfonic acid
MMA buffer	MOPS, MES, acetic acid (20 mM each)
MOPS	3-[N-Morpholino]propanesulfonic acid
NTA	Nitrilotriacetic acid
RNase A, RNase B	Ribonuclease A or B
SEC	Size-exclusion chromatography
SELDI TOF MS	Surface-enhanced laser desorption ionization time-of-flight mass spectrometry
t	Observed migration time of an analyte in electrophoresis
t_{EOF}	Migration time for an electroosmotic flow marker in electrophoresis
t_N	Normalized migration time for a solute in electrophoresis, where $t_N = t/t_{EOF}$
t_{No}	Normalized migration time for a solute during electrophoresis in the absence of an affinity ligand
TCAN	Triazacyclononane
TED	*Tris* (Carboxymethyl)ethylenediamine
TREN	*Tris* (2-Aminoethyl)amine
Δt_N	Change in the normalized migration time for a solute during electrophoresis in the absence versus presence of an affinity ligand, where $\Delta t_N = t_N - t_{No}$
Δt_{Nmax}	Maximum change in the normalized migration time for a solute during electrophoresis

REFERENCES

1. Gurd, F.R.N. and Wilcox, P.E., Complex formation between metallic cations and proteins, peptides and amino acids, *Adv. Prot. Chem.*, 11, 311–427, 1956.
2. Helfferich, F.J., Ligand-exchange chromatography, *Nature*, 189, 1001–1002, 1961.
3. Goldstein, G., Ligand-exchange chromatography of nucleotides, nucleosides and nucleic acid bases, *Anal. Biochem.*, 20, 477–483, 1967.
4. Porath, J., Carlsson, J., Olsson, I., and Belfrage, B., Metal chelate affinity chromatography, a new approach to protein fraction, *Nature*, 258, 598–599, 1975.
5. Vijayalakshmi, M.A., Pseudobiospecific ligand affinity chromatography, *Trends Biotechnol.*, 7, 71–76, 1989.
6. Winzerling, J.J., Berna, P., and Porath, J., How to use immobilized metal ion affinity chromatography, *Methods: Companion Methods Enzymol.*, 4, 4–13, 1992.
7. Chaga, G.S., Twenty-five years of immobilized metal ion affinity chromatography: past, present and future, *J. Biochem. Biophys. Methods*, 49, 313–334, 2001.
8. Porath, J. and Olin, B., Immobilized metal ion affinity adsorption and immobilized metal ion affinity chromatography of biomaterials: Serum protein affinities for gel-immobilized iron and nickel ions, *Biochemistry*, 22, 1621–1630, 1983.

9. Lönerdal, B. and Keen, C.L., Metal chelate affinity chromatography of proteins, *J. Appl. Biochem.*, 4, 203–208, 1982.

10. Majors, E. and Carr, P.W., Glossary of liquid-phase separation terms, *LC GC*, 19, 125–162, 2001.

11. Andersson, L. and Sulkowski, E., Evaluation of the interaction of protein α-amino groups with M(II) by immobilized metal ion affinity chromatography, *J. Chromatogr.*, 604, 13–17, 1992.

12. Sulkowski, E., Immobilized metal ion affinity chromatography of proteins, in *Protein Purification: Micro to Macro*, Burgess, R., Ed., Alan R. Liss, New York, 1987, pp. 149–162.

13. Sulkowski, E., The saga of IMAC and MIT, *BioEssays*, 10, 170–175, 1989.

14. Hubert, P. and Porath, J., Metal chelate affinity chromatography, I: Influence of various parameters on the retention of nucleotides and related compounds, *J. Chromatogr.*, 198, 247–255, 1980.

15. Hubert, P. and Porath, J., Metal chelate affinity chromatography, II: Group separation of mono and dinucleotides, *J. Chromatogr.*, 206, 164–168, 1981.

16. Fanou-Ayi, A. and Vijayalakshmi, M.A., Metal-chelate affinity chromatography as a separation tool, *Biochem. Eng. Ann. N.Y. Acad. Sci.*, 413, 300–306, 1983.

17. Dobrowolska, G., Muszunska, G., and Porath, J., Model studies on iron (III) ion affinity chromatography: Interaction of immobilized metal ions with nucleotides, *J. Chromatogr.*, 541, 331–339, 1991.

18. Willson, R.C., Murphy, J.C., and Jewell, D., Immobilized-metal affinity separation of nucleic acids, in *Proceedings of 14th Symposium on Affinity Technology and Biorecognition*, Cancun, Mexico, 2001, p. 218.

19. Hemdan, E.S. and Porath, J., Development of immobilized metal affinity chromatography, 2: Interaction of aminoacids with immobilized nickel iminodiacetate, *J. Chromatogr.*, 323, 255–264, 1985.

20. Yip, T.T., Nakagawa, Y., and Porath, J., Evaluation of interaction of peptides with Cu(II), Ni(II) and Zn(II) by high-performance immobilized metal ion affinity chromatography, *Anal. Biochem.*, 183, 159–171, 1989.

21. Sulkowski, E., Purification of proteins by IMAC, *Trends Biotechnol.*, 3, 1–7, 1985.

22. Hemdan, E.S., Zhao, Y.J., Sulkowski, E., and Porath, J., Surface topography of histidine residues: A facile probe by immobilized metal ion affinity chromatography, *Proc. Natl. Acad. Sci. U.S.A.*, 86, 1811–1815, 1989.

23. Mrabet, N.T., One-step purification of *Actinoplanes missouriensis* D-xylose isomerase by high-performance immobilized copper-affinity chromatography: Functional analysis of surface histidine residues by site-directed mutagenesis, *Biochemistry*, 31, 2690–2702, 1992.

24. Goubran-Botros, H. and Vijayalakshmi, M.A., Cell surface interactions with metal chelates, *J. Chromatogr.*, 495, 113–122, 1989.

25. Laboureau, E. and Vijayalakshmi, M.A., Concerning the separation of mammalian cells in immobilized metal ion partitioning systems: A matter of selectivity, *J. Mol. Recogn.*, 10, 262–268, 1997.

26. Hochuli, E., Bannwarth, W., Dobeli, H., Gentz, R., and Stuber, D., Genetic approach to facilitate purification of recombinant proteins with a novel metal chelate adsorbent, *Bio-Technology*, 6, 1321–1325, 1988.

27. Piatibratov, M., Hou, S., Brooun, A., Yang, J., Chen, H., and Alam, M., Expression and fast-flow purification of a polyhistidine-tagged myoglobin-like aerotaxis transducer, *Biochim. Biophys. Acta*, 1524, 149–154, 2000.

28. Xu, F., Jia, J., Jin, Y., and Wang, D.T., High-level expression and single-step purification of human tryptophanyl-tRNA synthetase, *Protein Exp. Purif.*, 23, 296–300, 2001.

29. Berna, P., Mrabet, N.T., Van Beeumen, J., Devreese, B., Porath, J., and Vijayalakshmi, M.A., Residue accessibility, hydrogen bonding, and molecular recognition: Metal-chelate probing of active site histidines in chymotrypsins, *Biochemistry*, 36, 6896–6905, 1997.

30. Boden, V., Rangeard, M.H., Mrabet, N., and Vijayalakshmi, M.A., Histidine mapping of serine proteases: A synergic study by IMAC and molecular modelling, *J. Mol. Recogn.*, 11, 32–39, 1998.

31. Goubran-Botros, H. and Vijayalakshmi, M.A., Immobilized metal ion affinity electrophoresis: A preliminary report, *Electrophoresis,* 12, 1028–1032, 1991.
32. Haupt, K., Roy, F., and Vijayalakshmi, M.A., Immobilized metal ion affinity capillary electrophoresis of proteins: A model for affinity capillary electrophoresis using soluble polymer-supported ligands, *Anal. Biochem.,* 234, 149–154, 1996.
33. Baek, W.O., Haupt, K., Colin, C., and Vijayalakshmi, M.A., Immobilized metal ion affinity gel electrophoresis: Quantification of protein affinity to transition metal chelates, *Electrophoresis,* 17, 489–492, 1996.
34. Baek, W.O. and Vijayalakshmi, M.A., Effect of chemical glycosylation of RNase A on the protein stability and surface histidines accessibility in immobilized metal ion affinity electrophoresis, *Biochim. Biophys. Acta,* 1336, 394–402, 1997.
35. Jiang, K.Y., Pitiot, O., Anissimova, M., Adenier H., and Vijayalakshmi, M.A., Structure-function relationship in glycosylated α-chymotrypsin as probed by IMAC and IMACE, *Biochim. Biophys. Acta,* 1433, 198–209, 1999.
36. Anissimova, M., Baek, W.O., Varlamov, V., Mrabet, N., and Vijayalakshmi, M.A., Study of oligomeric structures of bovine seminal and pancreatic ribonucleases and their functional relationship, *J. Mol. Recogn.,* submitted.
37. Varlamov, V., Anissimova, M., Baek, W.O., and Vijayalakshmi, M.A., Immobilized metal ion affinity gel electrophoresis of microbial ribonucleases and protein inhibitor Barstar, *Int. J. Bio/Chromatogr.,* 6, 109–120, 2001.
38. Tchaga, G.S., Hopp, J., and Nelson, P., Rapid one-step purification using TALON™ superflow, CLONTECHniques; available on-line at http://www.clontech.com, October 1998.
39. Hutchens, T.W., Yip, T.T., and Porath, J., Protein interaction with immobilized ligands: Quantitative analyses of equilibrium partition data and comparison with analytical chromatographic approaches using immobilized metal ion affinity adsorbents, *Anal. Biochem.,* 170, 168–182, 1988.
40. Millipore, Zip tip; available on-line at http://www.millipore.com/ziptip, accessed March 2005.
41. Crowe, J., Masone, B.S., and Ribbe, J., One-step purification of recombinant proteins with the 6xHis tag and Ni-NTA resin, *Mol. Biotechnol.,* 4, 247–258, 1995.
42. Boulis, Y., Adenier, H., and Vijayalakshmi, M.A., Purification of sweet potato sporamin by IMAC and its analysis by mass spectrometry, *J. Chromatogr. A,* submitted.
43. Nakagawa, Y., Yip, T.T., Belew, M., and Porath, J., High-performance immobilized ion affinity chromatography of peptides: Analytical separation of biologically active synthetic peptides, *Anal. Biochem.,* 168, 75–81, 1988.
44. Porath, J., High-performance immobilized metal-ion affinity chromatography of peptides and proteins, *J. Chromatogr.,* 443, 3–11, 1988.
45. Andersson, L., Sulkowski, E., and Porath, J., Purification of commercial human albumin on immobilized IDA-Ni(II), *J. Chromatogr.,* 421, 141–146, 1987.
46. Berna, P.P., Moraes, F.F., Barbotin, J.N., Thomas, D., and Vijayalakshmi, M.A., One-step affinity purification of a recombinant cyclodextrin glycosyl transferase by Cu(II), Zn(II) tandem column immobilized metal ion affinity chromatography, in *Advances in Molecular and Cell Biology,* Mosbach, K., Ed., JAI Press, Greenwich, NY, 1996, pp. 523–537.
47. Porath, J., From gel filtration to adsorptive size exclusion, *J. Prot. Chem.,* 16, 463–467, 1997.
48. Vijayalakshmi, M.A., Pitiot, O., Legallais, C., and Moriniere, P., Gel with adsorptive size exclusion chromatography AdSEC: Use for b2 microglobulin gel dialysis, French patent 98.13655, international extension PCT/FR99/02635.
49. Yang, L., Jia, L.Y., Zou, H.F., and Zhank, Y.K., Immobilized iminodiacetic acid (IDA)-type Cu(II) chelating membrane affinity chromatography for purification of bovine liver catalase, *Biomed. Chromatogr.,* 13, 229–234, 1999.
50. Ji, Z., Pinon, D.I., and Miller, L.J., Development of paramagnetic beads for rapid and efficient metal chelate affinity purification, *Anal. Biochem.,* 240, 197–201, 1996.
51. Gaberc-Porekar, V. and Menart, V., Perspectives of immobilized-metal affinity chromatography, *J. Biochem. Biophys. Methods,* 49, 335–360, 2001.

52. Reif, O.-W., Nier, V., Bahr, U., and Freitag, R., Immobilized metal affinity membrane adsorbents as stationary phases for metal interaction protein separation, *J. Chromatogr. A,* 664, 13–25, 1994.

53. Camperi, S.A., Grasselli, M., and Cascone, O., High-speed pectic enzyme fractionation by immobilized metal affinity membranes, *Bioseparation,* 9, 173–177, 2000.

54. Mantovaara, T., Pertoft, H., and Porath, J., Further characterization of carboxymethylated aspartic acid agarose: Purification of human alpha 2-macroglobulin and hemopexin, *Biotechnol. Appl. Biochem.,* 13, 371–379, 1991.

55. Jiang, W., Graham, B., Spiccia, L., and Hearn, M.T., Protein selectivity with immobilized metal ion-TACN sorbents: Chromatographic studies with human serum proteins and several other globular proteins, *Anal. Biochem.,* 255, 47–58, 1998.

56. Sundberg, L. and Porath, J., Preparation of adsorbents for biospecific affinity chromatography, I: Attachment of group-containing ligands to insoluble polymers by means of bifunctional oxiranes, *J. Chromatogr.,* 90, 87–98, 1974.

57. Sulkowski, E., Immobilized metal ion affinity chromatography of proteins on IDA-Fe(III), *Macromol. Chem. Macromol. Symp.,* 17, 335–348, 1988.

58. Boden, V., Colin, C., Barbet, J., Le Doussal, J.M., and Vijayalakshmi, M.A., Preliminary study of the metal binding site of an anti-DTPA-indium antibody by equilibrium binding immunoassays and immobilized metal ion affinity chromatography, *Bioconjug. Chem.,* 6, 373–379, 1995.

59. Boden, V., Winzerling, J.J., Vijayalakshmi, M.A., and Porath, J., Rapid one-step purification of goat immunoglobulins by immobilized metal ion affinity chromatography, *J. Immunol. Methods,* 181, 225–232, 1995.

60. Yip, T.T. and Hutchens, T.W., Immobilized metal ion affinity chromatography, *Mol. Biotechnol.,* 1, 151–164, 1994.

61. Pearson, R.G., Hard and soft acids and bases: HSAB, Part I, Fundamental principles, *J. Chem. Educ.,* 45, 581–587, 1968.

62. Mantovaara, T., Pertoft, H., and Porath, J., Purification of human serum amyloid P component (SAP) by calcium affinity chromatography, *Biotechnol. Appl. Biochem.,* 11, 564–570, 1989.

63. Porath, J., Olin, B,, and Grandstand, B., Immobilized metal affinity chromatography of serum proteins on gel-immobilized group III A metal ions, *Arch. Biochem. Biophys.,* 225, 543–547, 1983.

64. Vijayalakshmi, M.A., High-performance liquid chromatography with immobilized metal adsorbents, in *Affinity Chromatography and Biological Recognition,* Chaiken, I.M., Wilchek, M., and Parikh, I., Eds., Academic Press, Orlando, FL, 1983, pp. 269–272.

65. Andersson, L. and Porath, J., Isolations of phosphoproteins by immobilized metal (Fe(III)) affinity chromatography, *Anal. Biochem.,* 154, 250–254, 1986.

66. Muszynska, G., Andersson, L., and Porath, J., Selective adsorption of phosphoproteins on gel-immobilized ferric chelate, *Biochemistry,* 25, 6850–6853, 1986.

67. Zachariou, M. and Hearn, M.T.W., Protein selectivity in immobilized metal affinity chromatography based on the surface accessibility of aspartic and glutamic acid residues, *J. Prot. Chem.,* 14, 419–430, 1995.

68. Chaga, G.S., Ersson, B., and Porath, J., Isolation of calcium-binding proteins on selective adsorbents: Application to purification of bovine calmodulin, *J. Chromatogr. A,* 732, 261–269, 1996.

69. Zachariou, M. and Hearn, MTW., Adsorption and selectivity characteristics of several human serum proteins with immobilized hard Lewis metal ion-chelate adsorbents, *J. Chromatogr. A,* 890, 95–116, 2000.

70. Krishnan, S. and Vijayalakshmi, M.A., Purification and some properties of three serine carboxypeptidases from *Aspergillus niger, J. Chromatogr.,* 370, 315–326, 1986.

71. Mrabet, N., Physicochemical and structural implications for molecular recognition in immobilized metal ion affinity chromatography, *Methods: Companion Methods Enzymol.,* 4, 14–24, 1992.

72. Hutchens, T.W. and Yip, T.T., Protein interactions with surface-immobilized metal ions: Structure dependent variations in affinity and binding capacity constant with temperature and urea concentration, *J. Inorg. Biochem.*, 42, 105–118, 1991.

73. Sulkowski, E., Immobilized metal ion affinity chromatography: Imidazole proton pump and chromatographic sequelae, I: Proton pump, *J. Mol. Recogn.*, 9, 389–393, 1996a.

74. Sulkowski, E., Immobilized metal ion affinity chromatography: Imidazole proton pump and chromatographic sequelae, II: Chromatographic sequelae, *J. Mol. Recogn.*, 9, 494–498, 1996b.

75. Fitton, V. and Santarelli, X., Evaluation of immobilized metal affinity chromatography for purification of penicillin acylase, *J. Chromatogr. B*, 754, 135–140, 2001.

76. Krishnan, S., Geahel, I., and Vijayalakshmi, M.A., Semi-preparative scale isolation of carboxypeptidase isoenzymes from *Aspergillus niger* by a single metal chelate affinity chromatography step, *J. Chromatogr.*, 397, 339–346, 1987.

77. Ohkubo, I., Kondo, T., and Taniguchi, N., Purification of nucleosidediphosphatase of rat liver by metal-chelate affinity chromatography, *Biochim. Biophys. Acta*, 616, 89–93, 1980.

78. Kikuchi, H. and Watanabe, M., Significance of amino acids and histamine for the elution of non histone proteins in copper-chelate chromatography, *Anal. Biochem.*, 115, 109–112, 1981.

79. Krishnan, S. and Vijayalakshmi, M.A., Purification of an acid protease and a serine carboxy peptidase from *Aspergillus niger* using metal-chelate affinity chromatography, *J. Chromatogr.*, 329, 165–169, 1985.

80. Porath, J., Immobilized metal ion affinity chromatography, *Protein Express Purif.*, 3, 263–281, 1992.

81. Mrabet, N.T. and Vijayalakshmi, M.A., Immobilized metal-ion affinity chromatography: From phenomenological hallmarks to structure-based molecular insights, in *Biochromatography — Theory and Practice*, Vijayalakshmi, M., Ed., Taylor & Francis, London, 2002, pp. 272 –294.

82. Hochuli, E. and Piesecki, S., Interaction of hexahistidine fusion proteins with nitrilotriacetic acid-chelated Ni^{2+} ions, *Methods: Companion Methods Enzymol.*, 4, 68–72, 1992.

83. Smith, M.C., Furman, T.C., Ingolia, T.D., and Pidgeon, C., Chelating peptide-immobilized metal ion affinity chromatography: A new concept in affinity chromatography for recombinant proteins, *J. Biol. Chem.*, 263, 7211–7215, 1988.

84. Ljungquist, C., Breitholtz, A., Brink-Nilsson, H., Moks, T., Uhlen, M., and Nilsson, B., Immobilization and affinity purification of recombinant proteins using histidine peptide fusions, *Eur. J. Biochem.*, 186, 563–569, 1989.

85. Skerra, A., Pfitzinger, I., and Pluckthun, A., The functional expression of antibody Fv fragments in *Escherichia coli:* Improved vectors and a generally applicable purification technique, *Biotechnology*, 9, 273–278, 1991.

86. Sporeno, E., Barbato, G., Graziani, R., Pucci, P., Nitti, G., and Paonessa, G., Production and structural characterization of amino terminally histidine tagged human oncostatin M in *E. coli, Cytokine*, 6, 255–264, 1994.

87. Morganti, L., Huyer, M., Gout, P.W., and Bartolini, P., Production and characterization of biologically active Ala-Ser-(His)6-Ile-Glu-Gly-Arg-human prolactin (tag-hPRL) secreted in the periplasmic space of *Escherichia coli, Biotechnol. Appl. Biochem.*, 23, 67–75, 1996.

88. Dabrowski, S. and Kur, J., Recombinant His-tagged DNA polymerase, I: Cloning, purification and partial characterization of *Thermus thermophilus* recombinant DNA polymerase, *Acta Biochim. Pol.*, 45, 653–660, 1998.

89. Boulis, Y., Grenier-De March, G., Gomord, V., Adenier, H., Faye, L., and Vijayalakshmi, M.A., Analysis of sporamin forms expressed in different subcellular compartments of transgenic tobacco plants by IMAC and ESI-MS, *Plant Physiol. Biochem.*, 41, 215–221, 2003.

90. Todorova, D. and Vijayalakshmi, M.A., Study of amylolytic enzymes by IMAC, *Plant Physiol. Biochem.*, submitted.

91. Todorova, D. and Vijayalakshmi, M.A., Seasonal variations, effect of storage and identification of amylolytic enzymes from potato tubers cv. Bintje, *Plant Physiol. Biochem.*, submitted.

92. Zhao, Y.J., Sulkowski, E., and Porath, J., Surface topography of histidine residues in lysozymes, *Eur. J. Biochem.*, 202, 1115–1119, 1991.

93. Nanak, E., Abdul-Nour, J., and Vijayalakshmi, M.A., Metal affinity electrophoresis: An analytical tool, *Methods: Companion Methods Enzymol.,* 4, 97–102, 1992.

94. Heegaard, N.H. and Robey, F.A., Use of capillary zone electrophoresis to evaluate the binding of anionic carbohydrates to synthetic peptides derived from human serum amyloid component, *Anal. Chem.,* 64, 2479–2482, 1992.

95. Nanak, E., Chadha, K.C., and Vijayalakshmi, M.A., Segregation of normal and pathological human red blood cells, lymphocytes and fibroblasts by immobilized metal-ion affinity partitioning, *J. Mol. Recogn.,* 8, 77–84, 1995.

96. Goubran-Botros, H., Birkenmeier, G., Otto, A., Koperschlager, G., and Vijayalakshmi, M.A., Immobilized metal ion affinity partitioning of cells in aqueous two-phase systems: Erythrocytes as a model, *Biochim. Biophys. Acta,* 1074, 69–73, 1991.

97. Laboureau, E., Puech, C., Goubran-Botros, H., and Vijayalakshmi, M.A., Partition of erythrocytes by IMAP: From cellular to molecular level, *Int. J. Bio/Chromatogr.,* 4, 221–231, 1999.

98. Min, C. and Verdine, G.L., Immobilized metal affinity chromatography of DNA, *Nucleic Acids Res.,* 24, 3806–3810, 1996.

99. Smith, T.H., LaTour, J.V., Bochkariov, D., Chaga, G., and Nelson, P.S., Bifunctional phosphoramidite reagents for the introduction of histidyl and dihistidyl residues into oligonucleotides, *Bioconjugate Chem.,* 10, 647–652, 1999.

100. Posewitz, M.C. and Tempst, P., Immobilized gallium(III) affinity chromatography of phosphopeptides, *Anal. Chem.,* 71, 2883–2892, 1999.

101. Stesballe, A., Andersen, S., and Jensen, O.N., Characterization of phosphoproteins from electrophoretic gels by nano-scale Fe(III) affinity chromatography with off-line mass spectrometry analysis, *Proteomics,* 1, 207–222, 2001.

102. Ping, C. and Stults, J.T., Phosphopeptide analysis by on-line immobilized metal-ion affinity chromatography-capillary electrophoresis-electrospray ionization mass spectrometry, *J. Chromatogr. A,* 853, 225–235, 1999.

103. Riggs, L., Sioma, C., and Regnier, F.E., Automated signature peptide approach for proteomics, *J. Chromatogr. A,* 924, 359–368, 2001.

104. Merchant, M. and Weinberger, S.R., Recent advancements in surface-enhanced laser desorption/ ionisation-time of flight-mass spectrometry, *Electrophoresis,* 21, 1164–1177, 2000.

Section III

Preparative Applications

11

General Considerations in Preparative Affinity Chromatography

Anuradha Subramanian

Department of Chemical Engineering, University of Nebraska, Lincoln, NE

CONTENTS

11.1   INTRODUCTION

Historically, the main use of affinity chromatography has been in the isolation of biological agents. Examples given throughout this book include the use of affinity columns for the isolation of enzymes, antibodies, and recombinant proteins. This has made affinity chromatography an important preparative tool in biochemistry and biotechnology. But the successful use of affinity chromatography for large-scale isolations requires that several aspects of this method be considered and characterized. This chapter examines many of these factors and discusses their roles in affinity separations.

11.1.1   Challenges in Protein Purification

The key to commercial development of biotechnological manufacturing is undoubtedly downstream processing, which is used to separate and purify the product of interest [1]. It is estimated in most operations that downstream processing accounts for at least 50 to 80% of the total cost [2, 3]. Economics, efficiency, and practicality are some of the constraints dictating the search for novel chromatographic supports and methodologies that either offer novel selectivities or overcome the shortcomings of existing supports in the preparative-scale purification of proteins.

Along with these demands, there is increasing pressure on the health-care industry to provide ample amounts of protein-based therapeutics at a reduced cost [4, 5]. There is also pressure to have therapeutics be "specific-pathogen free" (SPF). Clinical validation is still an overriding cost to process restructuring and to the introduction of new protein-based drugs. In addition, the process costs for human plasma or tissue-derived products are frequently high due to the limited supply of SPF source material or the multiple steps that are needed to deal with low-concentration product [4]. Furthermore, the cost of recombinant products from sources like mammalian cell culture [6] and milk from transgenic livestock [5, 7] is still high in spite of their higher starting feed concentrations relative to most natural products.

The recovery of research expenses is a leading source of costs for many new products. However, process savings is still a major concern in light of the need to simultaneously increase both the volume and safety of therapeutic products. This is illustrated by recent shortages in the U.S. of immunoglobulin G (IgG) from human plasma. The unique specificity that antibodies like IgG can display for a target antigen makes these invaluable tools in diagnostics [8, 9], structural analysis [10, 11], histology [12], and immunotherapy [13]. For instance, purified immunoglobulins have been used to treat patients with inadequate immunoglobulin levels [14]. This and related applications have, in turn, given rise to an increased need for medical-grade immunoglobulins and better methods for their isolation from various sources.

To increase in process efficiency, it will be necessary to have prepurification steps and affinity adsorption processes that are selective for a desired protein subpopulation [2, 15]. Many proteins have complex functions that require a specific posttranslational structure. This creates the need for separation processes that can reproducibly select for a protein with a given structure and biological activity [5, 6]. Other problems with current adsorption processes include inefficiencies that lead to high buffer usage, a need for buffer-exchange steps, and slow volumetric throughput. They also may require multiple process steps for volume reduction [15], pathogen inactivation, or the removal of pathogen particles (e.g., HIV and HBV), but the addition of these steps also increases the process cost [16–19].

While recombinant products offer processing advantages with respect to SPF sources, there is still a need for variants of the same therapeutic protein. Examples include products from human blood plasma and tissue or natural products from the plasma and tissues of animals [20]. These variants are needed because there are frequently differences in their efficacy, such as might occur due to an individual's acquired immunity for a particular protein variant. This means that a battery of variants of the same protein or with a specific protein activity should be available to allow an alternative method of treatment.

Because of these factors, the plasma fractionation industry in the U.S. and other first-world countries is seeking to develop both improved and pathogen-safe products that are based on plasma-derived and recombinant therapeutic materials. Other nations are also trying to obtain plasma and recombinant products that employ leading-edge processing techniques and materials. The result is that there is also a demand for new techniques that can be used to process therapeutic proteins, especially for products that will soon be validated for clinical use.

11.1.2   Advantages of Affinity Methods for Protein Purification

Affinity and ion-exchange separations are both scalable adsorption methods that are important to protein purification [21–23]. Ideally, a new adsorption technology for a protein that employs such methods should have a high selectivity, high yield, and good throughput. The method should also be amenable to or provide improved specific pathogen removal, use minimal amounts of buffer, be easy to operate, and be less expensive than other available techniques. Separation methodologies based on affinity interactions are attractive for these applications because they offer exquisite selectivity and specificity and are relatively easy to operate.

To further improve selectivity and process stream throughput, new materials have been developed for protein purifications. Examples include hydrogel-filled zirconia beads such as Biosepra and perfusion-based matrixes like POROS. But these materials do have some accompanying disadvantages. For instance, these have increased matrix production costs due to their greater complexity versus other materials. They can also be less amenable to quality control restrictions and may require the use of complex equipment on a large scale (e.g., high-pressure pumps).

11.2   GOALS FOR PROTEIN PURIFICATION

The advances occurring in matrix technologies for therapeutic protein production are greatly constrained by the realities of clinical validation. Thus, any real advances in the processing methods for these proteins are most likely to consist of techniques that combine gains in process yield, throughput, and pathogen removal while also decreasing the complexity of the purification process [15, 22, 24]. It is also desirable to create a savings in production costs by developing techniques that reduce the number of process steps and amounts of buffer that are required.

An ideal affinity method for purification should combine an increase in target selectivity with an improved avidity for adsorbing the target from a crude sample. This technique should also have a high dynamic adsorption capacity at fast flow rates to reduce sample loading times [25]. In addition, the nonspecific adsorption of pathogens (as well as chemical agents used to inactivate these pathogens) after adsorption of the target should be negligible, or such agents should be easy to remove.

The flow in this ideal column should have a plug profile at a high length-to-diameter ratio for the column. This will reduce eluate volumes and the need for buffer changes or intermediate volume-reduction steps. If the purification can be performed at low pressures, then the equipment required for sample processing will also be simplified [25]. The means of solute transport within the matrix should also be considered. For instance, it was found that solute transport in hydrogel beads (prepared from cross-linked and derivatized cellulose and chitosan) was governed by a combination of convective and diffusive forces [26, 27]. By utilizing a convective component of mass transport in flow-through chromatography, it is possible to reduce retention times while maintaining good resolution and high sample throughput.

Real matrixes and affinity methods only approach some of these goals. Thus, a combined improvement in all of these process criteria requires a new engineering paradigm. This new paradigm should improve the placement of a ligand throughout the matrix, optimize void-flow fluid mechanics in a packed bed or other nonbatch-mode configuration for adsorption, and provide better insight into the manipulation of transport phenomena within a matrix [22, 25]. This combined understanding should seek to give higher densities of functional adsorption sites while allowing efficient access to these sites by intraparticle convection. This should also allow the minimization of pressure drops at low dispersion to maintain small product eluate volumes.

One factor that is important to consider in the design of an affinity purification process is the *time-dependent capacity* of the adsorption medium. This is also referred to as the "dynamic capacity" or "throughput" of the medium. The time-dependent capacity of an affinity matrix for large molecules like proteins, DNA, and RNA is chiefly limited by the speed of delivery to the adsorption site, along with its local accessibility and molecular structure. These latter two characteristics can be examined using classical, physical biochemical methods of analysis. Site accessibility is typically studied in the steady state using gel-permeation chromatography. However, more time-dependent analyses that emphasize transport phenomena and chemical kinetics can also be used to describe the accessibility and manipulation of adsorption-site structures [25].

Adsorption throughput is another factor to consider in affinity purification. This refers to the number of molecules that are captured per unit time per volume of matrix. For the capture of large biomolecules, the greatest gains in adsorption throughput have been made by employing sites at a low to moderate local density while also providing for rapid convective delivery of the target to these sites [22, 25]. This has provided a better understanding of the relationship between local site density and adsorption (affinity) site functionality, and of the average matrix capacity at a given average site density [22, 24]. These topics are discussed later in this chapter.

11.3 GENERAL CONSIDERATIONS IN PROTEIN PURIFICATION

Like other applications of affinity chromatography, preparative work requires that several important choices be made when developing the final separation system. Several of these items are discussed in the following subsections, including the selection of support material, ligand, and immobilization conditions, as well as the application and elution conditions.

11.3.1 Support Material

The selection of a solid-phase support for a chromatographic application is usually the first step in the development process. This is the case because the final performance of the overall sorbent will be dependent on the type of support to which the affinity ligand is immobilized. This performance will be determined by the material properties of the support, the derivatization chemistry used to modify this material, and the nature of the immobilized ligand [28]. The ideal support particle should be energetically homogeneous and have a high surface area for chemical modification. It should also be amenable to facile substitutions and have good physical and chemical stability over a wide range of pH, temperature, pressure, and solvent conditions [29–31]. It should also be available in a variety of particle diameters, pore sizes, and volumes.

Polymeric- or agarose-based supports are extensively used in chromatography. Both of these supports lend themselves to a variety of conjugation chemistries and offer reasonable physical and mechanical properties. They are also resistant to the various solvent systems used in chromatography [31, 32]. Supports based on metal oxides, silica, alumina, and zirconia are growing in use as well [32, 33]. Supports based on hydrogels also deserve consideration, since they may provide a convective mode of solute transport. A further detailed discussion of the various supports that can be used in affinity separations, and the properties of these materials, is provided in Chapter 2.

11.3.2 Affinity Ligand

The choice of an appropriate ligand is another critical step in the development of an affinity-based purification. In many cases, this choice will depend on the biochemical and biophysical properties of the target and its macroenvironment. For instance, if a process is needed for isolating a single protein or isoform of a target protein, then an affinity ligand with a high selectivity and specificity should be employed. This often requires the use of a monoclonal antibody (Mab) as the ligand.

Mabs are extremely selective for a given target. However, they are expensive to produce and are prone to fouling and denaturation. Immobilized Mabs have been used effectively in the separation and purification of medically relevant proteins, with some of these products having won FDA approval. If the levels of other proteins are high in such products, then it is often necessary to first use cation- or anion-exchange chromatography to reduce these levels.

Synthetic or biomimetic ligands can also be used to produce selective adsorbents, as discussed in Chapter 9. This is typically accomplished by using enzyme substrate analogs, inhibitors, or triazine scaffolds that resemble biochemical cofactors to generate a pseudo-affinity matrix. It is anticipated that with advances in combinatorial chemistry and molecular modeling, the efficient and economical generation of such ligands will cause these to be produced even more in the area of bioseparations.

11.3.3 Immobilization Conditions

In most cases, the support selected for an affinity column has to be chemically modified or derivatized to yield a group that can then react a ligand. This generally involves a reaction with primary amines, carboxyl groups, or sugar residues on a ligand to yield a stable covalent linkage. Methods based upon cyanogen bromide, tresyl chloride, N-hydroxysuccinimide, and epoxide groups are commercially available and can be used to immobilize proteins and antibodies onto a variety of supports. More information on these and other immobilization methods can be found in Chapter 3.

An activation chemistry that is stable over a wide range of pH and buffer conditions is often desirable for this process. The selected activation chemistry should result in negligible leakage of the ligand and be able to withstand the extremes in pH and temperatures that are encountered in routine sterilization procedures. Another factor that should be considered is the effect of immobilized ligand density on the performance of the affinity support. For instance, the effects of a high local antibody density and spatial distribution on antigen binding efficiency have been evaluated under conditions of dynamic loading and elution [22]. This is in contrast to previous studies that have emphasized the correlation of high-volume averaged antibody density with immunosorbent performance.

To illustrate the importance of ligand density for antibodies, the distribution of an antihuman Protein C monoclonal antibody was evaluated when this antibody was immobilized on an Emphaze AB1 support. This antibody was coupled through its amino groups with azalactone groups on the support at a pH of 7 or greater to give a stable covalent linkage. The distribution of the immobilized antibody was controlled by a two-step sequence of permeation and reaction, with the labeled antibody being visualized by immunofluorescence. At a low pH and low temperature in the presence of a competitor nucleophile, it was possible to depress the Thiele modulus for coupling to enable permeation of the antibody. The adsorption of the permeated antibody was enhanced by the presence of 0.75 M Na_2SO_4, with the pH then being raised to achieve rapid covalent coupling.

Based on this two-step immobilization approach, supports with bead-averaged antibody densities of 1 to 11 mg/ml hydrogel were examined. Immunosorbents that contained a more evenly distributed antibody gave a threefold greater binding efficiency for the target than supports with high local antibody densities. Under the same conditions, no appreciable changes in the mass-transfer characteristics were observed, as determined using breakthrough analysis, between the immunosorbents with distributed antibodies versus those with high local antibody densities.

11.3.4 Adsorption and Elution Conditions

Antigens and immobilized antibodies are bound to each other through a variety of forces that include ionic bonding, hydrophobic interactions, hydrogen bonding, and van der Waals attractions. The strength of an antibody-antigen complex depends on the relative affinity and avidity of the antibody. In addition, steric orientation, coupling density, and nonspecific interactions can also influence the binding by immobilized antibodies.

The objective of the elution step in an affinity method is to recover a specifically bound target with a high yield, purity, and stability. Elution conditions that might denature this product should be avoided in preparative work. An examination of the current literature shows that a wide variety of elution conditions are available, with the choice of these conditions often being empirical. However, the available elution strategies can be considered in a logical sequence when selecting an appropriate elution protocol. This sequence involves the consideration of (1) specific elution, (2) acid elution, (3) base elution, and (4) the use of chaotropic agents. Further details on these methods can be found in Chapter 4, but each will now be discussed briefly with regard to their use with immobilized antibodies.

Specific elution from immunosorbents makes use of the fact that certain antibodies can bind to their respective target at a high pH in the presence of metals like calcium or magnesium or in the presence of chelating agents like EDTA. The antigens that are bound by such antibodies can be eluted under relatively gentle conditions by either lowering the pH, adding EDTA to the elution buffer, or adding divalent metal ions that cause the antibody-antigen complex to disassociate.

Acid elution is the most widely used method for target desorption from an antibody column and is frequently effective. The most common acid eluants are pH 2.5 glycine-HCl,

0.02 M HCl, and pH 2.5 sodium citrate. Upon elution, any dissociated targets should be quickly neutralized to pH 7.0 with pH 8.5, 2 M Tris base to avoid acid-induced denaturation. In some cases, hydrophobic interactions between an adsorbed antigen and the immobilized antibodies can give rise to low recovery with acid elution. This can be solved by adding 1 M propionic acid, 10% dioxane, or ethylene glycol to the acid eluant to dissociate such complexes.

Base elution is less frequently employed than acid elution with immobilized antibodies. This is typically performed with 1 M NH$_4$OH or pH 11.5, 0.05 M diethylamine. This has been used to elute membrane proteins (i.e., targets with hydrophobic character) and antigens that precipitate in acid but are stable under basic conditions.

Chaotropic agents can also be used for elution. These agents disrupt the tertiary structure of proteins and, therefore, can be used to disrupt the antibody-antigen complexes. Chaotropic salts are particularly useful in disrupting ionic interactions and hydrogen bonding. In addition, they sometimes disrupt hydrophobic interactions. The relative effectiveness of chaotropic anions follows the order $SCN^- > ClO_4^- > I^- > Br^- > Cl^-$. The effectiveness of chaotropic cations follows the order guanidinium (Gu) $> Mg^{2+} > K^+ > Na^+$. This makes eluants like 8 M urea, 6 M guanidine-HCl, and 4 M NaSCN effective in disrupting most antigen-antibody (Ag:Ab) interactions. To avoid and minimize chaotropic-salt-induced protein denaturation, rapid desalting or dialysis of the eluant is advised.

11.4 OPTIMIZATION AND SCALE-UP FACTORS

Bioseparations often involve complex systems. Thus, developing a successful bioseparation technique will require that attention be given to several items. These items include the specific application being addressed and the fundamental nature of the separation process. To achieve high separation efficiency at a minimal cost, it is helpful to first obtain quantitative measures of the driving forces in this process and to identify the design and operating variables that will control performance [34–36]. Important design and operating parameters that should be considered include the adsorption rate constant, the rate of adsorption, and the number of transfer units.

It is useful to categorize separation techniques in terms of their "driving forces." This is because the knowledge obtained in generating process data, developing design models, and designing commercial equipment for one application will be useful in assessing feasibility and developing commercially viable technology for other applications involving the same driving force. Thus, chromatographic processes should be developed with an emphasis on full-scale design requirements throughout the development effort, beginning within the laboratory. A mathematical process model is needed to help guide these development and design efforts [37]. This can greatly reduce the number of experiments that will be required, as well as aid in optimization and scale-up.

To gain a better understanding of the interaction between a ligand and target protein, it is often necessary to quantitate such factors as the capacity of the affinity matrix and the dissociation constant of the corresponding protein-ligand interaction. Thus, studies aimed at understanding the adsorption mechanism and the nature of interactive forces between the immobilized ligand and target protein should also be undertaken.

11.4.1 Adsorption Isotherms

Identification of an adsorption isotherm that describes the adsorption process is an essential first step in the optimization of an affinity method. One thermodynamic model that can be used to describe the interactions of an antigen (Ag) with an antibody (Ab) is the Freundlich isotherm. This is represented by Equation 11.1,

$$q = bC^n \qquad (11.1)$$

where q is the amount of adsorbed target, C is its concentration in solution, and b and n are constants at a given temperature, with n being less than 1. This model is fairly accurate in describing adsorption on heterogeneous surfaces. However, it fails to predict the asymptotic nature of saturation and does not give a linear form at low concentrations. In addition, the applicability of this equation in modeling antigen-antibody interactions is rather limited.

Two other models that can be used are the Jovanovic isotherm and the Redlich-Peterson isotherm. The Jovanovic isotherm is represented by the equation,

$$q = \alpha[1 - \exp(-\beta C)] \qquad (11.2)$$

where α is the saturation capacity (or q_{max}) and β is the association constant. The Redlich-Peterson isotherm is described by the relationship

$$q = aC/[1 + bC^n] \qquad (11.3)$$

where a, b, and n are constants. Another possible model is the Langmuir-Freundlich isotherm, as given later in Equation 11.6.

As Equations 11.1 through 11.3 suggest, static and dynamic ligand-binding isotherms must be obtained by saturation experiments. This is accomplished by making a plot of the adsorbed solute as a function of its equilibrium concentration in the liquid phase, as shown in Figure 11.1. This yields an equilibrium curve or isotherm. The isotherm equation that best fits the data is then identified and the values for the corresponding parameters such as K_d (the dissociation equilibrium constant) and Q_{max} (the maximum static binding capacity) are then calculated [38].

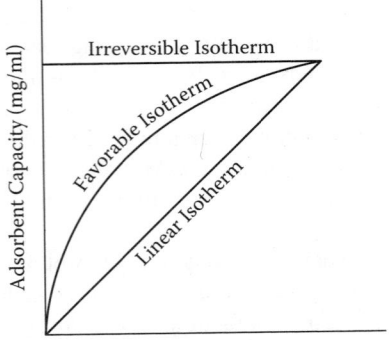

Figure 11.1   Some general isotherms that can be used for describing adsorption of target solutes to immobilized ligands in affinity chromatography (e.g., the binding of antigens to immobilized antibodies). The x-axis represents the concentration of solute that is in solution and in equilibrium with the affinity adsorbent. The y-axis represents the amount of solute that is bound per unit mass of the adsorbent.

11.4.2 Rate Equations

The selection of a rate equation to describe the binding of a target to an immobilized ligand is another component that should be considered early in the development of an affinity purification process. Based on molecular sorption to form a monolayer, Langmuir derived an equation that gives a good description for many cases of protein sorption [39]. The most general interaction between the unbound adsorbate (P) in the solution and the unoccupied immobilized ligand or matrix (L) is shown by Equation 11.4.

$$P + L \underset{k_2}{\overset{k_1}{\Leftrightarrow}} PL \tag{11.4}$$

This general relation can also be represented by a rate equation. The most common rate equation that is observed for such a reaction has a second-order rate for the forward process of adsorption and a first-order rate for the reversible dissociation of the adsorbed molecule. This rate equation is given below [40],

$$\frac{dq}{dt} = k_1 c (Q_{max} - q) - k_2 q \tag{11.5}$$

where Q_{max} is the maximum binding capacity of the support (in mg protein/ml support), q is the adsorbed protein concentration (in mg protein/ml support), and c is the protein concentration in the bulk phase. The dq/dt represents the change in the adsorbed protein concentration with time, and k_1 and k_2 are the forward and reverse rate constants for this reaction.

At equilibrium, the rate expression in Equation 11.5 reduces to the familiar relationship for a Langmuir isotherm,

$$q = \frac{Q_{max} c}{K_d + c} \tag{11.6}$$

where the dissociation equilibrium constant K_d is defined as being equal to the ratio k_2/k_1. An example of such an isotherm for the adsorption of human protein C to an immunosorbent containing immobilized antihuman protein C is shown in Figure 11.2. In the case of proteins, K_d depends on the isoelectric point of the protein, the pH of its surrounding solution, and the ionic strength of this solution. In practice, K_d and Q_{max} are often found by using the best-fit parameters obtained through nonlinear data-fitting programs.

An estimation of the rate of adsorption is also necessary for the design of an automated affinity purification. This rate is often expressed in terms of the amount of protein that is adsorbed per gram of support per unit time. To evaluate this, the relationship between the adsorption rate (AR) and the apparent adsorption rate constant (k_a) on the linear velocity (u) of the mobile phase passing through an affinity column can be evaluated by using a split-peak approach and empirical relationships, as given by Equations 11.7 and 11.8 [41],

$$k_a = a(u) + b \tag{11.7}$$

$$AR = m(u)^n \tag{11.8}$$

where a, b, m, and n are constants. Ideally, the dependence of adsorption rate (AR) on the linear velocity (u) should be evaluated at several feed concentrations. These equations can then be utilized to modify the method for use in process-scale chromatography.

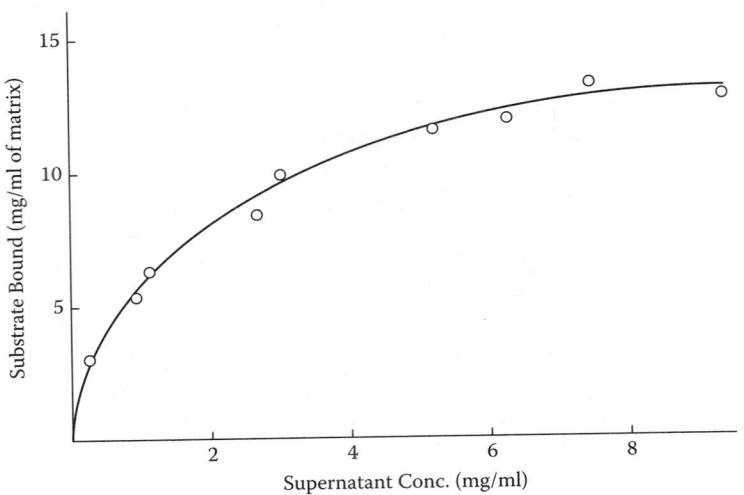

Figure 11.2 Isotherm for the binding of human protein C (hPC) to an Affiprep matrix containing immobilized anti-nPC monoclonal antibodies. These results were obtained in pH 6.5, 50 m*M* Tris-HCl containing 25 m*M* EDTA and 100 m*M* NaCl. This isotherm was generated by incubating a known amount of the immunosorbent with various concentrations of hPC for 24 h at 4°C. At the end of each experiment, the amount of hPC that remained in the supernatant was measured by absorbance measurements at 280 nm or by using an enzyme-linked immunosorbent assay (ELISA) for this substrate. (From Subramanian, A. and Velander, W.H., *Int. J. BioChromatogr.*, 5, 67–82, 2000. With permission.)

11.4.3 Breakthrough Curves

Most chromatographic separations on a preparative scale are carried out as packed-bed operations. But other configurations such as fluidized-bed, simulated moving-bed, and rotating annular chromatography are also possible (see Chapters 16 and 18). The adsorption and desorption of proteins on conventional beaded supports can be described using a combination of surface diffusion and pore diffusion with concurrent adsorption or desorption [36, 42–44]. The prediction and estimation of the underlying parameters that govern the transport of biomolecules in chromatographic supports is necessary for a valid scale-up and design strategy.

Most of the information needed to evaluate and model the performance of a column can be obtained from a breakthrough curve. This is a plot that shows the effluent concentration of a target as a function of time or throughput volume [28, 40, 45, 46]. Examples of such plots are given in the lower portion of Figure 11.3. The general properties of these curves and some of their uses are discussed in the following subsections.

11.4.3.1 *General Description of Breakthrough Curves*

This chapter will use the rate theory of chromatography rather than plate theory to explain the performance of packed-column affinity chromatography and the corresponding breakthrough curves. For a dynamic system, the change in solute concentration vs. time can be examined by using the following equation of continuity of the mobile phase [28, 38].

$$D_x \frac{\partial^2 c}{\partial x^2} - u \frac{\partial c}{\partial x} - R_i = \frac{\partial c}{\partial t} \qquad (11.9)$$

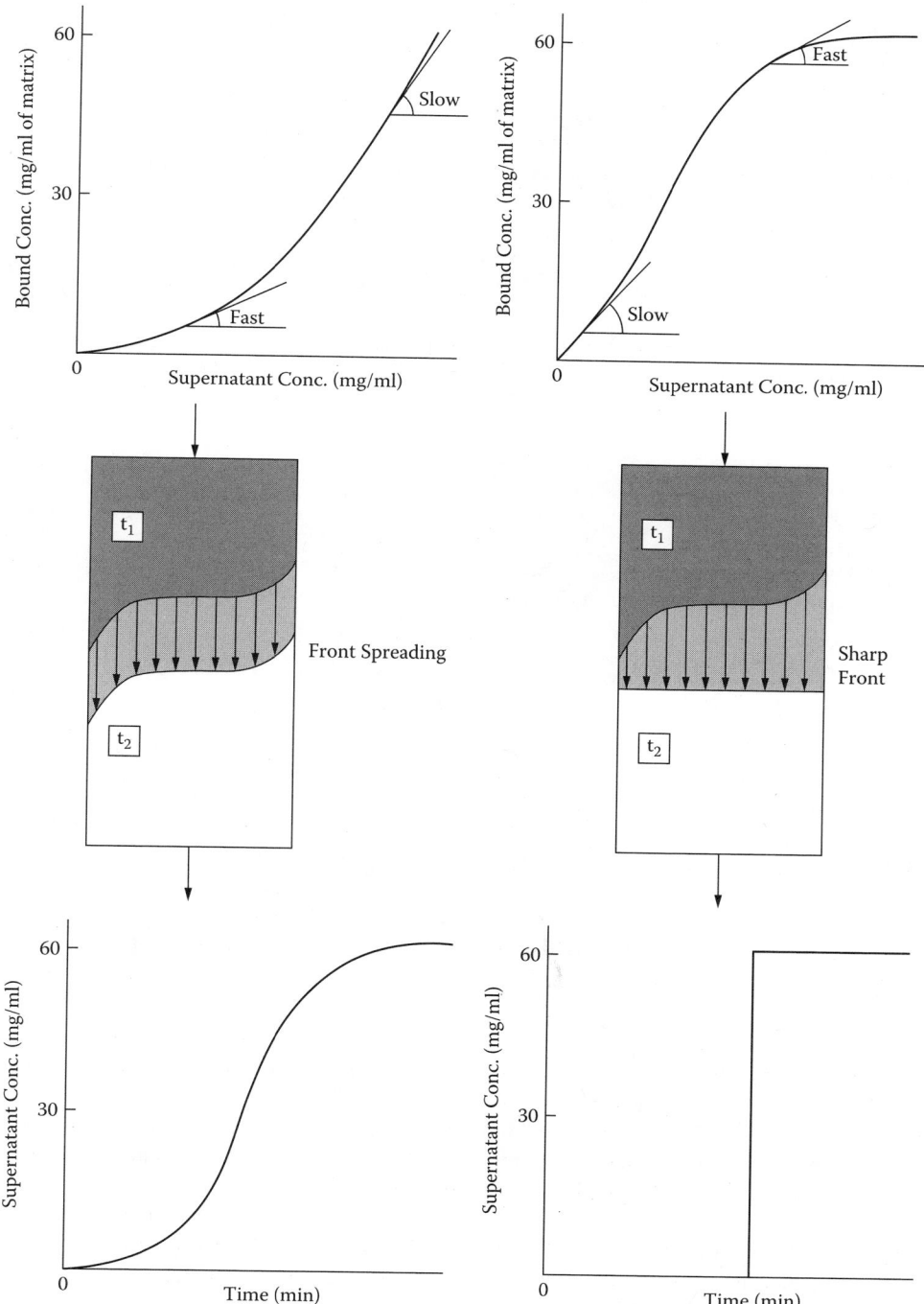

Figure 11.3 A schematic representation of the shape of the breakthrough curve and its impact on observed peak front.

In this relationship, D_x is the axial dispersion coefficient, u is the linear velocity, R_i is the rate of interface mass transfer (as given by Equation 11.5), x is the distance along the column bed, and t is the time.

In a breakthrough curve, the relative variation in the concentration of solute that is observed at the outlet of a column (as given by the dimensionless ratio C/C_0) at time t after the elution of a nonadsorbed component would have emerged from the column is described by Equation 11.10 [47],

$$\frac{C}{C_0} = \frac{J(n/r, nT)}{J(n/r, nT) + |1 - J(n, nT/r)| \exp|(1 - r^{-1})(n - nT)|} \tag{11.10}$$

where the terms r, n, and T are dimensionless operational parameters given by the following equations.

$$r = 1 + C_0/K_d \tag{11.11}$$

$$n = Q_{max} k_1 h A_c / F \tag{11.12}$$

$$T = Ft(K_d + C_0)/A_c Q_{max} h \tag{11.13}$$

Also, the function $J(\alpha, \beta)$ in Equation 11.10, where α equals n/r and β equals nT, can be approximated by using the relationship shown in Equation 11.14.

$$J(\alpha, \beta) \approx \frac{1}{2}\left[1 - erf\left(\sqrt{\alpha} - \sqrt{\beta}\right)\right] + \frac{\exp\left[1 - \left(\sqrt{\alpha} - \sqrt{\beta}\right)^2\right]}{2\pi^{\frac{1}{2}}\left[(\alpha\beta)^{\frac{1}{4}} + \beta^{\frac{1}{2}}\right]} \tag{11.14}$$

With Equation 11.10, it is possible to compute the value of C/C_0 (i.e., the response seen in the breakthrough curve) as a function of time for a wide variety of packed-bed affinity systems. In such work, small-scale batch experiments are often used to estimate the values of Q_{max}, K_d, k_1, and k_2 for these calculations.

Based on the use of a differential mass balance, the concentration propagation velocity ($\Delta l/\Delta t$) of a solute through an affinity column during a breakthrough analysis is given by the following equation,

$$\Delta l/\Delta t = v_s/[1 + \{(1 - \varepsilon)/\varepsilon\}\Delta q/\Delta c] \tag{11.15}$$

where l is the column length, v_s is the superficial velocity, ε is the interstitial porosity, and $\Delta q/\Delta c$ is the slope of the adsorption isotherm.

The shape of the resulting breakthrough curve will be a complex mix of the equilibrium and nonequilibrium processes involved in adsorption and will depend on the slope of the adsorption isotherm. This is demonstrated in Figure 11.3. A sharp breakthrough curve can be achieved if this isotherm is concave. This type of isotherm is considered "favorable" since it results in an efficient use of the sorbent within the column. However, convex isotherms cause front spreading and are considered unfavorable, since they give rise to less efficient use of the sorbent.

11.4.3.2 Use of Breakthrough Curves in Characterizing Separations

Breakthrough curves can be used to determine the rate-limiting step in solute transport and to model a separation process. A typical breakthrough profile for this type of study is shown in Figure 11.4. The area behind the curves in these plots corresponds to the

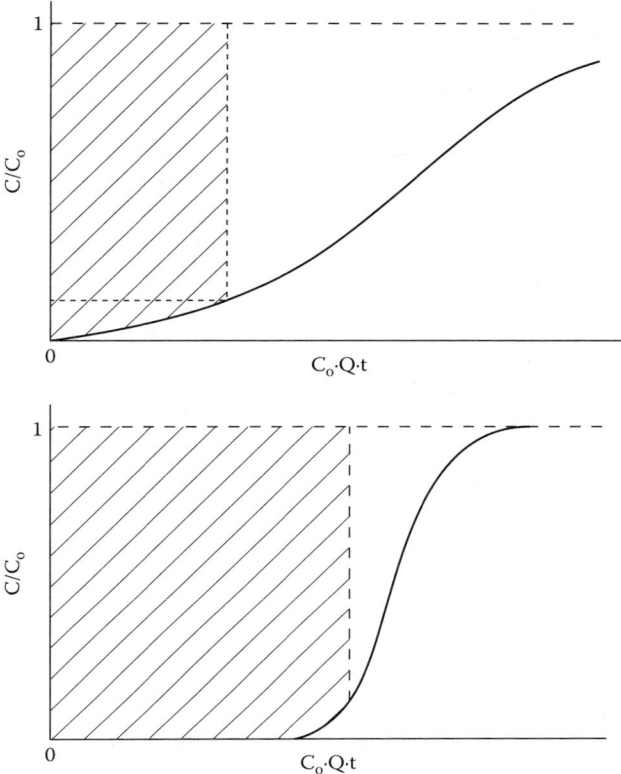

Figure 11.4 The effect of a breakthrough curve's profile on the performance of an affinity column. This figure shows a plot of the dimensionless concentration term C/C_0 versus the amount of solute or adsorbate that has been applied to the column at time t (C_0, Q, t). The term C_0 is the inlet feed concentration of the solute, C is the concentration that is observed at the column exit, and Q is the volumetric flow rate. As this figure indicates, either a shallow breakthrough profile (as shown at the top of this figure) or a sharper curve (shown at the bottom) can be obtained by altering the column operating conditions. The shaded areas represent the total amount of solute that has adsorbed to the column when the value of C/C_0 is 0.1 (i.e., 10% of the maximum breakthrough value has been attained).

maximum equilibrium capacity for a given feed concentration, and the area under the curve corresponds to the solute in the effluent. If the adsorptive step is stopped at its breakthrough concentration (C_{BT}), which is usually 10% of the feed or equilibrium concentration, the breakthrough curve can be used to determine such items as the (1) processing time, (2) percent of column capacity that is being utilized, and (3) concentration of protein in the effluent. This is precisely the information needed to optimize column performance.

Additional qualitative information about an adsorption process can be obtained from the profile of a breakthrough curve. Particularly, Equation 11.10 can be used in conjunction with computer simulations to determine the influence of varying a number of factors on the shape of breakthrough curves. A detailed description of this is provided elsewhere [35], but a few important features of the shape and position of a breakthrough curve will be discussed in this chapter. As an example, Figure 11.5 shows how the shape and position of a breakthrough curve varies with the length of a packed column when the column cross-sectional area, flow rate, and inlet feed concentration are all kept constant. This also shows

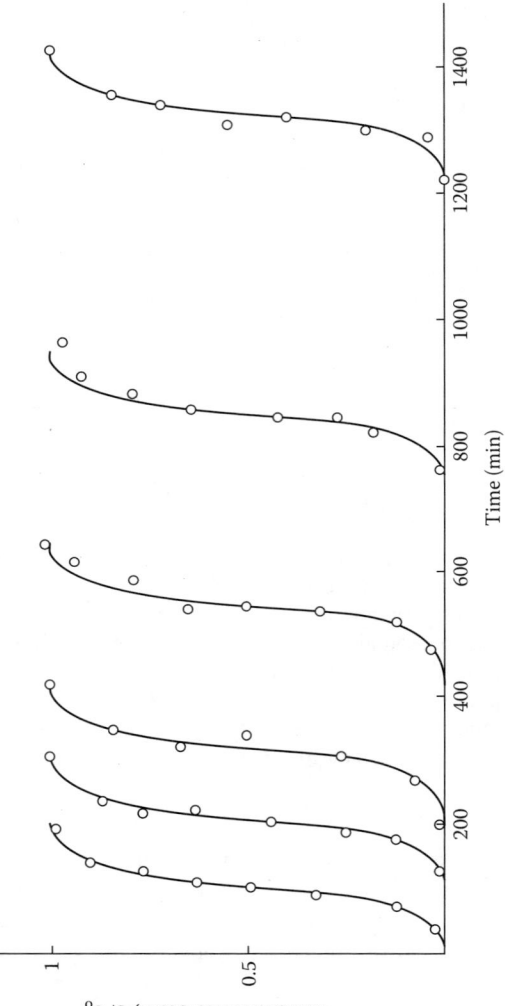

Figure 11.5 Variation of the shape and position of breakthrough profiles when using different column lengths. In this figure, the column diameter, the volumetric flow rate, and the inlet feed concentration of the solute (C_0) were kept constant while the column length was varied. The lengths of the columns used in this experiment increased in going from the left to the right. The open circles show the experimental data, and the solid lines are the predicted theoretical results.

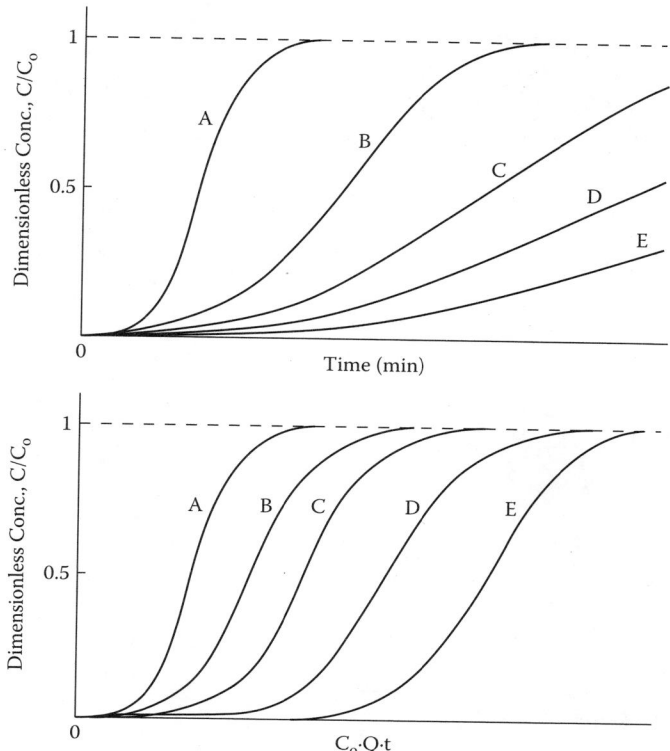

Figure 11.6 Breakthrough profiles obtained as a function of increasing flow rates through the column, where E is the fastest flow rate and A is the slowest flow rate tested. In the top figure, the relative response C/C_0 is plotted against time. In the bottom figure, C/C_0 is plotted against the amount of adsorbate or solute applied to the column, as represented by the combined term (C_0, Q, t).

that there can be excellent agreement between experimental breakthrough data and the breakthrough curves that are predicted by Equation 11.10. This also indicates that values for K_d and Q_{max} that are obtained from static ligand binding isotherms, as was used with Equation 11.10 in Figure 11.5, can be used to model packed-bed columns.

Figure 11.6 shows how the sharpness of a breakthrough profile changes as the volumetric flow rate is varied. In this figure, a significant improvement in performance can be observed when C/C_0 is plotted against the amount of adsorbate applied to the column, while a plot of C/C_0 vs. time shows little impact. In addition, the mean position of the breakthrough curve (i.e., the point at which $C/C_0 = 0.5$) is not affected by the flow rate.

As shown by Figure 11.7, varying the inlet feed concentration (C_0) for a column can have a marked impact on the shape and position of a breakthrough curve when C/C_0 is plotted against time. But when C/C_0 is replotted against the amount of adsorbate applied to the column, an effect is only noticed on the shape and position of the curve when $C_0 \leq K_d$. At $C_0 > K_d$ the shape and position of the curve become constant.

By using Equation 11.10, it is also possible to study the effect of varying the dissociation equilibrium constant (K_d) on the shape of a breakthrough curve. The same is true for studying the result of changing the amount of adsorbent in the packed column and the size of the forward rate constant (k_1) for the adsorption process.

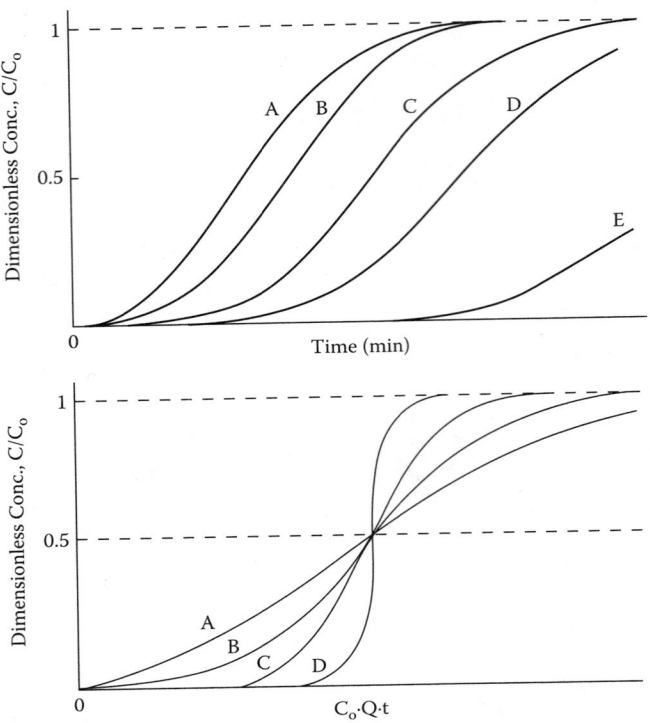

Figure 11.7   Breakthrough profiles obtained as a function of increasing inlet feed concentrations (C_0) through the column, where E is the lowest concentration and A is the highest concentration tested. In the top figure, the relative response C/C_0 is plotted against time. In the bottom figure, C/C_0 is plotted against the amount of adsorbate or solute applied to the column, as represented by the combined term (C_0, Q, t).

11.4.4   Number of Transfer Units

The eluant profiles for wash and elution steps are analogous to breakthrough curves, and the total recovery for a target solute can be obtained by integrating the area under these curves. Computer simulations are routinely used in this work to optimize and calculate processing times for a variety of product specifications. This involves design equations in terms of the number of transfer units (NTUs), which has been described in detail elsewhere for use in affinity chromatography [48, 49]. The models used to approximate and model protein breakthrough curves are also described elsewhere [35, 36].

11.4.4.1   General Description of NTUs

In immunoaffinity chromatography, there is usually strong binding between the immobilized antibodies and target (i.e., K_d is in the range of 10^{-8} to 10^{-10} M), and the kinetics of antigen-antibody interactions are often considered to be very rapid in comparison with diffusion steps. This gives rise to a system in which the binding of an antigen to an immobilized antibody can be considered irreversible during the loading step. This irreversible case can be represented in the general expression X = function (N,T), where X is the dimensionless concentration, N is the number of transfer units, and T is the dimensionless time or throughput. A brief description of the various values of N that are encountered in immunoaffinity chromatography is given in Table 11.1.

Table 11.1 Definitions for Theoretical Number of Transfer Units (N)

Term	Process	Description
N_d	Axial dispersion	$\dfrac{Pe_p L}{d_p}$
N_f	Film mass transfer	$\left(\dfrac{k_f a_p}{u_0}\right)L$
N_{pore}	Pore diffusion	$\left(\dfrac{k_{pore} a_p}{u_0}\right)L$
N_p	Solid homogeneous particle diffusion	$\psi_p \Gamma\left(\dfrac{k_p a_p}{u_0}\right)L$

Note: a_p = external surface area of sorbent particles per unit packed volume (cm^{-1}); Pe_p = packing Peclet number (≈ 0.5 for laminar flow); L = bed length (cm); d_p = particle diameter (cm); k or k_f = film mass-transfer coefficient (cm/sec); u_0 = superficial linear velocity (cm/sec); ψ = correction factor (0.59 for $R = 0$); Γ = distribution parameter, $= \rho_b(^*q_{feed})/c_{feed}$ (dimensionless); c_{feed} = concentration of feed solution (g/cm^3); q_{feed} = sorbate concentration in equilibrium with c_{feed} (g/g particles); ρ_b = particle bulk density $= (1-\varepsilon)\rho_p$ (g/cm^3 bed volume).

The expression that can be used to account for both film- and pore-diffusion limitations in a dimensionless breakthrough curve is given by Equation 11.16 [46],

$$T - 1 = \left(\frac{1}{N_{pore}} + \frac{1}{N_p}\right)\left\{\phi(X) + \frac{N_{pore}}{N_f}(\ln X + 1)\right\}\left(\frac{N_{pore}}{N_f} + 1\right)^{-1} \qquad (11.16)$$

where N_{pore} is the number of transfer units due to pore diffusion and N_f is the number of transfer units due to film diffusion. In Equation 11.16, the function $N(X)$ can be obtained through the use of the formula shown below,

$$\phi(X) \approx 2.44 - 3.66(1 - X)^{1/2} \qquad (11.17)$$

where X is the same as the term C/C_0 (the dimensionless concentration for the solute exiting the column), and T represents dimensionless time. This dimensionless time parameter is defined as follows,

$$T = \frac{V - \varepsilon v}{\Gamma v} \qquad (11.18)$$

where Γ is the partition coefficient for the separation process, as given in Equation 11.19, where ρ_p is the particle density.

$$\Gamma = \frac{\rho_p q_0}{c_0} \qquad (11.19)$$

For the case where pore diffusion alone is the rate-determining step in solute retention (i.e., $N_{pore}/N_f \approx 0$), Equation 11.17 simplifies to the form given below.

$$X = 1 - \left\{ \frac{2}{3} - 0.273 N_{pore}(T-1) \right\}^2 \qquad (11.20)$$

The breakthrough curve for solid homogeneous diffusion coupled with film mass transfer can be expressed as shown below.

$$N_f(T-1) = -mN_p(T-1) = \left\{ \begin{array}{l} \ln\dfrac{X}{\beta} + 1 + m;\ 0 \leq X \leq \beta \\[2mm] m\ln\dfrac{(1-X)}{(1-\beta)} + 1 + m;\ \beta \leq X \leq 1 \end{array} \right\} \qquad (11.21)$$

In Equation 11.21, the terms m and β are dimensionless groups given by the following relationships:

$$m = -N_f/N_p \qquad (11.22)$$

$$\beta = 1/(1 + N_f/N_p) \qquad (11.23)$$

For the case where homogeneous diffusion (or the quadratic driving force) in the solid phase is rate-limiting, the design equation takes the form shown in Equation 11.24.

$$N_p(T-1) = -1.69[\ln(1-X^2) + 0.61] \qquad (11.24)$$

This allows experimental breakthrough curves (plotted now as X vs. T) to be modeled with Equations 11.20, 11.21, and 11.24, providing the value of N_{pore} (or other NTU) that gives the best fit. This value can then, in turn, be used in the design and scale-up calculations for the separation method.

11.4.4.2   Use of NTUs in Characterizing Separations

NTUs can play an important role in characterizing an affinity purification process. For instance, the column capacity is the area behind the entire breakthrough curve, which is unity for the dimensionless curves written in terms of $X(C/C_0)$, T, and N. For a breakthrough function given by Equation 20, the column utilization ($Util_{col}$) can then be determined for the case where $T_f \geq 1 - (1.22/N_{pore})$ by using the following relationship.

$$Util_{col} = 100 - 36.1/N_{pore} + 44.4(T_f - 1) - 18N_{pore}(T_f - 1)^2 + 2.49N_{pore}^2(T_f - 1)^3 \qquad (11.25)$$

In the alternative case where $T_f \leq 1 - (1.22/N_{pore})$, the column utilization is instead determined by using Equation 11.26.

$$Util_{col} = 100 \times T_f \qquad (11.26)$$

Examples that illustrate the use of these and other equations for method evaluation are given in the next section.

11.5 CALCULATIONS FOR PROCESS EVALUATION

This section shows how the equations given earlier can be used to characterize a specific affinity purification method. In such studies, assumptions are usually made regarding the nature of the limiting mass-transfer kinetics and mechanism to further enable the approximation and modeling of the experimentally obtained breakthrough profiles. As an alternative, pulse analysis can be used as a method for evaluating and optimizing column performance. For example, pulse injection techniques have recently been used with classic plate-height equations to model and study the transport parameters in commercially available matrixes [47, 49]. The results obtained from these simulations can then be used to devise strategies for the design and optimization of the adsorptive, washing, and elution processes.

The use of design equations from Section 11.4 can be illustrated through an example involving immunoaffinity chromatography. For instance, consider a separation process that had a packing material (e.g., Emphaze) with a particle diameter (d_p) of 50 μm and a porosity (ε) of 0.45. To this support, monoclonal antibodies against human protein C (8861-Mab) were then immobilized, with the resulting immunosorbent being packed in a 1.0-cm I.D. × 10-cm column. A solution of human protein C (the antigen) was then introduced into the column and used as the feed.

The breakthrough curve that was obtained experimentally for this system is shown in Figure 11.8. Figure 11.8 also shows the predicted response that was obtained when using Equation 11.22. This fit gave a value of 8 for N_{pore} in this system and an area behind the breakthrough curve (T) of 8.0. This information was then used to further characterize this method.

11.5.1 Time Required for an Adsorption Process

The first calculation considered here is the time required to process a certain amount of feed. For instance, suppose that the column being used has a diameter of 80 cm, a length of 15 cm, and is operated at a superficial liquid velocity of 0.01 cm/sec. Also, suppose that the adsorption step is terminated when the effluent concentration reaches 10% of the feed concentration.

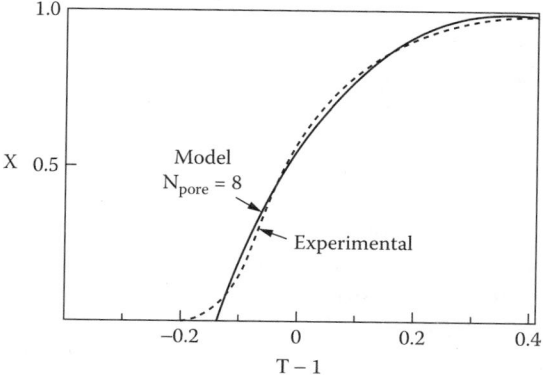

Figure 11.8 Breakthrough curve obtained on a 1.0-cm I.D. × 10.0-cm Affiprep column containing antihuman protein C monoclonal antibodies.

To determine the time required for this, it is possible to use the following values for superficial liquid velocity (u_0) and column cross section (A).

$$u_0 = 0.01 \text{ cm/sec}$$

$$A = \pi(80 \text{ cm})^2/4 = 5030 \text{ cm}^2$$

From these values, the volumetric flow rate (Q) is found as follows.

$$Q = u \times A = 181 \text{ l/h}$$

It is also known already for this system that $N_{pore} = 8.0$ (as determined by fitting Equation 11.26 to the breakthrough curve) and $X = 0.1$ (the defined point at which the operation is finished).

With this information, Equation 11.22 can be written as follows:

$$N_{pore}(T - 1) = 2.44 - 3.66(1 - X)^{1/2}$$

$$T_f = [2.44 - 3.66(0.9)^{1/2}]/8 + 1$$

$$= 0.871$$

The volume of feed processed during this adsorption step is then obtained from the definition of T provided by Equation 11.18,

$$T = \frac{V - \varepsilon v}{\Gamma v}$$

which can be rearranged to give the result shown below.

$$V - \varepsilon v = [0.871 \times 8.0 \times (15 \times 5030)]/1000$$

$$= 526 \text{ l}$$

Thus, the time required to complete this operation will be (526 l)/(181 l/h) = 2.90 h.

11.5.2 Utilization of Column Capacity

Another factor that can be determined for a chromatographic system is the degree to which the column capacity is being utilized during a separation ($Util_{col}$). For the case given in the previous subsection, this can be solved in the following manner. The term T_f is equal to 0.871 in this example, so $T_f > [1 - (1.22/N_{pore})]$. This means that Equation 11.25 should be used to give the value of column utilization. When the corresponding values for T_f and N_{pore} are placed into this formula, this gives a value for $Util_{col}$ of 86.99%.

11.5.3 Effect of Flow Rate on Column Performance

Another parameter that can be determined for a chromatographic purification system is the impact of the flow rate or the linear velocity on the column utilization and total process time. For example, suppose that the flow rate in the previous example is tripled, giving a new best-fit value for N_{pore} of 3. The volumetric flow rate (Q) would now be 543 l/h (i.e., three times its previous value). Equation 11.22 can then be rearranged to enable the calculation of T_f, as shown below.

$$T_f = [2.44 - 3.66(0.9)^{1/2}]/3 + 1$$

$$= 0.659$$

The volume of feed processed during this adsorption step can be obtained by rearranging Equation 11.18, as shown earlier, which gives the following result.

$$V - \varepsilon v = [0.659 \times 8.0 \times (15 \times 5030)]/1000$$
$$= 395 \text{ l}$$

This means that the time required to complete this operation will be (395 l)/(543 l/h) = 0.727 h. Since the value for T_f (i.e., 0.659) is greater than the term $[1 - (1.22/N_{pore})]$, Equation 11.25 is next used to give the column utilization ($Util_{col}$). By placing the appropriate values for N_{pore} and T_f into this equation, it is found that the column utilization is now 65.0% after the flow rate has been tripled.

11.5.4 Effects of Particle Size on Method Performance

It is also possible to consider how different diameters for the support particle will affect the separation. Larger particles will lead to lower pressure drops across a column, making them useful in large-scale processes, but such a change might also affect other factors in the separation. For instance, suppose that 50-μm support particles were replaced with 100-μm particles having an equivalent pore size and equilibrium capacity. Based on this information, it is possible to estimate what the new value for N_{pore} will be, thus allowing other system parameters to also be estimated.

First, the new value for N_{pore} can be determined by using the general equation N_{pore} = constant/d_p^2 between this term and the diameter of the support particles (d_p). From this, the change from the old value of 8.0 for N_{pore} to its new value can be found by using a simple proportionality expression.

$$(N_{pore})_{new} = (N_{pore})_{old} \times [(50 \text{ μm})^2/(100 \text{ μm})^2]$$
$$= 2.0$$

This factor can then be used with Equation 11.18 to give the new value for T_f, as shown below.

$$T_f = [2.44 - 3.66 \, (0.9)^{1/2}]/2.0 + 1$$
$$= 0.484$$

The volume of feed processed during this adsorption step can then be obtained by using the definition for T that is provided by Equation 11.18.

$$V - \varepsilon v = [0.484 \times 8.0 \times (15 \times 5030)]/1000$$
$$= 292 \text{ l}$$

This gives a new time for completing this operation of (526 l/h)/(292 l) = 1.60 h. Since the value for T_f (i.e., 0.659, as given previously for T_f in Section 11.5.3) is greater than the term $[1 - (1.22/N_{pore})]$, use of Equation 11.25 is next used to give the new column utilization ($Util_{col}$). By placing the appropriate values for N_{pore} and T_f into this equation, the percent column utilization is now estimated to be 59%.

11.6 SUMMARY AND CONCLUSIONS

The continued advances in sorbent materials, techniques for ligand immobilization, and activation chemistries should result in great improvements in downstream processing and preparative affinity chromatography. This, in turn, could give rise to huge advances in scaleable performance when process modeling is used to characterize and scale up such methods.

Predicting the behavior of affinity systems requires information on both the binding parameters and binding constants for these systems. For most proteins, the use of an appropriate adsorption isotherm to analyze solute interactions with a sorbent can usually be used to identify these items. In general, data obtained during the sample application step are more amenable to process modeling than data obtained during the washing and elution steps. Nevertheless, such information is quite valuable in both modeling and optimizing the performance of affinity purification techniques.

SYMBOLS AND ABBREVIATIONS

A	Column cross section
a_p	External surface area of sorbent particles per unit packed volume
AR	Adsorption rate
C	Concentration of liquid solute in the mobile phase or in the bulk
C_{BT}	Breakthrough concentration
C_0	Inlet or feed concentration of the adsorbate or solute
d_p	Particle diameter
D_i	Effective particle diffusion coefficient
DNA	Deoxyribonucleic acid
D_x	Axial dispersion coefficient
EDTA	Ethylenediaminetetraacetic acid
k or k_{film}	Film mass-transfer coefficient
k_1	Forward reaction rate constant
k_2	Backward reaction rate constant
k_a	Association rate constant
K_d	Dissociation equilibrium constant
L	Unoccupied immobilized ligand or matrix
L or l	Length of the column
N or **NTUs**	Number of transfer units
N_d	Number of transfer units due to axial dispersion
N_f	Number of transfer units due to film diffusion
N_p	Number of transfer units due to solid homogeneous particle diffusion
N_{pore}	Number of transfer units due to pore diffusion
P	Unbound adsorbate (P) in the solution
Pe_p	Packing Peclet number
R_i	Rate of interface mass transfer
RNA	Ribonucleic acid
q	Solid phase concentration of a solute under given conditions
Q	Volumetric flow rate
q_0 or Q_{max}	Maximum solid phase concentration of an adsorbed solute
R	Particle radius
t	Time
T_f	Dimensionless throughput parameter
u_0	Superficial liquid velocity
Util$_{col}$	Column utilization, percentage of initial capacity
v	Bed or total batch volume
V	Throughput volume

X	C/C_0, dimensionless bulk concentration
α	Saturation capacity (Q_{max}) in the Jovanovic isotherm
β	Association constant in the Jovanovic isotherm
ϵ	Column void fraction
ρ_p	Particle density
ψ	Correction factor

REFERENCES

1. Bonnerjea, J., Oh, S., Hoare, M., and Dunnhill, P., Protein purification: The right step at the right time, *Biotechnology,* 4, 954–958, 1986.
2. Labrou, N. and Clonis, Y., The affinity technology in downstream processing, *J. Biotechnol.,* 36, 95–119, 1994.
3. Ransohoff, T.C., Murphy, M.K., and Levine, H.L., Automation of biopharmaceutical purification processes, *BioPharm,* 3, 20–26, 1990.
4. Velander, W.H., Clark, D.B., Gee, D., Drohan, W.N., and Morcol, T., Technological challenges for large-scale purification of protein C, in *Protein C and Related Anticoagulants,* Bruley, D.F. and Drohan, W.N., Eds., Portfolio Publishing, The Woodlands, TX, 1990, pp. 11–27.
5. Paleyanda, R., Young, J., Velander, W., and Drohan, W., The expression of therapeutic proteins in transgenic animals, in *Recombinant Technology in Hemostasis and Thrombosis,* Hoyer, L.W. and Drohan, W.N., Eds., Plenum Press, New York, 1991, pp. 197–209.
6. Grinnell, B.W., Walls, J.D., Gerlitz, B., Berg, D.T., McClure, D.B., Ehrlich, H., Bang, N.U., and Yan, S.B., Native and modified recombinant human protein C: Function, secretion, and posttranslational modifications, in *Protein C and Related Anticoagulants,* Bruley, D.F. and Drohan, W.N., Eds., Gulf Publishing, Houston, 1990, pp. 29–63.
7. Velander, W.H., Johnson, J.L., Page, R.L., Russell, C.G., Subramanian, A., Wilkins, T.D., Gwazdauskas, F.C., Pittius, C., and Drohan, W.N., High level expression in the milk of transgenic swine using the cDNA encoding human protein C, *Proc. Natl. Acad. Sci. U.S.A.,* 89, 12003–12007, 1992.
8. Landon, J. and Akman, S., *Clinical Applications of Monoclonal Antibodies,* Plenum Press, New York, 1988.
9. Woodhead, J.S. and Aston, J.P., *Clinical Applications of Monoclonal Antibodies,* Plenum Press, New York, 1988.
10. Goding, J.W., Use of staphylococcal protein A as an immunological reagent, *J. Immunol. Methods,* 20, 241–253, 1978.
11. Abbas, A.K., Lichtman, A.H., and Jordan, S.P., *Cellular and Molecular Immunology,* Saunders, Philadelphia, 1997.
12. Johnstone, A.P. and Thrope, R., *Immunochemistry in Practice,* Blackwell, Oxford, U.K., 1996.
13. Harris, W.J., *Animal Cell Technology: Basics and Applied Aspects,* Kluwer, Dordrecht, Netherlands, 1994.
14. Greenbaum, B., Differences in immunoglobulin preparations for intravenous use: A comparison of six products, *Am. J. Pediatr. Hematol. Oncol.,* 12, 490–496, 1990.
15. Velander, W.H., Orthner, C.L., Tharakan, J.P., Madurawe, R.D., Ralston, A.H., Strickland, D.K., and Drohan, W.N., Process implications for metal-dependent immunoaffinity interactions, *Biotech. Prog.,* 5, 119–125, 1989.
16. Quinley, E.D., *Immunohematology: Principles and Practice,* Lippincott, Philadelphia, 1993, pp. 247–269.
17. Smith, T.F., *Transfusion-Transmitted Viruses: Epidemiology and Pathology,* in Insalca, S.J. and Menitove, J.E., Eds., American Association of Blood Banks, Bethesda, MD, 1987, pp. 11–18.

18. Diringer, H.H., Gelderblom, H., Hilmert, H., Ozel, M., Edelbluth, C., and Kimberlin, R.H., Scrapie infectivity, fibrils and low molecular weight protein, *Nature*, 306, 476–478, 1987.

19. Aronson, D.L. and Menache, D., Prevention of infectious disease transmission by blood and blood products, *Prog. Hematology*, 15, 221–223, 1987.

20. Brettler, D.B., Forsberg, A.D., Levine, P.H., Aledort, L.M., Hilgartner, M.W., Kasper, C.K., Lusher, J.M., McMillan, C., and Roberts, H., The use of porcine factor VIII concentrate in the treatment of patients with inhibitor antibodies to factor VIII, *Arch. Int. Med.*, 149, 1381–1392, 1989.

21. Velander, W.H., Subramanian, A., Madurawe, R.D., and Orthner, C.L., The use of Fab-masking antigens to enhance the activity of immobilized antibodies, *Biotech. Bioeng.*, 39, 1013–1023, 1992.

22. Subramanian, A., Van Cott, K.E., Milbrath, D.S., and Velander, W.H., Role of local antibody density effects on immunosorbent efficiency, *J. Chromatogr.*, 672, 11–24, 1994.

23. Orthner, C.L., Highsmith, F.A., Tharakan, J., Madurawe, R.D., Morcol, T., and Velander, W.H., Comparison of the performance of immunosorbents prepared by site-directed or random coupling of monoclonal antibodies, *J. Chromatogr.*, 558, 55–70, 1991.

24. Subramanian, A. and Velander, W.H., Ranking the factors impacting immunosorbent performance, *Intl. J. BioChromatogr.*, 5, 67–82, 2000.

25. Kaster, J.K., Oliveira, W., Glasser, W.G., and Velander, W.H., Optimization of pressure-flow limits, strength, intraparticle transport and dynamic capacity by hydrogel solids content and bead size in cellulose immunosorbents, *J. Chromatogr.*, 648, 79–90, 1993.

26. Subramanian, A. and Hommerding, J., The use of confocal laser scanning microscopy to study the transport of biomacromolecules in a macroporous support, *J. Chromatogr. B*, 819, 89, 2005.

27. Velander, W.H., Vancott, K.E., and Van Tasell, R.E., Inside-Out Crosslinked and Commercial Hydrogels, and Submacromolecular Selective Purification Using Hydrogels, U.S. patent 5,977,345, 1999.

28. Horstmann, B.J., Kenny, C.N., and Chase, H.A., Adsorption of proteins on Sepharose affinity adsorbents of varying particle size, *J. Chromatogr.*, 361, 179–190, 1986.

29. Leonard, M., New packing materials for protein chromatography, *J. Chromatogr.*, 699, 3–12, 1997.

30. Boschetti, E., Advanced sorbents for preparative protein separation purposes, *J. Chromatogr.*, 658, 207–213, 1994.

31. Narayan, S. and Crane, L., Affinity chromatography supports: A look at performance requirements, *Trends Biotechnol.*, 8, 12–16, 1990.

32. Nawrocki, J., Dunlap, C.J., Carr, P.W., and Blackwell, J.A., New materials for biotechnology: Chromatographic stationary phases based on zirconia, *Biotechnol. Prog.*, 10, 561–570, 1994.

33. Subramanian, A. and Sarkar, S., Use of a modified zirconia support in the preparation of immunoproteins, *J. Chromatogr.*, 944, 179–187, 2001.

34. Guiochon, G., Golshan Shirazi S., and Katti, A.M., *Fundamentals of Preparative and Nonlinear Chromatography*, Academic Press, New York, 1994.

35. Chase, H.A., Prediction of the performance of preparative affinity chromatography, *J. Chromatogr.*, 297, 179–185, 1984.

36. Arnold, F.H., Blanch, H.W., and Wilke, C.R., Analysis of affinity separations, I: Predicting the performance of affinity adsorbers, *Chem. Eng. J.*, 30, B9–B23, 1985.

37. Lenhoff, A.M., Significance and estimation of chromatographic parameters, *J. Chromatogr.*, 384, 285–299, 1987.

38. Belot, J.C. and Condoret, J.S., Modeling of liquid chromatography equilibria, *Proc. Biochem.*, 28, 365–372, 1993.

39. Langmuir, I., Constitution and fundamental properties of solids and liquids, *J. Am. Chem. Soc.*, 38, 2221–2295, 1916.

40. Horstmann, B.J. and Chase, H.A., Modeling the affinity adsorption of immunoglobulin G to protein A immobilized to agarose matrices, *Chem. Eng. Res. Des.*, 67, 243–254, 1989.

41. El-Kak, A., Manjini, S., and Vijayalakshmi, M., Interaction of immunoglobulin G with immobilized histidine: Mechanistic and kinetic aspects, *J. Chromatogr.,* 604, 29–37, 1992.

42. Nash, D.C. and Chase, H.A., Comparison of diffusion and diffusion-convection matrices for use in ion-exchange separations of proteins, *J. Chromatogr.,* 807, 185–207, 1998.

43. Carta, G., Exact analytical solution of a mathematical model for chromatographic operations, *Chem. Eng. Sci.,* 43, 2877–2883, 1988.

44. Rodrigues, A.E., Ramos, A.M.D., Loureiro, J.M., Diaz, M., and Lu, Z.P., Influence of adsorption-desorption kinetics on the performance of chromatographic processes using large-pore supports, *Chem. Eng. Sci.,* 47, 4405–4413, 1992.

45. Thomas, H.C., Heterogeneous ion exchange in a flowing system, *J. Am. Chem. Soc.,* 66, 1664–1671, 1994.

46. Arnold, F.H., Chalmers, J.J., Saunders, M.S., Croughan, M.S., Blanch, H.W., and Wilke, C.R., A rational approach to the scale-up of affinity chromatography, *Purif. Ferm. Prod.,* 271, 113–122, 1985.

47. Arnold, F.H. and Blanch, H.W., Analytical affinity chromatography, II: Rate theory and the measurement of biological binding kinetics, *J. Chromatogr.,* 355, 13–27, 1986.

48. Natarajan, V. and Cramer, S., A methodology for the characterization of ion-exchange resins, *Sep. Sci. Tech.,* 35, 1719–1742, 2000.

49. Sarkar, S. and Subramanian, A., Identification of the mass transfer mechanisms involved in the transport of human immunoglobulin-G in EDTPA modified zirconia, *J. Chromatogr.,* in press.

12

Affinity Chromatography of Enzymes

Felix Friedberg and Allen R. Rhoads

Department of Biochemistry and Molecular Biology,
Howard University, Washington, DC

CONTENTS

12.1   INTRODUCTION

The purification of enzymes can exploit either their general protein characteristics (e.g., when using a physical separation or a method based on relatively nonspecific interactions) or their specific biological interactions (e.g., as employed in bioselective separations). In practice, these two recognition modes are often used sequentially. This is the case because attempts at bioselective separations at an early stage in the purification process may encounter nonspecific binding, while physical separations or those based on nonspecific interactions are usually inadequate to distinguish between proteins that have similar physical properties.

Affinity separations for enzymes frequently employ agarose-based supports such as phenyl-Sepharose and heparin-Sepharose (i.e., materials that will be emphasized in this chapter). But other support materials besides agarose (as discussed in Chapter 2) can be used for this purpose. In addition, various types of affinity chromatography can be used in enzyme separations. For instance, immobilized metal-ion affinity chromatography (Chapter 10) or immunoaffinity chromatography (Chapter 6) can be utilized. Expression cloning and tagging has provided yet another tool for the isolation of enzymes, as is accomplished by inserting an affinity tag at the terminus of the desired protein (e.g., polyarginine for use in cation-exchange chromatography or polyphenylalanine for use in hydrophobic-interaction chromatography).

Enzymes can be purified by affinity chromatography through the use of either direct or subtractive methods. Affinity ligands used in the direct isolation of enzymes can be substrates, inhibitors, cofactors, or proteins associated with the biochemical pathways of the enzymes. Many examples of these ligands and methods are given in this chapter. Subtractive methods (e.g., those that use affinity columns to remove some proteins from a sample while allowing the enzyme of interest to pass through) can also be used in such separations.

12.2 STRUCTURE-BASED AFFINITY SEPARATIONS

The first group of affinity-based methods for enzymes are those in which the enzymes are isolated based on their protein structure. Several different types of ligands can be used for this purpose. The specific examples discussed here include Cibacron Blue F3GA, Procion Red HE3B, heparin, lectins, thiopropyl, and L-arginine or L-lysine. The use of hydrophobic-interaction chromatography and immobilized metal-ion affinity chromatography for enzyme purification is also discussed in this section.

12.2.1   Cibacron Blue F3GA
Cibacron Blue F3GA (also known as Procion Blue HB) is a monochlorotriazinyl dye that is often used in protein and enzyme purification. A detailed discussion of this and related dyes is provided in Chapter 9. This dye can be obtained in an immobilized form attached to Sepharose CL-6B from Amersham Pharmacia Biotech (see Figure 12.1). Other suppliers of such supports are given in Chapter 9. This dye has frequently been employed for the purification of dehydrogenases and finds continued use for this purpose [1].

When it is covalently coupled to a matrix, Cibacron Blue F3GA not only binds enzymes that interact with nucleotides, but it also exhibits a general affinity for enzymes that contain a nonpolar binding pocket next to a positive charge. Chlorotriazine rings are very reactive and can readily be coupled to a support for use in affinity chromatography. Examples of procedures for attaching Cibacron Blue F3GA and other triazine dyes to solid supports can be found in Baird et al. [2] or Dean and Watson [3], as well as in Chapter 9. These immobilized dyes can serve as substrates, competitive inhibitors, coenzymes, or effector analogs for a variety of enzymes. This is probably due to the fact that these dyes can assume the required geometry or polarity needed for these interactions.

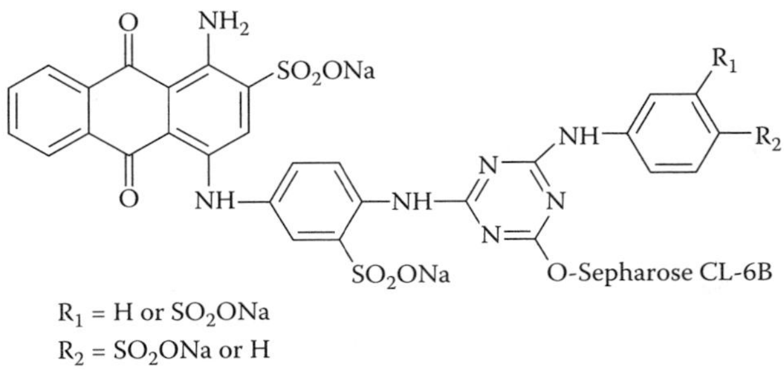

R_1 = H or SO$_2$ONa
R_2 = SO$_2$ONa or H

Figure 12.1 Partial structure of Blue Sepharose CL-6B.

Thus, the specificity of these dyes is broad, and nonspecific interactions can contribute significantly to their total binding energy.

Not only are nucleotide-requiring enzymes (e.g., dehydrogenases [4, 5], kinases [6, 7], and ligases [8]) bound to the Cibacron Blue F3GA columns, but other enzymes like organophosphate phosphatase [9] and proteins like albumin, α_2-macroglobulin, or interferon are retained. Another feature of Cibacron Blue F3GA and other immobilized triazine dyes is their high binding capacity compared with more specific nucleotide-affinity adsorbents [4]. To give specificity to these triazine dye supports, it is essential to utilize selective elution with a substrate or inhibitor. It should also be noted that the affinity of an enzyme for such a dye is considerably lessened in the presence of buffers that have an ionic strength in excess of 0.2 M, as is discussed in Chapter 9.

12.2.2 Procion Red HE3B

While Cibacron Blue F3GA is useful for isolating NAD^+-linked enzymes, Procion Red HE3B, another triazine dye, is often preferred for the purification of $NADP^+$-requiring dehydrogenases. When this dye is attached to Sepharose CL-6B, as much as 70 nmol of immobilized ligand can be attached per milliliter of settled resin [4]. Alcohol dehydrogenase from *Zymomonas mobilis* has been purified by using tandem columns containing Procion Green-HE4BD and Procion HB Blue-Sepharose CL-4B, with this enzyme being eluted with a buffer containing 1 mM NAD^+. This procedure resulted in a 36-fold purification of the enzyme and an 80% yield [5].

12.2.3 Heparin

Heparin (shown in Figure 12.2) is a ligand that has been of general value in the purification of restriction endonucleases and lipases. One support that contains this ligand is heparin-Sepharose CL-6B from Amersham Pharmacia Biotech. Because of its polyanionic character, heparin interacts with many cationic molecules, possibly competing with nucleic acid substrates. Whereas the ability of heparin to bind lipases is not well understood, its ability to release specific lipases from tissues and to stabilize their activities is well known. A method for preparing a heparin-agarose support can be found in Davison et al. [10]. Some specific applications for heparin-Sepharose are summarized in Table 12.1.

12.2.4 Lectins

Lectins are another group of ligands that have been used in enzyme purification. As discussed in Chapter 5, these ligands have the ability to react reversibly with specific sugar residues. Immobilized lectins are particularly useful for the isolation of extracellular components, detergent-solubilized cell membranes, subcellular particles, and glycoproteins (e.g., mucins).

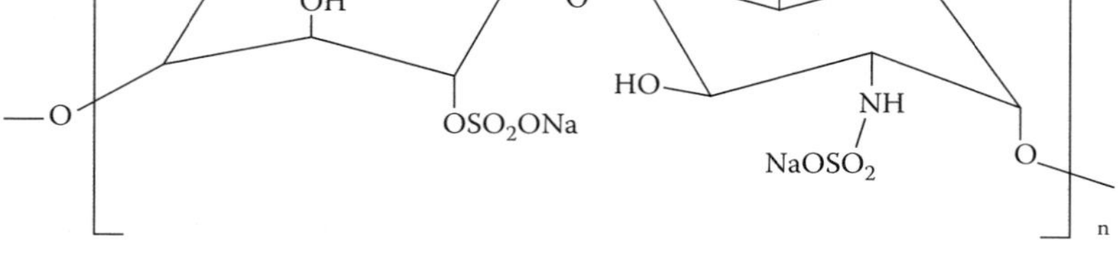

Figure 12.2 Partial structure of heparin.

Table 12.1 Enzymes Displaying Affinity for Immobilized Heparin

Enzyme	Elution Conditions	Degree of Purification[a]	References
Hamster liver lipase	0.05 M phosphate (pH 7.2), 0.2 M NaCl, 20% glycerol increased to 2 M NaCl	Not reported	[11]
Human plasma lipoprotein lipase	5 mM sodium barbital, 1.5 M NaCl	1500-fold	[12]
Yeast mitochondrial nuclease (has RNase, DNA endonuclease, and 5′-exonuclease activity)	0.25 to 0.6 M KCl in 0.01 M K_2PO_4-K_2HPO_4 (pH 7.4), 5 mM $MgCl_2$, 1 mM DTT, 15% glycerol, and 0.1% Nonidet P-40	20-fold	[13]
Yeast DNA exonuclease V	0.03 to 0.6 M $(NH_4)_2SO_4$ in 0.025 M K_2HPO_4-KH_2PO_4 (pH 7.0), 2 mM EDTA and 5 mM DTT, 2 mM benzamidine, 2 M pepstatin A, and 10% glycerol	7.5-fold	[14]
Bacteriophage N4-induced DNA polymerase (also has $3′ \rightarrow 5′$ exonuclease activity)	0.04 to 0.5 M NaCl in 0.1 M Tris-HCl (pH 8.0), 1 mM EDTA, 0.02 M 2-mercaptoethanol, and 10% glycerol	2-fold	[15]
Porcine liver mitochondrial DNA polymerase	0 to 0.8 M NaCl in 0.015 M Tris-HCl (pH 8.0), 5 mM 2-mercaptoethanol, 1 mM EDTA, and 10% glycerol	40-fold	[16]
Sulfolobus acidocaldarius DNA polymerase	0.2 to 1 M NH_4Cl in 0.05 M Tris (pH 7.5), 1 mM DTT (0.5 mM EGTA)	3-fold	[17]
Human lymphoblast DNA polymerase-α DNA primase	0.06 to 0.03 M K_2HPO_4-KH_2PO_4 (pH 7.8)	3-fold	[18]
Human liver ribonuclease	0.2 to 1.0 M NaCl in 0.02 M MES (pH 6.0), 0.2 M NaCl	40-fold	[19]
A431 cells phosphatidylinositol kinase	5 mM HEPES (pH 7.2), 0.25% Triton X-100, 0.05 M NaCl, 2 mM phosphatidyl inositol	11-fold	[20]
Yeast Cl tetrahydrofolate synthase (has 10-CHO-THF synthetase, 5,10-CH^+-THF cyclohydrolase, and 5,10-CH_2-THF dehydrogenase activity)	0.14 M KCl	66-fold	[21]
Bacterial 10-formyltetrahydrofolate synthetase	0.25 M KCl	19-fold	[21]
Tissue plasminogen activator	2% SDS, 5% glycerol in phosphate-buffered saline	Not reported	[22]

[a] A low recovery in the last stages of enzyme purification can frequently be very significant.

Wheat germ agglutinin (WGA) coupled to Sepharose 4B by the cyanogen bromide method is commercially available as Agarose Wheat Germ Lectin from Amersham Pharmacia Biotech. This lectin binds to terminal *N*-acetylglucosaminyl residues and glycoproteins that possess such residues. Similarly, concanavalin A (Con A) bound to agarose shows specificity for molecules that contain α-D-mannopyranosyl or α-D-glucopyranosyl residues. For this latter reaction, the binding sugar must have hydroxyl groups at positions C-3, C-4, and C-5 [23]. Con A-Sepharose (available from Amersham Pharmacia Biotech) can be prepared by coupling Con A to cyanogen bromide-activated Sepharose 4B.

Isoenzymes of adenosine deaminase have been found to interact differently with WGA and Con A affinity columns, suggesting that the various forms of this enzyme differ in the composition and accessibility of their sugar residues [24]. Because many extracellular carboxypeptidases (such as the N-type) contain covalently bound carbohydrate groups, they can readily be purified by a Con A-Sepharose column and eluted with a mobile phase containing 0.1 *M* α-methyl-D-mannoside [25]. Angiotensin-converting enzyme (ACE) has also been purified by a Con A-Sepharose column [26].

12.2.5 Thiopropyl Supports

A thiopropyl group is an additional type of ligand that can be used in enzyme purification. Thiopropyl-Sepharose 6B from Amersham Pharmacia Biotech has a short hydroxypropyl spacer arm and is an important tool for *covalent chromatography*. Ligands containing reactive thiol groups (cysteine residues) can be reversibly immobilized to such a matrix under relatively mild conditions. For instance, an acyl carrier protein (ACP)-Sepharose support was employed to confirm that the 2-acylglycerol phosphate ethanolamine acyl transferase and acyl ACP synthase activities of *Escherichia coli* are expressed by one enzyme and that this enzyme binds to ACP [27]. This was based on ACP's function as a carrier in fatty acid biosynthesis, in which it forms a thioester linkage between an acyl moiety and its 4'-phosphopantetheine prosthetic group.

ACP-Sepharose can be produced by first reducing ACP with dithiothreitol (DTT) and then treating this with 2,2'-dithiodipyridine to yield the 2-thiopyridine derivative of ACP. Next, activated thiopropyl-Sepharose 6B (shown in Figure 12.3) is reduced with dithiothreitol and washed to remove the 2-thiopyridine groups. This support and the ACP are then mixed in the presence of 0.2 *M* NaCl [28]. In the work described earlier, a crude enzyme preparation was applied to this ACP-Sepharose support, and the acyltransferase/synthetase was eluted as a single peak by using 0.5 *M* LiCl as the mobile phase, which significantly decreases the activity of the enzyme for ACP [27].

Another use of thiopropyl columns is in work with enzymes that have been modified through recombinant DNA techniques to contain specific proteins or polyamino acids that act as *affinity tags* or *affinity tails*. Often the selected tag or tail is a short homopolymer. For example, a synthetic DNA fragment encoding four additional cysteine residues was inserted at the 5' end of the *E. coli* galactokinase gene, *gal K*. After prepurification, the resulting recombinant enzyme was incubated with thiopropyl-Sepharose 6B and eluted with 10 m*M* dithiothreitol. Such treatment enhanced the specific activity of the enzyme

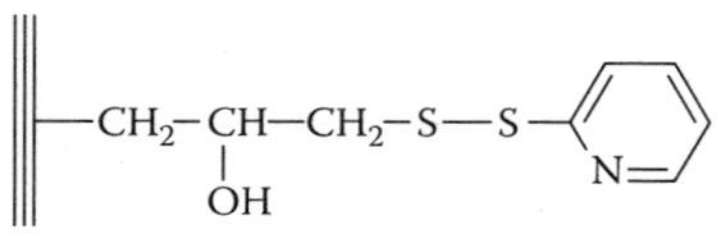

Figure 12.3 Partial structure of activated thiopropyl-Sepharose 6B.

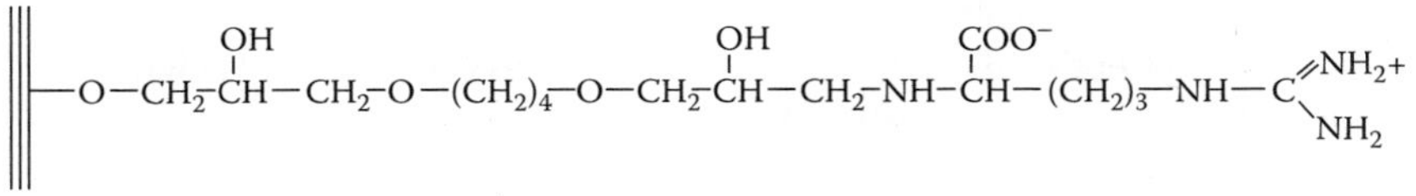

$$\underset{\text{\Large\textbar\textbar\textbar}}{\text{}} -\text{O}-\text{CH}_2\overset{\underset{\displaystyle |}{\text{OH}}}{\text{CH}}-\text{CH}_2-\text{O}-(\text{CH}_2)_4-\text{O}-\text{CH}_2\overset{\underset{\displaystyle |}{\text{OH}}}{\text{CH}}-\text{CH}_2-\text{NH}-\overset{\underset{\displaystyle |}{\text{COO}^-}}{\text{CH}}-(\text{CH}_2)_3-\text{NH}-\text{C}\overset{\displaystyle \nearrow \text{NH}_2+}{\underset{\displaystyle \searrow \text{NH}_2}{}}$$

Figure 12.4   Structure of L-arginine-agarose.

by approximately fivefold [29]. Other examples of the purification of affinity-tagged recombinant proteins are given in Section 12.2.8 of this chapter.

12.2.6   L-Arginine and L-Lysine

L-Arginine and L-lysine can be attached to agarose (as illustrated in Figure 12.4) and used as amphoteric derivatives that can interact through both electrostatic and stereospecific forces. For instance, L-arginine-Sepharose prepared according to Porath and Fornstedt [30] has been used in the purification of B-type carboxypeptidases (i.e., enzymes that recognize C-terminal arginine or lysine residues) [31, 32]. The matrix used in this work was prepared by converting agarose to an oxirane derivative by treatment with epichlorohydrin. The oxirane-agarose was then reacted with the amine groups of arginine or lysine and gave 14 to 20 µmol of ligand per milliliter of gel.

Carboxypeptidase B was eluted from this L-arginine matrix with a mobile phase that contained 1 mM guanidylthiopropionic acid (GEMSA), an inhibitor of this enzyme (note: 10 µM GEMSA completely inhibits the carboxypeptidase M in human placental microvilli) [32]. Carboxypeptidase B in human seminal plasma was purified nearly 1000-fold with a yield of 37% by this affinity matrix [31].

A slightly modified ligand, p-aminobenzoyl L-arginine, was immobilized on epoxy-activated Sepharose and used to purify carboxypeptidase H, a B-type carboxypeptidase [33]. When this matrix was employed as a terminal step and 1 mM GEMSA was used for elution, the enzyme was purified more than 100-fold with a yield of nearly 60%. Carboxypeptidase H from bovine pituitary glands was purified over 1300-fold by the same matrix, with a yield of 69% after elution using a 0 to 800 mM gradient of arginine [34].

L-Arginine- and L-lysine-agarose have also been used in the isolation of serine proteases that are specific for peptide bonds associated with basic amino acid residues. Examples include prothrombin [35], plasminogen activator [36], and prekallikrein [37], which have all been isolated using such matrices. Plasminogen activators (PAs) are trypsin-like serine proteases that are involved in pathological processes and are of value in thrombolytic therapy. Two important plasminogen activators are urokinase and tissue-type plasminogen activator (tPA). Lysine-Sepharose was used to isolate and purify tPA from murine mast cells [38]. Elution was promoted by using ε-aminocaproic acid, an inhibitor of plasminogen [39]. p-Aminobenzamidine is another strong inhibitor of serine proteinases and, as such, might also be suited for the elution of these agents [40].

12.2.7   Hydrophobic Interaction Chromatography

Hydrophobic interaction chromatography (HIC) is an additional technique that is valuable in enzyme purification. In this method, proteins are separated based on their surface hydrophobicity, where hydrophobic proteins are most strongly retained by the HIC column. If desired, once an enzyme or other type of protein has been eluted from the HIC column it can then be subjected to other separation methods, like traditional affinity chromatography.

The association energy needed for the interactions in HIC is a function of the ordered structure of water molecules that surround nonpolar regions on the surfaces of the protein and stationary phase. A phenyl-substituted resin is a good matrix for HIC, where the phenyl ligand is intermediate in hydrophobicity between an *n*-pentyl and *n*-butyl group. The aromatic amino acid residues of an enzyme bind to this matrix through π-π interactions.

In most instances, samples are loaded onto an HIC column at high ionic strength to enhance the hydrophobic interactions between a biomolecule and the column. Retained biomolecules like proteins can then be differentially eluted by a decrease in the mobile phase's salt concentration. The more hydrophobic the biomolecule, the less salt is needed to promote its binding to the HIC column, and the more strongly the biomolecule will be retained.

The addition of chaotropic molecules or miscible organic solvents (e.g., ethylene glycol) can be utilized in HIC to disrupt the hydrogen bonding of water or to change the polarity of the mobile phase to facilitate the release of biomolecules from the column. In the case of detergent-solubilized membrane proteins, the detergent is diluted for sample application (e.g., 0.2% cholate), while elution is promoted by increasing the detergent concentration of the mobile phase (e.g., 0.4% cholate). But it should be noted when doing this that a membrane-bound enzyme may behave in a hydrophobic manner merely because it holds onto other hydrophobic constituents.

HIC can be used with both native enzymes as well as those produced through recombinant DNA techniques. One example of the latter is work in which an affinity tail of 11 phenylalanine residues was attached to the amino terminus of β-galactosidase from *E. coli* and used to increase the hydrophobic character of this enzyme. A partially purified sample of this enzyme was adsorbed onto a column containing phenyl-Sepharose CL-4B and eluted using incremental amounts of ethylene glycol. A purification of approximately 14-fold was obtained by this step [29].

12.2.8 Immobilized Metal-Ion Affinity Chromatography

Metal ions are essential constituents of some protein structures. For example, Ca^{2+} forms stabilizing complexes with carboxyl and carbonyl oxygens, Zn^{2+} coordinates imidazole nitrogens and carboxyl oxygens and bonds covalently to thiols, and Fe^{3+} interacts with phosphoamino acids. As a result, metal-chelate columns can be used to isolate proteins with any of these groups, provided that these residues are favorably exposed in the native protein or have been genetically engineered into a terminal domain (i.e., as an affinity tail or tag).

The topic of immobilized metal-ion affinity chromatography (IMAC) is discussed in detail in Chapter 10. This method was originally described in 1975 by Porath et al. [41]. It is typically performed by first attaching a metal-chelating ligand (such as the disodium salt of iminodiacetic acid) to an epoxy-activated support like amino-Sepharose 6B. An example of such a matrix is shown in Figure 12.5. Next, a gel-bound chelate of the desired metal ion is prepared by passing a solution of this ion through the column. Using this type of column, a high-purity protein preparation (i.e., up to 95% pure) can be achieved.

The specific activity of a nucleoside diphosphatase sample increased from 47.5 to 185 units/mg when purified on a Ca^{2+}-chelate column. This same enzyme preparation was subsequently purified from 185 to 310 units/mg on a Zn^{2+} chelate column [42]. The enzyme in this case was eluted with a buffer that contained histidine. A Zn^{2+} chelate column (but not a Ca^{2+} column) has also been employed for the purification of collagenase, giving a

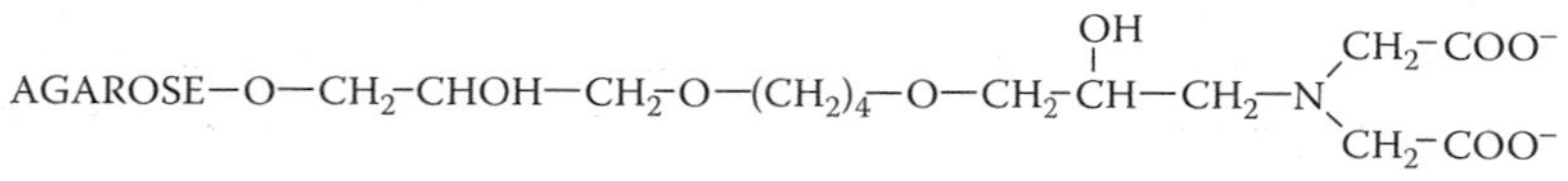

$$AGAROSE-O-CH_2-CHOH-CH_2-O-(CH_2)_4-O-CH_2-\overset{\overset{\displaystyle OH}{|}}{CH}-CH_2-N\overset{\diagup CH_2-COO^-}{\diagdown CH_2-COO^-}$$

Figure 12.5   Iminodiacetic acid derivative of oxirane-activated agarose.

twofold increase in specific activity [43]. In this latter example, the enzyme was eluted with a low-pH, high-ionic-strength buffer.

Recombinant proteins possessing multiple exposed histidine or cysteine residues can also bind to an IMAC column. Since such binding is pH dependent (as discussed in Chapter 10), the protein of interest can be eluted simply by reducing the pH of the buffer. To isolate mouse dihydrofolate reductase, a sequence coding for an affinity tag of six adjacent histidines at the amino terminus was placed into the gene for this enzyme [44]. After expression of this fusion protein in *E. coli*, the resulting cell extract was purified by IMAC using a Ni^{2+}-nitrilotriacetate support. The preparation of this adsorbent has been described by Hochuli et al. [45].

In the IMAC purification of an enzyme that has a histidine affinity tag, the concentration of imidazole in the wash and elution buffers can be adjusted to minimize copurification of nonspecifically bound proteins. After the IMAC purification step, the N-terminal histidine tag is then removed by treating the enzyme with carboxypeptidase A. Since recombinant fusion proteins may fail to fold properly to produce a native and active molecule, the removal of this added peptide tag is important. If, however, enzyme activity and substrate specificity appear unaltered, the modified enzyme can be used directly without treatment. In addition, by attaching two or more affinity tags to the same protein, immobilization and elution of the desired enzyme or protein can be performed under a variety of conditions [46].

The usefulness of tridentate iminodiacetic acid may not always be equal to that of tetradentate nitrilotriacetic acid when these are used as chelating agents in IMAC columns. For instance, one study compared these chelating agents when they were used in purifying a His-tagged recombinant form of the carboxyl half of HTLV-I surface envelope glycoprotein (overexpressed and secreted in *E. coli*). The conclusion of this study was that nitrilotriacetic acid was more suitable for the IMAC purification of this particular protein [47].

12.3 FUNCTION-BASED AFFINITY SEPARATIONS

A second group of affinity methods for purifying enzymes includes those based on the function, or binding activity, of the enzyme. Affinity columns that can be used for this purpose include those containing small or large nonprotein-based substrates or cosubstrates or ligands; columns that contain immobilized protein substrates or cosubstrates; and columns that make use of immobilized cofactors, enzyme inhibitors, bisubstrate inhibitors, or transition-state analogs. Each of these approaches are considered in the following subsections.

12.3.1 Small Immobilized Substrates or Cosubstrates

There are many examples in which small substrates or cosubstrates have been immobilized and used in enzyme purification. Some examples that will be considered in this section include the use of columns that contain diazonium, glutathione, coenzyme A, FMN, CDP, GM3, or boronate as affinity ligands.

12.3.1.1 Diazonium

The use of diazonium agarose as an affinity matrix for the purification of norsolorinic acid dehydrogenase has been described [48]. Norsolorinic acid dehydrogenase is an NADPH-requiring enzyme that is involved in aflatoxin biosynthesis. It catalyzes the conversion of norsolorinic acid to averantin, as shown in Figure 12.6. To purify this enzyme, ω-Aminohexylagarose (from Sigma) was treated with *p*-nitrobenzoyl azide in dimethylformamide to yield *p*-nitrobenzamidoalkylagarose. The *p*-nitro-benzamidoalkylagarose was

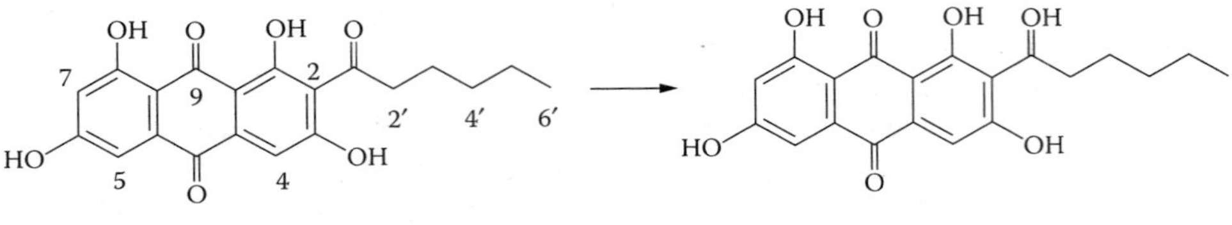

Figure 12.6 Conversion of norsolorinic acid to averantin by norsolorinic acid dehydrogenase.

then suspended in 0.2 M sodium dithionite and 0.5 M NaHCO$_3$, with sodium nitrite being added to produce the diazonium agarose derivative. This derivative was coupled to the hydroxyanthraquinone moiety of norsolorinic acid at positions 4, 5, or 7 (see Figure 12.7). This affinity matrix, however, exhibited little or no binding to the dehydrogenase. Since this enzyme is inactivated by high concentrations of the substrate (possibly because of the presence of the phenol groups), the investigators methylated the affinity matrix with diazomethane to block interactions with the more acidic phenol groups at positions 6 and 8 of norsolorinic acid. It was found that this methylated matrix allowed a one-step purification of the enzyme from a crude homogenate, resulting in a 138-fold purification [48].

12.3.1.2 Glutathione

Glutathione-S-transferase (GST) is another enzyme that has been purified by affinity chromatography. This is a drug-metabolizing enzyme that transfers a glutathione (GSH)

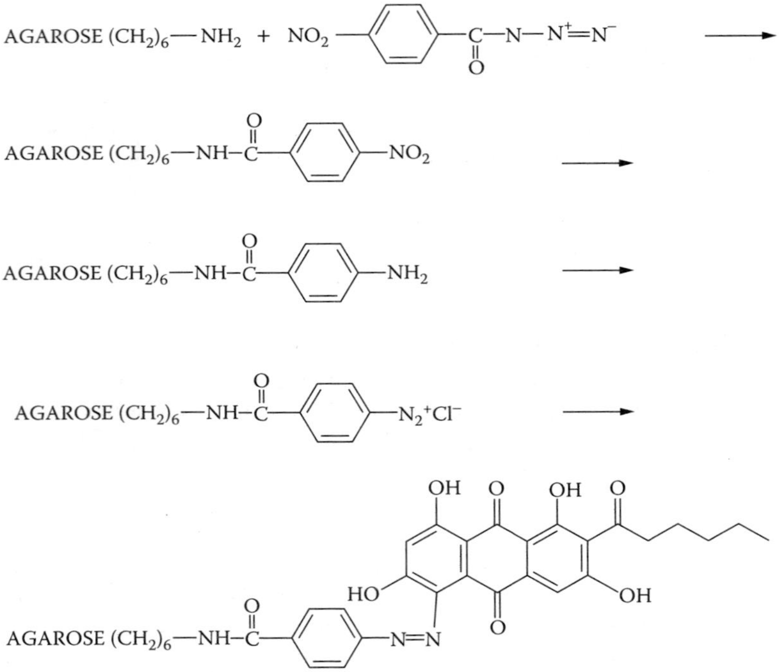

Figure 12.7 Matrix preparation for the isolation of norsolorinic acid dehydrogenase.

Figure 12.8   Formation of the LTA$_4$-GSH complex.

molecule (via its sulfur atom) to a wide spectrum of electrophiles, such as leukotriene A4 (LTA4). The reaction involved in this process is shown in Figure 12.8. Affinity matrixes commonly used for the purification of GST are glutathione-Sepharose 6B (also referred to as GSH-Sepharose) and *S*-hexylglutathione-Sepharose 7B. These are prepared by treating epoxy-activated Sepharose 6B with glutathione or *S*-hexylglutathione, respectively [49]. The GSH-Sepharose is prepared at pH 7, where the addition of GSH to epoxy-activated Sepharose presumably occurs by sulfhydryl addition to the epoxide groups on the support's surface (see Figure 12.9). At this pH value, the single amine group on GSH should be fully protonated and thus not available for reaction with the epoxide groups.

GSH can also be coupled to epoxy-activated Sepharose at a pH higher than 7, but the resulting support has poor affinity for transferases. At these higher pHs, the coupling of GSH is thought to proceed through the reaction of its amino group with the epoxide. Such a matrix, however, fails to bind an isoform of GST known as α-glutathione transferase subtype Ya1 Ya1.

It should be noted that the glutathione transferases actually represent a complex-multigene family, comprising many distinct classes of proteins, which are designated as alpha, mu, pi, theta, sigma, zeta, beta, phi, and omega. Initially, however, classification of GSTs was made simply by using the letter "Y." Whereas some GSTs show little enzymatic activity when bromosulfophthalein is used as a substrate, others like Ya1 Ya1 exhibit high activity. For the latter isoform, bromosulfophthalein-glutathione-Sepharose provides an effective matrix, and the adsorbed protein can be eluted with a buffer containing reduced glutathione [50]. More information on the employment of *S*-hexylglutathione-linked epoxy-activated Sepharose 6B for the purification of GST enzymes can be found in the literature [51].

Advantage can be taken of the interaction between GST and a GSH-affinity column for the isolation of enzymes or enzyme subunits. This can be accomplished by first overexpressing the enzyme (or enzyme subunit) as a fusion protein that has been modified through genetic engineering to contain GST. This fusion protein can then be isolated in a single step by means of a GSH-affinity column. This approach is particularly attractive in instances where discriminating, direct-affinity ligands are not readily available (e.g., as is the case for the many protein kinases) [52].

Figure 12.9   GSH coupled to epoxy-activated Sepharose 6B.

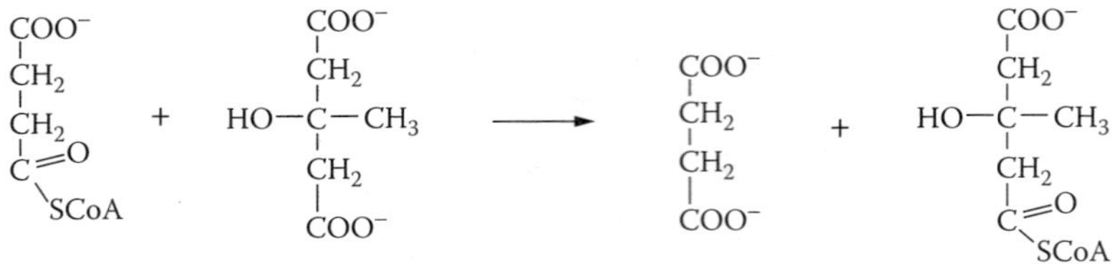

Figure 12.10 Reaction of succinyl CoA with 3-hydroxy-3-methylglutarate to form succinate and 3-hydroxy-3-methylglutaryl-CoA.

12.3.1.3   Coenzyme A

Coenzyme A transferases are a third group of enzymes that have been purified by using small immobilized substrates or cosubstrates. These enzymes catalyze the transfer of the coenzyme A (CoA) moiety from a CoA thiol ester to a carboxylic acid. An example of this type of enzyme is succinyl-CoA:3-hydroxyl-3methylglutarate CoA transferase, which catalyzes the transfer of CoA from succinic acid to 3-hydroxy-3-methyl glutaric acid (Figure 12.10). This enzyme can be purified on agarose-hexane-CoA, with a resulting increase in specific activity from 31.2 nmol/min/mg to 248.5 nmol/min/mg in one step [53]. Unfortunately, the amount of enzyme obtained by this type of affinity chromatography is limited, and the agarose-hexane-CoA support undergoes rapid degradation.

The CoA-agarose that is used in such a separation is prepared by first mixing 6-aminohexanoic acid with cyanogen bromide-activated Sepharose 4B to yield carboxyl-substituted Sepharose 4B (i.e., CH-Sepharose 4B, available from Amersham Pharmacia Biotech). The free carboxyl group at the end of the six-carbon spacer arm on this matrix is then activated with carbodiimide and coupled to N-hydroxysuccinimide ester. When this activated support is reacted with CoA, a thiol ester linkage is formed between the spacer and CoA, as illustrated in Figure 12.11.

12.3.1.4   FMN

Flavin adenine dinucleotide (FAD) synthetase (also known as ATP:FMN adenyltransferase) has been purified using a flavin mononucleotide (FMN) affinity matrix synthesized from CH-Sepharose 4B. FMN can be linked to CH-Sepharose 4B in the presence of water-soluble N-ethyl N'-(3-dimethylaminopropyl) carbodiimide, presumably by means of a mixed acid anhydride linkage to the phosphate group of FMN. To elute enzymes that are retained on this support, FMN can be added to the mobile phase.

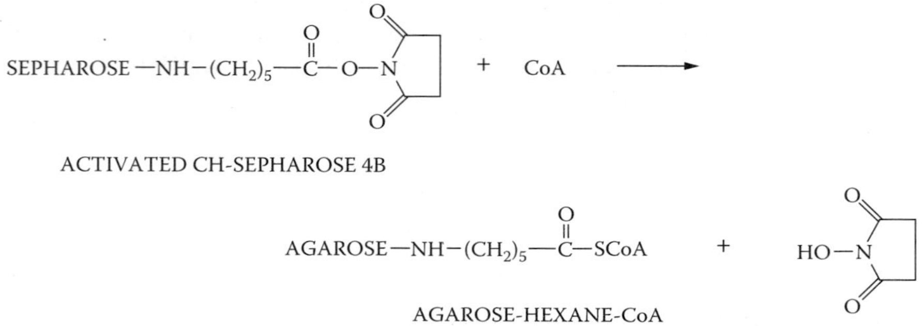

ACTIVATED CH-SEPHAROSE 4B

AGAROSE-HEXANE-CoA

Figure 12.11 Reaction of CoA with activated Sepharose.

When this FMN matrix was used in a one-step affinity chromatographic scheme, it increased the specific activity of FAD synthetase by approximately 130-fold (i.e., from 2.8 to 374 units/mg) [54]. Since the coupling of FMN to this support is through the terminal phosphate of FMN, enzymes which display an absolute requirement for such a group will not adhere to this affinity matrix. If FMN is coupled to Sepharose by using cyanogen bromide, the polyhydroxy ribityl side chain of FMN will also be attached to the support, thus further narrowing the ability of this ligand to be used in selective affinity separations.

12.3.1.5   CDP and GM3

Sialyl transferases are a group of glycosyl transferases that catalyze the transfer of a sialic acid (Neu Ac) residue to a nonreducing terminal sugar of a glycoprotein or glycolipid, as shown by the general reaction in Figure 12.12. Cytidine diphosphate (CDP)-Sepharose and ganglioside G_{M3} (GM3)-Sepharose were employed in the purification of CMP-Neu Ac:GM3 sialyl transferase and CMP-Neu Ac:lactosyl ceramide sialyl transferase from samples of developing rat brains. To separate these two enzymes from each other, a GM3 acid-Sepharose support was prepared by carrying out the nonspecific oxidation of GM3 with permanganate and coupling the resulting GM3 acid to aminohexyl-Sepharose 4B. For the GM3 sialyl transferase, a CDP-Sepharose column was used, leading to an increase in specific activity from 0.016 to 0.79 units/mg protein. This was then followed by the employment of a GM3-Sepharose affinity column, which raised the activity to 18.9 units/mg protein [55].

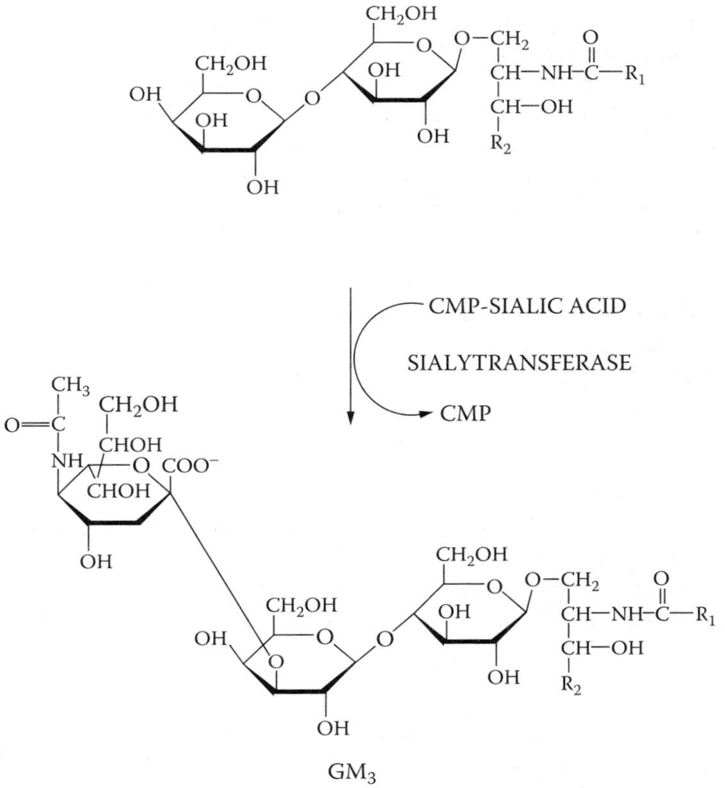

Figure 12.12   Reaction of CMP-Neu Ac with an HO acceptor to give CMP and Neu Ac-O-acceptor.

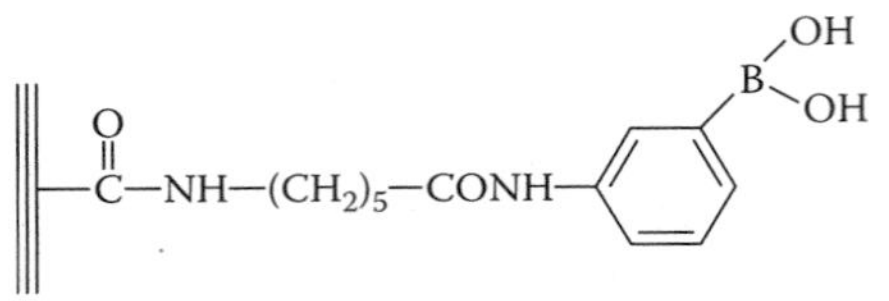

Figure 12.13   Partial structure of DHB-Sepharose.

12.3.1.6   Boronate

As discussed in Chapter 8, boronate affinity columns are valuable in separating low-molecular-weight compounds with coplanar *cis*-diol groups, including ribonucleotides, ribonucleosides, sugars, catecholamines, and coenzymes. But these columns can also be used to separate proteins based on their function. For instance, poly(ADP-ribose)glycohydrolase catalyzes the cleavage of ribosyl-ribose bonds, resulting in the liberation of ADP-ribose. An affinity matrix for the purification of this enzyme was prepared by allowing poly(ADP-ribose) to interact with dihydroxyboryl-Sepharose (DHB-Sepharose) [56]. This support (shown in Figure 12.13) was prepared by treating cyanogen bromide-activated Sepharose with 6-aminohexanoic acid, followed by coupling of the carboxy group of this spacer to *m*-aminophenylboronic acid hemisulfate [57].

12.3.2   Immobilized Large Nonprotein Substrates or Cosubstrates

Some large nonprotein-based substrates or cosubstrates can be used for enzyme purification by linking these ligands directly to a support like cyanogen bromide-activated agarose. As one example, DNA immobilized to agarose (e.g., DNA-agarose from Amersham Pharmacia Biotech) has been employed in the purification of DNA and RNA polymerases [58]. Similarly, Poly(U)-Sepharose 4B (also from Amersham Pharmacia Biotech) has been utilized in the isolation of recombinant HIV-1 reverse transcriptase [59]. In this second case, polyuridylic acid chains were coupled to the support via the tautomeric enolate form of the nucleotides. This resulted in a matrix that exhibited multipoint attachment of chains that were about 100 nucleotides long.

12.3.3   Immobilized Protein Substrates or Cosubstrates

Protein substrates or cosubstrates can also be utilized for the affinity isolation of enzymes. Examples of such ligands include myosin, fibrin, ferredoxin, actin, cytochrome c, ubiquitin, avidin, and enzyme subunits.

12.3.3.1   Myosin

AMP deaminase has been shown to bind to the myosin heavy chain *in vitro*. When coupled to activated Sepharose [60], myosin can thus be used as an affinity ligand for deaminase. Myosin-Sepharose has also been employed for isolating the protein phosphatase, which catalyzes the dephosphorylation of isolated myosin light chains [61].

12.3.3.2   Fibrin

A fibrin-Sepharose affinity matrix has been used to help define the mechanism of reaction between single-chain tissue plasminogen activator (sc-tPA) and human tPA inhibitor-1 (PAI 1) [62]. Isolation of a 1:1 complex of sc-tPA and PAI-1 by using fibrin-agarose indicated that cleavage of a "bait" peptide bond occurred during complex formation. The bait peptide

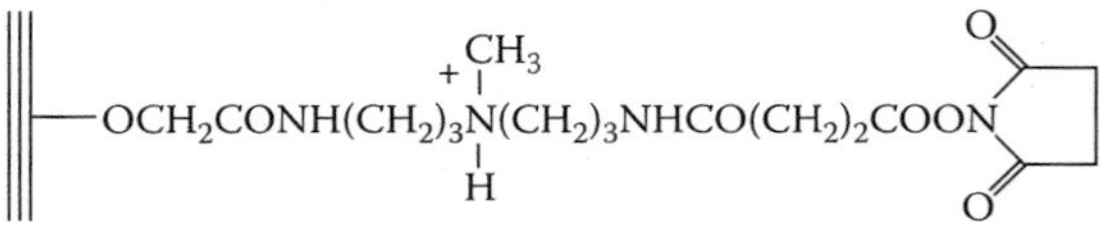

Figure 12.14 Partial structure of Affi-Gel 15.

Arg^{346}Met^{347}Ala-Pro-Glu-Glu- (which is located near the C terminus of inhibitor 1) resembles those found in other serine protease inhibitors known as serpins.

12.3.3.3 Ferredoxin

Ferredoxin:NADP$^+$ oxidoreductase (a flavoprotein) was isolated from plant tissue by employing a ferredoxin-Sepharose 4B affinity column. This enzyme was then eluted by using an increasing concentration of phosphate buffer. Such a separation gave an increase in the enzyme's specific activity from 7.1 to 30.5 units/mg [63].

12.3.3.4 Actin

Pancreatic deoxyribonuclease I (DNase I) forms a tight 1:1 complex with monomeric actin, giving a dissociation constant (K_d) of 10^{-10} M. But commercially available preparations of this enzyme also contain significant contamination from chymotrypsinogen, chymotrypsin, trypsin, and multiple ribonucleases. To purify this enzyme, an affinity column was prepared by adding 100 mg of G-actin to 25 ml of N-hydroxysuccinimide-substituted agarose, which was then used to treat the commercial DNase I. It should be noted that the specific matrix used in this work (Affi-Gel 15 from BioRad, shown in Figure 12.14) possesses a cationic charge in its 15-atom spacer arm. When working at a physiological pH, this property significantly enhances the coupling efficiency of this support for acidic proteins (i.e., proteins with a pI below 6.5).

In the work with DNase I, elution with a buffer that contained 10 M formamide yielded an enzyme with a specific activity of 2780 Kunitz units/mg, while the activity of the enzyme preparation before this chromatographic step was 2651 units/mg. While this change in activity might seem slight, the preparation was now essentially free of any protease or ribonuclease contamination [64].

12.3.3.5 Cytochrome C

Cytochrome c oxidase from heart mitochondria has been purified by linking cytochrome c directly to cyanogen bromide-activated Sepharose 4B. Cytochrome c can be attached to this support through multiple lysine residues, which are located mainly on one side of this protein (see Figure 12.15). Hydrolysis of some of the cyanogen bromide-activated groups on the support prior to coupling can minimize the multipoint attachment of cytochrome c through its lysine residues.

Alternatively, cytochrome c can be linked to activated thiol-Sepharose 4B (from Amersham Pharmacia Biotech). In this procedure, this protein is coupled through a single cysteine residue (see Figure 12.15). Since cytochrome c oxidase interacts with lysine residues that are located on the opposing side of cytochrome c, this particular method of immobilization is preferred for this cytochrome c.

Another way of immobilizing cytochrome c is to allow Affi-Gel 102 (i.e., an amino alkyl agarose gel from Bio-Rad with a six-atom hydrophilic spacer arm) to react with the

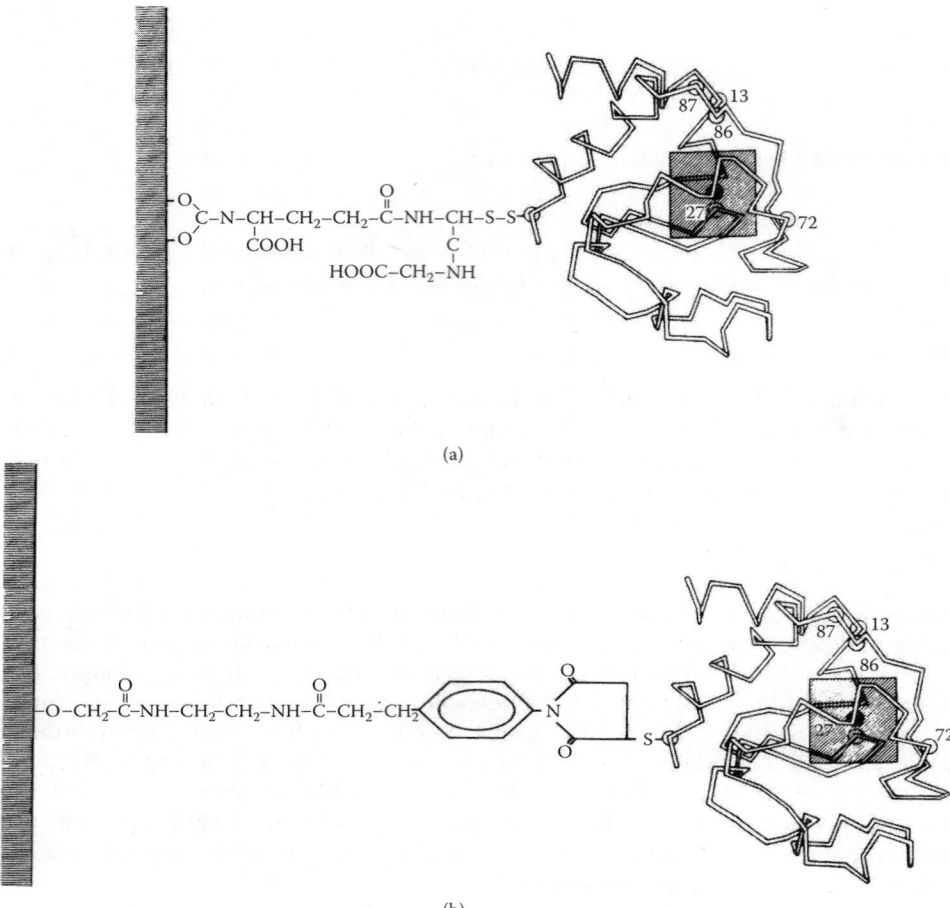

(a)

(b)

Figure 12.15 (a) Yeast cytochrome c linked to activated thiol-Sepharose 4B, and (b) yeast cytochrome c linked to Affi-Gel 102 through the heterobifunctional reagent SMPB. (Reproduced with permission from Broger, C., Bill, K., and Azzi, A., *Methods Enzymol.*, 126, 64–72, 1986.)

heterobifunctional reagent succinimidyl-4-(*p*-maleimidophenyl) butyrate (SMPB). Reduced cytochrome c is then covalently coupled to the extended spacer arm of this gel. Advantages of this matrix include its stability under reducing conditions and its high binding capacity for cytochrome c oxidase [65].

12.3.3.6   Ubiquitin

Restricting the number of attachment sites for a ligand increases the available binding domains on this ligand when it is immobilized to an affinity support. For instance, Hadas et al. [66] have used dimethylmaleic anhydride to reversibly protect some of the free amino groups of a protein, thus reducing the number of attachment points between the protein and a matrix. Duerksen-Hughes et al. [67] used the glyoxal derivative 4-(oxoacetyl) phenoxyacetic acid (OAPA) (Figure 12.16) to immobilize ubiquitin via its reactive arginine residues. To do this, the OAPA was first coupled with *N*-ethyl-*N′*-(3-dimethyl aminopropyl)-carbodiimide to amine-substituted polyacrylamide beads (i.e., Bio-Gel P-150 from Bio-Rad). This support was then allowed to bind to ubiquitin. In this binding step, phenyl glyoxal and its derivatives

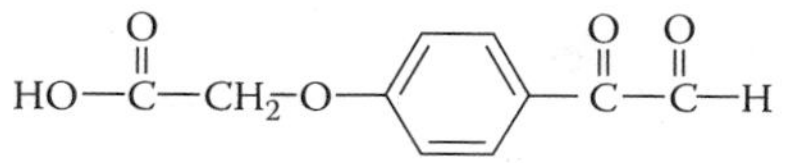

$$HO-\overset{O}{\underset{\|}{C}}-CH_2-O-\langle\bigcirc\rangle-\overset{O}{\underset{\|}{C}}-\overset{O}{\underset{\|}{C}}-H$$

Figure 12.16 The glyoxal derivative of OAPA.

react under mild conditions (pH 7.9) with mainly the guanidino group of the arginine residues on proteins. In this specific case, the immobilized OAPA bound primarily to three closely spaced arginine residues in the ubiquitin (arginines 42, 72, and 74, which are all within a few angstroms of each other). Ubiquitin carboxyl terminal hydrolase (an enzyme that hydrolyzes various amide- and ester-leaving groups from the carboxyl terminus) was specifically and reversibly bound by this affinity support. In addition, three other proteins exhibiting ubiquitin carboxyl-terminal hydrolytic activities were isolated. However, these same three proteins were not bound by a conventional matrix that contained ubiquitin immobilized through its lysine residues.

In eukaryotes, intracellular proteins targeted for degradation are first covalently linked to ubiquitin. A summary of this process is given in Figure 12.17. First, an enzyme (E_1) catalyzes adenyl group transfer from adenosine triphosphate (ATP) to the C-terminal glycine residue of ubiquitin, which is then moved to the sulfhydryl group of a cysteine residue within E_1. Next, ubiquitin is transferred from E_1 to another enzyme (E_2) by transthioesterification. A third enzyme (E_3) then binds the target protein at a specific protein-binding site and catalyzes the transfer of ubiquitin from E_2 to the ε-amino group of a lysine residue in the target protein.

Ubiquitin coupled to activated CH-Sepharose has not only allowed the isolation of ubiquitin-activating enzyme (E_1), but this has also been used to purify other components of the ubiquitin-protein ligase system (i.e., E_2 and E_3) [68]. It is important to note that there are hundreds of distinct E_3 proteins that recognize information in the amino acid sequences of proteins, making the latter targets for ubiquination. Sepharose must contain approximately 20 mg/ml of ubiquitin in the swollen gel if this matrix is to be used for the isolation of E_3; however, E_1 and E_2 can be completely bound by columns that contain much less ubiquitin (i.e., roughly 5 mg/ml). E_1 is covalently linked to such a column in the presence of ATP, but can be specifically eluted with adenosine monophosphate (AMP) and pyrophosphate. E_2 is bound when E_1 and ATP are present and is eluted when a high concentration of a thiol-containing compound is added to the mobile phase. E_3 is noncovalently adsorbed and is eluted at high salt concentrations or by increasing the pH.

12.3.3.7 Enzyme Subunits

In an earlier review [69], the possibility of using affinity chromatography for the isolation of enzyme subunits was mentioned. For instance, if aldolase that is covalently attached to

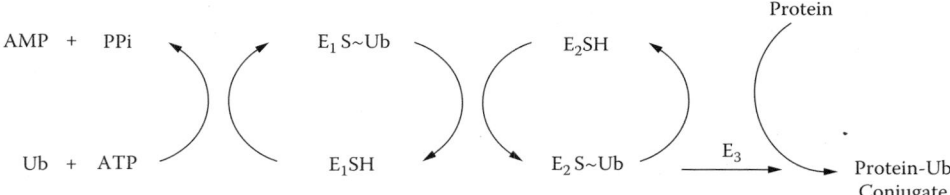

Figure 12.17 Suggested sequence of events in the ubiquitin-protein ligase system.

Sepharose is washed with 8 M urea, this enzyme will dissociate into its subunits. All subunits that are not covalently linked to the matrix can then be removed, leaving the remaining subunits behind for use as affinity ligands. It now appears that subunits of rabbit muscle aldolase are capable of enzyme activity and can be isolated for use in functional studies [70]. However, the approach of using these subunits as ligands has not been widely employed, even though it could accomplish the isolation of a pure oligomeric enzyme from a crude extract.

In one study, octameric porphobilinogen synthase was immobilized on Sepharose 4B and treated with 4 M urea. Four subunits per octamer were removed, which could be reassociated into a soluble octameric enzyme and gave a specific activity about one-half that of the immobilized octameric structure [71]. This is similar to the findings obtained with immobilized aldolase. However, for such experiments to succeed, the enzyme must be covalently linked to an affinity support through only a small fraction of its total subunits.

12.3.3.8   *Avidin*

The initial use of avidin affinity columns for the isolation of biotin-containing enzymes was unsuccessful because the affinity of native avidin for biotin is too high ($K_d = 10^{-15}$ M). In this case, the enzymes could be recovered from the column only under denaturing conditions. Tetrameric avidin, however, can be dissociated into monomers by urea or guanidine HCl, where the monomeric avidin exhibits a much lower affinity for biotin. A monomeric avidin affinity matrix can be prepared by treating cyanogen bromide-activated Sepharose 4B with avidin, followed by overnight exposure of this matrix to 6 M guanidine HCl.

A monomeric avidin support has been used to purify the biotin-containing bacterial enzyme propionyl-CoA carboxylase. The retained enzyme was eluted with 0.2 M biotin and 1 mM MgCl$_2$ in pH 7.0, 0.05 M potassium phosphate buffer containing 0.015 ng/ml DTT. This resulted in a sevenfold purification of the enzyme [72]. A monomeric avidin affinity column was also employed for the isolation of bacterial oxalacetate decarboxylase, giving a 15- to 20-fold purification [73].

12.3.4   Cofactors as Ligands

Immobilized cofactors have also been used as affinity ligands. One group of enzymes that can be isolated by this approach is the dehydrogenases. Because NADP$^+$ may undergo structural modification during coupling, it is usually not chosen for this purpose. However, 2′,5′ADP-Sepharose 4B (Amersham Pharmacia Biotech) has sufficient stability for use in a broad range of applications. This support can be prepared [74] from N^6-(6-amino-hexyl)adenosine 2′,5′-bisphosphate (see Figure 12.18).

Crude bacterial luciferase has been purified by using a cofactor as a ligand. Bacterial luciferase catalyzes the oxidation of reduced flavin mononucleotide and a long-chain

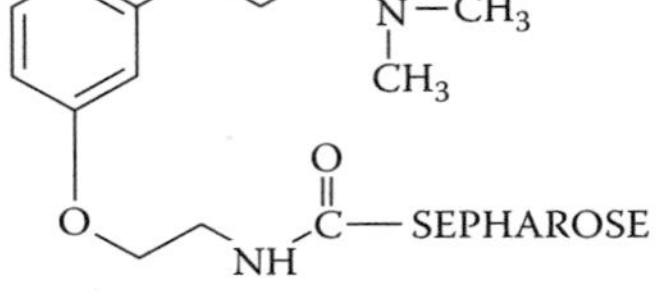

Figure 12.18   Structure of N-(6-aminohexyl)-adenosine 2′,5′-bisphosphate.

aldehyde by molecular oxygen. This enzyme was purified by first applying it to a Sepharose-hexanoate gel, which was used to remove proteins that bound nonspecifically to the support and its spacer arm. The enzyme and other nonretained components from the first column were then applied to a FMN-Sepharose column. The enzyme retained by this second column was later eluted by raising the concentration of either NaCl or FMN in the mobile phase. As a result of this purification, the activity of the luciferase preparation increased from 8×10^{12} to 1.4×10^{14} quanta/sec/mg [75].

One class of plant hydroxylases includes the nonheme iron-containing cytosolic dioxygenases. These enzymes require Fe^{2+} and ascorbate, as well as the cofactor α-ketoglutarate, for catalytic activity and are involved in the biosynthesis of abscissic acid, ethylene, and gibberellins. A 2-oxoglutarate-Sepharose affinity column has been utilized for purifying one such enzyme, flavonol 6-hydroxylase. In this example, the affinity column was used after ammonium sulfate precipitation, gel filtration, and ion-exchange chromatography [76].

12.3.5 Enzyme Inhibitors as Ligands

Enzyme inhibitors are yet another class of substances that can be used as ligands for enzyme purification. Examples of such ligands that will be discussed in this section include *p*-aminophenyl-β-D thiogalactopyranoside, CP-dMM, novobiocin, 2-[3-(2-ammonioethoxy)-benzoyl]ethyltrimethylammonium bromide, GTP, hadacidin, dATP, araUTP, amidines, cilastatin, pepstatins, peptide aldehyde inhibitors, *t*-butyloxycarbonyl-gly-phe-NHCH$_2$CN, and cycloheptaamylose.

12.3.5.1 *p-Aminophenyl-β-D Thiogalactopyranoside*

For the preparation of an affinity column specific for β-galactosidase, the competitive inhibitor *p*-aminophenyl-β-D thiogalactopyranoside was linked to the hydrocarbon arm of succinyl-agarose using *N*-cyclohexyl-*N'*-(2 morpholinoethyl)carbodiimide metho-*p*-toluenesulfonate as a coupling agent. A detailed description of this coupling method can be found in the literature [77]. When β-galactosidase was purified by this matrix, the specific activity of the enzyme preparation increased from 1.27 to 41 units/mg [78].

12.3.5.2 *CP-dMM*

Pig liver man$_9$-mannosidase (i.e., α-1,2-mannosidase) is an enzyme involved in the processing of *N*-linked oligosaccharides. This enzyme has been isolated on Con A-Sepharose columns, a general approach used for obtaining glycoproteins. However, this step was also preceded by the more efficient purification of this enzyme on a resin of synthetic *N*-5-carboxypentyl-l-deoxymannojirimycin (CP-dMM) that was linked to AH-Sepharose 4B (as obtained from Amersham Pharmacia Biotech). The structure of CP-dMM is given in Figure 12.19. The basic unit of the CP-dMM ligand is 1-deoxymannojirimycin, which is a basic sugar analog that is inhibitory to man$_9$-mannosidase's activity. With this ligand, it was possible to obtain a 280-fold increase in the specific activity of pig liver man$_9$-mannosidase [79].

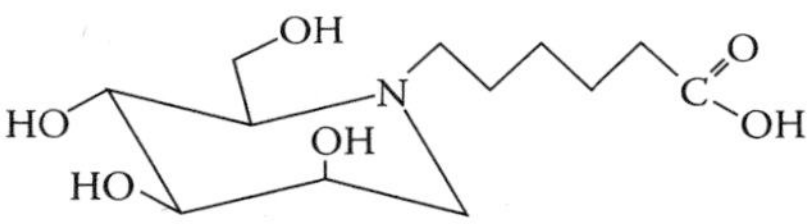

Figure 12.19 Structure of CP-dMM.

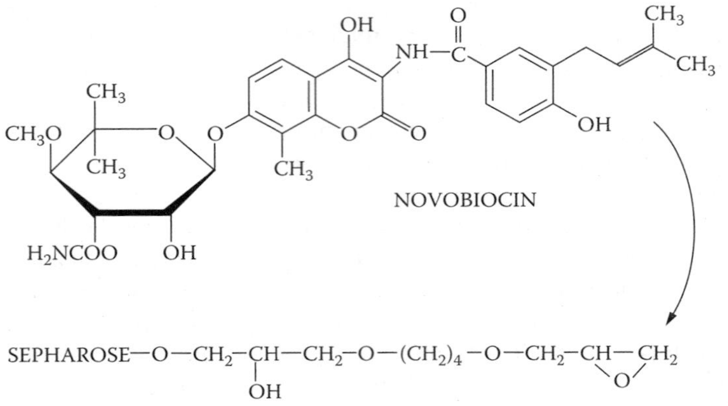

Figure 12.20   Mode of coupling novobiocin to epoxy-activated Sepharose.

12.3.5.3   Novobiocin

Novobiocin is a potent competitive inhibitor for the ATP-binding moiety of type II topoisomerases. These are enzymes that introduce negative superhelical turns into relaxed co-valently closed circular DNA. Novobiocin-Sepharose can be prepared by coupling of this antibiotic to epoxy-activated Sepharose 6B, as illustrated in Figure 12.20. This provides an efficient matrix for the purification of type II topoisomerase from *E. coli* [80]. More recently, mitochondrial type II topoisomerase was applied as a crude extract onto such a column and eluted with a buffer containing 20 mM ATP and 20 mM MgCl$_2$. This one step resulted in an increase in specific activity from 32 to 9600 units/mg [81].

12.3.5.4   2-[3-(2-Ammonioethoxy)benzoyl]ethyltrimethylammonium Bromide

Choline acetylase is an enzyme that catalyzes the synthesis of acetylcholine from acetyl CoA and choline. This has been isolated from the electric organ of *Torpedo marmorata*. This enzyme was first purified on a CoA-Sepharose affinity matrix prepared by coupling the CoA to Sepharose 4B via a six-carbon spacer. Next, a support containing the immobilized inhibitor 2-[3-(2-ammonioethoxy)benzoyl] ethyltrimethylammonium bromide was employed. The ligand used to make this second support contained a primary amino group that could react with the free carboxyl group of CH-Sepharose in the presence of *N*-ethyl-*N'*-(3-dimethylaminopropyl) carbodiimide at pH 4.8. This gave the structure shown in Figure 12.21.

In this separation, the material that was bound by the CoA-Sepharose 4B column was eluted with a buffer containing acetyl-CoA. The material retained by the second

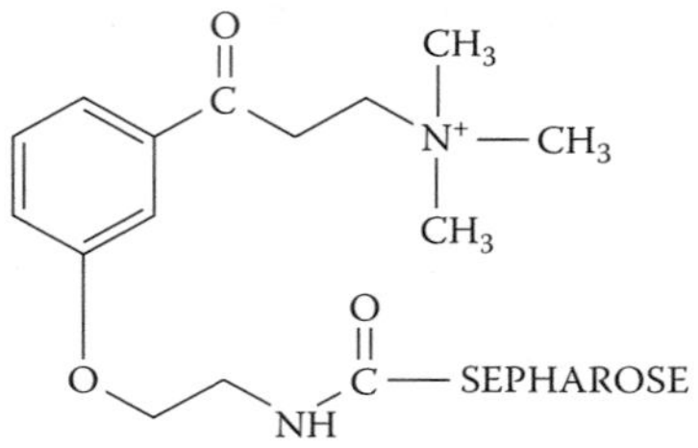

Figure 12.21   Structure of Sepharose-CONH-EtO(BzEt)-Me$_3$NBr.

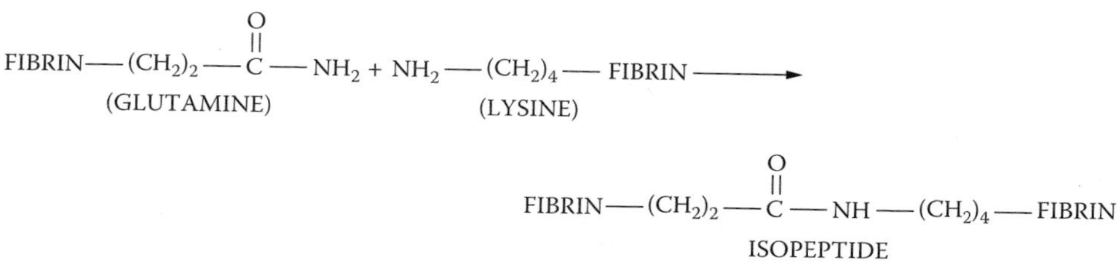

$$\text{FIBRIN}-(CH_2)_2-\overset{\overset{\textstyle O}{\|}}{C}-NH_2 + NH_2-(CH_2)_4-\text{FIBRIN}\longrightarrow$$

(GLUTAMINE)   (LYSINE)

$$\text{FIBRIN}-(CH_2)_2-\overset{\overset{\textstyle O}{\|}}{C}-NH-(CH_2)_4-\text{FIBRIN}$$

ISOPEPTIDE

Figure 12.22   Isopeptide bond formation in fibrin.

column, which contained the immobilized inhibitor, was eluted with a buffer that contained a soluble form of the inhibitor. Upon elution from the first column, the specific activity of the enzyme had risen from 0.02 to 0.11 unit/mg. But after elution from the second column, the specific activity was 73 units/mg [82].

12.3.5.5   GTP

The transglutaminases are a family of enzymes that catalyze the incorporation of primary amines into the glutamine residues of proteins. When the primary amine donor is a protein-bound lysine, an ε-(-glutamyl)lysine isopeptide is formed, as illustrated in Figure 12.22. These enzymes require Ca^{2+} and are differentially inhibited by guanosine triphosphate (GTP), where tissue-type transglutaminase is inhibited by GTP and epidermal transglutaminase is unaffected. This type of enzyme preparation was applied to a heparin-agarose column in a HEPES-EGTA buffer and eluted with the same buffer plus GTP, giving a purification of 10- to 17-fold. It may be assumed that the binding of GTP to this enzyme changes its conformation, thus promoting the enzyme's elution from a heparin column. After supplementing the mobile phase with $CaCl_2$, the same enzyme was adsorbed onto an α-casein-agarose column, from which it was later eluted through the addition of EGTA. This step allowed an additional fivefold purification to be obtained for this enzyme [83].

12.3.5.6   Hadacidin

Adenylosuccinate synthetase catalyzes the reaction where inosine monophosphate (IMP) and aspartate are combined with GTP to form adenylosuccinate, guanosine diphosphate (GDP), and phosphate. This enzyme was extracted from an ascites sarcoma sample by utilizing a hadacidin-Sepharose 4B support. Hadacidin (or N-formyl hydroxyaminoacetic acid) is an antibiotic that acts as a powerful competitive inhibitor of this enzyme. This antibiotic was coupled through its formyl group to aminohexyl-Sepharose 4B from Amersham Pharmacia Biotech. This support absorbs many kinds of proteins, but adenylosuccinate synthetase can be eluted specifically from this by adding aspartate to the mobile phase. This gives a 15-fold purification of this enzyme [84].

12.3.5.7   dATP

The allosteric inhibitor deoxyadenosine triphosphate (dATP) has been immobilized on Sepharose and used in isolating the anaerobic ribonucleoside triphosphate reductase of *E. coli*. This enzyme reduces cytidine triphosphate (CTP) to deoxycytidine triphosphate (dCTP) and requires NADPH, dithiothreitol, Mg^{2+}, and ATP. dATP-Sepharose was prepared by coupling aminophenyl-γ-dATP to cyanogen bromide-activated Sepharose [85], where the aminophenyl-γ-dATP was synthesized according to the procedure given by Knorre et al. [86].

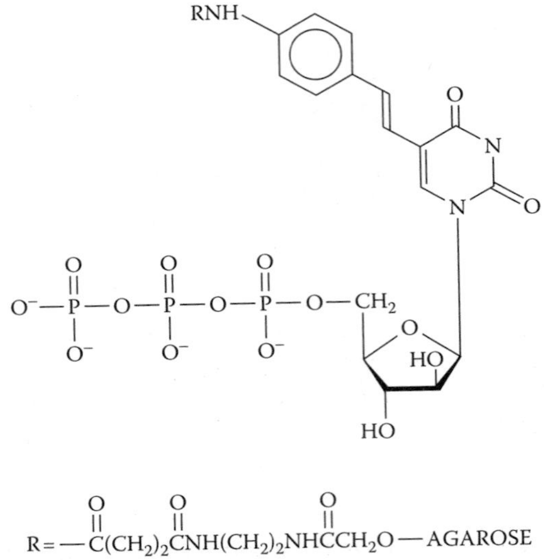

Figure 12.23   Partial structure of AraUTP-Affi-Gel 10.

12.3.5.8   *araUTP*

Affinity chromatography for isolation of the multiprotein complex DNA polymerase α-DNA primase from cherry salmon testes has been performed by using 5-(E)-(4 aminostyryl) araUTP as a ligand [87]. The structure of araUTP (arabinofuranosyluracil-5′-triphosphate) is given in Figure 12.23. This compound is a strong and specific inhibitor of DNA polymerase. This ligand can be coupled to Affi-Gel 10 (from Bio-Rad) by using a reactive *N*-hydroxysuccinimide ester spacer arm. The *N*-hydroxysuccinimide ester forms a stable covalent linkage with the primary amine of araUTP. But if the spacer of this support is omitted, enzymes will not be bound to this ligand. Furthermore, the desired enzymes will be retained by this affinity resin only in the presence of a template or a template primer.

12.3.5.9   *Amidines*

p-Aminobenzamidine is a strong inhibitor of trypsin and trypsinlike serine proteases, such as thrombin and urokinase. This compound was covalently coupled to CH-Sepharose 4B and used to isolate the high- and low-molecular-weight urokinases in urine. This gave a good yield when employed as a terminal step in a purification process [40, 88]. In addition, amidine derivatives coupled to silica beads have been coated with DEAE-dextran and used to purify serine proteases like thrombin, trypsin, and tPA by high-performance affinity chromatography [89].

12.3.5.10   *Cilastatin*

Renal dipeptidase, which also has β-lactamase activity, was affinity-purified using the antibiotic cilastatin. This ligand (shown in Figure 12.24) was attached to Sepharose 4B [90]. β-Lactamase activity has been shown to be selectively inhibited by cilastatin. The specific activity of the peptidase after purification by this ligand was increased from 1.04 to 65.7 μmol/min/mg.

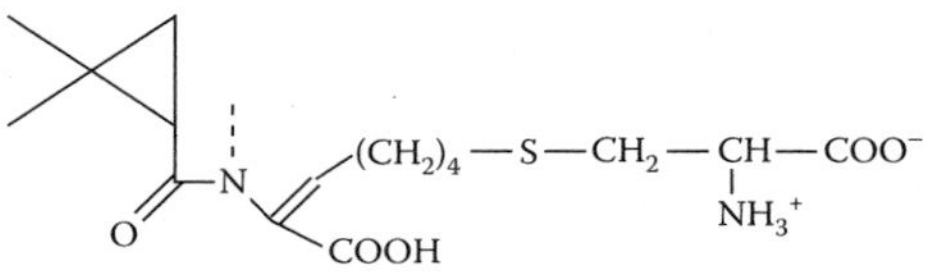

Figure 12.24 Structure of cilastatin.

12.3.5.11   Pepstatins

The pepstatins, of which pepstatin A is the most common example, are peptides that strongly inhibit certain proteases. These are produced by several *Streptomyces* species. Pepstatin A contains an unusual natural amino acid (4-amino-3-hydroxy-6-methyl heptanoic acid) that is essential for its inhibitory activity. Pepstatin is specific for the acid proteases. For instance, it strongly inhibits pepsin, renin, and cathepsin D at effective concentrations of 10^{-6} to 10^{-9} *M*, but does not inhibit trypsin, chymotrypsin, papain, or plasmin.

Pepstatin-agarose prepared according to Huang et al. [91] has been used to separate lysosomal procathepsin D from the active forms of cathepsin D [92]. Procathepsin D was bound by the pepstatin matrix (maintained at 4°C) when the pH of the mobile phase was 3.5. However, binding to this matrix did not occur at pH 5.3. This differed from the behavior noted for the active forms of the enzyme, which were bound at both pH values.

Active cathepsin was removed by the pepstatin agarose support at pH 5.3. The unbound procathepsin was then reapplied to a second pepstatin column at pH 3.5 and later eluted with a pH 8.0 buffer. The pure procathepsin D obtained by this technique appears to undergo an autocatalytic conversion to the active enzyme in a manner analogous to other carboxyl proteinases such as pepsinogen. When employed in combination with ion-exchange chromatography, pepstatin-agarose has also been used to obtain homogeneous recombinant HIV-1 and HIV-2 retroviral proteases from bacterial lysates [93].

12.3.5.12   Peptide Aldehyde Inhibitors

Lysosomal cysteine proteases (e.g., cathepsins B, H, and C) are important in cellular protein turnover. Peptide aldehyde inhibitors resembling leupeptin (Ac-leu-leu-argininal, where "argininal" refers to arginine aldehyde) are useful affinity ligands in the purification of cysteine proteases. An affinity matrix for the cysteine proteinase cathepsin B was produced by reacting gly-phe-glycinaldehyde semicarbazone with CH-Sepharose 4B [94]. Elution of the retained enzyme was effected by incubating the affinity gel with 1.5 m*M* 2,2′-dipyridyl disulfide (DDS) at pH 4. DDS reacts selectively with cysteine residues at the active site of the enzyme to produce an inactive and protected enzyme form that can later be rapidly reactivated by reducing agents under assay conditions. Cathepsin B was purified to near homogeneity by a two-step procedure that used acetone fractionation and this affinity matrix. Cathepsin L, papain, and actinidin were also bound by this affinity column, whereas other proteinases like cathepsin H and chymopapain were not [94]. A similar matrix was used to purify cathepsin B from purulent sputum [95].

12.3.5.13   t-Butyloxycarbonyl-gly-phe-NHCH₂CN

The agent *t*-butyloxycarbonyl-gly-phe-NHCH$_2$CN is a potent inhibitor of cysteine proteases like papain, chymopapain, and papaya proteinase III. After deprotection of its amino terminus, this inhibitor was reacted with CH-Sepharose 4B in the presence of carbodiimide to form Sepharose-6-aminohexanoyl-gly-phe-NHCH$_2$CN [96]. Hydroxyethyl disulfide was used to elute cysteine proteases that were retained by this support.

12.3.5.14 *Cycloheptaamylose*

The functional raw starch-binding protein domain of barley alpha-amylase, expressed in *E. coli*, was purified by affinity chromatography using its inhibitor cycloheptaamylose as a ligand. The cycloheptaamylose was cross-linked to epoxy-activated Sepharose 6B. The binding domain that was retained by this matrix was eluted with a buffer that contained 10 m*M* cycloheptaamylose [97].

12.3.6 Bisubstrate Inhibitors as Ligands

Protein kinases are enzymes that transfer a phosphoryl group from ATP to a serine/threonine or tyrosine on a protein. These represent the largest family of enzymes in eukaryotic cells. But using specific peptides as affinity ligands for these enzymes has led to disappointing results. This is due to the low binding effectiveness for protein kinases that are exhibited by such peptides, as well as the multiplicity of these enzymes and their substrates.

    An alternative approach that has been employed has been to use bisubstrate inhibitors as ligands. These are molecules that contain (1) a group that resembles a nucleotide (e.g., adenosine-5-carboxylic acid) and (2) amino-octanoic acid that is coupled with a peptide fragment by means of a linker group to epoxy-Sepharose 6B. An example of this is given in Figure 12.25. This type of support was used in a one-step purification to yield recombinant protein kinase A as a homogeneous protein from a crude cell extract [98]. Elution of the kinase from this affinity matrix was accomplished by using a buffer that contained increasing concentrations of L-arginine.

12.3.7 Immobilized Transition-State Analogs as Ligands

Certain tripeptides in which the C-terminal arginine is substituted with argininal (i.e., arginine aldehyde) resemble transition-state analogs for enzymes and are capable of forming a hemiacetal structure with the active-site serine of serine proteases [99]. Agarose containing the peptide D-phe-D-phe-argininal has been shown to be effective in binding tissue plasminogen activator (tPA) through hemiacetal formation. This tripeptide was formed sequentially on a

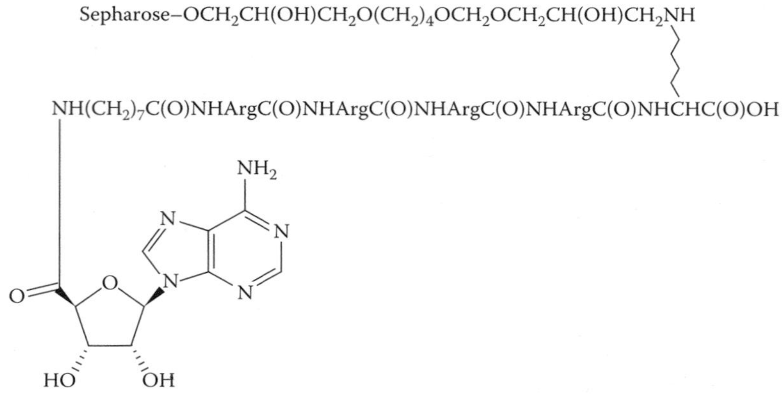

Sepharose–OCH$_2$CH(OH)CH$_2$O(CH$_2$)$_4$OCH$_2$OCH$_2$CH(OH)CH$_2$NH

NH(CH$_2$)$_7$C(O)NHArgC(O)NHArgC(O)NHArgC(O)NHArgC(O)NHCHC(O)OH

Figure 12.25 Epoxy-Sepharose amino-octanoic acid with peptide and adenosine-*S*-carboxylic acid attached.

carboxyl-substituted matrix by carbodiimide coupling. Alternatively, preformed dipeptides can be bound to succinimidyl-agarose, followed by carbodiimide coupling of this dipeptide with argininal semicarbazone and subsequent removal of the semicarbazide group.

A small percentage of the tPA that is bound by such a matrix can be eluted with 0.2 to 1.0 M NaCl, but complete desorption requires a low pH. For large-scale purification, recombinant tPA from conditioned media was captured on a Zn^{2+}-chelate column, applied to an argininal column and, after serial washing, eluted with 0.1 M acetic acid. Purity of the tPA in this case exceeded 95% and gave yields near 95%.

Kallikrein (a serine protease enzyme) has been purified using aldehyde peptide analogs that are unable to form a hemiacetal linkage. This was accomplished by using Leu-enkephalin-argininal semicarbazone that was covalently attached to succinimidyl-agarose. This was then used to purify plasma kallikrein. Unlike the immobilized aldehyde peptide analogs, this matrix interacted noncovalently and reversibly with kallikrein, giving a yield of 71% and a purification of about 350-fold [100]. The semicarbazones of peptide aldehyde inhibitors have also been used to purify cysteine proteases.

12.4 TANDEM METHODS

Several examples have already been given of tandem methods that employ affinity chromatography either alone or in combination with other methods. Immobilized-dye affinity columns used in tandem (where a specific positive-binding matrix is the last column in the chain) are particularly effective in the selective purification of proteins. As discussed in Chapter 9, a preliminary small-scale screening of these dye columns is recommended prior to their use in fractionation to assess the selectivity of these dyes for a given type of protein.

Thermostable DNA polymerase from the archaebacterium *Sulfolobus acidocaldarius* was fractionated in a tandem affinity method. The first support used in this process was phosphocellulose, followed by heparin-Sepharose and blue-Sepharose columns. This method gave an increase in specific activity for the enzyme from 4,000 to 11,800 units/mg during the heparin-Sepharose step and an increase to 200,000 units/mg after the blue-Sepharose step [17]. The DNA polymerase α-DNA primase complex (in which the two activities are tightly associated) from human lymphoblasts was similarly treated by phosphocellulose, heparin-Sepharose, and an antibody column [18].

Partially purified arginyl tRNA synthetase was applied to blue-Sephadex G-150 and eluted with 150 mM buffered KCl. This gave rise to at least a 16- to 18-fold increase in activity [101]. The purification of aminoacyl-tRNA provides an example of the success that can be obtained when general affinity chromatography is coupled to biospecific steps. For instance, during the isolation of yeast mitochondrial methionyl-tRNA synthetase, a partially purified extract was first placed onto a heparin-Ultrogel support, allowing for a rise in specific activity from 2.9 to 23 units/mg. This was next placed onto a tRNAmet-Sepharose column, giving an increase in activity to 500 units/mg, and finally onto an agarose-hexyl-AMP column, with a resulting activity of 1800 units/mg.

Glycosyltransferases from mammalian tissues have been purified using a support that contains a nucleoside diphosphate such as uridine diphosphate (UDP), which is linked via a spacer arm through the 5′-pyrophosphate group of ribose (e.g., UDP-hexanolamine-agarose) [102]. However, not all glycosyltransferases are retained by such a matrix. 5-Hg-UDP N-acetylglycosamine (linked through the 5-position of uracil via a mercaptide linkage to thiopropyl Sepharose 6B, as shown in Figure 12.26) has been used as a ligand for the purification of UDP N-acetylglucosamine α-D-mannoside β1→2 N-acetylglucosaminyl

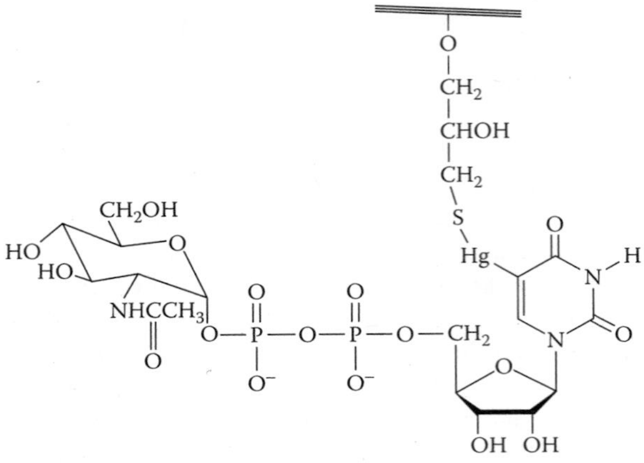

Figure 12.26   Chemical structure of 5-Hg-UDP-GlcNAc linked to thiopropyl-Sepharose.

transferase II. This purification step was repeated three times with the interphasing of a step that utilized Affi-Gel blue. The specific activity of the enzyme preparation increased over 440-fold during this process, from 0.0062 to 2.75 units/mg [103].

12.5   PURIFICATION OF PEPTIDES WITH IMMOBILIZED ENZYMES

12.5.1   Isolation of Inhibitory Peptides

Affinity chromatography of proteolytic enzymes often requires information concerning their naturally occurring inhibitors. Such inhibitors, in turn, have provided stable and effective ligands that have been useful in the purification of these enzymes. However, either immobilized enzymes or their apoenzymes can also be employed for inhibitor isolation. For instance, commercially available bovine carboxypeptidase A (without modification) was immobilized on cyanogen bromide-activated Sepharose and used to isolate carboxy-peptidase inhibitors from the parasitic nematode *Ascaris suum* [104]. Three isoforms of an 8-kDa carboxypeptidase inhibitor were later released from this column by applying a mobile phase that contained 1% formic acid.

Pepsin inhibitors from the parasite *A. suum* have been isolated by a two-step procedure that used native pepsin coupled to aminohexyl-Sepharose via carbodiimide [105]. Peptide extracts from this parasite were applied to this column at pH 1.95 and eluted at pH 10. Chromatofocusing yielded one major homogeneous inhibitor that was suitable for sequencing, along with three minor forms.

An immobilized apoenzyme of carboxypeptidase B has been used to isolate specific peptides [106]. Native porcine carboxypeptidase was coupled to cyanogen bromide-activated agarose and treated with *o*-phenanthroline and EDTA to produce a zinc-depleted, inactive apoenzyme that had the same affinity for peptide substrates as the catalytically active enzyme. Peptides with C-terminal basic residues (e.g., metenkephalin-ArgArg or dynorphin 1-13) could be separated from nonreactive peptides by using this apoenzyme column [89]. Bound peptides were released by decreasing the pH of the elution buffer to 4.0. Although the coupling efficiency was high (95%), only 2.7% of the immobilized carboxypeptidase (before conversion to the apoenzyme) retained activity.

12.5.2 Proteolytically Derived Peptides

The amino acid sequences of proteins that have been deduced from cDNA nucleotide sequences must be verified by direct protein sequencing of the N- and C-terminal regions to determine if posttranslational processing has occurred during the protein's biosynthesis. Anhydrotrypsin affinity chromatography has provided a simple and selective method for isolating the C-terminal fragment of proteins for sequencing [107].

Anhydrotrypsin is a catalytically inactive derivative of trypsin, in which serine 195 is converted to a dehydroalanine residue. This chemical conversion involves base elimination by alkaline treatment of phenylmethane-sulfonyl-trypsin. The affinity of the inactive anhydrotrypsin for peptides containing arginine or lysine at their C terminus is higher than that of the native enzyme [108]. Anhydrotrypsin has been immobilized to cyanogen bromide-activated Sepharose (giving a coverage of 7.6 mg/ml agarose), with the matrix then being treated with tosyl-L-lysine chloromethyl ketone to inhibit any residual tryptic activity. All of the peptides from tryptic digests of proteins without lysine or arginine at the C terminus were found to bind to this affinity matrix, with the exception of the C-terminal peptide, allowing this matrix to be used in a subtractive mode. In this approach, if a protein has a lysine or arginine residue at the C terminus, only the C-terminal peptide from a chymotryptic digest of this protein will bind to the column. Aside from its utility in isolating C-terminal peptides, two classes of soybean trypsin inhibitors were separated and purified on a preparative scale by an anhydrotrypsin matrix [92]. Inhibitors were applied at pH 7.5 and eluted with a pH 2.2, 0.05 *M* glycine-HCl buffer containing 0.2 *M* NaCl.

12.6 SUMMARY AND CONCLUSIONS

Various approaches for the isolation and purification of enzymes by affinity chromatography were examined. These approaches were grouped into two general categories: those that used the enzyme's protein structure and those that used its activity or functional properties. Ligands that can be used to separate an enzyme's structure include Cibacron Blue F3GA, Procion Red HE3B, heparin, lectins, thiopropyl groups, and L-arginine or L-lysine. The discussion also addressed the use of hydrophobic-interaction chromatography and immobilized metal-ion affinity chromatography as techniques for isolating and purifying enzymes.

Several ligands for function-based separations were examined as well. These included both small and large substrates or cosubstrates, cofactors, enzyme inhibitors, bisubstrate inhibitors, and transition-state analogs. Other procedures were also discussed, including the use of affinity columns in tandem methods for the purification of peptides with immobilized enzymes.

The discussion in this chapter has identified the relatively large number of affinity supports that are now commercially available for this type of work. However, as discussed in Chapter 2, not all support materials are equally suitable for enzyme purification, and the selection of supports with preattached ligands is not large. The sources for many of these affinity matrices have been mentioned throughout this chapter. These examples were provided without preference, and they may mirror the supplier's time of entrance into the market. Nevertheless, note that the real challenge and charm of affinity chromatography lies in the insight and cleverness of the investigator in designing and choosing a proper ligand and support.

SYMBOLS AND ABBREVIATIONS

ACE	Angiotensin-converting enzyme
ACP	Acyl carrier protein
AMP	Adenosine monophosphate

araUTP	Arabinofuranosyluracil-5′-triphosphate
ATP	Adenosine triphosphate
CDP	Cytidine diphosphate
CH-Sepharose 4B	Carboxyl-substituted Sepharose 4B
Con A	Concanavalin A
CP-dMM	N-5-Carboxypentyl-l-deoxymannojirimycin
CTP	Cytidine triphosphate
dATP	Deoxyadenosine triphosphate
dCTP	Deoxycytidine triphosphate
DHB-Sepharose	Dihydroxyboryl-Sepharose
DNase I	Deoxyribonuclease I
DTT	Dithiothreitol
EDTA	Ethylenediaminetetraacetic acid
EGTA	Ethylene glycol-*bis*(2-aminoethylether)-N,N,N',N'-tetraacetic acid
FAD	Flavin adenine dinucleotide
FMN	Flavin mononucleotide
GDP	Guanosine diphosphate
GM3	Ganglioside G_{M3}
GSH	Glutathione
GST	Glutathione-S-transferase
GTP	Guanosine triphosphate
HEPES	4-(2-Hydroxyethyl)piperazine-1-ethanesulfonic acid
HIC	Hydrophobic interaction chromatography
IMAC	Immobilized metal-ion affinity chromatography
IMP	Inosine monophosphate
K_d	Dissociation constant
MES	2-Morpholinoethanesulfonic acid
PA	Plasminogen activator
PAI-1	Human tPA inhibitor-1
sc-tPA	Single-chain tissue plasminogen activator
SDS	Sodium dodecyl sulfate
SMPB	Succinimidyl-4-(p-maleimidophenyl) butyrate
tPA	Tissue-type plasminogen activator
UDP	Uridine diphosphate
WGA	Wheat germ agglutinin

REFERENCES

1. Fischer, D., Ebenau-Jehle, C., and Grisebach, H., Phytoalexin synthesis in soybean: Purification and characterization of NADPH:2′-hydroxydaidzein oxidoreductase from elicitor-challenged soybean cell cultures, *Arch. Biochem. Biophys.*, 276, 390–395, 1990.
2. Baird, J.K., Sherwood, F., Carr, R.J.G., and Atkinson, A., Enzyme purification by substrate elution chromatography from procion dye-polysaccharide matrices, *FEBS Lett.*, 70, 61–66, 1976.
3. Dean, P.D.G. and Watson, D.H., Protein purification using immobilized triazine dyes, *J. Chromatogr.*, 165, 301–319, 1979.
4. Watson, D.H., Harvey, M.J., and Dean, P.D.G., The selective retardation of NADP⁺-dependent dehydrogenases by immobilized Procion Red HE-3B, *Biochem. J.*, 173, 591–596, 1978.

5. Neale, A.D., Scopes, R.K., Kelly, J.M., and Wettenhall, R.E., The two alcohol dehydrogenases of *Zymomonas mobilis:* Purification by differential dye ligand chromatography, molecular characterisation and physiological roles, *Eur. J. Biochem.,* 154, 119–124, 1986.

6. Baxter, A., Currie, L.M., and Durham, J.P., A general method for purification of deoxycytidine kinase, *Biochem. J.,* 173, 1005–1008, 1978.

7. Kasten, T.P., Naqui, D., Kruep, D., and Dunaway, G.A., Purification of homogeneous rat phosphofructokinase isozymes with high specific activities, *Biochem. Biophys. Res. Commun.,* 111, 462–469, 1983.

8. Sugiura, M., Purification of the T4 DNA ligase by blue Sepharose chromatography, *Anal. Biochem.,* 108, 227–229, 1980.

9. Pai, S.B., Purification of a bacterial organophosphate-hydrolyzing phosphatase by Cibacron 3GA-Sepharose affinity chromatography, *Biochem. Biophys. Res. Commun.,* 110, 412–416, 1983.

10. Davison, B.L., Leighton, T., and Rabinowitz, J.C., Purification of *Bacillus subtilis* RNA polymerase with heparin-agarose: *In vitro* transcription of phi 29 DNA, *J. Biol. Chem.,* 254, 9220–9226, 1979.

11. Jansen, H., Lammers, R., Baggen, M.G., Wouters, N.M., and Birkenhager, J.C., Circulating and liver-bound salt-resistant hepatic lipases in the golden hamster, *Biochim. Biophys. Acta,* 1001, 44–49, 1989.

12. Becht, I., Schrecker, O., Klose, G., and Greten, H., Purification of human plasma lipoprotein lipase, *Biochim. Biophys. Acta,* 620, 583–591, 1980.

13. Dake, E., Hofmann, T.J., McIntire, S., Hudson, A., and Zassenhaus, H.P., Purification and properties of the major nuclease from mitochondria of *Saccharomyces cerevisiae, J. Biol. Chem.,* 263, 7691–7702, 1988.

14. Burgers, P.M.J., Bauer, G.A., and Tam, L., Exonuclease V from *Saccharomyces cerevisiae:* A 5′–3′-deoxyribonuclease that produces dinucleotides in a sequential fashion, *J. Biol. Chem.,* 263, 8099–8105, 1988.

15. Lindberg, G.K., Rist, J.K., Kunkel, T.A., Sugino, A., and Rothman-Denes, L.-B., Purification and characterization of bacteriophage N4-induced DNA polymerase, *J. Biol. Chem.,* 263, 11319–11326, 1988.

16. Mosbaugh, D.W., Purification and characterization of porcine liver DNA polymerase gamma: Utilization of dUTP and dTTP during *in vitro* DNA synthesis, *Nucleic Acids Res.,* 16, 5645–5659, 1988.

17. Elie, C., DeRecondo, A.M., and Forterre, P., Thermostable DNA polymerase from the archaebacterium *Sulfolobus acidocaldarius:* Purification, characterization and immunological properties, *Eur. J. Biochem.,* 178, 619–626, 1989.

18. Bialek, G., Nasheuer, H.P., Goetz, H., Behnke, B., and Grosse, F., DNA polymerase alpha-DNA primase from human lymphoblasts, *Biochim. Biophys. Acta,* 951, 290–297, 1988.

19. Sorrentino, S., Tucker, G.K., and Glitz, D.G., Purification and characterization of a ribonuclease from human liver, *J. Biol. Chem.,* 263, 16125–16131, 1988.

20. Walker, D.H., Dougherty, N., and Pike, L.J., Purification and characterization of a phosphatidylinositol kinase from A 431 cells, *Biochemistry,* 27, 6504–6511, 1988.

21. Staben, C., Whitehead, T.R., and Rabinowitz, J.C., Heparin-agarose chromatography for the purification of tetrahydrofolate utilizing enzymes: C1-tetrahydrofolate synthase and 10-formyltetrahydrofolate synthetase, *Anal. Biochem.,* 162, 257–264, 1987.

22. Stein, P.L., van Zonneveld, A.-J., Pannekoek, H., and Strickland, S., Structural domains of human tissue-type plasminogen activator that confer stimulation by heparin, *J. Biol. Chem.,* 264, 15441–15444, 1989.

23. Kennedy, J.F. and Rosevear, A., An assessment of the fractionation of carbohydrates on concanavalin A-Sepharose 4B by affinity chromatography, *J. Chem. Soc. Perkin Trans.,* 19, 2041–2046, 1973.

24. Swallow, D.M., Evans, L., and Hopkinson, D.A., Several of the adenosine deaminase isozymes are glycoproteins, *Nature,* 269, 261–262, 1977.

25. Grimwood, B.G., Plummer, T.H., Jr., and Tarentino, A.L., Characterization of the carboxypeptidase N secreted by Hep G2 cells, *J. Biol. Chem.*, 263, 14397–14401, 1988.

26. Sharma, M. and Singh, U.S., Molecular and catalytic properties of angiotensin converting enzyme-I from bovine seminal plasma, *J. Biochem. (Tokyo)*, 104, 57–61, 1988.

27. Jackowski, S. and Rock, C.O., Ratio of active to inactive forms of acyl carrier protein in *Escherichia coli, J. Biol. Chem.*, 258, 15186–15191, 1983.

28. Cooper, C.L., Hsu, L., Jackowski, S., and Rock, C.O., 2-Acylglycerolphosphoethanolamine acyltransferase/acyl-acyl carrier protein synthetase is a membrane-associated acyl carrier protein binding protein, *J. Biol. Chem.*, 264, 7384–7389, 1989.

29. Persson, M., Bergstrand, M.G.S., Buelow, L., and Mosbach, K., Enzyme purification by genetically attached polycysteine and polyphenylalanine affinity tails, *Anal. Biochem.*, 172, 330–337, 1988.

30. Porath, J. and Fornstedt, N., Group fractionation of plasma proteins on dipolar ion exchangers, *J. Chromatogr.*, 51, 479–489, 1970.

31. Skidgel, R.A., Deddish, P.A., and Davis, R.M., Isolation and characterization of a basic carboxy-peptidase from human seminal plasma, *Arch. Biochem. Biophys.*, 267, 660–667, 1988.

32. Skidgel, R.A., Davis, R.M., and Tan, F., Human carboxypeptidase M: Purification and characterization of a membrane-bound carboxypeptidase that cleaves peptide hormones, *J. Biol. Chem.*, 264, 2236–2241, 1989.

33. Grimwood, B.G., Plummer, T.H., Jr., and Tarentino, A.L., Characterization of the carboxypeptidase N secreted by Hep G2 cells, *J. Biol. Chem.*, 263, 14397–14401, 1988.

34. Parkinson, D., Two soluble forms of bovine carboxypeptidase H have different amino-terminal sequences, *J. Biol. Chem.*, 265, 17101–17105, 1990.

35. Sakuragawa, N., Takahashi, K., and Ashizawa, T., Isolation of Prothrombin, *Acta Med. Biol.*, 25, 119–125, 1977.

36. Wu, M., Arimura, G.K., and Yunis, A.A., Purification and characterization of a plasminogen activator secreted by cultured human pancreatic carcinoma cells, *Biochemistry*, 16, 1908–1913, 1977.

37. Suzuki, T. and Takahashi, H., Purification of prekallikrein with arginine agarose, *Methods Enzymol.*, 34, 432–435, 1974.

38. Bartholomew, J.S. and Woolley, D.E., Plasminogen activator release from cultured murine mast cells, *Biochem. Biophys. Res. Commun.*, 153, 540–544, 1988.

39. Chibber, B.A.K., Deutch, D.G., and Mertz, E.T., Plasminogen, *Methods Enzymol.*, 34, 424–432, 1974.

40. Takahashi, R., Akiba, K., Koike, M., Noguchi, T., and Ezure, Y., Affinity chromatography for purification of two urokinases from human urine, *J. Chromatogr. B*, 742, 71–78, 2000.

41. Porath, J., Carlsson, J., Olsson, I., and Belfrage, G., Metal chelate affinity chromatography approach to protein fractionation, *Nature*, 258, 598–599, 1975.

42. Ohkubo, I., Kondo, T., and Taniguchi, N., Purification of nucleosidediphosphatase of rat liver by metal chelate affinity chromatography, *Biochim. Biophys. Acta*, 616, 89–93, 1980.

43. Cawston, T.E. and Tyler, J.A., Purification of pig synovial collagenase to high specific activity, *Biochem. J.*, 183, 647–656, 1979.

44. Crowe, J., Dobeli, H., Genz, R., Hochuli, E., Stuber, D., and Henco, K., 6xHis-Ni-NTA chromatography as a superior technique in recombinant protein expression/purification, *Methods Mol. Biol.*, 31, 371–387, 1994.

45. Hochuli, E., Dobeli, H., and Schacher, A., New metal chelate adsorbent selective for proteins and peptides containing neighboring histidine residues, *J. Chromatogr.*, 411, 177–184, 1987.

46. Panagiotidis, C.A. and Silverstein, S.J., pALEX, a dual-tag prokaryotic expression vector for the purification of full-length proteins, *Gene*, 164, 45–47, 1995.

47. Tallet, B., Astier-Gin, T., Castroviejo, M., and Santarelli, X., One-step chromatographic purification procedure of a His-tag recombinant carboxyl half part of the HTLV-I surface

envelope glycoprotein overexpressed in *Escherichia coli* as a secreted form, *J. Chromatogr. B,* 753, 17–22, 2001.

48. Chuturgoon, A.A., Dutton, M.F., and Berry, R.K., The preparation of an enzyme associated with aflatoxin biosynthesis by affinity chromatography, *Biochem. Biophys. Res. Commun.,* 166, 38–42, 1990.

49. Simons, P.C. and Vander Jagt, D.L., Purification of glutathione *S*-transferases from human liver by glutathione-affinity chromatography, *Anal. Biochem.,* 82, 334–341, 1977.

50. McLellan, L.I. and Hayes, J.D., Differential induction of class alpha glutathione *S*-transferases in mouse liver by the anticarcinogenic antioxidant butylated hydroxyanisole: Purification and characterization of glutathione *S*-transferase Ya1Ya1, *Biochem. J.,* 263, 393–402, 1989.

51. Anuradha, D., Reddy, K.V., Kumar, T., Neeraja, S., Reddy, P.R.K., and Reddanna, P., Purification and characterization of rat testicular glutathione *S*-transferases: Role in the synthesis of eicosanoids, *Asian J. Androl.,* 2, 277–282, 2000.

52. Sorol, M.R., Pastori, R.L., Muro, A., Moreno, S., and Rossi, S., Structural and functional analysis of the cAMP binding domain from the regulatory subunit of *Mucor rouxii* protein kinase A, *Arch. Biochem. Biophys.,* 382, 173–181, 2000.

53. Francesconi, M.A., Donella-Deana, A., Furlanetto, V., Cavallini, L., Palatini, P., and Deana, R., Further purification and characterization of the succinyl-CoA:3-hydroxy-3-methylglutarate coenzyme A transferase from rat-liver mitochondria, *Biochem. Biophys. Acta,* 999, 163–170, 1989.

54. Bowers-Komro, D.M., Yamada, Y., and McCormick, D.B., Substrate specificity and variables affecting efficiency of mammalian flavin adenine dinucleotide synthesis, *Biochemistry,* 28, 8439–8446, 1989.

55. Gu, X.B., Gu, T.J., and Yu, R.K., Purification to homogeneity of GD3 synthase and partial purification of GM3 synthase from rat brain, *Biochem. Biophys. Res. Commun.,* 166, 387–393, 1990.

56. Thomassin, H., Jacobson, M.K., Guay, J., Verreault, A., Aboul-ela, N., Menard, L., and Poirier, G.G., An affinity matrix for the purification of poly(ADP-ribose) glycohydrolase, *Nucleic Acids Res.,* 18, 4691–4694, 1990.

57. Jacobson, M.K., Payne, D.M., Alvarez-Gonzales, R., Jurez-Salinas, H., Sims, J.L., and Jacobson, E.L., Determination of *in vivo* levels of polymeric and monomeric ADP-ribose by fluorescence methods, *Methods Enzymol.,* 106, 483–494, 1984.

58. Greth, M. and Chevallier, M.R., Studies on deoxyribonucleases from Haemophilus influenzae on DNA agarose affinity chromatography: Two-step purification of ATP-dependent deoxyribonuclease, *Biochim. Biophys. Acta,* 390, 168–181, 1975.

59. Bhikhabhai, R., Joelson, T., Unge, T., Strandberg, B., Carlsson, T., and Loevgren, S., Purification, characterization and crystallization of recombinant HIV-1 reverse transcriptase, *J. Chromatogr.,* 604, 157–170, 1992.

60. Marquetant, R., Sabina, R.L., and Holmes, E.W., Identification of a noncatalytic domain in AMP deaminase that influences binding to myosin, *Biochemistry,* 28, 8744–8749, 1989.

61. Pato, M.D. and Kerc, E., Comparison of the properties of the protein phosphatases from avian and mammalian smooth muscles: Purification and characterization of rabbit uterine smooth muscle phosphatases, *Arch. Biochem. Biophys.,* 276, 116–124, 1990.

62. Lindahl, T.L., Ohlsson, P.-I., and Wiman, B., The mechanism of the reaction between human plasminogen-activator inhibitor 1 and tissue plasminogen activator, *Biochem. J.,* 265, 109–113, 1990.

63. Hirasawa, M., Chang, K.T., and Knaff, D.B., Characterization of a ferredoxin:NADP$^+$ oxidoreductase from a nonphotosynthetic plant tissue, *Arch. Biochem. Biophys.,* 276, 251–258, 1990.

64. Nefsky, B. and Bretscher, A., Preparation of immobilized monomeric actin and its use in the isolation of protease-free and ribonuclease-free pancreatic deoxyribonuclease I, *Eur. J. Biochem.,* 179, 215–219, 1989.

65. Broger, C., Bill, K., and Azzi, A., Affinity chromatography purification of cytochrome c oxidase from bovine heart mitochondria and other sources, *Methods Enzymol.,* 126, 64–72, 1986.

66. Hadas, E., Koppel, R., Schwartz, F., Raviv, O., and Fleminger, G., Enhanced activity of immobilized dimethylmaleic anhydride-protected poly- and monoclonal antibodies, *J. Chromatogr.,* 510, 303–309, 1990.

67. Duerksen-Hughes, P.J., Williamson, M.M., and Wilkinson, K.D., Affinity chromatography using proteins immobilized via arginine residues: Purification of ubiquitin carboxyl-terminal hydrolases, *Biochemistry,* 28, 8530–8536, 1989.

68. Hershko, A., Heller, H., Elias, S., and Ciechanover, A., Components of ubiquitin-protein ligase system: Resolution, affinity purification, and role in protein breakdown, *J. Biol. Chem.,* 258, 8206–8214, 1983.

69. Friedberg, F., Affinity chromatography and insoluble enzymes, *Chromatogr. Rev.,* 14, 121–131, 1971.

70. Chan, W.W., Matrix-bound protein subunits, *Biochem. Biophys. Res. Commun.,* 41, 1198–1204, 1970.

71. Gurne, D., Chen, J., and Shemin, D., Dissociation and reassociation of immobilized porphobilinogen synthase: Use of immobilized subunits for enzyme isolation, *Proc. Natl. Acad. Sci. U.S.A.,* 74, 1383–1387, 1977.

72. Henrikson, K.P., Allen, S.H., and Malloy, W.L., An avidin monomer affinity column for the purification of biotin-containing enzymes, *Anal. Biochem.,* 94, 366–370, 1979.

73. Dimroth, P., Purification of the sodium transport enzyme oxalacetate decarboxylase by affinity chromatography on avidin Sepharose, *FEBS Lett.,* 141, 59–62, 1982.

74. Brodelius, P., Larsson, P.-O., and Mosbach, K., Synthesis of three AMP-analogs: N^6-(6-aminohexyl)-adenosine 5′-monophosphate, N^6-(6-aminohexyl)-adenosine 2′,5′-bisphosphate, and N^6-(6-aminohexyl)-adenosine 3′,5′-bisphosphate and their application as general ligands in biospecific affinity chromatography, *Eur. J. Biochem.,* 47, 81–89, 1974.

75. Waters, C.A., Murphy, J.R., and Hastings, J.W., Flavine binding by bacterial luciferase: Affinity chromatography of bacterial luciferase, *Biochem. Biophys. Res. Commun.,* 57, 1152–1158, 1974.

76. Anzellotti, D. and Ibrahim, R.K., Thermostable beta-galactosidase from the archaebacterium *Sulfolobus solfataricus:* Purification and properties, *Arch. Biochem. Biophys.,* 382, 161–172, 2000.

77. Steers, E., Jr. and Cuatrecasas, P., Isolation of beta-galactosidase by chromatography, *Methods Enzymol.,* 34, 350–358, 1974.

78. Pisani, F.M., Rella, R., Raia, C.A., Rozzo, C., Nucci, R., Gambacorta, A., De Rosa, M., and Rossi, M., Thermostable beta-galactosidase from the archaebacterium *Sulfolobus solfataricus:* Purification and properties, *Eur. J. Biochem.,* 187, 321–328, 1990.

79. Schweden, J. and Bause, E., Characterization of trimming Man9-mannosidase from pig liver: Purification of a catalytically active fragment and evidence for the transmembrane nature of the intact 65 kDa enzyme, *Biochem. J.,* 264, 347–355, 1989.

80. Staudenbauer, W.L. and Orr, E., DNA gyrase: Affinity chromatography on novobiocin-Sepharose and catalytic properties, *Nucleic Acids Res.,* 9, 3589–3603, 1981.

81. Melendy, T. and Ray, D.S., Novobiocin affinity purification of a mitochondrial type II topoisomerase from the trypanosomatid *Crithidia fasciculata, J. Biol. Chem.,* 264, 1870–1876, 1989.

82. Raeber, A.J., Riggio, G., and Waser, P.G., Purification and isolation of choline acetyltransferase from the electric organ of *Torpedo marmorata* by affinity chromatography, *Eur. J. Biochem.,* 186, 487–492, 1989.

83. Dadabay, C.Y. and Pike, L.J., Purification and characterization of a cytosolic transglutaminase from a cultured human tumour-cell line, *Biochem. J.,* 264, 679–685, 1989.

84. Matsuda, Y., Shimura, K., Shiraki, H., and Nakagawa, H., Purification and properties of adenylosuccinate synthetase from Yoshida sarcoma ascites tumor cells, *Biochim. Biophys. Acta,* 616, 340–350, 1980.

85. Eliasson, R., Fontecave, M., Joernvall, H., Krook, M., Pontis, E., and Reichard, P., The anaerobic ribonucleoside triphosphate reductase from *Escherichia coli* requires *S*-adenosylmethionine as a cofactor, *Proc. Natl. Acad. Sci. U.S.A.,* 87, 3314–3318, 1990.

86. Knorre, D.G., Kurbatov, V.A., and Samukov, V.V., General method for the synthesis of ATP gamma-derivatives, *FEBS Lett.*, 70, 105–108, 1976.
87. Izuta, S. and Saneyoshi, M., AraUTP-Affi-Gel 10: A novel affinity absorbent for the specific purification of DNA polymerase alpha-primase, *Anal. Biochem.*, 174, 318–324, 1988.
88. Holmberg, L., Bladh, B., and Astedt, B., Purification of urokinase by affinity chromatography, *Biochim. Biophys. Acta*, 445, 215–222, 1976.
89. Khamlichi, S., Muller, D., Fuks, R., and Jozefonvicz, J., Specific adsorption of serine proteases on coated silica beads substituted with amidine derivatives, *J. Chromatogr.*, 510, 123–132, 1990.
90. Campbell, B.J., Forrester, L.J., Zahler, W.L., and Burks, M., Beta-lactamase activity of purified and partially characterized human renal dipeptidase, *J. Biol. Chem.*, 259, 14586–14590, 1984.
91. Huang, J.S., Huang, S.S., and Tang, J., Cathepsin D isozymes from porcine spleens: Large-scale purification and polypeptide chain arrangements, *J. Biol. Chem.*, 254, 11405–11417, 1979.
92. Conner, G.E., Isolation of procathepsin D from mature cathepsin D by pepstatin affinity chromatography: Autocatalytic proteolysis of the zymogen form of the enzyme, *Biochem. J.*, 263, 601–604, 1989.
93. Rittenhouse, J., Turon, M.C., Helfrich, R.J., Albrecht, K.S., Weigl, D., Simmer, R.L., Mordini, F., Erickson, J., and Kohlbrenner, W.E., Affinity purification of HIV-1 and HIV-2 proteases from recombinant *E. coli* strains using pepstatin-agarose, *Biochem. Biophys. Res. Commun.*, 171, 60–66, 1990.
94. Rich, D.H., Brown, M.A., and Barrett, A.J., Purification of cathepsin B by a new form of affinity chromatography, *Biochem. J.*, 235, 731–734, 1986.
95. Buttle, D.J., Bonner, B.C., Burnett, D., and Barrett, A.J., A catalytically active high-Mr form of human cathepsin B from sputum, *Biochem. J.*, 254, 693–699, 1988.
96. Buttle, D.J., Kembhavi, A.A., Sharp, S.L., Shute, R.E., Rich, D.H., and Barrett, A.J., Affinity purification of the novel cysteine proteinase papaya proteinase IV and papain from papaya latex, *Biochem. J.*, 261, 469–476, 1989.
97. Tibbot, B.K., Wong, D.W.S., and Robertson, G.H., A functional raw starch-binding domain of barley alpha-amylase expressed in *Escherichia coli*, *J. Protein Chem.*, 19, 663–639, 2000.
98. Loog, M., Uri, A., Jarv, J., and Ek, P., Bi-Substrate analogue ligands for affinity chromatography of protein kinases, *FEBS Lett.*, 480, 244–248, 2000.
99. Patel, A., O'Hara, M., Callaway, J.E., Greene, D., Martin, J., and Nishikawa, A.H., Affinity purification of tissue plasminogen activator using transition-state analogs, *J. Chromatogr.*, 510, 83–93, 1990.
100. Basak, A., Gong, Y.T., Cromlish, J.A., Paquin, J.A., Jean, F., Seidah, N.G., Lazure, C., and Chretien, M., Syntheses of argininal semicarbazone containing peptides and their applications in the affinity chromatography of serine proteinases, *Int. J. Peptide Protein Res.*, 36, 7–17, 1990.
101. Lin, S.X., Shi, J.P., Cheng, X.D., and Wang, Y.L., Arginyl-tRNA synthetase from *Escherichia coli*, purification by affinity chromatography, properties, and steady-state kinetics, *Biochemistry*, 27, 6343–6348, 1988.
102. Nishikawa, Y., Pegg, W., Paulsen, H., and Schachter, H., Control of glycoprotein synthesis: Purification and characterization of rabbit liver UDP-*N*-acetylglucosamine:alpha-3-D-mannoside beta-1,2-*N*-acetylglucosaminyltransferase I, *J. Biol. Chem.*, 263, 8270–8281, 1988.
103. Bendiak, B. and Schachter, H., Control of glycoprotein synthesis: Purification of UDP-*N*-acetylglucosamine:alpha-D-mannoside beta 1-2 *N*-acetylglucosaminyltransferase II from rat liver, *J. Biol. Chem.*, 262, 5775–5783, 1987.
104. Homandberg, G.A., Litwiller, R.D., and Peanasky, R.J., Carboxypeptidase inhibitors from *Ascaris suum*: The primary structure, *Arch. Biochem. Biophys.*, 270, 153–161, 1989.

105. Martzen, M.R., McMullen, B.A., Smith, N.E., Fujikawa, K., and Peanasky, R.J., Primary structure of the major pepsin inhibitor from the intestinal parasitic nematode *Ascaris suum, Biochemistry,* 29, 7366–7372, 1990.

106. Yasuhara, T. and Ohashi, A., Apocarboxypeptidase B-Sepharose: A specific adsorbent for peptides, *Biochem. Biophys. Res. Commun.,* 166, 330–335, 1990.

107. Kumazaki, T., Teraswa, K., and Ishii, S., Affinity chromatography on immobilized anhydrotrypsin: General utility for selective isolation of C-terminal peptides from protease digests of proteins, *J. Biochem.,* 102, 1539–1546, 1987.

108. Yokosawa, H. and Ishii, S., Anhydrotrypsin: New features in ligand interactions revealed by affinity chromatography and thionine replacement, *J. Biochem. (Tokyo),* 81, 647–656, 1977.

13

Isolation of Recombinant Proteins by Affinity Chromatography

Anuradha Subramanian

Department of Chemical Engineering, University of Nebraska, Lincoln, NE

CONTENTS

13.1   INTRODUCTION

Advances in recombinant DNA technology, genetics, and biotechnology have enabled the biosynthesis and production of complex therapeutic proteins in a variety of systems. Once they have been synthesized intracellularly in the host organism, these proteins are either exported into the culture medium or retained intracellularly as inclusion bodies. The purification method that is used for these proteins will depend on their concentrations and their final state in the raw sample. This chapter examines the use of antibodies and other affinity ligands for the purification of recombinant proteins from complex mixtures.

13.1.1 General Requirements for the Purification of Recombinant Proteins

The factors needed for the purification of recombinant proteins are similar for many source materials. This includes culture media used for bacteria, mammalian cells, and plant cells. It also includes the milk of transgenic animals. A sequence of six major unit operations is usually required for this purification. These steps are (1) the removal of solids, (2) volume reduction, (3) sample prepurification (e.g., as performed by ion-exchange), (4) viral inactivation, (5) high-resolution purification (e.g., as obtained by affinity chromatography), and (6) a final low-resolution cleanup [1, 2].

Another requirement for the purification procedure is that it must include methods to control proteolytic degradation of the recombinant protein. This is needed so that the purification process will ultimately yield a product in a biologically active form. Although meeting all of these requirements is not trivial, it is easier to accomplish in some expression systems than in others [3]. Various methods that can be used as part of this purification process are shown in Table 13.1. However, the emphasis in this current chapter will be on affinity chromatographic techniques.

The purification of recombinant proteins from biological sources is usually complicated by the presence of endogenous proteins [4]. Purification methods based on ion exchange or adsorption serve as excellent prepurification steps, but they usually fail to yield a single purified protein from a complex mixture [5]. Purification techniques based on more-specific interactions, such as those used in immunoaffinity chromatography, have also been developed through the use of a variety of biological and synthetic ligands [1, 5, 6]. Reviews on the topics of protein purification by dye-ligand and biomimetic affinity chromatography can be found in Chapters 9 and 18. In addition, Chapters 14 and 16 discuss the use of antibodies and receptors for product isolation. In this current chapter, the use of affinity chromatography with either protein-based or synthetic ligands will be considered for use in recombinant protein purifications.

Table 13.1   Methods for Recombinant Protein Purification

Method	Degree of Protein Resolution	Ease of Scale-Up
Ultrafiltration	Low	Easy
Precipitation	Low to medium	Easy
Molecular exclusion	Low to medium	Medium
Electrophoresis	Medium to high	Difficult
Electrofocusing	Medium to high	Difficult
Isotachophoresis	Medium	Difficult
Adsorption	Medium	Medium
Ion exchange	Medium	Medium
Affinity	High	Medium

Affinity chromatography was established as a practical technique for purification during the 1970s. Since that time, it has also grown in use as an analytical tool [7]. The exquisite selectivity and specificity of affinity-based separations has allowed the use of these methods in both laboratories and on a commercial scale [1, 5]. Affinity-based methods are now used in about 8 to 10% of the bioseparation procedures. However, with the introduction of new drugs and therapeutics based on monoclonal antibodies or humanized monoclonal antibodies, the use of affinity-based purifications is expected to grow.

13.1.2 Comparison of Expression Systems

Recent advances in recombinant DNA techniques and bioreactor technology have made it possible to make and modify proteins in a wide variety of hosts [8–11]. This ability has already delivered important technological innovations in the protection of humans, animals, and their environment through the discovery of lifesaving drugs, vaccines, and artificial organs [12, 13].

Current systems used for the expression of proteins include bioreactors based on cultures of bacteria, yeast, insect cells, or mammalian cells, as well as the use of bioreactors that are based on transgenic animals [3, 10]. Of these, the bacterial-expression system has been used extensively for heterologous protein production. In this approach, the bacteria often accumulate the recombinant protein product in the form of inclusion bodies. Isolating these inclusion bodies from bacterial cells by centrifugation represents a first major step in the purification of such recombinant proteins; these proteins are then solubilized with an organic solvent and, in some cases, treated so that they can properly refold.

In contrast to bacterial-expression systems, those that make use of yeast or insect cell cultures express and secrete recombinant proteins into an extracellular medium. Expression systems based on mammalian cells provide the most active product and one that has received the most proteolytic processing. However, the expression levels found in mammalian cells are typically lower than those found in yeast and insect cell cultures. In these systems, it may be necessary (depending on the vehicle for expression) to include a centrifugation step during protein purification to remove cells and other intracellular components. This would then be followed by a step that exchanges the buffer into a more useful form. The resulting sample would then be further processed by methods such as ion-exchange or affinity chromatography to isolate the desired recombinant protein from the treated cell culture medium.

13.2 IMMUNOAFFINITY CHROMATOGRAPHY

One approach that can be used to help isolate a recombinant protein is immunoaffinity chromatography (IAC). As discussed in Chapter 6, this is a method in which the binding between an antigen (i.e., the species of interest) and an immobilized antibody is utilized as the basis for the separation. In the isolation of a recombinant protein, this technique would involve obtaining an antibody that has specific binding for this protein and immobilizing this antibody to a solid support for use in a column. This immobilized antibody support is also known as an *immunosorbent*.

This support is used by passing the sample mixture over the immunosorbent under conditions that allow the antibodies to capture the protein of interest while allowing other proteins or sample components to wash through the column [14, 15]. An elution solvent is then passed through the immunosorbent to dissociate the retained protein and allow its release from the column. This elution generally makes use of a change in pH, added metal ions, or the use of reagents like sodium thiocyanate (NaSCN) and urea. The result is the appearance of a highly purified product in the column eluate [1].

13.2.1 Purification with Monoclonal Antibodies

Monoclonal antibodies (Mabs) have become indispensable tools for protein characterization and purification [15, 16], principally because they can be made with a customized avidity and specificity. The use of IAC and monoclonal antibodies to purify proteins is widespread, and a detailed description of the theoretical and practical aspects of this method can be found in earlier reports [17, 18]. Additional information on the use of immobilized antibodies in the isolation of chemicals can be found in Chapter 14.

The monoclonal antibodies used for purifying a given protein are usually made in hybridoma reactors or ascites fluid. These antibodies are then isolated from their cell culture and placed onto an immunosorbent. Monoclonal antibodies with dissociation constants of 10^{-7} to 10^{-8} M are usually preferred for this work, since these can allow a recombinant protein to be recovered under physiological conditions.

The high cost of producing monoclonal antibodies makes immunoaffinity chromatography a cost-intensive technique for protein purification. This is true both in the laboratory and on an industrial scale. Furthermore, large-scale applications of immunoaffinity chromatography have been limited by the low functional efficiency that results when antibodies are covalently anchored to supports. Thus, increasing the activity of this immunosorbent can reduce some of these costs. Using immobilized antibodies with a higher functional activity can also reduce buffer costs (which make up about 40% of the processing costs in large-scale chromatographic purifications) by decreasing the mobile-phase volumes required for column washing and product elution.

13.2.2 Optimizing Immunosorbent Performance

The performance of an immunosorbent is dependent on the physical characteristics of the matrix, the surface density of the immobilized antibodies, and the orientation of these antibodies [19]. For instance, a high ratio of reactive groups per unit area during the coupling of antibodies to a matrix can lead to the clustering of antibodies close to the surface of the matrix. This in, turn, can reduce the binding capacity of these antibodies for their targets due to steric hindrance. It is also imperative for the immobilized antibodies to have the correct orientation for binding with an antigen in solution. In addition, for an immunosorbent to obtain its maximum binding capacity, the binding sites on the antibodies should be protected during the immobilization process. Multipoint attachment of an antibody to the matrix can also lead to a loss in activity through blocking or distortion of the antibody's structure. Further details on these and other immobilization effects on the performance of affinity ligands can be found in Chapter 3.

Various methods have been described for enhancing the antigen-binding activity of immobilized antibodies [19]. Some of these involve the use of site-specific immobilization, as is described in Chapter 6. Another approach involves the protection of antigen-binding domains on monoclonal antibodies through the use of F_{ab}-masking antigens (FMAs). This has been illustrated in work with Emphaze AB1 and Affiprep supports that contained murine monoclonal antibodies against human protein C (hPC). These antibodies were immobilized in the presence of either synthetic or recombinant FMAs, where the synthetic FMAs consisted of peptides or polymer-peptide conjugates.

One FMA was prepared by modifying the lysine residues of recombinant human protein C (rhPC) with an acetic acid ester of N-hydroxysuccinimide. This FMA was then mixed with the corresponding antibodies, allowed to bind to their F_{ab} regions, and kept in the presence of these antibodies during their covalent immobilization. These antibodies were immobilized under conditions that gave low bead-averaged densities (i.e., 0.4 to 1.1 mg antibody/ml matrix) to minimize local density effects. Later, the remaining reactive groups on the support were inactivated, and the FMA was removed from the immobilized antibodies.

It was found that immunosorbents prepared with FMAs gave an antigen-binding efficiency (η_{Ag}) of 42 to 48% as compared with 18 to 22% for those attached without such agents. The theoretical antigen-binding efficiency was determined by using the following formula, which assumes two target antigens can bind to each antibody.

$$\eta_{Ag} = 100 \ (2/1) \ M \ V \ (MW_{Ag}/MW_{Mab}) \qquad (13.1)$$

In this equation, M is the density of the immobilized antibodies (in mg antibody/ml support), MW_{Mab} is the molecular weight of the antibody (150,000 g/mol, in this case), MW_{Ag} is the molecular weight of the antigen being purified (in g/mol), and V is the volume of the gel support (in ml). It was also found that after digestion with pepsin, immunosorbents that were prepared with FMA-masked antibodies gave three- to fourfold more released $F_{(ab)2}$ fragments than matrixes made with unmasked antibodies. This result gave correlation with the higher functional efficiency (η_{Ag}) of the masked immunosorbents and indicated that they also had antibodies with better orientation for antigen binding.

Another approach for maximizing the binding ability of an antibody is to immobilize the antibody through its F_c region. One way that this can be accomplished is by adsorbing antibodies to a secondary ligand, like protein A [20] or protein G, that interacts with this region. Further details on this approach can be found in Chapter 3. It is also possible to directly immobilize an antibody through its F_c region by using a hydrazide-based approach [21–26]. This makes use of the fact that antibodies belong to the family of glycoproteins, with carbo-hydrate groups distant from the F_{ab} domains and generally located in the F_c region. As a result, the chemical modification of an antibody at these groups will usually not affect its antigen-binding activity.

The immobilization of an antibody through its carbohydrate residues requires the use of a derivatized support that contains a hydrazido ($-NHNH_2$) group. This also requires that the carbohydrate residues on an antibody be previously oxidized with an agent like sodium metaperiodate to yield aldehyde groups. These aldehyde groups can then react with the activated support to yield stable hydrazone linkages [21, 22]. Further details on this reaction can be found in Chapter 6.

A variety of monoclonal and polyclonal antibodies have been immobilized through this hydrazide approach [21–25]. A two- to tenfold enhancement in immunosorbent performance versus random coupling methods was observed when this was used for monoclonal antibodies [21–25]. No significant improvement in the performance of immobilized polyclonal antibodies prepared by this approach was noted in one study [21]. One application of this method in the field of protein purification has been its use to prepare immobilized monoclonal antibody columns for the isolation of human factor IX from plasma fractions [26].

13.2.3   Development of an Immunoaffinity Purification

The use of immunoaffinity chromatography in protein purification can be illustrated by using recombinant human protein C (hPC) as an example. hPC is a vitamin K-dependent plasma protein that is important to the regulation of hemostasis and blood coagulation. It is a heterodimer composed of heavy and light chains and has an overall molecular weight of 62 kDa. A general representation of the structure of hPC is given in Figure 13.1. Some important items in the structure of hPC include (1) an activation peptide, which is critical to its anticoagulant function, (2) the presence of glycosylation at four N-linked sites, and

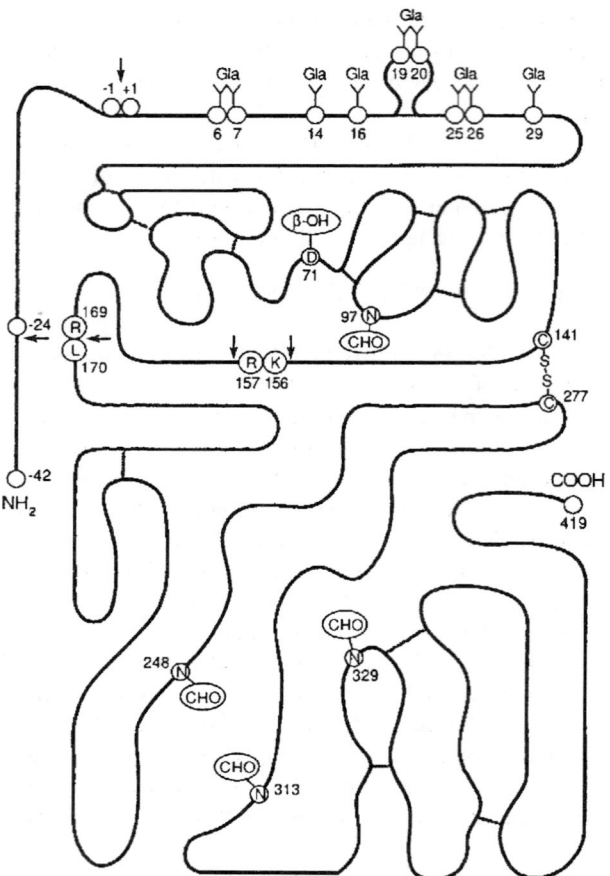

Figure 13.1   Schematic representation of human protein C. (From Yan, S.C.B., Razzano, P., Chao, Y.B., Walla, J.D., Berg, D.T., McClure, D.B., and Grinell, B.W., *BioTechnology*, 8, 655–661, 1992; and from Subramanian, A., Paleyanda, R.K., Lubon, H., Williams, B.L., Gwazdauskas, F.C., Knight, J., Drohan, W.N., and Velander, W.H., *Ann. N.Y. Acad. Sci.*, 782, 87–92, 1996. With permission.)

(3) the presence of γ-carboxylation at nine glutamic acid residues in the mature protein from plasma or in the recombinant protein product.

13.2.3.1   Importance of Protein Source and Subpopulations

The clinical value of hPC is well documented [12, 27], but the low concentration of hPC in plasma (4 mg/l) has resulted in hPC being replaced with recombinant human protein C (rhPC) for therapeutic purposes. Previous attempts to synthesize rhPC from various cell lines, such as Chinese hamster ovary (CHO), kidney (MK2 or HepG2), and mouse epi-thelioid, have resulted in mixtures of both mature and immature rhPC subpopulations [28, 29]. As an example, rhPC has been expressed in CHO cell bioreactors and in Syrian hamster cell lines (AV12-664) at levels ranging from 5 to 20 mg/l culture medium [28], with both mature and immature forms having been detected. Moreover, rhPC has also been expressed in the milk of transgenic mice and pigs at concentrations ranging from 0.25 to 2.5 g rhPC/l milk [30, 31].

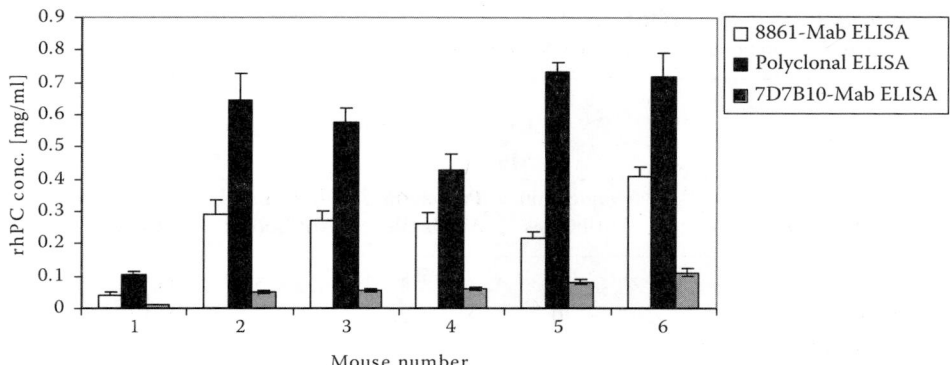

Figure 13.2 Detection of rhPC in the milk of transgenic mice or pigs by ELISA methods with monoclonal or polyclonal antibodies being used as reagents. In this study, the milk was collected on days 10 to 15 of lactation. The ELISAs were based on an EDTA-dependent monoclonal antibody (7D7B10-Mab), a pH-dependent monoclonal antibody (8861-Mab), or goat polyclonal antibodies against hPC. The values shown here represent an average of four to six dilutions performed in duplicate and in two independent assays.

In comparing these various sources, it is necessary to consider the immature populations of protein. For rhPC, these populations have incomplete protein processing, as indicated by their structures and decreased activity in coagulation assays. Thus, one challenge in the development of a downstream separation strategy for this protein lies in the creation of a procedure that can separate its subpopulations (e.g., forms of this protein with slight variations in glycosylation or altered proteolytic processing). The use of monoclonal antibodies and immunoaffinity chromatography is one way that this can be accomplished.

This presence of these immature subpopulations also requires specific assays that will enable the measurement of these various subpopulations in the cell culture supernatant or milk of transgenic animals. Monoclonal antibodies are also valuable for this type of work, since they can be used in methods such as an enzyme-linked immunosorbent assay (ELISA). For instance, Table 13.2 gives data for the production of rhPC in the milk of transgenic mice and pigs. In this case, hPC cDNA and hPC genomic DNA were used under the influence of a mammary gland specific promoter to express rhPC in the milk of transgenic animals (i.e., pigs and mice). The genomic DNA sequence contained the gene of interest along with all the introns and exons, while the cDNA contained the gene of interest without the introns, thus making it less complex.

As shown in Figure 13.2, an ELISA method using polyclonal antibodies gave an estimated level of expression for rhPC of 900 ± 100 μg/ml milk in pigs and 895 ± 20 μg/ml milk in mice when using the genomic DNA construct. This value represents the total amount of rhPC, including both mature and immature populations. This is much greater than the levels that are usually found in human plasma or that are expressed when using a CHO cell culture (see bottom entries in Table 13.2). But a similar assay that used monoclonal antibodies (7D7B10-Mab) against the γ-carboxyglutamic acid domain of the hPC light chain gave detected levels of 60 or 125 μg/ml, indicating that only 6 to 15% of the total rhPC was completely γ-carboxylated [30]. When the cDNA or genomic DNA of hPC was placed under the influence of the same promoter, the different lines yielded varying levels of expression.

Table 13.2 Detection of rhPC in Milk from Transgenic Pigs or Mice by ELISAs
Employing Monoclonal or Polyclonal Antibodies as Reagents

| | Expression Level of rhPC (μg/ml) | | | |
| | Mice | | Pigs | |
Type of Construct	Monoclonal Antibodies	Polyclonal Antibodies	Monoclonal Antibodies	Polyclonal Antibodies
CDNA	<1	11 ± 4	15 ± 5	350 ± 70
Genomic	125 ± 20	895 ± 20	60 ± 10	900 ± 100
Expression levels of rhPC in CHO cell culture	5–10 μg/ml			
Endogenous levels of hPC in plasma	0.4–2.0 μg/ml			

Note: Column 1 lists the gene construct used to create transgenic mouse and pig lines. Milk was collected on days 10 to 15 of lactation. Concentration of rhPC was assayed in whey by ELISA using EDTA-dependent 7D7B10-Mab or a goat polyclonal antibody to hPC. Values presented here are an average of four to six dilutions, performed in duplicate and in two independent assays.

In another example, detection was performed using a different monoclonal antibody for hPC (8861-Mab), which could bind near the activation peptide in this protein's heavy chain. In this case, the resulting ELISA values represented an overall average expression level and gave results that compared reasonably with those provided by the polyclonal ELISA method in Table 13.2. This information was then used to design a purification method for the various subpopulations of rhPC. This was accomplished by using these various monoclonal antibodies to make immunosorbents that could capture different types of rhPC. The separation strategy that was used in such work is shown in Figure 13.3.

13.2.3.2   Sample Pretreatment

Although it can be difficult to recover recombinant proteins from any system, some are easier to use than others [3]. One item to consider here is the nature of the components that are present with the recombinant protein in the initial sample. For example, milk is a complex biocolloid where the level and diversity of the background proteins presents unique challenges to the development of a purification methodology. This is illustrated by the electrophoretic profiles for various types of skim milk, as shown in Figure 13.4.

In addition to caseins, major whey proteins in such a sample include lactoglobulin (MW, 18 kDa) and lactalbumin (MW, 14 kDa). Nonprotein contaminants such as lipids, carbohydrates, and nucleic acids may also be present and must be removed during the purification of a recombinant product. In general, the level of difficulty in isolating a recombinant protein from milk is related to the total amount of protein that is present. Thus, as shown in Figure 13.4, it would be more technically challenging to work with milk from pigs as opposed to milk from cows, sheep, or goats [4].

The colloidal nature of milk adds a degree of complexity in isolating recombinant proteins that is not found with other expression systems. In such samples, the elimination of casein micelles is a top priority [32]. These casein micelles can be precipitated under acidic conditions (i.e., pH 4.6); however, this may also desialyate the recombinant protein product.

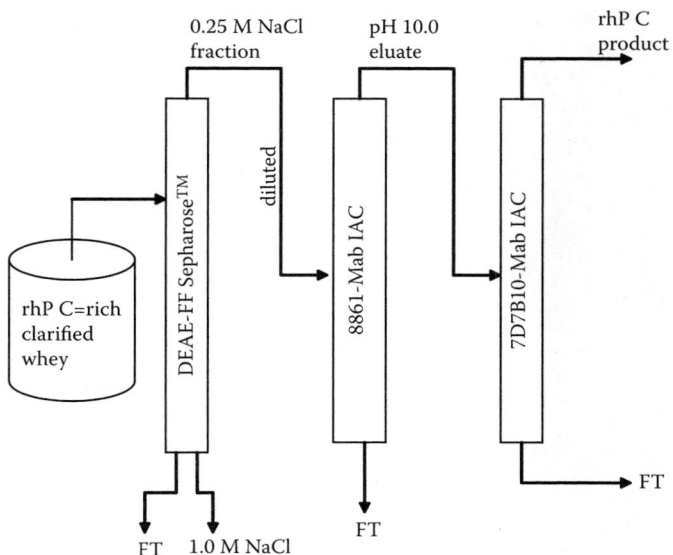

Figure 13.3 A method for the purification of recombinant human protein C (rhPC) from milk obtained from transgenic animals. Prior to this process, milk from transgenic pigs was treated with EDTA, centrifuged, and filtered to yield the rhPC-rich clarified whey. This clarified whey was applied to a DEAE fast-flow (FF) Sepharose ion-exchange column. The rhPC fraction was eluted with 0.25 M NaCl, diluted, and loaded onto an immunoaffinity column containing 8861-Mab antibodies as the ligands. The bound rhPC was later eluted with a pH 10.0, 0.1 M NaHCO$_3$ buffer containing 0.1 M NaCl. The eluted pH 10.0 fraction was then dialyzed against a pH 7.0, 10 mM Tris-HCl buffer containing 50 mM NaCl and 25 mM EDTA and loaded onto a second immunoaffinity column, which contained immobilized 7D7B10-Mab antibodies. The rhPC bound by this second IAC column was eluted with a pH 7.0, 10 mM Tris-HCl buffer containing 50 mM NaCl and 25 mM CaCl$_2$. The abbreviation "FT" in this figure refers to the column fall-through, or nonretained fraction. All collected fractions were assayed for their rhPC content by using ELISA assays, as described in the text.

An alternative route is to solubilize casein micelles by complexing them with EDTA or citrate. For transgenic milk containing rhPC, this can be accomplished by treating the milk with 200 mM EDTA solution in a 1:1 ratio. The resulting mixture is then filtered, and the milk solids are removed by centrifugation to yield clarified whey, with all of these steps being carried out at 4°C. Using this approach, 95% of the rhPC in transgenic milk will appear in the clarified whey product [12, 31].

After the casein micelles have been removed, the remaining whey can be subjected to classic chromatographic methods to recover and purify the desired recombinant protein. The most common technique used for this purpose is ion-exchange chromatography. But the presence of significant amounts of both cationic and anionic proteins in whey reduces the efficacy of this approach, causing it to be used mainly as a prepurification step to reduce the total amount of background or contaminating proteins. For the purification of rhPC from milk, an ion-exchange step using a DEAE-Sepharose fast-flow column has been used. In this case, a mobile phase containing 0.25 M NaCl was found to elute more than 80% of the rhPC from such a column. This eluate was then diluted to a lower ionic strength and applied to an immunoaffinity column, which was used for the further purification of rhPC.

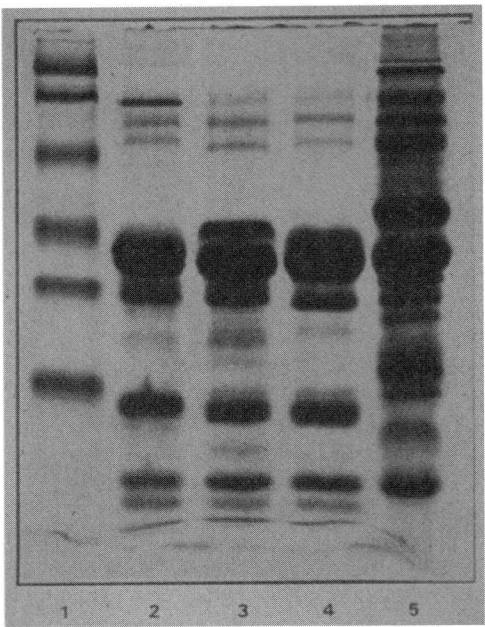

Figure 13.4 SDS-PAGE analysis of the protein in skim milk from various types of livestock. The proteins were detected by staining the final gel with Coomassie blue. Lane 1 contains molecular-weight markers with masses of 106, 80, 49.5, 32.5, and 27.5 kDa. Lanes 2 to 5 show the results for skim milk from a cow, goat, sheep, and pig, respectively.

13.2.3.3   Immunoaffinity Conditions

When purifying a recombinant protein by immunoaffinity chromatography, there are several properties that are desirable for the monoclonal antibodies that will be used in this method. First, they should give an immunosorbent that is able to recognize and bind variants of the recombinant protein (e.g., the different glycosylated forms seen in rhPC), as discussed previously. Second, the elution conditions needed for these antibodies should be gentle enough to retain the biological activity of the protein while also minimizing the coelution of contaminants and leaching of the antibody.

For hPC, immunoaffinity chromatography has been used with monoclonal antibodies that have allowed the elution of this protein under gentle elution conditions, making this a promising tool for the large-scale purification of hPC from plasma [1, 33]. In this case, an EDTA-dependent monoclonal antibody was used (7D7B10-Mab) that was able to bind hPC in the presence of EDTA but released this protein in the presence of 25 mM calcium. It was also possible to use a pH-dependent monoclonal antibody (8861-Mab) that could bind hPC through its activation peptide domain at pH 7.0 but released hPC at pH 10.

Both of these particular monoclonal antibodies were immobilized onto Affiprep beads according to instructions provided by the manufacturer. The amount of antibody remaining in solution after the coupling step was then determined by specific ELISA assays [34]. Immobilization efficiencies of 85 to 95% were obtained for these antibodies. The resulting immunosorbents were then packed onto 15 cm × 1 cm I.D. columns and used in separations performed at 4°C. Since the 8861-Mab column was able to bind a higher percentage of the rhPC populations in transgenic milk, the 0.25 M NaCl fraction

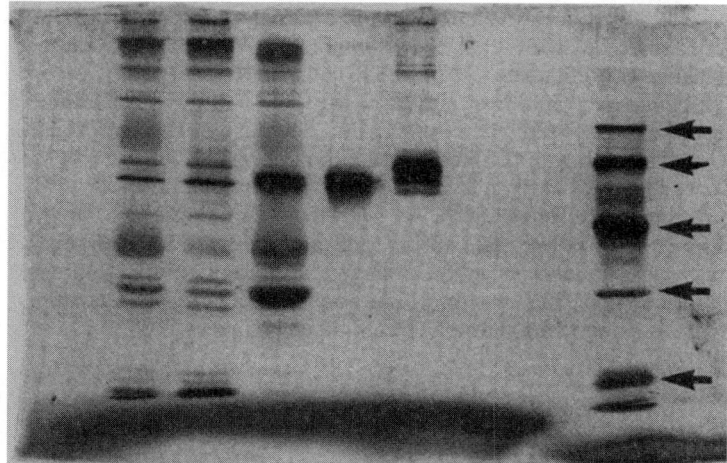

Figure 13.5 SDS-PAGE analysis of recombinant human protein C (rhPC) purified from milk. Lane 1 shows the clarified milk feed. Lane 2 is a 0.2 M retained portion from a DEAE ion-exchange column. Lane 3 shows the fall-through obtained when the sample was passed through an immunoaffinity column containing 8861-Mab antibodies. Lane 4 is the product eluted from the 8861-Mab column at pH 10.0. Lane 5 is a reference sample in which human protein C was present in plasma. Lane 6 shows the molecular-weight ladder for proteins with masses of 97, 66, 45, 31, and 22 kDa.

collected from an ion-exchange prepurification step (as described in the last section) was diluted and first loaded onto this column. The captured rhPC was then eluted from the 8861-Mab column with a pH 10.0 buffer. The typical yields for rhPC in this step were 80 to 90%, with the product being obtained having greater than 90% purity, as indicated in Figure 13.5 by sodium dodecyl sulfate polyacrylamide gel electrophoresis (SDS-PAGE).

In the next step, the rhPC obtained in the pH 10 elution buffer was dialyzed against pH 7.0, 10 mM Tris-HCl containing 50 mM NaCl and 25 mM EDTA. This sample was then loaded onto a column containing the second type of monoclonal antibody (7D7B10-Mab) to allow separation of the active versus inactive populations of rhPC. It was found that this column was only able to bind 33% of rhPC obtained in the pH 10.0 eluate from the first immunoaffinity column (Figure 13.6). However, the rhPC population that was purified by the 7D7B10-Mab column was the fully γ-carboxylated and most functionally active form of rhPC, as determined by amino-terminal sequence analysis. As shown in Table 13.3, the anticoagulation activity for the rhPC purified by this second immunoaffinity column was 67 to 160% of the reference activity for hPC. In contrast to this, the anticoagulation activity for rhPC that had been purified using only the 8861-Mab column was 13 to 35% of this reference activity.

13.3 PSEUDOBIOAFFINITY CHROMATOGRAPHY

In addition to antibodies, there are a variety of other biological ligands that can be used in protein purification. Examples discussed elsewhere in this book include the use of immobilized nucleic acids for the isolation of DNA-binding proteins (Chapter 7), the utilization

Table 13.3 Anticoagulant Activity for Recombinant Human Protein C (rhPC) Obtained from Various Constructs and Purified by Immunoaffinity Columns Containing 8861 Monoclonal Antibodies (Eluted at pH 10.0) or 7D7B10 Monoclonal Antibodies (Eluted with Calcium)

| | Anticoagulant Activity (U/mg) | | |
Construct	Activity of pH 10.0 Fraction	Activity of Calcium Fraction	Average rhPC Expression Level (μg/ml)
CDNA	87 ± 8	170 ± 20	90
CDNA	40 ± 8	400 ± 25	390
Genomic	70 ± 10	200 ± 15	280
Genomic	32 ± 4	190 ± 10	960
hPC	—	—	250

of immobilized cofactors or inhibitors for enzyme purification (Chapter 12), and the use of immobilized receptors for receptor-binding proteins (Chapter 16).

Although these natural ligands can have exquisite specificity, they also can have a high cost and can be labile. This has given rise to the development of alternative methods that make use of nonbiological ligands. This general area has been referred to as pseudo-bioaffinity chromatography (PAC) and is used to describe separations where the synthetic or biomimetic ligands exhibit specificities similar to those observed with natural-affinity ligands. Some specific examples of PAC are discussed in this section, including those that use histidine, biomimetic dyes, or anionic ligands.

13.3.1   Histidine PAC

The amino acid histidine has many unique properties. These properties include its mild hydrophobicity, its ability to undergo weak electron transfer due to its imidazole ring, its wide range of pK_a values, and its asymmetric carbon atom. In addition, histidine residues

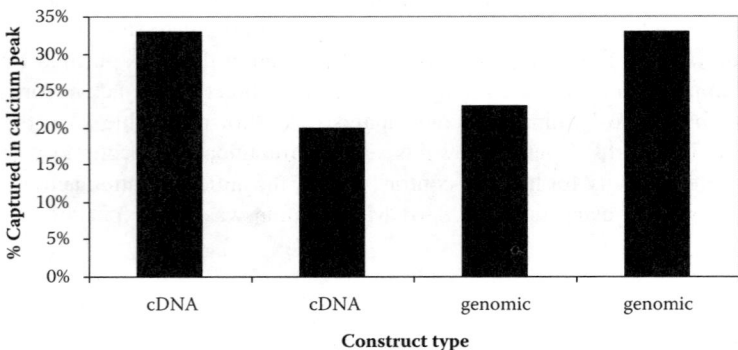

Figure 13.6 The percent of recombinant human protein C (rhPC) captured and eluted from an immunoaffinity column containing 7D7B10-Mab antibodies. The applied rhPC samples had been previously eluted at pH 10.0 from an 8861-Mab immunoaffinity column. The rhPC bound by the 7D7B10-Mab column was eluted with a pH 7.0, 10 mM Tris-HCl buffer containing 50 mM NaCl and 25 mM CaCl$_2$.

play a charge relay role in acid-base catalysis [35–38]. These properties enable histidine as a ligand to interact with proteins under various pH, temperature, and ionic-strength conditions. The interaction of histidine residues on proteins with immobilized metal-ion columns (Chapter 10) and dye-ligand systems suggests that histidine itself can be used as an affinity ligand.

Histidine has been immobilized on Sepharose, silica, and flat or hollow fiber membranes [35–38]. For instance, this has been performed on Sepharose gels by activating these with epichlorohydrin or diglycidyl ether and then reacting them with histidine. The ability of immobilized histidine to interact with many biological molecules has been demonstrated previously [38].

One reported application for immobilized histidine has been its use in purifying immunoglobulin G (IgG) and other antibodies [39, 40]. The binding between these proteins and the immobilized histidine is primarily based on a charge-charge interaction that occurs at or near the isoelectric point of the protein at low ionic strength. Depending on the geometry of the matrix (e.g., the use of beads versus flat membranes) and the activation chemistry that was used to immobilize histidine, varying binding parameters were obtained. For instance, IgG binding capacities of 10 to 20 mg/ml and 50 to 70 mg/ml with dissociation constants in the range of 10^{-6} to 10^{-8} M were reported for histidine immobilized to Sepharose and polyethylene vinyl alcohol, respectively. Immobilized histidine was then used to separate murine monoclonal antibodies from a cell culture supernatant [39] and to isolate IgG from untreated human serum. While moderate yields and purity have been reported for this method, it appears that this approach is not sufficiently specific for all antibody subspecies [40].

13.3.2   Biomimetic Affinity Chromatography

Biomimetic affinity chromatography (BAC) is another type of pseudobioaffinity chromatography. In this technique, a synthetic chemical moiety is used that mimics a natural ligand. Such agents have been investigated for use as sorbents in the purification of therapeutic proteins [40–43]. In addition, synthetic dyes based on triazines and related compounds have been employed as effective biomimetic ligands [44–46]. A more detailed discussion of such dyes and their properties can be found in Chapter 9.

Recently, biomimetic ligands aimed at particular binding sites on target proteins have been developed through the use of combinatorial chemistry and molecular modeling. An *a priori* knowledge of the binding site on a protein and structural information from x-ray crystallographic data combined with molecular modeling techniques has enabled the creation of diverse but focused libraries of biomimetic ligands. One selection criterion included the presence of an aliphatic or aromatic group on an agarose-immobilized triazine. In this work, amino group ligands obtained via molecular modeling were installed on an agarose matrix with cyanuric chloride to give a 1,3,5-trichloro-triazine scaffold. Networks of spatial arrangements were then obtained on this scaffold by substituting the three vulnerable chlorines with reactive amine substituents. Ligands capable of binding elastase, IgG, glycoproteins, and recombinant human factor VIIa were synthesized with this approach based on biomimetic design [47–50]. Affinity gels containing biomimetic ligands against elastase were shown to purify porcine pancreatic elastase and cod elastase from crude extracts.

To illustrate the efficacy and utility of BAC, an example is given in which this method was used to purify recombinant human factor VII [47]. A current method for isolating this factor uses a four-step sequential process based on filtration, ion-exchange chromatography with a mono-Q column, immunoaffinity chromatography with monoclonal antibodies, and a final polishing step on an ion-exchange column. This four-step procedure has a process yield of 40%. In an effort to eliminate the immunoaffinity step, a method based on BAC was developed to yield a durable and sterilizable affinity adsorbent that retained the Ca^{2+}-dependent on/off selectivity of the immunosorbent. Through the use of combinatorial chemistry and molecular modeling, solid-phase synthesis was carried out to give a directed library of 44 ligands on a Sepharose-CL 6B matrix. Immobilized ligands containing 3-aminobenzoic acid, 4-aminobenzoic acid, 2-amino-3-napthoic acid, 2-amino-benzimidazole, and 6-hydroxy-2-aminopurine on a triazine scaffold were particularly effective in giving a calcium-mediated interaction with factor VIIa [47].

Following their development, these synthetic ligands were used to purify recombinant factor VIIA expressed in baby hamster kidney (BHK) cells [47]. Yields and purity in the range of 95 to 99% were obtained when a partially pure, ion-exchanged cell culture medium was used as a feed. It was reported that an affinity matrix based on these synthetic ligands had a tenfold higher capacity than the previous immunoaffinity column, while also being more durable and able to withstand rigorous cleaning procedures.

13.3.3 Anionic Ligand Affinity Chromatography

Anionic ligands are other agents that have been used in pseudobioaffinity chromatography. In this work, chitosan beads with diameters of 400 to 600 μm and a solids content of 3.5% were derivatized with a difunctional epoxide (i.e., 1–4 diglycidyl ether butane diol) to yield reactive epoxide groups [51]. These epoxide groups were then end-titrated with a carboxylic

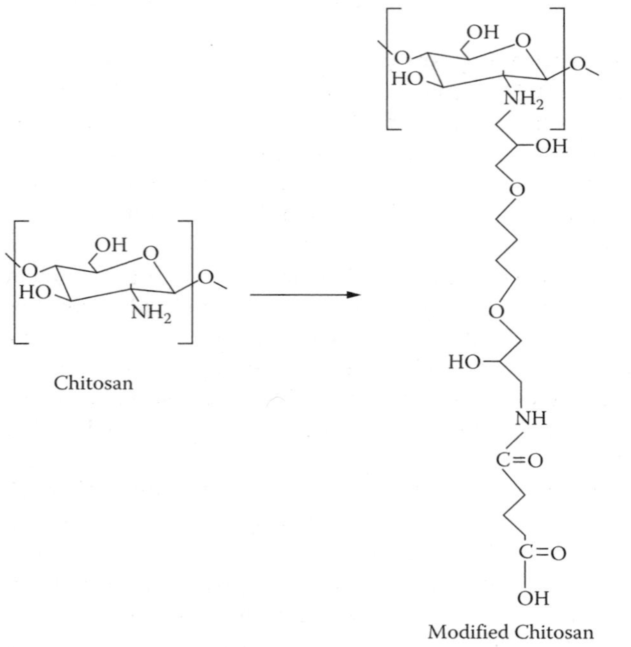

Figure 13.7 The structures of chitosan (left) and chitosan modified to contain an anionic ligand (right). The modified chitosan was prepared by derivatizing the primary amine groups of chitosan to give a reactive epoxide. This epoxide was then reacted with an anionic ligand (i.e., one containing a carboxylate group) that could be used as a pseudobioaffinity matrix for monoclonal antibodies. (From Roy, S.K., Todd, J.S., and Glasser, W.G., Crosslinked Hydrogel Beads from Chitosan, U.S. patent 5,770,712, 1998. With permission.)

acid-containing anionic ligand to yield a pseudobioaffinity matrix [52, 53]. The resulting ligand is shown in Figure 13.7. This matrix was then used in the separation of monoclonal antibodies from cell culture supernatants; to isolate IgG, IgA, and IgM from serum; and to obtain single-chain IgA from colostrum [53].

In one of these studies, a cell culture supernatant rich in monoclonal antibodies was obtained from a bioreactor and directly loaded onto a column packed with ligand-modified chitosan. It was found that this column could be used to give an effective isolation of the monoclonal antibodies in the presence of a salt gradient, as demonstrated in Figure 13.8. In this case, the separation of monoclonal antibodies from other proteins was possible through a differential in binding due to the pseudobioaffinity interactions. Using a similar protocol, this method has been used to separate and enrich subspecies of IgG_{2a}-, IgG_{2b}-, and IgG_3-class monoclonal antibodies from cell culture supernatants (data not included).

Static and dynamic IgG-binding capacities in the range of 55 to 75 mg IgG/ml gel and 10 to 20 mg IgG/ml gel, respectively, have been obtained with these chitosan supports. In addition, dissociation constants of 10^{-5} to 10^{-6} M for IgG have been obtained. Interestingly, anionic ligand-modified chitosan has been found to specifically interact with the F_{ab} region of IgG (see Table 13.4). Current efforts are underway to design other bioinspired ligand cages to selectively bind albumin, serine proteases, and hormones like human gonadotropin.

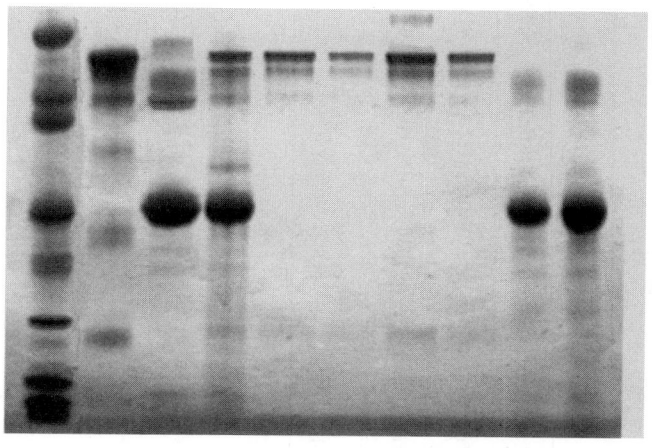

Figure 13.8 Results of an analysis by SDS-PAGE (using a 0.1% polyacrylamide gel and 4 to 12% gradient) of monoclonal antibodies that were purified from a cell culture supernatant using ligand-modified chitosan beads. Lane 1 shows the results for molecular-weight markers. Lanes 2 and 3 show the results for pure monoclonal antibodies and bovine serum albumin (BSA), respectively. Lane 4 shows a representative cell culture supernatant. Lanes 5 and 6 were obtained with an IgG_1 monoclonal antibody that had been purified with a ligand-modified chitosan bead column. Lanes 7 and 8 show a monoclonal antibody that had been purified with a protein A affinity column. Lanes 9 and 10 show the column fall-through.

Table 13.4 Total Recovery (TR) and Yield (Y) for Various Antibody Fragments Obtained on Columns Using Protein A or Ligand-Modified Chitosan as Affinity Ligands

Ligand		F_c Fragments			$F_{(ab)2}$ Fragments			F_{ab} Fragments		
		%TR	%Y	(%Y)/(%TR)	%TR	%Y	(%Y)/(%TR)	%TR	%Y	(%Y)/(%TR)
Protein A	Run 1	88.8	85.0	0.96	91.0	0.9	0.01	97.8	0.1	0.01
	Run 2	92.5	88.0	0.95	83.0	2.0	0.02	94.9	1.9	0.02
Ligand-modified chitosan[a]	Run 1	83.0	2.8	0.03	92.5	25.0	0.27	90.8	61.5	0.68
	Run 2	96.0	4.0	0.04	85.0	30.0	0.35	91.3	60.5	0.67

[a] The ligand-modified chitosan results were obtained with a Ligosep A column.

13.4 SUMMARY AND CONCLUSIONS

This chapter has discussed several affinity methods that have been used in the isolation of recombinant proteins. Methods based on immunoaffinity chromatography and monoclonal antibodies have played a significant role in this area due to the efficacy and the selectivity of monoclonal antibodies. In addition, several techniques have been reported that use pseudobioaffinity chromatography and biomimetic ligands for protein purification. Several examples were given to illustrate these various methods.

In the future, it appears that methods using monoclonal antibodies will continue to be important in these separations, especially as better techniques and tools are developed to enhance the performance of immobilized antibodies. While the design of synthetic affinity ligands has been impressive, the large-scale application of these ligands will depend on their availability and their facile synthesis. In addition, the stability of these ligands is an issue that must be considered, as well as the specificity and selectivity of these ligands for their target proteins. However, it is anticipated that affinity matrices based on biomimetic ligands will gain in commercial importance and be more widely employed in purification of medically relevant proteins.

SYMBOLS AND ABBREVIATIONS

BAC	Biomimetic affinity chromatography
BHK	Baby hamster kidney
BSA	Bovine serum albumin
cDNA	Complementary DNA
CHO	Chinese hamster ovary
DEAE	Diethylaminoethyl
DNA	Deoxyribonucleic acid
EDTA	Ethylenediaminetetraacetic acid
ELISA	Enzyme-linked immunosorbent assay
F_{ab}	Antigen-binding fragment of an antibody
F_c	Crystallizable fragment of an antibody
FMA	F_{ab}-Masking antigen
FT	Column fall-through
hPC	Human protein C
IAC	Immunoaffinity chromatography
IgA	Immunoglobulin A
IgG	Immunoglobulin G
IgM	Immunoglobulin M
M	Immobilized density of monoclonal antibodies (mg antibody/ml gel)
Mab	Monoclonal antibody
MOPS	3-Morpholinopropanesulfonic acid
MW_{Ag}	Molecular weight of an antigen
MW_{Mab}	Molecular weight of a monoclonal antibody
NaSCN	Sodium thiocyanate
PAC	Pseudobioaffinity chromatography
rhPC	Recombinant human protein C
SDS-PAGE	Sodium dodecyl sulfate polyacrylamide gel electrophoresis

TR	Total recovery
V	Volume of a gel (ml)
Y	Yield
η_{Ag}	Antigen binding efficiency

REFERENCES

1. Velander, W.H., Orthner, C.L., Tharakan, J.P., Madurawe, R.D., Ralston, A.H., Strickland, D.K., and Drohan, W.N., Process implications for metal-dependent immunoaffinity interactions, *Biotech. Prog.,* 5, 119–125, 1989.
2. Cartwright, T., Isolation and purification of products from animal cells, *TIBTECH,* 5, 25–30, 1987.
3. Marino, M.H., Expression systems for heterologous protein production, *BioPharm,* 7–8, 18–33, 1989.
4. Wilkins, T.D. and Velander, W.H., Isolation of recombinant proteins from milk, *J. Cell Biochem.,* 49, 333–338, 1992.
5. Chase, H.L., Affinity separations utilizing immobilized monoclonal antibodies, *Chem. Eng. Sci.,* 39, 1099–1125, 1983.
6. Weiss, A.M., Antonsen, K.P., Odde, D.L., and Yarmush, D.M., Immunoaffinity purification: Basic principles and operational considerations, *Biotech. Adv.,* 10, 413–416, 1992.
7. Dean, P.G.D., Johnson, W., and Middle, F., *Affinity Chromatography: A Practical Approach,* IRL Press, Oxford, U.K., 1985.
8. Betenbaugh, M.J., Lindsay, D.A., Juarbe-Osorio, L.G., Gorziglia, M., Vonderfecht, S., and Eiden, J.J., Genetically engineered viral antigens from insect cell cultures, *Ann. N.Y. Acad. Sci.,* 665, 210–215, 1992.
9. Clark, A.J., Simons, P., Wilmut, I., and Lathe, R., Pharmaceuticals from transgenic livestock, *TIBTECH,* 5, 20–24, 1987.
10. Hennighausen, L., The mammary gland as a bioreactor: Production of foreign proteins in milk, *Prot. Exp. Purif.,* 1, 3–8, 1990.
11. Osterrieder, N., Wagner, R., Pfeffer, M., and Kaaden, O.R., Expression of equine herpesvirus type 1 glycoprotein gp 14 in *E. coli* and in insect cells: A comparative study on protein processing and humoral immune responses, *J. Gen. Virology,* 75, 2041–2046, 1994.
12. Velander, W.H., Johnson, J.L., Page, R.L., Russell, C.G., Subramanian, A., Wilkins, T.D., Gwazdauskas, F.C., Pittius, C., and Drohan, W.N., High-level expression of a heterologous protein in the milk of transgenic swine using the cDNA encoding human protein C, *Proc. Nat. Acad. Sci. U.S.A.,* 89, 12003–12007, 1992.
13. Roy, S.N., Kudryk, B.J., and Redman, C.M., Secretion of biologically active recombinant fibrinogen by yeast, *J. Biol. Chem.,* 270, 23761–23767, 1995.
14. Clonis, Y.D., Process-scale affinity chromatography, in *Separation Processes in Biotechnology,* Asenjo, J., Ed., Marcel Dekker, New York, 1990, pp. 401–445.
15. Pfeiffer, N.E., Wylie, D.E., and Schuster, S.M., Immunoaffinity chromatography utilizing monoclonal antibodies, *J. Immunol. Methods,* 97, 1–9, 1987.
16. Milstein, C., Monoclonal antibodies, *Sci. Am.,* 234, 66–70, 1980.
17. Ehle, H. and Horn, A., Immunoaffinity chromatography of enzymes, *Bioseparations,* 3, 47–53, 1990.
18. Pepper, D.S., Selection of antibodies for immunoaffinity chromatography, in *Practical Protein Chromatography,* Kenney, A. and Fowell, S., Eds., Humana Press, Totowa, NJ, 1992, pp. 135–196.
19. Subramanian, A. and Velander, W.H., Ranking the factors impacting immunosorbent performance, *Int. J. BioChromatogr.,* 5, 67–82, 2000.
20. Schneider, C., Newman, R.A., Sutherland, R., Asser, U., and Greaves, M.F., A one-step purification of membrane proteins using a high efficiency immunomatrix, *J. Biol. Chem.,* 257, 10766–10769, 1982.

21. O'Shanessey, D.J. and Hoffman, W.L., Site-directed immobilization of glycoproteins on hydrazide-containing supports, *Biotech. Appl. Biochem.*, 9, 488–496, 1987.
22. Little, M.C., Siebert, C.J., and Matson, R.S., Strategy for the immobilization of monoclonal antibodies on solid-phase supports, *BioChromatography*, 3, 156–160, 1988.
23. Fleminger, G., Hadas, E., Wolf, T., and Solomon, B., Oriented immobilization of periodate-oxidized antibodies on amino and hydrazide derivatives of Eupergit C, *Appl. Biochem. Biotech.*, 23, 123–137, 1990.
24. Domen, P.L., Nevens, J.R., Mallia, A.K., Hermanson, G.T., and Klenk, D.C., Site directed immobilization of proteins, *J. Chromatogr.*, 510, 293–302, 1990.
25. Bonde, M., Pontoppidan, H., and Pepper, D.S., Direct dye binding: A quantitative assay for solid phase immobilized protein, *Prep. Chromatogr.*, 1, 269–277, 1991.
26. Orthner, C.L., Highsmith, F.A., Tharakan, J., Madurawe, R.D., Morcol, T.M., and Velander, W.H., Comparison of the performance of immunosorbents prepared by site-directed or random coupling of monoclonal antibodies, *J. Chromatogr.*, 558, 55–70, 1991.
27. Mann, K.G. and Bonvill. E.G., Protein C deficiency and thrombotic risk, in *Protein C and Related Anticoagulants*, Bruley, D.F. and Drohan, W.N., Eds., Gulf Publishing, Houston, TX, 1990, pp. 119–123.
28. Yan, S.C.B., Razzano, P., Chao, Y.B., Walla, J.D., Berg, D.T., McClure, D.B., and Grinell, B.W., Characterization and novel purification of recombinant protein C from three mammalian cell lines, *BioTechnology*, 8, 655–661, 1992.
29. Grinell, B.W., Walls, J.D., Gerlitz, B., Berg, D.T., McClure, D.B., Ehrlich, H., Bang, N.U., and Yan, S.B., Native and modified recombinant human protein C: Function, secretion, and posttranslational modifications, in *Protein C and Related Anticoagulants*, Bruley, D.F. and Drohan, W.N., Eds., Gulf Publishing, Houston, TX, 1990, pp. 29–63.
30. Van Cott, K.E., Lubon, H., Russell, C.G., Butler, S.P., Gwazdauskas, F.C., Knight, J., Drohan, W.N., and Velander, W.H., Phenotypic and genotypic stability of multiple lines of transgenic pigs expressing recombinant human protein C, *Transgenic Res.*, 6, 203–212, 1997.
31. Subramanian, A., Paleyanda, R.K., Lubon, H., Williams, B.L., Gwazdauskas, F.C., Knight, J., Drohan, W.N., and Velander, W.H., Rate limitations in posttranslational processing by the mammary gland of transgenic animals, *Ann. N.Y. Acad. Sci.*, 782, 87–92, 1996.
32. Ladisch, M.R., Rudge, S.R., Ruettimann, K.W., and Lin, L.K., Bioseparations of milk proteins, in *Bioproducts and Bioprocesses*, Fiechter, A., Okada, H., and Tanner, R.D., Eds., Springer-Verlag, Berlin, 1989, pp. 209–221.
33. Velander, W.H., Clark, D.B., Gee, D., Drohan, W.N., and Morcol, T., Technological challenges for large-scale purification of protein C, in *Protein C and Related Anticoagulants*, Bruley, D.F. and Drohan, W.N., Eds., Portfolio Publishing, The Woodlands, TX, 1990, pp. 11–27.
34. Subramanian, A., Van Cott, K.E., Milbrath, D.S., and Velander, W.H., Role of local antibody density effects on immunosorbent efficiency, *J. Chromatogr.*, 672, 11–24, 1994.
35. Wu, X., Haupt, K., and Vijayalakshmi, M., Separation of immunoglobulin G by high-performance pseudo-bioaffinity chromatography with immobilized histidine, *J. Chromatogr.*, 584, 35–41, 1992.
36. Bueno, S., Haupt, K., and Vijayalakshmi, M., *In vitro* removal of human IgG by pseudo-biospecific affinity membrane filtration on a large scale: A preliminary report, *Int. J. Artificial Organs*, 18, 392–398, 1995.
37. Vijayalakshmi, M.A., Pseudobiospecific ligand affinity chromatography, *TIBTECH*, 7, 71–76, 1989.
38. Mbida, A., Kannoun, S., and Vijayalakshmi, M.A., Purification of IgG1 subclass from human placenta by pseudo affinity chromatography, *Colloque INSERM*, 175, 237–244, 1989.
39. El-Kak, A., Manjini, S., and Vijayalakshmi, M., Interaction of immunoglobulin G with immobilized histidine: Mechanistic and kinetic aspects, *J. Chromatogr.*, 604, 29–37, 1992.
40. Li, R., Dowd, V., Stewart, D.J., Burton, S.L., and Lowe, C.R., Design, synthesis and application of a protein A mimetic, *Nat. Biotechnol.*, 16, 190–195, 1998.
41. Lowe, C.R. and Pearson, J.C., Affinity chromatography on immobilized dyes, *Methods Enzymol.*, 104, 97–113, 1984.

42. Clonis, Y.D., Stead, C.V., and Lowe, C.R., Novel cationic triazine dyes for protein purification, *Biotechnol. Bioeng.,* 30, 621–627, 1987.

43. Lowe, C.R., Burton, S.J., Burton, N.P., Alderton, W.K., Pitts, J.M., and Thomas, J.A., Designer dyes: Biomimetic ligands for the purification of pharmaceutical proteins by affinity chromatography, *TIBTECH,* 10, 442–448, 1992.

44. Burton, S.J., McLoughlin, S.B., Stead, C.V., and Lowe, C.R., Design and applications of biomimetic anthraquinone dyes, I: Synthesis and characterization of terminal ring isomers of CL reactive blue 2, *J. Chromatogr.,* 435, 127–137, 1988.

45. Burton, S.J., Stead, C.V., and Lowe, C.R., Design and applications of biomimetic anthraquinone dyes, II: The interaction of CL reactive blue 2 analogues with horse liver alcohol dehydrogenase, *J. Chromatogr.,* 455, 201–216, 1988.

46. Lindner, N.M., Jeffcoat, R., and Lowe, C.R., Design and applications of biomimetic anthraquinone dyes: Purification of calf intestinal alkaline phosphatase with immobilized terminal ring analogues of CL reactive blue 2, *J. Chromatogr.,* 473, 227–240, 1988.

47. Morill, P.R., Gupta, G., Sproule, K., Winzor, D., Christensen, J., Mollerup, I., and Lowe, C.R., Rational combinatorial chemistry based selection, synthesis and evaluation of an affinity adsorbent for recombinant human clotting factor VII, *J. Chromatogr.,* 774, 1–15, 2002.

48. Fillippusson, H., Erlendsson, L.S., and Lowe, C.R., Design, synthesis and evaluation of biomimetic affinity ligands for elastases, *J. Mol. Recognit.,* 13, 370–381, 2000.

49. Palaniswamy, U.D., Hussain, A., Iqbal, S., Sproule, K., and Lowe, C.R., Design, synthesis and characterization of affinity ligands for glycoproteins, *J. Mol. Recognit.,* 12, 57–66, 1999.

50. Teng, S.F., Sproule K., Hussain, A., and Lowe, C.R., A strategy for the generation of biomimetic ligands for affinity chromatography: Combinatorial synthesis and biological evaluation of an IgG binding ligand, *J. Mol. Recognit.,* 12, 67–75, 1999.

51. Roy, S.K., Todd, J.S., and Glasser, W.G., Crosslinked Hydrogel Beads from Chitosan, U.S. patent 5,770,712, 1998.

52. Subramanian, A., Mascoli, C., Roy, S.K., and Hommerding, J., The use of modified chitosan macrospheres in the selective removal of immunoglobulins, *J. Liq. Chromatogr.,* 24, 2649–2670, 2004.

53. Subramanian, A. and Hommerding, J., Interaction of hIgG with modified chitosan, *J. Liq. Chromatogr.,* 24, 2671–2688, 2004.

14

Affinity Chromatography in Antibody and Antigen Purification

Terry M. Phillips
Ultramicro Analytical Immunochemistry Resource, Division of Bioengineering and Physical Sciences, Office of Research Services, National Institutes of Health, Bethesda, MD

CONTENTS

14.1   INTRODUCTION

The use of immunological reagents such as antibodies and antigens has become popular in many biochemical, biotechnological, and medical fields. For instance, immunoaffinity separations are now widely used as tools for the preanalytical cleanup of samples, drug screening, toxicology, and clinical chemistry [1–7]. One of the major attractions of this approach is the specificity and selectivity of antibody-antigen interactions, which can be used to isolate a single analyte from both simple and complex biological matrices.

The use of antibodies in affinity columns, a technique referred to as *immunoaffinity chromatography*, is a common method used for isolating the corresponding antigens. This chapter examines this method for antigen purification and compares it with other available methods. However, affinity chromatography can also be used to purify antibodies. This may involve the use of immobilized antigens or other types of affinity ligands. These and other methods of antibody isolation are examined in this chapter. A more detailed description of immunoaffinity chromatography and other applications of this technique can be found in Chapter 6.

14.1.1   Antibodies

Antibodies belong to a group of proteins known as *immunoglobulins*. This group, in turn, can be divided into several different classes, as shown in Table 14.1. Antibodies can either be naturally occurring (i.e., raised through the immunization of an animal with a specific antigen) or selected and engineered to have specific characteristics.

Classical examples of naturally occurring antibodies are *polyclonal antibodies*. These are the types of antibodies obtained when an animal like a rabbit or goat is injected with a given antigen. The best-known example of an engineered antibody is a *monoclonal antibody* (Mab). A monoclonal antibody is constructed by fusing a cancer cell (myeloma) with an immune cell that has been sensitized to a specific antigen. Engineered antibodies

Table 14.1   Immunoglobulin Classes and Subclasses

Species	Immunoglobulin Classes	Immunoglobulin Subclasses
Human	IgG	IgG_1, IgG_2, IgG_3, IgG_4
	IgA	IgA_1, IgA_2
	IgM	—
	IgE	—
Mouse	IgG	IgG_1, IgG_{2a}, IgG_{2b}, IgG_3
	IgM	—
Rat	IgG	IgG_1, IgG_{2a}, IgG_{2b}, IgG_{2c}
	IgM	—
Rabbit	IgG	—
	IgM	—
Goat	IgG	IgG_1, IgG_2
	IgM	—
Sheep	IgG	IgG_1, IgG_2
	IgM	—
Horse	IgG	IgG_a, IgG_b, IgG_c, $IgG_{(T)}$

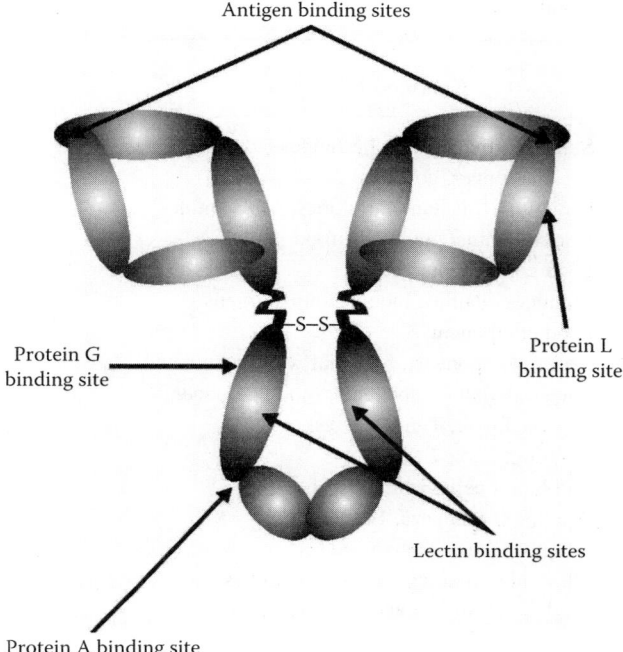

Antigen binding sites

—S–S—

Protein G
binding site

Protein L
binding site

Lectin binding sites

Protein A binding site

Figure 14.1 A general representation of an antibody, showing its various ligand-binding sites.

can also be made by employing techniques in molecular biology. It is possible through such means to produce not only monoclonal antibodies, but also *bifunctional antibodies* and *single-chain antibodies* with variable-region fragments (scF$_v$).

Antibodies exhibit a number of domains that bind a variety of molecules. These domains are illustrated in Figure 14.1 and allow a variety of techniques to be used for antibody purification. These methods range from physicochemical precipitation to affinity chromatography and immunoaffinity procedures. More details on these methods are provided in Section 14.2.

14.1.2 Antigens

The term *antigen* refers to any substance that is capable of eliciting the formation of a specific antibody directed against that substance. As shown in Table 14.2, an antigen can be a drug, peptide, protein, carbohydrate, or lipid. Antigens can also be a combination of these substances, such as a glycoprotein or lipoprotein.

The large number of substances that can act as antigens means that physicochemical techniques like traditional chromatography or electrophoresis can be used to isolate such substances. However, the most efficient approach for this work is to use immunoaffinity separations based on antibodies that are specific for these agents. Section 14.3 examines all of these methods and compares them in terms of their use in antigen purification.

14.2 ANTIBODY PURIFICATION

When isolating antibodies, the available purification procedures can be classified as being class-specific or antigen-specific. The choice of method depends on how broad or selective the final antibody preparation must be. It is also important for the investigator to consider

Table 14.2 Common Materials Classified as Antigens

Classes of Antigens	Common Examples
Inorganic and organic chemicals	Sulfur, formaldehyde, pesticides
Metals	Nickel, copper, gold
Pharmaceutical drugs	Penicillin, β-lactam antibiotics, sulfonimides
Peptides and proteins	Insulin, animal, and plant food proteins
Carbohydrates	Food components
Glycoproteins	Immunoglobulins, blood group antigens
Lipids	Food components
Lipoproteins	Serum components, bacterial wall components
Lipopolysaccharides	Bacterial wall components, food components
Pollens	Ragweed, oak, Timothy grass
Animal danders	Cat, dog, horse
Cells	Red blood cells, white blood cells
Viruses	Smallpox, influenza, HIV
Bacteria	Streptococcus, staphylococcus
Uni- and multicellular parasites	Plasmodia, filariae, tapeworms, flukes
Cell organelles	Nuclei, cell membranes, cell receptors

the source of the material to be used as the capture ligand and whether whole antibodies or antibody fragments are to be employed. Another important issue is the purity of the reagent. Each of these factors is considered in the following subsections with regard to antibody purification.

14.2.1 Considerations in Antibody Purification

Crude preparations of antibodies are usually present in plasma, serum, ascites fluid, or culture medium. However, all of these materials contain immunoglobulins plus other contaminating proteins. Initially, one has to decide whether to perform class-based isolation, such as one for all the immunoglobulin G-class antibodies (IgG) in a sample, or a more specific isolation for a specific type of antibody regardless of its immunoglobulin class.

Class-based antibody isolation can be performed by techniques like precipitation or traditional chromatographic methods that make use of differences in the physicochemical properties of the different types of antibodies [8, 9]. Class-specific isolations can also be performed using affinity techniques that employ bacterial proteins or lectins as general binding agents. However, antigen-specific isolations can only be performed by immunoaffinity procedures that use specific antigens as the capture ligands.

In addition to the class or fragmentation of antibodies, the method of antibody production and other solutes in the sample must also be considered when planning the isolation procedure. For instance, antibodies in ascites fluid are commonly contaminated with host proteins and must be regarded in the same light as antibodies present in serum or plasma. However, monoclonal antibodies produced in a cell culture will be contaminated only by the proteins used to feed the cells (i.e., fetal calf serum or synthetic serum).

14.2.2 Physicochemical Techniques for Antibody Purification

Some common physiochemical methods that are used for antibody purification are precipitation and chromatography. Antibody precipitation can be carried out by using salts, polyethylene glycol, or caprylic acid. Chromatographic techniques for antibody isolation (other than affinity methods) include ion-exchange and size-exclusion chromatography.

14.2.2.1   Antibody Precipitation

There are a number of techniques available for performing the class-specific precipitation of immunoglobulins. Antibody precipitation can be carried out using a variety of agents, ranging from distilled water to complex glycols and specialized acids. Perhaps the simplest technique is *euglobulin precipitation*. A euglobulin is a protein, such as IgM, that is soluble in salt solutions but insoluble in distilled water. IgM antibodies can easily be isolated from animal serum or plasma by simply dialyzing these samples overnight against water. Antigen-specific IgM antibodies cannot be isolated by this approach, but the technique has been used to obtain a starting material for further purification by chromatography [10]. However, not all IgM antibodies precipitate in water, and in such cases, these entities must be isolated by other techniques prior to further purification by chromatography. Water precipitation has also been applied to the preparation of IgG_3 and IgM monoclonal antibodies from mouse ascitic fluid, resulting in greater than 90% recovery with minimal contamination [11].

A more popular and widely applicable approach for immunoglobulin isolation is *salt precipitation*. Immunoglobulins precipitate out of saturated solutions of either sodium or ammonium sulfate. Of these two salts, ammonium sulfate is the more commonly applied reagent [12], especially when nonrefined fractions of immunoglobulin are required. In such cases, further refinement by ion exchange or other forms of chromatography are required to produce a pure class-specific preparation [13–15]. Figure 14.2 gives a flow diagram of the steps required for the ammonium sulfate precipitation of IgG.

Certain antibody classes can also be precipitated using polyethylene glycol (PEG) [16]. However, this has primarily been employed to isolate and analyze antibody/antigen

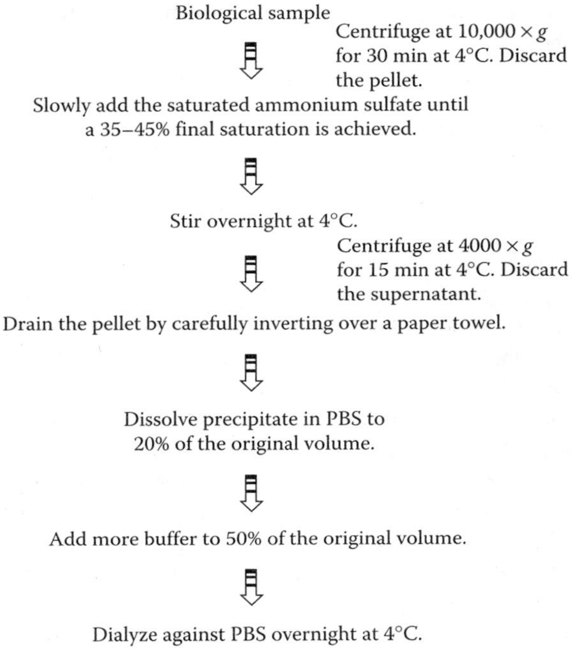

Biological sample
Centrifuge at $10,000 \times g$ for 30 min at 4°C. Discard the pellet.

Slowly add the saturated ammonium sulfate until a 35–45% final saturation is achieved.

Stir overnight at 4°C.
Centrifuge at $4000 \times g$ for 15 min at 4°C. Discard the supernatant.

Drain the pellet by carefully inverting over a paper towel.

Dissolve precipitate in PBS to 20% of the original volume.

Add more buffer to 50% of the original volume.

Dialyze against PBS overnight at 4°C.

Figure 14.2   Flowchart for the ammonium sulfate precipitation of immunoglobulins. (Abbreviations: PBS, phosphate buffered saline.)

complexes [17, 18]. PEG precipitation of IgG has been reported to be a useful preliminary step to the class-specific purification of immunoglobulins. Brooks et al. [19] used a combination of PEG and ammonium sulfate precipitation to purify monoclonal antibodies from cell culture supernatants. Perhaps the most useful application of PEG precipitation is in the purification of immunoglobulin Y from chicken eggs. Akita and Nakai [20] reported that PEG precipitation was superior to three other techniques for isolating this type of immunoglobulin from egg yolk protein. They estimated that up to 9.8 mg IgY per ml yolk could be extracted at a purity of 94%. Likewise, Romito et al. [21] reported that PEG precipitation was the most efficient way to isolate specific IgY from the yolks of eggs that had been laid by chickens immunized with plasmid DNA.

Caprylic acid is another agent that has been applied to the class-specific purification of antibodies [22, 23]. Caprylic acid is used to remove nonimmunoglobulin proteins from serum and plasma by precipitation, thus leaving a crude IgG fraction that can be purified further by chromatography [24]. This approach has gained popularity for preparing crude immunoglobulin fractions of hyperimmune serum for use as antivenom therapy [25]. However, it is necessary with this method to perform an additional ammonium sulfate precipitation step for the purification of class-specific IgG antibodies [26–28]. The efficiency of caprylic acid precipitation is debatable. Mohanty and Elazhary [29] reported that the caprylic acid–ammonium sulfate precipitation of IgG was not superior to straight ammonium sulfate treatment when isolating antibodies from bovine serum. Temponi et al. [30] found that caprylic acid was effective in precipitating some murine IgG antibody classes but not IgG_3- or IgA-type antibodies.

When considering precipitation as a means for antibody purification, one has to keep in mind that all of these techniques produce crude or class-specific immunoglobulins but never antigen-specific antibodies. Also, in almost every situation, further purification must be employed before immunologically pure reagents are obtained. However, if one does decide to use precipitation to produce starting materials for the further and more-refined purification of immunoglobulins, Table 14.3 can be used as a guide to the main features of the available precipitating reagents.

Table 14.3 Comparison of Immunoglobulin Precipitation Agents

Agent	Comments
Ammonium sulfate	Useful for polyclonal and monoclonal IgG isolation. A solution with 35% saturation will produce a pure IgG preparation, but not all IgG will be precipitated. A solution with 45% saturation will precipitate all IgG, but the sample will be contaminated with other proteins (e.g., albumin). Precipitation at 45% saturation produces an ideal starting material for other purification techniques.
Sodium sulfate	Useful only for rabbit or human polyclonal IgG. Often produces a purer preparation of IgG than other reagents, but is dependent on IgG concentration. The precipitation reaction is temperature sensitive and must be performed at 25°C.
Caprylic acid	Useful for polyclonal and monoclonal IgG. Purification is achieved by precipitation of non-IgG proteins. Different species require different final concentrations of caprylic acid: human and horse serum require 6.1%; goat serum requires 8.0%; and rabbit serum requires 8.2%.
Polyethylene glycol	Useful for both polyclonal and monoclonal antibodies. Good for IgM, but less efficient for IgG. This is a relatively mild procedure when using 20% PEG 6000.

14.2.2.2 General Chromatographic Methods

Since they are glycoproteins, antibodies exhibit properties that make them good candidates for isolation by a number of general chromatographic techniques. Two common techniques used for this purpose are ion-exchange and size-exclusion chromatography [31–33]. When considering which chromatographic approach should be used for immunoglobulin isolation, one needs to obtain a chromatographic system that is capable of isocratic and gradient elution at low to medium pressures. A popular commercial system for this type of work is a medium-pressure system for fast protein liquid chromatography (or FPLC), which is capable of performing immunoglobulin separations by a number of techniques (e.g., ion-exchange, size-exclusion, and affinity chromatography). However, other types of chromatographic systems can also be used for antibody purification.

The method of ion-exchange chromatography is based on the reversible adsorption of molecules to a charged matrix, followed by elution of the bound molecule in the presence of an increasing ionic-strength gradient. IgG-class antibodies can easily be isolated from hyperimmune serum through this approach by employing a simple batch technique (see Figure 14.3). This uses diethylaminoethyl (DEAE) as the ion-exchange group and either dextran beads (e.g., Sephadex) or cellulose as the support material and is extremely useful when small volumes of IgG must be isolated quickly. When DEAE Sephadex is mixed with a serum sample in 5 mM phosphate buffer (pH 6.5), IgG antibodies will be inhibited from adsorbing to the DEAE matrix, but other serum proteins will bind. This means that after removal of the ion-exchange material, the remaining supernatant will contain IgG with minimal contamination.

Column chromatography with a DEAE support, however, is the best way to isolate large quantities of IgG from serum or plasma. Under conditions similar to those used in Figure 14.3 for batch isolations, IgG can be eluted isocratically from a DEAE column as

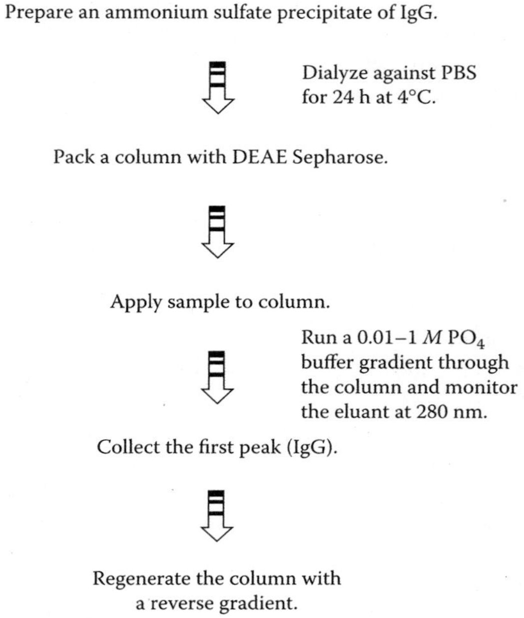

Prepare an ammonium sulfate precipitate of IgG.

 Dialyze against PBS
for 24 h at 4°C.

Pack a column with DEAE Sepharose.

Apply sample to column.

 Run a 0.01–1 M PO$_4$
buffer gradient through
the column and monitor
the eluant at 280 nm.

Collect the first peak (IgG).

Regenerate the column with
a reverse gradient.

Figure 14.3 Flowchart for a DEAE chromatographic technique used in the isolation of IgG-class immunoglobulins.

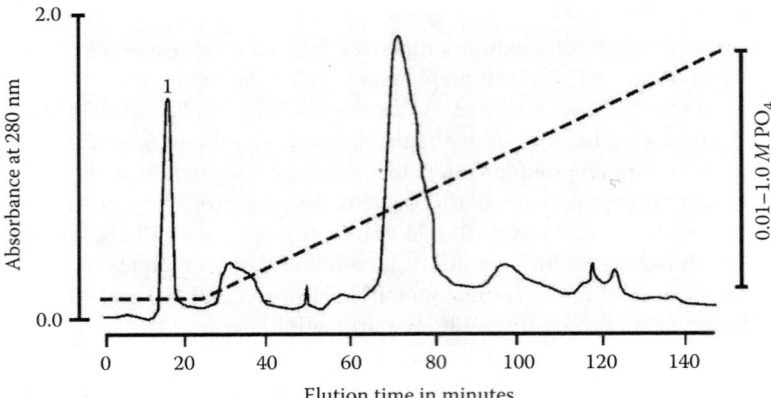

Figure 14.4 Chromatogram for a DEAE isolation of IgG. The first peak (1) contains pure IgG, although this material is class specific rather than antigen specific. The dashed line shows the phosphate gradient used to elute bound material from the DEAE column.

the first peak for a serum or plasma sample (see Figure 14.4). DEAE has also been used for the isolation of IgM Mabs by Burd and colleagues [34], who used a DEAE disk to isolate these immunoglobulins following elution with a sodium chloride step gradient.

Other ion-exchange materials have also been used for immunoglobulin purification. Gallo et al. [35] reported that anion exchange using Mono-Q columns combined with 0.05 to 0.3 M phosphate step-gradient elution could be used to isolate IgA, IgG, and IgM from human serum. Mono-Q anion-exchange chromatography followed by size-exclusion chromatography on Superose-6 has been used to purify IgM monoclonal antibodies from ascites fluid [36]. In addition, carboxymethyl (CM) Sephadex can be used to isolate F_{ab} fragments of digested immunoglobulins. For this latter procedure, the fragments of interest are eluted in the primary peak when applying a 0.1 M sodium acetate buffer. However, although ion-exchange chromatography is useful for preparing class-specific immunoglobulin fractions in a relatively pure form, it cannot be used to isolate antigen-specific antibodies or antibody fragments.

Size-exclusion chromatography is a useful technique for isolating different immunoglobulin classes according to their size. This process is achieved by passing the sample through a column packed with a porous support. Large molecules do not enter the pores and pass through the column with little or no retention. Smaller molecules are retarded as they enter pores and are temporarily retained, giving them longer elution times.

Depending on the size of the pores in the support, different sizes of molecules can be separated. Today, size-exclusion supports are available that possess separation ranges suitable for isolating all of the different immunoglobulins in animal and human serum, especially IgG [33]. A list of some commercial supports that can be used for this purpose is provided in Table 14.4, and a flowchart for the isolation of immunoglobulins by size-exclusion chromatography is given in Figure 14.5. Size-exclusion chromatography has also been used in the isolation of monoclonal antibodies from spent culture medium. Novales-Li [37] reported that IgM antibodies could easily be isolated by size-exclusion chromatography following salt precipitation.

One major disadvantage to size-exclusion chromatography is the time it requires to isolate class-specific antibodies and the high degree of technical skill that is often needed to produce good results. Although this method can be a powerful tool for antibody isolation,

Table 14.4 Selected Size-Exclusion Media for Immunoglobulin Isolation

Brand Name	Solute Size Range (Da)
Low- to Medium-Performance Supports	
Sepharose CL-6B	10,000–4,000,000
Sepharose CL-4B	60,000–20,000,000
Sepharose CL2B	70,000–40,000,000
Sephacryl S-300	10,000–1,500,000
Toyopearl HW-55F	1,000–700,000
Toyopearl HW-65F	40,000–5,000,000
High-Performance Supports	
TSK G3000SW	10,000–500,000
TSK G4000SW	20,000–10,000,000

it has lost its appeal to more modern approaches such as affinity and immunoaffinity techniques. However, with the increased interest in engineered antibodies (i.e., especially scF$_v$ antibodies produced by vector-expression systems), size exclusion may experience a revival, since it represents a relatively cheap approach to immunoglobulin purification [38].

Other chromatographic approaches can also be used to isolate specific classes of immunoglobulins. For instance, monoclonal antibodies can be purified according to their class by employing relatively simple chromatography on hydroxyapatite [39, 40]. However, this approach has not gained popularity, since many monoclonal antibodies are IgM-class

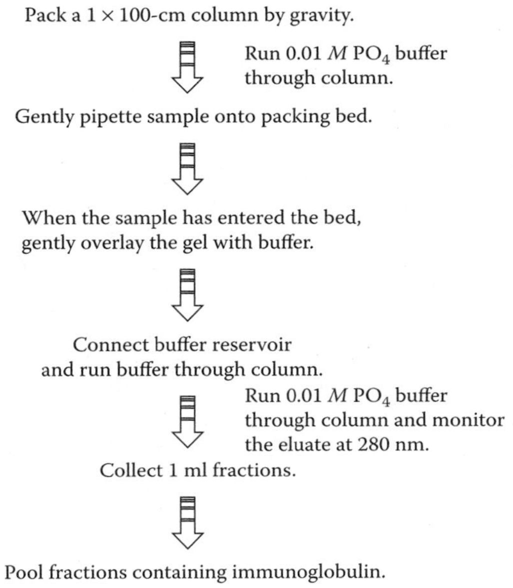

Pack a 1 × 100-cm column by gravity.

 Run 0.01 M PO$_4$ buffer through column.

Gently pipette sample onto packing bed.

When the sample has entered the bed, gently overlay the gel with buffer.

Connect buffer reservoir and run buffer through column.

 Run 0.01 M PO$_4$ buffer through column and monitor the eluate at 280 nm.

Collect 1 ml fractions.

Pool fractions containing immunoglobulin.

Figure 14.5 Flowchart for the isolation of immunoglobulins by size-exclusion chromatography.

immunoglobulins, which often precipitate in low-salt conditions such as those used during the commencement of this separation process.

14.2.3 Affinity Techniques for Antibody Purification

Many affinity techniques can be used for antibody purification. This can be performed either through the interactions of immunoglobulins with general immobilized ligands or immobilized antigens. In this section, both groups of ligands are considered as agents for antibody purification.

14.2.3.1 Immobilized Metal-Ion Affinity Chromatography

Immobilized metal-ion affinity chromatography (IMAC) is one affinity method that can be used for the general isolation of antibodies. A detailed discussion of IMAC and its applications can be found in Chapter 10. Boden et al. [41] reported that goat polyclonal antibodies could be easily isolated by IMAC using a column containing *tris*(2-aminoethyl)amine chelated with copper. With this procedure, they were able to recover an immunoglobulin fraction with greater than 95% homogeneity, as determined by silver-stained native gel electrophoresis and sodium dodecyl sulfate polyacrylamide gel electrophoresis.

14.2.3.2 Thiophilic Chromatography

Thiophilic chromatography is another affinity method that allows the isolation of class-specific immunoglobulins. As shown by the flowchart in Figure 14.6, this can be used for the isolation of monoclonal antibodies [42–45]. This involves the binding of proteins to

Pack 4 mL of gel into a 1 × 10 cm column.

 Equilibrate with 25 mL 0.1 M Tris-HCl/0.5 M K$_2$SO$_4$, pH 7.6.

Mix 1 mL of IgG with 2 mL of 0.1 M Tris-HCl/0.5 M K$_2$SO$_4$

 Wash with 20 mL 0.1 M Tris-HCl/0.5 M K$_2$SO$_4$, pH 7.6 and monitor the eluate at 280 nm until the baseline value returns.

Elute the bound IgG with 0.1 M ammonium bicarbonate.

Collect 1 ml fractions.

Pool immunoglobulin-containing fractions.

Dialyze against PBS overnight at 4°C.

Figure 14.6 Flowchart for the isolation of immunoglobulins by thiophilic chromatography. (Abbreviations: K$_2$SO$_4$, potassium sulfate; PBS, phosphate buffered saline.)

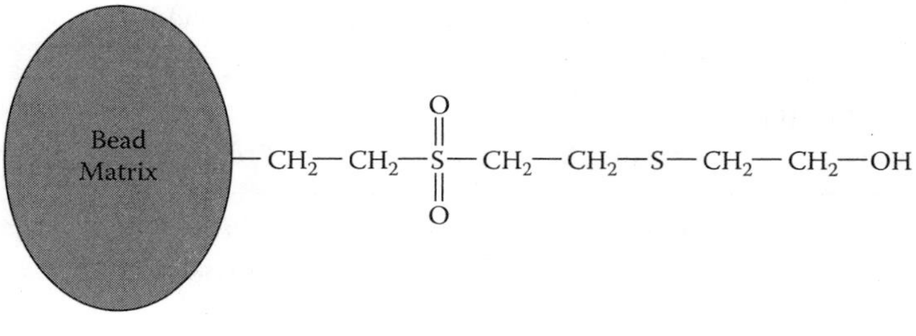

$$CH_2-CH_2-\overset{\overset{\displaystyle O}{\|}}{\underset{\underset{\displaystyle O}{\|}}{S}}-CH_2-CH_2-S-CH_2-CH_2-OH$$

Bead Matrix

Figure 14.7 A thiophilic matrix suitable for covalent affinity chromatography.

a matrix that has been formed by reacting β-mercaptoethanol with divinyl sulfone-activated agar to make a thiophilic or "T-gel," as shown in Figure 14.7 [46, 47].

Thiophilic chromatography has been used to isolate human serum proteins [48] and various immunoglobulins from a wide range of animals [49, 50]. This has also been reported to be a rapid technique for the purification of mouse $F_{(ab)'2}$ fragments from ascitic fluid [51] or recombinant sources [52]. Recovery of these fragments was performed using decreasing gradients of ammonium sulfate. Figure 14.8 shows a chromatogram for the isolation of IgG F_{ab} fragments on a thiophilic resin. Thiophilic adsorption chromatography has also been successfully applied to the recovery of a number of antibody fragments and specialized antibodies, such as recombinant single-chain antibodies [53], biotinylated F_{ab} [54], and bispecific antibodies [55].

14.2.3.3   Affinity Separations with Bacterial Coat Proteins

The isolation of immunoglobulins by bacterial Ig-binding receptors was first introduced in the early 1970s with the discovery of protein A, a bacterial coat protein from *Staphylococcus aureus* that was capable of selectively binding IgG [56–58]. Later, a more efficient Ig-binding protein, protein G, was isolated from G-148, a group-G streptococcal strain. This new protein was found to be more useful than protein A, since it could bind immunoglobulins from a wider variety of species [59, 60].

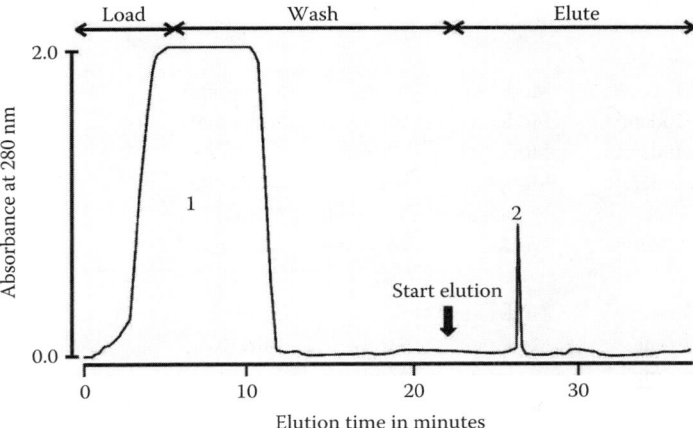

Figure 14.8 Chromatogram for the isolation of IgG F_{ab} fragments on a thiophilic resin. The F_{ab} fraction is contained in peak 2; peak 1 is the nonretained material. The procedure used in this separation is described in Figure 14.6.

Table 14.5 Binding of Different Animal Immunoglobulins to Proteins A, G, and L

Species	Immunoglobulin Class	Protein A	Protein G	Protein L
Chicken egg	IgY	No binding	No binding	Strong
Cow	IgG	No binding	Weak	No binding
Dog	IgG	Weak	Weak	—
	IgM	Weak	No binding	—
Goat	IgG	Weak	Strong	No binding
	IgM	No binding	No binding	No binding
Horse	IgG	Strong	Strong	No binding
Rabbit	IgG	Strong	Strong	Moderate
	IgM	No binding	No binding	Moderate
Rat	IgG	Weak	Moderate	Strong
	IgM	No binding	No binding	Strong
Sheep	IgG	Strong	Strong	No binding
	IgM	No binding	No binding	No binding

These bacterial proteins have since become popular as general affinity ligands for the isolation of immunoglobulins from many different animal species [61, 62]. The era of molecular biology has given rise to other related proteins produced by using recombinant varieties of Ig-binding proteins [63–65] or designing new molecules to more cheaply replace the naturally occurring forms [66]. A list of types of antibodies that bind to these ligands can be found in Table 14.5 and Table 14.6. More details on each of these ligands

Table 14.6 Binding of Murine and Human Antibodies to Proteins A, G, L, and Their Hybrids

Class	Protein A	Protein G	Protein L	A/G	G/L
Murine Monoclonal Antibodies					
IgG_1	Weak	Moderate	Moderate	Moderate	Moderate
IgG_{2a}	Moderate	Moderate	Moderate	Moderate	Moderate
IgG_{2b}	Moderate	Moderate	Moderate	Moderate	Moderate
IgG_3	Weak	Moderate	Moderate	Moderate	Moderate
IgM	Moderate	Weak	No binding	Moderate	Weak
IgA	Moderate	Moderate	Weak	Moderate	Moderate
Human Antibodies					
IgG_1	Strong	Strong	Moderate	Strong	Strong
IgG_2	Strong	Strong	Moderate	Strong	Strong
IgG_3	No binding	Strong	Moderate	Moderate	Strong
IgG_4	Strong	Strong	Moderate	Strong	Strong
IgA	Weak	No binding	Moderate	Weak	Moderate
IgM	Weak	No binding	Moderate	Weak	Moderate
IgE	Weak	No binding	Moderate	Weak	Weak

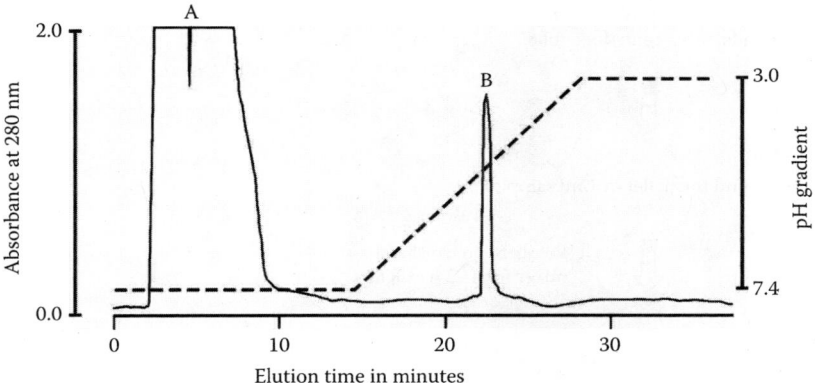

Figure 14.9 Chromatogram for the isolation of IgG by a protein A affinity column. Peak A is the nonretained material and peak B contains the class-specific IgG. The dashed line shows the pH gradient used to recover the IgG from this column.

and their use in antibody purification are given in the following subsections. Further information on these ligands can be found in Chapter 5.

14.2.3.3.1 Protein A. Since its introduction in the early 1970s, protein A has enjoyed great popularity as a quick and relatively simple approach for the isolation of IgG and some of its subclasses. Figure 14.9 shows a chromatogram for the isolation of IgG by a protein A affinity column. A flowchart for this type of application is given in Figure 14.10. For some species, protein A can also be used to isolate other immunoglobulin classes, such as IgA and IgM.

Protein A is covalently attached to the bacterial cell wall of *Staphylococcus aureus* and, when lightly fixed, whole bacteria have been used as a solid-phase matrix for this ligand [67]. However, protein A is also available in a pure form that is more suitable for use as an affinity ligand and in affinity chromatography [68, 69].

Recently, Denizli and Arica [70] described the design of a microporous membrane containing protein A for the class-specific purification of human IgG. Using this device, the authors claim that up to approximately 500 µg of IgG could be removed from human plasma. Protein A has also been used as a secondary ligand for attaching antibodies to solid supports for high-performance immunoaffinity chromatography [62, 71–73]. Solid-phase protein A can be used in batch techniques to isolate either certain IgG subclasses or F_c fragments, following the digestion of IgG by pepsin or papain.

A batch technique for the isolation of antibodies with solid-phase protein A is described in Figure 14.10. The recovery of absorbed antibodies from protein A columns can easily be achieved using mildly acid conditions (e.g., pH 3.1). However, Chalon et al. [74] reported that applying a prolonged amount of phosphate buffer at pH 7.0 to 7.4 to a protein A column could elute pure murine IgG_1. These investigators also reported that a simple acid-salt gradient (pH 7 to 3) could elute murine IgG_1, IgG_{2a}, IgG_{2b}, and IgG_3. However, a 0 to 3 M sodium thiocyanate gradient gave better recoveries [75]. Duhamel et al. [76] employed a citrate pH gradient to recover human IgG subclasses from a protein A column.

14.2.3.3.2 Protein G. Protein G from G-group streptococci has a more diverse binding range than protein A, making it a potentially more versatile reagent for antibody

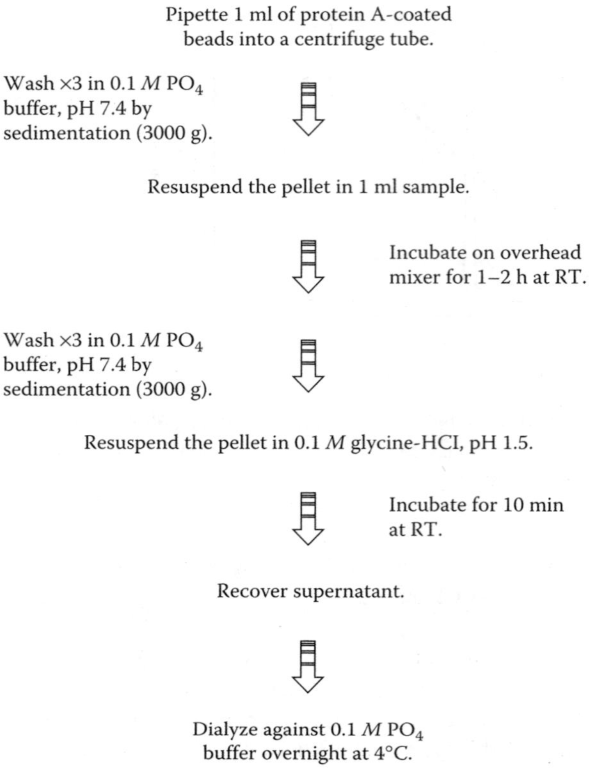

Pipette 1 ml of protein A-coated
beads into a centrifuge tube.

Wash ×3 in 0.1 M PO$_4$
buffer, pH 7.4 by
sedimentation (3000 g).

Resuspend the pellet in 1 ml sample.

    Incubate on overhead
mixer for 1–2 h at RT.

Wash ×3 in 0.1 M PO$_4$
buffer, pH 7.4 by
sedimentation (3000 g).

Resuspend the pellet in 0.1 M glycine-HCl, pH 1.5.

    Incubate for 10 min
at RT.

Recover supernatant.

Dialyze against 0.1 M PO$_4$
buffer overnight at 4°C.

Figure 14.10 Flowchart for the batch isolation of immunoglobulins using solid-phase protein A. (Abbreviations: PO$_4$, phosphate; RT, room temperature.)

purification. Protein G has been shown to bind to intact IgG as well as IgG fragments such as $F_{(ab)'2}$ and F_c, but with these binding events occurring at different sites [77, 78].

The ability of protein G to bind IgG from different species has elicited considerable interest from the scientific community. Recombinant protein G has been shown to be useful in isolating porcine IgG from whole serum and plasma [79]. Peng at al. [80] studied the binding of canine immunoglobulins to both proteins A and G. These investigators concluded that protein G was efficient for some immunoglobulin classes but that protein A was superior in isolating canine IgM and IgE, with the latter not binding to protein G. Fischer and Hlinak [81] reported that neither protein A nor protein G bound egg yolk antibodies produced by four different fowl species (i.e., turkey, duck, moskovy duck, and goose). Eliasson and colleagues [82] analyzed the different binding properties of protein A and G. They concluded that a chimeric recombinant form of these proteins (i.e., protein A/G) was required for their optimal use.

It has been shown that protein G can be effectively used as an affinity ligand for the class-specific purification of IgG. Pilcher et al. [83] compared the efficiency of three protein G affinity matrixes (CNBr-Sepharose, Tresyl-Sepharose, and Affigel-10) in isolating IgG from human, mouse, and goat serum. It was concluded that the optimum condition for IgG isolation involved the use of a pH 8.0 buffer containing 1.0 M sodium chloride. Other workers have used high-performance affinity chromatography and protein G as the affinity ligand for the purification of monoclonal antibodies [84, 85].

14.2.3.3.3   Protein L.   Protein L is a bacterial wall protein derived from *Peptostreptococcus magnus* and exhibits an affinity for human immunoglobulin light chains [86]. Studies have shown that this binding is mediated through five highly homologous domains, each comprising 72 to 76 amino acids, that interact with regions in the variable domain of immunoglobulin kappa light chains. This binding does not appear to interfere with the antigen-binding properties of the antibodies, indicating that an exterior binding site may be involved [87, 88]. Studies on the mechanism of this binding have demonstrated that specific residues are involved, especially the phenol group of tyrosine-53 [89].

Protein L has been reported to bind strongly to human IgG, IgA, IgM, and IgE [86]. In addition, protein L has been shown to bind to immunoglobulins from other species, including those commonly used to produce polyclonal and monoclonal antibodies (i.e., mice, rats, and rabbits).

Improvements in the binding properties of protein L have been made by engineering hybrids that combine this protein or its specific Ig-binding region with another Ig-binding protein. Svensson et al. [90] fused four protein L binding domains with the Ig-binding regions of protein A to produce a nonrestricted immunoglobulin-binding molecule. This engineered molecule was able to bind to IgG from a variety of animals, as well as different human immunoglobulins. This ligand was also useful in isolating kappa light-chain scF_v antibodies.

In similar work, several investigators have reported the development of protein L-protein G hybrids [91–93] capable of binding F_c fragments, F_{ab} fragments, and light chains, along with a large majority of intact human immunoglobulins. This binding was found to be specific with affinity constants of $5.9 \times 10^9\ M^{-1}$ for intact human IgG, $2.2 \times 10^9\ M^{-1}$ for IgG F_c fragments, and $2.0 \times 10^9\ M^{-1}$ for kappa light chains.

14.2.3.3.4   Other Bacterial Agents.   Other bacterial agents have also been found that bind immunoglobulins. For instance, Akessen et al. [94] described a protein derived from group A streptococcal strains that is capable of binding human IgG with little binding to other substances, except for a weak interaction with α_2-macroglobulin.

Similar findings were reported by Frick et al. [95], who described protein H derived from *Streptococcus pyogenes* and described its ability to bind to the F_c portion of human IgG. However, protein H was also found to express a strong affinity for human serum albumin. Later, this same group reported that protein H had a binding affinity for fibronectin type III and other matrix proteins containing the fibronectin type III motif [96]. These findings reduced the potential usefulness of this protein for antibody purification.

The group A streptococci, especially *Streptococcus pyogenes*, express a variety of other Ig-binding proteins, including protein M. This protein has been reported to bind human IgG subclasses but has not yet been used for the affinity purification of immunoglobulins [97–99].

14.2.3.4   Affinity Separations with Lectins

Naturally occurring immunoglobulins and some monoclonal antibodies are glycosylated, which means that they could potentially be isolated by using sugar-binding proteins or lectins [100]. A detailed discussion of lectins and their use in affinity chromatography can be found in Chapter 5. However, the use of lectins as binding agents has not received a lot of attention in antibody isolation. This is possibly due to their relatively nonspecific affinities, which results in the isolation of class-specific rather than antigen-specific immunoglobulins.

The most widely used lectin for antibody purification is *jacalin* [101]. This is an α-D-galactosyl (D-gal)-binding lectin derived from the seeds of the jackfruit (*Artocarpus integrifolia*). Although D-gal is a relatively common sugar, jacalin demonstrates an affinity for

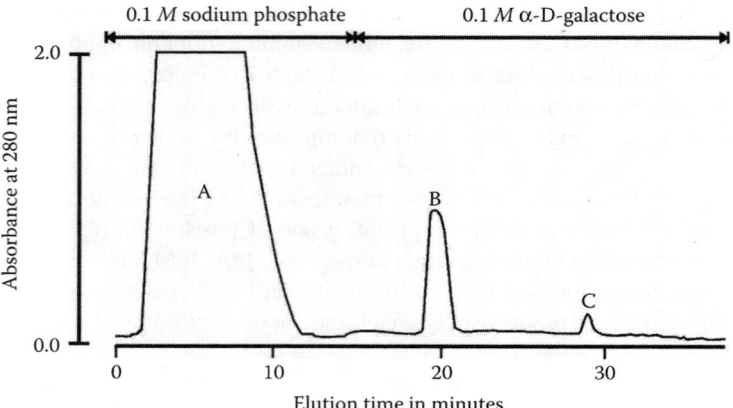

Figure 14.11 Chromatogram showing the isolation of IgA$_1$ on a jacalin lectin column. The non-retained material is represented by peak A, peak B is the class-specific IgA, and peak C is the small amount of IgD that is also retained on this column.

human IgA [102] but has little or no reactivity with pig, goat, horse, cow, or dog IgA [103]. Further, it has been reported that jacalin reacts only with IgA subclass 1, as demonstrated in Figure 14.11 [104]. Jacalin together with protein A can be used to isolate both IgA$_1$ and IgG from rabbit serum [104]. Other lectins isolated from different jackfruit, such as jacalin-H, have been reported to bind additional immunoglobulins as well as IgG aggregates, thus making these potentially useful for the purification of monomeric IgG by subtractive adsorption [105].

Lectins from additional sources have also been used to isolate immunoglobulin A. Irazoqui et al. [106] described the use of a lectin isolated from *Agaricus bisporus*. These authors reported that this lectin bound both the secretory IgA subclasses and subclass IgA$_2$. This indicated that this lectin had the ability to bind both *O*- and *N*-glycans. The recovery of these subclasses was achieved by eluting with ammonium hydroxide for IgA$_2$ and by adding sugars as competing agents for the other subclasses. IgA$_1$ has also been isolated using an extract from the albumin gland of the land snail. This material is specific for one IgA subclass, reacting with an *N*-acetyl glucosamine residue in these immunoglobulins [107].

Other lectins that have been used to isolate immunoglobulins in a class-specific manner are concanavalin A [108, 109], champedak lectin-M [110], castor bean agglutinin [111], snowdrop bulb lectin [112], and mannan-binding protein derived from a rabbit [113]. The specificities of these lectins are summarized in Table 14.7. A further discussion of their properties is given in Chapter 5.

14.2.3.5   Immunoaffinity Purification of Antibodies

The only way to purify specific and immunologically active antibodies of any class is to use immunoaffinity separations based on immobilized antigens. This is usually performed using column chromatography [114]. In most cases, the capture ligand is a chemical or biochemical antigen, although antibodies themselves can be used as ligands for isolating anti-immunoglobulin antibodies.

Immobilized antigen chromatography has successfully been used to isolate antigen-specific antibodies from a number of animal and human body fluids (e.g., serum or plasma) [115–121]. This has also been used to isolate monoclonal antibodies from ascites fluid and cell culture supernatants [122, 123]. An example of such a separation is shown in Figure 14.12.

Table 14.7 Specificities of Selected Lectins

Lectin	Sugar Specificity
Jacalin	α-D-galactosyl; -(1,3)-N-acetylgalactosamine
Concanavalin A	α-D-mannosyl; α-D-glycosyl
Champedak lectin-M	Man α1-3 Man
Rabbit mannan-binding protein	Specificity not assayed
Castor bean agglutinin	β-D-galactosyl
Snowdrop bulb lectin	Nonreducing terminus of α-D-mannosyl
Land snail lectin	N-acetyl glucosamine

The procedure for performing this type of isolation is relatively straightforward and is summarized by the flowchart in Figure 14.13.

Antibodies can also be isolated using mimetic ligands or mimotopes. A *mimetic ligand* is one that uses a synthetic chemical structure to resemble an antigen. For instance, a synthetic bifunctional ligand containing a triazine scaffold substituted with 3-aminophenol and 4-amino-1-naphthol has been immobilized on agarose to create a robust and highly selective affinity adsorbent for human IgG [124]. This adsorbent showed similar selectivity to immobilized protein A and was able to bind IgG from a number of species.

In a similar manner, monoclonal antibodies can be purified using a panel of epitope and mimotope adsorption columns [125]. A *mimotope* has a chemical structure that resembles an antigen's reactive site. For example, various synthetic peptides incorporating the immunodominant region of the MUC1 protein core were prepared and covalently linked to agarose beads for use as affinity matrices. Furthermore, an unrelated peptide was identified by phage-display methods as a mimotope for one of the anti-MUC1 antibodies, and this was evaluated for use as a potential affinity ligand.

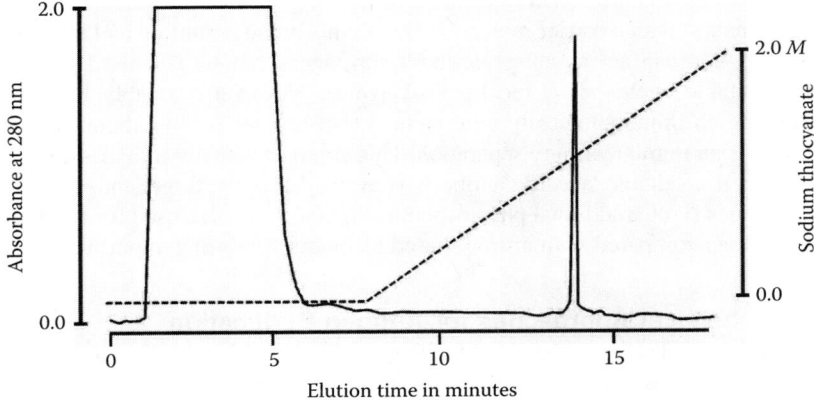

Figure 14.12 Chromatogram for the immunoaffinity purification of a specific antibody using an immobilized antigen. The first peak represents the nonreactive material in the sample, and the second peak represents the immunologically pure antibody. The dashed line shows the gradient of a chaotropic agent (sodium thiocyanate) that was used to elute the bound antibodies.

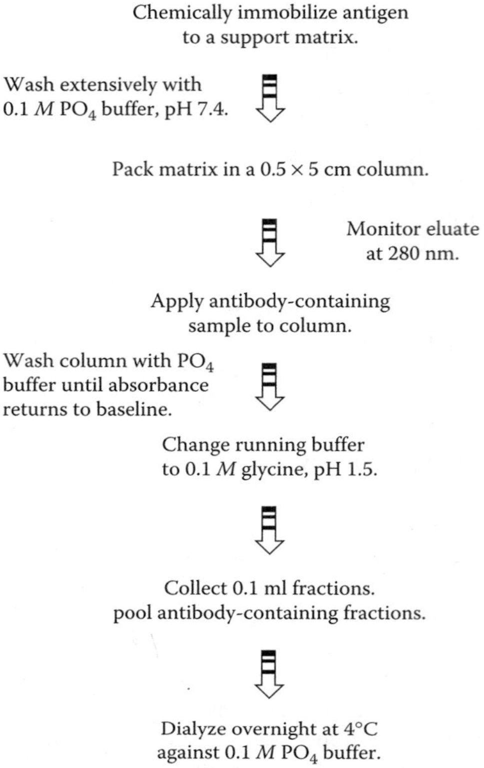

Chemically immobilize antigen
to a support matrix.

Wash extensively with
0.1 M PO$_4$ buffer, pH 7.4.

Pack matrix in a 0.5 × 5 cm column.

          Monitor eluate
at 280 nm.

Apply antibody-containing
sample to column.

Wash column with PO$_4$
buffer until absorbance
returns to baseline.

Change running buffer
to 0.1 M glycine, pH 1.5.

Collect 0.1 ml fractions.
pool antibody-containing fractions.

Dialyze overnight at 4°C
against 0.1 M PO$_4$ buffer.

Figure 14.13   Flowchart for the antigen-based affinity isolation of a specific antibody.

14.3   ANTIGEN PURIFICATION

As stated earlier, an antigen is any material capable of eliciting an immune response either by itself or in combination with a carrier molecule. Due to this broad definition and the wide range of compounds that can act as antigens, these substances can be isolated by many techniques using suitable chemical or biochemical agents. However, probably the only approach for isolating an immunologically pure antigen (but not necessarily chemically or biochemically pure) is an immunoaffinity separation. This approach usually provides biologically active antigens if a suitable, specific antibody is available for the target antigen. This section briefly examines some traditional physiochemical methods that are used for isolating antigens. These are then compared with affinity-based techniques for antigen purification.

14.3.1   Physicochemical Approaches for Antigen Purification

Antigens can often be isolated using general techniques like traditional chromatography or electrophoresis. Examples given in Table 14.8 include size-exclusion, ion-exchange, and reversed-phase chromatography, as well as isoelectric focusing and polyacrylamide gel electrophoresis. Although these methods can sometimes be used alone, there are many examples where combinations of such techniques are required before a sufficiently pure antigen preparation can be achieved [126, 127].

Table 14.8 Common Approaches for the Isolation and Purification of Antigens

Technique	Isolation Parameters
Physicochemical approaches	
Size-exclusion chromatography	Isolation by molecular weight
Ion-exchange chromatography	Isolation by charge
Reversed-phase chromatography	Isolation by hydrophobicity
Polyacrylamide-gel electrophoresis	Isolation by size in a gel matrix
Isoelectric focusing	Isolation by isoelectric point (pI) in a stabilized pH gradient
Two-dimensional electrophoresis	Isolation by pI in one dimension followed by molecular weight in the second dimension
Affinity-Based Approaches	
Affinity chromatography	Isolation by immunological reactivity
Immunoprecipitation	Isolation by immunoreactivity
Immunomagnetic separation	Isolation by immunoreactivity

14.3.1.1 General Chromatographic Methods

Antigens can be isolated and purified by a number of chromatographic techniques. Size-exclusion chromatography is perhaps the most useful of these methods, since one can isolate the molecule of interest on a size or molecular-weight basis. This technique is often combined with ion-exchange chromatography or reversed-phase chromatography for antigen isolation [128–130]. As discussed in Section 14.2, size exclusion is performed in columns that contain a porous support capable of retaining small molecules into their pores while larger molecules pass through. The advantage of this technique is that one does not need to know a vast amount of information about the antigen of interest. However, it must be kept in mind with this approach that molecules with similar molecular weights will not be fully resolved, thus leading to contamination in the final preparation. In addition, the resolving power of the column may not be able to separate the molecule of interest as a single peak, therefore requiring the use of additional purification methods.

Size-exclusion chromatography for antigens can be performed on low-, medium-, or high-performance systems using supports like those shown in Table 14.4. This wide selection allows a choice of many available separation systems, depending on the amount of starting material that is to be purified and the relative concentration of the antigen within this material. A large volume of starting material that contains a dilute antigen will usually give the best results when processed by low-performance (i.e., low pressure) chromatography using a large column (e.g., 1 cm I.D. × 30 to 60 cm long). Alternatively, a small sample containing a high concentration of antigen could be processed by FPLC or HPLC using a support such as Toyopearl or a TSK column (from Tosoh Bioscience).

Unfortunately, in most cases, antigen purification involves some trial and error, and the investigator is often forced to try several approaches before deciding on one method (or a series of methods) that is capable of producing a reasonably pure antigen preparation.

Other techniques that are often used for antigen isolation include ion-exchange and reversed-phase chromatography [131, 132]. These rely on differences in charge or hydrophobicity to separate an antigen from a sample. As was the case with size exclusion, the isolation of closely eluting molecules is often hard to obtain by these methods, although reversed-phase chromatography can provide high resolutions. An advantage of ion exchange is that it can be adapted for batch extraction once the adsorption and desorption conditions for the antigen on such a material have been established.

All of these chromatographic techniques are suitable for the separation of most classes of antigenic materials. However, there are difficulties encountered when using these to isolate and purify lipids, lipopolysaccharide, and lipoproteins. In these cases, alternative chromatographic supports and methods may have to be used, such as Sephadex LH (i.e., hydroxypropylated, cross-linked dextran beads) or thin-layer chromatography [133].

14.3.1.2   Electrophoretic Techniques

Electrophoresis is useful for antigen purification, but it usually does not yield an adequate amount of material for further studies. For instance, polyacrylamide gel electrophoresis (PAGE) is a good analytical tool, but it is not particularly helpful when it comes to isolating large quantities of a particular substance. Instead, this technique is better suited as a tool for analyzing materials that have been isolated by other techniques [134, 135].

The technique of isoelectric focusing (IEF) is more useful for antigen isolation and can be modified for work on a preparative scale. This technique isolates a molecule according to its isoelectric point (pI), as represented by the zone in a stabilized pH gradient where the molecule obtains a net neutral charge. This approach can be used with large electrophoretic systems, allowing the isolation of appreciable quantities of relatively pure antigens. A commercial version of such a device is called the Rotofor (from BioRad). This instrument allows the investigator to isolate proteins and other charged molecules by precipitating them at their pI into a large, rotating electrophoretic cell. The focused molecules are then harvested through a series of harvesting ports. When applied to proteins, peptides, glycoproteins, and polysaccharides, this instrument is capable of isolating up to gram amounts of the analyte of interest [136]. However, preliminary experiments have to be performed to first estimate the pI of the molecule of interest.

Two-dimensional electrophoresis is another tool that can be used to isolate minute amounts of a pure antigen. This procedure isolates substances based on their pI in the first dimension, followed by a second separation based on molecular weight. This second dimension is performed at 90° to the first, thus allowing the focused molecules to be analyzed on a gradient PAGE gel. Although a very powerful technique, two-dimensional electrophoresis

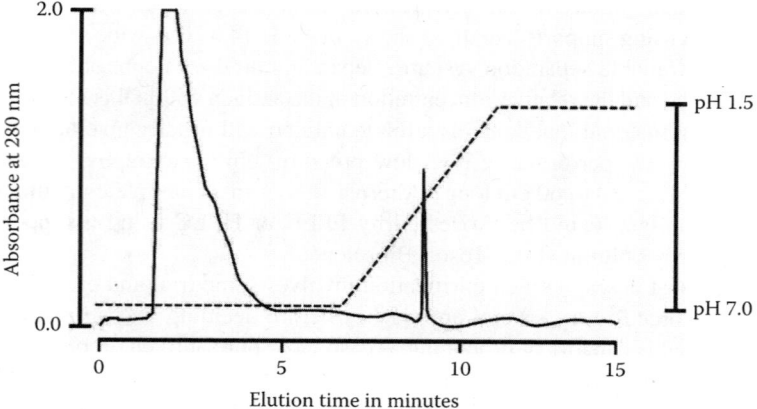

Figure 14.14   Chromatogram for the immunoaffinity isolation of a specific antigen using an immobilized-antibody column. The immunologically purified antigen is represented by the second peak, while all unreactive material is in the first peak. The dashed line represents the pH gradient that was used to elute the bound antigen in this example.

is more suited to analytical studies and is an excellent way to check the purity of an isolated analyte, especially if this is a protein, peptide, or glycoprotein [137, 138].

14.3.2 Affinity Methods for Antigen Isolation

As described earlier, immunoaffinity methods are probably the most practical approaches for isolating specific antigens. This makes use of the immunological activity of the antigens by binding them to immunoaffinity columns containing immobilized antibodies. More details on the field of immunoaffinity chromatography can be found in Chapter 6.

The isolation of antigens by this approach can employ polyclonal antibodies, affinity-purified antibodies, or specific monoclonal antibodies to isolate a single compound from essentially any matrix. An example of such a separation is given in Figure 14.14. Monoclonal

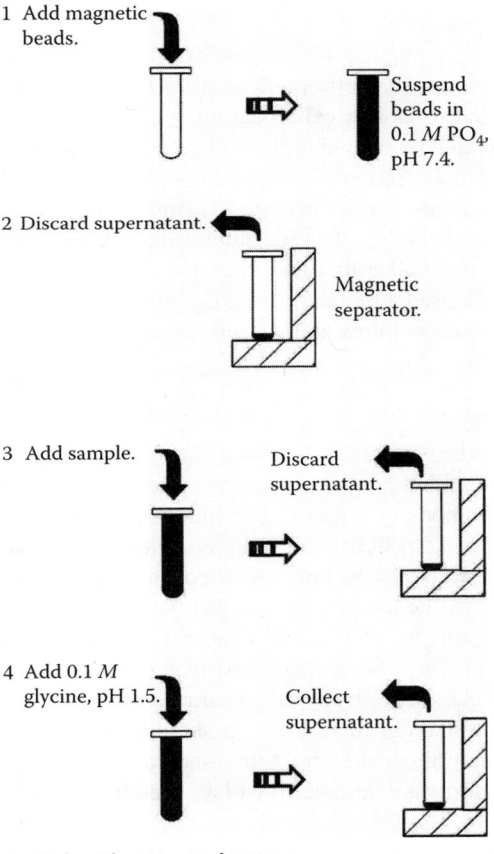

1 Add magnetic beads.

Suspend beads in 0.1 M PO$_4$, pH 7.4.

2 Discard supernatant.

Magnetic separator.

3 Add sample.

Discard supernatant.

4 Add 0.1 M glycine, pH 1.5.

Collect supernatant.

5 Dialyze the recovered antigen overnight at 4°C against PO$_4$ buffer.

Figure 14.15 Flowchart for the isolation of an antigen using magnetic antibody-coated beads. In Step 1, the antibody-coated beads are dispensed into a tube and suspended in buffer. In Step 2, a magnetic separator is used to sediment these beads and retain them while the supernatant is removed by decanting. In Step 3, the sample is added and the beads are suspended in this sample; it is at this time that the immobilized antibodies capture and retain the antigen. A magnetic separator is again used to sediment the beads. In Step 4, an elution buffer is added and the bound antigen released. A magnetic separator is again used to sediment the beads, allowing recovery of the released antigen in the buffer. In Step 5, the recovered antigen solution is dialyzed to remove the elution buffer.

antibodies have been used to isolate a large number of antigens and are often the reagents of choice when performing such isolations [139, 140].

Immunoaffinity separations have been successfully applied to many fields, including the isolation of allergens [141], the purification of parasite antigens for diagnostic purposes [142–145], and isolation of fungal [146] or viral components [147–150]. Specific antigens associated with receptors [151–155] and other cellular components [156–158] have been successfully isolated using immobilized-antibody columns. In addition, antibody-based systems have been used to purify recombinant proteins from expression systems such as bacterial cultures [159].

Along with whole antibodies, antibody fragments can be immobilized and used as immunoadsorbents for antigen isolation. As early as 1981, Kennedy and Barnes [160] used immobilized $F_{(ab)'2}$ fragments to isolate class-specific IgG from human serum. In the same manner, antibody F_{ab} fragments and scF$_v$s have been successfully used to isolate whole bacteria [161] and tagged recombinant proteins [162].

A technique that is becoming increasingly popular for immunoaffinity separations is a method performed with antibody-coated paramagnetic particles (see Figure 14.15). Although this approach has been mainly used to isolate cells bearing specific antigens, there is great potential for this to be used as a rapid technique for isolating immunologically specific antigens [73]. Immunomagnetic separations based on anti-immunoglobulin-coated particles have been employed in the recovery of muscle tropomyosin, vimentin, and myosin heavy chain [163]. Similarly, Kausch et al. [164] employed a procedure using streptavidin-coated particles to isolate a number of cell organelles that had been previously labeled with biotinylated antibodies. Angen et al. [165] used antibody-coated Dynabeads to isolate pleuropneumoniae serotype 2 bacteria from pure cultures and suspensions.

14.4 SUMMARY AND CONCLUSIONS

The isolation and purification of antibodies and antigens is critical to research involving these agents or their use within analytical methods. A variety of techniques can be used to isolate these substances, ranging from general physicochemical approaches to selective affinity-based methods. In choosing from these methods, one has to consider the nature and desired purity of the final target, as well as the amount of material that is available.

Physicochemical techniques often present the easiest and cheapest way to purify antibodies and antigens. However, it should always be remembered that these methods isolate a substance according to its general chemical or physical characteristics and not based on its ability to undergo a specific immunological reaction. It should also be kept in mind that only techniques employing immunological interactions, such as chromatography based on immobilized antibodies or antigens, are able to isolate immunologically active and specific reagents.

Many physicochemical techniques can be successfully applied to the isolation of monoclonal antibodies or engineered antibodies from culture media where only a single reagent is present at high levels. However, these techniques are less successful when isolating identical agents from complex biological fluids such as plasma or serum. Even nonimmunological affinity techniques, such as those that use bacterial proteins or lectins (e.g., protein A, protein G, and jacalin), cannot produce specific immunological reagents, even though these can be highly successful in purifying immunoglobulins on a class-specific basis.

The optimum approach for purifying specific antibodies or antigens from any sample is an immunoaffinity technique that utilizes an appropriate counteragent as the immunoaffinity ligand. The use of this approach results in the isolation of an immunologically specific agent, regardless of its chemical or biochemical nature. The disadvantage of this approach is that it is often expensive and time-consuming, making it prohibitive when production-scale purifications are required.

SYMBOLS AND ABBREVIATIONS

CM	Carboxymethyl
DEAE	Diethylaminoethyl
D-gal	α-D-Galactosyl
DNA	Deoxyribonucleic acid
F_{ab}	Fragment antibody
$F_{(ab)'2}$	Fragment antibody prime 2
F_c	Fragment crystalline
FPLC	Fast protein liquid chromatography
HPLC	High-performance liquid chromatography
IEF	Isoelectric focusing
IgA	Immunoglobulin A
IgA_1	Immunoglobulin A, subclass 1
IgA_2	Immunoglobulin A, subclass 2
Ig-binding	Immunoglobulin-binding
IgE	Immunoglobulin E
IgG	Immunoglobulin G
IgG_1	Immunoglobulin G, subclass 1
IgG_2	Immunoglobulin G, subclass 2
IgG_3	Immunoglobulin G, subclass 3
IgG_4	Immunoglobulin G, subclass 4
IgM	Immunoglobulin M
IgY	Immunoglobulin Y or egg yolk immunoglobulin
K_2SO_4	Potassium sulfate
M	Molar
Mab	Monoclonal antibody
mM	Millimolar
MUC1	Membrane-associated mucin 1
PAGE	Polyacrylamide gel electrophoresis
PBS	Phosphate buffered saline
PEG	Polyethylene glycol
pI	Isoelectric point
PO_4	Phosphate
RT	Room temperature
scF_v	Single-chain antibody
μg	Microgram

REFERENCES

1. Godfrey, M.A., Immunoaffinity extraction in veterinary residue analysis: A regulatory viewpoint, *Analyst,* 123, 2501–2506, 1998.
2. Shelver, W.L., Larsen, G.L., and Huwe, J.K., Use of an immunoaffinity column for tetra-chlorodibenzo-*p*-dioxin serum sample cleanup, *J. Chromatogr. B,* 705, 261–268, 1998.

3. Puech, L., Dragacci, S., Gleizes, E., and Fremy, J.M., Use of immunoaffinity columns for clean-up of diarrhetic toxins (okadaic acid and dinophysistoxins) extracts from shellfish prior to their analysis by HPLC/fluorimetry, *Food Addit. Contam.*, 16, 239–251, 1999.

4. Crabbe, P., Haasnoot, W., Kohen, F., Salden, M., and Van Peteghem, C., Production and characterization of polyclonal antibodies to sulfamethazine and their potential use in immunoaffinity chromatography for urine sample pre-treatment, *Analyst*, 124, 1569–1575, 1999.

5. Kondo, F., Matsumoto, H., Yamada, S., Tsuji, K., Ueno, K., and Harada, K., Immunoaffinity purification method for detection and quantification of microcystines in lake water, *Toxicon*, 38, 813–823, 2000.

6. Clarke, W., Chowdhuri, A.R., and Hage, D.S., Analysis of free drug fractions by ultrafast immunoaffinity chromatography, *Anal. Chem.*, 73, 2157–2164, 2001.

7. Hage, D.S., Affinity chromatography: A review of clinical applications, *Clin. Chem.*, 45, 593–615, 1999.

8. Page, M. and Thorpe, R., IgG purification, *Methods Mol. Biol.*, 80, 95–111, 1998.

9. Page, M. and Thorpe, R., Purification of monoclonal antibodies, *Methods Mol. Biol.*, 80, 113–119, 1998.

10. Hayzer, D.J. and Jaton, J.C., Immunoglobulin M (IgM), *Methods Enzymol.*, 116, 26–36, 1985.

11. Garcia-Gonzalez, M., Bettinger, S., Ott, S., Olivier, P., Kadouche, J., and Pouletty, P., Purification of murine IgG3 and IgM monoclonal antibodies by euglobulin precipitation, *J. Immunol. Methods*, 111, 17–23, 1988.

12. Page, M. and Thorpe, R., Purification of IgG by precipitation with sodium sulfate or ammonium sulfate, in *The Protein Protocols Handbook*, Walker, J.M., Ed., Humana Press, Totowa, NJ, 1996, pp. 721–722.

13. Saetang, T., Treamwattana, N., Suttijitpaisal, P., and Ratanabanangkoon, K., Quantitative comparison on the refinement of horse antivenom by salt fractionation and ion-exchange chromatography, *J. Chromatogr. B*, 700, 233–239, 1997.

14. Knutson, V.P., Buck, R.A., and Moreno, R.M., Purification of a murine monoclonal antibody of the IgM class, *J. Immunol. Methods*, 136, 151–157, 1991.

15. Zhou, W. and Cai, S., Production of monoclonal antibody in hollow fiber culture system with serum-free medium, *Chin. J. Biotechnol.*, 8, 41–49, 1992.

16. Page, M. and Thorpe, R., Purification of IgG by precipitation with polyethylene glycol, in *The Protein Protocols Handbook*, Walker, J.M., Ed., Humana Press, Totowa, NJ, 1996, p. 731.

17. Phillips, T.M., Immune complex assays: Diagnostic and clinical application, *Crit. Rev. Clin. Lab. Sci.*, 27, 237–264, 1989.

18. Telleman, P., Kingsbury, G.A., and Junghans, R.P., Direct immunoprecipitation of antigen with phage displaying immunoglobulin fragment, *BioTechniques*, 29, 1240–1248, 2000.

19. Brooks, D.A., Bradford, T.M., and Hopwood, J.J., An improved method for the purification of IgG monoclonal antibodies from culture supernatants, *J. Immunol. Methods*, 155, 129–132, 1992.

20. Akita, E.M. and Nakai, S., Comparison of four purification methods for the production of immunoglobulins from eggs laid by hens immunized with an enterotoxigenic *E. coli* strain, *J. Immunol. Methods*, 160, 207–214, 1993.

21. Romito, M., Viljoen, G.J., and Du Plessis, D.H., Eliciting antigen-specific egg-yolk IgY with naked DNA, *BioTechniques*, 31, 670–675, 2001.

22. Steinbuch, M. and Audran, R., The isolation of IgG from mammalian sera with the aid of caprylic acid, *Arch. Biochem. Biophys.*, 134, 279–284, 1969.

23. Page, M. and Thorpe, R., Purification of IgG using caprylic acid, in *The Protein Protocols Handbook*, Walker, J.M., Ed., Humana Press, Totowa, NJ, 1996, p. 723.

24. Habeeb, A.F. and Francis, R.D., Preparation of human immunoglobulin by caprylic acid precipitation, *Prep. Biochem.*, 14, 1–17, 1984.

25. Rojas, G., Jimenez, J.M., and Gutierrez, J.M., Caprylic acid fractionation of hyperimmune horse plasma: Description of a simple procedure for antivenom production, *Toxicon*, 32, 351–363, 1994.

26. McKinney, M.M. and Parkinson, A., A simple, non-chromatographic procedure to purify immunoglobulins from serum and ascites fluid, *J. Immunol. Methods,* 96, 271–278, 1987.

27. Reik, L.M., Maines, S.L., Ryan, D.E., Levin, W., Bandiera, S., and Thomas, P.E., A simple, non-chromatographic purification procedure for monoclonal antibodies: Isolation of monoclonal antibodies against cytochrome P450 isozymes, *J. Immunol. Methods,* 100, 123–130, 1987.

28. Perosa, F., Carbone, R., Ferrone, S., and Dammacco, F., Purification of human immunoglobulins by sequential precipitation with caprylic acid and ammonium sulphate, *J. Immunol. Methods,* 128, 9–16, 1990.

29. Mohanty, J.G. and Elazhary, Y., Purification of IgG from serum with caprylic acid and ammonium sulphate precipitation is not superior to ammonium sulphate precipitation alone, *Comp. Immunol. Microbiol. Infect. Dis.,* 12, 153–160, 1989.

30. Temponi, M., Kageshita, T., Perosa, F., Ono, R., Okada, H., and Ferrone, S., Purification of murine IgG monoclonal antibodies by precipitation with caprylic acid: Comparison with other methods of purification, *Hybridoma,* 8, 85–95, 1989.

31. Page, M. and Thorpe, R., Purification of IgG using ion-exchange HPLC/FPLC, in *The Protein Protocols Handbook,* Walker, J.M., Ed., Humana Press, Totowa, NJ, 1996, pp. 727–729.

32. Page, M. and Thorpe, R., Purification of IgG using DEAE-Sepharose chromatography, in *The Protein Protocols Handbook,* Walker, J.M., Ed., Humana Press, Totowa, NJ, 1996, pp. 725–726.

33. Page, M. and Thorpe, R., Purification of IgG using gel-filtration chromatography, in *The Protein Protocols Handbook,* Walker, J.M., Ed., Humana Press, Totowa, NJ, 1996, pp. 735–737.

34. Burd, R.S., Raymond, C.S., Ratz, C.A., and Dunn, D.L., A rapid procedure for purifying IgM monoclonal antibodies from murine ascites using a DEAE-disk, *Hybridoma,* 12, 135–142, 1993.

35. Gallo, P., Siden, A., and Tavolato, B., Anion-exchange chromatography of normal and monoclonal serum immunoglobulins, *J. Chromatogr.,* 416, 53–62, 1987.

36. Clezardin, P., Bougro, G., and McGregor, J.L., Tandem purification of IgM monoclonal antibodies from mouse ascites fluids by anion-exchange and gel fast protein liquid chromatography, *J. Chromatogr.,* 354, 425–433, 1986.

37. Novales-Li, P., Purification of murine immunoglobulin M from spent tissue culture supernatant by one-step gel filtration chromatography procedure, *Biomed. Chromatogr.,* 9, 42–47, 1995.

38. Laroche-Traineau, J., Clofent-Sanchez, G., and Santarelli, X., Three-step purification of bacterially expressed human single-chain Fv antibodies for clinical applications, *J. Chromatogr. B,* 737, 107–117, 2000.

39. Stanker, L.H., Vanderlaan, M., and Juarez-Salinas, H., One-step purification of mouse monoclonal antibodies from ascites fluid by hydroxylapatite chromatography, *J. Immunol. Methods,* 76, 157–169, 1985.

40. Josic, D., Loster, K., Kuhl, R., Noll, F., and Reusch, J., Purification of monoclonal antibodies by hydroxylapatite HPLC and size exclusion HPLC, *Biol. Chem. Hoppe Seyler,* 372, 149–156, 1991.

41. Boden, V., Winzerling, J.J., Vijayalakshmi, M., and Porath, J., Rapid one-step purification of goat immunoglobulins by immobilized metal ion affinity chromatography, *J. Immunol. Methods,* 181, 225–232, 1995.

42. Bog-Hansen, T.C., Separation of monoclonal antibodies from cell-culture supernatants and ascites fluid using thiophilic agarose, *Methods Mol. Biol.,* 45, 177–181, 1995.

43. Serres, A., Muller, D., and Jozefonvicz, J., Purification of monoclonal antibodies on dextran-coated silica support grafted by thiophilic ligand, *J. Chromatogr. A,* 711, 151–157, 1995.

44. Bog-Hansen, T.C., Separation of monoclonal antibodies from cell-culture supernatants and ascites fluid using thiophilic agarose, *Mol. Biotechnol.,* 8, 279–281, 1997.

45. Arguelles, M.E., Alonso, M., Garcia Suarez, M.D., Barneo, L., Sampedro, A., and de los Toyos, J.R., Performance of thiophilic adsorption chromatography in the purification of rat

IgG2b monoclonal antibodies from serum- and protein-free culture supernatants, *Biomed. Chromatogr.,* 13, 379–381, 1999.

46. Oscarsson, S. and Porath, J., Covalent chromatography and salt-promoted thiophilic adsorption, *Anal. Biochem.,* 176, 330–337, 1989.
47. Hutchens, T.W. and Porath, J., Thiophilic adsorption of immunoglobulins—analysis of conditions optimal for selective immobilization and purification, *Anal. Biochem.,* 159, 217–226, 1986.
48. Lihme, A. and Heegaard, P.M., Thiophilic adsorption chromatography: The separation of serum proteins, *Anal. Biochem.,* 192, 64–69, 1991.
49. Scholz, G.H., Vieweg, S., Leistner, S., Seissler, J., Scherbaum, W.A., and Huse, K., A simplified procedure for the isolation of immunoglobulins from human serum using a novel type of thiophilic gel at low salt concentration, *J. Immunol. Methods,* 219, 109–118, 1998.
50. Hansen, P., Scoble, J.A., Hanson, B., and Hoogenraad, N.J., Isolation and purification of immunoglobulins from chicken eggs using thiophilic interaction chromatography, *J. Immunol. Methods,* 215, 1–7, 1998.
51. Yurov, G.K., Neugodova, G.L., Verkhovsky, O.A., and Naroditsky, B.S., Thiophilic adsorption: Rapid purification of F(ab)2 and Fc fragments of IgG1 antibodies from murine ascitic fluid, *J. Immunol. Methods,* 177, 29–33, 1994.
52. Fiedler, M. and Skerra, A., Use of thiophilic adsorption chromatography for the one-step purification of a bacterially produced antibody F(ab) fragment without the need for an affinity tag, *Protein Expr. Purif.,* 17, 421–427, 1999.
53. Schulze, R.A., Kontermann, R.E., Queitsch, I., Dubel, S., and Bautz, E.K., Thiophilic adsorption chromatography of recombinant single-chain antibody fragments, *Anal. Biochem.,* 220, 212–214, 1994.
54. Lutomski, D., Joubert-Caron, R., Bourin, P., Bladier, D., and Caron, M., Use of thiophilic adsorption in the purification of biotinylated Fab fragments, *J. Chromatogr. B,* 664, 79–82, 1995.
55. Kreutz, F.T., Wishart, D.S., and Suresh, M.R., Efficient bispecific monoclonal antibody purification using gradient thiophilic affinity chromatography, *J. Chromatogr. B,* 714, 161–170, 1998.
56. Kronvall, G. and Frommel, D., Definition of staphylococcal protein A reactivity for human immunoglobulin G fragments, *Immunochemistry,* 7, 124–127, 1970.
57. Sjoquist, J., Structure and immunology of protein A, *Contrib. Microbiol. Immunol.,* 1, 83–92, 1973.
58. Lind, I., Occurrence of protein A in strains of *Staphylococcus aureus* demonstrated by means of labelled globulins, *Contrib. Microbiol. Immunol.,* 1, 93–97, 1973.
59. Bjorck, L. and Kronvall, G., Purification and some properties of streptococcal protein G, a novel IgG-binding reagent, *J. Immunol.,* 133, 969–974, 1984.
60. Akerstrom, B. and Bjorck, L., A physicochemical study of protein G, a molecule with unique immunoglobulin G-binding properties, *J. Biol. Chem.,* 261, 10240–10247, 1986.
61. Page, M. and Thorpe, R., Purification of IgG using protein A or protein G, in *The Protein Protocols Handbook,* Walker, J.M., Ed., Humana Press, Totowa, NJ, 1996, pp. 733–734.
62. Phillips, T.M., *Analytical Techniques in Immunochemistry,* Marcel Dekker, New York, 1992, pp. 75–86.
63. Stahl, S., Nygren, P.A., Sjolander, A., and Uhlen, M., Engineered bacterial receptors in immunology, *Curr. Opin. Immunol.,* 5, 272–277, 1993.
64. Tashiro, M. and Montelione, G.T., Structures of bacterial immunoglobulin-binding domains and their complexes with immunoglobulins, *Curr. Opin. Struct. Biol.,* 5, 471–481, 1995.
65. Fassina, G., Ruvo, M., Palombo, G., Verdoliva, A., and Marino, M., Novel ligands for the affinity-chromatographic purification of antibodies, *J. Biochem. Biophys. Methods,* 49, 481–490, 2001.
66. Teng, S.F., Sproule, K., Husain, A., and Lowe, C.R., Affinity chromatography on immobilized "biomimetic" ligands: Synthesis, immobilization and chromatographic assessment of an immunoglobulin G-binding ligand, *J. Chromatogr. B,* 740, 1–15, 2000.
67. Frohman, M.A., Frohman, L.A., Goldman, M.B., and Goldman, J.N., Use of protein A-containing *Staphylococcus aureus* as an immunoadsorbent in radioimmunoassays to separate antibody-bound from free antigen, *J. Lab. Clin. Med.,* 93, 614–621, 1979.

68. Goding, J.W., Use of staphylococcal protein A as an immunological reagent, *J. Immunol. Methods,* 20, 241–253, 1978.
69. Poiesi, C., Tamanini, A., Ghielmi, S., and Albertini, A., Protein A, hydroxyapatite and diethylaminoethyl: Evaluation of three procedures for the preparative purification of monoclonal antibodies by high-performance liquid chromatography, *J. Chromatogr. A,* 465, 101–111, 1989.
70. Denizli, A. and Arica, Y., Protein A-immobilized microporous polyhydroxyethyl-methacrylate affinity membranes for selective sorption of human-immunoglobulin-G from human plasma, *J. Biomater. Sci. Poly. Ed.,* 11, 367–382, 2000.
71. Phillips, T.M., Queen, W.D., More, N.S., and Thompson, A.M., Protein A-coated glass beads: Universal support medium for high-performance immunoaffinity chromatography, *J. Chromatogr.,* 327, 213–219, 1985.
72. Phillips, T.M., High-performance immunoaffinity chromatography, *Adv. Chromatogr.,* 29, 133–173, 1989.
73. Phillips, T.M. and Dickens, B.F., *Affinity and Immunoaffinity Purification Techniques,* Eaton Publishing, Natick, MA, 2000, pp. 111–137.
74. Chalon, M.P., Milne, R.W., and Vaerman, J.P., Interactions between mouse immunoglobulins and staphylococcal protein A, *Scand. J. Immunol.,* 9, 359–364, 1979.
75. Mackenzie, M.R., Warner, N.L., and Mitchell, G.F., The binding of murine immunoglobulins to staphylococcal protein A, *J. Immunol.,* 120, 1493–1496, 1978.
76. Duhamel, R.C., Schur, P.H., Brendel, K., and Meezan, E., pH gradient elution of human IgG1, IgG2 and IgG4 from protein A-Sepharose, *J. Immunol. Methods,* 31, 211–217, 1979.
77. Stone, G.C., Sjobring, U., Bjorck, L., Sjoquist, J., Barber, C.V., and Nardella, F.A., The Fc binding site for streptococcal protein G is in the C gamma 2-C gamma 3 interface region of IgG and is related to the sites that bind staphylococcal protein A and human rheumatoid factors, *J. Immunol.,* 143, 565–570, 1989.
78. Derrick, J.P. and Wigley, D.B., The third IgG-binding domain from streptococcal protein G: An analysis by X-ray crystallography of the structure alone and in a complex with Fab, *J. Mol. Biol.,* 243, 906–918, 1994.
79. Murphy, D.A., Van Alstin, W., Bowersock, T., and Burgos, C., Binding of a recombinant form of streptococcal protein G to porcine IgG, *J. Vet. Diagn. Invest.,* 4, 469–470, 1992.
80. Peng, Z.K., Simons, F.E., and Becker, A.B., Differential binding properties of protein A and protein G for dog immunoglobulins, *J. Immunol. Methods,* 145, 255–258, 1991.
81. Fischer, M. and Hlinak, A., The lack of binding ability of staphylococcal protein A and streptococcal protein G to egg yolk immunoglobulins of different fowl species, *Berl. Munch. Tierarztl. Wochenschr.,* 113, 94–96, 2000.
82. Eliasson, M., Andersson, R., Olsson, A., Wigzell, H., and Uhlen, M., Differential IgG-binding characteristics of staphylococcal protein A, streptococcal protein G, and a chimeric protein AG, *J. Immunol.,* 142, 575–581, 1989.
83. Pilcher, J.B., Tsang, V.C., Zhou, W., Black, C.M., and Sidman, C., Optimization of binding capacity and specificity of protein G on various solid matrices for immunoglobulins, *J. Immunol. Methods,* 136, 279–286, 1991.
84. Ohlson, S., Nilsson, R., Niss, U., Kjellberg, B.M., and Freiburghaus, C., A novel approach to monoclonal antibody separation (HPLAC) with SelectiSpher-10 protein G, *J. Immunol. Methods,* 114, 175–180, 1988.
85. Blank, G.S. and Vetterlein, D., Quantification of monoclonal antibodies in complex mixtures by protein G high-performance liquid affinity chromatography, *Anal. Biochem.,* 190, 317–320, 1990.
86. Akerstrom, B. and Bjorck, L., Protein L: An immunoglobulin light chain-binding bacterial protein: Characterization of binding and physicochemical properties, *J. Biol. Chem.,* 264, 19740–19746, 1989.
87. De Chateau, M., Nilson, B.H., Erntell, M., Myhre, E., Magnusson, C.G., Akerstrom, B., and Bjorck, L., On the interaction between protein L and immunoglobulins of various mammalian species, *Scand. J. Immunol.,* 37, 399–405, 1993.

88. Wikstrom, M., Sjobring, U., Drakenberg, T., Forsen, S., and Bjorck, L., Mapping of the immunoglobulin light chain-binding site of protein L, *J. Mol. Biol.,* 250, 128–133, 1995.

89. Beckingham, J.A., Housden, N.G., Muir, N.M., Bottomley, S.P., and Gore, M.G., Studies on a single immunoglobulin-binding domain of protein L from *Peptostreptococcus magnus*: The role of tyrosine-53 in the reaction with human IgG, *Biochem. J.,* 353, 395–401, 2001.

90. Svensson, H.G., Hoogenboom, H.R., and Sjobring, U., Protein LA, a novel hybrid protein with unique single-chain Fv antibody- and Fab-binding properties, *Eur. J. Biochem.,* 258, 890–896, 1998.

91. Kihlberg, B.M., Sjobring, U., Kastern, W., and Bjorck, L., Protein LG: A hybrid molecule with unique immunoglobulin binding properties, *J. Biol. Chem.,* 267, 25583–25588, 1992.

92. Vola, R., Lombardi, A., Tarditi, L., Zaccolo, M., Neri, D., Bjorck, L., and Mariani, M., Recombinant proteins L and LG: Two new tools for purification of murine antibody fragments, *Cell Biophys.,* pp. 24–25, 27–36, 1994.

93. Kihlberg, B.M., Sjoholm, A.G., Bjorck, L., and Sjobring, U., Characterization of the binding properties of protein LG, an immunoglobulin-binding hybrid protein, *Eur. J. Biochem.,* 240, 556–563, 1996.

94. Akesson, P., Cooney, J., Kishimoto, F., and Bjorck, L., Protein H: A novel IgG binding bacterial protein, *Mol. Immunol.,* 27, 523–531, 1990.

95. Frick, I.M., Akesson, P., Cooney, J., Sjobring, U., Schmidt, K.H., Gomi, H., Hattori, S., Tagawa, C., Kishimoto, F., and Bjorck, L., Protein H: A surface protein of *Streptococcus pyogenes* with separate binding sites for IgG and albumin, *Mol. Microbiol.,* 12, 143–151, 1994.

96. Frick, I.M., Crossin, K.L., Edelman, G.M., and Bjorck, L., Protein H: A bacterial surface protein with affinity for both immunoglobulin and fibronectin type III domains, *EMBO J.,* 14, 1674–1679, 1995.

97. Podbielski, A., Weber-Heynemann, J., and Cleary, P.P., Immunoglobulin-binding FcrA and Enn proteins and M proteins of group A streptococci evolved independently from a common ancestral protein, *Med. Microbiol. Immunol.,* 183, 33–42, 1994.

98. Katerov, V., Schalen, C., and Totolian, A.A., M-like, immunoglobulin-binding protein of *Streptococcus pyogenes* type M15, *Curr. Microbiol.,* 29, 31–36, 1994.

99. Krebs, B., Kaufhold, A., Boyle, M.D., and Podbielski, A., Different alleles of the fcrA/mrp gene of *Streptococcus pyogenes* encode M-related proteins exhibiting an identical immunoglobulin-binding pattern, *Med. Microbiol. Immunol.,* 185, 39–47, 1996.

100. Fassina, G., Ruvo, M., Palombo, G., Verdoliva, A., and Marino, M., Novel ligands for the affinity-chromatographic purification of antibodies, *J. Biochem. Biophys. Methods,* 49, 481–490, 2001.

101. Roque-Barreira, M.C. and Campos-Neto, A., Jacalin: An IgA-binding lectin, *J. Immunol.,* 134, 1740–1743, 1985.

102. Chui, S.H., Lam, C.W., Lewis, W.H., and Lai, K.N., High-performance liquid affinity chromatography for the purification of immunoglobulin A from human serum using jacalin, *J. Chromatogr.,* 514, 219–225, 1990.

103. Wilkinson, R. and Neville, S., Jacalin: Its binding reactivity with immunoglobulin A from various mammalian species, *Vet. Immunol. Immunopathol.,* 18, 195–198, 1988.

104. Kabir, S., Simultaneous isolation of intestinal IgA and IgG from rabbits infected intraduodenally with *Vibrio cholerae* 01 by combined lectin affinity chromatography involving jacalin and protein A, *Comp. Immunol. Microbiol. Infect. Dis.,* 16, 153–161, 1993.

105. Kondoh, H. and Kobayashi, K., Elimination of undesirable immunoglobulin contaminants including aggregated IgG from gamma-globulin preparations by jackfruit lectin affinity chromatography, *Clin. Chim. Acta,* 174, 15–23, 1988.

106. Irazoqui, F.J., Zalazar, F.E., Nores, G.A., and Vides, M.A., *Agaricus bisporus* lectin binds mainly O-glycans but also N-glycans of human IgA subclasses, *Glycoconj. J.,* 14, 313–319, 1997.

107. Booth, J.R., Munks, R., and Sokol, R.J., Isolation of IgA1 from human serum by affinity chromatography using an immobilized extract of the albumin gland of *Helix pomatia, Transfus. Med.,* 5, 117–121, 1995.

108. Weinstein, Y., Givol, D., and Strausbauch, P.H., The fractionation of immunoglobulins with insolubilized concanavalin A, *J. Immunol.,* 109, 1402–1404, 1972.

109. Peng, Z., Arthur, G., Simons, F.E., and Becker, A.B., Binding of dog immunoglobulins G, A, M, and E to concanavalin A, *Vet. Immunol. Immunopathol.,* 36, 83–88, 1993.

110. Lim, S.B., Chua, C.T., and Hashim, O.H., Isolation of a mannose-binding and IgE- and IgM-reactive lectin from the seeds of *Artocarpus integer, J. Immunol. Methods,* 209, 177–186, 1997.

111. Marches, R. and Ghetie, V., Interaction between human IgD and ricinus agglutinin, *Scand. J. Immunol.,* 24, 45–48, 1986.

112. Shibuya, N., Berry, J.E., and Goldstein, I.J., One-step purification of murine IgM and human alpha 2-macroglobulin by affinity chromatography on immobilized snowdrop bulb lectin, *Arch. Biochem. Biophys.,* 267, 676–680, 1988.

113. Nevens, J.R., Mallia, A.K., Wendt, M.W., and Smith, P.K., Affinity chromatographic purification of immunoglobulin M antibodies utilizing immobilized mannan binding protein, *J. Chromatogr.,* 597, 247–256, 1992.

114. Page, M. and Thorpe, R., Purification of IgG using affinity chromatography on antigen-ligand columns, in *The Protein Protocols Handbook,* Walker, J.M., Ed., Humana Press, Totowa, NJ, 1996, pp. 739–741.

115. Englebienne, P. and Doyen, G., An improved method for isolation of specific antibodies by affinity chromatography: Application to an antiserum to testosterone, *J. Immunol. Methods,* 62, 197–204, 1983.

116. Bird, P., Lowe, J., Stokes, R.P., Bird, A.G., Ling, N.R., and Jefferis, R., The separation of human serum IgG into subclass fractions by immunoaffinity chromatography and assessment of specific antibody activity, *J. Immunol. Methods,* 71, 97–105, 1984.

117. Tribbick, G., Triantafyllou, B., Lauricella, R., Rodda, S.J., Mason, T.J., and Geysen, H.M., Systematic fractionation of serum antibodies using multiple antigen homologous peptides as affinity ligands, *J. Immunol. Methods,* 139, 155–166, 1991.

118. Lambin, P., Bouzoumou, A., Murrieta, M., Debbia, M., Rouger, P., Leynadier, F., and Levy, D.A., Purification of human IgG4 subclass with allergen-specific blocking activity, *J. Immunol. Methods,* 165, 99–111, 1993.

119. Kannan, K., Lalitha, P., Rao, K.V., Narayanan, R.B., and Kaliraj, P., Optimisation of immunoaffinity purification of *Wuchereria bancrofti* specific antibodies from human sera, *Indian J. Exp. Biol.,* 35, 1076–1079, 1997.

120. Viani, E., Flamminio, G., Caruso, A., Foresti, I., De Francesco, M., Pollara, P., Balsari, A., and Turano, A., Purification of natural human IFN-gamma antibodies, *Immunol. Lett.,* 30, 53–58, 1991.

121. Beattie, R.E., Volsen, S.G., Smith, D., McCormack, A.L., Gillard, S.E., Burnett, J.P., Ellis, S.B., Gillespie, A., Harpold, M.M., and Smith, W., Preparation and purification of antibodies specific to human neuronal voltage-dependent calcium channel subunits, *Brain Res. Protoc.,* 1, 307–319, 1997.

122. Horstmann, B.J., Chase, H.A., and Kenney, C.N., Purification of anti-paraquat monoclonal antibodies by affinity chromatography on immobilised hapten, *J. Chromatogr.,* 516, 433–441, 1990.

123. Schwarz, A., Affinity purification of monoclonal antibodies, *Methods Mol. Biol.,* 147, 49–56, 2000.

124. Teng, S.F., Sproule, K., Husain, A., and Lowe, C.R., Affinity chromatography on immobilized "biomimetic" ligands: Synthesis, immobilization and chromatographic assessment of an immunoglobulin G-binding ligand, *J. Chromatogr. B,* 740, 1–15, 2000.

125. Murray, A., Sekowski, M., Spencer, D.I., Denton, G., and Price, M.R., Purification of monoclonal antibodies by epitope and mimotope affinity chromatography, *J. Chromatogr. A,* 782, 49–54, 1997.

126. Schletter, J., Kruger, C., Lottspeich, F., and Scutt, C., Improved method for preparation of lipopolysaccharide-binding protein from human serum by electrophoretic and chromatographic separation techniques, *J. Chromatogr. B,* 654, 25–34, 1994.

127. Tleugabulova, D., Sodium dodecylsulfate polyacrylamide gel electrophoresis of recombinant hepatitis B surface antigen particles, *J. Chromatogr. B,* 707, 267–273, 1998.

128. Gavrovic-Jankulovic, M., Cirkovic, T., Bukilica, M., Fahlbusch, B., Petrovic, S., and Jankov, R.M., Isolation and partial characterization of Fes p4 allergen, *J. Invest. Allergy Clin. Immunol.,* 10, 361–367, 2000.

129. Puri, N. and Saxena, R.K., Partial purification and characterization of a novel murine factor that augments the expression of class I MHC antigens on tumor cells, *Exp. Mol. Med.,* 30, 93–99, 1998.

130. Tejera, M.L., Villalba, M., Batanero, E., and Rodriguez, R., Identification, isolation, and characterization of Ole e7, a new allergen of olive tree pollen, *J. Allergy Clin. Immunol.,* 104, 797–802, 1999.

131. Fahlbusch, B., Rudeschko, O., Szilagyi, U., Schlott, B., Henzgen, M., Schlenvoigt, G., and Schubert, H., Purification and partial characterization of the major allergen, Cav p1, from guinea pig *Cavia poricellus, Allergy,* 57, 417–422, 2002.

132. Lee, H.J., Lee, C.S., Kim, B.S., Joo, K.H., Lee, J.S., Kim, T.S., and Kim, H.R., Purification and characterization of a 7-kDa protein from *Clonorchis sinensis* adult worms, *J. Parasitol.,* 88, 499–504, 2002.

133. Qureshi, N., Takayama, K., and Ribi, E., Purification and structural determination of nontoxic lipid A obtained from the lipopolysaccharide of *Salmonella typhimurium, J. Biol. Chem.,* 257, 11808–11815, 1982.

134. Nawata, S., Suminami, Y., Hirakawa, H., Murakami, A., Umayahara, K., Ogata, H., Numa, F., Nakamura, K., and Kato, H., Electrophoretic characterization of heat-stable squamous cell carcinoma antigen, *Electrophoresis,* 22, 3522–3526, 2001.

135. Bisht, V., Singh, B.P., Kumar, R., Arora, N., and Sridhara, S., Culture filtrate antigens and allergens of *Epicoccum nigrum* cultivated in modified semi-synthetic medium, *Med. Microbiol. Immunol.,* 191, 11–15, 2002.

136. Fisher, M.A., Bono, J.L., Abuodeh, R.O., Legendre, A.M., and Scalarone, G.M., Sensitivity and specificity of an isoelectric focusing fraction of *Blastomyces dermatitidis* yeast lysate antigen for the detection of canine blastomycosis, *Mycosis,* 38, 177–182, 1995.

137. Covert, B.A., Spencer, J.S., Orme, I.M., and Belisle, J.T., The application of proteomics in defining the T cell antigens of *Mycobacterium tuberculosis, Proteomics,* 1, 574–586, 2001.

138. Utt, M., Nilsson, I., Ljungh, A., and Wadstrom, T., Identification of novel immunogenic proteins of *Helicobacter pylori* by proteome technology, *J. Immunol. Methods,* 259, 1–10, 2002.

139. Brassfield, A.L., Antigen purification by monoclonal antibody immunoaffinity chromatography, *Methods Mol. Biol.,* 45, 195–203, 1995.

140. Cochet, S., Blancher, A., Roubinet, F., Hattab, C., Cartron, J.P., and Bertrand, O., Immunopurification of the blood group RhD protein from human erythrocyte membranes, *J. Chromatogr. B,* 735, 207–217, 1999.

141. Calabozo, B., Duffort, O., Carpizo, J.A., Barber, D., and Polo, F., Monoclonal antibodies against the major allergen of *Plantago lanceolata* pollen, Pla l 1: Affinity chromatography purification of the allergen and development of an ELISA method for Pla l 1 measurement, *Allergy,* 56, 429–435, 2001.

142. Ahn, M.H., Hyun, K.H., Kang, J.O., and Min, D.Y., Partially purified *Toxoplasma gondii* antigens by immunoaffinity chromatography, *Korean J. Parasitol.,* 35, 251–258, 1997.

143. Attallah, A.M., El Masry, S.A., Ismail, H., Attia, H., Abdel Aziz, M., Shehatta, A.S., Tabll, A., Soltan, A., and El Wassif, A., Immunochemical purification and characterization of a 74.0-kDa *Schistosoma mansoni* antigen, *J. Parasitol.,* 84, 301–306, 1998.

144. Ghosh, S., Khan, M.H., and Ahmed, N., Cross-bred cattle protected against *Hyalomma anatolicum anatolicum* by larval antigens purified by immunoaffinity chromatography, *Trop. Animal Health Prod.,* 31, 263–273, 1999.

145. Rodero, M., Jimenez, A., Chivato, T., Laguna, R., and Cuellar, C., Purification of *Anisakis simplex* antigen by affinity chromatography, *Parasitol. Res.,* 87, 736–740, 2001.

146. Gathumbi, J.K., Usleber, E., and Martlbauer, E., Production of ultrasensitive antibodies against aflatoxin B1, *Lett. Appl. Microbiol.,* 32, 349–351, 2001.

147. Cantalupo, P., Saenz-Robles, M.T., and Pipas, J.M., Expression of SV40 large T antigen in baculovirus systems and purification by immunoaffinity chromatography, *Methods Enzymol.,* 306, 297–307, 1999.

148. Bayry, J., Prabhudas, K., Bist, P., Reddy, G.R., and Suryanarayana, V.V., Immunoaffinity purification of foot and mouth disease virus type specific antibodies using recombinant protein adsorbed to polystyrene wells, *J. Virol. Methods,* 81, 21–30, 1999.

149. Soldan, S.S., Fogdell-Hahn, A., Brennan, M.B., Mittleman, B.B., Ballerini, C., Massacesi, L., Seya, T., McFarland, H.F., and Jacobson, S., Elevated serum and cerebrospinal fluid levels of soluble human herpesvirus type 6 cellular receptor, membrane cofactor protein, in patients with multiple sclerosis, *Ann. Neurol.,* 50, 486–493, 2001.

150. Hernandez, R., Chong, E., Morales, R., Perez, E., Amador, Y., Zubiaurrez, J.R., Valdes, R., Figueroa, A., Agraz, A., and Herrera, L., Stirrer tank: An appropriate technology to immobilize the CB.Hep-1 monoclonal antibody for immunoaffinity purification, *J. Chromatogr. B,* 754, 77–83, 2001.

151. Phillips, T.M., Isolation of an interleukin 2-binding receptor from activated lymphocytes by high-performance immunoaffinity chromatography, *J. Chromatogr.,* 550, 741–749, 1991.

152. Phillips, T.M., Isolation and measurement of membrane proteins by high-performance immunoaffinity chromatography, in *High-Performance Liquid Chromatography of Peptides and Proteins: Separation, Analysis, and Conformation,* Mant, C.T. and Hodges, R.S., Eds., CRC Press, Boca Raton, FL, 1991, pp. 517–527.

153. Eckard, C.P. and Beck-Sickinger, A.G., Characterisation of G-protein-coupled receptors by antibodies, *Curr. Med. Chem.,* 7, 897–910, 2000.

154. Chestnutt, K., Bird, J., and Carroll, K., Isolation and purification of an epidermal growth factor receptor-related inhibitor of cell growth from cultured rat astrocytes, *Neurosci. Lett.,* 294, 121–124, 2000.

155. Liitti, S., Matikainen, M.T., Scheinin, M., Glumoff, T., and Goldman, A., Immunoaffinity purification and reconstitution of human alpha(2)-adrenergic receptor subtype C2 into phospholipid vesicles, *Protein Expr. Purif.,* 22, 1–10, 2001.

156. Cochet, S., Blancher, A., Roubinet, F., Hattab, C., Cartron, J.P., and Bertrand, O., Immunopurification of the blood group RhD protein from human erythrocyte membranes, *J. Chromatogr. B,* 735, 207–217, 1999.

157. Fujiwara, H., Kikkawa, Y., Sanzen, N., and Sekiguchi, K., Purification and characterization of human laminin-8: Laminin-8 stimulates cell adhesion and migration through alpha3beta1 and alpha6beta1 integrins, *J. Biol. Chem.,* 276, 17550–17558, 2001.

158. Shin, K., Hayasawa, H., and Lonnerdal, B., Purification and quantification of lactoperoxidase in human milk with use of immunoadsorbents with antibodies against recombinant human lactoperoxidase, *Am. J. Clin. Nutr.,* 73, 984–989, 2001.

159. Hardy, E., Martinez, E., Diago, D., Diaz, R., Gonzalez, D., and Herrera, L., Large-scale production of recombinant hepatitis B surface antigen from *Pichia pastoris, J. Biotechnol.,* 77, 157–167, 2000.

160. Kennedy, J.F. and Barnes, J.A., Use of $F_{(ab')^2}$ antibody fragments in the synthesis of immunoadsorbents for preparing monospecific antigen, *J. Chromatogr.,* 212, 179–186, 1981.

161. Molloy, P., Brydon, L., Porter, A.J., and Harris, W.J., Separation and concentration of bacteria with immobilized antibody fragments, *J. Appl. Bacteriol.,* 78, 359–365, 1995.

162. Kipriyanov, S.M. and Little, M., Affinity purification of tagged recombinant proteins using immobilized single chain Fv fragments, *Anal. Biochem.,* 244, 189–191, 1997.

163. L'Ecuyer, T.J. and Fulton, A.B., Specific and quantitative immunoprecipitation of tropomyosin and other cytoskeletal proteins by magnetic separation, *BioTechniques,* 14, 436–441, 1993.

164. Kausch, A.P., Owen, T.P., Narayanswami, S., and Bruce, B.D., Organelle isolation by magnetic immunoabsorption, *BioTechniques,* 26, 336–343, 1999.

165. Angen, O., Heegaard, P.M., Lavritsen, D.T., and Sorensen, V., Isolation of *Actinobacillus pleuropneumoniae* serotype 2 by immunomagnetic separation, *Vet. Microbiol.,* 79, 19–29, 2001.

15

Affinity Chromatography of Regulatory and Signal-Transducing Proteins

Allen R. Rhoads and Felix Friedberg
Department of Biochemistry and Molecular Biology, College of Medicine,
Howard University, Washington, DC

CONTENTS

15.1   INTRODUCTION

15.1.1   Basic Principles and Applications

The field of signal transduction represents a wide range of systems and proteins. This chapter examines practical applications of affinity chromatography in selected areas of signal-transduction research. Signal transduction occurs when messages are communicated across the plasma membrane. In some cases, integral membrane receptor proteins are triggered, which causes protein phosphorylation or the production of a cellular second messenger, such as cyclic adenosine monophosphate (cyclic AMP) or inositol trisphosphate. Other trans-duction mechanisms include the direct interaction of hormones with cytosolic receptors, and interaction of extracellular factors with ion-channel receptors or with transmembrane receptors regulating cytoskeletal organization [1].

Enzymes and proteins of the transduction pathway may be relatively low in cellular concentration and readily susceptible to proteolysis, leading to a loss in regulatory function during isolation. The specificity and speed of affinity chromatography can be invaluable in these situations. Affinity chromatography also offers an inexpensive and rapid means for investigating the specific and sequential interactions between biomolecules in these pathways.

Signal-transducing proteins often possess strikingly different affinities for target proteins or associated subunits in the presence of a second messenger or when covalently modified by phosphorylation. This dependency can often be exploited in the development of an affinity chromatographic procedure. Proteins such as calmodulin or second messengers that interact with a variety of proteins in different signaling pathways can also serve as general affinity ligands for isolating multiple proteins.

15.1.2   General Immobilization and Elution Methods

The effectiveness of a particular affinity procedure often depends upon the method of immobilization. In contrast to small biomolecules, proteins can be immobilized through a wide variety of functional groups or fusion tags (see [2, 3] as well as Chapter 3). In many instances, the resulting ligands offer a higher degree of affinity and specificity. The stability of immobilized peptides and proteins can be increased through the incorporation of D-amino acids, cross-linking, or chemical modification, as long as the binding specificity is not compromised. Recent refinements in affinity chromatography that involve site-specific immobilization, the construction of fusion proteins for site-specific attachment, and the use of biotinylated ligands are advantageous for preserving ligand specificity, promoting high binding capacity, and providing ease of elution. For example, the immobilization of immunoglobulins by covalent coupling through their carbohydrate regions results in a more effective orientation of the antibody and preserves its antigen-binding ability [4, 5].

Immunoaffinity purification using immobilized monoclonal antibodies directed against specific peptide regions of regulatory proteins has been invaluable in elucidating and characterizing the protein components of signal-transduction pathways. Selective elution of a retained protein with a synthetic peptide antigen eliminates the need for elution by chaotropic agents that may denature proteins. The development of targeted monoclonal antibodies and these mild elution procedures have greatly enhanced the effectiveness of immunoaffinity chromatography for such applications. Other reviews are available for obtaining an even broader understanding of affinity chromatography for these and other applications [6–9].

15.2 CALCIUM REGULATION

15.2.1 Purification of Calmodulin Target Proteins

15.2.1.1 Structural Basis of Interaction

Calmodulin (CaM) can be divided into four homologous domains that each contain an EF-hand motif (see Figure 15.1). The level of homology is greatest when domain one (residues 8 to 40) is aligned with domain three (residues 81 to 113) and domain two (44 to 75) is aligned with domain four (residues 117 to 148) [10]. The four calcium-binding sites of CaM fall into two classes of high and low affinity. Regulation of CaM or troponin C is viewed as being controlled by the binding of an additional two calcium ions to domains 1 and 2 on the amino terminal half of the molecule. These sites are calcium-specific but have low affinity [11, 12].

The CaM target proteins that bind in a calcium-dependent manner to CaM are characterized by a basic amphipathic helix [13]. This helical region is composed of positively charged residues (mainly lysine and arginine) that are interspersed among hydrophobic residues in a sequence spanning approximately 20 amino acids. Polypeptide fragments of target proteins containing the CaM-recognition region [14–15] and certain analogous peptides from toxic components such as melittin [16] interact strongly with CaM. The CaM-recognition domain may occur near important regulatory or pseudosubstrate regions of target proteins. A 10-kDa CaM-binding domain that resulted from the cyanogen bromide (CNBr) cleavage of caldesmon was isolated by using a CaM-Sepharose column [17]. Cosedimentation with F-actin indicated that the retained peptide was bound by both actin and CaM and, thus, may be important in regulating the interaction of caldesmon with actin and CaM.

15.2.1.2 Applications of CaM-Affinity Chromatography

Watterson and Vanaman [18] and Miyake et al. [19] were the first to employ immobilized CaM for the purification of CaM target enzymes. Miyake [19] reported that the CaM-dependent phosphodiesterase of rat cerebrum was purified over 50-fold by a CaM-affinity column. These two groups used CNBr- and N-hydroxysuccinimide-activated Sepharose 4B to covalently couple CaM through its primary amine groups. Covalent attachment by either procedure could be expected at any seven of the eight lysine residues in CaM, since Lys 115 is trimethylated in most species and would thus be unreactive.

CaM-affinity chromatography has been used in the purification of numerous target proteins, as summarized in Table 15.1 [20–29]. Due to a high degree of evolutionary conservation [30], CaM is an effective ligand for target proteins from different species. The specificity of the CaM–target protein interaction has been exploited through recombinant technology by tagging proteins with a 4-kDa CaM-binding peptide. This peptide was derived from myosin light-chain kinase and permits rapid isolation of the protein product under mild conditions in one step when used with a CaM-affinity resin [31–33].

The value of the CaM-affinity matrix lies in the fact that, in the presence of calcium, most Ca^{2+}/CaM-dependent proteins bind with relatively high affinity to CaM (i.e., with dissociation constants in the nanomolar region). However, in the absence of calcium, these proteins have little affinity for CaM.

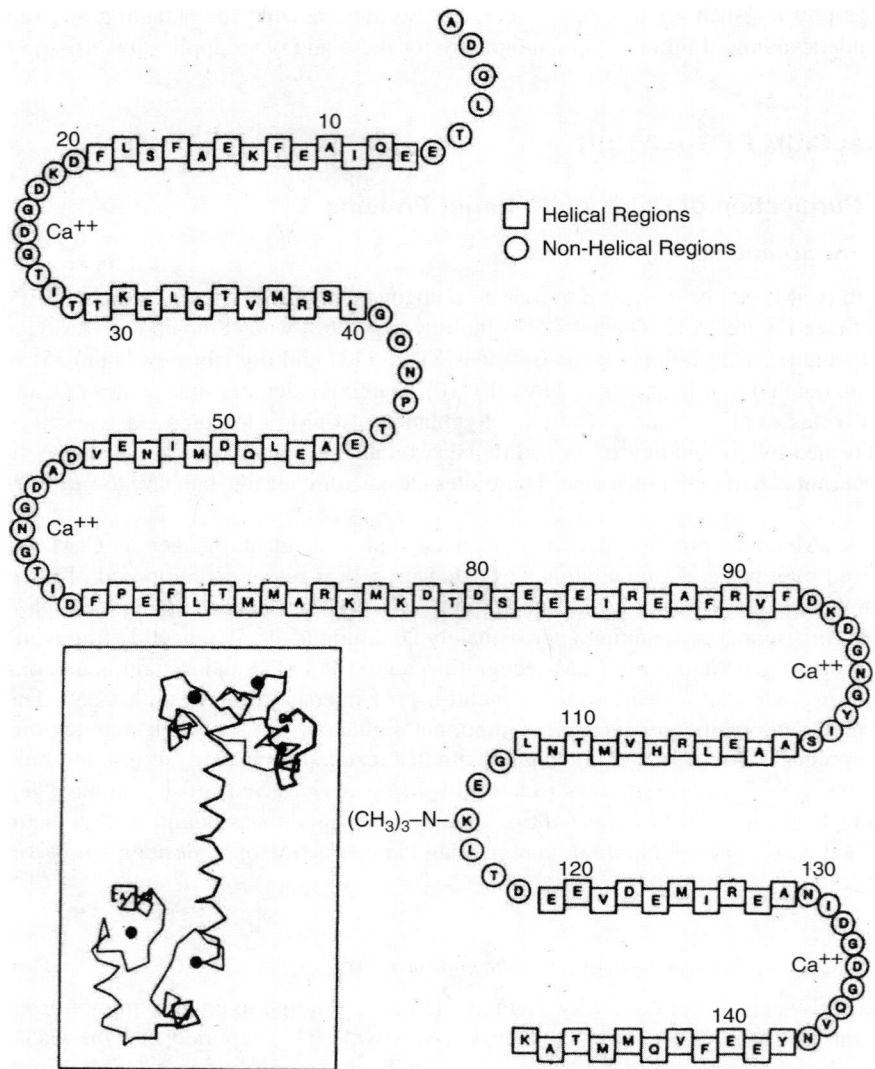

Figure 15.1 The primary amino acid sequence of vertebrate calmodulin (CaM). This sequence was derived from chicken CaM DNA, as given by Putkey et al. (*J. Biol. Chem.*, 258, 11864, 1983). The helical and nonhelical regions were assigned based on X-ray crystallographic data from Babu et al. (*Nature*, 315, 37, 1985, after Putkey et al., *J. Biol. Chem.*, 261, 9896, 1986). Calcium ions are shown in the vicinity of the four calcium-binding loops, or EF-hands. Each loop consists of the six residues that coordinate the calcium ion. These loops are flanked on both sides by alpha helical regions.

Other CaM-binding proteins include the Ca^{2+}-independent target proteins, such as neuromodulin and neurogranin. Neuromodulin binds specifically to CaM in the absence of calcium and can be released through the addition of calcium [24]. The structure responsible for this Ca^{2+}-independent interaction has been called the *IQ motif*. Neuromodulin is a primary protein of the growth cone associated with axonal growth. This is a major target for protein kinase C (PKC) phosphorylation and has been shown to stimulate the binding of guanosine 5′-*O*-3-thiotriphosphate (GTPγS) to the GTP-binding protein G_o in the growth

Table 15.1 Applications of CaM-Affinity Chromatography

Target Protein	Reference
Cyclic 3′,5′-nucleotide phosphodiesterase	[20]
Myosin light-chain kinase	[21]
CaM-dependent protein kinase	[22]
CaM-dependent protein phosphatase (2B)	[23]
Neuromodulin (GAP-43)	[24]
CaM-dependent histone 3-kinase	[25]
Inositol 1,4,5-trisphosphate 3-kinase	[26]
Major CaM-binding protein of cardiac muscle	[27]
CaM-dependent adenylyl cyclase	[28]
NO synthase	[29]

cone membrane [34]. CaM-affinity chromatography can also be valuable in the separation of CaM-sensitive and CaM-insensitive enzymes such as adenylyl cyclase [28].

Prior to the application of a sample to a CaM-affinity column, this column is usually equilibrated with Tris buffer containing 100 mM NaCl, 0.5 mM calcium, and calpain inhibitors. Protease inhibitors (e.g., leupeptin, soybean trypsin inhibitor, etc.) are also included to prevent the loss of CaM sensitivity from proteolysis. Retained components on the CaM matrix are later eluted with a buffer containing 5 mM EGTA (a specific calcium chelator).

Studies by Klee and others [35–37] on HPLC-purified CaM fragments have demonstrated that two fragments (one at the N-terminal end, residues 1 to 77, and the other at the C-terminal end, residues 78 to 148) can act as antagonists or agonists of target enzymes. By employing proteolytic fragments of CaM that were coupled to Sepharose 4B, Ni and Klee [38] developed affinity matrixes that exhibited some selectivity for target enzymes. These CaM fragments were coupled to CNBr-activated Sepharose through their lysine residues. Using this procedure, nearly all of the CaM-dependent enzymes could bind CaM fragment 78 to 148, but only phosphodiesterase and cyclic AMP-dependent protein kinase (PKA) were bound by fragment 1 to 77. A rapid procedure for the selective purification of CaM-stimulated phosphodiesterase was developed from these studies using CaM fragment 1 to 77 immobilized on Sepharose [39]. This technique required careful purification to obtain a homogeneous peptide that was free from intact CaM.

The extracellular adenylyl cyclase of *Bordetella pertussis* is activated by CaM in both the presence and absence of calcium [40, 41]. The affinity of CaM for this adenylyl cyclase is only slightly higher in the presence of Ca^{2+} than in its absence [42]. Because of this calcium-independent binding, elution of this enzyme from a normal CaM-affinity matrix requires the use of chaotropic agents. An engineered form of CaM (VU-8), in which three glutamic acid residues (positions 82 to 84) have been substituted with three lysines, has been found to possess a 1000-fold lower affinity for *B. pertussis* adenylyl cyclase than normal CaM in the absence of calcium. Thus, immobilized VU-8 can provide an affinity matrix for *B. pertussis* adenylyl cyclase that permits elution of this enzyme by simply removing calcium. A nearly homogeneous enzyme in high yield was obtained from a VU-8 matrix when it was eluted with 2 mM EGTA [42]. In addition, the holotoxin of *B. pertussis* has been purified to apparent homogeneity on CaM-Sepharose using a detergent-supplemented elution buffer containing 8 M urea [43]. This procedure effectively separated the proteolyzed low-molecular-weight catalytic subunits from the 216-kDa holotoxin.

15.2.1.3   Preparation of CaM for Immobilization

15.2.1.3.1   Conventional Purification.   CaM from the bovine brain is conveniently purified by the procedure of Dedman and Kaetzel [44], which uses the acetone powder of brain tissue. The purity of CaM can be assessed by a comparison of the total dry weight of protein and the amount of protein that is measured using an extinction coefficient of 0.18 at 276 nm for a reference protein concentration of 1 mg/ml [10]. The purified CaM should also be examined for its ability to activate bovine brain CaM-dependent phosphodiesterase [45] or other target enzymes. Furthermore, the purity of the CaM (within a range of 10 to 80 µg) can be assessed by SDS-PAGE to ensure that any column made with this ligand has good specificity [46].

15.2.1.3.2   Purification with Phenothiazine.   In the presence of calcium, CaM has two hydrophobic phenothiazine-binding sites on each half of this protein [47–49]. Because of this, fragments 1 to 77 and 78 to 148 of CaM, as well as intact CaM, can be purified by phenothiazine- and phenyl-agarose columns. Fluphenazine, a hydroxyphenothiazine, can be readily coupled to epoxy-activated Sepharose 4B and used to isolate CaM [50]. Elution of CaM from this matrix is accomplished by using EGTA-containing buffers of moderate ionic strength. Several other phenothiazines and CaM antagonists that differ in stability and capacity have also been employed as affinity ligands.

The phenothiazines are subject to photochemical degradation, but with proper precautions they can provide a rapid and effective means of purifying CaM from acidic proteins of similar molecular weight. Recently, phenothiazine-affinity columns have been used to isolate recombinant proteins fused to CaM [51]. Although possessing a lower capacity for CaM than a phenothiazine matrix, phenyl-Sepharose is a very stable support for the isolation of CaM, particularly when working with reactive protein extracts from plants [52].

15.2.1.3.3   Purification with Melittin.   Melittin is a model peptide that resembles the recognition sequence or CaM-binding region of many Ca^{2+}-dependent target proteins. Melittin is a basic amphipathic peptide, consisting of 26 amino acids, that is present in the bee venom of *Apis mellifera*. It forms a 1:1 Ca^{2+}-dependent, high-affinity complex with CaM, giving dissociation constants of 3 nM with calcium and 10 µM without calcium [16].

Melittin immobilized on CNBr-activated Sepharose displays high-affinity, Ca^{2+}-dependent binding to CaM [53]. This melittin matrix can be used to purify CaM and to measure the affinity of different peptides for radiolabeled CaM in a competitive-binding assay. Melittin-Sepharose has been employed to purify polyclonal antibodies against melittin [54]. These antibodies were then used to prepare antimellitin-Sepharose, which in turn was used to demonstrate that myosin light-chain kinase, CaM-dependent cyclic 3′,5′-nucleotide phosphodiesterase, and other CaM-binding proteins could interact with antibodies against this model peptide. Results from studies with this antimellitin Sepharose and with a monoclonal antibody [55] were the first to suggest that structural similarities exist among the CaM-recognition regions of several target proteins.

15.2.1.4   Preparation of CaM-Sepharose

CaM can be coupled through its primary amine groups to Sepharose 4B by the method of Klee et al. [56]. In this method, Sepharose 4B is activated by suspending 25 g of the moist agarose in 25 ml of 2 M sodium carbonate (pH 11.5), with this mixture then being stirred in a well-ventilated hood [57]. Next, a solution containing 3.45 g CNBr in 3 ml acetonitrile is added dropwise while the suspension is stirred. Caution should be exercised during this step, since CNBr is both volatile and toxic. After 5 min of additional stirring (at a constant pH), the mixture is placed on a 250-ml fritted glass funnel and washed with 500 ml of cold deionized water, followed by 1 l of cold 0.2 M sodium bicarbonate (pH 9.5). The activated Sepharose cake is then placed into 25 ml of a 0.1 M sodium borate buffer (pH 8.2), and 27 mg of CaM in 4.4 ml of borate buffer is added. After 10 to 18 h, this affinity gel is washed and capped with 0.5 M ethanolamine (pH 8.2) [56].

The final gel that is prepared by this method should contain approximately 0.8 mg CaM/ml gel. This is best determined by measuring the absorbance of the unreacted CaM in the filtrate at 276 nm. Alternatively, the amount of uncoupled CaM in the filtrate can be determined through its ability to activate CaM-deficient phosphodiesterase from bovine brain or by protein assays using either fluorescamine [58] or Coomassie blue [59] as reagents.

The effect of immobilizing CaM via its primary amine groups on its affinity for target enzymes is not clear. Chemical modification studies indicate that after carbamoylation of all seven lysines with isocyanate, CaM shows enhanced activation (152%) of brain adenylyl cyclase with no change in the apparent Michaelis constant (K_m) or concentration that is required for half-maximal stimulation [60]. Complete lysine modification of CaM has no effect on maximal activation of cyclic 3′,5′-nucleotide phosphodiesterase, but it increases the apparent K_m by two- to threefold above that of native CaM. However, monoacetylation of any of the lysines of CaM will reduce its affinity for calcineurin by five- to tenfold [61].

Although standard conditions for coupling CaM to CNBr-activated Sepharose do not involve the addition of calcium [56], the amount of calcium in laboratory reagents can approach a concentration of 1 μM unless special precautions are taken. Thus, under standard conditions, the coupling step probably takes place in the presence of a significant amount of calcium. Under these conditions, lysines 75 and 94 would be the most reactive of the seven lysines in CaM [62].

In the presence of calcium, CaM can be monomodified at lysine 75 by using carbamoylation with nitrosourea [63]. Phosphodiesterase requires a sevenfold higher amount of the monomodified CaM to reach normal maximal activation, while the activation profile of Ca^{2+}, Mg^{2+}-ATPase is unaffected. This suggests that stereochemical factors resulting from the modification of the flexible central helix in CaM may be important, depending upon the target enzyme. From these studies, it is difficult to predict the effect of immobilizing CaM through its lysines, but most target enzymes do have an affinity for immobilized CaM. However, the CNBr coupling of intact CaM to Sepharose via amine groups results in lower-than-calculated capacities. As a result, relatively large CaM columns are routinely used for the purification of milligram amounts of target proteins [64].

15.2.1.5   Preliminary Purification of CaM Target Proteins

CaM-binding proteins can be conveniently prepared according to the method of Hathaway et al. [65], as described by Ni and Klee [38]. Bovine brain cortex, a rich source of many CaM-dependent enzymes, should be obtained fresh and placed immediately on ice for transportation when it is used for such work.

The brain cortex should be homogenized for 1 min at 4°C in a Waring blender with 2.5 volumes of a 25 mM Tris-HCl buffer (pH 7.5) containing 2.5 mM dithiothreitol, 10 mM EDTA, 5 mM EGTA, and the following protease inhibitors: phenylmethylsulfonylfluoride (PMSF), 75 µg/ml; leupeptin, 1 µg/ml; soybean trypsin inhibitor, 50 µg/ml; alpha-N-benzoyl-L-arginine methyl ester, 100 µg/ml; and 1 mM benzamidine. Experience indicates that proteolysis occurs rapidly for highly labile CaM target proteins in the brain. Most CaM-binding proteins are sensitive targets for Ca^{2+}-dependent neutral proteases (i.e., calpains) [66].

Due to the close proximity or overlap between the autoinhibitory and CaM-binding domains of some target proteins, proteolysis often results in activation of enzyme activity with concomitant loss of CaM binding. Inclusion of cysteine protease inhibitors like leupeptin and antipain can minimize this proteolysis, which prevents binding of target proteins to the CaM-affinity matrix.

Before using CaM-affinity chromatography, endogenous CaM must be removed from the target enzymes by employing a diethylaminoethyl (DEAE)-cellulose column and NaCl gradient elution. This step must be performed with high resolution and care, or else the trace amounts of CaM that remain in the enzyme fraction will prevent target enzymes from binding to immobilized CaM. Often an additional preliminary step, such as gel-filtration or Blue Dextran chromatography [20, 28], can be used to ensure elimination of the endogenous CaM. The presence of endogenous CaM in a sample can be detected by measuring the activity of the CaM-dependent enzyme that is being purified in both the presence and absence of EGTA. Another approach that can be employed is to measure the endogenous CaM directly by an immunoassay. In addition, CaM can be measured by heating the sample to 100°C (which eliminates the activity of endogenous CaM-dependent enzymes) and then assaying the supernatant by looking at its degree of activation for a purified CaM-deficient reagent enzyme [45].

15.2.2   Purification of Other Calcium-Dependent Proteins

Grancalcin is a Ca^{2+}-binding protein that is abundant in human neutrophils. It belongs to the penta EF-hand subfamily (e.g., calpain, sorcin, peflin, and ALG-2). Like CaM, grancalcin undergoes conformational changes upon binding to Ca^{2+}. In one study, solubilized proteins from human neutrophils were applied to an immobilized grancalcin column, where the grancalcin was coupled to CNBr-activated Sepharose. It was demonstrated in this study that L-plastin, a leukocyte-specific actin-bundling protein, interacted with grancalcin in a Ca^{2+}-independent manner [67]. This interaction was confirmed by the observed co-immunoprecipitation of grancalcin with anti-L-plastin antibodies.

Binding studies have also been reported with another Ca^{2+}-binding protein, immobilized recoverin, that was isolated from bovine photoreceptors. Recoverin is related to neuronal-specific neurocalcin and hippocalcin. This ligand was used to identify rhodopsin kinase and tubulin as Ca^{2+}-dependent binding partners of recoverin. Recoverin was further shown to promote the Ca^{2+}-dependent inhibition of kinase-stimulated rhodopsin phosphorylation that mediates photoreceptor light adaptation [68].

Affinity isolation of the ryanodine receptor or calcium-release channel has been performed using a fusion protein that consisted of a 12-kDa FK506-binding protein (FKBP12) tagged with glutathione S-transferase (GST). The FKBP12 protein was employed, since it has a high affinity for the ryanodine receptor, while the GST tag has strong binding to glutathione [69]. In this study, a detergent-solubilized form of the ryanodine receptor was allowed to bind with the GST-tagged FKBP12 and passed through a glutathione-Sepharose column. The retained receptor (and GST-FKBP12) was then later eluted with a mobile phase that contained glutathione.

Several calcium-dependent phospholipid/membrane-binding proteins have been identified in tissues. These include calpactins, chromobindins, lipocortins, calcimedins, synexin, calelectrins, and endonexin, as well as others that are broadly referred to as annexins [70]. The cellular elements responsible for binding to these proteins (such as phospholipids, membranes, and actin) have all been immobilized on agarose and employed for the isolation of these proteins and for the study of their specific properties.

Glutaraldehyde-stabilized F-actin, prepared according to Herman and Pollard [71], has been employed to isolate annexins from bovine aorta [72], as shown in Table 15.2.

Table 15.2 Affinity Purification and Identification of Calcium-Dependent Binding Proteins

Protein	Ligand/Matrix	Binding/Elution Conditions	Results	Reference
Annexins of bovine aorta	Glutaraldehyde-stabilized F-actin	Tris buffer (pH 7.5), 100 mM NaCl, 0.5 mM dithiothreitol (DTT); eluted with 5 mM EGTA	Five proteins of 67, 36, 34, 32, and 10 kDa identified	[72]
Actin-binding proteins of chicken smooth muscle	F-actin (phalloidin-stabilized and suberimidate cross-linked actin)	HEPES buffer (pH 7.5), 0.05% Nonidet P40, 10% glycerol, protease inhibitors; batch elution with 0.2, 0.5, and 1.0 M KCl with 1 mM ATP/3 mM Mg^{2+}	Several proteins identified, including villin, fimbrin, spectrin, TW260/240, and filamin	[75]
Annexins of porcine intestine	Liposomes of dipalmitoyl-glycerophosphoserine entrapped in polyacrylamide gel	Applied sample at 2 mM Ca^{2+} and eluted at 8 mM EGTA	Proteolytic fragments of one or two annexins repeats have decreased (10- to 20-fold) affinity	[76]
Chromobindins of adrenal medullary cytosol	Immobilized chromaffin granule membrane or immobilized liposomes	Equilibrated with 240 mM sucrose, 30 mM KCL, 1 mM Mg^{2+} in HEPES buffer pH 7.3 at 37°C, and 4 mM Ca^{2+}/2 mM EGTA; batch elution with free Ca^{2+} at 40 and 0.1 µM	23 chromobindins identified, including CaM, PKC, caldesmon, synexin, calelectrins (32 and 68 kDa), and protein substrates for PKC and protein tyrosine kinases	[79–81]

Stabilization of F-actin by glutaraldehyde, phalloidin [73], or suberimidate cross-linking [74] prevents dissociation of the actin monomer from the immobilized filament during chromatography. Although the F-actin affinity matrix has been primarily used to study myosin interactions, a number of additional actin-binding proteins have been isolated by this support [75].

Liposomes entrapped within polyacrylamide particles or covalently linked to agarose have also been used to isolate members of the annexin family and to study their Ca^{2+}-binding properties (see Table 15.2). Proteolytic fragmentation of a 36-kDa annexin that normally contains four calcium-phospholipid binding regions has been found to give peptides that contain only one or two calcium-phospholipid binding regions. This leads to a 10- to 20-fold decrease in its Ca^{2+}-dependent affinity for the liposome matrix [76]. Recombinant annexin V has been purified to homogeneity by using its reversible Ca^{2+}-mediated binding to liposomes [77]. In addition, the specific and avid interaction of annexin V with phosphatidylserine has led to the use of annexin V as a reagent to detect cells in the early stages of cellular apoptosis, when phosphatidylserine undergoes externalization on the cell membrane [78].

Cytosolic proteins known as chromobindins can be bound to immobilized chromaffin granule membranes in the presence of calcium (see Table 15.2) and isolated by elution with EGTA [79–81]. The majority of isolated chromobindins are also bound by immobilized liposomes, which are prepared from total lipid extracts of chromaffin granule membranes by means of CNBr-activated Sepharose [79]. A total of 23 soluble chromobindins have been isolated by use of a matrix of chromaffin granule membranes. These proteins play an important role in membrane events such as exocytosis, membrane recycling, and cytoskeletal interactions.

Chromaffin granule proteins binding to microtubules in a calcium-independent manner have been isolated using taxol-stabilized microtubules. Taxol (shown in Figure 15.2) stabilizes microtubules by inhibiting their depolymerization. Severin et al. [82] employed taxol-stabilized microtubules that had been immobilized on agarose derivatized with N-hydroxysuccinimide ester. This was done to demonstrate that microtubule-associated protein 2 (MAP 2) was present in chromaffin granules and may be responsible for their interaction with microtubules. Microtubules bound to an immobilized derivative of colchicine have also been used to isolate microtubule-binding proteins [83].

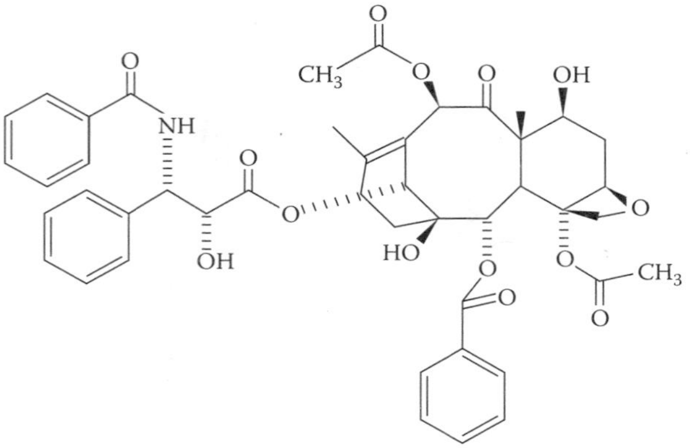

Taxol

Figure 15.2 Structure of taxol.

15.3 CYCLIC NUCLEOTIDE REGULATION

15.3.1 Adenylyl Cyclase

Signal transduction by receptors for hormones and neurotransmitters that are coupled to cyclic AMP production often involves mediation through a guanosine triphosphate (GTP)-binding protein that subsequently stimulates or inhibits the effector enzyme adenylyl cyclase [84]. The adenylyl cyclase regulated by GTP-binding proteins may be sensitive or insensitive to stimulation by CaM [85]. Classical GTP-binding proteins are heterotrimeric complexes consisting of α, β, and γ subunits. The binding of GTP or GTPγS to the α subunit of the $\alpha\beta\gamma$ complex, in exchange for GDP, leads to dissociation of the activated α subunit from the $\beta\gamma$ dimer. Activation of the α subunit can also occur through the interaction of AlF$_4^-$ with bound GDP, causing mimicry of GTP binding [86].

Due to its low concentration in membranes and its lability in the detergent-solubilized state, adenylyl cyclase was one component of this signal-transduction triad that proved difficult to purify by conventional techniques. Initially, adenylyl cyclase from cardiac muscle was purified over 40-fold using a N^6-adenosine triphosphate (ATP)-Sepharose column, but this enzyme was still associated with GTP-binding protein (G$_s$) even after extensive washing of the bound enzyme with 5 mM GTP [87]. However, GTP-Sepharose chromatography proved effective in separating G$_s$ from adenylyl cyclase [88] and demonstrated that membrane-associated GTP-binding and adenylyl cyclase activities were due to separate entities [89].

Forskolin, a major diterpene isolated from the roots of the Indian plant *Coleus forskohlii* (Figure 15.3), is known to be a potent activator of adenylyl cyclase from various sources. This activation explains the basis for many of forskolin's pharmacological actions [90, 91]. Forskolin, specifically the 7-O-hemisuccinyl-7-deacetyl derivative that is biologically active, has been useful as an affinity ligand for the purification and subsequent characterization and cloning of adenylyl cyclase [92–95]. In this case, deacetylforskolin provides a reactive derivative with an activation constant and maximum activation that are more closely related to the parent compound than that obtained by direct succinylation of the acetylated oligohydroxy derivative.

For the preparation of a forskolin-affinity matrix, forskolin is first deacetylated by alkaline hydrolysis and reacted with succinic anhydride to form 7-O-hemisuccinyl-7-deacetylforskolin [93]. This derivative is then coupled to aminoethyl-Sepharose that has been activated via the formation of an N-hydroxysuccinimide ester [96]. The degree of

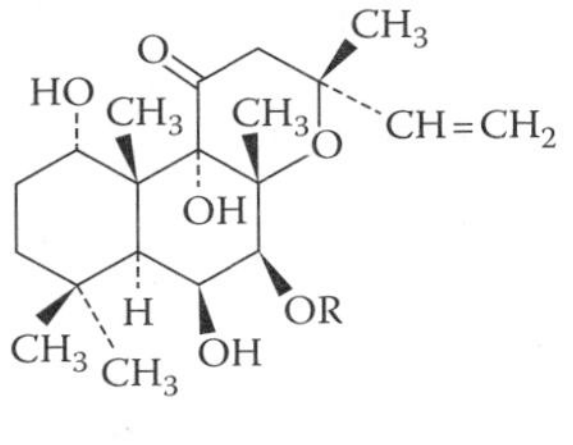

Forskolin: $R = \overset{\displaystyle O}{\overset{\|}{C}} - CH_3$

Forskolin 7-O-hemisuccinyl-7-deacetyl: $R = \overset{\displaystyle O}{\overset{\|}{C}} - CH_2 - CH_2 - COOH$

Figure 15.3 Structure of forskolin and its derivative.

immobilization of this ligand onto agarose can be measured by including radiolabeled forskolin in the coupling reaction, and there should be 0.4 to 0.8 μmol forskolin/ml packed resin.

Approximately 90% of Lubrol-solubilized adenylyl cyclase from cardiac muscle can be bound by this forskolin support, and 35 to 55% of the retained enzyme can be eluted with 100 μM forskolin in a buffer containing Tween 60. Detergent alone does not lead to enzyme release but is essential for elution with forskolin. A 300- to 500-fold enrichment of adenylyl cyclase was obtained by using this support in a single-step procedure [93].

With this forskolin-affinity support, adenylyl cyclase was readily separated from GTP-binding proteins, and thus was unresponsive to nonhydrolyzable derivatives of GTP or NaF. The isolated enzyme was depleted of bound forskolin by employing a Sephadex G-25 gel-filtration column. The forskolin-affinity column was regenerated by treating it with a buffer that contained urea. The ratio of basal to Mn^{2+} or forskolin-stimulated activity was essentially unchanged, indicating that the purified enzyme maintained its responsiveness to direct activators while losing responsiveness to guanine nucleotides and NaF that act via the GTP-binding protein. However, activation by GTP analogs and NaF can be fully restored by adding purified GTP-binding proteins to the purified enzyme [93, 94].

Under mild conditions (i.e., low $MgCl_2$ and NaCl concentrations), a GppNHp-activated adenylyl cyclase was purified over 3000-fold by a series of steps that involved a forskolin-Sepharose support [97]. The stoichiometry between the α_s, $\beta\gamma$, and catalytic subunits was near unity, suggesting that the $\beta\gamma$-dimer remains associated with the α-subunit during the activation cycle of adenylyl cyclase under physiological conditions. Forskolin-affinity chromatography has also been used to purify adenylyl cyclase to homogeneity from other sources [98, 99], including a novel 180-kDa form of the enzyme that was isolated from olfactory cilia [100].

Adenylyl cyclase from yeast has been purified by a recombinant tagging technique in which the expressed fusion protein of adenylyl cyclase was linked at the N-terminus to a peptide epitope (YPYDVPDYA) of hemagglutinin from the influenza virus [101]. This short peptide represents the complete epitope for a monoclonal antibody. The expressed adenylyl cyclase was adsorbed by this antibody in an immunoaffinity column and then released by using competitive elution with the peptide. This competitive-elution scheme avoided the deleterious extremes in pH that are normally employed to elute antigens from immunoaffinity columns.

A CaM-independent adenylyl cyclase from bovine brain was isolated in the unbound fraction by utilizing a CaM-Sepharose column [102]. This enzyme was then used to produce a monoclonal antibody, which was later adsorbed to protein A-Sepharose to produce an immunoaffinity matrix. The resulting antibody column specifically bound only the CaM-independent enzyme from brain samples. The N-terminal amino acid sequence of the purified protein later permitted synthesis of oligonucleotide probes that were employed to screen a cDNA library from the brain.

15.3.2 GTP-Binding Proteins

GTP binding proteins were first identified through their modulation of adenylyl cyclase activity. These proteins are now known to participate in the activation of other target systems, including ion channels and phospholipases [103, 104]. Isolation procedures for membrane-associated GTP-binding proteins have relied mainly on classical techniques and hydrophobic chromatography on heptyl- or phenyl-Sepharose [105]. However, in some cases GTP-Sepharose affinity chromatography has been employed [106]. To make GTP-Sepharose, phenylethylenediamine is coupled to the γ-phosphate of GTP. The resulting product is then immobilized on carboxypropylamino-Sepharose via carbodiimide coupling [87].

Affinity chromatography with immobilized βγ subunits has been used to isolate and purify a family of α subunits [107]. Because the different β and γ subunits are closely related in structure and function [108], βγ subunits were employed to isolate α subunits that bind to them in a reversible fashion. Binding was stabilized by GDP, and elution was promoted by adding GTP or AlF_4^- to the mobile phase. The βγ support bound more than 60% of the $α_s$, $α_{i1}$, $α_{i2}$, and $α_0$ subunits from Lubrol extracts of the brain and also retained a new 42-kDa α subunit protein (α42). This binding was found to be specific, since heat-inactivated βγ-agarose did not give the same results [96]. Sample extracts were added slowly to the βγ column to allow the α subunit to dissociate from endogenous βγ subunits and associate with the immobilized βγ subunits [109]. GTPγS (50 μM) was found to elute all α subunits except α42. To make use of this feature, the column was sequentially eluted with GTPγS and then with AlF_4^- to obtain a fraction enriched in the α42 subunit. This novel 42-kDa α subunit was not a substrate for ADP-ribosylation by the *B. pertussis* toxin. In addition, the new α subunit was shown to stimulate polyphosphoinositide-specific phospholipase C from bovine brain [110].

Large-scale affinity chromatography using immobilized Rab5 GTP-binding protein (which regulates eukaryotic intracellular transport) was used to identify effectors of Rab proteins [111]. A GST-Rab5 fusion protein was expressed in a pGEX vector (from Amersham Pharmacia) and was bound to glutathione-Sepharose 4B beads (see support shown in Figure 15.4). The resulting column contained 50 mg of GST-Rab5 protein per milliliter of Sepharose. Proteins that were retained by this support were later eluted by replacing GTPγS with GDP in the mobile phase. Over 20 proteins specific for the GTPγS form of Rab5 were identified and purified by this approach. This method was also used for cdc42 to better define its biological function [112].

The GTP-bound protein form of Rho family members (rho, cdc42, and rac) belonging to the Ras superfamily was used to identify potential effector targets for these molecular switches. Affinity chromatography was used in this situation by preparing GST fusion proteins of cdc42, rac1, rhoA, or H-ras. These fusion proteins were then adsorbed to glutathione-agarose in both their GDP- and GTPγS-bound states. A 195-kDa protein was purified on the immobilized GST-cdc42 column and found to be identical to IQGAP1, a protein reported previously to interact with Ras. Based on co-immunoprecipitation, CaM was found to be associated with IQGAP1 and with fragments containing the IQ motif that is responsible for binding to CaM.

To investigate the interactions of the importin nuclear transport receptors, a RanGTP protein was immobilized on IgG-Sepharose and used to identify nucleo-cytoplasmic

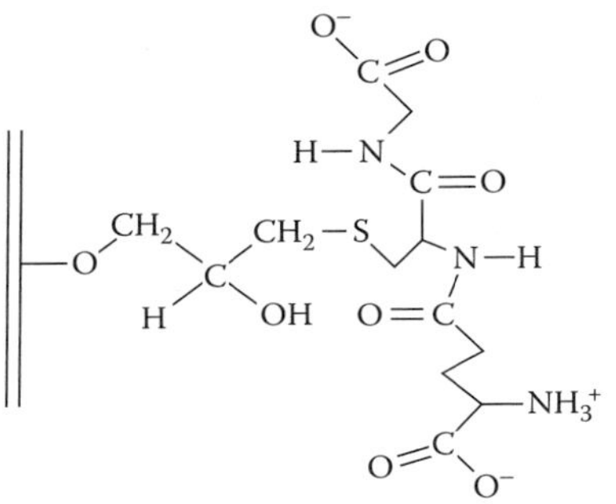

Figure 15.4 Structure of glutathione-Sepharose, where glutathione is coupled to epoxy-activated Sepharose.

transport substrates for Ran [113]. Based on this approach, new members of the importin β family were identified using wild-type RanGDP and a GTPase-deficient Ran mutant that was associated with bound GTP.

15.3.3   Guanylyl Cyclase

Cyclic GMP is formed by the action of guanylyl cyclase. Two major forms of this enzyme exist, one soluble and the other membrane-bound. The soluble form, which is composed of two subunits with masses of 82,000 and 70,000 kDa, is activated by free radicals, such as nitric oxide [114], and certain porphyrins [115]. The soluble enzyme has been isolated by immunoaffinity chromatography, based on an antibody that was developed against a synthetic peptide from the C-terminus of the 70-kDa subunit [116]. Competitive elution with a synthetic peptide was used to remove the retained enzyme from the immunoaffinity column, which successfully preserved the enzyme's activity (in contrast to elution with chaotropic agents). A soluble form of guanylyl cyclase was purified from rat lungs using monoclonal antibodies in this immunoaffinity procedure [117]. The 70-kDa subunit that was isolated through denaturing polyacrylamide gel electrophoresis was subjected to N-terminal amino acid sequencing and used to prepare an oligonucleotide probe for cDNA screening. Sequence information indicated that the 70-kDa subunit has a regulatory function rather than a catalytic function.

Membrane-bound guanylyl cyclase consists of only one polypeptide chain and can be stimulated by atriopeptins [118] and *Escherichia coli* enterotoxin [119]. GTP-agarose and wheat germ agglutinin-agarose have been used to purify this membrane-bound enzyme by approximately 600-fold [120]. This purified enzyme was then used to immunize mice for monoclonal antibody production. When the resulting monoclonal antibodies where immobilized on Sepharose, they bound a 180-kDa protein that exhibited both atrial natriuretic binding and guanylyl cyclase activity.

The extracellular domain of guanylyl cyclase was expressed in *E. coli* as a GST fusion protein and purified by glutathione-affinity chromatography [121]. The homogeneous extracellular domain was shown to bind the bacterial heat-stable enterotoxin peptide with an affinity comparable with the native receptor. Glycosylation was not required for binding. In other studies, guanylyl cyclase from the bovine photoreceptor outer segment was solubilized using dodecyl-β-D-maltoside and stabilized with phosphatidylcholine, glycerol, and dithiothreitol. This was then purified 250-fold by using GTP-affinity chromatography [122]. The photoreceptor membrane guanylyl cyclase that was obtained by this approach had properties similar to other membrane guanylyl cyclases.

15.3.4   Cyclic 3′,5′-Nucleotide Phosphodiesterase

Purification of cyclic 3′,5′-nucleotide phosphodiesterases has proved difficult. This is a result of the multiple forms of these enzymes that are present in the cytosol and in cell membranes, as well as the relatively low amounts of these enzymes. The multiplicity of these enzymes is due primarily to the expression of discrete enzymes that differ in substrate specificity, response to CaM, and allosteric modulation by cyclic GMP [123]. Proteolysis that leads to an alteration in enzyme properties during purification has also contributed to difficulties in clearly defining the physical and inhibitory properties of the different phosphodiesterases [124].

Cilostamide (see Figure 15.5a) is a quinoline compound that specifically inhibits an insulin-sensitive, low-K_m form of cyclic AMP phosphodiesterase that is found in fat cells and hepatocytes. This specific enzyme form has also been called a cyclic GMP-inhibited isozyme. To study this isozyme, the carboxylic acid analog of cilostamide (see Figure 15.5b) was converted to an *N*-(2-isothiocyanato)ethyl derivative and coupled to aminoethyl agarose

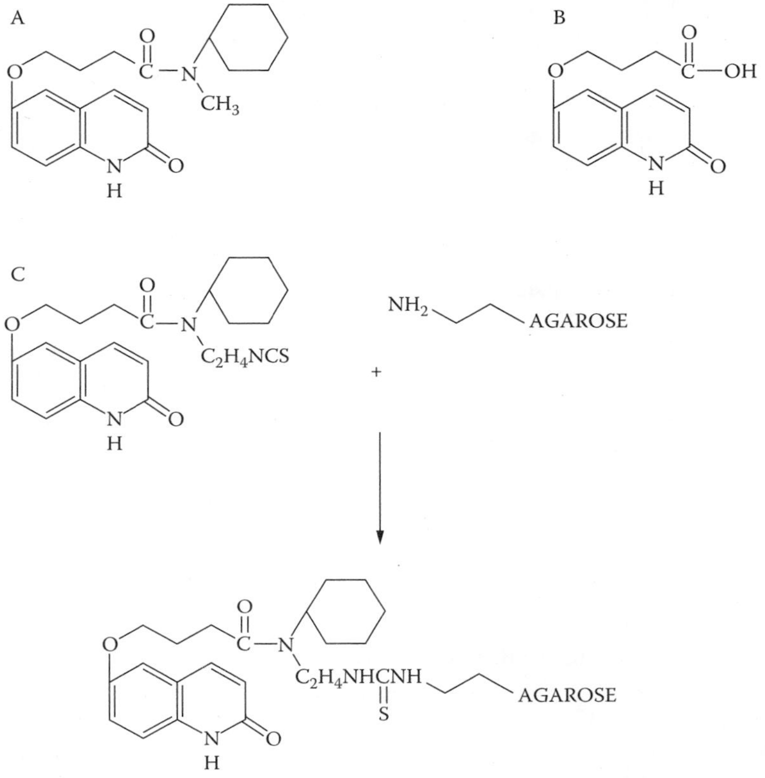

A

B

C

NH$_2$ AGAROSE

+

C$_2$H$_4$NHCNH AGAROSE

Figure 15.5 The structures of (a) cilostamide, (b) a carboxylic acid analog of cilostamide, and (c) an isothiocyanatoethyl analog of cilostamide coupled to amine-substituted agarose. (Reproduced with permission from Degerman, E., Belfrage, P., Newman, A.H., Rice, K.C., and Manganiello, V.C., *J. Biol. Chem.*, 262, 5797–5807, 1987.)

(Figure 15.5c) [125]. An enzyme preparation that was solubilized by 150 mM NaBr and a nonionic detergent was applied to the resulting affinity gel. This was later eluted with 100 mM NaBr and 50 mM cyclic AMP for a period of 5 h at 4°C. This resulted in the selective purification (300-fold) of the low-K_m isozyme with a 60% yield. A 63.8-kDa protein, representing the native form of cyclic GMP-inhibited phosphodiesterase from adipose tissue, was obtained with this affinity matrix.

A number of distinct isozymes of phosphodiesterase from the visual system have been identified by affinity chromatography. Phosphodiesterase from the visual transduction system of the cone photoreceptor was purified approximately 15,000-fold to homogeneity by a procedure using a cyclic GMP-Sepharose column [126]. This affinity column was prepared by reacting epoxy-activated Sepharose 6B (as prepared according to Sundberg and Porath [127]) with cyclic GMP, using the method of Martins et al. [128]. The coupling density of this matrix was 0.5 to 1.0 µmol GMP/ml agarose. The matrix was regenerated with 4 M guanidine, 2 M NaCl, and 1 mM EDTA. An 8-(6-aminohexylamino) cyclic AMP-Sepharose column was placed in tandem with this cyclic GMP-Sepharose column to eliminate minor contaminants due to the R-subunit of PKA and cyclic GMP-dependent protein kinase, which also bind to the cyclic AMP matrix. After applying a wash buffer to these columns, the cyclic GMP column was removed from the AMP column, and cyclic GMP phosphodiesterase was eluted by adding 5 mM cyclic GMP to the mobile phase.

A soluble rod phosphodiesterase was purified 2600-fold to homogeneity by using immunoaffinity chromatography with a monoclonal antibody (ROS 1a) affinity column [129]. The retained enzyme was eluted with a pH 10.7 buffer. Fractions were collected in tubes preloaded with a buffer below pH 7.0 to rapidly neutralize the eluted proteins. In another study, a distinct form of phosphodiesterase from *Neurospora crassa* was also purified by monoclonal antibodies in an immunoaffinity column [130]. Hybridomas that produced specific antibodies for this target were selected based on their capacity to inhibit cyclic AMP phosphodiesterase activity. The phosphodiesterase retained by these antibodies was eluted by increasing the ionic strength of the mobile phase.

A unique, soluble calcium-stimulated phosphodiesterase was isolated from the cytosolic fraction of bovine lungs [131]. This phosphodiesterase associates with CaM even in the presence of 0.1 mM EGTA. This enzyme was dissociated from CaM by first binding the holoenzyme to an immunoaffinity matrix that was prepared with an antibody against phosphodiesterase. This column was next washed with a buffer containing 0.1 mM EGTA and 0.5 M NaCl to remove CaM. The CaM-dependent enzyme was then eluted with 2.5 M MgCl$_2$. CaM-affinity chromatography was employed to further purify this enzyme. The enzyme obtained by this method was similar to *B. pertussis* toxin adenylyl cyclase, neuromodulin, and phosphorylase kinase in its ability to bind CaM at very low levels of calcium. *In vivo*, the enzyme can be expected to retain calcium sensitivity without being altered by acute changes in cellular CaM concentrations.

15.3.5 Cyclic Nucleotide-Dependent Protein Kinase

Regulation of proteins by cyclic AMP in cells occurs mainly through covalent modification by means of the phosphorylation of serine and threonine residues [132]. PKA is a tetramer consisting of two regulatory (R) and two catalytic (C) subunits. Cyclic AMP binds to the inhibitory R-subunit, leading to dissociation of an R-subunit dimer and two active C-subunits. Two different forms of the C-subunit and R-subunit have been identified [132].

Affinity chromatography of these protein kinases has exploited their interactions with ATP, allosteric modulators, pseudosubstrate domains, and the recognition sequences in the phosphorylation domains of target proteins. A synthetic peptide of 18 amino acids (which corresponds to the inhibitory domain of a heat-stable protein kinase inhibitor) [133] was used to construct an affinity column for purification of the catalytic subunit of PKA [134]. This heat-stable inhibitor is a highly specific inhibitor of PKA. The inhibitory peptide, which is homologous to the pseudosubstrate domain of the R-subunit of PKA, causes half-maximal inhibition at 200 nM. Immobilization of this peptide was accomplished by linking it to N-hydroxysuccinimide-agarose. Both the C_α and C_β catalytic subunits of PKA were isolated by this method from 3T-3 cells that had been transfected with expression vectors containing a metallothionein-promoter. These vectors encoded either the C_α or C_β isoforms of the catalytic subunit. A fraction containing the kinase activity of the transfected cells was eluted from the peptide-affinity column in the presence of 200 mM L-arginine, an inhibitor of kinases. Both purified isoforms had similar K_m values for ATP (4 µM) and Kemptide (5.6 µM), the latter being a synthetic phosphate-acceptor peptide that has the sequence LRRASLG.

In the yeast *S. cerevisiae*, three genes encode the catalytic subunits of PKA [135]. The holoenzyme (containing the regulatory and specific catalytic subunits of the TPK1 gene) was bound to an immunoaffinity column that had been prepared from a monoclonal antibody directed against the regulatory subunit of this yeast enzyme. This monoclonal antibody was adsorbed on protein A-Sepharose to make the immunoaffinity column. After the enzyme sample had been applied to this column, a homogeneous 52-kDa catalytic subunit was released by eluting it with 0.8 mM cyclic AMP.

Cyclic AMP-8-(6-aminohexylamino)-agarose has been used to isolate the regulatory subunit of PKA from *Trypanosoma cruzi* [136]. The nonretained fraction that passed through this support contained the catalytic subunit, while fractions eluted with cyclic AMP contained a 56-kDa regulatory component that cross-reacted with polyclonal antibodies against mammalian regulatory subunits type I or II. A rapid and efficient immunoaffinity method for purifying PKA holoenzyme was developed by coupling polyclonal antibodies (raised against the PKA regulatory subunit) to NHS-activated Sepharose [137]. The holoenzyme of PKA was purified 2430-fold with a 20% yield from a bivalve mollusk and was found to be fully active, as determined by its cAMP-binding activity, PKA activity, and ability to undergo autophosphorylation.

Cyclic GMP protein kinases have been purified using cyclic AMP-agarose [138, 139]. During the initial affinity isolation of this enzyme from bovine aorta on a cyclic GMP column [138], a major fraction of the Form I isozyme and some of Form II were eluted with 1 M NaCl. Subsequent elution with 0.1 mM cyclic GMP yielded mainly Form II. Upon performing cyclic AMP-affinity chromatography on the same enzyme preparation, Form I exchanged more rapidly than Form II with cyclic GMP during the elution step. Proteolytic treatment of the separated enzymes yielded unique peptides, suggesting differences in either their amino acid sequences or posttranslational modification. Cyclic AMP-agarose has also been employed in the purification of cyclic GMP-dependent protein kinase from *Ascaris suum* [139]. The yield in this step was 75%, with a purification of about tenfold.

To make a cyclic AMP column, 8-(6-aminohexylamino)-cyclic AMP can be synthesized according to the method of Jergil and Mosbach [140] by reacting 8-bromo-cyclic AMP with 1,6-diaminohexane. The resulting 8-(6-aminohexylamino)-cyclic AMP can be immobilized onto CNBr-activated Sepharose. Another useful immobilized cyclic AMP analog for the purification of protein kinases is *N*6-(2-aminoethyl) cyclic AMP, which can be synthesized from 6-chloropurine riboside 3′,5′-monophosphate and 1,2-aminoethane, as described in the literature [141].

15.3.6 Other Protein Kinases

Affinity chromatography of protein kinases can utilize distinct recognition sequences that occur within the phosphorylation domains of a target protein [142]. An example of this is work in which specific peptide sequences derived from glycogen synthase were synthesized and immobilized to provide an affinity support with a high density of binding sites for both glycogen synthase kinase-3 (GSK-3) and casein kinase II [143]. The synthetic peptide used in this study was coupled to *N*-hydroxysuccinimide-agarose. Within the recognition sequence, phosphorylation by casein kinase II was required for the binding of GSK-3, thus providing a highly selective separation procedure. These affinity columns were eluted by increasing the concentration of NaCl to 50 mM when using a phosphopeptide ligand or to 200 mM when using a dephosphopeptide ligand. The K_m value for the free peptides of the kinases was 10 μM or less.

Recently, affinity chromatography of GSK-3 was performed using an immobilized fragment of axin, a protein that interacts with β-catenin, phosphatase 2A, and adenomatous polyposis coli protein [144]. A polyhistidine-tagged axin peptide containing the complete GSK-3β and β-catenin binding domain was expressed in *E. coli*. This peptide was then immobilized on nickel-nitrilotriacetic acid (Ni-NTA) agarose and CNBr-activated Sepharose 4B. This Axin-His6 matrix was found to selectively bind recombinant rat GSK-3β and native GSK-3 from yeast, sea urchin embryos, and porcine brain. This single-step method may be useful for the large-scale purification of active recombinant or native GSK-3. This method might also be used to follow the phosphorylation state and kinase activity of this enzyme.

In other studies, histone-affinity chromatography was used to purify a histone-tubulin protein kinase from *L. donovani*. This gave a purification of approximately fivefold and a yield of 90% when the enzyme was eluted with 0.5 *M* NaCl [143]. The column used in this work was prepared by covalently attaching calf thymus histones to Sepharose 4B, as described by Wilchek et al. [145].

Peptide-affinity chromatography is another tool that has been quite effective in the identification of phosphorylated-dependent binding partners in signal transduction. The src homology 2 (SH2) and phosphotyrosine binding (PTB) domains of signaling proteins bind to phosphorylated tyrosines. SH2-containing proteins (e.g., Fyn, Csk, and phosphatidylinositol 3-kinase) have been purified on columns that contain immobilized synthetic phosphopeptides [146, 147]. Phosphopeptide matrixes have also been used to identify phosphorylation-dependent binding partners of EGF, IGF-1, and JAK proteins [148]. The peptides used in making these columns were synthesized by incorporating phosphotyrosine directly into the synthetic peptide. Control peptides were similarly made by using phenylalanine in place of phosphotyrosine. When a column was prepared that contained a 22-residue peptide derived from the S-6 protein of the 40 S-ribosomal subunit, the terminal mitogen-stimulated kinase p70[S6k] from rabbit liver was purified 6000- to 7000-fold in one step [149]. The isolated p70[S6k] kinase was then used for the screening of potential inhibitors for this enzyme.

15.3.7   Protein Phosphatases

Protein phosphatases have the reciprocal function to kinases in the regulation of proteins through phosphorylation. In addition, protein phosphatases interact with a wide range of proteins [150]. When protein phosphatase inhibitor-1 [151] and an analogous protein (DARPP-32) are phosphorylated by PKA, these are potent and specific inhibitors of protein phosphatase 1 (PPase1) [152, 153]. Ingebritsen and Ingebritsen [154] have used immobilized inhibitor-1 to study the mechanism of PPase1 inhibition (Table 15.3). It has been found that PPase1 does not bind to dephosphorylated inhibitor-1 or bovine serum albumin (BSA) coupled to Sepharose. Immobilized recombinant inhibitor-2 has also been used for a one-step affinity purification of PPase1 from rabbit muscle [155]. PPase1 from brain tissue has been purified by utilizing heparin-Sepharose [156], and protein phosphatase 2C has been isolated from muscle and liver by employing immobilized thiophosphorylated myosin light chains [157].

The major portion of PPase1 activity in mammalian tissues is tightly associated with glycogen. This fact allowed the use of β-cyclodextrin (immobilized on epoxy-activated Sepharose 6B) as a means for purifying both the catalytic and noncatalytic subunits of PPase1 in rat liver [158]. In this work, the protein phosphatase holoenzyme was transferred to an immobilized β-cyclodextrin column by the repeated cycling of isolated rat liver glycogen through this column. Selective elution resolved the 37-kDa catalytic subunit from the noncatalytic 161- and 54-kDa subunits, where the noncatalytic subunits act as the glycogen-binding domain of the hepatic protein phosphatase.

Recombinant rabbit muscle PPase1α was immobilized by reacting it with activated 6-amino hexanoic acid Sepharose 4B (i.e., carboxyl-substituted Sepharose 4B or CH-Sepharose 4B). This was then used to isolate the glycogen-binding regulatory subunit (R_G). This affinity gel contained 0.25 mg of PPase1 per milliliter of gel and allowed any retained protein to be eluted with 1 *M* NaCl. The R_G subunit was purified 4000- to 6000-fold by this column. It was found that this purified subunit could bind glycogen and be phosphorylated by PKA. The same affinity column was also effective in isolating a thymus nuclear inhibitor of PPase1 [159].

Table 15.3 Affinity Purification of Protein Phosphatases

Enzyme	Ligand/Matrix	Binding/Elution Conditions	Results	Reference
Protein phosphatase 1 (rabbit skeletal muscle)	Phosphorylated or dephosphorylated inhibitor 1 coupled to CNBr-activated Sepharose	Tris buffer (pH 7.0), 0.1 mM EGTA, 10% glycerol, 0.01% Brij 35, and protease inhibitors; elution with 3 M NaSCN or 1 M NaCl	Phosphatase activity (74%) was bound by the phosphorylated inhibitor matrix	[154]
Protein phosphatase 1 (bovine brain)	Heparin-Sepharose	Tris buffer (pH 7.0) containing 0.2 mM EGTA, 0.1% mercaptoethanol, and protease inhibitors; elution with 120 to 400 mM NaCl	Heparin-Sepharose was used at two steps in the purification; enzyme purified to near homogeneity	[156]
Protein phosphatase 2 C (rabbit muscle/liver)	Thiophosphorylated myosin-p-light-chain Sepharose	Triethanolamine buffer (pH 7.0), 1 mM EGTA, 0.1% mercaptoethanol, 5% glycerol, 10 mM Mg^{2+}, 25 mM NaCl; elution by replacing Mg^{2+} with 2 mM EDTA	Single activity peak of two isozymes with 44-fold purification and 81% yield was obtained	[157]
Protein phosphatase 1 (glycogen-associated, rat liver)	β-Cyclodextrin-Sepharose	Glycylglycine buffer (pH 7.4), 1 mM EGTA, 0.1% mercaptoethanol and protease inhibitors; elution with 2 M NaCl and the noncatalytic subunit with 1 mg/ml β-cyclodextrin	Catalytic subunit of 37 kDa and noncatalytic subunits of 161 and 54 kDa	[158]
Protein phosphatase 2A (mouse brain)	Microcystin LR-Sepharose	Tris buffer (pH 7.4), 10% glycerol, 2 mM EDTA, 2 mM EGTA; elution with 1 M NaCl	<50% protein phosphatase 1 bound; column has low binding capacity	[160]

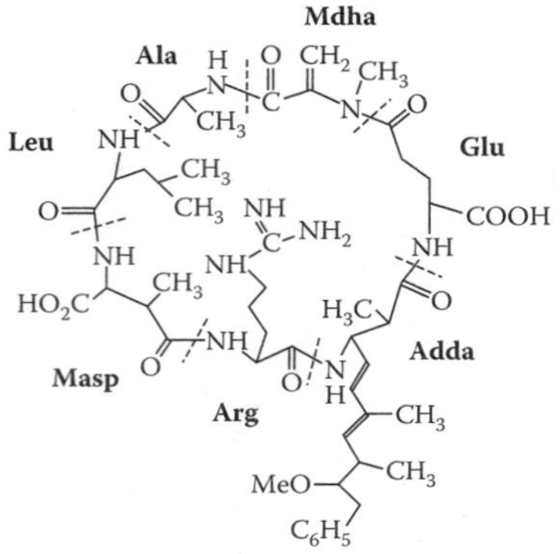

Figure 15.6 Structure of microcystin-LR. Abbreviations: Masp, β-methylaspartic acid; Mdha, *N*-methyldehydroalanine; Adda, 3-amino-9-methoxy-2,6,8-trimethyl-10-phenyldeca-4,6-dienoic acid. (Reproduced with permission from Nishiwaki, S., Fujika, H., Suganuma, M., Nishiwaki-Matsushima, R., and Sugimura, T., *FEBS Lett.*, 279, 115–118, 1991.)

The tumor-promoter okadaic acid inhibits protein phosphatases 1 and 2A. A photo-affinity probe (methyl 7-*O*-(4-azidobenzoyl)-okadaate) has been used to selectively label the catalytic subunit of the heterotrimeric protein phosphatase 2A [160]. Because of the functional significance of the carboxyl group of okadaic acid, another inhibitor of phosphatases 1 and 2A (i.e., microcystin LR, as shown in Figure 15.6) has been employed as an affinity ligand for the purification of protein phosphatase 2A [161]. Microcystin-Sepharose and gel-filtration chromatography were used to purify protein phosphatase 1 to near homogeneity. This enzyme was isolated as a dimer of catalytic subunits associated with the glycogen-binding subunit [162].

Biotinylated microcystin attached to avidin-Sepharose has also been used to purify phosphatase1 and 2A. The biotinylated derivative of microcystin is an effective inhibitor of these enzymes, with only nanomolar concentrations being needed for 50% inhibition. This was used in an affinity method by taking an extract of phosphatases (which had been precleared of endogenous biotinylated proteins by passage through an avidin-Sepharose column) and combining this with 10 μ*M* biotinylated microcystin. This mixture was then applied to an avidin-Sepharose column. After nonretained components had been washed from this column, the retained phosphatases were eluted by using a mobile phase that contained 1 m*M* biotin. An advantage of this method is that it employed relatively mild elution conditions, allowing the isolation of undissociated holoenzymes [163].

15.4 INOSITOL PHOSPHATE AND DIACYLGLYCEROL REGULATION

15.4.1 Phosphatidylinositol and Inositol Metabolism

Activation of phospholipase C by calcium-mobilizing hormones promotes the hydrolysis of phosphatidylinositol 4,5-bisphosphate to form inositol 1,4,5-trisphosphate [Ins(1,4,5)P$_3$ or IP$_3$] and diacylglycerol [164, 165]. The released IP$_3$ acts as a second messenger to mobilize intracellular calcium and can undergo phosphorylation to Ins (1,3,4,5) P$_4$ or IP$_4$ [166]. The diacylglycerol that is formed by phospholipase C action can activate protein kinase C [167] and serve as a source of arachidonic acid for the production of numerous icosanoid mediators [168].

An analog of IP_3 has been immobilized on agarose and employed in the isolation of inositol-metabolizing enzymes [169]. This support was made by reacting CH-Sepharose 4B (which contains the succinimide ester of aminocaproic acid as an activated spacer arm) with a primary amine group on tyramine. The diazotization product of 2-O-(4-aminobenzoyl)-1,4,5-tri-O-phosphono-myo-inositol was then coupled to the *ortho* position of the phenyl ring on the immobilized tyramine to form the final affinity matrix. In an alternative procedure, the aminocyclohexanecarbonyl analog of IP_3 was attached directly to an activated support via its primary amine group (i.e., without the use of tyramine and diazotization). These affinity supports were employed in the isolation of $Ins(1,4,5)P_3$ 5-phosphatase, $Ins(1,4,5)P_3$ 3-kinase, and $Ins(1,4,5)P_3$ binding protein [169]. The mobile phases used with these columns contained 2 mM EDTA to protect the immobilized inositol derivatives from Mg^{2+}-dependent IP_3 5-phosphatase activity. Enzymes that bound to these supports were eluted with 0.2 and 0.5 M KCl. IP_3 5-phosphatase was found to bind to the column that contained an IP_3 analog coupled via an azo linkage, but did not bind to the column containing the analog with an amide linker arm.

Four distinct groups (α, β, γ, and δ) of phospholipase C (PLC) enzymes occur in the cytosolic and membrane fractions of tissues [170]. Some of these groups have multiple isoforms. A specific PLC isozyme with a mass of 145 kDa (PLCγ) is a major target of epidermal growth factor-induced tyrosine phosphorylation. This isozyme was purified by immunoaffinity chromatography using an immobilized monoclonal antibody that was directed against phosphotyrosine residues [171] (see Table 15.4). PLCγ was eluted from this immunoaffinity column with phenylphosphate and was isolated by precipitation using an anti-PLCγ monoclonal antibody and cells of *Staphylococcus aureus*.

Phosphatidylinositol 4-phosphate kinase has been isolated from bovine brain membranes by using ATP-agarose as a final step in the purification [172]. A 55-kDa phosphatidylinositol 4-phosphate kinase from A431 cells was purified to near homogeneity by employing a combination of procedures that involved heparin and reactive green-agarose supports [173]. Quercetin, an oligohydroxy benzopyran inhibitor of the kinase, has also been used to purify phosphatidylinositol 4-phosphate kinase from rat liver plasma membranes [174].

Attenuation of the diacylglycerol signal can occur through phosphorylation of diacylglycerol by diacylglycerol kinase (DGK) or through lipase degradation of diacylglycerol, leading to the production of arachidonic acid. ATP-agarose, phenyl-Superose, and heparin-Sepharose have all been employed in the purification and separation of the different forms of diacylglycerol kinase, as indicated in Table 15.4 [175, 176].

A soluble CaM-dependent IP_3 kinase that converts IP_3 to $Ins(1,3,4,5)P_4$ has been isolated from several sources by procedures that employ CaM and adenosine 2,3-diphosphate agarose [26, 177–179]. IP_3 kinase from smooth muscle was purified about 25-fold with a 20% yield by using CaM-Sepharose chromatography [177]. The enzyme from rat brain was purified from cytosol to a specific activity of 2.3 μmol/min/mg protein (a 2700-fold purification) by chromatography on phosphocellulose, Orange A dye (from Millipore), CaM-agarose, and hydroxyapatite [179]. The enzyme from bovine brain has also been purified using CaM-Sepharose affinity chromatography [180]. Inclusion of calpain inhibitors during the purification from IP_3 kinase from rat brains is essential to obtain an intact 53-kDa form of this enzyme [178].

The ligand-activated PDGF-β receptor that associates with phospholipase C-γ, GAP, and phosphatidylinositol 3-kinase has been used in an affinity matrix to purify p85, a protein that links the kinase to the activated, tyrosine-phosphorylated receptor [181]. This unique approach may be useful with other PDGF receptor-associated proteins, as well as other tyrosine kinase receptors. A lithium-sensitive 29-kDa myoinositol monophosphate

Table 15.4 Affinity Purification of Enzymes in Inositol Metabolism

Enzyme	Ligand/Matrix	Binding/Elution Conditions	Results	Reference
Phospholipase C(γ-type)	Antiphosphotyrosine monoclonal antibody-Sepharose	Buffered elution with 0.075% Triton X-100 and 10 mM phenylphosphate	Large-scale immunoaffinity procedure for the isolation of phosphorylated phospholipase C (γ)	[171]
Phosphatidyl-inositol 4-phosphate kinase of brain	ATP-agarose	Kinase eluted with 300 to 400 mM NaCl in buffer containing 0.1% Triton X-100 and 0.11% PEG 20,000	ATP maintained at 50 M in binding buffer to elute diacylglycerol kinase	[172]
Phosphatidyl-inositol 4-kinase of A431 cells	Heparin- and reactive green-Sepharose	Elution buffers supplemented with 0.25% Triton X-100, heparin matrix eluted with salt gradient ± phosphoinositol; reactive green eluted with 1 M KCl	A nearly homogeneous 55-kDa protein was isolated	[173]
Phosphatidylinositol 4-kinase of rat liver	Quercetin-Sepharose	Elution performed with buffered 0.15 to 2.0 M NaCl containing 0.02% Triton X-100	The purification was 26-fold with 92% yield	[174]
Diacylglycerol kinase of rat brain	Affi-Gel blue and ATP-agarose	Affi-Gel blue enzyme eluted with 1.0 and 2.0 M NaCl; ATP-agarose: gradient of 0 to 0.5 M NaCl	Two isozymes were purified from cytosolic and membrane fractions	[175, 176]
Phosphatidyl-inositol 3-kinase and p85 of 3T3 cells	PDGF receptor-antibody complex immobilized on protein A-Sepharose	Immunocomplex washed extensively with buffers containing 1% Triton and 0.5 M LiCl	A 110-kDa kinase and an 85-kDa subunit that associates with PDGF receptor was isolated	[181]
Brain myoinositol monophosphate phosphatase	Phenyl Sepharose	Decreasing ammonium sulfate concentration	A 29-kDa protein was isolated	[182]

phosphatase that catalyzes the hydrolysis of Ins 1-P and Ins 4-P was purified to homogeneity from bovine brain by using phenyl-Superose chromatography [182] and a monoclonal immunoaffinity procedure [183]. In this method, low-ionic-strength buffers containing ethylene glycol were used to elute the enzyme in good yield (54%) and helped preserve the binding capacity of the immunoaffinity column. In contrast to this, elution with chaotropic agents (i.e., 4 M MgCl$_2$ or 3 M NaSCN) or a low-pH buffer caused a total loss of activity.

Affinity chromatography has been used to confirm findings of the yeast 2-hybrid system, which identified proteins interacting with the regulatory components of the multi-subunit NADPH oxidase. NADPH oxidase catalyzes superoxide anion formation for bactericidal activity and translocates to the membrane during its activation in neutrophils and B-lymphocytes [184]. The acidic C-terminal half of the oxidase (p67 containing an SH3 domain) was used to identify and clone a C2 domain-containing protein known as Jfc1. Jfc1 was shown to bind membrane phosphatidylinositol 3,4,5-trisphosphate (PIP$_3$) and phosphatidylinositol 3,4-bisphosphate. This protein also appeared to serve as an adaptor between phosphatidylinositol 3-kinase products and the cytosolic NADPH oxidase complex. Using an immobilized analog of PIP$_3$, several pleckstrin homology (PH) domain-containing proteins were identified [185]. These downstream targets of PI 3-kinase action appear to bind to PIP$_3$ and its analog through the PH domain.

15.4.2   Protein Kinase C

At least seven closely related phenotypes of protein kinase C (PKC) exist with different regional and cellular distributions [186]. Threonine-Sepharose [187] was used in combination with phenyl-Sepharose and high-resolution hydroxylapatite chromatography to resolve the isozymes of PKC [188, 189]. The threonine-Sepharose was prepared according to Kitano et al. [187]. To do this, L-threonine was first coupled to an amine-substituted agarose (AH-Sepharose 4B) through the addition of carbodiimide. When used in the purification of PKC, this L-threonine affinity matrix provided a 30-fold increase in specific activity with a 40% yield. PKC was eluted from this matrix with a linear gradient of NaCl (0 to 1.0 M).

PKC-ε expressed in insect cells via the baculovirus vector has been purified to homogeneity by a procedure involving serine-Sepharose chromatography and elution with a NaCl gradient (0.15 to 1.0 M) [190]. The serine-Sepharose column used in this work was prepared by the same procedure as described earlier for threonine-Sepharose [187].

In contrast to other protein kinases, PKC is inhibited by a number of specific agents, including melittin, polymyxin B, trifluoperazine, dibuccaine, compound W-7 (i.e., N-(6-aminohexyl)-5-chloro-1-naphthalenesulfonamide, as shown in Figure 15.7), and verapamil [167]. Most of these agents also inhibit CaM action by interacting with hydrophobic sites on this protein. With PKC, these agents may in some cases cause inhibition by preventing the enzyme from interacting with essential phospholipids.

The sulfonamide N-(2-aminoethyl)-5-isoquinolinesulfonamide (i.e., H-9, as shown in Figure 15.7) and other analogs (H-7 and H-8) have been shown to act as competitive inhibitors of PKC with respect to ATP [191]. PKC and other kinases bind immobilized H-9 and can be eluted from this matrix with 30 mM ATP or L-arginine [192]. This support can be prepared by coupling H-9 directly to CNBr-activated agarose, giving 10 to 12 μmol H-9/ml gel. PKC, cyclic GMP-, and cyclic AMP-dependent protein kinase can be eluted separately from this support by using a gradient of 0 to 1.5 M L-arginine, which is an inhibitor of all three kinases.

W-7-agarose has been used to investigate aspects of the interaction between W-7 and PKC [193]. For this work, PKC was bound by the immobilized W-7 and eluted with 0.1% Triton X-100 and MgATP. It was suggested that W-7 interacted mainly with the ATP-binding region of PKC in the absence of a phospholipid. In a similar study using melittin-agarose chromatography, PKC was bound and released by a buffer that contained Triton X-100 and

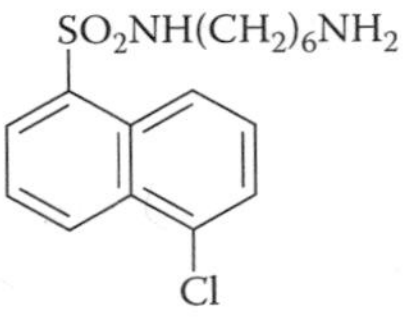

Figure 15.7   Structures of W-7, H-9, and tamoxifen.

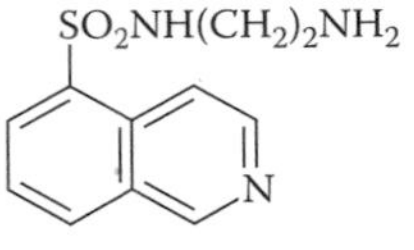

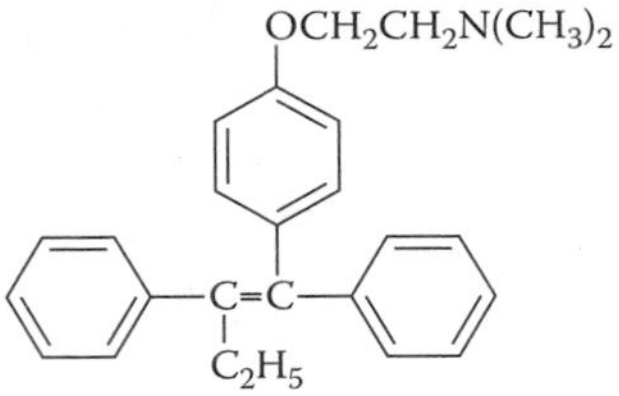

MgATP [194]. Based on ATP-sensitive binding and other results, melittin (like W-7) appears to interact specifically with the catalytic domain of PKC. The same group of investigators examined the binding of PKC to immobilized tamoxifen (shown in Figure 15.7), an anti-estrogen that is effective in the treatment of certain human breast cancers [195]. PKC was found to bind specifically and reversibly to N-didesmethyltamoxifen when this compound was coupled to CNBr-activated agarose through its primary amine group. A fraction with PKC activity was eluted from this tamoxifen matrix by using 0.2% (v/v) Triton X-100. A derivative with similar hydrophobicity (4-hydroxytamoxifen) was immobilized on epoxy-activated Sepharose but did not bind PKC, suggesting that PKC has specific binding to the didesmethyltamoxifen matrix.

Since PKC has an absolute requirement for calcium and acidic phospholipids, phosphatidylserine coupled to Affi-Gel 102 (an amine-substituted agarose from Bio-Rad) can also be used in the purification of PKC [196]. For instance, emulsified phosphatidylserine was coupled to amine-substituted agarose by using a water-soluble carbodiimide, 1-ethyl-3-(3-dimethylaminopropyl) carbodiimide. When bound to this support, PKC was eluted with a Tris buffer that contained 2 mM EDTA, 10 mM EGTA, 1 M NaCl, and 50 mM mercaptoethanolamine. In the last step of this procedure, PKC was purified by approximately fivefold but gave a recovery of only 5%. A Ca^{2+}-dependent phospholipid affinity column, consisting of phosphatidylserine immobilized to polyacrylamide, has also been used to purify PKC [197].

15.5   OTHER SIGNAL-TRANSDUCTION SYSTEMS

Myosins make up a superfamily of motor proteins. This superfamily contains more than 15 structurally distinct classes that have been linked to a number of diseases. In order to identify the function and protein cargo that associate with the unique tail regions of myosins, immobilized myosin VIIA was used as an affinity ligand [198]. Antimyosin VIIA polyclonal antibodies were coupled through their carbohydrate moieties to Affi-Gel Hz (from Bio-Rad), yielding 5 mg antimyosin VIIA antibodies/ml gel. Tissue homogenates were then applied to this immunoaffinity column, and the retained proteins were eluted with a pH 2.5, 100 mM glycine/HCl buffer. Fractions collected from this column were immediately neutralized, with their immunoreactivity then being examined by Western blotting with antimyosin VIIA antibodies. Proteins with masses of 17 and 55 kDa from mouse kidneys or cochlea were copurified with myosin VIIA by the affinity column. These proteins were identified as CaM and microtubule-associated protein 2B (MAP-2B). Both proteins were strongly associated with myosin, with dissociation constants for this binding being in the nanomolar range.

　　Using the SH-3 domain from different proteins that have been fused to GST, affinity chromatography has been performed with glutathione-Sepharose to isolate SH3-binding proteins from a human megakaryocytic cell line [199]. Proteins that bound to this column were released into an SDS-PAGE sample buffer. Microsequencing identified one of these proteins as WASp, a protein that is defective in Wiskott-Aldrich syndrome. A specific interaction was shown to occur between WASp and the SH-3 domain-containing Fyn protein-tyrosine kinase of human hematopoietic cells.

15.6　SUMMARY AND CONCLUSIONS

Affinity chromatography is a simple and inexpensive method for rapidly isolating specific proteins or groups of proteins under mild conditions. In this chapter, several examples of this were given in the field of regulatory and signal-transducing proteins. Specific types of proteins that were considered in this discussion included those involved in the regulation of calmodulin, cyclic nucleotides, and inositol phosphate/diacylglycerol, as well as other signal-transduction systems. Besides being a valuable tool for isolating proteins and other biological agents, affinity chromatography can also be used to identify protein-binding partners and to examine the effects of covalent modification and second messengers on binding. Several examples of these applications were also given in this chapter.

　　Fusion proteins and pharmacological agents with site-directed affinity tags such as biotin have greatly accelerated the application of affinity techniques to the delineation of protein interactions in functional genomics. Ligands containing expression or site-directed affinity tags ensure that the ligand attached to the matrix is homogeneous and specific and that bound proteins can be released from this ligand under relatively mild conditions. High accessibility, stabilization, and binding capacities for these matrix-attached ligands are important to consider in maintaining strong and specific matrix interactions. Proper controls to test interaction specificity and rule out nonspecific interaction are also important considerations in the effective design of affinity methods for regulatory or signal-transducing proteins.

SYMBOLS AND ABBREVIATIONS

AMP	Adenosine monophosphate
ATP	Adenosine triphosphate
BSA	Bovine serum albumin
CaM	Calmodulin
CH-Sepharose 4B	Carboxyl-substituted Sepharose 4B
CNBr	Cyanogen bromide
cyclic AMP	Cyclic adenosine monophosphate

DEAE	Diethylaminoethyl
DGK	Diacylglycerol kinase
DTT	Dithiothreitol
EDTA	Ethylenediaminetetraacetic acid
EGTA	Ethylene glycol-*bis*(2-aminoethylether)-*N*,*N*,*N'*,*N'*-tetraacetic acid
FKBP12	12-kDa FK506-binding protein
GAP	GTP-activating protein
GDP	Guanosine diphosphate
G_s	GTP-binding protein
GSK-3	Glycogen synthase kinase-3
GST	Glutathione *S*-transferase
GTP	Guanosine triphosphate
GTPγS	Guanosine 5'-*O*-3-thiotriphosphate
H-9	*N*-(2-Aminoethyl)-5-isoquinolinesulfonamide
HEPES	4-(2-Hydoxyethyl)piperazine-1-ethanesulfonic acid
HPLC	High-performance liquid chromatography
IP$_3$	Inositol 1,4,5-trisphosphate, or Ins(1,4,5)P$_3$
K_m	Michaelis constant
MAP 2	Microtubule-associated protein 2
MAP-2B	Microtubule-associated protein 2B
Ni-NTA	Nickel-nitrilotriacetic acid
PDGF	Platelet-derived growth factor
PH domain	Pleckstrin homology domain
PIP$_3$	Phosphatidylinositol 3,4,5-trisphosphate
PKA	Cyclic AMP-dependent protein kinase
PKC	Protein kinase C
PLC	Phospholipase C
PLCγ	145-kDa Isozyme of PLC
PMSF	Phenylmethylsulfonylfluoride
PPase1	Protein phosphatase 1
PTB	Phosphotyrosine binding
R_G	Glycogen-binding regulatory subunit
SDS-PAGE	Sodium dodecyl sulfate-polyacrylamide gel electrophoresis
SH2	Src homology 2
W-7	*N*-(6-Aminohexyl)-5-chloro-1-naphthalenesulfonamide

REFERENCES

1. Denhardt, D.T., in *The Molecular Basis of Cell Cycle and Growth Control,* Stein, G.S., Baserga, R., Giordano, A., and Denhardt, D.T., Eds., Wiley-Liss, New York, 1999, pp. 225–304.
2. Smith, D.B. and Johnson, K.S., Single-step purification of polypeptides expressed in *Escherichia coli* as fusions with glutathione S-transferase, *Gene,* 67, 31–40, 1988.
3. Van Dyke, M.W., Sirito, M., and Sawadogo, M., Single-step purification of bacterially expressed polypeptides containing an oligo-histidine domain, *Gene,* 111, 99–104, 1992.
4. Domen, P.L., Nevens, J.R., Mallia, A.K., Hermanson, G.T., and Klenk, D.C., Site-directed immobilization of proteins, *J. Chromatogr.,* 510, 293–302, 1990.
5. O'Shannessy, D.J., Hydrazido-derivatized supports in affinity chromatography, *J. Chromatogr.,* 510, 13–21, 1990.
6. Mohr, P. and Pommerening, K., *Affinity Chromatography: Practical and Theoretical Aspects,* Marcel Dekker, New York, 1985.

7. Dean, P.D.G., Johnson, W.S., and Middle, F.A., *Affinity Chromatography: A Practical Approach,* IRL Press, Oxford, U.K., 1985.

8. Bailon, P., Ehrlich, G.K., Fung, W.-J., and Berthold, W., Affinity chromatography: Methods and protocols, in *Methods in Molecular Biology,* Vol. 147, Humana Press, Totowa, NJ, 2000.

9. Hermanson, G.T., Mallia, A.K., and Smith,·P.K., *Immobilized Affinity Ligand Techniques,* Academic Press, New York, 1992.

10. Watterson, D.M., Sharief, F., and Vanaman, T.C., The complete amino acid sequence of the Ca^{2+}-dependent modulator protein (calmodulin) of bovine brain, *J. Biol. Chem.,* 255, 962–975, 1980.

11. Potter, J.D. and Gergely, J., The calcium and magnesium binding sites on troponin and their role in the regulation of myofibrillar adenosine triphosphatase, *J. Biol. Chem.,* 250, 4628–4633, 1975.

12. Yazawa, M., Yagi, K., Toda, H., Kondo, K., Narita, K., Yamazaki, R., Sobue, K., Kakiuchi, S., Nagao, S., and Nozawa, Y., The amino acid sequence of the *Tetrahymena* calmodulin which specifically interacts with guanylate cyclase, *Biochem. Biophys. Res. Commun.,* 99, 1051–1057, 1981.

13. O'Neil, K.T. and DeGrado, W.F., How calmodulin binds its targets: Sequence-independent recognition of amphiphilic alpha-helices, *Trends Biochem. Sci.,* 15, 59–64, 1990.

14. Blumenthal, D.K., Charbonneau, H., Edelman, A.M., Hinds, T.R., Rosenberg, G.B., Storm, D.R., Vincenzi, F.F., Beavo, J.A., and Krebs, E.G., Synthetic peptides based on the calmodulin-binding domain of myosin light chain kinase inhibit activation of other calmodulin-dependent enzymes, *Biochem. Biophys. Res. Commun.,* 156, 860–865, 1988.

15. Buschmeier, B., Meyer, H.E., and Mayr, G.W., Characterization of the calmodulin-binding sites of muscle phosphofructokinase and comparison with known calmodulin-binding domains, *J. Biol. Chem.,* 262, 9454–9462, 1987.

16. Comte, M., Maulet, Y., and Cox, J.A., Ca^{2+}-dependent high-affinity complex formation between calmodulin and melittin, *Biochem. J.,* 209, 269–272, 1983.

17. Bartegi, A., Fattoum, A., Derancourt, J., and Kassab, R., Characterization of the carboxyl-terminal 10 kDa cyanogen bromide fragment of caldesmon as an actin-calmodulin-binding region, *J. Biol. Chem.,* 265, 15231–15238, 1990.

18. Watterson, D.M. and Vanaman, T.C., Affinity chromatography purification of a cyclic nucle-otide phosphodiesterase using immobilized modulator protein, a troponin C-like protein from brain, *Biochem. Biophys. Res. Commun.,* 73, 40–46, 1976.

19. Miyake, M., Daly, J.W., and Creveling, C.R., Purification of calcium-dependent phospho-diesterases from rat cerebrum by affinity chromatography on activator protein-Sepharose, *Arch. Biochem. Biophys.,* 181, 39–45, 1977.

20. Sharma, R.K., Wang, T.H., Wirch, E., and Wang, J.H., Purification and properties of bovine brain calmodulin-dependent cyclic nucleotide phosphodiesterase, *J. Biol. Chem.,* 255, 5916–5923, 1980.

21. Adelstein, R.S. and Klee, C.B., Purification and characterization of smooth muscle myosin light chain kinase, *J. Biol. Chem.,* 256, 7501–7509, 1981.

22. Thiel, G., Czernik, A.J., Gorelick, F., Nairn, A.C., and Greengard, P., Ca^{2+}/calmodulin-dependent protein kinase II: Identification of threonine-286 as the autophosphorylation site in the alpha subunit associated with the generation of Ca^{2+}-independent activity, *Proc. Natl. Acad. Sci. U.S.A.,* 85, 6337–6341, 1988.

23. Tallant, E.A. and Wallace, R.W., Characterization of a calmodulin-dependent protein phos-phatase from human platelets, *J. Biol. Chem.,* 260, 7744–7751, 1985.

24. Andreasen, T.J., Leutje, C.W., Heideman, W., and Storm, D.R., Purification of a novel calmodulin binding protein from bovine cerebral cortex membranes, *Biochemistry,* 22, 4615–4618, 1983.

25. Wakim, B.T., Picken, M.M., and DeLange, R.J., Identification and partial purification of a chromatin bound calmodulin activated histone 3 kinase from calf thymus, *Biochem. Biophys. Res. Commun.,* 171, 84–90, 1990.

26. Biden, T.J., Comte, M., Cox, J.A., and Wollheim, C.B., Calcium-calmodulin stimulates inositol 1,4,5-trisphosphate kinase activity from insulin-secreting RINm5F cells, *J. Biol. Chem.,* 262, 9437–9440, 1987.

27. Sharma, R.K., Purification and characterization of novel calmodulin-binding protein from cardiac muscle, *J. Biol. Chem.*, 265, 1152–1157, 1990.

28. Toscano, W.A., Jr., Westcott, K.R., LaPorte, D.C., and Storm, D.R., Evidence for a dissociable protein subunit required for calmodulin stimulation of brain adenylate cyclase, *Proc. Natl. Acad. Sci. U.S.A.,* 76, 5582–5586, 1979.

29. Schmidt, H.H., Pollock, J.S., Nakane, M., Gorsky, L.D., Forstermann, U., and Murad, F., Purification of a soluble isoform of guanylyl cyclase-activating-factor synthase, *Proc. Natl. Acad. Sci. U.S.A.,* 88, 365–369, 1991.

30. Friedberg, F. and Rhoads, A.R., Evolutionary aspects of calmodulin, *IUBMB Life,* 51, 215–221, 2001.

31. Stofko-Hahn, R.E., Carr, D.W., and Scott, J.D., A single step purification for recombinant proteins: Characterization of a microtubule associated protein (MAP 2) fragment which associates with the type II cAMP-dependent protein kinase, *FEBS Lett.,* 302, 274–278, 1992.

32. Wyborski, D.L., Bauer, J.C., Zheng, C.F., Felts, K., and Vaillancourt, P., An *Escherichia coli* expression vector that allows recovery of proteins with native N-termini from purified calmodulin-binding peptide fusions, *Protein Expr. Purif.,* 16, 1–10, 1999.

33. Bonifácio, M.J., Vieira-Coelho, M.A., and Soares-da-Silva, P., Expression and characterization of rat soluble catechol-O-methyltransferase fusion protein, *Protein Expr. Purif.,* 23, 106–112, 2001.

34. Strittmatter, S.M., Valenzuela, D., Kennedy, T.E., Neer, E.J., and Fishman, M.C., G_0 is a major growth cone protein subject to regulation by GAP-43, *Nature,* 344, 836–841, 1990.

35. Newton, D.L., Oldewurtel, M.D., Krinks, M.H., Shiloach, J., and Klee, C.B., Agonist and antagonist properties of calmodulin fragments, *J. Biol. Chem.,* 259, 4419–4426, 1984.

36. Wolff, J., Newton, D.L., and Klee, C.B., Activation of *Bordetella pertussis* adenylate cyclase by the carboxyl-terminal tryptic fragment of calmodulin, *Biochemistry,* 25, 7950–7955, 1986.

37. Guerini, D., Krebs, J., and Carafoli, E., Stimulation of the purified erythrocyte Ca^{2+}-ATPase by tryptic fragments of calmodulin, *J. Biol. Chem.,* 259, 15172–15177, 1984.

38. Ni, W.-C. and Klee, C.B., Selective affinity chromatography with calmodulin fragments coupled to Sepharose, *J. Biol. Chem.,* 260, 6974–6981, 1985.

39. Draetta, G. and Klee, C.B., Purification of calmodulin-stimulated phosphodiesterase by affinity chromatography on calmodulin fragment 1-77 linked to Sepharose, *Methods Enzymol.,* 159, 573–581, 1988.

40. Wolff, L., Cook, H.G., Goldhammer, A.H., and Berkowitz, S.A., Calmodulin activates prokaryotic adenylate cyclase, *Proc. Natl. Acad. Sci. U.S.A.,* 77, 3841–3844, 1980.

41. Greenlee, D.V., Andreasen, T.J., and Storm, D.R., Calcium-independent stimulation of *Bordetella pertussis* adenylate cyclase by calmodulin, *Biochemistry,* 21, 2759–2764, 1982.

42. Haiech, J., Predeleanu, R.T., Watterson, D.M., Ladant, D., Bellalou, J., Ullmann, A., and Barzu, O., Affinity-based chromatography utilizing genetically engineered proteins: Interaction of *Bordetella pertussis* adenylate cyclase with calmodulin, *J. Biol. Chem.,* 263, 4259–4262, 1987.

43. Hewlett, E.L., Gordon, V.M., McCaffery, J.D., Sutherland, W.M., and Gray, M.C., Adenylate cyclase toxin from *Bordetella pertussis*: Identification and purification of the holotoxin molecule, *J. Biol. Chem.,* 264, 19379–19384, 1989.

44. Dedman, J.R. and Kaetzel, M.A., Calmodulin purification and fluorescent labeling, *Methods Enzymol.,* 102, 1–8, 1983.

45. Nibhanupudy, N., Jones, F., and Rhoads, A.R., Involvement of arginine residues in the activation of calmodulin-dependent 3′,5′-cyclic-nucleotide phosphodiesterase, *Biochemistry,* 27, 2212–2217, 1988.

46. Laemmli, U.K., Cleavage of structural proteins during the assembly of the head of bacteriophage T4, *Nature (London),* 227, 680–685, 1970.

47. Brzeska, H., Szynkiewicz, J., and Drabikowski, W., Localization of hydrophobic sites in calmodulin and skeletal muscle troponin C studied using tryptic fragments: A simple method of their preparation, *Biochem. Biophys. Res. Commun.,* 115, 87–93, 1983.

48. Head, J.F., Masure, H.R., and Kaminer, B., Identification and purification of a phenothiazine binding fragment from bovine brain calmodulin, *FEBS Lett.,* 137, 71–74, 1982.
49. Thulin, E., Andersson, A., Drakenberg, T., Forsen, S., and Vogel, H.J., Metal ion and drug binding to proteolytic fragments of calmodulin: Proteolytic, cadmium-113, and proton nuclear magnetic resonance studies, *Biochemistry,* 23, 1862–1870, 1984.
50. Charbonneau, H., Hice, R., Hart, R.C., and Cormier, M.J., Purification of calmodulin by Ca^{2+}-dependent affinity chromatography, *Methods Enzymol.,* 102, 17–39, 1983.
51. Schauer-Vukasinovic, V. and Daunert, S., Purification of recombinant proteins based on the interaction between a phenothiazine-derivatized column and a calmodulin fusion tail, *Biotechnol. Prog.,* 15, 513–516, 1999.
52. Anderson, J.M., Purification of plant calmodulin, *Methods Enzymol.,* 102, 9–17, 1983.
53. Cox, J.A., Comte, M., Fitton, A.J.E., and DeGrado, W.F., The interaction of calmodulin with amphiphilic peptides, *J. Biol. Chem.,* 260, 2527–2534, 1985.
54. Kaetzel, M.A. and Dedman, J.R., Affinity-purified melittin antibody recognizes the calmodulin-binding domain on calmodulin target proteins, *J. Biol. Chem.,* 262, 3726–3729, 1987.
55. Wang, K.C., Wong, H.Y., Wang, J.H., and Lam, H.-Y.P., A monoclonal antibody showing cross-reactivity toward three calmodulin-dependent enzymes, *J. Biol. Chem.,* 258, 12110–12113, 1983.
56. Klee, C.B., Krinks, M.H., Manalan, A.S., Cohen, P., and Stewart, A.A., Isolation and characterization of bovine brain calcineurin: A calmodulin-stimulated protein phosphatase, *Methods Enzymol.,* 102, 227–244, 1983.
57. March, S.C., Parikh, I., and Cuatrecasas, P., A simplified method for cyanogen bromide activation of agarose for affinity chromatography, *Anal. Biochem.,* 60, 149–152, 1974.
58. Udenfriend, S., Stein, S., Bohlen, P., Darma, W., Leimgruber, W., and Weigele, M., Fluorescamine: A reagent for assay of amino acids, peptides, proteins, and primary amines in the picomole range, *Science,* 178, 871–872, 1972.
59. Bradford, M., A rapid and sensitive method for the quantitation of microgram quantities of protein utilizing the principle of protein-dye binding, *Anal. Biochem.,* 72, 248–254, 1976.
60. Thiry, P., Vandermeers, A., Vandermeers-Piret, M.-C., Rather, J., and Christophe, J., The activation of brain adenylate cyclase and brain cyclic-nucleotide phosphodiesterase by seven calmodulin derivatives, *Eur. J. Biochem.,* 103, 409–414, 1980.
61. Manalan, A.S. and Klee, C.B., Affinity selection of chemically modified proteins: Role of lysyl residues in the binding of calmodulin to calcineurin, *Biochemistry,* 26, 1382–1390, 1987.
62. Giedroc, D.P., Sinha, S.K., Brew, K., and Puett, D., Differential trace labeling of calmodulin: Investigation of binding sites and conformational states by individual lysine reactivities: Effects of beta-endorphin, trifluoperazine, and ethylene glycol *bis*(beta-aminoethyl ether)-*N,N,N',N'*-tetraacetic acid, *J. Biol. Chem.,* 260, 13406–13413, 1985.
63. Mann, D.M. and Vanaman, T.C., Modification of calmodulin on Lys-75 by carbamoylating nitrosoureas, *J. Biol. Chem.,* 263, 11284–11290, 1988.
64. Sharma, R.K., Taylor, W.A., and Wang, J.H., Use of calmodulin affinity chromatography for purification of specific calmodulin-dependent enzymes, *Methods Enzymol.,* 102, 210–219, 1983.
65. Hathaway, D.R., Adelstein, R.S., and Klee, C.B., Interaction of calmodulin with myosin light chain kinase and cAMP-dependent protein kinase in bovine brain, *J. Biol. Chem.,* 256, 8183–8189, 1981.
66. Wang, K.K., Villalobo, A., and Roufogalis, B.D., Calmodulin-binding proteins as calpain substrates, *Biochem. J.,* 262, 693–706, 1989.
67. Lollike, K., Johnsen, A.H., Durussel, I., Borregaard, N., and Cox, J.A., Biochemical characterization of the penta-EF-hand protein grancalcin and identification of L-plastin as a binding partner, *J. Biol. Chem.,* 276, 17762–17769, 2001.
68. Chen, C.-K., Inglese, J., Lefkowitz, R.J., and Hurley, J.B., Ca^{2+}-dependent interaction of recoverin with rhodopsin kinase, *J. Biol. Chem.,* 270, 18060–18066, 1995.

69. Xin, H.B., Timerman, A.P., Onoue, H., Wiederrecht, G.J., and Fleishcher, S., Affinity purification of the ryanodine receptor/calcium release channel from fast-twitch skeletal muscle based on its tight association with FKBP12, *Biochem. Biophys. Res. Commun.*, 214, 263–270, 1995.

70. Hawkins, T.E., Merrifield, C.J., and Moss, S.E., Calcium signaling and annexins, *Cell Biochem. Biophys.*, 33, 275–296, 2000.

71. Herman, I.M. and Pollard, T.D., Comparison of purified anti-actin and fluorescent-heavy meromyosin staining patterns in dividing cells, *J. Cell Biol.*, 80, 509–520, 1979.

72. Martin, F., Derancourt, J., Capony, J.P., Watrin, A., and Cavadore, J.C., A 36 kDa monomeric protein and its complex with a 10 kDa protein both isolated from bovine aorta are calpactin-like proteins that differ in their Ca^{2+}-dependent calmodulin-binding and actin-severing properties, *Biochem. J.*, 251, 777–785, 1988.

73. Grandmont-Leblanc, A. and Gruda, J., Affinity chromatography of myosin, heavy meromyosin, and heavy meromyosin subfragment one on F-actin columns stabilized by phalloidin, *Can. J. Biochem.*, 55, 949–957, 1977.

74. Ohara, O., Takahashi, S., Ooi, T., and Fujiyoshi, T., Cross-linking study on skeletal muscle actin: Properties of suberimidate-treated actin, *J. Biochem.*, 91, 1999–2012, 1982.

75. Miller, K.G. and Alberts, B.M., F-actin affinity chromatography: Technique for isolating previously unidentified actin-binding proteins, *Proc. Natl. Acad. Sci. U.S.A.*, 86, 4808–4812, 1989.

76. Johnsson, N. and Weber, K., Structural analysis of p36, a Ca^{2+}/lipid-binding protein of the annexin family, by proteolysis and chemical fragmentation, *Eur. J. Biochem.*, 188, 1–7, 1990.

77. Burger, A., Berendes, R., Voges, D., Huber, R., and Demange, P., A rapid and efficient purification method for recombinant annexin V for biophysical studies, *FEBS Lett.*, 329, 25–28, 1993.

78. van Engeland, M., Nieland, L.J., Ramaekers, F.C., Schutte, B., and Reutelingsperger, C.P., Annexin V-affinity assay: A review on an apoptosis detection system based on phosphatidylserine exposure, *Cytometry*, 31, 1–9, 1998.

79. Martin, W.H. and Creutz, C.E., Interactions of the complex secretory vesicle binding protein chromobindin A with nucleotides, *J. Neurochem.*, 54, 612–619, 1990.

80. Creutz, C.E., Zaks, W.J., Hamman, H.C., Crane, S., Martin, W.H., Gould, K.L., Oddie, K.M., and Parsons, S.J., Identification of chromaffin granule-binding proteins: Relationship of the chromobindins to calelectrin, synhibin, and the tyrosine kinase substrates p35 and p36, *J. Biol. Chem.*, 262, 1860–1868, 1987.

81. Creutz, C.E., Dowling, L.G., Sando, J.J., Villar-Palasi, C., Whipple, J.H., and Zaks, W.J., Characterization of the chromobindins: Soluble proteins that bind to the chromaffin granule membrane in the presence of Ca^{2+}, *J. Biol. Chem.*, 258, 14664–14674, 1983.

82. Severin, F.F., Shanina, N.A., Kuznetsov, S.A., and Gelfand, V.I., MAP2-mediated binding of chromaffin granules to microtubules, *FEBS Lett.*, 282, 65–68, 1991.

83. Michalik, L., Vanier, M.T., and Launay, J.F., Microtubule affinity chromatography: A new technique for isolating microtubule-binding proteins from rat pancreas, *Cell Mol. Biol.*, 37, 805–811, 1991.

84. Hurley, J.H., Structure, mechanism, and regulation of mammalian adenylyl cyclase, *J. Biol. Chem.*, 274, 7599–7602, 1999.

85. Tang, W.-J., Krupinski, J., and Gilman, A.G., Expression and characterization of calmodulin-activated (type I) adenylylcyclase, *J. Biol. Chem.*, 266, 8595–8603, 1991.

86. Bigay, G., Deterre, P., Pfister, C., and Chabre, M., Fluoroaluminates activate transducin-GDP by mimicking the gamma-phosphate of GTP in its binding site, *FEBS Lett.*, 191, 181–185, 1985.

87. Homcy, C., Wrenn, S., and Haber, E., Affinity purification of cardiac adenylate cyclase: Dependence on prior hydrophobic resolution, *Proc. Natl. Acad. Sci. U.S.A.*, 75, 59–63, 1978.

88. Pfeuffer, T., GTP-binding proteins in membranes and the control of adenylate cyclase activity, *J. Biol. Chem.*, 252, 7224–7234, 1977.

89. Birnbaumer, L., Transduction of receptor signal into modulation of effector activity by G proteins: The first 20 years or so, *FASEB J.*, 4, 3178–3188, 1990.

90. Lindner, E. and Metzger, H., The action of forskolin on muscle cells is modified by hormones, calcium ions and calcium antagonists, *Arzneimittelforschung,* 33, 1436–1441, 1983.

91. Seamon, K.B., Padgett, W., and Daly, J.W., Forskolin: Unique diterpene activator of adenylate cyclase in membranes and in intact cells, *Proc. Natl. Acad. Sci. U.S.A.,* 78, 3363–3367, 1981.

92. Pfeuffer, T., Gaugler, B., and Metzger, H., Isolation of homologous and heterologous complexes between catalytic and regulatory components of adenylate cyclase by forskolin-Sepharose, *FEBS Lett.,* 164, 154–160, 1983.

93. Pfeuffer, T. and Metzger, H., 7-O-Hemisuccinyl-deacetyl forskolin-Sepharose: A novel affinity support for purification of adenylate cyclase, *FEBS Lett.,* 146, 369–375, 1982.

94. Pfeuffer, E., Dreher, R.-M., Metzger, H., and Pfeuffer, T., Catalytic unit of adenylate cyclase: Purification and identification by affinity crosslinking, *Proc. Natl. Acad. Sci. U.S.A.,* 82, 3086–3090, 1985.

95. Krupinski, J., Coussen, F., Bakalyar, H.A., Tang, W.-J., Feinstein, P.G., Orth, K., Slaughter, C., Reed, R.R., and Gilman, A.G., Adenylyl cyclase amino acid sequence: Possible channel-or transporter-like structure, *Science,* 244, 1558–1564, 1989.

96. Parikh, I., Sica, V., Nola, E., Puca, G.A., and Cuatrecasas, P., Affinity chromatography of estrogen receptors, *Methods Enzymol.,* 34, 670–688, 1974.

97. Marbach, I., Bar-Sinai, A., Minich, M., and Levitzki, A., Beta subunit copurifies with GppNHp-activated adenylyl cyclase, *J. Biol. Chem.,* 265, 9999–10004, 1990.

98. Coussen, F., Guermah, M., d'Alayer, J., Monneron, A., Haiech, J., and Cavadore, J.-C., Evidence for two distinct adenylate cyclase catalysts in rat brain, *FEBS Lett.,* 206, 213–217, 1986.

99. Minocherhomjee, M., Selfe, S., Flowers, N.J., and Storm, D.R., Direct interaction between the catalytic subunit of the calmodulin-sensitive adenylate cyclase from bovine brain with ^{125}I-labeled wheat germ agglutinin and ^{125}I-labeled calmodulin, *Biochemistry,* 26, 4444–4448, 1987.

100. Pfeuffer, E., Mollner, S., Lancet, D., and Pfeuffer, T., Olfactory adenylyl cyclase: Identification and purification of a novel enzyme form, *J. Biol. Chem.,* 264, 18803–18807, 1989.

101. Field, J., Nikawa, J., Broek, D., MacDonald, B., Rodgers, L., Wilson, I.A., Lemer, R.A., and Wigler, M., Purification of a RAS-responsive adenylyl cyclase complex from *Saccharomyces cerevisiae* by use of an epitope addition method, *Mol. Cell Biol.,* 8, 2159–2165, 1988.

102. Lipkin, V.M., Khramtsov, N.V., Andreeva, S.G., Moshnyakov, M.V., Petukhova, G.V., Rakitina, T.V., Feshchenko, E.A., Ishchenko, K.A., Mirzoeva, S.F., Chernova, M.N., and Dranytsyna, S.M., Calmodulin-independent bovine brain adenylate cyclase: Amino acid sequence and nucleotide sequence of the corresponding cDNA, *FEBS Lett.,* 254, 69–73, 1989.

103. Hamm, H.E., The many faces of G protein signaling, *J. Biol. Chem.,* 273, 669–672, 1998.

104. Freissmuth, M., Casey, P.J., and Gilman, A.G., G proteins control diverse pathways of transmembrane signaling, *FASEB J.,* 3, 2125–2131, 1989.

105. Sternweis, P.C. and Robishaw, J.D., Isolation of two proteins with high affinity for guanine nucleotides from membranes of bovine brain, *J. Biol. Chem.,* 259, 13806–13813, 1984.

106. Northup, J.K., Sternweis, P.C., Smigel, M.D., Schleifer, L.S., Ross, E.M., and Gilman, A.G., Purification of the regulatory component of adenylate cyclase, *Proc. Natl. Acad. Sci. U.S.A.,* 77, 6516–6520, 1980.

107. Pang, I.-H. and Sternweis, P.C., Isolation of the alpha subunits of GTP-binding regulatory proteins by affinity chromatography with immobilized beta gamma subunits, *Proc. Natl. Acad. Sci. U.S.A.,* 86, 7814–7818. 1989.

108. Birnbaumer, L., G proteins in signal transduction, *Ann. Rev. Pharmacol. Toxicol.,* 30, 675–705, 1990.

109. Pang, I.H. and Sternweis, P.C., Purification of unique alpha subunits of GTP-binding regulatory proteins (G proteins) by affinity chromatography with immobilized beta gamma subunits, *J. Biol. Chem.,* 265, 18707–18712, 1990.

110. Smrcka, A.V., Hepler, J.R., Brown, K.O., and Sternweis, P.C., Regulation of polyphosphoinositide-specific phospholipase C activity by purified Gq, *Science,* 251, 804–807, 1991.

111. Christoforidis, S. and Zerial, M., Purification and identification of novel Rab effectors using affinity chromatography, *Methods,* 20, 403–410, 2000.

112. Hart, M.J., Callow, M.G., Souza, B., and Polakis, P., IQGAP1, a calmodulin-binding protein with a rasGAP-related domain, is a potential effector for cdc42Hs, *EMBO J.,* 15, 2997–3005, 1996.

113. Kutay, U., Hartmann, E., Treichel, N., Calado, A., Carmo-Fonseca, M., Prehn, S., Kraft, R., Görlich, D., and Bischoff, R., Identification of two novel RanGTP-binding proteins belonging to the importin beta superfamily, *J. Biol. Chem.,* 275, 40163–40168, 2000.

114. Andreopoulos, S. and Papapetropoulos, A., Molecular aspects of soluble guanylyl cyclase regulation, *Gen. Pharmacol.,* 34, 147–157, 2000.

115. Koesling, D., Studying the structure and regulation of soluble guanylyl cyclase, *Methods,* 19, 485–493, 1999.

116. Humbert, P., Niroomand, F., Fischer, G., Koesling, D., Hinsch, D.-D., Gausepohl, H., Frank, R., Schultz, G., and Bohme, E., Purification of soluble guanylyl cyclase from bovine lung by a new immunoaffinity chromatographic method, *Eur. J. Biochem.,* 190, 273–278, 1990.

117. Nakane, M., Saheki, S., Kuno, T., Ishii, K., and Murad, F., Molecular cloning of a cDNA coding for 70 kilodalton subunit of soluble guanylate cyclase from rat lung, *Biochem. Biophys. Res. Commun.,* 157, 1139–1147, 1988.

118. Waldman, S.A., Rapoport, R.M., and Murad, F., Atrial natriuretic factor selectively activates particulate guanylate cyclase and elevates cyclic GMP in rat tissues, *J. Biol. Chem.,* 259, 14332–14334, 1984.

119. Field, M., Graf, L.H., Jr., Laird, W.J., and Smith, P.L., Heat-stable enterotoxin of *Escherichia coli*: *In vitro* effects on guanylate cyclase activity, cyclic GMP concentration, and ion transport in small intestine, *Proc. Natl. Acad. Sci. U.S.A.,* 75, 2800–2804, 1978.

120. Takada, M., Takeuchi, H., and Shino, M., Production and characterization of monoclonal antibodies against particulate guanylate cyclase in porcine kidney, *Biochem. Biophys. Res. Commun.,* 164, 653–663, 1989.

121. Nandi, A., Mathew, R., and Visweswariah, S.S., Expression of the extracellular domain of the human heat-stable enterotoxin receptor in *Escherichia coli* and generation of neutralizing antibodies, *Protein Expr. Purif.,* 8, 151–159, 1996.

122. Aparicio, J.G. and Applebury, M.L., The bovine photoreceptor outer segment guanylate cyclase: Purification, kinetic properties, and molecular size, *Protein Expr. Purif.,* 6, 501–511, 1995.

123. Degerman, E., Belfrage, P., and Manganiello, V.C., Structure, localization, and regulation of cGMP-inhibited phosphodiesterase (PDE3), *J. Biol. Chem.,* 272, 6823–6826, 1997.

124. Kincaid, R.L., Stith-Coleman, I.E., and Vaughan, M., Proteolytic activation of calmodulin-dependent cyclic nucleotide phosphodiesterase, *J. Biol. Chem.,* 260, 9009–9015, 1985.

125. Degerman, E., Belfrage, P., Newman, A.H., Rice, K.C., and Manganiello, V.C., Purification of the putative hormone-sensitive cyclic AMP phosphodiesterase from rat adipose tissue using a derivative of cilostamide as a novel affinity ligand, *J. Biol. Chem.,* 262, 5797–5807, 1987.

126. Gillespie, P.G. and Beavo, J.A., Characterization of a bovine cone photoreceptor phosphodiesterase purified by cyclic GMP-Sepharose chromatography, *J. Biol. Chem.,* 263, 8133–8141, 1988.

127. Sundberg, L. and Porath, J., Preparation of adsorbents for biospecific affinity chromatography: Attachment of group-containing ligands to insoluble polymers by means of bifunctional oxiranes, *J. Chromatogr.,* 90, 87–98, 1974.

128. Martins, T.J., Mumby, M.C., and Beavo, J.A., Purification and characterization of a cyclic GMP-stimulated cyclic nucleotide phosphodiesterase from bovine tissues, *J. Biol. Chem.,* 257, 1973–1979, 1982.

129. Gillespie, P.G., Prusti, R.K., Apel, E.D., and Beavo, J.A., A soluble form of bovine rod photoreceptor phosphodiesterase has a novel 15 kDa subunit, *J. Biol. Chem.,* 264, 12187–12193, 1989.

130. Ulloa, R.M., Rubinstein, C.P., Molinay Vedia, L., Torres, H.N., and Tellez-Inon, M.T., Cyclic nucleotide phosphodiesterase activity in *Neurospora crassa,* Purification by immunoaffinity chromatography and characterization, *FEBS Lett.,* 241, 219–222, 1988.

131. Sharma, R.K. and Wang, J.H., Purification and characterization of bovine lung calmodulin-dependent cyclic nucleotide phosphodiesterase: An enzyme containing calmodulin as a subunit, *J. Biol. Chem.,* 261, 14160–14166, 1986.

132. Taylor, S.S., cAMP-dependent protein kinase: Model for an enzyme family, *J. Biol. Chem.,* 264, 8443–8446, 1989.

133. Demaille, J.G., Peters, K.A., and Fischer, E.H., Isolation and properties of the rabbit skeletal muscle protein inhibitor of adenosine 3′,5′-monophosphate dependent protein kinases, *Biochemistry,* 16, 3080–3086, 1977.

134. Olsen, S.R. and Uhler, M.D., Affinity purification of the C alpha and C beta isoforms of the catalytic subunit of cAMP-dependent protein kinase, *J. Biol. Chem.,* 264, 18662–18666, 1989.

135. Zoller, M.J., Kuret, J., Cameron, S., Levin, L., and Johnson, K.E., Purification and characterization of C1, the catalytic subunit of *Saccharomyces cerevisiae* cAMP-dependent protein kinase encoded by TPK1, *J. Biol. Chem.,* 263, 9142–9148, 1988.

136. Ulloa, R.M., Mesri, E., Esteva, M., Torres, H.N., and Tellez-Inon, M.T., Cyclic AMP-dependent protein kinase activity in *Trypanosoma cruzi, Biochem. J.,* 255, 319–326, 1988.

137. Cao, J., Fernández, M., and Villamarin, J.A., A method for the purification of cAMP-dependent protein kinase using immunoaffinity chromatography, *Protein Expr. Purif.,* 14, 418–424, 1998.

138. Lincoln, T.M., Thompson, M., and Cornwell, T.L., Purification and characterization of two forms of cyclic GMP-dependent protein kinase from bovine aorta, *J. Biol. Chem.,* 263, 17632–17637, 1988.

139. Thalhofer, J.P. and Hofer, H.W., Purification and properties of cyclic-3′,5′-GMP-dependent protein kinase from the nematode *Ascaris suum, Arch. Biochem. Biophys.,* 273, 535–542, 1989.

140. Jergil, B. and Mosbach, K., Cyclic AMP: Purification of protamine kinase, *Methods Enzymol.,* 34, 261–264, 1974.

141. Dills, W.L., Goodwin, C.D., Lincoln, T.M., Beavo, J.A., Bechtel, P.J., Corbin, J.D., and Krebs, E.G., Purification of cyclic nucleotide receptor proteins by cyclic nucleotide affinity chromatography, *Adv. Cyclic Nucleotide Res.,* 10, 199–217, 1979.

142. Woodgett, J.R., Use of peptide substrates for affinity purification of protein-serine kinases, *Anal. Biochem.,* 180, 237–241, 1989.

143. Mukhopadyay, N.K., Shome, K., Saha, A.K., Hassell, J.R., and Glew, R.H., Heparin binds to *Leishmania donovani promastigotes* and inhibits protein phosphorylation, *Biochem. J.,* 264, 517–525, 1989.

144. Primot, A., Baratte, B., Gompel, M., Borgne, A., Liabeuf, S., Romette, J.-L., Jho, E.-H., Costantini, F., and Meijer, L., Purification of GSK-3 by affinity chromatography on immobilized axin, *Protein Expr. Purif.,* 20, 394–404, 2000.

145. Wilchek, M., Miron, T., and Kohn, J., Affinity chromatography, *Methods Enzymol.,* 104, 3–55, 1984.

146. Fry, M.J., Panayotou, G., Dhand, R., Ruiz-Larrea, F., Gout, I., Nguyen, O., Courtneidge, S.A., and Waterfield, M.D., Purification and characterization of a phosphatidylinositol 3-kinase complex from bovine brain by using phosphopeptide affinity columns, *Biochem. J.,* 288, 383–393, 1992.

147. Koegl, M., Kypta, R.M., Bergman, M., Alitalo, K., and Courtneidge, S.A., Rapid and efficient purification of Src homology 2 domain-containing proteins: Fyn, Csk and phosphatidylinositol 3-kinase p85, *Biochem. J.,* 302, 737–744, 1994.

148. Dumènil, G., Rubini, M., Dubois, G., Baserga, R., Fellous, M., and Pellegrini, S., Identification of signalling components in tyrosine kinase cascades using phosphopeptide affinity chromatography, *Biochem. Biophys. Res. Commun.,* 234, 748–753, 1997.

149. Ferrari, S., A rapid purification protocol for the mitogen-activated p70 S6 kinase, *Protein Expr. Purif.,* 13, 170–176, 1998.

150. Cohen, P. and Cohen, P.T., Protein phosphatases come of age, *J. Biol. Chem.,* 264, 21435–21438, 1989.

151. Ingebritsen, T.S. and Cohen, P., Protein phosphatases: Properties and role in cellular regulation, *Science,* 221, 331–338, 1983.

152. Hemmings, H.C., Jr., Nairn, A.C., and Greengard, P., DARPP-32, a dopamine- and adenosine 3′,5′-monophosphate-regulated neuronal phosphoprotein, II: Comparison of the kinetics of phosphorylation of DARPP-32 and phosphatase inhibitor 1, *J. Biol. Chem.,* 259, 14491–14497, 1984.

153. Hemmings, H.C., Jr., Greengard, P., Tung, H.Y., and Cohen, P., DARPP-32, a dopamine-regulated neuronal phosphoprotein, is a potent inhibitor of protein phosphatase-1, *Nature,* 310, 503–505, 1984.

154. Ingebritsen, V.M. and Ingebritsen, T.S., Immobilized inhibitor-1 binds and inhibits protein phosphatase 1, *Biochim. Biophys. Acta,* 1012, 1–4, 1989.

155. Zhang, Z., Zhao, S., Zirattu, S.D., Bai, G., and Lee, E.Y., Expression of recombinant inhibitor-2 in *E. coli* and its utilization for the affinity chromatography of protein phosphatase-1, *Arch. Biochem. Biophys.,* 308, 37–41, 1994.

156. Tung, H.Y. and Reed, L.J., Purification and characterization of protein phosphatase 1I activating kinase from bovine brain cytosolic and particulate fractions, *J. Biol. Chem.,* 264, 2985–2990, 1989.

157. Cohen, P., Klumpp, S., and Schelling, D.L., An improved procedure for identifying and quantitating protein phosphatases in mammalian tissues, *FEBS Lett.,* 250, 596–600, 1989.

158. Wera, S., Bollen, M., and Stalmans, W., Purification and characterization of the glycogen-bound protein phosphatase from rat liver, *J. Biol. Chem.,* 266, 339–345, 1991.

159. Zhao, S., Xia, W., and Lee, E.Y., Affinity chromatography of regulatory subunits of protein phosphatase-1, *Arch. Biochem. Biophys.,* 325, 82–90, 1996.

160. Nishiwaki, S., Fujiki, H., Suganuma, M., Ojika, M., Yamada, K., and Sugimura, T., Photo-affinity labeling of protein phosphatase 2A, the receptor for a tumor promoter okadaic acid, by [27-3H]methyl 7-O-(4-azidobenzoyl)okadaate, *Biochem. Biophys. Res. Commun.,* 170, 1359–1364, 1990.

161. Nishiwaki, S., Fujika, H., Suganuma, M., Nishiwaki-Matsushima, R., and Sugimura, T., Rapid purification of protein phosphatase 2A from mouse brain by microcystin-affinity chromatography, *FEBS Lett.,* 279, 115–118, 1991.

162. Moorhead, G., MacKintosh, C., Morrice, N., and Cohen, P., Purification of the hepatic glycogen-associated form of protein phosphatase-1 by microcystin-Sepharose affinity chromatography, *FEBS Lett.,* 362, 101–105, 1995.

163. Campos, M., Fadden, P., Alms, G., Qian, Z., and Haystead, T.A.J., Identification of protein phosphatase-1-binding proteins by microcystin-biotin affinity chromatography, *J. Biol. Chem.,* 271, 28478–28484, 1996.

164. Berridge, M.J., Inositol triphosphate and diacylglycerol: Two interacting second messengers, *Annu. Rev. Biochem.,* 56, 159–193, 1987.

165. Majerus, P.W., Ross, T.S., Cunningham, T.W., Caldwell, K.K., Jefferson, A.B., and Bansal, V.S., Recent insights in phosphatidylinositol signaling, *Cell,* 63, 459–465, 1990.

166. Berridge, M.J. and Irvine, R.F., Inositol phosphates and cell signalling, *Nature,* 341, 197–205, 1989.

167. Takai, Y., Kikkawa, U., Kaibuchi, K., and Nishizuka, Y., Membrane phospholipid metabolism and signal transduction for protein phosphorylation, *Adv. Cyclic Nucleotide Protein Phosphorylation Res.,* 18, 119–158, 1984.

168. Exton, J.H., Signaling through phosphatidylcholine breakdown, *J. Biol. Chem.,* 265, 1–4, 1990.

169. Hirata, M., Watanabe, Y., Ishimatsu, T., Yanaga, F., Koga, T., and Ozaki, S., Inositol 1,4,5-trisphosphate affinity chromatography, *Biochem. Biophys. Res. Commun.,* 168, 379–386, 1990.

170. Wahl, M. and Carpenter, G., Selective phospholipase C activation, *BioEssays,* 13, 107–113, 1991.

171. Wahl, M.I., Nishibe, S., Kim, J.W., Kim, H., Rhee, S.G., and Carpenter, G., Identification of two epidermal growth factor-sensitive tyrosine phosphorylation sites of phospholipase C-gamma in intact HSC-1 cells, *J. Biol. Chem.,* 265, 3944–3948, 1990.

172. Moritz, A., DeGraan, P.N.E., Ekhart, P.F., Gispen, W.H., and Wirtz, K.W.A., Purification of a phosphatidylinositol 4-phosphate kinase from bovine brain membranes, *J. Neurochem.,* 54, 351–354, 1990.

173. Walker, D.H., Dougherty, N., and Pike, L.J., Purification and characterization of a phosphatidylinositol kinase from A431 cells, *Biochemistry*, 27, 6504–6511, 1988.

174. Urumow, T. and Wieland, O.H., Purification and partial characterization of phosphatidylinositol-4-phosphate kinase from rat liver plasma membranes: Further evidence for a stimulatory G-protein, *Biochim. Biophys. Acta*, 1052, 152–158, 1990.

175. Kato, M. and Takenawa, T., Purification and characterization of membrane-bound and cytosolic forms of diacylglycerol kinase from rat brain, *J. Biol. Chem.*, 265, 794–800, 1990.

176. Yada, Y., Ozeki, T., Kanoh, H., and Nozawa, Y.G., Purification and characterization of cytosolic diacylglycerol kinases of human platelets, *J. Biol. Chem.*, 265, 19237–19243, 1990.

177. Yamaguchi, K., Hirata, M., and Kuriyama, H., Calmodulin activates inositol 1,4,5-trisphosphate 3-kinase activity in pig aortic smooth muscle, *Biochem. J.*, 244, 787–791, 1987.

178. Sim, S.S., Kim, J.W., and Rhee, S.G., Regulation of D-myo-inositol 1,4,5-trisphosphate 3-kinase by cAMP-dependent protein kinase and protein kinase C, *J. Biol. Chem.*, 265, 10367–10372, 1990.

179. Johanson, R.A., Hansen, C.A., and Williamson, J.R., Purification of D-myo-inositol 1,4,5-trisphosphate 3-kinase from rat brain, *J. Biol. Chem.*, 263, 7465–7471, 1988.

180. Takazawa, K., Passareiro, H., Dumont, J.E., and Erneux, C., Purification of bovine brain inositol 1,4,5-trisphosphate 3-kinase: Identification of the enzyme by sodium dodecyl sulphate polyacrylamide-gel electrophoresis, *Biochem. J.*, 261, 483–488, 1989.

181. Escobedo, J.A., Navankasattusas, S., Kavanaugh, W.M., Milfay, D., Fried, V.A., and Williams, L.T., cDNA cloning of a novel 85 kDa protein that has SH2 domains and regulates binding of PI3-kinase to the PDGF beta-receptor, *Cell*, 65, 75–82, 1991.

182. Gee, N.S., Ragan, C.I., Watling, K.J., Asplay, S., Jackson, R.G., Reid, G.G., Gani, D., and Shute, J.K., The purification and properties of myo-inositol monophosphatase from bovine brain, *Biochem. J.*, 249, 883–889, 1988.

183. Gee, N.S., Howell, S., Ryan, G., and Ragan, C.I., A monoclonal antibody to bovine brain inositol monophosphatase: Immunoaffinity purification of the brain and kidney enzymes and evidence for their structural identity, *Biochem. J.*, 264, 793–798, 1989.

184. McAdara-Berkowitz, J.K., Catz, S.D., Johnson, J.L., Ruedi, J.M., Thon, V., and Babior, B.M., JFC1, a novel tandem C2 domain-containing protein associated with the leukocyte NADPH oxidase, *J. Biol. Chem.*, 276, 18855–18862, 2001.

185. Shirai, T., Tanaka, K.-I., Terada, Y., Sawada, T., Shirai, R., Hashimoto, Y., Nagata, S., Iwamatsu, A., Okawa, K., Li, S., Hattori, S., Mano, H., and Fukui, Y., Specific detection of phosphatidylinositol 3,4,5-trisphosphate binding proteins by the PIP3 analogue beads: An application for rapid purification of the PIP3 binding proteins, *Biochim. Biophys. Acta*, 1402, 292–302, 1998.

186. Ohno, S. and Nishizuka, Y., Protein kinase C isotypes and their specific functions: Prologue, *J. Biochem. (Tokyo)*, 132, 509–511, 2002.

187. Kitano, T., Go, M., Kikkawa, U., and Nishizuki, Y., Assay and purification of protein kinase C, *Methods Enzymol.*, 124, 349–352, 1986.

188. Marais, R.M. and Parker, P.J., Purification and characterisation of bovine brain protein kinase C isotypes alpha, beta and gamma, *Eur. J. Biochem.*, 182, 129–137, 1989.

189. Wheeler, M.B. and Veldhuis, J.D., Purification of three forms of chromatographically distinct protein kinase C from the swine ovary, *Mol. Cell Endocrinol.*, 61, 117–122, 1989.

190. Schaap, D. and Parker, P.J., Expression, purification, and characterization of protein kinase C-epsilon, *J. Biol. Chem.*, 265, 7301–7307, 1990.

191. Hidaka, H., Inagaki, M., Kwamoto, S., and Sasaki, Y., Isoquinolinesulfonamides: Novel and potent inhibitors of cyclic nucleotide dependent protein kinase and protein kinase C, *Biochemistry*, 23, 5036–5041, 1984.

192. Inagaki, M., Watanabe, M., and Hidaka, H., *N*-(2-aminoethyl)-5-isoquinolinesulfonamide, a newly synthesized protein kinase inhibitor, functions as a ligand in affinity chromatography: Purification of Ca^{2+}-activated, phospholipid-dependent and other protein kinases, *J. Biol. Chem.*, 260, 2922–2925, 1985.

193. O'Brian, C.A. and Ward, N.E., Binding of protein kinase C to *N*-(6-aminohexyl)-5-chloro-1-naphthalenesulfonamide through its ATP binding site, *Biochem. Pharmacol.*, 38, 1737–1742, 1989.

194. O'Brian, C.A. and Ward, N.E., ATP-sensitive binding of melittin to the catalytic domain of protein kinase C, *Mol. Pharmacol.,* 36, 355–359, 1989.

195. O'Brian, C.A., Housey, G.M., and Weinstein, I.B., Specific and direct binding of protein kinase C to an immobilized tamoxifen analogue, *Cancer Res.,* 48, 3626–3629, 1988.

196. Wise, B.C., Raynor, R.L., and Kuo, J.F., Phospholipid-sensitive Ca^{2+}-dependent protein kinase from heart, I: Purification and general properties, *J. Biol. Chem.,* 257, 8481–8488, 1982.

197. Uchida, T. and Filburn, C.R., Affinity chromatography of protein kinase C-phorbol ester receptor on polyacrylamide-immobilized phosphatidylserine, *J. Biol. Chem.,* 259, 12311–12314, 1984.

198. Todorov, P.T., Hardisty, R.E., and Brown, S.D.M., Myosin VIIA is specifically associated with calmodulin and microtubule-associated protein-2B (MAP-2B), *Biochem. J.,* 354, 267–274, 2001.

199. Banin, S., Truong, O., Katz, D.R., Waterfield, M.D., Brickell, P.M., and Gout, I., Wiskott-Aldrich syndrome protein (WASp) is a binding partner c-Src family protein-tyrosine kinases, *Current Biol.,* 6, 981–988, 1996.

16

Receptor-Affinity Chromatography

Pascal Bailon, Michele Nachman-Clewner, Cheryl L. Spence,
and George K. Ehrlich

Hoffmann-La Roche Inc., Nutley, NJ

CONTENTS

16.1 INTRODUCTION

Affinity chromatography is a versatile method for the purification of various biological molecules, including enzymes, hormones, antibodies, and receptors [1–4]. In 1987, a technique known as receptor-affinity chromatography (RAC) was developed as an alternative to immunoaffinity chromatography [5]. Receptor-affinity chromatography makes use of the reversible interactions that occur between a matrix-bound receptor and a soluble protein target. In theory, receptor-affinity adsorbents are unique in their ability to bind only a fully active biomolecule in its native conformation. In contrast, immunoaffinity adsorbents often lack such specificity and bind to any form of a target that possesses the required epitope, irrespective of the target's degree of renaturation or biological activity. This difference has been illustrated in the use of both methods to purify recombinant human interleukin-2 [5, 6].

The selectivity of RAC for active biological molecules makes it an ideal method for the downstream purification of recombinant proteins. This includes proteins that are expressed as inclusion bodies in bacteria, as well as those expressed by mammalian cells grown in a tissue culture. This chapter examines the application of RAC as a purification tool for such samples. Using recombinant interleukin-2 (rIL-2) and soluble interleukin-2-receptor (IL-2R) as models, a systematic approach is presented for the development and optimization of an RAC system. This discussion covers both conventional-column RAC and the use of RAC with membranes or fluidized beds. Specific applications of RAC to be examined include its use in the purification of rIL-2, rIL-2 analogs, a rIL2-pseudomonas exotoxin fusion protein, and humanized anti-TAC antibodies.

16.2 BASIS OF RECEPTOR-AFFINITY CHROMATOGRAPHY

16.2.1 General Principles

Receptor-affinity chromatography is based on the selective interaction that occurs between a receptor and its target during the formation of a reversible receptor-target complex. RAC makes use of this specific and reversible binding by employing a recombinant receptor as the stationary phase. This stationary phase is then used to isolate and retain the corresponding target from applied samples.

Ideally there should be one-to-one binding between this immobilized receptor and its soluble target. However, this situation is seldom seen in RAC for various reasons. For instance, deviations from this expected behavior might occur if there is (1) reduced access of the target to the receptor (e.g., as might be caused by exclusion from the pores of the matrix), (2) decreased binding of the immobilized receptor due to improper orientation, or (3) impeded binding due to the close proximity of neighboring receptors. A more detailed discussion of these and other immobilization effects can be found in Chapter 3.

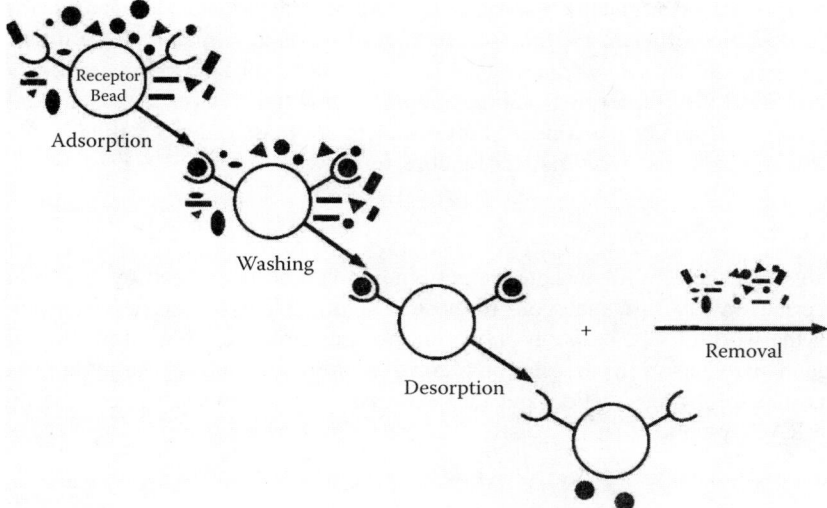

Figure 16.1 A typical scheme for receptor-affinity chromatography.

An illustration of how RAC is typically performed is shown in Figure 16.1. This requires that the receptor first be chemically attached to a solid support and packed into a column. This column is then equilibrated with an appropriate application buffer, which is later used to pass through a sample that contains the desired target. After nonretained sample materials have been washed from the column, the adsorbed target is eluted with a mild desorbing agent, such as a low-pH buffer.

The general scheme shown in Figure 16.1 is typical of many affinity methods, as discussed in Chapter 4. But there are several features that distinguish RAC from these other methods. These include (1) the ability of RAC to rapidly form a stable but reversible receptor-target complex, (2) its selective binding to a single target, (3) its ability to give a high recovery of biological activity for the isolated target, and (4) its ability to isolate biomolecules of all sizes.

16.2.2 Theory of RAC

Several theoretical issues are important to consider in the design of an RAC system. For instance, the two possible rate-limiting factors for target retention in RAC are mass-transfer and adsorption kinetics. To attain the maximum capabilities of RAC, it is important to optimize these two factors. Since the affinity of most receptor-target systems is high (i.e., implying a fast association rate or slow dissociation), the main factor that determines the binding rates in such systems is mass transfer within the chromatographic support. Another factor that should be considered during the use of RAC is the desorption rate of the target under the desired elution conditions. Each of these items is considered in greater detail in the following subsections.

16.2.2.1 Mass Transfer

Efficient mass transfer in RAC occurs when the diffusion time t_D (i.e., the time needed for a target to reach the immobilized receptor) is much shorter than the target's residence time in the column. The relation between t_D and the distance a target must diffuse is given by Equation 16.1.

$$t_D = L_D^2/D \qquad (16.1)$$

In this equation, t_D is expressed in units of seconds, L_D is the average diffusion distance (in cm), and D is the diffusion coefficient for the soluble target (i.e., approximately 10^{-7} cm²/sec for a typical protein). As shown by Equation 16.1, a small value for t_D is produced as the distance a target must diffuse becomes shorter. Thus, if mass transfer is the rate-limiting step in target retention, supports that have short distances for mass transfer will also give small values for t_D and better mass-transfer kinetics for RAC.

16.2.2.2 Adsorption Kinetics

A simple analysis of the rate for receptor-target binding can be performed by assuming that the interactions in RAC are analogous to those in solution. This can be accomplished by combining the work of Nishikawa et al. [7] on the equilibrium binding of a solute to an affinity ligand and Chase's [8] description of antigen-antibody binding. Based on these earlier reports, the following reaction and rate expressions can be used to describe the kinetics of an RAC system.

$$R + L \overset{k_1}{\underset{k_{-1}}{\longleftrightarrow}} R - L \tag{16.2}$$

$$d[R - L]/dt = k_1[R][L] - k_{-1}[R - L] \tag{16.3}$$

$$k_{de} = k_{-1}/k_1 = [R][L]/[R - L] = (q_m - q^*)C^*/q^* \tag{16.4}$$

In these equations, k_1 is the association rate constant for target L with an immobilized receptor R to form the complex $R - L$, k_{-1} is the dissociation rate constant for the target-receptor complex, and the molar concentration for each of these species is enclosed within brackets. The term k_{de} is the effective dissociation constant, which is related to the maximum binding capacity per unit volume of the affinity support (q_m), and the binding capacity per unit volume of the affinity support (q^*) when this support is in equilibrium with a target concentration of C^* in solution. In these relationships, the effective dissociation constant takes into account only the amount of receptor that remains functional after immobilization and may differ from the actual dissociation rate constant for the receptor when it is present in solution.

16.2.2.3 Desorption Kinetics

Another important aspect in RAC is the elution step. In this step, it is desirable to allow the target to be completely dissociated from the immobilized receptor. The rate of receptor-target dissociation during this step can be described in a manner analogous to that used for the dissociation of an antibody-antigen complex, as given by the following expression,

$$d[R - L]/dt = k_{-2}[R - L] \tag{16.5}$$

where k_{-2} is the dissociation rate constant for the receptor-ligand complex under the given elution conditions [9].

It is common in many affinity methods to use nonspecific solvents for solute desorption, such as low-pH buffers and mobile phases that contain chaotropic agents or denaturants. These tend to give rise to fast elution by breaking down a target-receptor complex rather than simply preventing rebinding of the target to the affinity sorbent. However, such nonspecific methods may also alter the three-dimensional structure of the receptor or target and can reduce or eliminate their binding ability. It is for this reason that gentler, more specific elution methods are employed in RAC. In general, the conditions chosen for dissociation should be selected based on the stability of the target, the strength of receptor-target binding, and the dominant forces involved in this binding (e.g., electrostatic and hydrophobic interactions, van der Waals forces, London dispersion forces, and hydrogen bonding).

16.3 DESIGN OF RAC SEPARATIONS

In the creation of any RAC method, it is necessary to first determine which conditions will promote the desired receptor-target interaction. The discussion in this section uses recombinant IL-2 as an example to highlight the general parameters that must be considered for the use of RAC in protein purification.

16.3.1 Selection of Receptors for Immobilization

The receptors chosen for immobilization should have high affinity during sample application but reversible binding to the target in the presence of mild desorbing agents, thus enabling the target's elution. The selection of suitable receptors is usually performed empirically by first performing a small-scale immobilization. These immobilized receptors are then examined to determine their binding capacities and selectivities for the desired target, as well as their ability to release the target in the presence of mild desorbing agents.

This process can be demonstrated using two soluble interleukin-2 receptors (IL-2Rs), which will be referred to as IL-2R-Δ-Nae and IL-2R-Δ-Mst. These receptors have been engineered and expressed in Chinese hamster ovary (CHO) cells [10, 11]. The transfected ovary cells were grown in a bioreactor, and the secreted IL-2Rs were collected from the conditioned CHO cell medium by filtration. These IL-2Rs were purified using target-affinity chromatography based on immobilized human rIL-2 [11]. After careful evaluation, it was determined that there was no significant difference in the affinity of these two receptors for human rIL-2. Thus, the IL-2R-Δ-Nae receptor, which from this point will be referred to as IL-2R, was arbitrarily chosen for use in further studies involving RAC.

16.3.2 Selection of a Support for RAC

The support used for receptor immobilization should have the following properties. First, it should be inert, hydrophilic, and chemically stable. It should also contain an optimum number of functional groups (e.g., hydroxyls, carboxyl groups, amines, or hydrazides) that can be easily activated for efficient protein coupling. In addition, it is desirable to have this support in a form with a high porosity, thereby allowing the rapid passage of potentially viscous fluids at moderate pressures. And finally, the support should provide a microenvironment that is biocompatible, hydrophilic, and conducive to receptor-target interactions. A further discussion of these and other desirable properties for an affinity support can be found in Chapter 2.

Several commercially available polymer supports were evaluated for use with IL-2R. These included Affi-Gel 10 from BioRad, Sepharose FF from Pharmacia, and glass beads, ProSep, and NuGel 500 Å from Separation Industries. Of these, the NuGel and glass beads were found to have the mechanical rigidity, flow-rate range, and flux that were best suited for industrial-scale operations. Thus, these two supports were chosen for the immobilization of IL-2R and its use in either conventional RAC or fluidized-bed RAC, as will be discussed later in this chapter.

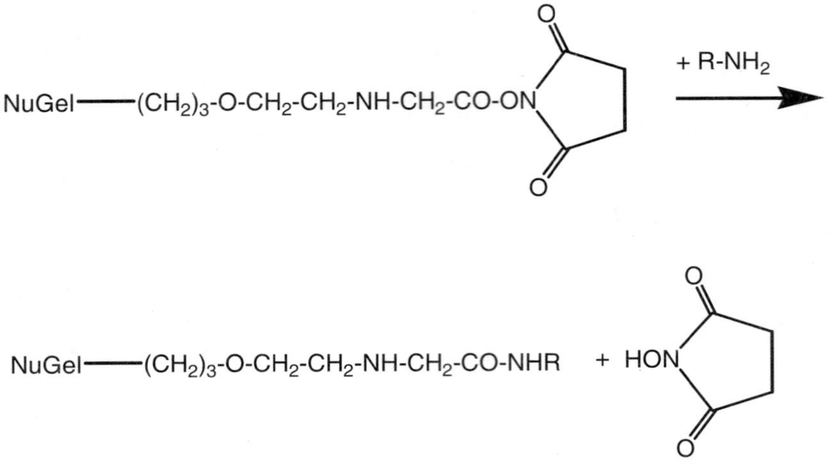

Figure 16.2 Coupling reaction between the activated support NuGel P-AF poly-*N*-hydroxysuccinimide and an amine-containing ligand, such as a protein.

16.3.3 Receptor Immobilization

The immobilization of a receptor for RAC is generally carried out with a chemically reactive support capable of forming a stable covalent bond with the receptor. Numerous immobilization methods are available for this purpose, as discussed in Chapter 3. In the case of IL-2R, this receptor was covalently coupled to a commercially available *N*-hydroxysuccinimide (NHS) ester form of a polyhydroxy silica gel (i.e., NuGel P-AF Poly-*N*-hydroxysuccinimide from Separation Industries). This immobilization method makes use of the reaction between nonprotonated amine groups on the receptor (which acted as nucleophiles) and the NHS groups on the support to form covalent bonds between the support and receptor. The mechanism for this reaction is shown in Figure 16.2. The excess NHS groups are then allowed to undergo hydrolysis to form carboxyl residues. Another immobilization method examined for IL-2R involved reductive amination between this receptor and aldehyde-activated glass beads, as will be discussed in Section 16.5 of this chapter.

16.3.4 Receptor-Affinity Chromatography

Once the receptor, support, and immobilization method have been selected, these can be combined for use in receptor-affinity chromatography. For the work with IL-2R, this receptor was immobilized by starting with a known weight of NuGel-NHS that had been washed quickly with three volumes of ice-cold water in a coarse sintered-glass funnel. This washed gel was transferred into a stoppered Erlenmeyer flask and suspended in an equal volume of cold coupling buffer (pH 7, 100 mM potassium phosphate buffer) that contained a known concentration of IL-2R. This slurry was shaken gently at 4°C for 16 h.

The uncoupled IL-2R was collected by filtering the slurry on a coarse sintered-glass funnel, followed by washing of the filtrate with two volumes of pH 7.4 phosphate buffered saline (PBS). The filtrate and washes were then combined, and the volume of this combined pool was measured. A small aliquot was dialyzed against PBS for later use in a protein assay. Any activated groups remaining on the NuGel support were neutralized by mixing this material with an equal volume of a pH 7.0, 100 mM ethanolamine-HCl solution and reacting for 1 h. The neutralized gel was washed with PBS and stored as a suspension in PBS at 4°C in the presence of 0.02% (w/w) sodium azide.

The protein concentration of the dialyzed aliquot was determined by the Lowry method [12] to give the amount of uncoupled IL-2R. The coupling density of the IL-2R affinity sorbent (expressed in mg IL-2R per ml support) was determined by taking the difference between the starting amount of IL-2R and the amount of uncoupled IL-2R that remained in solution. This later calculation made use of the fact that 1 g of NuGel has a volume of approximately 2 ml.

The binding capacity of the IL-2R adsorbent was also determined experimentally. To do this, a receptor-affinity column of known volume (generally 0.5 to 1.0 ml) was saturated with an excess of the desired target in either a purified or crude form. The nonadsorbed materials were washed away with the initial equilibration buffer, and the bound target was eluted with a mild desorbing agent. The amount of retained target (expressed in nmol) was calculated from the protein content of the eluate and molecular weight of the target protein. The amount of retained target per volume of gel (nmol target/ml gel) gave the binding capacity of the receptor-affinity adsorbent. From this value, it was then possible to determine how much crude media could be applied to a column containing this adsorbent for purification.

Several factors affect the coupling efficiency and binding capacity of an RAC support. These include the coupling pH, density of activated groups on the matrix, and the receptor coupling density [13]. For IL-2R, the optimal conditions for maximum retention of the target-binding capacity were found to be a coupling pH of 7 to 8, an activated-group density of 15 to 30 μmol/ml gel, and a receptor coupling density of 1 to 1.5 mg/ml gel (or 40 to 60 nmol/ml gel). In general, these conditions should be applicable to any soluble receptor; however, it may be necessary to perform some preliminary optimization studies with each new receptor.

16.3.5 Procedures for Receptor-Affinity Purification

After the RAC adsorbent has been prepared and characterized, it is ready for use in target purification. This process is illustrated in this section by using an IL-2R receptor-affinity column for the purification of some clinically important recombinant proteins: rIL-2 and rIL-2 analogs. The specific examples given are based on the isolation of these targets from *Escherichia coli* or mammalian sources.

The general scheme used for such samples when purifying them by RAC is shown in Figure 16.3. If required, the first step for the purification of recombinant proteins

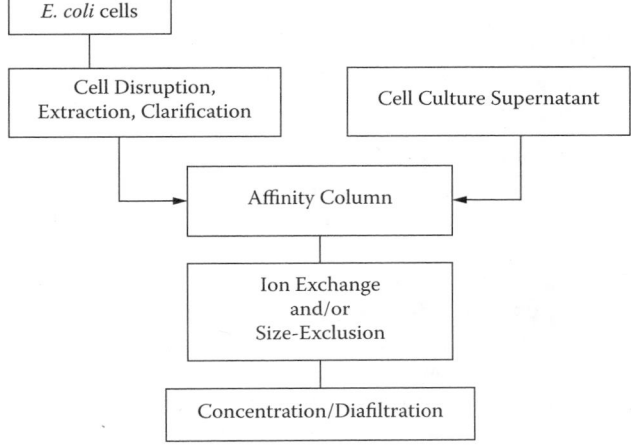

Figure 16.3 General scheme for recombinant protein purification by receptor-affinity chromatography.

expressed in microorganisms involves the extraction, solubilization, and renaturation of these proteins. This is followed by adsorption of the target of interest from the crude extract onto an RAC. After washing away nonadsorbed materials from the crude extract, the adsorbed target would then be desorbed using an appropriate eluant. Each of these steps is examined in greater detail in the following subsections.

16.3.5.1   Target Extraction, Solubilization, and Renaturation

Recombinant proteins that are expressed at high concentrations by E. coli are often present in an insoluble form as inclusion bodies. This occurred for rIL-2, which required the use of a strong denaturant (e.g., 7 M guanidine hydrochloride) for extraction and solubilization after its expression by E. coli. Previous studies have indicated that optimal refolding of recombinant proteins occurs when their concentrations are at or below the micromolar range [14]. Thus, prior to refolding, the relatively large concentration of rIL-2 that resulted from its extraction with 7 M guanidine hydrochloride required a 40-fold dilution and an aging period of 4 to 5 days [15]. When using recombinant proteins from another source, e.g., CHO cell culture supernatants, no special treatment prior to RAC is normally needed other than filtering through a 0.2-μm membrane.

16.3.5.2   Adsorption and Washing Steps

One of the most critical aspects of receptor-affinity chromatography is the sample adsorption step. During this step, the crude material should be in a buffer that facilitates maximum adsorption. Sufficient contact time should be maintained between the soluble target and the receptor adsorbent, through the use of carefully chosen flow rates, so that no product is lost or wasted during the adsorption phase. Previous studies have demonstrated that quantitative adsorption can be achieved at very high flow rates [13]. The high selectivity and specificity of the receptor adsorbent enables preferential binding of a fully renatured target from a heterogeneous population. Therefore, it is often advantageous to perform RAC at or above its saturation binding capacity.

    After target adsorption, the receptor column is immediately washed to remove contaminants that are within or surround the affinity adsorbent. This also serves to remove nonspecifically bound materials from the support and immobilized receptor. These nonspecifically bound contaminants can be removed by washing extensively with buffers that contain salts at a neutral or mildly alkaline pH or by using buffers that contain a low concentration of a nonionic detergent like Tween-20.

16.3.5.3   Elution, Polishing, and Storage

Elution of a specifically bound target from an RAC column requires dissociation of the receptor-target complex. This desorption is usually performed with nonspecific eluants such as low-pH buffers (i.e., pH < 3) or solutions that contain chaotropes (e.g., potassium thiocyanate, urea, or guanidine hydrochloride). These eluants should be removed after target desorption by using dialysis or gel permeation, which maintains the biological activity of the target. For IL-2R, no deleterious effects for the immobilized receptor are observed under these elution conditions.

    The most common steps used for the final polishing of recombinant proteins are gel permeation chromatography, filtration, and diafiltration. These provide a means for preparing a purified recombinant protein free of trace amounts of either high-molecular-weight contaminants (e.g., aggregates, pyrogens, and host-cell proteins) or low-molecular-weight contaminants (e.g., salts, detergents, and denaturants). Gel permeation is also a convenient way of exchanging purified proteins into the final formulation buffer. This step is carried out under aseptic conditions. The final step in polishing involves concentration of the target under sterile conditions through the use of a membrane with an appropriate molecular weight. This step can also be used to exchange the product into the final formulation buffer before storage.

16.3.5.4 Method Validation, Scale-Up, and Automation

The stability and longevity of a receptor column will determine how long it can be used in target purification. For the IL-2R column, this was determined by looking at the quality and quantitative recovery of its purified target proteins. This was accomplished by using such methods as SDS-PAGE, peptide mapping, and amino acid composition analysis. When studied by these techniques, no impairment in the performance of the RAC column was observed for at least 100 cycles of operation.

A microcomputer-based control system was used to perform unattended repetitive receptor-affinity purifications with the IL-2R column. Figure 16.4 shows the basic components of this system. The input of fluids through the RAC column and the column output were controlled by a collection of valves. Other major system components included a pump to control fluid flow, an ultraviolet monitor, a recorder for protein detection, and a fraction collector for collecting the column effluent. A column regeneration step using a solution of 3 M guanidine hydrochloride was included as part of the purification process.

Recombinant DNA technology has made possible the production of large quantities of soluble receptors. This, in turn, has enabled the construction of receptor columns suitable for large-scale purification, such as the one used here to isolate rIL-2 and related agents.

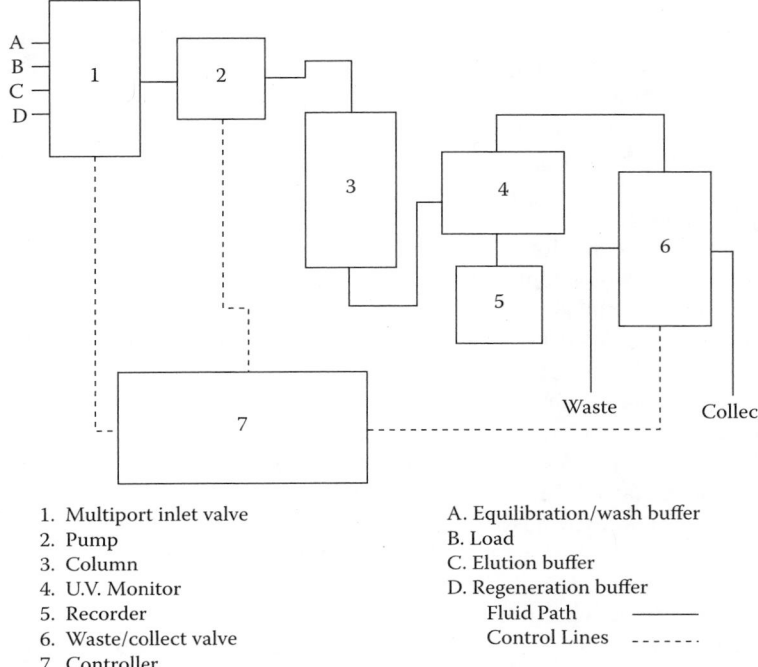

1. Multiport inlet valve	A. Equilibration/wash buffer
2. Pump	B. Load
3. Column	C. Elution buffer
4. U.V. Monitor	D. Regeneration buffer
5. Recorder	Fluid Path ————
6. Waste/collect valve	Control Lines - - - - - -
7. Controller	

Figure 16.4 Basic schematic of an automated receptor-affinity purification system.

444

Bailon et al.

The design of efficient large-scale receptor-affinity columns is based on the performance parameters determined from small-scale operations. Large-scale RAC purification involves processing large volumes of crude material that, in turn, will require a large receptor column and increased flow rates. This larger column size is best achieved by increasing the cross-sectional area of the column rather than its height, which provides a reasonably low resistance to flow and allows the use of high flow rates for sample application.

16.4 EXAMPLES OF RAC APPLICATIONS

Several applications have been reported for receptor-affinity chromatography. This section discusses some specific applications for the IL-2R adsorbent. These include the use of this adsorbent in the purification of human rIL-2, rIL-2 muteins, murine rIL-2, an interleukin-2 pseudomonas exotoxin fusion protein (IL-2-PE40), and a humanized monoclonal antibody (HAT).

16.4.1 Recombinant Human rIL-2

To obtain recombinant human rIL-2, a synthetic gene for IL-2 was constructed and introduced into *E. coli*. These modified bacteria were then grown in an appropriate medium in a large fermenter [16]. A Chinese hamster ovary cell line transfected with the IL-2 gene was used as the source of mammalian glycosylated rIL-2.

The bioactivity of the IL-2 was determined by a colorimetric assay; this assay measured the IL-2–dependent proliferation of murine CTLL cells, as determined through the production of lactic acid by these stimulated cells as a product of glucose metabolism [17]. A biologic response modifier protein (BRMP) reference agent, human rIL-2 (from Jurkat cells), was used as the reference standard. In this assay, one unit of activity was defined as the quantity of IL-2 needed to produce a half-maximal response in the assay.

The general RAC purification scheme for the production of rIL-2 from bacterial and mammalian sources was given earlier in Figure 16.3. The rIL-2 that was expressed in *E. coli* as an insoluble form within inclusion bodies was extracted with 7 *M* guanidine hydrochloride, as discussed in Section 16.3.5.1. The solubilized rIL-2 was diluted 40-fold with PBS and aged for 4 to 5 days, during which time the denatured rIL-2 refolded. The diluted and aged extract was clarified by centrifugation or by filtration prior to its application to an RAC column. The conditioned CHO media that contained rIL-2 was filtered through a 0.2-μm filter before being applied to the RAC column.

Adsorption, washing, and elution of the rIL-2 sample on the RAC column were performed by taking into account all of the general considerations described in Section 16.3.5. In this specific case, elution of rIL-2 from the column was achieved by dissociating the receptor-target complex with a low-pH buffer containing 0.2 *M* acetic acid and 0.2 *M* sodium chloride.

A final step to remove trace contaminants was performed by using gel permeation on a Sephadex G-50 superfine column. This column was equilibrated with a pH 3.5, 50 m*M* sodium acetate buffer that contained 5 mg/ml mannitol and 0.2 *M* sodium chloride. After gel permeation, the rIL-2 was concentrated to a level of roughly 5 mg/ml by using an Amicon stirred-cell concentrator with a 5000 molecular-weight-cutoff membrane. The rIL-2 obtained by this method was stored in the pH 3.5 acetate-mannitol buffer at 2 to 4°C without any significant loss of activity over at least one year.

The glycosylated mammalian rIL-2 was purified by RAC according to the same procedures as described for the rIL-2 obtained from *E. coli*. After RAC purification, both forms of rIL-2 had specific activities of approximately 2×10^7 U/mg.

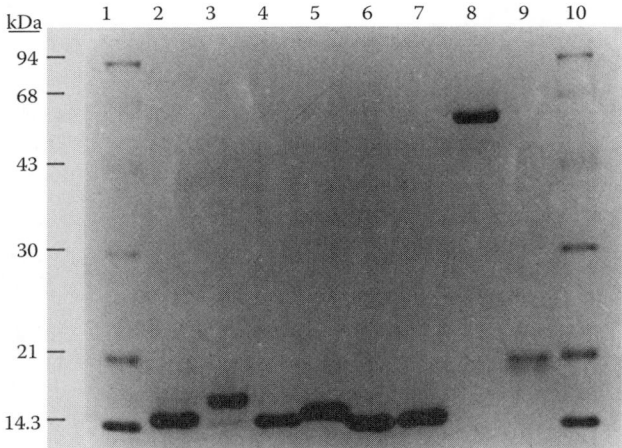

Figure 16.5 SDS-PAGE analysis of receptor-affinity purified recombinant interleukin-2 (rIL-2) and its analogs. Lanes 1 and 10 contain molecular-weight marker proteins. Lanes 2 through 9 represent microbial rIL-2, mammalian glycosylated rIL-2, yeast rIL-2, mutein Lys20, mutein Des-Ala(1)-Ser(125), seleno-Met-rIL-2, IL2-PE40, and a murine rIL-2 homolog, respectively.

16.4.2 Human rIL-2 Analogs

Using recombinant DNA technology, several analogs of human rIL-2, a murine homolog, and an interleukin-2 pseudomonas exotoxin fusion protein were synthesized in *E. coli* or in yeast (*Pichia pastoris*) and purified by RAC. The final purified rIL-2 and related molecules were subjected to SDS-PAGE analysis, with the results shown in Figure 16.5.

One of these agents was human rIL-2 expressed in yeast. A second was rIL-2 Lys^{20} mutein, in which the N-terminal aspartic acid at position 20 of human rIL-2 was replaced with a lysine residue [18], resulting in a 1000-fold reduction in bioactivity. This second analog has been used in the identification of amino acid residues involved in IL-2 receptor binding.

A third analog studied by RAC was rIL-2 $Des-Ala^1-Ser^{125}$ mutein, in which Ala^1 has been deleted and Cys^{125} has been substituted with a serine residue. The Cys^{125} substitution facilitates the refolding of rIL-2 that has been solubilized from inclusion bodies. RAC-purified rIL-2 mutein exhibits similar bioactivity to microbial rIL-2.

Seleno-methionine rIL-2 was the fourth mutein analog that was examined. This mutein was used in a structure determination of rIL-2 by X-ray crystallography. A fifth analog considered was murine rIL-2, which was used to conduct chronic studies in mice. All of these RAC-purified IL-2 analogs, with the exception of rIL-2 Lys^{20}, exhibited bioactivity similar to that of wild-type IL-2.

Another agent that was purified by RAC was the interleukin-2–pseudomonas exotoxin fusion protein (IL2-PE40). IL2-PE40 is a 54.4-kDa chimeric protein in which the recognition domain of the pseudomonas exotoxin has been replaced by IL-2. It is a cytotoxic agent to cells that bear IL-2 receptors, giving it potential clinical applications. These possible applications include the treatment of autoimmune diseases, organ-transplant rejection, and rheumatoid arthritis, which are conditions in which IL-2 receptors are thought to be involved. For this work, the IL2-PE40 was engineered and expressed in *E. coli* [19] and purified by RAC [20]. Its activity was determined by measuring its ability to inhibit IL-2–dependent murine CTLL proliferation [19].

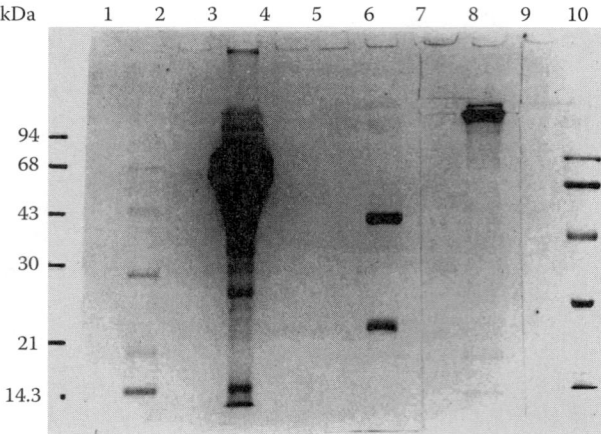

Figure 16.6 SDS-PAGE analysis of receptor-affinity purified anti-Tac (HAT). Lanes 1 and 10 contain molecular-weight marker proteins. Lane 3 represents crude CHO media, Lane 6 shows the results for purified HAT under reducing conditions, and Lane 8 shows the purified HAT under nonreducing conditions.

16.4.3 Humanized Anti-Tac (HAT)

Humanized anti-Tac (HAT) is yet another agent that has been purified by RAC. HAT is a humanized monoclonal antibody. This is produced by introducing into a human IgG framework selected amino acids from the variable region of a murine monoclonal antibody that binds to the low-affinity p55 subunit of the IL-2 receptor [21]. HAT is an immuno-suppressive drug that has been approved by the FDA for the prevention of kidney-transplantation rejection.

HAT has been expressed in a murine myeloma cell line (SP2/O) that was transfected with a gene encoding the chimeric antibody. The expressed HAT was then purified from its conditioned media by employing RAC. SDS-PAGE results for HAT that had been purified by RAC are shown in Figure 16.6, using both reducing and nonreducing conditions for the analysis. This preparation exhibited greater than 98% purity for the HAT, as determined by densitometric measurements.

16.5 MEMBRANE-BASED RECEPTOR-AFFINITY CHROMATOGRAPHY

Two advantages of receptor-affinity chromatography are its specificity and fast adsorption kinetics. However, the property of fast adsorption cannot be fully utilized in conventional particle-based RAC, since such columns are usually limited by the rate of mass transfer rather than adsorption. This results in less-than-optimum productivity.

In theory, the mass-transfer problems associated with conventional RAC could be circumvented if the receptor were instead immobilized within a thin microporous mem-brane. In such a membrane, the required distance for solute diffusion can be as short as 1 μm, instead of the typical distance of 100 μm in a gel-based particulate support. This shorter diffusion distance makes mass transfer much faster in these membranes. This, in turn, causes adsorption rather than mass transfer to become the rate-limiting step in solute retention and opens up the possibility of achieving faster and more optimum separations. It is for this reason that flat-sheet and hollow-fiber membranes have been considered for use in bioaffinity purifications [22–25]. This section examines the use of such membranes in RAC, giving rise to a method known as membrane-based receptor-affinity chromatography (MRAC) [23].

16.5.1 Theory of MRAC

A typical process used in MRAC is shown in Figure 16.7. In Figure 16.7a, a single hollow membrane is illustrated, while Figure 16.7b shows a membrane module that contains an array of hollow fibers. In either of these formats, a crude feed stream containing the target protein is pumped through the lumen of the hollow-fiber receptor-affinity membrane. The transmembrane pressure differential draws the applied fluid through the hollow fiber and results in binding between the target and the immobilized receptor on the membrane walls. The target-depleted filtrate that has been drawn through the membrane is then discarded, while the feed stream (which still contains some of the target protein) is circulated again through the membrane. After washing away nonretained contaminants, the bound target is eluted with an appropriate elution buffer and saved. A regeneration step is then used to prepare the system for the next cycle of operation.

In this type of membrane-based system, the diffusion time of the target is much smaller than the sample's residence time. Since receptor-target interactions usually have a low dissociation constant, k_{-1} is also quite small. Thus, according to the model presented in Section 16.2, the primary factors that determine the rate of solute retention in membrane RAC are the association rate constant (k_1) and the concentrations of the receptor and target ($[R]$ and $[L]$). Under these conditions, the adsorption kinetics approach those for a nondiffusion-controlled reaction in solution except that one of the reactants (i.e., the receptor) is now immobilized on a membrane. In this type of system, the value of k_1 is determined by the immobilized receptor density, and $[L]$ is given by the feed-stream concentration of the target.

Conceptually, membrane-based affinity purifications have several advantages over particle-based methods. For instance, the higher porosity and lower mass-transfer resistance of membranes allow them to have a high volumetric throughput and short sample-processing times. Also, the topography of membranes is such that the adsorbent molecule is situated on the inner surfaces of membrane pores and (for all practical purposes) along the flow path of the applied target. This minimizes the diffusion and accessibility problems associated with packed-bed columns. These features, plus the fast mass-transfer and adsorption kinetics of MRAC, allow for maximum utilization of the adsorbent molecule.

16.5.2 Construction of an IL-2R Affinity Membrane

The immobilization of a receptor onto an affinity membrane can be illustrated by using IL-2R as an example. Amine-based coupling methods, as discussed in Section 16.3, can often be used for this purpose. However, it is also possible to use the fact that IL-2R is a glycoprotein with N- and O-linked glycosylation sites. For instance, the sugar moieties on IL-2R can be oxidized with periodate to give aldehyde groups, which in turn can be covalently coupled to a hydrazide-containing hollow-fiber membrane to form a stable hydrazone bond.

To carry out this type of immobilization, IL-2R with a starting concentration of 2.5 to 10 mg/ml was dialyzed against a pH 5.5, 0.1 M potassium phosphate coupling buffer. This receptor solution was then mixed with a one-tenth volume of 0.1 M sodium metaperiodate in water and allowed to react for 1 h in the dark at room temperature. The oxidized IL-2R was desalted by passing it through a Sephadex G-25 Superfine column in the presence of the coupling buffer. The protein content of the mobile phase was monitored, and fractions were collected. Those fractions that contained the oxidized IL-2R were pooled and concentrated. The protein concentration of this preparation was determined by measuring its absorbance at 280 nm, using a reference absorbance value of 1.65 for a 1-mg/ml solution of IL-2R.

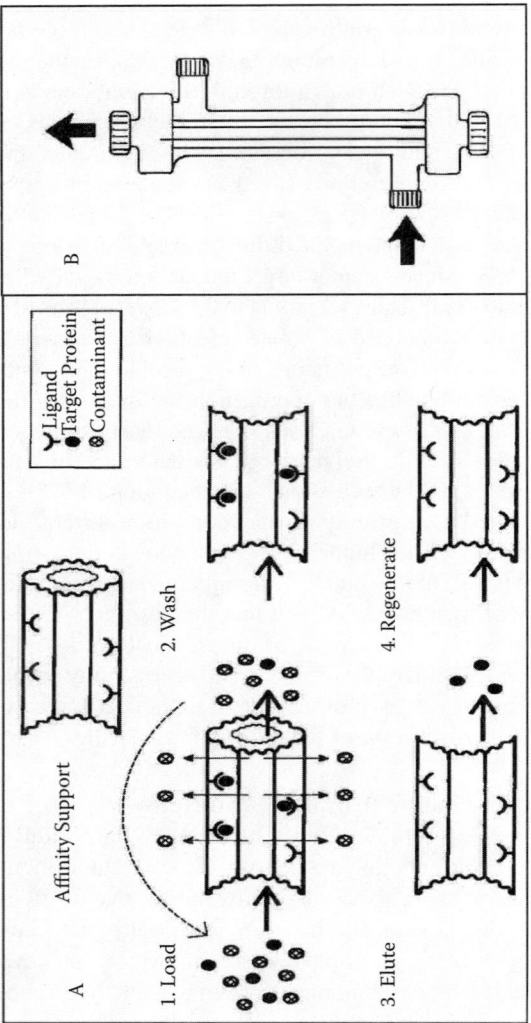

Figure 16.7 Operation and chromatographic bed used in membrane-based receptor-affinity chromatography (MRAC). In (a), the receptor is chemically attached to the hollow-fiber membrane in an oriented fashion, with a single hollow fiber being used here for illustrative purposes. The device in (b) shows a membrane module that contains an array of these hollow-fiber membranes.

The hydrazide derivative of a hollow-fiber membrane module (0.4-ml volume) was used for immobilizing the oxidized IL-2R. Approximately 4 mg of IL-2R in 6-ml coupling buffer was circulated overnight through this module at a flow rate of 2 ml/min. Thirty milliliters of PBS were later passed through this affinity membrane, followed by washings with 0.2 M sodium chloride in 0.2 M acetic acid. These washings were pooled, and the protein content (representing the uncoupled IL-2R) was determined by measuring its absorbance at 280 nm, as described in the previous paragraph.

16.5.3 Determining the Binding Capacity of an MRAC Membrane

The binding capacity of an MRAC membrane can be determined in a similar manner to that used for an affinity column. For example, the capacity of the IL-2R affinity membrane (0.4-ml volume) was determined by equilibrating this membrane with PBS and saturating it with an excess of purified or crude HAT (the soluble protein target), with the HAT being applied at a flow rate of 1 ml/min. Nonadsorbed materials were washed from the membrane at 2 ml/min by using PBS that contained 0.2 M sodium chloride. The adsorbed HAT was eluted by applying a solution of 0.2 M acetic acid and 0.2 M sodium chloride at a flow rate of 1 ml/min. Fractions of this eluant were collected every 2 min and monitored for the presence of protein, as described earlier. The protein-containing fractions were pooled, and their protein content was measured at 280 nm by using a reference absorbance of 1.4 for a 1-mg/ml HAT solution in PBS. This measured value gave the affinity membrane's observed binding capacity, which in turn determined the amount of crude HAT that could be purified by the membrane in one operation cycle.

The dynamic binding capacity of the affinity-membrane module was determined at a flow rate of 15 ml/min by applying solutions that contained HAT concentrations of 1.0, 2.5, 5.0, 10.0, or 25.0 nM. In all of these experiments, the total amount of HAT that passed through the membrane was 0.5 mg. The results indicated that the feed-stream concentrations had no significant effect on the capture efficiency of the affinity membrane.

16.5.4 Effects of Flow Rate and Coupling Density in MRAC

Two other factors that might affect the performance of MRAC are the flow rate used for sample application and the coupling density of the receptor. The effects of flow rate were considered for the IL-2R membrane (containing 0.73 mg IL-2R) by loading 500 μg HAT at a concentration of 3.6 μg/ml at 1, 5, 15, 25, 50, or 100 ml/min. The residence time (t_R, in sec) for the sample at each flow rate was determined by using the expression t_R = [membrane volume (in ml) × 60 sec/min]/[filtrate flow rate (in ml/min)]. The results indicated that the capture efficiency of the affinity membrane was unaffected by the flow rate, even when the residence time was as small as 2.6 sec.

The importance of coupling density in MRAC was examined by preparing five IL-2R affinity membranes with coupling densities of 0.55, 1.4, 1.82, 3.75, or 4.50 mg/ml. The binding capacity was 73% for the membrane with the lowest coupling density and dropped to 31% for the membranes with higher coupling densities, where 100% represents a perfect 1:1 stoichiometry for the binding reaction. These theoretical and expected binding capacities for the IL-2R affinity membranes were calculated from the coupling density by using a molar mass of 25 kDa for IL-2R.

16.5.5 Process-Scale MRAC

The use of MRAC for process-scale work can be illustrated through the purification of human rIL-2, IL2-PE40, and HAT from crude extracts by an IL-2R receptor-affinity membrane. This membrane had a total volume of 9.7 ml and contained 15.5 mg of IL-2R. To use this membrane, it was placed into a prototype automated purification system (i.e., Affinity-15 from Sepracor).

In this application, 500 ml of an *E. coli* extract containing roughly 5 μg/ml rIL-2 was circulated through the IL-2R affinity membrane at a loading flow rate of 1 l/min and a filtrate flow rate of 140 ml/min. Washing and elution of this membrane were performed as stated earlier for the purification of anti-Tac-H. The actual process time was approximately 5 min. The rIL-2 obtained by this method was later bioassayed and analyzed by SDS-PAGE.

A similar approach was used in the purification of IL2-PE40 and HAT by the IL-2R membrane. For the IL2-PE40, 500 ml of a crude *E. coli* extract containing about 10 μg/ml IL2-PE40 was circulated through the affinity membrane, with washing being performed as described for rIL-2 and elution being accomplished with a pH 6 buffer containing 1.5 *M* potassium thiocyanate and 50 m*M* potassium phosphate.

The HAT was purified by circulating through the IL-2R membrane 9 l of SP2/0 media that contained 4 μg/ml of this target. The membrane in this case was preequilibrated with PBS and operated at a loading flow rate of 2 l/min and a filtrate flow rate of 240 ml/min. Washing and elution were performed at a flow rate of 200 ml/min with PBS containing 0.25 *M* sodium chloride or 0.2 *M* acetic acid containing 0.2 *M* sodium chloride. One cycle of this operation was completed in 45 min. Both the purified IL2-PE40 and HAT were later examined by SDS-PAGE and a bioassay.

It was found in these applications that MRAC was both practical and efficient in purifying these three closely related recombinant proteins. It was further determined from this work that membrane-based affinity systems are well suited for the industrial-scale production of protein therapeutics. Some advantages of this approach include its use of less adsorbent than other methods and its short process times, which makes it possible to perform faster purifications.

16.6 FLUIDIZED-BED RECEPTOR-AFFINITY CHROMATOGRAPHY

Fluidized-bed chromatography is an alternative protein purification method to conventional packed-bed column chromatography [26, 27]. In a fluidized-bed recovery system, recombinant proteins are recovered directly from a crude feed that contains particulate matter, such as cell debris. In general, the clarification, concentration, and purification steps for recombinant proteins can be accomplished in a single step by fluidized-bed systems.

The differences between a fluidized-bed column and a conventional packed column are illustrated in Figure 16.8. In a packed-bed column, particulate matter from a crude feed stream can get trapped within the spaces between particles, resulting in clogging of the column. Solids such as cell debris can also settle as a cake at the column inlet and hinder the passage of fluids through the column bed (see Figure 16.8a). In a fluidized-bed system, the flow is upward and the force of the liquid flow causes the column bed to expand. This creates spaces between the particles and allows less-restricted passage of fluids through the column (see Figure 16.8b).

This section discusses the characterization and use of fluidized beds in RAC, giving rise to a method known as fluidized-bed receptor-affinity chromatography (FB-RAC). This approach is illustrated by using it in the purification of several targets based on the immobilized receptor IL-2R.

16.6.1 Immobilization of a Receptor for FB-RAC

A fluidized-bed containing the receptor IL-2R was prepared by mixing 125 ml of swelled Prosep-5CHO gel (which contains aldehyde groups) with 400 ml of 0.62-mg/ml IL-2R in

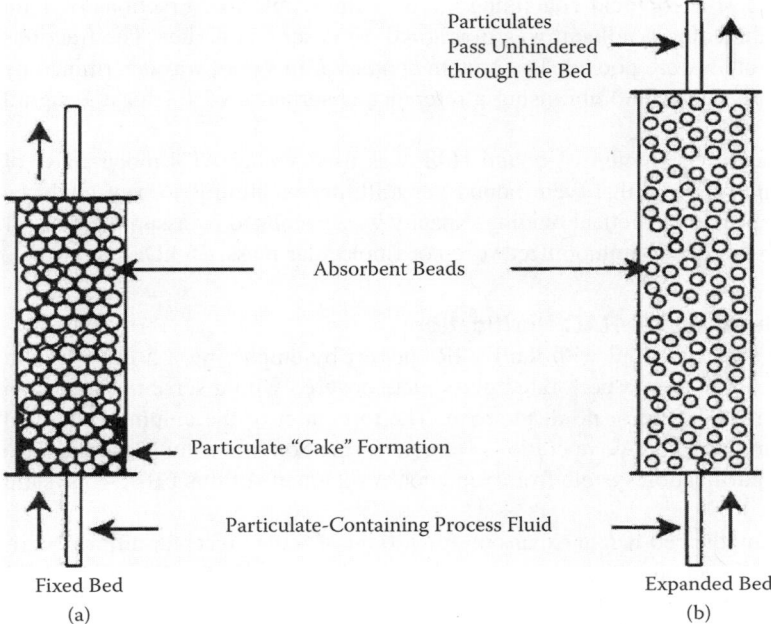

Figure 16.8   Illustration of the differences between (a) packed-bed and (b) fluidized-bed columns.

pH 7.2 PBS. This mixture was reacted at 4°C for 4 h. The gel was then allowed to settle, and the absorbance of the supernatant at 280 nm was determined, giving the amount of IL-2R remaining in solution. Solid glycine (7.88 g) and sodium cyanoborohydride (1.65 g) were added to this mixture (giving final concentrations of 0.2 M and 0.05 M, respectively) and allowed to react overnight to help remove any remaining aldehyde groups on the support. The support was then transferred to a coarse sintered-glass frit and washed three times with 625 ml PBS.

To remove additional aldehyde groups on the support, a 200-ml solution containing 0.1 M sodium borohydride was added and allowed to react with the support in a fume hood. The support was then washed with five volumes of PBS. This was next treated for 1 h with 200 ml of a 1% (w/v) solution of 20-kDa polyethylene glycol in PBS and washed three times with five volumes of PBS. The end product was stored in 0.04% sodium azide at 4°C.

By using the absorbance of the supernatant, the amount of uncoupled IL-2R in the immobilization reaction could be determined, as described in Section 16.5. The difference between the initial amount of IL-2R (400 ml × 0.62 mg/ml) and the remaining amount of uncoupled IL-2R gave the amount of immobilized receptor. From this, the coupling density was determined by dividing the amount of coupled receptor by the total volume of support (125 ml).

16.6.2   Determination of Binding Capacities for FB-RAC

The binding capacity of the IL-2R fluidized bed was found by taking 1 ml of the IL-2R affinity beads and packing this into an Amicon G 10 mm × 150 mm column. This column was equilibrated with PBS, and an excess of purified HAT was applied at a flow rate of 1 ml/min. The nonadsorbed material was washed away with PBS, while the adsorbed HAT was eluted with 0.2 *M* acetic acid containing 0.2 *M* sodium chloride. Fractions of 1 ml were collected as the column effluent was monitored, as described earlier. The fractions found to contain protein were pooled. The protein content of this pool was determined by measuring its absorbance at 280 nm, using a reference absorbance of 1.4 for a 1-mg/ml HAT in PBS.

When the measured amount of bound HAT was used with HAT's molar mass of 144.4 kDa, the nmol of HAT that were bound per milliliter of affinity sorbent could be determined. The expected theoretical binding capacity was calculated by assuming that 1:1 binding occurred between the immobilized receptor (molecular mass, 25 kDa) and HAT.

16.6.3 Example of an FB-RAC Purification

A fluidized-bed device was used with the IL-2R support by employing a 5 cm × 25 cm column that had at its bottom a perforated glass plate covered with a screen. The design of this inlet tube created little or no dead space. The top outlet of the column contained an adjustable piston. All FB-RAC operations were performed on this column using a Trio automated protein purification system from Sepracor. A schematic of this FB-RAC system is shown in Figure 16.9.

For this system, the bed-height expansion for 120 ml of settled receptor-affinity beads was determined as a function of flow rate. One liter of unclarified HAT cell culture media was used as the sample and was applied at flow rates of 15, 30, or 45 ml/min. The bed expansion at each flow rate was determined after equilibrating the column with PBS and applying 1 l of the cell culture medium in the upward-flow direction. The bed height expansion observed

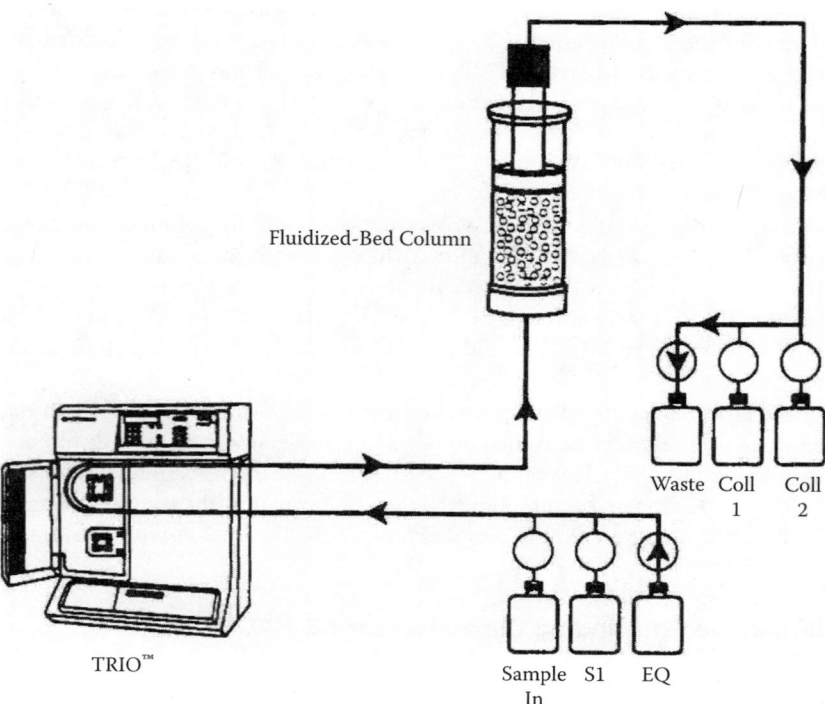

Figure 16.9 An example of an automated FB-RAC purification system.

Table 16.1 Bed-Height Expansion Versus
Flow Rate for an FB-RAC Column

Flow Rate (ml/min)	Bed Height (cm)[a]
0	6.2
15	10
30	14.6
45	19.5

[a] This gives the initial packed-bed height for 120 ml
of the controlled-pore glass bead IL-2R affinity sorbent. The bed-height expansion was determined
using 1 l of an unclarified cell culture supernatant.
Other details are given in the text.

at each flow rate is shown in Table 16.1. It can be seen from this that the initial bed height
of 6.2 cm for settled beads increased to 19.5 cm at a flow rate of 45 ml/min.

The fluidized-bed adsorption of HAT from increasing volumes of unclarified crude
media was used to determine the dynamic binding capacity of the FB-RAC system. This was
accomplished by equilibrating the FB-RAC column with 1.8 l of PBS for 1 h and applying
0.5 to 10 l of the crude media in an upward direction at a predetermined optimal flow rate
of 30 ml/min. After each application step, the nonadsorbed and loosely bound materials
were washed away from the fluidized bed through the use of PBS. The dynamic binding
capacities that were obtained are given in Table 16.2.

After the column was washed with PBS, the adsorbed HAT was eluted with 0.2 M
acetic acid containing 0.2 M sodium chloride. This was done in the column mode by
lowering the top piston. The column was then equilibrated again with PBS. Finally, the
top piston was raised, and the column was equilibrated with another 1.8 l of PBS before
the application of more crude HAT media.

Table 16.2 Dynamic Binding Capacities
for an FB-RAC Column

Feedstock Load (l)	HAT Adsorbed (mg)
0.5	12.2
1.0	30.0
2.0	66.3
5.0	145.1
10.0	166.0

Note: The FB-RAC column used in this study contained 120 ml of immobilized IL-2R affinity beads.
A fixed flow rate of 30 ml/min was used in all experiments. See the text for further details.

In the final purification method, the receptor beads were fluidized by applying PBS to them at an upward flow rate of 30 ml/min. Five to 10 l of unclarified HAT cell culture media were applied to these beads in the FB-RAC column. The nonadsorbed materials were washed away with PBS. The beads were allowed to rest, and the upper piston was lowered to meet the top of the settled bead bed. The adsorbed HAT was then eluted, as described in the previous paragraph. The column effluent was monitored, and the column was regenerated, also as described previously.

For the purification of rIL-2, 5 to 10 l of unclarified crude rIL-2 extract were applied to the receptor-affinity beads in the fluidized-bed mode. The rest of the purification procedure was the same as described for HAT, except elution was performed with a pH 5.3 buffer containing 0.2 M acetic acid and 0.2 M sodium chloride. Similarly, the anti-Tac(Fv)-PE38 was purified by this column by applying 5 to 10 l of unclarified crude extract. After washing away the nonadsorbed materials, the retained anti-Tac(fv)-PE38 was eluted in the conventional-column mode by using 3 M potassium thiocyanate in PBS. This eluted protein was later dialyzed again with three 2-l changes of buffer. The FB-RAC affinity column was regenerated with 300 ml of 2 M guanidine hydrochloride in PBS after each cycle of operation. This column was then reequilibrated with PBS. In each case, protein assays, SDS-PAGE, and bioassays were performed on the recovered products.

As can be seen in Figure 16.10, the IL-2R FB-RAC column was capable of purifying rIL-2, anti-Tac monoclonal antibodies, and anti-Tac(Fv)-PE38 with good yield and high purity. Since the column used unclarified crude extracts as the starting material, it avoided labor-intensive and time-consuming centrifugation and filtration steps. This makes this approach a highly productive and economical affinity purification method.

For clinical applications, proteins purified by FB-RAC may require additional polishing steps, such as ion-exchange or gel-permeation chromatography. FB-RAC columns also need extensive washing and regeneration to remove residual proteins and particulate matter after each cycle of operation.

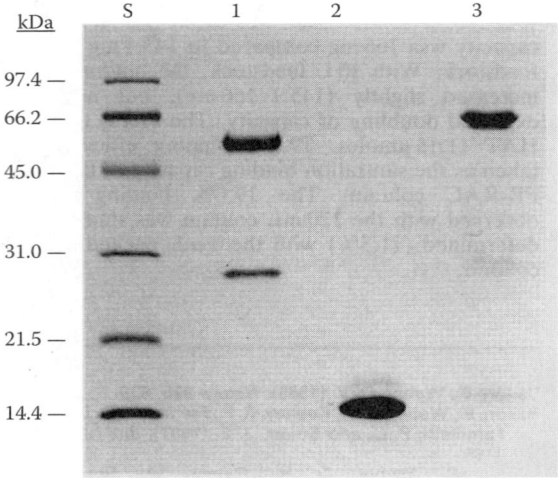

Figure 16.10 SDS-PAGE analysis of recombinant proteins that have been purified by FB-RAC. Lane S shows the molecular-weight marker proteins. Lane 2 shows purified HAT in a reduced form, Lane 2 gives the results for rIL-2, and Lane 3 shows purified anti-TAC-C3-(Fv)-PE38 in a nonreduced form.

16.7 SUMMARY AND CONCLUSIONS

This chapter discussed the method of receptor-affinity chromatography and showed how it exploits the specific and reversible interactions between an immobilized receptor and its corresponding soluble target. Three types of receptor-affinity chromatography were described: (1) conventional particle-based RAC, (2) membrane-based RAC, and (3) fluidized-bed RAC. Membrane-based RAC circumvents the mass-transfer and accessibility problems that are associated with conventional particle-based methods, while fluidized beds facilitate passage of cell debris and particulate matter in unclarified crude extracts. Each of these approaches was illustrated by using immobilized IL-2R receptors to purify recombinant human IL-2 and related proteins.

Overall, all three types of RAC were shown to be viable techniques that are scalable and productive as affinity purification methods. Since biotechnology has already produced or is on the verge of producing many soluble receptors of biomolecules such as IL-2, IL-1, γ-interferon, and tumor necrosis factor, it should be only a matter of time before RAC becomes an established method for the purification of high-value recombinant proteins.

SYMBOLS AND ABBREVIATIONS

BRMP	Biologic response modifier protein
C^*	Concentration of a target in solution when at equilibrium with an affinity support
CHO	Chinese hamster ovary cells
D	Diffusion coefficient
FB-RAC	Fluidized-bed receptor-affinity chromatography
HAT	Humanized anti-Tac antibodies
IL2-PE40	Interleukin-2-*Pseudomonas* exotoxin fusion protein
IL-2R	Interleukin-2 receptor
k_1	Association rate constant for the binding of a target to a receptor
k_{-1}	Dissociation rate constant for a receptor-target complex under application conditions
k_{-2}	Dissociation rate constant for a receptor-target complex under elution conditions
k_{de}	Effective dissociation constant, where $k_{de} = k_{-1}/k_1 = (q_m - q^*)C^*/q^*$
L	Soluble target
L_D	Average distance diffused by a solute in time t_D
MRAC	Membrane-based receptor-affinity chromatography
NHS	*N*-Hydroxysuccinimide ester
PBS	Phosphate-buffered saline
q^*	Binding capacity of an affinity support per unit volume when the support is in equilibrium with a concentration C^* of a solute
q_m	Maximum binding capacity of an affinity support per unit volume
R	Immobilized receptor
RAC	Receptor-affinity chromatography
r-IL-2	Recombinant interleukin-2
R-L	Receptor-target complex
SDS-PAGE	Sodium dodecyl sulfate polyacrylamide gel electrophoresis
t_D	Average diffusion time for a solute
t_R	Residence time of a sample in a chromatographic bed

REFERENCES

1. Cuatrecasas, P., Wilcheck, P., and Anfinsen, C.B., Selective enzyme purification by affinity chromatography, *Proc. Natl. Acad. Sci.,* 61, 636–643, 1968.
2. Chaiken, I., Wilcheck, M., and Parikh, I., *Affinity Chromatography and Biological Recognition,* Academic Press, New York, 1984.
3. Kline, T., Ed., *Handbook of Affinity Chromatography,* Marcel Dekker, New York, 1993.
4. Lowe, C.R., Affinity chromatography and related techniques: Perspectives and trends, *Adv. Mol. Cell Biol.,* 15B, 513–522, 1996.
5. Bailon, P., Weber, D., Keeney, R.F., Fredericks, J.E., Smith, C., Familletti, P.C., and Smart, J.E., Receptor-affinity chromatography: A one-step purification for recombinant interleukin-2, *BioTechnology,* 5, 1195–1198, 1987.
6. Bailon, P. and Weber, D., Receptor-affinity chromatography, *Nature,* 335, 839–840, 1988.
7. Nishikawa, A.H., Bailon, P., and Ramel, A.H., Design parameters in affinity chromatography, *J. Macromol. Sci. A,* 10, 149–190, 1976.
8. Chase, H.A., Affinity separations utilising immobilised monoclonal antibodies: A new tool for the biochemical engineer, *Chem. Eng. Sci.,* 39, 1099–1125, 1984.
9. Lowe, C.R. and Dean, P.D.G., *Affinity Chromatography,* Wiley, New York, 1974.
10. Hakimi, J., Seals, C., Anderson, L.E., Podlaski, F.J., Lin, P., Danho, W., and Jenson, J.C., Biochemical and functional analysis of soluble human interleukin-2 receptor produced in rodent cells: Solid phase reconstitution of a receptor-ligand binding reaction, *J. Biol. Chem.,* 262, 17336–17341, 1987.
11. Weber, D.V., Keeney, R.F., Familletti, P.C., and Bailon, P., Medium-scale ligand-affinity purification of two soluble forms of human interleukin-2 receptor, *J. Chromatogr.,* 431, 55–63, 1988.
12. Lowry, O.H., Protein measurement with Folin phenol reagent, *J. Biol. Chem.,* 193, 265–275, 1951.
13. Seetharam, R. and Sharma, S.K., Eds., *Purification and Analysis of Recombinant Proteins,* Marcel Dekker, New York, 1991, pp. 267–283.
14. Light, A., Protein solubility, protein modifications and protein folding, *BioTechniques,* 3, 298–304, 1985.
15. Weber, D.V. and Bailon, P., Application of receptor-affinity chromatography to bioaffinity purification, *J. Chromatogr.,* 510, 59–69, 1990.
16. Ju, G., Collins, L., Kaffke, K., Tsien, W.H., Chizzonite, R.A., Crowl, R., Bhatt, R., and Kilian, P., Structure-function analysis of human interleukin-2: Identification of amino acid residues required for biological activity, *J. Biol. Chem.,* 262, 5723–5731, 1987.
17. Familletti, P.C. and Wardwell, J.A., A novel approach to bioassays, *BioTechnology,* 6, 1169–1172, 1988.
18. Collins, L., Tsien, W.-H., Seals, C., Hakimi, J., Weber, D., Bailon, P., Hoskins, J., Green, W.C., Toome, V., and Ju, G., Identification of specific residues of human interleukin-2 that effect binding to the 70-kDa subunit (p70) of the interleukin-2 receptor, *Proc. Natl. Acad. Sci. U.S.A.,* 85, 7709–7713, 1988.
19. Loiberboum-Gaski, H., Fitzgerald, D., Chaudhary, V., Adhya, S., and Pastan, I., Cytotoxic activity of an interleukin-2-*Pseudomonas* exotoxin chimeric protein in *Escherichia coli, Proc. Natl. Acad. Sci. U.S.A.,* 85, 1922–1926, 1988.
20. Bailon, P., Weber, D., Gately, M., Smart, J.E., Loberboum-Galski, H., Fitzgerald, D., and Pastan, I., Purification and partial characterization of an interleukin-2-*Pseudomonas* exotoxin fusion protein, *BioTechnology,* 6, 1326–1329, 1988.
21. Queen, C., Schneider, W.P., Selick, H.E., Payne, P.W., Landolfi, N.F., Duncan, J.F., Avdalovick, N.M., Levitt, M., Junghans, R.P., and Waldman, T.A., A humanized antibody that binds to the interleukin-2 receptor, *Proc. Natl. Acad. Sci. U.S.A.,* 86, 10029–10033, 1989.
22. Brandt, S., Goffe, R.A., Kessler, S.B., O'Connor, J.L., and Zale, S.E., Membrane-based affinity technology for commercial-scale purifications, *BioTechnology,* 6, 779–782, 1988.

23. Nachman, M., Azad, A.R.M., and Bailon, P., Membrane-based receptor-affinity chromatography, *J. Chromatogr.,* 597, 155–166, 1992.
24. Nachman, M., Azad, A.R.M., and Bailon, P., Efficient recovery of recombinant proteins using membrane-based immunoaffinity chromatography (MIC), *Biotechnol. Bioeng.,* 40, 564–571, 1992.
25. Nachman, M., Kinetic aspects of membrane-based immunoaffinity chromatography, *J. Chromatogr.,* 597, 167–172, 1992.
26. Draeger, N. and Chase, H.A., Liquid fluidized bed adsorption of protein in the presence of cells, *Bioseparation,* 2, 67–80, 1991.
27. Spence, C.L., Schaffer, C.A., Kessler, S., and Bailon, P., Fluidized-bed receptor-affinity chromatography, *Biomed. Chromatogr.,* 8, 236–241, 1994.

Section IV

Analytical and Semipreparative Applications

17

Affinity Methods in Clinical and Pharmaceutical Analysis

Carrie A.C. Wolfe

Division of Science and Mathematics, Union College, Lincoln, NE

William Clarke

Department of Pathology, Johns Hopkins School of Medicine, Baltimore, MD

David S. Hage

Department of Chemistry, University of Nebraska, Lincoln, NE

CONTENTS

17.1   INTRODUCTION

There are few areas that compare with clinical chemistry in the variety of analytes it must consider. Substances that must be measured in this field range from small ions and organic compounds to proteins, viruses, and cells. There is also a wide range of samples that must be examined, such as urine, blood, cerebrospinal fluid, and tissue samples.

Affinity methods are valuable in dealing with these complex mixtures, since they can be used to selectively isolate or separate analytes in the presence of many other substances. A variety of ligands have been used for this purpose, such as antibodies, boronates, bacterial proteins, lectins, and chiral agents. In addition, numerous approaches have been performed with these ligands in the analysis of clinical samples. A few examples are affinity extraction, high-performance affinity chromatography (HPAC), and chromatographic immunoassays. These methods can also be used in combination with other analytical techniques, such as reversed-phase liquid chromatography (RPLC), gas chromatography (GC), capillary electrophoresis (CE), and mass spectrometry.

This chapter considers how affinity methods have been used for various classes of clinical analytes. The discussion covers the fields of enzymology, immunology, endocrinology, and toxicology. Others areas discussed include the use of affinity chromatography for monitoring therapeutic drugs, separating chiral drugs, and isolating lipids. The analysis of proteins, tumor markers, and proteomic libraries is considered as well. In each case, examples of applications are provided, and the advantages of using affinity ligands in these analyses are discussed.

17.2 ENZYMOLOGY

Enzymology was the earliest field in which affinity chromatography was employed. Some enzymes are useful for clinical testing because they are found in a high abundance in only a certain organ or tissue. This allows these enzymes to be used as selective markers for damage to this portion of the body. For instance, when disease or some other traumatic event occurs at the tissue or organ, some of these enzymes will leak out of cells and enter the blood. Thus, an elevated level of these enzymes in blood can be used as an indicator of this damage.

In the body, there are often several forms of an enzyme that catalyze the same reaction but that have slightly different amino acid sequences. These different forms are called isoenzymes or isozymes. It is common for a certain isoenzyme to be found in one organ, while a different isoenzyme may be found in a different portion of the body. For example, there are three forms of the enzyme creatine kinase (CK) in humans: CK-MM, CK-MB, and CK-BB. Of these, CK-MB is found mainly in heart muscle, while CK-MM is found in skeletal muscle. This means that the presence of an increase in CK-MB in blood can be used to determine if a patient has had a heart attack.

Another enzyme with a number of forms is alkaline phosphatase (ALP). ALP has several distinct isoenzymes that are found in the liver, bone, placenta, and intestinal cells [1]. One way liver and bone ALP isoenzymes can be separated and examined in serum is by using lectin affinity chromatography (see Figure 17.1). In this case, an HPAC column containing immobilized wheat germ agglutinin was used to distinguish between these two isoenzymes [2]. This approach was relatively rapid and gave good correlation when compared with a solid-phase immunoassay for ALP [2, 3].

Another enzyme with isoforms that can be monitored to determine the presence of disease is glutathione S-transferase. The level of this enzyme is elevated by many diseases that affect the liver, biliary tract, and pancreas [1]. Glutathione S-transferase isoenzymes in liver and human lung samples have been separated and analyzed using (S)-octylglutathione as a ligand in an affinity column [4, 5].

Other examples of enzymes that have been separated and analyzed by affinity chromatography are human plasminogen, angiotensinase A, and aminopeptidase M. Human plasminogen species were separated using immobilized p-aminobenzamidine and detected on-line with an immobilized urokinase column [6]. The purification and analysis of angiotensinase

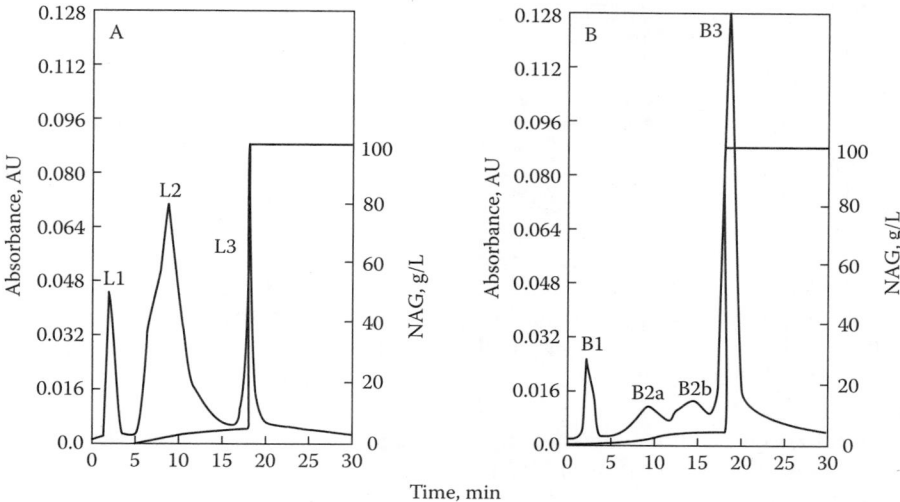

Figure 17.1 Separation of alkaline phosphatase (ALP) isoenzymes by lectin affinity chromatography. These results are for control samples containing (a) human liver ALP or (b) human bone ALP. The labeled peaks (L1–L3 and B1–B3) refer to the major enzyme fractions observed in these samples. The axis on the right gives the solvent conditions used during each step of this separation. (Reproduced with permission from Anderson, D.J., Branum, E.L., and O'Brien, J.F., *Clin. Chem.*, 36, 240–246, 1990.)

A and aminopeptidase M in urine and kidney samples have been achieved using a wheat germ agglutinin extraction column followed by HPLC [7].

Acetylcholinesterase (AChE) has been analyzed in amniotic fluid by combining size-exclusion chromatography with postcolumn immunodetection. In this system, an immuno-affinity column containing anti-AChE antibodies was used to capture AChE as it eluted from the size-exclusion column. A substrate for AChE was then passed through the immunoaffinity column, with the colored product being detected by an on-line absorbance detector [8].

Urokinases have been separated from human urine and detected using affinity-chromatography columns that contained 6-aminocaproic acid [9]. DNase II has been purified from human liver and urine by using murine antibodies against this enzyme, with the antibodies being conjugated to a formyl-cellulofine resin [10]. In addition, the interactions of enzymes with substrates and inhibitors have been investigated in numerous studies using the method of quantitative affinity chromatography [11, 12]. Further information on this last technique can be found in Chapters 22 and 23.

17.3 IMMUNOLOGY

Immunology is the second area of clinical chemistry in which affinity chromatography has been used. Much of this work has involved the use or detection of antibodies. Antibodies, or immunoglobulins, are soluble proteins that are part of the vertebrate immune system. Antibodies have a high selectivity and affinity for target substances known as antigens. The specificity of these interactions makes antibodies attractive for use as reagents in many analytical methods. Further information on the properties and production of antibodies can be found in Chapter 6 and in the literature [13].

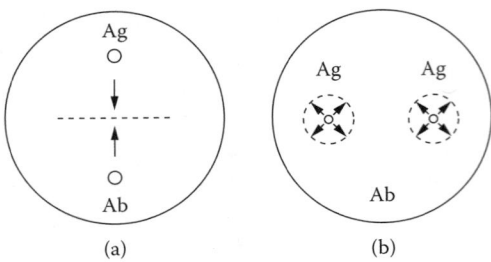

Figure 17.2 Basic approaches used in (a) the Ouchterlony assay and (b) radial immunodiffusion. The symbol Ab represents the antibodies and Ag is the antigen. The dashed lines represent the precipitation that forms as these two agents meet within the gel at appropriate concentrations for binding and cross-linking to occur.

Clinical assays can use antibody-antigen interactions for either qualitative or quantitative detection. For instance, in an Ouchterlony assay, solutions of an antibody and antigen are placed in wells within a gel and allowed to diffuse, as shown in Figure 17.2(a). As these solutions meet in the gel, the antibody and antigen will bind and, if present in the correct ratio, form a large cross-linked complex and visible precipitate [14]. This precipitate provides a signal that both an analyte and corresponding antibody were in the system. Although this approach is useful for qualitative work, it does have several limitations. For instance, it only works for antigens large enough to simultaneously bind more than one antibody. This method also requires the use of a polyclonal antibody or a mixture of monoclonal antibodies to allow cross-linking between antigen molecules.

A modified version of this approach is used in radial immunodiffusion. As illustrated in Figure 17.2(b), this again involves the diffusion of an antigen through a gel. However, this gel now contains a constant level of antibodies that are entrapped in this matrix. As the antigen diffuses from its sample well, it becomes diluted and interacts with the antibodies. When the antigen has reached an appropriate concentration, it will cross-link with the antibodies to form a visible precipitate that appears as a ring about the sample well. As a more concentrated antigen solution is analyzed by this approach, the size of this ring increases, with the area being proportional to the antigen's initial concentration. This allows quantitative information to be obtained on the antigen. However, this substance must again be large enough to bind more than one antibody simultaneously and must be present at a sufficiently high concentration to allow a visible precipitate to be observed.

A more sensitive means for detection can be accomplished by using an appropriate tag on the antibody or on an analog of the desired analyte. This gives rise to methods such as the competitive-binding immunoassay and sandwich immunoassay, as described in Chapter 29. The tags employed in these assays can be fluorescent molecules, radioisotopes, or enzymes, among others [14]. These methods are often performed through the use of a solid support that contains an immobilized antibody or antigen. This support can vary from a microtiter plate to a chromatographic column. Recent clinical applications that have used antibodies immobilized on chromatographic supports include methods for the isolation of MHC class II antigens from homogenized human intestine samples [15] and the isolation of immunoglobulins from diluted egg yolk [16]. Other examples can be found in Chapters 6 and 29.

Immobilized antibodies in HPLC columns have been used for the direct detection of numerous clinical analytes. These include immunoglobulin G (IgG) [17], immunoglobulin E (IgE) [18], anti-idiotypic antibodies [19, 20], human serum albumin (HSA) [21], granulocyte colony-stimulating factor [22], interferon [23, 24], tumor necrosis factor-α [24], interleukins

[24, 25], and β_2-microglobulin [26]. In this approach, an antibody or antibody fragment is used as an immobilized ligand in the column. This column is prepared by immobilizing or adsorbing these ligands to a support such as silica that can withstand the pressures and flow rates used in HPLC. Samples are applied to this column under physiological conditions, with the retained analytes later being desorbed by changing the pH or mobile-phase composition. The eluting analyte can be detected by various means, such as absorbance or fluorescence detection. Some advantages of this approach are its speed, precision, and ability to reuse the immobilized antibodies for several hundred assays [27]. The main disadvantage is the need for an analyte that has a sufficiently high concentration for detection. More details on the use of antibody columns for direct analyte detection can be found in Chapter 29.

Antibodies in clinical samples have often been measured by using protein A and protein G as ligands. For instance, human IgG-class antibodies have been analyzed in serum with HPLC columns containing protein A, protein G, or a mixture of these two proteins [28–30]. In one report, IgG and human serum albumin were measured simultaneously in serum by using two affinity columns, one containing protein A and the other containing anti-HSA antibodies (see Figure 17.3). These columns were connected in series during sample application and later eluted sequentially with a low-pH buffer. With this approach, it was possible to determine both compounds in only a few minutes, giving an albumin/IgG ratio that could be used as a general indicator of a patient's health [21].

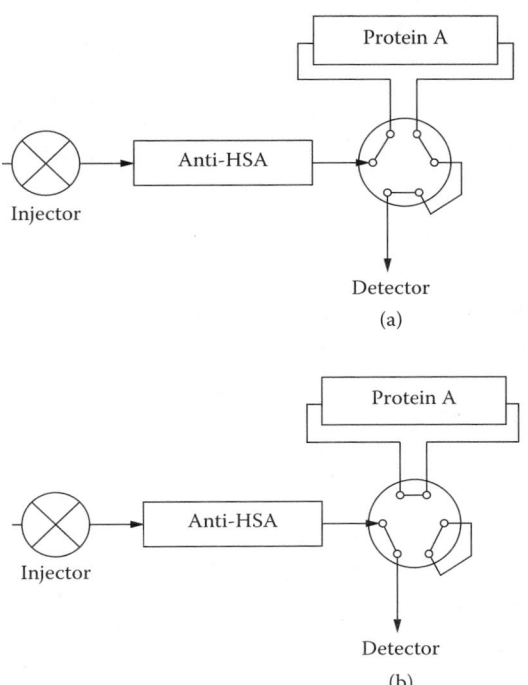

Figure 17.3 System used for the simultaneous determination of HSA and IgG in human serum by affinity columns containing anti-HSA antibodies and protein A. The diagram in (a) shows the configuration during sample injection, while (b) shows the configuration during the elution of analytes from the anti-HSA column. To elute substances from the protein A column, the system is set back to the mode shown in (a) while the elution buffer is passed through both columns. (From Hage, D.S. and Walters, R.R., *J. Chromatogr.*, 386, 37–49, 1987. With permission.)

Indirect detection has also been utilized with affinity columns in the measurement of clinical analytes. In the area of immunology, IgG has been examined using a chromatographic sandwich immunoassay [31] and purified using biotinylated chicken antibodies in conjunction with an immobilized avidin column [32]. IgE has been extracted from serum and detected by CE through the use of immobilized antibodies in microcapillary bundles or in laser-drilled glass rods [33]. In addition, chromatographic-based competitive-binding immunoassays have been used to determine IgG [34, 35], with similar methods being used for HSA [36, 37] and theophylline [38]. Further details on these and other formats for chromatographic immunoassays can be found in Chapter 29.

17.4   ENDOCRINOLOGY

A third area of clinical chemistry in which affinity ligands have been used is endocrinology. Endocrinology is the study of hormones and their effects on the body. Hormones are secreted by the endocrine glands, diffuse into the circulation, and are carried to their target tissue by the bloodstream. Upon reaching their target tissue, hormones can have a variety of effects, including the regulation of extracellular fluid volume, metabolism, smooth muscle and glandular contraction, immune-system functions, and blood clotting. Examples of hormones include thyroid hormone, adrenocortical steroids, gonadal steroids, and insulin. Severe consequences can result from the over- or underproduction of hormones. For instance, the lack of insulin production results in diabetes and leads to increased blood glucose concentrations. This, in turn, can lead to organ damage, mental impairment, and peripheral vascular damage.

The most popular use of affinity chromatography in the clinical laboratory is the determination of glycated hemoglobin for the assessment of long-term diabetes management [39]. The method is based on the use of boronate as a ligand in an affinity column [39–47]. Since glucose and many other sugars can interact with boronates, this agent can separate sugar-containing proteins like glycohemoglobin from normal hemoglobin [39, 41]. An example of such an analysis is given in Figure 17.4.

Boronate columns can also be used for other purposes in clinical testing. For instance, these have been employed in determining the relative amount of all glycated proteins in a sample [46]. Alternatively, a specific glycoprotein can be examined by combining a boronate column with a detection method selective for the protein of interest. Examples

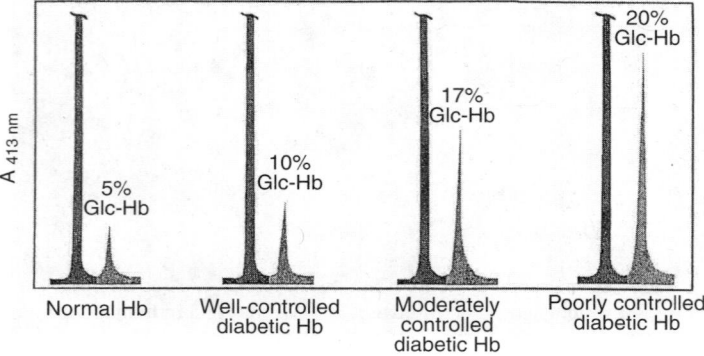

Figure 17.4   Determination of glycated hemoglobin (Glc-Hb) in blood by high-performance affinity chromatography. (Reproduced with permission from Singhal, R.P. and DeSilva, S.S.M., *Adv. Chromatogr.*, 31, 293–335, 1992.)

include the use of boronate columns followed by an immunoassay for the detection of glycated albumin in serum and urine [48] or for the determination of glycated apolipo-protein B in serum [49]. Another example is the determination of glycated albumin in serum by an HPLC boronate column preceded by an anion-exchange column [50].

Another group of compounds that have been analyzed with boronate columns are the catechols. Epinephrine, norepinephrine, dopamine [51–53], dihydroxyphenylalanine [54], dihydroxyphenylacetic acid [54, 55], 5-(S)-cysteinyldopa [56], and vanilmandelic acid [57] have all been examined by this approach. One example is given in Figure 17.5, where a boronate HPLC column was used in combination with ion-pair reversed-phase liquid chromatography to isolate and measure epinephrine, norepinephrine, and dopamine in urine.

The carbohydrate structures of follicle-stimulating hormone (FSH) and luteinizing hormone (LH) have been characterized under various clinical conditions using columns containing concanavalin A [58]. In another report, insulin was analyzed by CE after it

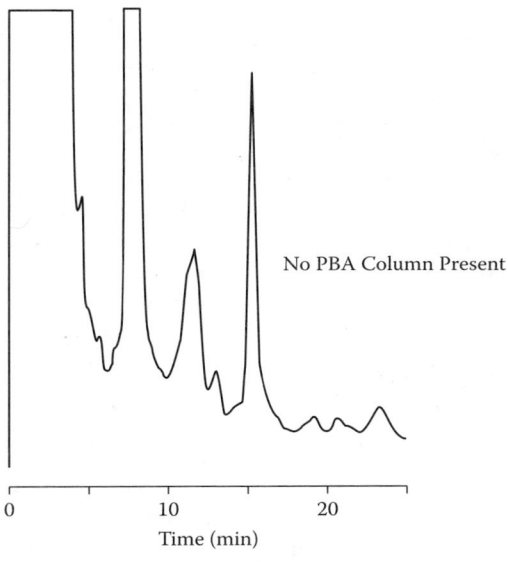

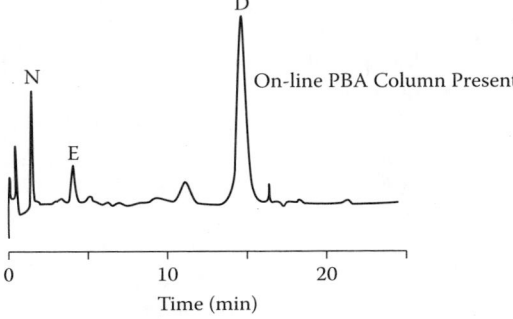

Figure 17.5 Separation of norepinephrine (N), epinephrine (E), and dopamine (D) by extraction with a phenylboronic acid (PBA) column on-line with ion-pair reversed-phase liquid chromatography. The top chromatogram shows the results obtained for a urine sample without the use of the PBA column, and the bottom shows the results for the same sample with extraction by the PBA column. (From Thomas, D.H. and Hage, D.S., unpublished results.)

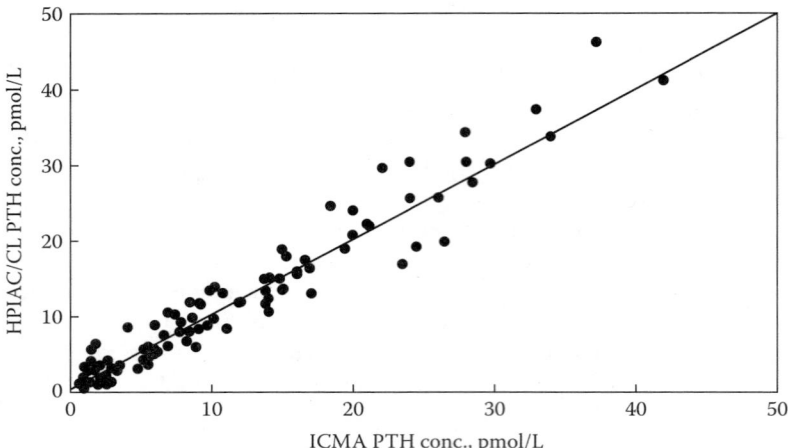

Figure 17.6 Correlation between results obtained for parathyrin (PTH) in plasma by high-performance immunoaffinity chromatography with chemiluminescence detection (HPIAC/CL) and a standard immunochemiluminometric assay (ICMA). The results shown in this graph are for 130 patients with various calcium-related disorders. (Reproduced with permission from Hage, D.S., Taylor, B., and Kao, P.C., *Clin. Chem.*, 38, 1494–1500, 1992.)

had been extracted from serum using a capillary packed with antibodies adsorbed onto protein G [59].

Chromatographic immunoassays have been used to measure a number of hormones in clinical samples [60–70]. For instance, thyroid-stimulating hormone [60], parathyrin [61, 62], and growth-hormone releasing factor [63] have been measured using chromatographic sandwich immunoassays (see Figure 17.6). Testosterone has been examined by a simultaneous-injection competitive-binding immunoassay, and thyroxine has been determined using a one-site immunometric assay [60]. Human chorionic gonadotropin (hCG) has been monitored in urine by combining immunoextraction with a reversed-phase liquid chromatography column [65]. hCG has also been determined in serum by a chromatographic sandwich immunoassay [66]. In addition, cortisol [67], estrogens [68, 69], and human epidermal growth factor [70] have been detected using on-line immunoextraction with reversed-phase liquid chromatography.

Many studies have reported the use of affinity columns for the off-line isolation and purification of clinical substances prior to their quantitation by a second method. In one study, melatonin was isolated from serum by low-performance columns containing polyclonal antibodies against melatonin, with the recovered analyte then being examined by HPLC and detected by electrospray ionization mass spectrometry [71]. Similarly, off-line immunoextraction has been used with reversed-phase liquid chromatography and gas chromatography to determine estrogens [72, 73], nortestosterone [74], and trenbolone [75] in urine and bile.

It is common in affinity extraction to employ immobilized antibodies that can cross-react with the desired target and related agents. For example, in one study, diethylstilbestrol, dienestrol, and hexestrol were extracted by the same antibody column for analysis by gas chromatography/mass spectrometry (GC-MS) [72]. This can be taken one step further by placing multiple ligands in the same column. This was demonstrated in a method described for nortestosterone, methyl testosterone, trenbolone, zeranol, estradiol, diethylstilbestrol, and related compounds in urine, in which samples were extracted off-line with an affinity column that contained seven types of antibodies [76].

17.5 TOXICOLOGY AND THERAPEUTIC DRUG MONITORING

The areas that have seen the greatest use of affinity columns in clinical testing are toxicology and therapeutic drug monitoring. In therapeutic drug monitoring, the amount of a drug in the circulation is examined to help control and adjust the dosing of this agent for treatment [77]. In toxicology, the goal is to identify and measure an unknown substance taken by a patient or to study the effects this agent has on the body.

There are many examples of affinity methods that have been developed for the determination of pharmaceutical drugs and their metabolites. Chloramphenicol [78], dexamethasone [79], flumethasone [80], ivermectin, avermectin [81], estrogens [72, 73], nortestosterone [74], and trenbolone [75] have all been detected using off-line immunoextraction for urine, serum, or tissue samples before the detection of these drugs with RPLC or GC. The determination of albuterol in plasma [82]; ochratoxin A in serum, plasma, or milk [83]; and antibiotic drug residues [84] has similarly been performed using immunoextraction with RPLC.

Some drugs have been measured by chromatographic immunoassays. For instance, theophylline has been quantitated in serum using fluorescein-impregnated liposomes as labels. This approach has been used in a liposome-based flow-injection immunoassay [38] and other flow-based methods [85–88].

Chiral ligands have become powerful tools for separating drugs and drug metabolites that contain more than one stereoisomer. Examples from clinical labs include the use of cyclodextrin columns for the analysis of chlorpheniramine [89], citalopram, desmethylcitalopram, didesmethylcitalopram [90], hexobarbital [89], and moguisteine metabolites [91]. For instance, propranolol is commonly used in the management of patients with cardiovascular disease. However, this drug has two enantiomers with different pharmacological properties. β-Cyclodextrin has been used in HPLC columns with fluorescence detection to separate and examine these enantiomers in both plasma and urine (see Figure 17.7) [92].

Immobilized proteins can also be used to separate chiral drugs. In one study, the enantiomers of pentazocine in serum were separated by using ovomucoid as a binding agent [93].

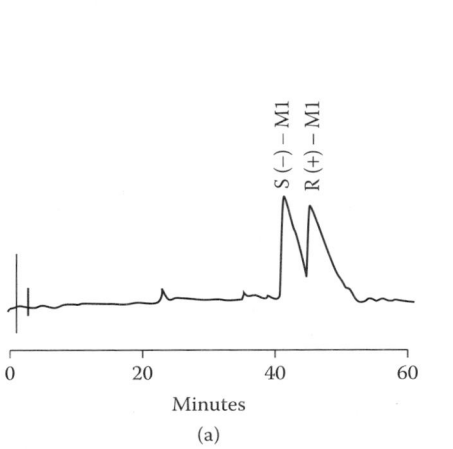
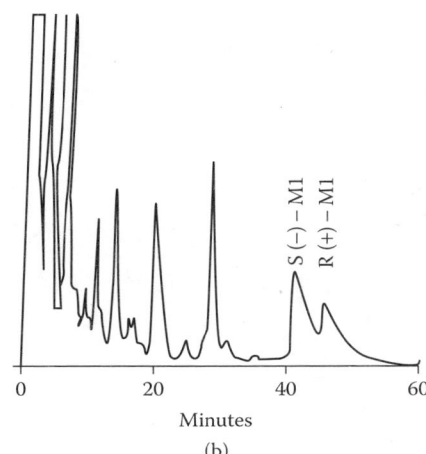

Figure 17.7 Separation of the $R(+)$- and $S(-)$-enantiomers of the M1 metabolite for moguisteine in (a) a standard solution of this metabolite and (b) urine from a patient treated with moguisteine. (Reproduced with permission from Castoldi, D., Oggioni, A., Renoldi, M.I., Ratti, E., DiGiovine, S., and Bernareggi, A., *J. Chromatogr. B,* 655, 243–252, 1994.)

Other chiral drugs in urine, serum, or plasma have been separated through the use of α_1-acid glycoprotein (AGP, or orosomucoid) as an immobilized ligand. These drugs have included bunolol [94], citalopram [95], fenoprofen [96], flurbiprofen [97], ibuprofen [96, 98], ketamine [99], ketoprofen [96], methadone [100–102], norketamine [99], norverapamil [103], pindolol [104], vamicamide [105], verapamil [103, 106], and thiopentone [107].

Immobilized antibody columns have frequently been utilized for off-line extraction in drug residue analysis [76, 88, 108–110]. As discussed in Chapter 6, one advantage of using an off-line antibody-extraction column is that substances collected from this column can be derivatized or placed into a different solvent before quantitation. This is particularly important when combining affinity extraction with GC, where solute derivatization is often required to improve volatility or detection. This also helps remove water from a sample before the analytes are injected onto a GC system. This has been demonstrated in the analysis of anabolic steroids [111] and Δ^9-tetrahydrocannabinol metabolites [112] by GC-MS.

Clinical applications of on-line immunoextraction and HPLC in drug analysis include methods reported for digoxin [113], lysergic acid diethylamide (LSD), LSD metabolites [114, 115], phenytoin [72], propranolol [114], and Δ^9-tetrahydrocannabinol [116]. All these examples used immunoaffinity columns combined with RPLC. One variation on this approach has been the coupling of on-line immunoextraction with CE. This was demonstrated in the analysis of cyclosporin A and its metabolites in tear samples from corneal transplant patients (see Figure 17.8) [117].

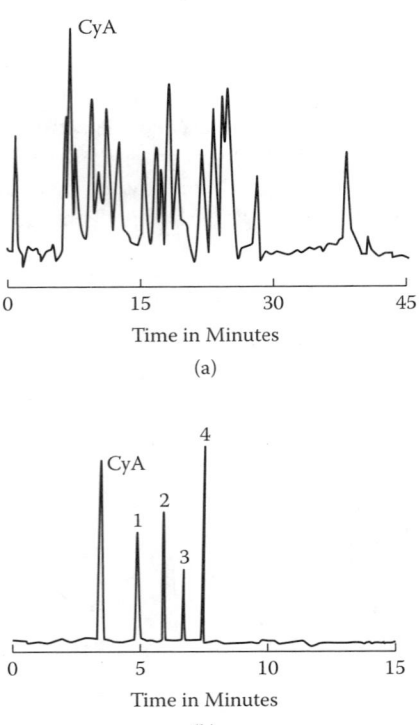

Figure 17.8 Analysis of cyclosporin in tear samples by (a) normal CE and (b) immunoextraction coupled on-line with CE. CyA represents cyclosporin A, and peaks 1 through 4 represent its metabolites. (Reproduced with permission from Phillips, T.M. and Chiemlinska, J.J., *Biomed. Chromatogr.*, 8, 242–246, 1994.)

Ligands besides antibodies can be employed for affinity extraction in toxicology or drug analysis. For instance, this has been used to measure 2-thioxothiazolidine-4-carboxylic acid in urine as an index of environmental exposure to carbon disulfide. This technique was performed by using an organomercurial column for off-line extraction, followed by analysis with RPLC [118].

17.6 LIPIDS AND LIPOPROTEINS

Lipids and lipoproteins are important in clinical testing due to their role in heart disease and atherosclerosis. Lipids include phospholipids, triglycerides, and sterols, which serve as components in cell membranes, the coating of nerve fibers, and as surfactants in the lungs. Lipids also are used as an energy source for the body and as precursors for steroid synthesis. Lipoproteins consist of a mixture of proteins and lipids in various proportions. These transport lipids in blood and are classified based on their density as being low-, intermediate-, or high-density particles, along with an alternative form known as a chylomicron.

There are several studies that report the use of affinity columns for the isolation or analysis of lipids and lipoproteins (see Figure 17.9). Immunoaffinity chromatography with antiapolipoprotein A-II antibodies has been used to separate apolipoprotein A-I and apolipoprotein A-II from lipoprotein particles. In the same study, Apo B-containing lipoproteins were also separated using immobilized antibodies [119]. Furthermore, low-performance columns containing concanavalin A have separated apolipoprotein A- and apolipoprotein B-containing lipoproteins from human plasma [120].

The glycated and nonglycated forms of apolipoprotein B in serum have been separated and quantitated through the use of an *m*-aminophenylboronate column followed by an immunoassay [49]. In addition, a boronate column and latex immunoagglutination have been used to determine glycated apolipoprotein A-I [121]. In other studies, prostaglandins and thromboxanes have been analyzed by GC and GC-MS following sample cleanup by affinity extraction [122–125].

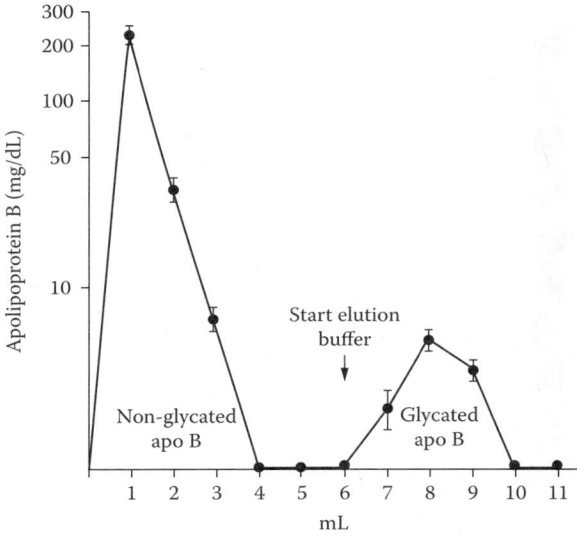

Figure 17.9 Elution of glycated and nonglycated apolipoprotein B (Apo B) on an aminophenyl-boronate column. (Reproduced with permission from Panteghini, M., Bonora, R., and Pagani, F., *Ann. Clin. Biochem.*, 31, 544–549, 1994.)

17.7 CANCER AND TUMOR MARKERS

The study and detection of cancer is another area that has made use of affinity chromatography. One such application has been the detection of tumor markers. A tumor marker is a substance produced by cells or tissues that indicates the presence of cancer. An ideal tumor marker should be specific for one type of cancer and never be present under normal conditions. Also, this marker should be released only in response to a tumor and allow for its early diagnosis and treatment. Most current tumor markers meet only some of these criteria. As a result, they are mainly used only after surgery to determine if a tumor has been completely removed or to follow the activity of a tumor during radiation treatment or chemotherapy.

Various biomolecules have been used as tumor markers, including enzymes, hormones, proteins, carbohydrates, oncofetal antigens, and modified DNA. As an example, exposure of the body to alkylating agents can result in covalently modified bases in DNA that, in turn, may give rise to a mutation and cause cancer. Monitoring these products in urine is one way to detect exposure to these alkylating agents. In one study, this was performed by using immobilized antibodies against alkylated adenine, which allowed the extraction of 3-alkyladenines and their quantitation by GC-MS [126, 127].

Protein-based tumor markers have also been studied using affinity chromatography. For instance, serial lectin columns containing concanavalin A and wheat germ agglutinin have been used to characterize prostatic acid phosphatase (PAP) with respect to its carbohydrate contents (see Figure 17.10) [128]. In another study, immunoaffinity chromatography was used to purify prostate-specific membrane antigen from prostate cancer cells [129]. A more detailed discussion on the use of affinity chromatography in the study of prostate cancer can be found in the literature [130].

Affinity chromatography has been used to monitor other types of cancer as well. Modified nucleosides in patients with gastrointestinal cancer have been analyzed with

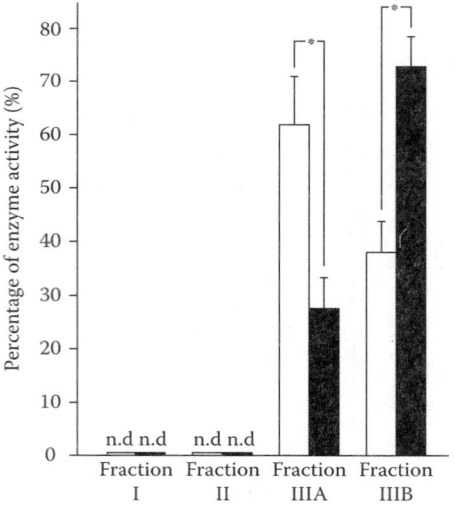

Figure 17.10   Relative amount of prostatic acid phosphatase detected in four fractions collected from wheat germ agglutinin and concanavalin A columns. The open bars are for a sample representing benign prostatic hyperplasia, and the closed bars are for a sample representing a prostate carcinoma. (Reproduced with permission from Yoshida, K.I., Honda, M., Arai, K., Hosoya, Y., Moriguchi, H., Sumi, S., Ueda, Y., and Kitahara, S., *J. Chromatogr. B*, 695, 439–443, 1997.)

boronate columns combined off-line with RPLC [131]. In addition, immobilized synthetic oligosaccharide tumor antigens have been used to assess the response of cancer patients to carbohydrate-based vaccines [132].

17.8 PROTEINS AND PROTEOMICS

Proteins are measured for a variety of reasons in clinical assays. For instance, they could be used to determine nutritional status, as is the case with albumin and transthyretin. They can also be used to explain clotting problems (e.g., fibrinogen), to test for iron deficiency (e.g., transferrin), or to detect cancer (e.g., prostate-specific antigen).

On-line immunoextraction with HPLC has been used to examine numerous proteins in biological fluids. This has been demonstrated in the analysis of α_1-antitrypsin [133], fibrinogen [134], HSA [21, 135], and transferrin [133, 136, 137]. The microheterogeneity of serum transferrin, which is a glycosylated protein, has been studied during alcoholic liver disease through the use of low-performance concanavalin A columns [138]. It has also been possible to separate and measure carbohydrate-deficient transferrin versus fully glycosylated transferrin by on-line immunoextraction coupled to electrospray-ionization mass spectrometry (see Figure 17.11) [139].

One specific application of lectin-affinity chromatography for proteins has been in the isolation of glycoprotein G from viral particles [140]. In this study, a rhabdovirus known as viral hemorrhagic septicemia virus was isolated and sonicated in polyethylene glycol (PEG) to liberate glycoprotein G. The samples were then passed over a concanavalin A column to separate glycoprotein G from the detergent-solubilized virion and virion-free PEG supernatants.

The determination of urinary albumin/creatinine ratios has been performed by combining high-performance immunoaffinity chromatography with flow-injection analysis [135]. In this method, an antialbumin antibody column was used for the capture and detection of HSA, while a Jaffe-based colorimetric reactor was used to quantitate creatinine in the nonretained portion of the sample (see Figure 17.12).

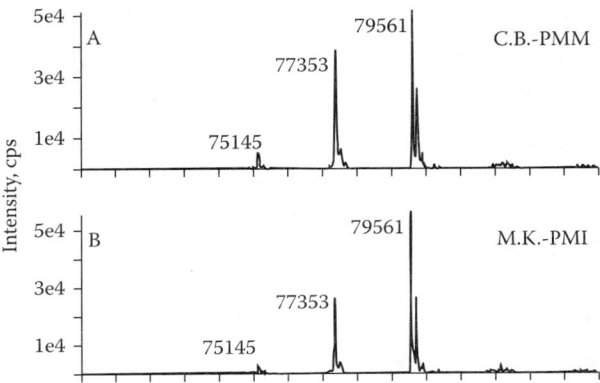

Figure 17.11 Mass spectra for transferrin isoforms in samples from patients with variant forms of (a) phosphomannomutase (PMM) or (b) phosphomannose isomerase (PMI), as determined by immunoaffinity chromatography coupled with electrospray-ionization mass spectrometry. The three peaks in each spectrum represent (from right to left) intact transferrin and transferrin that has lost one or two oligosaccharide groups. (Reproduced with permission from Lacey, J.M., Bergen, H.R., Magera, M.J., Naylor, S., and O'Brien, J.F., *Clin. Chem.*, 47, 513–518, 2001.)

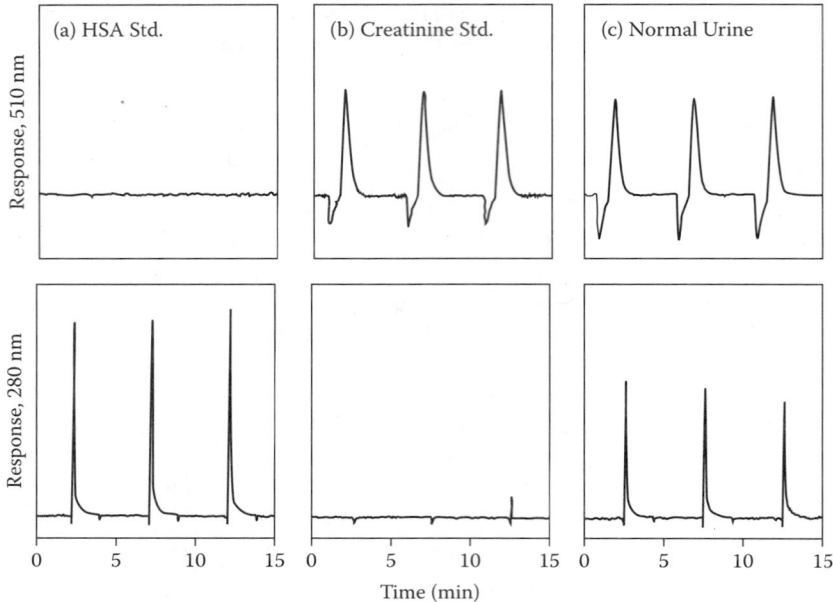

Figure 17.12 Typical responses obtained by high-performance immunoaffinity chromatography and flow-injection analysis (HPIAC/FIA) for the analysis of HSA and creatinine in urine. The results shown are for the injection of (a) HSA, (b) creatinine, and (c) normal urine. (Reproduced with permission from Ruhn, P.F., Taylor, J.D., and Hage, D.S., *Anal. Chem.*, 66, 4265–4271, 1994.)

Several groups have utilized chromatographic immunoassays for measuring proteins in biological fluids. Examples given earlier include direct-detection methods for acetyl-cholinesterase [8], IgG [17], IgE [18], HSA [21], and β_2-microglobulin [26]. Sandwich immunoassays have been used in chromatographic systems for the determination of IgG [31], thyroid-stimulating hormone [60], and parathyrin [61, 62]. Chromatographic competitive-binding assays have been described for HSA [36, 37].

Various blood-clotting proteins have been measured using affinity methods. For example, antithrombin III has been examined in plasma by employing an immobilized heparin column [141, 142]. In addition, human platelet glycocalicin has been purified with off-line boronate acid columns prior to analysis by anion-exchange HPLC [143].

The completion of the human genome project has led to even further clinical applications of affinity chromatography. This has mainly involved the use of this technique in proteomic studies [144]. The goal of such work is to characterize the proteins in a given cell or organelle under various conditions. This involves not only isolating the proteins but determining their structures, modifications, or locations in the system. Since many diseases are caused by the wrong structure, incorrect location, or abnormal concentration of a protein, a great deal of emphasis has been placed on developing faster methods for such work.

Two-dimensional gel electrophoresis is often used to separate proteins from cells or organelles. However, techniques based on HPLC can also be used. For instance, ion-exchange and affinity columns followed by RPLC can be coupled directly to a mass spectrometer for proteomic studies [145–147]. In these methods, the proteins are first digested with trypsin or some other proteolytic enzyme. The resulting peptides are then separated by an affinity or ion-exchange column, desalted, and further separated by RPLC, with the eluting peptides being passed onto a mass spectrometer for identification.

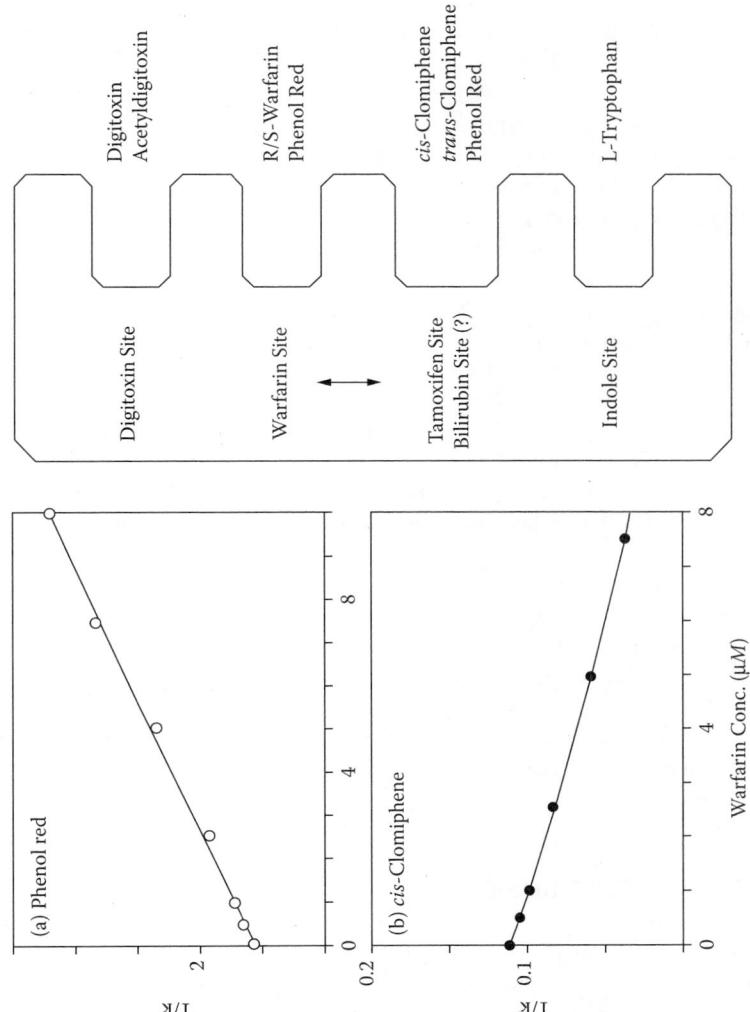

Figure 17.13 Competition between injected samples of (a) phenol red or (b) cis-clomiphene with warfarin in the mobile phase on an immobilized HSA column, and the use of these results to generate a map for the binding sites of HSA (right). (Reproduced with permission from Sengupta, A. and Hage, D.S., *Anal. Chem.*, 71, 3821–3827, 1999.)

One recent modification of this approach has involved the use of an on-line immobilized trypsin column with a mass spectrometer. This has decreased the total analysis time from more than 8 h to 2 h [148, 149]. Affinity columns that are specific for a given type of peptide are useful for such work. For instance, Ga(III) and Fe(III) chelate columns have both been employed for the isolation of phosphorylated peptides in these studies [148, 150]. More information on these and related applications can be found in Chapters 18 and 27.

Phosphorylated peptides are examples of signature peptides. These are peptides that are unique to a certain protein and can be used in its identification. Affinity chromatography is particularly useful in isolating such species. Besides the use of Ga(III) and Fe(III) chelate columns for phosphorylated peptides [148, 150], other applications of affinity ligands in this area include the isolation of glycosylated peptides by lectin [151–153] and boronate columns [143].

As discussed in Chapters 22 through 25, another application for affinity chromatography is as a tool to study solute-protein interactions. Reviews of this topic can be found in the literature [12, 154]. The study of solute-protein interactions involves the use of techniques like frontal analysis, zonal elution, and related methods, as described in Chapters 22 and 23. An example is given in Figure 17.13, where various probe compounds were used to characterize the drug-binding sites of HSA [155]. Another example is a frontal-analysis method employed for drug screening that used an affinity column containing a biotinylated enzyme adsorbed to immobilized streptavidin [156].

Affinity columns can also be used to capture a protein or peptide labeled with a particular tag [157–160]. This has been employed in quantitative proteomic studies by using an isotope-coded affinity tag (ICAT). This type of tag is described in detail in Chapters 18 and 27. It includes a biotin label for binding to streptavidin and a sulfhydryl reactive group for labeling free cysteine residues. This tag comes in two forms: a light form containing hydrogens and a heavy form that is labeled with deuterium. These two forms are used so that one can act as the internal standard for the other in a sample.

Proteins obtained under two different conditions (e.g., from normal cells or those representing a given disease state) are tagged with either the heavy or light form of the ICAT reagent. These proteins are then combined, digested with a protease like trypsin, and isolated by a streptavidin column. The tagged peptides obtained by this column are then eluted, separated by RPLC, and analyzed by mass spectrometry. The heavy and light peptide peaks in the mass spectrum are compared to identify proteins that have been expressed at different rates in the two original cell populations [157–159]. A similar approach can be used to label and analyze phosphoproteins [160].

17.9 SUMMARY AND CONCLUSIONS

It can be seen from the examples given in this chapter that affinity methods are valuable and effective tools for the isolation, purification, and detection of various clinical agents. Techniques used in this field include low- and high-performance affinity chromatography, chromatographic immunoassays, quantitative affinity chromatography, and affinity extraction. These techniques can be used either alone or in combination with other analytical methods. The selectivity of these tools has made them useful in a number of areas in clinical chemistry. Examples considered in this chapter included enzymology, immunology, endocrinology, toxicology, and therapeutic drug monitoring. Clinical applications in the areas of chiral separations, lipids or lipoprotein analysis, and the detection of tumor makers were considered as well. Finally, the use of affinity chromatography with proteins was discussed, along with its role in the growing field of proteomics.

SYMBOLS AND ABBREVIATIONS

Ab	Antibody
AChE	Acetylcholinesterase
Ag	Antigen
AGP	α_1-Acid glycoprotein
ALP	Alkaline phosphatase
Apo B	Apolipoprotein B
CE	Capillary electrophoresis
CK	Creatine kinase
CyA	Cyclosporin A
DNA	Deoxyribonucleic acid
FSH	Follicle-stimulating hormone
GC	Gas chromatography
GC-MS	Gas chromatography/mass spectrometry
Glc-Hb	Glycated hemoglobin
Hb	Hemoglobin
hCG	Human chorionic gonadotropin
HPAC	High-performance affinity chromatography
HPIAC/CL	High-performance immunoaffinity chromatography with chemiluminescence detection
HPIAC/FIA	High-performance immunoaffinity chromatography and flow-injection analysis
HPLC	High-performance liquid chromatography
HSA	Human serum albumin
ICAT	Isotope-coded affinity tag
ICMA	Immunochemiluminometric assay
IgE	Immunoglobulin E
IgG	Immunoglobulin G
LH	Luteinizing hormone
LSD	Lysergic acid diethylamide
MHC	Major histocompatibility complex
PAP	Prostatic acid phosphatase
PBA	Phenylboronic acid
PEG	Polyethylene glycol
PMI	Phosphomannose isomerase
PMM	Phosphomannomutase
PTH	Parathyrin
RPLC	Reversed-phase liquid chromatography

REFERENCES

1. Goldberg, D.M., Diagnostic enzymology, in *Applied Biochemistry of Clinical Disorders,* 2nd ed., Gornall, A.G., Ed., Lippincott, New York, 1986, pp. 37, 47.
2. Anderson, D.J., Branum, E.L., and O'Brien, J.F., Liver- and bone-derived isoenzymes of alkaline phosphatase in serum as determined by high-performance affinity chromatography, *Clin. Chem.,* 36, 240–246, 1990.
3. Gonchoroff, D.G., Branum, E.L., Cedel, S.L., Riggs, B.L., and O'Brien, J.F., Clinical evaluation of high-performance affinity chromatography for the separation of bone and liver alkaline phosphatase isoenzymes, *Clin. Chim. Acta,* 199, 43–50, 1991.

4. Wheatley, J.B., Kelley, M.K., Montali, J.A., Berry, C.O.A., and Schmidt, D.E., Jr., Examination of glutathione *S*-transferase isoenzyme profiles in human liver using high-performance affinity chromatography, *J. Chromatogr. A,* 663, 53–63, 1994.

5. Wheatley, J.B., Montali, J.A., and Schmidt, D.E., Jr., Coupled affinity-reversed-phase high-performance liquid chromatography systems for the measurement of glutathione *S*-transferase in human tissues, *J. Chromatogr. A,* 676, 65–79, 1994.

6. Abe, I., Ito, N., Noguchi, K., Kazama, M., and Kasai, K., Immobilized urokinase column as part of a specific detection system for plasminogen species separated by high-performance affinity chromatography, *J. Chromatogr.,* 565, 183–195, 1991.

7. Scherberich, J.E., Wiemer, J., Herzig, C., Fischer, P., and Schoeppe, W., Isolation and partial characterization of angiotensinase A and aminopeptidase M from urine and human kidney by lectin affinity chromatography and high-performance liquid chromatography, *J. Chromatogr.,* 521, 279–289, 1990.

8. Vanderlaan, M., Lotti, R., Siek, G., King, D., and Goldstein, M., Perfusion immunoassay for acetylcholinesterase: Analyte detection based on intrinsic activity, *J. Chromatogr. A,* 711, 23–31, 1995.

9. Takahashi, R., Akiba, K., Koike, M., Noguchi, T., and Ezure, Y., Affinity chromatography for purification of two urokinases from human urine, *J. Chromatogr. B,* 742, 71–78, 2000.

10. Nakajima, T., Yasuda, T., Takeshita, H., Mori, S., Mogi, K., Kaneko, Y., Nakazato, E., and Kishi, K., Production and characterization of murine monoclonal anti-human DNase II antibodies, and their use for immunoaffinity purification of DNase II from human liver and urine, *Biochim. Biophys. Acta,* 1570, 160–164, 2002.

11. Hage, D.S., Affinity chromatography, in *Handbook of HPLC,* Katz, E., Eksteen, R., Shoenmakers, P., and Miller, N., Eds., Marcel Dekker, New York, 1998, Chap. 13.

12. Chaiken, I.M., Ed., *Analytical Affinity Chromatography,* CRC Press, Boca Raton, FL, 1987.

13. Berg, J.M., Tymoczko, J.L., and Stryer, L., *Biochemistry,* 5th ed., W.H. Freeman and Co., New York, 2001, pp. 99–101.

14. Dubiski, S., Diagnostic immunology, in *Applied Biochemistry of Clinical Disorders,* 2nd ed., Gornall, A.G., Ed., Lippincott, New York, 1986, pp. 63, 65, 68.

15. Oshitani, N., Hato, F., Jinno, Y., Sawa, Y., Hara, J., Nakamura, S., Matsumoto, T., Arakawa, T., Kitano, A., Kitagawa, S., and Kuroki, T., Heterogeneity of HPLC profiles of human class II MHC-bound peptides isolated from intestine with inflammatory bowel disease, *Diagnostic Disease Sci.,* 47, 2088–2094, 2002.

16. Chen, C.C., Tu, Y.Y., Chen, T.L., and Chang, H.M., Isolation and characterization of immunoglobulin in yolk (IgY) specific against hen egg white lysozyme by immunoaffinity chromatography, *J. Agric. Food Chem.,* 50, 5424–5428, 2002.

17. Phillips, T.M., More, N.S., Queen, W.D., Holohan, T.V., Kramer, N.C., and Thompson, A.M., High-performance affinity chromatography: A rapid technique for the isolation and quantitation of IgG from cerebral spinal fluid, *J. Chromatogr.,* 317, 173–179, 1984.

18. Phillips, T.M., More, N.S., Queen, W.D., and Thompson, A.M., Isolation and quantitation of serum IgE levels by high-performance immunoaffinity chromatography, *J. Chromatogr.,* 327, 205–211, 1985.

19. Phillips, T.M., High-performance immunoaffinity chromatographic detection of immuno-regulatory anti-idiotypic antibodies in cancer patients receiving immunotherapy, *Clin. Chem.,* 34, 1689–1692, 1988.

20. Phillips, T.M. and Babashak, J.V., Isolation of anti-idiotypic antibodies by immunoaffinity chromatography on Affinichrom beads, *J. Chromatogr.,* 512, 387–394, 1990.

21. Hage, D.S. and Walters, R.R., Dual-column determination of albumin and immunoglobulin G in serum by high-performance affinity chromatography, *J. Chromatogr.,* 386, 37–49, 1987.

22. Phillips, T.M., Immunoaffinity measurement of recombinant granulocyte colony stimulating factor in patients with chemotherapy-induced neutropenia, *J. Chromatogr. B,* 662, 307–313, 1994.

23. Phillips, T.M., Measurement of recombinant interferon levels by high performance immunoaffinity chromatography in body fluids of cancer patients on interferon therapy, *Biomed. Chromatogr.,* 6, 287–290, 1992.

24. Phillips, T.M. and Krum, J.M., Recycling immunoaffinity chromatography for multiple analyte analysis in biological samples, *J. Chromatogr. B*, 715, 55–63, 1998.

25. Phillips, T.M., Measurement of total and bioactive interleukin-2 in tissue samples by immunoaffinity-receptor affinity chromatography, *Biomed. Chromatogr.*, 11, 200–204, 1997.

26. Mogi, M., Harada, M., Adachi, T., Kojima, K., and Nagatsu, T., Selective removal of β_2-microglobulin from human plasma by high-performance immunoaffinity chromatography, *J. Chromatogr.*, 496, 194–200, 1989.

27. Hage, D.S. and Nelson, M.A., Chromatographic immunoassays, *Anal. Chem.*, 73, 199A–205A, 2001.

28. Ohlson, S., High performance liquid affinity chromatography (HPLAC) with protein A-silica, in *Affinity Chromatography and Biological Recognition*, Chaiken, I.M., Wilchek, M., and Parikh, I., Eds., Academic Press, New York, 1983, pp. 255–256.

29. Crowley, S.C. and Walters, R.R., Determination of immunoglobulins in blood serum by high-performance affinity chromatography, *J. Chromatogr.*, 266, 157, 1983.

30. Cassulis, P., Magasic, M.V., and DeBari, V.A., Ligand affinity chromatographic separation of serum IgG on recombinant protein G-silica, *Clin. Chem.*, 37, 882–886, 1991.

31. Hacker, A., Hinterleitner, M., Shellum, C., and Gubitz, G., Development of an automated flow injection chemiluminescence immunoassay for human immunoglobulin G, *Frenz. J. Anal. Chem.*, 352, 793–796, 1995.

32. Kim, H.O., Durance, T.D., and Li-Chan, E.C., Reusability of avidin-biotinylated immunoglobulin Y columns in immunoaffinity chromatography, *Anal. Biochem.*, 268, 383–397, 1999.

33. Guzman, N.A., Biomedical applications of on-line preconcentration-capillary electrophoresis using an analyte concentrator: Investigation of design options, *J. Liq. Chromatogr.*, 18, 3751–3768, 1995.

34. de Alwis, U. and Wilson, G.S., Rapid heterogeneous competitive electrochemical immunoassay for IgG in the picomole range, *Anal. Chem.*, 59, 2786–2789, 1987.

35. Valencia-Gonzalez, M.H. and Diaz-Garcia, M.E., Flow-through fluorescent immunosensing of IgG, *Ciencia*, 4, 29–40, 1996.

36. Cassidy, S.A., Janis, L.J., and Regnier, F.E., Kinetic chromatographic sequential addition immunoassays using protein A affinity chromatography, *Anal. Chem.*, 64, 1973–1977, 1992.

37. Hage, D.S., Thomas, D.H., and Beck, M.S., Theory of a sequential addition competitive binding immunoassay based on high-performance immunoaffinity chromatography, *Anal. Chem.*, 65, 1622–1630, 1993.

38. Locascio-Brown, L., Plant, A.L., Chesler, T., Kroll, M., Ruddel, M., and Durst, R.A., Liposome-based flow-injection immunoassay for determining theophylline in serum, *Clin. Chem.*, 39, 386–391, 1993.

39. Mallia, A.K., Hermanson, G.T., Krohn, R.I., Fujimoto, E.K., and Smith, P.K., Preparation and use of a boronic acid affinity support for the separation and quantitation of glycosylated hemoglobins, *Anal. Lett.*, 14, 649–661, 1981.

40. Mayer, T.K. and Freedman, Z.R., Protein glycosylation in diabetes mellitus: A review of laboratory measurements and of their clinical utility, *Clin. Chim. Acta*, 127, 147–184, 1983.

41. Hjerten, S. and Li, J.P., High-performance liquid chromatography of proteins on deformed non-porous agarose beads: Fast boronate affinity chromatography of haemoglobin at neutral pH, *J. Chromatogr.*, 500, 543–553, 1990.

42. Fluckiger, R., Woodtli, T., and Berger, W., Quantitation of glycosylated hemoglobin by boronate affinity chromatography, *Diabetes*, 33, 73–76, 1984.

43. Gould, B.J., Hall, P.M., and Cook, J.G.H., Measurement of glycosylated hemoglobins using an affinity chromatography method, *Clin. Chim. Acta*, 125, 41–48, 1982.

44. Klenk, D.C., Hermanson, G.T., Krohn, R.I., Fujimoto, E.K., Mallia, A.K., Smith, P.K., England, J.D., Wiedmeyer, H.M., Little, R.R., and Goldstein, D.E., Determination of glycosylated hemoglobin by affinity chromatography: Comparison with colorimetric and ion-exchange methods, and effects of common interferences, *Clin. Chem.*, 28, 2088–2094, 1982.

45. Johnson, R.N. and Baker, J.R., Inaccuracy in measuring glycated albumin concentration by thiobarbituric acid colorimetry and by boronate chromatography, *Clin. Chem.*, 34, 1456–1459, 1988.

46. Singhal, R.P. and DeSilva, S.S.M., Boronate affinity chromatography, *Adv. Chromatogr.,* 31, 293–335, 1992.

47. Kitagawa, N. and Treat-Clemens, L.G., Chromatographic study of immobilized boronate stationary phases, *Anal. Sci.,* 7, 195–198, 1991.

48. Silver, A.C., Lamb, E., Cattell, W.R., and Dawnay, A.B., Investigation and validation of the affinity chromatography method for measuring glycated albumin in serum and urine, *Clin. Chim. Acta,* 202, 11–22, 1991.

49. Panteghini, M., Bonora, R., and Pagani, F., Determination of glycated apolipoprotein B in serum by a combination of affinity chromatography and immunonephelometry, *Ann. Clin. Biochem.,* 31, 544–549, 1994.

50. Yasukawa, K., Abe, F., Shida, N., Koizumi, Y., Uchida, T., Noguchi, K., and Shima, K., High-performance affinity chromatography system for the rapid, efficient assay of glycated albumin, *J. Chromatogr.,* 597, 271–275, 1992.

51. Edlund, P.O. and Westerlund, D., Direct injection of plasma and urine in automated analysis of catecholamines by coupled-column liquid chromatography with post-column derivatization, *J. Pharm. Biomed. Anal.,* 2, 315–333, 1984.

52. Ni, P., Guyon, F., Caude, M., and Rosset, R., Automated determination of catecholamines using on-column extraction of diphenylboronate-catecholamine complexes and high-performance liquid chromatography with electrochemical detection, *J. Liq. Chromatogr.,* 12, 1873–1888, 1989.

53. Boos, K.S., Wilmers, B., Sauerbrey, R., and Schlimme, E., Development and performance of an automated HPLC-analyzer for catecholamines, *Chromatographia,* 24, 363–370, 1987.

54. Edlund, P.O., Determination of dihydroxyphenylalanine and dihydroxyphenylacetic acid in biological samples by coupled-column liquid chromatography with dual coulometric-amperometric detection, *J. Pharm. Biomed. Anal.,* 4, 625–639, 1986.

55. Hansson, L., Glad, M., and Hansson, C., Boronic acid-silica: A new tool for the purification of catecholic compounds on-line with reversed-phase high-performance liquid chromatography, *J. Chromatogr.,* 265, 37–44, 1983.

56. Hansson, C., Kagedal, B., and Kallberg, M., Determination of 5-S-cysteinyldopa in human urine by direct injection in coupled-column high-performance liquid chromatography, *J. Chromatogr.,* 420, 146–151, 1987.

57. Eriksson, B.-M. and Wikstrom, M., Determination of vanilmandelic acid in urine by coupled-column liquid chromatography combining affinity to boronate and separation by anion exchange, *J. Chromatogr.,* 567, 1–9, 1991.

58. Papandreou, M.J., Asteria, C., Pettersson, K., Ronin, C., and Beck-Peccoz, P., Concanavalin A affinity chromatography of human serum gonadotropins: Evidence for changes in carbohydrate structure in different clinical conditions, *J. Clin. Endocrin. Metab.,* 76, 1008–1013, 1993.

59. Cole, L.J. and Kennedy, R.T., Selective preconcentration for capillary zone electrophoresis using protein G immunoaffinity capillary chromatography, *Electrophoresis,* 16, 549–556, 1995.

60. Hage, D.S., Chromatographic approaches to immunoassays, *J. Clin. Ligand Assay,* 20, 293–301, 1998.

61. Hage, D.S. and Kao, P.C., High-performance immunoaffinity chromatography and chemiluminescent detection in the automation of a parathyroid hormone sandwich immunoassay, *Anal. Chem.,* 63, 586–595, 1991.

62. Hage, D.S., Taylor, B., and Kao, P.C., Intact parathyroid hormone: Performance and clinical utility of an automated assay based on high-performance immunoaffinity chromatography and chemiluminescence detection, *Clin. Chem.,* 38, 1494–1500, 1992.

63. Cho, B.Y., Zou, H., Strong, R., Fisher, D.H., Nappier, J., and Krull, I.S., Immunochromatographic analysis of bovine growth hormone releasing factor involving reversed-phase high-performance liquid chromatography-immunodetection, *J. Chromatogr. A,* 743, 181–194, 1996.

64. Durst, R.A., Locascio-Brown, L., and Plant, A.L., in *Flow Injection Analysis (FIA) Based on Enzymes or Antibodies,* Schmid, R.D., Ed., VCH, New York, 1991, pp. 181–190.

65. Liu, C.L. and Bowers, L.D., Immunoaffinity trapping of urinary human chorionic gonadotropin and its high-performance liquid chromatographic-mass spectrometric determination, *J. Chromatogr. B,* 687, 213–220, 1996.

66. Johns, M.A., Rosengarten, L.K., Jackson, M., and Regnier, F.E., Enzyme-linked immunosorbent assays in a chromatographic format, *J. Chromatogr. A,* 743, 195–206, 1996.

67. Nilsson, B., Extraction and quantitation of cortisol by use of high-performance liquid affinity chromatography, *J. Chromatogr.,* 276, 413–417, 1983.

68. Farjam, A., Brugman, A.E., Lingeman, H., and Brinkman, U.A.T., On-line immunoaffinity sample pre-treatment for column liquid chromatography: Evaluation of desorption techniques and operating conditions using an anti-estrogen immuno-precolumn as a model system, *Analyst,* 116, 891–896, 1991.

69. Farjam, A., Brugman, A.E., Soldaat, A., Timmerman, P., Lingeman, H., de Jong, G.J., Frei, R.W., and Brinkman, U.A.T., Immunoaffinity precolumn for selective sample pretreatment in column liquid chromatography: Immunoselective desorption, *Chromatographia,* 31, 469–477, 1991.

70. Hayashi, T., Sakamoto, S., Wada, I., and Yoshida, H., HPLC analysis of human epidermal growth factor using immunoaffinity precolumn, II: Determination of hEGFs in biological fluids, *Chromatographia,* 27, 574–580, 1989.

71. Rolcik, J., Lenobel, R., Siglerova, V., and Strnad, M., Isolation of melatonin by immunoaffinity chromatography, *J. Chromatogr. B,* 775, 9–15, 2002.

72. Bagnati, R., Castelli, M.G., Airoldi, L., Oriundi, M.P., Ubaldi, A., and Fanelli, R., Analysis of diethylstilbestrol, dienestrol and hexestrol in biological samples by immunoaffinity extraction and gas-chromatography negative-ion chemical ionization mass spectrometry, *J. Chromatogr.,* 527, 267–278, 1990.

73. Bagnati, R., Oriundi, M.P., Russo, V., Danese, M., Berti, F., and Fanelli, R., Determination of zeranol and β-zeranol in calf urine by immunoaffinity extraction and gas chromatography-mass spectrometry after repeated administration of zeranol, *J. Chromatogr.,* 564, 493–502, 1991.

74. van Ginkel, L.A., Stephany, R.W., van Rossum, H.J., van Blitterswijk, H., Zoontjes, P.W., Hooijshuur, R.C.M., and Zuydendorp, J., Effective monitoring of residues of nortestosterone and its major metabolite in bovine urine and bile, *J. Chromatogr.,* 489, 95–104, 1989.

75. van Ginkel, L.A., van Blitterswijk, H., Zoontjes, P.W., van den Bosch, D., and Stephany, R.W., Assay for trenbolone and its metabolite 17α-trenbolone in bovine urine based on immunoaffinity chromatographic clean-up and off-line high-performance liquid chromatography-thin-layer chromatography, *J. Chromatogr.,* 445, 385–392, 1988.

76. van Ginkel, L.A., Immunoaffinity chromatography, its applicability and limitations in multiresidue analysis of anabolizing and doping agents, *J. Chromatogr.,* 564, 363–384, 1991.

77. Soldin, S.J., Therapeutic drug monitoring, in *Applied Biochemistry of Clinical Disorders,* 2nd ed., Gornall, A.G., Ed., Lippincott, New York, 1986, pp. 536, 537, 541.

78. Gude, T., Preiss, A., and Rubach, K., Determination of chloramphenicol in muscle, liver, kidney and urine of pigs by means of immunoaffinity chromatography and gas chromatography with electron-capture detection, *J. Chromatogr. B,* 673, 197–204, 1995.

79. Stanley, S.M.R., Wilhelmi, B.S., and Rodgers, J.P., Comparison of immunoaffinity chromatography combined with gas chromatography-negative ion chemical ionization mass spectrometry and radioimmunoassay for screening dexamethasone in equine urine, *J. Chromatogr.,* 620, 250–253, 1993.

80. Stanley, S.M.R., Wilhelmi, B.S., Rodgers, J.P., and Bertschinger, H., Immunoaffinity chromatography combined with gas chromatography-negative ion chemical ionisation mass spectrometry for the confirmation of flumethasone abuse in the equine, *J. Chromatogr.,* 614, 77–86, 1993.

81. Li, J. and Zhang, S., Immunoaffinity column cleanup and liquid chromatographic method for determining ivermectin in sheep serum, *J. AOAC Int.,* 79, 1300–1302, 1996.

82. Ong, H., Adam, A., Perreault, S., Marleau, S., Bellemare, M., and Du Souich, P., Analysis of albuterol in human plasma based on immunoaffinity chromatographic clean-up combined with high-performance liquid chromatography with fluorimetric detection, *J. Chromatogr.,* 497, 213–221, 1989.

83. Zimmerli, B. and Dick, R., Determination of ochratoxin A at the ppt level in human blood, serum, milk and some foodstuffs by high-performance liquid chromatography with enhanced fluorescence detection and immunoaffinity column cleanup: Methodology and Swiss data, *J. Chromatogr. B*, 666, 85–99, 1995.

84. Haagsma, N. and van de Water, C., Immunochemical methods in the analysis of veterinary drug residues, in *Analysis of Antibiotic Drug Residues in Food Products of Animal Origin*, Agarwal, V.K., Ed., Plenum Press, New York, 1992, pp. 81–97.

85. Yap, W.T., Locascio-Brown, L., Plant, A.L., Choquette, S.J., Horvath, V., and Durst, R.A., Liposome flow injection immunoassay: Model calculations of competitive immunoreactions involving univalent and multivalent ligands, *Anal. Chem.*, 63, 2007–2011, 1991.

86. Rico, C.M., Del Pilar Fernandez, M., Gutierrez, A.M., Conde, M.C.P., and Camara, C., Development of a flow fluoroimmunosensor for determination of theophylline, *Analyst*, 120, 2589–2591, 1995.

87. Palmer, D.A., Edmonds, T.E., and Seare, N.J., Flow injection immunosensor for theophylline, *Anal. Lett.*, 26, 1425–1439, 1993.

88. Palmer, D.A., Fernandez-Hernando, P., and Miller, J.N., A model on-line flow injection fluorescence immunoassay using a protein A immunoreactor and lucifer yellow, *Anal. Lett.*, 26, 2543–2553, 1993.

89. Haginaka, J. and Wakai, J., β-Cyclodextrin bonded silica for direct injection analysis of drug enantiomers in serum by liquid chromatography, *Anal. Chem.*, 62, 997–1000, 1990.

90. Rochat, B., Amey, M., and Baumann, P., Analysis of enantiomers of citalopram and its demethylated metabolites in plasma of depressive patients using chiral reverse-phase liquid chromatography, *Ther. Drug Monit.*, 17, 273–279, 1995.

91. Castoldi, D., Oggioni, A., Renoldi, M.I., Ratti, E., DiGiovine, S., and Bernareggi, A., Assay of moguisteine metabolites in human plasma and urine: Conventional and chiral high-performance liquid chromatographic methods, *J. Chromatogr. B*, 655, 243–252, 1994.

92. Pham-Huy, C., Radenen, B., Sahui-Gnassi, A., and Claude, J.R., High-performance liquid chromatographic determination of (*S*)- and (*R*)-propranolol in human plasma and urine with a chiral β-cyclodextrin bonded phase, *J. Chromatogr. B*, 665, 125–132, 1995.

93. Kelly, J.W., Stewart, J.T., and Blanton, C.D., HPLC separation of pentazocine enantiomers in serum using an ovomucoid chiral stationary phase, *Biomed. Chromatogr.*, 8, 255–257, 1994.

94. Li, F., Cooper, S.F., Cote, M., and Ayotte, C., Determination of the enantiomers of bunolol in human urine by high-performance liquid chromatography on a chiral AGP stationary phase and identification of their metabolites by gas chromatography-mass spectrometry, *J. Chromatogr. B*, 660, 327–339, 1994.

95. Haupt, D., Determination of citalopram enantiomers in human plasma by liquid chromatographic separation on a chiral-AGP column, *J. Chromatogr. B*, 685, 299–305, 1996.

96. Menzel-Soglowek, S., Geisslinger, G., and Brune, K., Stereoselective high-performance liquid chromatographic determination of ketoprofen, ibuprofen and fenoprofen in plasma using a chiral α₁-acid glycoprotein column, *J. Chromatogr.*, 532, 295–303, 1990.

97. Geisslinger, G., Menzel-Soglowek, S., Schuster, O., and Brune, K., Stereoselective high-performance liquid chromatographic determination of flurbiprofen in human plasma, *J. Chromatogr.*, 573, 163–167, 1992.

98. Pettersson, K.-J. and Olsson, A., Liquid chromatographic determination of the enantiomers of ibuprofen in plasma using a chiral AGP column, *J. Chromatogr.*, 563, 414–418, 1991.

99. Geisslinger, G., Menzel-Soglowek, S., Kamp, D.-H., and Brune, K., Stereoselective high-performance liquid chromatographic determination of the enantiomers of ketamine and norketamine in plasma, *J. Chromatogr.*, 568, 165–176, 1991.

100. Schmidt, N., Brune, K., and Geisslinger, G., Stereoselective determination of the enantiomers of methadone in plasma using high-performance liquid chromatography, *J. Chromatogr.*, 583, 195–200, 1992.

101. Beck, O., Boreus, L.O., LaFolie, P., and Jacobsson, G., Chiral analysis of methadone in plasma by high-performance liquid chromatography, *J. Chromatogr.*, 570, 198–202, 1991.

102. Kristensen, K., Angelo, H.R., and Blemmer, T., Enantioselective high-performance liquid chromatographic method for the determination of methadone in serum using an AGP and a CN column as chiral and analytical column, respectively, *J. Chromatogr. A,* 666, 283–287, 1994.

103. Chu, Y.-Q. and Wainer, I.W., Determination of the enantiomers of verapamil and norvera-pamil in serum using coupled achiral-chiral high-performance liquid chromatography, *J. Chromatogr.,* 497, 191–200, 1989.

104. Mangani, F., Luck, G., Fraudeau, C., and Verette, E., On-line column switching high-performance liquid chromatography analysis of cardiovascular drugs in serum with automated sample clean-up and zone cutting technique to perform chiral separation, *J. Chromatogr. A,* 762, 235–241, 1997.

105. Suzuki, A., Takagaki, S., Suzuki, H., and Noda, K., Determination of the *R,R-* and *S,S-* enantiomers of vamicamide in human serum and urine by high-performance liquid chromatography on a chiral-AGP column, *J. Chromatogr.,* 617, 279–284, 1993.

106. Fieger, H. and Blaschke, G., Direct determination of the enantiomeric ratio of verapamil: Its major metabolite norverapamil and gallopamil in plasma by chiral high-performance liquid chromatography, *J. Chromatogr.,* 575, 255–260, 1992.

107. Jones, D.J., Nguyen, K.T., McLeish, M.J., Crankshaw, D.P., and Morgan, D.J., Determination of *(R)*-(+)- and *(S)*-(−)-isomers of thiopentone in plasma by chiral high-performance liquid chromatography, *J. Chromatogr. B.* 675, 174–179, 1996.

108. Hage, D.S., Survey of recent advances in analytical applications of immunoaffinity chro-matography, *J. Chromatogr. B.,* 715, 3–28, 1998.

109. Katz, S.E. and Siewierski, M., Drug residue analysis using immunoaffinity chromatography, *J. Chromatogr.,* 624, 403–409, 1992.

110. Katz, S.E. and Brady, M.S., High-performance immunoaffinity chromatography for drug residue analysis, *J. Assoc. Off. Anal. Chem.,* 73, 557–560, 1990.

111. Dubois, M., Taillieu, X., Colemonts, Y., Lansival, B., De Graeve, J., and Delahaut, P., GC-MS determination of anabolic steroids after multi-immunoaffinity purification, *Analyst,* 123, 2611–2616, 1998.

112. Feng, S., El-Sohly, M.A., Salamone, S., and Salem, M.Y., Simultaneous analysis of Δ^9-THC and its major metabolites in urine, plasma, and meconium by GC-MS using an immunoaffinity extraction procedure, *J. Anal. Toxicol.,* 24, 395–402, 2000.

113. Reh, E., Determination of digoxin in serum by on-line immunoadsorptive clean-up high-performance liquid chromatographic separation and fluorescence-reaction detection, *J. Chromatogr.,* 433, 119–130, 1988.

114. Rule, G.S. and Henion, J.D., Determination of drugs from urine by on-line immunoaffinity chromatography-high-performance liquid chromatography-mass spectrometry, *J. Chro-matogr.,* 582, 103–112, 1992.

115. Cai, J. and Henion, J., On-line immunoaffinity extraction-coupled column capillary liquid chromatography/tandem mass spectrometry: Trace analysis of LSD analogs and metabolites in human urine, *Anal. Chem.,* 68, 72–78, 1996.

116. Kircher, V. and Parlar, H., Determination of Δ^9-tetrahydrocannabinol from human saliva by tandem immunoaffinity chromatography-high-performance liquid chromatography, *J. Chro-matogr. B,* 677, 245–255, 1996.

117. Phillips, T.M. and Chiemlinska, J.J., Immunoaffinity capillary electrophoresis analysis of cyclosporin in tears, *Biomed. Chromatogr.,* 8, 242–246, 1994.

118. Thienpont, L.M., Depourcq, G.C., Nelis, H.J., and De Leenheer, A.P., Liquid chromato-graphic determination of 2-thioxothiazolidine-4-carboxylic acid isolated from urine by affin-ity chromatography on organomercurial agarose gel, *Anal. Chem.,* 62, 2673–2675, 1990.

119. Alaupovic, P., Hodis, H.N., Knight-Gibson, C., Mack, W.J., LaBree, L., Cashin-Hemphill, L., Corder, C.N., Kramsch, D.M., and Blankenhorn, D.H., Effects of lovastatin on ApoA- and ApoB-containing lipoproteins: Families in a subpopulation of patients participating in the Monitored Atherosclerosis Regression Study (MARS), *Arterioscler. Thromb.,* 14, 1906–1913, 1994.

120. Tavella, M., Alaupovic, P., Knight-Gibson, C., Tournier, H., Schinella, G., and Mercuri, O., Separation of ApoA- and ApoB-containing lipoproteins of human plasma by affinity chromatography on concanavalin A, *Prog. Lipid Res.*, 30, 181–187, 1991.

121. Shishino, K., Murase, M., Makino, H., and Saheki, S., Glycated apolipoprotein A-I assay by combination of affinity chromatography and latex immunoagglutination, *Ann. Clin. Biochem.*, 37, 498–506, 2000.

122. Bachi, A., Zuccato, E., Baraldi, M., Fanelli, R., and Chiabrando, C., Measurement of urinary 8-epi-prostaglandin $F_{2\alpha}$, a novel index of lipid peroxidation *in vivo*, by immunoaffinity extraction/gas chromatography-mass spectrometry: Basal levels in smokers and non-smokers, *Free Radical Biol. Med.*, 20, 619–624, 1996.

123. Mackert, G., Reinke, M., Schweer, H., and Seyberth, H.W., Simultaneous determination of the primary prostanoids prostaglandin E_2, prostaglandin $F_{2\alpha}$ and 6-oxoprostaglandin $F_{1\alpha}$ by immunoaffinity chromatography in combination with negative ion chemical ionization gas chromatography-tandem mass spectrometry, *J. Chromatogr.*, 494, 13–22, 1989.

124. Chiabrando, C., Pinciroli, V., Campoleoni, A., Benigni, A., Piccinelli, A., and Fanelli, R., Quantitative profiling of 6-ketoprostaglandin $F_{1\alpha}$, 2,3-dinor-6-ketoprostaglandin $F_{1\alpha}$, thromboxane B_2 and 2,3-dinor-thromboxane B_2 in human and rat urine by immuno-affinity extraction with gas chromatography-mass spectrometry, *J. Chromatogr.*, 495, 1–11, 1989.

125. Ishibashi, M., Watanabe, K., Ohyama, Y., Mizugaki, M., Hayashi, Y., and Takasaki, W., Novel derivatization and immunoextraction to improve microanalysis of 11-dehydrothromboxane B_2 in human urine, *J. Chromatogr.*, 562, 613–624, 1991.

126. Prevost, V., Shuker, D.E.G., Friesen, M.D., Eberle, G., Rajewsky, M.F., and Bartsch, H., Immunoaffinity purification and gas chromatography-mass spectrometric quantitation of 3-alkyladenines in urine: Metabolism studies and basal excretion levels in man, *Carcinogenesis*, 14, 199–204, 1993.

127. Friesen, M.D., Garren, L., Prevost, V., and Shuker, D.E.G., Isolation of urinary 3-methyladenine using immunoaffinity columns prior to determination by low-resolution gas chromatography-mass spectrometry, *Chem. Res. Toxicol.*, 4, 102–106, 1991.

128. Yoshida, K.I., Honda, M., Arai, K., Hosoya, Y., Moriguchi, H., Sumi, S., Ueda, Y., and Kitahara, S., Serial lectin affinity chromatography with concanavalin A and wheat germ agglutinin demonstrates altered asparagine-linked sugar-chain structures of prostatic acid phosphatase in human prostate carcinoma, *J. Chromatogr. B*, 695, 439–443, 1997.

129. Sokoloff, R.L., Norton, K.C., Gasior, C.L., Marker, K.M., and Grauer, L.S., A dual-monoclonal sandwich assay for prostate-specific membrane antigen: Levels in tissues, seminal fluid and urine, *Prostate*, 43, 150–157, 2000.

130. Sumi, S., Arai, K., and Yoshida, K., Separation methods applicable to prostate cancer diagnosis and monitoring therapy, *J. Chromatogr. B*, 764, 445–455, 2001.

131. Nakano, K., Shindo, K., Yasaka, T., and Yamamoto, H., Reversed-phase liquid chromatographic investigation of nucleosides and bases in mucosa and modified nucleosides in urines from patients with gastrointestinal cancer, *J. Chromatogr.*, 332, 127–137, 1985.

132. Wang, Z.G., Williams, L.J., Zhang, X.F., Zatorski, A., Kudryashov, V., Ragupathi, G., Spassova, M., Bornmann, W., Slovin, S.F., Scher, H.I., Livingston, P.O., Lloyd, K.O., and Danishefsky, S.J., Polyclonal antibodies from patients immunized with a globo H-keyhole limpet hemocyanin vaccine: Isolation, quantification, and characterization of immune responses by using totally synthetic immobilized tumor antigens, *Proc. Natl. Acad. Sci. U.S.A.*, 97, 2719–2724, 2000.

133. Flurer, C.L. and Novotny, M., Dual microcolumn immunoaffinity liquid chromatography: An analytical application to human plasma proteins, *Anal. Chem.*, 65, 817–821, 1993.

134. McConnell, J.P. and Anderson, D.J., Determination of fibrinogen in plasma by high-performance immunoaffinity chromatography, *J. Chromatogr.*, 615, 67–75, 1993.

135. Ruhn, P.F., Taylor, J.D., and Hage, D.S., Determination of urinary albumin using high-performance immunoaffinity chromatography and flow injection analysis, *Anal. Chem.*, 66, 4265–4271, 1994.

136. Ohlson, S., Gudmundsson, B.-M., Wikstrom, P., and Larsson, P.-O., High-performance liquid affinity chromatography: Rapid immunoanalysis of transferrin in serum, *Clin. Chem.,* 34, 2039–2043, 1988.

137. Janis, L.J. and Regnier, F.E., Dual-column immunoassays using protein G affinity chromatography, *Anal. Chem.,* 61, 1901–1906, 1989.

138. Inoue, T., Yamauchi, M., Toda, G., and Ohkawa, K., Microheterogeneity with concanavalin A affinity of serum transferrin in patients with alcoholic liver disease, *Alcohol Clin. Exp. Res.,* 20, 363A–365A, 1996.

139. Lacey, J.M., Bergen, H.R., Magera, M.J., Naylor, S., and O'Brien, J.F., Rapid determination of transferrin isoforms by immunoaffinity liquid chromatography and electrospray mass spectrometry, *Clin. Chem.,* 47, 513–518, 2001.

140. Perez, L., Estepa, A., and Coll, J.M., Purification of the glycoprotein G from viral haemorrhagic septicaemia virus, a fish rhabdovirus, by lectin affinity chromatography, *J. Virol. Methods,* 76, 1–8, 1998.

141. Dawidowicz, A.L., Rauckyte, T., and Rogalski, J., The preparation of sorbents for the analysis of human antithrombin III by means of high-performance affinity chromatography, *Chromatographia,* 37, 168–172, 1993.

142. Dawidowicz, A.L., Rauckyte, T., and Rogalski, J., High-performance affinity chromatography for analysis of human antithrombin III, *J. Liq. Chromatogr.,* 17, 817–831, 1994.

143. DeCristofaro, R., Landolfi, R., Bizzi, B., and Castagnola, M., Human platelet glycocalicin purification by phenyl boronate affinity chromatography coupled to anion-exchange high-performance liquid chromatography, *J. Chromatogr.,* 426, 376–380, 1988.

144. Anonymous, What is genomics? The emergence of genomics as a discipline; available online at http://genomics.ucdavis.edu/what.html, accessed September 2003.

145. Opitek, G., Lewis, K., and Jorgenson, J., Comprehensive on-line LC/LC/MS of proteins, *Anal. Chem.,* 69, 1518–1524, 1997.

146. Davis, M.T., Beierle, J., Bures, E.T., McGinley, M.D., Mort, J., Robinson, J.A., Spahr, C.S., Yu, W., Luethy, R., and Patterson, S.D., Automated LC-LC-MS-MS platform using binary ion-exchange and gradient reversed-phased chromatography for improved proteomic analysis, *J. Chromatogr. B,* 752, 281–291, 2001.

147. Link, A., Eng, J., Schieltz, D., Carmack, E., Mize, G., Morris, D., Garvik, B., and Yates, J., Direct analysis of protein complexes using mass spectrometry, *Nat. Biotechnol.,* 17, 676–682, 1999.

148. Riggs, L., Sioma, C., and Regnier, F., Automated signature peptide approach for proteomics, *J. Chromatogr. A,* 924, 359–368, 2001.

149. Hsieh, Y., Wang, H., Elicone, C., Mark, J., Martin, S., and Regnier, F., Automated analytical system for the examination of protein primary structure, *Anal. Chem.,* 68, 455–462, 1996.

150. Stensballe, A., Andersen, S., and Jensen, O.N., Characterization of phosphoproteins from electrophoretic gels by nanoscale Fe(III) affinity chromatography with off-line mass spectrometry analysis, *Proteomics,* 1, 207–222, 2001.

151. Geng, M., Zhang, X., Bina, M., and Regnier, F., Proteomics of glycoproteins based on affinity selection of glycopeptides from tryptic digests, *J. Chromatogr. B,* 752, 293–306, 2001.

152. Ji, J., Chakraborty, A., Geng, M., Zhang, X., Amini, A., Bina, M., and Regnier, F., Strategy for qualitative and quantitative analysis in proteomics based on signature peptides, *J. Chromatogr. B,* 745, 197–210, 2000.

153. Geng, M., Ji, J., and Regnier, F., Signature-peptide approach to detecting proteins in complex mixtures, *J. Chromatogr. A,* 870, 295–313, 2000.

154. Hage, D.S., High-performance affinity chromatography: A powerful tool for studying serum protein binding, *J. Chromatogr. B,* 768, 3–30, 2002.

155. Sengupta, A. and Hage, D.S., Characterization of minor site probes for human serum albumin by high-performance affinity chromatography, *Anal. Chem.,* 71, 3821–3827, 1999.

156. Chan, N.W., Lewis, D.F., Hewko, S., Hindsgaul, O., and Schriemer, D.C., Frontal affinity chromatography for the screening of mixtures, *Comb. Chem. High Throughput Screen.,* 5, 395–406, 2002.

157. Regnier, F.E., Riggs, L., Zhang, R., Xiong, L., Liu, P., Chakraborty, A., Seeley, E., Sioma, C., and Thompson, R.A., Comparative proteomics based on stable isotope labeling and affinity selection, *J. Mass Spectrom.,* 37, 133–145, 2002.
158. Turecek, F., Mass spectrometry in coupling with affinity capture-release and isotope-coded affinity tags for quantitative protein analysis, *J. Mass Spectrom.,* 37, 1–14, 2002.
159. Gygi, S., Rist, B., Gerber, S., Turecek, F., Gelb, M., and Aebersold, R., Quantitative analysis of complex protein mixtures using isotope-coded affinity tags, *Nat. Biotechnol.,* 17, 994–999, 1999.
160. Goshe, M.B., Conrads, T.P., Panisko, E.A., Angell, N.H., Veenstra, T.D., and Smith, R.D., Phosphoprotein isotope-coded affinity tag approach for isolating and quantitating phospho-peptides in proteome-wide analyses, *Anal. Chem.,* 73, 2578–2586, 2001.

18

Affinity Chromatography in Biotechnology

Neil Jordan and Ira S. Krull

Department of Chemistry and Chemical Biology, Northeastern University, Boston, MA

CONTENTS

18.1 INTRODUCTION

Affinity chromatography is a separation technique that makes use of a ligand possessing a unique and specific affinity for an analyte [1]. Because of the extraordinary ability of biological ligands to discriminate between very similar compounds in their interactions, affinity separations that use such ligands have a high degree of selectivity. This is particularly advantageous in biotechnology, where a specific protein, usually present in a complex mixture, is often the desired target. Although other techniques are capable of complex separations, few if any are as rapid and selective as affinity chromatography. Other advantages of affinity chromatography for such work include its need for little or no sample preparation and its ability to select for biological activity.

This chapter discusses several ways in which affinity chromatography has been used in the field of biotechnology. One such area has been the use of affinity chromatography in industry for the downstream processing of fermentation products. Systems for this purpose have been designed to allow the continuous selection and concentration of product from raw feedstock. Process monitoring is another area in which affinity chromatography has been utilized. The rapid monitoring of biofermentors is critical to determining harvest times and maximizing yields. Being an on/off method, affinity chromatography easily lends itself to automation, since no qualitative interpretation of results is needed.

Affinity chromatography has also been used in biotechnology at the analytical level, where it is an extremely powerful tool in proteomics. For instance, it can be used to isolate a particular protein from a complex mixture or to characterize posttranslational modifications. Relative rates of protein expression can be quickly and accurately determined by this method, enabling comparisons of normal and abnormal cell states. Such information is vital in drug development and in the identification of disease markers.

Any disease in which a molecular marker has been identified lends itself to diagnosis through affinity chromatography. Quick and simple blood tests or urinalysis methods have already been developed for many agents. The sensitivity of affinity methods in detecting posttranslational modifications and the relative amounts of these modifications should be of interest, especially advances in the field of proteomics.

Affinity chromatography does require that a good ligand first be obtained for the compound of interest. Consequently, this chapter begins with a summary of the various types of ligands available for affinity chromatography. The discussion then focuses on the various applications that have been reported for affinity chromatography in the processing and analysis of agents in biotechnology.

18.2 AFFINITY TARGETS AND LIGAND SELECTION

The main requirement for affinity chromatography is that a good ligand must first be obtained for the compound of interest. Fortunately, there are many ligands that can be used for this purpose. This section briefly discusses the use of antibodies, enzymes, enzyme substrates, boronates, metal chelates, and other agents for the binding of such substances of interest in biotechnology. Ligands for particular types of groups are also discussed, including those that bind to phosphoryl, histidinyl, glycosyl, and cysteinyl residues. The use of engineered ligands and affinity tags is also examined.

18.2.1 Antibodies and Antigens

The ability to raise and purify antibodies to many biomolecules has made antibodies popular as ligands in affinity chromatography. In this application, antibodies are used to bind their respective target compound, or antigen. Another advantage of antibodies over nonbiological ligands is the ability to screen substances according to bioactivity and not just structure. Further details on the use of antibodies in affinity chromatography, a technique known as *immunoaffinity chromatography*, can be found in Chapter 6. The isolation of antibodies by affinity chromatography is discussed in Section 18.2.6 of this chapter, as well as in Chapter 14.

One example of the use of antibodies as affinity ligands is a method in which an HPLC (high-performance liquid chromatography)-based antibody column was used to screen plasma samples for parathyroid hormone [2], giving limits of detection in the attomole range with run times of 6 min. Sample preparation in this work consisted of a single filtration step and gave recoveries for this hormone of 99%.

Polyclonal antibodies were used as ligands to isolate O^6-butylguanine, a minor DNA adduct [3]. This was performed using samples of hydrolyzed DNA that were later analyzed by gas chromatography (GC). This provided a much simpler sample cleanup than traditional pretreatment methods while maintaining high sensitivity for the analysis.

Antibodies have also been used as affinity ligands in the isolation of prostate-specific antigen [4], insulin [5], human immunoglobulin G [6], and hemoglobin [7]. In these studies, limits of detection in the femtomole range and recoveries of 90% or better were commonly seen. Sample preparation in these reports typically consists of a single filtration or centrifugation step.

Johns et al. [8] reported the use of a bispecific antibody (i.e., an antibody with two different antigen-binding domains) in an affinity capillary to perform an on-line enzyme-linked immunosorbent assay (ELISA) in 20 min. This approach gave an increase in sensitivity versus a traditional ELISA. In other work, the use of bispecific antibodies has allowed the development of sandwich-type assays [2, 8], which significantly lowered the limits of detection for the analytes without adding a lengthy derivatization step.

A particularly novel application of antibodies has been their use in affinity columns for the detection of anti-idiotypic antibodies in cancer patients [9]. Tumor antigens were used to isolate antitumor antibodies produced by the patient's own immune system. These antitumor antibodies were then immobilized on an affinity column and used to determine the levels of anti-idiotypic antibodies in the same patient's blood.

18.2.2 Enzymes and Substrates

The specificity of an enzyme for its substrate, and the substrate for its enzyme, makes these agents useful as ligands for affinity chromatography. A more detailed discussion of enzyme and substrate-based ligands can be found in Chapters 5 and 12.

A specific example of how these ligands can be used in biotechnology is given by the purification of restriction endonucleases. For these enzymes, there is a tendency to break down the substrate when it is used as a ligand. A solution to this problem can be obtained by recognizing the need for Mg^{2+} for this process. By adding EDTA to the buffer in the affinity column, the magnesium ions will be chelated, and this enzymatic degradation will be inhibited. This can result in the purification of such enzymes at 75% yield when using DNA as a ligand [10]. A disadvantage to this approach is the need for a specific ligand for each enzyme to be purified. But by combining the recognition sequences of several enzymes on one oligonucleotide ligand, groups of restriction endonucleases can be isolated in a single separation [11]. Other examples of affinity purifications for DNA-binding enzymes and proteins can be found in Chapter 7.

18.2.3 Phosphoryl Groups

The phosphorylation of biomolecules is a ubiquitous feature of metabolism, where biological activity is often a function of a molecule's degree of phosphorylation. Indeed, the loss of regulation for this process has been implicated in many diseases. This makes it important to be able to determine both the number and location of phosphoryl groups on a biological substance. Since many metal ions have an affinity for the oxygen in phosphoryl groups, these can be used as ligands to help isolate such targets. More details on the use of metal ions for this purpose can be found in Chapter 10 on *immobilized metal-ion affinity chromatography* (IMAC). The use of IMAC for other types of functional groups (i.e., histidines) is discussed in Section 18.2.4 of this chapter.

Affinity columns containing chelates of iron(III) have been shown to selectively bind phosphoamino acids as well as phosphopeptides (see Figure 18.1) [12, 13]. A mixture of amino acids applied to a column containing an iminodiacetate-agarose gel and chelated Fe^{3+} ions showed little affinity for this material. But a mixture of phosphoamino acids applied to the same column was strongly retained. Elution of these compounds was accomplished through the addition of phosphate ions to the mobile phase. The same column allowed preparations of ovalbumin containing one, two, or no phosphoserines to be distinguished.

When comparing various metal chelates for the isolation of phosphopeptides prior to analysis by mass spectrometry, it has been shown that Ga(III) is preferred to Fe(III) [14].

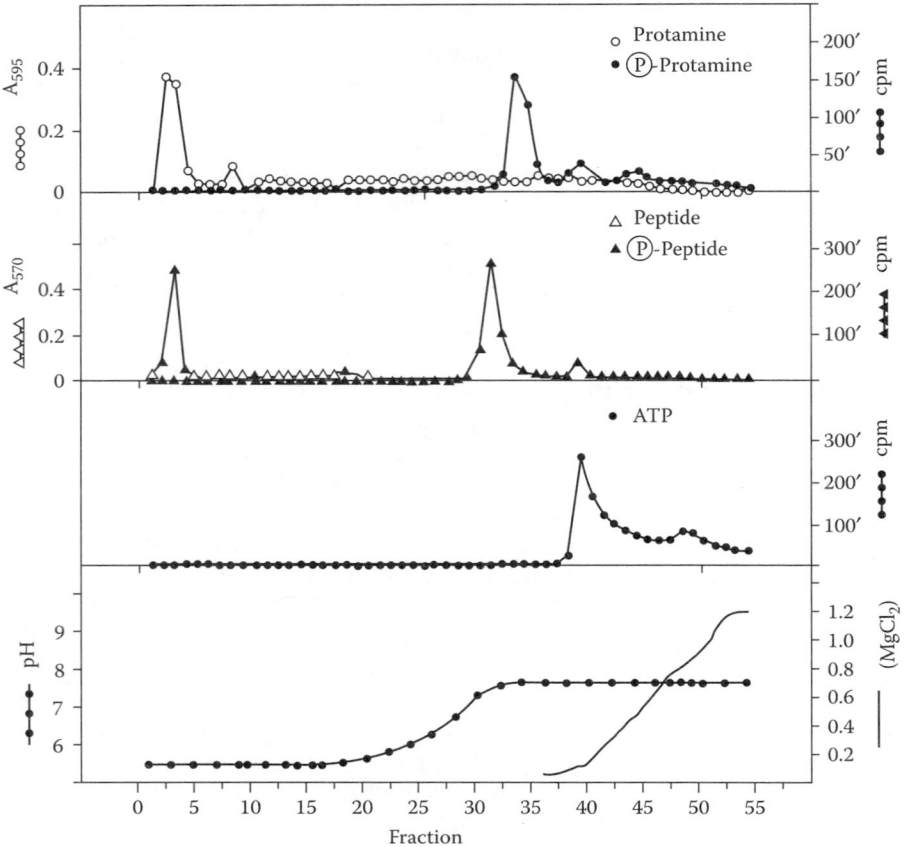

Figure 18.1   Separation of phosphopeptide and phosphoprotamine on Fe(III)-chelated Sepharose. (Reproduced with permission from Muszynska, G., Dobrowolska, G., Medin, A., Eckman, P., and Porath, J., *J. Chromatogr.*, 604, 19–28, 1992.)

Of the ions that were compared, Fe(III), Ga(III), and Zr(IV) showed the highest affinity for these compounds, giving greater than 80% retention for the phosphopeptides. In addition, Fe(III) and Ga(III) allowed the elution of these agents in about two column volumes and produced less dilution than Zr(IV), which required much larger elution volumes. When their eluants were analyzed by mass spectrometry, the Fe(III) resulted in much higher levels of interference than the Ga(III) or Zr(IV). Thus, in this particular application, Ga(III) appeared to be the best choice for use in analyte isolation. Other studies with immobilized Al(III) chelates have demonstrated a high selectivity for phosphoproteins but with apparently weaker binding [12, 15].

Riggs et al. used a Ga(III) chelate column to isolate phosphorylated peptides from tryptic digests of milk [16]. Following the isolation of these phosphopeptides, they were dephosphorylated and chromatographed on a reversed-phase column that was directly coupled to electrospray ionization mass spectrometry (ESI-MS). By comparing the known peptide signatures of milk proteins, it was possible to determine the phosphorylation sites on these proteins. This process, from digestion to final analysis, was completely automated and resulted in a 90% reduction in analysis time versus previous techniques, as well as providing greater reproducibility and fewer opportunities for sample loss.

Another type of ligand that has been used in protein analysis is a *phosphoprotein isotope coded affinity tag* (PhIAT). This is a commercially available affinity tag that recognizes phosphoryl groups on proteins [17]. The use of this ligand will be discussed in more detail in Section 18.2.10 of this chapter, which examines the topic of affinity tags.

18.2.4 Histidinyl Groups

Immobilized metal-ion affinity chromatography was originally used by Porath when he described a method for purifying proteins based on the affinity of surface imidazole and thiol groups for some metal ions [18, 19]. Some examples of these applications are given in Figure 18.2 and Table 18.1. The relative binding for the tested metal ions was Cu > Zn > Ni > Mn. Later studies [20, 21] indicated that Cu(II) and Zn(II) ions leach as histidinyl-metal complexes upon elution. This leaching of the metal ions led to a reduced recovery of activity for the enzyme glutathione-*S*-transferase (GST) by one-third when using Cu(II) and one-half with Zn(II), along with the formation of precipitates in the column. In contrast to this, GST retained 90% of its activity when purified on a Ni(II) chelate column.

Ni(II) has also been shown to be effective in resolving small differences in peptide affinities [22]. Consequently, Ni(II) chelates are now preferred as ligands for the isolation of histidine-containing peptides. There can be interferences from cysteine and tryptophan, which are also bound by immobilized metals. However, this can be controlled by reducing and alkylating the sulfhydryl groups on cysteine or acetylating a tryptophan-containing peptide [23]. More details on the use of Ni(II) in immobilized metal-ion affinity chromatography can be found in Chapter 10.

18.2.5 Glycosyl Groups and Sugar Residues

Electrophoresis, a common method for separating proteins, is not effective when applied to glycoproteins. The different glycoforms of a protein produce multiple masses for the same protein. This reduces the resolution and smears the protein bands in the final electropherogram. Affinity chromatography has the ability to separate the various glycoforms of a protein and is the preferred method for such separations.

Concanavalin A is a lectin that exhibits a broad affinity for glycosylated proteins and peptides, especially those containing *N*-type hybrid and high mannose oligosaccharides. Its affinity is reduced for diantennary oligosaccharides and is effectively absent for tri- and tetraantennary oligosaccharides. It has been used to detect changes in the carbohydrate

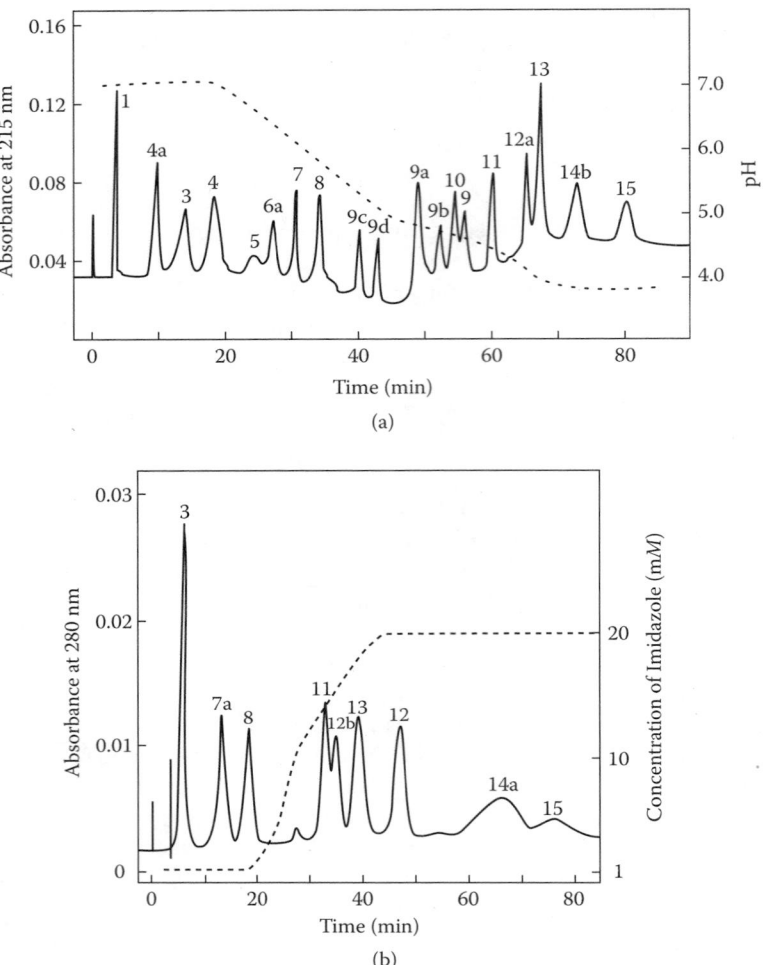

Figure 18.2 (a) Separation of 19 peptides on an immobilized Cu(II) column, and (b) separation of nine peptides on an immobilized Ni(II) column. A list of the compounds that are represented by peaks 1 to 15 is given in Table 18.1. (Reproduced with permission from Nakagawa, Y., Yip, T.T., Belew, M., and Porath, J., *Anal. Biochem.*, 168, 75–81, 1988.)

structures of human serum gonadotropins based on the changes in branching [24]. A summary of the properties of concanavalin A, as well as those for other lectins used in affinity chromatography, can be found in Chapter 5.

Bandeiraea simplicifolia lectin (BS-II) is another lectin commonly used in affinity methods. This lectin is highly selective for *N*-acetylglucosamine derivatives of glycoproteins. It has been used together with concanavalin A in serial lectin affinity chromatography to determine changes in prostate acid phosphatase that could not be detected with either column alone [25]. Wheat germ lectin is a third type of lectin common in affinity chromatography. It has been used to isolate the isoforms of alkaline phosphatase with better than 90% recovery.

As reviewed in Chapter 8, boronates can be used as yet another type of ligand for isolation of carbohydrate-containing compounds. Boronate derivatives can interact with the 1,2-*cis*-diol groups that appear on glucose and many other sugars. This ability gives these

Table 18.1   Identity of the Peptides in Figure 18.2

Peak No.	Peptide Name
1	Neurotensin
2	GRF
3	Oxytocin
3a	$[Asu^{1,6}]$ oxytocin
3b	$[Phe^4]$ oxytoxin
4	$[Leu^5]$ enkephalin
4a	Sulfated $[Leu^5]$ enkephalin
5	Mastoparan
6a	Tyr-bradykinin
7	Substance P
7a	$[Tyr^8]$ substance P
8	Somatostatin
9	Human calcitonin
9a	$[Asu^{1,7}]$ human calcitonin
9b	Human calcitonin [17–32]
9c	$[Asu^{1,7}]$ Eel calcitonin
9d	Eel calcitonin [11–32]
10	Bombesin
11	Angiotensin II
11a	*des*-Asp^1[Ile^8] angiotensin II
11b	$[\beta\text{-}Asp^1]$ angiotensin II
11c	$[D\text{-}His^6]$ angiotensin II
11d	$[Sar^1,Val^5,Ala^8]$ angiotensin II
11e	$[Sar^1,Ile^8]$ angiotensin II
12	Human GIP
12a	$[Trp(for)^{25,36}]$ human GIP [21–42]
12b	Bovine GIP
13	LH-RH
14a	Human PTH [1–34]
14b	Human PTH [13–34]
15	Angiotensin I
15a	$[Val^5]$ angiotensin I
15b	$[\beta\text{-}Asp^1]$ angiotensin I

Source: Data obtained from Nakagawa, Y., Yip, T.T., Belew, M., and Porath, J., *Anal. Biochem.*, 168, 75–81, 1988. With permission.

ligands a broad selectivity for glycoproteins. These ligands have stronger binding to galactose, mannose, and sialic acid than is seen for lectins, making boronates useful in situations where lectins are ineffective. For instance, phenylboronate-agarose (PBA) has been used to isolate glycocalicin (Gc), a subunit of the membrane platelet glycoprotein Ib where the sialic acid content has been reduced [26]. Unlike wheat germ lectin, which requires the presence of sialic acid to bind Gc, PBA instead binds through the galactose derivatives found on Gc.

18.2.6   Immunoglobulins

A variety of affinity ligands can be used for the isolation of antibodies (or immunoglobulins). A detailed discussion of such ligands can be found in Chapter 14. One example is
protein A. This is a bacterial cell wall protein found in *Staphylococcus aureus* that has
been shown to bind antibodies [27]. The binding occurs in the F_c region of antibodies.
This interaction is strongest at a neutral pH but weakens with decreasing pH. Protein A
has been used for both the industrial purification of recombinant antibodies in an expanded-
bed format [28] and in analytical separations of antibodies [29]. It has also been used as
a secondary ligand to temporarily bind antibodies to an affinity column (see Chapter 6).

Protein G is another bacterial protein that can bind to antibodies. This is produced by
group G streptococci and also binds to the constant region of immunoglobulins. Although it
has similar binding characteristics to protein A, its selectivity is different for antibodies from
various species and classes [30]. This has been used in the detection of immunoglobulins in
serum [31] and as a secondary ligand for the preparation of immunoaffinity supports [5, 32].

Another material that has been used in the separation of antibodies and other proteins
is thiophilic gel, or T-gel. It has been found to have a specific yet broad affinity for immunoglobulins [33]. The ligand in this gel has a sulfone group in close proximity to a thioether.
T-gel has been used in a one-step separation of serum proteins, giving a yield of 80% activity
[34]. But where T-gel may prove to be of greatest use is in the separation of bispecific antibodies.

Bispecific antibodies are becoming increasingly important in immunotherapy and immunochemistry but are especially difficult to purify because of the similarity of the production
by-products (for instance, see Figure 18.3). Other methods have problems in producing high
levels of yield or purity with these agents. However, T-gel is thought to bind unpaired cysteines.

Type of heavy (H) and light (L) chain combination	Diagrammatic representation
Monospecific (parental H-L associations)	
Bispecific (two homologous H-L and heterologous H-H associations)	
One non-homologous H-L association	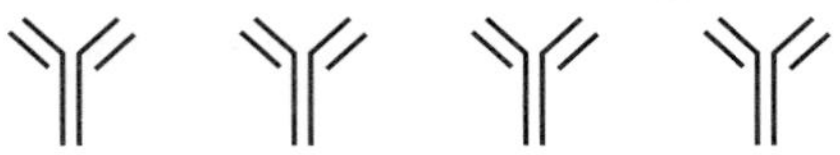
Two non-homologous H-L associations	

Figure 18.3   Ten possible types of antibody molecules that are produced by a hybrid-hybridoma.
(Reproduced with permission from Kreutz, F.T., Wishart, D.S., and Suresh, M.R., *J. Chromatogr. B*,
714, 161–170, 1998.)

As a result, the misalignment of cysteines in the different chains of bispecific antibodies leads to a greater affinity for these types of molecules. Based on this effect, T-gel has been used to purify bispecific antibodies in a single step with simple salt-gradient elution [35].

18.2.7 Cysteinyl Groups

Cysteine often forms disulfide bonds with other cysteines in the same protein. This can either link two chains together or form loops within the same chain. Not all cysteines are properly aligned within a protein to form such bonds, and so some are instead found in a reduced form. T-gel (as summarized in the last section) is thought to bind these available cysteine residues [35].

Immobilized metal-ion affinity chromatography is another tool that can be used with cysteine residues. Chelates of Cu(II), Zn(II), and Ni(II) are all known to bind histidinyl and thiol groups in proteins [18]. But the affinity of these chelates for such groups is not equal, and they can be separated by elution with a pH gradient. As discussed earlier, although Cu(II) and Zn(II) have higher affinities for proteins [18], they appear to experience some leaching upon elution [20, 21]. This means that Ni(II) is more commonly used for such separations.

A third way in which cysteine groups can be examined with affinity ligands is to use an *isotope-coded affinity tag*. This is a commercially available tag that recognizes cysteine residues in a protein or peptide [36]. Further details on this reagent are provided later in Section 18.2.10 on affinity tags.

18.2.8 Proteins

Besides the various ligands that have already been discussed, there are others that can also be used for the purification or separation of proteins. For instance, heparin is a highly sulfated glycosaminoglycan with a broad affinity for proteins. A partial structure for heparin is shown in Figure 18.4. The presence of several sulfate groups on heparin makes it useful as a cation-exchange agent.

Heparin appears to have a particular affinity for nucleic acid-binding proteins and has been used to separate polymerases [37, 38] and nucleases [39, 40]. It has also been shown to bind lipases [41, 42] and has been employed as an affinity ligand to separate antithrombin(III) from plasma, giving a purity of about 90% [43]. When used in an industrial separation of antithrombin, heparin provided better than 95% purity and activity for this substance [44]. Heparin has also been used in an enrichment and fractionation step for proteins from *Haemophilus influenzae* prior to the further analysis of these proteins by capillary electrophoresis and mass spectrometry [45].

Triazine compounds, which have long been used as textile dyes, have a broad affinity for proteins and have been used to purify proteins for many years [46]. One drawback to these ligands has been their nonspecific binding. The technique of *affinity elution*, where an eluant is chosen that specifically competes with the desired compound and preferentially

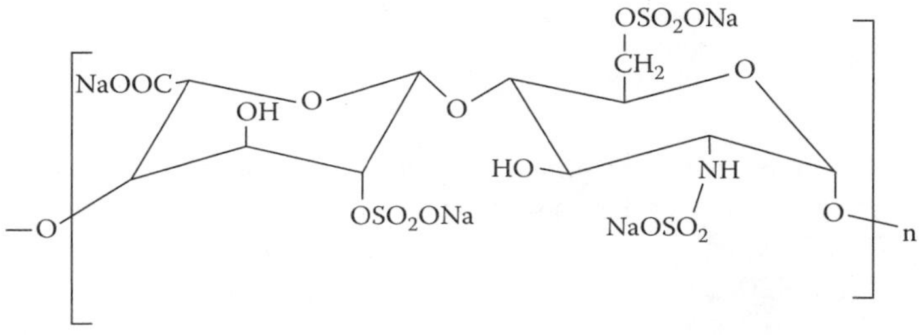

Figure 18.4 A partial structure of heparin.

elutes it, has improved this selectivity. The triazine dyes appear to mimic the structures of polynucleotides and have found use in separations of nucleotide-requiring enzymes, such as dehydrogenases [47–49] and ligases [50]. There also appears to be an electrostatic effect in their interactions, with elution from these ligands typically involving a salt gradient.

As discussed by Clonis et al. [46], triazine dyes have seen extensive use in the industrial purification of proteins. This is due to their low cost and high stability, which is of particular importance because of the harsh column cleanup reagents that are often used in commercial applications. The use of salts with these ligands to elute bound proteins allows for sufficiently mild conditions that provide a good retention of biological activity [47, 48, 51]. More information on triazine dyes and the related area of biomimetic ligands can be found in Chapter 9.

18.2.9   Engineered Ligands

Most of the ligands used in affinity chromatography are naturally occurring materials. But in many cases these natural ligands can have drawbacks, such as high cost, low stability, or the need for denaturing eluants. These disadvantages can limit the use of these ligands. To overcome some of these limitations, synthetic or altered ligands have been developed for affinity chromatography.

Biomimetic dyes make up one group of these alternative ligands (see Chapter 9). These ligands have been designed to mimic, with increasing accuracy, the substrates of various target enzymes. This makes use of the advantages of triazine dyes and related compounds while increasing their affinity and selectivity. As computer modeling software improves and the molecular structures of numerous proteins become better defined, these tailor-made ligands should become ever more powerful in their ability to selectively isolate biomolecules.

Molecularly imprinted polymers (MIPs) are a second group of synthetic ligands that are being investigated. A molecularly imprinted polymer is produced by polymerizing functional monomers around the compound of interest. Prior to polymerization, these functional monomers align themselves with complementary functional groups on the target molecule. After polymerization, the imprinted molecule is extracted, leaving behind a complementary binding site. MIPs have been used mostly as ligands for small molecules, but some work with protein and peptide separations has also been reported [52]. Further information on the preparation and use of MIPs can be found in Chapter 30.

As was discussed earlier, protein A is an effective ligand for isolating immunoglobulins, but it requires elution conditions that may destroy the biological activity of some retained antibodies. To overcome this problem, the binding domain of protein A has been altered slightly to reduce its binding strength and allow milder elution conditions [53]. With this modified protein A, immunoglobulins are less likely to be denatured, but there is also an accompanying loss of recovery for these agents.

A recombinant insulin precursor has been isolated by selecting an appropriate site on this target protein and constructing a complementary ligand with an MIP [54]. The result was an insulin precursor that was obtained with 90% recovery and more than 95% purity. This approach was aided by the production of *affinity fingerprints* [55, 56], which are essentially affinity maps based on the binding preferences of compounds to a set of reference proteins.

18.2.10   Affinity Tags

As discussed in Chapter 10, immobilized metal-ion affinity chromatography is commonly used for the purification of proteins and peptides containing histidine residues. This method has also been used along with polyhistidine tails that have been engineered into recombinant proteins for their isolation [20–22, 57, 58]. These histidine tails (or "His tags") are nonimmunogenic and can later be cleaved from the protein by also incorporating an encoded protease cleavage site into the protein's structure [57].

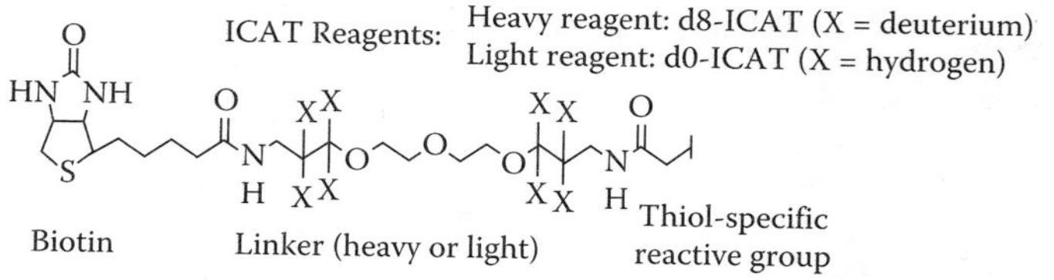

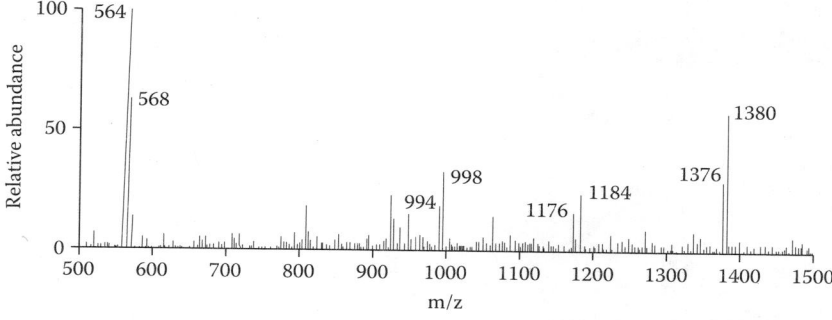

ICAT Reagents: Heavy reagent: d8-ICAT (X = deuterium)
Light reagent: d0-ICAT (X = hydrogen)

Biotin   Linker (heavy or light)   Thiol-specific reactive group

Figure 18.5 Structure of the ICAT reagent. (Reproduced with permission from Gygi, S.P., Rist, B., Gerber, S.A., Turecek, F., Gelb, M.H., and Aebersold, R., *Nat. Biotechnol.*, 17, 994–999, 1999.)

Another type of label used in affinity separations is the *isotope coded affinity tag* (ICAT). This tag labels sulfhydryl groups and possesses a biotin tail with a linker that may or may not be deuterated. The basic structure of this type of tag is shown in Figure 18.5 [36]. If this nondeuterated form of ICAT is used to label proteins expressed under normal conditions, and if the deuterated form is used to label the same proteins expressed under other conditions, then a comparison can be made by mass spectrometry between the levels of expression in these two groups. After labeling, the protein populations are combined, digested, and isolated by avidin affinity chromatography (i.e., using the biotin tag on the label). The isolated peptides are then analyzed by mass spectrometry. A typical mass spectrum obtained by the process is shown in Figure 18.6.

A *phosphoprotein isotope-coded affinity tag* (PhIAT) has similarly been used to selectively label phosphoseryl and phosphothreonyl residues [17]. This process, shown in Figure 18.7, involves the removal of a phosphate group on a protein and its replacement with an ethylene dithiol, which contains either only hydrogen or four deuterium atoms on it. This is followed by biotinylation of the molecule. Proteins present in cells under different conditions can be compared in terms of their levels of phosphorylation by differentially tagging them with the nondeuterated or deuterated PhIAT reagents and comparing these by mass spectrometry. One advantage of PhIAT over the conventional isolation of phosphoproteins by immobilized metal-ion affinity chromatography is the higher selectivity of the PhIAT process.

One problem with using an affinity tag for protein purification is that the tag must later be removed. This process can sometimes lead to degradation of the protein. The need for an additional step to separate the protein from the cleaved tag must also be addressed. A possible solution to these problems is to use *inteins*. Inteins are protein-splicing elements

Figure 18.6 Representative mass spectrum using the ICAT reagent. (Reproduced with permission from Gygi, S.P., Rist, B., Gerber, S.A., Turecek, F., Gelb, M.H., and Aebersold, R., *Nat. Biotechnol.*, 17, 994–999, 1999.)

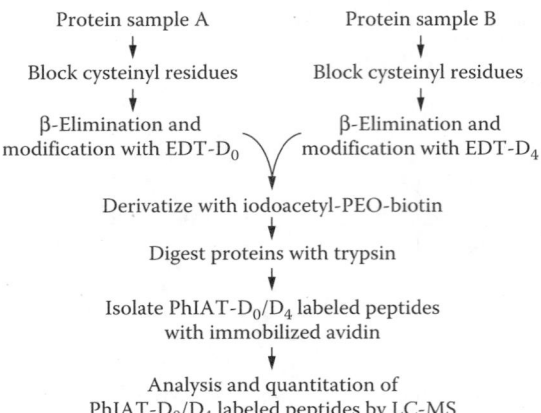

Protein sample A   Protein sample B
        ↓   ↓
Block cysteinyl residues   Block cysteinyl residues
        ↓   ↓
β-Elimination and   β-Elimination and
modification with EDT-D_0 modification with EDT-D_4

Derivatize with iodoacetyl-PEO-biotin
        ↓
Digest proteins with trypsin
        ↓
Isolate PhIAT-D_0/D_4 labeled peptides
with immobilized avidin
        ↓
Analysis and quantitation of
PhIAT-D_0/D_4 labeled peptides by LC-MS

Figure 18.7 Analytical scheme for the relative quantitation and isolation of phosphopeptides from two different samples using the PhIAT approach. (Reproduced with permission from Goshe, M.B., Conrads, T.P., Panisko, E.A., Angell, N.H., Veenstra, T.D., and Smith, R.D., *Anal. Chem.,* 73, 2578–2586, 2001.)

that cleave under conditions that leave the target protein unchanged [59]. If an affinity tag is attached to the intein and the intein is attached to the target molecule, the target can be cleaved from the intein, leaving the intein bound to the affinity column. The intein can then be eluted separately. This process is illustrated in Figure 18.8.

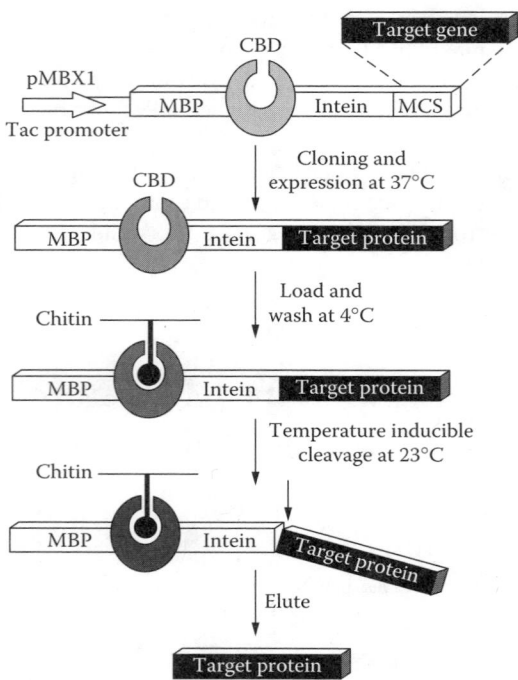

Figure 18.8 Purification scheme using an intein splice. (Reproduced with permission from Southworth, M.W., Amaya, K., Evans, T.C., Xu, M.Q., and Perler, F.B.P., *BioTechniques*, 27, 110–120, 1999.)

18.3 INDUSTRIAL APPLICATIONS

Affinity chromatography has played an important role in enzyme and protein purification for many decades (see Chapters 12 to 16). This section considers some specific applications for affinity methods in biotechnology, such as its use in downstream processing and process monitoring.

18.3.1 Downstream Processing

The downstream processing of recombinant proteins can account for over half the cost of their production [60]. Proteins destined for pharmaceutical use must be highly purified, and yields need to be as high as possible to maximize efficiency. The ideal production process would be fast, simple, and require only a minimum number of steps. Unfortunately, for most methods, these needs are often at odds with one another. However, affinity chromatography is almost ideally suited for downstream processing, since it can be fast, highly selective, and capable of high yields with the proper ligand selection.

18.3.1.1 Affinity Ligands for Downstream Processing

The demands of protein purification for the pharmaceutical market put additional demands on the selection criteria for affinity ligands. For instance, in addition to being selective, the ligand must be stable under the harsh conditions used for column cleanup and depyrogenation [47]. The ligand must also be readily available at a reasonable cost. In addition, the targeted molecules must be retrieved with little or no loss of bioactivity.

Most of the ligands used in affinity chromatography are selective but fail to meet such requirements as high stability and low cost. Because of these difficulties, ion-exchange techniques are often used instead for downstream processing. Although ion-exchange materials are stable, inexpensive, and allow elution with a salt gradient (which helps preserve bioactivity), they do lack the selectivity of affinity columns.

Biomimetic dyes have been used as affinity ligands in several applications with great success [47, 48, 51]. A review of such work can be found in Chapter 9. These dyes can be designed to mimic the active sites of many enzymes, rendering them very selective. As discussed earlier, they often make use of relatively general electrostatic interactions, but their selectivity can be improved through the use of affinity elution. These dyes are quite stable, even under the most extreme conditions used for column cleanup. They are also much less costly than biological ligands of comparable selectivity and provide recoveries that are often 90% or better. Furthermore, the elution conditions used with biomimetic dyes are sufficiently mild to allow the retention of activity for most biomolecules.

Immobilized metal-ion affinity chromatography with Ni(II) has been employed in conjunction with a genetically coded polyhistidine for the purification of glutathione-S-transferase [21]. Elution was performed with 500 mM imidazole in 100 mM sodium phosphate and 0.5 M sodium chloride. These conditions allowed for 85% of this enzyme's original activity to be retained. Although the polyhistidine tag did not seem to affect the structure or function of this protein, the tag could be enzymatically cleaved and removed by immobilized metal-ion affinity chromatography. Since Ni(II) is potentially harmful, additional steps would be necessary to ensure its complete removal from the finished product for use with living systems.

18.3.1.2 Methods for Downstream Processing

The processing of recombinant proteins by affinity chromatography is a well-understood process and makes use of established methods. However, the use of a packed chromatographic column for this purpose does require one or more preparative steps to first remove whole cells and cell fragments that would quickly foul the column. Every additional step added to the purification process results in lower yields and higher costs. Thus, the development of alternative methods that allow for the elimination of these pretreatment steps is desirable.

One such alternative method is *fluidized-bed adsorption*. This method allows the packing in a column to expand with the application of a sample during solvent flow in a reversed. (i.e., bottom-to-top) direction. This results in expansion of the packed bed and allows the passage of particulate matter through the column, eliminating the need for sample filtration or centrifugation prior to processing. Ideally, the adsorbents used in these columns should expand but maintain their relative positions in the column with little or no mixing, which would reduce the efficiency of the separation.

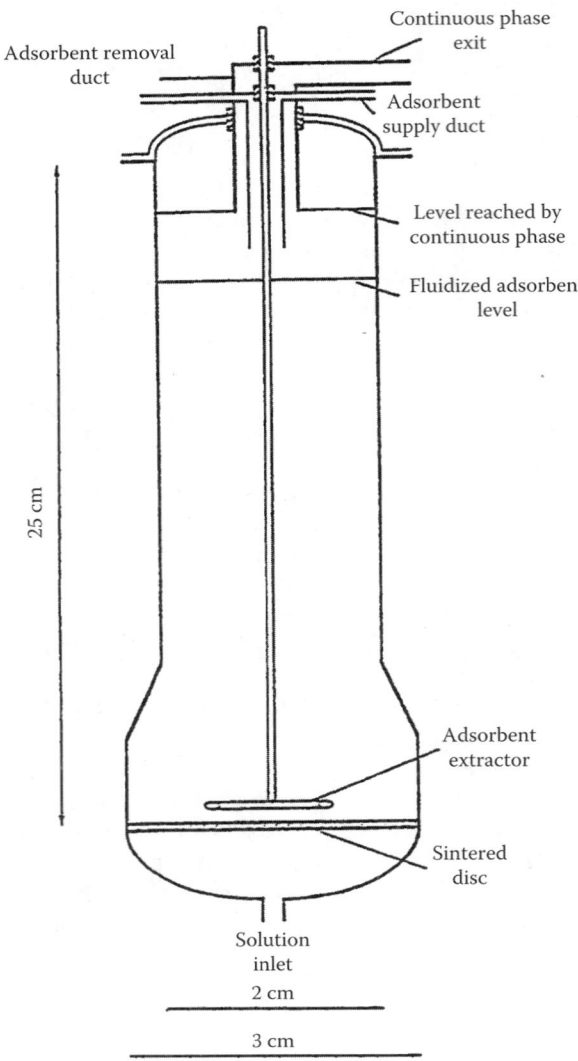

Figure 18.9   Illustration of a countercurrent contactor. (Reproduced with permission from Owen, R.O., McCreath, G.E., and Chase, H.A., *Biotechnol. Bioeng.,* 53, 427–441, 1997.)

An improvement on fluidized-bed adsorption can be obtained by using *expanded-bed adsorption* (EBA). This differs from fluidized-bed adsorption primarily in the degree of back-mixing that occurs. With careful control of the flow rates and particle density, and by incorporating a range of particles in the bed, it is possible in EBA to produce a static distribution of adsorbent particles in the column. Efficiencies using this technique can be attained that rival packed columns while allowing unclarified feedstocks to be processed [60, 61]. A further review of expanded-bed adsorption in industrial processing can be found in the literature [62].

Another alternative approach is to use a continuous method of processing. Continuous methods do not require the repeated loading, washing, elution, and regeneration steps needed with a batch process. This greatly increases the actual time spent performing separations versus these other steps and improves the overall efficiency.

Owen described a system in which an adsorbent was continuously cycled through the various stages [51]. This process made use of four contactors (see details given in Figure 18.9) arranged in a series that corresponded to the four steps of a batch process, as shown in Figure 18.10. In this system, the first contactor was used to load the feedstock, the second to wash, the third to elute the purified product, and the fourth to regenerate the adsorbent, all performed in a continuous manner.

A system based on *continuous annular chromatography* (CAC) has been described for processing a complex feed containing recombinant antibodies [28]. In CAC, an adsorbent is loaded between two concentric cylinders. The annulus is slowly rotated while the feed and mobile phase are continuously added from fixed inlets, as illustrated in Figure 18.11. The proteins in the feed are washed down through this bed in a helical band and, once an equilibrium state has been reached, they are eluted at some constant point along the cylinder. Results obtained with such a system are comparable with those obtained with a standard batch process.

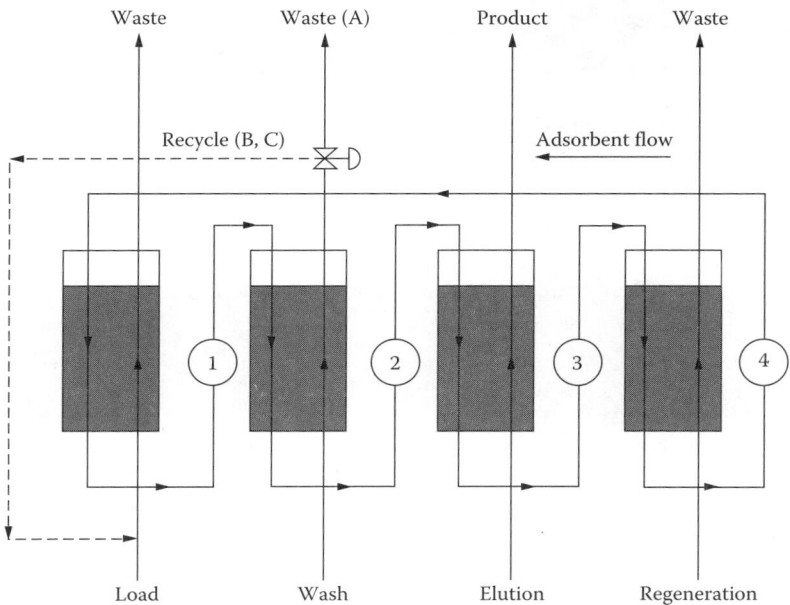

Figure 18.10 A four-stage countercurrent contactor arrangement. (Reproduced with permission from Owen, R.O., McCreath, G.E., and Chase, H.A., *Biotechnol. Bioeng.*, 53, 427–441, 1997.)

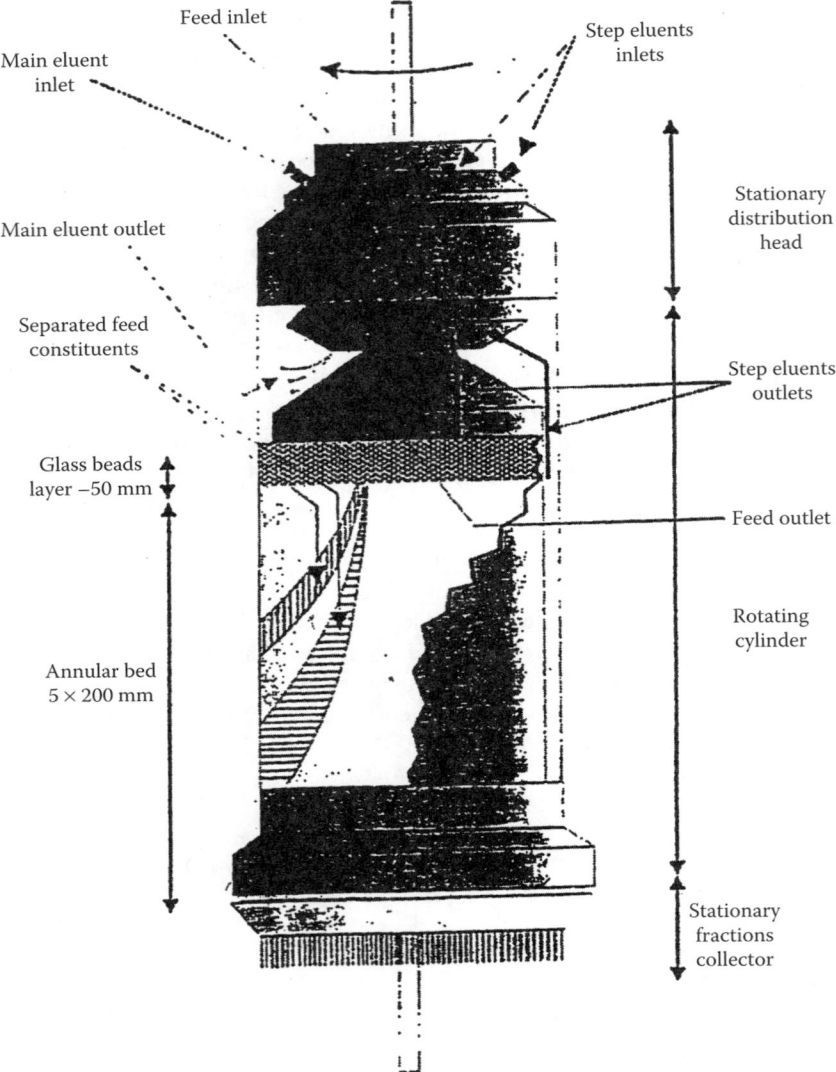

Figure 18.11   A continuous annular chromatographic system. (Reproduced with permission from Giovannini, R. and Freitag, R., *Biotechnol. Bioeng.,* 73, 522–529, 2001.)

18.3.2   Process Monitoring

The need for process monitoring in biotechnology is evident, since small problems in a fermentor can rapidly ruin any product. As an example, a single bacterium in a 100-l fermentor can completely overwhelm the system in less than 2 days [63]. Fermentation products will also start to degrade if not harvested. Living organisms will show variations in production rates from batch-to-batch and should be monitored to determine optimum harvest times. Thus, there is a need to monitor the fermentor to determine that production is proceeding within accepted limits.

To be effective, any monitoring system for a fermentor must be fast, selective, and sensitive. Moreover, the sampling that must be done while monitoring the process must be performed without introducing contamination to the product. Automation of this process is preferred. With the exception of the sampling technique, affinity chromatography can meet all these stated requirements. Affinity ligands can be chosen that have high levels of selectivity and sensitivity. Typical analysis times are just a few minutes, and affinity chromatographic methods have routinely been automated.

An automated system for monitoring the production of monoclonal antibodies using a protein A affinity column has been described [63]. Samples were taken automatically every 2 h, and the antibody level was determined within 3 min. The analysis was fast enough to allow the fermentor to be adjusted if necessary. Upon reaching a predetermined limit, the fermentor was automatically harvested. Since the analysis time was fast compared with the culture time, this same monitoring system could be applied to several fermentors. A similar approach could be used for other recombinant proteins by employing appropriate ligands for these targets.

18.4 ANALYTICAL APPLICATIONS

Besides preparative applications, affinity chromatography has also been used as an analytical tool in biotechnology. This section presents several examples, including its use in proteomics, drug development, and clinical diagnostics.

18.4.1 Proteomics

18.4.1.1 Analysis of Complex Mixtures

In proteomics, one of the most daunting challenges facing researchers is the sheer number of proteins found in an organism and even in an organelle. To try to isolate and identify every protein individually would be time consuming, inefficient, and expensive. Electrophoretic methods provide good resolution, but they are not capable of separating the hundreds and perhaps thousands of proteins that may be found in a single organelle. Two-dimensional gel electrophoresis has the required resolution but is time consuming and is not easily automated. Furthermore, low-abundance proteins are at risk of being lost in the large number of spots that need to be excised, eluted, and digested for identification by mass spectrometry.

The ability to fractionate a sample and concentrate minor proteins by affinity chromatography prior to electrophoresis can greatly increase the resolving power of such a method without greatly increasing its overall analysis time. For instance, Taylor et al. [64] used ion-exchange chromatography combined with two-dimensional gel electrophoresis to analyze the proteins of the rat liver Golgi complex. Almost 500 additional spots were observed in the final stained gel, and 22 previously unknown proteins were identified. Although an ion-exchange column was used in this particular method for sample pretreatment, affinity ligands could be used in the same manner.

The need to physically cut out and process the spots in a gel makes the isolation of minor proteins difficult and time-consuming. Methods that can be directly coupled to a mass spectrometer, even though they may possess lower resolving power, would be faster and more sensitive to the presence of minor proteins. Ion-exchange chromatography has been coupled with reversed-phase liquid chromatography (RPLC) and directly interfaced with a mass spectrometer to provide completely automated assays that can separate tryptic digests of complex protein mixtures in about 2 h [65–67]. The use of RPLC as the second dimension provided on-line desalting of the peptide mixtures before they were analyzed by mass spectrometry. Protein identification was achieved by matching the mass spectra obtained for the peptide fragments to computer libraries of known protein spectra.

An extension of this approach uses the idea that a protein can be identified by certain peptides obtained from a tryptic digest that is unique to that protein [68]. By using these signature peptides for identification, it is not necessary to separate and analyze all peptides from a digest, but only those that compose the signature. This allows the analysis to be significantly simplified. This approach was used to identify glycoproteins in human serum. In this method, the sample was reduced and alkylated, digested with trypsin, and applied to a lectin-affinity column to bind the glycopeptides. The retained peptides were eluted and directly transferred to a reversed-phase column using automated switching valves. The peptides were then deglycosylated and spotted onto a plate for analysis by matrix-assisted laser-desorption ionization (MALDI) mass spectrometry. Comparison of the resulting mass spectra with human protein databases allowed for the identification of these peptides and their corresponding proteins. The method was found to be fast and allowed for the isolation of proteins bearing a particular posttranslational modification. This particular method is selective for glycoproteins, but any type of protein could be isolated with the appropriate choice of ligand. A drawback to this method is the need for sample handling during the tryptic digestion and placing of the sample onto a MALDI plate.

Riggs et al. [16] described an automated system that reduced and alkylated proteins in an autosampler, and used an immobilized trypsin column for digestion. This system also included a direct electrospray ionization interface between the RPLC column used for digest separation and a mass spectrometer. A diagram of this system is shown in Figure 18.12. Using a Ga(III) chelate column to select for phosphoproteins in milk, a complete analysis was performed in 1 h and 35 min. This system also was able to process several samples simultaneously at different stages of the process, which further reduced the time needed to process multiple samples.

18.4.1.2   Structural Studies

18.4.1.2.1   Primary-Structure Variants.   The presence of primary-structure variants has been implicated in a variety of diseases and is thus of great medical importance. In biotechnology, structural variations in expressed proteins result in lower yields and by-products that might be immunogenic. Affinity chromatography is one approach that can be used to detect such variants.

A quick but effective affinity-chromatography method for detecting primary-structure variants in human hemoglobin has been described [7]. In this method, the target protein was first purified by immunoaffinity chromatography, desalted, and digested with an immobilized trypsin column. The digest was then separated by RPLC and monitored by ESI-MS. The immunoaffinity column allowed the isolation of hemoglobin from serum with greater than 98% purity and minimal interference from contaminant proteins. As this hemoglobin was eluted from the immunoaffinity column, it was transferred directly to an ion-exchange column that contained equal amounts of cation and anion exchangers. This was necessary to guarantee that all protein was retained when the pH was changed from the low value used for immunoaffinity elution to the higher pH that was used later to provide maximum activity for the trypsin column.

The setup for this system is shown in Figure 18.13. All of the columns were directly connected, and the entire process was automated, with the time from sample injection to final results being less than 2 h [7]. A comparison of the total ion current chromatograms indicated that a definite change had occurred in one of the detected peptides. Further analysis of the mass spectra for the variant peak and corresponding normal peak could indicate the exact location and type of modification that occurred for this peptide.

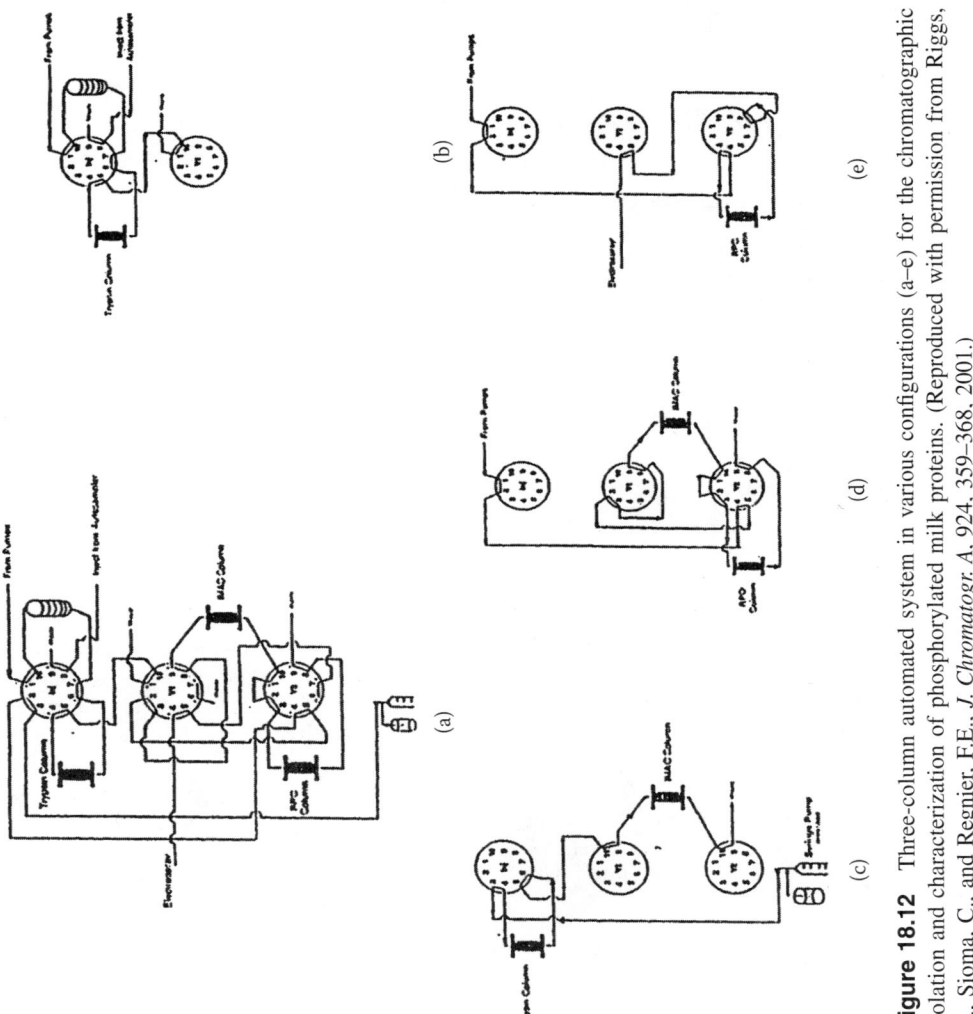

Figure 18.12 Three-column automated system in various configurations (a–e) for the chromatographic isolation and characterization of phosphorylated milk proteins. (Reproduced with permission from Riggs, L., Sioma, C., and Regnier, F.E., *J. Chromatogr. A*, 924, 359–368, 2001.)

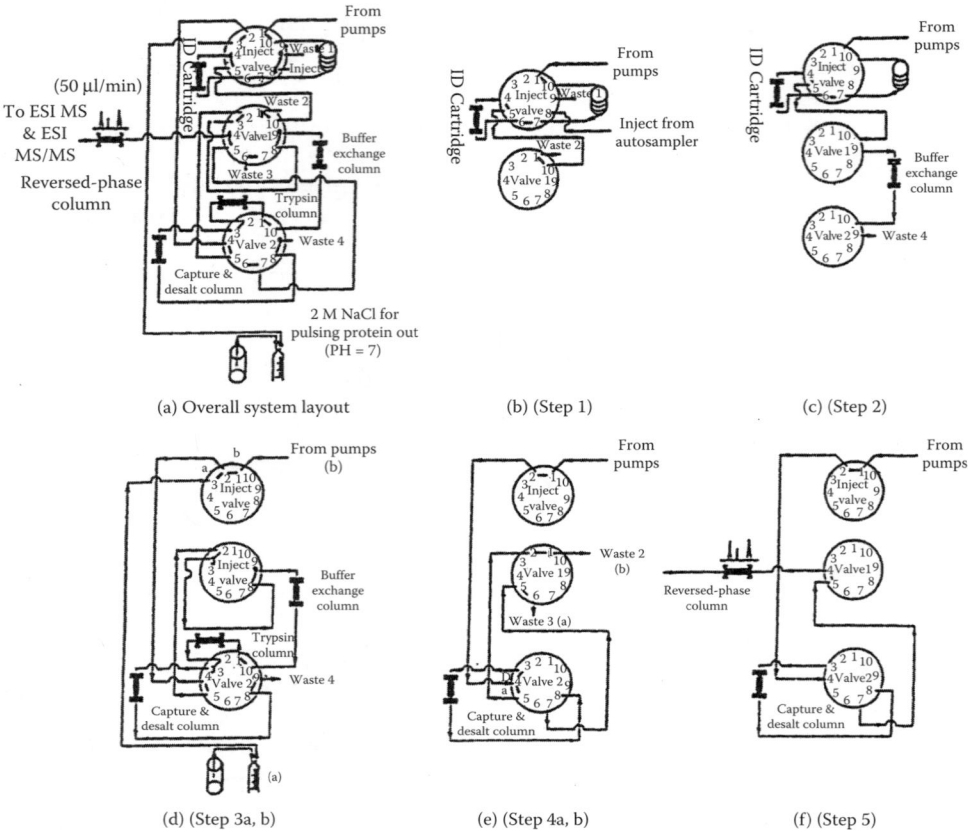

Figure 18.13 Five-column automated system for the isolation and characterization of human hemoglobin. The diagram in (a) shows the overall system. In Step 1 (b), immobilized antibodies are used in the ID cartridge to bind hemoglobin. In Step 2 (c), the bound hemoglobin is eluted to a buffer exchange column, where the pH is raised to 7.5. In Step 3 (d), 2 *M* NaCl is added to this column to elute the retained hemoglobin, which is then digested by immobilized trypsin and captured by a reversed-phase column. In Step 4 (e), the reversed-phase column is purged with 0.1% trifluoroacetic acid (TFA), and salts are removed from this column. Finally, in Step 5 (f), the hemoglobin digest is eluted onto a reversed-phase analytical column, separated, and monitored by mass spectrometry. (Reproduced with permission from Hsieh, Y.L., Wang, H., Elicone, C., Mark, J., Martin, S.A., and Regnier, F., *Anal. Chem.*, 68, 455–462, 1996.)

18.4.1.2.2 Posttranslational Modifications.

Phosphorylation and glycosylation are two common posttranslational modifications found in proteins. Such modifications are important in determining protein function, and as stated earlier, loss of control for these modifications may play a role in many diseases. Methods like amino acid analysis or nucleotide sequencing of a protein's gene can provide primary-structure information, but they do not provide information on the presence of these modifications. However, affinity chromatography with ligands specific for phosphoryl and glycosyl groups can be used for this purpose.

Immobilized metal-ion affinity columns, specifically those containing Ga(III) and Fe(III), can be used to select for phosphorylated peptides in a protein digest. Comparison of the peptides retained by such columns with the peptides present in the complete digest will indicate which sequences have been phosphorylated. In one such study, Posewitz and Tempst [14] isolated phosphopeptides of human β integrin using a Ga(III) column. The mass spectra that were obtained are shown in Figure 18.14. Analysis of these results can indicate not only which peptides have been phosphorylated, but also their degree of phosphorylation.

Modification by glycosylation can involve many different saccharides. Being able to distinguish the exact form of glycosylation is a problem that has yet to be resolved. As was discussed in the section on targeting glycosyl groups (Section 18.2.5), there are affinity ligands like lectins that have some selectivity for the various types of saccharides. For instance, the selection of N-type hybrid and high-mannose oligosaccharides by the lectin concanavalin A is based on the extent of branching [24, 69], and the lectin BS-II appears to preferentially bind O-linked over N-linked oligosaccharides [70]. Boronate ligands are much broader in their range of targets but have higher affinity for the 1,2-cis-diols of glucose and similar sugars. These latter ligands have been particularly useful in binding polysaccharides with low sialic acid content [26]. By combining different types of ligands, it is possible to determine in a general sense the type of sugar groups involved in glycosylation. Some examples include the use of serial lectin chromatography for differentiating the various isozymes of alkaline phosphatase [71], human serotransferrin [69], and prostatic acid phosphatase [25].

18.4.1.2.3 Metabolic Studies. The fact that different proteins are expressed at different rates in a cell seems obvious. However, a single cell may have different rates of expression for a protein under various conditions. The ability to determine how a change in a cell's state will affect its rate of protein expression can provide useful information on the cell's metabolism and the protein's function. Affinity methods have been developed that allow for either the direct quantitation of a protein in a given mixture or determination of the relative amounts of the same protein in two different mixtures.

Ji et al. [70] developed a method that involved the tryptic digestion of a protein mixture followed by acetylation of this digest with either N-acetoxysuccinimide or N-acetoxy-d₃-succinimide. This caused the free amino groups and lysines on each peptide to be labeled, with the two types of acetylation agents giving products that differed in mass by 3 to 6 Da. These labeled peptides were then affinity-selected using either immobilized metal-ion affinity chromatography or lectin columns. Following their elution from these affinity columns, the peptides were further fractionated and desalted by reversed-phase chromatography, which was interfaced with ESI-MS. The original protein content of the mixtures was determined using the signature-peptide approach, where the mass spectra obtained from the peptide mixtures were compared with the mass spectra of affinity-selected peptides from digests of known proteins.

By mixing these types of labeled peptides from cells that have been grown under different conditions, the relative rates of protein expression can be determined in these cell populations. Absolute quantitation can be obtained by mixing the labeled peptide digest with a digest of a known amount of the same protein mixture, which is used as an internal standard. Although this method, as described here, is not totally automated, it has been demonstrated elsewhere [16] that every step in this system can be linked in an automated fashion, thereby improving accuracy and reducing the analysis time.

A similar method for determining the relative rates of protein expression using the ICAT technique has been described [36]. In this work, the proteins that were expressed by *Saccharomyces cerevisiae* in the presence of two different carbon sources were analyzed. For this experiment, the proteins from one group of cells were tagged with the heavy ICAT reagent, and those from the other group were tagged with the light ICAT reagent. The two groups were then mixed, digested, and the ICAT tagged peptides (i.e., those originally

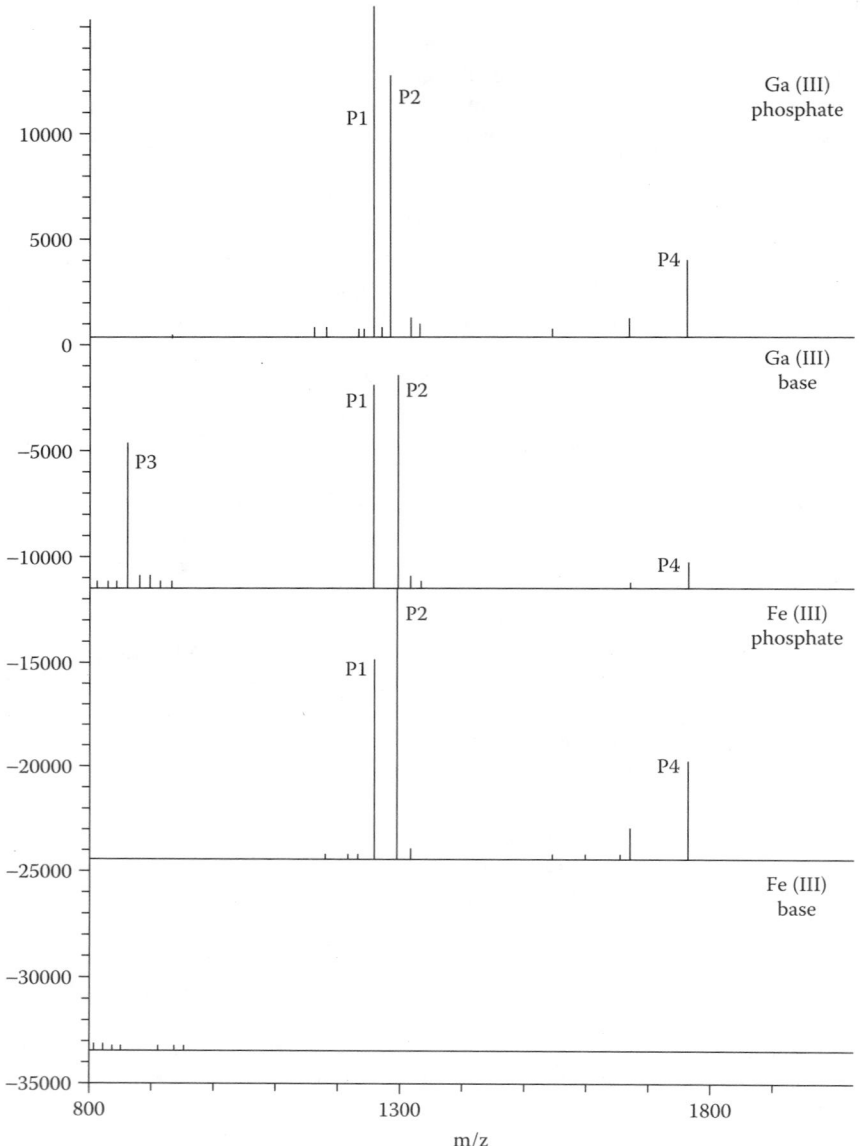

Figure 18.14 Mass spectra of four phosphopeptides that were captured and eluted from immobilized Ga(III) and Fe(III) supports. These peptides were eluted using either a phosphate-containing buffer (first and third mass spectra) or a mobile phase that contained ammonium hydroxide as a base (second and fourth mass spectra). (Reproduced with permission from Posewitz, M.C. and Tempst, P., *Anal. Chem.*, 71, 2883–2892, 1999.)

containing cysteine) were affinity-selected. The captured peptides were then eluted from the affinity column, fractionated, desalted by RPLC, and examined by MALDI-MS. These proteins were identified by the signature-peptide approach, with a comparison of the heavy and light peptide peaks in the mass spectra allowing for the relative quantitation.

Others have used the PhIAT reagent to isotopically tag and isolate phosphorylated peptides [17]. In this procedure, the proteins are digested and the phosphorylated peptides tagged with the PhIAT reagent. These peptides are then affinity-selected and analyzed by mass spectroscopy. By tagging two digests, one labeled with the heavy PhIAT reagent and the other with the light PhIAT reagent, the relative amounts of phosphorylated peptides in each group can be discerned. Additionally, any change in the phosphorylation of a protein will be indicated by the appearance or disappearance of a peak in the resulting mass spectra.

18.4.2 Drug Development

In the development of any drug, there are several criteria of importance. For instance, the drugs must be selective toward a given target and bind quickly and permanently to this target to be effective with minimal side effects. Affinity fingerprinting [55, 56] can be used to quickly screen potential candidates for their ability to meet these criteria. It may also be possible with this approach to predict cross-reactivity with other drugs that may produce unwanted side effects.

Hage and coworkers [72, 73] have performed extensive work in examining the binding characteristics of small molecules to human serum albumin. Using immobilized human serum albumin in affinity columns, they determined the binding kinetics and strengths of small molecules like warfarin and clomiphene. Binding ratios could also be calculated and, with the use of site-specific probes, binding sites could be identified. The competitive binding between drugs can also be analyzed in this manner.

Many drugs are known to bind to melanin, often with deleterious side effects [74]. This binding has been studied for several drugs using affinity chromatography. The binding constants for these drug-melanin interactions have been determined, along with the competitive binding between some drugs for this binding agent. Melanin also accumulates several carcinogenic substances. The existence of these various affinities suggests that melanin-mediated medications may be useful for the treatment of some cancers. This makes affinity chromatography attractive as a possible means for describing the interactions of such medicines with melanin.

As biotechnology continues to develop, the importance of immunotherapies and immunodiagnostics will become even greater. Several applications of bispecific antibodies in these areas have already been investigated [4, 75, 76]. As is the case with small-molecule drugs, the binding characteristics of the proteins and other macromolecules used in these areas will need to be investigated. Although the work described so far in this chapter has been performed with small molecules, the same approaches should be applicable for characterizing macromolecular binding. More information on approaches that could be used to characterize macromolecular binding is given in Chapters 22 and 23.

Phillips [77] used affinity chromatography to analyze samples taken from bone marrow aspirate fluid, cerebrospinal fluid, and plasma to determine their concentration of recombinant human granulocyte colony-stimulating factor. While the target colonies develop in the bone marrow, it is beneficial to know whether the administered drug is in fact reaching the target area. Any of this factor that is found in the cerebrospinal fluid will not achieve its intended purpose and might actually pose a risk by creating unwanted side effects. The results of this study demonstrate how affinity chromatography could be used to study and identify problems with the distribution of a drug in the body.

18.4.3 Diagnostics

The use of affinity chromatography in clinical diagnostics is another area that is related to biotechnology. A detailed review of this area can be found in Chapter 17, but a few examples are given in this section. One area in which affinity methods might be useful is, again, in the analysis of protein structure and posttranslational modifications. This is the case because changes in glycosylation have been observed in carcinoma [25], chronic uremia [24], osteoblastic bone disorder [71], and cholestatic disorders [71]. Changes in phosphorylation have also been implicated in certain types of cancers [14].

Variations in the levels of normal proteins have also been implicated in many diseases. For instance, abnormally high and low levels of fibrinogen are the cause of coagulopathies and cardiovascular disease [78]. Other examples include the problems associated with high levels of high-density lipoproteins in the blood, the use of DNA adducts as an early sign of cancer [3], and the production of an immunogenic response as the result of infections. In all of these instances, there will be some change produced in the amount or type of biological agent that, if detected, can be used for diagnostics. The sensitivity and selectivity of affinity chromatography, even for detection of differences in posttranslational modifications, makes it ideally suited for this purpose.

In one report, affinity chromatography was used to detect low levels of urinary albumin [79]. The presence of albumin in urine is indicative of the onset of kidney disease. But one of the problems in analyzing for any urinary component is the normal fluctuation of urine production and volume. This can be overcome by also testing for creatinine, which is released into the urine at a relatively constant rate. To do this, an antialbumin antibody was used to trap the albumin from serum on an affinity column. The nonretained fraction, which contained creatinine, was automatically directed to a reaction coil, where it was reacted with picric acid to form a colored complex. The absorbance of both this complex and the albumin that was later eluted from the antibody column was then measured. The system that was used for this purpose is shown in Figure 18.15. By determining the creatinine level in each sample, it was possible to correct for any dilution or concentration of the albumin due to changes in urine production. This entire system was fully automated and could process one sample every 5 min. This represents a significant decrease in analysis time over other available methods, while also providing more-precise results.

Another report used lectin columns to help detect osteoblastic and choleostatic hepatic disorders by measuring changes in the production of bone and liver alkaline phosphatase. Wheat germ lectin was used to first capture and then separate these two isozymes based on gradient elution [71]. Abnormal changes in the relative amounts of the two isozymes could be detected prior to any changes in total enzyme production. This, in turn, allowed clinicians to determine the likely source of the disorder. This method gave a limit of detection that was tenfold lower than that of traditional methods but with comparable accuracy.

A third study used affinity chromatography to measure fibrinogen [78]. Fibrinogen is an important blood protein that is involved in coagulation. Increased levels of this protein can result from myocardial infarction, burns, fractures, surgery, or pregnancy. In patients that have suffered a myocardial infarction, increased levels of fibrinogen lead to an increased risk of thromboembolism. Decreased levels can result from hereditary disorders, severe hemorrhaging, thrombolytic therapy, and other sources. Reduced levels of fibrinogen can result in excessive bleeding. Consequently, it is frequently necessary to quickly and accurately monitor fibrinogen levels in the blood. With antibody-based columns, it was possible to create a fast, simple, accurate, and sensitive method for this protein. The total analysis time was 15 min, and the recovery for fibrinogen was greater than 80%, with good reproducibility and no apparent interferences.

Numerous other diagnostic methods have been developed based on affinity chromatography. Examples include methods that have been reported for the analysis of prostatic acid phosphatase in serum for detecting human prostate carcinoma [25], human chorionic gonadotropin in urine as a measure of stimulated steroid production [80], and prostate-specific

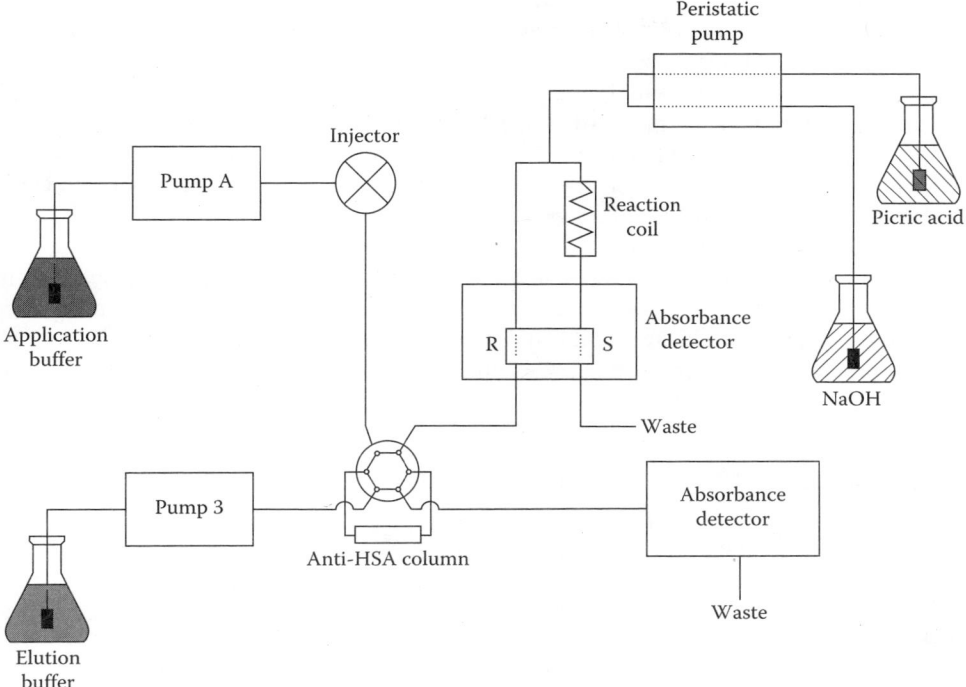

Figure 18.15 An automated system for the determination of urinary albumin based on immunoaffinity chromatography and flow-injection analysis. Samples were applied to the anti-HSA antibody column using a neutral-pH phosphate application buffer and were eluted using a lower pH buffer. Nonretained components that passed through this column were allowed to enter the on-line flow-injection analysis system, where the eluant was combined with sodium hydroxide and picric acid for the detection of creatinine. The symbols R and S refer to the reference and sample flow cells in the absorbance detector, respectively. (Reproduced with permission from Ruhn, P.F., Taylor, J.D., and Hage, D.S., *Anal. Chem.,* 66, 4265–4271, 1994.)

antigen as a marker for prostate cancer [4]. Affinity chromatography has also been used in a technique for measuring high-density lipoproteins (Zhong and Krull, unpublished results).

18.5 SUMMARY AND CONCLUSIONS

The vast number of ligands available for affinity chromatography has allowed this method to find applications in industrial, analytical, and clinical settings. With the ability to raise antibodies against a wide variety of compounds and the availability of new synthetic ligands, the possible applications seem boundless. With improvements in computer-aided molecular design and expanding affinity fingerprint libraries, it is not unreasonable to think that a ligand can be found or designed for any analyte of known structure. Increases in ligand selectivity and sensitivity should be realized along with reduced costs and increased stability.

Should the cost of materials reach sufficiently low levels, affinity chromatography could be applied to the purification of fluids and gases. Systems could be designed to eliminate microorganisms from the air and water supplies of sensitive buildings or as smaller portable units. Similar devices could be put to use as biohazard detectors in areas where biological contamination is suspected.

The process of expanded-bed adsorption, in a clinical setting, could permit the automated and rapid analysis of samples through the direct injection of whole biological fluids; this would eliminate much of the need for sample handling and keep the exposure of laboratory workers to biofluids at a minimum. Applications in blood purification would guarantee usable blood from less-than-ideal sources and greatly expand the donor pool. The ability to rapidly and effectively filter blood could lead to the use of affinity chromatography as a first-stage treatment in blood-borne diseases, lowering the levels of pathogens in blood and reducing the use of drugs in treatment or, in the case of untreatable diseases, extending the life of the patient.

Ultimately, the role of affinity chromatography in biotechnology should expand. It is an extraordinarily powerful separation method that will prove to become only more so in the future. As advances in structural and functional proteomics expand the use of biotechnology products, affinity chromatography will prove to be invaluable in the analysis of these products.

SYMBOLS AND ABBREVIATIONS

BS-II	*Bandeiraea simplicifolia* lectin
CAC	Continuous annular chromatography
DNA	Deoxyribonucleic acid
EBA	Expanded-bed adsorption
EDTA	Ethylene diamine tetraacetate
ELISA	Enzyme-linked immunosorbent assay
ESI-MS	Electrospray ionization mass spectrometry
FSH	Follicle-stimulating hormone
Gc	Glycocalicin
GC	Gas chromatography
GST	Glutathione-*S*-transferase
ICAT	Isotope-coded affinity tag
IRP	International Reference Preparations
IMAC	Immobilized metal-ion affinity chromatography
LH	Luteinizing hormone
MALDI	Matrix-assisted laser desorption ionization
MALDI-MS	Matrix-assisted laser desorption ionization mass spectrometry
PBA	Phenylborate agarose
PhIAT	Phosphoprotein isotope affinity tag
RPLC	Reversed-phase liquid chromatography
TBS	Tris-buffered saline
αMG	α-Methylglucopyranoside
αMM	α-Methylmannopyranoside

REFERENCES

1. Majors, R.E. and Carr, P., Glossary of liquid-phase separation terms, *LC/GC*, 19, 124–162, 2001.
2. Hage, D.S. and Kao, P.C., High-performance immunoaffinity chromatography and chemiluminescent detection in the automation of a parathyroid hormone sandwich immunoassay, *Anal. Chem.*, 63, 586–595, 1991.

3. Bonfanti, M., Magagnotti, A., Galli, A., Bagnati, R., Moret, M., Gariboldi, P., Fanelli, R., and Airoldi, L., Determination of O^6-butylguanine in DNA by immunoaffinity extraction/gas chromatography-mass spectrometry, *Cancer Res.,* 50, 6870–6875, 1990.

4. Kreutz, F.T. and Suresh, M.R., Novel bispecific immunoprobe for rapid and sensitive detection of prostate-specific antigen, *Clin. Chem.,* 43, 649–656, 1997.

5. Cole, L.J. and Kennedy, R.T., Selective preconcentration for capillary zone electrophoresis using protein G immunoaffinity capillary chromatography, *Electrophoresis,* 16, 549–556, 1995.

6. Hacker, A., Hinterleitner, M., Shellum, C., and Gubitz, G., Development of an automated flow injection chemiluminescence immunoassay for human immunoglobulin G, *Fresnius' J. Anal. Chem.,* 352, 793–796, 1995.

7. Hsieh, Y.L., Wang, H., Elicone, C., Mark, J., Martin, S.A., and Regnier, F., Automated analytical system for the examination of protein primary structure, *Anal. Chem.,* 68, 455–462, 1996.

8. Johns, M.A., Rosengarten, L.K., Jackson, M., and Regnier, F.E., Enzyme-linked immuno-sorbent assays in a chromatographic format, *J. Chromatogr. A,* 743, 195–206, 1996.

9. Phillips, T.M., High-performance immunoaffinity chromatographic detection of immuno-regulatory anti-idiotypic antibodies in cancer patients receiving immunotherapy, *Clin. Chem.,* 34, 1689–1692, 1988.

10. Vlatakis, G. and Bouriotis, V., Sequence-specific DNA affinity chromatography: Application to the purification of *EcoRI* and *SphI, Anal. Biochem.,* 195, 352–357, 1991.

11. Pozidis, C., Bouriotis, V., and Vlatakis, G., Sequence-specific DNA affinity chromatography: application of a group-specific adsorbent for the isolation of restriction endonucleases, *J. Chromatogr.,* 630, 151–157, 1993.

12. Andersson, L. and Porath, J., Isolation of phosphoproteins by immobilized metal (Fe^{3+}) affinity chromatography, *Anal. Biochem.,* 154, 250–254, 1986.

13. Muszynska, G., Dobrowolska, G., Medin, A., Eckman, P., and Porath, J., Model studies on iron(III) affinity chromatography, II: Interaction of immobilized iron (III) ions with phosphorylated amino acids, peptides and proteins, *J. Chromatogr.,* 604, 19–28, 1992.

14. Posewitz, M.C. and Tempst, P., Immobilized gallium(III) affinity chromatography of phosphopeptides, *Anal. Chem.,* 71, 2883–2892, 1999.

15. Andersson, L., Recognition of phosphate groups by immobilized aluminum(III) ions, *J. Chromatogr.,* 539, 327–334, 1991.

16. Riggs, L., Sioma, C., and Regnier, F.E., Automated signature peptide approach for proteomics, *J. Chromatogr. A,* 924, 359–368, 2001.

17. Goshe, M.B., Conrads, T.P., Panisko, E.A., Angell, N.H., Veenstra, T.D., and Smith, R.D., Phosphoprotein isotope-coded affinity tag approach for isolating and quantitating phospho-proteins in proteome-wide analyses, *Anal. Chem.,* 73, 2578–2586, 2001.

18. Porath, J., Carlsson, J., Olsson, I., and Belfrage, G., Metal chelate affinity chromatography, a new approach to protein fractionation, *Nature,* 258, 598–599, 1975.

19. Nakagawa, Y., Yip, T.T., Belew, M., and Porath, J., High-performance immobilized metal ion affinity chromatography of peptides: Analytical separation of biologically active synthetic peptides, *Anal. Biochem.,* 168, 75–81, 1988.

20. Carlsson, J., Mosbach, K., and Bulow, L., Affinity precipitation and site-specific immobili-zation of proteins carrying polyhistidine tails, *Biotechnol. Bioeng.,* 51, 221–228, 1996.

21. Clemmitt, R.H. and Chase, H.A., Facilitated downstream processing of a histidine-tagged protein from unclarified *E. coli* homogenates using immobilized metal affinity expanded-bed adsorption, *Biotechnol. Bioeng.,* 67, 206–216, 2000.

22. Smith, M.C., Furman, T.C., Ingolia, T.D., and Pidgeon, C., Chelating peptide-immobilized metal ion affinity chromatography: A new concept in affinity chromatography for recombi-nant proteins, *J. Biol. Chem.,* 263, 7211–7215, 1988.

23. Hansen, P., Lindeberg, G., and Andersson, L., Immobilized metal ion affinity chromatog-raphy of synthetic peptides binding via the α-amino group, *J. Chromatogr.,* 627, 125–135, 1992.

24. Papandreou, M.J., Asteria, C., Petterson, K., Ronin, C., and Beck-Peccoz, P., Concanavalin A affinity chromatography of human serum gonadotropins: Evidence for changes of carbohydrate structure in different clinical conditions, *J. Clin. Endocrinol. Metab.,* 76, 1008–1013, 1993.

25. Yoshida, K., Honda, M., Arai, K., Hosoya, Y., Moriguchi, H., Sumi, S., Ueda, Y., and Kitahara, S., Serial lectin affinity chromatography with concanavalin A and wheat germ agglutinin demonstrates altered asparagine-linked sugar-chain structures of prostatic acid phosphatase in human prostate carcinoma, *J. Chromatogr. B,* 695, 439–443, 1997.

26. De Cristofars, R., Landolfi, R., Bizzi, B., and Castagnola, M., Human platelet glycocalicin purification by phenyl boronate affinity chromatography coupled to anion-exchange high-performance liquid chromatography, *J. Chromatogr. B,* 426, 376–380, 1988.

27. Ey, P.W., Prowse, S.J., and Jenkin, C.R., Isolation of pure IgG, IgG_{2a} and IgG_{2b} immuno-globulins from mouse serum using protein A-Sepharose, *Immunochemistry,* 15, 429–436, 1978.

28. Giovannini, R. and Freitag, R., Isolation of a recombinant antibody from a cell culture supernatant: continuous annular versus batch and expanded-bed chromatography, *Biotechnol. Bioeng.,* 73, 522–529, 2001.

29. Ford, C.H., Osborne, P.O., Mathew, A., and Rego, B.G., Affinity purification of novel bispecific antibodies recognising carcinoembryonic antigen and deoxyrubicin, *J. Chromatogr. B,* 754, 427–435, 2001.

30. Eliasson, M., Olsson, A., Palmcrantz, E., Wibers, K., Inganas, M., Guss, B., Linberg, M., and Uhlen, M., Chimeric IgG-binding receptors engineered from staphylococcal protein A and streptococcal protein G, *J. Biol. Chem.,* 263, 4323–4327, 1988.

31. Cassulis, P., Magasic, M.V., and DeBari, V.A., Ligand affinity chromatographic separation of serum IgG on recombinant protein G-silica, *Clin. Chem.,* 37, 882–886, 1991.

32. Riggin, A., Sportsman, J.R., and Regnier, F., Immunochromatographic analysis of proteins: Identification, characterization and purity determination, *J. Chromatogr.,* 632, 37–44, 1993.

33. Hutchens, T.W. and Porath, J., Thiophilic adsorption of immunoglobulins-analysis of conditions optimal for selective immobilization and purification, *Anal. Biochem.,* 159, 217–226, 1986.

34. Lihme, A. and Heegaard, P.M.H., Thiophilic adsorption chromatography: The separation of serum proteins, *Anal. Biochem.,* 192, 64–69, 1991.

35. Kreutz, F.T., Wishart, D.S., and Suresh, M.R., Efficient bispecific monoclonal antibody purification using gradient thiophilic affinity chromatography, *J. Chromatogr. B,* 714, 161–170, 1998.

36. Gygi, S.P., Rist, B., Gerber, S.A., Turecek, F., Gelb, M.H., and Aebersold, R., Quantitative analysis of complex protein mixtures using isotope-coded affinity tags, *Nat. Biotechnol.,* 17, 994–999, 1999.

37. Elie, C., De Recondo, A.M., and Forterre, P., Thermostable DNA polymerase from the archaebacterium *Sulfolubu acidocaldarius:* Purification, characterization and immunological properties, *Eur. J. Biochem.,* 178, 619–626, 1989.

38. Bialek, G., Nashueuer, H., Goetz, H., Behnke, B., and Grosse, F., DNA polymerase α-DNA primase from human lymphoblasts, *Biochim. Biophys. Acta,* 951, 290–297, 1988.

39. Burgers, P.M.J., Bauer, G.A., and Tam, L., Exonuclease V from *Saccharomyces cerevisiae:* A 5′δ3′deoxyribonuclease that produces dinucleotides in a sequential fashion, *J. Biol. Chem.,* 263, 8099–8105, 1988.

40. Sorrentino, S., Tucker, G.R., and Glitz, D.G., Purification and characterization of a ribonu-clease from human liver, *J. Biol. Chem.,* 263, 16125–16131, 1988.

41. Jansen, H., Lammers, R., Baggen, M.G.A., Wouters, N.M.H., and Birkenhaeger, J.C., Circulating and liver-bound salt-resistant hepatic lipases in the golden hamster, *Biochim. Biophys. Acta,* 1001, 44–49, 1989.

42. Matsuoka, N., Shirai, K., and Jackson, R.L., Preparation and properties of immobilized lipoprotein lipase, *Biochim. Biophys. Acta,* 620, 308–316, 1980.

43. Dawidowicz, A.L., Rauckyte, T., and Rogalski, J., High performance affinity chromatography for analysis of human antithrombin III, *J. Liq. Chromatogr.,* 17, 817–831, 1994.

44. Lebing, W.R., Hammond, D.J., Wydick, J.E., III, and Baumbach, G.A., A highly purified antithrombin III concentrate prepared from human plasma fraction IV-1 by affinity chromatography, *Vox Sang.,* 67, 117–124, 1994.

45. Fountoulakis, M. and Takacs, B., Design of protein purification pathways: Application to the proteome of *Haemophilus influenzae* using heparin chromatography, *Prot. Exp. Pur.,* 14, 113–119, 1998.

46. Clonis, Y.D., Labrou, N.E., Kotsira, V.P., Mazitsos, C., Melissis, S., and Gogolas, G., Biomimetic dyes as affinity chromatography tools in enzyme purification, *J. Chromatogr. A,* 891, 33–44, 2000.

47. McCreath, G.E., Chase, H.A., and Owen, R.O., Expanded bed affinity chromatography of dehydrogenases from baker's yeast using dye-ligand perfluoropolymer supports, *Biotechnol. Bioeng.,* 48, 341–354, 1995.

48. Chang, Y.K., McCreath, G.E., and Chase, H.A., Development of an expanded bed technique for an affinity purification of G6PDH from unclarified yeast cell homogenates, *Biotechnol. Bioeng.,* 48, 355–366, 1995.

49. Watson, D.H., Harvey, M.J., Dean, P.D.G., The selective retardation of NADP$^+$-dependent dehydrogenase by immobilized Procion Red HE-3B, *Biochem. J.,* 173, 591–596, 1978.

50. Sugiura, M., Purification of the T4 DNA ligase by blue Sepharose chromatography, *Anal. Biochem.,* 108, 227–229, 1980.

51. Owen, R.O., McCreath, G.E., and Chase, H.A., A new approach to continuous counter-current protein chromatography: Direct purification of malate dehydrogenase from a *Saccharomyces cerevisiae* homogenate as a model system, *Biotechnol. Bioeng.,* 53, 427–441, 1997.

52. Kempe, M. and Mosbach, K., Separation of amino acids, peptides and proteins on molecularly imprinted stationary phases, *J. Chromatogr. A,* 691, 317–323, 1995.

53. Gulich, S., Uhlen, M., and Hober, S., Protein engineering of an IgG-binding domain allows milder elution conditions during affinity chromatography, *J. Biotechnol.,* 76, 233–243, 2000.

54. Sproule, K., Morrill, P., Pearson, J.C., Burton, S.J., Hejnaes, K.R., Valore, H., Ludvigsen, S., and Lowe, C.R., New strategy for the design of ligands for the purification of pharmaceutical proteins by affinity chromatography, *J. Chromatogr. B,* 740, 17–33, 2000.

55. Kauvar, L.M., Villar, H.O., Sportsman, J.R., Higgins, D.L., and Schmidt, D.E., Jr., Protein affinity map of chemical space, *J. Chromatogr. B,* 715, 93–102, 1998.

56. Kauvar, L.M., Affinity fingerprinting, *Biotech,* 13, 965–966, 1995.

57. Gentz, R., Chen, C.H., and Rosen, C.A., Bioassay for trans-activation using purified human immunodeficiency virus tat-encoded protein: Trans-activation requires mRNA synthesis, *Proc. Natl. Acad. Sci. U.S.A.,* 86, 821–824, 1989.

58. Kipriyanov, S.M., Little, M., Kropshofer, H., Breitling, F., Gotter, S., and Dubel, S., Affinity enhancement of recombinant antibody: Formation of complexes with multiple valency by a single-chain Fv fragment-core streptavidin fusion, *Protein Eng.,* 9, 203–211, 1996.

59. Southworth, M.W., Amaya, K., Evans, T.C., Xu, M.Q., and Perler, F.B.P., Purification of proteins fused to either the amino or carboxy terminus of the mycobacterium xenopi gyrase A intein, *BioTechniques,* 27, 110–120, 1999.

60. Labrou, N. and Clonis, Y.D., The affinity technology in downstream processing, *J. Biotech.,* 36, 95–119, 1994.

61. Chase, H.A. and Draeger, N.M., Expanded-bed adsorption of proteins using ion-exchangers, *Sep. Sci. Tech.,* 27, 2021–2039, 1992.

62. Hjorth, R., Expanded-bed adsorption in industrial processing: recent developments, *Trends Biotechnol.,* 15, 230–235, 1997.

63. Paliwal, S.K., Nadler, T.K., Wang, D.I.C., and Regnier, F.E., Automated process monitoring of monoclonal antibody production, *Anal. Chem.,* 65, 3363–3367, 1993.

64. Taylor, R.S., Wu, C.C., Hays, L.G., Eng, J.K., Yates, J.R., and Howell, K.E., Proteomics of rat liver Golgi complex: Minor proteins are identified through sequential fractionation, *Electrophoresis,* 21, 3441–3459, 2000.

65. Opitek, G.J., Lewis, K.C., and Jorgenson, J.W., Comprehensive on-line LC/LC/MS of proteins, *Anal. Chem.,* 69, 1518–1524, 1997.

66. Davis, M.T., Beierle, J., Bures, E.T., McGinley, M.D., Mort, J., Robinson, J.H., Spahr, C.S., Yu, W., Luethy, R., and Patterson, S.D., Automated LC-LC-MS-MS platform using binary ion-exchange and gradient reversed-phased chromatography for improved proteomic analysis, *J. Chromatogr. B,* 752, 281–291, 2001.

67. Link, A.J., Eng, J., Schieltz, D.M., Carmack, E., Mize, G.J., Morris, D.R., Garvik, B.M., and Yates, J.R., III, Direct analysis of protein complexes using mass spectrometry, *Nat. Biotechnol.,* 17, 676–682, 1999.

68. Geng, M., Ji, J., and Regnier, F., Signature-peptide approach to detecting proteins in complex mixtures, *J. Chromatogr. A,* 870, 295–313, 2000.

69. Geng, M., Zhang, X., Bina, M., and Regnier, F., Proteomics of glycoproteins based on affinity selection of glycopeptides from tryptic digests, *J. Chromatogr. B,* 752, 293–306, 2001.

70. Ji, J., Chakraborty, A., Geng, M., Zhang, X., Amini, A., Bina, M., and Regnier, F., Strategy for qualitative and quantitative analysis in proteomics based on signature peptides, *J. Chromatogr. B,* 745, 197–210, 2000.

71. Anderson, D.J., Branum, E.L., and O'Brien, J.F., Liver- and bone-derived isozymes of alkaline phosphatase in serum as determined by high-performance affinity chromatography, *Clin. Chem.,* 36, 240–246, 1990.

72. Hage, D.S. and Sengupta, A., Studies of protein binding to nonpolar solutes by using zonal elution and high-performance affinity chromatography: Interactions of *cis*- and *trans*-clomiphene with human serum albumin in the presence of β-cyclodextrin, *Anal. Chem.,* 70, 4602–4609, 1998.

73. Loun, B. and Hage, D.S., Chiral separation mechanism in protein-based HPLC columns, 2: Kinetic studies of (R)- and (S)-warfarin binding to immobilized human serum albumin, *Anal. Chem.,* 68, 1218–1225, 1996.

74. Knorle, R., Schniz, E., and Fuerstein, T.J., Drug accumulation in melanin: an affinity chromatography study, *J. Chromatogr. B,* 714, 171–179, 1998.

75. Warnaar, S.O., De Paus, V., Lardenoije, R., Machielse, B., De Graaf, J., Bregonje, M., and Van Haarlam, H., Purification of bispecific F(ab′)2 from murine trinoma OC/TR with specificity for CD3 and ovarian cancer, *Hybridoma,* 13, 519–526, 1994.

76. Bodey, B., Siegel, S.E., and Kaiser, H.E., Human cancer detection and immunotherapy with conjugated and non-conjugated monoclonal antibodies, *Anticancer Res.,* 16, 661–674, 1996.

77. Phillips, T.M., Immunoaffinity measurement of recombinant granulocyte colony stimulating factor in patients with chemotherapy-induced neutropenia, *J. Chromatogr. B,* 662, 307–313, 1994.

78. McConnell, J.P. and Anderson, D.J., Determination of fibrinogen in plasma by high-performance immunoaffinity chromatography, *J. Chromatogr.,* 615, 67–75, 1993.

79. Ruhn, P.F., Taylor, J.D., and Hage, D.S., Determination of urinary albumin using high-performance immunoaffinity chromatography and flow injection analysis, *Anal. Chem.,* 66, 4265–4271, 1994.

80. Liu, C.L. and Bowers, L.D., Immunoaffinity trapping of urinary human chorionic gonadotropin and its high-performance liquid chromatographic-mass spectrometric confirmation, *J. Chromatogr. B,* 687, 213–220, 1996.

19

Environmental Analysis by Affinity Chromatography

Mary Anne Nelson and David S. Hage
Department of Chemistry, University of Nebraska, Lincoln, NE

CONTENTS

19.1 INTRODUCTION

As awareness grows of the effects of even ultratrace contamination in the environment, there is an increasing demand for better analytical methods for studying environmental samples. For instance, new regulations by agencies in North America [1] and Europe [2] have forced laboratories to adopt methods with lower limits of detection, greater specificity, and higher precision for agents of environmental interest.

The analysis of environmental samples is complicated by the fact that the selected techniques must deal with a wide range of analytes and matrixes. Samples can range from simple matrixes like drinking water to more complex mixtures like soil extracts or foods. The analytes in these samples range from small molecules (e.g., organic pesticides) to large pharmaceuticals and even proteins. In many of these samples, the analyte is present at trace or ultratrace levels and in the presence of several interfering species. As an example, pesticides found in ground and surface water can occur as either the parent compound or as degradation products [3–5]. In a recent study performed by the U.S. Geological Survey, it was shown that more than 95% of sampled streams and almost 50% of wells had at least one pesticide, with low-level mixtures of several pesticides being the most common form of contamination [6]. These mixtures present a real challenge in the analysis of the water samples, since the substances they contain may have similar properties when detected by analytical techniques.

The most common methods used for environmental testing are gas chromatography (GC), gas chromatography/mass spectrometry (GC/MS) [7–10], high-performance liquid chromatography (HPLC) [11–15], and enzyme-linked immunosorbent assays (ELISAs) [16–19]. However, over the last decade there has also been growing interest in the use of affinity techniques for such work [20–26]. This has paralleled the increased use of ELISAs in environmental testing [27–29].

Affinity chromatography has several features that make it attractive for the study of pollutants and environmental agents. For instance, the ability to couple affinity chromatography with other methods (e.g., HPLC, GC, or capillary electrophoresis) can lead to the development of multidimensional methods that are easily automated, sensitive, and specific with excellent reproducibility [30]. In addition, the specific nature of many affinity ligands, such as antibodies [31], allows affinity chromatography to be used for the analysis of either a specific analyte or closely related group of analytes.

This chapter examines the types of affinity ligands that have been used in environmental testing. This includes a discussion of traditional ligands as well as those based on metal-ion systems [32–34] and molecularly imprinted polymers [35, 36]. The discussion then considers the various formats in which these ligands have been employed for environmental assays, such as direct detection methods, chromatographic immunoassays, affinity extraction, biosensors, and affinity capillary electrophoresis.

19.2   DIRECT DETECTION METHODS

The simplest type of environmental analysis that uses affinity chromatography is one based on the direct detection mode. This typically uses the traditional on/off format of affinity chromatography, as described in Chapter 4. This section discusses the general principles of this format and gives some specific examples of its use in environmental applications.

19.2.1   General Principles

The direct detection mode of affinity chromatography is based on the injection of a sample onto an affinity column under conditions that promote specific binding between the analytes and immobilized ligand. If the ligand has strong binding (i.e., an association constant greater than 10^5 to $10^6 \ M^{-1}$), then a change in column conditions will probably be needed to release the analytes for detection. This typically involves a change in the ionic strength or pH of the mobile phase, or the addition of a chaotropic agent or a displacer to this solvent [37]. If the ligand has weaker binding, then analyte elution and detection may be possible under isocratic conditions [38].

In addition to its simplicity, there are several other advantages to this approach. For instance, the use of a high-specificity ligand like an antibody can allow direct detection to be used for a particular analyte, while a more general ligand could be employed for the analysis of a broader class of compounds. Furthermore, when this approach is used as part of an HPLC system, measurements with precisions in the range of 1 to 5% can generally be produced in as little as a few minutes. Good limits of detection can also be obtained through this technique, although the exact values will vary with the analyte and type of detection being used [39].

Although the simplicity of this approach makes it attractive for environmental analysis, it does have some limitations. One limitation is the need for a reasonably high analyte concentration to allow the direct measurement of this compound as it exits the affinity column. The main challenge here is to discriminate between the eluting analyte and the change in background signal that often occurs when the elution buffer is applied to the affinity column. For ultraviolet or visible absorbance detection, this requires that the analyte have a high molar absorptivity. Alternatively, fluorescent detection can be used with analytes that contain natural fluorophores or that have been derivatized with fluorescent tags through either precolumn or postcolumn methods [30].

Another possible limitation of the direct detection mode is that it is based on the sequential injection of standards and samples. This makes this format best suited for situations where low or moderate numbers of samples must be processed. However, the use of sequential analysis does give direct detection other advantages. For instance, the sequential processing of samples creates a system that is easier to maintain and troubleshoot than batch analysis if a problem arises during the system's operation. The use of sequential injection also provides a faster turnaround time for individual samples than can be obtained by most batch techniques.

19.2.2   Examples of Applications

Most environmental applications of direct affinity detection have used this to measure metal ions or to study human exposure to such ions. Examples of methods that have been employed in this work include covalent affinity chromatography and immobilized metal-ion affinity chromatography.

*19.2.2.1   Analysis of Exposure to Heavy Metals
            by Covalent Affinity Chromatography*

One example of the direct detection mode in environmental analysis has been the use of this method to study human exposure to heavy metals. This was accomplished by determining the protein metallothionein (MT) by means of covalent affinity chromatography (or thiol-disulfide interchange chromatography). This made use of columns that separated proteins based on exchange between the sulfhydryl groups in proteins and sulfur-sulfur bridges immobilized in these columns.

This work was based on the fact that the exposure of humans to heavy metals leads to the appearance of certain amino acids and proteins in urine and plasma. One such agent is metalloprotein. This low-mass protein was discovered in 1957 [40] and is rich in sulfhydryl groups [41–43]. Metal ions like Cd^{2+}, Zn^{2+}, Cu^{2+}, Hg^{2+}, and Co^{2+} stimulate the biosynthesis of metallothionein and bind to this protein. For instance, it has been shown in several studies that exposure to heavy metals like Zn, Cd, and Hg results in an increased production of metallothionein [44–47]. Figure 19.1 shows the separation of a metallothionein Zn complex (i.e., Zn-thionein) from water, urine, plasma, and milk samples by covalent affinity chromatography [46]. The isolated protein was then measured and used as an indicator of heavy metals in blood and the kidneys or liver.

19.2.2.2   Analysis of Metal Ions and Metal Ligands by IMAC

Immobilized metal-ion affinity chromatography (IMAC) is another example of a direct detection method that has been used in environmental analysis. As discussed in Chapter 10,

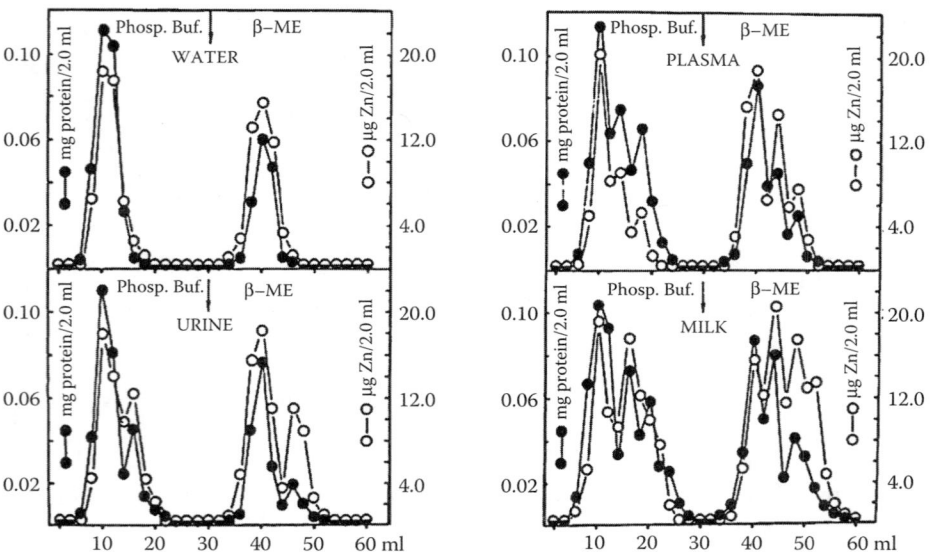

Figure 19.1   Separation of Zn-thionein from samples of water, human urine, plasma, and breast milk by covalent affinity chromatography. The support used in this column was Sepharose-DNTB. The application buffer was pH 7.0, 0.05 M phosphate, and the elution solvent was this same buffer plus 0.05 M β-mercaptoethanol (β-ME). The flow rate was 30 ml/h, and the sample volume was 5 ml. Fractions were collected every 2.0 ml. The arrow at the top of each chromatogram shows the point at which a switch was made between the application buffer and the elution buffer. (Reproduced with permission from Kabzinski, A.K.M., *Biomed. Chromatogr.*, 12, 281–290, 1998.)

the ligand in IMAC is a metal ion attached to an immobilized chelating agent. This type of support can be used to examine the interactions between the metal ion and other metal complexing ligands or to isolate such ligands for later use.

There have been several reports in which IMAC has been used for the extraction and cleanup of metal ligands in water samples [48–50]. In one report, IMAC was used to isolate copper ligands during the analysis of stream water [51]. This was performed with a gel column that contained iminodiacetate (IDA) as a chelating group. This chelating group was first loaded with copper ions by applying a solution of copper sulfate and rinsing with a pH 8.2 borate buffer. A sample was then applied to the column, followed by another rinsing step. The retained ligands (i.e., those complexed to copper ions) were later eluted using a pH step gradient. During their elution, these ligands were detected by monitoring their absorbance between 230 and 500 nm. Fractions of these ligands were then collected and examined using three-dimensional excitation/emission matrix fluorescence spectroscopy.

It is known that organic ligands play a significant role in trace-metal speciation, in the mobilization and transport of nutrients or pollutants in the environment, and in soil weathering [51]. In this work by Wu and Tanoue [51], two peaks were seen when stream water was applied to the copper-treated IMAC column. These peaks represented weak and strong ligands for copper. The ratio of the weak to strong ligand fractions ranged from 3.3 to 17.3, indicating that most of the isolated ligands had a weak affinity for copper ions. Although there is little information regarding the source of these ligands, a proteinlike fluorescence was observed for these substances, suggesting that they arose from a biogenic source [51].

19.3 INDIRECT DETECTION METHODS

Probably the most popular use of affinity chromatography in environmental analysis has been in the automation of immunoassays (i.e., methods that use antibodies as selective binding agents). The use of affinity chromatography in this manner is often referred to as a *chromatographic immunoassay* or *flow-injection immunoanalysis* (*FIIA*). A detailed description of this field can be found in Chapter 29. In addition, several authors have reviewed the use of such assays for environmental samples [52–54].

19.3.1 General Principles

Chromatographic immunoassays are based on the indirect detection of an analyte by observing how it reacts with a labeled binding agent or prevents a labeled analog of the analyte from interacting with an antibody [55]. In environmental testing, detection for these assays is usually accomplished by using an analyte analog that is labeled with a fluorescent group or enzymatic tag. This label is then monitored as the analog and analyte compete for antibody binding sites within a column or flow-injection chamber. This approach is particularly useful for trace analytes that do not themselves produce a signal that can be directly detected.

Two types of chromatographic immunoassays are the competitive-binding immunoassay and homogeneous immunoassay. The competitive-binding immunoassay can be used for either large or small analytes, while chromatographic homogeneous immunoassays involve a size separation that restricts their use to low-mass analytes. Another possible format is the two-site immunometric assay, which is used for macromolecules like proteins and peptides (see Chapter 29). Although this latter method has not yet been used in environmental testing, it could potentially be used for looking at protein- and peptide-based herbicides.

19.3.2   Competitive-Binding Immunoassays

A competitive-binding immunoassay is the most common type of chromatographic immunoassay in environmental testing. This can be conducted using either an immobilized analog
of the analyte or immobilized antibodies [52]. The first of these two approaches makes use
of a small amount of labeled antibodies in solution for which the analyte and immobilized
analog compete. The second method makes use of a labeled analog in solution that competes
with the analyte for a limited number of immobilized antibodies on a support. This latter format
is more cost effective with regard to the antibodies (generally the most expensive component
of the assay), since these can often be reused for multiple assays. Three specific examples
of this assay are given in the following subsections: the simultaneous-injection format,
the sequential-injection assay, and flow-injection liposome immunoanalysis.

19.3.2.1   Herbicide Detection by Simultaneous-Injection Assays

When using an immobilized antibody, there are two ways in which the sample and labeled
analog can be applied to the column. The first of these formats is the simultaneous-injection
assay. In this approach, the analyte and labeled analog are mixed off-line and injected
together onto the column, where they compete for a small number of antibody-binding
sites. The amount of analyte in the sample is then determined by using a calibration curve
that plots the amount of label that binds to or passes through the column versus the known
concentration of analyte in injected standards.

   The simultaneous-injection assay has been used in several environmental studies,
including a method developed for detecting the herbicide isoproturon. This made use of
a column that contained immobilized antibodies able to bind this herbicide and an analog
of isoproturon conjugated with the enzyme horseradish peroxidase. In this assay, samples
were combined off-line with a fixed concentration of enzyme-labeled isoproturon, and this
mixture was injected onto the antibody column. After nonretained agents had been washed
from the system, a substrate for the enzyme was applied for detection of the retained label.
The limit of detection reported for isoproturon was 0.12 µg/l, and the total run time was
approximately 25 min [56].

   A similar assay has been described for atrazine. In this method, a mixture of the
sample or standard was combined with an atrazine analog labeled with horseradish peroxidase. After injecting this mixture into a flow cell that contained immobilized antibodies
for atrazine, the nonretained species were washed from the column, and a substrate for
the enzyme label was applied. The resulting colored product was then detected at 405 nm,
giving a measure of the labeled analog that was bound to the column. Figure 19.2 shows
the response obtained in this assay at several atrazine concentrations. The run time was
15 min per sample, and the detection limit was 0.5 ng/l. However, this detection limit
could be reduced to 0.02 ng/l if the assay time was increased to 30 min per sample [57].

19.3.2.2   Herbicide Detection by Sequential-Injection Assays

A second format that can be used in a flow-based competitive-binding immunoassay is a
sequential-injection assay. This is performed by applying the sample first to the affinity
column, followed later by an injection of the label. One advantage of this format over the
simultaneous-injection method is that the label is not subject to matrix interferences, since
it is never in contact with the actual sample [58]. This is particularly important if a fluorescent
tag is used for the label, since it eliminates any quenching by sample components.

   Sequential injection has been used to develop an automated immunoassay for atrazine in water and soil samples. Figure 19.3 shows the system used to perform this assay.
In this approach, a flow cell containing immobilized antibodies was first rinsed with an

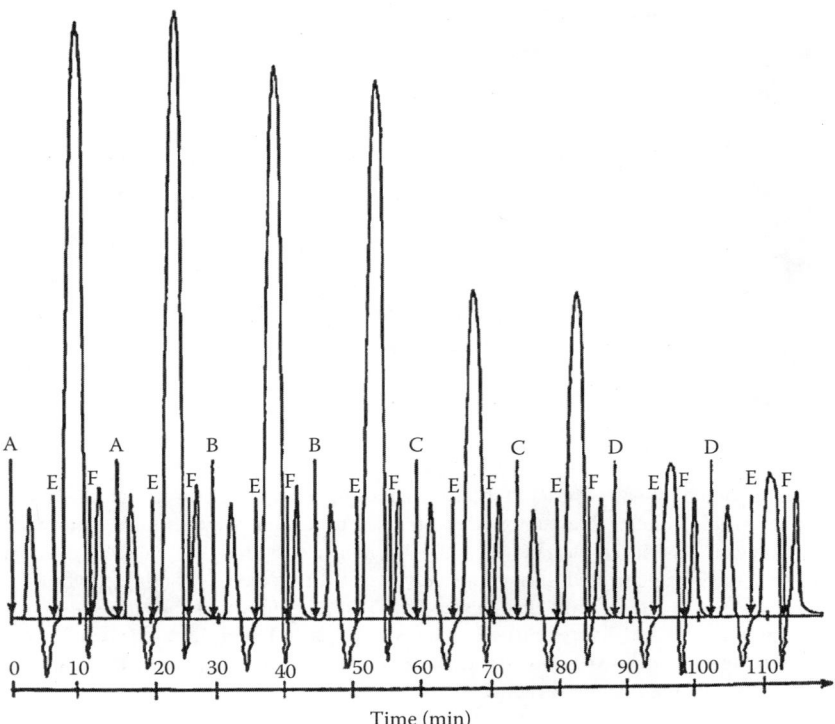

Figure 19.2 Responses obtained for duplicate injections of four atrazine standards in a flow-based competitive-binding immunoassay. The arrows at A, B, C, and D represent the injections of standards containing 0.001, 1, 10, and 100 ng/ml atrazine, respectively. The decrease in peak size at increasing atrazine concentrations is due to the decrease in the amount of bound labeled analog (i.e., the detected substance), as this analog is hindered by atrazine from binding to the immobilized antibodies. The arrows at E show the points at which a substrate was injected onto the system for measuring the amount of retained enzyme label. The arrows at F indicate when an elution buffer was applied to the system. (Reproduced with permission from Bjarnason, B., Bousios, N., Eremin, S., and Johansson, G., *Anal. Chim. Acta*, 347, 111–120, 1997.)

equilibration buffer. A standard or sample was then pumped into this flow cell, followed by injection of a fixed amount of peroxidase-labeled atrazine. After a short rinsing step, a substrate for the peroxidase label was added to the flow cell and allowed to react. The fluorescence of the enzyme-generated product was then detected downstream using an excitation wavelength of 320 nm and an emission wavelength of 404 nm.

Both the simultaneous- and sequential-injection formats can be used for either large or small analytes [59, 60]. However, they differ in their analytical characteristics. For instance, the sequential-injection assay provides a lower limit of detection and larger change in response than the simultaneous-injection assay at low analyte concentrations. However, the simultaneous-injection format is slightly easier to perform and has a calibration curve that covers a broader dynamic range. Further information on these assays and their properties can be found in Chapter 29.

19.3.2.3   *Environmental Testing by Flow-Injection Liposome Immunoanalysis*

An interesting variation of the flow-based competitive-binding immunoassay involves the use of liposomes for detection [61–63]. This has given rise to a method known as *flow-injection*

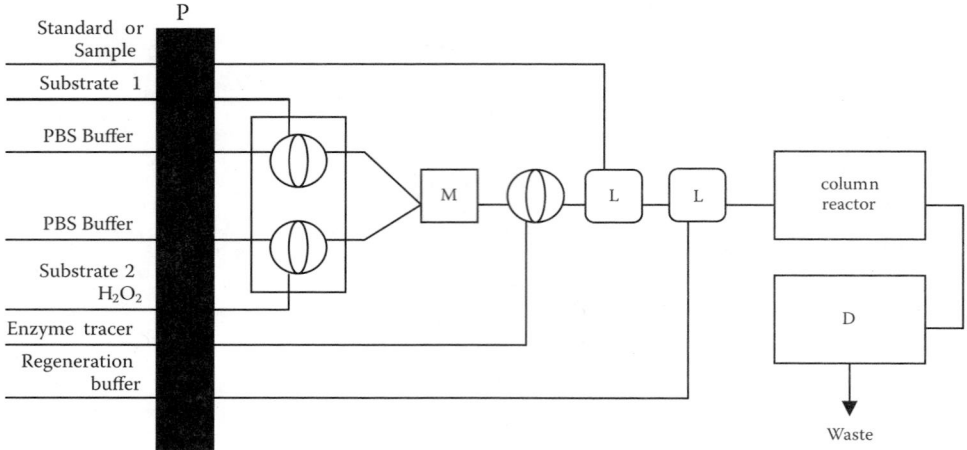

Figure 19.3 Design of a flow-based system for a sequential-injection competitive-binding immu-noassay for atrazine. In this system, five pumps (P) were used to deliver reagents in the following sequence: (1) the flow cell was rinsed with pH 7.2, 0.04 M phosphate buffered saline (PBS); (2) an atrazine standard or sample was injected through a Lee valve (L), followed by the injection of the atrazine-peroxidase tracer; (3) after a short rinsing step with PBS, the enzyme substrates (i.e., 3-(p-hydroxyphenyl)propionic acid and hydrogen peroxide, or substrates 1 and 2, respectively) were pumped into the mixing chamber (M); and (4) as these substrates passed through the antibody column, the fluorescence of the enzyme-generated product was measured downstream in a flow-through cell (D). (Reproduced with permission from Wittmann, C. and Schmid, R.D., *J. Agric. Food Chem.*, 42, 1041–1047, 1994.)

liposome immunoanalysis (FILIA). In this technique, a liposome containing a large amount of a fluorescent dye or other detectable marker is used to tag an analyte analog. This labeled analog is then allowed to compete with the analyte for immobilized antibodies in a flow cell. A detergent is later passed through the flow cell to lyse the retained liposomes and release their internal dye molecules or markers. These markers are then measured downstream by an on-line detector and provide a signal that is inversely related to the amount of analyte in the original sample.

This approach has been used for the determination of alachlor. The system employed in this study and some typical results are shown in Figure 19.4. A signal was generated in this assay by using liposomes that contained a fluorescent dye. These were used to tag an analog of alachlor and conduct a competitive-binding assay. This technique had an analysis time of 6 min when operated at a flow rate of 450 μl/min and gave a limit of detection for alachlor of 5 μg/l when 25 μg of antibody were used to prepare the immu-noaffinity column.

19.3.3 Homogeneous Immunoassays

Another type of chromatographic immunoassay used for environmental testing is the homogeneous immunoassay. This is a method in which the reaction between the antibody, analyte, and label occurs in solution. One way this can be performed by chromatography is to use a *restricted-access assay*. This employs a random-access-medium column that can separate a small, labeled analog from the larger antibody-analog complexes based on their differences in size.

Such an assay has been used to screen plasma and water samples for atrazine; for related *s*-triazines like simazine, propazine, and terbutylazine; and for atrazine degradation

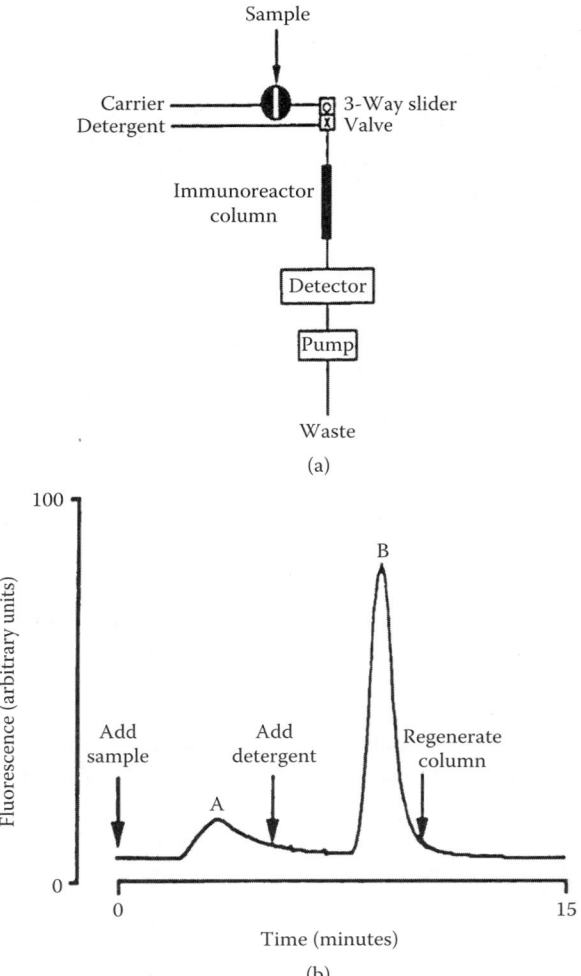

Figure 19.4 (a) System for performing a chromatographic immunoassay with liposome-based detection, and (b) a typical response for alachlor. In this method, samples were combined with alachlor-tagged liposomes and applied in a carrier buffer to an immunoreactor column that contained antibodies immobilized to glass beads. After nonbound liposomes were washed from the column (peak A in the lower figure), a detergent solution was passed through the column to lyse the retained liposomes and release their associated fluorescent dye. This released dye was measured by an on-line fluorescence detector, as indicated by peak B in the lower figure. Finally, the column was washed with Tris-buffered saline and allowed to regenerate before application of the next sample. (Reproduced with permission from Reeves, S.G., Rule, G.S., Roberts, M.A., Edwards, A.J., and Durst, R.A., *Talanta*, 41, 1747–1753, 1994.)

products such as deisopropyl-, deethyl-, and hydroxyatrazine. This method used a fluorescein-labeled analog of atrazine and off-line incubation of the sample with the labeled analog and a small amount of antiatrazine antibodies. This mixture was then introduced onto a restricted-access-medium column, as shown in Figure 19.5. This column contained a support with a reversed-phase stationary phase within its pores that retained the labeled analog but not the excluded analog-antibody complexes. The amount of antibody-bound analog in the nonretained

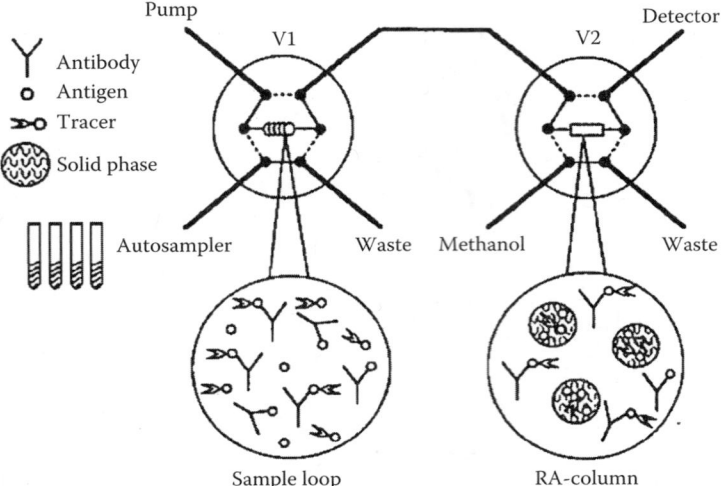

Figure 19.5 System for a restricted-access immunoassay. The RA column is the restricted-access support, which allows the antibody-bound fractions of the analyte and labeled analog (i.e., the antigen and tracer) to pass through nonretained, while the column binds to the free analyte or analog that remains in solution. V1 and V2 are the two valves used to inject the sample and pass this through the restricted-access column. (Reproduced with permission from Onnerfjord, P., Eremin, S.A., Emneus, J., and Marko-Varga, G., *J. Chromatogr. A*, 800, 219–230, 1998.)

peak was then used as an indirect measure of the analyte in the original sample. This gave a detection limit of 20 pg/ml for atrazine and a throughput of 80 samples per hour [64, 65].

19.4 AFFINITY EXTRACTION

The most popular use of affinity ligands in environmental testing has been in the field of affinity extraction. The term *affinity extraction* refers to the use of an affinity ligand for the isolation or concentration of a chemical prior to its detection or measurement by another method [58]. Immobilized antibodies are frequently used for this purpose, giving rise to a technique known as *immunoextraction*, but other ligands can also be employed. If immunoextraction is coupled directly with another method, this gives rise to a combined technique, such as *immunoaffinity/reversed-phase liquid chromatography* (IA-RPLC) or *immunoaffinity/gas chromatography* (IA-GC) [30].

19.4.1 General Principles

Combining affinity chromatography with other analytical techniques typically increases the selectivity of sample preparation while also providing a means for discriminating between several chemicals captured by the same affinity column. There are two basic approaches used for this purpose. The first is *off-line affinity extraction*, in which affinity chromatography is followed by fraction collection and subsequent analysis of the isolated substances by methods like GC, HPLC, or ELISA. The second approach is *on-line affinity extraction*, where the analyte is removed from a sample by an affinity column and passed directly onto a second method for measurement or detection. This latter approach often involves coupling affinity chromatography to reversed-phase liquid chromatography. However, there have been additional reports using on-line affinity extraction with GC [66, 67] or capillary electrophoresis [13, 27, 68].

The choice of off-line or on-line affinity extraction depends on the overall goal of the analysis and the compatibility of this extraction with the second analysis method. For example, when affinity extraction is combined with gas chromatography, an off-line mode is often used to go from the aqueous affinity effluent to a more volatile matrix that can be analyzed by GC. In this sense, the affinity column acts as a solid-phase extraction cartridge. However, when affinity extraction is used with reversed-phase liquid chromatography, an on-line method might be preferred because of its higher precision and greater ease of automation. In either situation, the use of affinity extraction gives higher selectivity and less interference than traditional solid-phase extraction [67]. This occurs because most common solid-phase extraction cartridges retain analytes based on their polarity [69], which will create problems if other sample or reagent components have similar polarities.

19.4.2   Sample Pretreatment by Off-Line Affinity Extraction

Off-line extraction is the easiest approach for combining affinity chromatography with other methods. This typically uses an affinity ligand immobilized onto a low-performance support (such as activated agarose) held within a disposable syringe or solid-phase extraction cartridge. After this affinity support has been conditioned with the necessary application buffer or solvents, the sample is applied, and nonbound sample components are washed through the support. An elution buffer is then passed through the support, and analyte fractions are collected.

Affinity extraction is popular for sample pretreatment, since it eliminates many of the extraction and derivatization steps required by traditional methods of sample preparation [3, 70]. This technique has been used in many environmental studies, including work with ochratoxins in baby food [71], deoxynivalenol in wheat [72], triazine biocides in seawater [73], and fumonisins in corn [74]. Typical limits of detection for these methods are in the range of nanograms per liter.

An important limitation of off-line affinity extraction is the need to transfer the extract to a second method. Because manual steps are generally used for both this transfer and the off-line extraction, the precision and recovery of the measurement may suffer. In addition, other steps (e.g., extraction into an organic solvent, and evaporation of the solvent) may be needed before the affinity extract can be analyzed by HPLC, GC, or ELISA. This not only increases the analysis time, but increases the cost of the assay and the amount of reagents needed for sample pretreatment.

One example of off-line affinity extraction has been its use in determining estrogens in wastewater [75]. In this method, wastewater was first treated by filtration and traditional solid-phase extraction. Next, the collected extracts were evaporated to dryness under nitrogen and reconstituted in a 5% solution of methanol in water. This reconstituted sample was then applied to an immobilized-antibody column. Estrogens from the sample were bound to this column and later eluted with a 70% solution of methanol in water. This eluant was collected and evaporated to dryness, followed by reconstitution in 25% acetonitrile in water. The reconstituted analytes were analyzed by HPLC and detected by electrospray-ionization mass spectrometry (ESI-MS) [75]. Figure 19.6 compares the chromatograms obtained for raw-sewage samples with and without the affinity extraction step. Without affinity extraction, 17β-estradiol (E2) and its degradation product estrone (E1) were masked by matrix components. But with the use of affinity extraction, both analytes gave well-defined peaks.

19.4.3   Environmental Testing by On-Line Affinity Extraction and HPLC

On-line affinity extraction couples the specificity of an affinity column with the ability to automatically transfer its retained components to a second technique for analysis. In environmental testing, the on-line coupling of immunoextraction with reversed-phase liquid chromatography has been used to create several methods with excellent

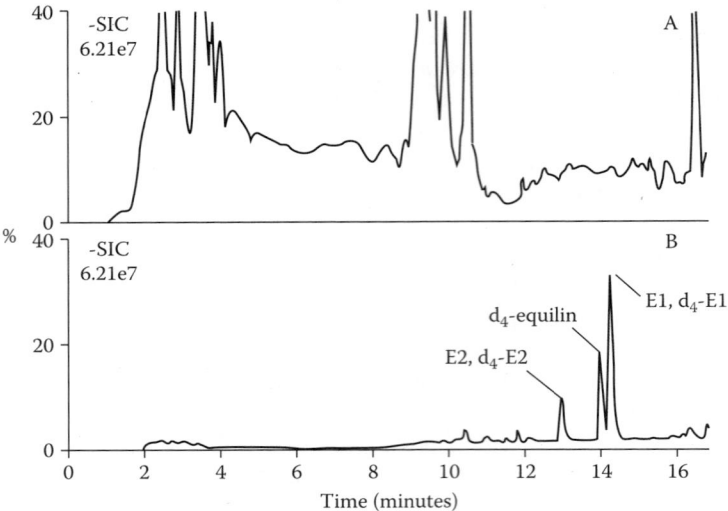

Figure 19.6 Summed-ion chromatograms (SICs) for the analysis of raw sewage by HPLC and electrospray-ionization mass spectrometry. The upper figure (a) shows the results obtained using only standard solid-phase extraction. The lower graph (b) shows the results for the same sample when immunoextraction was also used for sample pretreatment. The peaks labeled E1 and E2 represent estrone and 17β-estradiol, respectively, which were injected along with their deuterated forms (d_4-E1 and d_4-E2). Equilin was used as an internal standard and was also injected along with its deuterated form (d_4-equilin). (Reproduced with permission from Ferguson, P.L., Iden, C.R., McElroy, A.E., and Brownawell, B.J., *Anal. Chem.*, 73, 3890–3895, 2001.)

reproducibility and fast analysis times. For instance, Thomas et al. originally used this approach for the measurement of triazine herbicides and their degradation products at the parts-per-billion levels in 6 to 12 min [70]. This same approach has been used with carbofuran [76] and carbendazim [77]. Low limits of detection are possible with this method, since the affinity column is mass sensitive and responds to the moles, rather than the concentration, of a substance applied to the column. Thus, these columns can be optimized for work even at low parts-per-billion or parts-per-trillion levels [3].

A typical system used to perform on-line immunoextraction with reversed-phase liquid chromatography is shown in Figure 19.7. Three solvents are often used in such a system: an application buffer for the immunoaffinity column, an elution buffer for this column, and a mobile phase for the reversed-phase column. In this approach, the sample is first applied to the immunoaffinity column in the application buffer. Once binding of the analyte has occurred and nonretained components have been washed away, the antibody column is switched on-line with a C_{18} reversed-phase precolumn, and the elution buffer is applied. This causes the analytes to dissociate from the antibody column; however, since this buffer also acts as a weak mobile phase for the reversed-phase precolumn, the analytes are "trapped" in this precolumn. This precolumn is later switched on-line with a second, larger C_{18} column, and a mobile phase containing some organic modifier is applied, causing the elution and separation of the analytes based on their polarities.

A modified version of this system has been used for determining carbendazim in soil and lake water [77]. This was accomplished by using a high-performance protein G column coupled to a reversed-phase analytical column through the use of a restricted-access-medium trapping column. Prior to analysis, 20 μg of antibodies were loaded onto the protein G column to form an immunoaffinity support. A sample was then applied to

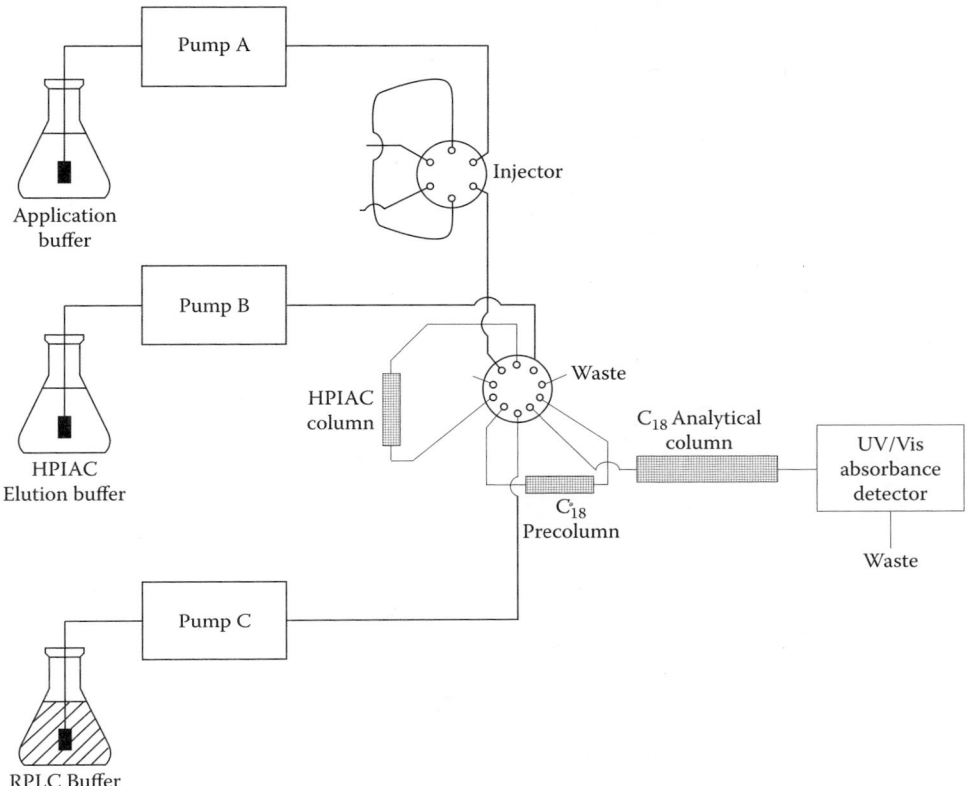

Figure 19.7 System for performing on-line immunoextraction with reversed-phase liquid chromatography. Details on the operation of this system are given in the text.

this column and the carbendazim extracted. A stripping buffer containing 2% acetic acid was next pumped through the affinity column, which caused elution of the adsorbed antibodies and carbendazim. The trapping column retained the carbendazim while the antibodies were washed through the system. The trapping column was then switched on-line with an analytical column and mass spectrometer. The limit of detection was 25 ng/l for carbendazim, and the throughput was three samples per hour [77].

Other reports of on-line affinity columns in HPLC have been described for estrogens [78] and polycyclic aromatic hydrocarbons [79]. The first of these two studies used estrogens and an antiestrogen antibody column to compare various elution techniques for immunoextraction, including pH step gradients, aqueous desorption with chaotropic or micellar additives, immunoselective desorption with analyte analogs in a 5% acetonitrile solution, and thermal desorption [78]. The second study used on-line immunoextraction with reversed-phase liquid chromatography and fluorescence detection to monitor 15 polycyclic aromatic hydrocarbons. The limits of detection for these compounds ranged from 10 to 20 ng/l for a 20-ml sample.

19.4.4 Analysis of Degradation Products by On-Line Affinity Extraction

A unique application of affinity extraction has been its use to examine both a parent compound and its degradation products in environmental samples. The analysis of degradation products for environmental contaminants is complicated by several factors.

Atrazine

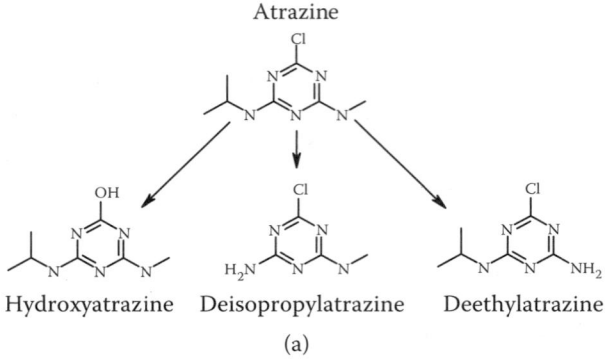

Hydroxyatrazine   Deisopropylatrazine   Deethylatrazine

(a)

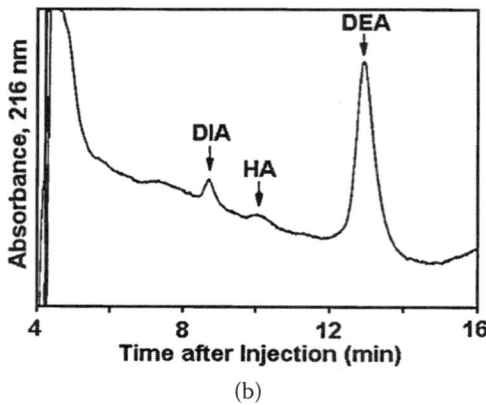

(b)

Figure 19.8 The major degradation products of atrazine (top) and a typical chromatogram obtained for the analysis of these degradation products by immunoextraction coupled on-line with reversed-phase liquid chromatography (bottom). The chromatogram shown at the bottom was obtained for a 45-ml sample of groundwater. The concentrations measured for deethylatrazine (DEA), hydroxyatrazine (HA), and deisopropylatrazine (DIA) in this sample were 210, 10, and 60 ng/l, respectively. (Reproduced with permission from Rollag, J.G., Beck-Westermeyer, M., and Hage, D.S., *Anal. Chem.,* 68, 3631–3637, 1996.)

First, these products generally occur at lower concentrations than the parent compound. Also, degradation of the parent compound often leads to a product that has a different polarity, activity, and ecological lifetime. Pathways for the degradation of herbicides can be of a chemical nature (e.g., hydrolysis or photodegradation) [4, 5, 80, 81] or of biological origin (e.g., due to microbial action) [82–84]. Examples of such reactions are shown in Figure 19.8.

Degradation analysis has been performed with on-line immunoextraction and reversed-phase HPLC in studies of atrazine in the environment. This was initially demonstrated by Rollag et al., who were able to detect the primary degradation products of atrazine (i.e., deethylatrazine, deisopropylatrazine, and hydroxyatrazine) at levels extending down to the low parts-per-trillion range [3]. A typical chromatogram showing the separation of these products is given in Figure 19.8.

This same method has been used to determine the adsorption isotherms for these degradation products on adsorbents used for water treatment [85, 86] and to examine the

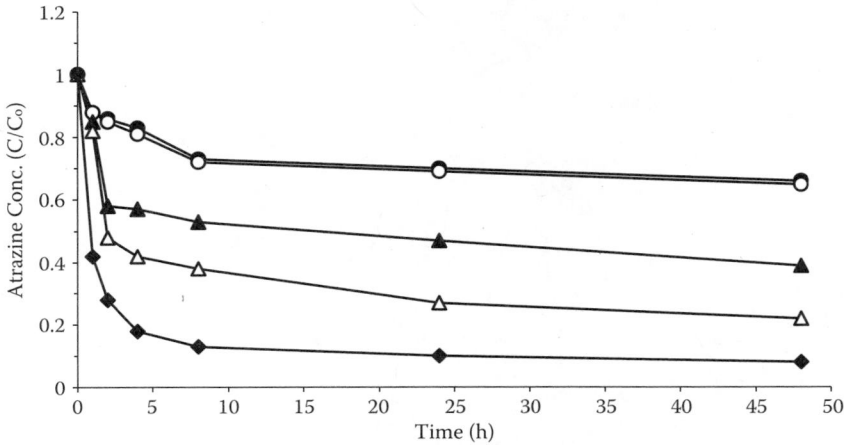

Figure 19.9 Relative amount of atrazine remaining in solution after treatment with zero-valent iron. This graph shows results obtained at (from top to bottom) 0.5, 1.0, 2.5, 5.0, and 10% (w/v) iron mixed with a water sample that initially contained 20 mg/l atrazine. (Reproduced with permission from Singh, J., Zhang, T.C., Shea, P.J., Comfort, S.D., Hundal, L.S., and Hage, D.S., *Proc. WEFTEC '96,* Dallas, TX, 1996, pp. 143–150.)

breakdown of atrazine by zero-valent iron (Fe^0) [87]. In this latter study, immunoextraction/RPLC was found to be a useful alternative to radiotracer studies. Some typical results are shown in Figure 19.9, where it was shown that the rate of atrazine removal was dependent upon the concentration of atrazine and the amount of added iron. For instance, immunoextraction indicated that a 20-µg/l solution of atrazine treated with 10% (w/v) Fe^0 caused the atrazine concentration to fall below 3 µg/l within 48 h [87].

19.4.5   On-Line Affinity Extraction and Field-Portable Systems

Another application reported for on-line affinity extraction has been its use in the development of field-portable systems for herbicide detection. This approach has recently been used in the creation of a system for measuring triazine herbicides in water [88]. A picture of this device and some typical results obtained with it are shown in Figure 19.10. This same system has recently been adapted for use with other classes of herbicides, such as 2,4-dichlorophenoxyacetic acid and related compounds. This system can be transported on a small cart by a single person and weighs 35 lb, with the power supply adding another 45 lb. Once at a site, this instrument takes 5 min to assemble and start. Since the solvent reservoirs are connected to the pumps during transport, it takes less than 10 min to establish a baseline and begin analysis.

One advantage of this device is its ability to analyze a small amount of sample with only limited pretreatment. For instance, ground- and surface-water samples can be examined by simply filtering them through a 0.2-µm syringe filter before injection. The speed of this method is also quite good, with an overall analysis time of 10 min per sample (in the case of triazine herbicides) and a throughput of 5 min per sample. The lower limit of detection reported for atrazine with a 2-ml sample was 0.3 µg/l, with a linear range extending up to 25 µg/l. However, this limit of detection and linear range could be adjusted by altering the amount of sample applied to the antibody column [88].

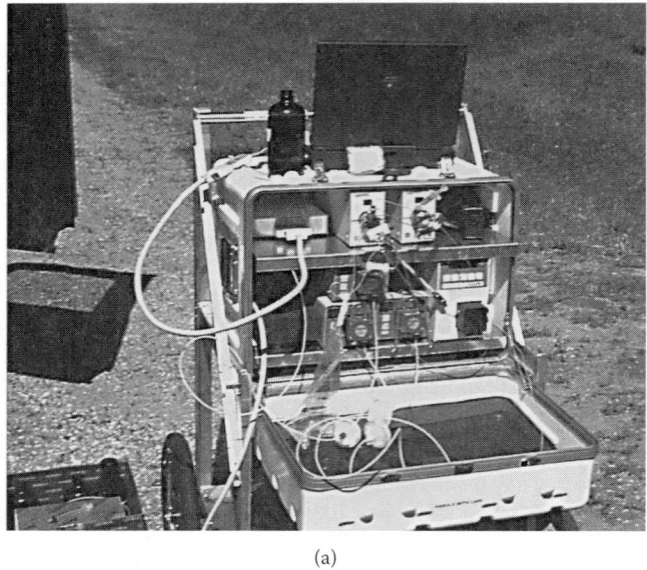

(a)

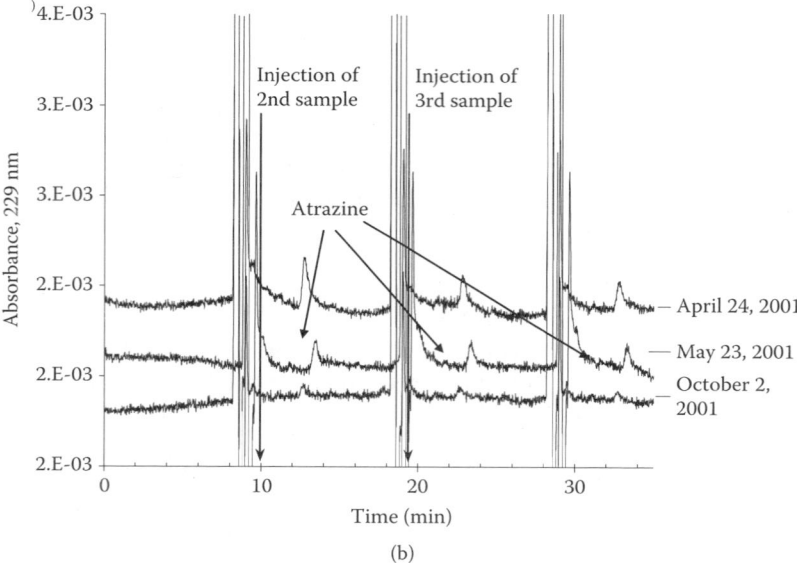

(b)

Figure 19.10 (a) A field-portable device based on immunoextraction and reversed-phase liquid chromatography for the analysis of herbicides in water samples, and (b) results obtained by this device for the triplicate analysis of triazines in river water at various times of the year. (Reproduced with permission from Nelson, M.A., Development of a Field Portable Immunoextraction/RPLC System for Environmental Analysis, Ph.D. dissertation, University of Nebraska-Lincoln, Lincoln, NE, 2003.)

19.4.6   On-Line Affinity Extraction with Other Methods

The on-line coupling of affinity columns is not limited to liquid chromatography but can also be performed with gas chromatography. As an example, Dalluge et al. coupled immunoextraction with gas chromatography for the measurement of atrazine in river water,

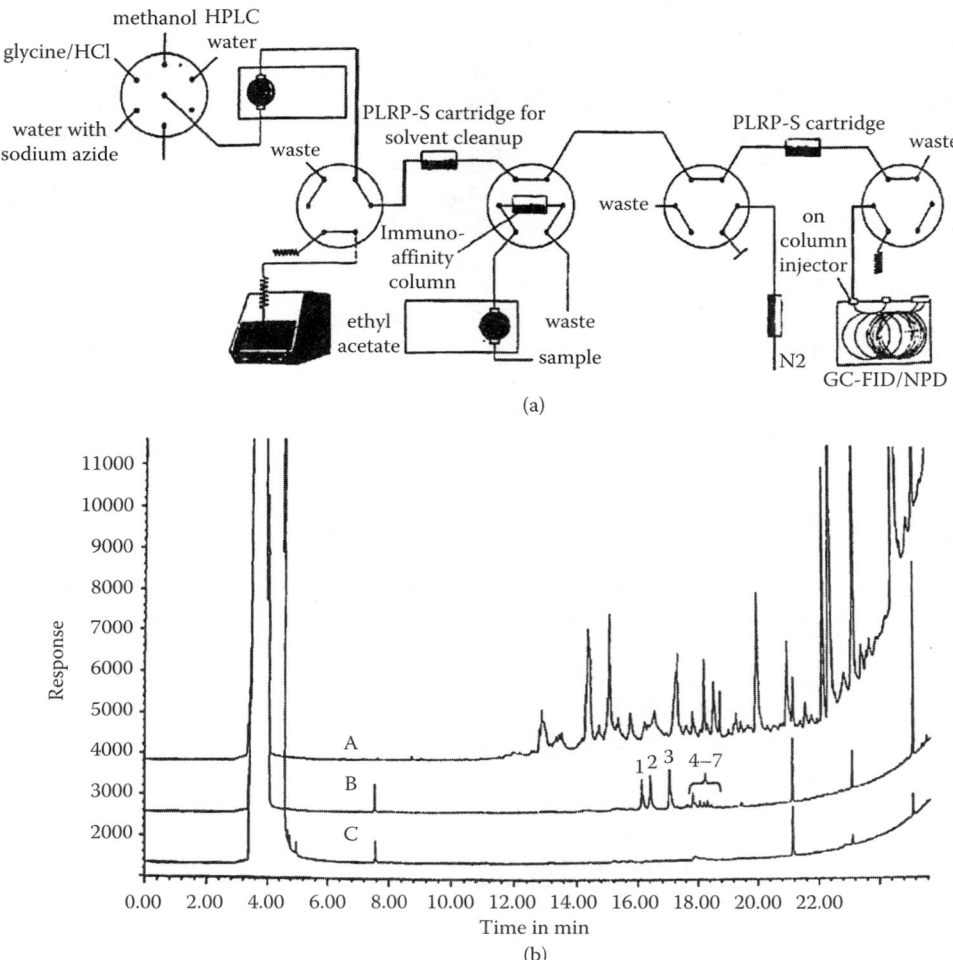

Figure 19.11 (a) System for combining immunoaffinity extraction on-line with gas chromatography and (b) chromatograms obtained with a system for the analysis of triazines. In this approach, styrene-divinyl benzene copolymer (PLRP-S) columns were used to (1) remove contaminant materials from the sample and (2) trap analytes as they eluted from the immunoaffinity column and place them in a GC-compatible solvent. The results given in (b) are for 10 ml of municipal wastewater treated by (A) traditional solid-phase extraction and (B) immunoaffinity extraction. A blank run obtained with 10 ml of HPLC-grade water is shown in (C) as a control. The spiked wastewater used in (A) and (B) contained 1 mg/l each of (1) atrazine, (2) terbuthylazine, (3) sebuthylazine, (4) simetryn, (5) prometryn, (6) terbutryn, and (7) dipropetryn. (Reproduced with permission from Dalluge, J., Hankemeier, T., Vreuls, R.J.J., and Brinkman, U.A.T., *J. Chromatogr. A*, 830, 377–386, 1999.)

wastewater, and orange juice [67]. Figure 19.11 shows the system developed for this work. In this method, samples were first injected onto an immunoaffinity column. The retained solutes were then desorbed and collected by an on-line cartridge containing a styrene-divinyl benzene copolymer (PLRP-S). This cartridge helped to reconcentrate the analytes into a narrow band. Next, these compounds were desorbed from the PLRP-S cartridge with ethyl acetate and fed directly into a GC system, where they were monitored by flame ionization or nitrogen phosphorus detectors. Prior to use on this system, all solvents were treated with a PLRP-S column to remove trace contaminants.

The recovery of triazines by this method was compared with that obtained using traditional solid-phase extraction. A spiked water sample containing 1 μg/l atrazine gave a recovery of 68% with traditional solid-phase extraction and 88% with immunoextraction. At 100 ng/l atrazine, immunoextraction gave 96% recovery for atrazine. The lower limit of detection for atrazine in this system was 170 pg when using a flame-ionization detector and 15 pg when using a nitrogen phosphorous detector.

Some results for this approach are shown in the lower portion of Figure 19.11 for a municipal wastewater sample spiked with seven s-triazine compounds. As noted earlier for immunoextraction and HPLC, the use of only traditional solid-phase extraction allows many sample components to make it to the GC column, where they can overwhelm and mask the peaks of interest. But immunoextraction provides more efficient sample cleanup and allows the analytes of interest to be more easily detected.

19.5 MOLECULARLY IMPRINTED POLYMERS

Molecularly imprinted polymers (or MIPs) are a relatively new class of affinity supports that have been examined for environmental analysis. A detailed discussion of MIPs and their preparation can be found in Chapter 30. These are created through the polymerization of a functional monomer (i.e., one that interacts with the target analyte) in the presence of the target and a cross-linking agent. However, more than one functional monomer can be used during this process to increase the possible interactions between the final binding site and target. The polymerization is carried out in the presence of a solvent that promotes the creation of pores, thus giving better access of analytes to the imprinted sites. After polymerization, the target is washed away and the polymer is ready for use. Mechanical crushing of the polymer is also often performed to reduce the particle size of this material for use as a column packing [89].

The interactions between the target and imprinted polymer usually rely heavily on the presence of hydrogen bonds during the polymerization process. Since water will tend to disrupt these interactions and decrease binding of the target, most MIPs are formed in and used with organic solvents. But there have been a few reports that have used aqueous mobile phases with MIPs for sample analysis. Examples include a method described by Haupt et al. [90–92] for 2,4-dichlorophenoxyacetic acid and an MIP produced by Matsui and coworkers [93] that gave selective binding for atrazine.

However, even MIPs that require an organic solvent can be used for environmental samples. This can be accomplished by utilizing a method known as the *six-S procedure* [94]. This is a multistep process that involves (1) the selective fractionation of a sample, (2) the use of an organic solvent to elute or obtain the analytes of interest, and (3) the use of a molecularly imprinted polymer to selectively bind the analyte of interest in the presence of the organic solvent.

A system for performing the six-S procedure is shown in Figure 19.12. In this system, the sample is first loaded onto a restricted-access-medium column; this separates the sample components based on size and their interactions with the stationary phase in this column. After high-mass components of the sample have been washed away, the smaller retained compounds are eluted with an organic solvent and applied to a second column that contains a molecularly imprinted polymer. After this second column has been washed to remove nonretained and weakly retained components, the retained solutes are eluted

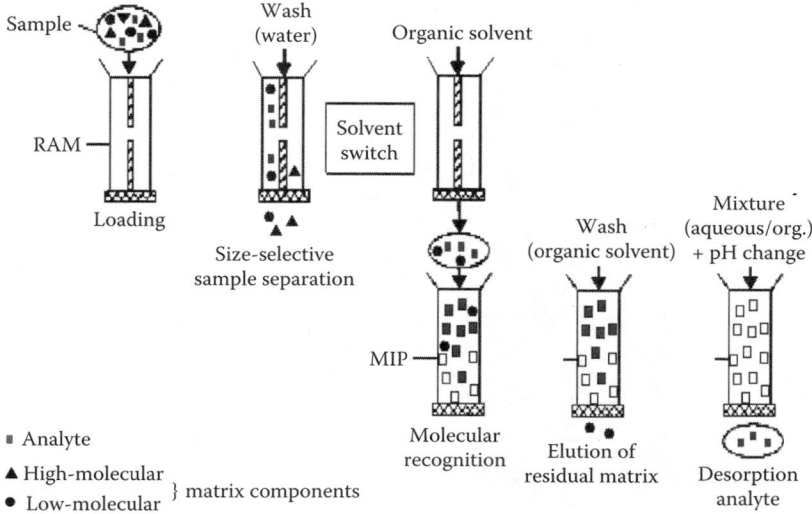

Figure 19.12 The six-*S* procedure for coupling a restricted-access media (RAM) column with a molecularly imprinted polymer (MIP). (Reproduced with permission from Koeber, R., Fleischer, C., Lanza, F., Boos, K.-S., Sellergren, B., and Barcelo, D., *Anal. Chem.*, 73, 2437–2444, 2001.)

by switching to an aqueous/organic mixture at a different pH than the original solvent. This acts to desorb the retained analytes and pass them onto an analytical HPLC column for separation and measurement. With this procedure, it is possible to detect small analytes without interferences from large matrix components, such as proteins, nucleic acids, or humic acids [94].

MIPs are attractive as alternatives to immunoaffinity columns when one must deal with nonaqueous samples or analytes against which antibodies cannot be easily obtained (e.g., toxins). However, further work is still needed in the characterization and optimization of MIPs. Fortunately, many groups are making such efforts. For instance, one group has examined the mechanism of atrazine binding to an MIP by using proton nuclear magnetic resonance spectroscopy [93]. This was performed with a molecularly imprinted polymer that used atrazine as the template and methacrylic acid as the monomer in deuterium-labeled chloroform. Additional work also needs to be conducted in optimizing the binding and dissociation of these polymers, as well as their mass-transfer properties.

19.6 STUDIES OF MOLECULAR INTERACTIONS

Research into the interactions between environmental contaminants and biological systems has recently become a priority within the U.S. Environmental Protection Agency [95]. Of particular concern is the appearance of endocrine disruptors that may affect humans and animals. One method that has been used to screen such agents (e.g., possible estrogen agonists and antagonists) is gel permeation chromatography with affinity ligands [96]. For instance, in one report, tritium-labeled estradiol and estradiol were incubated with a hepatic estrogen receptor in the presence of several suspected endocrine disruptors. Gel permeation chromatography was then used to separate the

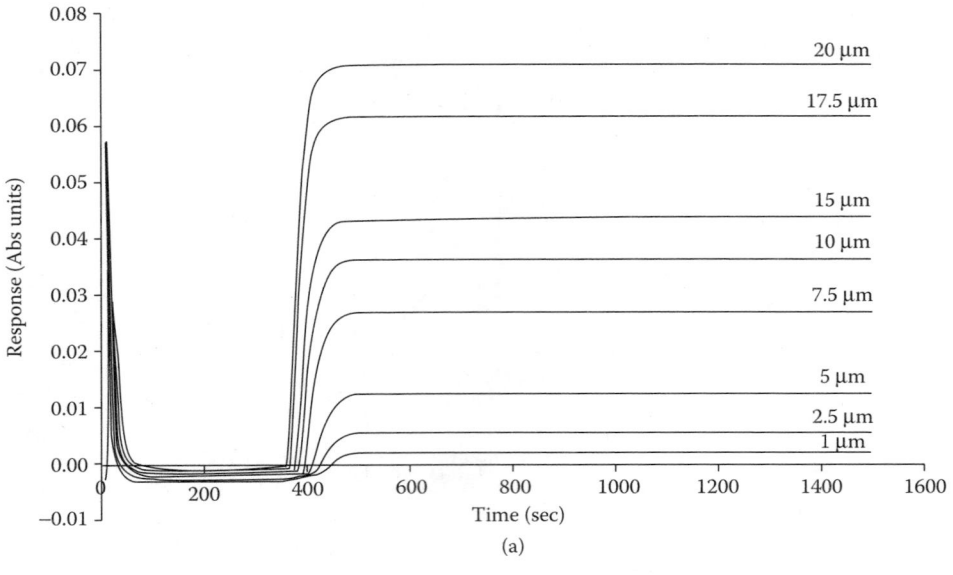

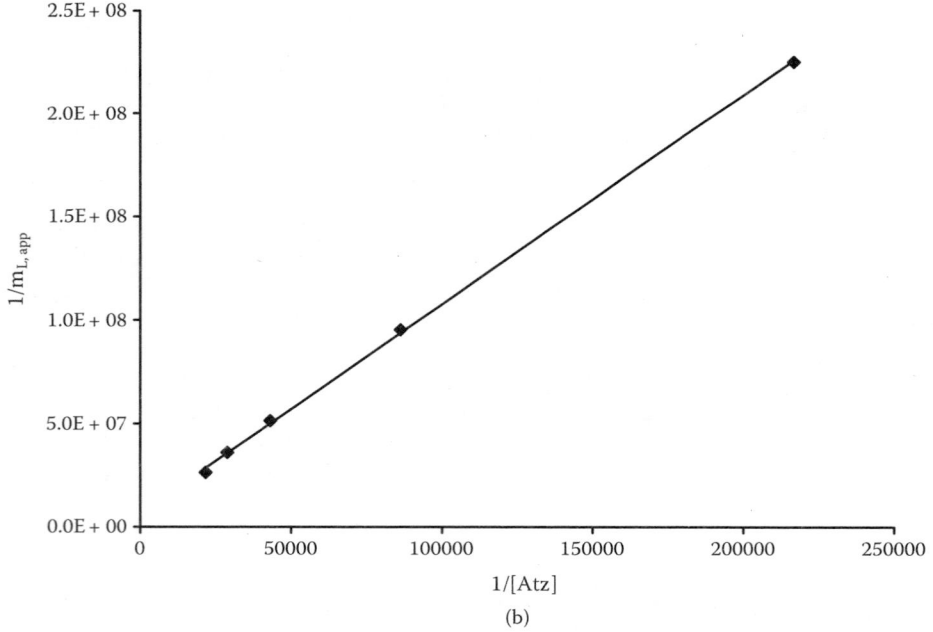

Figure 19.13 (a) Frontal-analysis curves for the binding of various concentrations of atrazine ([Atrazine]) to an immobilized human serum albumin (HSA) column, and (b) analysis of these data using a plot of $1/m_{L,app}$ vs. $1/$[Atrazine] to determine the binding constant for atrazine with HSA. In this latter plot, the intercept was used to determine the total number of binding sites for atrazine in the column, while the association equilibrium constant for atrazine with HSA was calculated using the ratio of the intercept vs. the slope.

low-affinity and high-affinity estrogen receptors and determine their saturation curves. Competitive assays using mixtures of estradiol and the suspected endocrine disruptors were also performed in the presence of each estrogen receptor. The results indicated that compounds like tetrachlorobiphenylol, *o,p*-DDT, and dieldrin had weak competition with estradiol, while stronger competition was noted for compounds like estrone and ethynyl-estradiol.

In recent work, affinity ligands immobilized within columns have been used to study the interactions of environmental agents with proteins. For example, atrazine, a suspected synthetic endocrine disruptor [95], has been examined for its binding to one of the major transport proteins in blood, human serum albumin (HSA). The affinity of this interaction was determined by frontal analysis [97]. In this study, HSA was immobilized onto a silica support and placed into a column, through which several known concentrations of atrazine were then applied in a continuous manner. Figure 19.13 shows some of the breakthrough curves that were obtained, where the mean position of each curve was used as a measure of the apparent moles of atrazine needed to saturate the column ($m_{L,app}$). This amount, in turn, was related to the concentration of applied atrazine and used to find both the binding capacity of the column and the affinity of atrazine for the immobilized HSA.

To obtain these last two items, the results were examined using a double reciprocal plot of $1/m_{L,app}$ vs. $1/$[Atrazine], according to methods described in Chapter 22. In this specific case, the result was a linear relationship, indicating that atrazine had only one major binding site on the immobilized HSA. It was also possible from this plot to determine the binding constant for atrazine to HSA, giving a value of $1 \times 10^4\ M^{-1}$ at pH 7.4 and 37°C. This value is comparable with that seen for the binding of HSA to endogenous compounds and drugs, suggesting that this protein may act as a carrier for atrazine in blood. These results also suggest that the presence of atrazine could affect the interactions of these other agents with HSA [97].

19.7   AFFINITY BIOSENSORS

Affinity biosensors represent another area in which affinity ligands have been used for environmental studies. The principles behind affinity-based optical biosensors are discussed in Chapter 25, while the theory behind these and other types of sensors is reviewed in Chapter 23. In general, these sensors monitor the reaction of a ligand with an immobilized analyte or examine the binding of an analyte with an immobilized ligand. This is often performed by measuring the change in optical properties at the sensor's surface. This gives a signal that is used to examine the rate of the analyte-ligand interaction or to estimate the binding constant for this interaction. This signal can also be used as a means for quantitating the amount of analyte in samples.

In one report, an optical biosensor was coupled on-line to an HPLC system [98]. For this work, samples were first applied to a reversed-phase chromatographic column and separated in the presence of a 37:63 methanol/water mixture. As solutes eluted from this column, they were split in a 1:3 ratio, with the smaller fraction being mixed with a solution of antibodies able to bind isoproturon. This mixture was passed through a reaction coil, where the solutes and antibodies were allowed to interact. The amount of free antibodies that remained was determined by passing this mixture through a biosensor containing an immobilized derivative of isoproturon at its surface. The adsorbed antibodies were then measured using reflectometric interference spectroscopy. As isoproturon or related

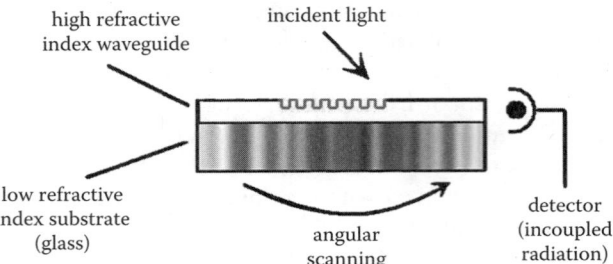

high refractive
index waveguide

incident light

low refractive
index substrate
(glass)

angular
scanning

detector
(incoupled
radiation)

Figure 19.14 Optical biosensor based on a film waveguide refractometer interfaced with a grating coupler. This design consists of a thin film of a high-refractive-index material deposited on glass with a lower refractive index. One item that affects the propagation of light through this system is the amount of analyte on top of the waveguide. Binding of substances to this surface decreases the propagation speed of light in this area and affects the signal. (Reproduced with permission from Brecht, A. and Gauglitz, G., *Anal. Chim. Acta*, 347, 219–233, 1997.)

compounds eluted from the HPLC column, these reacted with the antibodies and decreased the amount of retained antibodies in the biosensor. With this approach, limits of detection of 0.06, 0.09, 0.12, and 0.8 µg/ml were obtained for isoproturon, monuron, fenuron, and monolinuron, respectively [98].

In another study, four optical biosensors were compared for the analysis of atrazine and other *s*-triazine compounds [99]. This study involved the use of a competitive-binding immunoassay, in which antibodies were mixed with the sample before its application to the biosensor's surface. The antibodies were again allowed to interact with the analyte, while the remaining free antibodies were measured by using an immobilized analog of the analyte at the surface of the transducer. The four types of transducers evaluated in this study were a grating coupler, a channel waveguide interferometer, a waveguide for surface plasmon resonance, and a thin-film reflectometer (e.g., see Figure 19.14). All these transducers gave limits of detection for atrazine below 1 µg/l [99], with the thin-film reflectometer giving the lowest value (0.05 µg/l). The average analysis time was 20 min per sample, and reasonable correlation was noted versus an HPLC method. However, the results of these devices for real environmental samples were dependent on the sample matrix, with some irreversible binding being observed due, in part, to the presence of humic substances in water samples [99].

19.8   AFFINITY CAPILLARY ELECTROPHORESIS

Affinity capillary electrophoresis (ACE) is a type of capillary electrophoresis in which ligands are either added to a sample or placed within the electrophoretic system. As discussed in Chapter 26, this technique can be used to quantitate specific analytes, to separate solutes, or to examine the binding between a ligand and sample components.

Although ACE has not seen much use in environmental testing, there have been a few reports in this field. For instance, one report used ACE to investigate the binding of *s*-triazines to fulvic and humic acids from water and soil samples [100]. In this case, the investigators examined the partition coefficients for *s*-triazines between the dissolved humic substances and water. This was accomplished by adding increasing concentrations of humic acids to the running buffer while the migration time for an injected pesticide was measured (see Figure 19.15). It was found that the resulting partition coefficients gave good correlation with the degree of ionization for the *s*-triazines and humic acids, with these interactions

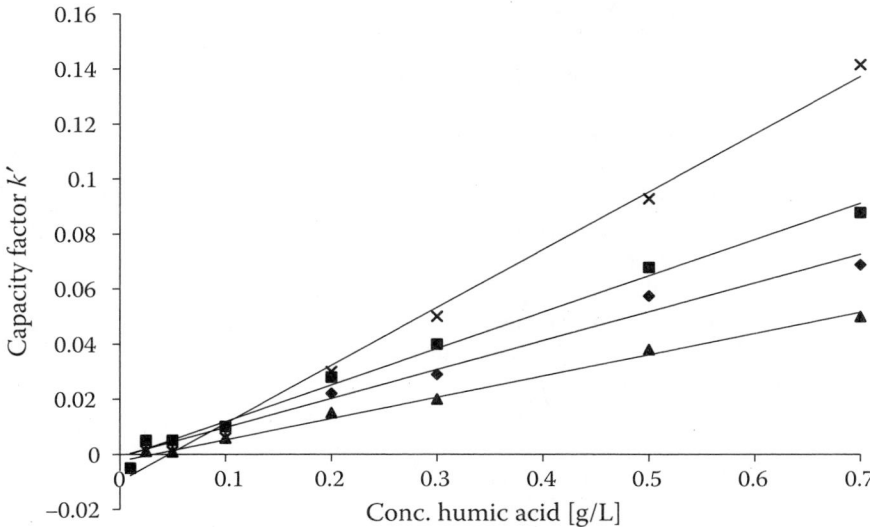

Figure 19.15 Measured retention factor (k') for four s-triazines in affinity capillary electrophoresis as various amounts of humic acid were placed into the running buffer (pH 4.6, 50 mM acetate). The results shown here are for hydroxyatrazine (X), atraton (■), ametryn (◆), and ameline (▲). (Adapted from Schmitt, P., Freitag, D., Trapp, I., Garrison, A.W., Schiavon, M., and Kettrup, A., *Chemosphere*, 35, 55–75, 1997. With permission.)

appearing to be based on hydrogen bonding and acid-base interactions. This same method shows promise for use in studying the binding between xenobiotics and natural dissolved organic matter.

19.9 SUMMARY AND CONCLUSIONS

Although affinity chromatography is a relatively new technique in the environmental area, there are many possible advantages of using this method for such work. These advantages include the specificity of affinity chromatography and its ability to be used directly with a wide variety of samples. Other potential benefits include the simplicity of this approach for sample pretreatment, its ability to be used for either selective measurements or the study of a group of related compounds, and the information it can provide on interactions between environmental agents and biological systems.

Some environmental applications that have already been reported for affinity chromatography include the use of thiol-disulfide interchange chromatography for the analysis of metallothioneins and the use of IMAC for the study of organic ligands and metal-ion transport. Affinity chromatography has also been used to perform immunoassays for environmental analytes using a variety of detection formats. Furthermore, off-line and on-line affinity extraction have been employed in work with food, water, and tissue samples.

Other emerging environmental methods include the use of affinity ligands in biosensors and in capillary electrophoresis. The study of endocrine disruptors by affinity methods and the development of molecular imprints for affinity columns are two additional areas that hold great promise. With such a range of potential capabilities, it is expected that affinity chromatography will play an even greater role in the future of environmental measurements.

SYMBOLS AND ABBREVIATIONS

ACE	Affinity capillary electrophoresis
DEA	Deethylatrazine
DIA	Deisopropylatrazine
E1	Estrone
E2	17β-Estradiol
ELISA	Enzyme-linked immunosorbent assay
ESI-MS	Electrospray-ionization mass spectrometry
Fe^0	Zero-valent iron
FIIA	Flow-injection immunoanalysis
FILIA	Flow-injection liposome immunoanalysis
GC	Gas chromatography
GC-MS	Gas chromatography/mass spectrometry
HA	Hydroxyatrazine
HPLC	High-performance liquid chromatography
HSA	Human serum albumin
IA-GC	Immunoaffinity/gas chromatography
IA-RPLC	Immunoaffinity/reversed-phase liquid chromatography
IMAC	Immobilized metal-ion affinity chromatography
k'	Retention factor
K_a	Association equilibrium constant
MIP	Molecularly imprinted polymer
MS	Mass spectrometry
MT	Metallothionein
PBS	Phosphate-buffered saline
PLRP-S	Styrene-divinyl benzene copolymer
RA	Restricted access
RAM	Restricted access media
RPLC	Reversed-phase liquid chromatography
SIM	Summed-ion chromatogram
β-ME	β-Mercaptoethanol

REFERENCES

1. U.S. Environmental Protection Agency, Agricultural Chemicals in Ground Water: Proposed Pesticide Strategy, U.S. EPA, Washington, DC, 1987.
2. Commission of the European Communities, Pesticides in Ground and Drinking Water, CEC, Brussels, Belgium, 1989.
3. Rollag, J.G., Beck-Westermeyer, M., and Hage, D.S., Analysis of pesticide degradation products by tandem high-performance immunoaffinity chromatography and reversed-phase liquid chromatography, Anal. Chem., 68, 3631–3637, 1996.
4. Korte, F., Konstantinova, T., Mansour, M., Ilieva, P., and Bogdanova, A., Photodegradation of some unsaturated triazine derivatives with herbicide and bactericide activity, Chemosphere, 35, 51–54, 1997.
5. Mansour, M., Feicht, E.A., Behechti, A., and Scheunert, I., Experimental approaches to studying the photostability of selected pesticides in water and soil, Chemosphere, 35, 39–50, 1997.
6. Gilliom, R., Barbash, J., Kolpin, D., and Larson, S., Testing water quality for pesticide pollution, Env. Sci. Tech., 33, 164A–169A, 1999.

7. Namiesnik, J. and Zygmunt, B., Selected concentration techniques for gas chromatographic analysis of environmental samples, *Chromatographia*, 56, S9–S18, 2002.
8. Zygmunt, B., Gas chromatographic determination of volatile environmental organic pollutants based on solvent-free extraction, *Chemia Inzynieria Ekologiczna*, 8, 973–980, 2001.
9. Eckenrode, B.A., Environmental and forensic applications of field-portable GC-MS: An overview, *J. Am. Soc. Mass Spectrom.*, 12, 683–693, 2001.
10. Clement, R.E., Yang, P.W., and Koester, C.J., Environmental analysis, *Anal. Chem.*, 73, 2761–2790, 2001.
11. Jedrzejczuk, A., Goralczyk, K., Czaja, K., Strucinski, P., and Ludwicki, J.K., High performance liquid chromatography: Application in pesticide residue analysis, *Roczniki Panstwowego Zakladu Higieny*, 52, 127–138, 2001.
12. Obrist, H., On-line solid phase extraction for HPLC analysis, *Chimia*, 55, 46–47, 2001.
13. Tribaldo, E.B., Residue analysis of carbamate pesticides in water, *Food Sci. Tech.*, 102, 537–570, 2000.
14. Boyd-Boland, A.A., SPME-HPLC of environmental pollutants, in *Applications of Solid Phase Microextraction*, Pawliszyn, J., Ed., Royal Society of Chemistry, Cambridge, U.K., 1999, pp. 327–332.
15. Pobozy, E., On-line preconcentration of trace elements for HPLC determination: A review, *Chemia Analityczna*, 44, 119–135, 1999.
16. Linde, C.D. and Goh, K.S., Immunoassays (ELISAs) for pesticide residues in environmental samples, *Pest Outlook*, 6, 18–23, 1995.
17. Niessner, R., Immunoassays in environmental analytical chemistry: Some thoughts on trends and status, *Anal. Meth. Instr.*, 1, 134–144, 1993.
18. Jeannot, R., Trends in analytical methods for determination of organic compounds in the environment: Application on waters and soils matrixes, *Spectra Analyse*, 26, 17–24, 1997.
19. Nunes, G.S., Toscano, I.A., and Barcelo, D., Analysis of pesticides in food and environmental samples by enzyme-linked immunosorbent assays, *Trends Anal. Chem.*, 17, 79–87, 1998.
20. Aga, D.S. and Thurman, E.M., Environmental immunoassays: Alternative techniques for soil and water analysis, in *Immunochemical Technology for Environmental Applications*, Aga, D.S. and Thurman, E.M., Eds., American Chemical Society, Washington, DC, 1997, pp. 1–20.
21. Bouzige, M. and Pichon, V., Immunoextraction of pesticides at the trace level in environmental matrixes, *Analusis*, 26, M112–M117, 1998.
22. Dankwardt, A. and Hock, B., Enzyme immunoassays for analysis of pesticides in water and food, *Food Tech. Biotech.*, 35, 165–174, 1997.
23. Fitzpatrick, J., Fanning, L., Hearty, S., Leonard, P., Manning, B.M., Quinn, J.G., and O'Kennedy, R., Applications and recent developments in the use of antibodies for analysis, *Anal. Lett.*, 33, 2563–2609, 2000.
24. Groopman, J.D. and Donahue, K.F., Aflatoxin, a human carcinogen: Determination in foods and biological samples by monoclonal antibody affinity chromatography, *J. AOAC*, 71, 861–867, 1988.
25. Harris, A.S., Wengatz, I., Wortberg, M., Kreissig, S.B., Gee, S.J., and Hammock, B.D., Development and application of immunoassays for biological and environmental monitoring, in *Multiple Stresses in Ecosystems*, Cech, J., Wilson, B., and Crosby, D., Eds., Lewis Publishers, Boca Raton, FL, 1998, pp. 135–153.
26. Hennion, M.C. and Barcelo, D., Strengths and limitations of immunoassays for effective and efficient use for pesticide analysis in water samples: A review, *Anal. Chim. Acta*, 362, 3–34, 1998.
27. Van Emon, J.M., Gerlach, C.L., and Bowman, K., Bioseparation and bioanalytical techniques in environmental monitoring, *J. Chromatogr. B*, 715, 211–228, 1998.
28. Newman, D.J. and Price, C.P., Future developments in immunoassay, in *Principles and Practice of Immunoassay*, Price, C.P. and Newman, D.J., Eds., Macmillan, London, U.K., 1997, pp. 649–656.
29. Unger, K.K. and Anspach, B., Trends in stationary phases in high-performance liquid chromatography, *Trends Anal. Chem.*, 6, 121–125, 1987.

30. Hage, D.S., Survey of recent advances in analytical applications of immunoaffinity chromatography, *J. Chromatogr. B,* 715, 3–28, 1998.

31. Stevenson, D., Immuno-affinity solid-phase extraction, *J. Chromatogr. B,* 745, 39–48, 2000.

32. Sharma, S., Desai, T.A., and Agarwal, G.P., Immobilized metal ion-protein interactions: Role of ionic strength and pH, in *Proceedings of 221st ACS Natl. Meeting,* San Diego, CA, 2001, p. 195.

33. Gonzalez-Vergara, E. and Vincent, J.B., Silica-conalbumin conjugate as an inexpensive alternative for metalloprotein affinity metal chromatography, *J. Environ. Sci. Health Part A,* A31, 2337–2347, 1996.

34. Sharma, S. and Agarwal, G.P., Interactions of proteins with immobilized metal ions, *J. Colloid Inter. Sci.,* 243, 61–72, 2001.

35. Jenkins, A.L., Yin, R., and Jensen, J.L., Molecularly imprinted polymer sensors for pesticide and insecticide detection in water, *Analyst,* 126, 798–802, 2001.

36. Jenkins, A.L., Yin, R., Jensen, J.L., and Durst, H.D., Molecularly imprinted polymers for the detection of chemical agents in water, *Polymeric Mats. Sci. Eng.,* 84, 76–77, 2001.

37. Green, T.M., Charles, P.T., and Anderson, G.P., Detection of 2,4,6-trinitrotoluene in seawater using a reversed-displacement immunosensor, *Anal. Biochem.,* 310, 36–41, 2002.

38. Pignatello, J.J., The measurement and interpretation of sorption and desorption rates for organic compounds in soil media, *Adv. Agron.,* 69, 1–73, 2000.

39. Piehler, J., Brandenburg, A., Brecht, A., Wagner, E., and Gauglitz, G., Characterization of grating couplers for affinity-based pesticide sensing, *Appl. Optics,* 36, 6554–6562, 1997.

40. Margoshes, M. and Vallee, B.L., A cadmium protein from equine kidney cortex, *J. Am. Chem. Soc.,* 79, 4813–4814, 1957.

41. Kagi, J.H.R. and Kojima, Y., Chemistry and biochemistry of metallothionein, *Experientia Suppl.,* 52, 25–61, 1987.

42. Kagi, J.H.R., Himmelhoch, S.R., Whanger, P.D., Bethune, J.L., and Vallee, B.L., Equine hepatic and renal metallothioneins: Purification, molecular weight, amino acid composition, and metal content, *J. Biol. Chem.,* 249, 3537–3542, 1974.

43. Buhler, R.H. and Kagi, J.H., Human hepatic metallothioneins, *FEBS Lett.,* 39, 229–234, 1974.

44. Kabzinski, A.K.M. and Paryjczak, T., Attempts of applying covalent affinity chromatography for determination of environmental exposition to heavy metals based on the quantitative isolation of metallothionein specific protein from human urine, *Chemia Analityczna,* 40, 831–846, 1995.

45. Kabzinski, A.K.M., The application of covalent affinity chromatography with thiol-disulfide interchange for quantitative determination of metallothionein-specific protein from human fluids, as indicator of environmental exposure to heavy metals, *Pol. J. Environ. Stud.,* 6, 61–67, 1997.

46. Kabzinski, A.K.M., Application of covalent affinity chromatography with thiol-disulfide interchange for determination of environmental exposition to heavy metals based on the quantitative determination of Zn-thionein from physiological human fluids by indirect method based on analysis of metal contents, *Biomed. Chromatogr.,* 12, 281–290, 1998.

47. Kabzinski, A.K.M., The determination of environmental and industrial exposure to heavy metals based on the quantitative isolation of metallothionein from human fluids, with application of covalent affinity chromatography with thiol-disulfide interchange gel as a solid-phase extraction support, *Talanta,* 46, 335–346, 1998.

48. Midorikawa, T. and Tanoue, E., Extraction and characterization of organic ligands from oceanic water columns by immobilized metal ion affinity chromatography, *Marine Chem.,* 52, 157–171, 1996.

49. Wu, F., Midorikawa, T., and Tanoue, E., Fluorescence properties of organic ligands for copper(II) in Lake Biwa and its rivers, *Geochem. J.,* 35, 333–346, 2001.

50. Wu, F.C., Evans, R.D., and Dillon, P.J., Fractionation and characterization of fulvic acid by immobilized metal ion affinity chromatography, *Anal. Chim. Acta,* 452, 85–93, 2002.

51. Wu, F. and Tanoue, E., Isolation and partial characterization of dissolved copper-complexing ligands in streamwaters, *Environ. Sci. Tech.,* 35, 3646–3652, 2001.

52. Zhi, Z.-L., Flow-injection immunoanalysis, a versatile and powerful tool for the automatic determination of environmental pollutants, *Lab. Rob. Automat.,* 11, 83–89, 1999.

53. Kramer, P.M., Franke, A., and Standfuss-Gabisch, C., Flow injection immunoaffinity analysis (FIIAA): A screening technology for atrazine and diuron in water samples, *Anal. Chim. Acta,* 399, 89–97, 1999.

54. Kraemer, P. and Schmid, R., Flow injection immunoanalysis (FIIA): A new format of immunoassay for the determination of pesticides in water, *GBF Monogr.,* 14, 243–246, 1991.

55. Hage, D.S. and Nelson, M.A., Chromatographic immunoassays, *Anal. Chem.,* 73, 198A–205A, 2001.

56. Katmeh, M.F., Godfrey, A.J.M., Stevenson, D., and Aherne, G.W., Enzyme immunoaffinity chromatography: A rapid semi-quantitative immunoassay technique for screening the presence of isoproturon in water samples, *Analyst,* 122, 481–486, 1997.

57. Bjarnason, B., Bousios, N., Eremin, S., and Johansson, G., Flow injection enzyme immunoassay of atrazine herbicide in water, *Anal. Chim. Acta,* 347, 111–120, 1997.

58. Hage, D.S., Affinity chromatography: A review of clinical applications, *Clin. Chem.,* 45, 593–615, 1999.

59. Wittmann, C. and Schmid, R.D., Development and application of an automated quasi-continuous immunoflow injection system to the analysis of pesticide residues in water and soil, *J. Agric. Food Chem.,* 42, 1041–1047, 1994.

60. Wittmann, C., Immunochemical techniques and immunosensors for the analysis of dealkylated degradation products of atrazine, *Intl. J. Environ. Anal. Chem.,* 65, 113–126, 1996.

61. Reeves, S.G., Rule, G.S., Roberts, M.A., Edwards, A.J., and Durst, R.A., Flow-injection liposome immunoanalysis (FILIA) for alachlor, *Talanta,* 41, 1747–1753, 1994.

62. Rule, G.S., Palmer, D.A., Reeves, S.G., and Durst, R.A., Use of protein A in a liposome-enhanced flow-injection immunoassay, *Anal. Proc.,* 31, 339–340, 1994.

63. Lee, M., Durst, R.A., and Wong, R.B., Comparison of liposome amplification and fluorophor detection in flow-injection immunoanalyses, *Anal. Chim. Acta,* 354, 23–28, 1997.

64. Onnerfjord, P., Eremin, S.A., Emneus, J., and Marko-Varga, G., A flow immunoassay for studies of human exposure and toxicity in biological samples, *J. Mol. Recogn.,* 11, 182–184, 1998.

65. Onnerfjord, P., Eremin, S.A., Emneus, J., and Marko-Varga, G., High sample throughput flow immunoassay utilising restricted access columns for the separation of bound and free label, *J. Chromatogr. A,* 800, 219–230, 1998.

66. Mackert, G., Reinke, M., Schweer, H., and Seyberth, H.W., Simultaneous determination of the primary prostanoids prostaglandin E2, prostaglandin F2 alpha and 6-oxoprostaglandin F1 alpha by immunoaffinity chromatography in combination with negative ion chemical ionization gas chromatography-tandem mass spectrometry, *J. Chromatogr.,* 494, 13–22, 1989.

67. Dalluge, J., Hankemeier, T., Vreuls, R.J.J., and Brinkman, U.A.T., Online coupling of immunoaffinity-based solid-phase extraction and gas chromatography for the determination of *s*-triazines in aqueous samples, *J. Chromatogr. A,* 830, 377–386, 1999.

68. Puchades, R. and Maquieira, A., Recent developments in flow injection immunoanalysis, *Crit. Rev. Anal. Chem.,* 26, 195–218, 1996.

69. Junk, G.A., Avery, M.A., and Richard, J.J., Interferences in solid-phase extraction using C-18 bonded porous silica cartridges, *Anal. Chem.,* 60, 1347–1350, 1988.

70. Thomas, D., Beck-Westermeyer, M., and Hage, D.S., Determination of atrazine in water using tandem high-performance immunoaffinity chromatography and reversed-phase liquid chromatography, *Anal. Chem.,* 66, 3823–3829, 1994.

71. Burdaspal, P., Legarda, T.M., Gilbert, J., Anklam, E., Apergi, E., Barreto, M., Brera, C., Carvalho, E., Chan, D., Felgueiras, I., Hald, B., Jorgensen, K., Langseth, W., MacDonald, S., Nuotio, K., Patel, S., Schuster, M., Solfrizzo, M., Stefanaki, I., Stroka, J., and Torgersen, T., Determination of ochratoxin A in baby food by immunoaffinity column cleanup with liquid chromatography: Interlaboratory study, *J. AOAC Intl.,* 84, 1445–1452, 2001.

72. Cahill, L.M., Kruger, S.C., McAlice, B.T., Ramsey, C.S., Prioli, R., and Kohn, B., Quantification of deoxynivalenol in wheat using an immunoaffinity column and liquid chromatography, *J. Chromatogr. A,* 859, 23–28, 1999.

73. Carrasco, P.B., Escola, R., Marco, M.P., and Bayona, J.M., Development and application of immunoaffinity chromatography for the determination of the triazinic biocides in seawater, *J. Chromatogr. A,* 909, 61–72, 2001.

74. De Girolamo, A., Solfrizzo, M., von Holst, C., and Visconti, A., Comparison of different extraction and clean-up procedures for the determination of fumonisins in maize and maize-based food products, *Food Add. Contam.,* 18, 59–67, 2001.

75. Ferguson, P.L., Iden, C.R., McElroy, A.E., and Brownawell, B.J., Determination of steroid estrogens in wastewater by immunoaffinity extraction coupled with HPLC-electrospray-MS, *Anal. Chem.,* 73, 3890–3895, 2001.

76. Rule, G.S., Mordehai, A.V., and Henion, J., Determination of carbofuran by online immunoaffinity chromatography with coupled-column liquid chromatography/mass spectrometry, *Anal. Chem.,* 66, 230–235, 1994.

77. Bean, K.A. and Henion, J.D., Determination of carbendazim in soil and lake water by immunoaffinity extraction and coupled-column liquid chromatography-tandem mass spectrometry, *J. Chromatogr. A,* 791, 119–126, 1997.

78. Farjam, A., Brugman, A.E., Lingeman, H., and Brinkman, U.A.T., On-line immunoaffinity sample pretreatment for column liquid chromatography: Evaluation of desorption techniques and operating conditions using an anti-estrogen immuno-precolumn as a model system, *Analyst,* 116, 891–896, 1991.

79. Bouzige, M., Pichon, V., and Hennion, M.C., Online coupling of immunosorbent and liquid chromatographic analysis for the selective extraction and determination of polycyclic aromatic hydrocarbons in water samples at the ng l^{-1} level, *J. Chromatogr. A,* 823, 197–210, 1998.

80. Stolpe, N.B. and Shea, P.J., Alachlor and atrazine degradation in a Nebraska soil and underlying sediments, *Soil Sci.,* 160, 359–370, 1995.

81. Nelieu, S., Kerhoas, L., and Einhorn, J., Atrazine degradation by ozonization in the presence of methanol as scavenger, *Int. J. Environ. Anal. Chem.,* 65, 297–311, 1996.

82. Boundy-Mills, H.L., de Souza, M.L., Mandelbaum, R.T., Wackett, L.P., and Sadowsky, M.J., The atzB gene of *Pseudomonas sp.* strain ADP encodes the second enzyme of a novel atrazine degradation pathway, *Appl. Environ. Microbiol.,* 63, 916–923, 1997.

83. Sadowsky, M.J., Wackett, L.P., De Souza, M.L., Boundy-Mills, K.L., and Mandelbaum, R.T., Genetics of atrazine degradation in *Pseudomonas sp.* strain ADP, *ACS Symp. Ser.,* 683, 88–94, 1998.

84. Shapir, N. and Mandelbaum, R.T., Atrazine degradation in subsurface soil by indigenous and introduced microorganisms, *J. Agric. Food Chem.,* 45, 4481–4486, 1997.

85. Hilts, B.A., Dvorak, B.I., Rodriguez-Fuentes, R., and Miller, J.A., GAC treatment of Lincoln's water: Implications of pretreatment and herbicides, *Proc. Ann. Conf. Am. Water Works Assoc.,* 2000, pp. 1060–1066.

86. Ashley, J.M., Dvorak, B.I., and Hage, D.S., Activated carbon treatment of *s*-triazines and their metabolites, *Water Res. Urban Environ-98, Proc. Nat. Conf. Environ. Eng.,* Chicago, 1998, pp. 338–343.

87. Singh, J., Zhang, T.C., Shea, P.J., Comfort, S.D., Hundal, L.S., and Hage, D.S., Transformation of atrazine and nitrate in contaminated water by iron-promoted processes, *Proc. WEFTEC '96,* Dallas, TX, 1996, pp. 143–150.

88. Nelson, M.A., Development of a Field Portable Immunoextraction/RPLC System for Environmental Analysis, Ph.D. dissertation, University of Nebraska-Lincoln, Lincoln, NE, 2003.

89. Mosbach, K., Yilmaz, E., and Haupt, K., Molecular imprinting, PCT Int. Appl. 0190228, 2001.

90. Haupt, K., Dzgoev, A., and Mosbach, K., Assay system for the herbicide 2,4-D using a molecularly imprinted polymer as an artificial recognition element, *Anal. Chem.,* 70, 628–631, 1998.

91. Haupt, K., Mayes, A.G., and Mosbach, K., Herbicide assay using an imprinted polymer-based system analogous to competitive fluoroimmunoassays, *Anal. Chem.,* 70, 3936–3939, 1998.

92. Haupt, K., Molecularly imprinted sorbent assays and the use of non-related probes, *React. Funct. Polym.,* 41, 125–131, 1999.

93. Matsui, J., Miyoshi, Y., Doblhoff-Dier, O., and Takeuchi, T., A molecularly imprinted synthetic polymer receptor selective for atrazine, *Anal. Chem.*, 67, 4404–4408, 1995.

94. Koeber, R., Fleischer, C., Lanza, F., Boos, K.-S., Sellergren, B., and Barcelo, D., Evaluation of a multidimensional solid-phase extraction platform for highly selective on-line cleanup and high-throughput LC-MS analysis of triazines in river water samples using molecularly imprinted polymers, *Anal. Chem.*, 73, 2437–2444, 2001.

95. Timm, G.E. and Maciorowski, A.F., Endocrine disruptor screening and testing: A consensus strategy, in *Analysis of Environmental Endocrine Disruptors,* Keith, L.H., Jones-Lepp, T.L., and Needham, L.L., Eds., American Chemical Society, Washington, DC, 2000, pp. 1–10.

96. Bunce, J.J., Cox, B.J., and Partridge, A.W., Method development and interspecies comparison of estrogen receptor binding assays for estrogen agonists and antagonists, in *Analysis of Environmental Endocrine Disruptors,* Keith, L.H., Jones-Lepp, T.L., and Neddham, L.L., Eds., American Chemical Society, Washington, DC, 2000, pp. 11–22.

97. Dodlinger, M., Chromatographic Studies of Atrazine Binding to Human Serum Albumin, M.S. thesis, University of Nebraska-Lincoln, Lincoln, NE, 2003.

98. Haake, H.-M., De Best, L., Irth, H., Abuknesha, R., and Brecht, A., Label-free biochemical detection coupled on-line to liquid chromatography, *Anal. Chem.*, 72, 3635–3641, 2000.

99. Brecht, A. and Gauglitz, G., Label-free optical immunoprobes for pesticide detection, *Anal. Chim. Acta,* 347, 219–233, 1997.

100. Schmitt, P., Freitag, D., Trapp, I., Garrison, A.W., Schiavon, M., and Kettrup, A., Binding of s-triazines to dissolved humic substances: Electrophoretic approaches using affinity capillary electrophoresis (ACE) and micellar electrokinetic chromatography (MEKC), *Chemosphere,* 35, 55–75, 1997.

20

Affinity Chromatography in Molecular Biology

Luis A. Jurado, Shilpa Oak, Himanshu Gadgil, Robert A. Moxley, and Harry W. Jarrett

Department of Biochemistry, University of Tennessee Health Science Center, Memphis, TN

CONTENTS

20.1   INTRODUCTION

Affinity chromatography encompasses a variety of techniques and is used in many fields. One of the fields in which it is widely used is molecular biology [1–53]. Although standard columns can be employed in such work, molecular biology often utilizes formats that are uncommon elsewhere. For example, packed columns containing protein A are often used in other fields for antibody purification. However, in molecular biology protein A is instead coupled to microbeads and used for immunoprecipitation. These beaded supports have also been combined with fluorescent dyes and flow cytometry to give a method known as a microsphere-bead assay, which can provide detection limits approaching or exceeding those of high-performance liquid chromatography (HPLC).

Affinity chromatography is most often used in molecular biology as a preparative method. Some of the examples that are discussed in this chapter include its use for the isolation of polynucleotides, recombinant proteins, antibodies or antigens, and cells. The analytical applications that are discussed include chromatin immunoprecipitation and multiplexed particle-based flow cytometric assays. Related methods, such as DNA or RNA microarrays and biosensors, will not be considered.

20.2   POLYNUCLEOTIDE ISOLATION

The isolation of polynucleotides is one important application of affinity chromatography in molecular biology. Affinity chromatography can be used to purify both RNA and DNA, but its use for the isolation of RNA is more common. The isolation of polynucleotides by affinity chromatography makes use of the natural pairing of nucleic acid bases. This can be performed using immobilized polynucleotides that bind to soluble polynucleotides with either a partial or full complementary sequence. Many methods for polynucleotide affinity chromatography exist, but all involve the use of a ligand that is either a homopolymer or heteropolymer. The use of both these classes of ligands is discussed in the following subsections.

20.2.1   Homopolymer Ligands

The homopolymer ligands used for polynucleotide purification consist of chains of a single type of nucleotide. Examples are polythymidylic acid (poly(dT)) and polyuridylic acid (poly(U)). These are used for isolating polyadenylic acid, or poly(A), mRNA from cell preparations. Poly(dT) and poly(U) selectively isolate mRNA in the presence of rRNA and tRNA.

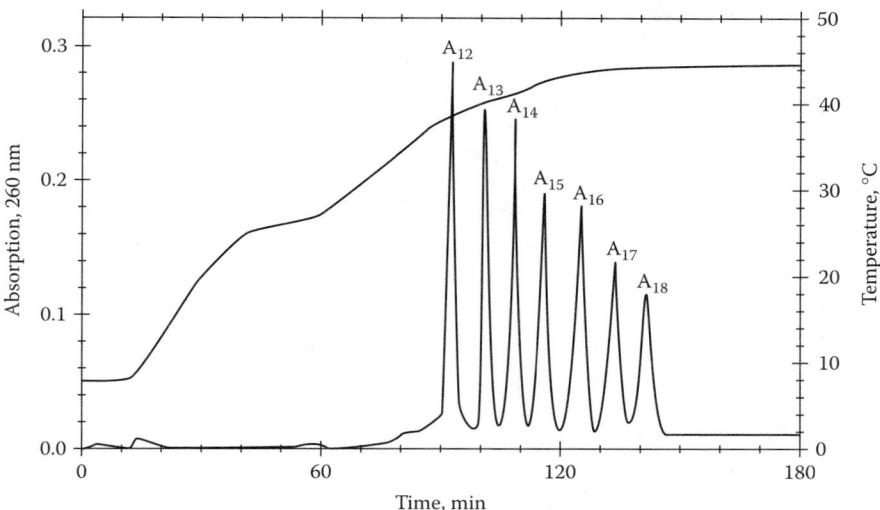

Figure 20.1 Separation of an adenylic acid mixture, $(dA)_{12-18}$, on a thymidylic acid $(dT)_{18}$-silica column using temperature-gradient elution. A major advance in increasing the resolution of $(dA)_n$ is demonstrated in this figure, where HPLC was used instead of conventional low-pressure chromatography. (Reproduced with permission from Goss, T.A., Bard, M., and Jarrett, H.W., *J. Chromatogr. A*, 508, 279–287, 1990).

To perform this type of separation, RNA in the sample is first chemically denatured to separate the intermingled ribonucleotides and to inactivate any RNases that are present. The denatured poly(A) mRNA is then applied to a column containing a ligand like immobilized poly(dT). The poly(A) tail of the mRNA then anneals to the immobilized poly(dT), while the other types of RNA wash through the column. The poly(A) mRNA is then eluted by removing salt from the buffer, thus disrupting interactions between the poly(dT) and poly(dA) by creating intrastrand charge repulsion. After the mRNA has been collected, additional steps are usually required to obtain a sufficient purity for use in the final application.

Immobilized homopolymers like poly(dT) and poly(U) are most frequently attached to chromatographic materials based on cellulose, agarose, or Sephadex. Poly(dT) has also been attached to silica by using *N*-hydroxysuccinimide activation, which has allowed polynucleotide purification by HPLC [1, 2]. Modern methods of homopolymer chromatography, like those shown in Figure 20.1 and Figure 20.2, can allow the separation of poly(dA) oligomers that differ by a single nucleotide base. In addition, these ligands can be used with paramagnetic beads for the isolation of mRNA from whole cells [3], since eukaryotic mRNA contains a poly(A) tail. Homopolymer libraries can also be used to construct a subtractive library, as discussed in the next section of this chapter.

20.2.2 Heteropolymer Ligands and Subtractive Hybridization

Heteropolymer ligands are sequences of RNA, DNA, or cDNA that are attached to a solid support. These are used primarily in *subtractive library construction*, which is also known as *subtractive hybridization*. Subtractive hybridization is a powerful method that can identify differentially expressed genes in closely related cell types of the same eukaryotic organism. By removing cDNA that binds a particular sequence, this method allows the construction of a cDNA library from selected and enriched populations of mRNA.

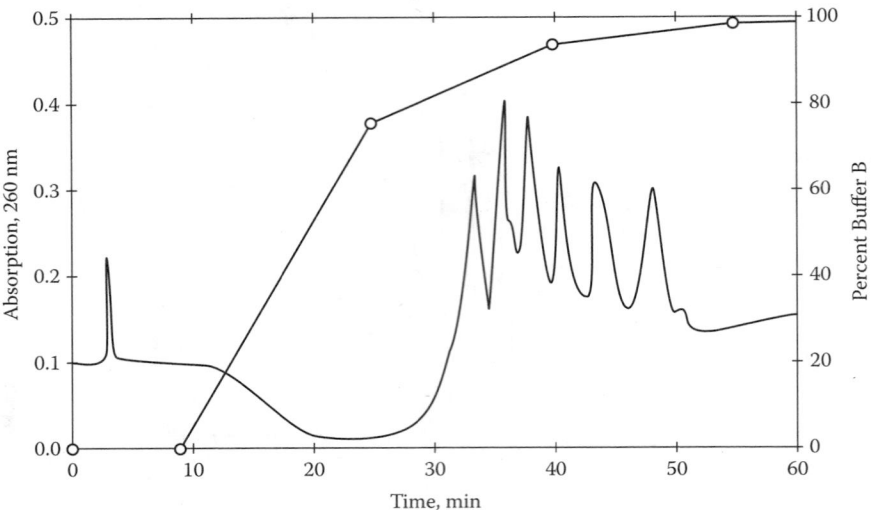

Figure 20.2 Separation of an adenylic acid mixture, $(dA)_{19-24}$, on a thymidylic acid $(dT)_{50}$-silica column using a formamide gradient. (Reproduced with permission from Goss, T.A., Bard, M., and Jarrett, H.W., *J. Chromatogr. A*, 508, 279–287, 1990.)

In this method, the tissue, cDNA library, or mRNA that is to be compared is first designated as the target [+] or driver [−]. The target is the cDNA containing the desired sequence. The driver is the polynucleotide that is complementary (or "antisense") to the target; this can be either mRNA or cDNA. The target cDNA is produced from a cDNA library or by using reverse transcription from the isolated tissue mRNA. Regardless of how they are produced, the target and driver must be complementary to one another.

In the hybridization step, the two populations (i.e., the target and driver) are first denatured and mixed. The target sequences then hybridize with the driver sequences. Any cDNA from the target that does not hybridize with the driver is recovered and saved, giving rise to the subtraction step in this procedure. Usually several hybridization and subtraction steps are performed to further deplete the target of any driver sequences. The resulting pool of enriched mRNA is then cloned into an appropriate library vector.

Subtractive library methods differ in the type of hybridization and subtraction that they employ. For example, hybridization can occur in solution using a free driver or it can make use of a driver bound to a support. When the hybridization is performed in solution, hydroxyapatite chromatography is usually used to separate the various double- and single-stranded polynucleotide fractions [4]. In this approach, the denatured target cDNA and driver mRNA are mixed in solution without a support and applied to a column that contains previously boiled hydroxyapatite. When the hybridization mixture is applied to this column, the unhybridized (i.e., single-stranded) target cDNA elutes under relatively mild conditions. This subtracted, target DNA can then be used for further rounds of subtraction and for later library construction. The double-stranded cDNA:mRNA and excess driver mRNA that are retained by the hydroxyapatite column can be eluted later, if desired, by applying a high-concentration phosphate buffer.

Subtraction can also be performed using immobilized polynucleotides. To do this, poly(dT) is attached to a support such as latex [5]. Denatured driver poly(A) mRNA is then added, and the poly(dT) and poly(A) are allowed to anneal. Next, the mRNA is reverse-transcribed to form an mRNA:cDNA-support complex, and the mRNA template is removed. This results in immobilized driver cDNA that can be mixed with the target mRNA for hybridization. Subtraction is performed by centrifuging the mixture and separating the unhybridized target mRNA in the supernatant from the solid support. After several rounds of this process to improve subtraction, the mRNA is used for cDNA library construction [5, 6]. In a recent variation of this method [6], driver mRNA was first converted to a sense strand of cDNA by using reverse transcription and the polymerase chain reaction. Since this resulted in amplified cDNA rather than RNA, which is less stable, this prevented loss of low-abundance sequences.

Paramagnetic beads can also be used for subtractive hybridization. In this case, a magnetic field is used for separating the support from the supernatant [3]. For instance, this can be accomplished by using Dynabeads oligo(dT), which contain a (dT)-oligomer covalently attached to paramagnetic beads. To make use of these in subtractive hybridization, the driver mRNA is mixed with these beads and annealed. The beads are then washed. Next, the annealed driver mRNA is reverse-transcribed to form a cDNA:mRNA complex, and the mRNA template is removed from the cDNA-beads by thermal denaturation. The resulting cDNA-poly(dT)-paramagnetic beads are then mixed with the target mRNA for hybridization. Finally, subtraction is performed by removing the cDNA:mRNA from the free target mRNA with a magnet that attracts and retains the paramagnetic beads.

20.3 PROTEIN ISOLATION

Another important use of affinity chromatography in molecular biology is in the isolation of recombinant proteins. Recombinant DNA technology has revolutionized the field of biology over the last few decades. Proteins that are present in low amounts in nature can now be overexpressed in a host such as baculovirus or *Escherichia coli*. Several expression systems (e.g., the T7 expression system, which utilizes a pET-based vector) are commercially available for such work. These systems also allow the tagging of proteins with small peptides or protein domains, which can be used for the detection and affinity purification of the tagged proteins.

For detection, *epitope tags* are often used to allow an antibody to interact with a recombinant protein. These epitope tags can be fused with either the *N*- or *C*-terminus of the recombinant protein [7, 8]. The Myc tag, HA tag, and FLAG tag are some common examples used for protein identification. Monoclonal and polyclonal antibodies against these tags are commercially available. This allows recombinant proteins containing these tags to be detected by using antibodies and one of the several immunological techniques, such as those described later in this chapter.

Affinity tags can be used for the purification of a recombinant protein [9, 10]. These tags can be a small peptide (e.g., a sequence of six histidines) [11] or an entire protein domain (e.g., glutathione *S*-transferase) [12]. Unlike epitope tags, which are primarily used for detection, the main function of an affinity tag is to facilitate purification. However, antibodies against affinity tags are also commercially available and are sometimes used for detecting affinity-tagged fusion proteins. Some common examples of affinity tags are presented in the following subsections.

20.3.1 Histidine Tags

A histidine tag (or His-Tag) is a common type of affinity tag [11]. An example is shown in Figure 20.3, which contains six histidine residues ($6 \times$ His). Polyhistidine tags like this have a strong affinity for transition metal ions such as Co^{2+}, Ni^{2+}, Cu^{2+}, and Zn^{2+}. Iminodiacetic acid (IDA) can be used as a chelating agent with these ions for the purification

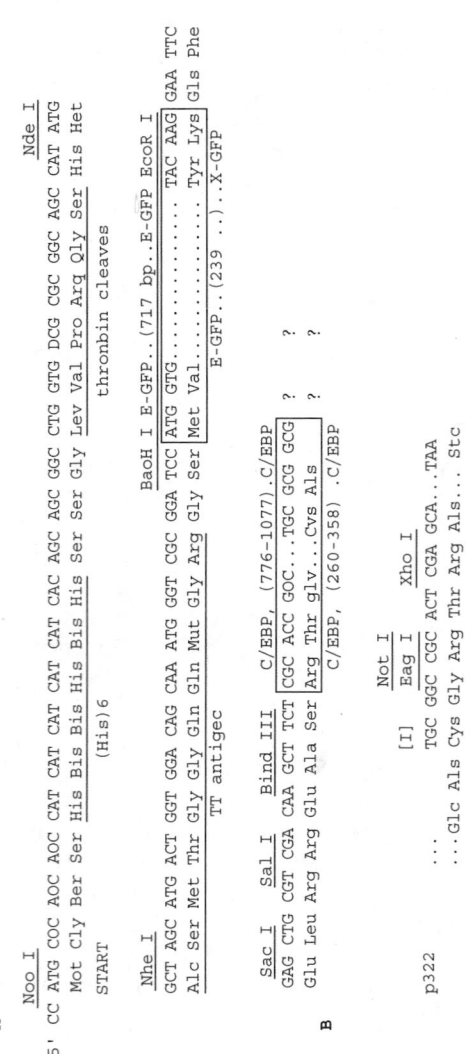
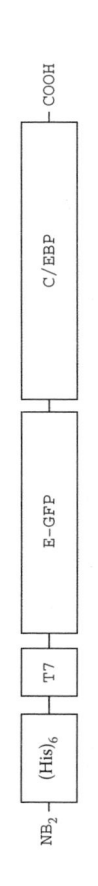

Figure 20.3 A polyhistidine tag chimeric protein. The region of the pI22-C/EBP plasmid that produces the chimeric fusion protein is diagrammed in (A). The position of unique restriction sites and other useful regions of the DNA and protein sequence are illustrated, with (B) showing the region of pI22 before modification with the C/EBP DNA insert, and (C) showing the GFP-C/EBP fusion protein that is produced. (Reproduced with permission from Jarrett, H.W. and Taylor, W.L., *J. Chromatogr. A*, 803, 131–139, 1998.)

of histidine-tagged proteins. For this, IDA is immobilized to a solid support and is allowed to bind to these transition metals by forming a coordination complex. The resulting complex can then be used to trap histidine-tagged proteins that also can bind these metal ions. More information on these and related columns can be found in Chapter 10 under the topic of immobilized metal-ion affinity chromatography (IMAC).

One problem with IDA is that it has a low affinity for metal ions. This can result in the leaching of a chelated metal from the IDA support during a protein's purification, giving a decreased yield for the protein. Modern supports, such as nickel-nitrilotriacetic acid (Ni^{2+}-NTA) and cobalt-carboxymethylasparate (Co^{2+}-CMA), have been developed that show minimal leaching of metal ions, making these more efficient for the purification of His-Tag proteins. These resins also have a high capacity and can bind up to 10 mg protein per milliliter support. To elute proteins from these materials, a mobile phase is used that contains either imidazole as a competing agent for the chelating group or a metal binding agent like ethylenediaminetetraacetic acid (EDTA).

Histidine tags have several advantages over other tags used for protein purification. For instance, the small size of a histidine tag ensures that it will give minimal interference with the folding of a fusion protein. Another advantage is that polyhistidine sequences can bind to metal ions even under denaturing conditions, which allows histidine-tagged proteins to be purified in the presence of agents like urea. This is important because expressed proteins are sometimes insoluble and found in "inclusion bodies" in a cell, requiring denaturation to make these proteins soluble.

A major disadvantage of histidine tags is that proteolytic degradation of even a few histidines in the tag can affect its affinity for nickel. This, in turn, will affect the yield of the tagged protein. Another disadvantage is that histidine tagged proteins are known to form dimers and tetramers in the presence of metal ions. This can lead to inaccurate estimates of the molecular mass for the fusion protein. A third disadvantage is that cellular proteins with two or more adjacent histidine residues can bind to the same immobilized metal ions as histidine-tagged proteins, causing these to coelute with a tagged protein and decrease its purity.

20.3.2 Calmodulin-Binding Peptide

A second type of affinity tag used in molecular biology is calmodulin-binding peptide (CBP). This is a 4-kDa peptide that binds strongly to calmodulin in the presence of calcium. pCAL expression systems contain the CBP encoding sequence, as shown in Figure 20.4. The resulting CBP fusion proteins can then be purified by binding them to calmodulin-agarose in the presence of Ca^{2+}, followed by their elution in the presence of 2 mM EDTA [13]. CBP tags are small compared with another affinity tag, glutathione S-transferase, and show lower interference with protein folding. The CBP tags are sometimes less sensitive to proteolysis than histidine tags. In addition, the chromatographic conditions required to bind a CBP tag are mild and help to better maintain the conformation of the tagged fusion protein.

20.3.3 Strep-Tag

Strep-Tag is a third type of affinity tag. This Strep-Tag contains nine amino acids with the sequence Ala-Trp-Arg-His-Pro-Gln-Phe-Gly-Gly. The name of this tag comes from its ability to strongly bind streptavidin. Strep-Tag fusion proteins can be purified using supports that contain streptavidin, StrepTactin, or monomeric avidin as ligands [14]. Biotin is a competitive inhibitor of the interaction between Strep-Tag and streptavidin and can be used for the elution of Strep-Tag proteins from streptavidin columns. These columns can later be regenerated and reused after biotin elution by washing them with 2(4′-hydroxy-azobenzene)benzoic acid (HABA).

554

Jurado et al.

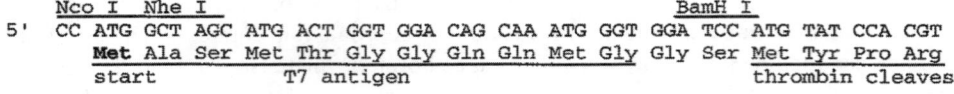

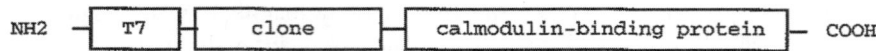

Figure 20.4 The pCAL-c plasmid for a calmodulin-binding peptide tag. The top figure shows the region of the plasmid that contains the multiple cloning restriction sites, the T7 antigen, and a diagram of the calmodulin-binding site. The lower diagram shows the protein produced by this plasmid.

20.3.4 Biotinylation of Fusion Proteins

Biotin protein ligase (BPL) is an enzyme that adds biotin to a lysine within a specific sequence in a protein. If this biotinylation sequence is introduced into a recombinant protein [15], the resulting fusion protein can be purified on a monomeric avidin column and later eluted in the presence of a low concentration of biotin. A major disadvantage to this approach is that all proteins that contain a biocytin (biotin) group will be isolated. However, *E. coli* contains only one such protein, the biotin carrier carboxyl protein (BCCP). Both Strep-Tag proteins and these biotinylated fusion proteins can be easily detected by using enzyme-conjugated streptavidin.

20.3.5 Glutathione *S*-Transferase

Smith and Johnson [16] first used glutathione *S*-transferase (GST) from *Schistosoma japonicum* as a tag for the purification of recombinant proteins. GST is a 28-kDa protein that catalyzes the transfer of the tripeptide glutathione to an electrophilic substrate. GST has a high affinity for glutathione, allowing immobilized glutathione to be used for the purification of GST-tagged proteins [16]. Proteins retained by glutathione-agarose can be eluted with a buffer that contains 10 mM glutathione as a competing agent [12]. Glutathione-agarose is relatively cheap and can be reused several times, making this a cost-effective approach for protein purification. In addition, GST-tagged proteins can be detected using commercially available anti-GST antibodies.

Since GST is a bacterial protein, GST fusion proteins show minimal formation of inclusion bodies when expressed in *E. coli*. But proteins purified through the use of glutathione-agarose are sometimes contaminated with a 60-kDa endogenous *E. coli* protein, thought to be one of the *E. coli* heat-shock induced chaperonins. However, this contaminant can be eliminated through the incubation of samples with 5 mM magnesium chloride and 5 mM ATP prior to their purification on the glutathione-agarose column.

20.3.6 Maltose-Binding Protein

Maltose-binding protein (MBP) is a 40-kDa protein encoded by the malE gene. MBP is a part of the maltose/maltodextrin system in *E. coli*, which is responsible for maltose uptake. MBP has a high affinity for amylose [17], making immobilized amylose useful for the purification of MBP fusion proteins [18]. Elution from an amylose column can be carried out by using a high concentration of maltose in the mobile phase. As was seen for GST, the fact that MBP is a bacterial protein means that MBP fusion proteins tend to resist the formation of inclusion bodies when they are expressed in *E. coli*.

20.3.7 S Tag Fusion Systems

Bovine pancreatic ribonuclease A can be cleaved to form two tightly associated fragments: the S peptide and S protein. The high affinity between the S peptide and S protein is exploited in the S Tag purification system [9]. In this method, the protein of interest is fused with the S peptide, with the resulting fusion protein later being purified on a column that contains immobilized S protein. Elution of the fusion protein is carried out with a buffer that contains 3 M guanidinium thiocyanate. One advantage of this system is that the interaction between S peptide and S protein is stable under moderately denaturing conditions (e.g., 2 M urea), such as those used to solubilize inclusion bodies.

20.3.8 Cellulose-Binding Domain Fusion Proteins

Cellulose-binding domains (CBDs) consist of the functional domains of several cellulolytic enzymes. These domains facilitate catalysis by bringing the active site and substrate for these enzymes in close proximity to each other. By using CBDs as affinity tags, CBD fusion proteins can be purified by cellulose affinity chromatography [19]. These columns make use of the fact that CBDs bind tightly to cellulose, with association equilibrium constants in the range of 10^5 to 10^7 M^{-1}. This interaction is also stable over a wide pH range. Because of this strong interaction, CBD fusion proteins have to be eluted from cellulose columns with a denaturing agent, such as urea or guanidium hydrochloride.

20.4 WORK WITH IMMOBILIZED ANTIBODIES AND ANTIGENS

Another group of affinity methods that are often used in molecular biology are those based on antibody-antigen interactions. Several examples are examined in this section, including immunoprecipitation techniques, chromatographic methods based on immobilized antibodies or antigens, and the use of immunological agents in the isolation of cells.

20.4.1 Immunoprecipitation

Immunoprecipitation (i.e., the formation of an insoluble complex as a result of an antibody-antigen interaction) is used for a variety of purposes in cellular and molecular biology. For instance, this method can be used to determine the molecular weight of a substance, detect the presence of a protein, or measure the quantity of a protein. Immunoprecipitation is also used to characterize the specificity of antibodies, determine binding constants for biological interactions, measure the rates of protein synthesis and degradation, examine the extent of a protein's posttranslational modification, and identify proteins that may be associated with a particular antigen. For instance, the isolation of proteins by immuno-precipitation is the method of choice for studies of biosynthesis and the transport of proteins in cultured cells. Recently, protein-protein interactions within a cell have also been widely studied through the use of this technique.

Table 20.1 General Steps in Immunoprecipitation

1. Prepare the sample or solubilize the antigen preparation
2. Preclear the sample of any insoluble proteins or proteins that bind nonspecifically to the protein A, protein G, or protein L support that will be used in Step 4
3. Incubate the sample with the primary antibodies against the target antigen of interest
4. Precipitate the primary antibody-antigen complex by using a solid support containing immobilized protein A, protein G, or protein L
5. Wash the immunoprecipitate
6. Elute the retained antibody-antigen complex and separate its components by electrophoresis

The basic steps in immunoprecipitation are shown in Table 20.1, with further details being given elsewhere [20–22]. As suggested by this table, immunoprecipitation is often associated with a high background signal. One source of this is the fact that during the immunoprecipitation of a specific antigen, it is possible to have coprecipitation of other proteins due to their direct absorption to the immobilized antibody-binding agent (e.g., proteins A, G, or L, as will be discussed in the next section) or through "nonspecific" interactions with the antibody. This background problem can be minimized by first treating (or "preclearing") the sample with several stringent washes or by adding detergents like Triton X-100 or IGEPAL. It is also usually desirable to use affinity-purified antibodies in the immunoprecipitation method, which helps further reduce the background and optimizes detection of the desired target.

In the study of protein-protein interactions within eukaryotic cells, experiments based on the co-immunoprecipitation of proteins from cellular extracts can provide convincing evidence that two proteins interact either directly or indirectly. However, the ability to perform this experiment is limited by the availability of appropriate antibodies and by the normal concentrations of the proteins being studied. As mentioned previously, another problem with this approach is the frequent occurrence of nonspecific binding between sample components and the antibodies or antibody-binding agents that are used for the immunoprecipitation. Thus, the detection of protein-protein interactions by co-immuno-precipitation should always be confirmed by an independent technique, such as cofrac-tionation. In addition, coprecipitation experiments should be performed with antibodies against all the proteins that are involved in the complex being studied.

Recently, co-immunoprecipitation was used to detect an association between syn-trophin, a component of the Duchenne muscular dystrophy complex, and Grb2, an SH2/SH3 adapter protein [20]. In this work, an antibody against syntrophin was used to precipitate syntrophin from rabbit skeletal muscle extract. This immunoprecipitate was then fractionated by sodium dodecyl sulfate polyacrylamide gel electrophoresis (SDS-PAGE) and transferred to a nitrocellulose membrane [21]. This blotted membrane was next incubated with anti-Grb2 antibodies, with any binding being detected through the use of enhanced chemiluminescence. To confirm the results of this work, Grb2 fusion proteins were shown to bind syntrophin by other methods, including surface plasmon resonance. Other techniques that can be used to detect protein-protein interactions by co-immunoprecipitation and dimerization are discussed in detail elsewhere [22].

20.4.2 Antibody Binding to Immobilized Bacterial Proteins
Although many techniques have been developed to purify antibodies, some of the most common methods are those that employ immobilized bacterial proteins. Examples of bacterial proteins that can bind to antibodies include proteins A, G, and L. A brief

discussion of these ligands and their properties is presented in this section. A more detailed description of these proteins and their use in antibody isolation can be found in Chapter 14.

Protein A is the bacterial protein used most for the isolation of immunoglobulin G (IgG)-class antibodies. This protein is a cell wall component of *Staphylococcus aureus* [23] and binds strongly to sites in the second and third constant regions of the F_c portion of immunoglobulins from many species [24]. Each molecule of IgG contains two binding sites for protein A. Since protein A has four potential binding sites for IgG [25], it is possible to form multimeric complexes between this agent and IgG. Although protein is often used for IgG-class antibodies, it generally shows much weaker binding or no interactions with other classes of immunoglobulins (see Table 20.2).

Protein G is another bacterial wall component that binds antibodies. This is a highly stable surface receptor from a species of *Streptococcus*. Protein G binds preferentially to the F_c portion of IgG but can also bind to IgG's F_{ab} region, making this useful as a ligand

Table 20.2 Binding of Proteins A, G, or L and Related Hybrids to Various Types of Immunoglobulins

	Protein A	Protein G	Protein G-plus	Protein L	Protein LA
Human immunoglobulins					
IgG$_1$	+++	+++	+++	++	+++
IgG$_2$	+++	+++	+++	++	+++
IgG$_3$	++	+++	+++	++	+++
IgG$_4$	++	+++	+	++	+++
IgA	+	−	−	+++	+++
IgD	+	−	−	+++	+++
IgE	+	−	−	+++	+++
IgM	+	−	−	+++	+++
F_{ab}	+	+	—	+++	+++
$F_{(ab)'2}$	+	+	—	+++	+++
K light chains	−	−	−	+++	+++
ScFv	+	−	−	+++	+++
Mouse immunoglobulins					
IgG$_1$	+/−	+++	+++	+++	+++
IgG$_{2a}$	+	+++	+	+++	+++
IgG$_{2b}$	++	+++	+	+++	+++
IgG$_3$	+++	+++	+++	+++	+++
IgM	—	—	−	+++	+
IgA	—	—	−	+++	—
Bovine immunoglobulins	+	+++	−	−	+
Sheep immunoglobulins	−	+	—	−	−
Goat immunoglobulins	−	+	—	−	−
Pig immunoglobulins	++	++	—	+++	+++
Rabbit immunoglobulins	++	+++	—	+/−	+++
Chicken immunoglobulins	−	+/−	—	+	+

Note: +++ = strong binding, ++ = moderately strong binding, + = moderate binding, +/− = weak binding, − = very weak or no binding. More details can be found in the literature [48–53].

for purifying $F_{(ab)'2}$ fragments. Each molecule of protein G can bind two molecules of IgG and can interact with immunoglobulins from a large number of species. Protein A and protein G differ in their ability to bind to some types of antibodies, as indicated in Table 20.2. However, both have quite strong binding to their target antibodies under physiological conditions. In addition, these interactions can easily be disrupted by decreasing the pH of the surrounding solution.

Protein G-plus is a form of streptococcal protein G that has been genetically engineered to increase its binding capacity for antibodies and to eliminate the albumin binding sites of native protein G. This protein is available commercially through companies like Novagen, Calbiochem, and Novabiochem. It is specific for IgG-class antibodies and shows the highest affinity for human IgG_1, IgG_2, and IgG_3, as well for mouse IgG_1 and IgG_3. It does not bind to mouse and human IgA, IgD, or IgM-class antibodies. It also does not bind to albumin.

Protein L is a cell wall protein from *Peptostreptococcus magnus* that binds immunoglobulins through their kappa light chains. The advantage of this is that such interactions do not interfere with an antibody's antigen-binding site. Protein L also binds to immunoglobulin fragments such as an scF_v (i.e., a single chain antibody variable region fragment) or a F_{ab} region. This ligand exhibits a high affinity for many classes of antibodies and has a broad range of species specificity. Because of these characteristics, protein L is more flexible than proteins A or G as an antibody-binding agent. In addition, since protein L does not bind to bovine, sheep, or goat immunoglobulins, background problems associated with the use of fetal bovine serum are avoided. Furthermore, protein L can be used for purifying humanized antibodies made in transgenic animals of these species, which is a route that has become a primary source for therapeutic antibody production.

Protein LA is a genetically engineered hybrid of proteins L and A. This combines the binding properties of these two proteins, making it useful for purifying immunoglobulins and antibody fragments from a wide variety of species. A comparison of the binding properties of protein LA with proteins A, G, G plus, and L can be found in Table 20.2. More details on these ligands and their binding properties can be found in Chapters 5 and 14.

20.4.3 Immobilized Antibodies and Antigens

As discussed in Chapter 14, columns and supports that use immobilized antibodies or antigens can be powerful tools for the isolation of biological substances. For instance, monoclonal antibodies are routinely used in molecular biology for purifying agents from complex mixtures like serum, tissue homogenates, and cell membranes. This makes use of the technique of immunoaffinity chromatography (see Chapters 6 and 14), in which the immobilized antibody acts as the stationary phase for a column.

In immunoaffinity chromatography, antibodies for the desired target are covalently coupled to a matrix like agarose and packed into a column. When a sample is passed through this column, the corresponding target is then retained. This target can often be recovered by changing the pH of the mobile phase, although harsher conditions are sometimes required. The eluted target is then restored to a neutral pH and, if necessary, dialyzed into an appropriate storage solvent. The antibody column can then be regenerated with a neutral buffer and prepared for the next sample.

The antibodies used within such a column should have stable binding to their target at a neutral pH but allow the dissociation of this target with a dilute acid or base. In addition, these antibodies should be able to survive several repeated cycles of binding and elution, thus allowing their use for multiple samples. The buffers used with these antibody columns are generally dictated by the nature of the sample. However, the use of a mobile phase that is at a slightly alkaline pH and that contains a detergent seems to result in less contamination for the final antibody-purified product [26].

Along with the use of immobilized antibodies for the isolation of target substances, it is also possible to use immobilized antigens for the purification of antibodies. Although the antisera obtained from animals can contain high concentrations of antibodies against a target antigen, these preparations almost always contain additional antibodies against other substances. To remove the desired antibodies from these other binding agents, the antiserum can be passed through an immobilized antigen column. In this column, the desired antigen is coupled to a support like Sepharose and will retain the antigen-specific antibodies while allowing other nonspecific antibodies to pass through. The bound antibodies can later be released and collected, typically by using a pH 2.5 or 11.5 elution buffer.

The first use of an immobilized antigen for antibody purification was by Campbell et al. in 1951, who used an antigen immobilized to p-aminobenzyl cellulose (see Chapter 6). In this work, nonspecific serum proteins were removed by washing the column with a 1% saline solution, while the retained antibodies were eluted under acidic conditions. This general method was later applied to other substances, including nonimmune system proteins, giving rise to some of the earliest examples of modern affinity chromatography [27]. It is possible with this approach to isolate pure and specific antibodies for a given agent, as would be required in quantitative studies of the binding between these antibodies and their target substances [28]. However, this method does require that a highly purified form of the antigen first be available for preparing the affinity column that will be used for antibody purification.

20.4.4   Cell Isolation

Yet another use for affinity ligands in molecular biology has been in cell isolation. This can be illustrated by work that has been performed with the hemopoietic system. This is a complex system where, at any given time, a blood sample may contain many types of cells or cells at several stages of development. These cells can be examined by using techniques like rosette fractionation, specific affinity fractionation, and flow cytometry. Affinity fractionation is another method that is useful for this type of work, since it is relatively simple and inexpensive to perform. The affinity ligand in this approach might be an antigen, lectin, or antibody that can recognize the surface properties of a cell and allow the cell's selective removal from a sample.

As an example, immunoaffinity chromatography has been used for more than 30 years to separate B cells and T cells for the study of lymphocyte function. The technique, first performed by Wigzell and Andersson in 1969 [29], makes use of unique cell surface markers. For instance, surface immunoglobulin (sIg) can be found on all B lymphocytes but is not found on other cells. Thus, this cell surface marker can be used for specific isolation of B lymphocytes. This can be performed by making an affinity column that contains immobilized antibodies against the surface marker for the cell type of interest. With this column, the desired cells will be retained, while other nonadhering cells can be removed by washing the column several times with a solution such as cold phosphate buffered saline. The retained cells can then be eluted by adding an excess of free antibody to the column or a soluble form of the marker as a competing agent [30].

It is also possible to use other ligands for cell isolation. For example, cells can first be incubated with antibodies against a specific cell surface marker. This mixture can then be applied to a protein A column, which will retain the antibodies and cells bound to these antibodies. Another example would include the use of a primary antibody (i.e., an antibody against a cell surface marker) that contains a tag, such as biotin or fluorescein, that can later be bound to an affinity column with a ligand that can bind to the tag (e.g., avidin or antifluorescein antibodies).

In this work with cell isolation, the strong and multiple binding of cells to immobilized antibodies (or any other type of ligand) is not desirable. If this does occur, it will give rise to retention for the cells that is nearly irreversible. It can also lead to stripping of the cell surface markers when the bound cells are forced to elute from the column. To avoid these problems, reducible disulfide bonds can be used for attaching the antibodies (or ligand) to the support, which can later be broken to detach the retained cells.

A major advantage of affinity chromatography over other cell-separation methods is that it allows for the possible enrichment of a particular cell type while also depleting a cellular subset. Collaborative workshops [31] have identified a large number of cell surface markers, known as *cluster of differentiation* (CD) *antigens*, that can be identified through the use of antibodies. Many CD antigens are specific to a given cell type or common to a given cell lineage. For example, CD19 is a marker for B lymphocytes, and the marker CD9 is found on T cells, B cells, and myeloid cells. Many of the antibodies for CD antigens are available commercially and can be used in affinity columns. One example is a report in which T cell depletion was performed by using the positive selection of CD34+ hematopoietic cells with a biotinylated monoclonal anti-CD34 antibody and an immunoaffinity column [32]. If an appropriate matrix is used for such work, this does not result in nonspecific stimulation or binding of the cells.

Affinity methods can also be applied to the large-scale separation of cells. However, the effectiveness of such a separation will be highly dependent on a number of factors. Two of these factors when using immobilized antibodies are (1) the number and distribution of antibody-reactive sites on the cell surface and (2) the activity of the immobilized antibody and its accessibility to the marker sites on the cells. These latter characteristics will, in turn, be affected by such items as the activity of the immobilized antibody, the orientation of the antibody on the support, and the number and types of bonds that hold the antibody within the column. A further discussion of these immobilization factors can be found in Chapter 3 and are important in the use of any immobilized ligand for cell separations.

Another item that should be considered in such work is the specificity of the ligand. For instance, protein A is of limited usefulness with B cells, since these cells may nonspecifically bind and be stimulated when using this ligand. Thus, the selection or adaptation of an immunoaffinity technique for the isolation of a given cellular subset always requires careful consideration of the relative advantages and disadvantages of the chosen procedure.

20.5 THE CHIP ASSAY

The *chromatin immunoprecipitation* (or ChIP) *assay* is another method used in molecular biology that involves immobilized ligands. This technique makes use of a nucleosome as a type of affinity support for binding transcription complexes. As a result, this technique is closely related to the method of DNA affinity chromatography (see Chapter 7). This section discusses the ChIP assay and provides the typical experimental conditions for such assays.

20.5.1 General Principles of the ChIP Assay

It is a challenging task to understand gene regulation and the numerous factors in multiprotein complexes that contribute to specific regulation events. The ability to investigate chromosome structure and function is one of the most important assets of the ChIP assay. This technique allows researchers to determine which proteins (e.g., transcription factors) bind to a specific region of DNA within a gene promoter. This method also provides information on whether the interaction of these proteins with DNA is direct or indirect in nature. It has been used to provide data on the chromatin structure of active genes at high resolution and at any given time, and it provides a means of analyzing low-abundance DNA-binding transcription factors and coactivators [33].

The ChIP assay uses the *in vivo* fixation of chromatin by formaldehyde and immuno-precipitation to provide information on the interaction of a given DNA site with nuclear proteins. This procedure begins with the cross-linking of chromosomal proteins with a DNA-binding site through the use of formaldehyde. Formaldehyde is a short-distance (2 Å) cross-linking agent in which the carbon atom acts as a nucleophilic center for the production of both protein-nucleic acid and protein-protein adducts. This agent reacts with amine groups on proteins (e.g., the amino terminus and lysine) or in DNA (e.g., as occurs on adenine, guanine, and cytosine) to form a Schiff base. This intermediate can then react further with a second amine group to produce the final DNA-protein or protein-protein complex [34].

Formaldehyde fixation usually takes place *in vivo* within 10 to 60 min; however, reaction times of several hours have also been used [35]. Further discussion about the experimental details of this procedure can be found in the next section. Since formaldehyde efficiently penetrates cells, it can be used to investigate higher-order chromatin structures while avoiding the risk of redistribution or reassociation of chromosomal proteins that can occur during the preparation of cellular or nuclear extracts. Other advantages of formal-dehyde include its high solubility in water and its ability to react under a wide range of buffer and temperature conditions.

After a sample has been treated with formaldehyde and sonicated, the next step in the ChIP assay is to immunoprecipitate the chromatin with a specific type of antibody, as illustrated in Figure 20.5. This is performed according to the general guidelines given in Section 20.4.1 for immunoprecipitation, where the antibody-target complex is later isolated

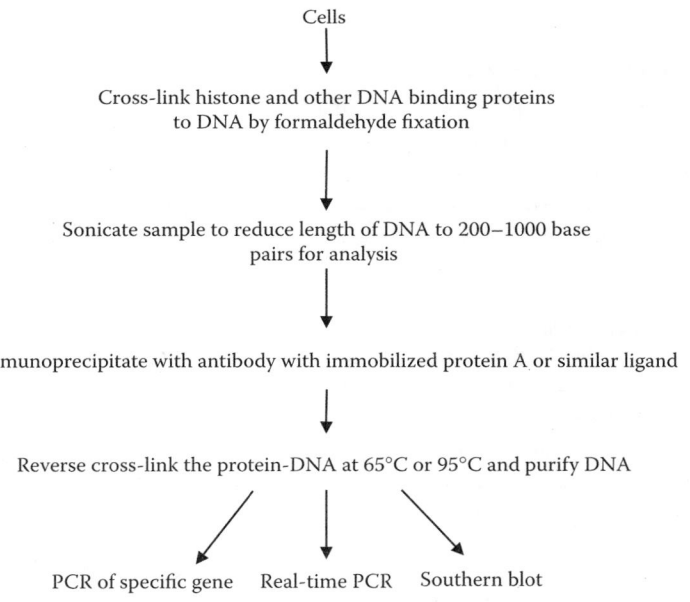

Cells

Cross-link histone and other DNA binding proteins
to DNA by formaldehyde fixation

Sonicate sample to reduce length of DNA to 200–1000 base
pairs for analysis

Immunoprecipitate with antibody with immobilized protein A or similar ligand

Reverse cross-link the protein-DNA at 65°C or 95°C and purify DNA

PCR of specific gene   Real-time PCR   Southern blot

Figure 20.5 The ChIP assay. Starting from live cells, DNA-binding proteins are cross-linked to DNA and other proteins by fixation with formaldehyde. The fixed cells are then lysed, and the DNA is sheared by sonication. Antibodies that are specific for acetyl-histones or the protein of interest are next added. This is followed by removal of these antibodies along with any substances that are bound to them through immunoprecipitation and adsorption onto a protein A support or similar antibody-binding matrix. The protein-DNA or protein-protein complexes are then eluted from this material, and these adducts are reverse-cross-linked to free the DNA. Finally, the DNA is analyzed using a Southern blot, a standard polymerase chain reaction (PCR), or real-time PCR.

using a support that contains an antibody-binding ligand like protein A or protein L. The retained target can then later be released from this support by using an appropriate elution solvent. After the immunoprecipitation has been completed, the cross-linking between the DNA and proteins is reversed. The DNA is then purified and analyzed by methods such as Southern-blot hybridization or the polymerase chain reaction (PCR).

The sensitivity of the ChIP assay mainly depends on the availability of specific, high-quality antibodies that not only react with the desired antigen, but also efficiently immunoprecipitate it under stringent conditions. Cross-linking with formaldehyde will lead to both protein-protein and DNA-protein adducts, but the reversal of this process at different temperatures can be used to selectively obtain each type of adduct. For instance, an extended incubation at 65°C breaks only protein-DNA bonds, while protein-protein adducts require temperatures close to boiling for reversal. As mentioned earlier, the DNA recovered from the immunoprecipitation is usually analyzed by PCR or hybridization. However, the recovered protein can also be analyzed if all of the original cell's protein were labeled with ^{35}S-methionine, which improves the detection of such proteins [35].

20.5.2   Experimental Procedures

As indicated in Figure 20.5, there are several steps involved in the ChIP assay. This section provides more details on each of these steps, including methods for the formaldehyde fixation of cells, the solubilization of chromatin, the immunoprecipitation of cross-linked chromatin, the reversal of cross-linking between DNA and proteins, the purification of the captured DNA, and the analysis of the immunoprecipitated DNA and identification of its binding sites.

20.5.2.1   Formaldehyde Fixation of Cells

For an *in vivo* chromatin immunoprecipitation experiment, the cells should first be treated with a 1% formaldehyde solution at 4°C. Cross-linking reaction times reported in the literature vary between 10 min and several hours for this reaction [35]. Glycine is later added to stop the cross-linking process. Next, the cells are usually washed and suspended in buffers containing protease inhibitors. These cells are then ready for chromatin solubilization, as described in the next section.

To optimize the cross-linking process, there are several parameters that should be considered. One of these is the reaction time. A long time for the cross-linking reaction will result in a loss of immunoprecipitable material and can result in a larger average size for the DNA fragments that are obtained after sonication (see the next section). For instance, a 70% loss of material and an average DNA fragment size of 20 kilobases was found to occur with a cross-linking time longer than 6 h [35]. During this reaction, it should also be kept in mind that formaldehyde is a moderate denaturing agent for proteins. As such, it may affect the structure of the regulatory complex under study.

20.5.2.2   Chromatin Solubilization

Since fixed cells are highly resistant to restriction-enzyme digestion or DNase I treatment, sonication is the only way to shear and solubilize the formaldehyde-fixed material and generate pieces of DNA with well-defined lengths. The conditions used for sonication and fixation are interrelated, since highly cross-linked material will not produce small chromatin fragments even after prolonged sonication [35]. The presence of short DNA fragments after sonication is desirable, since these will provide the highest mapping resolution for DNA.

Several variables can affect the efficiency of DNA shearing during sonication. These variables include the time allowed for sonication, the output control settings, the tip immersion depth, the presence of microglass beads (0.1 to 0.5 mm diameter), and the use of a microtip on the sonicator [36]. The sonication efficiency can be examined by using agarose gel electrophoresis to look at the size of resulting fragments. An average DNA fragment size of less than 600 base pairs is considered optimal for further analysis.

After sonication, additional chromatin purification can be performed by using isopycnic centrifugation. In this procedure, the nonchromatin-bound cross-linked proteins and naked DNA or RNA are eliminated by means of cesium chloride isopycnic centrifugation [34]. At this point, the isolated and fixed chromatin can be stored at –80°C until further use.

20.5.2.3 *Immunoprecipitation of Cross-Linked Chromatin*

The next step in the ChIP assay involves immunoprecipitation of the cross-linked chromatin by antibodies that are specific for the complex under study. As discussed in Section 20.4.1, this makes use of a protein A support or similar material that contains ligands capable of binding the antibodies and removing them from solution. The immunoprecipitation process should be performed under the most stringent conditions possible. Prior to this step, a preclearing step is also performed, in which the sample is passed through a protein A support to remove sample components that nonspecifically bind to this medium.

Affinity-purified polyclonal antibodies are normally used for immunoprecipitation in these studies [34]. In selecting the amount of antibody that should be used, an excess of antibody will result in a higher overall yield for the DNA, but it will also have less specific binding, giving a lower enrichment of the specifically immunoprecipitated DNA. In such work, it is recommended that control experiments be carried out with no antibodies or an unrelated antibody to account for nonspecific binding in the immunoprecipitated DNA. To release the immunoprecipitated material, the protein A beads used to capture the antibodies and associated target can be suspended in an elution buffer such as 0.1 *M* sodium bicarbonate plus 1% sodium dodecyl sulfate.

20.5.2.4 *Reversal of Cross-Linked DNA and DNA Purification*

An extended incubation at 65°C will break protein-DNA adducts, but protein-protein adducts require temperatures closer to boiling. For the purification of immunoprecipitated DNA, each cross-linked sample should be incubated at 65°C for 4 to 6 h. The proteins in the mixture can be digested with proteinase K at 37°C for several hours. This last step has been reported to ease the purification of the immunoprecipitated DNA [34]. The coprecipitation of RNA is also possible in the ChIP assay, but this can be avoided by treating the samples with RNase A for 30 min at 37°C. A typical ChIP assay results in 1 to 10 ng of purified DNA from a standard flask of culture cells. Recently, it has been shown that further fractionation of the purified samples on agarose gels and a selection of fragments in the range of 100 to 250 base pairs can increase the selectivity of this assay [36].

20.5.2.5 *Analysis of Immunoprecipitated DNA and Identification of Binding Sites*

Previous knowledge of a putative or suspected recognition DNA binding site usually guides the analysis of immunoprecipitated DNA in the ChIP assay. Several approaches can be used to identify the sequence of a DNA binding site that is recognized by a protein. In one approach, the immunoprecipitated DNA is allowed to hybridize with a radioactive probe for a suspected site in a slot blot experiment. This involves an estimation of the relative amount of immunoprecipitated DNA by immobilizing in adjacent slots equal amounts of the immunoprecipitated DNA and a control immunoprecipitate (obtained with no antibody or an unrelated antibody) to compare their radioactive hybridization signals.

A second approach for analyzing the immunoprecipitated DNA is to use PCR to amplify the DNA region containing the sequence of interest. In this approach, the presence of an expected or suspected DNA binding region is determined by using the immunoprecipitated DNA as a template for specific PCR amplification with gene-selected primers. The PCR products are then analyzed by polyacrylamide gel electrophoresis or by using agarose. A negative control for the PCR reaction should include the use of primers from an unrelated DNA sequence (e.g., upstream regions). A modified approach to detect the presence of specific genomic regions in the immunoprecipitate is to use multiplex PCR, in which several PCR products from different primer pairs are amplified simultaneously. Southern blot analysis can then be used to confirm the identity of each PCR product [37].

A third approach for verifying protein-DNA interactions at specific sites within a gene promoter is to use quantitative real-time PCR [38]. This method is based on the detection of a fluorescence signal that is proportional to the amplification of a PCR product and, thus, allows visualization of the exponential portion of the PCR reaction. The reporters used for real-time PCR can be the DNA-intercalating SYBR Green reagent or the TaqMan and Molecular Beacons probes, which are oligonucleotides that contain both reporter and quencher fluorophores for quantitation based on fluorescence resonance energy transfer. The probe and primers are selected to include the proximal promoter region where a suspected transcription response element resides.

Since the amount of immunoprecipitated DNA obtained from a ChIP assay is usually low (0.1 to 10 ng), a nested PCR strategy or a large number of cycles (e.g., 60) is usually needed for the real-time PCR [38]. When using a nested protocol, the first PCR step uses immunoprecipitated DNA as a template and primers flanking 5' to the real-time PCR products. This step increases the amount of template. A different primer set within the amplified sequence (3' to the first primer set) is then used for the real-time PCR. The specificity and sensitivity for the analysis of the immunoprecipitated DNA is increased by this approach [38]. Also, with the recent development of new fluorescent dyes for probe labeling, multiple targets in a single real-time PCR reaction can be examined. This expands the capacity of using ChIP and quantitative real-time PCR for examining *in vivo* protein-DNA interactions.

20.6 MULTIPLEXED PARTICLE-BASED FLOW CYTOMETRIC ASSAYS

Now that the sequence for the human genome has become available, there has been a growing need for high-throughput screening procedures to detect human genotypes on a large scale. Also, the ability to quantify different analytes from the same sample is becoming increasingly important. These types of measurements can be accomplished by applying the concepts of affinity chromatography to the detection of multiple analytes in a sample. It is for this purpose that a versatile technique known as the *microsphere-bead assay* (or Flow-Metrix system) has been developed. This can be used in such applications as nucleic acid testing, immunoassays, protein-protein interaction studies, and enzyme assays [39].

This system consists of the following elements: microspheres that contain an immobilized ligand and fluorescent markers, an additional fluorophore for analyte detection, and a standard bench flow cytometer [40]. This method first uses individual sets of 5.5-μm polystyrene microspheres that contain different combinations of orange- and red-emitting fluorophores. These microspheres often contain carboxylate groups to immobilize oligonucleotides, antibodies, enzyme substrates, or antigens that contain amine groups. Alternatively, avidin-coupled microspheres can be used to bind biotinylated molecules.

Next, a fluorescent reactant is prepared for each target. These fluorescent reactants can be oligonucleotides, antibodies, antigens, or other agents that bind to a target molecule on only one set of beads in the multiplexed assay [39].

After the detection parameters have been optimized separately for each desired analyte, microspheres against various targets can be mixed to perform a multiplexed assay. In this case, the various beads are mixed and added to a sample containing green fluorescent reactants. After a short incubation, the mixture of microspheres is analyzed with a flow cytometer, which can discriminate between the microspheres based on both their size and orange (585 nm) or red (>650 nm) fluorescence. In addition, the amount of bound analyte on each microsphere can be determined by examining the green fluorescence (530 nm) of the particle.

20.6.1 Nucleic Acid Hybridization

One way in which a microsphere-bead assay can be used is for genetic allelic sequence testing [39]. For instance, this has been performed by attaching to microsphere beads 16 different oligonucleotides containing a 5′-end amino linker and corresponding to allelic DNA sequences within the second exon of the HLA-DQA1 gene. After hybridizing a mixture of these microsphere beads with fluorescently labeled HLA-DQA1 allele-specific complementary sequences, a flow cytometer was used to separate each type of microsphere based on its orange and red fluorescence, followed by a separate analysis of their green fluorescence intensities. The results showed that the hybridization of a microsphere-bound oligonucleotide with its complementary probe was rapid, requiring no more than 15 min to reach completion. Hybridization at 55°C allowed this method to be used for differentiating between complementary oligonucleotide duplexes and mismatched complexes [39].

Specific oligonucleotide sequences from heterogeneous mixtures of DNA have also been detected using a multiplexed microsphere-bead assay [40]. In this work, several oligonucleotides (or "capture probes") were attached to the surface of microsphere beads. The analyte was a mixture of green-labeled cDNA sequences between 628 and 728 base pairs long that were obtained after the amplification of a 16S/23S intergenic spacer region in microorganisms collected from polluted groundwater. After mixing the microsphere beads in a single tube and exposing these to the same analyte, direct hybrid capture occurred between the matching capture probes and fluorescent cDNA sequences. Flow cytometry was then used to perform multiplexed detection based on measurements of the red, orange, and green emission intensities of the beads. A standard containing beads with a known number of fluorescent DNA molecules was developed and used to determine the resolution, sensitivity, and detection limit for this assay. The results showed that this method had better discrimination and more accurate quantitation than a DNA microarray procedure.

Simultaneous detection of wild-type and mutant DNA in the cystic fibrosis transmembrane conductance regulator (CFTR) gene was also performed using a multiplexed bead assay [41]. DNA samples from patients were used to amplify exon regions that contained the five most common mutations of the cystic fibrosis gene in North America. Each primer was labeled at its 5′ terminus with biotin. This process yielded amplicon sizes between 106 and 147 base pairs in length. Next, capture oligonucleotide probes specific for each wild-type and mutant sequence and complementary in sequence to the biotinylated amplicon were immobilized to different carboxylated microspheres. For the hybridization reaction, PCR amplification products were incubated at 55°C for 10 min with a mixture of the microspheres, each containing one type of immobilized probe. The microsphere beads were then removed through centrifugation, the hybridized amplicons were labeled with streptavidin-R-phycoerythrin, and the beads were analyzed by flow cytometry. It was found that this assay could simultaneously identify the genotype of the five selected CFTR mutations in patient DNA samples.

Another multiplexed bead assay was developed that could quantify the expression of genes based on hybridization of mRNA populations [42]. In this method, 20 probe oligo-nucleotides were derived from the genes of interest and immobilized to different color-coded microspheres in separate reaction tubes. For the analysis of a sample, the total RNA from *Arabidopsis thaliana* was converted to cDNA, followed by its *in vitro* transcription using T7 RNA polymerase in the presence of biotinylated UTP. The hybridization mixture contained target denatured cRNA samples and capture-probe microspheres that were incubated at 48°C for 1 h. The microspheres were then recovered by centrifugation, washed, and incubated with streptavidin-conjugated *R*-phycoerythrin. A flow cytometer was used to categorize these beads and measure the hybridization signal associated with each microsphere's surface. The specificity of hybridization did not change when using up to 20 target genes in the assay, and the smallest amount of detectable target was approximately 1.0 fmol.

20.6.2 Simultaneous Measurement of Cytokines

A more recent application of the microsphere-bead assay has been its use for the simultaneous measurement of several cytokines in the same sample [43]. This is based on a two-site sandwich immunoassay that makes use of two groups of anticytokine antibodies, two fluorescent dyes, and microspheres of uniform size. One of the two antibodies is used as the capture antibody and is coupled to beads with a particular color. The second type of antibody is conjugated with biotin and is used for analyte detection by reacting this labeled antibody with streptavidin that contains a conjugated fluorescent dye. In this method, the signal measured for this latter dye on each bead is proportional to the concentration of cytokines in the sample [43]. This has allowed a total of 15 different cytokines to be measured simultaneously in a 100-μl sample.

20.6.3 Analysis of Protein-Protein Interactions

A third application of microsphere-based assays has been their use in identifying and characterizing multiplexed interactions between nuclear receptors and peptides that represent the binding region of coactivator proteins [44]. To do this, several specific coactivator peptides were coupled to separate microsphere-bead populations. After these microspheres had been mixed, they were incubated with a nuclear receptor that had been coupled to a green fluorophore. Flow cytometry was then used to sort the microspheres and to examine the binding of the receptor to each type of coactivator peptide. Binding measurements were also made using the microspheres. The affinities determined by surface plasmon resonance and fluorescence gave good agreement with the results of other techniques. This approach was used to show that the interaction between the microsphere-immobilized peptides and receptor could be modulated by known agonists and antagonists for this system.

20.7 SUMMARY AND CONCLUSIONS

This chapter has examined the use of affinity ligands for the detection or the preparation of substances in molecular biology. The application of such ligands in this field is increasing and is becoming ever more inventive. Some examples of these applications include the use of homopolymers for polynucleotide isolation and heteropolymer ligands for subtractive hybridization assays. Another common example is the use of affinity tags for protein purification. The use of antibodies and antigens as reagents, such as in the methods of immunoprecipitation, immunoaffinity chromatography, and antibody purification, is another important area in this field.

A related technique that was discussed is the ChIP assay, which is used to detect endogenous transcription factors that are bound *in vivo* to promoter regions in DNA. This method can also be used to examine proteins that are not directly bound to DNA (e.g., coactivators and related proteins from the machinery of the transcription process) but that depend on other proteins for promoter binding. With the ChIP assay, it is possible to delineate the mechanisms by which nuclear receptors transmit a hormone-binding signal to the core transcription machinery [33]. This has led to the identification and characterization of several distinct multiprotein complexes containing coactivators and corepressors that directly interact with agonist-bound nuclear receptors. Another application of the ChIP assay has been its use to determine the spatial and temporal distribution of histone acetylation, a reversible post-translational modification that acts as a biochemical marker of active genes [45]. Future applications might include the identification of *in vivo* targets for transcription factors or related coactivators/corepressors or the combined use of chromatin immunoprecipitation with detection based on DNA microarrays [46] and real-time PCR [38].

The use of affinity ligands in particle-based flow cytometry was also examined. It was shown how assays based on this can be applied to virtually any type of molecular interaction, making these methods potentially useful in clinical diagnosis, environmental analysis, or agricultural testing [39]. Furthermore, this method can be applied to the multiplexed hybridization of nucleic acids for use in tissue typing, the diagnosis of genetic disease, and tests for mRNA/cDNA expression [39]. Another possible application would include the use of microspheres in multiplexed enzyme assays for analyzing both the substrate specificity and kinetics for multiple enzymes or multiple substrates. An example of such an assay might involve the preparation of microsphere beads that contain various fluorescent peptides. These microspheres could then be mixed and used to examine the specificity or relative activity of an endopeptidase for each peptide, while the enzyme activity could be measured by measuring the loss of fluorescence on the microspheres [39].

SYMBOLS AND ABBREVIATIONS

BCCP	Biotin carrier carboxyl protein
BPL	Biotin protein ligase
CBD	Cellulose-binding domain
CBP	Calmodulin-binding peptide
CD	Cluster of differentiation
cDNA	Complementary DNA
CF	Cystic fibrosis
CFTR	Cystic fibrosis transmembrane conductance regulator
ChIP	Chromatin immunoprecipitation
CMA	Carboxymethylasparate
DNA	Deoxyribonucleic acid
EDTA	Ethylenediaminetetraacetic acid
FITC	Fluorescein isothiocyanate
GST	Glutathione *S*-transferase
HABA	2(4′-Hydroxyazobenzene)benzoic acid
His-Tag	Histidine tag
HPLC	High-performance liquid chromatography

IDA	Iminodiacetic acid
IgA	Immunoglobulin A
IgD	Immunoglobulin D
IgG	Immunoglobulin G
IgM	Immunoglobulin M
IMAC	Immobilized metal-ion affinity chromatography
MBP	Maltose binding protein
mRNA	Messenger RNA
PCR	Polymerase chain reaction
Poly(A)	Polyadenylic acid
Poly(dT)	Polythymidylic acid
Poly(U)	Polyuridylic acid
RNA	Ribonucleic acid
ScFv	Single-chain antibody variable region fragment
SDS-PAGE	Sodium dodecyl sulfate polyacrylamide gel electrophoresis
sIg	Surface immunoglobulin

REFERENCES

1. Massom, L.R. and Jarrett, H.W., High-performance affinity chromatography of DNA: II. Porosity effects., *J. Chromatogr. A*, 600, 211–228, 1992.
2. Goss, T.A., Bard, M., and Jarrett, H.W., High-performance affinity chromatography of DNA, *J. Chromatogr. A*, 508, 279–287, 1990.
3. Rodriguez, I.R. and Chader, G.J., A novel method for the isolation of tissue-specific genes, *Nucleic Acids Res.*, 20, 3258, 1992.
4. Richards, E.J., Secondary separation of double- and single-stranded nucleic acids using hydroxylapatite chromatography, in *Current Protocols in Molecular Biology*, Wiley, New York, 1993, pp. 2.13.11–12.13.13.
5. Hara, E., Kato, T., Nakada, S., Sekiya, S., and Oda, K., Subtractive cDNA cloning using oligo(dT)30-latex and PCR: Isolation of cDNA clones specific to undifferentiated human embryonal carcinoma cells, *Nucleic Acids Res.*, 19, 7079–7104, 1991.
6. Hara, E., Yamaguchi, T., Tahara, H., Tsurui, H. Ide, T. Oda, K., DNA-DNA subtractive cDNA cloning using oligo(dT)30-Latex and PCR: Identification of cellular genes which are overexpressed in senescent human diploid fibroblasts, *Anal. Biochem.*, 214, 58–64, 1993.
7. Nasoff, M., Bergseid, M., Hoeffler, J.P. and Heyman, J.A., High-throughput expression of fusion proteins, *Methods Enzymol.*, 328, 515–529, 2000.
8. Fritze, C.E. and Anderson, T.R., Epitope tagging: General method for tracking recombinant proteins, *Methods Enzymol.*, 327, 3–16, 2000.
9. Raines, R.T., McCormick, M., Van Oosbree, T.R. and Mierendorf, R.C., The S.Tag fusion system for protein purification, *Methods Enzymol.*, 326, 362–376, 2000.
10. Skerra, A. and Schmidt, T.G.M., Use of the Strep-Tag and streptavidin for detection and purification of recombinant proteins, *Methods Enzymol.*, 326, 271–304, 2000.
11. Bornhorst, J.A. and Falke, J.J., Purification of proteins using polyhistidine affinity tags, *Methods Enzymol.*, 326, 245–254, 2000.
12. Smith, D.B., Generating fusions to glutathione S-transferase for protein studies, *Methods Enzymol.*, 326, 254–270, 2000.
13. Vaillancourt, P., Zheng, C.F., Hoang, D.Q. and Breister, L., Affinity purification of recombinant proteins fused to calmodulin or to calmodulin-binding peptides, *Methods Enzymol.*, 326, 340–362, 2000.
14. Skerra, A. and Schmidt, T.G., Use of the Strep-Tag and streptavidin for detection and purification of recombinant proteins, *Methods Enzymol.*, 326, 271–304, 2000.

15. Cronan, J.E. Jr and Reed, K.E., Biotinylation of proteins *in vivo*: A useful posttranslational modification for protein analysis, *Methods Enzymol.,* 326, 440–458, 2000.
16. Smith, D.B. and Johnson, K.S., Single-step purification of polypeptides expressed in Escherichia coli as fusions with glutathione S-transferase, *Gene,* 67, 31–40, 1988.
17. Chun, S.Y., Strobel, S., Bassford, P., Jr and Randall, L.L., Folding of maltose-binding protein. Evidence for the identity of the rate-determining step *in vivo* and *in vitro, J. Biol. Chem.,* 268, 20855–20862, 1993.
18. Sachdev, D. and Chirgwin, J.M., Fusions to maltose-binding protein: Control of folding and solubility in protein purification, *Methods Enzymol.,* 326, 312–321, 2000.
19. Park, J.S., Shin, H.S. and Doi, R.H., Fusion proteins containing cellulose-binding domains, *Methods Enzymol.,* 326, 418–429, 2000.
20. Oak, S.A. and Jarrett, H.W., The oligomerization of mouse a1-syntrophin and self-association of its pleckstrin homology domain 1, *Biochemistry,* 39, 8870–8877, 2000.
21. Towbin, H., Staehelin, T. and Gordon, J., Electrophoretic transfer of proteins from polyacrylamide gels to nitrocellulose sheets: Procedure and some applications, *Proc. Natl. Acad. Sci. U.S.A.,* 76, 4350–4354, 1979.
22. Ransone, L.J., Detection of protein-protein interactions by coimmunoprecipitation and dimerization, *Methods Enzymol.,* 254, 491–497, 1995.
23. Hjelm, H., Hjelm, K. and Sjoquist, J., Protein A from Staphylococcus aureus. Its isolation by affinity chromatography and its use as an immunosorbent for isolation of immunoglobulins, *FEBS Lett.,* 28, 73–76, 1972.
24. Deisenhofer, J., Crystallographic refinement and atomic models of a human Fc fragment and its complex with fragment B of protein A from Staphylococcus aureus at 2.9- and 2.8-A resolution, *Biochemistry,* 20, 2361–2370, 1981.
25. Sjodahl, J., Repetitive sequences in protein A from Staphylococcus aureus. Arrangement of five regions within the protein, four being highly homologous and Fc-binding, *Eur. J. Biochem.,* 73, 343–351, 1977.
26. Crumpton, M.J. and Parkhouse, R.M.E., Comparison of the effects of various detergents on antibody-antigen interactions, *FEBS Lett.,* 22, 210–212, 1975.
27. Wilchek, M., Miron, T. and Kohn, J., Affinity chromatography, *Methods Enzymol.,* 104, 3–55, 1984.
28. Louvard, D., Vannier, C., Maroux, S., Pages, J. and Lazdunski, C., A quantitative immunochemical technique for evaluation of the extent of integration of membrane proteins and of protein conformational changes and homologies, *Anal. Biochem.,* 76, 83–94, 1976.
29. Wigzell, H. and Andersson, B., Cell separation on antigen-coated columns. Elimination of high rate antibody-forming cells and immunological memory cells, *J. Exp. Med.,* 129, 23–36, 1969.
30. Rubin, B., Regulation of helper cell activity by specifically adsorbable T lymphocytes, *J. Immunol.,* 116, 80–85, 1976.
31. Schlossman, S., *Leukocyte Typing,* Oxford Press, Oxford, UK, 1995.
32. Cottler-Fox, M., Cipolone, K., Yu, M., Berenson, R., O'Shannessy, J. and Dunbar, C., Positive selection of CD34+ hematopoietic cells using an immunoaffinity column results in T cell-depletion equivalent to elutriation, *Exp. Hematol.,* 23, 320–322, 1995.
33. Shang, Y., Hu, X., DiRenzo, J., Lazar, M.A. and Brown, M., Cofactor dynamics and sufficiency in estrogen receptor-regulated transcription, *Cell,* 103, 843–852, 2000.
34. Orlando, V., Strutt, H., and Paro, R., Analysis of chromatin structure by *in vivo* formaldehyde cross-linking, *Methods,* 11, 205–214, 1997.
35. Orlando, V., Mapping chromosomal proteins *in vivo* by formaldehyde-crosslinked-chromatin immunoprecipitation, *Trends. Biochem. Sci.,* 25, 99–104, 2000.
36. Johnson, T.A., Wilson, H.L. and Roesler, W.J., Improvement of the chromatin immunoprecipitation (ChIP) assay by DNA fragment size fractionation, *Biotechniques,* 31, 740–742, 2001.
37. Hecht, A. and Grunstein, M., Mapping DNA interaction sites of chromosomal proteins using immunoprecipitation and polymerase chain reaction, *Methods Enzymol.,* 304, 399–414, 1999.

38. Christenson, L.K., Stouffer, R.L. and Strauss, J.F., 3rd., Quantitative analysis of the hormone-induced hyperacetylation of histone H3 associated with the steroidogenic acute regulatory protein gene promoter, *J. Biol. Chem.,* 276, 27392–27399, 2001.

39. Fulton, R.J., McDade, R.L., Smith, P.L., Kienker, L.J. and Kettman, J.R., Jr, Advanced multiplexed analysis with the FlowMetrix system, *Clin. Chem.,* 43, 1749–17456, 1997.

40. Spiro, A., Lowe, M. and Brown, D., A bead-based method for multiplexed identification and quantitation of DNA sequences using flow cytometry, *Appl. Environ. Microbiol.,* 66, 4258–4265, 2000.

41. Dunbar, S.A. and Jacobson, J.W., Application of the luminex LabMAP in rapid screening for mutations in the cystic fibrosis transmembrane conductance regulator gene: A pilot study, *Clin. Chem.,* 46, 1498–1500, 2000.

42. Yang, L., Tran, D.K. and Wang, X., BADGE, beadsarray for the detection of gene expression, a high-throughput diagnostic bioassay, *Genome Res.,* 11, 1888–1898, 2001.

43. Carson, R.T. and Vignali, D.A., Simultaneous quantitation of 15 cytokines using a multi-plexed flow cytometric assay, *J. Immunol. Methods,* 227, 41–52, 1999.

44. Iannone, M.A., Consler, T.G., Pearce, K.H., Stimmel, J.P., Parks, D.J. and Gray, J.G., Multiplexed molecular interactions of nuclear receptors using fluorescent microspheres, *Cytometry,* 44, 326–337, 2001.

45. Crane-Robinson, C., Myers, F.A., Hebbes, T.R., Clayton, A.L. and Thorne, A.W., Chromatin immunoprecipitation assays in acetylation mapping of higher eukaryotes, *Methods Enzymol.,* 304, 533–547, 1999.

46. Nal, B.E., Mohr, E. and Ferrier, P., Location analysis of DNA-bound proteins at the whole-genome level: Untangling transcriptional regulatory networks, *Bioessays,* 23, 473–476, 2001.

47. Jarrett, H.W. and Taylor, W.L., Transcription factor-green fluorescent protein chimeric fusion proteins and their use in studies of DNA affinity chromatography, *J. Chromatogr. A,* 803, 131–139, 1998.

48. Lindmark, R., Thoren-Tolling, K. and Sjoquist, J., Binding of immunoglobulins to protein A and immunoglobulin levels in mammalian sera, *J. Immunol. Methods,* 62, 1–13, 1983.

49. Surolia, A. and MIkhan, M., Protein A: Nature's universal anti-antibody, *Trends. Biochem. Sci.,* 7, 74–76, 1982.

50. Bjorck, L. and Protein L., A novel bacterial cell wall protein with affinity for Ig L chains, *J. Immunol.,* 140, 1194–1197, 1988.

51. Nilson, B.H., Logdberg, L., Kastern,W., Bjorck, L. and Akerstrom, B., Purification of antibodies using protein L-binding framework structures in the light chain variable domain, *J. Immunol. Methods,* 164, 33–40, 1993.

52. Akerstrom, B. and Bjorck, L., Protein L: An immunoglobulin light chain-binding bacterial protein. Characterization of binding and physicochemical properties, *J. Biol. Chem.,* 264, 19740–19746, 1989.

53. Langone, J.J., Applications of immobilized protein A in immunochemical techniques, *J. Immunol. Methods,* 55, 277–96, 1982.

21

Affinity-Based Chiral Stationary Phases

Sharvil Patel and Irving W. Wainer

Bioanalytical and Drug Discovery Unit, National Institute on Aging,
National Institutes of Health, Baltimore, MD

W. John Lough

Institute of Pharmacy and Chemistry, University of Sunderland,
Sunderland, U.K.

CONTENTS

21.1   INTRODUCTION

In 1848 Louis Pasteur used a hand lens and a pair of tweezers to separate the sodium ammonium salts of *p*-tartaric acid into their component enantiomers (D- and L-tartaric acid), thereby accomplishing the first enantioselective resolution [1]. Pasteur also accomplished the first chemical resolution of a racemic compound when he neutralized a solution of L-cinchonin with racemic tartaric acid [2]. With this experiment, he demonstrated that the transformation of enantiomers into diastereomers converted molecules with the same physicochemical properties (enantiomers) into compounds with different physicochemical properties (diastereomers). Thus, while the solubilities of D- and L-tartaric acids were identical, the L-tartrate-L-cinchonin salt was crystallized before the D-tartrate-L-cinchonin salt.

Enzymes, carrier proteins, and other biopolymers are chiral molecules that have the ability to form transient diastereomeric complexes with other enantiomeric compounds (ligands). The resulting diastereomeric complexes can differ in their relative stabilities, which can produce a physical or functional discrimination between the enantiomers of the ligand. The ability of enzymes and carrier proteins to act as chiral selectands has been extensively used in drug development. This property has also been adapted for use in the chromatographic sciences through the development of *chiral stationary phases* (CSPs). CSPs have been created through the immobilization of biopolymers on chromatographic supports and have been used primarily in liquid chromatographic systems. They have been particularly valuable in the quantification and separation of small chiral compounds on an analytical scale [3, 4].

The pharmacological aspects of biopolymer-based CSPs are addressed in Chapter 24. The current chapter addresses the analytical applications of these CSPs. Although the term *biopolymer-based CSP* can be extended to include CSPs derived from macrocyclic antibiotics and derivatized cellulose, the discussion here concentrates on those developed from proteins. The specific proteins discussed here include α_1-acid glycoprotein (AGP), serum albumins such as bovine serum albumin (BSA) and human serum albumin (HSA), and ovomucoid (OVM), as well as the enzymes α-chymotrypsin (ACHT), trypsin (TRYP), and cellobiohydrolase (CBH). Other albumins, binding proteins, and enzymes are also examined.

21.2   GENERAL EXPERIMENTAL APPROACHES

In general, two approaches have been used in the construction of biopolymer-based CSPs: covalent immobilization and entrapment/adsorption, as shown later in this chapter. The enzymes ACHT, TRYP, and CBH I have been covalently attached to silica supports; ACHT and TRYP have been entrapped in the interstitial spaces of an immobilized artificial membrane stationary phase (IAM-SP) [5, 6]; and ACHT has been adsorbed on LiChrospher phases [7]. The carrier-protein phases have been synthesized using entrapment, followed by cross-linking in the case of AGP [8] or covalent immobilization for the serum albumins [9, 10].

Zonal-elution chromatography is the most frequently used approach in the analytical separation of enantiomeric compounds on biopolymer-based CSPs. This technique can also be used to study the binding of drugs and other solutes on immobilized-protein columns, to determine drug-drug protein-binding interactions, and to develop quantitative structure-retention relationships that describe these binding processes (see Chapter 24). Information on the kinetics of solute-protein interactions can also be obtained if appropriate data are collected on the width and retention for solute peaks under various flow-rate conditions (see Chapters 22 and 23).

Frontal-analysis chromatography has also been used with immobilized-biopolymer CSPs [9]. Although frontal analysis generally requires a greater amount of compound than zonal elution, this technique does tend to provide binding constants that are more precise and accurate than those measured by zonal-elution methods and can be used to determine the number of active binding sites on the column (see Chapter 22).

Biopolymer-based CSPs are used in the reversed-phase mode, with mobile phases composed of buffers and various modifiers. On these phases, a change in buffer pH or ionic strength can dramatically alter chromatographic efficiency and enantioselectivity. The type and concentration of modifiers added to the buffer can also significantly change the chromatographic results. This has been demonstrated on an AGP-CSP, where the addition of an ion-pairing agent is necessary to produce an enantioselective separation [11, 12], or where the use of acetonitrile in place of 2-propanol can produce an enantiomeric separation [13]. In addition, competitive ligands can be used to reduce retention, with or without a reduction in enantioselectivity [14–17], or to increase both retention and enantioselectivity [18].

Since CSPs are composed of biologically active molecules, chemical and genetic methods can be used to alter their activity and chromatographic performance. For instance, specific amino acid residues on HSA-CSPs have been chemically modified, thereby changing the retention and enantioselectivities of these materials [19–20]. As an example, site-directed mutagenesis of CBH was used to produce D214N, which gave a chiral stationary phase (D214N-CSP) that had lower retention and enantioselectivities than CBH-CSP [21].

Temperature is another experimental variable that can be used to alter retention and enantioselectivity. For most biopolymer-based CSPs, increasing temperature produces decreased retention for both enantiomers, with a concomitant decrease in enantioselectivity. However, when the effect of temperature on the chiral separation of (R,S)-sotalol was investigated on a CBH-CSP, the elution order reversed between 17 and 28°C [22]. This phenomenon could be explained by the fact that the retention time of the second eluted enantiomer, (S)-sotalol, rapidly decreased with an increasing temperature, while the (R)-enantiomer exhibited an unusual endothermic adsorption behavior and gave an increase in retention with increasing temperature.

21.3 ENZYME-BASED CHIRAL STATIONARY PHASES

Several types of enzymes have been used as chiral stationary phases. The specific examples that are examined in this section include α-chymotrypsin, trypsin, and cellobiohydrolase I. The preparation of these chiral stationary phases is discussed, along with their use in chiral separations and the effects of various chromatographic conditions on their performance.

21.3.1 α-Chymotrypsin (ACHT)

α-Chymotrypsin (ACHT) is a serine protease that catalyzes the hydrolysis of peptide bonds at amino acid residues containing large apolar side chains. This enzyme is a chiral polymer composed of 245 amino acids with an ellipsoidal tertiary structure. The enzyme contains a very distinct active center composed of a hydrophobic pocket and a hydrolytically active

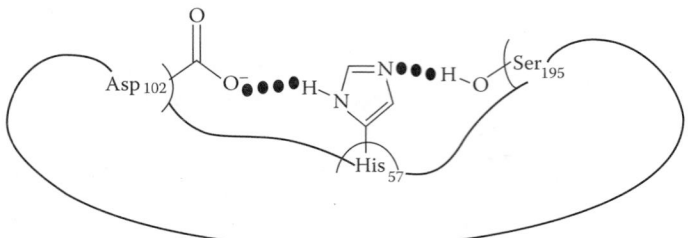

Figure 21.1 Schematic of the active site of α-chymotrypsin.

site that includes residues Ser-195, His-57, and Asp-102 (see Figure 21.1). The hydrolytic activity of ACHT involves the acylation of the hydroxyl group of Ser-195 by a carboxyl moiety on the substrate. Binding and hydrolysis of substrates at the ACHT active site involve a combination of hydrophobic, hydrogen bonding, and stereochemical interactions.

21.3.1.1 Preparation of an ACHT-CSP

Several methods have been used to prepare an ACHT chiral stationary phase (ACHT-CSP). ACHT was initially immobilized on a silica-based support containing covalently bonded glutaraldehyde [23]. However, immobilization via reactive epoxy (A), aldehyde (C), and tresyl (D) groups has also been studied [7]. The highest stereoselectivity and shortest retention times are generally observed when ACHT is immobilized on the epoxy silica, ACHT-A. This is due to a higher loading of the enzyme on this phase. However, ACHT-C has been preferred because of its greater stability and reproducibility than ACHT-A and the fact that it gives higher enantioselectivity than ACHT-D.

The physical adsorption of ACHT on silica phases has been studied as well [7]. Different types of silicas were investigated, and it was found that the amount of ACHT adsorbed was affected by both the pore diameter and mobile-phase pH. The pH used for these adsorption studies was lower than the isoelectric point of ACHT (pH 8.1 to 8.3), since this is beyond normal operating conditions for silica supports. However, even though there was significant adsorption of ACHT on LiChrospher silica and a significant enantioselectivity, a large amount of leakage of the ACHT from the coated silica phases led to a continuous decrease in retention factors. Accordingly, adsorbed ACHT-CSPs were not considered to be useful for the quantitative analyses of enantiomers.

ACHT has also been noncovalently entrapped on an IAM-SP [5, 6]. The IAM-SP was derived from the covalent immobilization of 1-myristoyl-2-[(13-carboxyl)tridecanoyl]-sn-3-glycerophosphocholine on aminopropyl silica, and resembles one-half of a cellular membrane. In the IAM-SP, the phosphatidylcholine head groups form the surface of the support, and the hydrocarbon side chains produce a hydrophobic interface that extends from the charged head group to the surface of the silica. With the IAM interphase, ACHT is embedded within the interphase surroundings.

The enzymatic activities of glutaraldehyde-immobilized ACHT and IAM-entrapped ACHT have been investigated using known ACHT substrates such as L-tryptophan amine (L-Trp-NH$_2$) and L-tryptophan methyl ester (L-Trp-OME) [5, 23]. The enzymatic activity observed when D- and L-Trp-NH$_2$ were injected onto the ACHT-CSP is shown in the resulting chromatogram in Figure 21.2a, which contains two peaks [23]. These peaks were collected and identified as L-Trp and D-Trp-NH$_2$.

Figure 21.2 illustrates how this immobilized enzyme reacts in the presence of inhibitors [23]. When treated with the reversible inhibitor 4-nitrophenyltrimethylacetate (4-NTA),

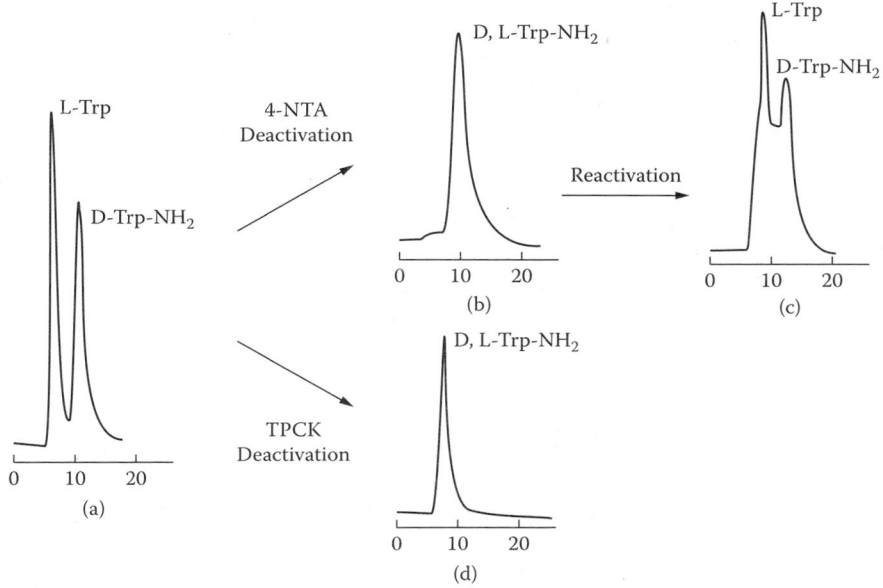

Figure 21.2 Chromatographic retention and resolution of L- and D-tryptophan amide (L- and D-Trp-NH$_2$) on chiral stationary phases based on α-chymotrypsin (ACHT-CSP) after treatment with 4-nitrophenyltrimethylacetate (4-NTA) or N-tosyl-L-phenylalanine chloromethyl ketone (TPCK). The results in (a) are for an untreated ACHT-CSP, (b) is for ACHT-CSP after treatment with 4-NTA, (c) shows the results after washing out of 4-NTA, and (d) gives the results obtained after treatment with TPCK. (Reprinted with permission from Jadaud, P., Thelohan, S., Schonbaum, G.R., and Wainer, I.W., *Chirality*, 1, 38–44, 1989.)

an injection of D- and L-Trp-NH$_2$ onto the ACHT-CSP column that produced two peaks in Figure 21.2a now gives only a single peak (see Figure 21.2b). After washing the 4-NTA from the column, the enzymatic activity was restored, as shown in Figure 21.2c. Enzymatic activity was also lost after the ACHT-CSP column was treated with N-tosyl-L-phenylalanine chloromethyl ketone (TPCK), an irreversible inhibitor of ACHT; the resulting chromatogram shown in Figure 21.2d again shows only a single peak for an injection of D,L-Trp-NH$_2$. The enzymatic activity of the immobilized ACHT in this last case could not be restored. The IAM form of the immobilized ACHT was also used to determine inhibition constants (K_i) as well as the form of inhibition for competitive and noncompetitive inhibitors of ACHT [5].

21.3.1.2 Chiral Separations Based on ACHT-CSPs

The ACHT-CSP has been of limited use in enantioselective separations. The primary substrates for ACHT are the aromatic amino acids tryptophan (Trp), phenylalanine (Phe), and tyrosine (Tyr). These compounds have been enantioselectively resolved as free amino acids as well as N- and O-derivatives (see Figure 21.3 and Table 21.1) [7]. The O-derivatives of Trp, Phe, and Tyr, such as methyl esters or amides, were enantioselectively resolved due to the enzymatic conversion of the L-derivative into the free amino acid. The D- and L-enantiomers of nonaromatic amino acids (e.g., leucine and alanine), which are not ACHT substrates, have also been stereochemically resolved on an ACHT-CSP as N-benzoyl or O-benzyl derivatives (see Table 21.1). The same is true for enantiomers of the nonsteroidal anti-inflammatory agent Naproxen (see Table 21.1) [7].

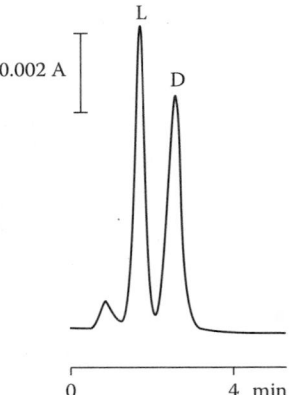

Figure 21.3 The enantioselective separation of D,L-*N*-acetyltryptophan on a chiral stationary phase based on α-chymotrypsin and using an aldehyde-activated silica support. (Reproduced with permission from Marle, I., Karlsson, A., and Pettersson, C., *J. Chromatogr.*, 604, 185–196, 1992.)

ACHT is an endopeptidase that requires at least four amino acid residues in both directions from the site of hydrolysis. Thus, this enzyme would not be expected to hydrolyze dipeptides. But a series of enantiomeric and diastereomeric dipeptides (as listed in Table 21.1) that contained Phe or Trp and a second amino acid have been stereochemically resolved on an ACHT-CSP [24]. This ACHT-CSP was also used to resolve the enantiomers of *N*-α-aspartyl-phenylalanine 1-methyl ester, with the L,L-form of this dipeptide being the same as the artificial sweetener aspartame [25].

Table 21.1 Enantioselective Separations of Small Molecules and Dipeptides on Immobilized α-Chymotrypsin Chiral Stationary Phases

Solute	Enantioselectivity Factor (α)
D,L-tryptophan	1.40[a], 1.16[b]
R,S-Naproxen	1.54[a]
D,L-*N*-carbobenzoxy-phenylalanine	1.62[a]
D-Phe-D-Leu and L-Phe-L-Leu	1.10[b]
D,L-Leu-D,L-Trp and L-Leu-L-Trp	1.83[b]
N-acetyl-D,L-tryptophan	1.07[c]
N-benzoyl-D,L-leucine	1.24[c]
N-benzoyl-D,L-phenylalanine	1.11[c]
D,L-alanine benzyl ester	1.74[c]

[a] Data obtained from Marle, I., Karlsson, A., and Pettersson, C., *J. Chromatogr.*, 604, 185–196, 1992.
[b] Data obtained from Jadaud, P. and Wainer, I.W., *J. Chromatogr.*, 476, 165–174, 1989.
[c] Data obtained from Jadaud, P., Thelohan, S., Schonbaum, G.R., and Wainer, I.W., *Chirality*, 1, 38–44, 1989.

21.3.1.3 Chromatographic Conditions for ACHT-CSPs

The effects of mobile-phase pH and ionic strength on retention and enantioselectivity on an ACHT-CSP have been investigated [7]. This study utilized mobile phases with pHs of 3.85, 5.28, and 6.40. It was found that pH had no dramatic effect on enantioselectivity, but increasing the pH did decrease the retention of acidic compounds and increased the retention of basic compounds. These results were probably due to changes in the nonspecific electrostatic interactions between the solutes and ACHT-CSP. The addition of charged modifiers to the mobile phase can also alter the chromatographic properties of this phase. Changing the ionic strength of the buffer from 0.025 to 0.10 had essentially the same effect.

The addition of octanesulfonate to the mobile phase produced an equivalent decrease in retention for (*R*)- and (*S*)-Naproxen and, therefore, had no effect on their observed enantioselectivity [7]. However, this modifier increased the chromatographic efficiency of the separation, probably through competitive binding to nonspecific binding sites. The addition of dimethyloctylamine to the mobile phase had no effect on the separation of anionic solutes, but it increased the retention and enantioselectivity of Trp and *N*-derivatives of Phe. These results suggest that the cationic modifier has an effect on the conformation of the immobilized ACHT.

21.3.2 Trypsin (TRYP)

Trypsin (TRYP) is a serine protease that preferentially catalyzes the hydrolysis of ester and peptide bonds involving the carboxyl group of arginine (Arg) and histidine (His). X-Ray crystallography indicates that this enzyme is a globular molecule consisting of two domains of approximately equal size [26]. The secondary structure of trypsin is predominantly a β structure, with each of the domains being present in a deformed β-barrel. The result of these two distorted β-barrel domains is that the loop regions, which protrude from the barrel ends, are almost symmetrically presented by each of the two folded domains. These loop structures combine to form a surface region of the enzyme that extends outward, above the catalytic site region of the S1 substrate-binding site.

21.3.2.1 Preparation of a TRYP-CSP

A few approaches have been used to prepare a trypsin chiral stationary phase (TRYP-CSP). For instance, TRYP has been immobilized through covalent binding to the glutaraldehyde moieties on a hydrophilic polymer that was first covalently bonded to 15-μm silica particles having 300-Å pores [27].

A TRYP-CSP can be synthesized in the same manner as described earlier for an ACHT-CSP. In this method, the support is first washed four times with pH 7.0, 0.1 *M* phosphate buffer. This support is then added to a pH 8.7, 0.1 *M* sodium borate solution in the ratio of 1 g support per 2.5 ml of the borate solution. Next, TRYP is added and the mixture is stored at 5°C for at least 12 h. This mixture is later filtered, and the solid phase is washed four times with pH 7.0, 0.1 *M* phosphate buffer prior to use.

The enzymatic activities of soluble and immobilized TRYP have been compared based on their ability to hydrolyze D,L-*N*-α-benzoyl-arginine *p*-nitroanalide (D,L-BAPNA) [27]. The results demonstrated that the hydrolytic activity of the immobilized enzyme was 72% of that seen for an equivalent molar amount of soluble TRYP.

21.3.2.2 Chiral Separations Based on TRYP-CSPs

The D- and L-isomers of arginine (Arg), lysine (Lys), and histidine (His) and some corresponding *N*- and *O*-derivatives were applied to the TRYP-CSP column made by Thelohan et al. [27]. The free amino acids gave low retention, and an accurate determination of enantioselectivity

Table 21.2 Effect of Temperature on the Enantioselective
Separations of Arginine, Histidine, and the *N*- or *O*-Derivatives
of These Amino Acids on an Immobilized Trypsin Chiral
Stationary Phase

	Enantioselectivity Factor (α)		
Solute	37°C	Ambient Temperature	5°C
D,L-Arg	5.33	1.20	1.11
N-Bz-D,L-Arg	1.01	1.00	1.01
D,L-Arg-Oet	1.78	2.40	3.00
D,L-Arg-Ome	1.33	2.00	2.90
D,L-BAPNA	11.41	10.77	8.28
D,L-His	1.09	1.07	1.00
N-Bz-D,L-His	1.02	1.00	1.03
D,L-His-Ome	1.05	1.06	2.32

Note: Arg = arginine; Bz = benzoyl; BAPNA = N-benzoylarginine-p-
nitroanilide; His = histidine; OEt = ethyl ester; OMe = methyl ester.

Source: Data obtained from Thelohan, S., Jadaud, P., and Wainer, I.W.,
Chromatographia, 28, 551–555, 1989.

was difficult. However, Arg and His were resolved with apparent enantioselectivity values (α) of 1.20 and 1.07, as shown in Table 21.2. *N*-derivatization increased the observed retention, where the retention factors (k') increased proportionally with the hydrophobicity of the *N*-substituted groups. However, no enantioselective separations were observed.

When the methyl and ethyl esters of Arg were chromatographed on the TRYP-CSP, an enantioselective separation was observed based on the stereospecific hydrolysis of the L-Arg esters (see Table 21.2). A similar separation of D,L-His-OMe was also observed. D,L-BAPNA produced the highest enantioselectivity (α = 10.77), representing the separation of N-Bz-L-Arg and D-BAPNA, as indicated in Table 21.2.

21.3.2.3 Chromatographic Conditions for TRYP-CSPs

The effect of temperature on the enantioselectivity of a TRYP-CSP is summarized in Table 21.2. Increasing the temperature from 5 to 37°C has no effect on the enantioselectivity for *N*-derivatized solutes. However, the retention times decreased with increasing temperature, which were the expected results and indicated that the primary retention mechanism involved nonspecific interactions with the stationary phase.

Under the same conditions, increasing temperature produced an increase in the observed enantioselectivities for Arg and His, with Arg experiencing the most dramatic change between ambient temperature and 37°C (i.e., α = 1.20 and 5.33, respectively). This change was primarily due to an increase in the retention of L-His, which was contrary to the expected effect. Since 37°C represents an optimum temperature for TRYP activity, this behavior could reflect a change in the active-site conformation of TRYP and indicates that the primary retention mechanism for these compounds involves interactions with the enzyme's active site.

With *O*-derivatized amino acids, increased temperature produced a decrease in the observed enantioselectivities, except for D,L-BAPNA. This effect reflected a temperature-associated increase in enzymatic activity, which produced the hydrolysis of D-Arg-OMe and D-His-OMe. At lower temperatures, the predominant substrates were the L-isomers, and the enantioselectivities reflected this. No hydrolysis of D-BAPNA was observed at any of the temperatures used in the study, with the observed increase in enantioselectivity being due to an increase in chromatographic efficiency.

21.3.3 Cellobiohydrolase I (CBH I)

Cellobiohydrolase I (CBH I) is the key enzyme in fungal cellulose degradation. The crystal structure of the CBH I catalytic domain reveals a large domain (434 residues) with the dimensions $60 \times 50 \times 40$ Å [28]. About one-third of this domain is folded into a β-sandwich composed of two antiparallel β-strands. CBH I contains extensive loops that are stabilized by disulfide bridges, together with the concave β-sandwich, to form a 40-Å-long tunnel in which the active site is enclosed.

21.3.3.1 Preparation of a CBH I-CSP

A CBH I chiral stationary phase (CBH I-CSP) can be prepared in the following manner. First, the alcohol moieties on diol silica were oxidized to form aldehydes by reacting this material with periodic acid. CBH I was then immobilized to this material through the formation of a Schiff base. The Schiff base was reduced with sodium cyanoborohydride to produce the final CBH I-CSP [29, 30].

21.3.3.2 Chiral Separations Based on CBH I-CSPs

Acidic and basic compounds have been enantioselectively separated on a CBH I-CSP [21, 31]. Representative enantioselectivities are presented in Table 21.3. The highest enantioselectivity was observed for the separation of β-blocking agents such as propranolol, alprenolol,

Table 21.3 Effect of pH on the Retention Factor for the Second Eluted Enantiomer (k'_2) and Enantioselectivity Factor (α) for Acidic and Basic Compounds on an Immobilized Cellobiohydrolase I Chiral Stationary Phase

Solute	pK$_a$	Parameter	pH					
			2.2	3.5	4.7	5.6	6.8	8.1
Acids								
Warfarin	5.0	k'_2	5.58	6.92	6.11	2.45	0.89	0.18
		α	1.27	1.31	1.26	1.14	1.00	1.00
Naproxen	4.2	k'_2	7.73	8.11	4.24	1.24	0.55	0.14
		α	1.01	1.00	1.00	1.00	1.00	1.00
Amines								
Prilocaine	7.9	k'_2	—	0.06	0.28	0.87	1.98	3.72
		α	—	1.00	1.00	1.00	1.22	1.19
Propranolol	9.5	k'_2	0.69	1.37	4.81	18.8	106	339
		α	1.00	1.05	1.42	2.34	4.31	3.83

Note: The pK$_a$ values given here describe the acid dissociation constants (K_a) for these solutes, where $pK_a = -\log K_a$.

Source: Data obtained from Marle, I., Erlandsson, P., Hansson, L., Isaksson, R., Pettersson, C., and Pettersson, G., *J. Chromatogr.*, 586, 233–248, 1991.

and metoprolol. The latter results are consistent with data from CBH I inhibition studies, which demonstrated that propranolol and alprenolol enantioselectively inhibit this enzyme [28]. In both cases, the (S)-enantiomers are stronger inhibitors than the (R)-enantiomers.

21.3.3.3 Chromatographic Conditions for CBH I-CSPs

The pH of the mobile phase is a key regulator of retention and enantioselectivity on a CBH I-CSP [21]. In general, with acidic solutes, increasing the pH reduces retention and enantioselectivity. The opposite effect is observed with amines. The results shown in Table 21.3 indicate that electrostatic interactions between the solutes and CBH I-CSP play a predominant role in retention and enantioselectivity.

The influence of ionic strength on the enantioseparation of (R,S)-sotalol showed that raising the buffer concentration from 10 to 20 mM strongly reduced the retention for both enantiomers while leaving the enantioselectivity unchanged [31]. When the injected solute was (R,S)-propranolol, an increase in ionic strength gave a decrease in retention and enantioselectivity; however, the retention of (R,S)-warfarin increased with an increase in ionic strength and gave no change in enantioselectivity [21]. These results are consistent with the theory that electrostatic interactions are the key driving force behind chromatographic retention and enantioselectivity for this stationary phase.

The effect of temperature on the retention and enantioselective separation of (R,S)-sotalol on CBH I-CSP has been investigated [22]. The data in Figure 21.4 indicate that the

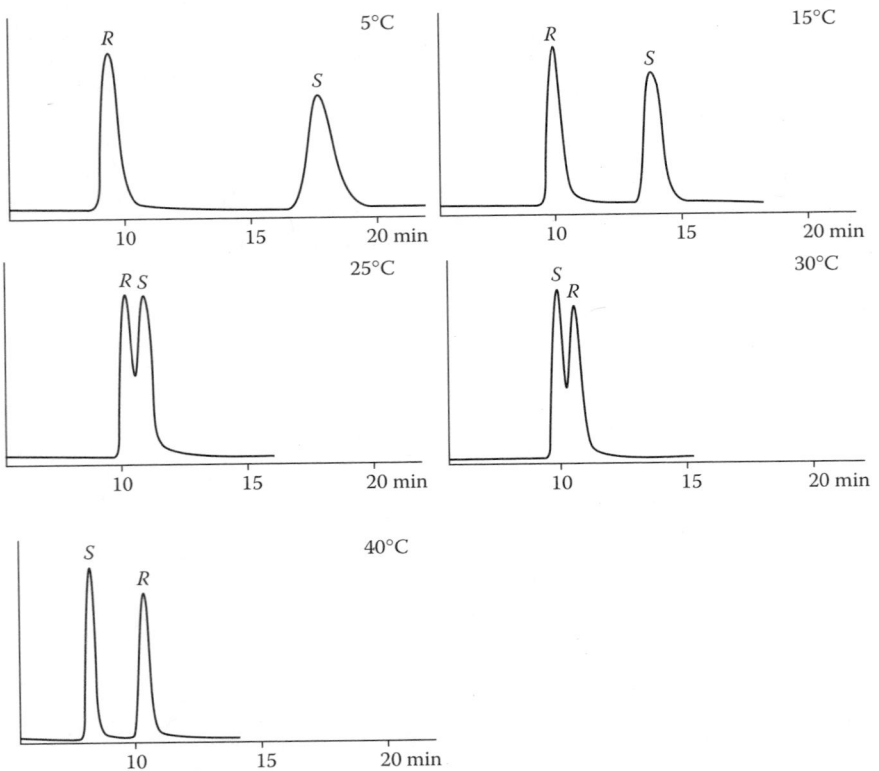

Figure 21.4 Effect of temperature on the retention and enantioselective separation of (R,S)-sotalol on a chiral stationary phase based on immobilized cellobiohydrolase I (CBH I-CSP). (Reproduced with permission from Fulde, K. and Frahm, A.W., *J. Chromatogr. A*, 858, 33–43, 1999.)

elution order of these enantiomers reversed between 25 and 30°C. This reversal was a result of the decrease in retention for the second eluting enantiomer, (S)-sotalol, with an increase in temperature, while the (R)-enantiomer exhibited an unusual endothermic adsorption behavior (i.e., a negative slope in its van't Hoff plot) and an increased retention time.

21.4   SERUM PROTEIN-BASED CSPS

Serum proteins are a large family of proteins including serum albumins, acid glycoproteins, and high- and low-density lipoproteins. Serum protein binding is an important pharmacological process that plays a role in determining the overall *in vivo* distribution, excretion, activity, and toxicity of a substance. These proteins have the ability to transport a wide variety of exogenous compounds throughout the body, such as fatty acids, metal ions, bilirubin, and several low-mass hormones [9].

Since binding of some drugs to these serum proteins can be stereoselective, it is possible that these interactions play a role in producing differences in the efficacy and toxicity of the enantiomers for a chiral drug [3]. The enantioselectivity of these proteins, particularly serum albumin (SA) and α_1-acid glycoprotein (AGP), have been used to produce a number of successful CSPs [4]. With these chiral stationary phases, the enantioselective affinities of the native proteins are retained after immobilization, making these useful in drug-protein-binding studies, as discussed in Chapter 24.

21.4.1   Serum Albumins (SAs)

Bovine serum albumin (BSA) and human serum albumin (HSA) are two closely related proteins with molecular weights of 66.2 and 66.5 kDa, respectively. They are both globular, hydrophobic proteins with 17 internal disulfide bridges that form nine double loops. BSA and HSA are relatively acidic proteins with isoelectric points of 4.7 and 4.9, respectively, and net charges of −18 and −20 at pH 7.5. The tertiary structures of these proteins contain several well-defined binding regions for solutes, including fatty acids, metal ions, and many small organic compounds.

21.4.1.1   *Preparation of SA-CSPs*

A variety of methods have been reported for the preparation of serum albumin chiral stationary phases (SA-CSPs). The immobilization of SAs on solid supports [9] can be carried out by covalently attaching SA to diol-bonded silica that has been activated with 1,1'-carbonyldiimidazole [10, 32], by attaching albumin to silica through a two-step [33] or three-step Schiff base method [33], or by using modified silica that has been activated with *N*-hydroxysuccinimide ester [34]. In addition, some studies have used albumin that is noncovalently adsorbed to ion-exchange columns [35] or immobilized on agarose [36] or hydroxyethylmethacrylate [37, 38].

21.4.1.2   *Chiral Separations Based on SA-CSPs*

A variety of acidic and neutral chiral compounds have been enantioselectively resolved on BSA and HSA-based chiral stationary phases. Examples are given in Table 21.4 [3, 4, 9, 39]. These compounds include many of clinical importance, such as coumarin derivatives (e.g., warfarin) [10], benzodiazepine derivatives [40, 41], barbiturates [39], benzothiadiazines [39], and α-arylpropionic acids (e.g., ibuprofen) [32, 33]. Of these compounds, the coumarin derivatives, benzodiazepines, and α-arylpropionic acids have received the most attention.

While SA-CSPs have been shown to enantioselectively resolve a wide range of compounds, these columns have not been extensively used in analytical or bioanalytical

Table 21.4 Representative Chiral Compounds Resolved by Bovine Serum Albumin and Human Serum Albumin Chiral Stationary Phases

Solute	BSA-CSP	HSA-CSP
Kynurenine	√	—
N-benzoyl	√	—
Dansyl	√	—
N-(2,4-dinitrophenyl)	√	—
Benzoin	√	—
Oxazepam	√	√
Warfarin	√	√
Leucovorin	√	√
Omeprazole	√	—
Ibuprofen	√	√
Lorazepam hemisuccinate	—	√
Tryptophan	—	√
Ketoprofen	√	√
Naproxen	√	√
Benoxaprofen	—	√
Fenoprofen	—	√
Suprofen	—	√
Pirprofen	√	√
Mandelic acid	√	—
Flurbiprofen	√	√

Source: Data obtained from Wainer, I.W., Ed., *Drug Stereochemistry: Analytical Methods and Pharmacology*, 2nd ed., Marcel Dekker, New York, 1993; Haginaka, J., *J. Chromatogr. A*, 906, 253–273, 2001; Hage, D.S. and Austin, J., *J. Chromatogr. B*, 739, 39–54, 2000; Allenmark, S. and Andersson, S., *Chirality*, 1, 154–160, 1989.

studies. The main problem with these materials is their low chromatographic efficiencies, which produce poor peak shapes. This problem can be overcome by using multidimensional chromatography in which the CSP is coupled to a more efficient achiral column. This approach was used to analyze the serum concentrations of (*R*)- and (*S*)-warfarin [42] and the stereoisomers of leucovorin [43] in samples from patients.

21.4.1.3 Chromatographic Conditions for SA-CSPs

For solutes that contain a carboxylic acid, there is an increase in retention with a decrease in the pH of the mobile phase on an SA-CSP [39, 44]. This has been attributed to the availability of a large number of positive charges on serum albumins, due to their protonated amino groups. This leads to increased electrostatic interactions with the carboxylate anion, causing the protein to act as a weak ion exchanger [44]. There is a similar pH effect on the retention of enantiomers for the acidic compound warfarin, but with an opposite effect on enantioselectivity [42]. There is an opposite effect of pH on basic compounds that contain an amine moiety [39, 44]. The retention of these compounds increases with an increase in pH, as does their observed enantioselectivity. The same effect occurs with aromatic amino acids like D,L-tryptophan and its metabolites.

The effect of ionic strength on the retention of charged solutes on serum albumin columns is more complex, since both electrostatic and hydrophobic interactions are influenced. The retention of carboxylic acids is increased dramatically by a decrease in buffer concentration from 50 to 10 mM [44]. This can be explained by the changes that occur in the electrostatic interactions between the charged solute and protein. However, at buffer concentrations greater than 200 mM, there was a reversal in retention for a series of N-benzoylamino acids, indicating that under these conditions, there is a greater influence of hydrophobic interactions [44].

The standard mobile phase used with a BSA-CSP and HSA-CSP consists of a phosphate buffer that contains 1-propanol [3, 4, 9]. The resolution and retention of solutes on these columns can be regulated by making small changes in the 1-propanol concentration (i.e., less than 8%). This has been shown for N-benzoyl-amino acids [39, 44] and warfarin [42]. In the latter case, an increase in 1-propanol concentration from 1 to 5% resulted in a decrease in the retention factor for (S)-warfarin from 41.1 to 12.0 and for (R)-warfarin from 54.6 to 13.8, with a corresponding decrease in enantioselectivity from 1.33 to 1.15.

Anionic modifiers can be used to reduce retention on serum albumin columns without having a significant effect on enantioselectivity. On a BSA-CSP, the addition of 5 mM trichloroacetic acid to the mobile phase reduced the retention of both (R)- and (S)-warfarin by almost 50%, while their enantioselectivity fell by only 4% [45].

21.4.2 Human α₁-Acid Glycoprotein (AGP)

Human α_1-acid glycoprotein (AGP) is also known as orosomucoid (ORM). This is a 41- to 43-kDa glycoprotein with a pI of 2.8 to 3.8 and a high carbohydrate content of 45% [46, 47]. The peptide portion of this protein has 183 amino acids in a single polypeptide chain with two disulfide bridges. The carbohydrate portion has five to six highly sialylated complex-type-N-linked glycans that are attached to the peptide chain. AGP has a high solubility in water and polar organic solvents.

AGP is one of the plasma proteins synthesized by the liver and is mainly secreted by the hepatocytes, although extrahepatic AGP gene expression has also been reported [46]. The hepatic production of these proteins, also known as "acute phase proteins," is increased in response to various types of stress, such as physical trauma or unspecific inflammatory stimuli. The biological function of AGP remains unknown. However, a number of possible physiological activities have been studied, such as immunomodulating effects. Due to its low pI value, AGP is the major plasma protein responsible for the binding of cationic drugs [4]. As will be seen later, this property has made AGP supports the most versatile of all the serum protein-based chiral stationary phases.

21.4.2.1 Preparation of AGP-CSPs

An AGP chiral stationary phase (AGP-CSP) was initially synthesized by coupling this protein to epoxide-activated silica [48]. The silica (LiChrospher Si 300) was activated using 3-glycidoxypropyltrimethylsilane and mixed with a solution of AGP dissolved in pH 8.5, 0.1 M borate buffer containing 0.2 M sodium chloride. The resulting AGP-CSP contained about 35 mg of AGP per gram silica.

The first commercially available form of AGP-CSP (i.e., EnantioPac) was prepared by first ionically binding AGP to diethylaminoethyl silica. This protein was then cross-linked using a process that involved the oxidation of the terminal alcohol groups to aldehydes, Schiff base formation, and reduction to form amine groups [8]. This process yielded

Table 21.5 Chiral Cationic and Anionic Drugs
Resolved on α_1-Acid Glycoprotein Chiral Stationary
Phases

Cationic Solutes	Anionic Solutes
Atropine	Ethotoin
Chlorpheniramine	Fenoprofen
Cocaine	Hexobarbital
Disopyramide	Ibuprofen
Methylphenidate	Ketoprofen
Metoprolol	Naproxen
Terbutaline	2-Phenylbutyric acid
Ketamine	3-Phenylbutyric acid
Methadone	2-Phenoxypropionic acid
Verapamil	2-Phenylpropionic acid

Source: Data obtained from Wainer, I.W., Ed., *Drug Stereo-
chemistry: Analytical Methods and Pharmacology*, 2nd ed.,
Marcel Dekker, New York, 1993; Haginaka, J., *J. Chromatogr.
A*, 906, 253–273, 2001; Wainer, I.W., in *Chiral Liquid Chro-
matography*, Lough, W.J., Ed., Blackie and Son, London,
1989, pp. 130–139; Hermansson, J. and Eriksson, M., *J. Liq.
Chromatogr.*, 9, 621–639, 1986; Schill, G., Wainer, I.W., and
Barkan, S.A., *J. Liq. Chromatogr.*, 9, 641–666, 1986; Enquist,
M. and Hermansson, J., *J. Chromatogr.*, 519, 271–283, 1990;
Enquist, M. and Hermansson, J., *J. Chromatogr.*, 519,
285–298, 1990.

an immobilized coating of approximately 180 mg protein per gram of silica. The resulting
column is a useful CSP but has a short lifetime. A second-generation column (i.e., Chiral
AGP) was later produced with better chromatographic performance and stability.

21.4.2.2 Chiral Separations Based on AGP-CSPs

AGP-CSPs have been the most widely used and effective chiral stationary phases. A wide
variety of basic, neutral, and acidic chiral compounds have been enantioselectively resolved
on AGP-CSPs, as indicated in Table 21.5 [3, 4, 8, 11–13, 17]. Due to its physicochemical
properties, AGP mainly binds to basic and neutral drugs from endogenous as well as
exogenous origins [46, 47]. This is reflected in the extensive use on AGP-CSPs in the
resolution of cationic drugs like verapamil and propranolol. However, through the use of
mobile-phase modifiers, including ion-pairing reagents, AGP-CSPs have been successfully
employed in the chiral separation of anionic drugs, such as ibuprofen, which do not bind
extensively to native AGP.

21.4.2.3 Bioanalytical Applications of AGP-CSPs

Because of the broad applicability of AGP-CSPs and their use with aqueous mobile phases,
columns based on these phases have been used in many bioanalytical studies. The AGP-
CSP can be used alone or in a multidimensional system. The latter approach involves the
use of an achiral column (usually as a precolumn) for the separation of compounds such

as a parent drug and its metabolites based upon structural differences; this is followed by an enantioselective separation on the AGP-CSP. However, the growing use of liquid chromatography/mass spectrometry (LC-MS) and selective ion monitoring has reduced the need for multidimensional systems, with the required chiral separations now often being accomplished using just the AGP-CSP.

Bioanalytical methods based on AGP-CSPs have been developed for disopyramide [49], atenolol [50], verapamil and norverapamil [51], citalopram [52], mepivacaine [53], and thiopentone [54]. Recently, a sensitive enantioselective liquid chromatographic assay with mass spectrometric detection has been developed and validated for simultaneous determination of the plasma concentrations for (R)- and (S)-ketamine, and (R)- and (S)-norketamine [54]. These compounds were extracted from human plasma using solid-phase extraction and then directly injected into an LC-MS system for detection and quantification. Enantioselective separations were achieved on an AGP-CSP column and did not require a coupled-column system. The mobile phase consisted of a 6:94 (v/v) 2-propanol/ammonium acetate buffer (10 mM, pH 7.6). The flow rate was 0.5 ml/min, and a temperature of 25°C was used during the analyses. Under these conditions, the analysis time was 20 min. The detection of ketamine, norketamine, and bromoketamine (the internal standard) was achieved using selected ion monitoring at mass-to-charge ratios of 238.1, 224.1, and 284, respectively. The extracted calibration curves were linear from 1 to 125 ng/ml per enantiomer for each analyte with correlation coefficients greater than 0.9993. The intraday and interday precisions were less than 8.0%. This method has been applied to the clinical study of ketamine in pain management. Figure 21.5 gives some representative chromatograms from this work.

21.4.2.4   Chromatographic Conditions for AGP-CSPs

The chromatographic retention and efficiency of an AGP-CSP, as well as its observed enantioselectivity, are highly sensitive to the chromatographic conditions. While there are many variables that can affect these parameters, the key factors are the pH of the mobile phase, the addition of mobile-phase modifiers, and the temperature at which the separation is conducted. These effects have been extensively reviewed [3, 4, 56] and are summarized in this section.

The pH effects for AGP-CSPs depend on the nature of the solute and the composition of the mobile phase [11, 12, 17, 48]. For example, a decrease in pH usually results in a drop in retention for cationic solutes [11] and an increase in retention for anionic solutes [12] (see Table 21.6). However, the effect of pH on the enantioselectivity can vary. In addition, these effects are extremely dependent on the composition of the mobile phase [12]. This is illustrated in the resolution of cyclopentolate. When the mobile phase contains sodium chloride and 2-propanol, a decrease in the pH of the mobile phase produces a decrease in α from 1.96 to 1.79; but when tetrabutylammonium bromide is used as the modifier, a corresponding decrease in pH produces an increase in α from 1.70 to 2.09 [12].

The pH of the mobile phase can also affect peak symmetry [12]. This effect is also dependent upon the structure of the solute. For example, when disopyramide is the solute, the first eluted enantiomer has a leading edge when the mobile-phase pH is 6.5, but it becomes nearly symmetrical when the pH is raised to 7.0 to 7.5. However, for propoxyphene, the peak corresponding to the second eluting enantiomer has extreme tailing when the mobile-phase pH is 7.0 or 7.5, but this peak becomes symmetrical when the pH is dropped to 6.5.

A large number of secondary, tertiary, and quaternary ammonium compounds have been used as mobile-phase modifiers with AGP-CSPs [3, 4, 8, 11, 12, 48]. Dimethyl octylamine (DMOA) has been the most extensively studied and appears to act via two mechanisms. With cationic solutes, DMOA competes for nonspecific and specific binding

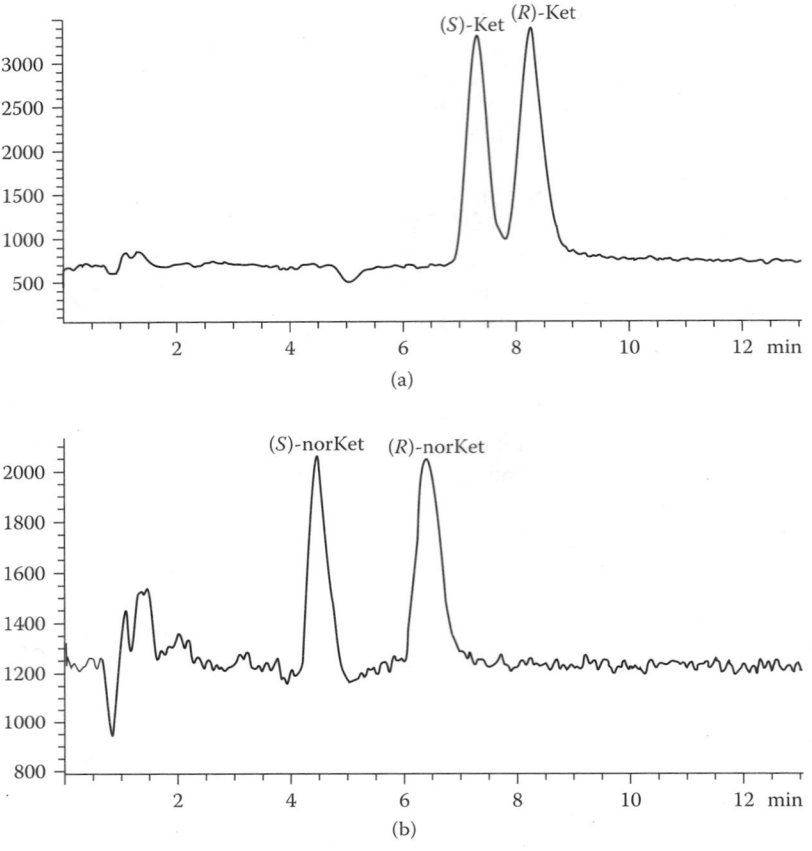

Figure 21.5 Separation and determination of the enantiomers of (a) ketamine (Ket) and (b) norketamine (norKet) in human plasma. This analysis was accomplished using an α_1-acid glycoprotein chiral stationary phase (Chiral AGP-CSP) and mass spectrometric detection. (Reproduced with permission from Rosas, M.E.R., Patel, S., and Wainer, I.W., *J. Chromatogr. B*, 794, 99–108, 2003.)

sites, reducing retention and increasing chromatographic efficiency [11, 12]. While some specific binding may be reduced, the increase in efficiency often outweighs any loss in selectivity and produces a net increase in the observed enantioselectivity. With anionic solutes, DMOA acts as an ion-pairing agent and increases the distribution of the solute to the stationary phase. This produces increased retention as well as higher α values [22].

Anionic modifiers such as butyric and octanoic acids have been successfully used in the enantioselective separation of cationic solutes [12]. For example, the addition of 0.05 *M* butyric acid or 0.01 *M* octanoic acid to a mobile phase containing pH 7, 0.02 *M* phosphate buffer resulted in a significant enantioselective separation of atropine ($\alpha = 1.23$ and 1.64, respectively), while no such selectivity was observed when the mobile phase contained only the phosphate buffer [12]. The primary effect of the anionic modifiers appears to be ion-pair formation with the cationic solutes. However, secondary effects due to modification of the immobilized protein may also occur.

A large variety of uncharged modifiers have been used to effect retention and enantioselectivity on AGP-CSPs. These include monovalent and divalent alcohols, acetonitrile, and amino acids [11–13, 17]. The addition of an uncharged modifier to the mobile

Table 21.6 Effect of pH on the Retention Factor (k')
and Enantioselectivity (α) of Anionic and Cationic
Compounds on an Immobilized α_1-Acid Glycoprotein
Chiral Stationary Phase

Solute	pH	$k'_1{}^a$	α
Anionic Compounds			
Mandelic acid[b]	4.95	10.2	1.0
	6.02	4.6	1.0
	7.02	2.1	1.0
Ibuprofen[b]	4.95	85.7	1.0
	6.02	27.3	1.14
	7.02	11.7	1.19
Cationic Compounds			
Disopyramide[c]	6.5	1.9	2.77
	7.0	2.7	2.70
	7.5	3.3	2.67
Propoxyphene[c]	6.5	3.2	1.52
	7.0	4.4	2.30
	7.5	5.1	2.40
Methadone[c]	6.5	5.0	1.51
	7.0	6.5	1.59
	7.5	7.4	1.57

[a] This value is the retention factor for the first eluted enantiomer.
[b] Data from Hermansson, J. and Eriksson, M., *J. Liq. Chromatogr.*,
9, 621–639, 1986.
[c] Data from Schill, G., Wainer, I.W., and Barkan, S.A., *J. Liq. Chromatogr.*, 9, 641–666, 1986.

phase usually reduces chromatographic retention and increases efficiency, but it is often accompanied by a decrease in enantioselectivity. For example, when the injected solute was alprenolol, the use of 0.7 M 1-propanol and 1.1 M 2-propanol as modifiers resulted in retention factors of 22.8 and 19.6 for the first eluted enantiomer, with enantioselectivities of 1.20 and 1.21, respectively. When the concentrations of the modifiers were increased to 1.06 M for 1-propanol and to 1.60 M for 2-propanol, the retention factor for the first eluting enantiomer was reduced to 12.9 or 12.8, and the corresponding enantioselectivity dropped to 1.0 and 1.07, respectively [17].

The most extensively used modifiers for AGP-CSPs are 1-propanol, 2-propanol, and acetonitrile. These modifiers often have vastly different effects on chromatographic retention and enantioselectivity, which depends on the structure of the solute being separated. For example, it has already been shown how the retention for alprenolol is affected by 1-propanol and 2-propanol. On the same AGP-CSP that was used to compare these additives, the use of 2.85 M acetonitrile as a mobile-phase modifier gave a retention factor for the first eluting enantiomer of 30.6 and an enantioselectivity of 1.83 [17].

Micellar chromatography has been used with charged mobile-phase modifiers to reduce retention and increase enantioselectivity in AGP-CSPs [52, 56]. Examples of this approach include the quantitative analysis of citalopram enantiomers in human plasma

[52] and the enantioselective separation of (R)- and (S)-Naproxen for the analysis of samples that have been incubated with human liver microsomes [55]. In the former assay, N-dodecyl-N,N-dimethylammonio-3-propanesulfonate was used as the surfactant, and hexanoic acid was the ion-pairing reagent. In the latter assay, Tween 20 was the surfactant and DMOA was the ion-pairing reagent.

21.5 SUMMARY AND CONCLUSIONS

Biopolymer-based chiral stationary phases represent a large class of successful analytical tools. However, CSPs based on enzymes and serum albumins have been essentially supplanted by CSPs based on derivatized cellulose and macrocyclic antibiotics. Only the AGP-CSP remains a viable analytical column. This is due, in part, to the broad enantioselectivity of AGP-CSPs and the advances in LC-MS technology that avoid time-consuming coupled-column systems.

Basic studies into the chiral recognition mechanisms that take place on enzyme and serum protein-based CSPs have demonstrated that the separations on these columns reflect the interactions that take place with the nonimmobilized biopolymer. Thus, studies with these CSPs can be used to provide invaluable insights into basic pharmacological and biochemical processes. A further discussion of this topic is provided in Chapter 24.

SYMBOLS AND ABBREVIATIONS

ACHT	α-Chymotrypsin
ACHT-SP	α-Chymotrypsin chiral stationary phase
AGP	α_1-Acid glycoprotein
AGP-SA	α_1-Acid glycoprotein chiral stationary phase
Arg	Arginine
BAPNA	N-Alpha-benzoyl-arginine p-nitroanalide
BSA	Bovine serum albumin
BSA-CSP	Bovine serum albumin chiral stationary phase
Bz	Benzoyl
CBH I	Cellobiohydrolase I
CBH I-CSP	Cellobiohydrolase I chiral stationary phase
CSP	Chiral stationary phase
DMOA	Dimethyloctylamine
His	Histidine
HSA	Human serum albumin
HSA-CSP	Human serum albumin chiral stationary phase
IAM-SP	Immobilized artificial membrane stationary phase
K_i	Inhibition constant
k'	Retention factor
LC-MS	Liquid chromatography/mass spectrometry
Lys	Lysine
4-NTA	4-Nitrophenyltrimetlylacetate
OEt	Ethyl ester
OMe	Methyl ester
ORM	Orosomucoid
OVM	Ovomucoid

Phe	Phenylalanine
SA	Serum albumin
TPCK	N-Tosyl-L-phenylalanine chloromethyl ketone
Trp	Tryptophan
Trp-NH$_2$	Tryptophan amine
Trp-OME	Tryptophan methyl ester
TRYP	Trypsin
TRYP-CSP	Trypsin chiral stationary phase
Tyr	Tyrosine
α	Enantioselectivity factor

REFERENCES

1. Pasteur, L., On the asymmetry of naturally occurring organic compounds, in *The Foundation of Stereochemistry: Memoirs by Pasteur, V. Hoff, L. Bel, and Wislicenus,* Richardson, G.M., Ed., American Book Co., New York, 1901, pp. 1–33.
2. Drayer, D., The early history of stereochemistry, in *Drug Stereochemistry, Analytical Methods and Pharmacology,* 2nd ed., Wainer, I.W., Ed., Marcel Dekker, New York, 1993, pp. 5–14.
3. Wainer, I.W., Ed., *Drug Stereochemistry: Analytical Methods and Pharmacology,* 2nd ed., Marcel Dekker, New York, 1993.
4. Haginaka, J., Protein-based chiral stationary phases for high-performance liquid chromatography enantioseparations, *J. Chromatogr. A,* 906, 253–273, 2001.
5. Chui, W.K. and Wainer, I.W., Enzyme-based HPLC supports as probes of enzyme activity and inhibition: The immobilization of trypsin and α-chymotrypsin on an immobilized artificial membrane HPLC support, *Anal. Biochem.,* 201, 237–245, 1992.
6. Alebic-Kolbah, T. and Wainer, I.W., Application of an enzyme-based stationary phase to the determination of enzyme kinetic constants and types of inhibition: A new HPLC approach utilizing an immobilized artificial membrane chromatographic support, *J. Chromatogr.,* 653, 122–129, 1993.
7. Marle, I., Karlsson, A., and Pettersson, C., Separation of enantiomers using α-chymotrypsin-silica as a chiral stationary phase, *J. Chromatogr.,* 604, 185–196, 1992.
8. Wainer, I.W., Immobilized proteins as HPLC chiral stationary phases, in *Chiral Liquid Chromatography,* Lough, W.J., Ed., Blackie and Son, Glasgow, 1989, pp. 130–139.
9. Hage, D.S. and Austin, J., High-performance affinity chromatography and immobilized serum albumin as probes for drug- and hormone-protein binding, *J. Chromatogr. B,* 739, 39–54, 2000.
10. Domenici, E., Bertucci, C., Salvadori, P., Felix, G., Cahagne, I., Motellier, S., and Wainer, I.W., Synthesis and chromatographic properties of an HPLC chiral stationary phase based upon human serum albumin, *Chromatographia,* 29, 170–176, 1990.
11. Hermansson, J. and Eriksson, M., Direct liquid chromatographic resolution of acidic drugs using a chiral α_1-acid glycoprotein column (EnantioPac), *J. Liq. Chromatogr.,* 9, 621–639, 1986.
12. Schill, G., Wainer, I.W., and Barkan, S.A., Chiral separation of cationic drugs on α_1-acid glycoprotein-bonded stationary phase, *J. Liq. Chromatogr.,* 9, 641–666, 1986.
13. Enquist, M. and Hermansson, J., Influence of uncharged mobile phase additives on retention and enantioselectivity of chiral drugs using an α_1-acid glycoprotein column, *J. Chromatogr.,* 519, 271–283, 1990.
14. Wainer, I.W. and Chu, Y.Q., Use of mobile phase modifiers to alter retention and stereoselectivity on a bovine serum albumin high-performance liquid chromatographic chiral stationary phase, *J. Chromatogr.,* 455, 316–322, 1988.

15. Noctor, T.A.G. and Wainer, I.W., The use of displacement chromatography to alter retention and enantioselectivity on a human serum albumin-based HPLC chiral stationary phase: A mini-review, *J. Liq. Chromatogr.*, 16, 783–800, 1993.

16. Noctor, T.A.G., Wainer, I.W., and Hage, D.S., Allosteric and competitive displacement of drugs from human serum albumin by octanoic acid, as revealed by high-performance liquid affinity chromatography, on a human serum albumin-based stationary phase, *J. Chromatogr.*, 577, 305–315, 1992.

17. Enquist, M. and Hermansson, J., Separation of the enantiomers of beta-receptor blocking agents and other cationic drugs using a CHIRAL-AGP column: Binding properties and characterization of immobilized alpha 1-acid glycoprotein, *J. Chromatogr.*, 519, 285–298, 1990.

18. Dominici, E., Bertucci, C., Salvadori, P., and Wainer, I.W., Use of a human serum albumin-based high-performance liquid chromatography chiral stationary phase for the investigation of protein binding: Detection of the allosteric interaction between warfarin and benzodiazepine binding sites, *J. Pharm. Sci.*, 80, 164–166, 1991.

19. Noctor, T.A.G. and Wainer, I.W., The *in situ* acetylation of an immobilized human serum albumin chiral stationary phase for high-performance liquid chromatography in the examination of drug-protein binding phenomena, *Pharma. Res.*, 9, 480–484, 1992.

20. Bertucci, C. and Wainer, I.W., Improved chromatographic performance of a modified human albumin-based stationary phase, *Chirality*, 9, 335–340, 1997.

21. Marle, I., Erlandsson, P., Hansson, L., Isaksson, R., Pettersson, C., and Pettersson, G., Separation of enantiomers using cellulase (CBH I) silica as a chiral stationary phase, *J. Chromatogr.*, 586, 233–248, 1991.

22. Fulde, K. and Frahm, A.W., Temperature-induced inversion of elution order in the enantioseparation of sotalol on a cellobiohydrolase I-based stationary phase, *J. Chromatogr. A*, 858, 33–43, 1999.

23. Jadaud, P., Thelohan, S., Schonbaum, G.R., and Wainer, I.W., The stereochemical resolution of enantiomeric free and derivatized amino acids using an HPLC chiral stationary phase based on immobilized α-chymotrypsin: Chiral separation due to solute structure or enzyme activity, *Chirality*, 1, 38–44, 1989.

24. Jadaud, P. and Wainer, I.W., Stereochemical recognition of enantiomeric and diastereomeric dipeptides by high-performance liquid chromatography on a chiral stationary phase based upon immobilized α-chymotrypsin, *J. Chromatogr.*, 476, 165–174, 1989.

25. Jadaud, P. and Wainer, I.W., The stereochemical resolution of the enantiomers of aspartame on an immobilized α-chymotrypsin HPLC chiral stationary phase: The effect of mobile-phase composition and enzyme activity, *Chirality*, 2, 32–37, 1990.

26. Schultz, R.M. and Liebman, M.N., Protein II: Structure-function relationship of protein families, in *Textbook of Biochemistry, with Clinical Correlations*, Wiley, New York, 1992, pp. 112–116.

27. Thelohan, S., Jadaud, P., and Wainer, I.W., Immobilized enzymes as chromatographic phases for HPLC: The chromatography of free and derivatized amino acids on immobilized trypsin, *Chromatographia*, 28, 551–555, 1989.

28. Henriksson, H., Stahlberg, J., Isaksson, R., and Pettersson, G., The active sites of cellulases are involved in chiral recognition: A comparison of cellobiohydrolase 1 and endoglucanase 1, *FEBS Lett.*, 390, 339–344, 1996.

29. Marle, I., Erlandsson, P., Hansson, L., Isaksson, R., Pettersson, C., and Pettersson, G., Separation of enantiomers using cellulase (CBH I) silica as a chiral stationary phase, *J. Chromatogr.*, 586, 233–248, 1991.

30. Henriksson, H., Joensson, S., Isaksson, R., and Pettersson, G., Chiral separation based on immobilized intact and fragmented cellobiohydrolase II (CBH II): A comparison with cellobiohydrolase I (CBH I), *Chirality*, 7, 415–424, 1995.

31. Hedeland, M., Holmin, S., Nygard, M., and Pettersson, C., Chromatographic evaluation of structure selective and enantioselective retention of amines and acids on cellobiohydrolase I wild type and its mutant D214N, *J. Chromatogr. A*, 864, 1–16, 1999.

32. Harada, K., Yuan, Q., Nakayama, M., and Sugii, A., Effects of organic modifiers on the chiral recognition by different types of silica-immobilized bovine serum albumin, *J. Chromatogr. A,* 740, 207–213, 1996.

33. Loun, B. and Hage, D.S., Characterization of thyroxine-albumin binding using high-performance affinity chromatography, I: Interactions at the warfarin and indole sites of albumin, *J. Chromatogr.,* 579, 225–235, 1992.

34. Millot, M.C., Sebille, B., and Mangin, C., Enantiomeric properties of human albumin immobilized on porous silica supports coated with polymethacryloyl chloride, *J. Chromatogr. A,* 776, 37–44, 1997.

35. Taleb, N.L., Millot, M.C., and Sebille, B., Enantioselectivity properties of human serum albumin immobilized on anion-exchangers based on polyvinylimidazole-coated silica: Effect of protein loading on separation properties, *J. Chromatogr. A,* 776, 45–53, 1997.

36. Fitos, I., Visy, J., Simonyi, M., and Hermansson, J., Chiral high-performance liquid chromatographic separations of vinca alkaloid analogues on α_1-acid glycoprotein and human serum albumin columns, *J. Chromatogr.,* 609, 163–171, 1992.

37. Simek, Z. and Vespalec, R., Interpretation of enantioselective activity of albumin used as the chiral selector in liquid chromatography and electrophoresis, *J. Chromatogr. A,* 685, 7–14, 1994.

38. Simek, Z. and Vespalec, R., Chromatographic properties of chemically bonded bovine serum albumin working as a chiral selector in alkaline mobile phases, *J. Chromatogr.,* 629, 153–160, 1993.

39. Allenmark, S. and Andersson, S., Optical resolution of some biologically active compounds by chiral liquid chromatography on BSA-silica (Resolvosil) columns, *Chirality,* 1, 154–160, 1989.

40. Noctor, T.A.G., Pham, C.D., Kaliszan, R., and Wainer, I.W., Stereochemical aspects of benzodiazepine binding to human serum albumin, I: Enantioselective high-performance liquid affinity chromatographic examination of chiral and achiral binding interactions between 1,4-benzodiazepines and human serum albumin, *Mol. Pharmacol.,* 42, 506–511, 1992.

41. Kaliszan, R., Noctor, T.A.G., and Wainer, I.W., Stereochemical aspects of benzodiazepine binding to human serum albumin, II: Quantitative relationships between structure and enantioselective retention in high-performance liquid affinity chromatography, *Mol. Pharmacol.,* 42, 512–517, 1992.

42. Chu, Y.Q. and Wainer, I.W., The measurement of warfarin enantiomers in serum using coupled achiral/chiral high-performance liquid chromatography (HPLC), *Pharm. Res.,* 5, 680–683, 1988.

43. Wainer, I.W. and Stiffin, R.M., Direct resolution of the stereoisomers of leucovorin and 5-methyltetrahydrofolate using a bovine serum albumin high-performance liquid chromatographic chiral stationary phase coupled to an achiral phenyl column, *J. Chromatogr.,* 424, 158–162, 1988.

44. Allenmark, S., Optical resolution by liquid chromatography on immobilized bovine serum albumin, *J. Liq. Chromatogr.,* 9, 425–442, 1986.

45. Wainer, I.W. and Chu, Y.Q., Use of mobile phase modifiers to alter retention and stereoselectivity on a bovine serum albumin high-performance liquid chromatographic chiral stationary phase, *J. Chromatogr.,* 455, 316–322, 1988.

46. Kremer, J.M.H., Wilting, J., and Janssen, L.H.M., Drug binding to human alpha-1-acid glycoprotein in health and disease, *Pharmacol. Rev.,* 40, 1–47, 1988.

47. Fournier, T., Medjoubi-N, N., and Porquet, D., Alpha-1-acid glycoprotein, *Biochim. Biophys. Acta,* 1482, 157–171, 2000.

48. Hermansson, J., Direct liquid chromatographic resolution of racemic drugs using α_1- 1-acid glycoprotein as the chiral stationary phase, *J. Chromatogr.,* 269, 71–80, 1983.

49. Hermansson, J., Eriksson, M., and Nyquist, O., Determination of (R)- and (S)-disopyramide in human plasma using a chiral α_1-acid glycoprotein column, *J. Chromatogr.,* 336, 321–328, 1984.

50. Enquist, M. and Hermansson, J., Separation and quantitation of (R)- and (S)-atenolol in human plasma and urine using an α_1-AGP column, *Chirality,* 1, 209–215, 1989.

51. Chu, Y.Q. and Wainer, I.W., Determination of the enantiomers of verapamil and norverapamil in serum using coupled achiral-chiral high-performance liquid chromatography, *J. Chromatogr.*, 497, 191–200, 1989.

52. Haupt, D., Determination of citalopram enantiomers in human plasma by liquid chromatographic separation on a chiral-AGP column, *J. Chromatogr. B*, 685, 299–305, 1996.

53. Vletter, A.A., Olieman, W., Burm, A.G.L., Groen, K., and Van Kleef, J.W., High-performance liquid chromatographic assay of mepivacaine enantiomers in human plasma in the nanogram per milliliter range, *J. Chromatogr. B*, 678, 369–372, 1996.

54. Jones, D.J., Nguyen, K.T., McLeish, M.J., Crankshaw, D.P., and Morgan, D.J., Determination of (*R*)-(+)- and (*S*)-(−)-isomers of thiopentone in plasma by chiral high-performance liquid chromatography, *J. Chromatogr. B*, 675, 174–179, 1996.

55. Rosas, M.E.R., Patel, S., and Wainer, I.W., Determination of the enantiomers of ketamine and norketamine in human plasma by enantioselective liquid chromatography-mass spectrometry, *J. Chromatogr. B*, 794, 99–108, 2003.

56. Haupt, D., Pettersson, C., and Westerlund, D., Separation of (*R*)- and (*S*)-naproxen using micellar chromatography and an α_1-acid-glycoprotein column: Application for chiral monitoring in human liver microsomes by coupled-column chromatography, *J. Biochem. Biophys. Methods*, 25, 273–284, 1992.

Section V

Biophysical Applications

22

Quantitative Affinity Chromatography: Practical Aspects

David S. Hage and Jianzhong Chen
Department of Chemistry, University of Nebraska, Lincoln, NE

CONTENTS

22.1 INTRODUCTION

Previous chapters have examined the use of affinity chromatography in the isolation of substances and in their analysis. However, this method can also be used to study the interactions between two or more chemicals. When employed for this purpose, this approach is often referred to as analytical affinity chromatography, quantitative affinity chromatography, or biochromatography [1–3]. One way this might be performed is by using the format given in Figure 22.1, in which the ligand of interest is immobilized in an affinity column while injections are made of a complementary solute in the mobile phase. By examining the elution time or volume of the solute after it has passed through the column, it is possible to obtain information on the equilibrium constants that describe the analyte's binding to the affinity ligand. If additional agents are also present in the mobile phase, data can be obtained on how these agents affect the solute-ligand interaction. Furthermore, information on the rates

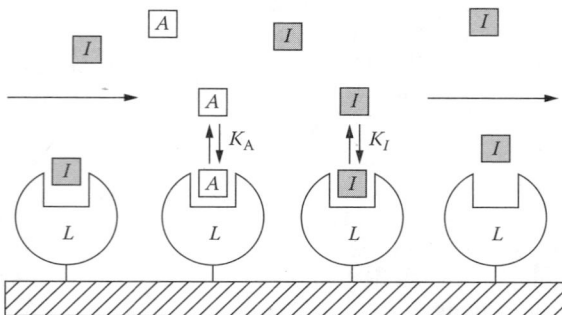

Figure 22.1 Reaction model for the study of solute-ligand binding by affinity chromatography, where A is the injected or applied analyte, I is a competing agent added in a known concentration to the mobile phase, and L is the immobilized ligand. The terms K_A and K_I are the association equilibrium constants for the binding of A and I to L, respectively. (Reproduced with permission from Hage, D.S., *J. Chromatogr. B*, 768, 3–20, 2002.)

of these binding processes can be acquired by examining the shape of the solute's elution profile. As will be seen later, all of these approaches have been used in the study of biological interactions.

One advantage of utilizing affinity chromatography is its ability to reuse the same ligand preparation for multiple experiments. For instance, columns containing immobilized serum proteins on supports for high-performance liquid chromatography (HPLC) have been used for 500 to 1000 injections [4–6]. This creates a situation in which only a relatively small amount of ligand is needed for a large number of studies. This helps give good precision by minimizing run-to-run variations. Other advantages include the ease with which affinity methods can be automated, especially when used as part of HPLC systems, and the relatively short periods of time required with such systems for solute-binding studies (i.e., often 5 to 15 min per analysis). The fact that the immobilized ligand is continuously washed with an applied solvent is another advantage, since this minimizes the effects produced by soluble contaminants in the initial ligand preparation [7].

Whenever an immobilized ligand is used to study a solute-ligand interaction, an important question to consider is, "How well does this ligand work as a model for the same ligand in solution?" The answer will depend on the ligand being considered and the way it has been immobilized. This currently needs to be evaluated on a case-by-case basis. For instance, in the case of human serum albumin (HSA) and bovine serum albumin (BSA), there is a large body of evidence indicating that the immobilized forms of these proteins can provide good qualitative and quantitative agreement with the behavior seen for the same proteins in solution [5–16]. However, this is not the case for columns with cross-linked albumin [17] or columns containing cross-linked α_1-acid glycoprotein (AGP) [18, 19]. Although the reasons for such differences are not fully understood, they do indicate the importance of using model compounds with known binding properties to assess any immobilized protein before it is used to examine other substances. As an alternative, experiments can be performed in which both soluble and immobilized ligands are used.

This chapter discusses the most common formats for performing quantitative studies of solute-ligand interactions by affinity chromatography. The techniques considered include zonal elution, frontal analysis, and various kinetic techniques. The emphasis in this chapter is on traditional chromatographic systems; related flow-based methods that employ biosensors are discussed in Chapters 23 and 25. The general principles of each method are examined, and their general applications are presented. In addition, several practical considerations in the design and use of these methods are given.

22.2 ZONAL ELUTION

The most popular affinity method used to examine solute-ligand binding is zonal elution. This is performed in the same mode used for most analytical applications of chromatography, in which a narrow plug of solute is injected onto a column while the solute's elution time or volume is monitored [3]. A typical system for performing zonal elution by HPLC is shown in Figure 22.2.

Zonal elution was first used for quantitative affinity chromatography in 1974 by Dunn and Chaiken, who examined the retention of staphylococcal nuclease on a low-performance column containing immobilized thymidine-5′-phosphate-3′-aminophenylphosphate [20]. By the late 1980s and early 1990s, reports also began to appear in which this approach was used with HPLC [7, 9–11, 21–23]. It has since been used for a variety of applications, as will be seen later in this chapter.

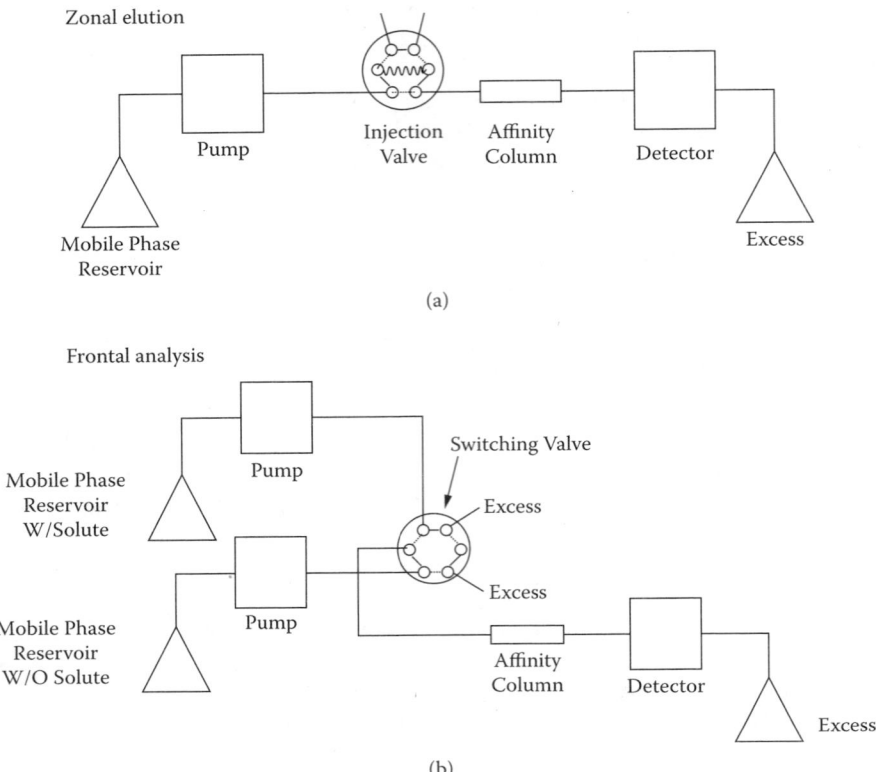

Figure 22.2 Typical experimental systems for performing (a) zonal elution and (b) frontal analysis.

22.2.1   General Principles

Zonal elution is generally performed by injecting a small amount of analyte through a column under linear elution conditions while elution of the analyte is monitored by an on-line detector. A larger amount of analyte and nonlinear conditions have also been employed in some cases [24, 25]. Furthermore, fraction collection and off-line detection can be used, especially when working with low-performance affinity systems [3]. The mobile phase in these studies has a known composition (usually a buffer with a physiological pH) and often contains a fixed concentration of an additive or competing agent. Information on the interactions between this analyte and the ligand within the column is obtained by looking at how the analyte retention changes with the mobile-phase composition or other conditions (e.g., changes in the temperature or solute structure).

A typical zonal-elution experiment is shown in Figure 22.3. In this example, binding of L-tryptophan to immobilized HSA is being examined in the presence of various mobile-phase concentrations of the drug phenytoin as a competing agent. As can be seen, the retention of L-tryptophan shifts to lower values as the phenytoin concentration is increased, indicating that direct or negative allosteric competition is occurring between these two solutes on HSA [26]. Besides adding a competing agent to the mobile phase, other factors that can be altered in these experiments include the pH, ionic strength and polarity of the mobile phase, the column temperature, and the type of solute or ligand in the column.

An important advantage of zonal elution is that it requires only a small amount of solute per injection. It is also possible to examine more than one compound per injection, provided that these substances can be detected separately or that adequate resolution is

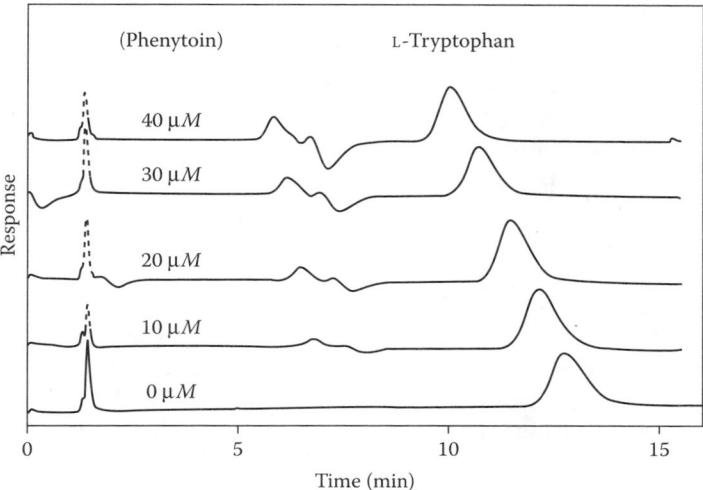

Figure 22.3 Chromatograms obtained in a zonal-elution study for the injection of L-tryptophan into the presence of phenytoin as a mobile-phase additive. These studies were performed at 37°C in a pH 7.4, 0.067 M phosphate buffer.

obtained between their corresponding peaks. Another advantage of zonal elution is that it can easily be performed with standard HPLC equipment. The only modification required is the addition of temperature control for the column and mobile phase. If performed properly, changes in retention of only a few seconds can easily be detected by this approach [27].

22.2.2 Common Applications of Zonal Elution

There are a variety of ways in which zonal elution has been used to obtain information on the binding of solutes to ligand. These include measurements of the degree and affinity of solute-protein binding, studies examining changes in binding as the mobile-phase composition or temperature is varied, and experiments that consider how alterations in solute or protein structure affect the interactions of these species. More details on each application are provided later in this section.

Each of the applications relies on the fact that the retention observed for an injected analyte is a direct measure of that analyte's interactions within the column. This is described by Equation 22.1, which shows how the analyte's overall retention factor (k) is related to the number of binding sites it has in the column and to the equilibrium constants for the analyte at these sites [1].

$$k = (K_{A1}\, n_1 + \ldots + K_{An}\, n_n)\, m_L/V_M \qquad (22.1)$$

In this equation, the retention factor is calculated by using $k = (t_R - t_M)/t_M$ or $k = (V_R - V_M)/V_M$, where t_R is the retention time of the injected compound, V_R is the corresponding retention volume, t_M is the column void time, and V_M is the void volume. Other terms in Equation 22.1 include the total moles of binding sites in the column (m_L), the association equilibrium constants for the analyte at the individual sites in this population (K_{A1} through K_{An}), and the fraction of each type of site in the column (n_1 through n_n). From this equation, it can be seen that a change in the strength of binding, the number of binding sites, or the relative distribution of these sites can result in a shift in analyte retention. It will be shown later how this shift can be used to provide both qualitative and quantitative information on the interactions between a solute and an immobilized ligand.

22.2.2.1   Estimation of Relative Binding

One way zonal elution has been employed is as a means to measure the average extent of binding between a solute and an immobilized ligand. This is based on the fact that the retention factor, when measured at true equilibrium, is equal to the fraction of an injected solute that is bound to the ligand (b) divided by the fraction of solute that remains free in the mobile phase (f), or $k = b/f$. By using the fact that the bound plus free fractions must equal 1, it is possible to rearrange this relationship into the following form, which allows the bound fraction of the solute to be calculated from its retention factor [28].

$$b = k/(1 + k) \tag{22.2}$$

Another way in which the relative binding of two solutes can be compared is by taking the ratio of their retention factors on the same affinity column. According to Equation 22.1, if both solutes have a single, common binding site on the ligand, the ratio of their retention factors should equal the ratio of their association constants at this site. However, caution must be exercised when using this approach with solutes that have multisite binding or different binding sites on a ligand, since these sites may have different susceptibilities to a loss of activity during immobilization. For instance, it is known that the warfarin-azapropazone and indole-benzodiazepine sites on HSA are affected to different degrees by the immobilization of this protein [7, 29]. Unless it can be ensured that such sites have equal levels of activity, the relative binding determined by this method should be viewed as only an approximation of the behavior expected for the same ligand in solution.

22.2.2.2   Competition and Displacement Studies

The most common use for zonal elution in quantitative affinity chromatography has been in competition and displacement studies. This is performed by injecting the analyte while a fixed concentration of a potential competing agent is passed through the column in the mobile phase. An example of this type of experiment was given in Figure 22.3.

It is relatively easy from such work to determine qualitatively whether or not two compounds interact as they bind to the same immobilized ligand. But to obtain further information, such as the nature of this competition and the number of sites involved, it is necessary to compare the zonal elution data to the responses expected for various inter-action models. The equations used for this purpose are shown in Table 22.1. These describe models that range from direct competition at a single binding site to those that involve multisite interactions, simple allosteric effects, or noncompetitive binding.

In analyzing zonal elution data, it is often helpful to first make a plot of $1/k$ vs. the concentration of the competing agent. If no competition is occurring between the analyte and mobile-phase additive, only random variations should be observed, as demonstrated in Figure 22.4a. If a linear relationship with a positive slope is seen, as illustrated in Figure 22.4b, there is probably direct competition between the analyte and additive at a single common binding site on the ligand. A nonlinear response with a positive slope indicates that multisite competition or negative allosteric interactions are present, as shown in Figure 22.4c. And, as illustrated in Figure 22.4d, nonlinear behavior with a negative slope is an indication of positive allosteric interactions.

After the general model that fits the zonal elution data has been identified, the results can be fit to the equations for this model to obtain information on the strength of this interaction. This is accomplished by using the equations in Table 22.1 to determine the binding parameters for the system. For instance, the equations in Table 22.1 predict a linear relationship for a plot of $1/k$ vs. $[I]$ for an analyte and competing agent with single-site competition on an immobilized ligand, as illustrated in Figure 22.4b. By determining the

Table 22.1 Models and Equations Used in Zonal Elution

Type of System [References]	Model	Predicted Response
Self competition of an injected analyte (A) with itself as a competing agent at a single type of binding site (L) [24, 30]	$A + L \overset{K_A}{\longleftrightarrow} A - L$	$\dfrac{1}{k_A} = \dfrac{[A]}{C_L} + \dfrac{1}{K_A C_L}$
Direct competition of injected analyte (A) and competing agent (I) at a single common binding site (L), where A has no other binding sites and I has no other interactive binding sites with A or itself [1, 31]	$I + L \overset{K_I}{\longleftrightarrow} I - L$ $A + L \overset{K_A}{\longleftrightarrow} A - L$	$\dfrac{1}{k_A} = \dfrac{K_I[I]}{K_A C_L} + \dfrac{1}{K_A C_L}$
Direct competition of injected analyte (A) and competing agent (I) at a single common binding site (L), where A has one or more additional sites ($L_{N1}...L_{Nn}$) that produce no competition, and A is present in a small amount versus L; I has no other common binding sites with A or itself [10]	$A + L \overset{K_A}{\longleftrightarrow} A - L$ $I + L \overset{K_I}{\longleftrightarrow} I - L$ $A + L_{N1} \overset{K_{A_{N1}}}{\longleftrightarrow} A - L_{N1}$ $A + L_{Nn} \overset{K_{A_{Nn}}}{\longleftrightarrow} A - L_{Nn}$	$\dfrac{1}{k_A - X} = \dfrac{K_I[I]}{K_A C_L} + \dfrac{1}{K_A C_L}$ where $X = \dfrac{K_{AL_{N1}} mL_{N1} + ... + K_{AL_{Nn}} mL_{Nn}}{V_M}$
Self-competition of an injected analyte (A) with itself as a competing agent at two types of binding sites (L_1 and L_2) [24, 30, 32]	$A + L_1 \overset{K_{A_1}}{\longleftrightarrow} A - L_1$ $A + L_2 \overset{K_{A_2}}{\longleftrightarrow} A - L_2$	$k_A = \dfrac{K_{A_1}}{1 + K_{A_1}[A]} C_{L_1}$ $+ \dfrac{K_{A_2}}{1 + K_{A_2}[A]} C_{L_2}$
Direct competition of injected analyte (A) and competing agent (I) at two common binding sites (L_1 and L_2), where there are no other interactions between the binding sites for A and I	$A + L_1 \overset{K_{A_1}}{\longleftrightarrow} A - L_1$ $A + L_2 \overset{K_{A_2}}{\longleftrightarrow} A - L_2$ $I + L_1 \overset{K_{I_1}}{\longleftrightarrow} I - L_1$ $I + L_2 \overset{K_{I_2}}{\longleftrightarrow} I - L_2$	$k_A = \dfrac{K_{A_1}}{1 + K_{I_1}[I]} C_{L_1}$ $+ \dfrac{K_{A_2}}{1 + K_{I_2}[I]} C_{L_2}$
Direct competition between a soluble ligand (S) and immobilized ligand (L) for an injected analyte (A), where there is no interaction between S and L, and $K_{AS}[A] \ll 1$ [33]	$A + S \overset{K_{AS}}{\longleftrightarrow} A - S$ $A + L \overset{K_A}{\longleftrightarrow} A - L$	$\dfrac{1}{k_A} = \dfrac{K_{AS}[S]}{K_A C_L} + \dfrac{1}{K_A C_L}$

(*continued*)

Table 22.1   Models and Equations Used in Zonal Elution (Continued)

Type of System [References]	Model	Predicted Response
Self-competition of an injected analyte (A) with itself as a competing agent at a single type of binding site (L) when a solubilizing agent (S) is present and there is no interaction between S and L [34]	$A + S \overset{K_{AS}}{\longleftrightarrow} A - S$ $A + L \overset{K_A}{\longleftrightarrow} A - L$	$\dfrac{1}{k_A} = \dfrac{C_A}{C_L} + \dfrac{1 + K_{AS}C_S}{K_A C_L}$
Direct competition of injected analyte (A) and competing agent (I) at a single common binding site (L) when a solubilizing agent (S) is present, where A has no other binding sites and I has no other interactive binding sites with A or itself [35]	$A + L \overset{K_A}{\longleftrightarrow} A - L$ $I + L \overset{K_I}{\longleftrightarrow} I - L$ $A + S \overset{K_{AS}}{\longleftrightarrow} A - S$ $I + S \overset{K_{IS}}{\longleftrightarrow} I - S$	$\dfrac{1}{k_A} = \dfrac{(1 + K_{AS}C_S)K_I C_I}{(1 + K_{IS}C_S)K_A C_L}$ $+ \dfrac{1 + K_{AS}C_S}{K_A C_L}$
Allosteric interaction between an injected analyte (A) at one site on a ligand (L_1) when another solute (B) binds at a second site on the same ligand (L_2) [36]	$A + L_1 \overset{K_A}{\longleftrightarrow} A - L_1$ $B + L_2 \overset{K_B}{\longleftrightarrow} B - L$ $A + L - B \overset{K'_A}{\longleftrightarrow} A - L - B$	$\dfrac{k_{A,0}}{k - k_{A,0}} = \dfrac{1}{\left(\dfrac{K'_A}{K_A} - 1\right)}\left(\dfrac{1}{K_B C_B} + 1\right)$
Allosteric interaction between an injected analyte (A) at one site on a ligand (L_1) when another solute (B) binds at a second site on the same ligand (L_2) in the presence of a solubilizing agent (S), where there is no interaction between S and L [34]	$A + L_1 \overset{K_{A_1}}{\longleftrightarrow} A - L_1$ $B + L_2 \overset{K_B}{\longleftrightarrow} B - L$ $A + L - B \overset{K'_A}{\longleftrightarrow} A - L - B$ $A + S \overset{K_{AS}}{\longleftrightarrow} A - S$ $B + S \overset{K_{BS}}{\longleftrightarrow} B - S$	$\dfrac{k_{A,0}}{k - k_{A,0}} = \dfrac{1 + K_{BS}C_S}{\left(\dfrac{K'_A}{K_A} - 1\right)}\left(\dfrac{1}{K_B C_B} + 1\right)$

best-fit slope and intercept for this plot and taking the ratio of the slope to intercept, the result should give the association constant for I at the site of competition (K_I). In this situation, the association constant for the analyte (K_A) can also be obtained by using the intercept and a separate measurement for the concentration of ligand sites (m_L/V_M). A similar approach can be used with the other equations in Table 22.1 to examine cases where the analytes and additives have multiple sites of competition or independent binding regions.

Some advantages of using zonal elution for this work are its speed, its need for only a small amount of analyte, and its possible use with multiple analytes per injection. In addition, the association constants measured by this approach have good precision

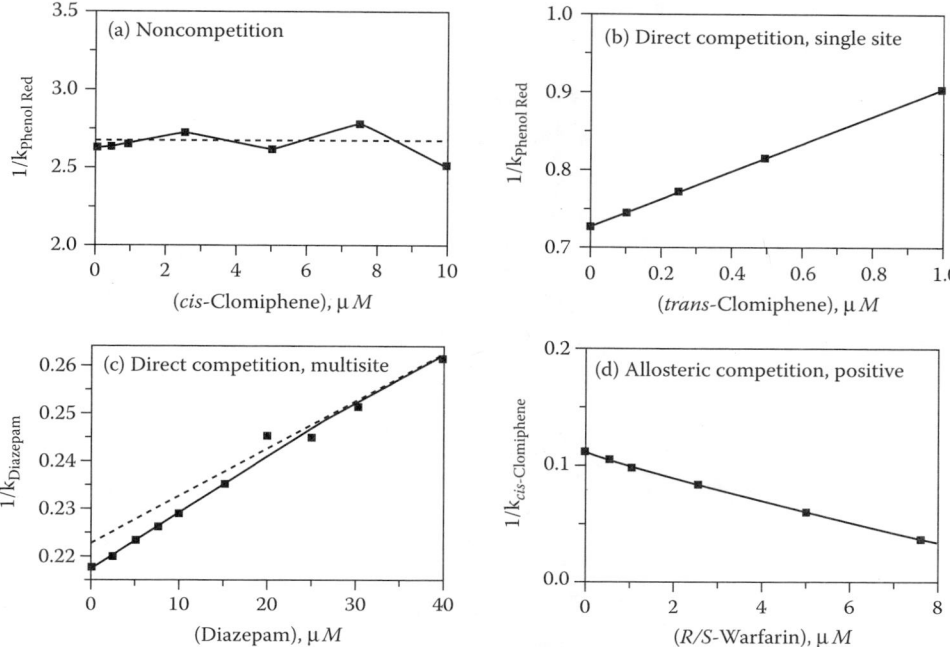

Figure 22.4 Reciprocal plots prepared for analyte and competing agents with various types of competition on immobilized HSA columns. (Reproduced with permission from Hage, D.S., *J. Chromatogr. B*, 768, 3–20, 2002.)

(typically 5 to 10%) when performed on an HPLC system. The use of slope/intercept ratios in generating such numbers has the added benefit of giving values that show little variation with changes in the ligand content. For instance, it has been shown that a long-term change in activity of 20% for an immobilized HSA support gave a random change of only ±2% in association constants measured by this strategy [7].

22.2.2.3 Solvent and Temperature Studies

A third way that zonal elution can be used is to consider how changes in the reaction conditions affect solute-ligand binding. For instance, this can be examined by varying the pH, ionic strength, or general content of the mobile phase. This is valuable in helping estimate the relative contributions of various forces to the formation and stabilization of a solute ligand. As one example, changing the pH can affect the interactions between a ligand and solute by changing their conformations, net charges, or coulombic interactions. An increase in ionic strength tends to decrease coulombic interactions through a shielding effect, but at the same time, this may cause an increase in nonpolar solute adsorption. Adjusting the solvent's polarity by adding a small amount of organic modifier can alter solute-ligand binding by disrupting nonpolar interactions or causing a change in their structures [17, 37].

Figure 22.5 shows the results of a solvent study in which changes in the pH, ionic strength, and organic modifier content of the mobile phase were used to examine the interactions of carbamazepine with HSA [38]. It was concluded in this case that nonpolar interactions were particularly important in this binding process, as indicated by the decrease in binding with the addition of propanol and its increase with phosphate concentration. The

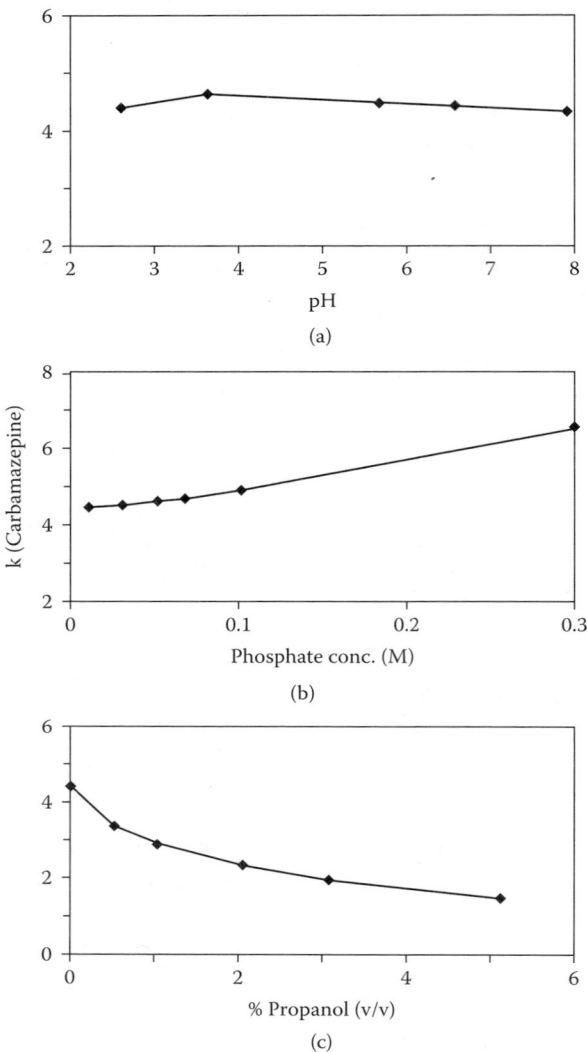

Figure 22.5 Change in retention with pH, ionic strength, and mobile-phase organic modifier content for the binding of carbamazepine to immobilized HSA. (Reproduced with permission from Kim, H. and Hage, D.S., *J. Chromatogr. B*, 816, 57–66, 2005.)

use of different buffer salts, chaotropic agents, and other additives have also been employed in such experiments.

Temperature is another factor that can be varied during zonal-elution studies. An example is given in Figure 22.6 for the binding of D- and L-tryptophan to HSA [14]. Besides providing qualitative data on the effects of temperature on binding, zonal elution can be used to determine thermodynamic constants. For instance, the following equation can be used for a system with 1:1 binding,

$$\ln k = -(\Delta H/R\ T) + \Delta S/R + \ln(m_L/V_M) \qquad (22.3)$$

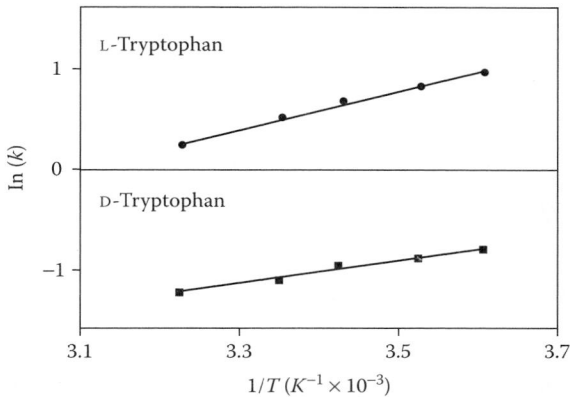

Figure 22.6 Effects of temperature on the binding of D- and L-tryptophan to HSA. (Reproduced with permission from Yang, J. and Hage, D.S., *J. Chromatogr. A*, 725, 273–285, 1996.)

where T is absolute temperature at which the retention factor is measured, R is the ideal gas law constant, ΔH is the change in enthalpy for the reaction, ΔS is the change in entropy, and other terms are as defined previously.

Equation 22.3 has frequently been employed in thermodynamic studies that involve zonal-elution experiments [29, 39–42] and in studies that consider how temperature alters the selectivity of protein columns [43, 44]. If it is known that there is no temperature dependence in the number of binding sites (m_L) for a ligand, the slope of a linear plot of ln k vs. $1/T$ can be used to determine the value of ΔH for a solute-ligand system [14, 29].

One precaution that needs to be exercised in interpreting the results of zonal-elution temperature and solvent studies is that the observed changes in retention may be due to variations in either the number of active sites on the ligand or in the degree of solute binding at these sites. For instance, it has been shown that changing the temperature, pH, or organic modifier content of the mobile phase can cause changes in the retention factors for L-tryptophan on immobilized HSA that are related to alterations in both the association constants and number of binding sites for this analyte. However, for D-tryptophan, the change in retention seen under identical conditions is related only to a change in its association constant; a similar observation has been made for both D- and L-tryptophan when varying the concentration of phosphate buffer in the mobile phase [6].

22.2.2.4   *Determining the Location and Structure of Binding Sites*

Yet another application of zonal elution in affinity chromatography has been its use in determining the location and structures of binding regions on a ligand. For instance, if it is known where one agent interacts with a ligand, competition studies with this agent can be used to determine if other compounds bind at the same site. This approach has been used in many reports, including work with HSA and other albumins in their binding to nonsteroidal anti-inflammatory drugs [45, 46], *R*- and *S*-ibuprofen [30], *cis*- and *trans*-clomiphene [35], digitoxin or acetyldigitoxin [27], and benzodiazepines [46]. It is even possible to generate maps that show the relationship between the binding regions on a ligand. An example is shown in Figure 22.7 based on competition studies performed between carbamazepine and various probes for the major and minor binding regions of HSA [38].

Another approach for learning about binding sites is to study how changes in the structure of a solute or ligand affect their interactions. This is the principle behind the use

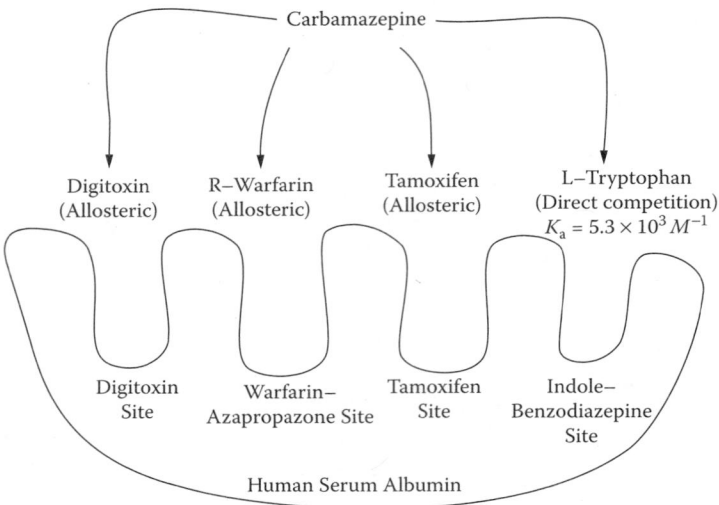

Figure 22.7 Interactions of carbamazepine with the major and minor binding regions of HSA. (Reproduced with permission from Kim, H. and Hage, D.S., *J. Chromatogr. B,* 816, 57–66, 2005.)

of zonal elution to develop a quantitative structure-retention relationship (QSRR) that describes the binding of drugs and their analogs to a protein column. This involves measuring retention factors for a large set of structurally related compounds under constant temperature and mobile-phase conditions. The resulting data are then compared with factors that describe various structural features of the solutes. This is done to determine which of these factors are most important in the retention and binding of the tested compounds. From these data, information can be obtained on the forces involved in drug-protein binding, and an approximate description can be developed for the sites involved in these interactions [23, 47].

A complementary approach to QSRRs is to use zonal elution to investigate how solute retention changes as alterations are made to its binding sites on a ligand. An example is work that involved the acetylation of HSA with *p*-nitrophenyl acetate. This reagent is thought to mainly modify the Tyr-411 residue of HSA, which is located at the indole-benzodiazepine site of this protein. This modification was shown to change the retention for a variety of solutes injected onto normal versus modified HSA columns [48]. A similar study used o-nitrophenylsulphenyl chloride to modify the lone tryptophan residue on HSA (Trp-214), which is located within the warfarin-azapropazone site of HSA [49]. This latter modification did not change the moles of binding sites, but it did result in a complete loss of HSA's stereoselectivity for *R*- and *S*-warfarin. As a result, it was concluded that Trp-214 or its neighboring residues played an important role in determining the chiral recognition of these compounds by HSA. Similar studies have modified the lone free cysteine residue on HSA with ethacrynic acid [50] or used BSA fragments for the chiral separation of benzoin and other drugs [50, 51].

22.2.3 Alternative Applications

Besides the applications already presented, there are a number of alternative uses for zonal elution that have been reported. This section examines some of these applications, including those involving normal-role affinity chromatography, experiments with low-solubility compounds, work with heterogeneous systems, and quantitative studies of allosteric interactions.

22.2.3.1   Normal-Role Affinity Chromatography

Up to this point, all the competition studies described for zonal elution have made use of a small, soluble agent that competes with the analyte for sites on an immobilized ligand. This format is referred to as reversed-role affinity chromatography. However, it is also possible to examine the competition between a soluble ligand and immobilized ligand for the analyte. This latter mode is known as normal-role affinity chromatography. Although not as common as the reversed-role format, normal-role affinity chromatography was the first of these techniques to be developed for use in quantitative studies of solute-ligand interactions [20, 33, 52].

One advantage of this method is that it provides information on the binding of a solute with both soluble and immobilized ligands. Thus, a direct comparison can be made between these two forms of the ligand, allowing immobilization effects to be avoided or evaluated. A recent example of this approach is work that has examined the binding between drugs and β-cyclodextrin using an immobilized HSA column (see Figure 22.8). The model and equation used to analyze the data in this study are given in Table 22.1. This approach made it possible to directly determine the stability constant for drug-cyclodextrin complexes in solution, giving results that were consistent with other solution-phase methods [33].

22.2.3.2   Work with Low-Solubility Compounds

One requirement of zonal elution is that enough analyte must be injected to give an observable peak. If a competing agent is used, this agent must also have a sufficient solubility in

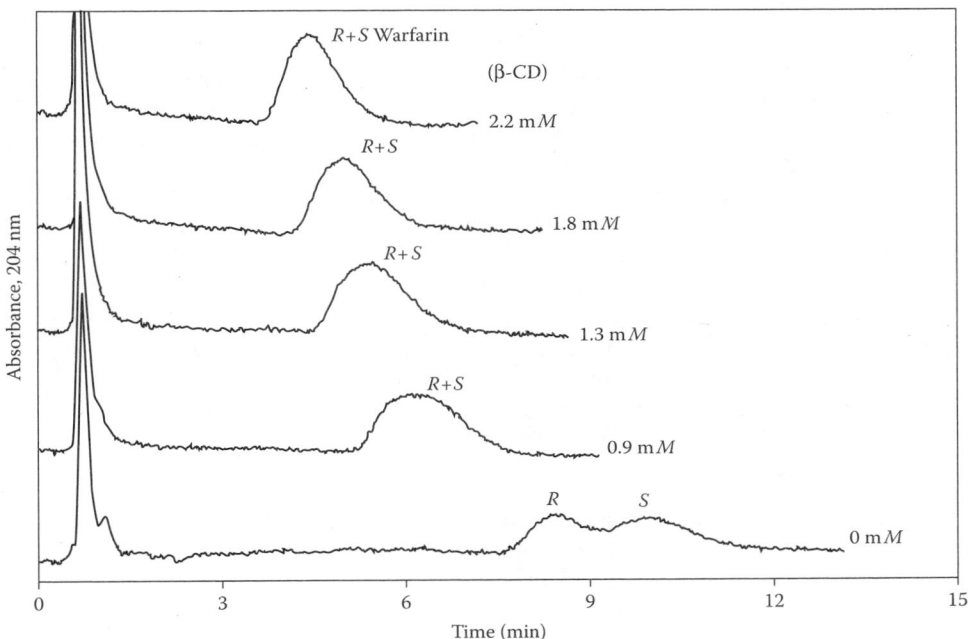

Figure 22.8 Chromatograms obtained in a zonal-elution study for the injection of warfarin into the presence of β-cyclodextrin (β-CD) as a mobile-phase additive. The mobile-phase concentrations of β-CD (from bottom to top) were 0, 0.9, 1.3, 1.8, and 2.2 mM. These studies were performed at 37°C in a pH 7.4, 0.067 M phosphate buffer. (Reproduced with permission from Chen, J., Ohnmacht, C.M., and Hage, D.S., *J. Chromatogr. A*, 1033, 115–126, 2004.)

the mobile phase to give a measurable shift in the analyte's peak. However, a problem can arise with one or both of these requirements when dealing with low-solubility compounds. Since the mobile phase for quantitative affinity chromatography is usually an aqueous buffer, one approach often used to increase solubility is to add an organic modifier. But this may also change the binding between the solute and ligand. A similar effect can occur when alterations are made in the pH, ionic strength, or temperature of the system.

Another approach that can sometimes be used to increase a compound's solubility is to use a secondary binding agent in the mobile phase. For example, cyclodextrins have been used for this purpose in zonal-elution studies involving a number of nonpolar drugs [13, 27, 35]. This was originally performed by making plots of $1/k$ vs. $[I]$ at several cyclodextrin concentrations. The intercepts and slopes obtained from these plots were then used to prepare a separate graph of the intercept/slope ratio vs. the concentration of cyclodextrin. In this final plot, the reciprocal of the intercept gave the association constant between the competing agent and ligand, while the ratio of the slope and intercept gave the stability constant for the competing agent with the cyclodextrin.

An alternative approach has recently been developed in which the stability constant between the competing agent and cyclodextrin is first determined by normal-role affinity chromatography, as discussed in the previous section. Next, zonal elution with the analyte is performed in the presence of various amounts of the competing agent, but using only one concentration of cyclodextrin. A plot is again made of $1/k$ vs. $[I]$, but now the intercept, slope, and previously measured stability constant are used together to give the association constant between the desired agent and ligand [34].

22.2.3.3 Studies Involving Binding Heterogeneity

Another group of recent studies have considered the effects of ligand heterogeneity in zonal-elution experiments. For instance, if the injected analyte has binding sites that are not affected by the presence of a mobile-phase additive, the competition between these agents can still be examined by plotting $1/(k - X)$ vs. $[I]$, where X is a constant that represents the analyte retention due to sites not involved in the competition of A with I [10].

In the case where the analyte and competing agent both have more than one binding site, the situation becomes more complicated. When the amount of these binding sites is known, the association constants for each site can be obtained using nonlinear regression. This can be accomplished by using the multisite equations given in Table 22.1. As an alternative, it may be possible to use other compounds to more selectively examine the interactions at these sites. For instance, this approach has been used to independently examine the binding of L-thyroxine and related compounds at the warfarin-azapropazone and indole-benzodiazepine sites of HSA [7].

22.2.3.4 Studies of Allosteric Interactions

It has recently been shown that quantitative information on allosteric interactions can also be obtained through zonal-elution experiments [36]. An example is shown in Figure 22.9, where S-lorazepam acetate was applied to an immobilized HSA column in the presence of R- or S-ibuprofen in the mobile phase. For both competing agents, a positive allosteric interaction was noted as the retention of S-lorazepam acetate increased with their mobile-phase concentrations. By analyzing these plots according to the equation given in Table 22.1 for allosteric systems, it was found that there was a 3- to 5.8-fold increase in the apparent binding constant for S-lorazepam acetate in the presence of these other drugs. A similar approach can be used for nonpolar compounds by using a cyclodextrin or other solubilizing agent in this system.

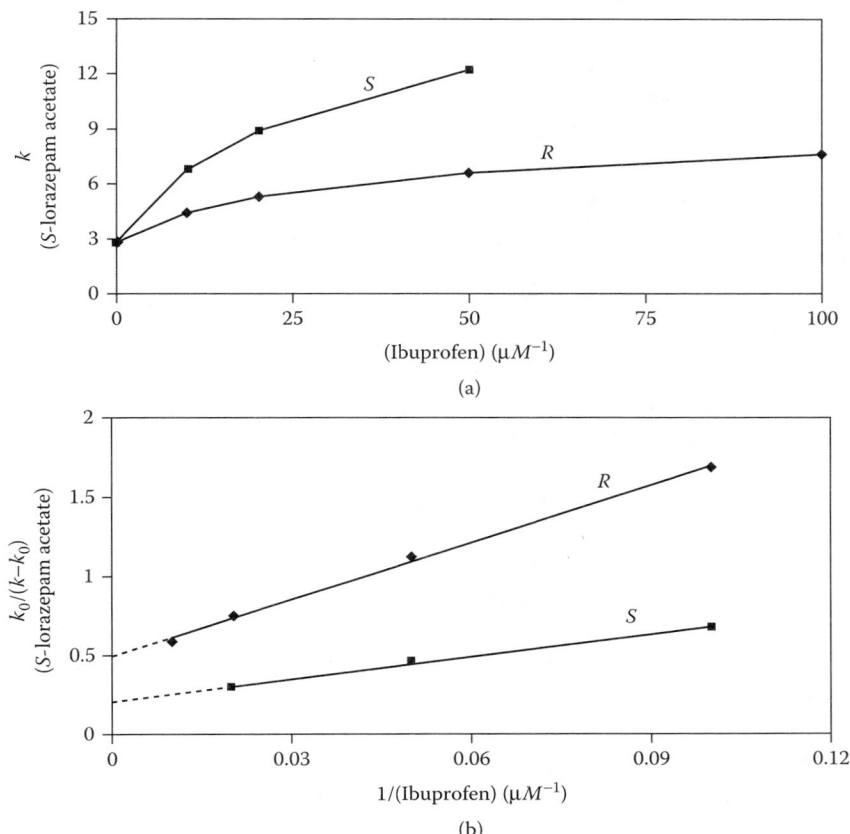

Figure 22.9 Retention of *S*-lorazepam acetate on an HSA column in the presence of *R*- or *S*-ibuprofen. (Based on data from Fitos, I., Visy, J., Simonyi, M., and Hermansson, J., *Chirality,* 11, 115–120, 1999.)

22.2.4 Practical Considerations

Although zonal elution is relatively easy to perform, the proper use of this method does require that several experimental factors be considered. Examples include the choice of affinity column, the need for an accurate measurement of analyte retention, the selection of mobile-phase conditions, and the use of an appropriate sample size. These and other items are discussed in the following subsections.

22.2.4.1 Choice of Affinity Column

When using affinity chromatography for zonal elution, the amount and type of ligand in the column should be carefully considered. For a solute with weak binding, a high-capacity column is preferred to increase retention and give more easily detected peak shifts. However, for compounds with high retention, a short or low-capacity column may be preferred to provide more reasonable analysis times.

If the zonal-elution experiments are based on an immobilized ligand, this ligand should be immobilized under mild conditions to help retain its activity. It is also wise to first characterize this ligand with compounds that have known binding properties before it is used to study the binding of other substances. These items are less crucial when using normal-role affinity chromatography, since the role of the immobilized ligand in this format is to compete with a soluble ligand in binding to the analyte.

The nature of the chromatographic support should be considered as well. For some supports, nonspecific binding may be appreciable and account for a significant part of the analyte's retention. This can be examined by using a control column to detect and correct for such binding. This control column should be made in the same way as the affinity column, but with no ligand being present. If nonspecific interactions are detected, an adjustment can be made for this by using the techniques mentioned earlier for heterogeneous binding systems.

Whenever zonal elution is performed, such items as the length and diameter of the column, the type of support (including pore size and particle size), and type of ligand should be reported. Ideally, the total ligand content of the column should also be provided, along with the amount of ligand that was active. Such information is useful in comparing the results of different studies and in selecting conditions for a zonal-elution experiment or troubleshooting such a system. Yet another experimental factor that should be reported and monitored is the pressure across the column. Although very high pressures are needed to affect protein structure, recent studies have shown that at least some solute-ligand systems do have slight changes in binding near the upper-pressure limits of HPLC systems [54]. To test for such effects, initial zonal-elution studies should be performed at several flow rates and back-pressures to identify conditions in which minimal variations in retention are produced.

22.2.4.2   Choice of Additive Concentrations

When examining binding constants or changes in retention due to a mobile-phase additive, it is necessary to have a shift in analyte retention that can be measured. The conditions required for this can be selected by considering how the retention factor k changes as it moves between its maximum and minimum values, k_{max} and k_{min}. In a system where the injected analyte and additive have direct competition at a single type of site, the following relationship describes the relative shift in retention, $(k - k_{min})/(k_{max} - k_{min})$, that will be observed at various additive concentrations [55].

$$\frac{k - k_{min}}{k_{max} - k_{min}} = \frac{1}{1 + K_I[I]} \qquad (22.4)$$

According to Equation 22.4, the size of this shift depends on the association constant for the competing agent (K_I) and this agent's concentration in the mobile phase ($[I]$). Another way to view this ratio is as the fraction of binding sites that remain unoccupied and able to bind to the analyte as it contacts the stationary phase. Based on this model, the analyte's retention factor should equal k_{max} when the concentration of competing agent is zero, with the retention factor approaching k_{min} as $[I]$ approaches infinity.

Equation 22.4 shows that a minimum shift in retention of 10% is produced when the term $K_I[I]$ is between 0.1 and 9.0. This provides the range of competing-agent concentrations that are required for zonal-elution studies at this level of precision. For a minimum shift of 5%, competing agents with concentrations that give $K_I[I]$ values of 0.05 to 19 should be used. This indicates that more precise retention measurements will also allow a greater range of competing-agent concentrations to be employed in the study.

Although Equation 22.4 provides the range of competing-agent concentrations that should ideally be used for a particular solute-ligand system, other factors may prevent this full range from being tested. For instance, the competing agent may have a limited solubility in the mobile phase, or it may have a high background signal that prevents its use at high concentrations. In some cases it is possible to switch to a different competing agent with a lower background or better solubility. Another approach is to increase solubility by adding an organic modifier or changing the pH of the mobile phase, but this can also affect binding between the immobilized ligand and analyte or competing agent. A third alternative is to employ a solubilizing agent, as discussed earlier in this chapter [27, 35]. One disadvantage of using a solubilizing agent is that it complicates the data analysis, since both the concentration of the solubilizing agent and competing agent must now be controlled or varied. Also, the binding of an analyte to a solubilizing agent produces lower retention, which decreases the range of retention factors that can be used during the zonal-elution experiments [27].

Since many zonal-elution studies involve the use of several mobile phases, it is crucial to ensure that the chromatographic system be equilibrated with each new solvent before any retention measurements are made. For this reason, it is recommended that several injections be made under each set of conditions to ensure that reproducible results are obtained. When changing the pH, buffer composition, or organic modifier content of the mobile phase, it is usually sufficient to wash the column with 10 to 20 void volumes of the new solvent. However, an even larger volume may be needed when altering the competing agent's concentration.

22.2.4.3   *Selection of Sample Size*

Although it is possible to work under nonlinear conditions through the use of computer modeling and nonlinear regression [24, 25], most zonal-elution studies assume that the amount of injected analyte is small compared with the amount of active ligand in the column. This creates a problem with columns containing large ligands (e.g., proteins), because they usually have a much lower capacity than traditional columns, such as those used for reversed-phase chromatography. Fortunately, it is easy to test for linear elution conditions by injecting the analyte at several concentrations and seeing if the measured retention factor is a consistent value.

An example of such an experiment is shown in Figure 22.10 for the injection of D- and L-tryptophan on an HSA column [14]. As can be seen in this figure, the sample-size

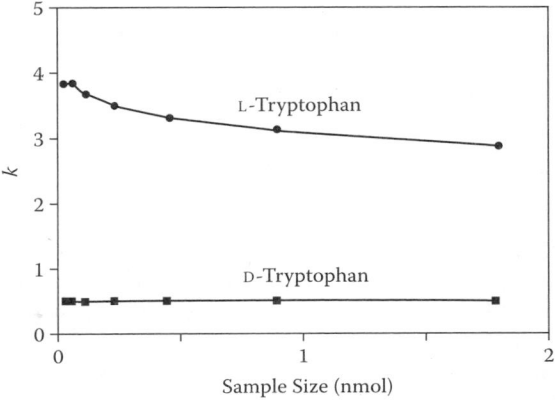

Figure 22.10   Effect of sample size on the retention factors measured for D- and L-tryptophan on an immobilized HSA column. The column size was 10 cm × 4.1 mm I.D. and contained 500 nmol HSA/g silica. (Reproduced with permission from Yang, J. and Hage, D.S., *J. Chromatogr.*, 645, 241–250, 1993.)

dependence of the retention factor can vary significantly between two different solutes on the same column. Thus, this needs to be examined on a case-by-case basis. However, even solutes with large sample-size effects tend to give satisfactory results with sample loads of 0.01 to 0.2% versus the total column binding capacity.

Two other factors to consider when selecting sample concentrations for zonal-elution studies are the solubility and detectability of the analyte. Solubility places an upper limit on the amount of analyte that can be applied with each injection. For most compounds, work in the micromolar range, as is often needed to obtain linear elution conditions, provides sufficient solubility for injection. However, this can be difficult to accomplish when working with nonpolar compounds that are only sparingly soluble in aqueous buffer. One possibility is to use a small amount of organic modifier in the sample solvent. But caution must be followed in doing this, since it may create background peaks that make it difficult to examine solute retention or cause shifts in retention due to differences between the mobile phase and injection solvent. Another possibility is to add a solubilizing agent to the injection solvent or mobile phase. Although this greatly expands the types of compounds that can be examined in zonal-elution experiments, it does add an extra variable to the study (i.e., the solubilizing agent's concentration).

The detectability of the analyte depends on the detection scheme being employed, the properties of the analyte, and the background signal due to the mobile phase. Detection in zonal-elution studies is most often accomplished by absorbance detection, but other approaches like liquid chromatography-mass spectrometry (LC/MS) have been employed [56]. Sample concentrations in the micromolar range are compatible with most of these methods. If an additive is in the mobile phase, which also gives a response to the detector, an increase in background will be seen as greater amounts of this additive are used. This may result in greater noise and give a lower signal-to-noise ratio for the analyte, requiring the use of higher sample concentrations for detection. Also, a change in the mobile phase's composition (e.g., its pH or organic modifier content) may increase or decrease the response for the analyte by changing its physicochemical properties. Thus, detectability of the analyte should be evaluated in each zonal-elution study by using the full range of mobile phases that are to be examined.

If the analyte cannot be placed onto the column at a level suitable for detection, it might be possible to use a labeled analog that can be monitored at lower concentrations than the analyte. An example is a work in which radiolabels were used in zonal-elution studies to examine the binding of various solutes to low-performance albumin columns [57]. Alternative labels, such as those based on fluorescent tags, might also be employed. However, precautions must be taken in this approach to ensure that the labeled analogs are mimicking the behavior expected for the analyte of interest.

22.2.4.4   *Determination of Analyte Retention*

An essential requirement for any zonal-elution experiment is that the analyte must have reproducible and observable retention. As pointed out earlier, this retention will be related to the association constants for solute-ligand binding and the moles of binding sites within the column. Although the amount of ligand in an affinity column may only be in the nmol-mol range, the relatively strong binding that occurs in many solute-ligand systems can result in high retention factors during zonal-elution experiments.

Having large retention factors makes it easy to detect small shifts in retention, but it also leads to long analysis times. One way to reduce the retention time for a strongly retained compound is to increase the flow rate; however, care must be taken to ensure that this does not affect the solute's retention factor. Another option is to change the ligand content or size of the column, with a reduction in either item leading to a proportional decrease in retention. Mobile-phase additives can also be used to adjust retention, but caution must be taken with this option to avoid altering the nature of the solute-ligand interaction being studied.

One assumption made in most zonal-elution studies is that the center of an analyte's peak represents a point of local equilibrium within the column. This assumption should be tested for each new analyte by performing a few zonal-elution studies at several flow rates. If the assumption is not valid, a shift in retention factor will result. This indicates that a slower flow rate or a longer column should be used.

To obtain an accurate measure of retention, it is recommended that the true center of a solute's peak be employed. It is important to note that this is not the same as the peak maximum, since many solutes on affinity columns produce tailing peaks. This tailing occurs even under linear elution conditions and is a result of relatively slow association and dissociation kinetics. For chromatograms that are collected by a computer, the position of an analyte's true retention time can be determined by calculating the central moment of its peak [3]. During this process, it is recommended that a correction first be made for any shifts in the baseline in the region of the peak of interest. An alternative approach is to use one of several empirical expressions that allow t_R to be calculated from a peak's elution time at its maximum and other factors, such as its width at half-height and its asymmetry factor [58, 59].

22.2.4.5   Other Considerations

There are a number of additional items to consider during zonal-elution studies. For instance, it is essential to control and report the pH and composition of the mobile phase and samples. This includes the type of buffer and its concentration as well as any additives that are present, since even small changes in these factors can alter solute-ligand binding. In addition, the temperature should be controlled by a column heater or a water jacket attached to a circulating water bath. A water jacket is highly recommended for such studies due its accuracy and greater flexibility in controlling the temperature.

22.3   FRONTAL ANALYSIS

The second-most-common method in affinity chromatography for studying solute-ligand binding is frontal analysis. This differs from zonal elution in that it involves the continuous, rather than discrete, application of an analyte to a column. The result is essentially a titration of the active sites within the column [1, 3]. Figure 22.2b shows a typical system used for such an experiment when performed by HPLC.

Frontal analysis was first employed in the investigation of solute-ligand binding in 1975 by Kasai and Ishii, who used this method with low-performance affinity columns [60]. In 1978, Nakano et al. [61] also used this technique with low-performance columns to study the binding of immobilized BSA with salicylate, and Lagercrantz and coworkers [57] employed a similar method in 1979 to examine the binding of salicylate with HSA. In 1992 Loun and Hage [7] reported the use of frontal analysis and HPLC as a means to characterize the binding of immobilized HSA for various solutes. The following section discusses the underlying principles of frontal analysis and various applications of this method. The general theory of this technique is given, and a number of practical factors are provided that need to be considered when using this approach.

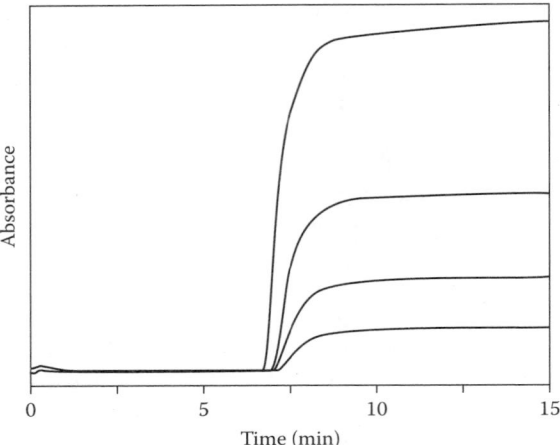

Figure 22.11 Typical frontal-analysis experiments performed with L-tryptophan on an immobilized HSA column. The concentrations of applied L-tryptophan (from left to right) were 100, 50, 25, and 12.5 μM. The flow rate was 0.25 ml/min, and the column void time was 3.6 min. (Reproduced with permission from Yang, J. and Hage, D.S., *J. Chromatogr. A*, 725, 273–285, 1996.)

22.3.1 General Basis of Method

In frontal analysis, a solution containing a known concentration of solute is continuously applied to a column with an immobilized ligand. As the solute binds to this ligand, the column becomes saturated, and the amount of solute eluting from the column gradually increases, forming a characteristic breakthrough curve. An example of such a curve is shown in Figure 22.11. If fast association and dissociation kinetics are present in the system, the mean position of the breakthrough curve can be related to the concentration of applied solute, the amount of ligand in the column, and the association constants for solute-ligand binding. As will be shown later, this provides a means for quantitating the active ligand in the column and the affinity of this ligand for the applied solute.

Like zonal elution, frontal analysis can be performed with either low-performance or HPLC columns. Although frontal analysis does require more analyte than zonal elution, it provides more information per experiment. Its main advantage is its ability to separately measure both the association constants and number of binding sites within a column. This makes this approach valuable in characterizing the properties of a column and in obtaining careful measurements of binding affinity and activity.

22.3.2 Applications

Like zonal elution, frontal analysis can be used to provide information regarding many aspects of a solute-ligand system. Applications that have been reported include the use of this method to determine the affinity and number of binding sites for a solute in a column, the nature of this binding (e.g., single site or multisite), the effects of temperature or solvent on this binding, and the changes that occur in the presence of a competing agent.

22.3.2.1 Measurement of Affinity and Number of Binding Sites

The main application of frontal analysis has been to provide quantitative data on the affinity and amount of ligand in a column. This is accomplished by measuring the breakthrough times for a solute at several concentrations and fitting the results to equations based on

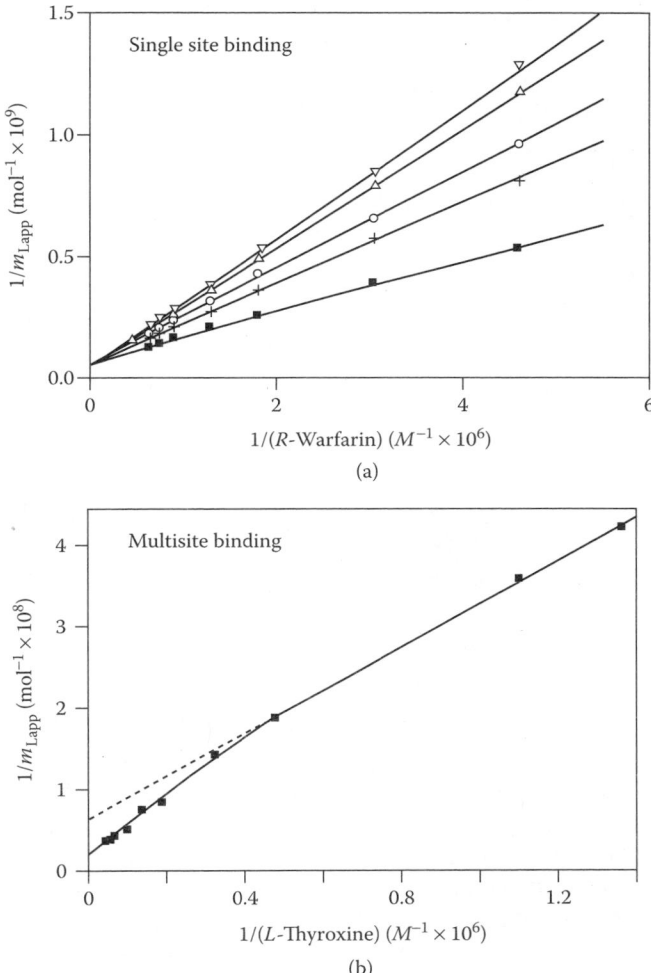

Figure 22.12 Examples of double-reciprocal frontal analysis plots for systems with (a) single-site binding and (b) multisite binding. (Reproduced with permission from Loun, B. and Hage, D.S., *Anal. Chem.*, 66, 3814–3822, 1994; and Tweed, S.A., Loun, B., and Hage, D.S., *Anal. Chem.*, 69, 4790–4798, 1997.)

various reaction models. Examples of such fits are shown in Figure 22.12. Double-reciprocal plots are particularly useful for this purpose, where $1/m_{Lapp}$ (i.e., the apparent moles of analyte required to saturate the column) is plotted vs. $1/[A]$ (i.e., the inverse of the applied analyte's concentration). According to the equations given in Table 22.2, the result for this plot should be a linear relationship if the analyte has a single type of binding site in the column. If more than one type of binding site is present, negative deviations will be seen at high analyte levels (i.e., low values for $1/[A]$) [62, 63].

Once it has been determined which reaction model describes the solute-ligand system, the affinity and number of binding sites in the column can be determined from the best-fit parameters for the experimental data. For instance, the equations in Table 22.2 for a system with single-site binding predict that a plot of $1/m_{Lapp}$ versus $1/[A]$ will give a linear response

Table 22.2   Models and Equations Used in Frontal Analysis

Type of System [References]	Model	Predicted Response
Binding of analyte (A) to a single type of ligand site (L) [7]	$A + L \underset{}{\overset{K_A}{\longleftrightarrow}} A - L$	$m_{\text{Lapp}} = \dfrac{m_L K_A [A]}{1 + K_A [A]}$ or $\dfrac{1}{m_{\text{Lapp}}} = \dfrac{1}{K_A m_L [A]} + \dfrac{1}{m_L}$
Binding of analyte (A) at two types of sites (L_1 and L_2) [63]	$A + L_1 \underset{}{\overset{K_{A_1}}{\longleftrightarrow}} A - L_1$ $A + L_2 \underset{}{\overset{K_{A_2}}{\longleftrightarrow}} A - L_2$	$m_{\text{Lapp}} = \dfrac{m_{L_1} K_{A_1}}{1 + K_{A_1}[A]} + \dfrac{m_{L_1} K_{A_2}}{1 + K_{A_2}[A]}$ or $\dfrac{1}{m_{\text{Lapp}}} = \dfrac{1 + K_{A_1}[A] + \beta_2 K_{A_1}[A] + \beta_2 K_{A_1}{}^2 [A]^2}{m_L \{ (\alpha_1 + \beta_2 - \alpha_1 \beta_2) K_{A_1}[A] + \beta_2 K_{A_1}{}^2 [A]^2 \}}$ where $\alpha_1 = m_{L_1}/m_{L_2}$ and $\beta_2 = K_{A_2}/K_{A_1}$

with a slope equal to $1/(K_A \, m_L)$ and an intercept of $1/m_L$. This means the total binding capacity of the column can be obtained from the inverse of the intercept, while the ligand's association constant can be determined by dividing the intercept by the slope. A similar approach, but using a combination of both nonlinear and linear fits, can be used for more complex systems that involve multisite interactions [62, 63]. If desired, equivalent information can be obtained by analyzing the data with Scatchard plots [57, 61, 63, 64].

The main advantage of frontal analysis over zonal elution is that it can simultaneously provide information on both the association constant for a solute and its number of binding sites. This makes frontal analysis the method of choice when information is needed on the binding capacity. Frontal analysis is also preferred for accurate measurements of association constants, since the values that it provides for K_A can be determined independent from the binding capacity [16].

22.3.2.2   Competition and Displacement Studies

A second application of frontal analysis has been as a tool to examine the competition between solutes for an immobilized ligand. This has been used to examine the competition of sulfamethizole with salicylic acid for HSA [64], and salicylate with clofibric acid, octanoic acid, or estradiol for sites on BSA [57]. This experiment is performed in a similar manner to that described for zonal elution, in which the change in analyte retention is measured as a function of the competing agent's concentration in the mobile phase. In frontal analysis, direct competition between the analyte and competing agent leads to a smaller breakthrough time for the analyte as the competing agent's level is increased. Positive or negative allosteric effects can also be observed, which lead to a shift to higher or lower breakthrough times, respectively, with an increase in the competing agent's concentration.

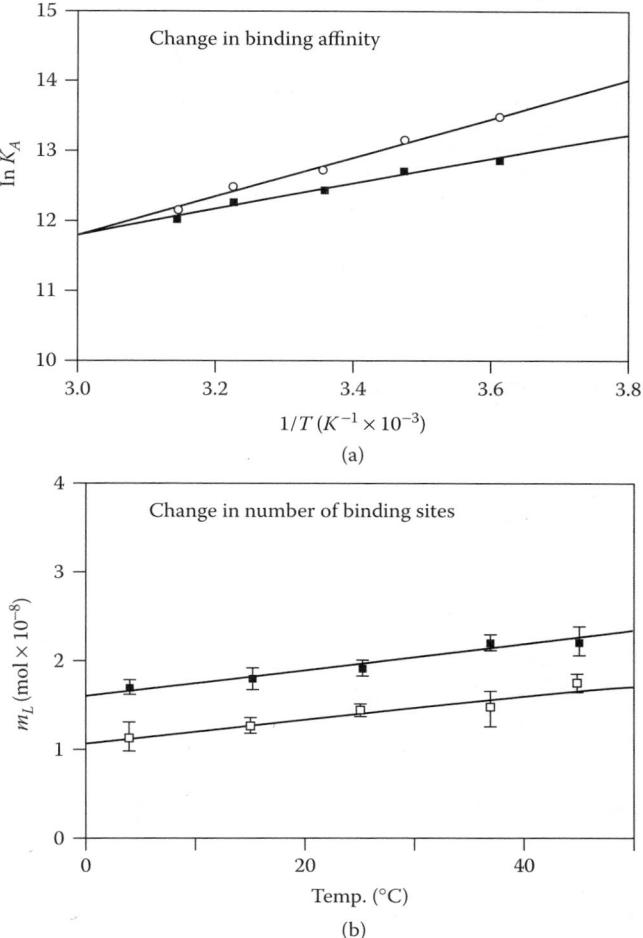

Figure 22.13 Use of frontal analysis to examine the change in affinity and number of binding sites for *R*-warfarin (■) and *S*-warfarin (○, □) on an immobilized HSA column as a function of temperature. (Reproduced with permission from Loun, B. and Hage, D.S., *Anal. Chem.*, 66, 3814–3822, 1994.)

22.3.2.3 Solvent and Temperature Studies

It was mentioned for zonal elution that caution must always be used in interpreting solvent and temperature studies performed by this approach, since the observed retention shifts may be due to alterations in either the affinity or number of binding sites. But this is not an issue in frontal analysis, since data on both affinity and activity are provided in the same experiment. For instance, Figure 22.13 shows how frontal analysis was used to measure m_L and K_A for *R*- and *S*-warfarin at various temperatures on an immobilized HSA column. Based on this information, it was possible to predict how these analytes would behave during zonal elution. Frontal analysis has also been used to examine the binding and separation of D- and L-tryptophan on immobilized HSA under a variety of temperature and mobile-phase conditions, many of which affected both the strength and number of binding sites for these agents [14].

22.3.2.4   Studies with Modified Ligands

Like zonal elution, frontal analysis has been used to examine the binding of solutes to modified proteins. In one case, frontal analysis was used to compare the binding capacities for monomeric versus dimeric HSA in their interactions with various solutes [64]. In addition, this approach has been used to study HSA that had been modified with o-nitrophenylsul-phenyl chloride. In this study, it was determined that this modification did not change the amount of warfarin-azapropazone sites on HSA, but it did lower the association constant of this protein for *R*-warfarin. Similar studies with L-tryptophan indicated that the number of moles of indole-benzodiazepine sites was not affected; however, an allosteric decrease in affinity was detected on the modified HSA [49].

22.3.3   Practical Considerations

As was true for zonal elution, frontal analysis has a number of factors that need to be considered and optimized for the proper use of this method. Some of these items include the choice of affinity column, the selection of analyte concentrations, and correct determination of breakthrough times.

22.3.3.1   Choice of Affinity Column

The same general factors concerning the selection of a column that were noted in zonal elution are also important in frontal analysis. For instance, the size of the column and the amount of ligand in this column can be increased or decreased to adjust the breakthrough time for an applied solute. The ligand should again be immobilized in a manner that will allow it to retain its activity. And an evaluation should be made of the nonspecific binding that occurs between the analyte and support in this column.

This last item can be examined by using a control column of the same size and containing the same support material as the affinity column but with no affinity ligand being present. If significant binding is seen for the analyte on this control column, it is sometimes possible to correct for this by subtracting the breakthrough times of the control column from the affinity column, assuming that the nonspecific binding in these is the same. Alternatively, the frontal-analysis data can be analyzed using equations that allow for heterogeneous binding sites, as given in Table 22.2.

22.3.3.2   Choice of Analyte Concentrations

The main variable in frontal analysis is the concentration of analyte applied to the column. A key factor in choosing this concentration is the size of the analyte's association constant. This is demonstrated in Equation 22.5, where the fraction of active column sites that are bound to the analyte (m_{Lapp}/m_L) is dependent only on the applied analyte's concentration and the association constant for the analyte at these sites.

$$m_{Lapp}/m_L = K_A[A]/(1 + K_A[A]) \tag{22.5}$$

As indicated by Equation 22.5, the value of m_{Lapp}/m_L will equal zero when no analyte is applied and will approach 1 as the analyte concentration approaches infinity. Between these two extremes is a range of intermediate concentrations in which the ratio of m_{Lapp}/m_L gives a measurable shift in the mean position of the breakthrough curve.

For a minimum shift of ±10%, the analyte concentrations that will be needed to produce this shift must provide a product $K_A[A]$ that is equal to 0.1 to 10 (i.e., [A] is equal to 0.1 to 10 times $1/K_A$). Similarly, a minimum shift of 5% in the breakthrough curve will require that $K_A[A]$ be in the range of 0.05 to 20.

Detectability and solubility are other issues to consider when choosing analyte conditions for frontal analysis. In this case, the applied analyte solutions should provide a measurable increase in signal versus the baseline, and the maximum change in signal should have a linear dependence on analyte concentration. Whether this is the actual case for a given solute can be determined by passing the selected analyte solutions through the chromatographic system when no column is present, with the resulting signal then being examined to see if it is both measurable and proportional to the analyte's concentration. If analyte solubility is an issue, this can be improved by using the various methods described earlier for zonal elution.

Once an appropriate range of analyte concentrations has been identified, it is necessary to select other conditions that will help make shifts in binding easier to observe. Factors that can be adjusted for this include the size of the column, the amount of ligand in the column, and the application flow rate. This can be demonstrated with a modified version of Equation 22.5 (as shown in Equation 22.6), which shows how the difference in time between the mean position of the breakthrough curve (t_{mean}) and the column void time will vary with flow rate (F), column size, and amount of ligand (m_L).

$$t_{mean} - t_M = \frac{m_L K_A[A]}{F[A](1 + K_A[A])} \tag{22.6}$$

As can be seen from this equation, the difference ($t_{mean} - t_M$) will increase by using a column with a greater amount of immobilized ligand or by using a slower flow rate for analyte application. The disadvantage of this is that it will increase the analysis time and require more ligand and analyte for each experiment.

22.3.3.3 Determination of Breakthrough Times

The determination of binding capacities and association constants by frontal analysis requires that careful measurements be made of the analyte's average breakthrough time or volume. For a symmetric breakthrough curve, this will be equal to the point halfway between the baseline and upper plateau. However, most of these curves are not symmetric, so an alternative way of determining this mean position is needed. One approach is to integrate below the front portion of the curve and above the latter part until a point is reached at which these two areas are equal. This is the equivalent of converting the frontal-analysis data into a step function, where the mean breakthrough time will be the point at which the step function changes its value. The same goal can be reached by taking the first derivative of the breakthrough curve and determining the central moment of this derivative.

When using the equations in Table 22.2 for measuring association constants, it is assumed that the mean point of the breakthrough curve represents a point of equilibrium between the analyte and ligand. The validity of this assumption can be tested by performing studies at several flow rates and seeing if consistent breakthrough volumes are obtained. If a shift in this volume is noted, a lower concentration of the analyte, a higher ligand capacity column, or a slower flow rate is needed to allow a local equilibrium to be achieved within the column. The same study can be employed to determine whether the breakthrough volume is independent of the pressure across the column.

When acquiring data for a frontal-analysis experiment, it is best to have a well-defined difference between the void time of the column and beginning of the breakthrough curve. It is also necessary to ensure that the analyte is applied for a sufficient amount of time to reach and form the upper plateau of the curve. This often involves applying the analyte solution well past the mean breakthrough time to achieve this result.

22.3.3.4   Other Considerations

Additional factors that should be reported and considered in frontal analysis include the pH and composition of the mobile phase and the temperature used during the study. The flow rates and the back-pressures that were present should also be given. And finally, the size of the column, the support it contains, the column's ligand content, and the means by which this ligand was immobilized should be described. As mentioned earlier, this information is useful in comparing data from different studies and in troubleshooting affinity chromatographic experiments.

22.4   KINETIC METHODS

The methods described up to this point have focused on solute-ligand interactions at a local equilibrium. However, affinity chromatography can also be used to study the kinetics of these processes. Examples of methods that allow this information to be obtained are band-broadening studies, the peak decay method, and free fraction measurements.

22.4.1   Band-Broadening Studies

The earliest chromatographic approach used for kinetic studies was the band-broadening method. This makes use of the peak broadening that occurs during zonal-elution experiments. The method is performed by injecting the analyte at several flow rates while determining its corresponding peak widths and plate-height values. This information is then used to prepare a van Deemter-type plot of plate height vs. flow rate or linear velocity. These studies are performed both on the affinity column and on an identical inert control column that contains the same support but no immobilized ligand. Work on this second column is used to determine the plate-height contributions due to processes other than analyte–stationary-phase interactions. Finally, the plate-height contribution due to the stationary-phase interactions are calculated by taking the difference between the total plate height measured for the affinity column and the plate height estimated for the nonstationary-phase processes.

Once the plate-height contribution due to the stationary-phase interaction has been determined, this can be related to the kinetics of analyte dissociation from the ligand, as shown by Equation 22.7.

$$H_s = \frac{2uk}{k_d(1+k)^2} \tag{22.7}$$

In this equation, u is the linear velocity of mobile phase in the column, k is the retention factor of the injected solute, H_s is the plate height due to the stationary-phase interaction, and k_d is the dissociation rate constant between the analyte and immobilized ligand. Based on Equation 22.7, a plot of H_s vs. $uk/(1 + k)^2$ should give a slope of $2/k_d$ and an intercept of zero. By using the k_d values obtained from this plot along with independent estimates of the equilibrium constants for the system, the association rate constants for the solute and ligand can also be obtained [5, 6].

An example of a band-broadening study is shown in Figure 22.14 for D-tryptophan injected onto an immobilized HSA column [6]. As can be seen from this figure, this system gave good agreement with the behavior predicted by Equation 22.7. Similar studies have been performed with R- or S-warfarin [5] and L-tryptophan [6] on immobilized HSA columns. From these experiments, it has been possible to obtain the association and dissociation rate constants for these solutes on their affinity columns.

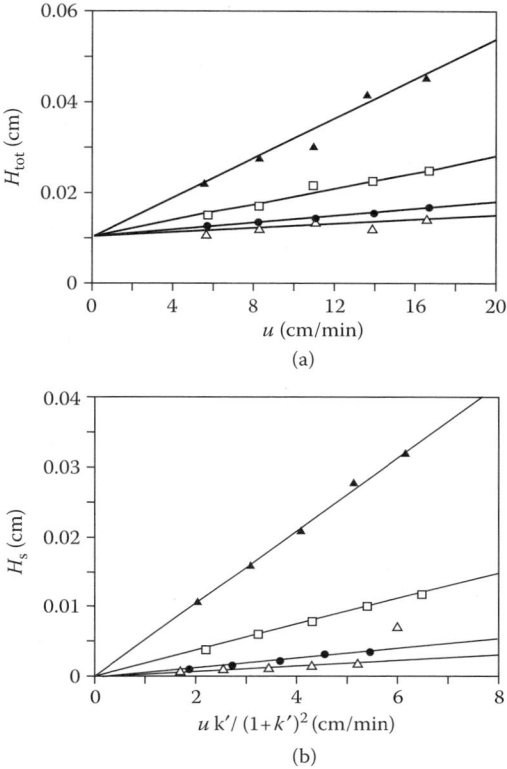

Figure 22.14 (a) Total plate height (H_{tot}) vs. linear velocity (u) for the injection of D-tryptophan onto an immobilized HSA column, and (b) the plate-height contribution due to stationary-phase mass transfer (H_s) versus $[uk/(1 + k)^2]$ for the same system. (Reproduced with permission from Yang, J. and Hage, D.S., *J. Chromatogr. A*, 766, 15–25, 1997.)

The resulting data have provided useful insights into the energetics of these binding processes and on how they are affected by changes in temperature or solvent. For instance, by combining the results of frontal analysis and band-broadening measurements for the interactions of D- and L-tryptophan with HSA, it was possible to show that the change in binding with pH for these solutes was the result of alterations in both their association and dissociation rates. The same types of experiments have been used to examine how these rates are affected by changes in the buffer composition, temperature, and organic-modifier content of the mobile phase [5, 6].

22.4.2 Peak Decay Analysis

Another way to determine dissociation rate constants for a solute-ligand system is peak decay analysis. In this technique, the affinity column is first saturated with the analyte. The excess analyte is then washed away, and the release of the adsorbed analyte is observed over time. If this is performed under conditions that minimize rebinding of solute to the column, the release of adsorbed analyte should follow a first-order decay curve, as shown in Figure 22.15.

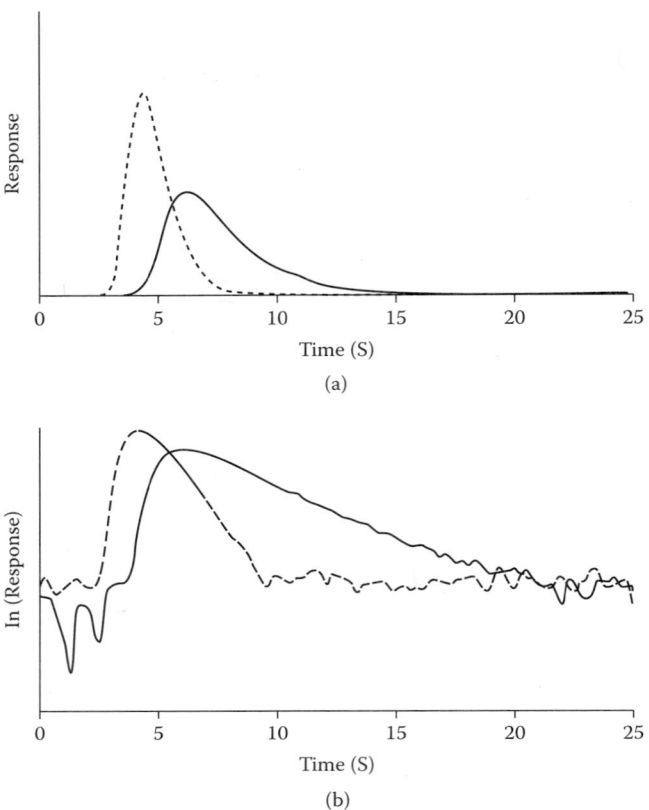

Figure 22.15 Peak decay experiments performed with *R*-warfarin on an inert control column (dashed lines) and a column containing immobilized HSA (solid lines). The plots in (a) are the original chromatograms obtained for the injection of a warfarin sample onto columns measuring 2.5 mm × 2.1 mm I.D., and the results in (b) show these same data when plotted on a logarithmic scale. (Reproduced with permission from Chen, J. and Hage, D.S., *J. Chromatogr. A*, submitted.)

This approach was originally developed in 1987 for the study of systems with essentially irreversible dissociation [65]. The method has been modified since then to include the presence of weak to moderate interactions between the analyte and immobilized ligand [66]. When reassociation of the analyte with the ligand is negligible and analyte dissociation is the rate-limiting step in its release from the column, this leads to the following relationship for the observed response [65],

$$\ln(dm_{Ae}/dt) = \ln(k_d\, m_{A0}) - k_d\, t \qquad (22.8)$$

where m_{A0} is the initial moles of analyte bound to the ligand, m_{Ae} is the moles of analyte in the flowing mobile phase at time t during the elution process, and k_d is the dissociation rate constant for the analyte from the ligand. This equation predicts a linear relationship between the natural logarithm of the detector response, $\ln(dm_{Ae}/dt)$, and time. Furthermore, from the slope of this curve it is possible to obtain the dissociation rate constant for the system. Examples of such plots are also given in Figure 22.15.

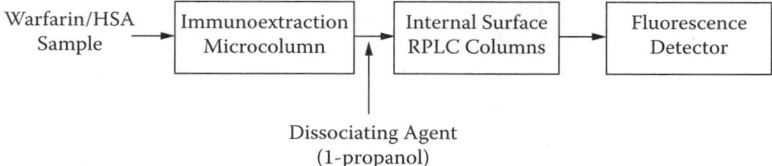

Figure 22.16 System used to determine the free fraction of warfarin in warfarin/HSA mixtures by using a sandwich microcolumn containing antiwarfarin antibodies. The dissociating agent is used to promote the release of HSA-bound warfarin after this has passed through the immunoextraction column. The released warfarin is then separated from HSA by using internal-surface reversed-phase liquid chromatography (RPLC) columns. (Reproduced with permission from Clarke, W., Chowdhuri, A.R., and Hage, D.S., *Anal. Chem.*, 73, 2157–2164, 2001.)

22.4.3 Free-Fraction Analysis

The final method that will be discussed for kinetic studies is free-fraction analysis. This technique uses small columns with antibodies that bind the solute of interest and are capable of extracting this solute in very short periods of time (i.e., 80 to 200 msec). With such a column, it is possible to isolate the nonbound fraction of a solute from a solution in which most of this compound is bound to a soluble ligand, even when dissociation of the solute from this ligand occurs on the time scale of a few seconds.

This approach has already been demonstrated by using the binding of *R*- and *S*-warfarin to HSA as a model system [67]. In this case, antiwarfarin antibodies were used in an immuno-affinity microcolumn, and warfarin/HSA samples were injected onto this column with only 180 msec being allowed for extraction. Under these conditions, it was predicted that the amount of extracted warfarin should have very little interference from any warfarin that was initially bound to HSA and later released as the sample passed through the immunoaffinity support (see Figure 22.16). The amount of free warfarin in each sample was then determined by using fluorescence measurements to examine the amount of warfarin that eluted nonretained and comparing this with the total amount of warfarin in the original sample. One advantage of this approach is that it is extremely fast. In addition, it examines the binding of solutes to ligands that are in solution rather than immobilized to a solid support.

22.5 SUMMARY AND CONCLUSIONS

This chapter examined a variety of methods for the study of solute-ligand interactions by affinity chromatography. Particular attention was given to the techniques of zonal elution and frontal analysis, which are the most popular formats for this type of work. Various applications of these methods were described, including their use to determine the extent of solute-ligand binding, the number of sites involved in these interactions, and the equilibrium constants for these processes. It was also shown how these methods could be used to investigate the ability of a solute to be displaced by other compounds, the effects of temperature or solvent composition on a solute-ligand reaction, and the structure and location of the binding sites on an immobilized ligand.

Numerous practical issues were discussed for the design and use of these methods, such as the concentration of analyte or competing agents that should be used and factors that affect the accurate measurement of analyte retention. Some newer approaches for the study of solute-ligand systems were described as well, including kinetic methods based on band-broadening measurements, peak decay analysis, and free-fraction measurements.

The variety of ways in which affinity chromatography can be used for this work and the wealth of information it can provide have resulted in its being utilized in an increasing number of applications. It is expected that this trend will continue as affinity chromatography becomes more common in clinical and pharmaceutical research as a means for studying the interactions of drugs, hormones, and other solutes with biological agents.

SYMBOLS AND ABBREVIATIONS

A	Injected or applied analyte
AGP	α_1-Acid glycoprotein
b	Fraction of a solute that is bound to a ligand
B	Substance in the mobile phase that acts as an allosteric agent
BSA	Bovine serum albumin
C_B	Total concentration of a solute that binds at a different site on L from A
C_L	Total concentration of an immobilized ligand
C_I	Total concentration of a competing agent (I) in the mobile phase
C_S	Total concentration of a solubilizing agent
f	Free fraction of a solute
F	Flow rate
$HPLC$	High-performance liquid chromatography
H_s	Plate-height term due to stationary-phase mass transfer
HSA	Human serum albumin
I	Competing agent
k	Retention factor
k_A	Retention factor for analyte A
$k_{A,0}$	Retention factor for analyte A in the absence of a competing agent
k_d	Dissociation rate constant
K_A	Association equilibrium constant for an analyte (A) with an immobilized ligand (L)
$K_A{}'$	Apparent equilibrium association constant for an analyte (A) with an immobilized ligand (L) due to allosteric effects by another agent (B) binding to the same ligand
K_{AS}	Association equilibrium constant for an analyte (A) with a solubilizing agent (S)
K_B	Association equilibrium constant for an allosteric agent (B) binding with an immobilized ligand (L)
K_{BS}	Association equilibrium constant for an allosteric agent (B) with a solubilizing agent (S)
K_I	Association equilibrium constant for a competing agent (I) in the mobile phase with an immobilized ligand (L)
K_{IS}	Association equilibrium constant for a competing agent (I) in the mobile phase with a solubilizing agent (S)
k_{max}	Maximum retention factor observed for analyte A during a competition study
k_{min}	Minimum retention factor observed for analyte A during a competition study
L	Immobilized ligand
LC/MS	Liquid chromatography-mass spectrometry
m_{A0}	Initial moles of analyte bound to the ligand
m_{Ae}	Moles of analyte in the flowing mobile phase at time t during elution
m_L	Total moles of active ligand in a column

m_{Lapp}	Apparent moles of active ligand in a column
n	Fraction of binding sites in a given category
$QSRR$	Quantitative structure retention relationship
R	Ideal gas law constant
S	Solubilizing agent
t	Time
T	Absolute temperature
t_M	Column void time
t_{mean}	Mean breakthrough time for a frontal-analysis experiment
t_R	Retention time
u	Linear velocity
V_M	Column void time
V_R	Column retention time
X	Retention factor for an analyte due to nonspecific binding
β-CD	β-Cyclodextrin
ΔH	Change in enthalpy
ΔS	Change in entropy

REFERENCES

1. Hage, D.S. and Tweed, S.A., Recent advances in chromatographic and electrophoretic methods for the study of drug-protein interactions, *J. Chromatogr. B,* 699, 499–525, 1997.
2. Hage, D.S., Affinity chromatography, in *Handbook of HPLC,* Katz, E., Eksteen, R., and Miller, N., Eds., Marcel Dekker, New York, 1998, chap. 13.
3. Chaiken, I.M., Ed., *Analytical Affinity Chromatography,* CRC Press, Boca Raton, FL, 1987.
4. Loun, B. and Hage, D.S., Characterization of thyroxine-albumin binding using high-performance affinity chromatography, II: Comparison of the binding of thyroxine, triiodo-thyronines and related compounds at the warfarin and indole sites of human serum albumin, *J. Chromatogr. B,* 665, 303–314, 1995.
5. Loun, B. and Hage, D.S., Chiral separation mechanisms in protein-based HPLC columns, 2: Kinetic studies of (*R*)- and (*S*)-warfarin binding to immobilized human serum albumin, *Anal. Chem.,* 68, 1218–1225, 1996.
6. Yang, J. and Hage, D.S., Effect of mobile phase composition on the binding kinetics of chiral solutes on a protein-based high-performance liquid chromatography column: Interactions of D- and L-tryptophan with immobilized human serum albumin, *J. Chromatogr. A,* 766, 15–25, 1997.
7. Loun, B. and Hage, D.S., Characterization of thyroxine-albumin binding using high-performance affinity chromatography, I: Interactions at the warfarin and indole sites of albumin, *J. Chromatogr.,* 579, 225–235, 1992.
8. Domenici, E., Bertucci, C., Salvadori, P., Felix, G., Cahagne, I., Motellier, S., and Wainer, I.W., Synthesis and chromatographic properties of an HPLC chiral stationary phase based upon human serum albumin, *Chromatographia,* 29, 170–176, 1990.
9. Noctor, T.A.G., Pham, C.D., Kaliszan, R., and Wainer, I.W., Stereochemical aspects of benzodiazepine binding to human serum albumin, I: Enantioselective high-performance liquid affinity chromatographic examination of chiral and achiral binding interactions between 1,4-benzodiazepines and human serum albumin, *Mol. Pharmacol.,* 42, 506–511, 1992.
10. Noctor, T.A.G., Wainer, I.W., and Hage, D.S., Allosteric and competitive displacement of drugs from human serum albumin by octanoic acid, as revealed by high-performance liquid chromatography, on a human serum albumin-based stationary phase, *J. Chromatogr.,* 577, 305–315, 1992.

11. Domenici, E., Bertucci, C., Salvadori, P., Motellier, S., and Wainer, I.W., Immobilized serum albumin: Rapid HPLC probe of stereoselective protein-binding interactions, *Chirality,* 2, 263–268, 1990.

12. Domenici, E., Bertucci, C., Salvadori, P., and Wainer, I.W., Use of a human serum albumin-based high-performance liquid chromatography chiral stationary phase for the investigation of protein binding: Detection of the allosteric interaction between warfarin and benzodiazepine binding sites, *J. Pharmaceut. Sci.,* 80, 164–166, 1991.

13. Sengupta, A. and Hage, D.S., Characterization of minor site probes for human serum albumin by high-performance affinity chromatography, *Anal. Chem.,* 71, 3821–3827, 1999.

14. Yang, J. and Hage, D.S., Role of binding capacity versus binding strength in the separation of chiral compounds on protein-based high-performance liquid chromatography columns: Interactions of D- and L-tryptophan with human serum albumin, *J. Chromatogr. A,* 725, 273–285, 1996.

15. Miller, J.H.M. and Smail, G.A., Interaction of the enantiomers of warfarin with human serum albumin, peptides and amino acids, *J. Pharm. Pharmacol.,* 29, 33–33, 1977.

16. Loun, B. and Hage, D.S., Chiral separation mechanisms in protein-based HPLC columns, 1: Thermodynamic studies of (*R*)- and (*S*)-warfarin binding to immobilized human serum albumin, *Anal. Chem.,* 66, 3814–3822, 1994.

17. Allenmark, S., *Chromatographic Enantioseparation: Methods and Applications,* 2nd ed., Ellis Horwood, New York, 1991, chap. 7.

18. Jewell, R.C., Brouwer, K.L.R., and McNamara, P.J., α1-Acid glycoprotein high-performance liquid chromatography column (Enantiopac) as a screening tool for protein binding, *J. Chromatogr.,* 487, 257–264, 1989.

19. Schill, G., Wainer, I.W., and Barkan, S.A., Chiral separations of cationic and anionic drugs on an α1-acid glycoprotein-bonded stationary phase (Enantio-Pac), II: Influence of mobile phase additives and pH on chiral resolution and retention, *J. Chromatogr.,* 365, 73–88, 1986.

20. Dunn, B.M. and Chaiken, I.M., Quantitative affinity chromatography: Determination of binding constants by elution with competitive inhibitors, *Proc. Natl. Acad. Sci. U.S.A.,* 71, 2382–2385, 1974.

21. Fitos, I., Visy, J., Simonyi, M., and Hermansson, J., Chiral high-performance liquid chromatographic separations of vinca alkaloid analogs on α1-acid glycoprotein and human serum albumin columns, *J. Chromatogr.,* 609, 163–171, 1992.

22. Dalgaard, L., Hansen, J.J., and Pedersen, J.L., Resolution and binding site determination of DL-thyronine by high-performance liquid chromatography using immobilized albumin as chiral stationary phase: Determination of the optical purity of thyroxine in tablets, *J. Pharmaceut. Biomed. Anal.,* 7, 361–368, 1989.

23. Kaliszan, R., Noctor, T.A.G., and Wainer, I.W., Stereochemical aspects of benzodiazepine binding to human serum albumin, II: Quantitative relationships between structure and enantioselective retention in high-performance liquid affinity chromatography, *Mol. Pharmacol.,* 42, 512–517, 1992.

24. Vidal-Madjar, C., Jaulmes, A., Racine, M., and Sebille, B., Determination of binding equilibrium constants by numerical simulation in zonal high-performance affinity chromatography, *J. Chromatogr.,* 458, 13–25, 1988.

25. Arnold, F.H., Schofield, S.A., and Blanch, H.W., Analytical affinity chromatography, I: Local equilibrium theory and the measurement of association and inhibition constants, *J. Chromatogr.,* 355, 1–12, 1986.

26. Chen, J., Ohmacht, C.M., and Hage, D.S., Studies of phenytoin binding to human serum albumin by high-performance affinity chromatography, *J. Chromatogr. B,* 809, 137–145, 2004.

27. Hage, D.S. and Sengupta, A., Characterization of the binding of digitoxin and acetyldigitoxin to human serum-albumin by high-performance affinity chromatography, *J. Chromatogr. B,* 724, 91–100, 1999.

28. Noctor, T.A.G., Diaz-Perez, M.J., and Wainer, I.W., Use of a human serum albumin-based stationary phase for high-performance liquid chromatography as a tool for the rapid determination of drug-plasma protein binding, *J. Pharmaceut. Sci.,* 82, 675–676, 1993.

29. Yang, J. and Hage, D.S., Characterization of the binding and chiral separation of D- and L-tryptophan on a high-performance immobilized human serum albumin column, *J. Chromatogr.*, 645, 241–250, 1993.

30. Hage, D.S., Noctor, T.A.G., and Wainer, I.W., Characterization of the protein binding of chiral drugs by high-performance affinity chromatography: Interactions of *R*- and *S*-ibuprofen with human serum albumin, *J. Chromatogr. A*, 693, 23–32, 1995.

31. Nakano, N.I., Shimamori, Y., and Yamaguchi, S., Binding capacities of human serum albumin monomer and dimer by continuous frontal affinity chromatography, *J. Chromatogr.*, 237, 225–232, 1982.

32. Zhivkova, Z. and Russeva, V., New mathematical approach for the evaluation of drug binding to human serum albumin by high-performance liquid affinity chromatography, *J. Chromatogr. B*, 707, 143–149, 1998.

33. Chen, J., Ohnmacht, C.M., and Hage, D.S., Characterization of drug interactions with soluble beta-cyclodextrin by high-performance affinity chromatography, *J. Chromatogr. A*, 1033, 115–126, 2004.

34. Chen, J. and Hage, D.S., Affinity chromatographic studies of allosteric interactions: Analysis of albumin binding to non-polar drugs in the presence of a solubilizing agent, *Anal. Chem.*, submitted.

35. Hage, D.S. and Sengupta, A., Studies of protein binding to nonpolar solutes by using zonal elution and high-performance affinity chromatography: Interactions of *cis*- and *trans*-clomiphene with human serum albumin in the presence of β-cyclodextrin, *Anal. Chem.*, 70, 4602–4609, 1998.

36. Chen, J. and Hage, D.S., Quantitative studies of allosteric drug-protein interactions by high-performance affinity chromatography, *Nat. Biotechnol.*, 22, 1445–1448, 2004.

37. Hage, D.S., Chromatographic and electrophoretic studies of protein binding to chiral solutes, *J. Chromatogr. A*, 906, 549–481, 2001.

38. Kim, H. and Hage, D.S., Chromatographic analysis of carbamazepine binding to human serum albumin, *J. Chromatogr. B*, 816, 57–66, 2005.

39. Gilpin, R.K., Ehtesham, S.E., and Gregory, R.B., Liquid chromatographic studies of the effect of temperature on the chiral recognition of tryptophan by silica-immobilized bovine albumin, *Anal. Chem.*, 63, 2825–2828, 1991.

40. Gilpin, R.K., Ehtesham, S.B., Gilpin, C.S., and Liao, S.T., Liquid chromatographic studies of memory effects of silica-immobilized bovine serum albumin, I: Influence of methanol on solute retention, *J. Liq. Chromatogr.*, 19, 3023–3035, 1996.

41. Peyrin, E., Guillaume, Y.C., Morin, N., and Guinchard, C., Sucrose dependence of solute retention on human serum albumin stationary phase: Hydrophobic effect and surface tension considerations, *Anal. Chem.*, 70, 2812–2818, 1998.

42. Su, W., Gregory, R.B., and Gilpin, R.K., Liquid chromatographic studies of silica-immobilized HEW lysozyme, *J. Chromatogr. Sci.*, 31, 285–280, 1993.

43. Peyrin, E., Guillaume, Y.C., and Guinchard, C., Peculiarities of dansyl amino acid enantioselectivity using human serum albumin as a chiral selector, *J. Chromatogr. Sci.*, 36, 97–103, 1998.

44. Peyrin, E. and Guillaume, Y.C., Chiral discrimination of *N*-(dansyl)-DL-amino acids on human serum albumin stationary phase: Effect of a mobile phase modifier, *Chromatographia*, 48, 431–435, 1998.

45. Rahim, S. and Anne-Francoise, A., Location of binding sites on immobilized human serum albumin for some nonsteroidal anti-inflammatory drugs, *J. Pharmaceut. Sci.*, 84, 949–952, 1995.

46. Aubry, A.-F., Markoglou, N., and McGann, A., Comparison of drug binding interactions on human, rat and rabbit serum albumin using high-performance displacement chromatography, *Comp. Biochem. Physiol. C: Pharmacol. Toxicol. Endocrinol.*, 112C, 257–266, 1995.

47. Wainer, I.W., Enantioselective high-performance liquid affinity chromatography as a probe of ligand-biopolymer interactions: An overview of a different use for high-performance liquid chromatographic chiral stationary phases, *J. Chromatogr. A*, 666, 221–234, 1994.

48. Noctor, T.A.G. and Wainer, I.W., The *in situ* acetylation of an immobilized human serum albumin chiral stationary phase for high-performance liquid chromatography in the examination of drug-protein binding phenomena, *Pharmaceut. Res.,* 9, 480–484, 1992.

49. Chattopadhyay, A., Tian, T., Kortum, L., and Hage, D.S., Development of tryptophan-modified human serum albumin columns for site-specific studies of drug-protein interactions by high-performance affinity chromatography, *J. Chromatogr. B,* 715, 183–190, 1998.

50. Haginaka, J. and Kanasugi, N., Enantioselectivity of bovine serum albumin-bonded columns produced with isolated protein fragments, II: Characterization of protein fragments and chiral binding sites, *J. Chromatogr. A,* 769, 215–223, 1997.

51. Haginaka, J. and Kanasugi, N., Enantioselectivity of bovine serum albumin-bonded columns produced with isolated protein fragments, *J. Chromatogr. A,* 694, 71–80, 1995.

52. Andrews, P., Kitchen, B.J., and Winzor, D.J., Affinity chromatography for the quantitative study of acceptor-ligand interactions: The lactose synthetase system, *Biochem. J.,* 135, 897–900, 1973.

53. Fitos, I., Visy, J., Simonyi, M., and Hermansson, J., Stereoselective allosteric interaction on human serum albumin between ibuprofen and lorazepam acetate, *Chirality,* 11, 115–120, 1999.

54. Ringo, M.C., Hage, D.S., and Evans, C.E., Pressure effects in the binding of solutes to immobilized protein columns, *Anal. Chem.,* submitted.

55. Hage, D.S., High-performance affinity chromatography: A powerful tool for studying serum protein binding, *J. Chromatogr. B,* 768, 3–30, 2002.

56. Hage, D.S. and Austin, J., High-performance affinity chromatography and immobilized serum albumin as probes for drug- and hormone-protein binding, *J. Chromatogr. B,* 739, 39–54, 2000.

57. Lagercrantz, C., Larsson, T., and Karlsson, H., Binding of some fatty acids and drugs to immobilized bovine serum albumin studied by column affinity chromatography, *Anal. Biochem.,* 99, 352–364, 1979.

58. Foley, J.P. and Dorsey, J.G., A review of the exponentially modified Gaussian (EMG) function: evaluation and subsequent calculation of universal data, *J. Chromatogr. Sci.,* 22, 40–46, 1984.

59. Anderson, D.J. and Walters, R.R., Effect of baseline errors on the calculation of statistical moments of tailed chromatographic peaks, *J. Chromatogr. Sci.,* 22, 353–359, 1984.

60. Kasai, K. and Ishii, S., Affinity chromatography of trypsin and related enzymes, I: Preparation and characteristics of an affinity adsorbent containing tryptic peptides from protamine as ligands, *J. Biochem. (Tokyo),* 78, 653–662, 1975.

61. Nakano, N.I., Oshio, T., Fujimoto, Y., and Amiya, T., Study of drug-protein binding by affinity chromatography: Interaction of bovine serum albumin and salicylic acid, *J. Pharmaceut. Sci.,* 67, 1005–1008, 1978.

62. Jacobson, S.C., Andersson, S., Allenmark, S.G., and Guiochon, G., Estimation of the number of enantioselective sites of bovine serum albumin using frontal chromatography, *Chirality,* 5, 513–515, 1993.

63. Tweed, S.A., Loun, B., and Hage, D.S., Effects of ligand heterogeneity in the characterization of affinity columns by frontal analysis, *Anal. Chem.,* 69, 4790–4798, 1997.

64. Nakano, N.I., Shimamori, Y., and Yamaguchi, S., Mutual displacement interactions in the binding of two drugs to human serum albumin by frontal affinity chromatography, *J. Chromatogr.,* 188, 347–356, 1980.

65. Moore, R.M. and Walters, R.R., Peak-decay method for the measurement of dissociation rate constants by high-performance affinity chromatography, *J. Chromatogr.,* 384, 91–103, 1987.

66. Chen, J. and Hage, D.S., Peak decay analysis of drug-protein dissociation rates, *J. Chromatogr. A,* submitted.

67. Clarke, W., Chowdhuri, A.R., and Hage, D.S., Analysis of free drug fractions by ultrafast immunoaffinity chromatography, *Anal. Chem.,* 73, 2157–2164, 2001.

23

Quantitative Affinity Chromatography: Recent Theoretical Developments

Donald J. Winzor

Department of Biochemistry, University of Queensland,
Brisbane, Queensland, Australia

CONTENTS

23.1   INTRODUCTION

Quantitative affinity chromatography refers to the characterization of biospecific interactions by methods that are essentially adaptations of preparative affinity techniques. This use of affinity chromatography was given extensive coverage in the previous edition of this handbook [1], which remains a valid summary of the first 20 years in this field. The basic principles behind quantitative affinity chromatography are described here in Chapter 22 of this current book, with the main emphasis in this chapter being recent theoretical developments that have occurred in this area.

The most notable change in quantitative affinity chromatography in recent years has been the introduction of biosensor technology [2, 3]. This has provided a means of monitoring the concentration of a partitioning solute that is bound to immobilized affinity sites—a parameter that previously had to be inferred from the extent of solute depletion in the mobile phase. Other interesting applications of quantitative affinity chromatography that have been reported include the immobilization of human red blood cells or related membranes to study the interactions of cytochalasin B and glucose with transmembrane glucose and nucleoside transporters [4, 5]. In addition, work has been performed in which quantitative affinity chromatography has been used to characterize the kinetics, as well as the thermodynamics, of solute-ligand binding [6, 7].

Other new developments have included the discovery of a rectangular hyperbolic relationship that describes the interaction of a multivalent partitioning solute with immobilized affinity sites [8] and a simple algebraic derivation of the multivalent Scatchard expression [9] for the case of solute bivalence [10]. By supplementing the already impressive potential of quantitative affinity chromatography, these advances reinforce the assertions made in the previous edition of this review [1] about the power and scope of this method for characterizing biospecific interactions.

23.2   SUMMARY OF BASIC EXPRESSIONS

Quantitative affinity chromatography is generally based on the characterization of reactions involving (1) a partitioning solute (A), (2) an immobilized affinity ligand (X), and (3) a soluble ligand (S) that interacts either with A or X. Several possible combinations for conducting these studies are shown in Figure 23.1. The most common combination involves

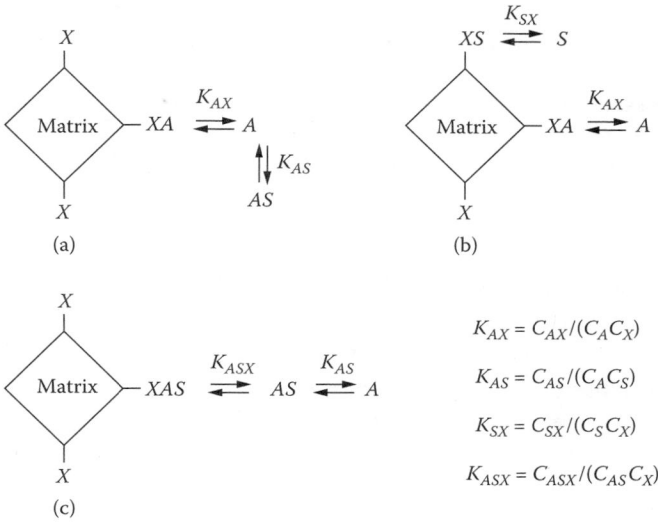

$$K_{AX} = C_{AX}/(C_A C_X)$$

$$K_{AS} = C_{AS}/(C_A C_S)$$

$$K_{SX} = C_{SX}/(C_S C_X)$$

$$K_{ASX} = C_{ASX}/(C_{AS} C_X)$$

Figure 23.1 Schematic representation of the interactions encountered in quantitative affinity chromatography. (a) Competition between a soluble ligand, S, and an immobilized affinity site, X, for a partitioning solute, A. (b) Competition between S and A for immobilized affinity sites. (c) Ligand-retarded desorption of a partitioning solute, resulting from ternary complex formation between the immobilized affinity sites and the solute-ligand complex, AS.

the ligand-facilitated elution of a solute as a result of competition between the soluble ligand and immobilized affinity sites for the solute. This situation, shown in Figure 23.1a, occurs when X is an immobilized form of the soluble ligand [11, 12].

Ligand-facilitated elution can also be created by the competition between a soluble ligand and solute for immobilized affinity sites, as shown in Figure 23.1b. An example of this situation is the competition that occurs between two saccharides (A and S) as they both bind to a lectin-affinity column [13, 14].

A third possible situation is one with ligand-retarded elution. This can be produced when the interaction of a solute with immobilized affinity sites requires that the solute first have complex formation with a soluble ligand. This is illustrated by the reaction scheme in Figure 23.1c. One example of this situation is work that has been performed with the lactose synthetase system, which involved interactions between galactosyltransferase and N-acetylglucosamine on an affinity column that contained α-lactalbumin-linked Sepharose [15]. Another example is research with lactate dehydrogenase-NADH interactions that has been conducted using an oxamate-Sepharose column [16].

Although the quantity that is usually measured in quantitative affinity chromatography is a retention time or elution volume, the thermodynamic parameter that these monitor is a partition coefficient for the distribution of an applied solute between its soluble and adsorbed states. Thus, theoretical expressions for the description of data in quantitative affinity chromatography tend to be derived on the basis of species concentrations and the law of mass action. This means that writing the corresponding chromatographic relationships merely involves relating the experimental data to the term $(\bar{C}_A/\overline{C}_A)$, which represents the ratio of the total liquid-phase concentration of solute $(\bar{C}_A)$ to its overall total concentration $(\overline{C}_A)$. This is usually accomplished by replacing $(\bar{C}_A/\overline{C}_A)$ with the ratio $(V_A^*/\bar{V}_A)$, where $\bar{V}_A$ is the measured elution volume for the solute in the presence of the solute-matrix interaction, and V_A^* is the elution volume of the solute in the absence of this interaction [17, 18].

However, this substitution can also be performed using the corresponding retention time ratio $(t_A^*/\bar{t}_A)$ instead of $(V_A^*/\bar{V}_A)$, since the retention time and elution volume are related to each other through the flow rate.

For systems that involve solute-ligand competition (e.g., as described by the reaction schemes in Figure 23.1a and Figure 23.1b), the value of V_A^* can be obtained in two ways. In the first, V_A^* is given by the term $\bar{V}_A$ in the presence of a saturating concentration of ligand S. In the second approach, V_A^* is given by $\bar{V}_A$, as determined from an experiment performed with a column that is identical to the affinity column but that is devoid of any immobilized affinity sites, X [11, 19, 20]. For the reaction system that is shown in Figure 23.1c, V_A^* is simply obtained by using the elution volume for the solute in the absence of ligand (S).

23.2.1 Equations for Univalent Partitioning Solutes

To illustrate how expressions can be derived for quantitative affinity chromatography, let us begin by considering the concentration of each solute species that is present in the liquid phase or in an adsorbed state. For a situation that involves a competition between X and S for a univalent solute A (the case shown in Figure 23.1a), the concentration of solute in the liquid phase ($\bar{C}_A$) will depend on both the concentration of solute-ligand complex AS and the concentration of free A. This gives rise to the relationship shown in Equation 23.1,

$$\bar{C}_A = C_A + K_{AS}C_AC_S = C_A(1 + K_{AS}C_S) \tag{23.1}$$

where the concentration of complex AS is found by using the product of its association equilibrium constant (K_{AS}) and the free concentrations of the solute and soluble ligand (C_A and C_S). In this same situation, the concentration of the adsorbed solute, as given by ($\bar{C}_A - C_A$), is used to reflect the concentration of the solute-matrix complex. For this species, we can write the following equation,

$$(\bar{C}_A - C_A) = K_{AX}C_AC_X \tag{23.2}$$

where the concentration of the adsorbed solute is also described in terms of the association equilibrium constant for solute-matrix binding (K_{AX}) and the free concentrations of the solute and matrix sites (C_A and C_X).

By combining Equations 23.1 and 23.2, the ratio of solute concentrations in the two phases of the system becomes the ratio shown in Equation 23.3.

$$(\bar{C}_A - C_A)/\bar{C}_A = [(\bar{C}_A/C_A) - 1] = K_{AX}C_X/(1 + K_{AS}C_S) \tag{23.3}$$

Although the concentration of free affinity sites (C_X) cannot be measured, it can be calculated by using the difference between the total concentration of affinity sites, $\bar{C}_X$ (a constant parameter whose magnitude evolves from the analysis) and the concentration of occupied affinity sites, ($\bar{C}_A - C_A$). Making this substitution into Equation 23.3 gives rise to the following new relationship

$$[(\bar{C}_A/C_A) - 1] = [K_{AX}/(1 + K_{AS}C_S)]\bar{C}_X - [K_{AX}/(1 + K_{AS}C_S)](\bar{C}_A - C_A) \tag{23.4}$$

or its column chromatographic counterpart, as given by Equation 23.5 in terms of elution volumes.

$$\left[(\bar{V}_A/V_A^*) - 1\right] = [K_{AX}/(1 + K_{AS}C_S)]\bar{C}_X - [K_{AX}/(1 + K_{AS}C_S)]\bar{C}_A\left[(\bar{V}_A/V_A^*) - 1\right] \tag{23.5}$$

The quantity $[K_{AX}/(1 + K_{AS}C_S)]$ retains a constant magnitude for solute solutions with a range of $\bar{C}_A$ values but a constant free ligand concentration C_S. As a result of this, the linear dependence of $[(\overline{C}_A/\overline{C}_A) - 1]$ vs. $(\overline{C}_A - \overline{C}_A)$, or of $[(\overline{V}_A/V_A^*) - 1]$ vs. $\overline{C}_A[(\overline{V}_A/V_A^*) - 1]$, will allow for the evaluation of $\overline{C}_X$ and the parameter $[K_{AX}/(1 + K_{AS}C_S)]$, which is called the *constitutive solute-matrix affinity constant* $(\overline{K}_{AX})$ [1]. Such measurements are made at a series of fixed free-ligand concentrations (C_S). This makes it possible to determine K_{AX} as the value of $\overline{K}_{AX}$ measured in the absence of ligand $(C_S = 0)$ and to determine K_{AS} as the slope obtained for the dependence of $K_{AX}/\overline{K}_{AX}$ upon C_S. A detailed account of the experiments involved in this approach is given in the literature [21].

Adoption of this solute-distribution approach has led to an expression that makes use of a Scatchard linear transform [22]. But this equation is open to criticism because it fails to achieve a separation of variables. However, this problem can be readily avoided by instead fitting the data for a given value of C_S to the transform's original rectangular hyperbolic relationship. Specifically, the results for $(\overline{C}_A, \overline{C}_A)$ or $(\overline{V}_A, \overline{C}_A)$ at a given value for C_S are fitted to Equation 23.6.

$$(\overline{C}_A - \overline{C}_A) = \overline{C}_A[(\overline{V}_A/V_A^*) - 1] = \frac{[K_{AX}/(1 + K_{AS}C_S)]\overline{C}_X\overline{C}_A}{1 + [K_{AX}/(1 + K_{AS}C_S)]\overline{C}_A} \qquad (23.6)$$

This allows one to obtain the effective total concentration of affinity sites $(\overline{C}_X)$ and the constitutive affinity constant, $\overline{K}_{AX} = [K_{AX}/(1 + K_{AS}C_S)]$, which are the two curve-fitting parameters that result from a nonlinear least-squares analysis of such data.

Expressions for the analysis of data for affinity systems that follow the reaction schemes in Figure 23.1b and Figure 23.1c are provided in Equations 23.7 and 23.8, respectively.

$$(\overline{C}_A - \overline{C}_A) = \overline{C}_A[(\overline{V}_A/V_A^*) - 1] = \frac{[K_{AX}/(1 + K_{SX}C_S)]\overline{C}_X\overline{C}_A}{1 + [K_{AX}/(1 + K_{SX}C_S)]\overline{C}_A} \qquad (23.7)$$

$$(\overline{C}_A - \overline{C}_A) = \overline{C}_A[(\overline{V}_A/V_A^*) - 1] = \frac{[K_{AS}K_{ASX}C_S/(1 + K_{AS}C_S)]\overline{C}_X\overline{C}_A}{1 + [K_{AS}K_{ASX}C_S/(1 + K_{AS}C_S)]\overline{C}_A}$$

$$= \frac{K_{AS}K_{ASX}C_S\overline{C}_X\overline{C}_A}{1 + K_{AS}C_S(1 + K_{ASX}\overline{C}_A)} \qquad (23.8)$$

Most characterizations of the interactions discussed thus far have used linear transforms of these dependencies. However, the widespread availability of nonlinear curve-fitting programs now makes it easy to analyze these results in terms of an untransformed relationship. This avoids potential problems with error distortion in the transformed experimental results [23]. Indeed, some researchers may even opt to analyze all results simultaneously by using a global fit of the combined $(\overline{C}_A, \overline{C}_A, C_S)$ or $(\overline{V}_A, \overline{C}_A, C_S)$ data to the appropriate form of Equation 23.6.

The relationships presented up to this point have presumed that we have knowledge and control of C_S, the free concentration of soluble ligand. In the initial studies performed with quantitative affinity chromatography [11, 15, 19], this condition was not restrictive, since the relatively weak strength of the solute-ligand interactions under investigation required the use of soluble ligand concentrations that were much greater than $\overline{C}_A$. It was therefore justifiable to make the approximation that the free and total concentrations of the soluble ligand were approximately equal (or $C_S \approx \overline{C}_S$). In other studies [24, 25], the value of C_S for small soluble ligands has been established directly by dialysis. When results are not being processed on the basis of a constant free-ligand concentration, advantage can be taken of the relationships in Equations 23.6 to 23.8 by incorporating the following substitution [14, 26],

$$C_S = \overline{C}_S - (Q - 1)\overline{C}_A/Q \tag{23.9}$$

in which Q represents the ratio shown in Equation 23.10.

$$Q = K_{AX}/\overline{K}_{AX} \tag{23.10}$$

23.2.2 Equations for Multivalent Solutes

For a partitioning solute that has multiple binding regions for an immobilized affinity site X, the reaction of A with X is not necessarily restricted to 1:1 complex formation, an assumption that has been inherent in the approaches discussed up to this point. If a single intrinsic binding constant K_{AX} governs all solute-matrix interactions [27], then the general counterparts of Equations 23.4 and 23.5 for a solute with f binding regions (i.e., an f-valent partitioning solute) will take the forms shown in Equations 23.11 and 23.12 [9].

$$\left[(\overline{C}_A/\overline{C}_A)^{1/f} - 1\right] = \overline{K}_{AX}\overline{C}_X - f\overline{K}_{AX}\overline{C}_A^{(f-1)/f}\left(\overline{C}_A^{1/f} - \overline{C}_A^{1/f}\right) \tag{23.11}$$

$$\left[(\overline{V}_A/V_A^*)^{1/f} - 1\right] = \overline{K}_{AX}\overline{C}_X - f\overline{K}_{AX}\left(\overline{V}_A/V_A^*\right)^{(f-1)/f}\overline{C}_A\left[\left(\overline{V}_A/V_A^*\right)^{1/f} - 1\right] \tag{23.12}$$

These expressions predict a linear dependence of $[(\overline{C}_A/\overline{C}_A)^{1/f} - 1]$ upon $\overline{C}_A^{(f-1)/f}(\overline{C}_A^{1/f} - \overline{C}_A^{1/f})$, or of $[(\overline{V}_A/V_A^*)^{1/f} - 1]$ upon $(\overline{V}_A/V_A^*)^{(f-1)/f}\overline{C}_A[(\overline{V}_A/V_A^*)^{1/f} - 1]$. Based on such plots, results obtained at a constant C_S value can be used to evaluate $\overline{K}_{AX}$ and $\overline{C}_X$ by using the slope $(-f\overline{K}_{AX})$ and ordinate intercept $(\overline{K}_{AX}\overline{C}_X)$ of Equations 23.11 and 23.12.

This approach is illustrated in Figure 23.2a, using a study of the NADH interaction with lactate dehydrogenase on a trinitrophenyl-Sepharose column as an example. In this

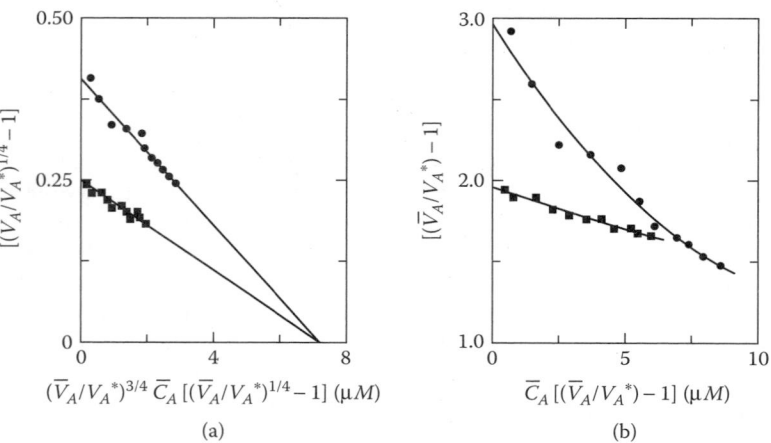

Figure 23.2 Effect of the valence assigned to a partitioning solute on the Scatchard analysis of frontal-affinity chromatography data. This example is based on the interaction of rabbit muscle lactate dehydrogenase with trinitrophenyl-Sepharose in the absence (•) and presence (■) of NADH (5 μM). (a) Plots of results according to Equation 23.12 with $f = 4$ for the tetrameric enzyme. (b) Conventional Scatchard plots ($f = 1$) of the same results. (Data taken from Table 1 of Ref. 21.)

case, the common abscissa intercept $(\bar{C}_X/4)$ for the results obtained in the absence (•) and presence (■) of 5 μM NADH gives a value of 28.7 μM for the effective total concentration of immobilized ligand sites. In addition, a binding constant (K_{AX}) of $1.4 \times 10^4\ M^{-1}$ can be found for the interaction of lactate dehydrogenase with the trinitrophenyl matrix by utilizing the slope of the plot in the absence of NADH. Also, interpretation of the decreased slope $(0.84 \times 10^4\ M^{-1})$ in the presence of 5 μM NADH as $K_{AX}/(1 + K_{AS}C_S)$ yields an intrinsic binding constant of $1.3 \times 10^5\ M^{-1}$ for the interaction of NADH with the tetrameric lactate dehydrogenase. For studies in which $\bar{C}_S$ rather than C_S is the ligand concentration of known magnitude [14], the expression that corresponds to Equation 23.9 for C_S is shown in Equation 23.13.

$$C_S = \bar{C}_S - (Q - 1)f\,\bar{C}_A/Q \qquad (23.13)$$

Analysis of the results in Figure 23.2a by a classic Scatchard approach is illustrated in Figure 23.2b. The curvature in this second group of plots seemingly suggests that some form of NADH-dependent negative cooperativity is present for the immobilized ligand sites. In reality, this curvature simply reflects the effects of solute tetravalence, as already indicated in Figure 23.2a. This example indicates the importance of taking into account the effects of solute multivalence in situations where the partitioning solute is likely to possess more than one binding site for an immobilized ligand. Although some reluctance to assign a magnitude to f is understandable, it should be realized that resorting to a conventional Scatchard analysis does not avoid the problem. Instead, this merely means that the experimenter has chosen, or used by default, $f = 1$ as the value for the solute valence.

23.3   EXPERIMENTAL PROTOCOLS

Quantitative affinity chromatography was initially developed as a column chromatographic procedure [11, 12, 15]. However, it was rapidly adapted to accommodate the results of partition-equilibrium experiments [6, 19, 21, 28], including those obtained by solid-phase immunoassays [29, 30]. The advent of biosensor technology has created additional interest in the partition-equilibrium variant of quantitative affinity chromatography by providing a means for the direct monitoring of complex formation between a solute and immobilized ligand sites [31, 32]. Detailed descriptions of the procedures that are used in this field are available in the literature [1, 10, 21, 33–35], as well as in Chapters 22 and 24 of this book. Thus, the emphasis in this current chapter will be on the underlying principles rather than the finer experimental details.

23.3.1   Column Chromatography

As noted previously [1], a decision has to be made whether to use frontal-affinity chromatography [20, 36] or zonal-affinity chromatography [11, 15] when performing column-based measurements. These methods are also often referred to as *frontal analysis* and *zonal elution*, respectively. Frontal analysis is the procedure of choice from a theoretical viewpoint, while zonal elution allows the use of a simplifying combination of circumstances for the characterization of a solute-ligand or matrix-ligand interaction. These approaches differ only in terms of the volume of analyte solution that is applied to the column. Frontal analysis requires that a sufficient volume of solution be used to ensure that the elution profile contains a plateau region with the same composition as the initial applied solution (see Figure 23.3a), while in zonal elution a much smaller sample volume can be used (Figure 23.3b).

An important factor to consider when using zonal elution is the extent of dilution that occurs for the applied analyte A during the chromatographic run. Such dilution means

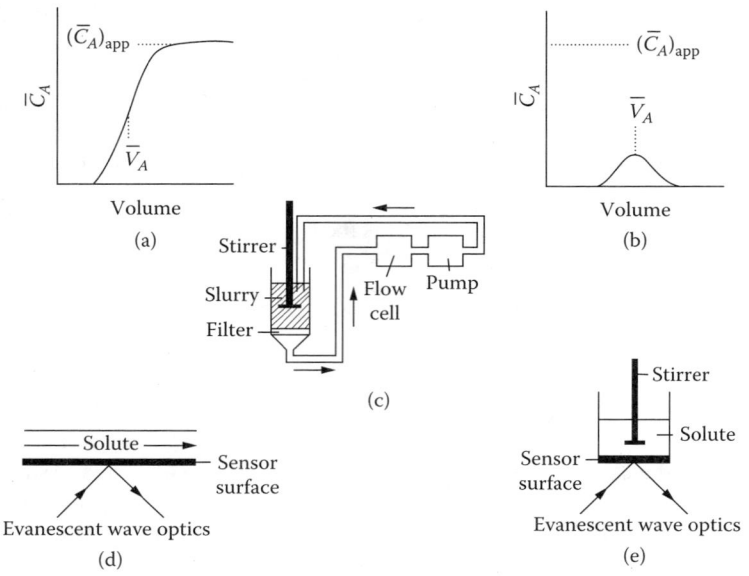

Figure 23.3 Aspects of the different experimental protocols for conducting quantitative affinity chromatography. (a) Form of the elution profile in frontal-affinity chromatography for a solution with applied concentration $\bar{C}_A$. (b) The problem of solute dilution in zonal-affinity chromatography. (c) Apparatus for the recycling partition variant of quantitative affinity chromatography. (d) The closed microfluidics feature of the Biacore biosensor. (e) The general design of cuvette-based biosensors, such as IAsys and IBIS.

that an exact value cannot be assigned to $\bar{C}_A$, as is required for the expressions presented in Section 23.2. Thus, the researcher is instead forced to assume that $\bar{C}_A$ contributes negligibly to a relationship like the one shown in Equation 23.12, which then simplifies to the following approximate expression.

$$\left[\left(\bar{V}_A/V_A^*\right)^{1/f} - 1\right] \approx K_{AX}\bar{C}_X/(1 + K_{AS}C_S) \qquad (23.14)$$

In this situation, the elution volume ($\bar{V}_A$) of solute A is obtained by applying small zones of this solute to an affinity column that has been preequilibrated with mobile phases that contain several known concentrations (C_S) of a soluble ligand S. The resulting data are then used with Equation 23.14 to determine K_{AS} for a system that involves competition between the soluble ligand and matrix for the partitioning solute.

Although zonal-elution chromatography has been strongly recommended for the characterization of biospecific interactions [37, 38], there has until recently [39] been no specific mention of the inherent assumption that $C_X \approx \bar{C}_X$ [24]. Instead, it has been suggested [37] that the loading concentration of partitioning solute should be low relative to $1/K_{AX}$ to create a situation where one is working under linear elution conditions (i.e., elution behavior that is independent of the amount of applied solute) [39]. Such action is certainly a way of making C_X approximately equal to $\bar{C}_X$, but this implies that quantitative affinity chromatography is only useful for the study of relatively weak interactions. In practice, however, this technique has been used to characterize binding constants that range from 10^3 to $10^9\ M^{-1}$ [1, 10].

Thus, although zonal elution has the potential to play an extremely useful role in the characterization of biospecific interactions, precautions need to be taken to ensure the validity of the approximations that are inherent in this method. Frontal analysis is a more rigorous procedure in that it entails no such assumptions. However, the requirement for a much greater amount of solute does set a practical limit on the usefulness of this latter technique in the study of some biological systems.

23.3.2 Partition-Equilibrium Methods

Because the equations for quantitative affinity chromatography were initially derived in terms of a solute's distribution between a solid matrix and a liquid phase, these expressions can also be employed in simple partition experiments to characterize biospecific interactions [19, 35]. For instance, this approach has been used to study the interactions of glycolytic enzymes with muscle myofibrils [40–42].

A problem encountered in simple partition-equilibrium experiments is the need to dispense an accurate amount of affinity matrix into each reaction mixture. One way of overcoming this difficulty is to perform a series of experiments with the same slurry of affinity matrix. This method, known as the *recycling partition technique* [26, 28], can be used in situations where the concentration of a partitioning solute is suitable for continuous measurement by a flow-through monitor (see Figure 23.3c). In this technique, a set of $(\bar{\bar{C}}_A, \bar{C}_A)$ results is generated through the successive addition of solute aliquots to the stirred slurry. This is followed by successive aliquots of a competing ligand solution to provide the $(\bar{\bar{C}}_A, \bar{C}_A, \bar{C}_S)$ data set that is required for determining K_{AS} or K_{SX} [26].

Although the amount of affinity matrix is fixed in this method, it is necessary to consider the decrease in $\bar{C}_X$ that occurs with each addition of the solute or competing ligand solution. In doing this, it is helpful to remember that the amount of affinity matrix in the initial slurry, $(V_A^*)_0(\bar{C}_X)_0$, must describe the corresponding product, $V_A^*\bar{C}_X$, at all stages of the experiment. Thus, the results of this experiment can be related to a system with an initial affinity site concentration of $(\bar{C}_X)_0$ by multiplying Equation 23.4 by $V_A^*/(V_A^*)_0$. This provides the following Scatchard relationship,

$$\left[V_A^*\Big/\left(V_A^*\right)_0\right][(\bar{\bar{C}}_A/\bar{C}_A)-1] = \bar{K}_{AX}(\bar{C}_X)_0 - \bar{K}_{AX}\left[V_A^*\Big/\left(V_A^*\right)_0\right](\bar{\bar{C}}_A - \bar{C}_A) \qquad (23.15)$$

which, in turn, is a linear transform of the rectangular hyperbolic expression shown in Equation 23.16.

$$\left[V_A^*\Big/\left(V_A^*\right)_0\right](\bar{\bar{C}}_A - \bar{C}_A) = \bar{K}_{AX}(\bar{C}_X)_0\bar{C}_A/(1 + \bar{K}_{AX}\bar{C}_A) \qquad (23.16)$$

This correction for dilution effects is illustrated in Figure 23.4 for the evaluation of K_{AX} and K_{AS} for a univalent partitioning solute system. In this particular example, recycling partition studies were used to determine the binding constant between antithrombin and high-affinity heparin on a heparin-Sepharose affinity column [26]. This example also illustrates the use of Equation 23.9 for C_S when the total concentration of the competing ligand $(\bar{C}_S)$ is the parameter of known magnitude. Further experimental details on the recycling partition technique can be found in the literature [26], as well as in earlier reviews of the topic [21, 35].

23.3.3 Biosensor Technology

In the previous section on partition-equilibrium experiments, the concentration of solute that was bound to the immobilized affinity sites was determined by taking the difference between the total solute concentration $(\bar{\bar{C}}_A)$ and its concentration in the liquid phase $(\bar{C}_A)$. But this

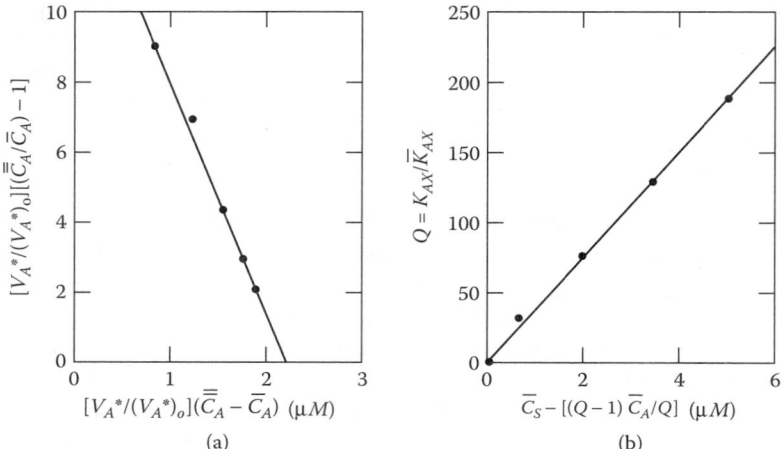

Figure 23.4 Characterization of the interaction between high-affinity heparin (S) and antithrombin (A) by recycling partition-equilibrium studies with heparin-Sepharose as the affinity matrix. (a) Scatchard plot for the evaluation of K_{AX} from data for antithrombin alone (Equation 23.15 with $C_S = 0$). (b) Evaluation of K_{AS} by the application of Equation 23.9 to $\overline{K}_{AX}$ values obtained for heparin-antithrombin mixtures. (Data taken from Hogg, P.J., Jackson, C.M., and Winzor, D.J., *Anal. Biochem.*, 192, 303–311, 1991.)

approach does not work for high-affinity interactions, where $\overline{C}_A$ tends to be indistinguishable from zero. In this situation, it would be desirable to have a more direct assessment of complex formation. This can now be accomplished through biosensor technology [2, 3]. Previously, direct complex monitoring had been performed through the use of radioimmunoassays or ELISA (enzyme-linked immunosorbent assay) techniques. However, these earlier approaches also relied on the approximation that the measured concentration of the adsorbed solute (after removal of all solute from the liquid phase) represented the composition of adsorbed solute in the original and unperturbed equilibrium mixture [29, 30].

One commercial biosensor instrument that is used for direct complex monitoring is Biacore [2, 31, 32]. In this instrument, the affinity matrix is located on a sensor surface that forms the base of a capillary channel. A solution of partitioning solute with a concentration $\overline{C}_A$ is then passed through this channel, as is shown in Figure 23.3d. Evanescent-wave technology is used to produce a response that is directly related to the concentration of the adsorbed solute, $(\overline{\overline{C}}_A - \overline{C}_A)$. This is accomplished by using the increased refractive index of the affinity-matrix layer due to solute binding. Because there is a continuous flow of solute across the affinity surface, the concentration of solute in the liquid phase at equilibrium should be the same as the applied concentration, $\overline{C}_A$. Further details on this instrument can be found in Chapter 24.

Two other biosensor instruments that are available commercially are IAsys [3, 43, 44] and IBIS [45]. These also allow the direct detection of a bound solute through evanescent-wave technology. However, they differ from Biacore in that they use a cuvette-based system rather than a flow-through capillary, as is illustrated in Figure 23.3e.

As was true for the partition-equilibrium methods that were described in Section 23.3.2, complex formation in the IAsys and IBIS biosensors occurs on the affinity surface at the expense of the solute's concentration in the liquid phase. However, these approaches differ in that the liquid-phase concentration $\overline{C}_A$ is the parameter that must be deduced with the biosensors, rather than the concentration of the adsorbed solute. This liquid-phase concentration can be determined based on the known concentration of solute that was placed in the flow cell or cuvette ($\bar{C}_A$) and the measured concentration of the matrix-bound solute ($\bar{C}_A - \bar{\bar{C}}_A$). Concentrations of bound solute equivalent to the nanomolar range can be monitored by using surface-plasmon resonance in the Biacore and IBIS instruments or by using a resonant mirror detection system in the IAsys biosensor.

Because of the direct proportionality between the equilibrium biosensor response (R_e) and the concentration of matrix-bound solute [44, 46], the counterpart to Equation 23.6 for a univalent partitioning solute analyzed by the Biacore instrument is as follows,

$$R_e = R_m \bar{K}_{AX} \bar{C}_A / (1 + \bar{K}_{AX} \bar{C}_A) \quad (23.17)$$

where R_m is the response that is equivalent to $\bar{C}_X$. This expression applies directly to the ($R_e, \bar{C}_A$) data generated by Biacore. For cuvette-based biosensors, the corresponding relationship takes the form shown in Equation 23.18 [47, 48],

$$R_e = R_m \bar{K}_{AX}[\bar{\bar{C}}_A - (R_e/P)]/\{1 + \bar{K}_{AX}[\bar{\bar{C}}_A - (R_e/P)]\} \quad (23.18)$$

in which P is a proportionality constant that relates the instrument's response to the bound ligand concentration. This latter expression is used to account for the ligand depletion that occurs in the liquid phase of cuvette-based biosensors.

The need to consider the effects of ligand depletion with the cuvette-based instruments is a bit of a disadvantage. However, this is offset in thermodynamic studies by their ability to conduct stepwise titrations [47], a feature shared with the recycling partition technique that was discussed in Section 23.3.2. An example of such an experiment is given in Figure 23.5. A desire to conduct stepwise titrations has also led to the conversion of the Biacore-X microfluidics system into a recycling variant of a cuvette-based biosensor [49]. A comparison of these and other features of flow cell versus cuvette-based biosensors was given in a recent review [50].

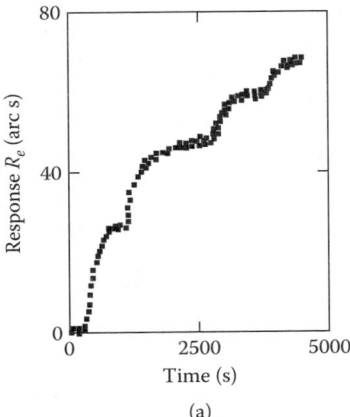
(a)

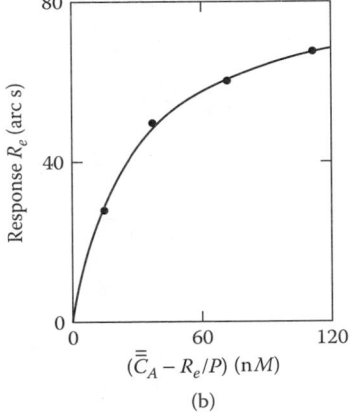
(b)

Figure 23.5 Thermodynamic characterization of the interaction between apocarboxypeptidase A and a specific monoclonal antibody on the sensor surface of an IAsys cuvette. (a) Stepwise titration obtained by supplementing the liquid phase with successive aliquots of antigen. (b) Rectangular hyperbolic dependence of the equilibrium response (R_e) plotted according to Equation 23.18, together with the best-fit line ($K_{AX} = 3.3 \times 10^7 \, M^{-1}$, $R_m = 86$ arc sec). (Data taken from Hall, D.R. and Winzor, D.J., *Anal. Biochem.*, 244, 152–160, 1997.)

23.4   RECENT THEORETICAL DEVELOPMENTS

Given that quantitative affinity chromatography has been in existence for over 30 years, it is not surprising that there have been relatively few new theoretical developments over the past decade. Furthermore, the developments that have occurred have tended to be consolidating rather than innovative in nature. Examples of theoretical work that are discussed in this section include an alternative derivation of the Scatchard equation for a multivalent solute, a demonstration of how multivalent solutes can give rise to linear Scatchard plots, and the development of a rectangular hyperbolic relationship for multivalent partitioning solutes.

23.4.1   Alternative Derivation of the Scatchard Expression for a Multivalent Solute

The general counterpart of the Scatchard equation for an f-valent partitioning solute (see Equations 23.11 and 23.12) has been obtained [9]. This was found by rearranging an expression that was derived by applying reacted-site probability theory [51] to the problem of determining the intrinsic binding constant for solute-matrix interactions in quantitative affinity chromatography [28]. Because an understanding of the logic involved in this process requires a level of math beyond that of many researchers in biology, a simple algebraic derivation has recently been devised [10] for the particular case of a bivalent solute ($f = 2$). This approach has now been extended to the situation where a partitioning solute with f equivalent and independent sites interacts with immobilized affinity sites.

In the absence of a competing soluble ligand, the total concentration of a partitioning solute can be described by a series of stoichiometric binding constants $K_i = K_1 \ldots K_f$, where K_i is the product of the intrinsic binding constant and the number of ways of forming the complex AX_i. The relationship between the total solute concentration and these binding constants is given in Equation 23.19.

$$\bar{C}_A = \bar{C}_A + K_1 \bar{C}_A C_X + K_1 K_2 \bar{C}_A C_X^2 + \ldots + K_1 K_2 \ldots K_f \bar{C}_A C_X^f \tag{23.19}$$

In this relationship, the stoichiometric constant K_i (where $f \geq i \geq 1$) is related to the intrinsic binding constant K_{AX} through Equation 23.20 [27].

$$K_i = [(f - i + 1)/i!] K_{AX} \tag{23.20}$$

By substituting Equation 23.20 into Equation 23.19, one obtains the following expression,

$$\bar{C}_A = \bar{C}_A + f K_{AX} \bar{C}_A C_X + f(f-1)/(2!) K_{AX}^2 \bar{C}_A C_X^2 + \ldots + K_{AX}^f \bar{C}_A C_X^f$$

$$= \bar{C}_A (1 + K_{AX} C_X)^f \tag{23.21}$$

which can be rearranged into the form shown in Equation 23.22.

$$[(\bar{C}_A/\bar{C}_A)^{1/f} - 1] = K_{AX} C_X \tag{23.22}$$

Similarly, the total concentration of immobilized affinity sites ($\bar{C}_X$) is described by using Equation 23.23.

$$\bar{C}_X = C_X + f K_{AX} \bar{C}_A C_X + \{2f(f-1)/(2!)\} K_{AX}^2 \bar{C}_A C_X^2 + \ldots + f K_{AX} \bar{C}_A C_X^f$$

$$= C_X + f K_{AX} \bar{C}_A C_X [1 + (f-1) K_{AX} C_X + \{(f-1)(f-2)/(2!)\} K_{AX}^2 C_X^2 + K_{AX}^{(f-1)} C_X^{(f-1)}]$$

$$\tag{23.23}$$

The free concentration of immobilized affinity sites now becomes

$$C_X = \bar{C}_X - fK_{AX}\bar{C}_A C_X(1 + K_{AX}C_X)^{f-1} \tag{23.24}$$

so Equation 23.22 can be rewritten into the following form.

$$[(\bar{C}_A/\bar{C}_A)^{1/f} - 1] = K_{AX}\bar{C}_X - fK_{AX}^2\bar{C}_A C_X(1 + K_{AX}C_X)^{f-1} \tag{23.25}$$

Substitution of the left-hand side of Equation 23.22 for the product $K_{AX}C_X$ then leads to the relationship,

$$\left[(\bar{C}_A/\bar{C}_A)^{1/f} - 1\right] = K_{AX}\bar{C}_X - fK_{AX}\bar{\bar{C}}_A^{(f-1)/f}\left(\bar{C}_A^{1/f} - \bar{C}_A^{1/f}\right) \tag{23.26}$$

which is the multivalent Scatchard expression that was deduced previously (see Equation 23.11) [9]. One application of this equation has been its use in determining a binding constant for the interaction of tetrameric enzyme aldolase ($f = 4$) with muscle myofibrils [8, 40]. An example of a plot from this study is provided in Figure 23.6a.

An advantage of this last derivation is that it obtains expressions for $\bar{C}_A$ and $\bar{C}_X$ by employing a method that follows the standard approach initiated by Klotz in the original studies of ligand binding [27]. This should create greater confidence in the validity of using Equation 23.26 as the Scatchard expression for a multivalent partitioning solute, an analysis that has been used only rarely despite its relevance to a large number of experimental systems.

23.4.2 Linearity of the Conventional Scatchard Plot for a Multivalent Solute

Using an incorrect value (usually unity) for f in Equations 23.11 or 23.12 should lead to a curved Scatchard plot when one is working with a multivalent solute. This was demonstrated

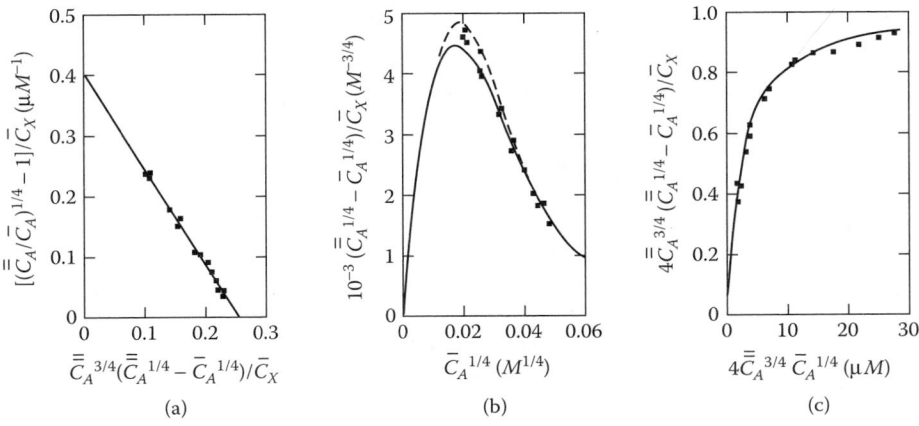

(a) (b) (c)

Figure 23.6 Analysis of binding data for the interaction of aldolase with muscle myofibrils. (a) Multivalent Scatchard analysis (Equation 23.11) on the basis of enzyme tetravalence, giving a K_{AX} of 3.8×10^5 M^{-1}. (b) Plot of the results according to Equation 23.31, together with theoretical relationships for systems with K_{AX} and $\bar{C}_X$ values of 2.1 μM (—) and 1.5 μM (---) (i.e., the experimental limits of $\bar{C}_X$ in the partition-equilibrium study) [40]. (c) Rectangular hyperbolic dependence obtained by analyzing the results in terms of Equation 23.32; the best-fit curve again corresponds to a binding constant (K_{AX}) of 3.8×10^5 M^{-1}. (Data taken from Harris, S.J., Jackson, C.M., and Winzor, D.J., *Arch. Biochem. Biophys.*, 316, 20–23, 1995.)

earlier in Figure 23.2b. But situations can be encountered in which the conventional Scatchard plot (for example, as based on Equations 23.4 or 23.26 with $f = 1$) is linear despite the fact that the solute is multivalent [16]. This experimental observation had been predicted [19] as a situation that might occur when steric restrictions preclude the multiple interactions of a solute with immobilized binding sites. This intuitive reasoning has now been replaced by an explicit analysis of the circumstances required for this to occur. Specifically, $(\overline{\overline{C}}_A - \overline{C}_A)$ must be much less than $\overline{C}_A$ for a multivalent solute to exhibit such behavior [52].

If the concentration of the matrix-bound solute $(\overline{\overline{C}}_A - \overline{C}_A)$ is small relative to the concentration of its liquid-phase counterpart $(\overline{C}_A)$, the expression for the total concentration of solute can be written as shown below.

$$\overline{\overline{C}}_A = \overline{C}_A + (\overline{\overline{C}}_A - \overline{C}_A) = \overline{C}_A(1 + \delta) \tag{23.27}$$

where:

$$\delta = (\overline{\overline{C}}_A - \overline{C}_A)/\overline{C}_A \ll 1 \tag{23.28}$$

By using the binomial theorem, this means that the following statement must be true.

$$\overline{\overline{C}}_A^{1/f} = \overline{C}_A^{1/f}(1 + \delta/f + \cdots) \tag{23.29}$$

A reasonable approximation for Equation 23.26 can then be given by Equation 23.30 [52].

$$(\overline{\overline{C}}_A - \overline{C}_A)/\overline{C}_A = fK_{AX}\overline{C}_X - fK_{AX}(\overline{\overline{C}}_A - \overline{C}_A) \tag{23.30}$$

Under these circumstances, the conventional Scatchard plot should be linear, implying that the bound solute exists solely in the form of the 1:1 complex AX. This linearity also confirms that the affinity parameter determined from the plot is the stoichiometric constant for 1:1 complex formation (fK_{AX}), which reflects the f ways of forming complex AX with an intrinsic binding constant K_{AX} [19, 28].

23.4.3 The Rectangular Hyperbolic Relationship for a Multivalent Partitioning Solute

As noted in Section 23.2.1, the use of Scatchard analysis for evaluating equilibrium constants and reaction stoichiometry is open to criticism because of (1) its reliance on transformed experimental parameters and (2) the statistical distortion of a data distribution that results from such transformations [23, 53]. However, the multivalent counterpart of the Scatchard equation has been derived without recourse to the rectangular hyperbolic expression of which it must be the linear transform. Indeed, ten years elapsed between the first report [9] of Equation 23.26 and its corresponding rectangular hyperbolic relationship [8].

In an attempt to obtain this type of untransformed relationship, Equation 23.26 was initially rearranged as shown in Equation 23.31.

$$r_f = \left(\overline{\overline{C}}_A^{1/f} - \overline{C}_A^{1/f} \right)/\overline{C}_X = K_{AX}\,\overline{C}_A^{1/f} \left/ \left[1 + fK_{AX}\overline{\overline{C}}_A^{(f-1)/f}\overline{C}_A^{1/f} \right] \right. \tag{23.31}$$

This achieves the desired separation of variables in the sense that the multivalent binding function (r_f, the dependent variable) is expressed as a function of the liquid-phase concentration of the partitioning solute raised to the appropriate power ($\overline{C}_A^{1/f}$, the independent variable). However, the presence of the total-solute concentration term, $\overline{\overline{C}}_A^{(f-1)/f}$, in the denominator of Equation 23.31 does give rise to some surprising features in the dependence of r_f upon $\bar{C}_A^{1/f}$. This is illustrated in Figure 23.6b, where the results for the myofibrillar interaction with aldolase [8, 40] are replotted in this format. For instance, the appearance of a maximum in this plot and a dependence of the binding function r_f upon the concentration of the immobilized affinity sites ($\bar{C}_X$) clearly places this form of data representation at variance with the usual concept of a binding curve. Nevertheless, global fitting of the data for $\bar{C}_A$, C_A, and $\bar{C}_X$ to Equation 23.31 can still be used to deduce the value of K_{AX} [8].

Independence of the ordinate parameter upon $\bar{C}_X$ can be achieved by multiplying Equation 23.31 by $f\bar{C}_A^{(f-1)/f}$ to obtain the relationship given in Equation 23.32.

$$f\bar{C}_A^{(f-1)/f} r_f = f\bar{C}_A^{(f-1)/f} K_{AX} \bar{C}_A^{1/f} \bigg/ \left[1 + f\bar{C}_A^{(f-1)/f} K_{AX} \bar{C}_A^{1/f} \right] \qquad (23.32)$$

This relationship describes the rectangular hyperbolic dependence of ($f\bar{C}_A^{(f-1)/f} r_f$) upon ($f\bar{C}_A^{(f-1)/f} \bar{C}_A^{1/f}$), as illustrated in Figure 23.6c. Based on this equation, the binding constant K_{AX} (or its reciprocal, the dissociation constant) can be obtained by simply substituting ($f\bar{C}_A^{f-1)/f} r_f$) for the binding function (or initial velocity) and by substituting ($f\bar{C}_A^{(f-1)/f} \bar{C}_A^{1/f}$) for the solute (substrate) concentration in any commercial computer program that is used to evaluate binding or Michaelis-Menten parameters.

In many studies, the capacity of the affinity matrix for a partitioning solute is also a parameter that is evaluated. Under these circumstances, $\bar{C}_X$ can be replaced by $\bar{c}_X/M_X$. This new term represents the ratio of the weight-concentration of the affinity matrix ($\bar{c}_X$) to the mass of matrix that binds a mol of solute (M_X, the reciprocal of the matrix capacity for solute). The use of this approach to evaluate the matrix capacity for a solute (expressed as mol solute/g matrix) along with K_{AX} is illustrated in recent studies that have examined the interactions between muscle myofibrils and glycolytic enzymes [8, 41].

23.5 BIOMEMBRANE-AFFINITY CHROMATOGRAPHY

An interesting development in quantitative affinity chromatography has been the use of frontal analysis with immobilized cell membranes and proteoliposomes to study the interactions of transmembrane proteins such as glucose transporter Glut 1 [4] and nucleoside transporter [5, 54]. In some applications of this technique, referred to as biomembrane-affinity chromatography [5, 55, 56], immobilized erythrocytes have been used as the affinity matrix [5, 57] to provide a situation that bears an even greater similarity to glucose/nucleoside transport *in vivo*. One specific application [57] is described in the next section to illustrate various quantitative aspects of this work.

23.5.1 Studies with Erythrocyte Membranes as the Affinity Matrix

In biomembrane-affinity chromatography, the interaction of interest involves an immobilized affinity site (X) for which two soluble reactants (solute A and soluble ligand S) compete. This type of reaction is represented by the model in Figure 23.1b. For example, in studies with immobilized glucose transporter Glut 1, the binding constant of this transporter for glucose (K_{SX}) was determined using the glucose-facilitated elution of either cytochalasin B [4, 55, 57] or forskolin [55, 57], two competitive inhibitors of glucose transport. For such work, a freeze-thaw technique has been used to immobilize either proteoliposomes enriched with Glut 1 [58] or red blood cell membrane vesicles [4] on Superdex 200 gel beads. Furthermore, red blood cells have been immobilized by mixing these with a gel synthesized from derivatized acrylamide monomers and containing positively charged polymer aggregates with big channels [57, 59].

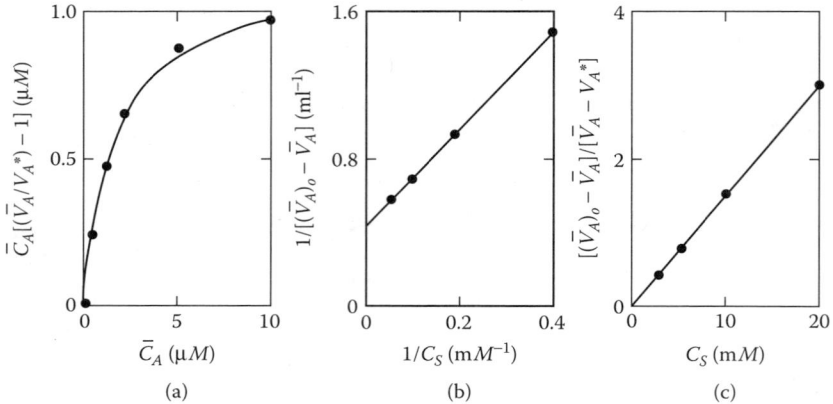

(a) (b) (c)

Figure 23.7 Characterization of the interaction between glucose and the red-cell glucose transporter (Glut 1) by the glucose-facilitated elution of forskolin in biomembrane-affinity chromatography on immobilized erythrocytes. (a) Evaluation of K_{AX} for the interaction of forskolin with
immobilized red-cell Glut 1 sites in the absence of glucose (Equation 23.6 with $C_S = 0$). (b) Analysis
of results for mixtures of forskolin (2 nM) and glucose in terms of Equation 23.33. (c) Corresponding
analysis in terms of Equation 23.34. (Data taken from Zeng, C.-M., Zhang, Y., Lu, L., Brekkan, E.,
Lundqvist, A., and Lundahl, P., *Biochim. Biophys. Acta*, 1325, 91–98, 1997.)

Data are shown in Figure 23.7a for the determination of K_{AX} in the interaction of
forskolin with Glut 1 sites on immobilized erythrocytes. In this case, an analysis based on
Equation 23.6 with $C_S = 0$ yields a binding constant of 5.6×10^5 M^{-1}. Experimental details for
the determination of K_{SX} in this approach differ from the protocol described in Section 23.2.1.
In this previous section, the analysis was based on an interpretation of the dependence of
$(\bar{V}_A - V_A^*)$ upon $\bar{C}_A$ at various fixed ligand concentrations (C_S), while the approach used
with the glucose-facilitated elution of forskolin was monitored via the dependence of
$(\bar{V}_A - V_A^*)$ upon C_S for a single forskolin concentration ($\bar{C}_A$). The form in which these
results were presented (as shown in Figure 23.7b) reflects an analysis in terms of the
following relationship [57],

$$\frac{1}{(\bar{V}_A)_0 - \bar{V}_A} = \frac{1 + K_{AX}\bar{C}_A}{V_A^* K_{AX}\bar{C}_X} + \frac{(1 + K_{AX}\bar{C}_A)^2}{V_A^* K_{AX}\bar{C}_X K_{SX}C_S} \quad (23.33)$$

as recommended [19, 33] for this earlier experimental design, in which $(\bar{V}_A)_0$ is the value
of $\bar{V}_A$ for the fixed concentration of forskolin (2 nM) in the absence of glucose. The ratio
of the slope to the ordinate intercept from Figure 23.7b yields a value of 6.8 mM for
$(1 + K_{AX}\bar{C}_A)/K_{SX}$. This is essentially the same as the dissociation constant ($1/K_{SX}$), since
this system has a K_{AX} value of 5.6×10^5 M^{-1} (see Figure 23.7a) and $\bar{C}_A = 2$ nM, which
makes $(1 + K_{AX}\bar{C}_A) = 1.001$. Based on these results, K_{SX} for this system is equal to 150 M^{-1}.

As noted in the literature [34], a preferred way of analyzing such data involves the
substitution of $[(\bar{V}_A)_0 - V_A^*](1 + K_{AX}\bar{C}_A)$ for $V_A^* K_{AX}\bar{C}_X$, a replacement that follows from
Equation 23.5. This gives results that are then amenable to direct analysis, as can be shown
by rewriting Equation 23.33 in the form shown in Equation 23.34.

$$[(\bar{V}_A)_0 - \bar{V}_A]/(\bar{V}_A - V_A^*) = K_{SX}C_S/(1 + K_{AX}\bar{C}_A) \quad (23.34)$$

Equation 23.34 has the statistical advantage of having a dependent variable, $[(\bar{V}_A)_0 - \bar{V}_A]/$
$(\bar{V}_A - V_A^*)$, that is expressed as a function of the independent variable, C_S, as is illustrated in

Figure 23.7c. The value of $160\ M^{-1}$ that is obtained for K_{SX} through this analysis compares favorably with the estimate that was deduced earlier through double-reciprocal analysis (as represented by Figure 23.7b).

Because erythrocytes already comprise a separate phase, these do not have to be immobilized when they are used to characterize the interactions of Glut 1 with glucose and competitive inhibitors of glucose transport by quantitative affinity chromatography. Indeed, the simple partition-equilibrium form of quantitative affinity chromatography has provided a means of characterizing the interactions of glycolytic enzymes with erythrocyte membranes [60, 61], as well as the myofibrillar matrix of muscle [40–42]. Another such example involving erythrocyte membranes is described below.

23.5.2 Erythrocyte Ghosts as an Affinity Matrix

The characterization of interactions between long-chain fatty acids and albumin has posed a difficult challenge [62, 63] because of the small proportion (0.01%) of unbound fatty acid molecules that occur in albumin/fatty acid mixtures [64]. Although most quantitative studies of this system have examined the effect of albumin on the partitioning of fatty acids between aqueous and n-heptane phases [64], work has also been performed that makes use of the specific binding and incorporation of fatty acids into erythrocyte ghosts [65–67]. This system is represented schematically by the reaction scheme in Figure 23.1a, in which the fatty acid would be the partitioning solute (A), and albumin would be the soluble ligand (S) that competes with erythrocyte membrane sites (X).

Experimentally, these studies have followed the simple partition-equilibrium protocol. This has involved making suspensions of erythrocyte ghosts containing albumin and tritium-labeled palmitate that are then allowed to equilibrate. The distribution of the fatty acid is then determined by measuring the [^{3}H]palmitate concentration in the liquid phase ($\overline{C}_A$) of a mixture that has a known total fatty acid concentration ($\overline{C}_A$). Despite the fact that fatty acid incorporation into erythrocyte ghosts is a specific binding phenomenon [66], the dependence of ($\overline{C}_A - \overline{C}_A$) upon $\overline{C}_A$ in the absence of albumin (where $\overline{C}_A$ is synonymous with the free concentration C_A) has been found to be essentially linear over the limited concentration range in which palmitate remains nonmicellar (Figure 23.8a). This is a consequence of having a situation where $K_{AX}\overline{C}_A \ll 1$. Under those conditions, the form of Equation 23.6 with $C_S = 0$ becomes

$$(\overline{C}_A - \overline{C}_A) = K_{AX}\overline{C}_X\,\overline{C}_A/(1 + K_{AX}\overline{C}_A)$$

$$\approx K_{AX}\overline{C}_X\,\overline{C}_A \tag{23.35}$$

which is a linear relationship with a slope of $K_{AX}\overline{C}_X$.

An inability to assign a value to K_{AX} has made it necessary to use a calibration plot like Figure 23.8a to determine the concentration of free palmitate (C_A) in albumin-containing suspensions based on the measured concentration of palmitate that is associated with erythrocyte ghosts. By doing this, the resulting binding curve that is obtained for the interaction of palmitate with bovine serum albumin is essentially linear (see Figure 23.8b). Based on an application of the binomial theorem, the rectangular hyperbolic expression for palmitate binding to n equivalent and independent albumin sites can be written in the form given in Equation 23.36.

$$(\overline{C}_A - \overline{C}_A)/\overline{C}_S = nK_{AS}C_A/(1 + K_{AS}C_A)$$

$$\approx nK_{AS}C_A - nK_{AS}^2 C_A^2 + \dots \tag{23.36}$$

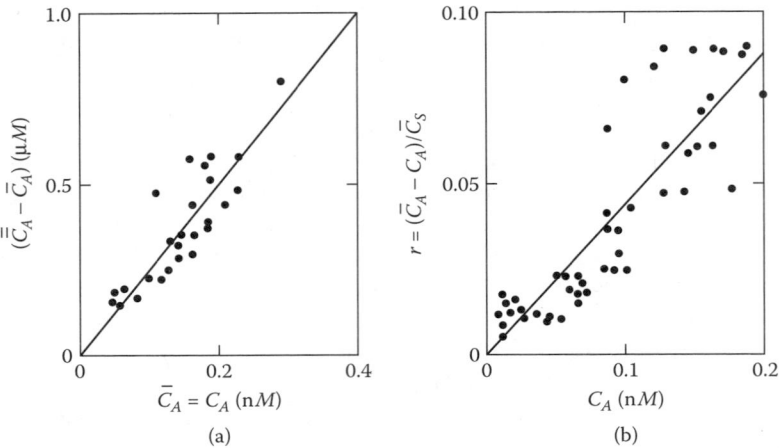

Figure 23.8 Characterization of the interaction between palmitate (A) and human serum albumin (S) by partition-equilibrium studies with erythrocyte ghosts as the affinity matrix. (a) Calibration plot for relating the concentrations of matrix-bound [³H]palmitate to its aqueous-phase concentration $(\bar{C}_A)$ in the absence of albumin. (b) Binding curve for the albumin-palmitate interaction deduced from measurements of $(\bar{C}_A - C_A)$ for mixtures with known composition $(\bar{C}_A, \bar{C}_S)$. (Data taken from Pond, S.M., Gordon, R.A., Simi, A.L., and Winzor, D.J., *Anal. Biochem.*, 237, 232–238, 1996.)

From this result, the slope of Figure 23.8b can be used to assign a value of $4.4 \times 10^8 \ M^{-1}$ to the product of the number of sites on albumin and the intrinsic binding constant for palmitate at these sites [67].

23.6 EVALUATION OF RATE CONSTANTS BY AFFINITY CHROMATOGRAPHY

Quantitative affinity chromatography has traditionally been used to determine equilibrium constants. This has been done by either using the elution volume or retention time for solute *A* or by using the solute's distribution $(\bar{C}_A/C_A)$ to characterize the thermodynamics of its interaction with immobilized affinity sites in the column (*X*). However, affinity chromatography can also be used to estimate the association (k_a) and dissociation (k_d) rate constants that determine the magnitude of the solute's binding constant (K_{AX}), where $K_{AX} = k_a/k_d$. Although initial attempts to achieve this goal by zonal-elution chromatography were disappointing [13, 68–72], a later attempt based on the theoretical plate-height concept has seemingly met with success [73]. Frontal analysis [6, 7, 74] allows the same information to be derived from the flow-rate dependence of the variance (or spread) of the eluted boundary, while the mean volume of this boundary $(\bar{V}_A)$ remains constant—a mandatory requirement for the measurement of an equilibrium parameter.

23.6.1 Theoretical Considerations

For situations in which the elution profile reflects contributions from partition kinetics as well as the kinetics of a solute-matrix interaction, DeLisi and Hethcote [75] have provided the following expression for the dependence of the boundary variance (ρ_A^2)

on the mobile-phase flow rate, F.

$$d\rho_A^2/dF = 2\left(V_A^* - V_o\right)\left[(1 + K_{AX}C_X)^2/k_{-p} + K_{AX}C_X/k_d\right]$$ (23.37)

The effects of partition kinetics (i.e., mobile-phase mass transfer) in this equation are reflected by the term k_{-p}, which is the rate constant for the movement of a solute from a stationary-phase volume $(V_A^* - V_0)$ to the mobile-phase volume (V_0), where V_A^* is the total column volume accessible to the solute. Introduction of $\bar{V}_A$, the elution volume of the partitioning solute, leads to elimination of the terms containing C_X (i.e., the free concentration of immobilized affinity sites) by means of the substitution shown below [7].

$$K_{AX}C_X = \left(\bar{V}_A - V_A^*\right)/\left(V_A^* - V_0\right)$$ (23.38)

The flow-rate dependence of the variance for the solute's elution profile then takes on the form in Equation 23.39,

$$d\rho_A^2/dF = 2\left\{(\bar{V}_A - V_0)^2/\left[k_{-p}\left(V_A^* - V_o\right)\right] + \left(\bar{V}_A - V_A^*\right)/k_d\right\}$$ (23.39)

in which k_{-p} and k_d are the parameters whose values must be determined.

The first of these two parameters (k_{-p}) can be evaluated by using the corresponding flow-rate dependence of ρ_A^2 from an experiment that is conducted on a control column of identical size to the column of interest, but which contains no immobilized affinity sites. The behavior of this control column is then examined by means of the following expression to obtain k_{-p}.

$$k_{-p} = 2\left(V_A^* - V_0\right)/\left(d\rho_A^2/dF\right)$$ (23.40)

Once k_{-p} has been estimated, k_d becomes the only parameter of unknown magnitude in Equation 23.37. To find this parameter, it is helpful to know that Equation 23.37 is based on a pseudo-first-order kinetic assumption that C_X, the concentration of free affinity sites, remains essentially unchanged during complex formation. Since the validity of this improves as $\bar{C}_A$ is decreased, the apparent rate constant $(k_d)_{app}$ that is determined from Equation 23.39 is extrapolated to zero concentration of the partitioning solute to fulfill this inherent assumption.

23.6.2   Experimental Applications

Several experimental aspects of the determination of rate constants by quantitative affinity chromatography can be illustrated by means of published frontal-analysis studies [6, 7]. This example entails the elution of p-nitrophenylmannoside from an affinity column that contained concanavalin A immobilized to controlled-pore glass beads (glyceryl-CPG 3000).

Figure 23.9a shows the effect of column flow rate on the variance of the advancing elution profiles for 40 μM p-nitrophenylmannoside on a concanavalin A-CPG 3000 column (■), as obtained during frontal analysis. This figure also shows the results obtained on an identical column that contained underivatized glyceryl-CPG 3000 (□). This second plot provided an estimate of 4.79 sec^{-1} for k_{-p}, based on Equation 23.40 and the use of a value for $(V_A^* - V_0)$ of 1.58 ml [6]. When the results of this experiment were extrapolated to $\bar{C}_A = 0$ for $(k_d)_{app}$ (with this latter value being obtained by applying Equation 23.39 to the variances of advancing elution profiles), this gave a rate constant of 0.40 sec^{-1} for the dissociation of p-nitrophenylmannoside from its complex with the immobilized

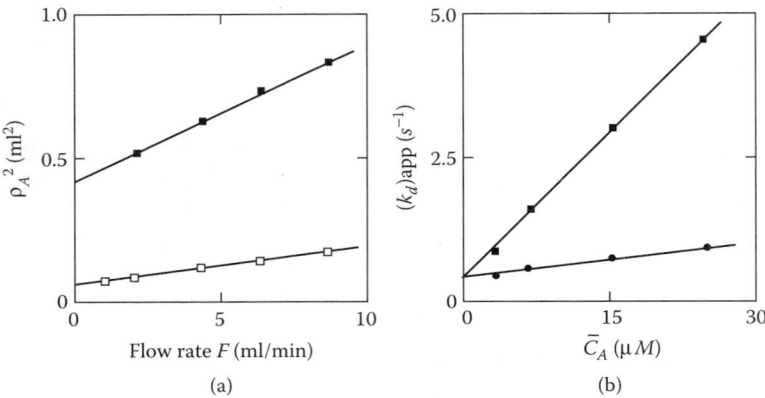

Figure 23.9 Use of high-performance affinity chromatography to evaluate the rate constant for desorption of p-nitrophenylmannoside from concanavalin A immobilized on CPG 3000. (a) Effect of column flow rate on the variance of the advancing elution profiles for p-nitrophenylmannoside on derivatized (■) or underivatized (□) CPG 3000. (b) Determination of the dissociation rate constant (k_d) as the limiting magnitude ($\bar{C}_A \to 0$) of apparent values obtained by applying Equation 23.39 to the flow-rate dependence of variances for advancing (■) and trailing (●) elution profiles. (The data in (a) were taken from Munro, P.D., Winzor, D.J., and Cann, J.R., *J. Chromatogr.*, 646, 3–15, 1993; data in (b) were taken from Munro, P.D., Winzor, D.J., and Cann, J.R., *J. Chromatogr.*, 659, 267–273, 1994.)

concanavalin A (■, Figure 23.9b). This rate constant can also be obtained by performing an analysis of boundary spreading in the trailing elution profiles (●, Figure 23.9b). By combining this information with a binding constant (K_{AX}) for the system of $2.4 \times 10^4\ M^{-1}$ [6], the association rate constant for the same system can be calculated by using the relationship $k_a = K_{AX}/k_d$. This gives a value of $6.0 \times 10^5\ M^{-1}\mathrm{sec}^{-1}$ for k_a.

These and other studies have established the feasibility of characterizing the kinetics of solute interactions with immobilized binding sites by using the flow-rate dependence of boundary spreading in frontal-analysis and zonal-elution chromatography. However, the advent of biosensor technology has afforded a simpler and more direct means for determining the association and dissociation rate constants between a soluble solute and immobilized affinity sites. Further information on this alternative approach is given in the next section.

23.7 IMPACT OF BIOSENSOR TECHNOLOGY

As noted in Section 23.3.3, the use of biosensors in the study of solute binding to immobilized affinity sites has the advantage of allowing complex formation to be monitored directly. This is an important consideration in the characterization of high-affinity interactions. Although initially seen as a way of quantifying the kinetics of biospecific adsorption/desorption phenomena [31, 32, 43, 44], biosensor technology also has much to offer as a means for characterizing the thermodynamics of an interaction [48, 50] (e.g., see Figure 23.10). Since most applications of quantitative affinity chromatography have involved the determination of binding constants, the use of biosensors in thermodynamic studies is discussed first in this section.

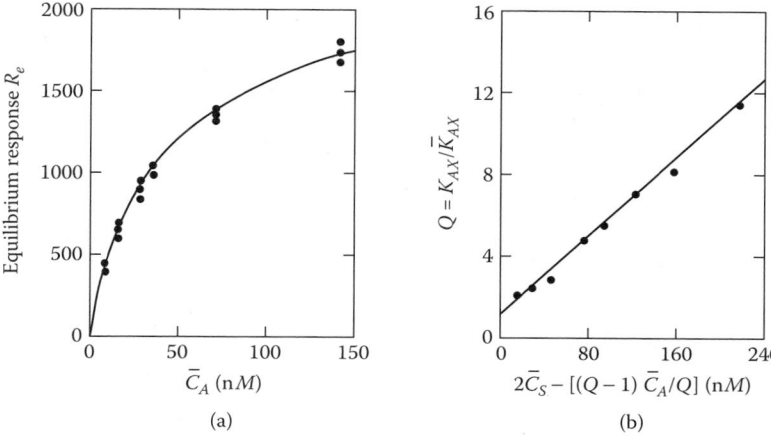

Figure 23.10 Thermodynamic characterization of the interaction between interleukin-6 (S) and the soluble form of its biospecific receptor (A) by means of a Biacore study with interleukin-6 also being used as the immobilized ligand (X). (a) Evaluation of K_{AX} by analysis of the equilibrium response (R_e) in terms of Equation 23.17. (b) Application of a modified form of Equation 23.9 that accounts for interleukin-6 bivalency. (Data taken from Ward, L.D., Howlett, G.J., Hammacher, A., Weinstock, J., Yasukawa, K., Simpson, R.J., and Winzor, D.J., *Biochemistry*, 34, 2901–2907, 1995.)

23.7.1 Biosensors in Thermodynamic Studies

23.7.1.1 Interaction between Interleukin-6 and a Soluble Form of Its Biospecific Receptor

Consideration of biosensor technology as another way of obtaining quantitative affinity chromatography data began in a Biacore study of the interaction between interleukin-6 and a soluble form of its biospecific receptor [76]. This study entailed immobilization of the dimeric recombinant cytokine to avoid the problems of solute bivalence. Application of Equation 23.17 to determine K_{AX} from equilibrium (time-independent) biosensor responses for a series of receptor concentrations is summarized in Figure 23.10a, which signifies a maximal response R_m of 2200 units and a binding constant of $2.4 \times 10^7 \ M^{-1}$. This information was then used in conjunction with competition experiments involving mixtures of receptor and soluble interleukin-6 in the liquid phase to obtain K_{AS}. Figure 23.10b plots these results according to an adapted form of Equation 23.9 that takes into account the bivalence of the competing cytokine (ligand). The slope of Figure 23.10b represents an intrinsic binding constant of $4.8 \times 10^7 \ M^{-1}$ for the interaction of the receptor with two equivalent and independent sites on the recombinant interleukin-6 [76].

23.7.1.2 Allowance for Solute Depletion in Experiments with Cuvette-Based Biosensors

A rigorous thermodynamic analysis of the results obtained with a cuvette-based biosensor will rely on Equation 23.18 to take into account the depletion of solute in the liquid phase as a complex forms on the sensor's surface. The importance of this consideration is illustrated in Figure 23.11. This figure is based on a study with an IBIS instrument [45] that examined the interaction between tyrosine kinase lck SH2 and an immobilized phosphotyrosine peptide. The solid symbols in this figure represent the experimental points

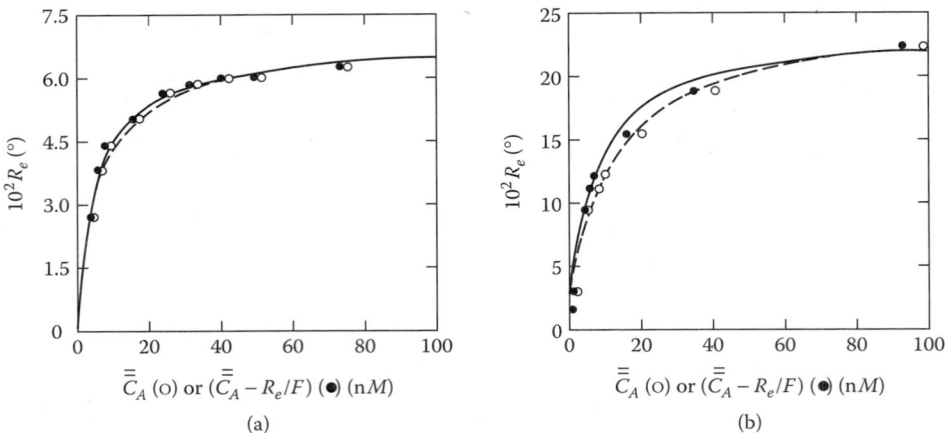

Figure 23.11 Effect of solute depletion in the liquid phase on the analysis of equilibrium responses for the interaction of tyrosine kinase lck SH2 domain with a phosphotyrosine peptide immobilized on the sensor surface of an IBIS cuvette. (a) Binding isotherms obtained by inclusion ($\bullet$) and neglect ($\circ$) of the abscissa R_e term (Equation 23.18) in the analysis of data obtained with a lightly derivatized sensor surface. (b) Corresponding analysis of data for a highly substituted sensor surface. (Data taken from de Mol, N.J., Plomp, E., Fischer, M.J.E., and Ruijtenbeek, R., *Anal. Biochem.*, 279, 61–70, 2000.)

with the abscissa plotted correctly, and the open symbols reflect the result of neglecting the R_e/P factor in Equation 23.18. In an experiment with a lightly substituted biosensor surface (illustrated in Figure 23.11a), the extent of solute depletion is of little consequence. However, solute depletion can become significant in a series of experiments conducted with a highly substituted sensor surface, as is the case in Figure 23.11b. In this particular example, the binding constant of 9.8 (± 0.8) $\times 10^7$ M^{-1} that was obtained from the incorrectly plotted data ($\circ$) increased by almost two-fold to 1.7 (± 0.2) $\times 10^8$ M^{-1} when the correct estimates of the abscissa parameter were used [45].

23.7.1.3   Contribution of Biosensor Studies to an Understanding of RNA Trafficking

The relatively broad range of information that can be gained by employing biosensors for simple thermodynamic measurements is illustrated by a recent IAsys study [77]. In this example, the interaction being studied was between a short segment of myelin basic protein mRNA and heterogeneous nuclear ribonucleoprotein A2 (hnRNP A2), a member of a family of ribonucleoproteins involved in RNA packaging. The mRNA examined in this work was a 21-nucleotide response element (A2RE) that has been shown to be necessary and sufficient for the transport of the message-encoding myelin basic protein in the cytoplasm of oligodendrocytes [78–80]. Because of its fundamental role in RNA trafficking, the specificity of this interaction between hnRNP A2 and A2RE is of great interest. The strength and stoichiometry of this interaction are also of interest.

    The first question addressed for this system through the use of biosensors was the specificity that is exhibited by hnRNP A2 toward 21-residue polyribonucleotides. To determine this, an identical amount of hnRNP A2 was immobilized on the two sensor surfaces of a double-compartment cuvette. This was done to allow a comparison to be made between the time-course response for a solution of A2RE and the response for a polyribonucleotide

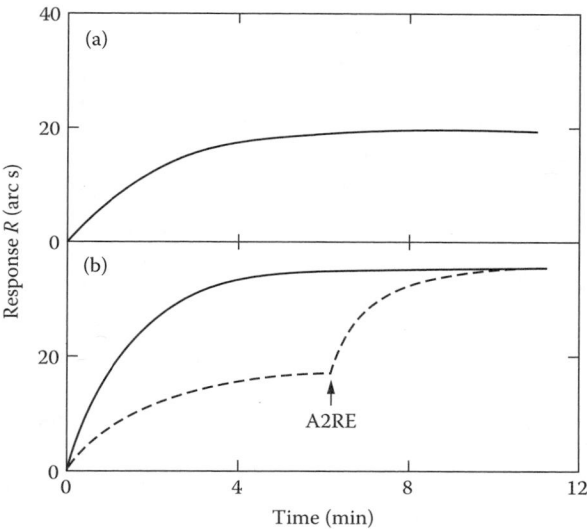

Figure 23.12 Examination of the specificity of hnRNP A2 for the 21-ribonucleotide response element A2RE by IAsys studies with the heterogeneous ribonucleoprotein as immobilized ligand. (a) Time-course of response obtained for a 10 μM solution of a 21-residue ribonucleotide with a scrambled sequence. (b) Comparison of the corresponding time course for 0.5 μM A2RE (——) with that (---) obtained for the sequential addition of scrambled nucleotide (10 μM) and A2RE (0.5 μM) to the cuvette. (Data taken from Shan, J., Moran-Jones, K., Munro, T.P., Kidd, G.J., Winzor, D.J., Hoek, K.S., and Smith, R., *J. Biol. Chem.*, 275, 38286–38296, 2000.)

that contained a scrambled version of the A2RE sequence. From the time-course data that were generated for the latter polyribonucleotide (see Figure 23.12a), it was found that an affinity for hnRNP A2 was not restricted to the A2RE sequence. However, a twofold greater amount of complex formation occurred in the presence of a polyribonucleotide that had the sequence for which specificity was anticipated (Figure 23.12b, solid line). This indicated that hnRNP A2 must possess two classes of sites with affinities for polyribonucleotides.

To assess the relationship between these two site classes, an experiment was next performed that involved the sequential bathing of the sensor surface with solutions of the scrambled ribonucleotide and A2RE. A logical conclusion that can be drawn from the resulting time-course data (see the broken line in Figure 23.12b) is that the binding of hnRNP A2 to nonspecific RNA is restricted to one class of sites, while the specific response element A2RE exhibits an affinity for both classes of sites.

The number of hnRNP A2 sites with an affinity for A2RE was determined by first establishing a binding isotherm for this specific polyribonucleotide (see Figure 23.13a). This calibration plot of response as a function of free ribonucleotide concentration also allowed the equilibrium response to be expressed in terms of the concentration of uncomplexed A2RE (C_A) that occurred in competition experiments performed with mixtures of hnRNP A2 and A2RE in the liquid phase (having concentrations of $\bar{C}_S$ and $\bar{C}_A$, respectively). The resulting plot is shown in Figure 23.13b. From the Scatchard plot for these results (see Figure 23.13c), the abscissa intercept indicated that there was a total of two binding sites for A2RE. It can therefore be concluded that hnRNP A2 possesses a single site with specificity for the myelin basic protein response element (with K_{AS} for this site being greater than $2 \times 10^7 \, M^{-1}$) along with a nonspecific site ($K_{AS} = 3$ to $4 \times 10^6 \, M^{-1}$) that has an affinity for any polyribonucleotide (as well as a charged polysaccharide such as heparin) [77].

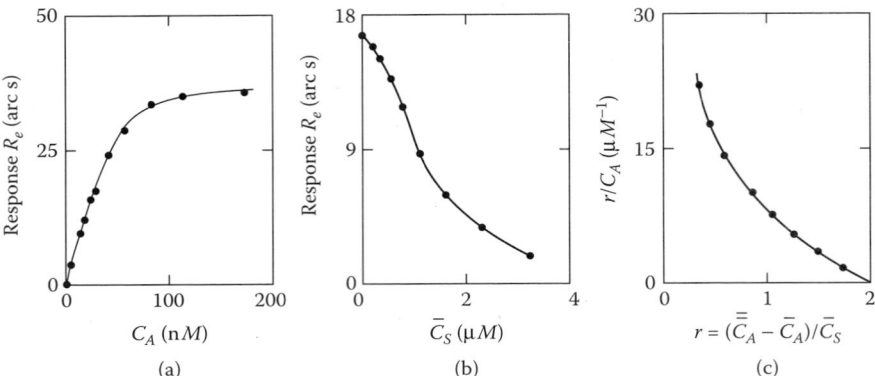

Figure 23.13 Use of equilibrium biosensor responses to determine the stoichiometry for the interaction between hnRNP A2 and the response element A2RE. (a) Calibration curve for the dependence of the IAsys biosensor response on the concentration of A2RE ($\bar{C}_A$) in equilibrium with immobilized hnRNP A2. (b) Dependence of R_e on the concentration of hnRNP A2 (C_S) in mixtures containing a fixed concentration of A2RE ($\bar{C}_A = 1\ \mu M$). (c) Scatchard plot of the consequent binding data obtained on the basis of $\bar{C}_A$ values deduced from R_e and the given calibration plot. (Data taken from Shan, J., Moran-Jones, K., Munro, T.P., Kidd, G.J., Winzor, D.J., Hoek, K.S., and Smith, R., *J. Biol. Chem.*, 275, 38286–38296, 2000.)

23.7.2 Kinetic Analysis of Biosensor Data

While the thermodynamic interpretation of biosensor data merely entails a slight adaptation of existing theory for quantitative affinity chromatography, it has been necessary to develop new expressions for extracting rate constants from such data. In past work, considerations seem to have been confined to the interaction of a solute with immobilized affinity sites in the belief (albeit mistaken) that K_{AX} accurately describes the corresponding reaction between the same two species in solution. For consistency within this chapter, $\bar{C}_A$ will still be used to represent the concentration of solute in the liquid phase. However, the overbar notation used earlier in this chapter now becomes redundant, since the concentration $\bar{C}_A$ invariably refers to the free solute in the kinetic expressions that will be presented.

23.7.2.1 Pseudo-First-Order Kinetics

The standard kinetic approach [31, 32] for characterizing the interaction between a solute and immobilized affinity site is based on a 1:1 solute-matrix interaction (see Figure 23.1a and Figure 23.1b). In this situation, the binding constant K_{AX} can be expressed as the ratio of the association and dissociation rate constants (k_a and k_d), where $K_{AX} = k_a/k_d$. Based on this relationship, standard kinetic considerations yield the following differential rate equation for such a system,

$$dC_{AX}/dt = k_a\bar{C}_A C_X - k_d C_{AX} \tag{23.41}$$

where C_X, the free concentration of immobilized affinity sites, can be replaced by ($\bar{C}_X - C_{AX}$). By substituting the instrumental response R for the concentration of bound solute (C_{AX}) at time t and by substituting R_m for $\bar{C}_X$, Equation 23.41 takes on the form shown in Equation 23.42.

$$dR/dt = k_a R_m \bar{C}_A - (k_a\bar{C}_A + k_d)R \tag{23.42}$$

In the Biacore instrument, the continual flow of solute across the sensor surface is considered to ensure a constant free-solute concentration in the liquid phase $(\overline{C}_A)$. Under these conditions, Equation 23.42 becomes a pseudo-first-order kinetic expression [32],

$$R = [R_m k_a \overline{C}_A / (k_a \overline{C}_A + k_d)][1 - \exp\{-(k_a \overline{C}_A + k_d)t\}] \qquad (23.43)$$

where the term $R_m k_a \overline{C}_A / (k_a \overline{C}_A + k_d)$ can be replaced by the equilibrium response, R_e [52]. During the adsorption stage of an experiment on this instrument, the time dependence of the response is thus described by a limiting (equilibrium) asymptote of R_e and a pseudo-first-order rate constant, $(k_a \overline{C}_A + k_d)$.

The desorption stage of a biosensor experiment is initiated by a switch from the solute solution to a buffer that contains no solute. If it is assumed that the flow of this buffer across the sensor's surface makes $\overline{C}_A$ equal to zero, an analysis of the desorption time-course data by using the relationship

$$R/R_0 = \exp(-k_d t) \qquad (23.44)$$

will provide an estimate of the dissociation rate constant. This is accomplished by using the exponential decay of the biosensor response R relative to its value R_0 at the beginning of the analysis of the desorption step. In practice, better agreement with Equation 23.44 is often obtained by employing a solution with a competing ligand in place of only buffer. This occurs because the ligand helps complex the free solute and ensures closer compliance with the assumption in Equation 23.44 that $\overline{C}_A$ is equal to zero [45, 81, 82].

23.7.2.2   Second-Order Kinetics

Although Equation 23.43 has also been used for analyzing results obtained with cuvette-based biosensors [43, 44], the assumption that $\overline{C}_A$ is constant is clearly a more questionable approximation in an experiment where the adsorption of solute occurs at the expense of its liquid-phase concentration. Elimination of this assumption from the derivation of the differential rate equation for a 1:1 interaction leads to the following relationship [47],

$$dC_{AX}/dt = k_a \overline{\overline{C}}_A \overline{C}_X + k_a \left[C^2_{AX} - C_{AX}(\overline{C}_A + \overline{C}_X) \right] - k_d C_{AX} \qquad (23.45)$$

where $\overline{\overline{C}}_A$ again represents the total solute concentration in the cuvette. This second-order differential rate equation can be solved either analytically [83] or through numerical integration [47]. When expressed in terms of the biosensor's response, the resulting integrated expression is given in Equation 23.46.

$$R = R_e[1 - (p - R_e)/\{R_e \exp[(k_a/P)(p - R_e)t] - R_e\}] \qquad (23.46)$$

In this expression, R_e (the equilibrium response) and p are the two roots of the quadratic equation

$$R^2 - (P\overline{\overline{C}}_A + R_m + Pk_d/k_a)R + P\overline{\overline{C}}_A R_m = 0 \qquad (23.47)$$

and P is the concentration-response proportionality constant (as described earlier for Equation 23.18).

A second-order kinetic analysis is also required in Biacore studies with bivalent solutes [84, 85]. For a system in which a bivalent solute interacts with immobilized affinity sites according to the scheme shown in Equation 23.48,

$$A + X \overset{(k_a)_1}{\underset{(k_d)_1}{\longleftrightarrow}} AX + X \overset{(k_a)_2}{\underset{(k_d)_2}{\longleftrightarrow}} AX_2 \tag{23.48}$$

the differential rate equations for complex formation take on the forms shown in Equations 23.49 and 23.50.

$$dC_{AX}/dt = 2(k_a)_1 \bar{C}_A C_X - (k_d)_1 C_{AX} - (k_a)^2 C_{AX} C_X + 2(k_d)_2 C_{AX_2} \tag{23.49}$$

$$dC_{AX_2}/dt = (k_a)_2 C_{AX} C_X - 2(k_d)_2 C_{AX} \tag{23.50}$$

After substituting the instrumental responses for the complexed ligand concentrations, these differential expressions then become

$$dR_1/dt = 2(k_a)_1 \bar{C}_A (R_m - R_1 - R_2) - (k_d)_1 R_1 - (k_a)_2 R_1 (R_m - R_1 - R_2) + 2(k_d)_2 R_2 \tag{23.51}$$

$$dR_2/dt = (k_a)_2 R_1 (R_m - R_1 - R_2) - 2(k_d)_2 R_2 \tag{23.52}$$

where both R_1 and R_2 (the responses that reflect concentrations AX and AX_2, respectively) contribute to the observed instrument response (R). Numerical integration has been used to solve these two differential rate equations, in conjunction with the constraint that $R = (R_1 + R_2)$; this has been performed through the use of BIAeval 3.0 global analysis software [84, 85].

23.7.2.3 Biacore Analysis of a 1:1 Interaction

An example of a pseudo-first-order kinetic analysis by the Biacore instrument is a study described by O'Shannessy et al. [32]. This work examined the binding of an F_{ab} fragment from an anti-CD4 monoclonal antibody (10 nM) with its antigen, with the latter being immobilized on the sensor surface. The resulting desorption curve is shown in Figure 23.14a.

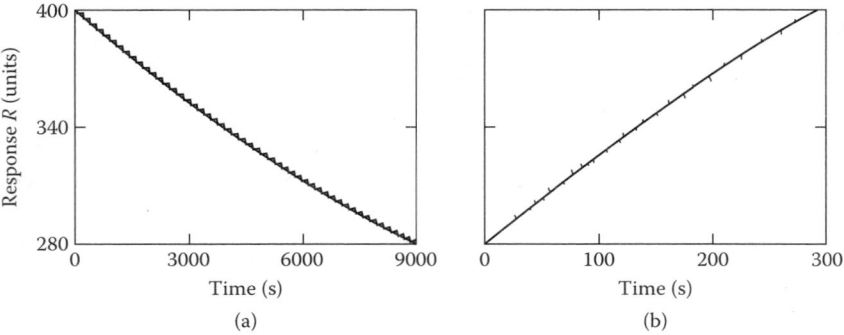

Figure 23.14 Kinetic analysis of Biacore data for the interaction between an anti-CD4 Fab fragment and its immobilized antigen. (a) Determination of k_d by fitting Equation 23.44 to the desorption data for an experiment performed with 10-nM Fab fragment. (b) Analysis of the adsorption stage of the same experiment (Equation 23.43) to obtain k_a and R_m as the remaining parameters of unknown magnitude. (Data taken from O'Shannessy, D.J., Brigham-Burke, M., Soneson, K.K., Hensley, P., and Brooks, I., *Anal. Biochem.*, 212, 457–468, 1993.)

Analysis of this plot according to Equation 23.44 (with k_d as the sole curve-fitting parameter) gave a dissociation rate constant of 3.98×10^{-5} sec^{-1}. The adsorption time-course data for the same concentration of F_{ab} fragments (Figure 23.14b) were then fit to Equation 23.43, with k_d being fixed at its previously measured value; this gave an association rate constant (k_a) of 1.04×10^5 M^{-1}sec^{-1}. A binding constant (K_{AX}) of 2.6×10^9 M^{-1} was then calculated from these rate constants [32].

Global analysis [86, 87] by means of CLAMP software [88] or BIAeval 3.0 software can be used to obtain a unique set of best-fit estimates for k_a, k_d, and R_m by the simultaneous curve-fitting of all time-course data. However, such analysis places stringent demands on the data quality, which must conform rigorously to the pseudo-first-order kinetic behavior that is dictated by a simple 1:1 interaction model.

Regarding this last item, it is disturbing that deviations from pseudo-first-order behavior are exhibited by interactions for which the inherent requirement for 1:1 stoichiometry should be a reasonable assumption. Some explanations that have been suggested for this deviant behavior have included heterogeneity of the immobilized affinity sites [89], isomerization of the AX complex [90], and the restricted access of solute to affinity sites as a result of mass-transport limitations [91–95]. The temporary unavailability of unoccupied sites that arises from the inefficient location of bound solute molecules (i.e., the parking problem) can also give rise to this behavior in instances where the partitioning solute is larger than the immobilized ligand [96, 97].

Although it is certainly possible to develop an analysis technique that encompasses a single complicating factor like limited mass transport [45, 98], the general problem of devising a kinetic analysis that accommodates potential contributions from a combination of such factors is seemingly intractable. When deviations from pseudo-first-order behavior are encountered in the kinetic analysis of Biacore data, it may be prudent to instead revert to a thermodynamic characterization. This is because the validity of this latter approach only comes into question if the time-independent response might be reflecting a transient steady state prior to elimination of the parking problem [99].

23.8   SUMMARY AND CONCLUSIONS

The most significant developments in quantitative affinity chromatography over the past decade have involved its application in the area of biosensor technology, a field still in its infancy at the time of the previous edition of this book [100]. Biosensor technology has the advantage of allowing the study of interactions that cannot be characterized by other types of quantitative affinity chromatography (i.e., those that monitor complex formation through the extent of solute depletion in the liquid phase). The ability of biosensor technology to directly examine complex formation thus adds another dimension to the versatility of quantitative affinity chromatography by extending the range of measurable binding constants. This is especially true for interactions that were originally too strong to study by other techniques.

Many studies have been based on the mistaken belief that the major contribution of biosensor technology would be the provision of information on the kinetics and mechanism of interaction between a partitioning solute and immobilized affinity sites. But this approach merely provides an alternative means for obtaining the kinetic parameters that pertain to a thermodynamic description of a reaction confined to a single phase [101, 102]. Furthermore, because this characterization is based on the interaction of a solute with a chemically modified (immobilized) affinity partner, it does not necessarily describe the corresponding interaction in solution (i.e., the reaction that is usually of interest). However, this latter system can be characterized by instead using a competitive-binding assay in which the interaction of a solute with immobilized affinity sites is used to assess its free concentration in a known mixture of the solute and a competing ligand [47, 49, 76, 77, 82]. In this regard, the necessity to conduct competition experiments has always been a central theme in the development of quantitative affinity chromatography for the evaluation of binding constants. It is therefore surprising that so little attention has been directed toward this aspect in the characterization of interactions by biosensor technology.

Both this review and its predecessor [1] have indicated the power of quantitative affinity chromatography in its various forms as a means for studying the biological interactions. This work involves not only an identification of reactants, but also a characterization of equilibrium constants, thus allowing a complete understanding of physiological systems at the molecular level. Of the many methods that are available for such endeavors, quantitative affinity chromatography is certainly one of the most versatile. There is no restriction on the size of ligand that can be examined, and a vast range of binding constants (e.g., less than $10^3 \ M^{-1}$ to greater than $10^9 \ M^{-1}$) can be measured by this technique [10, 21].

SYMBOLS AND ABBREVIATIONS

A	Partitioning solute
AS	Complex between a partitioning solute and soluble ligand
ASX	Ternary complex between a solute (A), a soluble ligand (S), and an immobilized affinity site (X)
AX	Complex between a partitioning solute and immobilized affinity site
C_i	Free molar concentration of species i
$\overline{C}_i$	Total molar concentration of constituent i in a liquid phase ($i = A$ or S)
$\overline{C}_X$	Effective total concentration of immobilized affinity sites
$\overline{\overline{C}}_i$	Total molar concentration of species i (A or S) in a two-phase system
f	Number of sites on a partitioning solute (i.e., the solute valence)
F	Column flow rate
k_a	Association rate constant for the interaction of a solute with immobilized affinity sites
k_d	Dissociation rate constant for the interaction of a solute with immobilized affinity sites
k_{-p}	Rate constant for the efflux of a partitioning solute from a solid phase
K_{AS}	Intrinsic binding constant (M^{-1}) for the interaction of a partitioning solute with a soluble ligand
K_{ASX}	Binding constant (M^{-1}) for ternary complex formation between a solute-ligand complex (AS) and immobilized affinity sites
K_{AX}	Intrinsic binding constant (M^{-1}) for interaction of a partitioning solute with an immobilized affinity site (i.e., the solute-matrix interaction)
$\overline{K}_{AX}$	Constitutive equilibrium constant (M^{-1}) for the solute-matrix interaction in competitive-binding studies
K_{SX}	Binding constant (M^{-1}) for interaction between a soluble ligand and immobilized affinity sites
P	Proportionality constant relating a biosensor's response to a bound-solute's concentration
Q	Ratio of solute-matrix binding constants ($K_{AX}/\overline{K}_{AX}$) measured in the absence and presence of a competing soluble ligand
r_f	Binding function for an f-valent solute, where $r_f = (\overline{\overline{C}}_A{}^{1/f} - \overline{C}_A{}^{1/f})/\overline{C}_X$
R	Biosensor response

R_e	Equilibrium biosensor response for a concentration $\bar{C}_A$ of a solute in the liquid phase
R_m	Maximum biosensor response (i.e., the equivalent of $\bar{C}_X$)
S	Soluble ligand that interacts either with a partitioning solute (A) or with an immobilized affinity site (X)
t	Time (in biosensor studies)
t_A^*	Retention time for a partitioning solute in the absence of an interaction with immobilized affinity sites
$\bar{t}_A$	Measured (constituent) retention time pertaining to a constituent concentration $\bar{C}_A$ for a partitioning solute
V_A^*	Elution volume of a solute in the absence of any interaction with immobilized affinity sites
$\bar{V}_A$	Measured (constituent) elution volume pertaining to a constituent concentration $\bar{C}_A$ for a partitioning solute
V_0	Volume of mobile phase in an affinity column (i.e., the void volume)
X	Immobilized affinity (ligand) site
σ_A^2	Variance (spread) of a boundary in frontal affinity chromatography

REFERENCES

1. Winzor, D.J. and Jackson, C.M., Determination of binding constants by quantitative affinity chromatography: Current and future applications, in *Handbook of Affinity Chromatography*, Kline, T., Ed., Marcel Dekker, New York, 1993, pp. 258–298.
2. Jönsson, U., Fägerstam, L., Ivarsson, B., Johnsson, B., Karlsson, R., Lundh, K., Löfäs, S., Persson, B., Roos, H., Rönnberg, I., Sjölander, S., Stenberg, E., Ståhlberg, R., Urbaniczky, C., Östlin, H., and Malmqvist, M., Real-time biospecific interaction analysis using surface plasmon resonance and a sensor chip technology, *BioTechniques,* 11, 620–628, 1991.
3. Cush, R., Cronin, J.M., Stewart, W.J., Maule, C.H., Molloy, J., and Goddard, N.J., The resonant mirror: A novel optical biosensor for direct sensing of biomolecular interactions, I: Principle of operation, *Biosens. Bioelectron.,* 8, 347–354, 1993.
4. Brekkan, E., Lundqvist, A., and Lundahl, P., Immobilized membrane vesicle or proteoliposome chromatography: Frontal analysis of interactions of cytochalasin *B* and *D*-glucose with the human red cell glucose transporter, *Biochemistry,* 35, 12141–12145, 1996.
5. Haneskog, L., Zeng, C.-M., Lundqvist, A., and Lundahl, P., Biomembrane affinity chromatographic analysis of inhibitor binding to the human red cell nucleoside transporter in immobilized cells, vesicles and proteoliposomes, *Biochim. Biophys. Acta,* 1371, 1–4, 1998.
6. Munro, P.D., Winzor, D.J., and Cann, J.R., Experimental and theoretical studies of rate constant evaluation by affinity chromatography: Determination of rate constants for the interaction of saccharides with concanavalin A, *J. Chromatogr.,* 646, 3–15, 1993.
7. Munro, P.D., Winzor, D.J., and Cann, J.R., Allowance for kinetics of partition in the determination of rate constants by affinity chromatography, *J. Chromatogr.,* 659, 267–273, 1994.
8. Harris, S.J., Jackson, C.M., and Winzor, D.J., The rectangular hyperbolic binding equation for multivalent ligands, *Arch. Biochem. Biophys.,* 316, 20–23, 1995.
9. Hogg, P.J. and Winzor, D.J., Effects of ligand multivalency in binding studies: A general counterpart of the Scatchard analysis, *Biochim. Biophys. Acta,* 843, 159–163, 1985.
10. Winzor, D.J., Quantitative affinity chromatography, *J. Biochem. Biophys. Methods,* 49, 99–121, 2001.
11. Dunn, B.M. and Chaiken, I.M., Quantitative affinity chromatography: Determination of binding constants by elution with competitive inhibitors, *Proc. Natl. Acad. Sci. U.S.A.,* 71, 2382–2385, 1974.

12. Dunn, B.M., Danner-Rabovsky, J., and Cambias, J.S., Application of quantitative affinity chromatography to the study of protein-ligand interactions, in *Affinity Chromatography and Biological Recognition,* Chaiken, I.M., Wilchek, M., and Parikh, I., Eds., Academic Press, New York, 1983, pp. 93–102.

13. Muller, A.J. and Carr, P.W., Chromatographic study of the thermodynamic and kinetic characteristics of silica-bound concanavalin A, *J. Chromatogr.,* 284, 33–51, 1984.

14. Winzor, D.J., Munro, P.D., and Jackson, C.M., The study of high-affinity interactions by quantitative affinity chromatography: Analytical expressions in terms of total ligand concentration, *J. Chromatogr.,* 597, 57–66, 1992.

15. Andrews, P., Kitchen, B.J., and Winzor, D.J., Use of affinity chromatography for the quantitative study of acceptor-ligand interactions: The lactose synthetase system, *Biochem. J.,* 135, 897–900, 1973.

16. Brinkworth, R.I., Masters, C.J., and Winzor, D.J., Evaluation of equilibrium constants for the interaction of lactate dehydrogenase with reduced nicotinamide adenine dinucleotide by affinity chromatography, *Biochem. J.,* 151, 631–636, 1975.

17. Gilbert, G.A., Elution volume versus reciprocal elution volume in the interpretation of gel filtration experiments, *Nature,* 210, 299–300, 1966.

18. Hogg, P.J. and Winzor, D.J., Quantitative affinity chromatography: Further developments in the analysis of experimental results from column chromatographic and partition equilibrium experiments, *Arch. Biochem. Biophys.,* 234, 55–63, 1984.

19. Nichol, L.W., Ogston, A.G., Winzor, D.J., and Sawyer, W.H., Evaluation of equilibrium constants by affinity chromatography, *Biochem. J.,* 143, 435–443, 1974.

20. Kyprianou, P. and Yon, R.J., A quantitative study of biospecific desorption of rat liver (M4) lactate dehydrogenase from 10-carboxydecylamino-Sepharose: Determination of the number of ligand-binding sites blocked on adsorption, *Biochem. J.,* 207, 549–556, 1982.

21. Winzor, D.J., Quantitative affinity chromatography, in *Affinity Separations: A Practical Approach,* Matejtschuk, P., Ed., IRL Press, Oxford, U.K., 1997, pp. 39–60.

22. Scatchard, G., The attraction of proteins for small molecules and ions, *Ann. N.Y. Acad. Sci.,* 51, 660–672, 1949.

23. Klotz, I.M., Ligand-receptor interactions: What we can and cannot learn from binding measurements, *Trends Pharmacol. Sci.,* 4, 253–255, 1983.

24. Bergman, D.A. and Winzor, D.J., Quantitative affinity chromatography: Increased versatility of the technique for studies of ligand binding, *Anal. Biochem.,* 153, 380–386, 1986.

25. Waltham, M.C., Holland, J.W., Nixon, P.F., and Winzor, D.J., Thermodynamic characterization of the interactions of methotrexate with dihydrofolate reductase by quantitative affinity chromatography, *Biochem. Pharmacol.,* 37, 541–545, 1988.

26. Hogg, P.J., Jackson, C.M., and Winzor, D.J., Use of quantitative affinity chromatography for characterizing high-affinity interactions: Binding of heparin to antithrombin III, *Anal. Biochem.,* 192, 303–311, 1991.

27. Klotz, I.M., The application of the law of mass action to binding by proteins: Interactions with calcium, *Arch. Biochem.,* 9, 109–117, 1946.

28. Nichol, L.W., Ward, L.D., and Winzor, D.J., Multivalency of the partitioning species in quantitative affinity chromatography: Evaluation of the site-binding constant for the aldolase-phosphate interaction from studies with cellulose phosphate as the affinity matrix, *Biochemistry,* 20, 4856–4860, 1981.

29. Hogg, P.J. and Winzor, D.J., Further probes into quantitative aspects of competitive binding assays: Allowance for effects of antigen multivalency in immunoassays, *Arch. Biochem. Biophys.,* 254, 92–101, 1987.

30. Hogg, P.J., Johnston, S.C., Bowles, M.R., Pond, S.M., and Winzor, D.J., Evaluation of equilibrium constants by solid-phase immunoassay: The binding of paraquat to its elicited mouse monoclonal antibody, *Mol. Immunol.,* 24, 797–801, 1987.

31. Karlsson, R., Michäelsson, A., and Mattsson, L., Kinetic analysis of monoclonal antibody-antigen interactions with a new biosensor-based analytical system, *J. Immunol. Methods,* 145, 229–240, 1991.

32. O'Shannessy, D.J., Brigham-Burke, M., Soneson, K.K., Hensley, P., and Brooks, I., Determination of rate constants for macromolecular interactions using surface plasmon resonance: Use of nonlinear least-squares analysis methods, *Anal. Biochem.,* 212, 457–468, 1993.
33. Winzor, D.J., Quantitative characterization of interactions by affinity chromatography, in *Affinity Chromatography: A Practical Approach,* Dean, P.D.G., Johnson, W.S., and Middle, F.A., Eds., IRL Press, Oxford, U.K., 1985, pp. 149–168.
34. Winzor, D.J., Measurement of binding constants by frontal affinity chromatography, in *Quantitative Analysis of Biospecific Interactions,* Lundahl, P., Lundqvist, A., and Greijer, E., Eds., Harwood Academic Publishers, Amsterdam, 1998, pp. 35–53.
35. Winzor, D.J., Determination of affinity constants by partition equilibrium methods, in *Quantitative Analysis of Biospecific Interactions,* Lundahl, P., Lundqvist, A., and Greijer, E., Eds., Harwood Academic Publishers, Amsterdam, 1998, pp. 55–77.
36. Kasai, K. and Ishii, S., Affinity chromatography of trypsin and related enzymes, V: Basic studies of quantitative affinity chromatography, *J. Biochem. (Tokyo),* 84, 1051–1060, 1978.
37. Abercrombie, D.M. and Chaiken, I.M., Zonal elution quantitative affinity chromatography and analysis of molecular interactions, in *Affinity Chromatography: A Practical Approach,* Dean, P.D.G., Johnson, WS., and Middle, F.A., Eds., IRL Press, Oxford, U.K., 1985, pp. 169–189.
38. Swaisgood, H.E. and Chaiken, I.M., Analytical affinity chromatography and the characterization of biomolecular interactions, in *Analytical Affinity Chromatography,* Chaiken, I.M., Ed., CRC Press, Boca Raton, FL, 1987, pp. 65–115.
39. Hage, D.S., High-performance affinity chromatography: A powerful tool for studying serum protein binding, *J. Chromatogr. B,* 768, 3–30, 2002.
40. Kuter, M.R., Masters, C.J., and Winzor, D.J., Equilibrium partition studies of the interaction between aldolase and myofibrils, *Arch. Biochem. Biophys.,* 225, 384–389, 1983.
41. Harris, S.J. and Winzor, D.J., Equilibrium partition studies of the myofibrillar interactions of glycolytic enzymes, *Arch. Biochem. Biophys.,* 275, 185–191, 1989.
42. Harris, S.J. and Winzor, D.J., Effect of calcium ion on the interaction of aldolase with rabbit muscle myofibrils, *Biochim. Biophys. Acta,* 999, 95–99, 1989.
43. Buckle, P.E., Davies, R.J., Kinning, T., Yeung, D., Edwards, P.R., Pollard-Knight, D., and Lowe, C.R., The resonant mirror biosensor: A novel optical biosensor for direct sensing of biomolecular interactions, Part II: Applications, *Biosens. Bioelectron.,* 8, 355–363, 1993.
44. Edwards, P.R., Gill, A., Pollard-Knight, D.V., Hoare, M., Buckle, P.E., Lowe, P.A., and Leatherbarrow, R.J., Kinetics of protein-protein interactions at the surface of an optical biosensor, *Anal. Biochem.,* 231, 210–217, 1995.
45. de Mol, N.J., Plomp, E., Fischer, M.J.E., and Ruijtenbeek, R., Kinetic analysis of the mass transport limited interaction between the tyrosine kinase lck SH2 domain and a phosphorylated peptide studied by a new cuvette-based surface plasmon resonance instrument, *Anal. Biochem.,* 279, 61–70, 2000.
46. Stenberg, E., Persson, B., Roos, H., and Urbaniczky, C., Quantitative determination of surface concentration of protein with surface plasmon resonance using radiolabeled proteins, *J. Colloid Interface Sci.,* 143, 513–526, 1991.
47. Hall, D.R. and Winzor, D.J., Use of a resonant mirror biosensor to characterize the interaction of carboxypeptidase A with an elicited monoclonal antibody, *Anal. Biochem.,* 244, 152–160, 1997.
48. Hall, D.R. and Winzor, D.J., Potential of biosensor technology for the characterization of interactions by quantitative affinity chromatography, *J. Chromatogr. B,* 715, 163–181, 1998.
49. Schuck, P., Millar, D.B., and Korṛt, A., Determination of binding constants by equilibrium titration with a circulating sample in a surface plasmon resonance biosensor, *Anal. Biochem.,* 265, 79–91, 1998.
50. Ward, L.D. and Winzor, D.J., Relative merits of optical biosensors based on flow-cell and cuvette designs, *Anal. Biochem.,* 285, 179–193, 2000.
51. Flory, P.J., Molecular size distribution in three-dimensional polymers, I: Gelation, *J. Am. Chem. Soc.,* 63, 3083–3090, 1941.

52. Kalinin, N.L., Ward, L.D., and Winzor, D.J., Effects of solute multivalence on the evaluation of binding constants by biosensor technology: Studies with concanavalin A and interleukin-6 as partitioning proteins, *Anal. Biochem.,* 228, 238–244, 1995.

53. Thompson, C.J. and Klotz, I.M., Macromolecule–small-molecule interactions: Analytical and graphical reexamination, *Arch. Biochem. Biophys.,* 147, 178–185, 1971.

54. Haneskog, L., Lundqvist, A., and Lundahl, P., Biomembrane affinity chromatographic analysis of nitrobenzylthioinosine binding to the reconstituted red cell nucleoside transporter, *J. Mol. Recognit.,* 11, 58–61, 1998.

55. Lu, L., Lundqvist, A., Zeng, C.-M., Lagerquist, C., and Lundahl, P., D-Glucose, forskolin and cytochalasin B affinities for the glucose transporter Glut 1, *J. Chromatogr. A,* 776, 81–86, 1997.

56. Lundqvist, A., Brekkan, E., Haneskog, L., Yang, Q., Miyaki, J., and Lundahl, P., Determination of transmembrane protein affinities for solutes by frontal chromatography, in *Quantitative Analysis of Biospecific Interactions,* Lundahl, P., Lundqvist, A., and Greijer, E., Eds., Harwood Academic Publishers, Amsterdam, 1998, pp. 79–93.

57. Zeng, C.-M., Zhang, Y., Lu, L., Brekkan, E., Lundqvist, A., and Lundahl, P., Immobilization of human red cells in gel particles for chromatographic activity studies of the glucose transporter Glut 1, *Biochim. Biophys. Acta,* 1325, 91–98, 1997.

58. Yang, Q. and Lundahl, P., Immobilized proteoliposome affinity chromatography for quantitative analysis of specific interactions between solutes and membrane proteins: Interaction of cytochalasin B with the glucose transporter Glut 1, *Biochemistry,* 34, 7289–7294, 1995.

59. Hjertén, S., Liao, J.-L., and Zhang, R., High-performance liquid chromatography on continuous polymer beds, *J. Chromatogr.,* 473, 273–275, 1989.

60. Kelley, G.E. and Winzor, D.J., Quantitative characterization of the interactions of aldolase and glyceraldehyde-3-phosphate dehydrogenase with erythrocyte membranes, *Biochim. Biophys. Acta,* 778, 67–73, 1984.

61. Harris, S.J. and Winzor, D.J., Interactions of glycolytic enzymes with erythrocyte membranes, *Biochim. Biophys. Acta,* 1038, 306–314, 1990.

62. Pedersen, A.O. and Brodersen, R., Myristic acid binding to human serum albumin investigated by dialytic exchange rate, *J. Biol. Chem.,* 263, 10236–10239, 1988.

63. Burczynski, F.J., Pond, S.M., Davis, C.K., Johnson, L.P., and Weisiger, R.A., Calibration of albumin-fatty acid binding constants measured by heptane-water partition, *Am. J. Physiol. G,* 265, 555–563, 1993.

64. Goodman, D.S., The interaction of human serum albumin with long-chain fatty acid anions, *J. Am. Chem. Soc.,* 80, 3892–3898, 1958.

65. Bojesen, I.N. and Bojesen, E., Palmitate binding to and efflux kinetics from human erythrocyte ghosts, *Biochim. Biophys. Acta,* 1064, 297–307, 1991.

66. Bojesen, I.N. and Bojesen, E., Exchange efflux of [³H]palmitate from human red cell ghosts to bovine serum albumin in buffer: Effects of medium volume and concentration of bovine serum albumin, *Biochim. Biophys. Acta,* 1111, 185–196, 1992.

67. Pond, S.M., Gordon, R.A., Simi, A.L., and Winzor, D.J., Further observations on the measurement of fatty acid incorporation by erythrocyte ghosts to quantify unbound palmitate concentration in albumin-palmitate mixtures, *Anal. Biochem.,* 237, 232–238, 1996.

68. Chaiken, I.M., Quantitative uses of affinity chromatography, *Anal. Biochem.,* 97, 1–10, 1979.

69. Kasche, V., Bucholz, K., and Galunsky, B., Resolution of high-performance liquid affinity chromatography: Dependence on eluate diffusion into the stationary phase, *J. Chromatogr.,* 216, 169–174, 1981.

70. Nilsson, K. and Larsson, P.-O., High-performance liquid affinity chromatography on silica-bound alcohol dehydrogenase, *Anal. Biochem.,* 134, 60–72, 1983.

71. Anderson, D.J. and Walters, R.R., Equilibrium and rate constants of immobilized concanavalin A determined by high-performance affinity chromatography, *J. Chromatogr.,* 376, 69–85, 1986.

72. Walters, R.R., Practical approaches for the measurement of rate constants by affinity chromatography, in *Analytical Affinity Chromatography,* Chaiken, I.M., Ed., CRC Press, Boca Raton, FL, 1987, pp. 117–156.

73. Loun, B. and Hage, D.S., Chiral separation mechanisms in protein-based HPLC columns, 2: Kinetic studies of (*R*)- and (*S*)-warfarin to immobilized human serum albumin, *Anal. Chem.,* 68, 1218–1225, 1996.

74. Winzor, D.J., Munro, P.D., and Cann, J.R., Experimental and theoretical studies of rate constant evaluation for the solute-matrix interaction in affinity chromatography, *Anal. Biochem.,* 194, 54–63, 1991.

75. DeLisi, C. and Hethcote, H.W., Chromatographic theory and application to quantitative affinity chromatography, in *Analytical Affinity Chromatography,* Chaiken, I.M., Ed., CRC Press, Boca Raton, FL, 1987, pp. 1–63.

76. Ward, L.D., Howlett, G.J., Hammacher, A., Weinstock, J., Yasukawa, K., Simpson, R.J., and Winzor, D.J., Use of a biosensor with surface plasmon resonance detection for the determination of binding constants: Measurement of interleukin-6 binding to the soluble interleukin-6 receptor, *Biochemistry,* 34, 2901–2907, 1995.

77. Shan, J., Moran-Jones, K., Munro, T.P., Kidd, G.J., Winzor, D.J., Hoek, K.S., and Smith, R., Binding of an RNA trafficking response element to heterogeneous nuclear ribonucleoproteins A1 and A2, *J. Biol. Chem.,* 275, 38286–38296, 2000.

78. Ainger, K., Arossa, D., Diana, A.S., Barbarese, E., and Carson, J.H., Transport and localization of elements in myelin basic protein mRNA, *J. Cell Biol.,* 138, 1077–1087, 1997.

79. Hoek, K.S., Kidd, G.J., and Smith, R., hnRNP A2 selectively binds the cytoplasmic transport sequence of myelin basic protein mRNA, *Biochemistry,* 37, 7021–7029, 1998.

80. Munro, T.P., Magee, R.J., Kidd, G.J., Carson, J.H., Barbarese, E., Smith, L.M., and Smith, R., Mutational analysis of a heterogeneous nuclear ribonucleoprotein A2 response element for RNA trafficking, *J. Biol. Chem.,* 274, 34389–34395, 1999.

81. Panayotou, G., Gish, G., End, P., Truong, O., Dhand, R., Fry, M.J., Hiles, I., Pawson, T., and Waterfield, M.D., Interactions between SH2 domains and tyrosine-phosphorylated platelet-derived growth factor beta-receptor sequences: analysis of kinetic parameters by a novel biosensor approach, *Mol. Cell. Biol.,* 13, 3567–3576, 1993.

82. Nieba, L., Krebber, A., and Pluckthün, A., Competition BIAcore for measuring true affinities: Large differences from values determined from binding kinetics, *Anal. Biochem.,* 234, 155–165, 1996.

83. Edwards, P.R., Maule, C.H., Leatherbarrow, R.J., and Winzor, D.J., Second-order kinetic analysis of IAsys biosensor data: Its use and applicability, *Anal. Biochem.,* 263, 1–12, 1998.

84. Müller, K.M., Arndt, K.M., and Pluckthün, A., Model and simulations of multivalent binding to fixed ligands, *Anal. Biochem.,* 261, 149–158, 1998.

85. Cooper, M.A. and Williams, D.H., Kinetic analysis of antibody-antigen interactions at a supported lipid monolayer, *Anal. Biochem.,* 276, 36–47, 1999.

86. Morton, T.A., Myszka, D.G., and Chaiken, I.M., Interpreting complex binding kinetics from optical biosensors: A comparison of analysis by linearization, the integrated rate equation, and numerical integration, *Anal. Biochem.,* 227, 176–185, 1995.

87. Roden, L.D. and Myszka, D.G., Global analysis of a macromolecular interaction measured on BIAcore, *Biochem. Biophys. Res. Commun.,* 225, 1073–1077, 1996.

88. Myszka, D.G. and Morton, T.A., CLAMP: A biosensor kinetic data analysis program, *Trends Biochem. Sci.,* 23, 149–150, 1998.

89. Corr, M., Slanetz, A.E., Boyd, L.F., Jelonek, M.T., Khilko, S., Al-Rhamadi, B.K., Lim, Y.S., Maher, S.E., Bothwell, A.L., and Marulies, D.H., T cell receptor–MHC class 1 peptide interactions: Affinity kinetics and specificity, *Science,* 265, 946–949, 1994.

90. Fischer, R.J., Fivash, M., Casas-Finet, J., Erickson, J.W., Kondoh, A., Bladen, S.V., Fisher, C., Watson, D.K., and Papas, T., Real-time DNA binding measurements of the ETS1 recombinant oncoproteins reveal significant kinetic differences between the p42 and p51 isoforms, *Protein Sci.,* 3, 257–266, 1994.

91. Glaser, R.W., Antigen-antibody binding and mass transport by convection and diffusion to a surface: A two-dimensional computer model of binding and dissociation kinetics, *Anal. Biochem.,* 213, 152–161, 1993.

92. Karlsson, R., Roos, H., Fägerstam, L., and Persson, B., Kinetic and concentration analysis using BIA technology, *Methods,* 6, 99–110, 1994.

93. Schuck, P., Kinetics of ligand binding to receptor immobilized on a polymer matrix, as detected with an evanescent wave biosensor, 1: A computer simulation of the influence of mass transport, *Biophys. J.,* 70, 1230–1249, 1996.

94. Schuck, P. and Minton, A.P., Kinetics of ligand binding to receptor immobilized in a polymer matrix, as determined with an evanescent wave biosensor, 2: Analysis of mass transport-limited binding kinetics in evanescent wave biosensors, *Anal. Biochem.,* 240, 262–272, 1996.

95. Myszka, D.G., Morton, T.A., Doyle, M.L., and Chaiken, I.M., Kinetic analysis of a protein-antibody interaction limited by mass transport on an optical biosensor, *Biophys. Chem.,* 64, 127–137, 1997.

96. O'Shannessy, D.J. and Winzor, D.J., Interpretation of deviations from pseudo-first-order kinetic behavior in the characterization of ligand binding by biosensor technology, *Anal. Biochem.,* 236, 275–283, 1996.

97. Hall, D.R. and Winzor, D.J., Parking problems as potential sources of deviation from predicted kinetic behavior in biosensor studies with small immobilized antigens, *Int. J. BioChromatogr.,* 4, 175–186, 1999.

98. Myszka, D.G., He, X., Demko, M., Morton, T.A., and Goldstein, B., Extending the range of rate constants available from BIAcore: Interpreting mass-transport-influenced data, *Biophys. J.,* 75, 583–594, 1998.

99. Munro, P.D., Jackson, C.M., and Winzor, D.J., On the need to consider kinetic as well as thermodynamic consequences of the parking problem in quantitative studies of nonspecific binding between proteins and linear polymer chains, *Biophys. Chem.,* 71, 185–198, 1998.

100. Fägerstam, L.G. and O'Shannessy, D.J., Surface plasmon resonance detection in affinity technologies: BIAcore, in *Handbook of Affinity Chromatography,* Kline, T., Ed., Marcel Dekker, New York, 1993, pp. 229–252.

101. Hall, D.R., Gorgani, N.N., Altin, J.G., and Winzor, D.J., Theoretical and experimental considerations of the pseudo-first-order approximation in conventional kinetic analysis of IAsys biosensor data, *Anal. Biochem.,* 253, 145–155, 1997.

102. Winzor, D.J., Biphasic thermodynamic characterization of interactions by quantitative affinity chromatography and biosensor technology, *J. Chromatogr. A,* 803, 291–297, 1998.

24

Chromatographic Studies of Molecular Recognition and Solute Binding to Enzymes and Plasma Proteins

Sharvil Patel and Irving W. Wainer

Bioanalytical and Drug Discovery Unit, National Institute on Aging,
National Institutes of Health, Baltimore, MD

W. John Lough

Institute of Pharmacy and Chemistry, University of Sunderland, Sunderland, U.K.

CONTENTS

24.1 INTRODUCTION

Biopolymers are primarily composed of chiral subunits such as L-amino acids and D-carbo-hydrates. Examples of these biopolymers include enzymes, receptors, and carrier proteins (e.g., albumin). These molecules can act as chiral binding agents due to their individual chiral subunits and secondary or tertiary structures. A biopolymer's ability to discriminate between three-dimensional structures during enzyme-substrate or ligand-receptor interactions is a complicated but basic aspect of biological and pharmacological processes. The initial discovery of this process can be attributed to Louis Pasteur, who in 1858 reported that D-ammonium tartrate was more rapidly destroyed by the mold *Penicillium glaucum* than L-ammonium tartrate [1, 2].

Since this first report of chiral recognition, the pharmacological aspects of chirality have been extensively reviewed (e.g., [3, 4]). In these systems, the key step is the formation of transient diastereomeric complexes. This mechanism is based upon the capacity of one chiral molecule (the "selector") to interact with the enantiomers of a second (the "selectand"). The differentiation between these enantiomers during the chiral recognition process is a result of energy differences between the diastereomeric selector-selectand complexes.

Chiral stationary phases (CSPs) based on enzymes and carrier proteins have been useful in the study of protein-solute interactions, including the extent of these interactions and their underlying mechanisms. Examples include the use of enzyme-based stationary phases to determine Michaelis-Menten kinetics [5, 6] and the use of serum albumin-based stationary phases (SA-CSPs) to study drug or solute binding processes [7–15]. In this latter case, CSPs have been used to determine the extent and enantioselectivity of solute binding to a variety of serum albumins [7, 8], to identify the sites at which this binding occurs [7, 9–11], to investigate competitive drug-drug interactions [7, 12–14], and to identify allosteric drug-drug interactions [14, 15]. This chapter discusses the basic mechanisms behind the chiral recognition of solutes by proteins and provides examples of how affinity columns and CSPs have been used in such work.

24.2 CHIRAL RECOGNITION MECHANISMS

Enantioselective separations performed on small molecule-based CSPs have been attributed to a three-point chiral recognition mechanism [16]. In this model, the "points" refer to defined atomic and molecular structures on the chiral selector and complementary sites on the enantiomeric solutes. These structures include groups capable of participating in attractive interactions like hydrogen bond formation, π-π charge transfer complexes, and dipole-dipole interactions. In addition, it is possible to have repulsive steric interactions, where any molecular entity with a significant size can be included as one of the "points" in this model.

While this mechanism may be applicable to small molecule-based CSPs, it does not appear to describe enantioselective separations on biopolymer-based CSPs. Instead, solute-protein interactions on CSPs appear to take place in large but sterically defined sites on a protein. These sites include hydrophobic pockets, catalytic pockets, and areas with electrostatic interactions. Proteins like serum albumins possess multiple binding sites [9–11], and these sites may have different enantioselectivities. Thus, a simple extension of the three-point mechanism is not useful for these binding agents. Instead, the enantioselective resolution of protein-based CSPs can be described in terms of a multistep, interconnected process based on kinetic- and thermodynamic-driven chiral recognition [17]. The steps in this mechanism are tethering, conformational adjustments, activation, and stabilization [18–21].

Tethering refers to the formation of the selector-selectand complex. In this step, the solute (or selectand) distributes from the mobile phase to the chiral stationary phase (the selector) through an initial interaction involving electrostatic forces, hydrogen bonding, dipole-dipole forces, or other processes. Since the physicochemical properties of the enantiomers are identical, this interaction tethers the selectand to the selector but does not in itself produce energetically different diastereomeric complexes.

Once their initial complex has formed, the selectand and selector adjust to each other to allow secondary interactions to occur. This involves conformational adjustments that may range from simple rotational changes in the conformation of the selectand and selector to more significant molecular adjustments. If there are major differences in the conformational energies of the selectand or selector in their binding conformations relative to their native states, then kinetic-driven chiral recognition will contribute to any observed enantioselectivity.

As the selectand and selector make conformational adjustments to each other, secondary interactions occur that determine the position of these two molecules relative to one another. This step, referred to as "activation," is also a process that occurs in stages. If significant conformational adjustments occur, they will contribute to kinetic-driven chiral recognition.

As secondary interactions occur, stabilization takes place in the selectand-selector complex by means of one or more attractive interactions. This may involve electrostatic, hydrogen bonding, π-π, and hydrophobic interactions. It is at this time that the selector and selectand are brought closer to each other. This can destabilize the complex through repulsive van der Waals interactions, which raises the free energy of the system. The magnitude of the stabilizing and destabilizing interactions will reflect the three-dimensional structures of the enantiomeric selectands and the chiral selector, which may result in differences in the relative stabilities of the diastereomeric complexes. If this process is the key source of the observed enantioselectivity, then the chiral recognition mechanism is driven by thermodynamics.

24.3 CHIRAL RECOGNITION BY ENZYMES

Many of the same immobilized enzymes that are used for chiral separations (see Chapter 21) have also been used in studies of chiral recognition. Examples include α-chymotrypsin and cellobiohydrolase I. The use and properties of these enzymes in chiral recognition studies are examined in the following subsections.

24.3.1 α-Chymotrypsin

One enzyme that has been used in chromatographic work is α-chymotrypsin (ACHT). The chiral recognition mechanisms that take place on an α-chymotrypsin chiral stationary phase (ACHT-CSP) have been studied by using a series of N- and O-derivatized amino acids on active and inactive forms of this enzyme [19]. ACHT contains a distinct active site that is composed of a hydrophobic pocket and a hydrolytic cavity. It has been found that the chromatographic retention of an ACHT-CSP for N- and O-derivatized amino acids is a function of specific and nonspecific interactions between these solutes and the stationary phase. However, the observed enantioselective interactions with the D- and L-forms of these amino acids are due to specific interactions with ACHT. It has also been determined that the primary region for these specific interactions is the active site of the enzyme.

It has been found that two recognition mechanisms appear to operate at the ACHT active site. These are (1) a chiral process based solely on molecular fit and (2) chiral resolution due to preferential hydrolysis. In a work by Jadaud et al. [19], the enantioselective separation of the N-acetyl-derivatives of D- and L-phenylalanine, which are not ACHT substrates, was attributed to the molecular-fit mechanism. In contrast to this, the observed

Table 24.1 Retention Factors for *N*- and *O*-Derivatized Amino Acids on ACHT-CSP with and without Treatment by *N*-Tosyl-L-Phenylalanine Chloromethyl Ketone (TPCK)

Solutes	Retention Factor for Untreated, ActiveACHT-CSP (k'_{active})	Retention Factor for TPCK-Deactivated ACHT-CSP (k'_{TPCK})	Change in Retention Factor for the Active vs. Deactivated ACHT-CSP (%) [a]
N-Acetyl-L-phenylalanine	0.92	0.80	−13
N-Acetyl-D-phenylalanine	1.24	0.75	−40
N-Benzoyl-L-leucine	3.64	3.12	−14
N-Benzoyl-D-leucine	4.52	3.12	−31
L-Phenylalaninamide	0.39	0.56	+44
D-Phenylalaninamide	0.49	0.60	+22
L-Tryptophanamide	0.96	1.10	+15
D-Tryptophanamide	2.19	1.10	−50
L-Tyrosine methyl ester	0.41	0.38	−7
D-Tyrosine methyl ester	0.42	0.72	+71

Note: The mobile phase in these studies was pH 6.0 with 0.13 M phosphate buffer ($I = 0.140$).

[a] This value was calculated using the formula $100 \cdot (k'_{active} - k'_{TPCK})/k'_{active}$.

Source: Data obtained from Jadaud, P., Thelohan, S., Schonbaum, G.R., and Wainer, I.W., *Chirality*, 1, 38–44, 1989.

enantiomeric resolution for the *O*-derivatives of L-tyrosine and L-tryptophan were found to be due to preferential hydrolysis [19].

These conclusions are supported by data from studies with deactivated ACHT-CSP (see Table 24.1). When the hydrophobic pocket of this immobilized enzyme is blocked with the ACHT inhibitors 4-NTA and *N*-tosyl-L-phenylalanine chloromethyl ketone (TPCK), this produces a loss of enantioselectivity and, in some cases, a large decrease in chromatographic retention. This indicates that for these solutes, the initial step in chiral recognition by ACHT is the formation of a solute-CSP complex by insertion of the solute's hydrophobic portion into the hydrophobic cavity of ACHT.

However, formation of this complex does not directly result in chiral recognition. The second step in this process involves electrostatic interactions and hydrogen bonding within the hydrolytic cavity. In this case, differences in the three-dimensional structures of solutes will lead to different interactions with the amino acids of ACHT comprised within its hydrolytic site. This produces different stabilities for the resulting complexes or, in the case of substrates, enantioselective hydrolysis.

For *N*-derivatized amino acids like *N*-acetyl-D,L-phenylalanine and *N*-benzoyl-D,L-leucine, TPCK inactivation essentially eliminates their enantioselective interactions with ACHT and reduces their retention on an ACHT column (see Table 24.1). This demonstrates for these solutes that binding at the active site of ACHT is indeed the source of the observed enantioselectivity and contributes significantly to their retention.

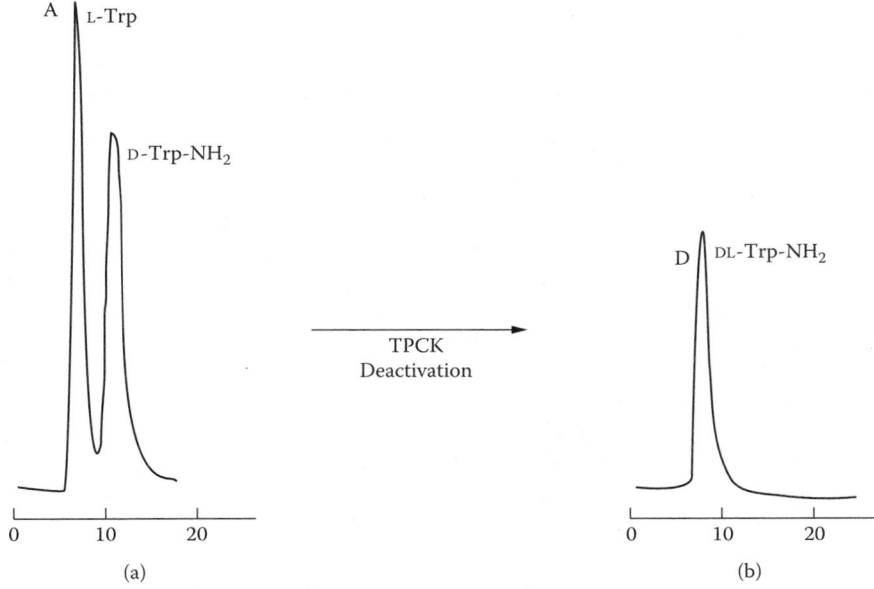

Figure 24.1 Retention of D- and L-tryptophan amides on (a) ACHT and (b) deactivated ACHT. (Reproduced with permission from Jadaud, P., Thelohan, S., Schonbaum, G.R., and Wainer, I.W., *Chirality*, 1, 38–44, 1989.)

The effect of blocking the active site of ACHT on its enantioselectivity is consistent with data obtained for D- and L-tryptophan amide and D- and L-phenylalanine amide. For these solutes, enantioselectivity is lost after the inactivation of ACHT by TPCK. In this case, the enantioselectivity of ACHT is due to enzymatic hydrolysis of the L-substrates, which gives rise to a separation of the L-acids and D-amides, as shown in Figure 24.1a. The loss of ACHT activity after treatment with TPCK means that the TPCK-ACHT chiral stationary phase is unable to separate the D- and L-amides (see Figure 24.1b).

However, these latter solutes show different behavior from *N*-derivatized amino acids. For instance, as shown in Table 24.1, the observed retention for three amides increased on a TPCK-ACHT column and only decreased for D-tryptophan amide. Since there are no enantioselective separations for either the *N*-derivatized amino acids or amino acid amides of a TPCK-ACHT column, it is tempting to assign the residual retention of these substances to nonspecific interactions with the chromatographic support or protein. But, there is a different outcome when the injected solutes are ester derivatives. For instance, both D- and L-tyrosine methyl ester (Tyr-OMe) are hydrolyzed on the active ACHT-CSP and give an observable enantioselective separation, as indicated in Table 24.1, while the TPCK-ACHT column gives no hydrolysis for these esters. On this second column, the retention for L-Tyr-OMe decreases by 7%, and the retention for D-Tyr-OMe increases by 71% relative to the active column, producing an enantioselective separation.

These results suggest that enantioselective interactions on ACHT occur at a site other than ACHT's active site. Since ACHT requires at least four amino acid residues on its substrates in both directions from the site of hydrolysis, it would be logical to assume that this requirement reflects a second or third binding site that tethers the peptide to this enzyme. Although these additional sites have not been characterized, the results obtained with D- and L-tyrosine methyl esters on a TPCK-ACHT column indicate that a second binding site does exist on ACHT and that this site has the ability to enantioselectively bind chiral solutes.

24.3.2   Cellobiohydrolase I

The chiral recognition of cellobiohydrolase I (CBH I) has been studied by using this protein as a stationary phase. This was accomplished by comparing the chromatographic results obtained on this support with those obtained on columns containing closely related enzymes [22–25]. Two of the enzymes used in this comparison were endoglucanase I (EG I), which has a 40% sequence identity with CBH I [23–25], and D214N, which is a single site-specific mutant form of CBH I [22, 24, 25].

In these studies, the enzymatic inhibition of CBH I and EG I by the enantiomers of propranolol and alprenolol were compared with the separation of these compounds on columns containing CBH I and EG I [23]. These experiments confirmed that the active sites for these enzymes were involved in chiral recognition. For the tested solutes, the inhibition constants (K_I) for the (S)-enantiomers were lower than those for the (R)-enantiomers on CHB I and EG I, and the (S)-enantiomers were more strongly retained on the CHB I and EG I columns. In addition, the K_I values for the enantiomers of propranolol and alprenolol were lower for the inhibition of CBH I than for EG I, and the corresponding enantioselectivities on the CBH I-CSP were greater than those on the EG I-CSP [23].

The main difference between these enzymes is that the active site on CBH I is located inside a tunnel, while in EG 1 it is located in an open groove. The results obtained by Henriksson et al. [23] demonstrated that the amino alcohols used in the inhibition experiments interacted more strongly with CBH I than EG I. The presence of the outer wall in CBH I was the most probable explanation for why this enzyme was more efficient in separating the given enantiomers. Also, the environment in the tunnel of CBH I would be shielded from the bulky solvent, which could favor the stereospecific binding of small solutes through hydrophobic or charge-transfer interactions to aromatic residues. This would also favor strong electrostatic interactions due to the presence of low dielectric constants in the tunnel.

In the active site of CHB I, residues Glu-212, Asp-214, and Glu-217 have been proposed to play important roles in this enzyme's hydrolytic activity and chiral recognition of enantiomers [23]. For instance, the electrostatic interactions of these three carboxylic residues are responsible for hydrolysis at the active site of CHB I. Glu-212 and Glu-217 are positioned on opposite sides of a glycosidic linkage, with their carboxylate groups being roughly 6 Å apart. This is concordant with the distance between the carboxylate groups of a nucleophile and the general acid/base catalyst observed in other retaining enzymes.

This proposed mechanism has been further examined in studies that compared the enantioselective retention of amines and acids on columns containing immobilized CBH I or D214N [22]. The enzyme D214N that was used in this second CSP was created by site-directed mutagenesis of CBH I, where the carboxylic acid group at Asp-214 was replaced with an amide [26]. This mutation was found to affect the retention, structure selectivity, and enantioselectivity of the resulting chiral stationary phase for amino alcohols. However, no change in enantioselectivity for prilocaine was observed. This indicated that the chiral discrimination of prilocaine was different on these columns from that seen for amino alcohols. In addition, no enantioselectivity was observed for mono- or divalent acids on these chiral stationary phases, although their retention was affected by the mutation. This suggested that chiral recognition was occurring at a site (or sites) that overlapped the catalytic site of CBH I.

24.4   CHIRAL RECOGNITION BY SERUM PROTEINS

Affinity chromatography has also been used to examine the binding of chiral solutes to immobilized serum proteins. Specific examples that will be considered in this section are studies that have been performed with the proteins human serum albumin (HSA) and

α_1-acid glycoprotein (AGP). More information on the use of these proteins as ligands in chiral separations can be found in Chapter 21.

24.4.1 Human Serum Albumin

The three-dimensional structure of HSA has been determined by X-ray crystallography at a resolution of 2.8 Å [27]. This protein is composed of three homologous domains, with each domain being the product of two subdomains possessing common structural motifs. The principle regions for solute binding to HSA occur in hydrophobic cavities in subdomains IIA and IIIA. These regions are known as "site I" and "site II," respectively. These binding areas have similar features, including an area near the surface of the cavity that is available for electrostatic interactions or hydrogen bonding with solutes.

Semiquantitative structure-activity relationships have been used to describe the binding of HSA to chiral 2-arylpropionic acid nonsteroidal anti-inflammatory agents (NSAIDs). This allowed a model to be developed for the indole-benzodiazepine binding site of this protein (i.e., site II) [28]. This study suggested that site II is a hydrophobic cleft about 16 Å long and 8 Å wide, with one or more cationic groups located near the surface.

The results obtained in chromatographic work with a series of achiral and chiral benzodiazepines on a human serum albumin chiral stationary phase (HSA-CSP) are consistent with the proposed structure of site II [11]. These studies demonstrated that there were both specific binding regions (i.e., site II) and nonspecific regions on HSA. For the chiral benzodiazepines, the observed enantioselectivity was a function of differential binding to site II.

Chromatographic data and molecular descriptors were used to construct quantitative structure-enantioselective retention relationships (QSERRs) for these interactions. This indicated that the binding of benzodiazepines to the nonspecific site of HSA could be estimated using structural descriptors related to the hydrophobicity of these compounds and their width in the direction of their phenyl substituent, as shown in Figure 24.2 [11].

However, an additional structural descriptor was needed to describe the retention of benzodiazepines at site II. This descriptor was referred to as the "polarity parameter" (P_{SM}) and was defined by the following equation,

$$P_{SM} = \Delta \times D_{HRN} \tag{24.1}$$

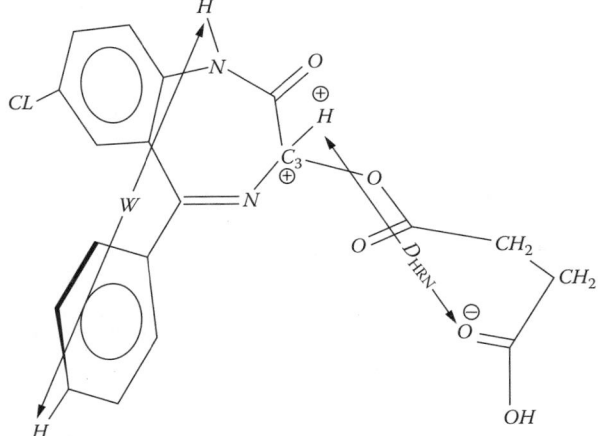

Figure 24.2 Structural descriptors used in a QSERR analysis of the binding of benzodiazepines to human serum albumin. (Reproduced with permission from Kalizan, R., Noctor, T.A.G., and Wainer, I.W., *Mol. Pharmacol.*, 42, 512–517, 1992.)

where Δ is the excess charge on the hydrogen atom attached to the chiral center at C_3 (i.e., the most negatively charged atom on the substituent attached to C_3) and D_{HRN} is the distance between specific atoms in this structure, as shown in Figure 24.2. The need for this additional descriptor was consistent with the involvement of electrostatic interactions in the binding of benzodiazepines and amino acids at site II of HSA.

The results obtained for the enantioselective separation of a series of chiral arylcarboxylic acids on an HSA-CSP are consistent with the site II model developed for chiral benzodiazepines [29]. These data indicate that hydrophobicity and molecular volume play key roles in chromatographic retention and enantioselectivity when using immobilized HSA. These results are also consistent with a stepwise chiral-recognition mechanism, as discussed in Section 24.2, in which the ability of a bound solute to make conformational adjustments to a chiral cavity affects the magnitude of the observed enantioselectivity.

In a work by Adrisano et al. [29], the relationship between chromatographic retention (expressed as the retention factor, k') and hydrophobicity (described using the logarithm of the partition coefficient, log P) was examined using regression analysis. A significant correlation was observed for nonresolved compounds and the first eluting enantiomer of resolved compounds ($r^2 = 0.6080$ and $p = 0.0017$) and for the second eluting enantiomer of the resolved compounds ($r^2 = 0.4933$ and $p = 0.0109$). Significant correlations were also obtained when molecular volumes were substituted for log P values ($r^2 = 0.7745$ and $p = 0.0001$, or $r^2 = 0.6046$ and $p = 0.0026$, respectively). When molecular volume and log P values were compared, a correlation of $r^2 = 0.7756$ ($p < 0.0001$) was observed, suggesting that these two variables were interchangeable. Thus, it was found for this series of compounds that retention on the HSA-CSP was primarily a function of solute hydrophobicity.

However, when the relationship between enantioselectivity (described using the enantioselectivity factor, α) and log P was examined for these same solutes, no significant correlation was observed. When α and the molecular volume were compared, a bimodal distribution in the data was detected. For compounds with molecular volumes between 121.7 and 136.2 Å^3, α increased with the molecular volume ($r^2 = 0.7526$ and $p = 0.05$). However, for compounds with molecular volumes ranging from 136.2 to 164.4 Å^3, α decreased with an increase in molecular volume ($r^2 = 0.8914$ and $p = 0.0046$).

From this, it was concluded that the enantioselectivity of HSA for these compounds was mainly a function of solute binding at a chiral cavity with a defined structure. It was also concluded that an optimum fit between the solute and chiral selector occurred when the solute volume was close to 136 Å^3. When the molecular volume of the solute was below this value, enantiomeric solutes may have had the ability to adjust to the cavity and optimize interactions that stabilized the complex or minimize interactions that destabilized it. When the molecular volume exceeded 136 Å^3, solutes may have only fit partially into the cavity, thereby diminishing the ability of HSA to discriminate between enantiomers.

The data from Adrisano et al. [29] were used to create QSERR equations to describe the retention of enantiomers on HSA columns. These equations were consistent with the observation that hydrophobicity is the primary factor in chromatographic retention. Furthermore, the equation for the retention of the second eluting enantiomer contained a term related to the positive contribution of the solute's electrostatic potential. This suggested that additional electrostatic interactions or hydrogen bonding contributed to the increase in retention and, therefore, to the enantioselectivity.

The results for the retention and enantioselective separation of the *erythro-* and *threo-* forms of 2,3-substituted 3-hydroxypropionic acids on an HSA-CSP were also consistent with the mechanisms for retention and enantioselectivity presented in Section 24.2 [18]. In this case, the lipophilicity of the solutes was measured by using their retention on a C_{18} column to produce partition coefficients expressed as log k'_w. It was found that their

Table 24.2 Structures, Conformations, and Enantioselectivities of 2,3-Substituted 3-Hydroxypropionic Acids (See Figure 24.3) on an Immobilized Human Serum Albumin Chiral Stationary Phase

Compound	Structure (R)	Conformation	Enantioselectivity (α)
1E	CH_3	Folded	1.08
1T	CH_3	Linear	1.41
2E	Ph	Linear	1.65
2T	Ph	Folded	1.42
3E	$PhCH_3\,(p)$	Linear	1.54
3T	$PhCH_3\,(p)$	Folded	1.45
4E	$PhCl\,(p)$	Linear	1.18
4T	$PhCl\,(p)$	Folded	1.08
5E	$PhBr\,(p)$	Linear	1.17
5T	$PhBr\,(p)$	Folded	1.0

Note: In the names of compounds, E represents the *erythro*-isomer, and T is the *threo*-isomer.

Source: Data obtained from Andrisano, V., Bertucci, C., Cavrini, V., Recanatini, M., Cavalli, A., Varoli, L., Felix, G., and Wainer, I.W., *J. Chromatogr. A*, 876, 75–86, 2000.

retention on the HSA column SAH (expressed as log k') was predominately a function of log k'_w, where the correlation between these two factors was $r^2 = 0.969$ for the first eluting enantiomer and $r^2 = 0.986$ for the second eluting enantiomer.

In this work, the lipophilicity of the solutes did not correlate with the observed enantioselectivity. In fact, the use of QSERRs could not identify a molecular descriptor for the observed enantioselective separations. However, a trend in the size of α was observed when molecular conformations were considered. As shown in Table 24.2, it was found through calculations that the most stable conformation for the 4/5 *erythro*-isomers was linear, while a folded conformation was present in the 4/5 *threo*-isomers. In each case, the observed enantioselectivity was greater between enantiomers in the linear conformation than for those in the folded conformation. This suggested that these compounds interacted with the hydrophobic cavity at site II of HSA in a predominantly linear conformation, where compounds that have a folded conformation expend more energy to assume a conformation capable of interacting with this cavity. When more energy is required to create the diastereomeric solute-site II complexes, smaller differences in energy will be seen between the complexes. This results

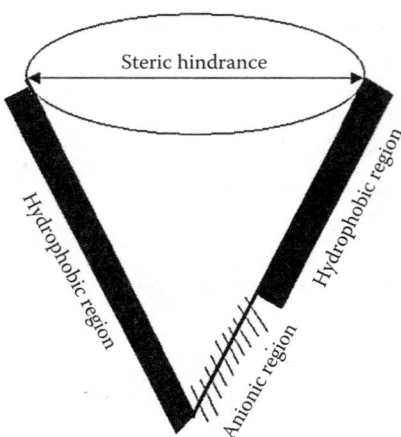

Figure 24.3 Schematic of the active site of AGP, as determined by QSERR studies.

in a smaller observed enantioselectivity. For the solutes tested by Andrisano et al. [18], the average difference in free-energy change (i.e., $\Delta\Delta G$, as calculated from α) for the folded conformations was 60 cal lower than for the linear conformations. This produced a decrease in the average observed enantioselectivity from 1.39 to 1.21.

While conformational adjustments appear to affect the magnitude of enantioselectivity for such solutes, the data from Andrisano et al. [18] suggest that the actual sources of this selectivity are the steric interactions between the solute and the chiral regions of site II. These interactions occur during insertion of the solute into this cavity. This is suggested by the enantiomeric elution order that was observed by Andrisano et al. [18]. For instance, in both the *erythro* and *threo* series of compounds, the more weakly retained enantiomers had an absolute configuration of R at carbon-2, the carbon atom adjacent to the carboxylic acid moiety.

In summary, the model for chiral recognition suggested by these studies with benzodiazepines, arylcarboxylic acids, and 2,3-substituted 3-hydroxypropionic acids is consistent with a stepwise process. This process involves (1) tethering of the solute to site II on HSA through electrostatic interactions or hydrogen bonding; (2) conformational adjustments that bring the chirality of the solute into position to interact with the chirality of the binding site; and (3) insertion of the solute into the hydrophobic cavity of the binding site, which drives the overall enantioselectivity.

24.4.2 α_1-Acid Glycoprotein

The binding of basic drugs to α_1-acid glycoprotein (AGP) has been studied extensively [30–36]. Initial *in vitro* displacement studies with antiarrhythmic drugs such as propafenone indicated that there were two classes of specific, saturable sites on AGP [32]. However, a reevaluation of propafenone displacement using the separate (R)- and (S)-enantiomers as well as the racemate has suggested that the curvilinear Scatchard plots obtained in the earlier work may have been due to the different binding properties of the propafenone enantiomers [33]. At the present, the prevalent view is that AGP has only one site that binds drugs through hydrophobic and electrostatic interactions [31, 34, 35].

The binding of a series of antihistamine drugs [34, 35], H_2 histamine receptor antagonists [34], and β-adrenoceptor antagonists to an AGP chiral stationary phase (AGP-CSP) has been investigated using the QSERR approach. The three major descriptors that have been found to describe the retention of these compounds on an AGP-CSP are a size parameter, hydrophobicity, and the electron excess charge on aliphatic nitrogens.

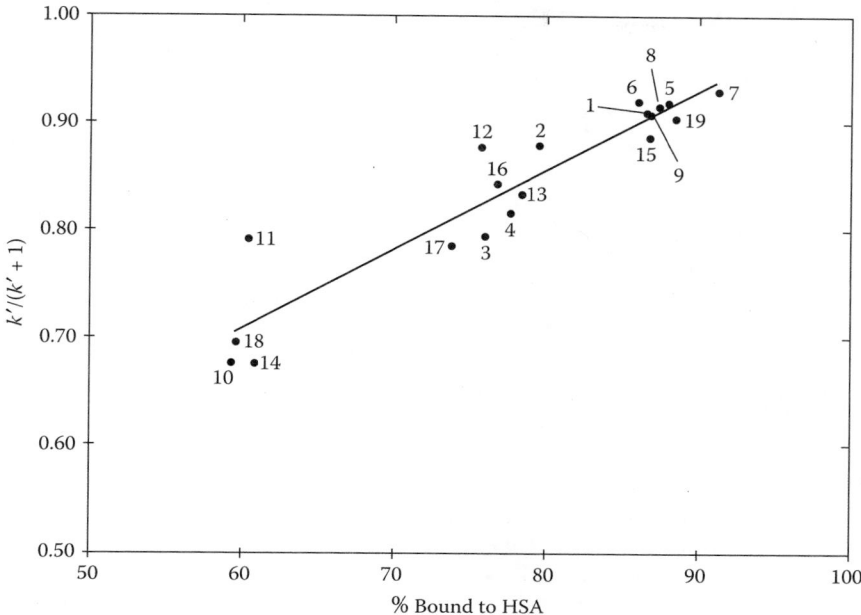

Figure 24.4 The relationship between chromatographic retention (expressed as k') on a chiral stationary phase containing immobilized human serum albumin (HSA-CSP) and the percent binding of soluble HSA to a series of benzodiazepines (points 1 through 18). (Reprinted with permission from Noctor, T.A.G., Diaz-Perez, M.J., and Wainer, I.W., *J. Pharm. Sci.*, 82, 675–676, 1993.)

The results from these studies suggest that the binding site of AGP is a conical pocket with lipophilic regions at the mouth and an anionic region close to the spike of the cone, as shown in Figure 24.4. The binding of solutes to this site appears to be driven by protonated aliphatic nitrogens, which can guide a drug toward the anionic region of this binding site. The hydrophobic aryl moieties of a drug will provide anchoring for this molecule in the lipophilic regions of the binding pocket, while steric restriction will prevent the molecule from plunging deep into the binding site. Chiral recognition by AGP appears to be based on its asymmetric negative charge distribution. Thus, there should be a differential fit between enantiomers that produce differences in electrostatic interactions and, therefore, different stabilities for their diastereomeric complexes with AGP.

However, experimental data obtained with an AGP-CSP indicate that AGP may have multiple binding sites [30, 36]. For instance, frontal chromatography performed with (−)-terodiline indicated that this molecule was adsorbed to one site with high affinity and to at least one other low-affinity site [30]. It was also observed that the enantiomers of amines, acids, and nonproteolytic compounds competed with (−)-terodiline for binding to these sites.

Another study that examined the effect of dimethyloctylamine (DMOA) on the enantioselective separation of chloroquine, mefloquine, and enpiroline indicated that these compounds were binding to at least two sites on an AGP-CSP [36]. These data indicated that the site at which DMOA competed with the enantiomers of mefloquine represented two-thirds of the total retention for these enantiomers. This study also suggested that while DMOA competitively displaced enpiroline and chloroquine at one site, the binding of DMOA to the AGP-CSP produced cooperative allosteric interactions at one or more additional sites, increasing the observed enantioselectivity for these compounds.

24.5 PROTEIN BINDING STUDIES

As has already been indicated in this chapter, immobilized serum proteins and enzymes have been used in many studies to characterize and study solute-protein interactions. Data that can be obtained through such work include estimates of the percent binding of a solute to a protein and measures of the binding affinity for this process. These materials can also be used to characterize binding regions on a protein, to study drug-drug interactions, and to help in the approval process for new drugs.

24.5.1 Determination of Percent Binding and Affinity

Zonal elution has been used as a chromatographic technique with an HSA-CSP to determine the percent binding of solutes to HSA [8]. This was performed using a series of benzodiazepines and coumarin derivatives applied to an HSA-CSP column, with the retention factor (k') for each compound then being obtained. These k' values were compared with ultrafiltration data by plotting $k'/(k' + 1)$ vs. the percent of solute that was found to bind to HSA by ultrafiltration. This gave correlation factors of 0.999 for both series of compounds (see Figure 24.5). The binding of HSA to triazole derivatives was also studied by this approach, but no correlation was observed in this case. This reflected the fact that members of this series did not bind significantly to HSA (i.e., the percent binding for all triazoles was less than 40%) and gave low retention on the HSA-CSP (i.e., all retention factors were less than 0.70). These data indicate that an HSA-CSP can be used to determine whether compounds bind or do not bind to HSA and can provide an estimate of the extent of this binding.

Frontal analysis has also been used with an HSA-CSP to determine the extent of binding between HSA and substances like (R)-warfarin and L-tryptophan [37]. The association equilibrium constants measured by this approach for (R)-warfarin and L-tryptophan were

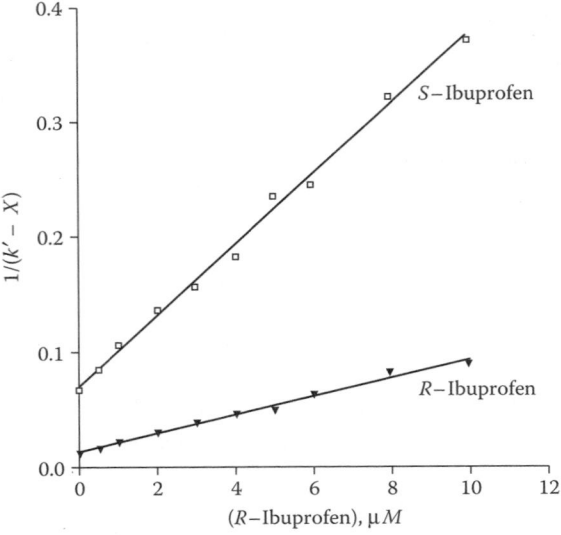

Figure 24.5 The effect of the mobile-phase concentration of (R)-ibuprofen on the retention of (R)- and (S)-ibuprofen on a human serum albumin column. (Reproduced with permission from Hage, D.S., Noctor, T.A.G., and Wainer, I.W., *J. Chromatogr. A*, 693, 23–32, 1995.)

$2.47 \times 10^5 \, M^{-1}$ and $1.10 \times 10^4 \, M^{-1}$, respectively, at pH 7.4 and 37°C, equivalent to results obtained by equilibrium dialysis or zonal elution.

24.5.2 Characterization of Binding Sites on HSA

As mentioned earlier, HSA contains two major drug-binding sites, known as site I (the warfarin-azapropazone site) and site II (the indole-benzodiazepine site) [38]. Warfarin is commonly used as a probe compound for binding at site I. Enantioselectivity in the binding of (*R*)- and (*S*)-warfarin to HSA has been reported by a number of authors. For example the association constants for these compounds with HSA have been reported as $1.5 \times 10^5 \, M^{-1}$ and $2.4 \times 10^5 \, M^{-1}$, respectively [38]. Other chiral coumarin drugs like acenocoumarol also bind enantioselectively to HSA, with the association constant for (*R*)-acenocoumarol being 2.5-times greater than for (*S*)-acenocoumarol. However, a related compound, phenprocoumon, had a twofold higher affinity for its (*S*)-enantiomer [38].

Compounds that are often used to probe site II on HSA are L-tryptophan and the 1,4-benzodiazepines (BDZs) [38]. The enantioselective binding of BDZs to site II has been extensively studied using HSA-CSPs and quantitative structure activity relationships [10, 11]. The results from these studies have raised the possibility that site II is a broad binding area rather than a clearly defined site. Because of this, it has been suggested that the terms *type I* and *type II* binding be used rather than *site I* and *site II* to characterize the interactions of drugs with HSA [10].

This latter conclusion has been supported by chromatographic studies with ibuprofen-like nonsteroidal anti-inflammatory agents [10, 14, 37]. Examples include reports in which competitive-displacement studies have looked at the ability of (*R*)- and (*S*)-ibuprofen to displace each other [39] or to act as displacers of BDZs [10]. In another study, octanoic acid has been used as a mobile-phase additive with NSAIDs as the solutes [14]. The results obtained with (*R*)-ibuprofen and (*S*)-ibuprofen alone are illustrated in Figure 24.6 [39]. These data

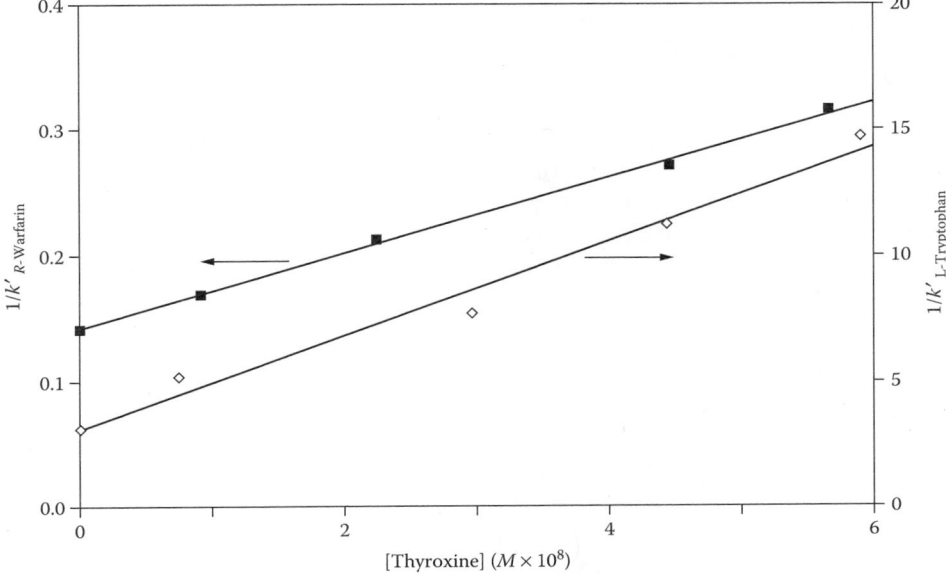

Figure 24.6 Competitive binding of L-thyroxine with (*R*)-warfarin (■) and L-tryptophan (◇) on an immobilized human serum albumin column. (Reproduced with permission from Loun, B. and Hage, D., *J. Chromatogr. B*, 579, 225–235, 1992.)

indicate that these two enantiomers have one common binding site on immobilized HSA, but that (S)-ibuprofen has at least one other major binding site. The association equilibrium constant estimated in this study for (R)-ibuprofen was $5.3 \times 10^5 \ M^{-1}$ at pH 6.9 and 25°C, while the association constants for (S)-ibuprofen at its two binding sites were estimated to be $1.1 \times 10^5 \ M^{-1}$ and $1.2 \times 10^5 \ M^{-1}$. These results are in good agreement with other association constants reported for ibuprofen on HSA. Data from other studies also suggest that there are at least two binding regions for ibuprofen on HSA that have been referred to as IBU_S and IBU_R. In addition, while NSAIDs like suprofen and ketoprofen primarily bind to site II of HSA, their enantioselective binding appears to stem from interactions at site I [39].

Chromatographic studies have also been used to investigate the binding of L-thyroxine to sites I and II of HSA [37]. Frontal analysis using (R)-warfarin and L-tryptophan as probes for these sites demonstrated that immobilized HSA had binding behavior similar to that observed for HSA in solution. The injection of (R)-warfarin or L-tryptophan in the presence of excess L-thyroxine was then used to independently examine the binding of L-thyroxine to sites I and II. This gave association constants for L-thyroxine at these sites of $1.4 \times 10^5 \ M^{-1}$ and $5.7 \times 10^5 \ M^{-1}$, respectively, at pH 7.4 and 37°C.

Using a van't Hoff approach, it has been determined that an HSA-CSP can be used to study the thermodynamics of solute-HSA interactions. In the case of L-thyroxine, the change in total free energy (ΔG) for its binding to sites I and II of HSA ranged from -7 to -8 kcal/mol and had a significant entropy component [37]. The effect of temperature on the competitive binding of thyroxine at sites I and II of HSA was also investigated, as shown in Figure 24.6. This type of study could play a valuable role in examining the site-specific binding of other drugs and hormones to HSA.

24.5.3 Characterizing Binding Sites Using Chemically Modified HSA

HSA contains two amino acid residues that are particularly susceptible to chemical modification. The first of these is a tryptophan residue (Trp-214), which resides in site I. The second is a reactive tyrosine residue (Tyr-411), which is located within site II. Both of these residues have been modified, and the resulting effects on the enantioselectivity and retention of compounds on HSA have been determined. The results of these studies have provided insight into the interactions of various compounds with this protein.

The effect of modifying Trp-214 with *o*-nitrophenylsulphenyl chloride has been studied using (R)- and (S)-warfarin as marker compounds [40]. For instance, frontal-analysis studies on Trp-214-modified HSA gave the same number of binding sites as normal HSA for (R)-warfarin but a lower association constant for this solute. It is believed that this is due to the direct blocking of HSA's warfarin-binding site by the modified Trp-214 residue. Furthermore, zonal-elution studies with racemic warfarin on columns containing normal or modified HSA indicated that Trp-214 and its neighboring residues played an important role in determining the affinity and enantioselectivity of HSA for (R)- and (S)-warfarin.

The reagent *p*-nitrophenyl acetate has been used to modify Tyr-411, and the effect of this modification on the retention and enantioselectivity of HSA for a variety of site II ligands has been investigated [41]. It was expected that this modification would decrease the binding of site II solutes to HSA. However, the binding of several solutes was enhanced. These results are consistent with the description of site II as a large, flexible area in which the precise topography is determined not only by the interacting solutes, but also by the presence of other molecules that may be binding at other, possibly remote sites.

Cysteine-34 (Cys-34) of HSA has been modified using ethacrynic acid [42]. Significant differences were observed in the binding of drugs known to interact at site I, such as (R,S)-warfarin and phenylbutazone, and those known to bind at site II, such as BDZs and NSAIDs. There was a significant decrease in the chromatographic retention for most of

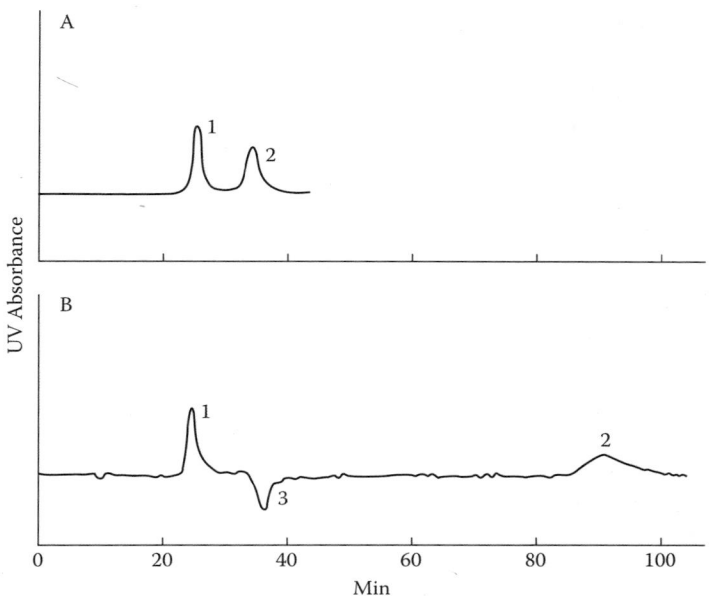

Figure 24.7 Effect of the presence of (*S*)-warfarin in the mobile phase on the retention and enantioselective separation of (*R*)- and (*S*)-lorazepam, where (A) no (*S*)-warfarin is in the mobile phase; (B) 10 mM (*S*)-warfarin is added to the mobile phase. In this figure, peak 1 is the (*R*)-lorazepam hemisuccinate, peak 2 is the (*S*)-lorazepam hemisuccinate, and peak 3 is a system peak due to warfarin. (Reproduced with permission from Domenici, E., Bertucci, C., Salvadori, P., and Wainer, I.W., *J. Pharm. Sci.*, 80, 164–166, 1991.)

these drugs on a column containing the modified HSA. In the case of the chiral compounds, significant differences in enantioselectivities were also observed. Some representative results from this study are given in Table 24.3. As indicated by this table, ethacrynic acid affected both of the most important binding sites on the HSA, leading to a marked increase in enantioselectivity for warfarin and a large decrease in enantioselectivity for the chiral benzodiazepines.

24.5.4 Studies of Drug-Drug Interactions

A number of studies with soluble and immobilized HSA have demonstrated that the simultaneous binding of two compounds at the same site on this protein will reduce the observed percent binding for one or both compounds [7, 43]. This effect is often viewed in terms of direct competition and requires the existence of limited and fairly rigid binding sites.

However, there is another view of this process in which the binding sites or areas are seen as being conformationally mobile, with multisolute interactions being a combination of direct and indirect displacement [44]. In this model, the simultaneous binding of two compounds to a protein such as HSA can be categorized as being independent, noncooperative (i.e., competitive displacement), anticooperative, or cooperative. The first two of these possibilities can, on the surface, be determined in a straightforward manner by standard protein-binding techniques. The second two possibilities, often referred to as *allosteric interactions*, are more difficult to examine by standard techniques, but these can be readily identified by chromatographic experiments on a support such as one containing immobilized HSA.

Cooperative interactions occur when the binding of one compound to a protein increases the affinity of a second compound. Such allosteric enhancements have been

Table 24.3 Effect of Ethacrynic Acid on the Interactions
of Drugs with a Chiral Stationary Phase Based on Human
Serum Albumin

Solute	$\Delta k'_1$ (%)	$\Delta k'_2$ (%)	$\Delta\alpha$ (%)
Temazepam	−19.1	−16.4	2.7
Oxazepam	−5.4	−1.9	3.7
Lorazepam	−20.8	−20.2	0.9
Warfarin	−15.6	−6.2	11.4
Ketoprofen	−28.1	−30.8	−3.6
Fenoprofen	−9.6	−7.6	2.3

Note: The values for $\Delta k'_1$ and $\Delta k'_2$ are the percent differences in the retention
factors for the first and second eluting enantiomers before and after the
modification of Cys_{34} on HSA with ethacrynic acid. The term is the
percent difference in enantioselectivity before and after this modification.

Source: Data obtained from Bertucci, C. and Wainer, I.W., *Chirality*, 9,
335–340, 1997.

noted for HSA in the effect of warfarin on the binding of BDZs to an HSA-CSP [45].
These studies investigated the effect of adding (R)- and (S)-warfarin to the mobile phase
while examining the retention and selectivity for the (R)- and (S)-enantiomers of lorazepam
hemisuccinate (LOH). The addition of (R)-warfarin had no effect; however, when 10 mM
(S)-warfarin was added to the mobile phase, the retention for (S)-LOH increased by 72%,
while the retention for (R)-LOH was unaffected (Figure 24.7). The net effect was a 76%
increase in the observed enantioselectivity. These results indicated that while (S)-LOH binds
to site II on HSA, (R)-LOH does not bind at this site. In addition, the binding of (S)-
warfarin to site II or a region adjacent to site II was proposed to cause a conformational
adjustment in this site, increasing the affinity for (S)-LOH.

Anticooperative interactions occur when the binding of one compound decreases
the affinity for a second solute. Such interactions have been identified for site II of HSA
in the effect of octanoic acid on the binding of NSAIDs [14, 46]. The addition of small
amounts of octanoic acid (less than 0.25 mM) to the mobile phase resulted in a significant
reduction in retention for NSAIDs on the HSA-CSP [46]. The magnitude of this reduction
was proportional to the initial affinities of the NSAIDs, where those compounds with the
strongest binding experienced the greatest amount of displacement. If this decrease in
retention were due to competitive displacement, then an inverse relationship between
affinity and retention would have been observed. Thus, the displacement of NSAIDs by
octanoic acid appears to have been anticooperative in nature.

Studies with suprofen, ketoprofen, warfarin, and phenylbutazone have also demonstrated that octanoic acid binds at site I on HSA [14]. At this site, the interaction can be
best described as a competitive interaction. This means that octanoic acid as well as
ketoprofen and suprofen appear to have multisite binding in their interactions with HSA.

24.5.5 Application of HSA-CSPs in New Drug Approval
Part of the drug approval process involves an investigation of the effect of commonly coadministered therapeutic agents on the protein binding of an applicant drug. The use of
immobilized HSA columns in this process has been demonstrated with oxaprozin [47].

Displacement studies using marker compounds for the major binding sites of HSA have shown that oxaprozin has a high affinity for site II and a significantly lower affinity for site I. Chromatographic studies on an HSA-CSP and ultrafiltration studies with HSA in solution were used to screen oxaprozin for possible competitive and allosteric interactions with potential coadministered drugs like NSAIDs, antipyretics, hypoglycemics, inhibitors of angiotensin-converting enzyme, anesthetics, metal ions, and anticancer agents. Competitive interactions occurred with drugs that were bound at site II, but none of these interactions were of clinical significance. Warfarin also displaced oxaprozin, but the reverse process did not occur, and this interaction was not clinically significant.

24.5.6 Solute Binding to Other Serum Albumins

The extent of solute binding and enantioselectivity of serum albumins (SAs) differs between mammalian species [38]. Serum albumin chiral stationary phases (SA-CSPs) produced by using albumins from various sources can be used to rapidly screen for interspecies differences. This was demonstrated using SA-CSPs developed with rat serum albumin (RtSA), rabbit serum albumin (RbSA), and HSA [48]. A series of compounds binding to sites I and II on these serum albumins were applied to columns containing these proteins, and their retentions and enantioselectivities were determined, as shown in Table 24.4. These experiments were able to rapidly distinguish differences in the binding affinities between these three species. For example, (R)-warfarin was more tightly bound to RtSA and HSA than (S)-warfarin, while the reverse was true with RbSA. This suggests that a rabbit would not be a good model for a human in pharmacokinetic studies.

Table 24.4 Enantioselectivity (α) and Elution Order for Enantiomers of Several Representative Compounds on Stationary Phases Based on Rat Serum Albumin (RtSA), Rabbit Serum Albumin (RbSA), and Human Serum Albumin (HSA)

Solute	Enantioselectivity and Elution Order [a]		
	RtSA	RbSA	HSA
Suprofen [b]	1.00[c]	1.85[c]	3.96[c]
Ibuprofen	2.41[c] (S,R)	2.18[d] (S,R)	1.60[c] (S,R)
Flurbiprofen	1.43[e] (S,R)	1.75[d] (S,R)	2.09[c] (S,R)
Tryptophan	7.40[f] (D,L)	2.14[f] (D,L)	2.20[f] (D,L)
Warfarin	1.47[c] (R,S)	1.57[c] (R,S)	2.56[c] (R,S)

[a] The elution order is given in the parentheses.

[b] The enantiomer elution order for suprofen was not determined due to unavailability of its single enantiomers.

[c] Mobile phase used in this separation: a 94:6 mixture of pH 7, 100 mM phosphate buffer and 1-propanol.

[d] Mobile phase used in this separation: 1 mM octanoic acid in a 94:6 mixture of pH 7, 100 mM phosphate buffer and 1-propanol.

[e] Mobile phase used in this separation: 0.3125 mM octanoic acid in a 94:6 mixture of pH 7, 100 mM phosphate buffer and 1-propanol.

[f] Mobile phase used in this separation: pH 7, 100 mM phosphate buffer.

Source: Data obtained from Massolini, G., Aubry, A.F., McGann, A., and Wainer, I.W., *Biochem. Pharmacol.*, 46, 1285–1293, 1993.

A series of NSAIDs were also injected onto each of these phases, with the percent binding to the individual SAs being determined for these compounds using ultrafiltration. Correlation coefficients of 0.999 were found between the chromatographic retention seen on the RtSA, RbSA, and HSA supports versus the results from the ultrafiltration studies. Thus, it was concluded that serum albumin columns could be used to rapidly determine interspecies affinities for such compounds.

24.6 SUMMARY AND CONCLUSIONS

The use of immobilized proteins as stationary phases and probes of molecular interactions has been an important application of affinity chromatography. The data presented in this chapter have shown that information about solute binding can be rapidly and accurately obtained using this approach. As demonstrated by the identification of a second binding site on ACHT or the investigation of cooperative and anticooperative drug-drug interactions, this technique can provide information that is difficult to obtain by other methods. The rapid developments occurring in immobilization, detection, and miniaturization suggest that this approach should play an even greater future role in drug discovery and development programs.

SYMBOLS AND ABBREVIATIONS

ACHT	α-Chymotrypsin
ACHT	α-Chymotrypsin chiral stationary phase
AGP	α_1-Acid glycoprotein
AGP-CSP	α_1-Acid glycoprotein chiral stationary phase
Asp	Aspartic acid
BDZs	1,4-Benzodiazepines
CBH I	Cellobiohydrolase I
CBH I-CSP	Cellobiohydrolase I chiral stationary phase
CSP	Chiral stationary phase
Cys	Cysteine
D_{HRN}	Distance parameter used in QSERR studies with benzodiazepines and HSA
DMOA	Dimethyloctylamine
EG I	Endoglucanase I
EG I-CSP	Endoglucanase I chiral stationary phase
Glu	Glutamic acid
HSA	Human serum albumin
HSA-CSP	Human serum albumin chiral stationary phase
IBU$_R$, IBU$_S$	Two binding sites for ibuprofen on HSA
k'	Retention factor
LOH	Lorazepam hemisuccinate
NSAIDs	Nonsteroidal anti-inflammatory agents
OMe	Methyl ester
P	Partition coefficient
Phe	Phenylalanine
P_{SM}	Polarity parameter
QSERR	Quantitative structure-enantioselective retention relationship
RbSA	Rabbit serum albumin

RtSA Rat serum albumin
SA Serum albumin
SA-CSP Serum albumin chiral stationary phase
TPCK N-Tosyl-L-phenylalanine chloromethyl ketone
TPCK-ACHT α-Chymotrypsin modified with TPCK
Trp Tryptophan
Tyr Tyrosine
α Enantioselectivity factor
ΔG Change in total free energy
$\Delta\Delta G$ Difference in change in the total free energy
$\Delta k'$ Change in retention factor for a solute

REFERENCES

1. Pasteur, L., On the asymmetry of naturally occurring organic compounds, in *The Foundation of Stereochemistry: Memoirs by Pasteur, V. Hoff, L. Bel, and Wislicenus,* Richardson, G.M., Ed., American Book Co., New York, 1901, pp. 1–33.
2. Drayer, D., The early history of stereochemistry, in *Drug Stereochemistry: Analytical Methods and Pharmacology,* 2nd ed., Wainer, I.W., Ed., Marcel Dekker, New York, 1993, pp. 5–24.
3. Lough, W.J. and Wainer, I.W., Eds., *Chirality in Natural and Applied Science,* Blackwell Science, Oxford, U.K., 2002.
4. Wainer, I.W., *Drug Stereochemistry: Analytical Methods and Pharmacology,* 2nd ed., Marcel Dekker, New York, 1993.
5. Chui, W.K. and Wainer, I.W., Enzyme based HPLC supports as probes of enzyme activity and inhibition: The immobilization of trypsin and α-chymotrypsin on an immobilized artificial membrane HPLC support, *Anal. Biochem.,* 201, 237–245, 1992.
6. Alebic-Kolbah, T. and Wainer, I.W., Application of an enzyme-based stationary phase to the determination of enzyme kinetic constants and types of inhibition: A new HPLC approach utilizing an immobilized artificial membrane chromatographic support, *J. Chromatogr.,* 653, 122–129, 1993.
7. Hage, D. and Austin, J., High-performance affinity chromatography and immobilized serum albumin as probes for drug- and hormone-protein binding, *J. Chromatogr. B,* 739, 39–54, 2000.
8. Noctor, T.A.G., Diaz-Perez, M.J., and Wainer, I.W., Use of a human serum albumin-based stationary phase for high-performance liquid chromatography as a tool for the rapid determination of drug-plasma protein binding, *J. Pharm. Sci.,* 82, 675–676, 1993.
9. Domenici, E., Bertucci, C., Salvadori, P., Felix, G., Cahagne, I., Motellier, S., and Wainer, I.W., Synthesis and chromatographic properties of an HPLC chiral stationary phase based upon human serum albumin, *Chromatographia,* 29, 170–176, 1990.
10. Noctor, T.A.G., Pham, C.D., Kalizan, R., and Wainer, I.W., Stereochemical aspects of benzodiazepine binding to human serum albumin, I: Enantioselective high-performance liquid affinity chromatographic examination of chiral and achiral binding interactions between 1,4-benzodiazepines and human serum albumin, *Mol. Pharmacol.,* 42, 506–511, 1992.
11. Kalizan, R., Noctor, T.A.G., and Wainer, I.W., Stereochemical aspects of benzodiazepine binding to human serum albumin, II: Quantitative relationship between structure and enantioselective retention in high-performance liquid affinity chromatography, *Mol. Pharmacol.,* 42, 512–517, 1992.
12. Wainer, I.W. and Chu, Y.Q., Use of mobile phase modifiers to alter retention and stereoselectivity on a bovine serum albumin high-performance liquid chromatographic chiral stationary phase, *J. Chromatogr.,* 455, 316–322, 1988.
13. Noctor, T.A.G. and Wainer, I.W., The use of displacement chromatography to alter retention and enantioselectivity on a human serum albumin-based HPLC chiral stationary phase: A mini-review, *J. Liq. Chromatogr.,* 16, 783–800, 1993.

14. Noctor, T.A.G., Wainer, I.W., and Hage, D.S., Allosteric and competitive displacement of drugs from human serum albumin by octanoic acid, as revealed by high-performance liquid affinity chromatography on a human serum albumin-based stationary phase, *J. Chromatogr.,* 577, 305–315, 1992.

15. Dominici, E., Bertucci, C., Salvadori, P., and Wainer, I.W., Use of·a human serum albumin-based high-performance liquid chromatography chiral stationary phase for the investigation of protein binding: Detection of the allosteric interaction between warfarin and benzodiazepine binding sites, *J. Pharm. Sci.,* 80, 164–166, 1991.

16. Pirkle, W.H., Hyun, M.H., and Bank, B., A rational approach to the design of highly effective chiral stationary phases, *J. Chromatogr.,* 316, 585–604, 1984.

17. Booth, T.D., Wahnon, D., and Wainer, I.W., Is chiral recognition a three-point process? *Chirality,* 9, 96–98, 1997.

18. Andrisano, V., Bertucci, C., Cavrini, V., Recanatini, M., Cavalli, A., Varoli, L., Felix, G., and Wainer, I.W., Stereoselective binding of 2,3-substituted-3-hydroxy-propionic acids on an immobilized human serum albumin chiral stationary phase: Stereochemical characterization and QSSR study, *J. Chromatogr. A,* 876, 75–86, 2000.

19. Jadaud, P., Thelohan, S., Schonbaum, G.R., and Wainer, I.W., The stereochemical resolution of enantiomeric free and derivatized amino acids using an HPLC chiral stationary phase based on immobilized α-chymotrypsin, *Chirality,* 1, 38–44, 1989.

20. Massolini, G., De Lorenzi, E., Calleri, E., Bertucci, C., Monaco, H.L., Perduca, M., Caccialanza, G., and Wainer, I.W., Properties of a stationary phase based on immobilized chicken liver basic fatty acid-binding protein, *J. Chromatogr. B,* 751, 117–130, 2001.

21. Greer, J. and Wainer, I.W., The molecular basis of chiral recognition, in *Chirality in Natural and Applied Science,* Lough, W.J. and Wainer, I.W., Eds., Blackwell Science, Oxford, U.K., 2002, pp. 87–108.

22. Marle, I., Erlandsson, P., Hansson, L., Isaksson, R., Pettersson, C., and Pettersson, G., Separation of enantiomers using cellulase (CBH I) silica as a chiral stationary phase, *J. Chromatogr.,* 589, 233–248, 1991.

23. Henriksson, H., Stahlberg, J., Isaksson, R., and Pettersson, G., The active sites of cellulase are involved in chiral recognition: A comparison of cellobiohydrolase 1 and endoglucanase 1, *FEBS Lett.,* 390, 339–344, 1996.

24. Hedeland, M., Holmin, S., Nygard, M., and Pettersson, G., Chromatographic evaluation of structure selective and enantioselective retention of amines and acids on cellobiohydrolase I wild type and its mutant D214N, *J. Chromatogr. A,* 864, 1–16, 1999.

25. Henriksson, H., Stahlberg, J., Koivula, A., Pettersson, G., Divne, C., Valtcheva, L., and Isaksson, R., The catalytic amino acid residues in the active site of cellobiohydrolase 1 are involved in chiral recognition, *J. Biotech.,* 57, 11–125, 1997.

26. Stahlberg, J., Divne, C., Koivula, A., Piens, K., Claeyssens, M., Teeri, T.T., and Jones, T.A., Activity studies and crystal structures of catalytically deficient mutants of cellobiohydrolase I from *Trichoderma reesei, J. Mol. Biol.,* 264, 337–349, 1996.

27. Wainmimolruk, S., Birkett, D.J., and Brooks, P.M., Structural requirements for drug binding to site II on human serum albumin, *Mol. Pharmacol.,* 24, 458–463, 1983.

28. He, X.M. and Carter, D.C., Atomic structure and chemistry of human serum albumin, *Nature,* 358, 209–215, 1992.

29. Adrisano, V., Booth, T.D., Cavrini, V., and Wainer, I.W., Enantioselective separation of chiral arylcarboxylic acids on an immobilized human serum albumin chiral stationary phase, *Chirality,* 9, 178–183, 1997.

30. Enquist, M. and Hermansson, J., Separation of the enantiomers of beta-receptor blocking agents and other cationic drugs using a CHIRAL-AGP column: Binding properties and characterization of immobilized alpha 1-acid glycoprotein, *J. Chromatogr.,* 519, 285–298, 1990.

31. Fournier, T., Medjoubi, N., and Proquet, D., Alpha-1-acid glycoprotein, *Biochim. Biophys. Acta,* 1482, 157–171, 2000.

32. Gillis, A.M., Yee, Y.G., and Kates, R.E., Binding of antiarrhythmic drugs to purified human α_1-acid glycoprotein, *Biochem. Pharmacol.,* 34, 4279–4282, 1985.

33. Šoltés, L., Sébille, B., and Szalay, P., Propafenone binding interaction with human α_1-acid glycoprotein: Assessing experimental design and data evaluation, *J. Pharmaceut. Biomed. Anal.*, 12, 1295–1302, 1994.

34. Kaliszan, R. and Wainer, I.W., Combination of biochromatography and chemometrics: A potential new research strategy in molecular pharmacology and drug design, in *Chromatographic Separations Based on Molecular Recognition*, Jinno, K., Ed., Wiley-VCH, New York, 1996, pp. 294–299.

35. Kaliszan, R., Nasal, A., and Turowski, M., Quantitative structure-retention relationships in the examination of the topography of the binding site of antihistamine drugs on α-acid glycoprotein, *J. Chromatogr. A*, 722, 25–32, 1996.

36. Aubry, A.F., Gimenez, F., Farinotti R., and Wainer, I.W., Enantioselective chromatography of the antimalarial agents chloroquine, mefloquine and enpiroline on a α_1-acid glycoprotein chiral stationary phase: Evidence for a multiple-site chiral recognition mechanism, *Chirality*, 4, 30–35, 1992.

37. Loun, B. and Hage, D., Characterization of thyroxine-albumin binding using high-performance affinity chromatography, *J. Chromatogr. B*, 579, 225–235, 1992.

38. Noctor, T.A.G., Enantioselective binding of drug top plasma proteins, in *Drug Stereochemistry: Analytical Methods and Pharmacology*, 2nd ed., Wainer, I.W., Ed., Marcel Dekker, New York, 1993, pp. 337–364.

39. Hage, D.S., Noctor, T.A.G., and Wainer, I.W., Characterization of the protein binding of chiral drugs by high-performance affinity chromatography interactions of *R*- and *S*-ibuprofen with human serum albumin, *J. Chromatogr. A*, 693, 23–32, 1995.

40. Chattopadhyay, A., Tian, T., Kortum, L., and Hage, D.S., Development of tryptophan-modified human serum albumin columns for site-specific studies of drug-protein interactions by high-performance affinity chromatography, *J. Chromatogr. B*, 715, 183–190, 1998.

41. Noctor, T.A.G. and Wainer, I.W., The *in situ* acetylation of an immobilized human serum albumin chiral stationary phase for high-performance liquid chromatography in the examination of drug-protein binding phenomena, *Pharmaceut. Res.*, 9, 480–484, 1992.

42. Bertucci, C. and Wainer, I.W., Improved chromatographic performance of a modified human serum albumin based stationary phase, *Chirality*, 9, 335–340, 1997.

43. Sjoholm, I., The specificity of binding sites on plasma proteins, in *Drug Protein Binding*, Reidenberg, M.M. and Eril, S., Eds., Prager Publishers, Philadelphia, 1988, pp. 36–45.

44. Honore, B., Conformational changes in human serum albumin induced by ligand binding, *Pharmacol. Toxicol.*, 66(Suppl. 2), 7–26, 1990.

45. Domenici, E., Bertucci, C., Salvadori, P., and Wainer, I.W., The use of a human serum albumin-based HPLC chiral stationary phase for the investigation of protein binding: The detection of the allosteric interaction between warfarin and benzodiazepine binding sites, *J. Pharm. Sci.*, 80, 164–166, 1991.

46. Noctor, T.A.G., Felix, G., and Wainer, I.W., Stereochemical resolution of enantiomeric 2-aryl propionic acid non-steroidal anti-inflammatory drugs on a human serum albumin based high-performance liquid chromatographic chiral stationary phase, *Chromatographia*, 31, 55–59, 1991.

47. Aubry, A.F., Markoglou, N., Adams, M.H., Longstreth, J., and Wainer, I.W., The effect on oxaprozin binding to human serum albumin of coadministered drugs, *J. Pharm. Pharmacol.*, 47, 937–944, 1995.

48. Massolini, G., Aubry, A.F., McGann, A., and Wainer, I.W., Determination of the magnitude and enantioselectivity of ligand binding to rat and rabbit serum albumins using immobilized-protein high-performance liquid chromatography stationary phases, *Biochem. Pharmacol.*, 46, 1285–1293, 1993.

25

Affinity-Based Optical Biosensors

Sheree D. Long
Biacore Inc., Piscataway, NJ

David G. Myszka
Center for Biomolecular Interaction Analysis, University of Utah, Salt Lake City, UT

CONTENTS

25.1 INTRODUCTION

Affinity-based optical biosensors represent a highly evolved state of analytical affinity chromatography. In this approach, packed columns are replaced with thin-layer surfaces. Moreover, postcolumn detection based on radioactivity, ultraviolet or visible absorbance, fluorescence, or light scattering is replaced by a single surface-integrated sensor. Instead of monitoring a binding process indirectly through the use of elution times, as is employed in analytical affinity chromatography, biosensors monitor formation of complexes directly on a surface. This is done in real time and without the need for labeling. These properties—plus the speed and flexibility of optical biosensors—have made these devices indispensable tools for the life sciences and pharmaceutical research.

The most versatile and widely used type of optical biosensor is that based on *surface plasmon resonance* (SPR) [1]. SPR detectors monitor changes in the refractive index at a sensor's surface as complexes form and break during a reaction. Since these instruments are essentially responding to the mass of molecules at the surface, no labeling is required. This greatly simplifies the experiment.

The ability to measure molecular interactions as they occur in real time makes it possible for optical biosensors to determine rate constants and binding affinities for nearly any noncovalent reaction. As a result, biosensors have been used to monitor interactions as diverse as the field of biology itself. For instance, in addition to protein-protein interactions, biosensors have been used to study reactions involving nucleic acids, lipid membranes, carbohydrates, and small molecules [2–4]. Biosensor technology is also now used routinely for protein identification, purification, and production. This chapter focuses on the principles and basic applications of optical biosensors, with a comparison being made between these and traditional affinity chromatographic-based methods.

25.2 PRINCIPLES AND USE OF BIOSENSORS

25.2.1 Commercial Biosensors

There are a number of commercial biosensors available for real-time binding studies. These instruments vary in sensitivity, function, and cost. The manual models are extremely flexible and suitable for use at the research bench, while highly automated platforms are amenable to drug-discovery applications. A recent review of commercial instruments has compared key features of the models that are available from current manufacturers [5].

Although these instruments vary in their design and intended applications, all contain three essential components: a detector, a sensor surface, and a sample delivery system. Surface plasmon resonance represents the predominant label-free detection method in these devices. SPR detectors can be set up in various configurations [1]. In these instruments, the sensor surface is typically a thin metal layer (usually gold) deposited on a glass support. The most common method for delivering samples to the sensor surface is through a flow cell, but cuvette-based and probe-based systems have also been used. A flow-cell configuration, which promotes rapid and uniform sample delivery to the surface, is suggested for kinetic applications.

25.2.2 Experimental Design for Biosensors

As with affinity chromatography, biosensor experiments require that one of the molecules of interest be attached to a surface, as seen in Figure 25.1(A). The advantage of a biosensor over traditional affinity chromatography is that it can directly monitor the attachment of a ligand to the surface, making it easy to control and quantitate the level of immobilization. In addition, the biosensor surface area is quite small (i.e., usually 0.1 to 1 mm^2), so only 0.1 to 1 μg of a protein-based ligand would be required for this immobilization.

The majority of applications for biosensors employ surfaces that are coated with a thin layer of carboxymethyl dextran. Dextran, which has been historically used as a chromatographic resin, creates a flexible and noncrosslinked hydrogel layer that extends 100 nm

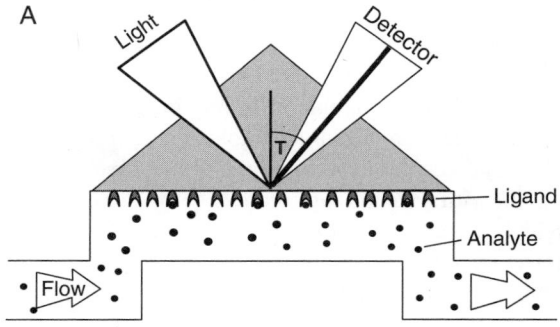

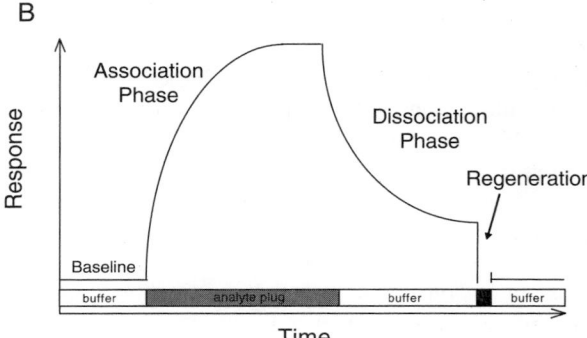

Figure 25.1 (A) Configuration of a standard SPR biosensor system and (B) the biosensor response during a typical binding cycle. In (A), the ligand is immobilized onto one wall of the flow cell, which is in proximity to the detector. Samples containing the analyte are introduced into the flow stream during the association phase of the measurement, and the flow cell is rapidly washed during the dissociation phase. In (B), the response is stable during the initial injection of buffer, since there is no change in the refractive index of the biosensor. However, when a sample plug is introduced, the response increases as the analyte accumulates at the surface. During the dissociation phase, the sensor surface is washed with buffer, and the signal goes toward its original baseline level as the analyte-ligand complexes break down. A short regeneration step may then be used to remove any remaining bound material. Once the system has been reequilibrated with the buffer, the signal returns to its initial baseline.

from the sensor surface in the presence of a physiological buffer. Immobilizing a ligand onto a dextran matrix offers several advantages over immobilization on a flat surface. For example, dextran gives reduced nonspecific adsorption for proteins on the gold surface, and it can be used to covalently attach ligands through well-defined coupling chemistries. Dextran also provides more-specific interaction sites per unit surface area, giving it an increased immobilization capacity. Furthermore, the flexibility of the dextran chain helps give functional activity to the immobilized molecule that is similar to what is observed in solution. Most manufacturers of biosensors supply premade sensors with various surface chemistries. For example, sensor surfaces are available for the construction of self-assembled monolayers as well as for other specialized applications [6].

25.2.3 Ligand Immobilization for Biosensors

The same coupling chemistries used in affinity chromatography are often applied to biosensors. The principal requirement for a ligand immobilized on the sensor surface is that it retain its functional interaction with its binding partner. Amine coupling is the most widely used method for linking proteins to a sensor surface. However, random amine-coupling methods may produce a heterogeneous surface if the reactive amine is in or near the ligand's binding site. If a homogeneous ligand population is required, site-specific coupling methods can be used, such as immobilization via unique thiols or aldehyde groups.

Alternatively, ligand attachment can be accomplished through a specific capturing molecule, such as an antibody for a ligand with a fusion tag, protein A for immunoglobulins, or streptavidin for biotinylated ligands. Ligand capturing has the advantage of creating a chemically homogeneous surface. However, for high-affinity interactions, regeneration of the sensor surface often removes the ligand, which must be recaptured prior to each binding cycle. Although recapturing new ligand for each binding cycle consumes more of this ligand, it also ensures a more consistent activity for the ligand throughout the analysis.

25.2.4 Typical Biosensor Cycle

Once a ligand has been immobilized on a sensor's surface, binding experiments can be performed to characterize this ligand's interactions with an analyte. The major steps in this experiment are illustrated in Figure 25.1(B). This cycle includes an association phase, a dissociation phase, and a regeneration step.

During the association phase, a sample is delivered to the biosensor's surface. This is typically performed by injecting a sample plug. As the analyte-ligand complexes form, the accumulation of mass at the sensor's surface changes the refractive index of the solvent layer near this surface, and an increase in response is recorded. If the binding reaction is permitted enough time to reach an equilibrium, the signal will reach a plateau. At this point, an equal number of complexes are forming and breaking at any given moment. A key benefit of biosensors is that they are able to quantitate the amount of complex that has been formed in the presence of the free analyte. Unlike a filter-binding assay or bead assay, no separation is needed for the reactants and products, thereby avoiding a perturbation of the reaction. In addition, formation of complexes can be monitored by biosensors for very weak or rapidly dissociating complexes.

During the wash-out step (or dissociation phase) of the experiment, the surface is flushed with a buffer, and the complexes at the surface are allowed to dissociate. As this occurs, the refractive index near the sensor's surface returns to its original value, and the signal decreases. Complexes that involve weak interactions may give a signal that decays back to the original baseline within minutes. However, tightly bound complexes often require a chemical-disruption step to release the analyte and regenerate the ligand for the next binding cycle. Typical protein-protein interactions can be disrupted by briefly washing the sensor with a dilute acid or base, which removes the bound material but retains the functional properties of the immobilized ligand. Again, a significant advantage of biosensors over traditional affinity chromatography is that the amount of material remaining on the surface can be quantitated directly.

Once an analyte has completely dissociated from the ligand, the experimental cycle can be repeated with a new analyte or new experimental conditions (e.g., a different temperature, pH, salt concentration, or buffer composition). A typical experimental cycle takes between 5 and 15 min. In addition, fully automated systems like the Biacore 2000 or 3000 can analyze up to 200 samples without interruption. One advantage of these automated systems is that they are exceptionally reproducible. The sensor surfaces themselves are chemically stable and can be used for several hundred binding cycles without any appreciable loss of capacity. However, the overall precision of a biosensor depends on the quality and stability of the biological reagents that are used with this device.

25.3 APPLICATIONS OF BIOSENSORS

25.3.1 Kinetic Measurements

While the power of affinity chromatography lies in its use for separation and purification, the calling card for optical biosensors is their use for the precise determination of both affinity constants and rate constants. While affinity constants provide information on the strength of an interaction, they do not contain any detailed information on how fast a complex is forming or falling apart. However, this information can be provided by optical biosensors through the measurement of rate constants for an interaction. To obtain accurate rate constants, careful experimental design and accurate interpretation of binding data are required. Recent advances and improvements in these areas have eliminated many of the experimental artifacts that can complicate biosensor-based measurements (e.g., instrument drift or mass transport effects) [6, 7].

One limitation of biosensors when they are used to study very fast association rates is created by the phenomenon of mass transport [8]. Mass transport is related to the speed at which molecules can be transported to and from a surface. Since one of the reactants in a biosensor is tethered to a surface, the other binding partner must diffuse from the bulk sample to this surface to bind. If the binding rate is faster than the transport rate, the apparent reaction rate will be limited by the transport step. This transport rate is determined by the diffusion rate of the analyte, the flow rate of the applied buffer, and the size and shape of the flow cell. The use of fast flow rates and low-capacity surfaces decreases the demand for analyte at the surface, making it possible to significantly reduce mass-transport effects [7]. In cases where the effects of mass transport cannot be completely eliminated, a simple model can be used during data analysis to accurately account for its effects [9].

An example of a kinetic analysis by a biosensor is shown in Figure 25.2. In this case, acetazolamide at several different concentrations was injected over a surface containing the immobilized enzyme carbonic anhydrase II. Figure 25.2 shows that as the concentration of acetazolamide was increased, the observed binding rate increased. This would be expected for a bimolecular reaction, since this requires that two molecules come together to form a complex. As the concentration of one of these agents is increased, the rate of collision increases, which also increases the observed binding rate.

The instrumental responses in Figure 25.2 were analyzed with a global fit to determine the values of the rate constants for the given reaction. In this process, the entire data set was examined simultaneously, where data from both the association and dissociation phases of the experiment were analyzed according to the same reaction model. This approach has the advantage of being nonsubjective and provides a stringent test for the reaction mechanism.

For the data in Figure 25.2, the association and dissociation rate constants for the interaction between acetazolamide and the immobilized carbonic anhydrase II were found by global fitting to be 2.93 (± 0.02) $\times$ 10^6 $M^{-1}sec^{-1}$ and 0.056 (± 0.001) sec^{-1}, respectively. The equilibrium dissociation constant (K_D, or the affinity constant) for this reaction was determined by calculating the ratio of these rate constants, since $K_D = k_d/k_a$. This gave an affinity constant of 19.0 (± 0.1) nM. Under standard operating conditions, SPR biosensors can measure association rate constants that range from 10^2 to 10^8 $M^{-1}sec^{-1}$ and dissociation rate constants that vary from 10^{-6} to 1 sec^{-1}. This provides these biosensors with the ability to determine dissociation equilibrium constants that range from millimolar to picomolar values (10^{-3} to 10^{-12} M).

25.3.2 Determination of Active Sample Concentrations

An important but often underutilized application of biosensors is their use for the determination of active sample concentrations. Quantitating the amount of an active protein in a

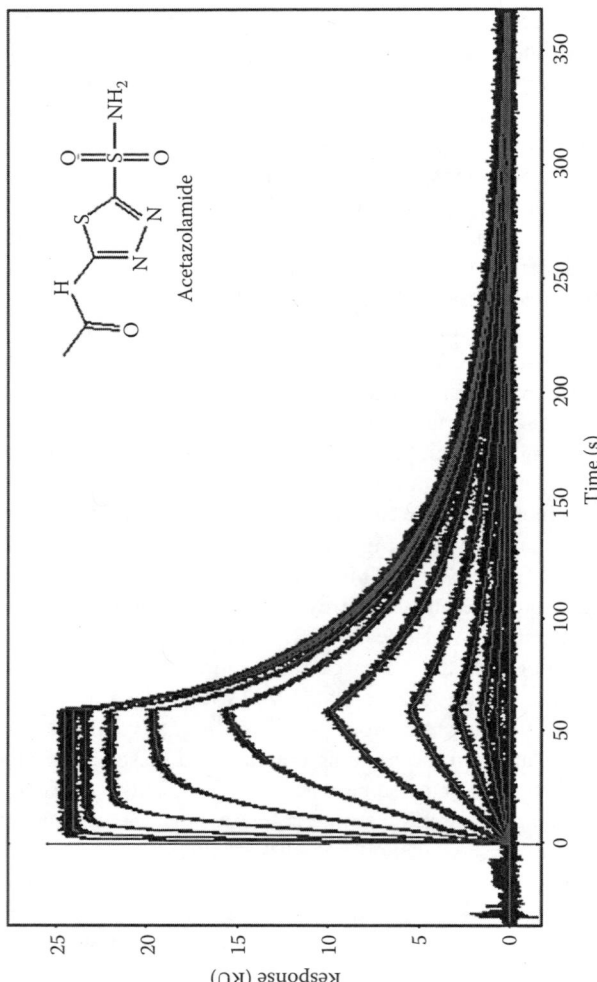

Figure 25.2 SPR kinetic analysis of the binding between acetazolamide and the enzyme carbonic anhydrase II, where the enzyme is immobilized on the biosensor's surface. In this example, the acetazolamide was injected in a twofold concentration series going from 20 μM to 20 nM (top to bottom). Each concentration was injected five times to demonstrate that the binding behavior was reproducible. The solid lines superimposed on the data represent a global fit to a 1:1 interaction model. The association and dissociation rate constants found to describe these data were $2.93 \pm 0.02 \times 10^6$ $M^{-1}\text{sec}^{-1}$ and 0.056 ± 0.001 sec^{-1}, respectively.

sample is essential for an accurate kinetic analysis. While techniques like spectroscopy and gel electrophoresis may provide information on how much protein is in a sample or its level of purity, they do not provide any information on the amount of functional protein in the sample.

By taking advantage of mass transport, active concentration assays on a biosensor can be set up under conditions where the reactions are either partially [10] or completely controlled by diffusion. In addition, when used for quality control studies, standard binding curves for known samples can be used to determine the activity of unknown samples. In fact, Biacore AB (Uppsala, Sweden) recently released an instrument and associated software that can be used for concentration analysis in a setting that must follow good manufacturing practices (GMPs). This instrument, known as the Biacore C, represents a trend in biosensor technology where specific instrument platforms and support software are being designed for dedicated applications.

25.3.3 Purification and Identification of Biochemicals

Biosensors have also become important for screening because they can act as micropurification systems when dealing with complex biological mixtures. The ability to detect and quantitate biospecific interactions for mixtures like cell lysates and conditioned media makes biosensors useful for identifying the ligands of orphan receptors. For example, Lackmann et al. recently used a biosensor to monitor the purification of an active component from such crude mixtures [11]. In this case, SPR biosensor technology was combined with traditional separation chromatography.

The integration of biosensors with mass spectrometry can produce a powerful platform that can isolate and identify surface-bound material without requiring further sample manipulation. In the last decade, several groups have been actively involved in interfacing SPR with mass spectrometry for the functional and structural characterization of proteins. Numerous studies have demonstrated that these two technologies can be interlinked without modification of existing software and hardware. One specific application has involved the direct identification of adsorbed molecules on a sensor chip using matrix-assisted laser desorption/ionization time-of-flight mass spectrometry (MALDI-TOF MS) [12]. In this work, sensor surfaces were removed from a biosensor instrument and placed directly onto a MALDI platform.

Alternatively, adsorbed material in a biosensor can be eluted from a sensor's surface while the surface remains in the biosensor device, and the eluted substances can be identified by electrospray mass spectrometry [13]. This on-line recovery of the analyte offers several advantages over chip-based analysis. These advantages include (1) the ability to concentrate samples prior to their mass spectrometric analysis, and (2) the ability to employ more than one approach in mass spectrometry for identifying the sample.

The value of using SPR and mass spectrometry in tandem, giving rise to a method know as SPR/MS, is best shown in "ligand fishing" studies. In these studies, the analyte is captured by a sensor surface from a complex biological mixture and is then identified using mass spectrometry. In a study performed by Zhukov et al. [14], SPR/MS gave functional as well as structural data on calmodulin and its binding partners. In this study, calmodulin was captured from a bovine brain extract using a sensor chip coated with a peptide fragment of myosin light-chain kinase. The captured material was then eluted using a microrecovery method, as shown in Figure 25.3. The recovered protein was next concentrated, desalted, and identified by MALDI-TOF MS. This report also demonstrated that the captured protein could be identified by proteolysis, followed by peptide mapping using mass spectrometry. The ability to incorporate postcapture enzymatic digestion of a

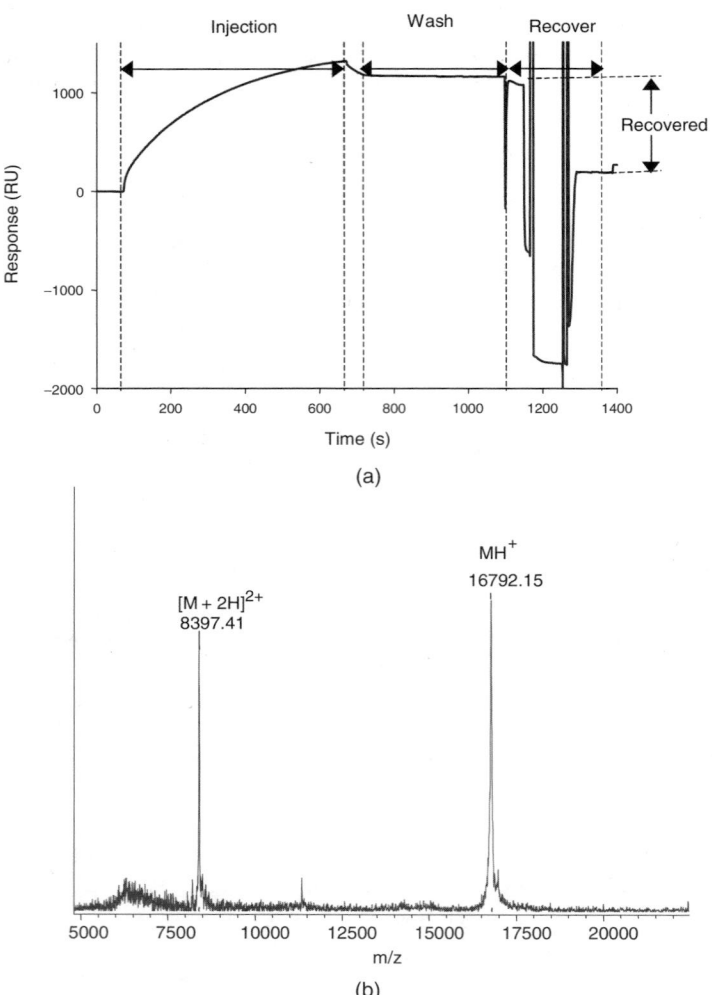

(a)

(b)

Figure 25.3 (a) A sensorgram showing the binding and microrecovery of an M13-specific binding agent that was initially present in a bovine brain extract, and (b) a MALDI mass spectrum for an SPR eluate of a calmodulin eluate that was isolated using the M13 sensor surface. The analyte in (a) was injected over the immobilized M13 surface, with this surface then being washed to remove any nonspecifically bound material. The wash step was followed by recovery of the M13-bound material. The amount of recovered agent was determined directly from the sensorgram. The mass spectrum shown in (b) for the material recovered from the M13 surface produced a spectrum that contained only the single- and double-charged calmodulin molecular ions. (These data were obtained from Zhukov, A., Suckau, D., Buis, J., and Jansson, O., *Biacore J.*, 2, 9–11, 2002.)

protein into this process expands the capabilities of SPR/MS as a tool for functional proteomic studies.

    Another approach for identifying a captured ligand is to use on-chip proteolytic digestion in which the ligand is affinity-captured on the sensor's surface and digested on a separate surface that has been derivatized with a proteolytic enzyme [15]. MALDI-TOF MS is then used to provide the molecular masses of both the native protein and its resulting peptide fragments. As both biosensors and mass spectrometry continue to evolve, the near future should see the production of a hybrid instrument in which the identification of an analyte and a characterization of its binding properties will occur within a single platform.

25.3.4 Characterization of Membrane-Associated Systems

Another recent use of biosensors has been their application in the characterization of membrane-associated systems. Historically, most biophysical analyses of interacting systems have avoided membrane-associated receptors because maintaining the membrane's functional activity is difficult. One advantage of a biosensor is that its surface can provide a pseudomembrane environment that can be used to study the interaction of proteins, drugs, and receptors within a lipid bilayer. For instance, Biacore AB has developed specialized surfaces that are ideal for creating membrane environments. One of these surfaces, referred to as HPA, consists of a self-assembled alkane monolayer that can support a hybrid lipid bilayer [16]. Such surfaces have been used to analyze how proteins interact with specific phospholipid head groups on a membrane surface [17]. Another surface provided by Biacore, known as L1, contains a matrix with an alkane linker that is capable of capturing liposome vesicles. Once captured, these liposomes are stable but can be easily removed from the surface with a detergent wash. Such liposome surfaces have been used to model drug-lipid interactions [18, 19].

To more closely mimic the native environment of membrane-associated receptors, Karlsson and Lofas immobilized rhodopsin on an L1 sensor chip and then reconstituted a bilayer around the receptor by injecting synthetic liposomes over the surface [20]. The process for building the bilayer on this surface is illustrated in Figure 25.4. This work demonstrated that a specific G-coupled protein could be bound to the reconstituted surface and that light induced dissociation of this protein from the surface, thereby confirming that the immobilized rhodopsin was functional. These results hint at the opportunities for using biosensor technology to characterize previously unexamined binding events that occur with membrane-associated receptors.

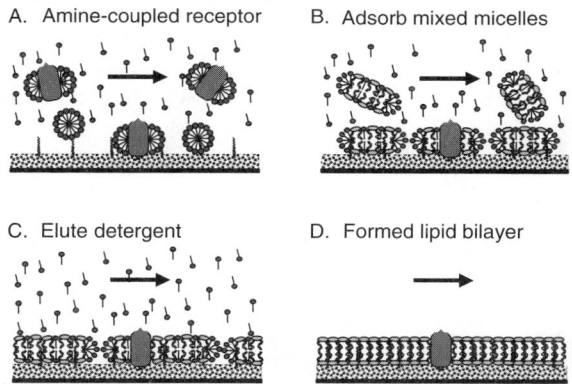

Figure 25.4 An approach for the on-surface reconstitution of lipid bilayers. This approach involves the following steps: (A) amine-coupled receptors are added to the sensor surface in the presence of a detergent; (B) the detergent-lipid mixed micelles are deposited on the surface immediately after receptor immobilization; (C) while detergent-free buffer is passed through the flow cell, the mixed micelles remain adhered to the surface, but the detergent is rapidly depleted from the surface; and (D) as the micelles become detergent-free, they fuse on the surface and build a continuous lipid bilayer around the receptor. (Reproduced with permission from Karlsson, O.P. and Lofas, S., *Anal. Biochem.*, 300, 132–138, 2002.)

25.4 RECENT APPLICATIONS OF BIOSENSORS

25.4.1 Drug Discovery

The hardware for SPR biosensors continues to evolve. Recently, Biacore AB released a new system (Biacore S51) that has a design intended for drug-discovery and drug-binding studies for compounds as small as 100 Da to macromolecular targets. With this instrument, lead compounds that have been identified in high-throughput screens can be confirmed and optimized through inhibition studies, affinity determinations, or detailed kinetic evaluation. This device can also be used in ADME (adsorption, distribution, metabolism, and excretion) studies, giving it a range of drug-assay formats that span from target identification to clinical development.

The flow cell of this instrument permits the simultaneous detection of multiple spots, which, in turn, increases sensitivity and improves referencing between surfaces. This also makes this device well suited for studying small molecules and resolving rapid kinetics. The flow cell has three detection spots arranged transversely across the direction of flow. These spots can be addressed separately during ligand immobilization, as shown in Figure 25.5a. The design of the flow cell is based on the principle of hydrodynamic addressing, which allows one reference and two ligand surfaces to be created within a single flow cell. By adjusting relative flow rates across the flow cell, buffer can be directed to one or the other of the addressable detection spots. This provides enhanced flexibility in the assay design and data output.

The other components of the Biacore S51 consist of a fluidics system, low-pulsation pumps, and an autosampler for the direct injection of analytes to the flow cell. As a result, sample consumption and carryover between cycles are reduced. In addition, the incorporation of 384-well microtiter plates and faster cycling times can permit high-throughput analysis and the fast evaluation of bimolecular interactions.

25.4.2 Array Systems

With the proteomics revolution in full swing, there is an obvious need for high-throughput systems to address questions about protein identification, cell signaling pathways, and translation profiling. An approach based on array SPR is being created through a collaboration of HTS Biosystems (Hopkinton, MA) and Biacore AB (Uppsala, Sweden). These groups are developing a system capable of monitoring up to about 400 ligand samples at one time. As shown in Figure 25.5b, ligands can be spotted (using available nanoliter

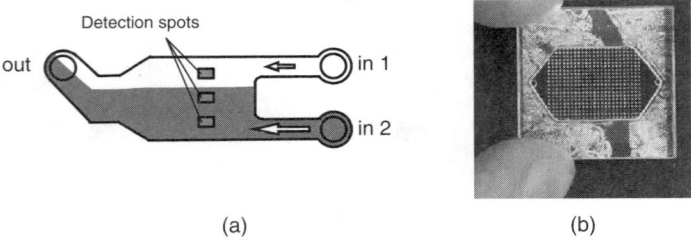

(a)   (b)

Figure 25.5 Two new biosensor designs: (a) the Biacore S51 hydrodynamic addressing flow cell, and (b) the HTS Biosystems and Biacore Flexchip. For the system shown in (a), it is possible to immobilize ligands on each side of the flow cell while maintaining the center as a reference surface by controlling the flow rate from the two inlets. For the array illustrated in (b), a 20 × 20 grid of 200-μm protein spots was prepared to collect binding data throughout the entire flow path of the biosensor.

spotting technology) in a grid format within a 1.2 × 2.2-cm area on a sensor's surface. This array sensor can simultaneously monitor the binding of a single analyte to all the spotted ligands. Analyte samples are delivered in a flow-cell system with rapid buffer exchange, allowing for a full kinetic analysis. Potential applications of this technology include antibody screening and characterization, protein profiling using antibody surfaces, and the analysis of defined nucleic acid and peptide libraries.

25.5 SUMMARY AND CONCLUSIONS

The availability of biosensors and their ease of use have resulted in a rapid application of this technology to diverse areas of research. Before biosensor technology was commercially available, real-time molecular interaction studies were limited to molecules that possessed suitable spectroscopic properties or labels for their detection. In addition, the detailed analysis of molecular interactions, as well as multimolecular complex assembly, was difficult. However, the high information content available from biosensors has significantly advanced our understanding of molecular function, just as high-resolution crystallography has advanced our knowledge of molecular structure.

New areas like the rapid advancement of membrane-associated receptor systems and the coupling of biosensors with mass spectrometry will permit the in-depth characterization of biomolecules in complex environments and at physiological concentrations. While traditional affinity chromatography remains a standard technique for purification and separation, the methods and experience developed over the years with this method have been adapted to optical biosensors, allowing these to provide a high-resolution approach for directly monitoring molecular interactions.

SYMBOLS AND ABBREVIATIONS

ADME	Adsorption, distribution, metabolism, and excretion
GMPs	Good manufacturing practices
k_a	Association rate constant
k_d	Dissociation rate constant
K_D	Dissociation equilibrium constant
MALDI	Matrix-assisted laser desorption/ionization
MALDI-TOF MS	Matrix-assisted laser desorption/ionization time-of-flight mass spectrometry
SPR	Surface plasmon resonance
SPR/MS	Surface plasmon resonance-mass spectrometry

REFERENCES

1. Homola, J., Yee, S.S., and Myszka, D.G., Surface plasmon resonance biosensors, in *Optical Biosensors: Present and Future,* Ligler, C.A., Ed., Elsevier, Amsterdam, 2002, pp. 207–251.
2. Myszka, D.G., Survey of the 1998 optical biosensor literature, *J. Mol. Recognit.,* 12, 390–408, 1999.
3. Rich, R.L. and Myszka, D.G., Survey of the 1999 surface plasmon resonance biosensor literature, *J. Mol. Recognit.,* 13, 388–407, 2000.
4. Rich, R.L. and Myszka, D.G., Survey of the year 2001 commercial optical biosensor literature, *J. Mol. Recognit.,* 15, 352–376, 2002.

5. Baird, C.L. and Myszka, D.G., Current and emerging commercial optical biosensors, *J. Mol. Recognit.*, 14, 261–268, 2001.

6. Rich, R.L. and Myszka, D.G., Advances in surface plasmon resonance biosensor analysis, *Curr. Opin. Biotechnol.*, 11, 54–61, 2000.

7. Myszka, D.G., Improving biosensor analysis, *J. Mol. Recognit.*, 12, 279–284, 1999.

8. Myszka, D.G., Morton, T.A., Doyle, M.L., and Chaiken, I.M., Kinetic analysis of a protein antigen-antibody interaction limited by mass transport on an optical biosensor, *Biophys. Chem.*, 64, 127–137, 1997.

9. Myszka, D.G., He, X., Dembo, M., Morton, T.A., and Goldstein, B., Extending the range of rate constants available from BIACORE: Interpreting mass transport-influenced binding data, *Biophys. J.*, 75, 583–594, 1998.

10. Sikavitsas, V., Nitsche, J.M., and Mountziaris, T.J., Transport and kinetic processes underlying biomolecular interactions in the BIACORE optical biosensor, *Biotechnol. Prog.*, 18, 885–897, 2002.

11. Lackmann, M., Bucci, T., Mann, R.J., Kravets, L.A., Viney, E., Smith, F., Moritz, R.L., Carter, W., Simpson, R.J., Nicola, N.A., Mackwell, K., Nice, E.C., Wilks, A.F., and Boyd, A.W., Purification of a ligand for the EPH-like receptor HEK using a biosensor-based affinity detection approach, *Proc. Natl. Acad. Sci. U.S.A.*, 93, 2523–2527, 1996.

12. Nelson, R.W. and Krone, J.R., Advances in surface plasmon resonance biomolecular interaction analysis mass spectrometry (BIA/MS), *J. Mol. Recognit.*, 12, 77–93, 1999.

13. Natsume, T., Nakayama, H., Jansson, O., Isobe, T., Takio, K., and Mikoshiba, K., Combination of biomolecular interaction analysis and mass spectrometric amino acid sequencing, *Anal. Chem.*, 72, 4193–4198, 2000.

14. Zhukov, A., Suckau, D., Buis, J., and Jansson, O., Ligand fishing with Biacore 3000: Selective binding, recovery and identification by MALDI-MS, *Biacore J.*, 2, 9–11, 2002.

15. Nelson, R., Nedelkov, D., and Tubbs, K., Biosensor chip mass spectrometry: A chip based proteomics approach, *Electrophoresis*, 21, 1155–1163, 2000.

16. Nakajima, H., Kiyokawa, N., Katagiri, Y.U., Taguchi, T., Suzuki, T., Sekino, T., Mimori, K., Ebata, T., Saito, M., Nakao, H., Takeda, T., and Fujimoto, J., Kinetic analysis of binding between Shiga toxin and receptor glycolipid Gb3Cer by surface plasmon resonance, *J. Biol. Chem.*, 276, 42915–42922, 2001.

17. Santagata, S., Boggon, T.J., Baird, C.L., Gomez, C.A., Zhao, J., Shan, W.S., Myszka, D.G., and Shapiro, L., G-Protein signaling through tubby proteins, *Science*, 292, 2041–2050, 2001.

18. Danelian, E., Karlen, A., Karlsson, R., Winiwarter, S., Hansson, A., Lofas, S., Lennernas, H., and Hamalainen, M.D., SPR biosensor studies of the direct interaction between 27 drugs and a liposome surface: Correlation with fraction absorbed in humans, *J. Med. Chem.*, 43, 2083–2086, 2000.

19. Baird, C.L., Courtenay, E.S., and Myszka, D.G., Surface plasmon resonance characterization of drug/liposome interactions, *Anal. Biochem.*, 310, 93–99, 2002.

20. Karlsson, O.P. and Lofas, S., Flow-mediated on-surface reconstitution of G-protein coupled receptors for applications in surface plasmon resonance biosensors, *Anal. Biochem.*, 300, 132–138, 2002.

Section VI

Recent Developments

26

Affinity Ligands in Capillary Electrophoresis

Niels H. H. Heegaard and Christian Schou

Department of Autoimmunology, Statens Serum Institute,
Copenhagen, Denmark

CONTENTS

26.1   INTRODUCTION

Methods for separating and categorizing the components of complex mixtures are crucial for the biological sciences, as is illustrated by the numerous projects that now deal with genomic or proteomic analysis. This has accelerated the development of high-resolution and high-throughput separation methods, such as capillary electrophoresis for DNA sequencing and two-dimensional gel electrophoresis for the mapping of proteomic libraries. The outcome of such work has been the creation of molecular catalogs that map genes in genomes and that characterize proteins with respect to their size, charge, and abundance. This information can, in turn, be used to help our understanding of the biological processes in health and disease.

To completely understand how biological systems work, the vast information in these molecular catalogs must be matched with data on molecular function. Clues on molecular function can be provided by the up-and-down regulation of genes and proteins in disease, the link between pathology and mutations or structural abnormalities in genes and proteins, and animal knock-out or knock-in models. However, the identification of ligands that bind to a specific analyte and measurements of the strength of these interactions are indispensable to a true understanding of biological function.

Unfortunately, most high-resolution separation methods cannot work under conditions that allow the survival of functions such as enzyme activity and ligand binding. For instance, two-dimensional gel electrophoresis requires the strong anionic detergent sodium dodecyl sulfate (SDS) to separate polypeptides according to their molecular weights. Another example of this problem is found in reversed-phase HPLC (high-performance liquid chromatography), which generally uses organic solvents and a low pH to achieve high-resolution separations for proteins and peptides. A notable exception to these other techniques is capillary electrophoresis (CE). This is a separation method in which molecules migrate through 25- to 100-μm I.D. quartz capillaries that are subjected to 100- to 500-V/cm electrical fields [1–3]. In CE, it is not only possible to use solvents that mimic physiological conditions, but it is also possible to perform experiments that examine the interactions of biological ligands under such conditions.

Affinity electrophoresis [4] is a technique in which the migration pattern of a compound in an electrical field is used to examine the binding of this compound with other agents or to estimate the binding constants for these interactions [5–10]. The first use of affinity ligands in electrophoresis was probably work that investigated ovalbumin-nucleic acid interactions in 1942 [11]. Besides providing a means for examining interactions, affinity electrophoresis can also be used as a tool to evaluate sample homogeneity and to measure the concentration of a specific analyte.

Capillary electrophoresis has been used for both affinity electrophoresis and functional applications. Advantages of CE include its quantitative capabilities, good reproducibility, low sample consumption, speed, versatile analytical range, and high resolution. The first applications of affinity ligands in CE were published in the early 1990s [12–22]. The resulting method is now often referred to as *affinity capillary electrophoresis* (ACE). Several recent reviews have covered the theory, methodology, and use of ACE [12, 23–48]. This chapter examines the types of affinity ligands that have been employed in CE and discusses various applications of this method. Of particular interest is the way in which ligands can be combined with CE to increase its selectivity, to perform quantitative measurements, and to characterize noncovalent analyte-ligand interactions.

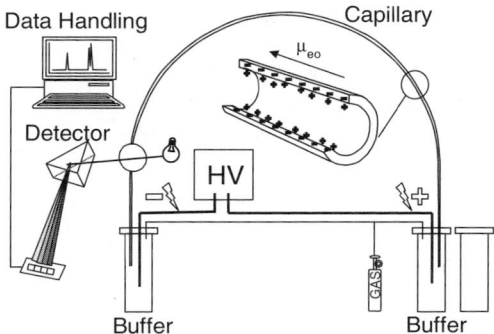

Figure 26.1 Schematic drawing of a capillary electrophoresis system. Abbreviations: HV, high-voltage power supply; μ_{eo}, mobility due to electroosmotic flow.

26.2 GENERAL USE OF LIGANDS IN CE

Tiselius first used free-solution electrophoresis in the 1930s. By performing protein analysis with moving-boundary electrophoresis in glass tubes, he noted the same fundamental features that 50 years later gave rise to capillary zone electrophoresis. These features included the need for strong electrical fields to achieve high efficiencies, the need for small dimensions to avoid the zone-broadening effects of Joule heating, and the advantages of using quartz tubes for UV detection [49, 50].

The key elements of a modern CE system are shown in Figure 26.1. This system provides CE with many advantages and differences when compared with traditional slab gel electrophoresis, as demonstrated in Table 26.1. These same features have led to the

Table 26.1 Comparison of Conventional Gel Electrophoresis and Capillary Electrophoresis

Feature	Gel Electrophoresis	Capillary Electrophoresis
Analyte detection	After separation	During separation
Analyte quantitation	Indirect	Direct
Theoretical plates	10^3–10^4	10^5–10^6
Detection limit (proteins)	nM	μM [a]
Typical sample volume	μl	nl
Parallel sample analysis	Yes	No[b]
Speed	Hours	Minutes
Migration units	Distance	Time
Separation field strength	2–20 V/cm	200–500 V/cm
Preparative capability	Good	Poor
Types of analytes	Macromolecules	Small molecules and macromolecules
Physiological buffer	Usually not compatible	Often compatible conditions

[a] Using UV-based detection systems.
[b] Capillary arrays have been devised for DNA sequencing.

Source: Heegaard, N.H.H., in *Protein-Ligand Interactions: Hydrodynamics and Calorimetry*, Harding, S.E. and Chowdhry, B.Z., Eds., Oxford University Press, Oxford, U.K., 2001, pp. 171–195.

unique ability of CE to be used as a tool for studying and exploiting various binding processes.

The selectivity of any separation method can be increased by introducing a specific affinity ligand into the system, so long as the interactions of this ligand are not inhibited by the separation conditions and the complexed molecules are distinguishable from those that remain free in solution [31, 32, 51]. When affinity ligands are used in CE, this produces a multimode separation. This combination was first utilized in the early 1990s as a way to enhance the selectivity of CE separations and to use CE as a tool for binding studies [12–16, 19, 20, 52–57].

In CE, there are basically two ways to use an affinity ligand. For instance, the ligand can be included in the sample containing the analytes, or the ligand can be included in the electrophoresis system (e.g., as a buffer additive). Differences exist with regards to the state of the ligand (e.g., soluble, derivatized, or immobilized) and the amount of analyte and ligand that are used in such methods. However, the fundamental idea of using electrophoresis to separate the complexed and noncomplexed forms of the analyte, thereby making it possible to visualize this binding process, remains the same.

An understanding of reversible interactions and the purpose of the desired analysis are helpful in deciding which CE approach to use. Weak interactions will generally be examined by or used in an on-line approach. In this technique, the ligand is employed as a buffer additive to increase selectivity, to identify interacting molecules, or to calculate binding constants [46]. The reason for adding the ligand to the running buffer is that any weakly bound complexes in preincubated samples will quickly dissociate when entering the CE system, thus making them undetectable. But by instead adding the ligand to the buffer, these same complexes can form, dissociate, and re-form throughout the separation, making them now possible to observe.

This problem does not occur with stronger complexes, where a stable separation and direct quantitation of the bound and free analyte in preincubated samples may be possible. This particular situation is exploited in a number of competitive and noncompetitive CE immunoassays to measure specific analytes [58–61], as will be discussed in Section 26.5.

A third possible case occurs when the system of interest has an intermediate binding strength. This often means that the complex of interest has a dissociation rate constant on the order of the separation time scale. This can give undesired behavior for the injected sample, including peak broadening and asymmetric peaks. This makes it difficult to use CE to quantitatively analyze such systems except in special cases [62].

26.3   THEORY OF CE WITH AFFINITY LIGANDS

The resolution (R) of an analytical separation technique can be defined as the degree of overlap between two neighboring zones or peaks. This is a product of the efficiency (Ef) and the selectivity (S) for a particular separation, as given by the relationship $R \approx Ef \times S$. Efficiency in CE is often expressed as the number of theoretical plates (N) for the system, which is a measure of narrowness of the individual zones or peaks. For instance, the low zone dispersion and small degree of diffusion usually encountered in CE favor high efficiencies or narrow peaks. The term *selectivity* in CE refers to the relative velocity difference of two analyte zones. In this case, a high selectivity means that there is a large velocity difference between two analytes, thus making it easy to separate them in a capillary of a given length [63].

The electrophoretic mobility (μ) of an analyte is another important factor to consider in CE. This term can be related to several experimental parameters, as illustrated by Equation 26.1.

$$\mu_{exp} = L/Et = L_{tot}L/Vt = \mu + \mu_{eo} \tag{26.1}$$

In this equation, L is the length of the capillary from the point of sample entry to the detector, E is the electric field strength, V is the applied voltage, L_{tot} is the total capillary length, and t is the time of appearance for the analyte peak at the detector. Notice here that the experimentally observed mobility (μ_{exp}) is actually the sum of the inherent mobility of the analyte (μ) and the electroosmotic mobility of the system (μ_{eo}). Also note that μ (or actually, $\mu + \mu_{eo}$) is proportional to $1/t$, which means that high electrophoretic mobilities correspond to short migration times.

The value of μ for an analyte will depend on two main factors [26, 46]. The first of these is the charge-to-size/shape ratio of the analyte, where size and shape are closely related to the analyte's mass. This ratio of charge to size/shape represents the balance between the ionic attraction and drag (or molecular friction) that is acting on the analyte. It is the balance between these forces that makes a solute migrate at a constant velocity in the electric field. The second factor affecting the value of μ is the presence of other molecules in the environment of the analyte. This is a chemical factor. If another substance (such as an additive) interacts with the analyte, the charge-to-size/shape ratio for the analyte-additive complex will almost certainly be different from the free analyte and thus will change the measured value of μ.

The mobility of this complex is not always predictable because the complex's size and shape may be the same, larger, or smaller than that of the analyte. This, in turn, will depend on the complex's conformation. However, the high efficiency of CE makes it possible to detect even interactions that lead to a small change in mobility.

Some fundamental parameters that can be used to describe these binding interactions are the reaction's equilibrium constant and stoichiometry (e.g., see Figure 26.2). Both of these parameters can be estimated by affinity CE [29, 64]. This can be done through ligand-binding studies that are based on the direct quantitation of analytes (e.g., peak-area measurements) in binding mixtures, as might be used in work with strong interactions in preincubated samples. This might also be accomplished through an analysis of mobility

$$A + L \underset{k_{off}}{\overset{k_{on}}{\rightleftharpoons}} AL \qquad\qquad K_D = k_{off}/k_{on}$$

$$K_D = ([A] \bullet [L])/[AL] = 1/K_A$$

$$[AL] = ([A]_{tot} \bullet [L])/K_D + [L]$$

Figure 26.2 A simple bimolecular interaction involving an analyte A and ligand L. In this model, $[A]_{tot}$ is total concentration of binding sites, $[A]$ is the concentration of free analyte, $[L]$ is the concentration of free ligand, and $[AL]$ is the concentration of analyte-ligand complex. The terms K_D and K_A are the dissociation and association equilibrium constants for this system, and k_{off} and k_{on} are the dissociation and association rate constants.

shifts, as would be used for systems with weak interactions, by employing a ligand as a buffer additive.

The theory behind using peak-area measurements for binding studies is the same as for any binding assay in which the equilibrium concentrations for the analyte, ligand, and analyte-ligand complex are determined at different analyte-ligand ratios. Binding curves that plot the bound concentration of ligand ($[B]$) as a function of the free ligand concentration ($[L]$) will show saturability and can yield binding constants. This is preferably accomplished by employing nonlinear curve fitting and data that cover 10 to 90% of the saturation range [65, 66]. Such data can also be analyzed by using a linear transform, as is obtained by plotting $[B]/[L]$ as a function of $[B]$, but such plots can be misleading (e.g., as occurs when the data range does not include saturation conditions) [66]. Although other, more reliable methods, for the linear transformation of binding graphs can be used [29, 67], these are not really necessary if computer programs are available for obtaining nonlinear fits.

The theory behind converting mobility shifts to binding constants in CE has its origins in similar work that has been performed in affinity chromatography, Tiselius electrophoresis, and gel affinity electrophoresis [8, 67–76]. This theory is simply an application of the law of mass balance to a chemical equilibrium and to the method (CE, in this case) that is being used to study the reaction [29, 77–79].

For instance, one possible situation that might occur is when the binding and dissociation rates for an analyte-ligand complex are fast compared with the individual migration rates of the analyte and complex (i.e., creating a local equilibrium in the system). Under these conditions, the two possible forms of the analyte — the free analyte and the analyte-ligand complex — will appear to migrate as one peak in the capillary. The observed mobility of this peak will be a weighted average of the mobilities of the two analyte species, with this weighted average being determined by the equilibrium constant for the interaction and by the ligand's concentration [26, 29, 51, 80–82]. Thus, the position of this peak is a measure of the average time the analyte spends in its free and complexed states.

In traditional slab gel electrophoresis, the electrophoretic mobility μ is directly proportional to the relative migration distance (d) of the analyte at a fixed time (i.e., the time at which the electrophoretic run is terminated). To correct for nonspecific migration shifts, a nonreacting internal marker can be added to the sample to adjust the measured distances. An experiment is then performed in which the migration distance of the analyte is measured relative to this reference distance (d_{ref}) at various ligand concentrations (c). This system can be described by the following linear equation [8] if the analyte-ligand complex has zero mobility, which links the association constant for the complex to the experimentally determined distance values.

$$1/d = 1/d_0[1 + K_A c] \qquad (26.2)$$

In Equation 26.2, d_0 is the relative migration distance of the analyte in the absence of ligand, and (K_A) is the association equilibrium constant for the analyte-ligand complex.

In CE, a similar approach can be used, but it is necessary to now consider the mobility of the analyte-ligand complex, since this may no longer be equal to zero. For this type of system, the relationship between K_A and migration becomes

$$\Delta\mu = (\Delta\mu_{max}K_A c)/(1 + K_A c) = \Delta\mu_{max}c/(K_D + c) \qquad (26.3)$$

in which $\Delta\mu$ is the difference in electrophoretic mobility between the analyte in the absence and presence of ligand with a given concentration c, and $\Delta\mu_{max}$ is the maximum difference between the mobility of the free and complexed forms of the analyte.

Since μ is directly proportional to $1/t$ (see Equation 26.1), it follows that $1/t$ can be substituted for μ if a correction is made for variations in the electroosmotic flow by using a noninteracting internal marker. Sources of these variations include fluctuations in the temperature of the system and in the viscosity or dielectric constant of the running buffer at different additive or ligand concentrations. If the distance to the detection window and the field strength are the same in each experiment, Equation 26.3 can then be written in any of the forms shown below,

$$\Delta(1/t) = (1/t - 1/t_m)$$
$$= [(1/t - 1/t_m)_{max} K_A c]/(1 + K_A c)$$
$$= [(1/t - 1/t_m)_{max} c/K_D]/(1 + c/K_D)$$
$$= \Delta 1/t_{max} c/(K_D + c) \tag{26.4}$$

where t_m is the migration time for the internal marker and $(1/t - 1/t_m)_{max}$, or $\Delta 1/t_{max}$, represents the maximum possible difference in inverse migration times that can be obtained between the analyte and internal marker (i.e., the difference between a 100% complexed analyte and the internal marker). It follows from these relationships that a plot of $\Delta(1/t)$ versus c can yield the dissociation constant K_D (or $1/K_A$) through nonlinear curve fitting, as illustrated in Figure 26.3b. As was mentioned earlier for traditional binding assays, this binding curve should show saturation (an indicator of specificity), and its data points should cover a range of 10 to 90% saturation [83, 84]. Again, there are several ways of linearizing such data, including a Scatchard-type plot, but these may introduce errors and should be used with caution [28, 46, 81].

There are several assumptions, limitations, and requirements that must be met for the proper use of Equations 26.2 to 26.4. For instance, these equations require that the analyte and analyte-ligand complex have different mobilities under the conditions used in the study. It is also assumed that the experiments are carried out with a ligand concentration that is much higher (i.e., 10 to 500 times greater) than the analyte's concentration. In addition, these particular equations assume that the ligand has a single type of binding site, is homogeneously distributed throughout the system, and has a precisely known concentration. Furthermore, it is assumed that the analyte and ligand have 1:1 binding stoichiometry, fast interaction kinetics, and no interactions with the capillary wall or unwanted buffer components. As mentioned earlier, it is also necessary to use an internal marker in these experiments to correct for nonspecific migration shifts [23–27, 29, 33, 37, 44, 82, 83]. This same approach can be extended to the use of more than one ligand, as is described in previous reports [79, 84, 85].

One inherent assumption made during CE binding studies is that the binding constant or interaction being measured is not influenced by the electric field. Another general assumption is that analyte stacking effects [86] do not invalidate the underlying requirement for a ligand excess.

26.4 GENERAL TYPES OF LIGANDS IN CE

There are several forms that the ligand can take in CE. Figure 26.4 and Figure 26.5 show some of the typical states of ligands that are employed in affinity CE. Depending on the requirements of the experiment, this ligand may be present either free in solution or as an immobilized form. The ligand may also be specifically modified to endow it with a customized mobility. All of these approaches are discussed in this section.

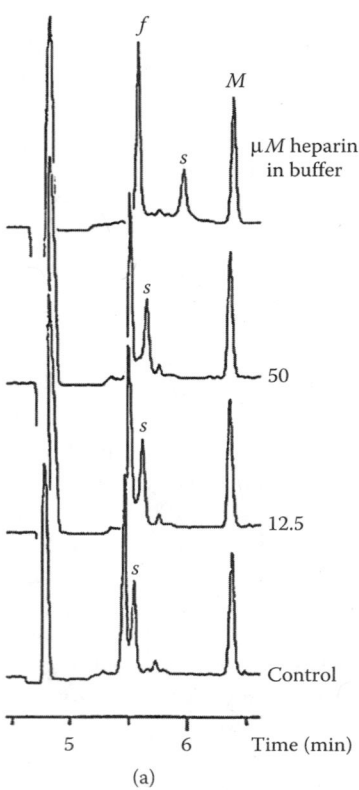

μM heparin in buffer

50

12.5

Control

5   6   Time (min)

(a)

Figure 26.3 Characterization of the interaction between LMW heparin and cleaved (Lys^{58}-β_2m) or wild-type β_2-microglobulin (β_2m) by affinity CE. In (a), heparin at various concentrations was added to the electrophoresis buffer. This was used for analyzing 4-sec injections of a mixture that contained 0.27 mg/ml Lys^{58}-β_2m (23 μM, as prepared from a stock solution of 1.3 mg/ml diluted in water) and 0.03 mg/ml of a marker peptide (M). The two species of Lys^{58}-β_2m are indicated in this electropherogram as being f (fast) or s (slow). This CE study was performed at 10 kV in pH 7.65, 0.2 M Tricine/NaOH. The results in (b) show a plot of the resulting migration-shift data. The peak appearance times (t) were corrected for electroosmotic flow variations by using the internal marker peptide. The data in these plots represent the means and standard deviations of triplicate experiments. The fitted curves were obtained by using a single site-binding hyperbola ($R^2 > 0.99$ for all fitted plots). Linearization of these data (see inset) was performed according to the equation $\Delta(1/t) = \Delta(1/t)_{max} - K_d[\Delta(1/t)/c]$, where c is the ligand concentration (i.e., [LMW heparin] in this study) and the slopes of the fitted lines would be equal to $-K_D$. (Reproduced with permission from Heegaard, N.H.H., Roepstorff, P., Melberg, S.G., and Nissen, M.H., *J. Biol. Chem.*, 277, 11184–11189, 2002.)

26.4.1 Free Ligands

The use of a nonmodified, soluble ligand (as shown in Figure 26.4) is usually preferred in quantitative applications of affinity CE. This is the case because it is easy to control the concentration of an unmodified ligand. This also makes it possible to rule out any changes in binding that may result from ligand immobilization or derivatization (see Chapter 3). The quantitative use of affinity CE with a buffer additive requires that the whole separation path be filled with the ligand. In contrast to this, partial filling (PF) of a ligand in a capillary [87]

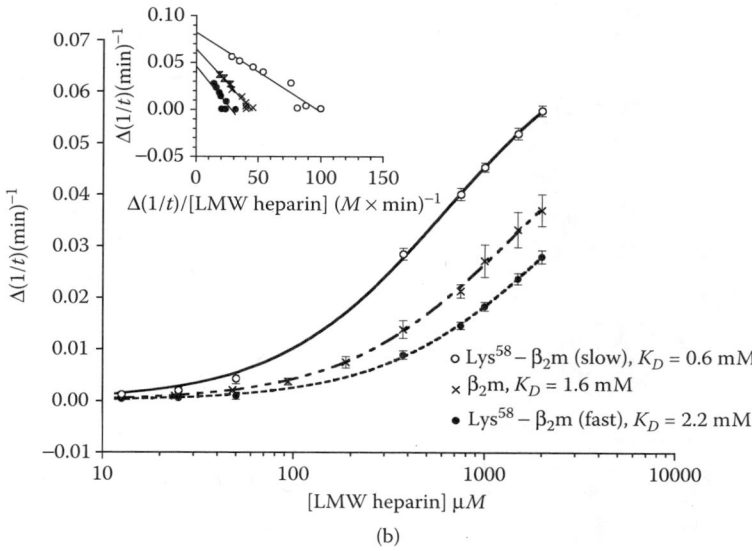

Figure 26.3 (*Continued*)

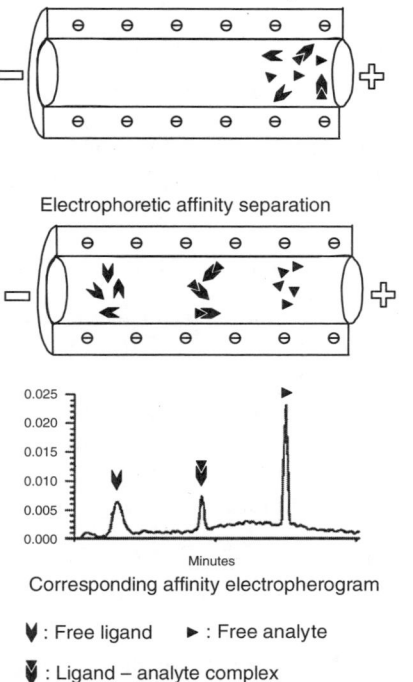

Corresponding affinity electropherogram

▼ : Free ligand   ▶ : Free analyte

▼ : Ligand – analyte complex

Figure 26.4 Separation of free and ligand-bound analyte by affinity CE based on a soluble ligand and strong analyte-ligand complex.

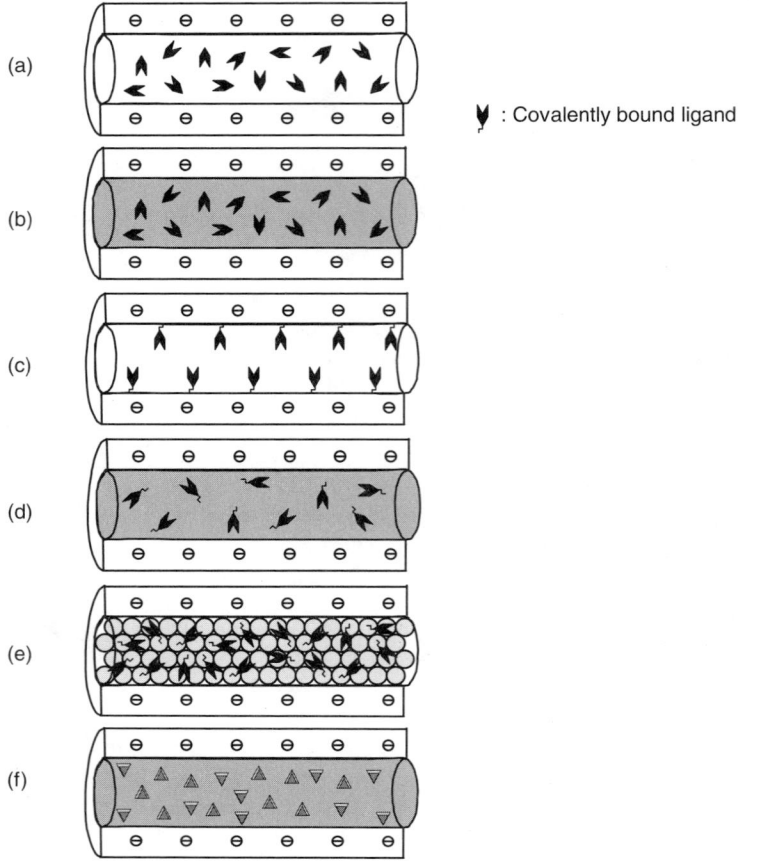

: Covalently bound ligand

Figure 26.5   Various types of ligand states in affinity CE: (a) free in solution, (b) entangled in a gel, (c) covalently coupled to the capillary wall, (d) coupled to polymers, (e) coupled to beads, or (f) formed as a molecular imprint of the analyte in a cross-linked polymer.

can be employed to create an increase in selectivity while avoiding detector interferences from the ligand and while minimizing any waste of the ligand.

In cases where charged additives are utilized, the capillary should always be filled with a vial that is separate from the one used during the electrophoretic run. If this is not done, the concentration of the ligand will change with time. When additives migrate faster than the analyte, it is also necessary to have the additive present in the electrophoresis run buffer vial, with this solution being prepared fresh prior to each run. More slowly moving additives need to be present only in the capillary.

26.4.2   Soluble Modified Ligands and Ligand Constructs

Because of poor limits of detection, it is often not possible to characterize strong binding events with unlabeled ligands when UV/Vis absorbance detection is used. Instead, it may be necessary to label the ligand or analyte with a fluorescent tag. When used in combination with laser-induced fluorescence detection (LIF), this will lower detection limits considerably [88].

A number of approaches exist for placing a fluorescent label on ligands or solutes, some of which are specific for a particular region or functional group (e.g., the carbohydrate groups on glycoproteins). It is therefore prudent, if possible, to label regions on the ligand or analyte that are not likely to participate in binding. An example of this is a study of peptide binding by a recombinant major histocompatibility complex HLA-DR4 molecule, where the carbohydrate moiety of the DR4 molecule (known to not participate in peptide binding) was labeled using a fluorescein-thiosemicarbazide tag [89].

Derivatizing the ligand is also a useful alternative in cases where it is not possible by electrophoresis to directly discern between the complexed and free forms of the ligand. For instance, this can occur if the ligand is a small neutral molecule. In this situation, an approach using a labeled and mobile ligand construct known as an *affinophore* or *affinity probe* may be considered [60, 90, 91]. The resulting method is known as affinity-probe capillary electrophoresis (APCE). Examples of such an approach are given in Table 26.2.

26.4.3 Immobilized Ligands

Ligands can also be immobilized in CE capillaries. This can be done physically, such as by entangling the ligands in gels. It can also be done chemically, as is accomplished by covalently bonding the ligand to the capillary wall [92, 93], to polymers [17, 94–98], or to frits or beads [99–102] inside the capillary. This is illustrated in Figure 26.5.

Yet another approach is to make an imprint [103] of a specific molecule that is incorporated into a crosslinked polymer during its synthesis and later removed. This provides a stationary phase for an affinity capillary that is now selective for the molecule, group of molecules, or functional group that was originally imprinted in the polymer [41, 96, 104–106].

The affinity material in these approaches may occupy a part of the capillary or the entire length of the separation path. The principle behind the resulting ligand-based separations is similar to that for traditional column chromatography, except that the driving forces now include electrophoresis and electroosmosis. Therefore, the use of ligands in such methods can be considered a type of capillary electrochromatography (CEC) [107].

Migration shifts are maximized in experiments that use immobilized ligands, since the mobility of the analyte-ligand complex is zero. This approach can also be utilized for preparative purposes. For instance, this can be used to improve detection limits by concentrating specific analytes in on-line capillary sections that contain enclosed immobilized antibodies or imprinted polymers [99, 108].

26.5 APPLICATIONS OF LIGANDS IN CE

There are three main applications for ligands in CE. The first of these is their use to enhance the selectivity of an analytical or preparative separation. The second is their utilization in the quantitative measurement or enrichment of specific analytes (i.e., assays). And the third general use of ligands in CE is in binding studies to identify and characterize interacting molecules. Examples for each of these applications are given in this section.

26.5.1 Selectivity Enhancement

This application uses ligands to improve resolution by increasing selectivity. This, in turn, makes it possible to visualize and quantitate the different components of the separated mixture. Examples of this application are given in Table 26.2. Within this area, the separation of chiral compounds has developed into an important field and is especially important for

Table 26.2 Examples of Affinity Ligands in Capillary Electrophoresis

Analyte	Interacting Molecules (Ligand Analyte or Analytes)	CE Mode	References
Ions and ion-binding compounds	Ca^{2+}- and C-reactive proteins	ACE	[55]
	Ca^{2+}- and Zn^{2+}-binding proteins	ACE	[56]
	[28]ane-N_6O_2-Fe$(CN)_6^{3-}$ and -Fe$(CN)_6^{4-}$	ACE with immobilized ligand	[92]
	CD: tetraphenylborate and tetraphenylphosphonium ions	ACE	[205]
	Zn^{2+}-HIV nucleocapsid protein, NCp7	ACE	[206]
	Fe^{2+}/Fe^{3+}: nicotinamine	—	[207]
	Cl^-, ClO_4^-: BSA	ACE	[208]
	Cu(II)-iminodiacetate: proteins	ACE with immobilized ligand	[97]
Drugs	α_1-Acid glycoprotein: basic drugs	PF-ACE	[87]
	α_1-Acid glycoprotein: disopyramide and remoxipride	PF-ACE	[166]
	Cortisol: Ab	ACE-CIA incl. chip format	[131, 209, 210]
	Serum proteins: β-blockers	FA-CE	[154]
	BSA: warfarin	ACE, FA-, and VP-CE	[12]
	HSA and CD: verapamil	FA-CE	[156, 157]
	BSA: leucovorin	ACE coated	[18]
	CD: leucovorin metabolites	ACE	[211]
	CD: model drugs	ACE	[212]
	Hydroxypropyl-β-CD: tioconazole stereoisomers	ACE	[213]
	Cholesteryl-10-undecenoate: benzodiazepine	ACE with immobilized ligand	[93]
	Charged surfactants: cephalosporins	ACE	[214]
	HDL, LDL, oxidized LDL: verapamil and nilvadipine	FA-CE	[151, 153]
	Methyl-β-CD: orciprenaline	PF-ACE	[215]
	Digoxin: F_{ab} Ab	ACE-CIA	[121]

(continued)

	Imprinted chiral stationary phase: propranolol and metoprolol	ACE	[96, 104]
	Digoxin: scFv antidigoxin	APCE-NCIA	[60]
	Dorzolamide: fluorescein-carbonic anhydrase	APCE	[123]
	Drugs: HSA, dye	Competitive Dye-APCE	[216]
	Theophylline: Mab and mouse Ig, mAb	ACE-CIA and NCIA-chip	[129, 130]
	Transthyretin: flufenamic acid and flurbiprofen	ACE and FA-CE	[152]
	Estradiol: Mab	ACE-NCIA-multichannel chip	[127]
	Ovalbumin: Mab		
	Cyclosporine: Mab	ACE-CIA-FLPO	[217]
	Morphine, PCP, methadone, cocaine, amphetamines: Mabs	ACE-CIA	[119, 218]
Small organics	Achiral crown ether + CD: stereoisomers of racemates with primary amines	ACE	[219]
	CD + SDS: Neutral hydrophobic compounds	MEKC	[220]
	TM-β-CD: phenoxy acid herbicides	ACE-CE-MS	[221]
	Thyroxine: Ab	Chip ACE-CIA	[222]
	Fumonisin B_1: Mab	ACE-CIA	[223]
Carbohydrates	Labeled aminophenyl-mannopyranoside; Con A, unlabeled monosaccharide	APCE	[90]
	Vancomycin: peptidoglycan precursors	ACE	[224]
	Methylglucose lipopolysaccharides: palmitoyl-CoA	ACE-ESI-MS	[170]
	Polyiodides: dextrins	ACE	[225]
	mAbs: α- and β-maltose	CAGE	[98]
DNA or RNA	9-Vinyladenine: oligodeoxynucleotides	CAGE	[17]
	Ethidium bromide: DNA	CAGE	[19]
	Bromodeoxyuridine: Mab + Ab	ACE-NCIA	[226]

(continued)

Table 26.2 Examples of Affinity Ligands in Capillary Electrophoresis (Continued)

Analyte	Interacting Molecules (Ligand Analyte or Analytes)	CE Mode	References
Peptides	CD: dipeptide stereoisomers	ACE with urea	[227]
	Carboxymethyl-β-CD: tripeptide stereoisomers	ACE	[228]
	SDS: cyclic peptide and its oligomers	MEKC	[229]
	Heparin: synthetic peptides	ACE	[13]
	Heparin: proteolytic peptides	ACE	[145, 146, 149, 230]
	Latex: Mab + hCG	CE	[100]
	GnRH: Ab F_{ab} fragments	IA-CE-MS	[99]
	Vancomycin: peptide libraries, dipeptide	ACE	[15, 53, 231]
	Vancomycin: peptide libraries	ACE-MS	[144, 148]
	Glucagon: Mab	RPLC-CE-CIA	[126]
	Scarpie-peptide: Mab	ACE-CIA	[214]
Proteins	Benzene sulfonamides: bovine carbonic anhydrase	ACE	[16, 62, 64, 231–233]
	TFE: β₂-microglobulin	On-line mixing CE	[163]
	Congo red: β₂-microglobulin	ACE	[162–164]
	ANS: β₂-microglobulin	ACE	[163]
	DNA aptamer: IgE	APCE	[125]
	DNA aptamer: HIV-1 RT	APCE	[234]
	Phosphorthioate oligo: HIV-1 gp120	APCE and competitive	[235]
	Anti-inflammatory drugs: HSA and BSA	ACE	[236]
	Porphyrins: HSA	ACE	[237]
	Polysuccinylcarbohydrate: pea lectin	APCE	[165, 238]
	DNA: MyoD/E47 DNA-binding proteins	ACE	[239]
	HLA class II: peptide	ACE	[89]
	Ds and ssDNA: Mab	ACE	[240]

(continued)

GFP-cyclophilin: HIV-1 p24	ACE	[241]
Peptides: erythropoietin receptor	ACE	[242]
Heparin: antithrombin	ACE	[243]
Streptavidin: biotin	ACE	[64]
F-ds-oligonucleotide: SpP3A2 transcription factor	ACE	[244]
Deoxyspegualin: Hsc70 heat shock protein	ACE	[245]
Fucose-1-phosphate: pea lectin	ACE	[246]
Trypsin and CT inhibitors: trypsin and CT and mAb	ACE	[116]
mAb: troponin I	ACE with immobilized mAb	[140]
IgA: succinyl-$F_{(ab')2}$	APCE-NCIA	[122]
Human growth hormone: Mab	APCE	[118]
Human growth hormone: Mab	ACE-NCIA	[52]
Human growth hormone: Mab F_{ab} fragment	APCE-IEF	[112]
Human serum albumin: Mab	ACE	[64]
Insulin: Mab	ACE-CIA	[59, 247–250]
Insulin-glucagon: Mabs	ACE-CIA	[58]
Morphine: Mab	ACE-CIA	[120]
Hapten: Mab	FA-CE	[117]
BSA: Mab and Mab F_{ab}	ACE-CIA	[251]
IgG: protein G	APCE	[252]
Enzyme-substrate systems		
Glucose-6-phosphate dehydrogenase: glucose-6-phosphate	On-line mixing CE	[158]
Ethanol: alcohol dehydrogenase	On-line mixing CE	[158]
tRNA^{trp}: Bacterial tryptophanyl-tRNA synthetase	APCE	[253]
Supramolecular assemblies		
HRV2: VLDLR-MBP	ACE	[254]
HRV: Mab and receptor fragments	ACE	[255, 256]

characterizing drug enantiomers. A few examples of this are given in Table 26.2; however, more complete coverage can found in recent reviews [78, 109, 110].

26.5.2 Quantitative Assays

One analytical issue in CE is its poor concentration limit of detection. This is a direct consequence of the narrow separation path in CE with which the detector must operate. Electrophoretic techniques—stacking [86], isotachophoresis [111], and isoelectric focusing [112]—will narrow analyte zones, resulting in taller, narrower peaks and improved detection limits. Off- and on-line concentration can also help with this problem [108]. This is often accomplished by using an adsorbing material of low specificity (such as a reversed-phase matrix) in front of the buffer-containing capillary or by using a charged pseudostationary phase in the electrophoresis buffer [113]. More specific enrichment of trace analytes can be accomplished by employing immobilized ligands, as has been demonstrated in reports using immobilized antibodies [39, 101, 114, 115]. When combined with sensitive detection systems like laser-induced fluorescence, these ligand-aided approaches can yield very low limits of detection. However, in this approach, attention must be paid to the conductivity of the sample versus the conductivity of the running buffer, as well as to other technical difficulties in placing and retaining inside the capillary a solid matrix with appropriate flow characteristics [114].

Quantitative measurements of analytes by CE typically use immunoreagents (i.e., antibodies), where CE simply provides a means for discriminating between the free and antibody-bound forms of the analyte [61]. Examples are given in Table 26.2. Since the initial work that demonstrated the possibility of separating immunocomplexes by CE [52, 116–118], practically all CE-based immunoassays have used more sensitive detection than UV/Vis absorption. Typically, laser-induced fluorescence detection is utilized for this purpose [59, 119–122].

Another means for increasing the detection limits of quantitative CE is to employ specific mobile affinity ligands (i.e., affinophores or affinity probes). These consist of an affinity ligand plus a mobilizing (or charged) group. When used in conjunction with the pronounced concentration effect that is achievable by isoelectric focusing, this approach has achieved impressive limits of detection [42, 122]. For instance, a detection limit of 5×10^{-12} M has been reported with this technique for recombinant human growth hormone [112], and a detection limit of approximately 35×10^{-12} M has been reported for α_1-antitrypsin [91]. The main problem with the use of these synthesized probes is the need to obtain affinity probes that are electrophoretically homogeneous or nearly homogeneous [60].

A competitive immunoassay (CIA) in CE (as illustrated in Figure 26.6) has the advantage of eliminating the need for a labeled antibody or antibody construct. But the sensitivity of these assays is limited by the dissociation constant of the antibody or capture reagent [123]. This means that high-affinity antibodies will increase the sensitivity of these assays and lower their limits of detection.

A noncompetitive immunoassay (NCIA) in CE is one that uses antibodies or antibody fragments with a fluorescent label. This is a direct assay that, as a rule, is more sensitive than a competitive immunoassay, since its sensitivity is not limited by the antibody's binding constant [124]. In addition, these assays are linear over a larger range (i.e., almost three orders of magnitude in response) [91]. For these reasons, noncompetitive immunoassays are preferred for trace analysis. But these do require a labeled antibody [122], an antibody fragment, or some other type of analyte-binding molecule, such as an aptamer [125]. In addition, the generation of an immunoreagent that contains a fluorescent label and that is charge-optimized, soluble, and homogeneous is not a trivial task [60, 91].

Traditional solid-phase sandwich immunoassays [124] cannot usually be used for small analytes because they require two types of antibodies that bind simultaneously to the

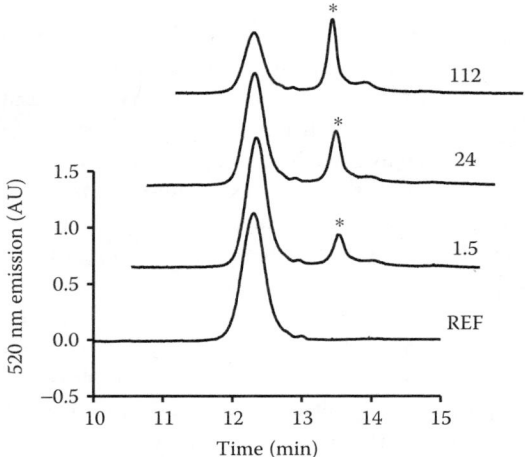

Figure 26.6 Example of a CE competitive immunoassay. This was performed using an in-house monoclonal antibody against human β_2-microglobulin. This antibody was mixed with fluorescein-labeled, purified β_2-microglobulin and analyzed by CE using laser-induced fluorescence detection. This analysis was conducted in pH 7.7, 0.1 M phosphate buffer by using a 75-μm I.D. capillary developed at 12 kV. An excess of unlabeled β_2-microglobulin was added to the sample, and repeated analyses after this addition gave a gradual displacement of the labeled β_2-microglobulin from antibodies to give a free fraction (marked with *) that was clearly separated from the peak for the antibody complex. Detection in this study was based on emission at 520 nm. The time (in min) after the addition of the unlabeled β_2-microglobulin is shown on the right.

same target. This is not the case in the formats used for CE immunoassays, which only require one type of antibody and have no limitations with regard to analyte size. This means that such assays can be used for small analytes. For instance, a CE immunoassay has provided 50- to 1000-pM detection limits in biological samples for drugs like dorzolamide and digoxin in 1 to 10 min [60, 123].

Charge variants of analytes can be quantitated simultaneously by CE because of their electrophoretic separation. This was shown in an analysis of recombinant human growth hormone [112].

Trace-enrichment techniques can be used to further lower detection limits in these assays. One example is the digoxin assay mentioned earlier that, when combined with solid-phase extraction, gave detection limits of 400 fM [60]. Another example is a CE immunoassay that was performed on-line with reversed-phase HPLC, giving a limit of detection of 20 pM for glucagon when used in a competitive format [126].

Several other developments have recently been reported in this field. For instance, some work has examined the lack of parallel processing of samples in CE to help further improve this approach [127]. In addition, work with chip immunoassays promises even faster analysis times and less reagent consumption [127–132]. Advances in alternative detection systems (e.g., fluorescence polarization) [133] may further improve the limits of detection in noncompetitive CE immunoassays and provide a way for directly identifying peaks that represent complexes. This is also the case when using CE with mass spectrometry [134–138], as might be performed on-line with immunoaffinity CE [99, 139]. Finally, ligand-mediated capture of specific analytes is growing in use for trace enrichment in procedures that are integrated with CE [99, 114, 126, 140, 141].

26.5.3   CE Binding Studies

There are several reasons why CE is often used for binding studies. For instance, this approach is valuable in situations where the amount of sample is limiting or where purified analyte or an analyte preparation with homogeneous binding activity is not available. CE is also useful in cases where complex formation is difficult to detect by other binding assays, especially when changes in electrophoretic mobility are expected as a result of complex formation. Two other situations in which CE binding assays are useful include cases where the analyte or ligand cannot be analyzed by gel electrophoresis, and experiments in which many ligands are to be screened for binding activity or that deal with systems that have weak binding (dissociation constants in the μM to M range) [44].

The separation time scale and estimated complex stability both determine the design of a quantitative CE binding experiment. In practical terms, the equilibrium that is being studied should be sufficiently fast to give clean peak shifts when the ligand is employed as a buffer additive (i.e., no peak broadening, splitting, or disappearance should be observed). This occurs when the complex dissociation half-time ($\ln2/k_{off}$) is less than or equal to 1% the time required to separate the free and bound molecules [51, 80, 142]. Thus, very long runs would, in theory, permit the analysis of slowly dissociating interactions by this type of affinity CE [80]. Conversely, if the time needed for the separation and sample introduction is fast (as in chip-based analysis) [131], peak broadening may occur, even with fast interactions; in these situations, a preequilibration approach is better suited for the estimation of binding constants [143]. Thus, the interplay between the dissociation rate constant (k_{off}) and separation time scale (for intermediate-affinity binding) and the lower limit of detection (for high-affinity binding) determines the experimental approach. A further discussion of the importance of rate constants in determining a system's compatibility with a given set of CE separation parameters is given in the literature [51, 62].

High-affinity interactions will normally be analyzed after preequilibration. In these situations, interference from molecules that dissociate during sample introduction and electrophoresis into an empty buffer is avoided if their dissociation rate constant is less than $0.105/t$, where t is the time required to separate the associated peaks. This ensures that no more than 10% of the bound ligand dissociates during the separation [65]. The analyte does not necessarily have to be pure in such an experiment, since it is electrophoretically separated from other components in the sample. Quantitation, however, does require that the analyte concentration be known.

An advantage of using a buffer additive in affinity CE is that knowledge of the exact concentration of the analyte is not needed for calculating binding constants. Many examples of this approach are listed in Table 26.2. The combination of affinity CE with mass spectrometry is extremely beneficial for this work, as illustrated by a study that gave binding constants for the interactions of 19 different peptides with one ligand (vancomycin) while the individual peptides were identified by MS [144].

A standard approach [44] for examining an interaction by affinity CE is to perform a *migration-shift experiment*. This is an experiment in which the ligand is added to the electrophoresis buffer. Both strong and weak interactions can be revealed by this method through the changes they produce in analyte peak profiles. Subsequently, any changes in the separation pattern should be checked for ligand dependency, as can be accomplished by performing runs in the presence of varying concentrations of the ligand. If no changes are observed in the absence or presence of ligand, there may be no binding, or one of several factors might be present that obscure or inhibit binding. Such factors include cases where the mobilities of the complexed and free forms of the analyte are the same (e.g., as occurs when a small uncharged ligand binds to a large molecule). Other obscuring factors include (1) the preexistence of a strong complex in the sample; (2) blocking of the ligand through its interactions with the capillary wall; (3) the presence of buffer components, other molecules, or buffer conditions that are not conducive to binding; and (4) the blocking of binding by components that bind to the analyte but that do not change its mobility.

The buffer additive approach is useful for discovering analytes in complex mixtures that bind to a specific ligand [15, 53]. This is especially true when the sample material is available in only scarce amounts, as happens when examining the binding of peptides in enzymatic digests of purified biomolecules [145–147]. Structure-function studies, like the mapping of specific binding activities to the fragments of a protein, often require subsequent identification of these fragments by another technique, such as mass spectrometry [148, 149].

Some modified versions of the buffer additive approach are useful for selected applications. One example is the use of affinity CE to determine the stoichiometry of binding, which provides a determination of how many ligands are bound to one analyte molecule [12]. This number can be derived by adapting the Hummel-Dreyer method [150] to affinity CE. In this approach, the ligand is added to both the running buffer and sample. One requirement here is that the ligand itself must give a detectable signal. When the components of an equilibrated sample plug (i.e., a mixture containing the analyte-ligand complex as well as the free analyte and ligand) begin to separate, the concentration of ligand in the zone corresponding to free ligand will always be lower than the buffer ligand concentration if some of this ligand was bound to the analyte. This localized ligand deficiency gives a negative peak that has the same mobility as the free analyte. Conversely, for the zone where the ligand is bound to the analyte, there will be an increased response, giving a positive peak. Since this approach is an equilibrium method (where the analyte is in equilibrium with free ligand during the run), it is well suited for the study of weak binding systems. When measuring stoichiometries by this method, the ligand should be used at saturating concentrations (i.e., at concentrations much greater than K_D). A plot of the free ligand peak areas vs. the total ligand concentration should then yield a value of total ligand concentration that gives a zero peak area (i.e., the peak area where the free ligand concentration is the same as the ligand concentration in the buffer). From this value, the bound ligand concentration can be derived and, since the analyte concentration is known, the binding stoichiometry can be determined [64].

A related approach is the *vacancy peak method (VP)*. In this technique, the running buffer contains the analyte and ligand, while the injected sample contains only buffer. This results in negative peaks that represent the free and bound ligand [12].

Another method for examining weak interactions is *frontal analysis (FA)* (see Figure 26.7). In this technique, the approach is the same as for strongly interacting analyte-ligand pairs, where the interacting molecules are preequilibrated and then injected. However, in frontal analysis a very large sample plug is now injected (i.e., typically an injection time of 50 to 100 sec, which corresponds to approximately 5 to 15% of the total volume of the capillary). This produces a peak plateau of analyte plus ligand, where the edge of the plateau will have a trailing or leading plateau that corresponds to the unbound ligand. By comparing this plateau height with the plateau heights of the ligand at the same total concentrations but in the absence of the analyte, a direct measure of the bound material is obtained. Frontal analysis has been used in many drug-protein binding studies (see Table 26.2). This is especially useful because protein interactions with the capillary are less important here than in other methods, making frontal analysis valuable in cases where the material being studied is available in relatively large amounts [151–157].

The use of a capillary as a microreactor, where enzymatic reactions can be followed after on-line mixing, is another means by which molecular function can be studied by CE [158, 159]. Protein folding is an example of intramolecular reversible reactions for which CE has proved to be a promising tool because of its speed, resolution, and wide range of possible buffers [160–163]. This approach can use buffers with ligand additives to characterize

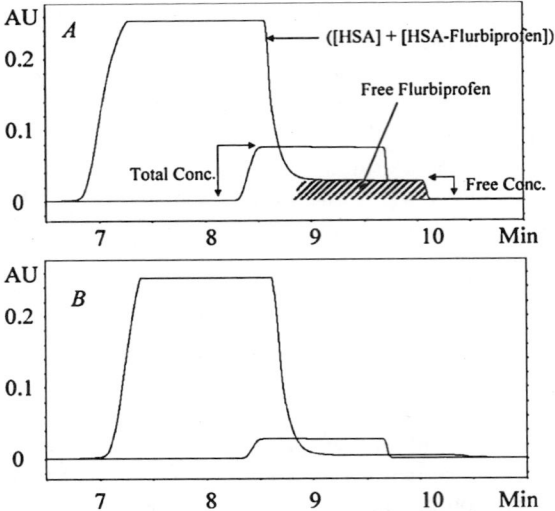

Figure 26.7 Frontal-analysis study examining the binding between flurbiprofen and human serum albumin (HSA). The electrophoresis buffer was pH 7.4, 67 mM phosphate. Samples were injected for 99 sec at 0.5 psi into a 50-μm I.D. uncoated capillary with a total length of 57 cm (50 cm to the detector). This corresponded to a sample volume of approximately 90 nl. The electrophoresis took place at 15 kV and 37°C. Detection was performed by using UV absorbance at 200 nm. In (a) is given the results for a sample containing [flurbiprofen]$_{total}$ = 607 μM and [HSA] = 56 μM. This is overlaid with the results for flurbiprofen alone at 607 μM. The plateau height of flurbiprofen in the HSA sample (i.e., the tailing plateau of the has plug) corresponds to the free flurbiprofen in the sample (shaded area = 211 μM). From this region, the amount of bound drug in the flurbiprofen-HSA mixture can be calculated. In (b), a sample was applied that contained [flurbiprofen]$_{total}$ = 203 μM and [HSA] = 56 μM. The overlaid trace in this case is for a sample of flurbiprofen alone at 203 μM. From these data, the free flurbiprofen in the second flurbiprofen-HSA sample was determined to be 25 μM. (Based on results published in Østergaard, J., Schou, C., Larsen, C., and Heegaard, N.H.H., *Electrophoresis*, 23, 2842–2853, 2002.)

functional differences between protein conformers [164] and allows on-line, time-controlled mixing with specific solvents that migrate at different velocities from the analyte [163].

26.6 SPECIFIC LIGANDS USED IN CE

There are many types of ligands that have been used in CE. Table 26.2 lists some of the pioneering applications along with recent and illustrative work. This table is divided into several sections based on the molecular nature of the ligand and analyte. It is obvious from this list that there are several classes of molecules that cannot be analyzed by gel electrophoresis but that can be analyzed by CE. Important examples are small organic drugs, small peptides, and ions. Also, these applications reflect the fact that affinity CE, in contrast to many other methods, is especially well suited for the characterization of weak binding. Examples of this include lectin-carbohydrate interactions [38, 165] and drug binding to plasma proteins [166].

A number of recent reviews provide further information on this topic. For instance, quantitative aspects of CE, such as its use in the estimation of binding constants and reaction stoichiometries, are discussed in several references [29, 32, 35, 37, 38, 46]. A focus on affinity probes is found in other references [42], along with additional tables of affinity CE applications [30, 35, 38]. A recent review has covered the use of affinity CE in both soluble and solid-phase techniques [40]. Other reports have placed a specific emphasis on the use of affinity CE in pharmaceutical applications [167] and in the analysis of weak interactions [38, 46].

26.7 TRENDS AND CHALLENGES IN AFFINITY CE

The information available from UV/Vis absorbance detection is not sufficient to identify and characterize the structure of the molecules and complexes that are separated by CE. Approaches that might overcome this limitation include the interfacing of CE with mass spectrometry (MS), nuclear magnetic resonance (NMR) spectroscopy, or biosensors. Examples are listed in Table 26.2. Some issues that must be overcome in the application of these schemes concern the development of appropriate interfaces, and the production of adequate detection limits and sampling rates.

26.7.1 CE-MS

The interfacing of CE with MS (giving rise to the method of *capillary electrophoresis-mass spectrometry*, or CE-MS) provides the ability to both separate chemicals and identify their structures, as well as the ability to verify ligand complexation and directly measure reaction stoichiometries. However, the introduction of liquid-phase analytes that have been separated by CE into the gas phase and interior vacuum of a mass analyzer is a challenge. A solution [134, 168] came when *electrospray-ionization* (ESI or EI) interfaces for mass spectrometers were introduced in the late 1980s. Electrospray ionization generates molecular ions with multiple charges, which facilitates the detection of even large molecules in the limited mass range of most mass spectrometers. Molecular masses are then deconvoluted from the charge state series (or charge envelopes) that are produced by each detected species [169].

An important factor in the use of mass spectrometry for binding studies is the fact that electrospray ionization is a relatively mild method that allows the survival and subsequent characterization of noncovalent molecular complexes in the gas phase [137, 170]. The mechanism of ion formation in ESI is still under investigation. However, it is known that the main items that affect electrospray-ion formation are the electrical field; the electrospray capillary diameter; the capillary heat capacity; the solvent's ionic strength, pH, and flow rate; the solvent saturation of ambient gas; and the analyte's distribution of ionic and hydrophobic groups, proton affinity, pK_a, and solvation energy [171, 172].

Designs for CE-MS interfaces all face the problem of how to replace the missing exit buffer reservoir from the CE system while still ensuring electrical continuity in both the CE and MS systems. Several interface designs are illustrated in Figure 26.8. The two most common of these are the liquid sheath flow system [173, 174] and the sheathless system [135, 175, 176].

In the *liquid sheath flow system* [148, 177], a metal sheath completes the electrical circuit of the CE capillary with the electrospray capillary. An MS-compatible sheath liquid acts as the CE terminal buffer reservoir, and a sheath gas at the nebulizer tip helps droplet formation. The liquid flow rate for this interface is typically a few μl/min, but some applications can go down to 250 nl/min [177]. This type of interface only causes limited band broadening.

A *sheathless interface* [171, 175, 178, 179] uses a CE capillary that is sleeved into a metal tip or a capillary that is pulled to a fine tip and coated with metal or that contains an electrode to complete the electrical circuit. Since no sheath liquid is required here, the elution of analytes is undisturbed. This results in high sensitivity. However, electroosmotic flow is needed to transport analytes into the sprayer, and there may be problems with incompatibility of the CE buffer with the electrospray process [180]. Newer developments in the area of liquid junctions have been made, where sample transport from the capillary exit to the

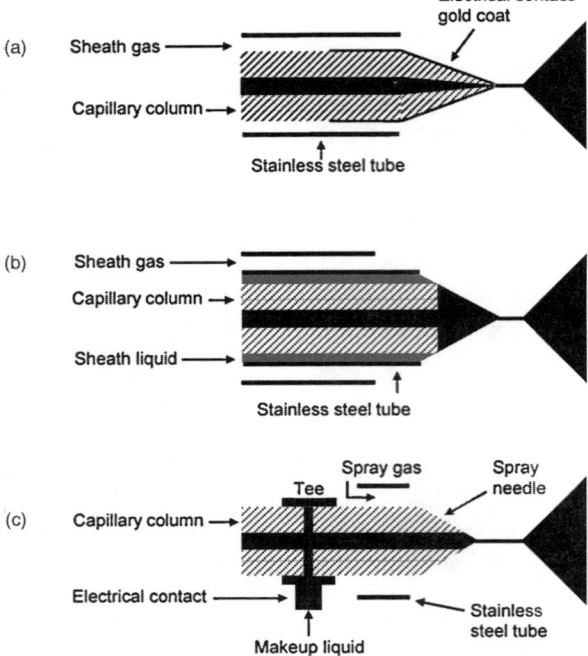

Figure 26.8   Several interfaces for coupling of CE with electrospray ionization-mass spectrometry: (a) sheathless interface, (b) sheath flow interface, and (c) liquid junction interface. (Based on a figure published in Banks, J.F., Recent advances in capillary electrophoresis/electrospray/mass spectrometry, *Electrophoresis*, 18, 2255–2266, 1997.)

electrospray needle is achieved by pressure. For instance, a method using a subatmospheric chamber enclosing a micro-electrospray tip [180] appears to be promising. This ensures simple and rugged operation and allows coupling to microfabricated devices [181].

With respect to the mass analyzer, the issue in interfacing CE with MS is the compatibility of the MS data acquisition with the extremely narrow peaks that appear from the CE instrument. Even though newer mass analyzers have high sampling rates, most of these systems are too slow to preserve the peak efficiencies that are typical of CE separations. This is especially true for mass analyzers where changes in magnetic or electric fields are used to selectively detect ions. For this reason, the most compatible type of mass analyzer for CE peak detection is a time-of-flight (TOF) instrument. In addition, Fourier transform ion cyclotron mass analyzers, although expensive, offer both high mass resolution and sensitivity and can be used in higher order MS analyses [178].

As indicated by Table 26.2 [137, 144, 148, 170, 182–185], there are still relatively few reported applications of mass spectrometry with affinity CE or CE that employs specific analyte enrichment. However, the number of these applications should increase. Reasons for this include several reports that noncovalent interactions can be studied by electrospray-ionization mass spectrometry, and the recent development of relatively simple interfaces for CE and MS instruments.

26.7.2   CE-NMR

NMR is one of the most comprehensive analytical detection techniques in existence. This makes the possible combination of this method with CE (known as *capillary electrophoresis-nuclear magnetic resonance spectroscopy*, or CE-NMR) of great interest. But the use of NMR as an on-line technique with trace-level separation methods like CE has been difficult. NMR measures the resonance spin states of various naturally occurring isotopes (e.g., ^1H, ^{13}C, ^{15}N, ^{32}P, etc.) in an external magnetic field. The main problem with this originates from the inherently poor sensitivity of NMR. This is due to the limited difference in spin-state populations that are present in even high magnetic fields and large sample volumes. Much effort has been put into improving NMR instruments to overcome this problem. However, even though larger magnets will continue to improve the sensitivity of NMR, the resulting instruments are unique, costly, and not designed for interfacing with liquid-separation methods.

The most likely avenues of improved sensitivity in CE-NMR are represented by an increase in computational power, minimization of the detection flow cell, and optimization of the solvent [186–188]. The low solvent consumption of CE limits the cost of using fully deuterated solvents, as opposed to HPLC-NMR systems [189]. For the study of biomolecules and biomolecular complexes, high-resolution and multidimensional NMR techniques are also required. Initial studies in this area have been performed using microcoil-based on-line sample cells [190, 191]. The challenge in data acquisition for such cells is illustrated by the fact that their volume is more than 10,000-fold lower than conventional NMR sample cells [191]. Also, the limited residence time of analytes that pass through this coil reduces sensitivity. This means that the high efficiency of CE is not easily made compatible or preserved with NMR detection. In addition, high concentrations of analytes are necessary to get useful signals [192]. No examples of affinity CE-NMR have yet been published. However, when CE-NMR does become further developed, it is anticipated that this will be an extremely informative tool with regard to molecular complexation.

26.7.3 CE with Biosensors

CE can also be coupled to devices that directly measure analyte-ligand binding. This is accomplished by using various biosensor systems, including cell-based sensors [193–195]. These biosensors include an immobilized element (often of biological origin) that acts as the sensing region. This, in turn, is connected to a transducer for translation of the chemical signal.

Most biosensors that are based on an immobilized ligand will respond to the association and dissociation of an analyte-ligand interaction as a function of time. The specificity of the biosensor reflects the affinity constants that it measures for a specific analyte or class of analytes [196, 197]. The direct combination of an analytical separation with measurements of the interaction kinetics for the separated compounds has been reported in on-line applications of other separation methods [198, 199]. But this is especially attractive with CE because of the high resolution and low sample consumption of this technique. One example is the use of patch clamp detection, where detector cells containing specific receptors yielded a measurable current when combined with CE [200].

The total number of reported applications for biosensors in CE is still low [193–195], but this combination has much to offer. In addition, the use of specifically tailored bioreporters with CE could possibly be extended to include other methods like mass spectrometry.

26.7.4 Other Detection Schemes

Further specialized detection schemes in CE may have applications for both analyte measurement and characterization. For example, electrochemical detection has been employed in CE by using Cu^{2+} as a complexing agent for peptides [201]. Another example is the use of fluorescence polarization to directly verify DNA-protein and protein-protein complex formation [133, 202].

26.8   SUMMARY AND CONCLUSIONS

The high resolution and high specificity that are possible in affinity CE have been helpful in characterizing many biomolecular systems. There are three ways that affinity ligands can be used in this technique. First, they can be used as tools to increase the selectivity of a separation, as might be needed to improve resolution. Specific interactions with ligands like antibodies may also be used to measure analytes. And finally, affinity CE can be a valuable method for identifying and characterizing reversible molecular interactions.

The theoretical basis of affinity CE for quantitative measurements is derived from the theory of traditional binding assays and from the various modes of affinity operations that have been used in conventional electrophoresis. The success of affinity CE is primarily due to the inherent and unique features of CE as a separation technique, including its analytical versatility, reproducibility, precision, and resolving power. The use of ligands in CE can provide information on molecules that are not analyzable by conventional electrophoresis (e.g., peptides and drug molecules). Depending on the application, this may transform CE into a two-dimensional or multimode separation technique.

Developments in detector technology, like the creation of interfaces for CE with information-rich methods like NMR and MS (along with the optimization of other existing detectors), should increase the versatility of affinity CE. This should ensure that this approach continues to play a unique role in the functional characterization of biomolecules and genome products.

ACKNOWLEDGMENTS

Dr. Jesper Østergaard performed the experiments shown in Figure 26.7. Part of the work reported herein was supported by the Danish Medical Research Council, Fonden til Lægevidenslkabens Fremme, Apotekerfonden af 1991, Lundbeckfonden, and the M. L. Jørgensen og Gunnar Hansens Fond.

SYMBOLS AND ABBREVIATIONS

[A]	Free-analyte concentration
Ab	Antibody (polyclonal)
ACE	Affinity capillary electrophoresis
[AL]	Analyte-ligand complex concentration
ANS	8-Anilino-1-naphthalene sulfonic acid
APCE	Affinity probe capillary electrophoresis
[B]	Bound-ligand concentration
BSA	Bovine serum albumin
c	Ligand (additive) concentration
CAGE	Capillary affinity gel electrophoresis
CD	Cyclodextrin
CE	Capillary electrophoresis
CEC	Capillary electrochromatography
CE-MS	Capillary electrophoresis-mass spectrometry
CE-NMR	Capillary electrophoresis-nuclear magnetic resonance spectroscopy
CIA	Competitive immunoassay
Con A	Concanavalin A

CT	Chymotrypsin
d	Relative migration distance for an analyte
d_0	Relative migration distance for an analyte without ligand
d_{ref}	Relative migration distance for an internal marker
ds	Double-stranded
E	Electric field strength
Ef	Efficiency
ESI (or EI)	Electrospray ionization
F	Fluorescein
FA	Frontal analysis
FLPO	Fluorescence polarization
GFP	Green fluorescent protein
GnRH	Gonadotropin-releasing hormone
hCG	Human chorionic gonadotropin
HDL	High-density lipoprotein
HIV	Human immunodeficiency virus
HPLC	High-performance liquid chromatography
HRV	Human rhinovirus
HSA	Human serum albumin
IA	Immunoaffinity (solid phase)
IEF	Isoelectric focusing
K_A	Association equilibrium constant
K_D	Dissociation equilibrium constant
k_{off}	Dissociation rate constant
k_{on}	Association rate constant
$[L]$	Free-ligand concentration
L	Capillary length-to-detector window
L_{tot}	Total capillary length
LDL	Low-density lipoprotein
LIF	Laser-induced fluorescence detection
mAb	Monoclonal antibody
MBP	Maltose binding protein
MEKC	Micellar electrokinetic chromatography
MS	Mass spectrometry
N	Number of theoretical plates
NCIA	Noncompetitive immunoassay
NMR	Nuclear magnetic resonance
PCP	Phencyclidine
PF	Partial filling
R	Resolution
RPLC	Reversed-phase liquid chromatography
RT	Reverse transcriptase
S	Selectivity
SDS	Sodium dodecyl sulfate
ss	Single-stranded
t	Peak appearance time
t_{eo}	Electroosmotic flow marker appearance time
t_m	Migration time for an internal marker
TFE	Trifluoroethanol
TM	Heptakis(2,3,6,-tri-O-methyl)

V	Voltage
VLDLR	Very-low-density lipoprotein receptor
VP	Vacancy peak method
$\Delta 1/t_{max}$	Maximum possible difference in the inverse migration times between the analyte and internal marker (i.e., the difference that occurs when the analyte is 100% complexed)
$\Delta\mu$	Difference in electrophoretic mobility of the analyte in the absence versus presence of ligand
$\Delta\mu_{max}$	Maximum mobility difference for an analyte
μ	Electrophoretic mobility
μ_{exp}	Experimentally observed mobility
μ_{peo}	Mobility due to electroosmotic flow

REFERENCES

1. FEP Mikkers, FM Everaerts, TPEM Verheggen. High performance zone electrophoresis. J Chromatogr 169:11–20, 1979.
2. JW Jorgenson, KD Lukacs. Zone electrophoresis in open-tubular glass capillaries. Anal Chem 53:1298–1302, 1981.
3. JW Jorgenson, KD Lukacs. Capillary zone electrophoresis. Science 222:266–272, 1983.
4. TC Bøg-Hansen. Crossed immuno-affinoelectrophoresis. An analytical method to predict the result of affinity chromatography. Anal Biochem 56:480–488, 1973.
5. NHH Heegaard. Electrophoretic analysis of reversible interactions. In: P Lundahl, A Lundqvist, E Greijer, eds. Quantitative Analysis of Biospecific Interactions. New York: Harwood Academic, 1998, pp 1–13.
6. S Nakamura, T Wakeyama. An attempt to demonstrate the distribution of trypsin inhibitors in the sera of various animals. J Biochem Tokyo 49:733–741, 1961.
7. M Szylit. Electrophorèse avec adsorption specifique: Interactions amylase-amidon en gel mixte di polyacrylamide-amidon. Ann Biol Clin 29:215–227, 1971.
8. K Takeo, S Nakamura. Dissociation constants of glucan phosphorylases of rabbit tissues studied by polyacrylamide gel disc electrophoresis. Arch Biochem Biophys 153:1–7, 1972.
9. SJ Gerbrandy, A Doorgeest. Potato phosphorylase isoenzymes. Phytochemistry 11:2403–2407, 1972.
10. NHH Heegaard. A history of the use and measurement of affinity interactions in electrophoresis. In: HJ Issaq, ed. A Century of Separation Science. New York: Marcel Dekker, 2002, pp 527–554.
11. LG Longsworth, DA MacInnes. An electrophoretic study of mixtures of ovalbumin and yeast nucleic acid. J Gen Physiol 25:507–516, 1942.
12. JC Kraak, S Busch, H Poppe. Study of protein-drug binding using capillary zone electrophoresis. J Chromatogr 608:257–264, 1992.
13. NHH Heegaard, FA Robey. Use of capillary zone electrophoresis to evaluate the binding of anionic carbohydrates to synthetic peptides derived from serum amyloid P component. Anal Chem 64:2479–2482, 1992.
14. H Kajiwara, H Hirano, K Oono. Binding shift assay of parvalbumin, calmodulin and carbonic anhydrase by high-performance capillary electrophoresis. J Biochem Biophys Methods 22:263–268, 1991.
15. Y-H Chu, GM Whitesides. Affinity capillary electrophoresis can simultaneously measure binding constants of multiple peptides to vancomycin. J Org Chem 57:3524–3525, 1992.
16. Y-H Chu, LZ Avila, HA Biebuyck, GM Whitesides. Use of affinity capillary electrophoresis to measure binding constants of ligands to proteins. J Med Chem 35:2915–2917, 1992.

17. Y Baba, M Tsuhako, T Sawa, M Akashi, E Yashima. Specific base recognition of oligode-oxynucleotides by capillary affinity gel electrophoresis using polyacrylamide-poly(9-vinyladenine) conjugated gel. Anal Chem 64:1920–1925, 1992.

18. GE Barker, P Russo, RA Hartwick. Chiral separation of leucovorin with bovine serum albumin using affinity capillary electrophoresis. Anal Chem 64:3024–3028, 1992.

19. A Guttman, N Cooke. Capillary gel affinity electrophoresis of DNA fragments. Anal Chem 63:2038–2042, 1991.

20. JL Carpenter, P Camilleri, D Dhanak, D Goodall. A study of the binding of vancomycin to dipeptides using capillary electrophoresis. J Chem Soc Chem Commun 11:804–806, 1992.

21. S Honda, A Taga, K Suzuki, S Suzuki, K Kakehi. Determination of the association constant of monovalent mode protein-sugar interaction by capillary zone electrophoresis. J Chromatogr 597:377–382, 1992.

22. SAC Wren, RC Rowe. Theoretical aspects of chiral separation in capillary electrophoresis. J Chromatogr 603:235–241, 1992.

23. KL Rundlett, DW Armstrong. Examination of the origin, variation, and proper use of expressions for the estimation of association constants by capillary electrophoresis. J Chromatogr A 721:173–186, 1996.

24. S Bose, J Yang, DS Hage. Guidelines in selecting ligand concentrations for the determination of binding constants by affinity capillary electrophoresis. J Chromatogr B 697:77–88, 1997.

25. DJ Winzor. Measurement of binding constants by capillary electrophoresis. J Chromatogr A 696:160–163, 1995.

26. X Peng, MT Bowser, P Britz-McKibbin, GM Bebault, JR Morris, DDY Chen. Quantitative description of analyte migration behavior based on dynamic complexation in capillary electrophoresis with one or more additives. Electrophoresis 18:706–716, 1997.

27. H Hoppe. System peaks and non-linearity in capillary electrophoresis and high-performance liquid chromatography. J Chromatogr A 831:105–121, 1999.

28. KL Rundlett, DW Armstrong. Methods for the determination of binding constants by capillary electrophoresis. Electrophoresis 22:1419–1427, 2001.

29. KL Rundlett, DW Armstrong. Methods for the estimation of binding constants by capillary electrophoresis. Electrophoresis 18:2194–2202, 1997.

30. Y-H Chu, CC Cheng. Affinity capillary electrophoresis in biomolecular recognition. Cell Mol Life Sci 54:663–683, 1998.

31. NHH Heegaard. Biospecific interactions measured by capillary electrophoresis. In: W Ens, KG Standing, IV Chernushevich, eds. New Methods for the Study of Molecular Complexes. Dordrecht: Kluwer Academic, 1998, pp 305–318.

32. IJ Colton, JD Carbeck, J Rao, GM Whitesides. Affinity capillary electrophoresis: A physical-organic tool for studying interactions in biomolecular recognition. Electrophoresis 19:367–382, 1998.

33. NHH Heegaard. Capillary electrophoresis for the study of affinity interactions. J Mol Recogn 11:141–148, 1998.

34. NHH Heegaard. Paper symposium: Affinity capillary electrophoresis. Electrophoresis 19:367–464, 1998.

35. Y-H Chu, LZ Avila, J Gao, GM Whitesides. Affinity capillary electrophoresis. Acc Chem Res 28:461–468, 1995.

36. K Shimura, K Kasai. Affinophoresis: Selective electrophoretic separation of proteins using specific carriers. Methods Enzymol 271:203–218, 1996.

37. NHH Heegaard, K Shimura. Determination of affinity constants by capillary electrophoresis. In: P Lundahl, A Lundqvist, E Greijer, eds. Quantitative Analysis of Biospecific Interactions. New York: Harwood Academic, 1998, pp 15–34.

38. K Shimura, K-I Kasai. Affinity capillary electrophoresis: A sensitive tool for the study of molecular interactions and its use in microscale analyses. Anal Biochem 251:1–16, 1997.

39. NA Guzman. On-line bioaffinity, molecular recognition, and preconcentration in CE technology. LC-GC 17:19–27, 1999.

40. NHH Heegaard, S Nilsson, NA Guzman. Affinity capillary electrophoresis: Important application areas and some recent developments. J Chromatogr B 715:29–54, 1998.

41. T Takeuchi, J Haginaka. Separation and sensing based on molecular recognition using molecularly imprinted polymers. J Chromatogr B 728:1–20, 1999.

42. K Shimura, K Kasai. Capillary affinophoresis as a versatile tool for the study of biomolecular interactions: A mini-review. J Mol Recogn 11:134–140, 1998.

43. NHH Heegaard. Paper symposium: Functional electrophoresis on chips and in capillaries. Electrophoresis 23:813, 2002.

44. NHH Heegaard. Capillary electrophoresis. In: SE Harding, BZ Chowdhry, eds. Protein-Ligand Interactions: Hydrodynamics and Calorimetry. Oxford, UK: Oxford University Press, 2001, pp 171–195.

45. NHH Heegaard, RT Kennedy. Identification, quantitation, and characterization of biomolecules by capillary electrophoretic analysis of binding interactions. Electrophoresis 20:3122–3133, 1999.

46. NHH Heegaard, MH Nissen, DDY Chen. Applications of on-line weak affinity interactions in free solution capillary electrophoresis. Electrophoresis 23:815–822, 2002.

47. G Rippel, H Corstjens, HAH Billiet, J Frank. Affinity capillary electrophoresis. Electrophoresis 18:2175–2183, 1997.

48. RT Kennedy, N Schultz. Studies of antibody-antigen interactions by capillary electrophoresis. Abstract HPCE V. 1993.

49. A Tiselius. The moving boundary method of studying the electrophoresis of proteins. Nova acta regiae societatis scientiarum upsaliensis Ser. IV, 7, No.4:1–107, 1930.

50. A Tiselius. A new apparatus for electrophoretic analysis of colloidal mixtures. Trans Faraday Soc 33:524–531, 1937.

51. M Mammen, FA Gomez, GM Whitesides. Determination of the binding of ligands containing the N-2,4-dinitrophenyl group to bivalent monoclonal rat anti-DNP antibody using affinity capillary electrophoresis. Anal Chem 67:3526–3535, 1995.

52. RG Nielsen, EC Rickard, PF Santa, DA Sharknas, GS Sittaampalam. Separation of antibody-antigen complexes by capillary zone electrophoresis, isoelectric focusing and high-performance size-exclusion chromatography. J Chromatogr 539:177–185, 1991.

53. Y-H Chu, LZ Avila, HA Biebuyck, GM Whitesides. Using affinity capillary electrophoresis to identify the peptide in a peptide library that binds most tightly to vancomycin. J Org Chem 58:648–652, 1993.

54. NHH Heegaard, FA Robey. Use of capillary zone electrophoresis for the analysis of DNA-binding to a peptide derived from amyloid P component. J Liq Chromatogr 16:1923–1939, 1993.

55. NHH Heegaard, FA Robey. A capillary electrophoresis-based assay for the binding of Ca^{2+} and phosphorylcholine to human C-reactive protein. J Immunol Methods 166:103–110, 1993.

56. H Kajiwara. Application of high-performance capillary electrophoresis to the analysis of conformation and interaction of metal-binding proteins. J Chromatogr 559:345–356, 1991.

57. S Honda, S Suzuki, A Nose, K Yamamoto, K Kakehi. Capillary zone electrophoresis of reducing mono- and oligo-saccharides as the borate complexes of their 3-methyl-1-phenyl-pyrazolin-5-one derivatives. Carbohydrate Res 215:193–198, 1991.

58. I German, RT Kennedy. Rapid simultaneous determination of glucagon and insulin by capillary electrophoresis immunoassays. J Chromatogr B 742:353–362, 2000.

59. NM Schultz, RT Kennedy. Rapid Immunoassays using capillary electrophoresis with fluorescence detection. Anal Chem 65:3161–3165, 1993.

60. FT Hafner, RA Kautz, BL Iverson, RC Tim, BL Karger. Noncompetitive immunoassay of small analytes at the femtomolar level by affinity probe capillary electrophoresis: Direct analysis of digoxin using a uniform-labeled scFv immunoreagent. Anal Chem 72:5779–5786, 2000.

61. NHH Heegaard, RT Kennedy. Antigen-antibody interactions in capillary electrophoresis. J Chromatogr B 768:93–103, 2002.

62. LZ Avila, Y-H Chu, EC Blossey, GM Whitesides. Use of affinity capillary electrophoresis to determine kinetic and equilibrium constants for binding of arylsulfonamides to bovine carbonic anhydrase. J Med Chem 36:126–133, 1993.

63. JC Giddings. Zone formation and resolution. In: JC Giddings. Unified Separation Science. New York: Wiley, 1991, pp 86–111.
64. Y-H Chu, WJ Lees, A Stassinopoulus, CT Walsh. Using affinity capillary electrophoresis to determine binding stoichiometries of protein-ligand interactions. Biochemistry 33:10616–10621, 1994.
65. Hulme EC, ed. Receptor-Ligand Interactions. Oxford: IRL, Oxford University Press, 1992.
66. IM Klotz. Ligand-protein binding affinities. In: TE Creighton, ed. Protein Function. A Practical Approach. Oxford: IRL Press, 1989, pp 25–54.
67. JN Wilson. A theory of chromatography. J Am Chem Soc 62:1583–1591, 1940.
68. J Weiss. On the theory of chromatography. J Chem Soc 81:297–303, 1943.
69. D DeVault. The theory of chromatography. J Am Chem Soc 65:532–540, 1943.
70. RA Alberty, HHJr Marvin. Study of protein-ion interaction by the moving boundary method. J Phys Colloid Chem 54:47–55, 1950.
71. LG Longsworth. Moving boundary electrophoresis—theory. In: M Bier, ed. Electrophoresis. Theory, Methods and Applications. New York: Academic Press, 1959, pp 91–136.
72. RA Alberty, HH Marvin, Jr. Study of protein-ion interaction by the moving boundary method. The combination of bovine serum albumin with chloride ion. J Am Chem Soc 73:3220–3223, 1951.
73. K Takeo. Affinity electrophoresis. In: A Chrambach, MJ Dunn, BJ Radola, eds. Advances in Electrophoresis. Vol. I. Weinheim: VCH Verlag, 1987, pp 229–279.
74. TC Bøg-Hansen, K Takeo. Determination of dissociation constants by affinity electrophoresis: Complexes between human serum proteins and concanavalin A. Electrophoresis 1:67–71, 1980.
75. V Horejsí. Some theoretical aspects of affinity electrophoresis. J Chromatogr 178:1–13, 1979.
76. K Takeo. Detection of phosphorylase on disc electrophoresis using polyacrylamide as supporting medium. Ann Rep Soc Protein Chem Yamaguchi Univ School Med 4:41–48, 1970.
77. Giddings JC. Unified Separation Science. New York: John Wiley, 1991.
78. A Rizzi. Fundamental aspects of chiral separations by capillary electrophoresis. Electrophoresis 22:3079–3106, 2001.
79. MT Bowser, GM Bebault, X Peng, DD Chen. Redefining the separation factor: A potential pathway to a unified separation science. Electrophoresis 18:2928–2934, 1997.
80. V Matousek, V Horejsí. Affinity electrophoresis: A theoretical study of the effects of the kinetics of protein-ligand complex formation and dissociation reactions. J Chromatogr 245:271–290, 1982.
81. DS Hage, SA Tweed. Recent advances in chromatographic and electrophoretic methods for the study of drug-protein interactions. J Chromatogr B 699:499–525, 1997.
82. J Heintz, M Hernandez, FA Gomez. Use of a partial-filling technique in affinity capillary electrophoresis for determining binding constants of ligands to receptors. J Chromatogr A 840:261–268, 1999.
83. J Kawaoka, FA Gomez. Use of mobility ratios to estimate binding constants of ligands to proteins in affinity capillary electrophoresis. J Chromatogr B 715:203–210, 1998.
84. MT Bowser, AR Kranack, DDY Chen. Properties of multivariate binding isotherms in capillary electrophoresis. Anal Chem 70:1076–1084, 1998.
85. MT Bowser, DDY Chen. Dynamic complexation of solutes in capillary electrophoresis. Electrophoresis 19:383–387, 1998.
86. DS Burgi, R-L Chien. Optimization in sample stacking for high-performance capillary electrophoresis. Anal Chem 63:2042–2047, 1991.
87. Y Tanaka, S Terabe. Separation of the enantiomers of basic drugs by affinity capillary electrophoresis using a partial filling technique and α_1-acid glycoprotein as chiral selector. Chromatographia 44:119–128, 1997.
88. Y-F Cheng, NJ Dovichi. Subattomole amino acid analysis by capillary zone electrophoresis and laser-induced fluorescence. Science 242:562–564, 1988.

89. NHH Heegaard, BE Hansen, A Svejgaard, LH Fugger. Interactions of the human class II major histocompatibility complex protein HLA-DR4 with a peptide ligand demonstrated by affinity capillary electrophoresis. J Chromatogr A 781:91–97, 1997.

90. K Shimura, K Kasai. Determination of the affinity constants of Concanavalin A for monosaccharides by fluorescence affinity probe capillary electrophoresis. Anal Biochem 227:186–194, 1995.

91. K Shimura, M Hoshimo, K Kamiya, K Katoh, S Hisada, H Matsumoto, K Kasai. Immunoassay of serum α_1-antitrypsin by affinity-probe capillary isoelectric focusing using a fluorescence-labeled recombinant antibody fragment. Electrophoresis 23:909–917, 2002.

92. C-Y Liu, W-H Chen. Electrophoretic separation of inorganic anions with an anion complex one-modified capillary column. J Chromatogr A 815:251–263, 1998.

93. AP Catabay, H Sawada, K Jinno, JJ Pesek, MT Matyska. Separation of benzodiazepines using cholesterol-modified fused-silica capillaries in capillary electrochromatography. J Cap Electrophor 5:89–95, 1998.

94. A Guttman, A Paulus, AS Cohen, N Grinberg, BL Karger. Use of complexing agents for selective separation in high-performance capillary electrophoresis. Chiral resolution via cyclodextrins incorporated within polyacrylamide gel columns. J Chromatogr 448:41–53, 1988.

95. H Ljungberg, S Nilsson. Protein-based capillary affinity gel electrophoresis for chiral separation of β-adrenergic blockers. J Liq Chromatogr 18:3685–3698, 1995.

96. S Nilsson, L Schweitz, M Petersson. Three approaches to enantiomer separation of β-adrenergic antagonists by capillary electrochromatography. Electrophoresis 18:884–890, 1997.

97. K Haupt, F Roy, MA Vijayalakshmi. Immobilized metal ion affinity capillary electrophoresis of proteins—a model for affinity capillary electrophoresis using soluble polymer-supported ligands. Anal Biochem 234:149–154, 1996.

98. H Ljungberg, S Ohlson, S Nilsson. Exploitation of a monoclonal antibody for weak affinity-based separation in capillary gel electrophoresis. Electrophoresis 19:461–464, 1998.

99. NA Guzman. Determination of immunoreactive gonadotropin-releasing hormone in serum and urine by on-line immunoaffinity capillary electrophoresis coupled to mass spectrometry. J Chromatogr B 749:197–213, 2000.

100. H Nilsson, M Wiklund, T Johansson, HM Hertz, S Nilsson. Microparticles for selective protein determination in capillary electrophoresis. Electrophoresis 22:2384–2390, 2001.

101. LG Rashkovetsky, YV Lyubarskaya, F Foret, DE Hughes, BL Karger. Automated microanalysis using magnetic beads with commercial capillary electrophoretic instrumentation. J Chromatogr 781:197–204, 1997.

102. M Quaglia, E De Lorenzi, C Sulitzky, G Massolini, B Sellergren. Surface initiated molecularly imprinted polymer films: A new approach in chiral capillary electrochromatography. Analyst 126:1495–1498, 2001.

103. G Vlatakis, LI Andersson, R Müller, K Mosbach. Drug assay using antibody mimics made by molecular imprinting. Nature 361:645–647, 1993.

104. L Schweitz, LI Andersson, S Nilsson. Capillary electrochromatography with predetermined selectivity obtained through molecular imprinting. Anal Chem 69:1179–1183, 1997.

105. JM Lin, T Nakagama, K Uchiyama, T Hobo. Temperature effect on chiral recognition of some amino acids with molecularly imprinted polymer filled capillary electrochromatography. Biomed Chromatogr 11:298–302, 1997.

106. JM Lin, T Nakagama, K Uchiyama, T Hobo. Capillary electrochromatographic separation of amino acid enantiomers using on-column prepared molecularly imprinted polymer. J Pharm Biomed Anal 15:1351–1358, 1997.

107. B Behnke, E Bayer. Pressurized gradient electro-high-performance liquid-chromatography. J Chromatogr A 680:93–98, 1994.

108. T Stroink, E Paarlberg, JCM Waterval, A Bult, WJM Underberg. On-line sample preconcentration in capillary electrophoresis, focused on the determination of proteins and peptides. Electrophoresis 22:2375–2383, 2001.

109. A Amini. Recent developments in chiral capillary electrophoresis and applications of this technique to pharmaceutical and biomedical analysis. Electrophoresis 22:3107–3130, 2001.

110. K Verleysen, P Sandra. Separation of chiral compounds by capillary electrophoresis. Electrophoresis 19:2798–2833, 1998.
111. TJ Thompson, F Foret, P Vouros, BL Karger. Capillary electrophoresis/electrospray ionization mass spectrometry: Improvement of protein detection limits using on-column transient isotachophoretic sample preconcentration. Anal Chem 65:900–906, 1993.
112. K Shimura, BL Karger. Affinity probe capillary electrophoresis: Analysis of recombinant human growth hormone with a fluorescent labeled antibody fragment. Anal Chem 66:9–15, 1994.
113. JP Quirino, S Terabe. Exceeding 5000-fold concentration of dilute analytes in micellar electrokinetic chromatography. Science 282:465–468, 1998.
114. DH Thomas, DJ Rakestraw, JS Schoeniger, V Lopez-Avila, J Van Emon. Selective trace enrichment by immunoaffinity capillary electrochromatography on-line with capillary zone electrophoresis-laser-induced fluorescence. Electrophoresis 20:57–66, 1999.
115. LJ Cole, RT Kennedy. Selective preconcentration for capillary zone electrophoresis using protein G immunoaffinity capillary chromatography. Electrophoresis 16:549–556, 1995.
116. AM Arentoft, H Frokiaer, S Michaelsen, H Sorensen, S Sorensen. High-performance capillary electrophoresis for the determination of trypsin and chymotrypsin inhibitors and their association with trypsin, chymotrypsin and monoclonal antibodies. J Chromatogr A 652:189–198, 1993.
117. MHA Busch, HFM Boelens, JC Kraak, H Poppe, AAP Meekel, M Resmini. Critical evaluation of the applicability of capillary zone electrophoresis for the study of hapten-antibody complex formation. J Chromatogr 744:195–203, 1996.
118. PD Grossman, JC Colburn, HH Lauer, RG Nielsen, RM Riggin, GS Sittampalam, EC Rickard. Application of free-solution capillary electrophoresis to the analytical scale separation of proteins and peptides. Anal Chem 61:1186–1194, 1989.
119. F Chen, RA Evangelista. Feasibility studies for simultaneous immunochemical multianalyte drug assay by capillary electrophoresis with laser-induced fluorescence. Clin Chem 40:1819–1822, 1994.
120. RA Evangelista, FT Chen. Analysis of structural specificity in antibody-antigen reactions by capillary electrophoresis with laser-induced fluorescence detection. J Chromatogr A 680:587–591, 1994.
121. FT Chen, SL Pentoney, Jr. Characterization of digoxigenin-labeled B-phycoerythrin by capillary electrophoresis with laser-induced fluorescence. Application to homogeneous digoxin immunoassay. J Chromatogr A 680:425–430, 1994.
122. F-TA Chen. Characterization of charge-modified and fluorescein-labeled antibody by capillary electrophoresis using laser-induced fluorescence. Application to immunoassay of low level immunoglobulin A. J Chromatogr A 680:419–423, 1994.
123. RC Tim, RA Kautz, BL Karger. Ultratrace analysis of drugs in biological fluids using affinity probe capillary electrophoresis: Analysis of dorzolamide with fluorescently labeled carbonic anhydrase. Electrophoresis 21:220–226, 2000.
124. Wild D, ed. The Immunoassay Handbook. London: Nature Publishing Group, 2001.
125. I German, DD Buchanan, RT Kennedy. Aptamers as ligands in affinity probe capillary electrophoresis. Anal Chem 70:4540–4545, 1998.
126. I German, RT Kennedy. Reversed-phase capillary liquid chromatography coupled on-line to capillary electrophoresis immunoassays. Anal Chem 72:5365–5372, 2000.
127. SB Cheng, CD Skinner, J Taylor, S Attiya, WE Lee, G Picelli, DJ Harrison. Development of a multichannel microfluidic analysis system employing affinity capillary electrophoresis for immunoassay. Anal Chem 73:1472–1479, 2001.
128. CS Effenhauser, GJM Bruin, A Paulus. Integrated chip-based capillary electrophoresis. Electrophoresis 18:2203–2213, 1997.
129. N Chiem, DJ Harrison. Microchip systems for immunoassay: An integrated immunoreactor with electrophoretic separation for serum theophylline determination. Clin Chem 44:591–598, 1998.
130. N Chiem, DJ Harrison. Microchip-based capillary electrophoresis for immunoassays: analysis of monoclonal antibodies and theophylline. Anal Chem 69:373–378, 1997.

131. LB Koutny, D Schmalzing, TA Taylor, M Fuchs. Microchip electrophoretic immunoassay for serum cortisol. Anal Chem 68:18–22, 1996.

132. RM Guijt, E Baltussen, GWK van Dedem. Use of bio-affinity interactions in electrokinetically controlled separations on microfabricated devices. Electrophoresis 23: 2002.

133. QH Wan, XC Le. Fluorescence polarization studies of affinity interactions in capillary electrophoresis. Anal Chem 71:4183–4189, 1999.

134. RD Smith, HR Udseth. Capillary zone electrophoresis-MS. Nature 331:639–640, 1988.

135. GA Valaskovic, NL Kelleher, FW McLafferty. Attomole protein characterization by capillary electrophoresis-mass spectrometry. Science 273:1199–1202, 1996.

136. RD Smith, HR Udseth, JH Wahl, DR Goodlett, SA Hofstadler. Capillary electrophoresis-mass spectrometry. Methods Enzymol 271:448–486, 1996.

137. PJ Vollmerhaus, FWA Tempels, JJ Kettenes-van den Bosch, AJR Heck. Molecular interactions of glycopeptide antibiotics investigated by affinity capillary electrophoresis and bioaffinity electrospray ionization mass spectrometry. Electrophoresis 23:868–879, 2002.

138. JS Rossier, F Reymond, PE Michel. Polymer microfluidic chips for electrochemical and biochemical analyses. Electrophoresis 23:858–867, 2002.

139. J Cai, J Henion. On-line immunoaffinity extraction-coupled column capillary liquid chromatography/tandem mass spectrometry: Trace analysis of LSD analogs and metabolites in human urine. Anal Chem 68:72–78, 1996.

140. JJ Dalluge, LC Sander. Precolumn affinity capillary electrophoresis for the identification of clinically relevant proteins in human serum: Application to human cardiac troponin I. Anal Chem 70:5339–5343, 1998.

141. TM Phillips, JJ Chiemlinska. Immunoaffinity capillary electrophoretic analysis of cyclosporin in tears. Biomed Chromatogr 8:242–246, 1994.

142. WA Lim, RT Sauer, AD Lander. Analysis of DNA-protein interactions by affinity coelectrophoresis. Methods Enzymol 208:196–210, 1991.

143. NH Chiem, DJ Harrison. Monoclonal antibody binding affinity determined by microchip-based capillary electrophoresis. Electrophoresis 19:3040–3044, 1998.

144. YM Dunayevskiy, YV Lyubarskaya, YH Chu, P Vouros, BL Karger. Simultanous measurement of nineteen binding constants of peptides to vancomycin using affinity capillary electrophoresis-mass spectrometry. J Med Chem 41:1201–1204, 1998.

145. NHH Heegaard, PMH Heegaard, P Roepstorff, FA Robey. Ligand binding sites in human serum amyloid P component. Eur J Biochem 239:850–856, 1996.

146. NHH Heegaard, HD Mortensen, P Roepstorff. Demonstration of a heparin-binding site in serum amyloid P component using affinity capillary electrophoresis as an adjunct technique. J Chromatogr 717:83–90, 1995.

147. NHH Heegaard. Microscale characterization of the structure-activity relationship of a heparin-binding glycopeptide using affinity capillary electrophoresis and immobilized enzymes. J Chromatogr A 853:189–195, 1999.

148. Y-H Chu, YM Dunayevskiy, DP Kirby, P Vouros, BL Karger. Affinity capillary electrophoresis-mass spectrometry for screening combinatorial libraries. J Am Chem Soc 118:7827–7835, 1996.

149. NHH Heegaard, P Roepstorff. Preparative capillary electrophoresis and mass spectrometry for the identification of a putative heparin-binding site in amyloid P component. J Capillary Electrophor 2:219–223, 1995.

150. JP Hummel, WJ Dreyer. Measurement of protein-binding phenomena by gel filtration. Biochim Biophys Acta 63:530–532, 1962.

151. Y Kuroda, B Cao, A Shibukawa, T Nakagawa. Effect of oxidation of low-density lipoprotein on drug binding affinity studied by high performance frontal analysis-capillary electrophoresis. Electrophoresis 22:3401–3407, 2001.

152. E De Lorenzi, C Galbusera, V Bellotti, P Mangione, G Massolini, E Tabolotti, A Andreola, G Caccialanza. Affinity capillary electrophoresis is a powerful tool to identify transthyretin binding drugs for potential therapeutic use in amyloidosis. Electrophoresis 21:3280–3289, 2000.

153. NA Mohamed, Y Kuroda, A Shibukawa, T Naakagawa, SE Gizawy, HF Askal, MEE Kommons. Binding analysis of nilvadipine to plasma lipoproteins by capillary electrophoresis-frontal analysis. J Pharm Biomed Anal 21:1037–1043, 1999.

154. PA McDonnell, GW Caldwell, JA Masucci. Using capillary electrophoresis/frontal analysis to screen drugs interacting with human serum proteins. Electrophoresis 19:448–454, 1998.

155. PA McDonnell, GW Caldwell, JA Masucci. Determination of the binding constants of drugs to human serum proteins and human serum using capillary electrophoresis frontal analysis. Ninth International Symposium on High Performance Capillary Electrophoresis and Related Microscale Techniques. 1997, p 629.

156. T Ohara, A Shibukawa, T Nakagawa. Capillary electrophoresis/frontal analysis for microanalysis of enantioselective protein binding of a basic drug. Anal Chem 67:3520–3525, 1995.

157. A Shibukawa, Y Yoshimoto, T Ohara, T Nakagawa. High-performance capillary electrophoresis/frontal analysis for the study of protein binding of a basic drug. Anal Chem 83:616–619, 1994.

158. LZ Avila, GM Whitesides. Catalytic activity of native enzymes during capillary electrophoresis: An enzymatic microreactor. J Org Chem 58:5508–5512, 1993.

159. S Van Dyck, A Van Schepdael, J Hoogmartens. Michaelis-Menten analysis of bovine plasma amine oxidase by capillary electrophoresis using electrophoretically mediated microanalysis in a partially filled capillary. Electrophoresis 22:1436–1442, 2001.

160. VJ Hilser, E Freire. Qunatitative analysis of conformational equilibrium using capillary electrophoresis: Applications to protein folding. Anal Biochem 224:465–485, 1995.

161. B Verzola, F Chiti, G Manao, PG Righetti. Monitoring equilibria and kinetics of protein folding/unfolding reactions by capillary zone electrophoresis. Anal Biochem 282:239–244, 2000.

162. NHH Heegaard, P Roepstorff, SG Melberg, MH Nissen. Cleaved β_2-microglobulin partially attains a conformation that has amyloidogenic features. J Biol Chem 277:11184–11189, 2002.

163. NHH Heegaard, JW Sen, NC Kaarsholm, MH Nissen. Conformational intermediate of the amyloidogenic protein β_2-microglobulin at neutral pH. J Biol Chem 276:32657–32662, 2001.

164. NHH Heegaard, JW Sen, MH Nissen. Congophilicity (Congo red affinity) of different β_2-microglobulin conformations characterized by dye affinity capillary electrophoresis. J Chromatogr A 894:319–327, 2000.

165. K Shimura, K-I Kasai. Capillary affinophoresis of pea lectin with polyliganded affinophores: A model study of divalent-polyvalent interactions. Electrophoresis 19:397–402, 1998.

166. A Amini, D Westerlund. Evaluation of association constants between drug enantiomers and human α_1-acid glycoprotein by applying a partial-filling technique in affinity capillary electrophoresis. Anal Chem 70:1425–1430, 1998.

167. RH Neubert, MA Schwarz, Y Mrestani, M Platzer, K Raith. Affinity capillary electrophoresis in pharmaceutics. Pharm Res 16:1663–1673, 1999.

168. JB Fenn, M Mann, CK Meng, SF Wong, CM Whitehouse. Electrospray ionization for mass spectrometry of large biomolecules. Science 246:64–71, 1989.

169. M Labowski, C Whitehouse, J Fenn. 3-Dimensional deconvolution of multiply charged spectra. Rapid Commun Mass Spectrom 7:71–84, 1993.

170. G Tuffal, A Tuong, C Dhers, F Uzabiaga, M Riviere, C Picard, G Puzo. Direct evidence of methylglucose lipopolysaccharides/palmitoyl-CoA noncovalent complexes by capillary zone electrophoresis-electrospray/mass spectrometry. Anal Chem 70:1853–1858, 1998.

171. RD Smith, JA Olivares, NT Nguyen, HR Udseth. Capillary zone electrophoresis-mass spectrometry using an electrospray ionization interface. Anal Chem 60:436–441, 1988.

172. R King, R Bonfiglio, C Fernandez-Metzler, C Miller-Stein, T Olah. Mechanistic investigation of ionization suppression in electrospray ionization. J Am Soc Mass Spectrom 11:942–950, 2000.

173. TG Huggins, JD Henion. Capillary electrophoresis mass-spectrometry determination of inorganic-ions using an ion spray-sheath flow interface. Electrophoresis 14:531–539, 1993.

174. F Garcia, J Henion. Fast capillary electrophoresis ion spray mass-spectrometric determination of sulfonylureas. J Chromatogr 606:237–247, 1992.

175. JA Olivares, NT Nguyen, CR Yonker, RD Smith. On-line mass spectrometric detection for capillary zone electrophoresis. Anal Chem 59:1230–1232, 1987.

176. GA Valaskovic, FW McLafferty. Sampling error in small-bore sheathless capillary electrophoresis/electrospray-ionization mass spectrometry. Rapid Commun Mass Spectrom 10:825–828, 1996.

177. DP Kirby, JM Thorne, WK Gotzinger, BL Karger. A CE/ESI-MS interface for stable, low-flow operation. Anal Chem 68:4451–4457, 1996.

178. SA Hofstadler, JH Wahl, JE Bruce, RD Smith. On-line capillary electrophoresis with Fourier ttransform ion cyclotron resonance mass spectrometry. J Am Chem Soc 115:6983–6984, 1993.

179. P Cao, M Moini. Capillary electrophoresis/electrospray ionization high mass accuracy time-of-flight mass spectrometry for protein identification using peptide mapping. Rapid Commun Mass Spectrom 12:864–870, 1998.

180. F Foret, H Zhou, E Gangl, BL Karger. Subatmospheric electrospray interface for coupling of microcolumn separations with mass spectrometry. Electrophoresis 21:1363–1371, 2000.

181. B Zhang, H Liu, BL Karger, F Foret. Microfabricated devices for capillary electrophoresis-electrospray mass spectrometry. Anal Chem 71:3258–3264, 1999.

182. J Ding, P Vouros. Advances in CE/MS. Anal Chem 71:378A–385A, 1999.

183. A von Brocke, G Nicholson, E Bayer. Recent advances in capillary electrophoresis/electrospray-mass spectrometry. Electrophoresis 22:1251–1266, 2001.

184. AJ Tomlinson, LM Benson, WD Braddock, RP Oda, S Naylor. On-line preconcentration-capillary electrophoresis-mass spectrometry (PC-CE-MS). J High Res Chromatogr 17:729–731, 1994.

185. AJ Tomlinson, S Naylor. Systematic development of on-line membrane preconcentration-capillary electrophoresis-mass spectrometry for the analysis of peptide mixtures. J Capillary Electrophor 2:225–233, 1995.

186. AK Malik, W Faubel. A review of capillary electrophoretic separations and their studies by nuclear magnetic resonance. J Capillary Electrophor 6:97–108, 1999.

187. P Gfrorer, J Schewitz, K Pusecker, LH Tseng, K Albert, E Bayer. Gradient elution capillary electrochromatography and hyphenation with nuclear magnetic resonance. Electrophoresis 20:3–8, 1999.

188. B Behnke, G Schlotterbeck, U Tallarek, S Strohschein, LH Tseng, T Keller, K Albert, E Bayer. Capillary HPLC-NMR coupling: High resolution 1H NMR spectrometry in the nanoliter scale. Anal Chem 68:1110–1115, 1996.

189. DL Olson, ME Lacey, AG Webb, JV Sweedler. Nanoliter-volume 1H NMR detection using periodic stopped-flow capillary electrophoresis. Anal Chem 71:3070–3076, 1999.

190. N Wu, TL Peck, AG Webb, RL Margin, JV Sweedler. Nanoliter volume sample cells for ^{1}H NMR: Application to on-line detection in capillary electrophoresis. J Am Chem Soc 116:7929–7930, 1994.

191. DL Olson, TL Peck, AG Webb, JV Sweedler. On-line NMR detection for capillary electrophoresis applied to peptide analysis. In: PTP Kaumaya, RS Hodges, eds. Peptides: Chemistry, Structure and Biology. Kingswinford: Mayflower Scientific, 1996, pp 730–731.

192. RA Kautz, ME Lacey, AM Woiters, F Foret, AG Webb, BL Karger, JV Sweedler. Sample concentration and separation for nanoliter-volume NMR spectroscopy using capillary isotachophoresis. J Am Chem Soc 123:3159–3160, 2001.

193. L Castelletti, SA Piletsky, AP Turner, PG Righetti, A Bossi. Development of an integrated capillary electrophoresis/sensor for L-ascorbic acid detection. Electrophoresis 23:209–214, 2002.

194. A Bossi, SA Piletsky, PG Righetti, AP Turner. Capillary electrophoresis coupled to biosensor detection. J Chromatogr A 892:143–153, 2000.

195. HA Fishman, O Orwar, RH Scheller, RN Zare. Identification of receptor ligands and receptor subtypes using antagonists in a capillary electrophoresis single-cell biosensor separation system. Proc Natl Acad Sci USA 92:7877–7881, 1995.

196. NHH Heegaard, DT Olsen, KL Larsen. Immuno-capillary electrophoresis for the characterization of a monoclonal antibody against DNA. J Chromatogr A 744:285–294, 1996.
197. M Strandh, M Ohlin, CA Borrebaeck, S Ohlson. New approach to steroid separation based on a low affinity IgM antibody. J Immunol Methods 214:73–79, 1998.
198. L Leickt, A Grubb, S Ohlson. Screening for weak monoclonal antibodies in hybridoma technology. J Mol Recognit 11:114–116, 1998.
199. TP Obrenovitch, E Zilkha. Microdialysis coupled to online enzymatic assays. Methods 23:63–71, 2001.
200. C Farre, A Sjoberg, K Jardemark, I Jacobson, O Orwar. Screening of ion channel receptor agonists using capillary electrophoresis-patch clamp detection with resensitized detector cells. Anal Chem 73:1228–1233, 2001.
201. AJ Gawron, SM Lunte. Optimization of the conditions for biuret complex formation for the determination of peptides by capillary electrophoresis with ultraviolet detection. Electrophoresis 21:2067–2073, 2000.
202. Q-H Wan, XC Le. Studies of protein-DNA interactions by capillary electrophoresis/laser-induced fluorescence polarization. Anal Chem 72:5583–5589, 2000.
203. J Østergaard, C Schou, C Larsen, NHH Heegaard. Evaluation of capillary electrophoresis frontal analysis for the study of low molecular weight drug-human serum albumin interactions. Electrophoresis 23:2842–2853, 2002.
204. JF Banks. Recent advances in capillary electrophoresis/electrospray/mass spectrometry. Electrophoresis 18:2255–2266, 1997.
205. T Nhujak, DM Goodall. Comparison of binding of tetraphenylborate and tetraphenylphosphonium ions to cyclodextrins studied by capillary electrophoresis. Electrophoresis 22:117–122, 2001.
206. T Guszczynski, TD Copeland. A binding shift assay for the zinc-bound and zinc-free HIV-1 nucleocapsid protein by capillary electrophoresis. Anal Biochem 260:212–217, 1998.
207. N von Wiren, S Klair, S Bansal, JF Briat, H Khodr, T Shiori, RA Leigh, RC Hider. Nicotinamine chelates both FeIII and FEII. Implications for metal transport in plants. Plant Physiol 119:1107–1114, 1999.
208. MK Menon, AL Zydney. Measurement of protein charge and ion binding using capillary electrophoresis. Anal Chem 70:1581–1584, 1998.
209. D Schmalzing, W Nashabeh, M Fuchs. Solution-phase immunoassay for determination of cortisol in serum by capillary electrophoresis. Clin Chem 41:1403–1406, 1995.
210. D Schmalzing, W Nashabeh, XW Yao, R Mhatre, FE Regnier, NB Ayefan, M Fuchs. Capillary electrophoresis-based immunoassay for cortisol in serum. Anal Chem 67:606–612, 1995.
211. A Shibukawa, DK Lloyd, IW Wainer. Simultaneous chiral separation of leucovorin and its major metabolite 5-methyl-tetrahydrofolate by capillary electrophoresis using cyclodextrins as chiral selectors: Estimation of the formation constant and mobility of the solute-cyclodextrin complexes. Chromatographia 35:419–429, 1993.
212. M Platzer, MA Schwarz, RHH Neubert. Determination of formation constants of cyclodextrin inclusion complexes using affinity capillary electrophoresis. J Microcolumn Sep 11:215–222, 1999.
213. SG Penn, DM Goodall, JS Loran. Differential binding of tioconazole enantiomers to hydroxypropyl-β-cyclodextrin studied by capillary electrophoresis. J Chromatogr 636:149–152, 1993.
214. MJ Schmerr, A Jenny. A diagnostic test for scrapie-infected sheep using a capillary electrophoresis immunoassay with fluorescent-labeled peptides. Electrophoresis 19:409–414, 1998.
215. A Amini, N Merclin, S Bastami, D Westerlund. Determination of association constants between enantiomers of orciprenaline and methyl-β-cyclodextrin as chiral selector by capillary zone electrophoresis using a partial filling technique. Electrophoresis 20:180–188, 1999.
216. J Sowell, JC Mason, L Strekowski, G Patonay. Binding constant determination of drugs toward subdomain IIIA of human serum albumin by near-infrared dye-displacement capillary electrophoresis. Electrophoresis 22:2512–2517, 2001.

217. L Ye, XC Le, JZ Xing, M Ma, R Yatscoff. Competitive immunoassay for cyclosporine using capillary electrophoresis with laser induced fluorescence polarization detection. J Chromatogr B 714:59–67, 1998.

218. J Caslavska, D Allemann, W Thormann. Analysis of urinary drugs of abuse by a multianalyte capillary electrophoretic immunoassay. J Chromatogr A 838:197–211, 1999.

219. DW Armstrong, LW Chang, SSC Chang. Mechanism of capillary electrophoresis enantioseparations using a combination of an achiral crown ether plus cyclodextrins. J Chromatogr A 793:115–134, 1998.

220. S Terabe, Y Miyashita, O Shibata, ER Barnhart, LR Alexander, DG Patterson, BL Karger, K Hosoya, N Tanaka. Separation of highly hydrophobic compounds by cyclodextrin-modified electrokinetic chromatography. J Chromatogr 516:23–31, 1990.

221. K Otsuka, LJ Smith, J Grainger, JR Barr, DG Patterson, N Tanaka, S Terabe. Stereoselective separation and detection of phenoxy acid herbicide enantiomers by cyclodextrin-modified capillary zone electrophoresis-electrospray ionization mass spectrometry. J Chromatogr A 817:75–81, 1998.

222. D Schmalzing, LB Koutny, TA Taylor, W Nashabeh, M Fuchs. Immunoassay for thyroxine (T4) in serum using capillary electrophoresis and micromachined devices. J Chromatogr B 697:175–180, 1997.

223. CM Maragos. Detection of the mycotoxin fumonisin B_1 by a combination of immunofluorescence and capillary electrophoresis. Food Agric Immunol 9:147–157, 1997.

224. J Liu, KJ Volk, MS Lee, M Pucci, S Handwerger. Binding studies of vancomycin to the cytoplasmic peptidoglycan precursors by affinity capillary electrophoresis. Anal Chem 66:2412–2416, 1994.

225. M Hong, H Soini, MV Novotny. Affinity capillary electrophoretic studies of complexation between dextrin oligomers and polyiodides. Electrophoresis 21:1513–1520, 2000.

226. XC Le, JZ Xing, J Lee, SA Leadon, M Weinfeld. Inducible repair of thymine glycol detected by an ultrasensitive assay for DNA damage. Science 280:1066–1069, 1998.

227. J Li, KC Waldron. Estimation of the pH-independent binding constants of alanylphenylalanine and leucylphenylalanine stereoisomers with β-cyclodextrin in the presence of urea. Electrophoresis 20:171–179, 1999.

228. S Sabah, GKE Scriba. pH-Dependent reversal of the chiral recognition of tripeptide enantiomers by carboxymethyl-β-cyclodextrin. J Chromatogr A 833:261–266, 1999.

229. M Petersson, K Walhagen, A Nilsson, K-G Wahlund, S Nilsson. Separation of a bioactive cyclic peptide and its oligomeric forms by micellar electrokinetic chromatography. J Chromatogr A 769:301–306, 1997.

230. NHH Heegaard. A heparin-binding peptide from human serum amyloid P component characterized by affinity capillary electrophoresis. Electrophoresis 19:442–447, 1998.

231. E Mito, Y Zhang, S Esquivel, FA Gomez. Estimation of receptor-ligand interactions by the use of a two-marker system in affinity capillary electrophoresis. Anal Biochem 280:209–215, 2000.

232. J Gao, M Mammen, GM Whitesides. Evaluating electrostatic contributions to binding with the use of protein charge ladders. Science 272:535–537, 1996.

233. J Gao, FA Gomez, R Harter, GM Whitesides. Determination of the effective charge of a protein in solution by capillary electrophoresis. Proc Natl Acad Sci USA 91:12027–12030, 1994.

234. V Pavski, XC Le. Detection of human immunodeficiency virus type 1 reverse transcriptase using aptamers as probes in affinity capillary electrophoresis. Anal Chem 73:6070–6076, 2001.

235. W Zhou, KB Tomer, MG Khaledi. Evaluation of the binding between potential anti-HIV DNA-based drugs and viral envelope glycoprotein gp120 by capillary electrophoresis with laser-induced fluorescence detection. Anal Biochem 284:334–341, 2000.

236. P Sun, A Hoops, RA Hartwick. Enhanced albumin protein separations and protein-drug binding constant measurements using anti-inflammatory drugs as run buffer additives in affinity capillary electrophoresis. J Chromatogr B 661:335–340, 1994.

237. Y Ding, B Lin, CW Huie. Binding studies of porphyrins to human serum albumin using affinity capillary electrophoresis. Electrophoresis 22:2210–2216, 2001.

238. K Shimura, K Kasai. Determination of the affinity constants of pea lectin for neutral sugars by capillary affinophoresis with a monoligand affinophore. J Biochem 120:1146–1152, 1996.

239. GJ Foulds, FA Etzkorn. DNA-binding affinities of MyoD and E47 homo- and hetero-dimers by capillary electrophoresis mobility shift assay. J Chromatogr A 862:231–236, 1999.

240. NHH Heegaard, DT Olsen, K-LP Larsen. Immuno-capillary electrophoresis for the characterization of a monoclonal antibody against DNA. J Chromatogr 744:285–294, 1996.

241. S Kiessig, J Reissmann, C Rascher, G Kullertz, A Fischer, F Thunecke. Application of a green fluorescent fusion protein to study protein-protein interactions by electrophoretic methods. Electrophoresis 22:1428–1435, 2001.

242. GW Caldwell, PA McDonnell, JA Masucci, DL Johnson, LK Jolliffe. Using affinity capillary electrophoresis to study the interaction of the extracellular binding domain of erythropoietin receptor with peptides. J Biochem Biophys Methods 40:17–25, 1999.

243. K Gunnarsson, L Valtcheva, S Hjertén. Capillary zone electrophoresis for the study of the binding of antithrombin to low-affinity heparin. Glycoconjugate J 14:859–862, 1997.

244. J Xian, MG Harrington, EH Davidson. DNA-protein binding assays from a single sea urchin egg: A high-sensitivity capillary electrophoresis method. Proc Natl Acad Sci USA 93:86–90, 1996.

245. K Nadeau, SG Nadler, M Saulnier, MA Tepper, CT Walsh. Quantitation of the interaction of the immunosuppressant deoxyspergualin and analogs with Hsc70 and Hsp90. Biochemistry 33:2561–2567, 1994.

246. R Kuhn, R Frei, M Christen. Use of capillary affinity electrophoresis for the determination of lectin-sugar interactions. Anal Biochem 218:131–135, 1994.

247. L Tao, RT Kennedy. Measurement of antibody-antigen dissociation constants using fast capillary electrophoresis with laser-induced fluorescence detection. Electrophoresis 18:112–117, 1997.

248. NM Schultz, L Huang, RT Kennedy. Capillary electrophoresis-based immunoassay to determine insulin content and insulin secretion from single islets of langerhans. Anal Chem 67:924–929, 1995.

249. L Tao, CA Aspinwall, RT Kennedy. On-line competitive immunoassay based on capillary electrophoresis applied to monitoring insulin secretion from single islets of Langerhans. Electrophoresis 19:403–408, 1998.

250. L Tao, RT Kennedy. On-line competitive immunoassay for insulin based on capillary electrophoresis with laser-induced fluorescence detection. Anal Chem 68:3899–3906, 1996.

251. JP Ou, QG Wang, TM Cheung, ST Chan, WS Yeung. Use of capillary electrophoresis-based competitive immunoassay for a large molecule. J Chromatogr B 727:63–71, 1999.

252. OW Reif, R Lausch, T Scheper, R Freitag. Fluorescein isothiocyanate-labeled protein G as an affinity ligand in affinity/immunocapillary electrophoresis with fluorescence detection. Anal Chem 66:4027–4033, 1994.

253. B Zhang, H Xue, B Lin. Application of capillary zone electrophoresis to the study of interactions between *Bacillus subtilis* tryptophanyl-tRNA synthetase and tRNA^{Trp}. Chromatographia 48:268–272, 1998.

254. VM Okun, R Moser, B Ronacher, E Kenndler, D Blaas. VLDL receptor fragments of different lengths bind to human rhinovirus HRV2 with different stoichiometry—an analysis of virus-receptor complexes by capillary electrophoresis. J Biol Chem 276:1057–1062, 2001.

255. VM Okun, R Moser, D Blaas, E Kenndler. Complexes between monoclonal antibodies and receptor fragments with a common cold virus: Determination of stoichiometry by capillary electrophoresis. Anal Chem 73:3900–3906, 2001.

256. VM Okun, B Ronacher, D Blaas, E Kenndler. Affinity capillary electrophoresis for the assessment of complex formation between viruses and monoclonal antibodies. Anal Chem 72:4634–4639, 2000.

27

Affinity Mass Spectrometry

Chad J. Briscoe
MDS Pharma Services, Lincoln, NE

William Clarke
Department of Pathology, Johns Hopkins School of Medicine, Baltimore, MD

David S. Hage
Department of Chemistry, University of Nebraska, Lincoln, NE

CONTENTS

27.1   INTRODUCTION

In recent years there has been a growing interest in combining affinity methods with mass spectrometry. This is shown by Figure 27.1, where an almost-exponential increase in such methods has been reported over the last decade. Part of the reason for this growth has been the need for more-sophisticated methods for the analysis of complex systems like combinatorial and proteomic libraries. Another factor is the advances that have recently been made in mass spectrometric methods for proteins and other biological agents.

The combined use of affinity ligands with mass spectrometry is referred to in this chapter as *affinity mass spectrometry*. As will be shown in this chapter, there are many formats in which affinity ligands can be used for such work. The earliest examples were in the late 1970s, when affinity supports were used for off-line sample preparation prior to gas chromatography-mass spectrometry [1, 2]. But it was not until the early 1990s that methods like electrospray ionization allowed the on-line use of affinity ligands with mass spectrometry [3].

This chapter reviews the various ways in which affinity ligands have been combined with mass spectrometry. The discussion includes off-line methods as well as techniques

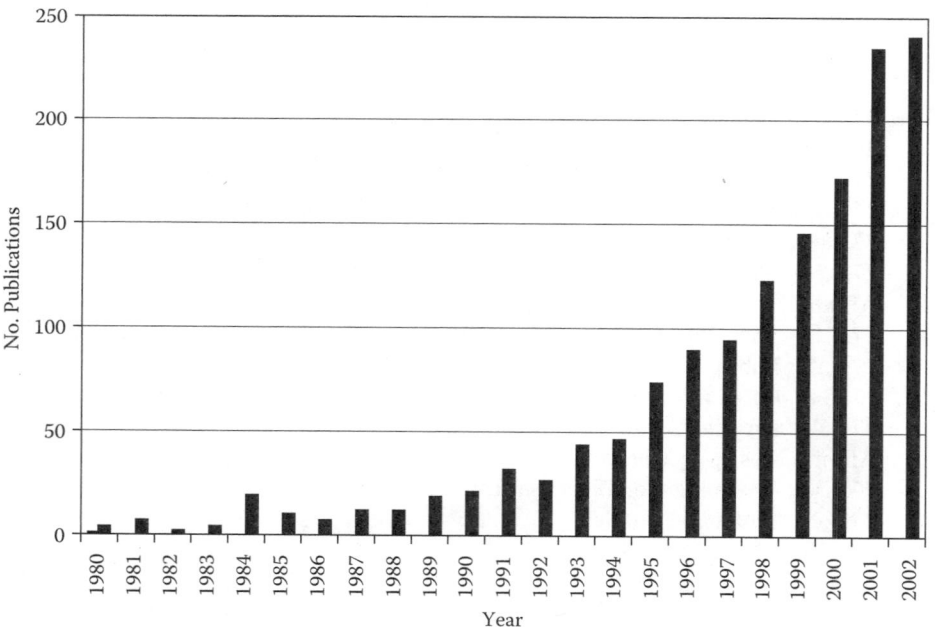

Figure 27.1   Publications appearing between 1980 and 2002 on the use of affinity ligands in mass spectrometry. These data were obtained by searching *Chemical Abstracts* for combined use of the terms *ligand* or *affinity* and *mass spectrometry* with a filter to remove unrelated topics, such as "proton affinity mass spectrometry."

for directly coupling affinity ligands with electrospray ionization, matrix-assisted laser desorption ionization, and other types of mass spectrometry. Several applications of these approaches are also examined, including their use for sample extraction, the screening of combinatorial or proteomic mixtures, the study of biological interactions, and the use of mass spectrometry for detection in affinity-based analysis methods.

27.2 TYPES OF AFFINITY MASS SPECTROMETRY

A large number of approaches have been used to combine affinity ligands with mass spectrometry. The earliest of these involved the off-line utilization of affinity ligands, but several on-line methods are also now available. These latter approaches include the use of affinity ligands with electrospray ionization, matrix assisted laser desorption ionization, fast atom bombardment, and inductively coupled plasma mass spectrometry. This section provides an overview of these methods, along with their advantages and disadvantages.

27.2.1 Off-Line Methods

The simplest way of combining an affinity ligand with mass spectrometry is to use the ligand as part of an off-line method. This can be used in situations where the affinity ligand or sample matrix is not directly compatible with mass spectrometry. This can also be employed when sample preparation must take place at a location separate from the site of analysis. A review of off-line methods that use antibodies for sample pretreatment has been given in the literature [4] for such techniques as gas chromatography-mass spectrometry (GC-MS) and liquid chromatography-mass spectrometry (LC-MS). The use of off-line methods with matrix-assisted laser desorption ionization is discussed in Section 27.2.3.

Many applications of off-line affinity chromatography with GC-MS have been reported. Examples include the use of this method for such compounds as urinary 8-epiprostaglandin F2a [5], flumethasone [6], dexamethasone [7], 3-alkyladenines [8], 3-methyladenine [9], 11-dehydrothromboxane B_2 [10], thromboxane B_2 [11], and O^6-butylguanine in DNA [12]. Additional examples can be found in the literature [13].

GC-MS with off-line affinity chromatography has often been used to identify and quantitate analytes in the parts-per-trillion range [14–18]. Typically, this method requires the derivatization of analytes after their affinity isolation to make them more volatile or stable for GC analysis. One key advantage to using off-line affinity ligands with GC-MS is that this allows the direct analysis of aqueous-based samples, since analytes that are captured and eluted from the affinity column can be later dried and reconstituted into a more volatile solvent prior to injection onto a GC system [4].

Off-line affinity purification has also been used with LC-MS. In this case, it is desirable to have a sample solution free of interfering salts, especially when electrospray ionization is being employed. The analytes should also be in an aqueous solution or a solvent like methanol or acetonitrile that is highly miscible with water. One example of off-line affinity extraction in LC-MS is a study in which immunoaffinity ultrafiltration and LC-MS were used for screening small combinatorial libraries [19].

The greater ease of combining affinity ligands with off-line LC-MS has made this approach more popular in recent years than off-line affinity separations in GC-MS, especially in work with polar compounds and aqueous samples. However, there are still many advantages to using affinity ligands with GC-MS methods. These include the greater efficiency of GC versus LC, the better speed and resolution of GC for volatile chemicals, and the superior performance of electron impact and chemical ionization versus LC-MS ionization methods for many low-mass substances.

27.2.2 Electrospray Ionization Mass Spectrometry

Currently, the most common method for directly using affinity ligands in mass spectrometry is electrospray ionization (ESI). In such work, the electrospray source can be used with several types of mass analyzers, including quadrupole, ion trap, time-of-flight, or Fourier transform instruments [20].

Quadrupoles and ion traps are commonly employed for selected-ion monitoring (SIM) or selected-reaction monitoring (SRM) to achieve high selectivity and sensitivity [21]. In selected-ion monitoring, the mass filter is set at a specified mass-to-charge ratio (m/z), where only compounds with the same m/z ratio as the analyte will be detected. In selected-reaction monitoring, a specific group of parent ions is chosen for further fragmentation, and their product ions are detected, thereby adding a second dimension of information to the analysis.

Time-of-flight and Fourier transform instruments can achieve high selectivity by means of their resolving power. This means that only compounds with a specific molecular formula will give a signal under the selected detection conditions. More sophisticated instruments that make use of a quadrupole time-of-flight (QqTOF) or tandem time-of-flight analyzers can be used to achieve even higher selectivity and add more diversity to the types of experiments that can be performed on a sample [22].

The development of atmospheric pressure ESI is one reason why affinity methods can be combined on-line with mass spectrometry. Figure 27.2 shows how this type of ionization is performed. In this technique, the sample is placed into a solvent and sprayed from a highly charged needle (3 to 5 kV). The solvent in the charged droplets produced

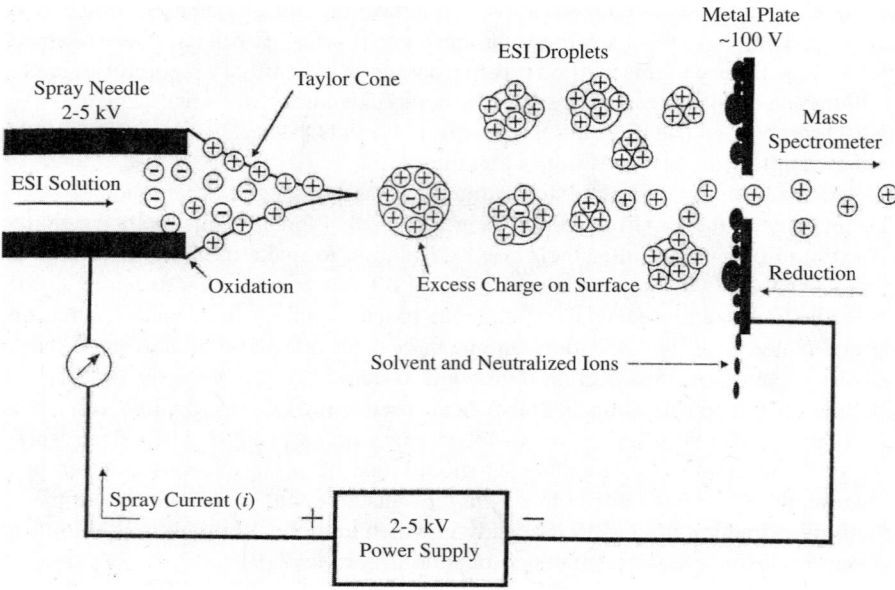

Figure 27.2 The electrospray ionization process. In this method, a sample solution is pumped through a needle held at a positive voltage. This results in the production of tiny droplets with an excess of positive charge. As these droplets travel through space, their charge begins to concentrate, and the droplet begins to break apart. Eventually, some of the molecules they contain are ejected as positively charged ions that are sent into a mass analyzer. (Reproduced with permission from Cech, N.B. and Enke, C.G., *Mass Spectrom. Rev.*, 20, 362–387, 2001.)

by this spray evaporate away quickly, giving smaller droplets with an excess of positive or negative charge. At some critical point, the coulombic forces in this droplet will exceed the cohesive forces, causing the droplet to divide. Eventually, the molecules in this droplet are desorbed as ions and enter the gas phase, where they can be analyzed by the mass spectrometer.

Most electrospray sources accept solution flow rates that range from 20 nl/min to 4 ml/min [23]. This makes it possible to directly couple almost any LC method to this type of device. In addition, ESI can be used with a variety of substances, ranging from small polar compounds to proteins. However, for this to work, a few requirements must be met. For instance, the solvent used to spray the sample must be volatile and free of any strong ion-pairing agents. In addition, this solvent must have the ability to carry some charge. These requirements are needed for the production of a suitably large number of sample ions.

When coupling ESI with affinity chromatography, the elution buffer should contain only volatile buffer salts like ammonium acetate and ammonium formate. Nonvolatile salts, such as potassium phosphate, are typically avoided because they lower the ionization efficiency. One reason this occurs is because nonvolatile components in the mobile phase can accumulate near the vacuum interface, causing a decrease in sensitivity over time. In addition, substances often present in these solvents, like phosphate and trifluoroacetic acid, can act as strong ion-pairing agents, which neutralize the electrospray droplets and prevent the transfer of charge to molecules within these droplets [24].

Another factor to consider with ESI is that its sensitivity tends to be highest in mobile phases that contain more than 50% methanol or acetonitrile in water. Thus, if the mobile phase is a simple aqueous buffer, this may need to be combined with methanol or acetonitrile before it enters the electrospray source. The solubility of the analytes must also be considered when using this technique, since substances like proteins may precipitate in the presence of solutions that have a high content of an organic solvent.

ESI mass spectrometry is extremely useful in the analysis of proteins and peptides. Although the mass-to-charge ratio for a singly charged protein or peptide may be outside the range of most mass analyzers, in ESI there are usually many charges placed on each of these biomolecules. This gives a mass-to-charge ratio that is now in the range of common mass spectrometers. For instance, a protein with a molecular mass of 16 kDa may carry 16 positive charges after ESI, giving it a mass-to-charge ratio of 1000. One minor difficulty associated with this process is that a single protein or peptide can give rise to many molecular ions, as shown in Figure 27.3 [25]. However, it is possible to convert these data to a simpler form through deconvolution.

27.2.3 Matrix Assisted Laser Desorption Ionization Mass Spectrometry

Matrix assisted laser desorption ionization (MALDI) was developed at about the same time as ESI in the late 1980s. MALDI is a relatively straightforward technique that involves two steps [26, 27]. First, the sample is dissolved in a matrix such as cinnamic acid that can absorb ultraviolet light. This solution is then crystallized on a target surface. In the second step of this procedure, the target surface is shot with short, intense pulses of light from a laser. As these shots are made, the matrix absorbs some of the light and dissipates this energy, causing the sample to be vaporized and released into the gas phase as ions. These ions are then separated and detected using a time-of-flight mass analyzer.

One advantage of MALDI is that it can be used for ions of both small substances (e.g., peptides with masses of 300 to 500 Da) as well as for large biomolecules (e.g., proteins with masses of 1 million Da), even when these contain only single charges.

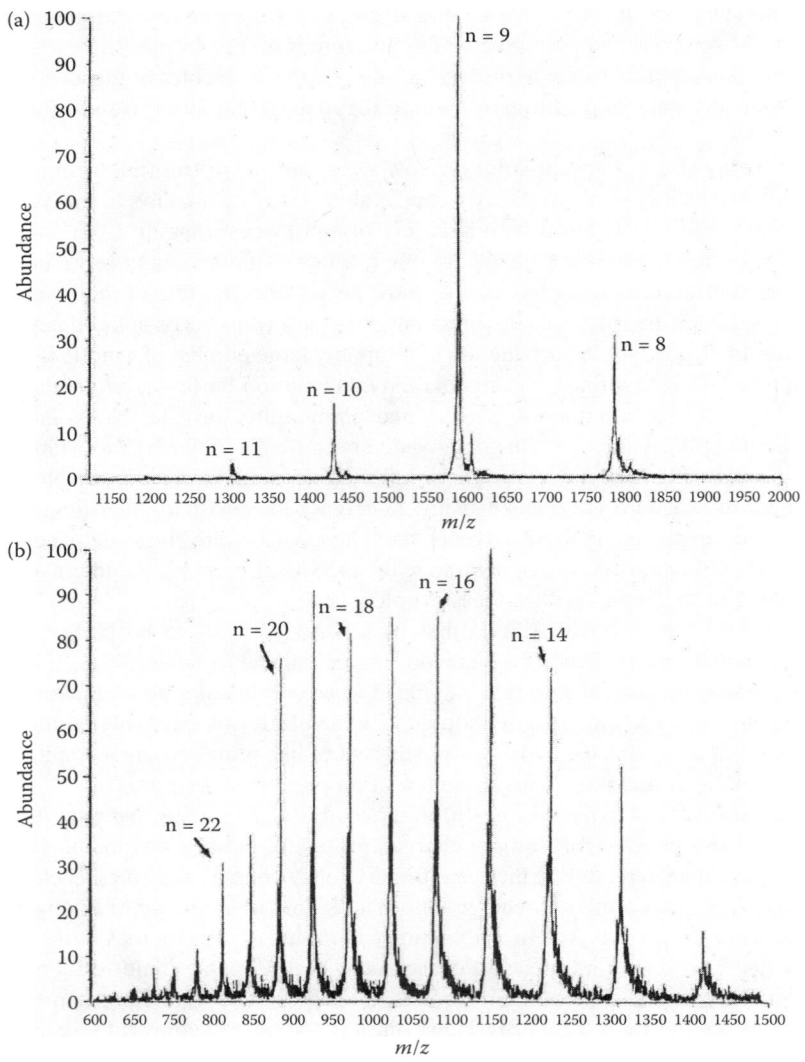

Figure 27.3 The charge envelopes obtained by ESI for (a) hen egg white lysozyme and (b) equine myoglobin. The peaks in each of these spectra are the ions for a single type of protein with different numbers of positive charges, n. The protein's mass can be obtained from these spectra by using the formula $m/z = [M + n \cdot H]/n$, where M is the protein's true mass and H is the mass of a proton. (Reproduced with permission from Cody, R.B., Tamura, J., and Musselman, B.D., *Anal. Chem.*, 64, 1561–1570, 1992.)

MALDI can also be performed very quickly. For instance, a 96-well plate can be prepared with commercially available robotic stations and analyzed in under 5 min by this approach.

MALDI does have some drawbacks, especially in its sample preparation and high background signals. As an example, the matrix used for MALDI can show a large variation in how it affects the ionization of different analytes. Matrix selection will depend on both the type of laser being employed and the classes of compounds being analyzed. The high background signal of MALDI is due to the organic and inorganic contaminants in the sample and matrix. This can be dealt with in several ways. One of these involves the use of a method known as surface-enhanced laser desorption ionization, as will be discussed in Section 27.3.4.

Affinity ligands can be combined with MALDI in either an off-line or on-line manner. Off-line applications generally involve using an affinity method to separate the components of a sample, which are then collected as fractions and combined with an appropriate MALDI matrix. Figure 27.4 shows an example in which a small antibody column was used to isolate an analyte for a MALDI-based immunoassay. In on-line techniques, an affinity ligand is immobilized to a MALDI target and used to capture specific chemicals from a sample.

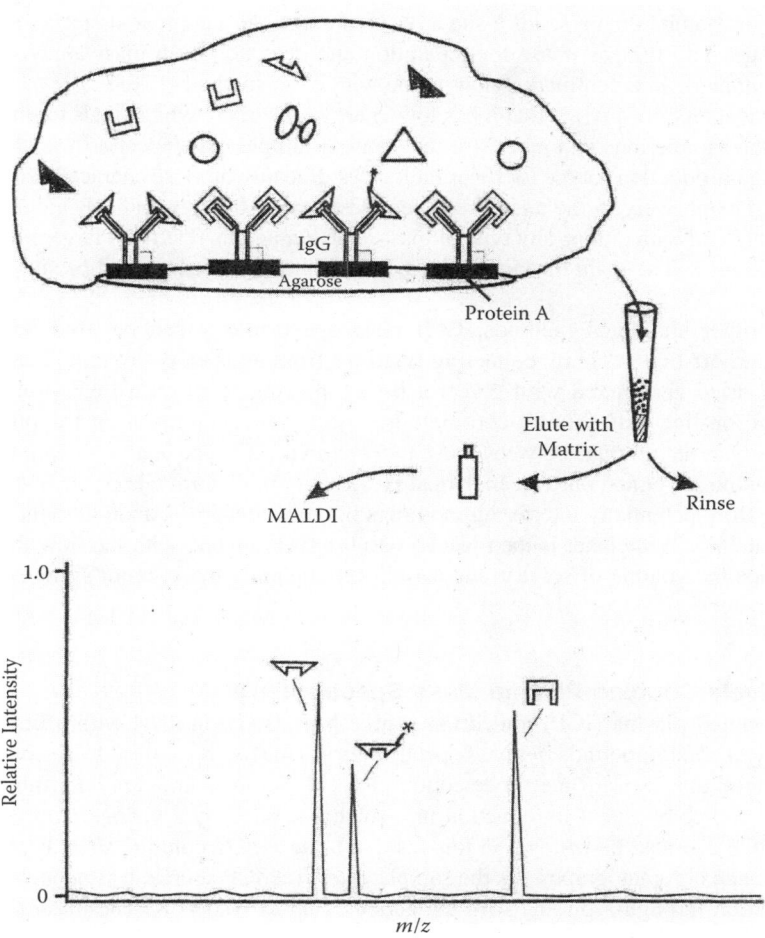

Figure 27.4 Use of off-line immunoextraction with MALDI mass spectrometry to perform an immunoassay. In this method, analytes were first extracted from samples by using antibodies adsorbed to protein A agarose in a pipette tip. The retained compounds were eluted from this support and combined with a matrix for analysis by MALDI mass spectrometry. This gave rise to a mass spectrum in which the analyte, related substances, and labeled analogs of the analyte could all be examined simultaneously based on their different mass-to-charge ratios. (Reproduced with permission from Niederkofler, E.E., Tubbs, K.A., Gruber, K., Nedelkov, D., Kiernan, U.A., Williams, P., and Nelson, R.W., *Anal. Chem.*, 73, 3294–3299, 2001.)

The contents of the target are later ionized and sent into a mass analyzer for study. Further details on both of these techniques are given in Sections 27.3.3 and 27.3.4.

27.2.4 Fast Atom Bombardment Mass Spectrometry

Fast atom bombardment (FAB) is another ionization method for mass spectrometry that has been used with affinity ligands [28]. FAB was one of the first ionization techniques used for analyzing nonvolatile and thermally labile compounds by mass spectrometry [20, 29, 30]. In this method, the sample is dissolved in a nonvolatile matrix such as glycerol. This mixture is then placed at the end of a probe and inserted into the vacuum chamber of a mass spectrometer. Next, the sample mixture is bombarded with neutral atoms like argon (as is used in traditional FAB) or ions like Cs^+ (as is used in *liquid secondary-ion mass spectrometry*, or LSIMS) [21]. This bombardment sends a shock wave through the sample and ejects ions along with the matrix into the gas phase for separation and detection by a mass analyzer. FAB can be used alone or in a continuous-flow mode with other methods [31, 32].

One of the advantages of FAB is that it has low-energy ionization, which leads to little fragmentation of the sample ions and gives rise to relatively simple mass spectra. It is also an excellent sample introduction source for polar molecules. Because of these characteristics, FAB has been used extensively in the analysis of peptides. Other strengths of FAB include its ability to be interfaced with almost any type of mass analyzer and its relatively large mass range (up to 10,000 m/z). The main disadvantage of FAB is the high background produced by its matrix.

Like many other analytical methods, FAB mass spectrometry can be used with affinity ligands in an off-line mode by collecting fractions from an affinity column. These fractions are then dried and mixed with glycerol before placing them on a FAB probe. When FAB is used on-line with affinity chromatography, a preconcentration or trapping column must be used as an interface between these two methods [20]. The analytes captured on the trapping column are eluted onto an analytical HPLC column in a mobile phase that is as free as possible from potentially interfering modifiers but that contains a small amount of glycerol (i.e., about 1%). This eluant is then placed into the FAB source, with the flow first being split to reduce the volume of solvent and avoid overwhelming the vacuum system of the mass spectrometer.

27.2.5 Inductively Coupled Plasma Mass Spectrometry

An inductively coupled plasma (ICP) ionization source has also been used with affinity ligands in the analysis of compounds by mass spectrometry. This has been used to perform an immunoassay with mass spectrometric detection [33–35]. Samples are generally introduced into an ICP source by direct introduction in a solvent applied by a syringe pump or peristaltic pump. ICP can also be coupled in this fashion to an HPLC column, such as one based on ion-exchange chromatography. As the sample enters the ICP source, it is nebulized and passed with argon through a plasma with temperatures up to 10,000 K. The plasma is maintained by the induced movement of argon ions in the presence of an electromagnetic field produced by high radio frequency coils. As the sample enters this plasma, its components are atomized and converted into ions [20, 21, 36].

Because of the destructive nature of this technique, only atomic-state ions or simple molecular ions can be analyzed by ICP. However, this method is very sensitive and has a large dynamic range. For this reason, antibodies tagged with metals have been used as a means for detection in ICP mass spectrometry for several types of immunoassays [33–35]. Specific examples are described in Section 27.6.2.

27.3 AFFINITY EXTRACTION

One way in which affinity ligands have been used in mass spectrometry is as a means for selectively isolating a given set of analytes from a sample prior to their mass analysis. In this approach, the affinity ligand is used as a way of reducing the complexity of the sample before its measurement. This technique is known as affinity extraction, and it can be accomplished by either off-line or on-line methods. Both approaches are examined in this section, with particular attention being given to antibody-based extraction, lab-on-valve systems, and methods involving MALDI mass spectrometry.

27.3.1 Mass Spectrometry with Immunoextraction

A relatively popular application of affinity ligands in mass spectrometry has been in the area of *immunoextraction* [37–42]. The theory and basis of immunoextraction is discussed in detail in Chapter 6. This current section focuses on the specific use of this technique with mass spectrometry. Advantages of this combination include its low limits of detection and its ability to reduce the preparation time needed for complex samples.

An example of immunoextraction in mass spectrometry is a method reported for the measurement of LSD in human urine [37]. The system used for this analysis is shown in Figure 27.5. This contained a column loaded with anti-LSD antibodies. The sample was first passed through this antibody column in the presence of phosphate-buffered saline. The retained drug was later eluted, while the protein *G* column was kept in line with a trapping column. The LSD captured by the trapping column was then back-flushed onto a 0.3 mm I.D. × 150 mm C_{18} analytical column coupled with a triple quadrupole mass spectrometer.

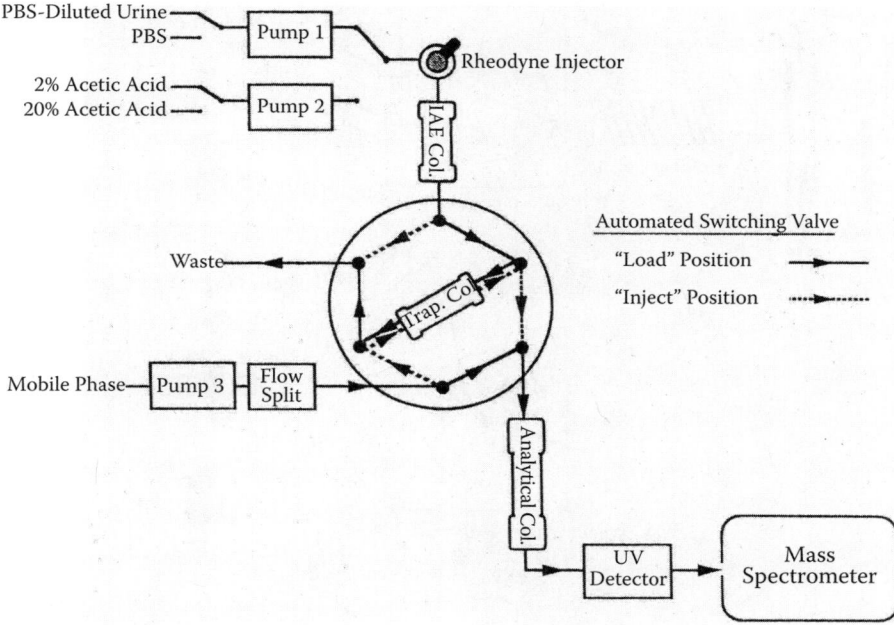

Figure 27.5 An LC-MS system with on-line immunoextraction for the analysis of LSD in urine samples. (Reproduced with permission from Cai, J. and Henion, J., *Anal. Chem.*, 68, 72–78, 1996.)

    Other examples of on-line immunoextraction in LC-MS are given in Chapter 6. With
this approach, it is possible to automate sample pretreatment while decreasing the overall
analysis time of a method and improving its precision. In addition, this avoids the use of
manual extraction steps and makes use of the high reproducibility of modern HPLC injection
systems and solvent delivery devices.

27.3.2 Lab-on-Valve Systems

Another area in which mass spectrometry and affinity chromatography have been combined
is in *affinity capture-release electrospray ionization mass spectrometry* (AC-ESI-MS) [43].
This is actually a subset of the field of on-line affinity extractions, as discussed earlier in
this chapter.

    In one recent application of this approach [44], a lab-on-valve system was created for
the automated capture and release of biotin-containing conjugates. The system used in this
work is shown in Figure 27.6. This contained a multiposition switching valve and a packed
flow cell in line with two of the valve's ports. Affinity beads were placed in this cell to allow
its use as both an affinity column and sample cell for absorbance detection.

    In one application, this system was used to measure the dissociation kinetics of biotin-
containing conjugates from streptavidin beads. This was studied by using geometry 2 in
Figure 27.6 along with two biotin conjugates: (1) 1-biotinyl-10-(4′nitrophenylthiomethyl-
carbonyl)-1,10-diaza-4,7-dioxadecane), or BPN, and (2) 1-(*N*-biotinyl)sarcosinyl-15-
(4′nitrophenylthiomethylcarbonyl)-1,15-diaza-5,8,11-trioxapenta-decane), or BSN. Once
these conjugates had been loaded separately onto the streptavidin beads, they were eluted
using biotin solutions. Absorbance measurements of the eluting BSN or BPN gave a decay

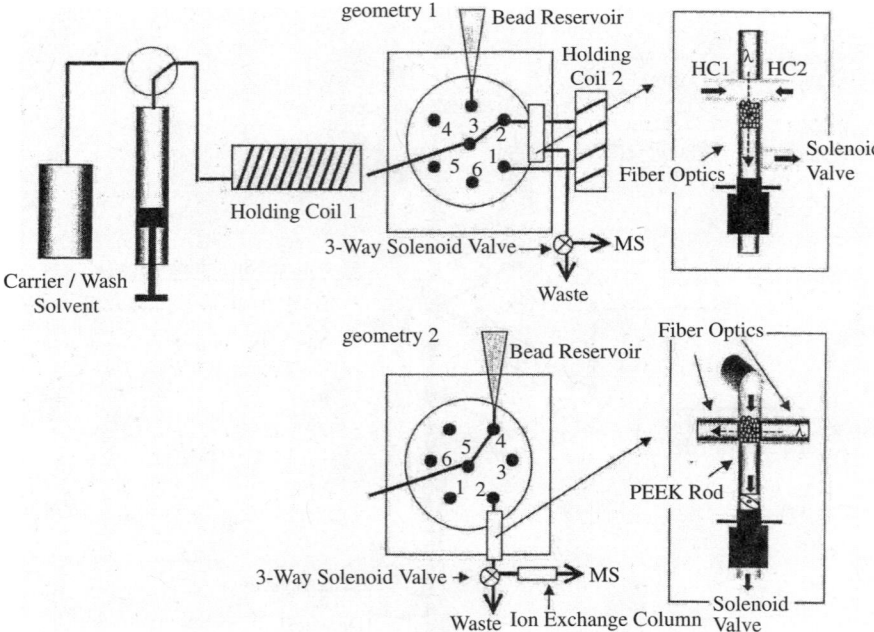

Figure 27.6 A lab-on-valve apparatus using affinity beads in combination with mass spectrometric
detection. The use of geometry 1 and 2 is discussed in the text. (Reproduced with permission from
Ogata, Y., Scampavia, L., Ruzicka, J., Scott, C.R., Gelb, M.H., and Turecek, F., *Anal. Chem.*, 74,
4702–4708, 2002.)

curve that showed the decrease in concentration for the bound BPN or BSN, while ESI mass spectrometry gave the actual concentration being displaced by biotin.

This lab-on-valve system was also used to develop an assay for β-galactosidase, which was accomplished using geometry 1 in Figure 27.6. This was based on an enzyme assay in which β-galactosidase cleaved galactose, giving a shift of 162 units in its mass-to-charge ratio. In this assay, a cell homogenate sample containing a biotin-substrate conjugate, enzyme reaction product, and deuterated internal standard was injected onto the streptavidin beads. These beads were then washed with a 50% methanol solution. The biotin conjugate and internal standard were released from the beads by applying a biotin solution in 0.03% acetic acid. This gave a method with a total analysis time of 4.5 min.

27.3.3 Off-Line Affinity MALDI

The use of off-line affinity ligands with MALDI is another area that has seen appreciable growth in recent years. An example is a mass spectrometric immunoassay that was created for the determination of β-2 microglobulin in human plasma [45]. This made use of a 96-well robotic workstation and pipette tips that had been modified to contain antibodies directed against β-2 microglobulin (see Figure 27.4). In this work, β-2 microglobulin was extracted from plasma using the immunoaffinity tips and later eluted onto the MALDI target. The matrix was then applied and data collected using an automated mass analyzer, making it possible to process 100 samples per hour.

Another application of off-line affinity MALDI has been in the characterization of antibody-antigen interactions. In one example [46], monoclonal antibodies against a known protein were immobilized to a nitrocellulose membrane. The protein of interest was then digested with trypsin, and its peptide fragments were applied to the immunoaffinity membrane. After the unbound fragments were washed away, the antibody-peptide complexes were applied to a MALDI target and analyzed to identify the retained peptides.

Off-line affinity MALDI has also been used to verify protein-protein interactions [47, 48] and to confirm the immunoaffinity purification of α-solanine, α-chaconine [49], lysozyme [50], and mycotoxin [51]. In addition, it has been used in conjunction with avidin-biotin methods to detect bradykinin and insulin [52].

27.3.4 On-Line Affinity MALDI

Affinity ligands can be used on-line with MALDI by modifying the MALDI target for use in affinity extraction. This approach is part of a larger set of techniques known as *surface-enhanced laser desorption ionization* (SELDI) mass spectrometry [53]. This is accomplished by placing a chromatographic stationary phase onto the MALDI target surface. This stationary phase might be reversed-phase coating, an ion-exchange group, or an affinity ligand, such as an antibody, dye, or nucleic acid.

The method of SELDI mass spectrometry has recently seen extensive use in the field of proteomics. For instance, one study used this method to screen serum samples for potential biomarkers related to breast cancer [54]. In this work, the affinity extraction of biomarkers was carried out directly on a MALDI target using a Ni^{2+} chelate as a ligand. After washing this surface, it was examined with MALDI and a time-of-flight mass analyzer. The resulting peaks were identified and explored for use as biomarker candidates using an advanced bioinformatics approach. This gave rise to a panel of three biomarkers that yielded a clinical sensitivity of 93% and specificity of 91%. A similar approach has been used to study ovarian cancer [55] and prostate cancer [56].

On-line affinity MALDI has also been used to characterize microorganisms. As an example, lectins and carbohydrates have been placed on MALDI targets to detect bacteria and viruses from a variety of samples, including urine, milk, and chicken [57].

The immobilized lectins were used to capture microbes through their interactions with cell wall glycoproteins, while the immobilized carbohydrates were used to interact with lectins expressed on the surfaces of microbes.

A variety of other applications have been reported for SELDI mass spectrometry. This technique was recently used to study the binding of anaplastic lymphoma kinase receptor to pleiotrophin growth factor [58]. Protein phosphorylation has also been studied by this method [59]. Furthermore, this approach has been used to examine the conjugation of oligosaccharides to proteins [60] and for screening transcription factors [61].

27.4 AFFINITY MASS SPECTROMETRY IN DRUG DISCOVERY

An area in which affinity mass spectrometry has become particularly important is in drug discovery [47]. For instance, the development and mapping of the human proteome by affinity mass spectrometry has played a critical role in the identification of new lead compounds. In addition, the combination of affinity chromatography with mass spectrometry has been used to screen combinatorial libraries. Examples of these and related applications are discussed in this section.

27.4.1 Screening Combinatorial Libraries

The screening of combinatorial libraries can be accomplished by either using on-line or off-line methods in affinity mass spectrometry [20, 62–65]. The use of LC with multiple stages of filtering by mass spectrometry (MS/MS) is particularly valuable in this work, since it allows more compounds to be examined in a shorter period of time.

In one application, ESI mass spectrometry was coupled directly with an affinity column to screen a library of 361 peptides [64]. A small trap was used to remove interfering salts from the peptides, with the retained peptides being eluted with a pH gradient into a mass spectrometer. In another application, immobilized vancomycin was used as an affinity ligand in high-performance affinity chromatography for screening a library of peptides [65]. An absorbance detector monitored the elution of peptides from this column, with the peaks of interest being trapped in a sample loop through the use of a two-way valve. This valve was then switched to inject the contents of each desired peak into an ion-trap mass spectrometer for identification.

Mass spectrometry has played a key role in the identification of noncovalent receptor-ligand interactions. In most of these approaches, mixtures of proteins and analytes are incubated for a given period of time. Following this incubation, one of several routes can be taken for introducing this mixture into a mass spectrometer. In the technique of *bioaffinity characterization mass spectrometry* (BAC MS), this mixture is infused directly into a Fourier transform mass spectrometer. In this device, the samples undergo quadrupolar excitation, and the protein target is released (see Figure 27.7). The analytes are then identified by high-resolution mass spectrometry and, if necessary, by MS/MS [66].

Another approach that can be used to examine noncovalent interactions in protein/solute mixtures involves the selection of receptor-ligand complexes by size-exclusion chromatography followed by an on-line desalting column and analysis by ESI mass spectrometry. In this method, the receptor-ligand complexes are compared with a control sample containing the ligand. This method has proved useful as a quick prescreening method to select lead compounds [67]. On-line ultrafiltration has been used in a similar manner [68]. In this case, the samples are loaded onto an ultrafiltration disk with a high molecular weight cutoff and washed thoroughly with water. The filtered complex is then dissociated and the analyte eluted with a pulse of a 50:50 methanol:water solution for study by mass spectrometry.

(a)

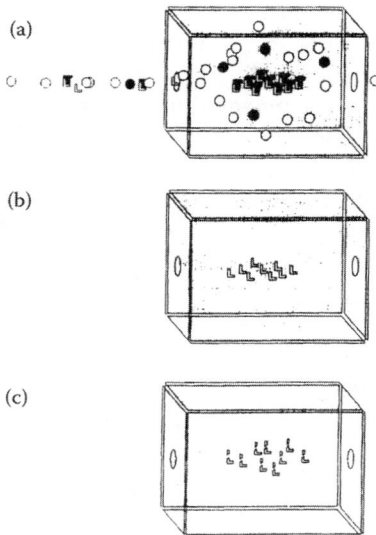

(b)

(c)

Figure 27.7 Principle of bioaffinity characterization mass spectrometry (BAC MS), as performed using a Fourier transform ICR trap. In this approach, (a) a noncovalent biological complex is accumulated in the trap, (b) these complexes are dissociated through heating, and (c) the ligands remaining in the trap are analyzed and identified. (Reproduced with permission from Bruce, J.E., Anderson, G.A., Chen, R., Cheng, X., Gale, D.C., Hofstadler, S.A., Schwartz, B.L., and Smith, R.D., *Rapid Commun. Mass Spectrom.*, 9, 644–650, 1995.)

27.4.2 Proteomic Studies

It has been said that mass spectrometry is the most important tool in proteomics [69]. Affinity chromatography has played a key role in helping make this possible. Even the highest resolution mass analyzer would have significant limitations if affinity chromatography or some other method were not first used for sample pretreatment and complexity reduction. The affinity ligands used for this purpose usually take advantage of some common feature on specific peptides and proteins in the sample. For example, lectins can be used to retain glycopeptides or glycoproteins in such studies [70, 71].

Immobilized metal-ion affinity chromatography (IMAC) is one affinity technique that has often been used to enhance the detection of proteins by mass spectrometry. In this approach, metal ions like Cu^{2+}, Ni^{2+}, or Ga^{2+} are chelated with a support containing immobilized iminodiacetic acid or nitrilotriacetic acid [72–76]. For example, when a mixture of digested proteins is passed through a Ga^{2+} chelate column, this gives rise to selective enrichment of phosphopeptides. Similarly, the use of IMAC columns containing Ni^{2+} or Cu^{2+} can be used to enrich peptides that have acidic amino acids or histidine [76, 77]. Examples illustrating the combination of IMAC with mass spectrometry are given in the literature [78–80]. Further information on the general field of IMAC can be found in Chapter 10.

An example of this is a study in which IMAC was used in a single experiment with mass spectrometry to characterize nearly all of the phosphoproteins in a whole-cell lysate [72]. In this study, the cell was lysed and its proteins were digested with trypsin. The resulting phosphopeptides were then enriched with an IMAC column, followed by their separation through reversed-phase liquid chromatography and analysis by ESI mass spectrometry. More than 1000 phosphopeptides were detected when this method was used to examine a whole-cell lysate from *S. cerevisiae* [81].

Labeling a protein with an affinity tag and expressing it in cell culture is another way in which affinity ligands can be used with mass spectrometry to investigate protein-protein interactions. After the tagged proteins and associated binding partners have been isolated on an affinity column for the tag, the retained protein complexes can be eluted and digested with trypsin. Methods like Fourier transform mass spectrometry are then used to identity the original interacting proteins based on their peptide fragments [82, 83].

One affinity tag that has recently become popular is the *isotopically coded affinity tag* (ICAT) [82–88]. This is used to compare the protein content in two samples. The structure of the ICAT reagent is shown in Figure 27.8a. At one end of this reagent is a biotin label that will be retained by an immobilized avidin or streptavidin column. The other end contains an iodoacetamide group that will react with free cysteine residues on proteins or peptides. This reagent is used in two forms: (1) a heavy form containing deuterium atoms, and (2) a light form containing hydrogen atoms.

Figure 27.8b shows how these reagents are utilized. First, the heavy and light forms of the reagent are added separately to a control cell lysate or cell lysate from the condition

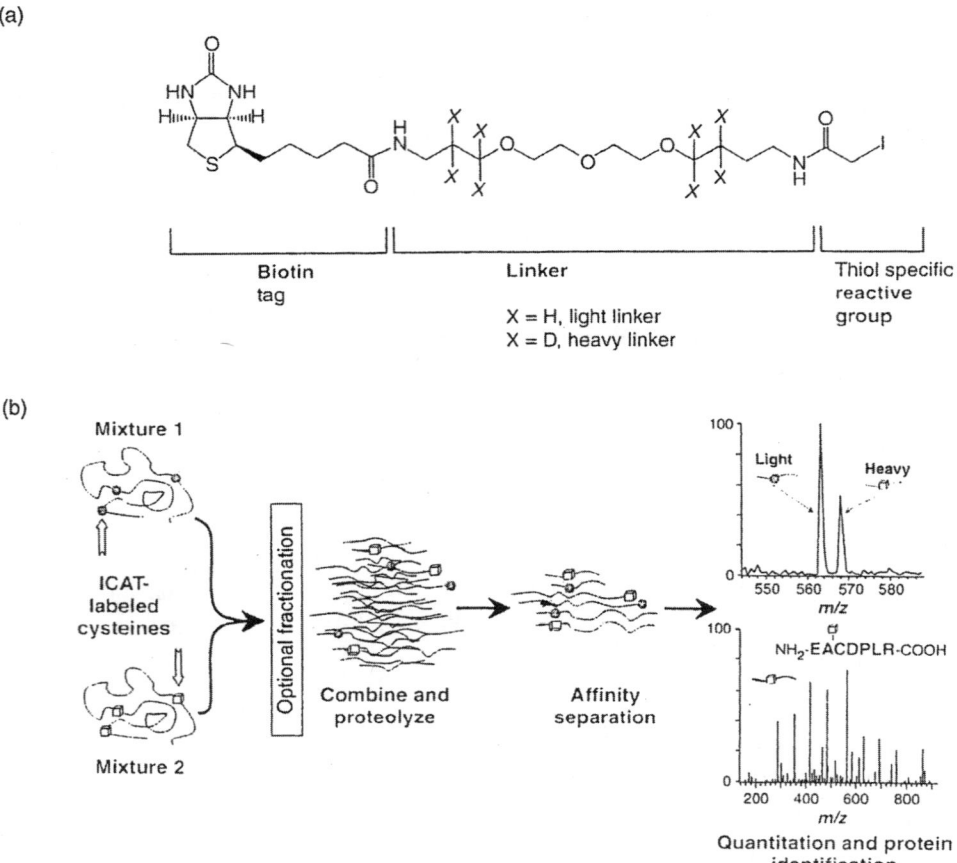

Figure 27.8   (a) Structure of a cysteine-selective ICAT reagent, and (b) the use of such a reagent in quantitative proteomic studies. (Reproduced with permission from Turecek, F., *J. Mass Spectrom.*, 37, 1–14, 2002.)

of interest. As this reagent reacts with the samples, it will covalently bind to the free cysteines on proteins. The cell lysate mixtures are then combined, digested with trypsin, and isolated on a streptavidin column, which captures the modified peptides through the biotin tag. After the captured peptides have been eluted, a second chromatographic step is used to further separate these before they are introduced into a mass analyzer. As they are examined by the mass spectrometer, the relative amount of each peptide in the cell lysate of interest can be compared with the same peptide (but with a slightly different mass tag) from the control sample. This provides each peptide with its own internal standard and allows the relative amount of each peptide to be determined in the cells of interest versus the control.

27.5 STUDIES OF BIOLOGICAL INTERACTIONS

Yet another use of affinity mass spectrometry has been in fundamental studies of biological interactions. This has involved both gas-phase and solution-phase studies. Gas-phase studies of biological interactions are unique to mass spectrometry and can be conducted quickly without significant requirements for sample preparation. Mass spectrometry can also be used in solution-phase studies to increase the power of traditional techniques for binding studies. For instance, with mass spectrometry such studies can often be conducted with multiple analytes simultaneously and with the use of only small amounts of analyte or ligand. Both gas-phase and solution-phase applications of affinity mass spectrometry are examined in this section.

27.5.1 Gas-Phase Studies

One of the unique features of ESI mass spectrometry is its ability to detect intact noncovalent complexes. An example is the production of observable peaks for both myoglobin and apomyoglobin by this method, as shown in Figure 27.9. In this case, a difference in mass of 615 units separates the reconstructed molecular ion peaks for these proteins. This mass difference represents the heme subunit, which is present in the intact form of myoglobin.

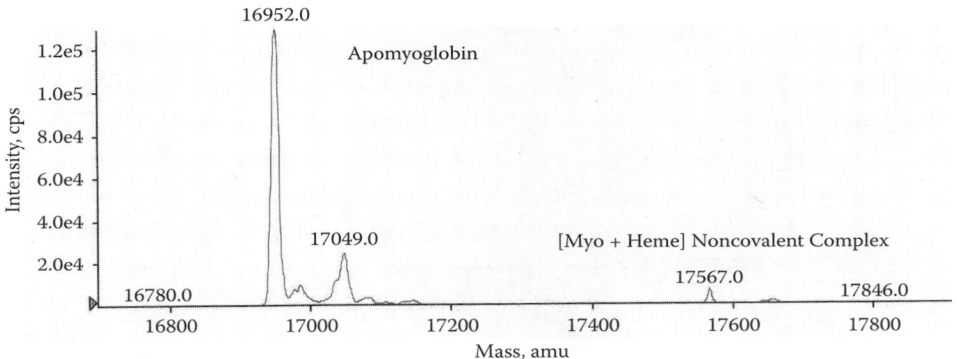

Figure 27.9 Reconstructed mass spectrum for horse heart myoglobin, showing the noncovalently bound heme complex.

Other examples of noncovalent interactions that have been examined by ESI include the binding of solutes to albumin or other proteins and the interactions of substances with DNA [89–97]. This work generally involves the off-line mixing of the ligand (e.g., a protein) and the binding partner of interest. This solution is then infused into an ESI mass spectrometer by using a syringe pump at a low flow rate (e.g., 5 µl/min). Low entrance voltages are used in the ESI source to avoid breaking the noncovalent interactions.

One report utilized this method to examine the binding of suramin, an anticancer drug, to albumin [93]. It was found that this drug bound to albumin in a 4:1 ratio at therapeutic levels and even higher levels in the presence of excess suramin. This approach makes it possible to quickly get an estimate of the number of potential binding sites for a solute on albumin. However, it is difficult to currently say whether the gas-phase results really mimic those seen in solution. For instance, it was pointed out in these studies that there was probably a conformational change in albumin in the gas phase that made it able to bind 20 suramin molecules [93].

27.5.2   Solution-Phase Studies

Mass spectrometry has been used as a detection method for frontal-affinity chromatography (FAC). A detailed discussion of the principles behind frontal analysis can be found in Chapter 22. The use of mass spectrometry with FAC is useful, since it can use small columns and low flow rates [98]. This makes this approach valuable in work with small amounts of ligands and solutes, as is required in many proteomic studies. Furthermore, mass spectrometry can allow the simultaneous detection of several analytes eluting from the same affinity column [62].

A specific example of such an application is a study in which a mixture of enzyme inhibitors was characterized with regard to their effects on enzyme substrate kinetics [99]. This was performed with an immobilized streptavidin column that was bound to a biotin tag on the enzyme N-acetylglucosaminyltransferase V (GnT-V). The mixture of inhibitors that was examined contained eight trisaccharides. These could be studied simultaneously, since the mass spectrometer was operated in the SIM mode and could differentiate between these based on their masses. The elution order of these compounds from the enzyme column was used as an indication of their strength as inhibitors (see Figure 27.10).

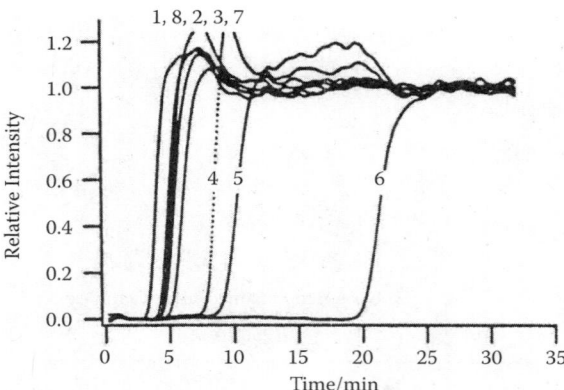

Figure 27.10   FAC mass spectrometry screening for a mixture of eight trisaccharide compounds. (Reproduced with permission from Zhang, B., Palcic, M.M., Schriemer, D.C., Alvarez-Manilla, G., Pierce, M., and Hindsgaul, O., *Anal. Biochem.*, 299, 173–182, 2001.)

By using a separate measure of the column's binding capacity, the dissociation constant for each of the trisaccharides could also be estimated.

27.6 MASS SPECTROMETRIC DETECTION IN AFFINITY METHODS

The final way in which mass spectrometry has been used with affinity ligands is as a detection device for techniques like capillary electrophoresis and immunoassays. Both methods are considered in this section.

27.6.1 Affinity Capillary Electrophoresis

The use of affinity ligands in capillary electrophoresis is discussed in detail in Chapter 26. This overcomes many of the selectivity limitations of ordinary capillary electrophoresis by using an affinity ligand as a buffer additive or immobilized binding agent. But there are several problems with the traditional methods of detection utilized in these techniques. For instance, UV/Vis absorbance in capillary electrophoresis has a concentration-based detection limit of only about 10^{-6} M. Laser-induced fluorescence allows for much lower detection limits but requires that the test compounds be fluorescent or derivatized with a fluorescent label. The use of ESI mass spectrometry as an alternative means of detection is appealing for these applications, since it does not require derivatization but can still reach detection limits in the 10^{-9} M range [100]. Special interfaces are often required for this purpose [101]. Figure 27.11 illustrates a few of these devices [102].

Several studies have used mass spectrometry with affinity capillary electrophoresis to examine the binding of peptides to vancomycin [103–106]. In one case, vancomycin was used as a model ligand to screen a combinatorial library of 361 peptides [103, 104]. Advantages of this approach include (1) its use of a solution-phase system, (2) its need for only small amounts of sample and receptor, and (3) the ability of mass spectrometry to sequence peptide candidates as they migrate from the capillary. In a related application, mass spectrometry and affinity capillary electrophoresis were used to simultaneously measure the binding constants for 19 solutes with vancomycin [105].

Several other uses for affinity capillary electrophoresis with mass spectrometry have recently been developed. For instance, this has been used as a tool for antibody epitope mapping [107]. In addition, the technique has been used for the concentration of samples and the determination of solute concentrations in biological fluids [108]. Finally, affinity capillary electrophoresis and mass spectrometry have been used on-line with IMAC in proteomic studies to identify phosphorylation sites on proteins [109, 110].

27.6.2 Mass Spectrometric Immunoassays

Immunoassays are another area in which mass spectrometry has been used for detection in an affinity method. By employing antibody fragments tagged with a metal, an immunoassay has been developed that can be directly coupled with ICP mass spectrometry [33–35]. This makes use of antibody F'_{ab} fragments that contain embedded gold nanoclusters. ICP mass spectrometry is then used to detect and measure these labeled fragments. ICP is used over ESI for this purpose because it is less sensitive to the matrix, gives equivalent sensitivity for all tagged antibody fragments, has a wide linear range, and gives low detection limits.

An example of this method is an off-line filter immunoassay that used centrifugal filters to separate bound from unbound antigen-antibody complexes. The labeled antibody fragments in these fractions were then measured by ICP mass spectrometry. Another reported

(a)

(b)

(c)

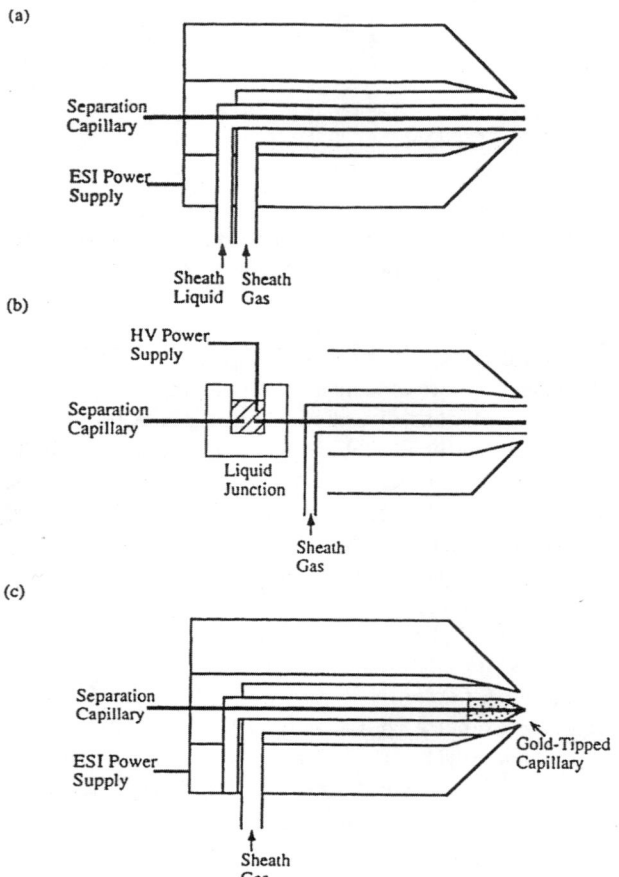

Figure 27.11 Three interfaces for coupling capillary electrophoresis with mass spectrometry: (a) a coaxial sheath-flow interface, (b) a liquid-junction interface, and (c) a sheathless interface. (Reproduced with permission from Cole, R.B., *Electrospray Ionization Mass Spectrometry*, John Wiley, New York, 1997.)

application is an immunoprecipitation technique that used protein A Sepharose to capture analytes along with their gold-labeled F'_{ab} fragments [33–35].

27.7 SUMMARY AND CONCLUSIONS

It has been shown in this chapter that affinity ligands are being used in a variety of ways with mass spectrometry. The first such methods used affinity chromatography or affinity extraction for off-line sample preparation in mass spectrometry. More recently, the combination of affinity ligands with mass spectrometry has become an integral part of drug discovery and development. This has been made possible by the development of methods like ESI and MALDI and through the creation of new formats for the use of ligands with these techniques. As a result, the use of affinity mass spectrometry in fields like proteomics is becoming more common. As further work is performed in this area, more applications should be seen.

SYMBOLS AND ABBREVIATIONS

AC-ESI-MS	Affinity capture-release electrospray ionization mass spectrometry
BAC MS	Bioaffinity characterization mass spectrometry
BPN	1-Biotinyl-10-(4′nitrophenylthiomethylcarbonyl)-1,10-diaza-4,7-ioxadecane
BSN	1-(N-Biotinyl)sarcosinyl-15-(4′nitrophenylthiomethylcarbonyl)-1,15-diaza-5,8,11-trioxapentadecane
ESI	Electrospray ionization
FAB	Fast atom bombardment
FAC	Frontal affinity chromatography
GC	Gas chromatography
GC-MS	Gas chromatography/mass spectrometry
GnT-V	N-Acetylglucosaminyltransferase V
HPLC	High-performance liquid chromatography
ICAT	Isotopically coded affinity tags
ICP	Inductively coupled plasma
IMAC	Immobilized metal-ion affinity chromatography
LC	Liquid chromatography
LC-MS	Liquid chromatography/mass spectrometry
LSD	Lysergic acid diethyl amide
LSIMS	Liquid secondary-ion mass spectrometry
MALDI	Matrix assisted laser desorption ionization
MS/MS	Mass spectrometry with two stages of filtering
m/z	Mass-to-charge ratio
QqTOF	Quadrupole time-of-flight
SELDI	Surface-enhanced laser desorption ionization
SIM	Selected-ion monitoring
SRM	Selected-reaction monitoring
TOF-TOF	Tandem time-of-flight analyzer

ACKNOWLEDGMENTS

This work was supported in part by MDS Pharma Services and by the National Institutes of Health under grant no. RO1 GM44931.

REFERENCES

1. Devreaux, M., Esnault, D., and Gastal, F., Essential tobacco oil: Study of the fatty acid fraction, *Ann. Tabac.,* 14, 111–117, 1976.
2. Grossman, S., Shahin, I., and Sredni, B., Rat testis lipoxyenase-like enzyme: Characterization of products from linoleic acid, *Biochim. Biophys. Acta,* 572, 293–297, 1979.
3. Rule, G.S. and Henion, J.D., Determination of drugs from urine by on-line immunoaffinity chromatography-high-performance liquid chromatography-mass spectrometry, *J. Chromatogr.,* 582, 103–112, 1992.
4. Hage, D., Survey of recent advances in analytical applications of immunoaffinity chromatography, *J. Chromatogr. B,* 715, 28 Mar, 1998.
5. Bachi, A., Zuccato, E., Baraldi, M., Fanelli, R., and Chiabrando, C., Measurement of urinary 8-epiprostaglandin F2a: A novel index of lipid peroxidation *in vivo* by immunoaffinity extraction/gas chromatography-mass spectrometry basal levels in smokers and nonsmokers, *Free Rad. Biol. Med.,* 290, 619–624, 1996.

6. Stanley, S.M.R., Wilhelmi, B.S., Rodgers, J.P., and Bertschinger, H., Immunoaffinity chromatography combined with gas chromatography-negative ion chemical ionization mass spectrometry for the confirmation of flumethasone abuse in the equine, *J. Chromatogr.,* 614, 77–86, 1993.

7. Stanley, S.M.R., Wilhelmi, B.S., and Rodgers, J.P., Comparison of immunoaffinity chromatography combined with gas chromatography-negative ion chemical ionization mass spectrometry and radioimmunoassay for screening dexamethasone in equine urine, *J. Chromatogr.,* 620, 250–253, 1993.

8. Prevost, A., Shuker, D.E.G., Friesen, M.D., Eberle, G., Rajewsky, M.F., and Bartsch, H., Immunoaffinity purification and gas chromatography-mass spectrometric quantification of 3-alkyladenines in urine metabolism studies and basal excretion levels in man, *Carcinogenesis,* 14, 199–204, 1993.

9. Friesen, M.D., Garren, L., Prevost, V., and Shuker, D.E.G., Isolation of urinary 3-methyladenine using immunoaffinity columns prior to determination by low-resolution gas chromatography-mass spectrometry, *Chem. Res. Toxicol.,* 4, 102–106, 1991.

10. Ishibashi, M., Watanabe, K., Ohyama, Y., Mizugaka, M., Hayashi, Y., and Takasaki, W., Novel derivatization and immunoextraction to improve microanalysis of 11-dehydrothromboxane B2 in human urine, *J. Chromatogr.,* 562, 613–624, 1991.

11. Chiabrando, C., Benigni, A., Piccinelli, A., Carminati, C., Cozzi, E., Remuzzi, G., and Finely, R., Antibody-mediated extraction/negative-ion chemical ionization mass spectrometric measurement of thromboxane B2 and 2,3-dinor-thromboxane B2 in human and rat urine, *Anal. Biochem.,* 163, 255–262, 1987.

12. Bonfanti, M., Magagnotti, C., Galli, A., Bagnati, R., Moret, M., Gariboldi, P., Finely, R., and Airoldi, L., Determination of O-6-butylguanine in DNA by immunoaffinity extraction/gas chromatography-mass spectrometry, *Cancer Res.,* 50, 6870–6875, 1990.

13. Tsikas, D., Affinity chromatography as a method for sample preparation in gas chromatography/mass spectrometry, *J. Biochem. Biophys. Methods,* 49, 705–731, 2001.

14. Chiabrando, C., Pinciroli, V., Campoleoni, A., Benigni, A., Piccinelli, A., and Fanelli, R., Quantitative profiling of 6-ketoprostaglandin F1a, 2,3-dinor-6-ketoprostaglandin F1a, thromboxane B2 and 2,3-dinor-thromboxane B2 in human and rat urine by immunoaffinity extraction with gas chromatography-mass spectrometry, *J. Chromatogr.,* 495, 1–11, 1989.

15. Mackert, G., Reinke, M., Schweer, H., and Seyberth, H., Simultaneous determination of the primary prostanoids prostaglandin E2, prostaglandin F2a and 6-oxoprostaglandin F1a by immunoaffinity chromatography in combination with negative ion chemical ionization gas chromatography-tandem mass spectrometry, *J. Chromatogr.,* 494, 13–22, 1989.

16. Van Ginkel, L.A., Stephany, R.W., Van Rossum, H.U., Steinbuch, H.M., Zomer, G., Van de Heeft, E., and De Jong, A.P., Multi-immunoaffinity chromatography: A simple and highly selective clean-up method for multi-anabolic residue analysis of meat, *J. Chromatogr.,* 489, 111–120, 1989.

17. Bagnati, R., Castelli, M.G., Airoldi, L., Oriundi, M.P., Uvalde, A., and Finely, R., Analysis of diethylstilbestrol, dienestrol and hexestrol in biological samples by immunoaffinity extraction and gas chromatography-negative-ion chemical ionization mass spectrometry, *J. Chromatogr.,* 527, 267–278, 1990.

18. Bagnati, R., Oriundi, M.P., Russo, V., Danese, M., Berti, F., and Fanelli, R., Determination of zeranol and b-zearalanol in calf urine by immunoaffinity extraction and gas chromatography-mass spectrometry after repeated administration of zeranol, *J. Chromatogr.,* 564, 493–502, 1991.

19. Wieboldt, R., Zweigenbaum, J., and Henion, J., Immunoaffinity ultrafiltration with ion spray HPLC/MS for screening small-molecule libraries, *Anal. Chem.,* 69, 1683–1691, 1997.

20. Watson, J.T., *Introduction to Mass Spectrometry,* 3rd ed., Lippincott-Raven, Philadelphia, 1997.

21. de Hoffmann, E. and Stroobant, V., *Mass Spectrometry: Principles and Applications,* 2nd ed., John Wiley, New York, 2001.

22. Chernusivech, I., Loboda, A.V., and Thomson, B.A., An introduction to quadrupole-time-of-flight mass spectrometry, *J. Mass Spectrom.,* 36, 849–865, 2001.

23. Cole, R.B., Some tenets pertaining to electrospray ionization mass spectrometry, *J. Mass Spectrom.,* 35, 763–772, 2000.

24. Cech, N.B. and Enke, C.G., Practical implications of some recent studies in electrospray ionization fundamentals, *Mass Spectrom. Rev.,* 20, 362–387, 2001.

25. Cody, R.B., Tamura, J., and Musselman, B.D., Electrospray ionization/magnetic sector mass spectrometry: Calibration, resolution, and accurate mass measurement, *Anal. Chem.,* 64, 1561–1570, 1992.

26. Gluckmann, M. and Karas, M., The initial ion velocity and its dependence on matrix, analyte and preparation method in ultraviolet matrix-assisted laser desorption/ionization, *J. Mass Spectrom.,* 34, 467–477, 1999.

27. Chen, X.J., Carroll, J.A., and Beavis, R.C., Near-ultraviolet-induced matrix-assisted laser desorption/ionization as a function of wavelength, *J. Am. Soc. Mass Spectrom.,* 9, 885–891, 1998.

28. Davoli, E., Fanelli, R., and Bagnati, R., Purification and analysis of drug residues in urine samples by on-line immunoaffinity chromatography/high performance liquid chromatography/continuous-flow fast atom bombardment mass spectrometry, *Anal. Chem.,* 65, 2679–2685, 1993.

29. Barber, M., Bordoli, R.S., Sedgwick, R.D., and Fan, A.N., Fast atom bombardment of solids (FAB): A new ion source for mass spectrometry, *J. Chem. Soc. Chem. Commun.,* 7, 325–327, 1981.

30. Barber, M., Bordoli, R.S., Elliott, G.J., Sedgwick, R.D., and Tyler, A.N., Fast atom bombardment mass spectrometry, *Anal. Chem.,* 54, 645A–657A, 1982.

31. Caprioli, R.M., Continuous flow sample probe for FAB MS, *Anal. Chem.,* 58, 2949–2954, 1986.

32. Caprioli, R.M., *Continuous-Flow FAB-MS,* John Wiley, Chichester, U.K., 1990.

33. Quinn, Z.A., Baranov, V.I., Tanner, S.D., and Wrana, J.L., Simultaneous determination of proteins using an element-tagged immunoassay coupled with ICP-MS detection, *J. Anal. Atomic Spectrom.,* 17, 892–896, 2002.

34. Baranov, V., Quinn, Z., Bandura, D., and Tanner, S.D., A sensitive and quantitative element-tagged immunoassay with ICPMS detection, *Anal. Chem.,* 74, 1629–1636, 2002.

35. Baranov, V.I., Quinn, Z.A., Bandura, D.R., and Tanner, S.D., The potential for elemental analysis in biotechnology, *J. Anal. Atomic Spectrom.,* 17, 1148–1152, 2002.

36. Houk, R.S., Elemental isotopic analysis by ICP-MS, *Acc. Chem. Res.,* 27, 333–339, 1994.

37. Cai, J. and Henion, J., On-line immunoaffinity extraction-coupled column capillary liquid chromatography/tandem mass spectrometry: Trace analysis of LSD analogs and metabolites in human urine, *Anal. Chem.,* 68, 72–78, 1996.

38. Rule, G.S., Mordehai, A.V., and Henion, J., Determination of carbofuran by on-line immunoaffinity chromatography with coupled-column liquid chromatography/mass spectrometry, *Anal. Chem.,* 66, 230–235, 1994.

39. Creaser, C.S., Feely, S.J., Houghton, E., Seymour, M., and Teale, P., On-line immunoaffinity chromatography-high-performance liquid chromatography-mass spectrometry for the determination of dexamethasone, *Anal. Commun.,* 33, 5–8, 1996.

40. Cai, J. and Henion, J., Quantitative multi-residue determination of β-agonists in bovine urine using on-line immunoaffinity extraction-coupled packed capillary liquid chromatography-tandem mass spectrometry, *J. Chromatogr. B,* 691, 357–370, 1997.

41. Onorato, J. and Henion, J.D., Evaluation of triterpine glycoside estrogenic activity using LC/MS and immunoaffinity extraction, *Anal. Chem.,* 73, 4704–4710, 2001.

42. Bergen, H.R., Lacey, J.M., O'Brien, J.F., and Naylor, S., Online single-step analysis of blood proteins: The transferrin story, *Anal. Biochem.,* 296, 122–129, 2001.

43. Gerber, S.A., Scott, C.R., Turecek, F., and Gelb, M.H., Direct profiling of multiple enzyme activities in human cell lysates by affinity chromatography/electrospray ionization mass spectrometry, Application to clinical enzymology, *Anal. Chem.,* 73, 1651–1657, 2001.

44. Ogata, Y., Scampavia, L., Ruzicka, J., Scott, C.R., Gelb, M.H., and Turecek, F., Automated affinity capture-release of biotin-containing conjugates using a lab-on-valve apparatus coupled to UV/visible and electrospray ionization mass spectrometry, *Anal. Chem.,* 74, 4702–4708, 2002.

45. Niederkofler, E.E., Tubbs, K.A., Gruber, K., Nedelkov, D., Kiernan, U.A., Williams, P., and Nelson, R.W., Determination of β-2 microglobulin levels in plasma using a high-throughput mass spectrometric immunoassay system, *Anal. Chem.*, 73, 3294–3299, 2001.

46. Sun, S., Mo, W., Ji, Y., and Liu, S., Use of nitrocellulose films for affinity-directed mass spectrometry for the analysis of antibody/antigen interactions, *Rapid Commun. Mass Spectrom.*, 15, 1743–1746, 2001.

47. Kelly, M.A., McLellan, T.J., and Rosner, P.J., Strategic use of affinity-based mass spectrometry techniques in the drug discovery process, *Anal. Chem.*, 74, 1–9, 2002.

48. Rudiger, A.H., Rudiger, M., Carl, U.D., Chakraborty, T., Roepstorff, P., and Wehland, J., Affinity mass spectrometry-based approaches for the analysis of protein-protein interaction and complex mixtures of peptide-ligands, *Anal. Biochem.*, 275, 162–170, 1999.

49. Driedger, D.R. and Sporns, P., Immunoaffinity sample purification and MALDI-TOF MS analysis of α-solanine and α-chaconine in serum, *J. Agric. Food Chem.*, 49, 543–548, 2001.

50. Brockman, A.H. and Orlando, R., Probe-immobilized affinity chromatography/mass spectrometry, *Anal. Chem.*, 67, 4581–4585, 1995.

51. Nelson, R.W., Krone, J.R., Bieber, A.L., and Williams, P., Mass spectrometric immunoassay, *Anal. Chem.*, 67, 1153–1158, 1995.

52. Schriemer, D.C. and Li, L., Combining avidin-biotin chemistry with matrix-assisted laser desorption/ionization mass spectrometry, *Anal. Chem.*, 68, 3382–3387, 1996.

53. Issaq, H.J., Conrads, T.P., Prieto, D.A., Tirumalai, R., and Veenstra, T.D., SELDI-TOF MS for diagnostic proteomics, *Anal. Chem.*, 75, 149A–155A, 2003.

54. Li, J., Zheng, Z., Rosenzweig, J., Wang, Y.Y., and Chan, D.W., Proteomics and bioinformatics approaches for identification of serum biomarkers to detect breast cancer, *Clin. Chem.*, 48, 1296–1304, 2002.

55. Petricoin, E.F., Ardekani, A.M., Hitt, B.A., Levine, P.J., Fusaro, V.A., Steinberg, S.M., Mills, G.B., Simone, C., Fishman, D.A., Kohn, E.C., and Liotta, L.A., Use of proteomic patterns in serum to identify ovarian cancer, *Lancet*, 359, 572–577, 2002.

56. Qu, Y., Adam, B.L., Yasui, Y., Ward, M.D., Cazares, L.H., Schellhammer, P.F., Feng, Z., Semmes, O.J., and Wright, G.L., Jr., Boosted decision tree analysis of surface-enhanced laser desorption/ionization mass spectral serum profiles discriminates prostate cancer from non-cancer patients, *Clin. Chem.*, 48, 1835–1843, 2002.

57. Bundy, J.L. and Fenselau, C., Lectin and carbohydrate affinity capture surfaces for mass spectrometric analysis of microorganisms, *Anal. Chem.*, 73, 751–757, 2001.

58. Stoica, G.E., Kuo, A., Aigner, A., Sunitha, I., Souttou, B., Malerczyk, C., Caughey, D.J., Wen, D., Karavanov, A., Riegel, A.T., and Wellstein, A., Identification of anaplastic lymphoma kinase as a receptor of the growth factor pleiotrophin, *J. Biol. Chem.*, 276, 16772–16779, 2001.

59. Cardone, M.H., Roy, N., Stennicke, H.R., Salvesen, G.S., Franke, T.F., Stanbridge, E., Frisch, S., and Reed, J.C., Regulation of cell death protease caspase-9 by phosphorylation, *Science*, 282, 1318–1321, 1998.

60. Chernyak, A., Karavanov, A., Ogawa, Y., and Kovac, P., Conjugating oligosaccharides to proteins by squaric acid diester chemistry: Rapid monitoring of the progress of conjugation, and recovery of the unused ligand, *Carbohydr. Res.*, 330, 479–486, 2001.

61. Forde, C.E., Gonzales, A.D., Smessaert, J.M., Murphy, G.A., Shields, S.J., Fitch, J.P., McCutchen-Maloney, S.L., A rapid method to capture and screen for transcription factors by SELDI mass spectrometry, *Biochem. Biophys. Res.*, 290, 1328–1335, 2002.

62. Tiller, P.R., Mutton, I.M., Lane, S.J., and Bevan, C.D., Immobilized human serum albumin: Liquid chromatography/mass spectrometry as a method of determining drug-protein binding, *Rapid Commun. Mass Spectrom.*, 9, 261–263, 1995.

63. Shin, Y.G. and van Breemen, R.B., Analysis and screening of combinatorial libraries using mass spectrometry, *Biopharm. Drug Dispos.*, 22, 353–372, 2001.

64. Kelly, M.A., Liang, H.B., Sytwu, I.I., Vlattas, I., Lyons, N.L., Bowen, B.R., and Wennogle, L.P., Characterization of SH2-ligand interactions via library affinity selection with mass spectrometric detection, *Biochemistry*, 35, 11747–11755, 1996.

65. Lynen, F., Borremans, F., and Sandra, P., Affinity chromatography on vancomycin coupled to reversed phase liquid chromatography/electrospray-ion trap mass spectrometry for the screening of combinatorial libraries, *Chromatographia,* 54, 433–437, 2001.

66. Bruce, J.E., Anderson, G.A., Chen, R., Cheng, X., Gale, D.C., Hofstadler, S.A., Schwartz, B.L., and Smith, R.D., Bio-affinity characterization mass spectrometry, *Rapid Commun. Mass Spectrom.,* 9, 644–650, 1995.

67. Kaur, S., McGuire, L., Tang, D., Dollinger, G., and Huebner, V., Affinity selection and mass spectrometry-based strategies to identify lead compounds in combinatorial libraries, *J. Protein Chem.,* 16, 505–511, 1997.

68. van Breemen, R.B., Huang, C.R., Nikolic, D., Woodbury, C.P., Zhao, Y.Z., and Venton, D.L., Pulsed ultrafiltration mass spectrometry: A new method for screening combinatorial libraries, *Anal. Chem.,* 69, 2159–2164, 1997.

69. Aebersold, R. and Goodlett, D.R., Mass spectrometry in proteomics, *Chem. Rev.,* 101, 269–295, 2001.

70. Geng, M., Ji, J., and Regnier, F.E., Signature-peptide approach to detecting proteins in complex mixtures, *J. Chromatogr. A,* 870, 295–313, 2000.

71. Geng, M., Zhang, X., Bina, M., and Regnier, F.E., Proteomics of glycoproteins based on affinity selection of glycopeptides from tryptic digests, *J. Chromatogr. B,* 752, 293–306, 2001.

72. Hunt, D.F., Personal commentary on proteomics, *J. Proteome Res.,* 1, 15–19, 2002.

73. Neville, D.C., Rozanas, D.R., Price, E.M., Gruis, D.B., Verkman, A.S., and Townsend, R.R., Evidence for phosphorylation of serine 753 in CFTR using a novel metal-ion affinity resin and matrix-assisted laser desorption mass spectrometry, *Protein Sci.,* 6, 2436–2445, 1997.

74. Nuwaysir, L.M. and Stults, J.T., Electrospray ionization mass spectrometry of phosphopeptides isolated by on-line immobilized metal-ion affinity chromatography, *J. Am. Soc. Mass Spectrom.,* 4, 662–669, 1993.

75. Tempst, P., Link, A.J., Riviere, L.R., Fleming, M., and Elicone, C., Internal sequence analysis of proteins separated on polyacrylamide gels at the submicrogram level: Improved methods, applications and gene cloning strategies, *Electrophoresis,* 11, 537–445, 1990.

76. Zhou, W., Merrick, B.A., Khaledi, M.G., and Tomer, K.B., Detection and sequencing of phosphopeptides affinity bound to immobilized metal ion beads by matrix-assisted laser desorption/ionization mass spectrometry, *J. Am. Soc. Mass Spectrom.,* 11, 273–282, 2000.

77. Stensballe, A., Andersen, S., and Jensen, O.N., Characterization of phosphoproteins from electrophoretic gels by nanoscale Fe(III) affinity chromatography with off-line mass spectrometry analysis, *Proteomics,* 2, 207–222, 2001.

78. Porath, J., Immobilized metal ion affinity chromatography, *Protein Express. Purif.,* 3, 263–281, 1992.

79. Gallis, B., Corthals, G.L., Goodlett, D.R., Ueba, H., Kim, F., Presnell, S.R., Figeys, D., Harrison, D.G., Berk, B.C., Aebersold, R., and Corson, M.A., Identification of flow-dependent endothelial nitric-oxide synthase phosphorylation sites by mass spectrometry and regulation of phosphorylation and nitric oxide production by the phosphatidylinositol 3-kinase inhibitor LY294002, *J. Biol. Chem.,* 274, 30101–30108, 1999.

80. Riggs, L., Sioma, C., and Regnier, F.E., Automated signature peptide approach for proteomics, *J. Chromatogr. A,* 924, 359–368, 2001.

81. Ficarro, S.B., McCleland, M.L., Stukenberg, P.T., Burke, D.J., Ross, M.M., Shabanowitz, J., Hunt, D.F., and White, F.M., Phosphoproteome analysis by mass spectrometry and its application to *Saccharomyces cerevisiae, Nat. Biotech.,* 20, 301–305, 2002.

82. Gavin, A.C., Bosch, M., Krause, R., Grandi, P., Marzioch, M., Bauer, A., Schultz, J., Rick, J.M., Michon, A.M., Crusiat, C.M., Remor, M., Hofert, C., Schelder, M., Brajenovic, M., Ruffner, H., Merino, A., Klein, K., Hudak, M., Dickson, D., and Rudi, T., Functional organization of the yeast proteome by systematic analysis of protein complexes, *Nature,* 415, 141–147, 2002.

83. Ho, Y., Gruhler, A., Hellbut, A., Bader, G., Moore, L., Adams, S.U., Millar, A., Taylor, P., Bennett, K., Boutiller, K., Yang, L., Wolting, C., Donaldson, I., Schandorff, S., Shewnarane, J., Vo, M., Taggartt, J., Goudreault, M., Muskat, B., and Alfarano, C., Systematic identification of protein complexes in *Saccharomyces cerevisiae* by mass spectrometry, *Nature,* 415, 180–183, 2002.

84. Turecek, F., Mass spectrometry in coupling with affinity capture-release and isotope-coded affinity tags for quantitative protein analysis, *J. Mass Spectrom.*, 37, 1–14, 2002.

85. Regnier, F.E., Riggs, L., Zhang, R., Xiong, L., Liu, P., Chakraborty, A., Seeley, E., Sioma, C., and Thompson, R.A., Comparative proteomics based on stable isotope labeling and affinity selection, *J. Mass Spectrom.*, 37, 133–145, 2002.

86. Wang, S. and Regnier, F.E., Proteomics based on selecting and quantifying cysteine containing peptides by covalent chromatography, *J. Chromatogr. A*, 924, 345–357, 2001.

87. Wang, S., Zhang, X., and Regnier, F.E., Quantitative proteomics strategy involving the selection of peptides containing both cysteine and histidine from tryptic digests of cell lysates, *J. Chromatogr. A*, 949, 153–162, 2002.

88. Ji, J., Chakraborty, A., Geng, M., Zhang, X., Amini, A., Bina, M., and Regnier, F.E., Strategy for qualitative and quantitative analysis in proteomics based on signature peptides, *J. Chromatogr. B*, 745, 197–210, 2000.

89. Ganem, B., Li, Y.T., and Henion, J.D., The use of ion-spray mass spectrometry to directly detect protein-ligand interactions, *Chemtracts Org. Chem.*, 5, 386–388, 1992.

90. Ganem, B. and Henion, J.D., Detecting non-covalent complexes of biological macromolecules: New applications of ion-spray mass spectrometry, *Chemtracts Org. Chem.*, 6, 1–22, 1993.

91. Bakhtiar, R. and Stearns, R.A., Studies on non-covalent associations of immunosuppressive drugs with serum albumin using pneumatically assisted electrospray ionization mass spectrometry, *Rapid Commun. Mass Spectrom.*, 9, 240–244, 1995.

92. Baczynskyj, L., Bronson, G.E., and Kubiak, T.M., Application of thermally assisted electrospray ionization mass spectrometry for detection of no covalent complexes of bovine serum albumin with growth hormone releasing factor and other biologically active peptides, *Rapid Commun. Mass Spectrom.*, 8, 280–286, 1994.

93. Roboz, J., Deng, L., Ma, L., and Holland, J.F., Investigation of suramin-albumin binding by electrospray mass spectrometry, *Rapid Commun. Mass Spectrom.*, 12, 1319–1322, 1998.

94. Triolo, A., Arcamone, F.M., Raffaelli, A., and Salvadori, P., Non-covalent complexes between DNA-binding drugs and double-stranded deoxyoligonucleotides: A study by ionspray mass spectrometry, *J. Mass Spectrom.*, 32, 1186–1194, 1997.

95. Greig, M.J., Gaus, H., Cummins, L.L., Sasmor, H., and Griffey, R.H., Measurement of macromolecular binding using electrospray mass spectrometry: Determination of dissociation constants for oligonucleotide-serum albumin complexes, *J. Am. Chem. Soc.*, 117, 10765–10766, 1995.

96. Sannes-Lowery, K.A., Griffey, R.H., and Hofstadler, S.A., Measuring dissociation constants of RNA and aminoglycoside antibiotics by electrospray ionization mass spectrometry, *Anal. Biochem.*, 280, 264–271, 2000.

97. Siuzdak, G., Krebs, J.F., Benkovic, S.J., and Dyson, H.J., Binding of hapten to a single-chain catalytic antibody demonstrated by electrospray mass spectrometry, *J. Am. Chem. Soc.*, 116, 7937–7938, 1994.

98. Schriemer, D.C., Bundle, D.R., Li, L., and Hindsgaul, O., Micro-scale frontal affinity chromatography with mass spectrometric detection: A new method for the screening of compound libraries, *Angew. Chem. Int. Ed.*, 37, 3383–3387, 1998.

99. Zhang, B., Palcic, M.M., Schriemer, D.C., Alvarez-Manilla, G., Pierce, M., and Hindsgaul, O., Frontal affinity chromatography coupled to mass spectrometry for screening mixtures of enzyme inhibitors, *Anal. Biochem.*, 299, 173–182, 2001.

100. Guit-van Duijn, R.M., Frank, J., Van Dedem, G.W.K., and Baltussen, E., Recent advances in affinity capillary electrophoresis, *Electrophoresis*, 21, 3905–3918, 2000.

101. von Brocke, A., Nicholson, G., and Bayer, E., Recent advances in capillary electrophoresis/electrospray mass spectrometry, *Electrophoresis*, 22, 1251–1266, 2001.

102. Cole, R.B., *Electrospray Ionization Mass Spectrometry*, John Wiley, New York, 1997.

103. Chu, Y.H., Kirby, D.P., and Karger, B.L., Free solution identification of candidate peptides from combinatorial libraries by affinity capillary electrophoresis/mass spectrometry, *J. Am. Chem. Soc.*, 117, 5419–5420, 1995.

104. Chu, Y.H., Dunayevskiy, Y.M., Kirby, D.P., Vouros, P., and Karger, B.L., Affinity capillary electrophoresis-mass spectrometry for screening combinatorial libraries, *J. Am. Chem. Soc.,* 118, 7827–7835, 1996.

105. Dunayevskiy, Y.M., Lyubarskaya, Y.V., Chu, Y.H., Vouros, P., and Karger, B.L., Simultaneous measurement of nineteen binding constants of peptides to vancomycin using affinity capillary electrophoresis-mass spectrometry, *J. Med. Chem.,* 41, 1201–1204, 1998.

106. Lynen, F., Zhao, Y., Becu, C., Borremans, F., and Sandra, P., Considerations concerning interaction characterization of oligopeptide mixtures with vancomycin using affinity capillary electrophoresis-electrospray mass spectrometry, *Electrophoresis,* 20, 2462–2474, 1999.

107. Lyubarskaya, Y.V., Dunayevskiy, Y.M., Vouros, P., and Karger, B.L., Microscale epitope mapping by affinity capillary electrophoresis-mass spectrometry, *Anal. Chem.,* 69, 3008–3014, 1997.

108. Guzman, N.A. and Stubbs, R.J., The use of selective adsorbents in capillary electrophoresis-mass spectrometry for analyte preconcentration and microreactions: A powerful three-dimensional tool for multiple chemical and biological applications, *Electrophoresis,* 22, 3602–3628, 2001.

109. Cao, P. and Stults, J.T., Phosphopeptide analysis by on-line immobilized metal-ion affinity chromatography-capillary electrophoresis-electrospray ionization mass spectrometry, *J. Chromatogr. A,* 853, 225–235, 1999.

110. Cao, P. and Stults, J.T., Mapping the phosphorylation sites of proteins using on-line immobilized metal affinity chromatography/capillary electrophoresis/electrospray ionization multiple stage tandem mass spectrometry, *Rapid Commun. Mass Spectrom.,* 14, 1600–1606, 2000.

28

Microanalytical Methods Based on Affinity Chromatography

Terry M. Phillips

Ultramicro Analytical Immunochemistry Resource, Division of Bioengineering
and Physical Sciences, Office of Research Services, National Institutes
of Health, Bethesda, MD

CONTENTS

28.1 INTRODUCTION

The development of micro- and ultramicro (nano) analytical systems has been envisioned for some time, but only in recent years have such systems become a reality. The marrying of analytical science with microfabrication technologies, developed in the silicon electronics sector, have made it possible to manufacture or fabricate electrophoretic and chromatographic devices on an extremely small scale, thus creating the new fields of microtechnology and nanotechnology.

One of the most appealing advantages of microfabricated devices is their low consumption of reagents and (in the biomedical sciences) samples. Additionally, a reduction in size is often accompanied by a considerable reduction in the time required to perform analyses. These characteristics have made these devices useful not only in the analytical field, but also in medical diagnostics, where such devices can be produced in relatively large numbers at low cost, making them disposable and lowering the risk of contamination.

Nanotechnology has opened up a new frontier for analytical sciences, leading to the development of micro-total analytical systems (μTAS). These systems are fabricated on glass, silicon, or polymer wafers and fulfill the concept envisioned for the creation of a complete "laboratory-on-a-chip" [1–6]. Table 28.1 gives a list of terms commonly used in the microfabrication of these systems and provides a brief explanation of their meaning. The emerging field of proteomics has introduced an additional requirement for these systems in which multiple analytes must be simultaneously measured within the same sample. This has resulted in the development of *static affinity-based microdevices* or *arrays*.

The continuing demand for ultramicroanalytical procedures has led to a growth in applications for instruments capable of analyzing sample volumes that were previously unreachable, such as nanoliter (nl = 10^{-9} l) or picoliter (pl = 10^{-12} l) volumes and smaller (see Table 28.2). As instruments get smaller and their resolving power increases, the application of affinity and immunoaffinity techniques to the analytical process is becoming a logical and even an essential approach. Although not yet fully exploited, the application of affinity techniques to microanalysis has been reported in areas such as genomic and proteomic arrays [7–9], chip-based analytical systems [10–12], and biosensors (see Chapter 25). This chapter examines some of these applications and provides some illustrations of their use.

28.2 MANUFACTURING MICROFABRICATED DEVICES

Reasonably small microdevices, such as miniature chromatographic systems, have been around for several years and can be quite easily assembled from commercially available components. However, true micro- and nanodevices are fabricated in relatively specialized facilities that require expensive lithography and machining tools as well as high-quality clean-room facilities. Microfabrication was originally (and still is) performed in materials such as glass and silica, but recently other polymer-based materials have also become popular, such as polydimethylsiloxene (PDMS), polymethyl-methacrylate (PMMA), polycarbonate, polyethylene terephthalate (PET), Teflon, polystyrene, and poly ether-ether ketone (PEEK) [13, 14].

The fabrication of microdevices is performed using modifications of techniques that have been developed in the microelectronics industry. Although a complete coverage of these procedures is outside the scope of this chapter, a brief introduction to the most common techniques is included here to give the reader a better understanding of this field. Figure 28.1 outlines the micromachining and photolithographic procedures that are most commonly used in the microfabrication of chromatographic devices. A further discussion of these procedures is provided in the following subsections.

Table 28.1 Common Terms Used in Microfabrication

Term	Meaning or Explanation
Electrokinetic	Driving fluids through microchannels and microdevice patterns by electricity
Embossing	Procedure for pressing or stamping a pattern into a wafer or chip
Imprinting	Stamping a pattern onto a chip or wafer
Injection molding	Formation of a microdevice by injection of molding material into a solid master mold
Lab-on-a-chip	A complete system, incorporating sample preparation, injection, and analysis within a microfabricated device
LIGA	A process combining three different techniques, lithography, electroplating, and molding, to fabricate complex structures incorporating multiple dimensions and electroconductive surfaces
Lithography	A procedure used to transfer a defined pattern from one medium to another, such as transferring an image on a mask to a glass or silicon wafer
MicroElectro Mechanical Systems	Microelectromechanical systems, or MEMS; also popularly "mechanical" referred to as micromachines, nanomachines, or transducers; these are characterized by being less than a square millimeter in size; in the most general form, MEMS consist of mechanical microstructures, microsensors, microactuators, and electronics, all integrated onto the same chip
Microfabrication	A collective name for the processes through which microdevices are manufactured
Microfluidics-based chips	These chips, which contain tiny channels in which the movement-based chips of fluids is controlled, allow the integration and miniaturization of a range of laboratory processes
Micromachining	A process using laser, chemical etching, or ultrasonic etching to machine or cut channels in a substrate or chip
Photolithography	A process utilizing light (usually UV radiation) to transfer a pattern from a mask to a silicon wafer or chip
Pressure stamping	Imprinting or impressing a pattern into a wafer or chip by applying pressure to a stamp
Rapid prototyping	A rapid procedure for mass-producing microdevices by contact photolithography
Replica molding	A technique for the mass reproduction of microdevices by casting a polymer against a master pattern
Reynolds number	A dimensionless number used to characterize turbulence and laminar flow of a fluid within a cylinder; this number is a product of the diameter plus the density and velocity of the fluid traveling through the cylinder divided by the fluid velocity
Soft lithography	Lithographic processes performed in a polymer, usually PDMS; this group of fabrication processes use a PDMS "stamp" to print or mold microdevice patterns
Total Analysis System (TAS)	A complete analytical system, designed to completely process a sample from injection to detection; a micro-TAS is the chip-based version of this
X-ray lithography	Technique where an X-ray source is used to transfer a pattern from a mask to a silicon wafer

Table 28.2 Units of Mass and Volume in Microtechniques

Prefix	Symbol	Use in Mass	Use in Volume
Milli-	m	10^{-3} grams	10^{-3} liters
Micro-	μ	10^{-6} grams	10^{-6} liters
Nano-	n	10^{-9} grams	10^{-9} liters
Pico-	p	10^{-12} grams	10^{-12} liters
Femto-	f	10^{-15} grams	10^{-15} liters
Atto-	a	10^{-18} grams	10^{-18} liters
Zepto-	z	10^{-21} grams	10^{-21} liters
Yocto-	y	10^{-24} grams	10^{-24} liters

28.2.1 Micromachining, Embossing, and Imprinting

Micromachining, embossing, and imprinting are commonly used to fabricate microdevices. Micromachining can involve chemical and ultrasonic etching or laser-beam cutting. The last of these is probably the most common approach and requires specialized facilities and

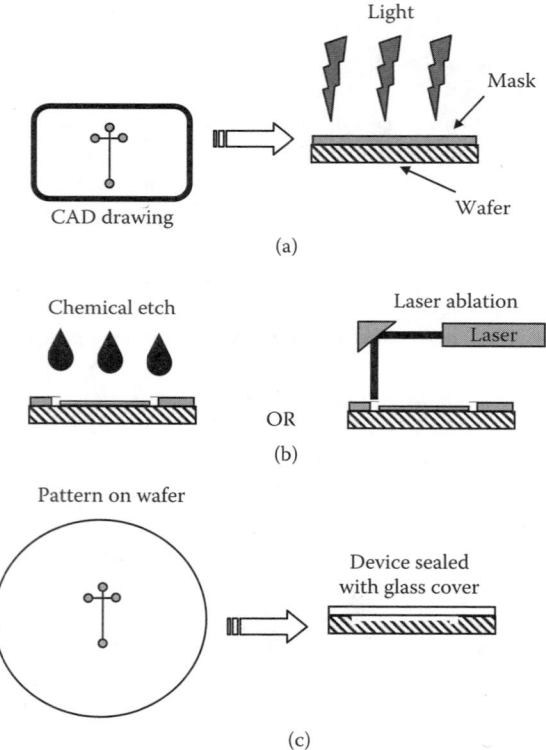

Figure 28.1 The most common techniques used in the microfabrication of microchromatographic devices. In (a) a computer-assisted design (CAD) drawing is made of the basic structure and transferred to a mask by photolithography. In (b) the channels and reservoirs are cut from the wafer surface by using either chemical etching or laser ablation. In (c) the mask is removed, thus exposing the surface. The open channels are sealed by annealing with a glass or silica cover.

instrumentation. In this method, a computer-assisted design (CAD) drawing is made of the desired device, scaled accordingly, and fed into a computer-controlled laser-milling instrument, which produces the desired pattern in a silicon or glass wafer.

Although most micromachining is performed at dedicated facilities, a laboratory-built system employing a scanning near-field optical microscope and a 244-nm UV laser has been described by Sun et al. [15] that is capable of creating structures at approximately 40-nm resolution. Other approaches have used focused ion beams for milling microcavities [16] and X-ray lithography for micromachining devices in PMMA [17]. The latter process produced channels that were 20 μm in width and 50 μm deep, depending upon the length of X-ray exposure and the final resolution of the design pattern. The channels produced by this approach are not square in cross-section but are often elliptical due to the focusing shape of the cutting beam (see Figure 28.2); this can cause some problems when calculating flow dynamics and rates.

Embossing and imprinting involve the construction of a master pattern and the creation of negative "stamps" from which positive patterns can be pressed into or stamped onto suitable materials and surfaces. This technology is suitable for manufacturing devices from polymeric materials like PDMS or PMMA.

28.2.2 Lithography

Lithography has become a popular approach to microfabrication due mainly to the numerous reports generated by the Whitesides group in Boston [18–21]. This is especially true for soft lithography, which can be performed in a number of different materials, the major one of which is PDMS. The appeal of this approach is that a relatively small amount of capital investment is required, and fabrication can be performed under regular laboratory conditions. In addition, fabrication costs are low when compared with other techniques and easily within most laboratory budgets.

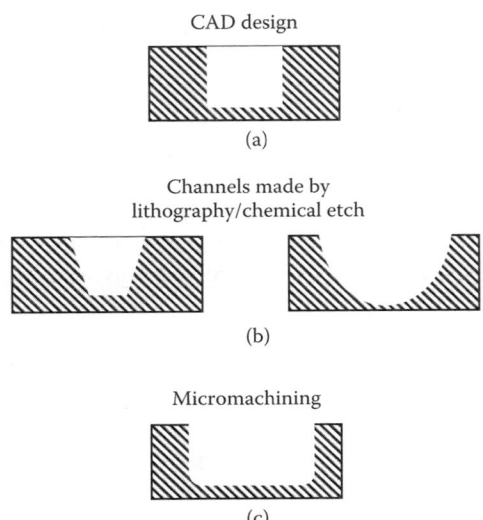

Figure 28.2 Channel shapes produced by microfabrication techniques: (a) an ideal channel visualized by a computer-assisted drawing program; (b) the shape of channels produced by lithography or chemical etching; and (c) channels produced by micromachining through chemical or light ablation.

A major starting component for this type of microfabrication is the design and manufacture of a mask from which the fabricated device will be made. The manufacturing of these masks can be performed at a specialized facility, or a procedure can be used in which a regular office printer is used to fabricate both masks and molds [22].

PDMS is a suitable material for fabricating structures in the 1-μm to 50-nm range. However, a serious limitation of this material up to the present has been its inability to be used for the fabrication of multidimensional structures and devices. This is because PDMS is prone to distortion, thus making the alignment of fabricated features difficult or even impossible. Additionally, there are a number of areas in which both the novice and experienced investigator may incur defects, such as dust contamination, bubble formation, and poor adhesion to the substrate. Further information on soft lithography and its applications can be found in two excellent and informative reviews [23, 24].

28.2.3 Molding

As its name suggests, molding is procedure in which devices are cast in micromolds or stamps. In this technique, stamps consisting of PDMS or alternative polymers are made from a lithographically designed master and used to mold patterns in other polymeric materials. This can be achieved by replica molding, in which polyurethane is molded against the stamp to fabricate patterns such as channels or reaction chambers [23, 25].

28.2.4 Other Techniques

Microlamination, pressure stamping, and *photochemical machining* (or a combination of these techniques) can be used to make metal reaction chambers and mixers. *LIGA* (which is derived from the German acronym for *Lithographie, Galvoformung*, and *Abformung*) is a fairly new process that combines lithography, electroplating, and molding. LIGA is relatively difficult and requires specialized equipment or facilities. It has been used to fabricate a number of three-dimensional microdevices, including reaction chambers and microreactors. An advantage of this process is that it is capable of fabricating devices in a wide variety of materials, and it can apply conductive surface materials to electroformed structures. This latter characteristic is useful for building control units for electrokinetic microfluidics.

28.3 MICRO- AND NANODEVICES

The increasing demand for reduction in reagent costs and analysis time, as well as the conservation of precious samples, has led to a demand for miniaturization and the development of new technologies. Affinity and immunoaffinity separation techniques are well suited to integration into these microsystems due to their specificity and fast reactions with the analyte of interest. General-affinity ligands are useful for isolating class-specific analytes from complex mixtures, while immunoaffinity matrices (as discussed in Chapter 6) are valuable in the isolation of a single analyte from a complex biological matrix.

Affinity- and immunoaffinity-based systems range from static devices (e.g., nucleic acid microarrays) to the newly developed protein and antibody arrays. In addition, affinity and immunoaffinity methods have been incorporated into microfluidic devices that have been used for capillary electrophoresis (see Chapter 26) and chromatographic devices.

28.3.1 Static Devices

As its name suggests, *array technology* involves placing patterns or arrays of ligands onto a solid matrix for use in detecting molecules that interact with these ligands. This is illustrated in Figure 28.3. The term *microarray* is generally used to refer to an ordered pattern of nucleic acids, proteins, small molecules, cells, or other substances that enable parallel analysis of

(a)

(b)

(c)

(d)

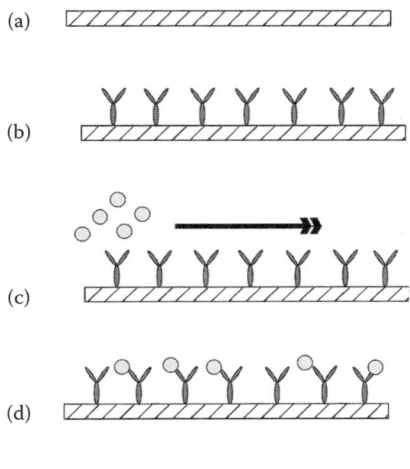

(e)   Scan array with a confocal
           fluorescence microscope

Figure 28.3   The basics of static-array technology. (a) A wafer possessing an activated surface
is used as the base support. (b) The capture ligand (Y) is immobilized to the activated surface.
(c) The fluorescent-labeled sample (•) is introduced and allowed to flow over or cover the ligand-
coated surface. (d) The sample is selectively captured by the ligand, and nonreactive materials are
removed. (e) The array is scanned by a fluorescence detector, and the positions of the bound sample
bands are recorded and analyzed.

complex biochemical samples. The advantages of arrays and microarrays are that they allow
an investigator to examine many interactions simultaneously and to determine the origin of
the molecules that are involved in these reactions.

The first arrays were developed by the Brown Laboratory at Stanford University
using multiple cDNAs as the immobilized ligands [26], with these ligands later being
replaced by oligonucleotides. Based on this approach, thousands of individual genes can
be spotted onto a 3×1-in. slide and used to analyze the messenger RNA present in such
samples as lysed cells, homogenized tissue, or biological fluids. An excellent book on
nucleic acid array technology has been edited by Schena [27] and is recommended reading
for those who wish to know more about this subject.

Just as nucleic acid-based microarrays have revolutionized and advanced genomic
studies, protein arrays (especially antibody arrays) hold the same potential for proteomics.
Once they have been fully developed and standardized, protein arrays could be employed in
a number of fields. Possible examples of applications are the study of molecular interactions
between known proteins and other proteins or nucleic acids, the study of antibody-antigen
reactions, and work examining protein interactions with cells and small molecules [28–31].

However, a number of technical difficulties have to be overcome before high-density
protein arrays become a reality. For example, high-sensitivity detection is required along
with the development of efficient protein-expression strategies that are capable of retaining
a protein's specific activity. Methods are also needed for optimizing protein immobilization. In the case of enzymes, antibodies, and receptors, maintaining bioactivity becomes
a major issue, especially following the immobilization of these ligands. These problems
not only plague static-array technology, but all areas of microseparations.

Affinity-based static arrays have been modified to accommodate the performance
of multianalyte immunoassays in microfabricated devices. Moody et al. [29] used a

miniarray to measure seven cytokines by patterning immobilized antibodies to the bottom of a 96-well polystyrene plate. The captured analytes were visualized by using a second preparation of antibodies that contained enzymes as labels. The resulting reaction pattern was then recorded using a CCD (charge-coupled device) camera.

Jones et al. [32] described a system that utilized atomic-force microscopy (AFM) to measure the degree of antibody binding to immobilized rabbit IgG. Although it was claimed that this was a relatively simple and straightforward technique, the use of AFM introduces a number of problems, the least of which is the need to acquire an atomic-force microscope.

Another approach has used chemically derivatized beads in micromachined cavities as capture devices in immunoassays [33]. This silicone chip-based sensor was capable of performing up to 100 individual assays in pyramidal cavities, which allowed the use of adequate fluidics and provided a suitable optical interface for detection.

28.3.2 Devices Based on Microfluidics

True microfluidic devices use some form of physical force to drive liquids through micro-channels and into a reaction chamber or separation chamber. In several systems, microfluidics have also been used to elute the separated analytes into detection chambers or flow cells. A schematic of a generalized microfluidic device is given in Figure 28.4.

Microfluidics can be performed in several ways, but two approaches have dominated the field: electrokinetically or chromatographically pumped devices (see Table 28.3). Both types of devices are discussed in this section. There are pros and cons to each of these approaches, although both have been successfully applied to affinity-based microanalytical systems. Even though few such devices have been described to date, this field is rapidly expanding, and commercially available devices that combine electrokinetic pumping with other fluid-control systems should soon be available for both analytical and clinical purposes.

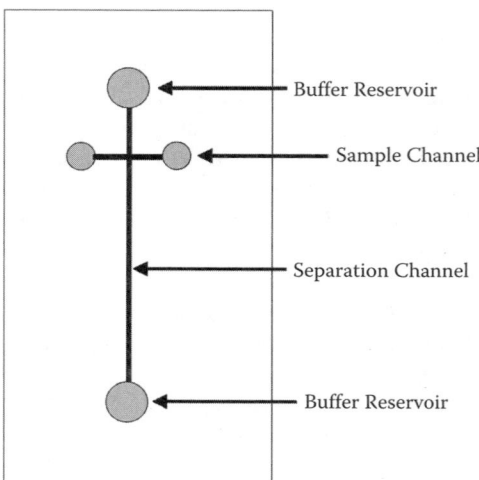

Figure 28.4 General schematic for an electrokinetic microfluidics device used for microchromatography. The running buffer is placed in the top and bottom reservoirs, and the separation channel is filled with a packing material. Samples are electrokinetically injected by introducing a low current between the electrodes placed in the two sample wells. A chromatographic separation is achieved by passing a current between the electrodes placed into top and bottom buffer reservoirs, driving the sample through the packing medium. On-line detection is achieved by positioning a detector at the lower end of the separation channel.

Table 28.3 Microfluidic Devices and Their Applications

Device Type	Application
Electrokinetic flow	Capillary electrophoresis, affinity/immunoaffinity CE, chip-based CE
	Capillary electrochromatography
	Affinity electrochromatography
	Chip-based immunoassays
	μ-TAS
Pressure-induced flow	Microcolumn chromatography
	Affinity/immunoaffinity chromatography
	Immunoaffinity flow-through arrays
	μ-TAS
Combination of electrokinetic and pressure-induced flow	Flow-through affinity/immunoaffinity arrays
	μ-TAS

28.3.2.1   Electrokinetic Devices

A large number of microfabricated devices have been based on capillary electrophoresis (e.g., see Chapter 26). This is mainly due to the relative ease with which fluids can be electrokinetically manipulated in such devices. Bruin [34] has reviewed the rapid developments in the field of electrokinetically driven microfluidic separation devices, providing examples of integrated systems for sample preconcentration, filtering, DNA amplification, and on-chip detection.

Electrokinetic force is perhaps the most controllable way to drive microfluidic devices, since this can provide flow rates in the nano- and picoliter range [35], thus enabling the stopped-flow approaches that are often essential to incubation steps and reagent mixing. Dodge et al. [12] used electrokinetic manipulations to develop an affinity assay for IgG based on picoliter-sized reaction chambers. This device was able to measure analyte concentrations down to 50 nM in less than 5 min.

Linder et al. [36] modified the internal surface of a PDMS device with Neutravidin to reduce the hydrophobicity of this material and provide a suitable surface for immobilizing biotinylated ligands. Using this approach, they made an electrokinetically driven device for performing immunoassays in a chip format.

Guijt et al. [37] reviewed the application of electrokinetically controlled microdevices based on affinity interactions, and they concluded that this approach was suitable for developing devices capable of high-throughput screening and routine biomedical analysis. Today, several commercial enterprises (e.g., Agilent Technologies and Caliper) are applying electrokinetic technology to the development of analytical instruments.

28.3.2.2   Chromatographic Devices

Chromatographic microdevices for analysis of proteins, peptides, and other biochemical entities are becoming increasingly popular. This is especially true for devices based on a chip design [38]. However, investigators have been building microchromatography systems for some years, and such systems have been applied to a number of different fields. This section provides a summary of these applications.

28.3.2.2.1 Applications of Microchromatography. In 1989, Kennedy and colleagues [39] reported the first application of microcolumn technology for the quantitative analysis of multiple analytes in single cells. These authors used this approach to analyze amino acids and neurotransmitters in single neurons by using nanoliter and subnanoliter samples. Ewing [40] described another open-tube liquid chromatographic technique for analyzing cellular and subcellular components in the picoliter- and femtoliter-volume range. Later, the Kennedy group reported the application of a microcolumn immunoassay to the determination of insulin secretion from a single islet of Langerhans [41]. This system used a 150-μm I.D. capillary column with protein G-immobilized anti-insulin antibodies as the capture ligands. The eluate from this column was analyzed by reversed-phase chromatography in a capillary column.

Other affinity techniques have also employed microcolumns. For instance, Zhang et al. [42] immobilized a mushroom lectin onto porous glass beads and packed these into approximately 10-μl columns. In the Phillips laboratory, a recycling immunoaffinity array consisting of 30 linked 10 mm × 150 μm I.D. fused capillary columns (packed with 10-μm-diameter antibody-coated glass beads) was used to assess the immunological status of clinical subjects using multianalyte analysis [43].

Wang et al. [44] employed a plastic microfluidic system containing microchannels. In this system, bovine serum albumin was adsorbed onto a porous poly(vinylidene fluoride) (PVDF) membrane and sandwiched between the microchannel plates for affinity-based chiral separations of racemic tryptophan and thiopental mixtures.

Microchannels packed with microspheres have also been used for bioaffinity separations. The developers of this device claim that it has the potential to be applied to a number of affinity assays, including the study of protein complexes and multiple analytes [45].

A similar microchannel device employing 50-μm-I.D. channels and coupled with laser-induced fluorescence detection is being developed in the Phillips laboratory, as shown in Figure 28.5. In this device, each channel has a different antibody F_{ab} fragment immobilized directly onto the internal walls of the channel. This makes it possible to simultaneously analyze 20 different analytes in a 20-pl sample.

28.3.2.2.2 Design of Components for Microchromatography. Recent advances in applying microfabrication techniques to the design and manufacture of miniature chromatography components have greatly advanced the field of microchromatography. However, many of these devices have not been applied to affinity and immunoaffinity separations.

Figeys and Aebersold [46] described a functional solvent-gradient system capable of delivering nanoliter/minute flow rates in microchannel assemblies. These gradients were generated through computer control of differential electroosmotic pumping between the solvent reservoirs. Although this has not been used for affinity separations, this device holds great potential for future developments in that field. As an example, in this report the authors demonstrated that this device was capable of determining protein and peptide digests at fM concentrations or even lower when using a reversed-phase microcolumn and electrospray ionization with an ion-trap mass spectrometer. Another report described the use of soft lithography to fabricate multilayer devices that could be used as valves and pumps in micro- and nanochromatography systems [47].

One major obstacle to the success of microchromatography systems is the gradient mixing that occurs in microchannels. This has become a serious impairment to successful separation in microdevices. However, an ingenious technique that could potentially resolve this issue was reported by the Whitesides group [48]. These investigators used base-relief structures fabricated into the floor of microchannels to produce a passive mixing technique that could revolutionize the efficiency of gradient microfluidic devices.

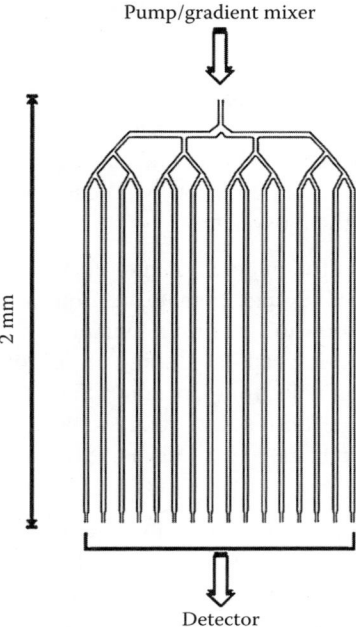

Pump/gradient mixer

2 mm

Detector

Figure 28.5 A 20-column microchannel (50-μm wide × 50-μm deep) chromatographic device employed for multianalyte affinity or immunoaffinity chromatography.

The Regnier group at Purdue has described microfabricated columns for electrochromatography, although their design also has potential for affinity applications [49]. These columns were initially produced by the *in situ* micromachining of a quartz wafer. This wafer contained a stationary phase that was bonded to $5 \times 5 \times 10$ μm collocated monolith support structures separated by 1.5-μm wide rectangular channels. The authors of this report estimated that the volume of a 150 μm × 4.6 cm column made by this approach was 18 nl. Recently, the same group fabricated columns in PDMS to form a relatively cheap, disposable chromatography system. When used in capillary electrochromatography, these columns gave efficiencies of 4×10^5 plates/m [50].

For investigators interested in investing in true micro- or nanochromatography systems, the most productive approach is to become involved in collaborative studies with investigators who have access to microfabrication facilities. As an alternative, there are a number of university-based microfabrication units that are prepared to manufacture microdevices on a contractual basis. The National Nanofabrication Users Network (www.nnun.org) is a useful reference site for investigators interested in obtaining information regarding microfabrication and registered sites where such technology exists.

Apart from true nanofabricated devices, most laboratories can develop and construct useful microassemblies that consist of microcolumns, mixers, pumps, and detectors from commercial sources. One of the most useful of these sources is Upchurch Scientific (Oak Harbor, WA). As an alternative to an on-line detector in these devices, sample fractions could also be collected and read in a microspectrophotometer or spectrometer.

28.3.2.2.3 Pumps. Syringe pumps are adequate for most microtechniques. This is especially true for those devices equipped with serial ports for computer control, such as those used for microdialysis or stopped-flow analysis. This feature also allows the system to be automated.

Dual pumps are required if gradient elution is to be incorporated into the system. In such cases, a mixing tee can be constructed using a microtee or a preassembled micromixer called a Nanomixer that is commercially available from Upchurch Scientific. In the Phillips laboratory, dual microdialysis syringe pumps from CMA Microdialysis (Chelmsford, MA) and dual infusion syringe pumps from Harvard Apparatus, Inc. (Holliston, MA) have been successfully used as pumps for gradient microchromatography systems.

28.3.2.2.4   Injection Systems.   Simple injectors for microchromatography systems can be constructed using low-pressure switching valves and loops made from capillary tubing. However, such loops will require calibration to determine their injection volume. A convenient way to make injection loops is by using capillary tubing made from PEEK (386 μm O.D., 50 to 100 μm I.D.), which is available from Upchurch Scientific and compatible in size to fused-silica capillary tubing. For those not wishing to manufacture their own injectors, one is available from Upchurch; however, this is reasonably expensive, and one still has to make the injection loop.

28.3.2.2.5   Microcolumn Construction.   Microcolumns can easily be made by packing laboratory-made affinity matrices into small-bore guard columns, especially those available for capillary columns. Microcolumns can also be constructed from polyamide-coated fused-silica capillary tubing, which is available in a wide range of internal diameters [43, 51]. End frits for these columns are usually made by annealing packed silica or glass particles at one end of the column (see Figure 28.6). However, this procedure can be easily

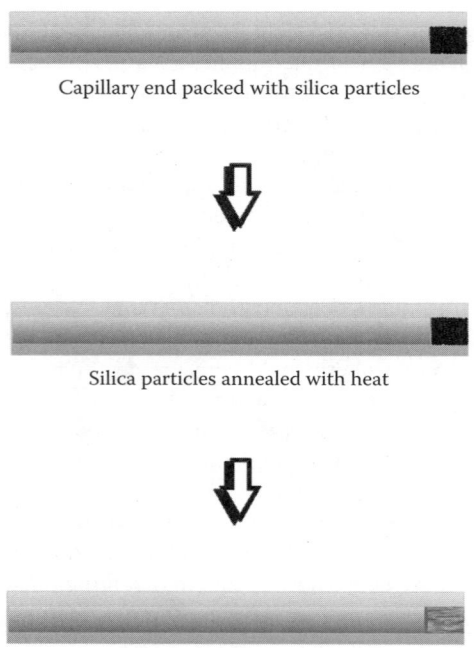

Capillary end packed with silica particles

Silica particles annealed with heat

Annealed particles form a porous frit

Figure 28.6   Preparation of microaffinity column frits by annealing silica particles packed into one end of a fused-silica capillary.

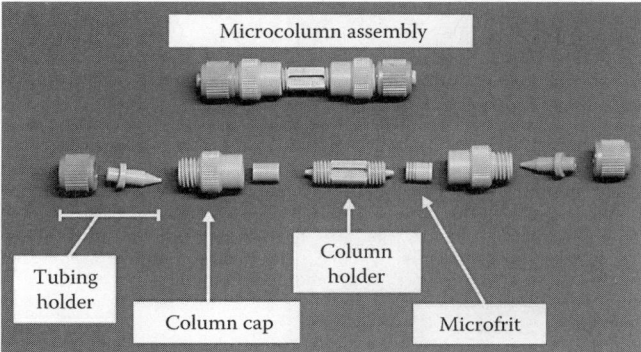

Figure 28.7 An assembled Upchurch Scientific microcolumn (top of photograph) and the components used in its assembly (middle of photograph).

performed by buying a capillary column holder capable of compressing a ready-made frit onto the end of the column. Such devices are available from Upchurch, complete with end fittings (Figure 28.7).

28.3.2.2.6 Building a Microchromatography System. Detection of the separated analytes is a constant challenge in microseparations, but there are commercially available instruments capable of performing this task. Most detectors equipped to perform on-line measurements in CE are adequate for microanalytical systems. However, there are also specialized detectors that can be used, such as a fluorescence detector that is available from CMA Microdialysis and a laser-induced fluorescence detector (the ZetaLIF) from Picometrics (available through ESA, Chelmsford, MA). Alternatively, as mentioned earlier, fractions of the column eluant can be collected and measured in an off-line detector.

In the Phillips laboratory, two microanalytical devices are in operation: a microimmunoaffinity chromatography system (Figure 28.8a) and an immunoaffinity array (Figure 28.8b). Both instruments were built with components from Upchurch Scientific, using capillary PEEK tubing throughout the first system and a combination of PEEK and fused-silica capillary tubing in the second [43]. In the first system, the neuropeptide calcitonin gene-related peptide can be isolated from 5 nl of cerebral spinal fluid (CSF) samples and detected in less than 2 min using postcolumn fluorescence detection (see Figure 28.9). The application of the microcolumn array to multiple analytes is illustrated in Figure 28.10. In this case, a patient's immunological status was assessed using a 20-column array and a scanning fluorescence detector. The detector in this was a converted DNA chip reader that was adapted to read the 1 cm × 2.5 cm column array. With this device, the captured analytes were detected *in situ*, and the array was regenerated using acid elution [43].

28.3.2.3 Lab-on-a-Chip

One of the ideals of microanalysis is the design of a complete laboratory system on a microfabricated platform or "chip." In the past 2 to 3 years, such a platform has become known as a "lab-on-a-chip," and these range from simple electrokinetically driven devices to complete chromatographic systems (see Figure 28.11). The Regnier group at Purdue University was perhaps the first to review the idea that complete chromatography and electrophoresis systems could be integrated into a microchip format [11]. Since then, several investigators have elaborated on the potential of this technology in both the analytical and medical sciences [3, 52–54].

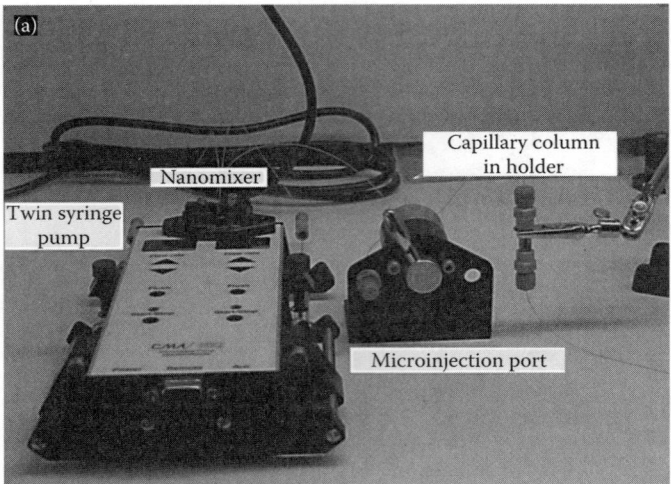

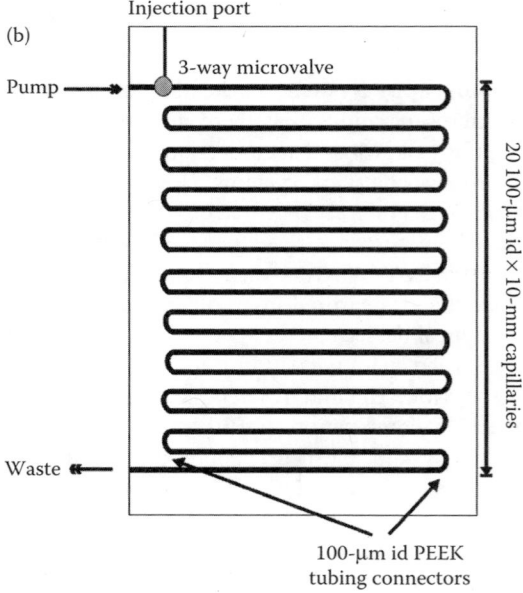

Figure 28.8 (a) A microimmunoaffinity chromatographic system consisting of a twin-microdialysis pump, a nanomixer, microinjection port, and capillary column in its holder. All parts shown here are commercially available from Upchurch Scientific. (b) A microcolumn immunoaffinity array containing 20 10-mm capillaries connected by PEEK capillary tubing to form a continuous serpentine pattern. Each capillary contains a different antibody immobilized to its internal surface. Samples, following injection, are pumped through the array, allowing each column to capture and remove its specific analyte. Detection is achieved by using a scanning fluorescence monitor.

Throughout the development of these devices, affinity and immunoaffinity capture systems have been central to obtaining good selectivity and specificity. In one example, Chiem and Harrison [10] developed a glass-based chip for performing a competitive immunoassay of the drug theophylline in human serum samples. This device handled the

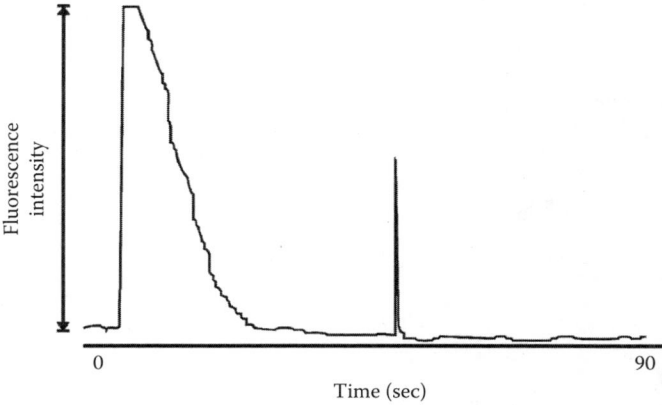

Figure 28.9 Immunoaffinity chromatogram of calcitonin gene-related peptide isolation from a 5-nl sample of cerebral spinal fluid using the system illustrated in Figure 28.8a. The first peak represents the nonreactive material, and the second peak represents the retained analyte following acid elution. This system used LIF detection at 633 nm and an application solvent that consisted of a pH 7.4, 0.01 M phosphate buffer.

complete analytical process from reagent mixing to fluorescence detection and required only approximately 50 nl of sample.

Even though affinity and immunoaffinity lab-on-a-chip technology has not yet gained its full potential, this situation should change in the foreseeable future. One of the major obstacles to this is the need for techniques that can immobilize affinity reagents in microfabricated devices and maintain their bioactivity. Annealing at high temperatures is the usual approach for sealing microdevices, but in most cases this damages or destroys the bioactivity of the immobilized ligand. Thus, the development of gentler sealing

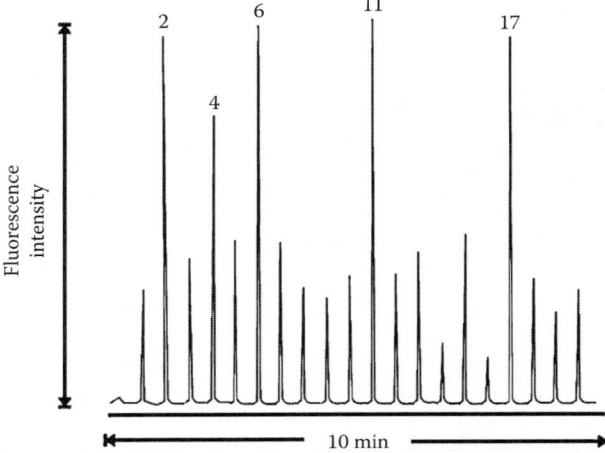

Figure 28.10 Assessment of the immune status of a human subject through the use of a 20-microcolumn anticytokine immunoaffinity array similar to the system illustrated in Figure 28.8b. Peaks 2, 4, 6, 11, and 17 in this result represent elevations in interleukin (IL)-2, -3, -12, -15, and gamma interferon, generally indicating activation of a cell-mediated immune response.

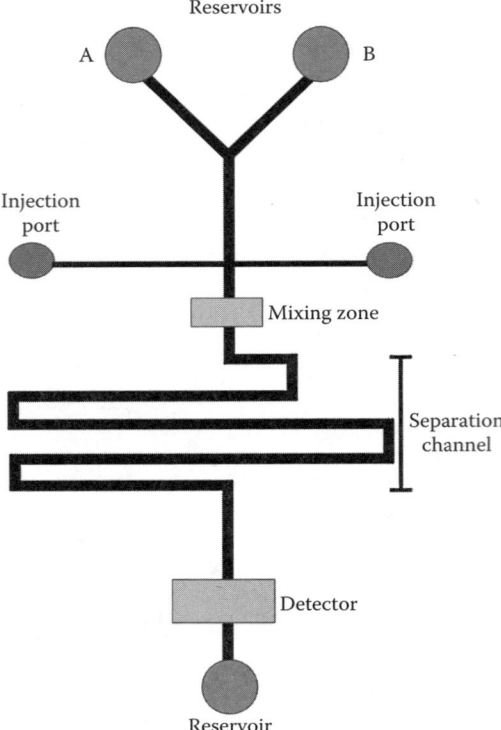

Figure 28.11 Schematic of an affinity chromatographic system based on a "lab-on-a-chip" format. Buffer solutions are placed in reservoirs A and B so that gradients can be formed using the mixing zone prior to the introduction of a sample via the injection ports. The separation channel is packed with lectin-coated glass beads secured by a glass frit that is placed immediately before the detector flow cell. The sample is injected by osmotic pumping between the two ports with the running buffers, with the separation and elution gradient being controlled by a current alternating between reservoirs A and B and terminating in the lower reservoir.

processes is required before affinity techniques can advance. One such approach is the use of cements and light-annealing glues, similar to those used in the optical industry.

There are, however, several reports of immunoassays being performed in microfabricated devices. Wang et al. [55] described a microfluidic device capable of performing electrochemical enzyme immunoassays. This lab-on-a-chip integrated (1) precolumn mixing of an enzyme-labeled antibody with its appropriate antigen, (2) electrophoretic separation of the reaction products, and (3) amperometric detection of 4-aminophenol, following the postseparation reaction of the enzyme label with 4-aminophenyl phosphate.

28.4   DETECTION SYSTEMS

A major problem associated with the diminishing size of micro- and nanodevices is the need to detect the analytes once they are separated. Lessons on how a variety of molecules can be measured on-line in small volumes have been learned from capillary electrophoresis (CE). Even though this information does not always apply to nanodevices, it can be of great help in solving detection problems in microdevices. Basically, there are four major on-line detection systems in CE: absorbance, fluorescence, chemiluminescence, and electrochemical detection, each of which is suitable for different types of analysis [56]. This section considers each of these detection modes and looks at how they have been used in micro- or nanoscale devices.

28.4.1 Ultraviolet Absorption

Ultraviolet absorption is useful for monitoring relatively high concentrations of proteins, peptides, and other molecules in HPLC and CE. However, its limits of detection have been found to be generally inadequate for most micro- and nanoscale analysis. Thus, further consideration will instead be given to the other common modes of detection in CE.

28.4.2 Fluorescence

Perhaps the most popular approaches for detection in microanalytical systems are fluorescence [57] and laser-induced fluorescence (LIF) [58–62]. Fluorescence detection can be performed with fixed-wavelength or scanning detectors, but LIF is becoming more popular, especially when using near-infrared diode lasers as the excitation source [63–66]. These latter lasers are relatively cheap and have quite long lifetimes. Simple LIF detectors can be built in any laboratory based on small (5- to 12-mW) diode lasers with emission lines at 633 and 650 nm. Figure 28.12 illustrates the basic design of a laboratory-built LIF detector. The use of such lasers has been supported by the availability of a number of fluorochromes that are capable of being excited by these lasers (e.g., Cyanine-5 or Cy-5 from Amersham Pharmacia Biotechnology and Alexa 633 from Molecular Probes; see Table 28.4).

One of the appeals of fluorescence detection is its versatility and the potential for incorporating a LIF detector directly into a chip-based device. Oldenburg et al. [67] developed a wavelength-resolved fluorescence detector using a postcolumn sheath detector and a charge-injection-device array detector. This provided a limit of detection (LOD) of 4.8×10^{-11} M for fluorescein isothiocyanate. Legendre et al. [63] coupled a near-infrared diode laser with a single-photon avalanche diode as the detector. This combination enabled the investigators to detect the dye IR-132 and various amino acids in the low zM range. In addition, both the groups of Harrison and Phillips have used a scanning fluorescence detector to measure multiple analytes in a capillary-based immunoassay system [43, 66, 68].

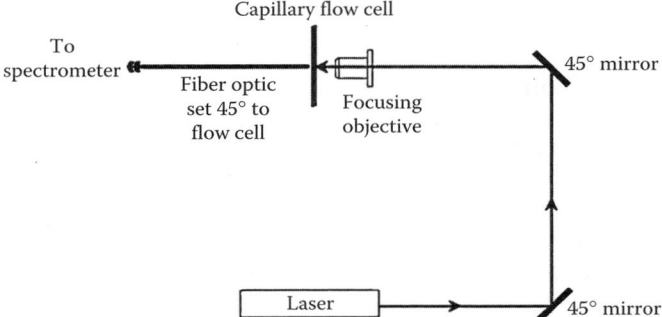

Figure 28.12   Schematic of a simple laboratory-built LIF detector consisting of a 633-nm, 10-mW laser focused onto a capillary flow cell by a 10× objective and deflected by two 45° mirrors. The mirrors are used to aid in accurate focusing of the laser beam onto the capillary flow cell. Light emitting from the labeled analyte is collected by a 600-μm fiber placed at 45° to the flow cell. The light is transferred via the fiber to either a spectrometer or photomultiplier tube.

Table 28.4 Fluorophores Useful for Detection in Microfabricated Devices

Fluorophore	Absorbance Wavelength (nm)	Emission Wavelength (nm)	Supplier
Alexa 633	632	649	Molecular Probes
Alexa 647	650	668	Molecular Probes
Alexa 660	664	689	Molecular Probes
Bodipy 630	625	642	Molecular Probes
Cy 5	649	670	Amersham-Pharmacia Biotechnology
Cy 5.5	675	694	Amersham-Pharmacia Biotechnology
SYTO 17	621	634	Molecular Probes
Texas red-X	595	615	Molecular Probes Research Organics
TO-PRO-3	613	629	Molecular Probes

28.4.3 Integrated Detection Systems

One of the advantages of microfabrication is its ability to integrate optical components directly into a chip. Such components can range from a microlens to avalanche microphotodiode detectors and microlasers or other light-emitting sources. Chabinyc et al. [69] fabricated an integrated fluorescence detector into a PDMS microfluidic device. This used an optical fiber to deliver the light for fluorescence excitation and an avalanche microphotodiode as the detector.

Another system [70] utilized integrated photodiodes in a silicone substrate to fabricate an on-chip fluorescence detector. This device was capable of detection at the femtogram level using SYBR green I dye for DNA restriction fragments. Although many of these detectors have not been applied to affinity-based separations, they are good examples of where the field is headed and the different types of devices that can be fabricated.

28.4.4 Chemiluminescence

Chemiluminescence is another detection system suitable for micro- and nanodevices. This approach has been applied to CE for a number of different analyses. However, no chemiluminescence detector has yet been integrated into a working microdevice.

28.4.5 Electrochemical Detection

Electrochemical detection has also enjoyed success in CE, with these techniques usually employing a sheath or modifications to the CE outlet. Electrochemical detection has the advantage of giving a signal that is proportional to the concentration of analytes even in extremely small volumes, a characteristic that has made this form of detection compatible with CE and other microtechniques. One particularly interesting application of electrochemical detection has been its use for measuring analytes in single cells [71, 72].

Electrochemical detection can be used in conjunction with immunoassays. For instance, the product 4-aminophenyl can be measured electrochemically following reduction of a 4-aminophenyl phosphate by the enzyme alkaline phosphatase. An example of this is an electrochemical enzyme-based immunoassay that has been described for the measurement of macrolide antibiotics in animal meats [73]. With this assay, it was reported that the system exhibited excellent specificity for two drugs and exhibited a limit of detection of 400 μg/ml for each analyte.

Electrochemical detectors have been integrated into chip formats, usually in conjunction with CE. The Lunte group has developed a PDMS-based device incorporating a dual-electrode system [74] for measuring catechol in test substrates. This device was capable of detecting catechol down to levels of 4 μM. The same group later reported another dual-electrode PDMS chip-based CE system using carbon paste to construct the electrodes [75]. The advantage of this design was that catalysts could be incorporated into the electrode, thus allowing for adjustable selectivity.

Wang et al. [76] reported the fabrication of a disposable PMMA chip-based system equipped with a thick-film electrochemical detector integrated into the system. This suggested that the production of this low-cost apparatus could lead to disposable analytical tools. Previously, these researchers reported the development of a chip-based electrochemical enzyme immunoassay using a postcolumn reactor to measure mouse IgG. This approach was capable of detecting 1.7×10^{-18} M (or 1.7 ag/ml) of IgG in a model system [55]. A further review concerning the use of electrochemical detectors in microfluidic devices has been written by Rossier et al. [24].

28.4.6 Mass Spectrometry

The coupling of mass spectrometry (MS) to microanalytical devices has been available for some time, and several groups have developed devices for coupling CE to MS. One of the most prolific advocates of immuno-concentrator-CE analysis coupled to MS is the Guzman group, which has successfully integrated immunoaffinity separations on a microscale with CE analysis and MS detection [51, 77, 78].

Microfabricated interfaces have also been described that use either a capillary to introduce a sample from the chip to a mass spectrometer or direct integration of the chip with MS. The Karger group [79] has designed a microfabricated and miniaturized subatmospheric electrospray interface for coupling a chip-based CE with an MS. Likewise, Kim and Knapp [80, 81] have described MS electrospray emitters fabricated by soft lithography in PDMS. This device was shown to provide a limit of detection of 1 μM at a signal-to-noise ratio (S/N) of 18. Further information on the coupling of various scales of liquid-phase separations with MS can be found in a review by Gelpi [82].

The eventual incorporation of MS as an integral part of an analytical chip is in the foreseeable future. Mass spectrometers are also becoming miniaturized, especially iontrap and time-of-flight spectrometers [83–86]. However, at the present time, these are at the micro level rather than the ultramicro or nano stage. These current instruments are being developed for field applications (e.g., for use on space shuttles and space stations where laboratory space is limited) rather than for general analytical use. A drawback to these instruments is their sensitivity, which at present has not been fully resolved.

28.5 SUMMARY AND CONCLUSIONS

This chapter has discussed the new horizons in nanotechnology that are actively being developed and has considered their possible use with affinity chromatographic techniques. Improvements in the manufacturing of microdevices on both silica wafers and "disposable" plastic substrates have greatly advanced the introduction of these devices into analytical science. In addition, developments in small-scale columns, mixers, pumps, and detectors have enhanced the availability and use of micro- and nanoscale devices.

Basic and clinical research has shown a growing need for the miniaturization of analytical techniques. This includes a need for lower sample consumption, lower reagent costs, and increased detection sensitivity. Micro- and nanoscale systems can be applied not only to situations where standard analytical techniques are employed, but also to the analysis of samples too small for conventional methods. Examples of such applications are the analysis of single cells and their secretions, studies involving precious archival samples, and the analysis of samples from neonatal patients.

The availability of commercial components for the construction of micro-HPLC systems and the rise of custom service microfabrication centers has opened the door for this technology to be employed in any laboratory. The chapter also described some practical issues regarding the construction of laboratory-built systems and discussed the types of available detection systems.

The application of affinity and immunoaffinity ligands to both micro- and nanoscale chromatographic systems has been slow to emerge, but several groups are now reporting both basic research and clinical studies that have been performed on microscale equipment or devices. The introduction of these miniature devices holds great promise and could revolutionize affinity and immunoaffinity chromatography in the near future.

SYMBOLS AND ABBREVIATIONS

a	Atto (10^{-18} units)
AFM	Atomic force microscopy
ag	Attogram
CAD	Computer-assisted design
CCD	Charge-coupled device
CE	Capillary electrophoresis
CSF	Cerebral spinal fluid
Cy-5	Cyanine-5 dye
f	Femto (10^{-15} units)
F_{ab}	Fragment antibody
fl	Femtoliter
fM	Femtomolar
I.D.	Internal diameter
IgG	Immunoglobulin G
LIF	Laser-induced fluorescence
LIGA	*Lithographie, Galvoformung*, and *Abformung*
LOD	Limit of detection
m	Milli (10^{-3} units)
MEMS	Microelectromechanical system
MS	Mass spectrometry
mW	Milliwatt
n	Nano (10^{-9} units)
nl	Nanoliter
nM	Nanomolar
nm	Nanometer
O.D.	Outer diameter
p	Pico (10^{-12} units)
PDMS	Polydimethylsiloxene
PEEK	Poly ether-ether ketone
PET	Polyethylene terephthalate
pl	Picoliter
PMMA	Polymethylmethacrylate

PVDF	Poly(vinylidene fluoride)
S/N	Signal-to-noise ratio
UV	Ultraviolet
y	Yocto (10^{-24} units)
z	Zepto (10^{-21} units)
zM	Zeptomolar
μ	Micro (10^{-6} units)
μl	Microliter
μM	Micromolar
μm	Micrometer
μTAS	Micro-total analytical system

REFERENCES

1. Voldman, J., Gray, M.L., and Schmidt, M.A., Microfabrication in biology and medicine, *Ann. Rev. Biomed. Eng.,* 1, 401–425, 1999.
2. Jakeway, S.C., de Mello, A.J., and Russell, E.L., Miniaturized total analysis systems for biological analysis, *Fresenius J. Anal. Chem.,* 366, 525–539, 2000.
3. Kricka, L.J., Microchips, microarrays, biochips and nanochips: Personal laboratories for the 21st century, *Clin. Chim. Acta,* 307, 219–223, 2001.
4. Chovan, T. and Guttman, A., Microfabricated devices in biotechnology and biochemical processing, *Trends Biotechnol.,* 20, 116–122, 2002.
5. Verpoorte, E., Microfluidic chips for clinical and forensic analysis, *Electrophoresis,* 23, 677–712, 2002.
6. Mouradian, S., Lab-on-a-chip: Applications in proteomics, *Curr. Opin. Chem. Biol.,* 6, 51–56, 2002.
7. Knezevic, V., Leethanakul, C., Bichsel, V.E., Worth, J.M., Prabhu, V.V., Gutkind, J.S., Liotta, L.A., Munson, P.J., Petricoin, E.F., and Krizman, D.B., Proteomic profiling of the cancer microenvironment by antibody arrays, *Proteomics,* 1, 1271–1278, 2001.
8. Cahill, D.J., Protein and antibody arrays and their medical applications, *J. Immunol. Methods,* 250, 81–91, 2001.
9. de Wildt, R.M., Mundy, C.R., Gorick, B.D., and Tomlinson, I.M., Antibody arrays for high-throughput screening of antibody-antigen interactions, *Nature Biotech.,* 18, 989–994, 2000.
10. Chiem, N. and Harrison, D.J., Microchip-based capillary electrophoresis for immunoassays: Analysis of monoclonal antibodies and theophylline, *Anal. Chem.,* 69, 373–378, 1997.
11. Regnier, F.E., He, B., Lin, S., and Busse, J., Chromatography and electrophoresis on chips: Critical elements of future integrated, microfluidic analytical systems for life science, *Trends Biotechnol.,* 17, 101–106, 1999.
12. Dodge, A., Fluri, K., Verpoorte, E., and de Rooij, N.F., Electrokinetically driven microfluidic chips with surface-modified chambers for heterogeneous immunoassays, *Anal. Chem.,* 73, 3400–3409, 2001.
13. Becker, H. and Gartner, C., Polymer microfabrication methods for microfluidic analytical applications, *Electrophoresis,* 21, 12–26, 2000.
14. Chen, Y. and Pepin, A., Nanofabrication: conventional and nonconventional methods, *Electrophoresis,* 22, 187–207, 2001.
15. Sun, S., Chong, K.S., and Leggett, G.J., Nanoscale molecular patterns fabricated by using scanning near-field optical lithography, *J. Am. Chem. Soc.,* 124, 2414–2415, 2002.
16. Vasile, M.J., Nassar, R., Xie, J., and Guo, H., Microfabrication techniques using focused ion beams and emergent applications, *Micron,* 30, 235–244, 1999.
17. Ford, S.M., Davies, J., Kar, B., Qi, S.D., McWhorter, S., Soper, S.A., and Malek, C.K., Micromachining in plastics using X-ray lithography for the fabrication of microelectro-phoresis devices, *J. Biomech. Eng.,* 121, 13–21, 1999.

18. Anderson, J.R., Chiu, D.T., Jackman, R.J., Cherniavskaya, O., McDonald, J.C., Wu, H., Whitesides, S.H., and Whitesides, G.M., Fabrication of topologically complex three-dimensional microfluidic systems in PDMS by rapid prototyping, *Anal. Chem.*, 72, 3158–3164, 2000.

19. McDonald, J.C., Duffy, D.C., Anderson, J.R., Chiu, D.T., Wu, H., Schueller, O.J., and Whitesides, G.M., Fabrication of microfluidic systems in poly(dimethylsiloxane), *Electrophoresis*, 21, 27–40, 2000.

20. Whitesides, G.M., Ostuni, E., Takayama, S., Jiang, X., and Ingber, D.E., Soft lithography in biology and biochemistry, *Ann. Rev. Biomed. Eng.*, 3, 335–373, 2001.

21. McDonald, J.C., Metallo, S.J., and Whitesides, G.M., Fabrication of a configurable, single-use microfluidic device, *Anal. Chem.*, 73, 5645–5650, 2001.

22. Deng, T., Wu, H., Brittain, S.T., and Whitesides, G.M., Prototyping of masks, masters, and stamps/molds for soft lithography using an office printer and photographic reduction, *Anal. Chem.*, 72, 3176–3180, 2000.

23. Xia, Y. and Whitesides, G.M., Extending microcontact printing as a microlithographic technique, *Langmuir*, 13, 2059–2067, 1997.

24. Rossier, J., Reymond, F., and Michel, P.E., Polymer microfluidic chips for electrochemical and biochemical analysis, *Electrophoresis*, 23, 858–867, 2002.

25. Zhao, X.-M., Xia, Y., and Whitesides, G.M., Fabrication of three-dimensional micro-structures: Microtransfer molding, *Adv. Mater.*, 8, 837–840, 1996.

26. Brown, P.O. and Botstein, D., Exploring the new world of the genome with DNA microarrays, *Nat. Genet.*, 21 Suppl., 33–37, 1999.

27. Schena, M., Ed., *Microarray Biochip Technology*, Eaton Publishing, Natick, MA, 2000.

28. Adam, B.L., Vlahou, A., Semmes, O.J., and Wright, G.L., Proteomic approaches to biomarker discovery in prostate and bladder cancers, *Proteomics*, 1, 1264–1270, 2001.

29. Moody, M.D., Van Arsdell, S.W., Murphy, K.P., Orencole, S.F., and Burns, C., Array-based ELISAs for high-throughput analysis of human cytokines, *BioTechniques*, 31, 186–194, 2001.

30. Huang, R.P., Simultaneous detection of multiple proteins with an array-based enzyme-linked immunosorbent assay (ELISA) and enhanced chemiluminescence (ECL), *Clin. Chem. Lab. Med.*, 39, 209–214, 2001.

31. Lin, S.C., Tseng, F.G., Huang, H.M., Huang, C.Y., and Chieng, C.C., Microsized 2D protein arrays immobilized by micro-stamps and micro-wells for disease diagnosis and drug screening, *Fresenius J. Anal. Chem.*, 371, 202–208, 2001.

32. Jones, V.W., Kenseth, J.R., Porter, M.D., Mosher, C.L., and Henderson, E., Microminiatur-ized immunoassays using atomic force microscopy and compositionally patterned antigen arrays, *Anal. Chem.*, 70, 1233–1241, 1998.

33. Goodey, A., Lavigne, J.J., Savoy, S.M., Rodriguez, M.D., Curey, T., Tsao, A., Simmons, V.G., Wright, J., Yoo, S.J., Sohn, Y., Anslyn, E.V., Shear, J.B., Neikirk, D.P., and McDevitt, J.T., Development of multianalyte sensor arrays composed of chemically derivatized poly-meric microspheres localized in micromachined cavities, *J. Am. Chem. Soc.*, 123, 2559–2570, 2001.

34. Bruin, G.J., Recent developments in electrokinetically driven analysis on microfabricated devices, *Electrophoresis*, 21, 3931–3951, 2000.

35. Bousse, L., Cohen, C., Nikiforov, T., Chow, A., Kopf-Sill, A.R., Dubrow, R., and Parce, J.W., Electrokinetically controlled microfluidic analysis systems, *Annu. Rev. Biophys. Bio-mol. Struct.*, 29, 155–181, 2000.

36. Linder, V., Verpoorte, E., Thormann, W., de Rooij, N.F., and Sigrist, H., Surface biopassi-vation of replicated poly(dimethylsiloxane) microfluidic channels and application to heter-ogeneous immunoreaction with on-chip fluorescence detection, *Anal. Chem.*, 73, 4181–4189, 2001.

37. Guijt, R.M., Baltussen, E., and van Dedem, G.W., Use of bioaffinity interactions in electro-kinetically controlled assays on microfabricated devices, *Electrophoresis*, 23, 823–835, 2002.

38. Regnier, F.E., He, B., Lin, S., and Busse, J., Chromatography and electrophoresis on chips: Critical elements of future integrated, microfluidic analytical systems for life science, *Trends Biotechnol.*, 17, 101–106, 1999.

39. Kennedy, R.T., Oates, M.D., Cooper, B.R., Nickerson, B., and Jorgenson, J.W., Microcolumn separations and the analysis of single cells, *Science,* 246, 57–63, 1989.

40. Ewing, A.G., Microcolumn separations of single nerve cell components, *J. Neurosci. Methods,* 48, 215–224, 1993.

41. Shen, H., Aspinwall, C.A., and Kennedy, R.T., Dual microcolumn immunoassay applied to determination of insulin secretion from single islets of Langerhans and insulin in serum, *J. Chromatogr. B,* 689, 295–303, 1997.

42. Zhang, B., Palcic, M.M., Mo, H., Goldstein, I.J., and Hindsgaul, O., Rapid determination of the binding affinity and specificity of the mushroom *Polyporus squamosus* lectin using frontal affinity chromatography coupled to electrospray mass spectrometry, *Glycobiology,* 11, 141–147, 2001.

43. Phillips, T.M., Multi-analyte analysis of biological fluids with a recycling immunoaffinity column array, *J. Biochem. Biophys. Methods,* 49, 253–262, 2001.

44. Wang, P.C., Gao, J., and Lee, C.S., High-resolution chiral separation using microfluidics-based membrane chromatography, *J. Chromatogr. A,* 942, 115–122, 2002.

45. Buranda, T., Huang, J., Perez-Luna, V.H., Schreyer, B., Sklar, L.A., and Lopez, G.P., Biomolecular recognition on well-characterized beads packed in microfluidic channels, *Anal. Chem.,* 74, 1149–1156, 2002.

46. Figeys, D. and Aebersold, R., Nanoflow solvent gradient delivery from a microfabricated device for protein identifications by electrospray ionization mass spectrometry, *Anal. Chem.,* 70, 3721–3727, 1998.

47. Unger, M.A., Chou, H.P., Thorsen, T., Scherer, A., and Quake, S.R., Monolithic microfabricated valves and pumps by multilayer soft lithography, *Science,* 288, 113–116, 2000.

48. Stroock, A.D., Dertinger, S.K., Ajdari, A., Mezic, I., Stone, H.A., and Whitesides, G.M., Chaotic mixer for microchannels, *Science,* 295, 647–651, 2002.

49. He, B., Tait, N., and Regnier, F.E., Fabrication of nanocolumns for liquid chromatography, *Anal. Chem.,* 70, 3790–3797, 1998.

50. Slentz, B.E., Penner, N.A., Lugowska, E., and Regnier, F.E., Nanoliter capillary electro-chromatography columns based on collocated monolithic support structures molded in poly(dimethyl siloxane), *Electrophoresis,* 22, 3736–3743, 2001.

51. Guzman, N.A. and Stubbs, R.J., The use of selective adsorbents in capillary electrophoresis-mass spectrometry for analyte preconcentration and microreactions: A powerful three-dimensional tool for multiple chemical and biological applications, *Electrophoresis,* 22, 3602–3628, 2001.

52. Stephenson, J., Lab-on-a-chip shows promise in defining and diagnosing cancers, *J.A.M.A.,* 282, 1801–1802, 1999.

53. Figeys, D. and Pino, D., Lab-on-a-chip: A revolution in biological and medical sciences, *Anal. Chem.,* 72, 330A–335A, 2000.

54. Krishnan, M., Namasivayam, V., Lin, R., Pal, R., and Burns, M.A., Microfabricated reaction and separation systems, *Curr. Opin. Biotechnol.,* 12, 92–98, 2001.

55. Wang, J., Ibanez, A., Chatrathi, M.P., and Escarpa, A., Electrochemical enzyme immunoassays on microchip platforms, *Anal. Chem.,* 73, 5323–5327, 2001.

56. Swinney, K. and Bornhop, D.J., Detection in capillary electrophoresis, *Electrophoresis,* 21, 1239–1250, 2000.

57. Landers, J.P., Oda, R.P., Spelsberg, T.C., Nolan, J.A., and Ulfelder, K.J., Capillary electrophoresis: A powerful microanalytical technique for biologically active molecules, *BioTechniques,* 14, 98–111, 1993.

58. MacTaylor, C.E. and Ewing, A.G., Critical review of recent developments in fluorescence detection for capillary electrophoresis, *Electrophoresis,* 18, 2279–2290, 1997.

59. Bergquist, J., Gilman, S.D., Ewing, A.G., and Ekman, R., Analysis of human cerebrospinal fluid by capillary electrophoresis with laser-induced fluorescence detection, *Anal. Chem.,* 66, 3512–3518, 1994.

60. Gilman, S.D. and Ewing, A.G., Analysis of single cells by capillary electrophoresis with on-column derivatization and laser-induced fluorescence detection, *Anal. Chem.,* 67, 58–64, 1995.

61. Tao, L. and Kennedy, R.T., On-line competitive immunoassay for insulin based on capillary electrophoresis with laser-induced fluorescence detection, *Anal. Chem.*, 68, 3899–3906, 1996.

62. Ferrance, J. and Landers, J.P., Exploiting sensitive laser-induced fluorescence detection on electrophoretic microchips for executing rapid clinical diagnostics, *Luminescence*, 16, 79–88, 2001.

63. Legendre, B.L., Moberg, D.L., Williams, D.C., and Soper, S.A., Ultrasensitive near-infrared laser-induced fluorescence detection in capillary electrophoresis using a diode laser and avalanche photodiode, *J. Chromatogr. A*, 779, 185–194, 1997.

64. Jiang, G., Attiya, S., Ocvirk, G., Lee, W.E., and Harrison, D.J., Red diode laser induced fluorescence detection with a confocal microscope on a microchip for capillary electrophoresis, *Biosens. Bioelectron.*, 14, 861–869, 2000.

65. McWhorter, S. and Soper, S.A., Near-infrared laser-induced fluorescence detection in capillary electrophoresis, *Electrophoresis*, 21, 1267–1280, 2000.

66. Phillips, T.M., Analysis of single-cell cultures by immunoaffinity capillary electrophoresis with laser-induced fluorescence detection, *Luminescence*, 16, 145–152, 2001.

67. Oldenburg, K.E., Xi, X., and Sweedler, J.V., High resolution multichannel fluorescence detection for capillary electrophoresis: Application to multicomponent analysis, *J. Chromatogr. A*, 788, 173–183, 1997.

68. Cheng, S.B., Skinner, C.D., Taylor, J., Attiya, S., Lee, W.E., Picelli, G., and Harrison, D.J., Development of a multichannel microfluidic analysis system employing affinity capillary electrophoresis for immunoassay, *Anal. Chem.*, 73, 1472–1479, 2001.

69. Chabinyc, M.L., Chiu, D.T., McDonald, J.C., Stroock, A.D., Christian, J.F., Karger, A.M., and Whitesides, G.M., An integrated fluorescence detection system in poly(dimethylsiloxane) for microfluidic applications, *Anal. Chem.*, 73, 4491–4498, 2001.

70. Webster, J.R., Burns, M.A., Burke, D.T., and Mastrangelo, C.H., Monolithic capillary electrophoresis device with integrated fluorescence detector, *Anal. Chem.*, 73, 1622–1626, 2001.

71. Bergquist, J., Josefsson, E., Tarkowski, A., Ekman, R., and Ewing, A.G., Measurements of catecholamine-mediated apoptosis of immunocompetent cells by capillary electrophoresis, *Electrophoresis*, 18, 1760–1766, 1997.

72. Cannon, D.M., Winograd, N., and Ewing, A.G., Quantitative chemical analysis of single cells, *Annu. Rev. Biophys. Biomol. Struct.*, 29, 239–263, 2000.

73. Draisci, R., delli Quadri, F., Achene, L., Volpe, G., Palleschi, L., and Palleschi, G., A new electrochemical enzyme-linked immunosorbent assay for the screening of macrolide antibiotic residues in bovine meat, *Analyst*, 126, 1942–1946, 2001.

74. Martin, R.S., Gawron, A.J., and Lunte, S.M., Dual-electrode electrochemical detection for poly(dimethylsiloxane)-fabricated capillary electrophoresis microchips, *Anal. Chem.*, 72, 3196–3202, 2000.

75. Martin, R.S., Gawron, A.J., Fogarty, B.A., Regan, F.B., Dempsey, E., and Lunte, S.M., Carbon paste-based electrochemical detectors for microchip capillary electrophoresis/electrochemistry, *Analyst*, 126, 277–280, 2001.

76. Wang, J., Pumera, M., Chatrathi, M.P., Escarpa, A., Konrad, R., Griebel, A., Dorner, W., and Lowe, H., Towards disposable lab-on-a-chip: Poly(methylmethacrylate) microchip electrophoresis device with electrochemical detection, *Electrophoresis*, 23, 596–601, 2002.

77. Tomlinson, A.J., Guzman, N.A., and Naylor, S., Enhancement of concentration limits of detection in CE and CE-MS: a review of on-line sample extraction, cleanup, analyte preconcentration, and microreactor technology, *J. Capillary Electrophor.*, 2, 247–266, 1995.

78. Guzman, N.A., Determination of immunoreactive gonadotropin-releasing hormone in serum and urine by on-line immunoaffinity capillary electrophoresis coupled to mass spectrometry, *J. Chromatogr. B*, 749, 197–213, 2000.

79. Zhang, B., Foret, F., and Karger, B.L., High-throughput microfabricated CE/ESI-MS: automated sampling from a microwell plate, *Anal. Chem.*, 73, 2675–2681, 2001.

80. Kim, J.S. and Knapp, D.R., Microfabrication of polydimethylsiloxane electrospray ionization emitters, *J. Chromatogr. A*, 924, 137–145, 2001.

81. Kim, J.S. and Knapp, D.R., Microfabricated PDMS multichannel emitter for electrospray ionization mass spectrometry, *J. Am. Soc. Mass Spectrom.*, 12, 463–469, 2001.

82. Gelpi, E., Interfaces for coupling liquid-phase separation/mass spectrometry techniques: An update on recent developments, *J. Mass Spectrom.*, 37, 241–253, 2002.

83. Badman, E.R. and Cooks, R.G., A parallel miniature cylindrical ion trap array, *Anal. Chem.*, 72, 3291–3297, 2000.

84. Cornish, T.J., Ecelberger, S., and Brinckerhoff, W., Miniature time-of-flight mass spectrometer using a flexible circuitboard reflector, *Rapid Commun. Mass Spectrom.*, 14, 2408–2411, 2000.

85. Berkout, V.D., Cotter, R.J., and Segers, D.P., Miniaturized EI/Q/oa TOF mass spectrometer, *J. Am. Soc. Mass Spectrom.*, 12, 641–647, 2001.

86. Moxom, J., Reilly, P.T., Whitten, W.B., and Ramsey, J.M., Double resonance ejection in a micro ion trap mass spectrometer, *Rapid Commun. Mass Spectrom.*, 16, 755–760, 2002.

29

Chromatographic Immunoassays

Annette C. Moser and David S. Hage

Department of Chemistry, University of Nebraska, Lincoln, NE

CONTENTS

29.1   INTRODUCTION

A chromatographic immunoassay is a flow-based technique in which antibodies or antibody-related substances are used as selective binding agents for chemical detection. This is also known as a flow-injection immunoassay (FIIA), but it is more appropriately classified as a chromatographic method, since it makes use of a stationary phase (generally based on an immobilized antibody or antigen) for sample separation and analysis.

Antibodies are glycoproteins produced by the body's immune system in response to a foreign agent, or antigen. Even though antibodies interact with antigens through noncovalent interactions, the large variety of such interactions for even a single antigen makes antibodies quite specific for their targets and gives them association equilibrium constants in the range of 10^5 to 10^{12} M^{-1}. As a result of this high specificity, antibodies are often used as analytical reagents for complex mixtures such as blood, plasma, urine, and food. More information on antibodies and their properties can be found in Chapter 6.

Chromatographic immunoassays have been around since 1977, when this method was used to measure human serum albumin (HSA) [1]. Since that time, a variety of detection schemes and formats have been created for these methods. For instance, chromatographic immunoassays can either detect an analyte directly or use a chemical tag for indirect detection. The emphasis in this chapter is placed on heterogeneous assay methods, although some homogeneous techniques have also been performed with chromatographic systems. For each of these techniques, this chapter provides examples of applications and examines several theoretical and practical factors to consider in the development and use of such assays.

29.2 LABELING AND DETECTION METHODS

Like traditional immunoassays, the detection schemes used in chromatographic immunoassays and related methods (e.g., immunoassays based on capillary electrophoresis) are quite varied. As indicated in Table 29.1, these methods can include the use of absorbance, fluorescence, radioactivity, electrochemical reactions, and thermometric detection. Along with these basic types of detection, chemical labels can also be employed. These labels can be based on enzymes, fluorescent tags, chemiluminescent agents, liposomes, or radioisotopes.

29.2.1 Absorbance Detection

One of the easiest means for detection in chromatographic immunoassays is UV/Vis absorbance. Absorbance detectors are common in high-performance liquid chromatography (HPLC) and provide a nondestructive but precise means for detecting many types of analytes. The one limitation of this approach is that it has detection limits of only 10^{-8} to 10^{-7} M for a standard HPLC system, which limits it to the detection of moderate or relatively high concentrations of an analyte or labeled analog. This also requires that the analyte or label have a chromophore with a reasonably high molar absorptivity at the detection wavelength. For the detection of protein and peptides, absorbance measurements at 210 to 215 nm or 280 nm are most commonly used. However, alternative wavelengths can be selected for other analytes.

For some applications of chromatographic immunoassays, limits of detection based on direct UV/Vis detection have been reported in the nM range. This is possible because the high affinity of most antibody-antigen interactions tends to give the immobilized antibodies (or antigens) a response related to the moles of applied solute rather than to its initial sample concentration [2]. As a result, only a small sample volume is required to detect an analyte at intermediate or high concentrations, but a larger volume can be used for more dilute solutions. However, caution must be used with this approach to minimize the nonspecific adsorption of other sample components, which could lead to an increased background signal.

29.2.2 Fluorescence

Some of the most common labels used in chromatographic immunoassays are fluorescent tags. This is due to their good limits of detection and the ease with which they can be employed.

Table 29.1 Common Labels Used in Chromatographic Immunoassays

Detection Method	Labels or Detected Substances
Absorbance	Native absorbance or enzymatic products
Chemiluminescence	Acridinium ester, luminol
Electrochemical activity	Redox-active products generated by alkaline phosphatase, β-galactosidase, horseradish peroxidase, glucose oxidase, or adenosine deaminase
Fluorescence	Fluorescein, Lucifer yellow, Texas red, fluorescent products generated by alkaline phosphatase or fluorescein, liposomes with fluorescent markers
Radioactivity	Iodine-125
Thermal measurements	Heat produced as a result of catalysis by alkaline phosphatase, β-galactosidase, or catalase

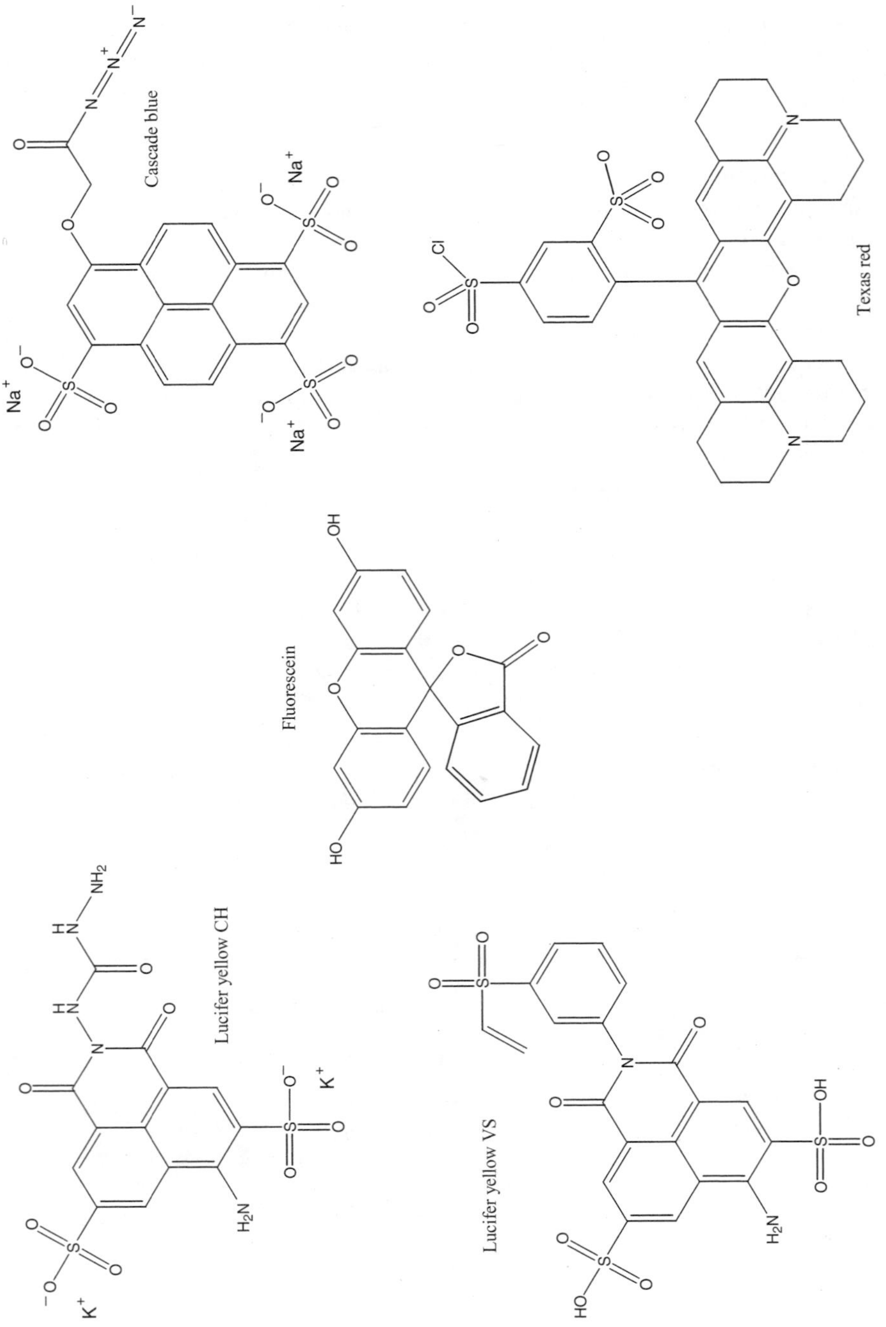

Figure 29.1 Structures of several labels used for fluorescence detection in chromatographic immunoassays.

These agents can be used either for direct analyte detection or for indirect detection by placing them on an antibody or an analyte analog. Several types of fluorescent labels have been used in chromatographic immunoassays. These include Lucifer yellow [3–5], Texas red [6], Cascade blue [7], and fluorescein [7–32], which are shown in Figure 29.1.

To obtain adequate limits of detection, a fluorescent label for a chromatographic immunoassay should have several key characteristics. First, it should have a high fluorescence quantum yield (i.e., up to 1.0). Second, conjugation of the label should require mild conditions and not significantly disturb the conformation of the molecule to which it is being linked. In addition, the label should be stable, and its conjugation should not adversely affect the binding of the antibody to the analyte or labeled analog in the immunoassay.

Fluorescein is the most common fluorescent label used in chromatographic immunoassays. Part of its popularity is due to its excitation wavelengths of 488 to 495 nm, which closely match the emission wavelength of an argon laser (488 nm) [33]. The quantum yield for fluorescein can be as high as 0.75 under ideal conditions, but its intensity fades when this compound is dissolved in a buffer, exposed to light, or stored for extended periods of time [33]. Furthermore, the pH of fluorescein's surrounding solution can greatly influence its emission intensity. For instance, at a pH below 7, the fluorescence of this agent is quenched by 50% [33].

Fluorescein isothiocyanate (FITC) is the most common reagent used to form fluorescein conjugates. FITC has excitation and emission wavelengths of 494 nm and 520 nm, respectively [33]. Since isothiocyanates are capable of forming stable products when reacted with primary amines, FITC is selective for modifying ε- and N-terminal amines in proteins and peptides. This is based on the reaction shown in Figure 29.2, which can give up to 8 to 10 labels per protein, with 4 to 5 labels usually being optimal.

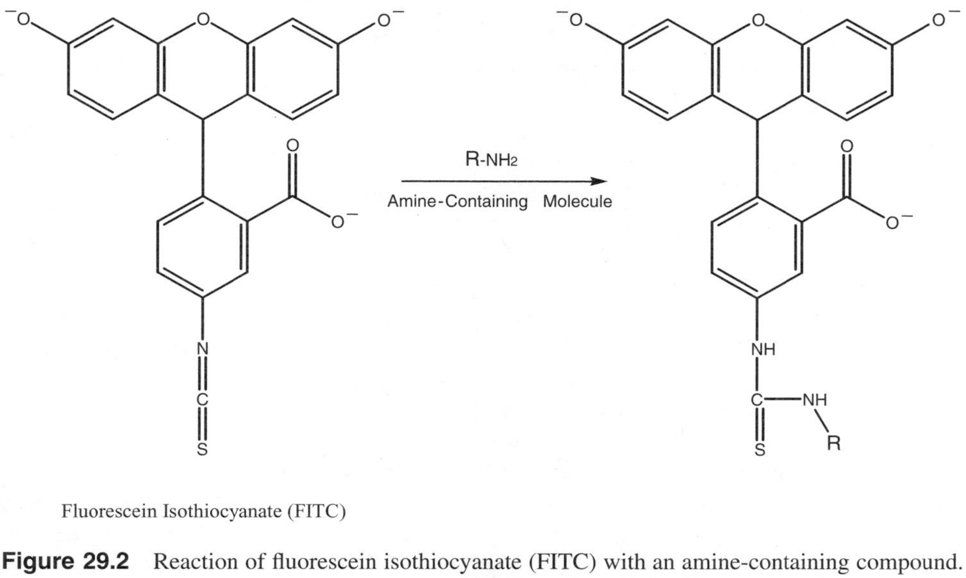

Fluorescein Isothiocyanate (FITC)

Figure 29.2 Reaction of fluorescein isothiocyanate (FITC) with an amine-containing compound. The only stable product is formed via reactions with primary amines, which makes FITC selective for ε- and N-terminal amines on proteins or peptides. In this reaction, a nucleophilic amine attacks the central carbon on the isothiocyanate group, resulting in a thiourea linkage between FITC and the protein or peptide [33].

Lucifer yellow is another label used for fluorescence detection in chromatographic immunoassays. It is a water-soluble dye available in several forms, including Lucifer yellow VS (LyVS) and Lucifer yellow CH (LyCH). The excitation wavelength for Lucifer yellow is 428 nm, and the emission wavelength is 540 nm, with a quantum yield of 0.25 [34]. LyVS contains a vinyl sulfone group for attachment to amine-containing compounds. LyCH has a hydrazide group that can react with aldehyde-containing compounds to form a hydrazone bond. Experimental details for these reactions can be found in previous reports [3–6, 34, 35]. After conjugation, the fluorescence quantum yield for Lucifer yellow is generally between 0.1 and 0.2. One advantage of using this as a label is that its fluorescence intensity is unchanged over a wide pH range (i.e., pH 2–10) [36].

Texas red is a third label employed for fluorescence detection in chromatographic immunoassays. This agent is a derivative of rhodamine and is also referred to as sulforhodamine 101. Texas red has a fluorescence quantum yield of approximately 0.25, an excitation wavelength of 589 nm, and an emission wavelength of 615 nm [33]. Although Texas red has a lower fluorescence quantum yield than fluorescein, its emission is more long-lived when it is stored or exposed to light [33]. In its active halogen form (i.e., Texas red sulfonyl chloride), Texas red can label amine-containing compounds to form stable and highly fluorescent conjugates.

Cascade blue is another fluorescent label used for detection in chromatographic immunoassays. This label has excitation wavelengths at 375 and 400 nm and an emission wavelength at 410 nm [33]. Several derivatives of Cascade blue exist for use in labeling. For amine-containing compounds, Cascade blue acetyl azide is a common choice. If two labels are to be used in an immunoassay, Cascade blue can be used with LyCH, since these can be simultaneously excited at 400 nm but detected separately at 410 and 530 nm [33].

29.2.3 Enzyme Labels

Enzymes are another group of labels often employed in chromatographic immunoassays. These are detected by adding substrates and examining the resulting products. A big advantage of using enzymes as labels is their ability to act as catalysts. This means that many product molecules can be generated by a single copy of the enzyme. This gives an amplified signal that is determined by the amount of available substrate, the reaction time, and the number of enzyme labels present. With this approach, the detection limit can be lowered by increasing the amount of substrate or increasing the reaction time, but this can also lead to an increase in the background signal or analysis time.

The detection method used to monitor an enzymatic product will be determined by the substrate being used, the required limits of detection, and interferences in the sample. Many techniques can be used for this purpose, including electrochemical, thermometric, fluorescence, absorbance, and chemiluminescence detection. Further information on each of these detection schemes is given later in this chapter.

When choosing an enzyme label for a chromatographic immunoassay, one must consider several items. For instance, enzymes with high turnover numbers tend to give the most sensitive results. In addition, it is necessary to have a substrate for the enzyme that is relatively stable, making it possible to maximize the sensitivity of the method and the amount of generated product. The enzyme label should be easy to detect, be stable under the assay conditions, and not be present in the original sample. It should also not contain any groups that will attach it to the analyte, antibodies, or other assay components. Finally, it should be possible to store the label for long periods of time, and it should be available at a low cost in a pure form.

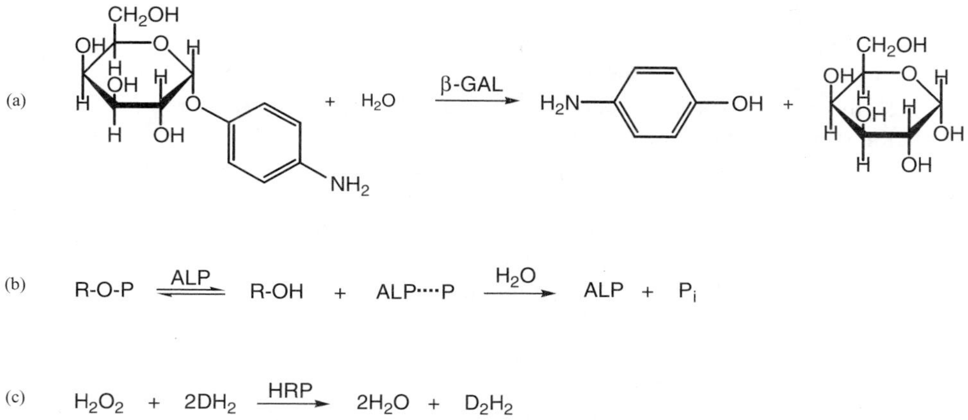

(a)

(b)   R-O-P $\xrightleftharpoons{\text{ALP}}$ R-OH + ALP····P $\xrightarrow{\text{H}_2\text{O}}$ ALP + P$_i$

(c) H_2O_2 + 2DH_2 $\xrightarrow{\text{HRP}}$ $2\text{H}_2\text{O}$ + D_2H_2

Figure 29.3 Typical reactions catalyzed by (a) β-galactosidase (β-GAL), (b) alkaline phosphatase (ALP), and (c) horseradish peroxidase (HRP).

There are only a few enzymes that have been used as labels in chromatographic immunoassays (see Figure 29.3). One of these is β-galactosidase (β-GAL), which is also known as lactase. β-GAL is a multimeric enzyme with a molecular weight of approximately 500 kDa. It catalyzes the hydrolysis of terminal nonreducing β-D-galactose residues in β-galactosides and takes part in transgalactosidation (i.e., the transfer of a galactosyl moiety from a substrate to an acceptor other than water). The conjugation of β-GAL to other molecules is generally accomplished through its ε-amine groups or free cysteine residues [37].

Several substrates and detection schemes have been used with β-GAL in flow-based immunoassays. For example, electrochemical detection has been used with the substrate aminophenylgalactopyranoside, which is converted by β-GAL to galactose and aminophenol. In this scheme, aminophenol is detected as it is oxidized to iminoquinone [38]. The conversion of lactose into glucose and galactose by β-GAL has been detected thermometrically [39, 40]. In addition, absorbance detection at 405 nm has been used to monitor the conversion of o-nitrophenylgalactoside into 2-nitrophenol and galactose in the presence of this enzyme [41].

Alkaline phosphatase (ALP) is a second enzyme that has been employed as a label in flow-based immunoassays. This is a metalloenzyme with a molecular weight of 84 to 150 kDa. ALP is stable at room temperature in neutral or slightly alkaline solutions. It catalyzes the hydrolysis of P–F, P–O–C, P–O–P, P–S, and P–N bonds and takes part in transphosphorylation reactions. This enzyme can be conjugated to other molecules in several ways, such as the use of a heterobifunctional crosslinking agent like succinimidyl 4-(N-maleimidolmethyl)-cyclohexanecarboxylate [37].

There are many substrates available for ALP making it possible to use a variety of detection schemes with this enzyme. These schemes include electrochemical detection of the conversion of p-aminophenyl phosphate to p-aminophenol [42–46], absorbance and electrochemical detection of the conversion of p-nitrophenyl phosphate to p-nitrophenol [47, 48], thermometric detection of the dephosphorylation of phosphoenol pyruvate [49], laser-induced fluorescence detection of the dephosphorylation of fluorescein diphosphate [50], and fluorescence detection of the dephosphorylation of 4-methylumbelliferyl phosphate [28, 46, 51].

The third enzyme label used in chromatographic immunoassays is horseradish peroxidase (HRP). This is a holoenzyme with a molecular weight of 40 kDa. It catalyzes the oxidation of various compounds by hydrogen peroxide or related agents. HRP can be conjugated to both large proteins (i.e., antibodies) and small molecules (i.e., antigens) by numerous methods [37].

Many substrates and detection schemes can be used with HRP. For example, the oxidation of *p*-hydroxyphenylpropionic acid by HRP has been detected by fluorescence measurements [28, 52–57]. The oxidation of 3,3,5,5-tetramethylbenzidine [58] and *o*-phenylene diamine dihydrochloride [59] by HRP has been measured using absorbance detection. In addition, the conversion of phenol by HRP has been monitored by electrochemical [60] and absorbance detection [61], and the action of HRP upon luminol has been examined through the use of chemiluminescence [62, 63].

A few other enzymes have been utilized as labels in chromatographic immunoassays. The decomposition of peroxide by catalase has been detected thermometrically [64]. The use of hemin as a label has allowed the reduction of hydrogen peroxide to be monitored through electrochemical detection [65]. Glucose oxidase catalyzes a reaction between glucose and diatomic oxygen to create gluconolactone and hydrogen peroxide, which has been detected electrochemically [66, 67]. In addition, adenosine deaminase, which converts the amine group on adenosine to a carbonyl, has been used as a label in chromatographic immunoassays [68]. In this last case, the resulting ammonium ions were detected using an ion-selective electrode.

29.2.4 Chemiluminescence

Chemiluminescence is the production of light as a result of a chemical reaction. This is yet another approach for detection in chromatographic immunoassays. In this method, the analyte, antibody, or analyte analog is tagged with a compound capable of producing light or giving rise to a light-generating reaction. This reaction is later initiated as the labeled compound elutes from the chromatographic system. Agents utilized for such detection in flow-based immunoassays include luminol and acridinium ester.

The use of chemiluminescent labels in chromatographic immunoassays has several advantages. For instance, their production of light can be quite fast, which allows for rapid detection. These reactions also provide good sensitivities and have low background signals, which can result in limits of detection in the femtomole or attomole range. The main limitation of this approach is that it requires the use of a postcolumn reactor for on-line detection. Furthermore, it is necessary to carefully optimize the reagents for the chemiluminescent reaction to achieve good sensitivity and reproducibility.

Luminol (i.e., 5-amino-2,3-dihydro-1,4-phthalazinedione) can be used as a chemiluminescent agent in reactions catalyzed by HRP. The reactions involved in this process are shown in Figure 29.4a. In this process, an excited molecule of 3-aminophthalate is formed through the reaction of luminol with hydrogen peroxide under basic conditions. This molecule releases visible light as it falls to the ground state. To improve this response, an enhancer such as *p*-iodophenol can be added, which increases the sensitivity by 1000-fold [69]. Examples of chromatographic immunoassays in which HRP has been used along with luminol for detection can be found in the literature [62, 63].

Acridinium ester is another agent employed for chemiluminescence detection in chromatographic immunoassays [63, 70–77]. Figure 29.4b shows how light is produced by this agent. The reaction occurs under basic conditions when hydrogen peroxide dissociates to form the hydroperoxyl anion, HO_2^-. This anion attacks the aromatic ring of acridinium ester to form a peroxide, which then reacts with a hydroxide anion to form an excited-state molecule of *N*-methylacridone. The excited molecule then releases light in the visible range as it falls to the ground state.

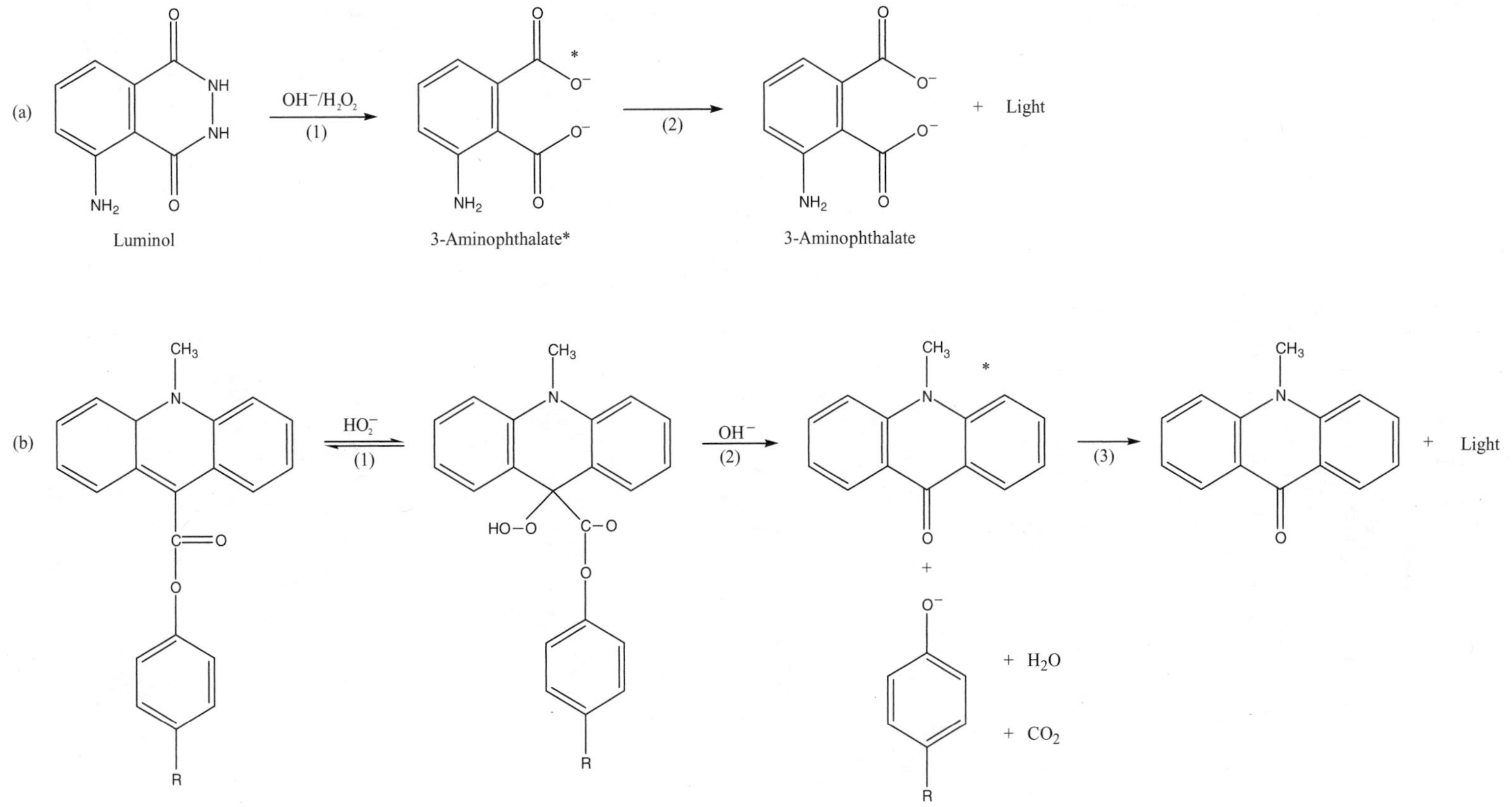

Figure 29.4 The chemiluminescent reactions of (a) luminol and (b) acridinium ester. The asterisk represents a molecule in the excited state.

Acridinium ester gives a lower limit of detection than luminol due to its faster rate of light production. In addition, its excited-state species (N-methylacridone) is released into solution before it emits light, which minimizes the effect of conjugation on this process [70]. However, the fast rate of acridinium ester chemiluminescence makes it crucial to carefully optimize the flow rates and reagent concentrations that are used with this agent to initiate light production. In addition, it is essential to minimize the time between the addition of these reagents and the point at which the light is measured [71]. Acridinium ester labels have been used in a variety of chromatographic immunoassays, including both competitive-binding and immunometric methods [63, 70–77]. Examples of these assays are given later in this chapter.

29.2.5   Liposome Labels

One way the signal in a chromatographic immunoassay can be improved is through the use of liposome labels. Liposomes form spontaneously when phospholipids are dispersed in water [78]. The liposomes used as labels in chromatographic immunoassays contain a lipid bilayer that surrounds an aqueous center with easily detectable marker compounds. These markers are usually fluorescent dyes, but other agents can also be used. One liposome can contain many copies of these markers, with up to 10^3 markers per liposome being reported in some studies [79].

Liposomes can be created by an injection method [78] in which water-soluble markers are placed in solution and combined with phospholipids. To create liposomes that interact with a particular antibody, modified phospholipids having the desired antigen linked to their polar head groups are used. These modified phospholipids are typically combined with unmodified phospholipids at a level of 1 mol% prior to liposome formation [79].

The markers are detected by disrupting the liposome and releasing them into solution. This disruption can be accomplished by several methods, including the addition of a detergent, the use of shear forces, or the promotion of lipid loss. These last two factors must be considered in the use of liposomes with chromatographic systems, since they can also result in the premature leakage of markers [80]. Liposomes have been used in several chromatographic immunoassays, including methods for theophylline [81–83], caffeine, antitheophylline [82], and imazethapyr [19, 84, 85]. An example of such a method is shown in Figure 29.5 for the detection of 17-estradiol by a competitive-binding immunoassay [86].

29.2.6   Radiolabels

One of the first approaches for detection in immunoassays involved the use of radioactive labels. These have also been used in a few chromatographic immunoassays. The main advantage of radiolabels is their low limits of detection. For example, Kusterbeck et al. developed a chromatographic immunoassay for 2,4-dinitrophenol that used iodine-125 as a label. This gave a limit of detection for 2,4-dinitrophenol of 140 nM with a linear range of 570 to 4600 nM [23]. Iodine-125 has also been used as a label for [Tyr^0]1-34 in the purification of parathyroid hormone-related peptide [87] and in the radiometric detection of methotrexate [88].

There are several disadvantages to using radiolabels in flow-based immunoassays. One obvious problem is the need to handle radioisotopes and the associated cost and safety of dealing with such materials. In chromatographic immunoassays, radiolabels also require the use of an instrument that is solely dedicated to their use. Another disadvantage is the need to routinely prepare or obtain fresh batches of the radiolabels, since they will continue to decay regardless of whether they are actually being used in an assay.

Step 1. Estradiol immobilized on particle

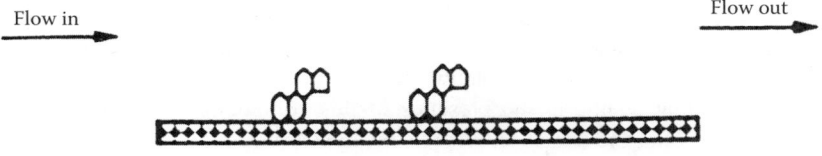

Step 2. Anti-Estradiol liposome binding to solid support

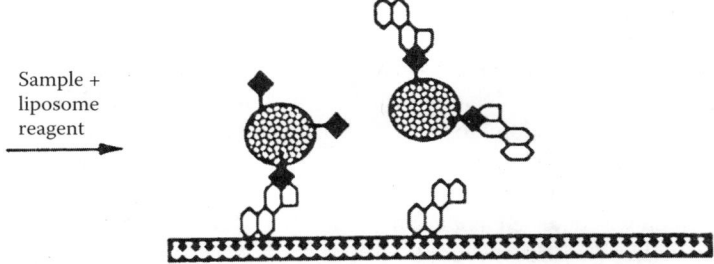

Step 3. Lysis of bound liposomes and detection

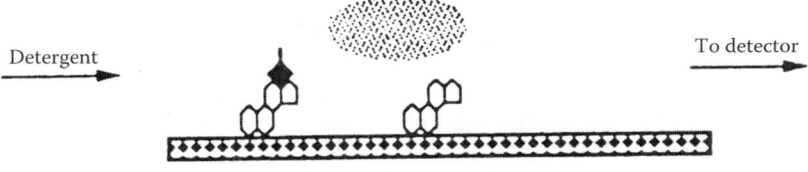

Step 4. Disruption of antigen-antibody complexes

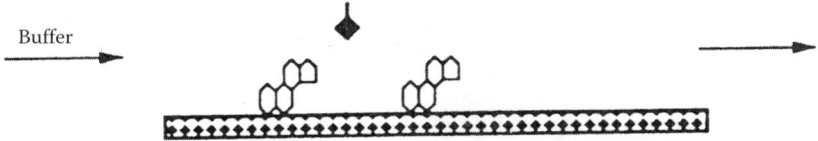

Figure 29.5 General scheme for the detection of 17-β-estradiol by a flow-based competitive-binding immunoassay with liposomes as labels. In this method, estradiol was placed within a column (step 1). Liposomes containing antiestradiol antibodies were then allowed to bind with soluble 17-β-estradiol in a sample and later passed through the immobilized estradiol column (step 2). After washing away the nonretained components, liposomes bound to the column were lysed (step 3) by adding a detergent, which released the entrapped fluorescent markers and allowed these to pass through an on-line detector. This gave a signal that was inversely related to the amount of soluble 17-β-estradiol in the original sample. In step 4, the remaining bound antibodies were removed from the column by passing through an elution buffer, and the column was allowed to regenerate prior to the injection of the next sample. (Reproduced with permission from Locascio-Brown, L. and Choquette, S.J., *Talanta*, 40, 1899–1904, 1993.)

29.2.7 Electrochemical Detection

Electrochemical detection has been employed in some work with flow-based immunoassays. This involves the use of a compound that can undergo either oxidation or reduction and is generally performed by measuring the change in current that occurs when this compound is oxidized or reduced at a constant applied potential. However, it is also possible to monitor the change in potential that results in the presence of this compound at essentially zero current (e.g., as occurs during the use of an ion-selective electrode) [68].

Electrochemical detection is often used for enzymatic reactions, since some of these reactions can be made to produce an electrochemically active product. For instance, an electrochemical enzyme immunoassay was used by Kaneki et al. for determining digoxin. In this assay, p-aminophenol was formed by means of the enzymatic reaction of ALP on p-aminophenyl phosphate. This product was measured through the change in current it created when oxidized at +300 mV versus a silver/silver chloride reference electrode [43]. Similar schemes based on p-aminophenol detection can be found in other studies [39, 49].

The use of electrochemical detection has been reported in other flow-based assays. For instance, immunoglobulin G (IgG) has been measured by a sandwich immunoassay using glucose oxidase as a label. This label was used to catalyze the formation of hydrogen peroxide from glucose, which was detected amperometrically [67]. Electrochemical detection has also been combined with capillary electrophoresis to develop a competitive-binding immunoassay (see Figure 29.6). In this case, HRP was used as a label to catalyze the oxidation of 3,3′,5,5′-tetramethylbenzidine, which was monitored through its reduction at a carbon-fiber microdisc-bundle electrode [89].

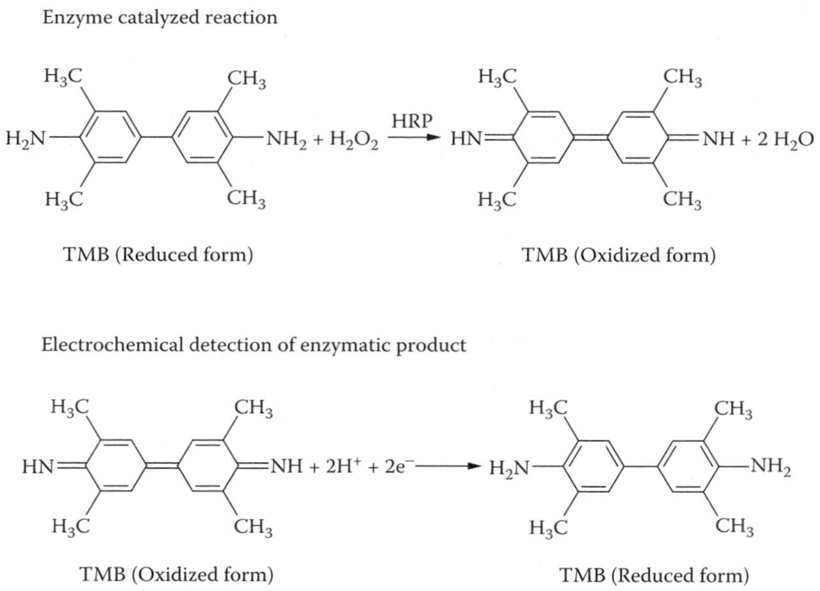

Enzyme catalyzed reaction

TMB (Reduced form)   TMB (Oxidized form)

Electrochemical detection of enzymatic product

TMB (Oxidized form)   TMB (Reduced form)

Figure 29.6 An electrochemical detection scheme for use with a horseradish peroxidase (HRP) label. In this method, 3,3′,5,5′-tetramethylbenzidine (TMB) in its reduced form is used as the substrate. This form is first converted into its oxidized form by HRP in the presence of hydrogen peroxide. The oxidized TMB is then detected by reducing it back to its initial state at a carbon electrode. (Reproduced with permission from He, Z. and Jin, W., *Anal. Biochem.*, 313, 34–40, 2003.)

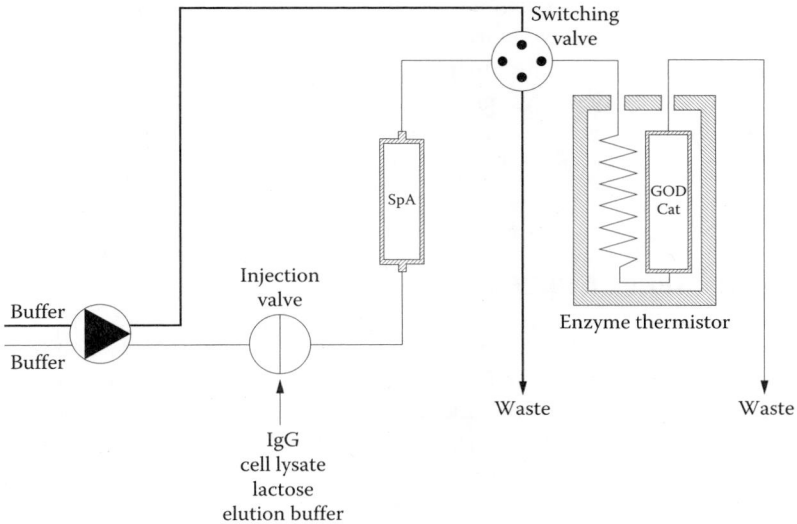

Figure 29.7 Instrument for measuring IgG based on a sandwich-type assay and thermometric detection. The affinity column used to bind the IgG contained immobilized staphylococcal protein A (SpA). Following the injection of IgG onto this column, SpA labeled with β-galactosidase was also injected for formation of a sandwich complex. Lactose was then passed through this column as a substrate for the signal of the β-galactosidase enzyme label. This produced glucose, which then proceeded to a second column (GOD cat) that contained immobilized glucose oxidase and catalase. As the glucose and resulting galactose were acted upon by this second column, the change in heat was monitored and used as a measure of the IgG in the first column. (Reproduced with permission from Brandes, W., Maschke, H.E., and Scheper, T., *Anal. Chem.*, 65, 3368–3371, 1993.)

29.2.8 Thermometric Detection

Thermometric detection measures the heat generated by a reaction. Since almost all enzymatic reactions generate heat, this method of detection can be used with enzyme labels. An example of such a system is shown in Figure 29.7. Like other enzyme-based methods of detection, this approach gives rise to an amplified signal that is related to the amount of product generated by the enzyme. This signal, in turn, is directly proportional to the amount of enzyme label in the measured fraction.

Thermometric detection was first used in a flow-based system by Mattiasson et al. [1]. This resulted in a method referred to as *thermometric enzyme-linked immunosorbent assay (TELISA)*. In this assay, catalase was used as the enzyme label, and hydrogen peroxide was its substrate. With this approach, it was possible to measure HSA at a concentration of 10^{-10} *M*. Other examples of thermometric detection have also been reported, including a sandwich assay for the detection of IgG [39] and a competitive-binding immunoassay for gentamicin [64].

29.2.9 Miscellaneous Methods

In addition to using labeled components for detection in flow-based immunoassays, the precolumn derivatization of analytes is possible. Examples of this include the use of fluorescent tags on recombinant granulocyte-stimulating factor [90] and interleukin-2 [91] for the analysis of these substances by immunoaffinity chromatography. Similar work with radiolabels has been used for the measurement of oligosaccharides by weak-affinity antibody columns [92, 93].

Off-line detection methods can also be implemented. This can be performed by collecting fractions of the column eluant and analyzing its contents by a second technique. One detection method used for this purpose is a standard immunoassay, as has been reported for serum IgE [94] and β2-microglobulin [95]. Bioassays can be used for off-line detection as well. This has been demonstrated for glutamine synthetase [96] and in a technique used to measure both the total and active amounts of interleukin-2 in tissue samples [91].

29.3   ASSAYS WITH DIRECT DETECTION

Flow-based immunoassays can be classified into two groups based on their format: direct- and indirect-detection methods. In a direct-detection method, the measured signal is generated by the analyte. This requires that the analyte be present at a reasonably high concentration and have good detection properties. If a substance does not meet these criteria, then indirect detection can be used, as discussed in Section 29.4.

29.3.1   General Principles

Direct detection is the simplest format for a chromatographic immunoassay. This is illustrated in Figure 29.8. In this method, a sample is injected onto an immobilized antibody column in the presence of an application buffer. Under these conditions, the analyte is allowed to bind with the antibodies while other sample components pass through the column as a nonretained peak. An elution buffer is used to disrupt the analyte's binding to the column, causing this substance to exit the column. The analyte is then passed through an on-line detector for measurement. The application buffer is later reapplied to the column, allowing the antibodies to regenerate before application of the next sample.

Step 1: Inject sample onto antibody column

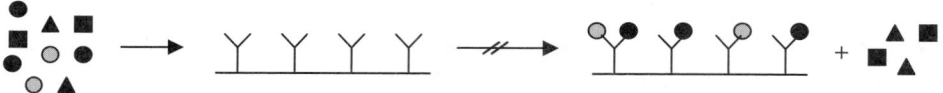

Step 2: Elution and detection of retained analyte

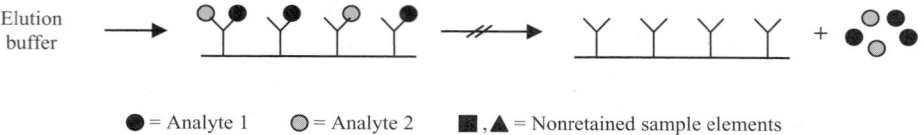

● = Analyte 1   ○ = Analyte 2   ■,▲ = Nonretained sample elements

Figure 29.8   Scheme for a direct-detection chromatographic immunoassay. Sample is applied to antibody column under physiological conditions. Compounds with affinity for the immobilized antibodies bind. Elution is performed by disruption of the noncovalent forces that hold the analytes on the antibody column. After elution, the column is regenerated by reapplying the application buffer to renature the antibodies. If necessary, a separation step to separate the analytes can be performed after elution from the antibody column.

Two advantages of direct detection are its speed and the ease with which it can be set up in an affinity system. However, it is difficult to use this method for trace substances unless they have been concentrated or labeled with a suitable tag prior to analysis. For this reason, direct detection is typically used to monitor substances with moderate to high concentrations in samples.

29.3.2 Applications

Several applications for the direct-detection method are given in Table 29.2. Further examples can be found in Chapter 6. A large number of monitoring techniques have been employed for this purpose, including UV/Vis absorbance, fluorescence, and mass spectrometry. Direct detection has been used with a wide range of substances, such as proteins, peptides, herbicides, carcinogens, and bacteria. It has also been used with a variety of samples, including serum, plasma, urine, saliva, tissue, bacterial or cell culture extracts, and fungal isolates (see Chapter 6).

An illustration of this format is given by a study in which recombinant tissue-type plasminogen activator (rt-PA) and recombinant antithrombin III (r-AT III) were measured [97]. In this work, anti-rt-PA or anti-r-AT III antibodies were immobilized onto Sepharose and used to extract the desired analytes from cell culture samples. The rt-PA and r-AT III were then eluted and detected with a spectrofluorometer. This method had a range of 2 to 500 μg/ml and could analyze one sample every 6 to 8 min [97].

Table 29.2 Examples of Direct Detection in Chromatographic Immunoassays

Analyte	Detection Method and Assay Characteristics	References
Human serum albumin	UV absorbance; LOD, 0.02 mg/ml	[16]
rt-PA and r-AT III	Fluorescence; range, 2–500 μg/ml	[97]
Albumin and IgG	UV absorbance; LOD, 3.5 μg (HSA)	[98]
E. coli	Fluorescence; LOD, 5×10^7 CFU/ml	[99]
rt-PA, monoclonal antibodies	Fluorescence; range, 1–1000 mg/l	[100]
Isoproturon and phenylurea herbicides	Diode array absorbance	[101]
Benzidine, dichlorobenzidine, aminoazobenzene and azo dyes	UV diode array absorbance; LOD, 0.05–0.1 μg/l; range, 0.1–5 μg/l	[102]
Triazine and phenylurea herbicides	CI-MS; LOD, 0.001–0.005 μg/l	[103]
Diethylstilbestrol	UV absorbance and cf-FAB-MS; LOD, 2 ng/ml	[104]
Cy5-labeled antirabbit IgG	Evanescent field-based fluorescence; linear range, 3–30 nM	[105]
Human cardiac troponin I	CE with UV absorbance; LOD, 2 nM	[106]
IgE	CE with UV absorbance	[107]
Polycyclic aromatic hydrocarbons	UV absorbance; LOD, 1–40 μg/l	[108]
Acetylcholinesterase	UV absorbance	[109]
IgG	Fluorescence; linear range, <75 μg/ml	[110]
Transferrin	UV absorbance; linear range, 5×10^1–1×10^5 ng	[111]
Insulin	CE with UV absorbance	[112]

Direct detection can also be performed by combining an immunoaffinity column with a second separation method, such as high-performance liquid chromatography (HPLC), capillary electrophoresis (CE), or gas chromatography (GC). The use of antibody-based columns in this fashion is known as *immunoextraction*. Some examples of this are given in Table 29.2 [16, 97–112], with further examples being provided in Chapter 6. In this approach, the immunoaffinity column is used to extract a class of analytes from the sample, which are then resolved by the second separation step.

A recent example of immunoextraction coupled with HPLC was reported in the analysis of benzidine and azo dyes in surface water and industrial effluents (see Figure 29.9). This analysis made use of antibodies that could cross-react with these agents and extract them as a class of structurally related molecules. The retained compounds were then eluted, separated by a C_{18} reversed-phase HPLC column, and measured using a diode array absorbance detector. Based on this method, it was possible to measure benzidine and related dyes at 0.1 to 1 µg/l [102].

Direct-detection immunoassays have also been used after other methods to monitor the presence of a given class of compounds. When this approach is used as part of an HPLC system, it is referred to as *postcolumn immunodetection* (see Chapter 6). An example is the use of an antibody column and direct detection to monitor the elution of acetylcholinesterase in amniotic fluid applied to a size-exclusion column [100].

The main advantage of direct postcolumn immunodetection is its specificity. However, it does require that the analyte be present at moderate to high concentrations and have a sharp, well-defined peak. An additional consideration is the need to ensure that the eluant from the first separation step is compatible with the immunodetection

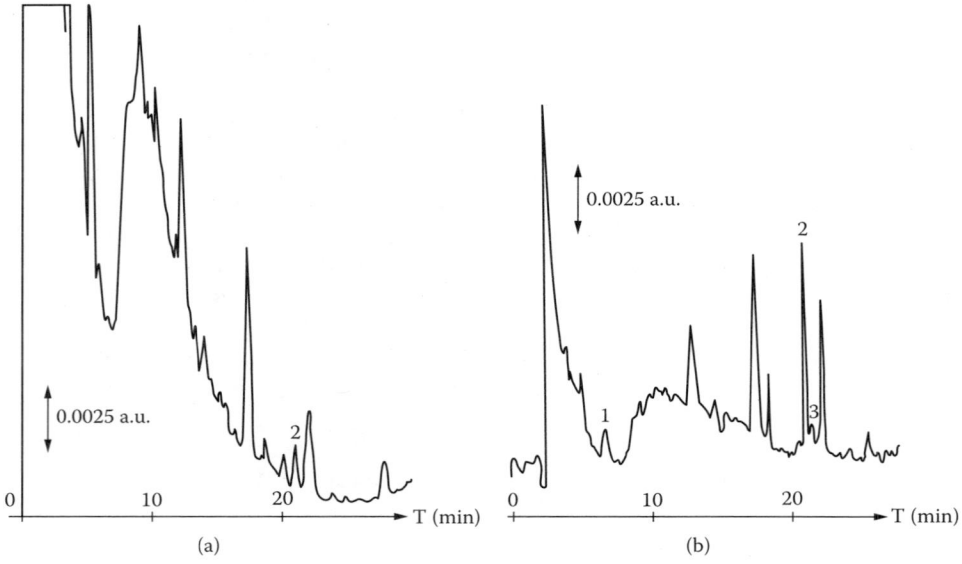

Figure 29.9 On-line analysis of a 2.5-ml textile effluent sample using immunoextraction and reversed-phase liquid chromatography. The result in (a) was obtained using a standard reversed-phase extraction cartridge for sample pretreatment, while the result in (b) made use of an antibenzidine antibody support. Peaks 1, 2, and 3 are benzidine, 3,3′-dichlorobenzidine, and 4-aminoazobenzene, respectively. (Reproduced with permission from Bouzige, M., Legeay, P., Pichon, V., and Hennion, M.C., *J. Chromatogr. A*, 846, 317–329, 1999.)

column. This usually involves adjusting the pH, ionic strength, or amount of organic modifiers to ensure that the immobilized antibodies will be capable of binding the eluting analytes.

29.3.3 Theory

In the direct-detection format, a high-affinity antibody is generally used to retain the analyte while other sample components pass through the system. If the association equilibrium constant for this interaction is sufficiently high and the time spent by the sample within the antibody column is relatively short, the binding of the analyte will be essentially irreversible under the application conditions. This is represented by the following reaction, in which k_a is the association rate constant for binding of the analyte (A) with the immobilized antibody (Ab).

$$A + Ab \xrightarrow{\ k_a\ } A - Ab \tag{29.1}$$

This model assumes that the chromatographic support has fast mass-transfer properties (e.g., an HPLC-grade material), as would be desired for use in an analytical method. For this type of system, the fraction (B) of analyte bound by the immunoaffinity column is given by Equation 29.2 [113, 114].

$$B = 1 - \{S_0/\text{Load } A\} \ln\left[1 + \left(e^{(\text{Load } A/S_0)} - 1\right)e^{-1/S_0}\right] \tag{29.2}$$

In this expression, Load A is the relative amount of analyte applied to the column versus the amount of active antibody present (m_L) (i.e., Load A = mol A/m_L). The term S_0 is a factor that depends on the association rate constant k_a and flow rate (F) of the injected analyte, as well as the amount of active antibody in the column, where $S_0 = F/(k_a m_L)$.

Based on Equation 29.2, the relative response of a direct-detection assay can be described by using the product ($B \cdot \text{Load } A$). This product, in turn, can be estimated by Equation 29.3.

$$\text{Relative response} = (B \cdot \text{Load } A)$$
$$= \text{Load } A - S_0 \ln\left[1 + \left(e^{(\text{Load } A/S_0)} - 1\right)e^{-1/S_0}\right] \tag{29.3}$$

This relative response will have a minimum value of zero at low analyte levels and a maximum value of 1 at high analyte loads. A similar expression can be used to give the absolute response by multiplying both sides of Equation 29.3 by a proportionality constant that is determined by the detectability of the analyte and the type of monitoring system being employed.

Figure 29.10 shows the response predicted by Equation 29.3 for a direct-detection assay under various flow-rate and sample-load conditions. In general, this response increases with the amount of applied analyte until the column reaches a saturation point. As this suggests, one important factor in determining the response of a direct-detection assay is the amount of active antibody in the column. This helps set the linear and dynamic ranges of the calibration curve. The application conditions are also important, since a small S_0 term (i.e., a slow injection flow rate or fast adsorption rate) will give the highest capture efficiency and the best response at low analyte levels. If a higher dynamic range is desired, then a faster flow and larger S_0 value should be used for the assay.

One interesting feature of Equation 29.3 is that it reflects the moles of analyte rather than its concentration. This is a result of the essentially irreversible binding that takes place between the analyte and antibodies during sample application. This means that the

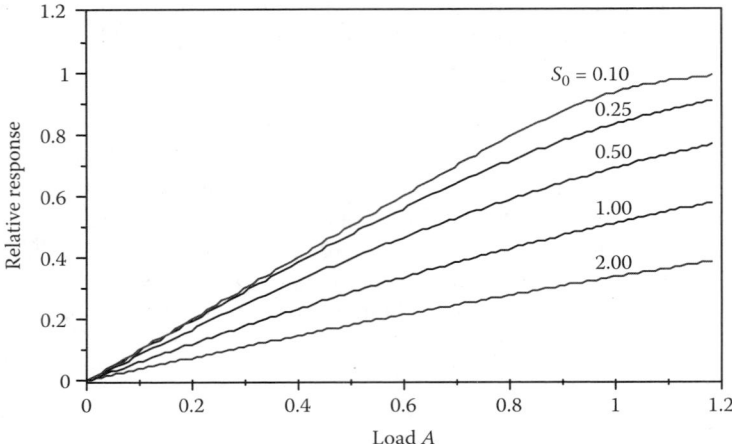

Figure 29.10 Theoretical calibration curves for a direct-detection assay, as determined by using Equation 29.3. This plot shows that as the application flow rate decreases or as the adsorption kinetics become faster (i.e., as S_0 becomes smaller), the slopes of the curves increase, providing a lower limit of detection. With faster flow rates or slower adsorption kinetics (i.e., larger S_0 values), the dynamic range of the assay increases.

concentration-based limit of detection for this assay can be adjusted by altering the volume of applied sample.

29.3.4   Practical Considerations

A number of factors should be considered when designing or optimizing a direct-detection assay. For example, a detection format should be used that will give a good response to even relatively small amounts of the analyte. The expected concentration of analyte in typical samples should be considered as well, since this will determine whether direct detection is feasible with the available amount of sample. This can be determined experimentally by injecting several standards onto the immunoaffinity column and estimating the lower limit of detection at a given sample volume. The sample volume can then be increased in proportion to the decrease in detection limit that is required. If it is not possible to obtain the desired detection range with a reasonable sample size, then an alternative approach should be used, such as one of the various methods of indirect detection that are described in Section 29.4.

The strong binding of most immunoaffinity columns means that elution in the direct-detection mode is generally accomplished by a step-gradient or linear-elution scheme. In an analytical application, this usually involves the use of a solvent that will disrupt noncovalent interactions between the antibody and analyte or alter their conformations. This is generally accomplished by altering the pH of the mobile phase, but the addition of a detergent, chaotropic salt, organic solvent, or denaturing agent is also possible. In work with weak- to moderate-strength antibodies (i.e., those with association constants less than $10^6 \, M^{-1}$), isocratic elution has even been employed [92, 93, 115].

Since the columns in chromatographic immunoassays are often reused for multiple samples, it is necessary to choose elution conditions that will not cause irreversible damage to the antibodies or support. If this is not done, the column may lose activity or have an increase in nonspecific binding over time. This can be monitored by routinely injecting control samples onto the column. If stored properly and used with relatively mild elution conditions, it is often possible to use an immobilized antibody column for several hundred elution cycles. In some cases, column lifetimes of over 500 to 1000 cycles have been reported.

If an analyte has slow dissociation from the immunoaffinity column, a second column may be required for the reconcentration of this compound before it is analyzed. This is usually performed through the use of a small reversed-phase precolumn, as is common when using immunoextraction prior to HPLC, CE, or GC. A further discussion of this technique can be found in Chapter 6. However, one precaution that should be exercised in this approach is that the second column must have sufficiently strong interactions with the analytes to keep them within this column as they elute from the immunoaffinity support. This can be confirmed by measuring the retention factors for these solutes on the precolumn at or near the desired elution conditions.

If an antibody column is to be used to examine several analytes, it is desirable for all these compounds to be quantitatively extracted during the application step. As indicated by Equations 29.2 and 29.3, this is best accomplished by using a slow flow rate for sample application (i.e., a small value for S_0) and a large-capacity column (i.e., a large m_L value, which gives small values for S_0 and Load A). With mixed samples, it is especially important that the total amount of all analytes be much less than the column capacity [116], thus preventing competition between the analytes and avoiding column-overloading effects.

29.4 COMPETITIVE-BINDING ASSAYS

The second type of chromatographic immunoassay is one that makes use of indirect detection for the analyte. This provides lower limits of detection than direct-detection methods, as is accomplished by using easy-to-measure tags on the antibody or on an analog of the analyte. One method that uses indirect detection is a competitive-binding immunoassay. This is the most common type of chromatographic immunoassay. It is based on competition between the analyte and a fixed amount of a labeled analog for binding sites on immobilized antibodies. This section considers three types of competitive-binding immunoassays for flow-based systems: the simultaneous-injection, sequential-injection, and displacement methods.

29.4.1 Simultaneous-Injection Method

29.4.1.1 General Principles

The most common competitive-binding format in a chromatographic immunoassay is the simultaneous-injection method. The general scheme for this method is illustrated in Figure 29.11. In this approach, a sample is first mixed with a constant amount of a labeled analyte analog (i.e., the label). This mixture is then applied to a column containing a small amount of immobilized antibodies (Ab). As the sample passes through this column, the analyte (A) and label (A^*) compete for binding sites on the antibodies. This competition is represented by the following reactions, where $A - Ab$ and $A^* - Ab$ are the antibody complexes formed by the analyte and label.

$$A + Ab \leftrightarrow A - Ab \tag{29.4}$$
$$A^* + Ab \leftrightarrow A^* - Ab \tag{29.5}$$

As a result of this competition, the amount of label in both the bound and the nonretained fractions will be affected by the presence of A. When the amount of label in these

Step 1: Inject sample and label onto antibody column

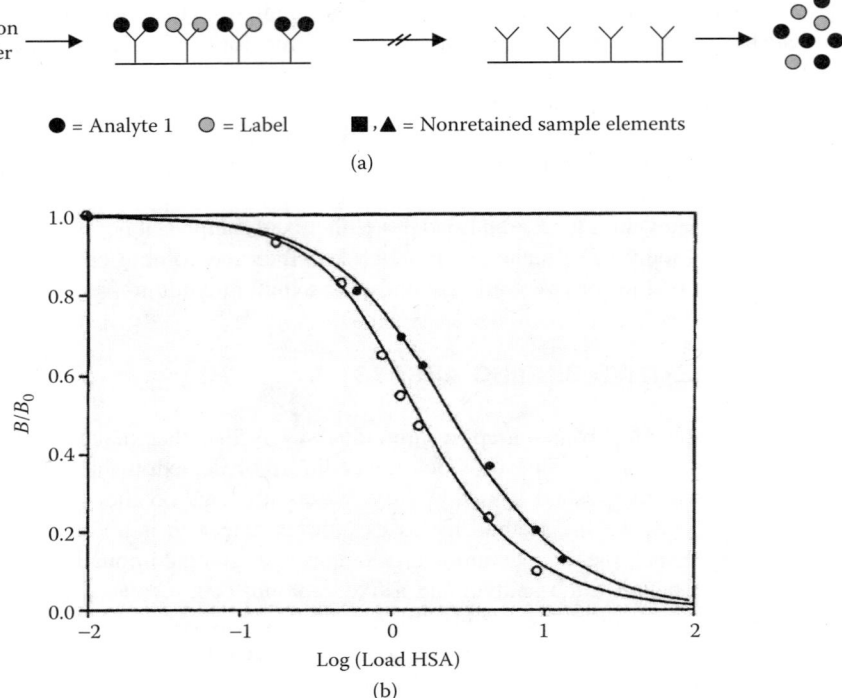

Step 2: Elution and detection of retained analyte and label

● = Analyte 1   ○ = Label   ■, ▲ = Nonretained sample elements

(a)

(b)

Figure 29.11 (a) Scheme for a simultaneous-injection chromatographic immunoassay and (b) results obtained for the injection of HSA as both the analyte and label onto an anti-HSA antibody column at a flow rate of 0.5 ml/min. The points in (b) represent calibration curves collected at Load A* values of 0.86 (○) and 1.7 (●). The lines are theoretical results predicted by Equation 29.6 under these conditions. (Reproduced with permission from Hage, D.S., Thomas, D.H., Chowdhuri, A.R., and Clarke, W., *Anal. Chem.*, 71, 2965–2975, 1999.)

fractions is determined, this gives an indirect measure of the analyte in the original sample.

A calibration curve for this assay is typically prepared by using a relative response that is given by the ratio B/B_0, where B is the amount of label bound in the presence of a given amount of analyte and B_0 is the amount of label bound in the absence of any analyte. This ratio is plotted as a function of the analyte's concentration or as the logarithm of this concentration, with the latter being used to examine a broader range of concentrations with a single graph. An example of such a calibration curve is shown in Figure 29.11. This type of graph will have a maximum value for B/B_0 of 1 when no analyte is present (i.e., $[A] = 0$). At high analyte concentrations, B/B_0 should approach zero in the absence of any nonspecific binding between the label and column.

29.4.1.2 Applications

Table 29.3 lists some analytes that have been measured using simultaneous-injection chromatographic immunoassays. This can be accomplished by using either covalently immobilized antibodies or antibodies adsorbed to protein A or protein G supports. In the first of these approaches, the antibodies remain on the column and can be used for multiple analyses, while in the second approach, a new layer of antibodies is used for each sample. One advantage of the first technique is that it requires a smaller amount of antibodies. However, the use of protein A or protein G to adsorb antibodies has the advantage of allowing the same column to be used for many analytes. In addition, this approach can provide better reproducibility, since it uses a fresh portion of antibodies for each injection.

An example of a method that uses simultaneous injection with immobilized antibodies is a technique reported for gentamicin, a potent antibiotic with a narrow therapeutic range. In this work, antigentamicin antibodies were immobilized onto a sol gel mesoporous material to create an affinity column. A mixture of serum samples containing gentamicin and a fixed amount of fluorescein-labeled gentamicin was then injected onto this column. Fluorescence detection of the retained label was used to determine the amount of gentamicin in the samples. The results obtained were comparable with those of a traditional fluorescence immunoassay and fluorescence-polarization immunoassay [14].

A simultaneous-injection method using a protein A column for antibody adsorption has been described for the measurement of atrazine, a common herbicide [9]. In this assay, the sample and a fixed amount of fluorescein-labeled atrazine were mixed and injected onto a protein A column after it had been coated with anti-atrazine antibodies. Once the excess label had been washed from the column, the bound fraction was eluted and its fluorescence measured. Using this method, it was possible to measure atrazine at levels of 2.1 to 50 µg/l [9].

29.4.1.3 Theory

The theory of a flow-based, simultaneous-injection, competitive-binding immunoassay has been described in detail by Hage et al. [113]. For this type of assay, it has been shown that the response for a system with homogeneous antibodies and adsorption-limited kinetics can be described by the following equation [113].

Relative response = B/B_0

$$= \frac{1 - \{S_0/(\text{Load } A + \text{Load } A^*)\} \ln\left[1 + \left(e^{(\text{Load } A + \text{Load } A^*)/S_0} - 1\right)e^{-1/S_0}\right]}{1 - \{S_0/\text{Load } A^*\} \ln\left[1 + \left\{e^{\text{Load } A^*/S_0} - 1\right\}e^{-1/S_0}\right]} \quad (29.6)$$

In Equation 29.6, Load A^* is the relative amount of label injected onto the column (Load $A^* = \text{mol } A^*/m_L$); Load A is the relative amount of analyte injected (Load $A = \text{mol } A/m_L$); and S_0 reflects the flow rate and adsorption kinetics of the system, as given by the relationship $S_0 = F/(k_a \, m_L)$.

A typical calibration curve for a simultaneous-injection assay is shown in Figure 29.11, along with the response predicted by Equation 29.6. Although Equation 29.6 assumes that homogeneous ligands are present in the column, it can be seen that good

Table 29.3 Examples of Simultaneous-Injection Chromatographic Immunoassays

Analyte	Detection Method and Tag	Assay Characteristics	References
Methods Based on Antibodies Adsorbed to Protein A or Protein G			
Human transferrin	Fluorescence; LyVS	Range, <500 µg/ml	[3]
Transferrin	Fluorescence; LyVS	LOD, 25 µg/ml	[4]
Adrenocortico-tropic hormone	Fluorescence; LyVS	Range, 0.2–10 mg/l	[5]
Testosterone	Fluorescence; Texas red	LOD, 0.5 µg/ml	[6]
Theophylline	Fluorescence; FITC	LOD, 0.3 ng/ml; range, <500 µg/l	[8]
Atrazine	Fluorescence; FITC	LOD, 2.1 µg/l; range, 2.1–50 µg/l	[9]
Cephalexin	Electrochemical; alkaline phosphatase	LOD, 1 µg/l	[42]
IgG	Absorbance; alkaline phosphatase	LOD, 333 zmol	[47]
Theophylline	Electrochemical; alkaline phosphatase	LOD, 25 ng/ml	[48]
Anti-BSA	Fluorescence; Cy5	LOD, 0.2 nM; linear range, 0.4–8 nM	[124]
Methods Based on Immobilized Antibodies			
HSA	Thermometric; catalase	LOD, 10^{-10} M	[1]
Theophylline	Fluorescence; FITC	LOD, 3 µg/l; range, 3–75 µg/l	[10]
Pullonsase, IgG, antithrombin III, rt-PA	Fluorescence; FITC	Range, µg/ml	[11]
IgG	Fluorescence; FITC	LOD, 155 ng/ml	[12]
IgG	Fluorescence; FITC	LOD, 4×10^{-9} M	[13]
Gentamicin	Fluorescence; FITC	LOD, 200 ng/ml; working range, 250–5000 ng/ml	[14]
IgG	Fluorescence; fluorescein	Linear range, 1–5 µg/ml	[18]
Insulin	Thermometric; alkaline phosphatase	LOD, 0.025 µg/ml; working range, 0.05–2 µg/ml	[49]
Carbaryl	Fluorescence; HRP	LOD, 26 ng/l	[52]
Atrazine	Fluorescence; HRP	LOD, 75 ng/l	[53]
Isoproturon	Absorbance; HRP	LOD, 0.09 µg/l	[58]
Gentamicin	Thermometric; catalase	Range, 10–400 µg/l	[64]
IgG	Electrochemical; glucose oxidase	—	[66]
Digoxin	Chemiluminescence; acridinium ester	LOD, 0.2 ng/ml	[72]
IgG	Chemiluminescence; acridinium ester	LOD, 7 fmol	[73]

(continued)

Table 29.3 Examples of Simultaneous-Injection Chromatographic Immunoassays (Continued)

Analyte	Detection Method and Tag	Assay Characteristics	References
Theophylline	Fluorescence; liposomes with carboxyfluorescein	—	[81]
Theophylline	Fluorescence; liposomes with carboxyfluorescein	Range, 0.025–0.4 mg/l	[83]
Theophylline, caffeine	Fluorescence; liposomes with carboxyfluorescein	Range, $3 \times 10^{-5} - 3 \times 10^{-8}\,M$	[84]
Antitheophylline	Fluorescence; liposomes with carboxyfluorescein	Range, $4 \times 10^{-7} - 6 \times 10^{-9}\,M$	[84]
Methotrexate	Radioactivity; ^{125}I	Range, 1–100 µg/l	[88]
Thyroxine	Electrochemical; HRP	LOD 25 µg/l; linear range 25–50 µg/l	[89]
Cortisol	Fluorescence; fluorescein	Range, 1–60 µg/dl	[125]

agreement is obtained even for systems with polyclonal antibodies, as is the case in Figure 29.11. Similar fits with Equation 29.6 have been noted under a variety of other experimental conditions [113].

Equation 29.6 indicates that the response of a simultaneous-injection assay depends on three parameters. The first of these is the relative amount of analyte that has been injected (Load A). In addition, the response will depend on the flow rate and adsorption kinetics of the system, as represented by S_0. Another important factor is the relative amount of injected label (Load A^*) [113]. The following subsection discusses how each of these parameters affects a simultaneous-injection immunoassay and will discuss their roles in assay development.

29.4.1.4 Practical Considerations

The factors that affect the behavior of a simultaneous-injection assay have been examined in several studies [113, 114]. As indicated by Equation 29.6, one factor to consider in such an assay is the relative amount of analyte that is applied to the column. This, in turn, depends on the moles of injected analyte and the binding capacity of the column. For optimum performance, it is recommended that the column binding capacity be approximately equal to the average moles of analyte present in a typical sample. This will give an assay in which an intermediate response is obtained for such a sample, thus allowing both higher and lower amounts of analyte to also be measured.

Two other factors that can be adjusted in this assay are the application flow rate, which affects S_0 in Equation 29.6, and the amount of label added to the sample, which determines the size of Load A^* [113, 117]. For the lowest limit of detection, a slow injection flow rate and small amount of label should be used. The best combination occurs at S_0 values of 0.1 to 1.0 and Load $A^* = 1.0$ for most columns, with slightly larger limits of detection being obtained at smaller values for these terms [113]. If desired, higher S_0 or Load A^* values can be used to adjust the calibration curve to higher analyte concentrations, but this also results in a system that is more sensitive to changes in the flow rate and amount of label [113].

The label used in this assay should give a signal that is not produced by any sample components. For instance, a fluorescent tag can be used as long as its excitation and emission wavelengths are different from those for any proteins or other molecules in the sample. It is also important to consider how the sample matrix will affect the label's signal. This is especially true for fluorescent tags or chemiluminescent agents, which can be either quenched or enhanced by the sample matrix.

The typical dynamic range for a simultaneous-injection assay extends over a 100-fold range when using lower and upper limits for B/B_0 of 0.95 and 0.05. However, the actual range will vary slightly with the experimental conditions and can be made larger through the use of high-precision measurements and low background signals [117]. In this format, it is also necessary to make a correction for nonspecific binding between the label and column. This can be performed by injecting the label in the presence of a large amount of analyte and measuring the corresponding value of B/B_0. This value is then subtracted from the response for every sample and standard [113].

29.4.2 Sequential-Injection Method

29.4.2.1 General Principles

The second possible format for a flow-based competitive-binding immunoassay is the sequential-injection method. As shown in Figure 29.12, this technique differs from the simultaneous-injection method in that the sample is now injected first, followed later by the label. The bound analyte and label are then eluted together, and the system is regenerated prior to the injection of the next sample [114].

Like the simultaneous-injection method, a calibration curve for the sequential-injection assay is usually prepared by plotting the relative response B/B_0 vs. the concentration of the analyte of the logarithm of this concentration. This graph will again have a maximum value of 1 for B/B_0 when no analyte is present and a value that approaches zero at high analyte concentrations in the absence of any nonspecific binding.

Although the sequential-injection method involves an additional chromatographic step, it has several advantages over the simultaneous-injection format. One advantage is that the label never actually comes in contact with the sample, since these pass at different times through the system. This makes it possible to eliminate matrix effects during label detection and provides improved reproducibility. The sequential injection of the sample and label also means that a wider range of detection formats can be used than are possible with simultaneous injection. For instance, it has been shown that even unlabeled proteins and simple absorbance measurements can be used for detection in this format [114]. Such an approach would not be possible in the analysis of real samples by simultaneous injection, where the sample matrix would usually give rise to a large background signal.

29.4.2.2 Applications

Several applications have been reported for the sequential-injection format, as shown in Table 29.4. One of the earliest examples is illustrated in Figure 29.12, in which HSA was detected by using unlabeled HSA as the label [114]. This particular method had an analysis time of less than 10 min at a flow rate of 0.5 ml/min, but this could be reduced to less than 2 min when using injection flow rates up to 4 ml/min.

A more recent example of this approach has been reported for imazethapyr, the active ingredient in the herbicide formulation PURSUIT [19]. This assay was performed using a column that contained immobilized anti-imazethapyr antibodies. The sample was injected first onto this column, followed by an injection of either a liposome conjugate of imazethapyr

Step 1: Inject sample onto antibody column

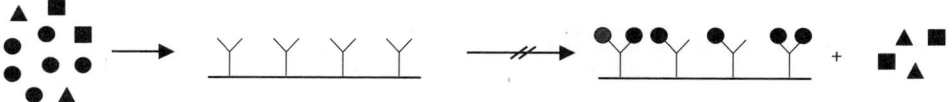

Step 2: Inject label onto antibody support

Step 3: Detect excess label and/or elute label for detection (column regeneration)

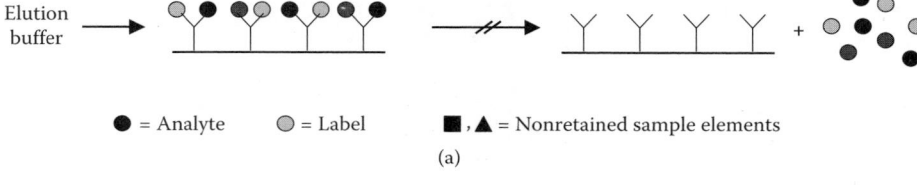

● = Analyte   ◉ = Label   ■, ▲ = Nonretained sample elements

(a)

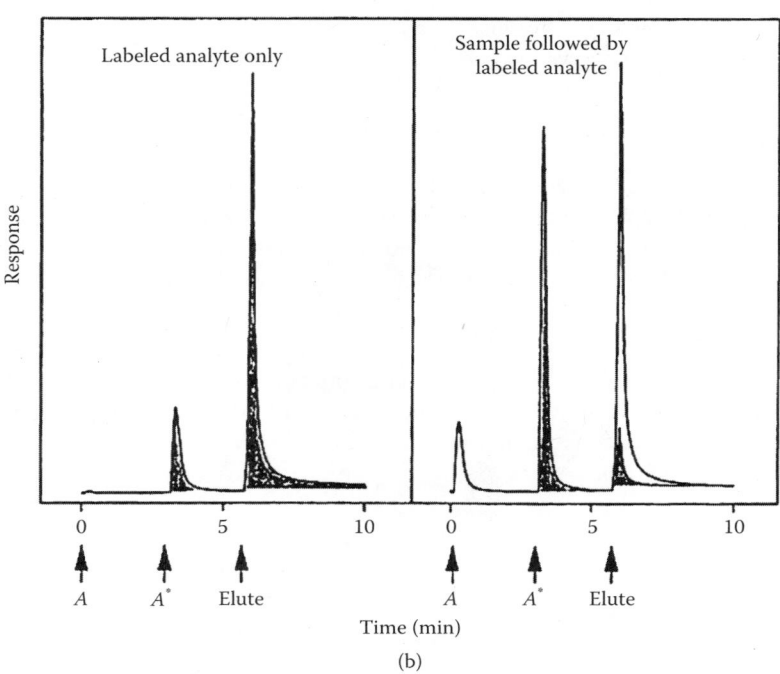

(b)

Figure 29.12 (a) Scheme for a sequential-injection chromatographic immunoassay and (b) a typical chromatogram for this assay. The results in (b) were obtained for the injection of HSA, with both the analyte and label being applied to an anti-HSA antibody column at a flow rate of 0.5 ml/min. (The lower portion of this figure is reproduced with permission from Hage, D.S., Thomas, D.H., and Beck, M.S., *Anal. Chem.*, 65, 1622–1630, 1993.)

Table 29.4 Examples of Sequential-Injection Chromatographic Immunoassays

Analyte	Detection Method and Tag	Assay Characteristics	References
Imazethapyr	Fluorescence; fluorescein	LOD, 500 ppb	[19]
Imazethapyr	Fluorescence; liposomes with carboxyfluorescein	LOD, 0.5 ppb	[19]
IgG	Thermometric; β-galactosidase	Range, 10–400 µg/ml	[39]
IgG	Thermometric; β-galactosidase	LOD, 33 pmol	[40]
Digoxin	Electrochemical; alkaline phosphatase	LOD, 10 pg/ml; range, 10–1000 pg/ml	[43]
Atrazine	Fluorescence; HRP	Range, 0.02–0.3 µg/l	[54]
Atrazine, propazine	Fluorescence; HRP	Range, 0.03–0.5 µg/l	[57]
HSA	Electrochemical; HRP	—	[60]
α-Amylase	Absorbance; HRP	—	[61]
Anti-IgG	Chemiluminescence; HRP	LOD, 1 fmol	[62]
Imazethapyr	Fluorescence; liposomes with sulforhodamine B or carboxyfluorescein	LOD, 0.1 ppb	[84]
Imazethapyr	Fluorescence; liposomes with carboxyfluorescein	Range, 0.1–100 ng/l	[85]

or fluorescein-labeled imazethapyr. In both cases, the amount of label bound to the column was determined by fluorescence detection. When using the liposome conjugate as a label, a limit of detection of 0.5 ppb was achieved. This was approximately 1000-fold better than the results obtained when using fluorescein-labeled imazethapyr [19].

29.4.2.3   Theory

An equation for the response of a sequential-injection assay (as given by B/B_0) has been derived for conditions in which adsorption-limited kinetics are present and the column contains a homogeneous population of antibodies [114]. The resulting expression is shown in Equation 29.7.

Relative response = B/B_0

$$= \frac{1 - \left\{ S_0 / \text{Load } A^* \right\} \ln \left[\dfrac{1 + \{ e^{(\text{Load } A + \text{Load } A^*)/S_0} - 1 \} e^{-1/S_0}}{1 + \{ e^{(\text{Load } A)/S_0} - 1 \} e^{-1/S_0}} \right]}{1 - \left\{ S_0 / \text{Load } A^* \right\} \ln \left[1 + \left\{ e^{(\text{Load } A^*)/S_0} - 1 \right\} e^{-1/S_0} \right]} \qquad (29.7)$$

All terms in this equation have the same meaning as in Equation 29.6. The only real difference between this expression and the one given in Equation 29.6 for a simultaneous-injection assay is in the number of terms that are present on the right-hand side. This difference reflects the fact that the analyte and label are now applied in separate steps to the antibody column rather than as a single mixture.

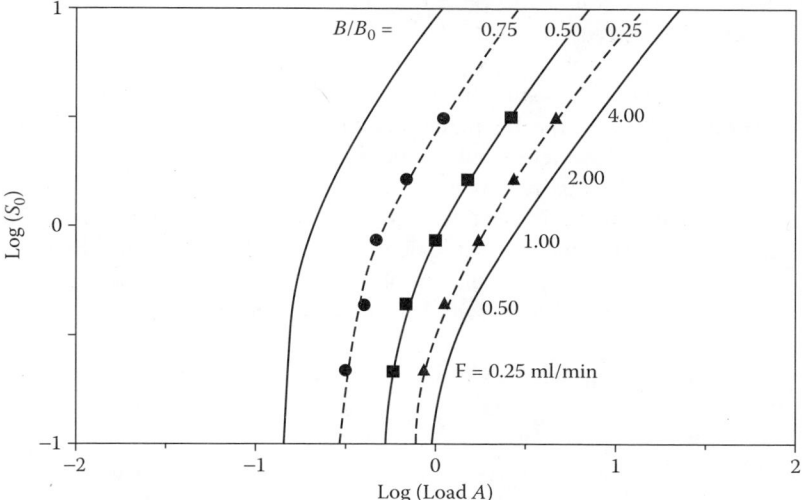

Figure 29.13 Example of a calibration curve for a sequential-injection chromatographic immunoassay. These results were obtained for the injection of HSA, with both the analyte and label being applied to an anti-HSA antibody column at flow rates of 0.5 to 4.00 ml/min and with Load $A^* = 1.0$. The lines are the theoretical results predicted by Equation 29.7. (Reproduced with permission from Hage, D.S., Thomas, D.H., Chowdhuri, A.R., and Clarke, W., *Anal. Chem.*, 71, 2965–2975, 1999.)

The response predicted by Equation 29.7 is illustrated in Figure 29.13. Like the simultaneous-injection assay and the direct-detection format, the sequential-injection assay gives a response that is related to the moles of injected analyte rather than to its concentration [114]. One consequence of this behavior is that lower concentration detection limits can be obtained by applying larger sample volumes to the column. Other parameters that can be varied to control the assay response include the flow rate, which affects S_0, and the amount of applied label, which affects Load A^*.

29.4.2.4   *Practical Considerations*

Like the simultaneous-injection format, a sequential-injection immunoassay is affected by changes in the application flow rate. It has been shown through the use of Equation 29.7 that the lowest limit of detection in this assay will typically be obtained at low flow rates, with faster flow rates giving a calibration curve that shifts to higher analyte levels [114]. However, changing the flow rate has different effects on the sensitivities of the sequential- and simultaneous-injection assays. While the sensitivity (i.e., the slope of the calibration curve) is constant over a wide range of flow rates for a simultaneous-injection assay, lowering the flow rate in a sequential-injection assay can give an increase in sensitivity. This occurs because the use of a lower flow rate in the sequential-injection format provides additional time for the label to bind any remaining antibody sites within the column [117].

A sequential-injection assay also differs from the simultaneous-injection format in that the amount of label now has little effect on the relative position of the calibration curve. This occurs because, in the sequential-injection method, the analyte is always allowed to bind the antibodies before the label. One benefit of this is that it provides the sequential-injection assay with a lower limit of detection than the simultaneous-injection method. However, the simultaneous-injection format does have a higher upper limit of detection and a wider dynamic range [117].

The best response for a sequential-injection method tends to occur when the binding capacity of the column is about the same as the average moles of analyte in a typical sample. The typical dynamic range for this method is a tenfold range in analyte concentration from $B/B_0 = 0.95$ to 0.05, which is much smaller than seen for the simultaneous-injection format [117]. However, this range can be broadened slightly by increasing the precision of the assay. It is also again necessary to make a correction for nonspecific binding. As was discussed for the simultaneous-injection method, this can be performed by injecting the label in the presence of a large amount of analyte and using the corresponding value of B/B_0 to correct the response of all samples and standards [114].

29.4.3   Displacement Methods

29.4.3.1   General Principles

The third possible format for a competitive-binding immunoassay in chromatography is the displacement immunoassay method. This format employs an immobilized antibody column that is saturated with a labeled analog of the analyte. As a sample is applied to this column, any of the labeled analog that is momentarily free in solution will be displaced and eluted from the column (see Figure 29.14). This displaced label is then detected, giving a response proportional to the amount of analyte. In many cases, several samples can be injected onto the column before it is regenerated, provided that enough label remains on the column for the production of a consistent and measurable signal.

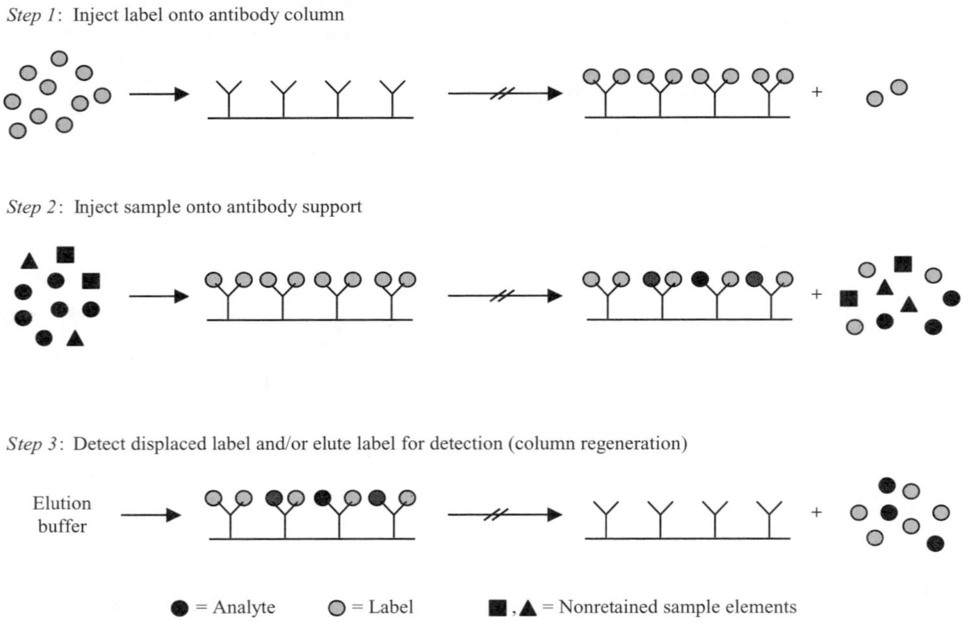

Figure 29.14   Typical scheme for a displacement immunoassay. A column containing immobilized antibodies is first loaded with a labeled analog of the analyte. When a sample is applied to this column, the analyte tends to displace the retained labeled analog, resulting in a displacement peak. The size of this peak can then be used to determine the amount of analyte in the original sample.

Table 29.5 Examples of Chromatographic Displacement Immunoassays

Analyte	Detection Method and Tag	Assay Characteristics	References
Cocaine, benzoylecgonine	Fluorescence; fluorescein	—	[20]
TNT, DNT	Fluorescence; fluorescein	LOD, 2.5 ng/ml; range, 20–1200 ng/ml	[21]
Cocaine, benzoylecgonine	Fluorescence; fluorescein	—	[22]
2,4-Dinitrophenol	Radioactivity; ^{125}I	LOD, 140 nM; linear range, 570–4600 nM	[23]
2,4-Dinitrophenol	Fluorescence; fluorescein	Linear range, 290–2300 nM	[23]
Cortisol	Fluorescence; alkaline phosphatase	Dynamic range, 12.5–1250 pmol	[46]
Polychlorinated biphenyls	Fluorescence; Cy5.29	LOD, 4 ppm; linear range, 4–20 µg/ml	[118]
Transferrin, HSA	Absorbance; transferrin and HSA	—	[126]

The ability to use a single application of label for the injection of multiple samples is one attractive feature of this approach. Another is its speed, since the displacement peak appears near the column void volume. The fact that the signal increases with analyte concentration is an advantage compared with other competitive-binding formats. However, care must be taken in selecting the proper conditions for this approach to work properly. This includes the choice of labeled analog, column size, and flow rate.

29.4.3.2   Applications

There have been a large number of applications reported for the displacement immunoassay, as indicated in Table 29.5. One example is an assay developed for the detection of polychlorinated biphenyls (PCBs). In this study, anti-PCB antibodies were immobilized onto Emphaze beads and placed into a column. This column was then loaded with a fluorescent PCB derivative, 2,3,5-trichlorophenoxypropyl-Cy5. Samples were next injected onto the column, resulting in displacement of some of the label. This label was measured with a fluorescence detector using an excitation wavelength of 635 nm and an emission wavelength of 661 nm. The resulting calibration curve gave an increase in signal as the amount of PCBs was increased. The limit of detection was 4 parts-per-billion, and the linear range extended up to 20 µg/ml [118].

Another example of a displacement immunoassay is a technique that was created for the detection of free thyroxine in serum [119]. In this approach, a small column containing immobilized protein G was loaded with anti-thyroxine antibodies that had been preincubated with an acridinium ester conjugate of triiodothyronine (i.e., a compound related to thyroxine). After the excess antibodies and labeled conjugate had been washed from the column, a sample containing thyroxine was applied. This gave a displacement peak that appeared within 25 sec of sample injection. One unique feature of this assay was that the flow rate and column size employed created a condition in which the sample spent only 60 to 100 msec in the affinity column. This was much shorter than the time required for thyroxine to be released from its binding proteins in serum, providing an assay that could directly measure the free form of this hormone.

29.4.3.3   Theory

The net reaction involved in the displacement of a labeled analog (A^*) from an antibody column by an analyte (A) has been described using the following reaction [120].

$$A^* - Ab + A \leftrightarrow A - Ab + A^* \tag{29.8}$$

Based on this model, a modified Langmuir isotherm has been used to describe the behavior of a displacement immunoassay. The expression employed for this purpose is shown below,

$$1/r_{im} = 1 + \{[A^* - Ab]/(K_{disp} [A_{loaded}])\}^{1/2} \tag{29.9}$$

where the terms K_{disp} and r_{im} are given by the following relationships.

$$K_{disp} = ([A^*][A - Ab])/([A][A^* - Ab]) \tag{29.10}$$

$$r_{im} = \text{displaced } A^*/\text{displaceable } A^* \tag{29.11}$$

In these equations, r_{im} is the relative amount of displaced label, K_{disp} is the displacement equilibrium constant for the reaction in Equation 29.9, and $[A_{loaded}]$ is the equilibrium concentration of the analyte. Other terms in these equations include displaced A^*, the amount of label displaced per sample injection, and displaceable A^*, the amount of label bound prior to sample injection [120].

A typical calibration curve for a displacement immunoassay is given in Figure 29.15. As stated earlier, this gives a response that increases with the amount of analyte in the sample.

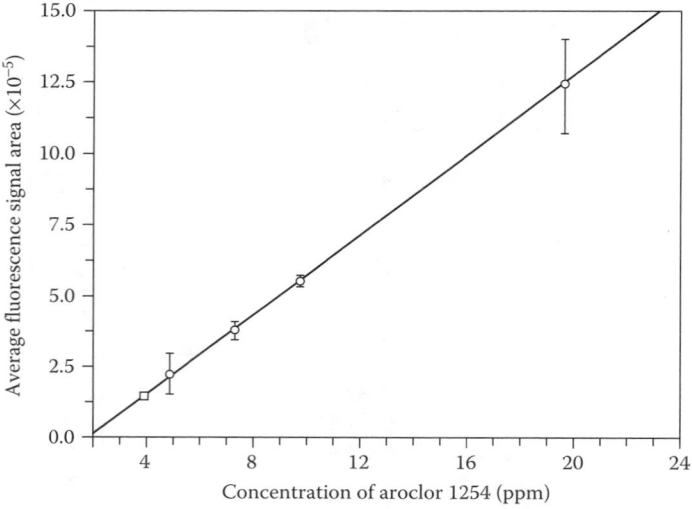

Figure 29.15   Calibration curve for a displacement immunoassay, using the analysis of Aroclor 1254 as an example. The label in this case was an analog of Aroclor 1254 that was conjugated with a fluorescent dye. (Reproduced with permission from Charles, P.T., Conrad, D.W., Jacobs, M.S., Bart, J.C., and Kusterbeck, A.W., *Bioconj. Chem.*, 6, 691–694, 1995.)

Although Equation 29.9 assumes that a local equilibrium is present within the column during such an assay (a situation that may not actually occur when using small columns and fast flow rates), this relationship is still useful in indicating the general factors that will affect the assay's behavior. For instance, Equation 29.9 shows that the behavior of a displacement immunoassay depends on the amount of antibody in the column, the amount of displaceable label, and the amount of injected analyte. However, it also depends on the relative extent to which the analyte can displace this label from the column.

29.4.3.4 Practical Considerations

An important advantage of the displacement immunoassay is its ability to be used for multiple sample injections [23]. This is possible as long as the amount of retained label is sufficiently stable to provide a reproducible and measurable signal. The stability of this signal depends on the rate of dissociation of the labeled analog from the immobilized antibodies [21, 23]. A labeled analog with a relatively fast rate of dissociation provides a large signal for the first few samples, but this signal swiftly decreases for subsequent injections. On the other hand, an analog with a slow rate of dissociation tends to give a broad displacement peak and may be difficult to detect. This means that a compromise must be reached between the rate of dissociation and the desired stability of the retained analog signal.

The rate of analog dissociation must currently be determined experimentally for each analog and antibody combination. However, it can be predicted that low-affinity analogs will give faster dissociation rates than high-affinity analogs, since the dissociation rate constant for a solute-antibody system is a major factor in determining the binding constant for this interaction. This also means that analogs whose structures differ the most from the antigen used to raise the immobilized antibodies tend to give the highest rates of dissociation.

Ideally, the analyte should have a large binding constant for the immobilized antibodies in a displacement immunoassay and a fast rate of binding to these antibodies. This helps to promote displacement of the labeled analog as a well-defined peak. It is also useful in a displacement assay to have a relatively high-capacity column. This allows the use of multiple injections between regeneration steps. Finally, the displacement effect tends to increase as slow flow rates are used, since this provides additional time for label dissociation [21, 23].

29.5 IMMUNOMETRIC ASSAYS

Another group of techniques that use indirect detection are the immunometric methods. These are also known as *noncompetitive immunoassays*, since there is no competition between the analyte and other substances in these methods. There are two types of immunometric methods used in flow-based systems: (1) the sandwich immunoassay, or two-site immunometric assay, and (2) the one-site immunometric assay. Both formats are examined in this section.

29.5.1 Sandwich Immunoassays

A sandwich immunoassay uses two different antibodies to bind to the same analyte. The first of these antibodies is immobilized onto a solid support and used for extracting the analyte from a sample. In the case of a chromatographic immunoassay, this antibody is immobilized or adsorbed to a support within a column. The second antibody contains an easily measured tag (e.g., an enzyme or fluorescent label) and is added in solution to the analyte either before or after the sample has been injected onto the immunoaffinity column. It is through the use

Step 1: Inject sample onto antibody column

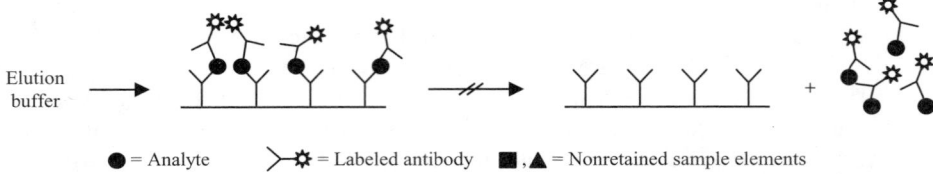

Step 2: Inject labeled antibody onto column

Step 3: Elute and detect bound labeled antibodies

● = Analyte   ⟩—☆ = Labeled antibody   ■,▲ = Nonretained sample elements

Figure 29.16 Typical scheme for a chromatographic sandwich immunoassay. In this particular example, the sample is first injected onto an antibody column, allowing the analyte of interest to bind while other nonretained compounds are eluted. Another antibody that is labeled with an easily detectable tag is then applied to the column and allowed to bind the analyte, creating a sandwich immunocomplex. The analyte-labeled antibody complex is then eluted by changing the buffer and monitored to determine the amount of analyte in the original sample.

of this second antibody that a tag is placed onto the analyte for its measurement. The result is an assay that has an increase in signal as the amount of analyte is increased.

29.5.1.1 General Principles

The most common approach for performing a sandwich immunoassay in chromatography is shown in Figure 29.16. In this scheme, the sample and labeled antibody are mixed and allowed to bind with each other before they are injected onto the immobilized antibody column. Alternatively, the sample and labeled antibody can be injected sequentially onto this column. This latter method eliminates the preincubation step but also tends to give worse limits of detection and requires more labeled antibodies.

An advantage of a sandwich immunoassay is that its response is directly proportional to the amount of analyte in the sample. This is demonstrated in Figure 29.17, where the signal for the retained peak increases with the concentration of analyte. This direct relationship is different from the response seen in most competitive-binding immunoassays (e.g., the simultaneous- and sequential-injection methods), which tend to have a decrease in signal as the amount of analyte is increased. This gives sandwich immunoassays better signal-to-noise ratios and lower limits of detection than competitive-binding immunoassays.

Another advantage of a sandwich immunoassay is its good selectivity. Because two types of antibodies are used for each analyte, only those chemicals that are able to bind both antibodies at the same time will be detected. This makes sandwich immunoassays especially useful in the analysis of substances that may have larger numbers of interferences present [76].

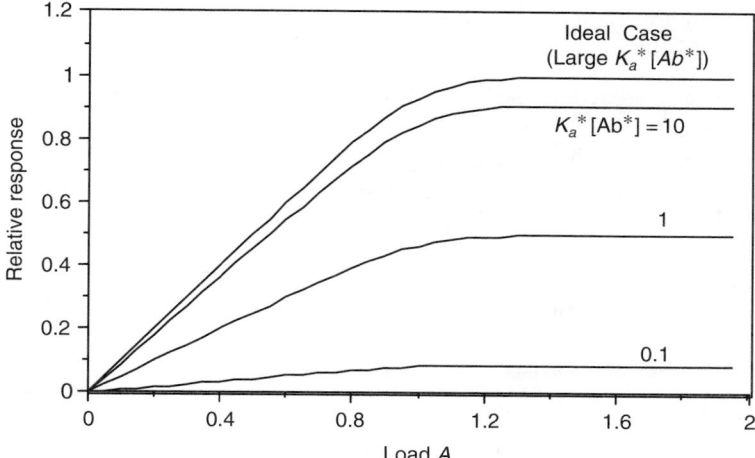

Figure 29.17 Theoretical calibration curves for a chromatographic sandwich immunoassay, as predicted by Equation 29.15. These results were obtained using a value of 0.1 for S_0. The top curve at $K_a^*[Ab^*]$ shows the ideal case, where all of the analyte is completely bound by the labeled antibodies prior to injection. As the value of $K_a^*[Ab^*]$ decreases, a smaller fraction of analyte is complexed with these antibodies prior to injection, leading to a small response.

A disadvantage of the sandwich immunoassay is that it can only be used for analytes that are large enough to simultaneously bind two antibodies. This generally limits this technique to large peptides, proteins, and other biomacromolecules. The need for two antibodies per analyte can also make this method more difficult to create and more expensive to perform than a competitive-binding assay.

29.5.1.2   Applications

Table 29.6 lists various analytes that have been examined using flow-based sandwich immunoassays. One early example was described by Hage and Kao [75], in which this method was used to measure parathyroid hormone (PTH) in human plasma. In this work, antibodies against one end of PTH were immobilized onto a silica-based support and packed into a column. A second antibody, labeled with acridinium ester and directed against a different portion of PTH, was then combined with the sample and allowed to bind to PTH prior to injection. The resulting complex was then captured by the immobilized antibodies. After the other sample components and excess labeled antibodies had been washed from the antibody column, the labeled antibodies and PTH were eluted. This fraction was then passed through a postcolumn detector, where the chemiluminescence of the acridinium ester tags was monitored. This method could detect 10^{-17} mol of PTH in human plasma with a throughput of one sample every 6 min.

Chromatographic sandwich immunoassays have also been used for postcolumn immunodetection in reversed-phase liquid chromatography (RPLC). This was demonstrated in the analysis of bovine growth-hormone releasing factor (GHRF) [121]. To do this, a column containing anti-GHRF antibodies was placed after a RPLC column and used to capture any eluting analyte. Next, an excess of fluorescein-labeled anti-GHRF antibodies was passed through this antibody column to form a sandwich immune complex. The GHRF and associated labeled antibodies were then eluted and monitored by an on-line fluorescence

Table 29.6   Examples of Chromatographic Sandwich Immunoassays

Analyte	Detection Method and Tag	Assay Characteristics	References
TSH	Absorbance; HRP	Range 0–0.29 nM	[15]
hCG	Fluorescence; FITC	Range, 0–66.6 ng/ml	[15]
HSA	Fluorescence; FITC	LOD, 0.001 mg/ml	[16]
PTH, interleukin-5	Fluorescence; FITC	LOD, 10 μM; linear range, <250 μM	[17]
IgG	Absorbance; alkaline phosphatase	LOD, 3 fmol; range, 3.33–133 fmol	[47]
IgG	Fluorescence; alkaline phosphatase	Linear range, 0.5–50 pmol/l	[50]
HSA	Electrochemical; hemin	Range, 1–10 mg/ml	[65]
Anti-IgG	Electrochemical; glucose oxidase	Range, 3–225 fmol	[67]
IgG	Electrochemical; adenosine deaminase	Range, 5–400 ng/ml	[68]
IgG	Chemiluminescence; acridinium ester	Range, 0.2–20 fmol	[74]
Parathyroid hormone	Chemiluminescence; acridinium ester	LOD, 0.24 pM (16 amol); linear range, 0.24–67 pM	[75]

detector. Although this approach was able to detect GHRF at reasonably low levels, it was noted that several measures had to be taken to minimize the background caused by nonspecific adsorption of the labeled antibodies. In addition, it was necessary to modify the eluant from the RPLC column to allow binding of the analyte to the immobilized antibody column.

29.5.1.3   Theory

Although the theory of a chromatographic sandwich immunoassay has not been described in previous studies, it is possible to obtain equations for this method by adapting the expressions given earlier in this chapter. For instance, the general reaction of an analyte (A) with a soluble and labeled antibody (Ab^*) can be represented by Equation 29.12,

$$A + Ab^* \xleftrightarrow{K_a^*} A - Ab^* \tag{29.12}$$

where K_a^* is the association equilibrium constant for this process. If the labeled antibodies are present in a large excess and allowed to reach an equilibrium with A, then the fraction of analyte bound to these antibodies (α_{A-Ab^*}) is given by Equation 29.13,

$$\alpha_{A-Ab^*} = 1/(1 + K_a^*[Ab^*]) \tag{29.13}$$

in which $[Ab^*]$ is the concentration of labeled antibodies in solution.

The second step in this process involves the adsorption of $A - Ab^*$ to the immobilized antibody (Ab). This is described by Equation 29.14.

$$A - Ab^* + Ab \xleftrightarrow{k_a^*} Ab^* - A - Ab \tag{29.14}$$

In this reaction, k_a^* is the association rate constant for the binding of $A - Ab^*$ to Ab. As was true for the methods examined earlier in this chapter, it is assumed in this reaction that the release of any $A - Ab^*$ from the column during the application step is negligible, making this an essentially irreversible binding process.

Since the reaction in Equation 29.14 is essentially the same model used to describe a direct-detection assay, the response for a sandwich immunoassay can be obtained by combining Equations 29.3 and 29.13. This gives rise to a relative response equal to (α_{A-Ab^*} · B · Load A), as shown in Equation 29.15.

Relative response $= (\alpha_{A-Ab^*} \cdot B \cdot \text{Load } A)$

$$= \left\{ \text{Load } A - S_0 \ln \left[1 + \left(e^{(\text{Load } A/S_0)} - 1 \right) e^{-1/S_0} \right] \right\} \Big/ \left(1 + K_a^* \left[Ab^* \right] \right) \quad (29.15)$$

According to Equation 29.15, the same factors that affect the response of a direct-detection assay will also affect a sandwich immunoassay (i.e., S_0 and Load A). In addition, the amount of labeled antibody that is combined with the sample and the association constant for the binding of this antibody with the analyte will determine the amount of analyte that is eventually present in the sandwich complex. The amount of time allowed for the reaction between A and Ab^* is also important, since a maximum response will be obtained only when these two agents are allowed to approach an equilibrium.

29.5.1.4   Practical Considerations

A calibration curve for a sandwich immunoassay is typically constructed by plotting the relative response versus the amount of injected analyte. The effects of various experimental factors on this response are illustrated in Figure 29.17. For instance, to obtain good sensitivity, low flow rates should be used, since this allows more time for the analyte and analyte-antibody complex to bind to the column. This is represented by the term S_0 in Equation 29.15 and has the same effect on the response as seen earlier in Figure 29.10 for a direct-detection assay.

Equation 29.15 indicates that the degree of binding between the analyte and labeled antibodies also needs to be optimized. This is usually accomplished by using a large excess of these antibodies, which increases the product $K_a[Ab^*]$ and produces more of the complex $A - Ab^*$. Although this results in a bigger signal, it can also give rise to a greater background due to nonspecific binding of the labeled antibodies in the column. Thus, some intermediate concentration of labeled antibodies is usually optimal for this assay.

Another factor to consider in the design of a chromatographic sandwich immunoassay is the size of the immunoaffinity column. It is desirable to have a large excess of such antibodies to give efficient capture of the analyte. However, as the size of this column is increased, there will also be a larger surface area available for nonspecific binding. Thus, a column with a high capture efficiency but low surface area is best for such work.

29.5.2   One-Site Immunometric Assays

29.5.2.1   General Principles

The second type of noncompetitive immunoassay is a one-site immunometric assay. The basic format for this method is shown in Figure 29.18. First, the sample is incubated with a known excess of labeled antibodies or F_{ab} fragments that can bind to the analyte. This mixture, which now contains analytes bound to the labeled agents as well as excess labeled

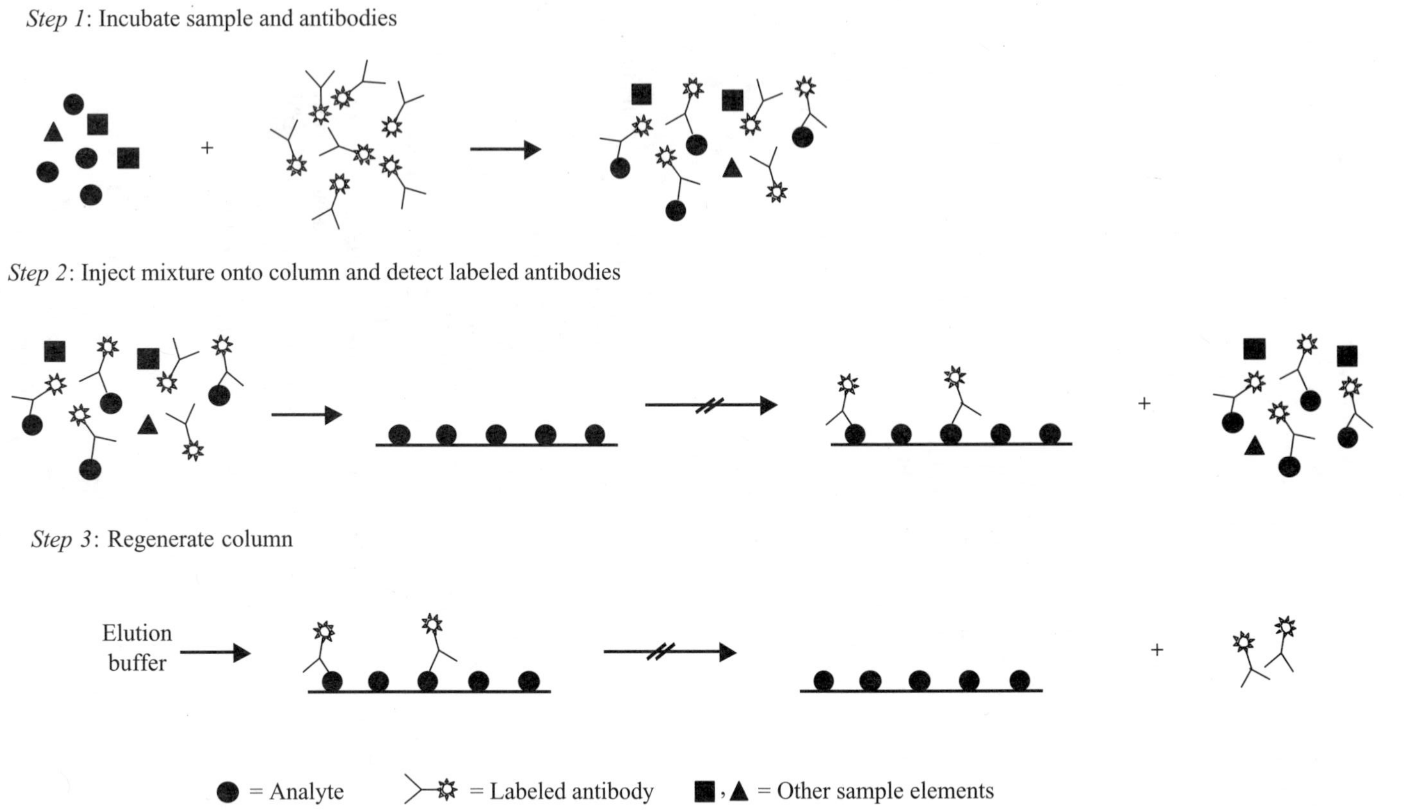

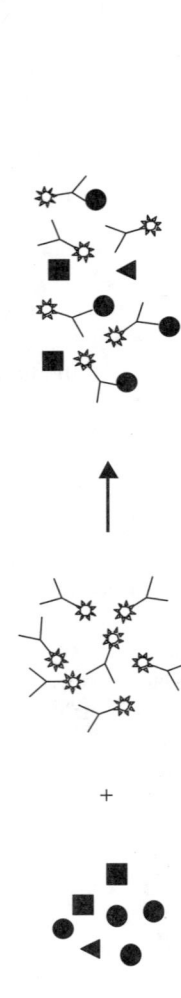

Step 1: Incubate sample and antibodies

Step 2: Inject mixture onto column and detect labeled antibodies

Step 3: Regenerate column

Elution
buffer

● = Analyte   = Labeled antibody   ■, ▲ = Other sample elements

Figure 29.18   Typical scheme for a flow-based one-site immunometric assay. In this method, the sample and labeled antibodies or F_{ab} fragments are first mixed and allowed to bind. The excess labeled binding agent is then removed by injecting this mixture onto a column that contains an immobilized analog of the analyte. The amount of labeled binding agent that elutes in either the retained or nonretained peak is then determined, providing a measure of the analyte in the original sample.

fragments and antibodies, is injected onto a column in which an analog of the analyte has been immobilized. This column is used to extract the excess antibodies or F_{ab} fragments, allowing those that are bound to the soluble analyte to pass through nonretained. The labeled antibodies or F_{ab} fragments in this nonretained peak give a signal proportional to the amount of analyte in the original sample.

A typical calibration curve for a one-site immunometric chromatographic immunoassay is constructed by plotting the amount of analyte vs. the response measured for the nonretained fraction of the labeled binding agent. As an alternative, the difference between the total and retained fraction of the labeled binding agent can be measured, resulting in a curve with an inverse relationship.

Although one-site immunometric assays have been used in only a few studies, they have shown several potential advantages over other chromatographic immunoassays. For instance, like competitive-binding immunoassays, these methods can detect both small and large solutes. They are also similar to sandwich immunoassays, since they can give a signal directly proportional to the amount of analyte. In addition, the fact that an immobilized analyte analog is used rather than an immobilized antibody creates the possibility of employing a wider range of elution conditions for column regeneration.

One disadvantage of this approach is that a different immobilized analog column must be used for each analyte. In addition, highly active and pure labeled antibodies or F_{ab} fragments are needed to provide a low background signal when monitoring the nonretained fraction. This means that special precautions must be taken when purifying these labeled agents to ensure that they do not lose a significant amount of their activity. It is also necessary to carefully monitor the stability of these agents during their storage and use.

29.5.2.2 Applications

A variety of analytes have been examined by one-site immunometric assays, as indicated in Table 29.7. A recent example is a method developed for thyroxine [122]. In this assay, thyroxine samples were mixed with an excess of acridinium ester-labeled F_{ab} fragments and applied to a column in which thyroxine had been immobilized through its amine group. This column was shown to extract the excess F_{ab} fragments while allowing the soluble thyroxine-F_{ab} complexes to pass through nonretained. These nonretained complexes were then passed into a postcolumn reactor for chemiluminescence detection. This method allowed one sample to be injected every 1.5 min, following a 20- to 45-min preincubation [122].

One-site immunometric assays have also been used for postcolumn immunodetection. This generally involves taking the eluant from an HPLC column and combining it with a solution of labeled antibodies or F_{ab} fragments that will bind the analyte. The mixture is then allowed to react in a mixing coil and is passed through an immunodetection column that contains an immobilized analog of the analyte. Examples of this approach have been reported for digoxin and related substances after their separation by RPLC [27] or a restricted-access column [26]. A similar approach has been used for human methionyl granulocyte colony-stimulating factor [24].

29.5.2.3 Theory

Little work has appeared on the theory of one-site immunometric assays. However, some general equations can again be obtained by using expressions similar to those employed for other chromatographic immunoassays. Like a sandwich immunoassay, a one-site

Table 29.7 Examples of Chromatographic One-Site Immunometric Assays

Analyte	Detection Method and Tag	Assay Characteristics	References
GCSF	Fluorescence; fluorescein	LOD, 1.5 ng/120 µl	[24]
PEG-GCSF	Fluorescence; fluorescein	LOD, 7.5 ng/120 µl	[24]
Digoxin and metabolites	Fluorescence; fluorescein	LOD, 2×10^{-10} M	[25]
Digoxin and metabolites	Fluorescence; fluorescein	LOD, 160 pg/ml; linear range, 0.2–2 nmol/l	[26]
Digoxin	Fluorescence; fluorescein	LOD, 200 fmol	[27]
Digoxigenin	Fluorescence; fluorescein	LOD, 50 fmol; linear range, 50–1000 fmol	[27]
Digoxin	Fluorescence; HRP, alkaline phosphatase, fluorescein	Fluorescence detection and enzyme label; LOD, 0.025 nM	[28]
Digoxigenin	Fluorescence; HRP, alkaline phosphatase, fluorescein	LOD, 0.01 nM	[28]
Interleukin-10	Fluorescence; fluorescein	LOD, 40 fmol	[29]
Digoxin	Absorbance; β-galactosidase	LOD, 0.2 µg/l	[41]
2,4-D	Electrochemical; alkaline phosphatase	LOD, 0.25 µg/l	[44]
Digoxigenin	Electrochemical; alkaline phosphatase	LOD, 0.5 amol; linear range, 0.38–7.7 fmol	[51]
DMFO	Fluorescence; HRP	LOD, 200 amol; linear range, 5×10^{-11} – 2.5×10^{-9} M	[55]
Fatty acid binding protein	Absorbance; HRP	Range, 2–12 µg/l and 12–2000 µg/l	[59]
Thyroxine	Chemiluminescence; HRP	LOD, 10^{-11} M	[63]
4-Amino-L- and D-phenylalanine	Chemiluminescence; acridinium ester	LOD, 1.76 pmol/ml	[77]
17-Estradiol	Fluorescence; liposome with carboxyfluorescein	—	[86]
Alpha-fetoprotein	Fluorescence; FITC	LOD, 0.1 ng/ml; linear range 0.5–60 ng/ml	[127]
Terbutryn	Grating coupler	LOD, 15 µg/l; linear range, 20–200 µg/l	[128]

immunometric assay involves two types of reactions. The first of these is the solution-phase reaction between the analyte and labeled antibodies or F_{ab} fragments (represented here by Ab^*). This reaction is summarized in Equation 29.16.

$$A + Ab^* \xleftrightarrow{K_a^*} A - Ab^* \qquad (29.16)$$

If this reaction is allowed to reach equilibrium and the concentration of Ab^* is much greater than that of A, then the fraction of analyte that is bound by Ab^* (α_{A-Ab^*}) can be obtained by using the following equation.

$$\alpha_{A-Ab^*} = 1 \big/ \left(1 + K_a^*\left[Ab^*\right]\right) \qquad (29.17)$$

It is this nonretained fraction of the analyte (i.e., the portion of A in the complex $A - Ab^*$) that is generally used for detection in a one-site immunometric assay.

If it is assumed that the column containing the immobilized analog has a high capture efficiency for the labeled binding agent and that there is an excess of this agent versus the analyte, then the relative response generated by the nonretained peak is given by Equation 29.18.

$$\text{Relative response} = \alpha_{A-Ab^*} \, \text{Load } A$$

$$= \text{Load } A/(1 + K_{a^*}[Ab^*]) \qquad (29.18)$$

Equation 29.18 indicates that the size of the nonretained signal in this method depends on the degree to which A has complexed with the labeled binding agent and the overall amount of A in the sample. This is illustrated by the curves shown in Figure 29.19. For the sake of convenience, the relationship in Equation 29.18 is given in terms of the relative moles of A that are applied to the column, but equivalent expressions can be written in terms of the concentration of A in the sample.

29.5.2.4 Practical Considerations

There are many factors to consider in the design of a one-site immunometric assay. For instance, Equation 29.18 indicates that the signal for this assay, as obtained for

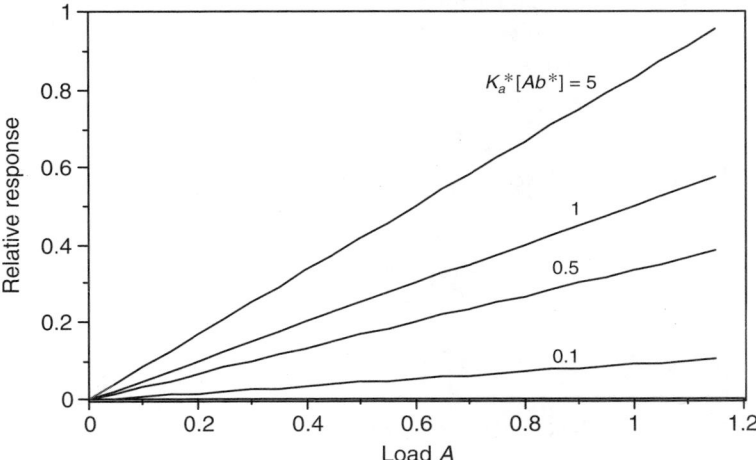

Figure 29.19 Theoretical calibration curves for a chromatographic one-site immunometric assay, as predicted by Equation 29.18. These curves assume that there is an excess of labeled binding agents versus the analyte and that there is 100% capture of the nonbound form of these labeled binding agents by the affinity column. Under these conditions, the response given by the nonretained peak will increase in direct proportion to the amount of analyte in the sample. As the degree of binding between the labeled binding agents and the analyte increases (i.e., as $K_a^*[Ab^*]$ becomes larger), the relative signal also increases.

the nonretained peak, is determined by the concentration of labeled binding agents that are combined with the analyte and the association constant between these reactants. The incubation time allowed for this reaction is also important, since a maximum response is obtained only when it has reached equilibrium, as is assumed in Equation 29.18.

To obtain effective capture of the excess binding agents, it is necessary to consider their flow rate and adsorption kinetics on the affinity column [122]. In addition, it is essential to verify that the column capacity is much larger than the amount of binding agent that is applied between regeneration steps. Finally, it is critical to have a pure and active preparation of the labeled binding agents. This is needed to create an assay in which all of the excess agent binds to the affinity column, giving a low background signal [122].

29.6   SUMMARY AND CONCLUSIONS

This chapter has examined the technique of chromatographic immunoassays and has shown how this approach can be used for chemical analysis. A large number of detection modes were described for these assays. This included absorbance and fluorescence detection, as well as enzymatic methods, liposome tags, and electrochemical labels. It was also shown how it is possible to use these approaches in either direct or indirect methods for analyte quantification. The direct-detection mode works best for moderate- to high-concentration analytes, while indirect methods like competitive-binding assays or immunometric methods are best for the determination of trace compounds.

Examples were given for these various assay formats, and several practical and theoretical aspects of each were considered. This included the presentation of equations that can be used to predict their response under certain conditions. Such methods have already been used in a broad range of applications [115, 123], including clinical analysis, environmental testing, and biotechnology. As more is learned about these assays, it is expected that further applications will appear.

SYMBOLS AND ABBREVIATIONS

A	Analyte
A^*	Labeled analog of the analyte (i.e., the label)
Ab	Antibody
Ab^*	Labeled antibody
$[A^*_{loaded}]$	Concentration of a labeled analog loaded onto an antibody column
ALP	Alkaline phosphatase
B	Fraction of bound label in the presence of analyte from a sample
B_0	Fraction of bound label in the absence of any analyte
B/B_0	Relative response of a competitive-binding immunoassay
BSA	Bovine serum albumin
CE	Capillary electrophoresis
cf-FAB-MS	Continuous-flow fast-atom-bombardment mass spectrometry
CFU	Colony-forming units
CI-MS	Chemical-ionization mass spectrometry
DMFO	α-(Difluoromethyl)ornithine

DNT	Dinitrotoluene
F	Flow rate
F_{ab}	Antigen-binding region of antibody
FIIA	Flow-injection immunoassay
FITC	Fluorescein isothiocyanate
GC	Gas chromatography
GCSF	Granulocyte colony-stimulating factor
GHRF	Growth-hormone releasing factor
hCG	Human chorionic gonadotropin
HPLC	High-performance liquid chromatography
HRP	Horseradish peroxidase
HSA	Human serum albumin
IgE	Immunoglobulin E
IgG	Immunoglobulin G
k_a	Association rate constant between an analyte and antibody
k_a^*	Association rate constant between $A - Ab^*$ and Ab
K_a^*	Association equilibrium constant between A and Ab^*
K_{disp}	Net equilibrium constant between A and A^* in a displacement immunoassay
Load A	Relative amount of A applied to a column versus the amount of ligand present (i.e., Load A = mol A/m_L)
Load A^*	Relative amount of A^* applied to a column versus the amount of ligand present (i.e., Load A^* = mol A^*/m_L)
LOD	Limit of detection
LyCH	Lucifer yellow CH
LyVS	Lucifer yellow vinyl sulfone
m_L	Moles of ligand in a column
PCB	Polychlorinated biphenyl
PEG-GCSF	Poly(ethylene glycol)-modified granulocyte colony-stimulating factor
PTH	Parathyroid hormone
r-AT III	Recombinant antithrombin III
r_{im}	Relative amount of displaced label in a displacement immunoassay
RPLC	Reversed-phase liquid chromatography
rt-PA	Recombinant tissue-type plasminogen activator
S_0	Term reflecting the flow rate and absorption kinetics of a chromatographic system, where $S_0 = F/(k_a m_L)$ for a system with simple adsorption-limited kinetics
SpA	Staphylococcal protein A
TELISA	Thermometric enzyme-linked immunosorbent assay
TMB	3,3′,5,5′-Tetramethylbenzidine
TNT	Trinitrotoluene
TSH	Thyroid-stimulating hormone
α_{A-Ab^*}	Fraction of analyte bound to labeled antibody
β-GAL	β-Galactosidase
2,4-D	2,4-Dichlorophenoxyacetic acid.

ACKNOWLEDGMENT

This work was supported by the National Institutes of Health under grant RO1 GM99431.

REFERENCES

1. Mattiasson, B., Borrebaeck, C., Sanfridson, B., and Mosbach, K., Thermometric enzyme linked immunosorbent assay: TELISA, *Biochim. Biophys. Acta,* 483, 221–227, 1977.

2. Thomas, D.H., Beck-Westermeyer, M., and Hage, D.S., Determination of atrazine in water using tandem high-performance immunoaffinity chromatography and reversed-phase liquid chromatography, *Anal. Chem.,* 66, 3823–3829, 1994.

3. Ren, X., Flow injection fluoroimmunoassay for human transferrin using a protein A immunoreactor, *Anal. Lett.,* 27, 1067–1074, 1994.

4. Palmer, D.A., Fernandez-Hernando, P., and Miller, J.N., A model online flow injection fluorescence immunoassay using a protein A immunoreactor and Lucifer yellow, *Anal. Lett.,* 26, 2543–2553, 1993.

5. Martin-Esteban, A., Fernandez, P., Perez-Conde, C., Gutierrez, A.M., and Camara, C., New fluorescence immunoassay for adrenocorticotropic hormone determination using flow injection analysis, *Anal. Quim. Int. Ed.,* 92, 37–40, 1996.

6. Palmer, D.A., Evans, M., Miller, J.N., and French, M.T., Rapid fluorescence flow injection immunoassay using a novel perfusion chromatographic material, *Analyst,* 119, 943–947, 1994.

7. Guo, J.C., Miller, J.N., Evans, M., and Palmer, D.A., Dual analyte flow injection fluorescence immunoassays using thiophilic gel reactors and synchronous scanning detection, *Analyst,* 125, 1707–1708, 2000.

8. Rico, C.M., Fernandez, M.D., Gutierrez, A.M., Conde, M.C.P., and Camara, C., Development of a flow fluoroimmunosensor for determination of theophylline, *Analyst,* 120, 2589–2591, 1995.

9. Turiel, E., Fernandez, P., Perez-Conde, C., Gutierrez, A.M., and Camara, C., Flow-through fluorescence immunosensor for atrazine determination, *Talanta,* 47, 1255–1261, 1998.

10. Garcinuno, R.M., Fernandez, P., Perez-Conde, C., Gutierrez, A.M., and Camara, C., Development of a fluoroimmunosensor for theophylline using immobilized antibody, *Talanta,* 52, 825–832, 2000.

11. Middendorf, C., Schulze, B., Freitag, R., Scheper, T., Howaldt, M., and Hoffmann, H., Online immunoanalysis for bioprocess control, *J. Biotechnol.,* 31, 395–403, 1993.

12. Pollema, C.H., Ruzicka, J., Christian, G.D., and Lernmark, A., Sequential injection immunoassay utilizing immunomagnetic beads, *Anal. Chem.,* 64, 1356–1361, 1992.

13. Valencia-Gonzalez, M.J. and Diaz-Garcia, M.E., Flow-through fluorescent immunosensing of IgG, *Ciencia,* 4, 29–40, 1996.

14. Yang, H.-H., Zhu, Q.-Z., Qu, H.-Y., Chen, X.-L., Ding, M.-T., and Xu, J.-G., Flow injection fluorescence immunoassay for gentamicin using sol-gel-derived mesoporous biomaterial, *Anal. Biochem.,* 308, 71–76, 2002.

15. Johns, M.A., Rosengarten, L.K., Jackson, M., and Regnier, F.E., Enzyme-linked immunosorbent assays in a chromatographic format, *J. Chromatogr. A,* 743, 195–206, 1996.

16. Yoshikawa, T., Terashima, M., and Katoh, S., Immunoassay using HPLAC and fluorescence-labeled antibodies, *J. Ferm. Bioeng.,* 80, 200–203, 1995.

17. Hayes, M.A., Polson, N.A., Phayre, A.N., and Garcia, A.A., Flow-based microimmunoassay, *Anal. Chem.,* 73, 5896–5902, 2001.

18. Pollema, C.H. and Ruzicka, J., Flow injection renewable surface immunoassay: A new approach to immunoanalysis with fluorescence detection, *Anal. Chem.,* 66, 1825–1831, 1994.

19. Lee, M., Durst, R.A., and Wong, R.B., Comparison of liposome amplification and fluorophore detection in flow-injection immunoanalyses, *Anal. Chim. Acta,* 354, 23–28, 1997.

20. Rabbany, S.Y., Kusterbeck, A.W., Bredehorst, R., and Ligler, F.S., Effect of antibody density on the displacement kinetics of a flow immunoassay, *J. Immunol. Methods,* 168, 227–234, 1994.

21. Whelan, J.P., Kusterbeck, A.W., Wemhoff, G.A., Bredehorst, R., and Ligler, F.S., Continuous-flow immunosensor for detection of explosives, *Anal. Chem.,* 65, 3561–3565, 1993.

22. Wemhoff, G.A., Rabbany, S.Y., Kusterbeck, A.W., Ogert, R.A., Bredehorst, R., and Ligler, F.S., Kinetics of antibody binding at solid-liquid interfaces in flow, *J. Immunol. Methods,* 156, 223–230, 1992.

23. Kusterbeck, A.W., Wemhoff, G.A., Charles, P.T., Yeager, D.A., Bredehorst, R., Vogel, C.W., and Ligler, F.S., A continuous flow immunoassay for rapid and sensitive detection of small molecules, *J. Immunol. Methods,* 135, 191–197, 1990.

24. Miller, K.J. and Herman, A.C., Affinity chromatography with immunochemical detection applied to the analysis of human methionyl granulocyte colony stimulating factor in serum, *Anal. Chem.,* 68, 3077–3082, 1996.

25. Irth, H., Oosterkamp, A.J., Tjaden, U.R., and van der Greef, J., Strategies for online coupling of immunoassays to HPLC, *Trends Anal. Chem.,* 14, 355–361, 1995.

26. Oosterkamp, A.J., Irth, H., Beth, M., Unger, K.K., Tjaden, U.R., and van de Greef, J., Bioanalysis of digoxin and its metabolites using direct serum injection combined with liquid chromatography and online immunochemical detection, *J. Chromatogr. B,* 653, 55–61, 1994.

27. Irth, H., Oosterkamp, A.J., van der Welle, W., Tjaden, U.R., and van der Greef, J., Online immunochemical detection in liquid chromatography using fluorescein-labeled antibodies, *J. Chromatogr.,* 633, 65–72, 1993.

28. Lindgren, A., Emneus, J., Marko-Varga, G., Irth, H., Oosterkamp, A., and Eremin, S., Optimization of a heterogeneous non-competitive flow immunoassay comparing fluorescein, peroxidase and alkaline phosphatase as labels, *J. Immunol. Methods,* 211, 33–42, 1998.

29. Kjellstrom, S., Emneus, J., and Marko-Varga, G., Flow immunochemical bio-recognition detection for the determination of interleukin-10 in cell samples, *J. Immunol. Methods,* 246, 119–130, 2000.

30. Onnerfjord, P. and Marko-Varga, G., Development of fluorescence based flow immunoassays utilising restricted access columns, *Chromatographia,* 51, 199–204, 2000.

31. Locascio-Brown, L., Martynova, L., Christensen, R.G., and Horvai, G., Flow immunoassay using solid-phase entrapment, *Anal. Chem.,* 68, 1665–1670, 1996.

32. Oosterkamp, A.J., Irth, H., Tjaden, U.R., and van der Greef, J., Online coupling of liquid chromatography to biochemical assays based on fluorescent-labeled ligands, *Anal. Chem.,* 66, 4295–4301, 1994.

33. Hermanson, G.T., *Bioconjugate Techniques*, Academic Press, New York, 1996.

34. Stewart, W.W., Lucifer dyes: Highly fluorescent dyes for biological tracing, *Nature,* 292, 17–21, 1981.

35. Keener, C.R., Wolfe, C.A.C., and Hage, D.S., Optimization of oxidized antibody labeling with Lucifer yellow CH, *BioTechniques,* 16, 894–897, 1994.

36. Stewart, W.W., Functional connections between cells as revealed by dye-coupling with a highly fluorescent naphthalimide tracer, *Cell,* 14, 741–759, 1978.

37. Butler, J.E., *Immunochemistry of Solid-Phase Immunoassay,* CRC Press, Boca Raton, FL, 1991.

38. Nistor, C., Rose, A., Wollenberger, U., Pfeiffer, D., and Emneus, J., A glucose dehydrogenase biosensor as an additional signal amplification step in an enzyme-flow immunoassay, *Analyst,* 127, 1076–1081, 2002.

39. Scheper, T., Brandes, W., Maschke, H., Ploetz, F., and Mueller, C., Two FIA-based biosensor systems studied for bioprocess monitoring, *J. Biotech.,* 31, 345–356, 1993.

40. Brandes, W., Maschke, H.E., and Scheper, T., Specific flow injection sandwich binding assay for IgG using protein A and a fusion protein, *Anal. Chem.,* 65, 3368–3371, 1993.

41. Freytag, J.W., Lau, H.P., and Wadsley, J.J., Affinity-column-mediated immunoenzymometric assays: Influence of affinity-column ligand and valency of antibody-enzyme conjugates, *Clin. Chem.,* 30, 1494–1498, 1984.

42. Meyer, U.J., Zhi, Z.-L., Meusel, M., Spener, F., and Loomans, E., Automated stand-alone flow injection immunoanalysis system for the determination of cephalexin in milk, *Analyst,* 124, 1605–1610, 1999.

43. Kaneki, N., Xu, Y., Kumari, A., Halsall, H.B., Heineman, W.R., and Kissinger, P.T., Electro-chemical enzyme immunoassay using sequential saturation technique in a 20-mL capillary: Digoxin as a model analyte, *Anal. Chim. Acta,* 287, 253–258, 1994.

44. Wilmer, M., Trau, D., Renneberg, R., and Spener, F., Amperometric immunosensor for the detection of 2,4-dichlorophenoxyacetic acid (2,4-D) in water, *Anal. Lett.,* 30, 515–525, 1997.

45. Palmer, D.A., Edmonds, T.E., and Seare, N.J., Flow injection electrochemical enzyme immunoassay for theophylline using a protein A immunoreceptor and *p*-aminophenyl phosphate *p*-aminophenol as the detection system, *Analyst,* 117, 1679–1682, 1992.

46. Kronkvist, K., Loevgren, U., Svenson, J., Edholm, L.-E., and Johansson, G., Competitive flow injection enzyme immunoassay for steroids using a post-column reaction technique, *J. Immunol. Methods,* 200, 145–153, 1997.

47. de Frutos, M., Paliwal, S.K., and Regnier, F.E., Liquid chromatography based enzyme-amplified immunological assays in fused-silica capillaries at the zeptomole level, *Anal. Chem.,* 65, 2159–2163, 1993.

48. Palmer, D.A., Edmonds, T.E., and Seare, N.J., Flow-injection immunosensor for theophylline, *Anal. Lett.,* 26, 1425–1439, 1993.

49. Mecklenburg, M., Lindbladh, C., Li, H., Mosbach, K., and Danielsson, B., Enzymic amplification of a flow-injected thermometric enzyme-linked immunoassay for human insulin, *Anal. Biochem.,* 212, 388–393, 1993.

50. Wang, Q., Wang, Y., Luo, G., and Yeung, W.S.B., Feasibility study of enzyme-amplified sandwich immunoassay using protein G capillary affinity chromatography and laser induced fluorescence detection, *J. Liq. Chromatogr.,* 24, 1953–1963, 2001.

51. Lovgren, U., Kronkvist, K., Backstrom, B., Edholm, L.-E., and Johansson, G., Design of non-competitive flow injection enzyme immunoassays for determination of haptens: Application to digoxigenin, *J. Immunol. Methods,* 208, 159–168, 1997.

52. Gonzalez-Martinez, M.A., Morais, S., Puchades, R., Maquieira, A., Abad, A., and Montoya, A., Development of an automated controlled-pore glass flow-through immunosensor for carbaryl, *Anal. Chim. Acta,* 347, 199–205, 1997.

53. Gascon, J., Oubina, A., Ballesteros, B., Barcelo, D., Camps, F., Marco, M.-P., Angel Gonzalez-Martinez, M., Morais, S., Puchades, R., and Maquieira, A., Development of a highly sensitive enzyme-linked immunosorbent assay for atrazine: Performance evaluation by flow injection immunoassay, *Anal. Chim. Acta,* 347, 149–162, 1997.

54. Kramer, P. and Schmid, R., Flow injection immunoanalysis (FIIA): A new immunoassay format for the determination of pesticides in water, *Biosens. Bioelectron.,* 6, 239–243, 1991.

55. Gunaratna, P.C. and Wilson, G.S., Noncompetitive flow injection immunoassay for a hapten, α-(difluoromethyl)ornithine, *Anal. Chem.,* 65, 1152–1157, 1993.

56. Gonzalez-Martinez, M.A., Penalva, J., Puchades, R., Maquieira, A., Ballesteros, B., Marco, M.P., and Barcelo, D., An immunosensor for the automatic determination of the antifouling agent Irgarol 1051 in natural waters, *Environ. Sci. Technol.,* 32, 3442–3447, 1998.

57. Kramer, P.M. and Schmid, R.D., Automated quasi-continuous immunoanalysis of pesticides with a flow injection system, *Pestic. Sci.,* 32, 451–462, 1991.

58. Katmeh, M.F., Godfrey, A.J.M., Stevenson, D., and Aherne, G.W., Enzyme immunoaffinity chromatography: A rapid semi-quantitative immunoassay technique for screening the presence of isoproturon in water samples, *Analyst,* 122, 481–486, 1997.

59. Kaptein, W.A., Korf, J., Cheng, S., Yang, M., Glatz, J.F.C., and Renneberg, R., Online flow displacement immunoassay for fatty acid-binding protein, *J. Immunol. Methods,* 217, 103–111, 1998.

60. Nilsson, M., Haakanson, H., and Mattiasson, B., Process monitoring by flow-injection immunoassay: Evaluation of a sequential competitive binding assay, *J. Chromatogr.,* 597, 383–389, 1992.

61. Nilsson, M., Mattiasson, G., and Mattiasson, B., Automated immunochemical binding assay (flow-ELISA) based on repeated use of an antibody column placed in a flow-injection system, *J. Biotechnol.,* 31, 381–394, 1993.

62. Liu, H., Yu, J.C., Bindra, D.S., Givens, R.S., and Wilson, G.S., Flow injection solid-phase chemiluminescent immunoassay using a membrane-based reactor, *Anal. Chem.,* 63, 666–669, 1991.

63. Aref'ev, A.A., Vlasenko, S.B., Eremin, S.A., Osipov, A.P., and Egorov, A.M., Flow-injection enzyme immunoassay of haptens with enhanced chemiluminescence detection, *Anal. Chim. Acta,* 237, 285–289, 1990.

64. Mattiasson, B., Svensson, K., Borrebaeck, C., Jonsson, S., and Kronvall, G., Non-equilibrium enzyme immunoassay of gentamicin, *Clin. Chem.,* 24, 1770–1773, 1978.

65. Karube, I., Matsunaga, T., Satoh, T., and Suzuki, S., A catalytic immunoreactor for the amperometric determination of human serum albumin, *Anal. Chim. Acta,* 156, 283–287, 1984.

66. De Alwis, U. and Wilson, G.S., Rapid heterogeneous competitive electrochemical immunoassay for IgG in the picomole range, *Anal. Chem.,* 59, 2786–2789, 1987.

67. de Alwis, W.U. and Wilson, G.S., Rapid sub-picomole electrochemical enzyme immunoassay for immunoglobulin G, *Anal. Chem.,* 57, 2754–2756, 1985.

68. Lee, I.H. and Meyerhoff, M.E., Rapid flow-injection sandwich-type immunoassays of proteins using an immobilized antibody reactor and adenosine deaminase-antibody conjugates, *Anal. Chim. Acta,* 229, 47–55, 1990.

69. Thorpe, G.H.G., Kricka, L.J., Moseley, S.B., and Whitehead, T.P., Phenols as enhancers of the chemiluminescent horseradish peroxidase-luminol-hydrogen peroxide reaction: Application in luminescence-monitored enzyme immunoassays, *Clin. Chem.,* 31, 1335–1341, 1985.

70. Butt, W., *Practical Immunoassay: The State of the Art,* Marcel Dekker, New York, 1984.

71. Rollag, J.G., Liu, T., and Hage, D.S., Optimization of post-column chemiluminescent detection for low-molecular-mass conjugates of acridinium esters, *J. Chromatogr. A,* 765, 145–155, 1997.

72. Dreveny, D., Seidl, R., Gubitz, G., and Michalowski, J., Development of solid-phase chemiluminescence immunoassays for digoxin comparing flow injection and sequential injection techniques, *Analyst,* 123, 2271–2276, 1998.

73. Hacker, A., Hinterleitner, M., Shellum, C., and Guebitz, G., Development of an automated flow injection chemiluminescence immunoassay for human immunoglobulin G, *Fresnius J. Anal. Chem.,* 352, 793–796, 1995.

74. Shellum, C. and Guebitz, G., Flow-injection immunoassays with acridinium ester-based chemiluminescence detection, *Anal. Chim. Acta,* 227, 97–107, 1989.

75. Hage, D.S. and Kao, P.C., High-performance immunoaffinity chromatography and chemiluminescent detection in the automation of a parathyroid hormone sandwich immunoassay, *Anal. Chem.,* 63, 586–595, 1991.

76. Hage, D.S., Taylor, B., and Kao, P.C., Intact parathyroid hormone: Performance and clinical utility of an automated assay based on high-performance immunoaffinity chromatography and chemiluminescence detection, *Clin. Chem.,* 38, 1494–1500, 1992.

77. Silvaieh, H., Schmid, M.G., Hofstetter, O., Schurig, V., and Gubitz, G., Development of enantioselective chemiluminescence flow- and sequential-injection immunoassays for α-amino acids, *J. Biochem. Biophys. Methods,* 53, 1–14, 2002.

78. Batzri, S. and Korn, E.D., Single bilayer liposomes prepared without sonication, *Biochim. Biophys. Acta,* 298, 1015–1019, 1973.

79. Locascio-Brown, L., Plant, A.L., and Durst, R.A., Liposome-based flow injection immunoassay system, *J. Res. Nat. Inst. Stand. Technol.,* 93, 663–665, 1988.

80. Plant, A.L., Locascio-Brown, L., Brizgys, M.V., and Durst, R.A., Liposome-enhanced flow injection immunoanalysis, *BioTechnology,* 6, 266–269, 1988.

81. Yap, W.T., Locascio-Brown, L., Plant, A.L., Choquette, S.J., Horvath, V., and Durst, R.A., Liposome flow injection immunoassay: Model calculations of competitive immunoreactions involving univalent and multivalent ligands, *Anal. Chem.,* 63, 2007–2011, 1991.

82. Locascio-Brown, L., Plant, A.L., Horvath, V., and Durst, R.A., Liposome flow injection immunoassay: Implications for sensitivity, dynamic range, and antibody regeneration, *Anal. Chem.,* 62, 2587–2593, 1990.

83. Locascio-Brown, L., Plant, A.L., Chesler, R., Kroll, M., Ruddel, M., and Durst, R.A., Liposome-based flow-injection immunoassay for determining theophylline in serum, *Clin. Chem.,* 39, 386–391, 1993.

84. Lee, M., Durst, R.A., and Wong, R.B., Development of flow-injection liposome immunoanalysis (FILIA) for imazethapyr, *Talanta,* 46, 851–859, 1998.

85. Lee, M. and Durst, R.A., Determination of imazethapyr using capillary column flow injection liposome immunoanalysis, *J. Agr. Food Chem.,* 44, 4032–4036, 1996.

86. Locascio-Brown, L. and Choquette, S.J., Measuring estrogens using flow injection immunoanalysis with liposome amplification, *Talanta,* 40, 1899–1904, 1993.

87. Hage, D.S., Taylor, R.L., and Kao, P.C., Improved recovery of a radiolabeled peptide with an albumin-treated reversed-phase HPLC column, *Clin. Chem.,* 38, 303–304, 1992.

88. Kamel, R., Landon, J., and Forrest, G.C., A fully automated, continuous-flow radioimmunoassay for methotrexate, *Clin. Chem.,* 26, 97–100, 1980.

89. He, Z. and Jin, W., Capillary electrophoretic enzyme immunoassay with electrochemical detection for thyroxine, *Anal. Biochem.,* 313, 34–40, 2003.

90. Phillips, T.M., Immunoaffinity measurement of recombinant granulocyte colony stimulating factor in patients with chemotherapy-induced neutropenia, *J. Chromatogr. B,* 662, 307–313, 1994.

91. Phillips, T.M., Measurement of total and bioactive interleukin-2 in tissue samples by immunoaffinity-receptor affinity chromatography, *Biomed. Chromatogr.,* 11, 200–204, 1997.

92. Zopf, D., Ohlson, S., Dakour, J., Wang, W., and Lundblad, A., Analysis and purification of oligosaccharides by high-performance liquid affinity chromatography, *Methods Enzymol.,* 179, 55–64, 1989.

93. Dakour, J., Lundblad, A., and Zopf, D., Separation of blood group A-active oligosaccharides by high-pressure liquid affinity chromatography using a monoclonal antibody bound to concanavalin A silica, *Anal. Biochem.,* 161, 140–143, 1987.

94. Phillips, T.M., More, N.S., Queen, W.D., and Thompson, A.M., Isolation and quantitation of serum IgE levels by high-performance immunoaffinity chromatography, *J. Chromatogr.,* 327, 205–211, 1985.

95. Mogi, M., Harada, M., Adachi, T., Kojima, K., and Nagatsu, T., Selective removal of β2-microglobulin from human plasma by high-performance immunoaffinity chromatography, *J. Chromatogr.,* 496, 194–200, 1989.

96. Alhama, J., Lopez-Barea, J., Toribio, F., and Roldan, J.M., Purification and determination of glutamine synthetase by high-performance immunoaffinity chromatography, *J. Chromatogr.,* 589, 121–126, 1992.

97. Beyer, K., Reinecke, M., Noe, W., and Scheper, T., Immunobased elution assay for process control, *Anal. Chim. Acta,* 309, 301–305, 1995.

98. Hage, D.S. and Walters, R.R., Dual-column determination of albumin and immunoglobulin G in serum by high-performance affinity chromatography, *J. Chromatogr.,* 386, 37–49, 1987.

99. Bouvrette, P. and Luong, J.H.T., Development of a flow injection analysis (FIA) immunosensor for the detection of *Escherichia coli, Int. J. Food Microbiol.,* 27, 129–137, 1995.

100. Schulze, B., Middendorf, C., Reinecke, M., Scheper, T., Noe, W., and Howaldt, M., Automated immunoanalysis systems for monitoring mammalian cell cultivation processes, *Cytotechnology,* 15, 259–269, 1994.

101. Delaunay-Bertoncini, N., Pichon, V., and Hennion, M.C., Comparison of immunoextraction sorbents prepared from monoclonal and polyclonal anti-isoproturon antibodies and optimization of the appropriate monoclonal antibody-based sorbent for environmental and biological applications, *Chromatographia,* 53, S224–S230, 2001.

102. Bouzige, M., Legeay, P., Pichon, V., and Hennion, M.C., Selective on-line immunoextraction coupled to liquid chromatography for the trace determination of benzidine, congeners and related azo dyes in surface water and industrial effluents, *J. Chromatogr. A,* 846, 317–329, 1999.

103. Ferrer, I., Hennion, M.C., and Barcelo, D., Immunosorbents coupled online with liquid chromatography/atmospheric pressure chemical ionization/mass spectrometry for the part per trillion level determination of pesticides in sediments and natural waters using low preconcentration volumes, *Anal. Chem.,* 69, 4508–4514, 1997.

104. Davoli, E., Fanelli, R., and Bagnati, R., Purification and analysis of drug residues in urine samples by on-line immunoaffinity chromatography/high-performance liquid chromatography/continuous-flow fast-atom-bombardment mass spectrometry, *Anal. Chem.,* 65, 2679–2685, 1993.

105. Hofmann, O., Voirin, G., Niedermann, P., and Manz, A., Three-dimensional microfluidic confinement for efficient sample delivery to biosensor surfaces: Application to immunoassays on planar optical waveguides, *Anal. Chem.,* 74, 5243–5250, 2002.

106. Dalluge, J.J. and Sander, L.C., Precolumn affinity capillary electrophoresis for the identification of clinically relevant proteins in human serum: Application to human cardiac troponin I, *Anal. Chem.,* 70, 5339–5343, 1998.

107. Guzman, N.A., Biomedical applications of online preconcentration-capillary electrophoresis using an analyte concentrator: Investigation of design options, *J. Liq. Chromatogr.,* 18, 3751–3768, 1995.

108. Perez, S., Ferrer, I., Hennion, M.C., and Barcelo, D., Isolation of priority polycyclic aromatic hydrocarbons from natural sediments and sludge reference materials by an anti-fluorene immunosorbent followed by liquid chromatography and diode array detection, *Anal. Chem.,* 70, 4996–5001, 1998.

109. Vanderlaan, M., Lotti, R., Siek, G., King, D., and Goldstein, M., Perfusion immunoassay for acetylcholinesterase: Analyte detection based on intrinsic activity, *J. Chromatogr. A,* 711, 23–31, 1995.

110. Stoecklein, W., Jaeger, V., and Schmid, R.D., Monitoring of mouse immunoglobulin G by flow-injection analytical affinity chromatography, *Anal. Chim. Acta,* 245, 1–6, 1991.

111. Janis, L.J. and Regnier, F.E., Dual-column immunoassays using protein G affinity chromatography, *Anal. Chem.,* 61, 1901–1906, 1989.

112. Cole, L.J. and Kennedy, R.T., Selective preconcentration for capillary zone electrophoresis using protein G immunoaffinity capillary chromatography, *Electrophoresis,* 16, 549–556, 1995.

113. Hage, D.S., Thomas, D.H., Chowdhuri, A.R., and Clarke, W., Development of a theoretical model for chromatographic-based competitive binding immunoassays with simultaneous injection of sample and label, *Anal. Chem.,* 71, 2965–2975, 1999.

114. Hage, D.S., Thomas, D.H., and Beck, M.S., Theory of a sequential addition competitive binding immunoassay based on high-performance immunoaffinity chromatography, *Anal. Chem.,* 65, 1622–1630, 1993.

115. Puchades, R. and Maquieira, A., Recent developments in flow injection immunoanalysis, *Crit. Rev. Anal. Chem.,* 26, 195–218, 1996.

116. Pichon, V., Bouzige, M., and Hennion, M.C., New trends in environmental trace-analysis of organic pollutants: Class-selective immunoextraction and clean-up in one step using immunosorbents, *Anal. Chim. Acta,* 376, 21–35, 1998.

117. Nelson, M.A., Reiter, W.S., and Hage, D.S., Chromatographic competitive binding immunoassays: A comparison of the sequential and simultaneous injection methods, *Biomed. Chromatogr.,* 17, 188–200, 2003.

118. Charles, P.T., Conrad, D.W., Jacobs, M.S., Bart, J.C., and Kusterbeck, A.W., Synthesis of a fluorescent analog of polychlorinated biphenyls for use in a continuous flow immunosensor assay, *Bioconj. Chem.,* 6, 691–694, 1995.

119. Clarke, W., Schiel, J.E., Moser, A., and Hage, D.S., Analysis of free hormone fractions by ultrafast immunoextraction/displacement immunoassay: Studies using free thyroxine as a model system, *Anal. Chem.,* 76, 1859–1866, 2005.

120. Rabbany, S.Y., Piervincenzi, R., Judd, L., Kusterbeck, A.W., Bredehorst, R., Hakansson, K., and Ligler, F.S., Assessment of heterogeneity in antibody-antigen displacement reactions, *Anal. Chem.,* 69, 175–182, 1997.

121. Cho, B.-Y., Zou, H., Strong, R., Fisher, D.H., Nappier, J., and Krull, I.S., Immunochromatographic analysis of bovine growth hormone releasing factor involving reversed-phase high-performance liquid chromatography-immunodetection, *J. Chromatogr. A,* 743, 181–194, 1996.

122. Oates, M.R., Clarke, W., Zimlich, A., II, and Hage, D.S., Optimization and development of a high-performance liquid chromatography-based one-site immunometric assay with chemiluminescence detection, *Anal. Chim. Acta,* 470, 37–50, 2002.

123. Hage, D.S., Survey of recent advances in analytical applications of immunoaffinity chromatography, *J. Chromatogr. B,* 715, 3–28, 1998.

124. Wang, Q., Luo, G., Wang, Y., and Yeung, W.S.B., Sandwich immunoassay for monoclonal antibody using protein G immunoaffinity capillary chromatography and diode laser induced fluorescence detection, *J. Liq. Chromatogr.,* 23, 1489–1498, 2000.

125. Schmalzing, D., Nashabeh, W., and Fuchs, M., Solution-phase immunoassay for determination of cortisol in serum by capillary electrophoresis, *Clin. Chem.,* 41, 1403–1406, 1995.

126. Cassidy, S.A., Janis, L.J., and Regnier, F.E., Kinetic chromatographic sequential addition immunoassays using protein A affinity chromatography, *Anal. Chem.,* 64, 1973–1977, 1992.

127. Wang, R., Lu, X., and Ma, W., Non-competitive immunoassay for α-fetoprotein using micellar electrokinetic capillary chromatography and laser-induced fluorescence detection, *J. Chromatogr. B,* 779, 157–162, 2002.

128. Bier, F.F., Jockers, R., and Schmid, R.D., Integrated optical immunosensor for *S*-triazine determination: Regeneration, calibration and limitations, *Analyst,* 119, 437–441, 1994.

30

Molecularly Imprinted Polymers:
Artificial Receptors for
Affinity Separations

Karsten Haupt

Compiègne University of Technology, Compiègne, France

CONTENTS

30.1 INTRODUCTION

The technique of molecular imprinting allows for the formation of specific recognition sites in synthetic polymers through the use of templates or imprint molecules. These recognition sites mimic the binding sites of biological receptors like antibodies and enzymes. The result is known as a molecularly imprinted polymer (MIP) or imprinted phase. An MIP can potentially be used in any application that requires a specific binding event, such as an affinity separation, a binding assay, or a biosensor. MIPs also have possible uses in organic synthesis and catalysis. The stability, ease of preparation, and low cost of MIPs make them particularly attractive for these applications. This chapter focuses on recent developments in molecular imprinting and the applications of MIPs in affinity separations and related techniques.

30.2 MOLECULARLY IMPRINTED POLYMERS

30.2.1 General Principle of Molecular Imprinting

The design and synthesis of biomimetic receptors that are capable of binding a target with similar affinity and specificity to antibodies has been a long-term goal of bioorganic chemistry. One technique that has been increasingly adopted for the generation of artificial, macromolecular receptors is the molecular imprinting of synthetic polymers [1, 2]. Molecular imprinting is a process in which functional monomers and cross-linking agents are copolymerized in the presence of a target (i.e., the *imprint molecule*), which acts as a molecular template. The functional monomers initially form a complex with the imprint molecule. Following polymerization, these functional groups are held in position by the highly cross-linked polymeric structure. Subsequent removal of the imprint molecule gives binding sites that are complementary in size and shape to the desired target or analyte. In this way, a molecular memory is introduced into the polymer, making it now capable of selectively rebinding this target (see Figure 30.1).

The complex formed between the monomers and imprint molecule can be produced through reversible covalent bonds or noncovalent interactions. Examples of the latter include hydrogen bonding, ionic bonds, hydrophobic interactions, and van der Waals forces. A combination of these two types of interactions can also be used.

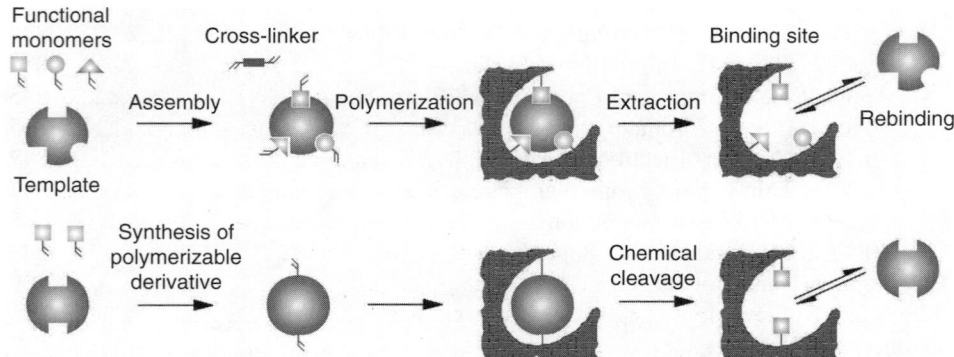

Figure 30.1   The molecular imprinting principle based on schemes for both the noncovalent (top) and covalent (bottom) methods.

Several items have to be considered when comparing covalent and noncovalent imprinting. The noncovalent approach, which was pioneered by Arshady and Mosbach [3], is more flexible in the possible choice of functional monomers, target molecules, and use of the imprinted materials. After polymerization by this method, the imprinted molecule can be removed by simple solvent extraction. However, it should be noted that the prepolymerization complex in this approach is a reversible system at equilibrium, with a stability that depends on the affinity constant between the imprint molecule and functional monomers. This may give rise to some heterogeneity in the imprinted binding sites.

Covalent imprinting was developed primarily by Wulff and Sarhan [4]. In this approach, a polymerizable derivative of the imprint molecule is synthesized. After creation of the polymer, the imprint molecule is removed through chemical cleavage. One problem with this method is that the association kinetics may be slow if the covalent bonds with the target have to be re-formed upon use of the polymer. On the other hand, the stability of these covalent bonds should yield a more homogeneous population of binding sites in the polymer. Moreover, the yield in binding sites relative to the amount of imprint molecule that is used (referred to as the *imprinting efficiency*) should be higher in this approach than it is for noncovalent protocols.

A simple demonstration of the molecular imprinting effect is shown in Figure 30.2. In this example, a nonimprinted copolymer of trifluoromethylacrylic acid and divinylbenzene was first synthesized. Increasing amounts of a template molecule (theophylline) were added to this monomer mixture before polymerization. When the resulting polymers were checked for their ability to bind a labeled analog of the template (i.e., ^{3}H-labeled theophylline), even a small quantity of template (i.e., a 1:5000 mole ratio versus the amount of functional monomer, trifluoromethylacrylic acid) was found to double the binding capacity of the polymer versus the same polymer in a nonimprinted form. Furthermore, it can be seen that as more template was used during polymerization, a higher capacity was obtained for the desired target in the imprinted polymer. In this particular case, a maximum capacity was reached at a mole ratio of 1:12 for the template versus functional monomer [5].

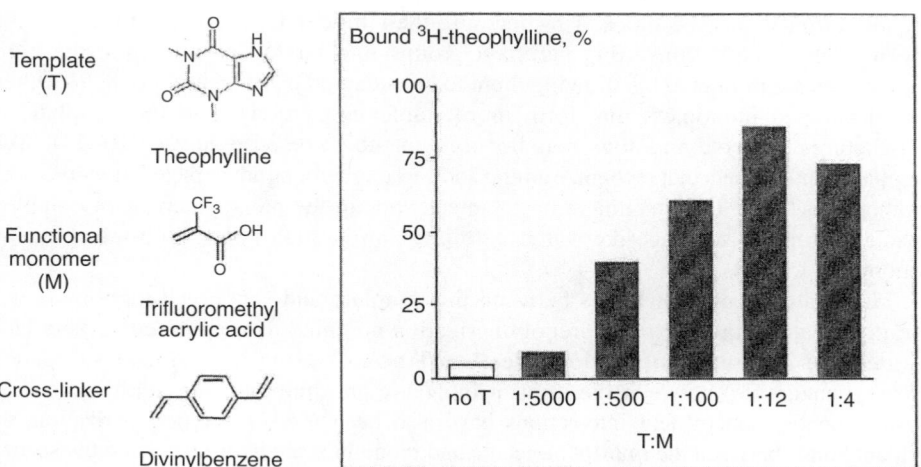

Figure 30.2 The imprinting effect. The left of this figure shows the ingredients for a theophylline-imprinted polymer. The plot on the right shows the measured binding capacity for radiolabeled theophylline on polymers that were prepared from these ingredients using different amounts of the template. (Data from Yilmaz, E., Mosbach, K., and Haupt, K., *Anal. Commun.*, 36, 167–170, 1999.)

30.2.2 Imprinting Matrix

Various materials can be used as the imprinting matrix in the creation of MIPs. Some examples that are considered in this section include acrylic and vinyl polymers, alternative organic polymers, and imprinting matrices such as silica and titanium dioxide.

30.2.2.1 *Acrylic and Vinyl Polymers*

Up to the present, the majority of reports on molecularly imprinted polymers have described organic polymers synthesized by radical polymerization of functional and cross-linking monomers that have vinyl or acrylic groups. Most have also used noncovalent interactions between monomers and the template. This can be attributed to the rather straightforward synthesis of these materials and to the large number of available monomers. These monomers can be basic (e.g., vinylpyridine) or acidic (e.g., methacrylic acid). They may also be permanently charged (3-acrylamidopropyl trimethylammonium chloride), hydrogen bonding agents (acrylamide), or hydrophobic substances (styrene).

These rather simple monomers normally have association constants with templates that are too low to form stable complexes. However, in the final polymer, the formation of several simultaneous interactions with the target, as well as the presence of a favorable entropy term, help assure tight binding between the imprint and target. These monomers have to be used in excess to shift the equilibrium toward complex formation. Association between the polymer and template is governed by an equilibrium that results in the formation of many different complexes. This results in a binding-site population that can be rather heterogeneous in terms of the number of functional groups that are incorporated. Some of these functional groups will even be randomly incorporated into the polymer without forming a binding site.

When using a Langmuir model to estimate the affinity constants for these imprinted polymers, models based on two or more sites usually fit the experimental data better than a one-site model [6]. In reality, however, the Langmuir model does not adequately describe most MIPs [7].

As a remedy to this problem of binding site heterogeneity, some have suggested that low-affinity sites be blocked by reacting their functional groups with a chemical reagent (for example converting carboxyl groups into methyl ester), while the high-affinity sites are protected by allowing them to be occupied by the template [8, 9]. Others have developed monomers that form more stable interactions with the template, or substructures thereof, and thus can be used in stoichiometric levels [10–14]. Two examples of monomers that recognize amino and carboxyl groups are depicted in Figure 30.3. Another possibility for obtaining stronger interactions in the prepolymerization complex, especially in polar solvents like water, is to use coordination bonds with metal chelate monomers [15–17].

Ultimately, covalent bonds between the template and functional monomers will give the greatest stability for a prepolymerization complex. Klein and coworkers have reported the imprinting of a tripeptide (Lys-Trp-Asp) using a sacrificial spacer (*O*-hydroxybenzamide) between the imprint molecule and monomer. In addition to these covalent bonds, noncovalent interactions have also been used. After polymerization, the covalent bonds between the imprint molecule and monomers are hydrolyzed and the spacers are eliminated, giving room and suitable functional groups for noncovalent interactions with the target molecule (Figure 30.4). In this example, the peptide rebinds to the polymer only through noncovalent interactions, which takes advantage of their faster association/dissociation kinetics versus covalent processes [18].

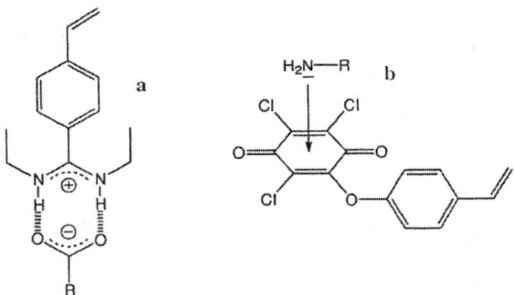

Figure 30.3 (a) An amidine functional monomer binding to a carboxyl group, and (b) a tetra-chloroquinone monomer complexing with an amino group. (Reproduced with permission from Wulff, G., Gross, T., and Schönfeld, R., *Angew. Chem. Int. Ed.*, 36, 1962–1964, 1997; and from Lübke, C., Lübke, M., Whitcombe, M.J., and Vulfson, E.N., *Macromolecules*, 33, 5098–5105, 2000.)

To obtain an optimized polymer for a given target analyte, combinatorial approaches to MIP synthesis have been used [5, 19, 20]. In this format, the ingredients for the imprinting recipe, such as the type and mole ratio of the functional monomers, are varied. This can be performed using automated procedures [19].

This combinatorial approach is illustrated by an MIP that was developed for the triazine herbicide terbutylazine. This MIP was optimized and created from a number of different MIPs that were synthesized on a small scale (i.e., approximately 55 mg) [20]. The functional monomer used in the final MIP was selected from a library of six candidates: methacrylic acid, methylmethacrylate, hydroxyethyl methacrylate, trifluoromethylacrylic acid, 4-vinylpyridine, and *N*-vinyl-α-pyrrolidone. Initial screening of these candidates was performed by determining which functional monomer retained the template most strongly. Among the six monomers tested, methyl methacrylate, 4-vinylpyridine, and *N*-vinyl-α-pyrrolidone led to polymers that allowed the imprint molecule to be rapidly and quantitatively extracted, while methacrylic acid and trifluoromethylacrylic acid led to polymers that strongly retained the template. Using these last two monomers, a secondary screening for selectivity was performed. For this screening, nonimprinted control polymers were also

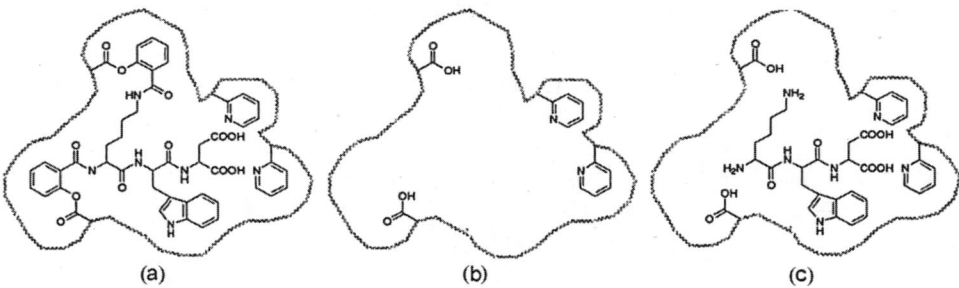

| (a) | (b) | (c) |

Figure 30.4 Molecular imprinting of the tripeptide Lys-Trp-Asp through covalent and noncovalent interactions. The figure in (a) shows a binding site with a covalently bound imprint molecule; (b) shows the binding site after chemical cleavage and extraction of the imprint molecule; and (c) shows rebinding of the imprint molecule through noncovalent interactions. (Reproduced with permission from Klein, J.U., Whitcombe, M.J., Mulholland, F., and Vulfson, E.N., *Angew. Chem. Int. Ed.*, 38, 2057–2060, 1999.)

prepared. Binding of the analyte to the MIPs and control polymers was next evaluated in the batch mode. In this study, the polymer that gave the highest selectivity for the desired analyte was found to be the one based on methacrylic acid [20].

30.2.2.2   Other Organic Polymers

In recent years, other polymers have appeared for use in preparing MIPs that are either better suited for a specific application or easier to synthesize in the desired form. For example, polymers such as polyphenols [21], poly(aminophenyl boronate) [22], poly(phenylene diamine) [23], poly(phenylene diamine-co-aniline) [24], polyurethanes [25], and overoxidized polypyrrole [26] have been used. Compared with polymers based on acrylic and vinyl monomers, the use of these other polymers seems to be somewhat restricted due to their limited choice of available functional monomers.

30.2.2.3   Other Imprinting Matrices

Sol gels like silica and titanium dioxide are now gaining in importance as imprinting matrices. Silica has been used as an imprinting matrix for inorganic ions [27] and organic molecules [28–32]. With this matrix, either the bulk material can be imprinted by the sol gel method, thus creating microporous materials with specifically arranged functional groups [27, 29, 32, 33], or an imprinted polysiloxane layer can be deposited at the silica's surface [28, 31, 34–36].

Recently, Katz and Davis have reported the molecular imprinting of bulk amorphous silica with single aromatic molecules by using a covalent monomer template complex, thereby creating shape-selective catalysts [33]. Through the use of physical adsorption experiments, they have been able to directly observe molecular-imprint-generated microporosity, with additional porosity being created in the silica upon template removal.

Another material that has been imprinted using the sol gel technique is titanium oxide [37–39]. For example, Willner and coworkers have functionalized the silica gate of an ion-sensitive field-effect transistor (ISFET) with a titanium dioxide film that included molecularly imprinted sites for 4-chlorophenoxyacetic acid or 2,4-dichlorophenoxyacetic acid. A titanium(IV) butoxide solution was reacted with the respective carboxylic acid, and the resulting mixture that included the titanium(IV) butoxidecarboxylate complex was deposited onto the ISFET gate. The sol gel polymerization of this mixture on the silica gate interface resulted in a titanium dioxide film with the embedded carboxylate. Treatment of this film with an ammonia solution resulted in the elimination of the carboxylate and the formation of imprinted molecular sites for the respective acid in the titanium dioxide film. The functionalized devices with the imprinted surfaces were used as chemical sensors and selectively sensed the sodium salts of the imprinted substrates [37–39]. A similar imprinting recipe was used by Kunitake and coworkers, who synthesized thin titanium dioxide films imprinted with carbobenzyloxy-L-alanine. The imprinted layers were used as the recognition element in an acoustic sensor (a quartz-crystal microbalance), which was able to selectively recognize carbobenzyloxy-L-amino acids [37–39].

30.2.3   Target Molecules

One of the attractive features of molecular imprinting is that it can be applied to a wide range of targets. The imprinting of small organic compounds such as pharmaceuticals, pesticides, amino acids, peptides, nucleotide bases, steroids, and sugars is now well established and considered almost routine. Metal ions and other ions have also been used as templates to induce the specific arrangement of functional groups in an imprinting matrix [27, 40–42].

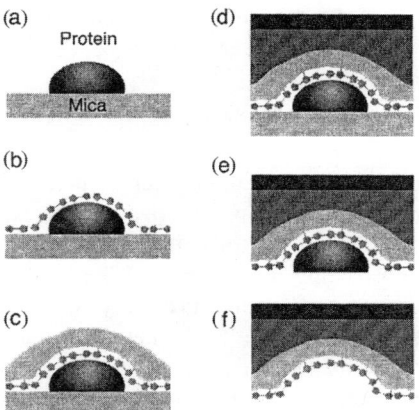

Figure 30.5 The generation of a protein imprint. In this example, the protein is (a) adsorbed to mica, (b) coated with disaccharide, and (c) overlaid with a fluoropolymer layer by plasma deposition. The polymer is then (d) glued to a glass substrate, (e) the mica is peeled off, and (f) the protein is removed, leaving a binding site. (Adapted with permission from Shi, H.Q., Tsai, W.B., Garrison, M.D., Ferrari, S., and Ratner, B.D., *Nature*, 398, 593–597, 1999.)

Larger organic compounds like peptides can be imprinted through similar approaches, but the imprinting of much larger structures is still a challenge. Specially adapted protocols have been proposed to create imprints of proteins in a thin layer of acrylic polymer on a silica surface [17]. Related work has examined the creation of imprints of cells using a lithographic technique [43] and imprints of the surface structure of mineral crystals [44].

Figure 30.5 shows one interesting approach that has been used to create imprints of proteins on a surface [45]. In this method, the protein of interest is first adsorbed onto an atomically flat mica surface. It is then spin-coated with a disaccharide solution, which upon drying forms a thin layer (1 to 5 nm) that is attached through multiple hydrogen bonds to the protein. This protective disaccharide shell is next covered with a fluoropolymer layer via glow-discharge plasma deposition, which covalently incorporates the sugar molecules. Finally, the polymer layer is attached to a glass substrate using an epoxy glue. After peeling off the mica, the protein is removed by treatment with an aqueous solution of sodium hydroxide and sodium hypochlorite (NaClO). This leaves nanocavities, as revealed by tapping-mode atomic force microscopy.

For these surface protein imprints, it has been reported that the resulting cavities are complementary in size and, to some extent, the functionality of the template protein. For example, it has been shown that a surface imprinted with bovine serum albumin (BSA) preferentially adsorbs this protein from a mixture that also contains immunoglobulin G. Moreover, an RNase A imprint preferentially adsorbed RNase A over lysozyme, which is similar to RNase A in its size and isoelectric point, and vice versa. This approach, although somewhat complex, might be applicable to the creation of other protein imprints.

30.2.4 Physical Forms and Preparation Methods for MIPs

Traditionally MIPs have been prepared as bulk polymer monoliths, which are then treated by mechanical grinding to obtain smaller micron-sized particles. The materials obtained through this somewhat inelegant method are useful for many applications. However, some applications require MIPs with a better-defined form. During the past few years, several aspects have been addressed in the development of such MIPs. These factors have included (1) the synthesis of small, spherical particles with sizes below the micron range, (2) the synthesis of thin layers, and (3) the creation of surface imprints.

30.2.4.1   Imprinted Particles

Imprinted particles like MIP nanobeads can be synthesized by methods such as precipitation polymerization and emulsion polymerization. Precipitation polymerization can be performed with prepolymerization mixtures similar to those used for bulk polymers, but with the relative amount of solvent in the mixture now being much higher. When polymerization progresses, imprinted nano- or microspheres precipitate instead of polymerizing together to form a macroporous polymer monolith. One drawback of this method is that, due to the dilution factor, higher amounts of the imprint molecule are needed than in traditional techniques. However, this is partly compensated for by the method's higher yields. This approach was used by Ye at al. to prepare imprinted particles for binding assays [46, 47]. It has been shown in some applications that these particles perform better than those obtained through mechanical grinding [48].

Wulff's group has used an approach that is similar to precipitation polymerization [49]. However, they adjust polymerization conditions so that soluble polymer microgels are produced. These microgels have molecular weights in the range of 10^6 g/mol, which places them close to proteins with respect to size. Although microgels could be synthesized using a monomer mixture adapted to the imprinting process, this method appears to be less straightforward as a means for obtaining selective imprinted materials, and more work in its optimization still needs to be performed.

Ishi-i et al. [50, 51] created "imprints" at the surface of fullerenes. This was accomplished by introducing two boronic acid groups into [60]fullerene using saccharides as template molecules. The resulting material gave regioselective and stereoselective rebinding of the saccharide [50, 51]. Later, Zimmerman et al. published a report on molecular imprinting inside dendrimers [52]. Their method involved the covalent attachment of dendrons to a porphyrin core (the template), cross-linking the end groups of the dendrons, and removal of the porphyrin template by hydrolysis. This technique appeared to yield homogeneous binding sites, allowed quantitative template removal, and produced only one binding site per polymer molecule. In addition, the materials were soluble in common organic solvents.

30.2.4.2   Thin-Imprinted Polymer Films

When a polymer is needed in the form of a thin film at a surface, one can choose between several standard techniques for polymer synthesis, such as spin coating or spray coating. In the MIP field, surface-bound films are often required (e.g., as in the construction of sensors). Several protocols have been used for this, most of which are based on *in situ* polymer synthesis. One elegant way of accomplishing this is to apply soft lithography [53]. This technique can create patterned surfaces with MIPs that are useful in multianalyte sensors and high-throughput screening systems [54]. Unfortunately, current imprinting recipes are not always compatible with the poly(dimethylsiloxane) stamps used for soft lithography.

MIPs can be synthesized at an electrode surface by electropolymerization [21, 23, 24, 26] or at a nonconducting surface by chemical grafting [22, 25]. Another recent development is the growth of an imprinted polymer from a surface by using polymerization initiators that are chemically bound [55, 56] or physically adsorbed [57] to the surface. This still results in binding sites that are contained in the bulk of the MIP layer. However, if a highly porous starting material is used and if the grafted layer is relatively thin, fast mass transfer will result. For example, Ulbricht and coworkers have photografted

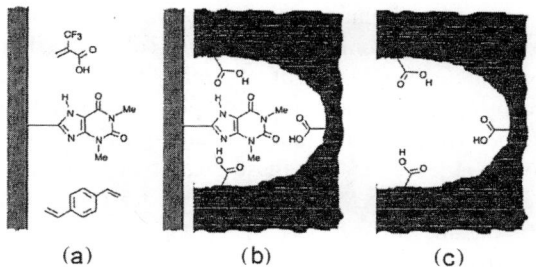

Figure 30.6 Molecular imprinting of theophylline immobilized onto a solid support. This figure shows (a) the immobilized template with monomers, (b) the composite material after polymerization, and (c) the imprinted polymer after dissolution of the support. (Reproduced with permission from Yilmaz, E., Haupt, K., and Mosbach, K., *Angew. Chem. Int. Ed.*, 39, 2115–2118, 2000.)

an MIP layer with a 10-nm thickness (in the dry state) onto a polypropylene membrane [55]. Sellergren's group has developed a method to synthesize MIP layers at the pore surface of porous silica particles. They were able to control the layer thickness to values between 0.8 and 7 nm [56].

30.2.4.3   Imprinting at Surfaces

Imprinted materials with binding sites situated at or close to the surface of the imprinting matrix have many advantages. For instance, the imprinted sites in these materials are more accessible and mass transfer is faster. The binding kinetics may also be faster, and target molecules conjugated with bulky labels can still bind. The reason that these materials are not universally used is because their preparation is less straightforward than for bulk polymers, and it requires specially adapted protocols.

Pérez and coworkers have developed a technique for imprinting at surfaces based on emulsion polymerization. In this method, small beads are created in an oil-in-water biphasic system that is stabilized by a surfactant. The imprint molecule (e.g., cholesterol) is part of the surfactant (i.e., pyridinium 12-(cholesteryloxycarbonyloxy)dodecane sulfate) [58]. As a result, all binding sites are situated at the particle surface, as was demonstrated by flocculation experiments using PEG-*bis*-cholesterol.

Another protocol for the creation of surface binding sites was described by Yilmaz, Haupt, and Mosbach. Here, the imprint molecule was immobilized onto a solid support, e.g., porous silica beads, prior to polymerization [59]. The pores were then filled with the monomer mixture, and the polymerization was initiated. The silica was removed by chemical dissolution, which left behind a porous polymeric structure that was a negative image of the original bead. The resulting binding sites were all situated at the surface of the polymer and were uniformly oriented (see Figure 30.6).

30.3   APPLICATIONS OF IMPRINTED POLYMERS IN AFFINITY SEPARATIONS

30.3.1   Liquid Chromatography

The first application for MIPs was their use as stationary phases in affinity chromatography. In particular, these were used for the separation of racemic mixtures of chiral compounds. This is possible because the imprinting process introduces enantioselectivity into polymers that are synthesized (in most cases) from nonchiral monomers.

A unique feature of MIPs versus conventional chiral stationary phases is that they are tailor-made for a specific target molecule, giving them a predetermined selectivity. For example, when a polymer is imprinted with the L-enantiomer of an amino acid, a column for high-performance liquid chromatography (HPLC) that contains this polymer will retain the L-enantiomer more than the D-enantiomer. However, a column containing a chemically identical nonimprinted polymer will not be able to separate these enantiomers. Typical values for the separation factor (α) of enantiomers on MIPs are between 1.5 and 5, but in some cases much higher values can be obtained.

A pronounced chiral selectivity has been observed with an MIP for the cinchona alkaloids cinchonidine and cinchonine, which gave a separation factor up to 31 [60]. It is even possible to prepare chromatographic supports selective for compounds that contain several chiral centers. For instance, a polymer imprinted with the dipeptide Ac-L-Phe-L-Trp-OMe was able to specifically recognize this imprint isomer over three other stereoisomers, where the LL form was more retained than the DD, DL, or LD forms [61].

If the molecule of interest contains more than two chiral centers (e.g., a carbohydrate), these properties of MIPs become even more important. As an example, in one study where polymers were imprinted against a glucose derivative, high selectivities were recorded between the various stereoisomers and anomers of glucose [62].

These examples suggest that good enantioseparations are achievable with MIPs. But in reality, the resolutions that have been obtained with MIPs in these separations are rather low. The same is true for the efficiencies of MIP-based columns, which are typically only 2000 to 5000 plates/m. The low resolution and efficiencies are both due to the severe peak broadening and tailing that is often seen for MIP supports, especially for the more strongly retained enantiomer. This, in turn, can be attributed to a heterogeneous population of binding sites in the MIP (with respect to their affinities and accessibilities) and to the low functional capacity of these materials [63].

Another problem with MIPs is that a fraction of the template molecules often cannot be extracted from the imprint after polymerization. This occurs because some of the template is deeply buried in the polymer. Even if extraction is possible, part of the resulting imprint sites have such a low accessibility that they are useless in chromatographic applications.

To obtain a mechanically stable material that is suitable for chromatography, a large percentage of the added monomers must be cross-linkers (i.e., typically 80 to 90% for bifunctional cross-linkers). This limits the amount of functional monomers and template molecule that can be added. Moreover, if a noncovalent imprinting protocol is used, the functional monomer has to be present in excess to shift the equilibrium toward complex formation. This inevitably results in some monomers being randomly distributed in the polymer rather than situated in binding sites, which leads to the creation of weak or nonspecific binding sites.

Numerous attempts have been made to improve the performance of MIPs and to avoid these problems. The easiest approach is to optimize the conditions for the selected MIP material. This can be done by optimizing the separation protocol used with this material, including the separation temperature, mobile-phase composition, and use of competing agents or gradient elution protocols [64] to improve peak shapes. It is also possible to chemically block the nonspecific or low-affinity sites in MIPs [8]. However, these same sites may be the ones that actually come into play in chromatography because of their faster kinetics than high-affinity sites. Thus, this approach tends to give only limited improvements in chromatographic behavior. Probably the best approach to overcome these problems is a "preventive" one in which efforts are made to synthesize or select better imprints.

During the last few years, work with MIPs in chromatography has focused on two key aspects. The first of these is the synthesis of uniformly shaped and sized particles with narrow pore-size distributions and improved mass-transfer properties. The second is the development of MIPs with better quality binding sites, ideally using stoichiometric ratios of the template and functional monomer.

Uniformly sized spherical MIP particles for chromatography can be synthesized in a variety of ways. These include organic-in-water suspension polymerization, suspension polymerization using perfluorocarbon liquids as the dispersing phase [65], and multistep swelling procedures [66]. These materials should have better chromatographic behavior than the more common ground bulk polymers. For example, it has been shown that a 25 cm × 4 mm I.D. column filled with MIP beads prepared through suspension polymerization in perfluorocarbon could resolve 1 mg of Boc-DL-Phe at flow rates up to 5 ml/min, a result not easily obtained with ground bulk polymer [65].

30.3.2 Capillary Electrochromatography

Capillary electrochromatography (CEC) might be one of the more promising chromatographic techniques to be used in combination with MIPs, in particular for chiral separations [67, 68]. *MIP-capillary electrochromatography* (MIP-CEC) profits from the inherent separation power of this method. Compared with MIP-based HPLC, better resolutions (due to efficiencies in CEC of more than 100,000 plates/m [67]) and larger separation factors can be achieved.

In one study, a chiral separation of the β-blockers propranolol and metoprolol was achieved with MIP-CEC. The polymer for this separation was cast *in situ* in the capillary in the form of a macroporous monolith attached to the inner wall. This capillary was prepared and conditioned within a few hours [69]. When this capillary was used in CEC, the components of racemic propranolol were resolved within only 120 sec, as shown in Figure 30.7. When samples were injected that contained mainly the *R*-enantiomer of propranolol, small amounts (1%) of the *S*-enantiomer could also be distinguished by this approach. Other possible uses for CEC and capillary electrophoresis include MIPs that are in the form of continuous polymer rods [70], as particles included in a gel matrix [71], or as small particles suspended in the carrier electrolyte [72].

30.3.3 Thin-Layer Chromatography

MIPs have also been used as stationary phases in thin-layer chromatography (TLC). Although the number of publications in this area is limited, the speed, simplicity, and ability of TLC to analyze parallel samples may make MIPs useful in TLC for determining the enantiomeric purity of compounds such as chiral drugs.

The first use of MIPs in TLC was by Mosbach and coworkers [73]. They performed chiral TLC using a finely ground imprinted polymer coated onto an inert support. They found that several racemic amino acids could be resolved. But problems were encountered due to band broadening, which led to the formation of zones rather than small spots or thin bands. This, in turn, led to band overlap and poor resolution, making it difficult to measure the retention factors. Further work in the optimization of particle shape, size, and porosity could result in an improved band shape and better resolution in such a method.

Recently, Suedee et al. reported similar work in which they used imprinted polymer particles of submicron size for TLC [74, 75]. The polymer in this study was mixed with calcium sulfate and wet-coated onto microscope slides. With this approach, the separation of enantiomers (e.g., ephedrine) and diastereomers (e.g., quinine/quinidine) was possible on a polymer imprinted with quinine, with separation factors typically below 1.5 [74]. However, they also reported that some of the quinine imprint molecule remained in the MIP after processing of the particles. It is therefore not certain whether the separations

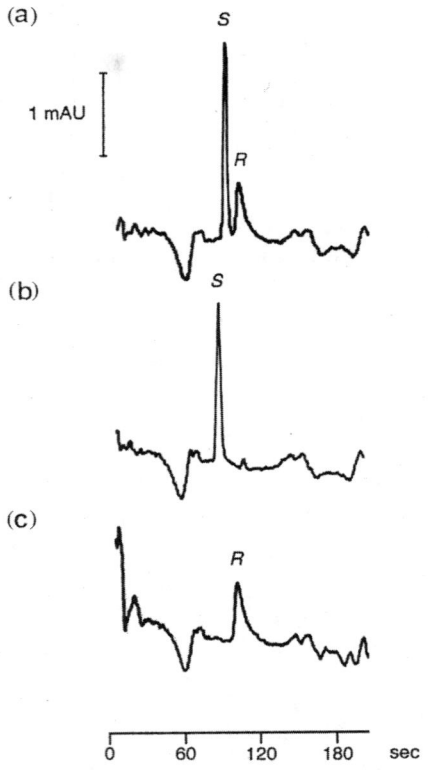

Figure 30.7 Chromatograms obtained by capillary electrochromatography for (a) racemic pro-pranolol, (b) *S*-propranolol, and (c) *R*-propranolol on an *R*-propranolol-imprinted polymer. (Adapted with permission from Schweitz, L., Andersson, L.I., and Nilsson, S., *Anal. Chem.*, 69, 1179–1183, 1997.)

observed with this material were solely due to the imprinted sites or also due to the chiral template remaining in the polymer.

30.3.4 Solid-Phase Extraction

In recent years, the separation technique that has been considered most for use with MIPs is solid-phase extraction (SPE) [76–82]. This is due to the ongoing need for efficient methods that can concentrate and clean up samples for medical, food, or environmental analyses.

Compared with liquid-liquid extraction, SPE is faster, more reproducible, and provides cleaner extracts. In SPE, emulsion formation is also not an issue, solvent consumption is reduced, and smaller sample sizes can be used. Moreover, SPE can be easily incorporated into automated procedures. Advantages of using molecular imprinting with SPE include the low price and stability of MIPs in different environments. In addition, some of the disadvantages of MIPs (e.g., their low efficiency and resolution) are less important in SPE than in other separation techniques, since SPE works in an on/off retention mode.

MIPs are not only more selective than traditional SPE materials, such as C_{18} or ion-exchange phases, but they are also more stable (and can be more selective) than immunoextraction matrices [83]. Since MIPs are compatible with organic solvents, *MIP solid-phase extraction* (MIP-SPE) can be performed directly after a solvent-extraction step.

These characteristics make SPE the most promising application for MIPs. This is also the application closest to commercialization, as is reflected in the large number of reports that have used MIP-SPE with real samples. For instance, MIP-SPE has been used to extract analytes from plasma and serum [80, 84], urine [85], bile [80], liver extracts [77], chewing gum [78], environmental water and sediment samples [86], and plant tissues [82].

A good example of the utility of imprinted polymers in SPE is given by their use in quantitating the herbicide atrazine in beef liver [77]. In this analysis, atrazine was first extracted from liver tissue with chloroform. An imprinted polymer was then used to clean this chloroform extract and to further concentrate the atrazine prior to measurement. In this example, the binding capacity of the polymer for atrazine in chloroform was found to be 19 µmol/g. The atrazine was eluted from this polymer with acetonitrile containing 10% acetic acid. It was then dried and reconstituted in acetonitrile or buffer and analyzed by reversed-phase HPLC or an enzyme-linked immunosorbent assay (ELISA). When the purified and nonpurified chloroform extracts were compared by reversed-phase HPLC, it was found that the MIP-SPE step considerably improved the accuracy and the precision of the HPLC method while lowering the detection limit for atrazine from 20 ppb to 5 ppb. This was made possible through the removal of interfering components in the sample, resulting in baseline resolution for the atrazine peak. Furthermore, when the MIP-SPE step was included the recovery for atrazine increased from 60.9% to 88.7% (as measured by HPLC) and from 79.6 to 92.8% (as determined by ELISA).

When MIPs are used as SPE materials, one of their more troublesome features is template leakage. Generally, once an MIP has been synthesized, it is subjected to exhaustive solvent extraction to remove the template from the polymer matrix. The difficulty in extracting 100% of the template molecule from an imprinted polymer has long been recognized, although it was widely believed that the few percent of template remaining within the polymer was permanently entrapped. However, recent work has demonstrated that this is not necessarily the case. What can and does occur is slow leakage of a portion of the remaining template from the polymer matrix over a period of time, even after exhaustive extraction of the polymer. This can have serious implications when the polymer is to be used as an SPE sorbent in trace and ultratrace analyses.

Some have tried, with some success, to find suitable washing procedures that allow for complete template removal from MIPs [87]. One possible method for circumventing this bleeding problem is to use a target molecule analog during the imprinting step rather than the target molecule itself [79, 88, 89]. This approach was first described by Andersson et al. [79]. In their report, MIPs were used for concentrating the drug sameridine from human plasma prior to the measurement of this drug by gas chromatography (GC). At the nanomolar drug levels used in the study, leakage of template from the polymer matrix during sample handling was easily detected by GC, which led to large errors in the drug's measurement. To fix this problem, a close structural analog of sameridine (see Figure 30.8) was used as the template molecule in the imprinting step, which gave an imprinted polymer that still displayed a strong affinity for sameridine. Following solid-phase extraction of sameridine from human plasma with this polymer, leakage of the analog from the polymer matrix did occur, but this analog could now be resolved from sameridine by the GC system. The results were comparable with those obtained through liquid-liquid extraction, with the added advantage that the injected sample now contained fewer matrix contaminants.

30.3.5 Membrane-Based Separations

Chromatographic separations are well established and widely used. However, they do have some limitations, especially in their scale-up. For large-scale work, they are often replaced

(a)

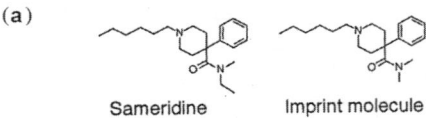

Sameridine   Imprint molecule

(b)

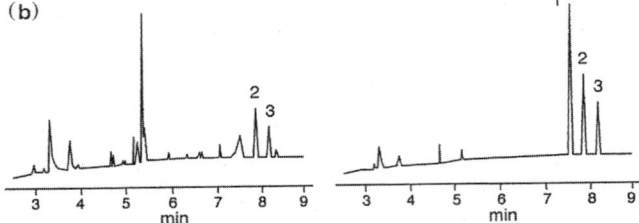

Figure 30.8 Solid-phase extraction of sameridine using an MIP. The figure in (a) shows the structures of sameridine and an analog that was used as the imprint molecule. The figures in (b) are GC chromatograms for a solvent extract (left) and an MIP extract (right) of a plasma sample. Peaks 1 to 3 in these chromatograms are (1) the imprint molecule, (2) sameridine, and (3) an internal standard. (Adapted with permission from Andersson, L.I., Paprica, A., and Arvidsson, T., *Chromatographia*, 46, 57–62, 1997.)

by membrane-based techniques that, in most cases, can be used in continuous mode rather than the typical batchwise operation of chromatography.

Imprinted membranes have been prepared in several ways. For instance, they can be cast directly as a thin layer on a flat surface [90] or placed between two surfaces [91]. Another way to obtain MIP membranes has been proposed by Ulbricht and coworkers, who photografted an MIP layer of 10 nm (in the dry state) onto a polypropylene membrane by physically adsorbing the photoinitiator onto the membrane [55]. Alternatively, MIP membranes can be prepared by a phase-inversion precipitation technique [92].

Imprinted membranes have great potential in chiral separations but can also be used as recognition elements in biomimetic sensors [93, 94]. Depending on the structure of the membrane, the target molecule can be selectively retained by the membrane [95]. If pore flux is limited, selective transport through the membrane may take place. For example, a freestanding membrane imprinted with 9-ethyladenine gave faster transport for adenosine than guanosine [90]. Also, a theophylline-imprinted membrane cast on the surface of a porous aluminum membrane transported theophylline faster than caffeine, which has a structure related to that of theophylline [96]. Such membranes could be useful in continuous separation processes.

30.4 SUMMARY AND CONCLUSIONS

MIPs have already been used in several applications, and they are close to commercialization in areas such as solid-phase extraction. However, more work needs to be performed to make these alternatives to affinity supports that use biomolecules as ligands. In particular, work is needed in the development of MIPs that contain a more homogeneous binding-site population, that have a higher affinity for targets, and that can be routinely used in aqueous solvents. Much of the current research with MIPs is already dealing with these problems. Meanwhile, the stability of MIPs, their low cost, and the fact that they can be tailor-made for analytes for which a biological receptor cannot be found are all properties that make them attractive and suitable for many possible applications in affinity separations.

SYMBOLS AND ABBREVIATIONS

BSA	Bovine serum albumin
CEC	Capillary electrochromatography
ELISA	Enzyme-linked immunosorbent assay
GC	Gas chromatography
HPLC	High-performance liquid chromatography
ISFET	Ion-sensitive field-effect transistor
MIP	Molecularly imprinted polymer
MIP-CEC	Molecularly imprinted polymer-capillary electrochromatography
MIP-SPE	Molecularly imprinted polymer solid-phase extraction
SPE	Solid-phase extraction
TLC	Thin-layer chromatography
α	Separation factor

REFERENCES

1. Sellergren, B., *Molecularly Imprinted Polymers: Man-Made Mimics of Antibodies and Their Applications in Analytical Chemistry,* Elsevier, Amsterdam, 2001.
2. Komiyama, M., Takeuchi, T., Mukawa, T., and Asanuma, H., *Molecular Imprinting: from Fundamentals to Applications,* Wiley-VCH, Weinheim, Germany, 2002.
3. Arshady, R. and Mosbach, K., Synthesis of substrate-selective polymers by host-guest polymerization, *Makromol. Chem.,* 182, 687–692, 1981.
4. Wulff, G. and Sarhan, A., The use of polymers with enzyme-analogous structures for the resolution of racemates, *Angew. Chem. Int. Ed.,* 11, 341, 1972.
5. Yilmaz, E., Mosbach, K., and Haupt, K., Influence of functional and cross-linking monomers and the amount of template on the performance of molecularly imprinted polymers in binding assays, *Anal. Commun.,* 36, 167–170, 1999.
6. Vlatakis, G., Andersson, L.I., Müller, R., and Mosbach, K., Drug assay using antibody mimics made by molecular imprinting, *Nature,* 361, 645–647, 1993.
7. Umpleby, R.J., Bode, M., and Shimizu, K.D., Measurement of the continuous distribution of binding sites in molecularly imprinted polymers, *Analyst,* 125, 1261–1265, 2000.
8. McNiven, S., Yokobayashi, Y., Cheong, S.H., and Karube, I., Enhancing the selectivity of molecularly imprinted polymers, *Chem. Lett.,* 12, 1297–1298, 1997.
9. Umpleby, R.J., Rushton, G.T., Shah, R.N., Rampey, A.M., Bradshaw, J.C., Berch, J.K., and Shimizu, K.D., Recognition directed site-selective chemical modification of molecularly imprinted polymers, *Macromolecules,* 34, 8446–8452, 2001.
10. Steinke, J.H.G., Dunkin, I.R., and Sherrington, D.C., A simple polymerisable carboxylic acid receptor: 2-acrylamido pyridine, *Trends Anal. Chem.,* 18, 159–164, 1999.
11. Spivak, D. and Shea, K.J., Molecular imprinting of carboxylic acids employing novel functional macroporous polymers, *J. Org. Chem.,* 64, 4627–4634, 1999.
12. Wulff, G., Gross, T., and Schönfeld, R., Enzyme models based on molecularly imprinted polymers with strong esterase activity, *Angew. Chem. Int. Ed.,* 36, 1962–1964, 1997.
13. Yano, K., Tanabe, K., Takeuchi, T., Matsui, J., Ikebukuro, K., and Karube, I., Molecularly imprinted polymers which mimic multiple hydrogen bonds between nucleotide bases, *Anal. Chim. Acta,* 363, 111–117, 1998.
14. Lübke, C., Lübke, M., Whitcombe, M.J., and Vulfson, E.N., Imprinted polymers prepared with stoichiometric template-monomer complexes: Efficient binding of ampicillin from aqueous solutions, *Macromolecules,* 33, 5098–5105, 2000.
15. Mallik, S., Johnson, R.D., and Arnold, F.H., Synthetic bis-metal ion receptors for bis-imidazole "protein analogs," *J. Am. Chem. Soc.,* 116, 8902–8911, 1994.

16. Hart, B.R. and Shea, K.J., Synthetic peptide receptors: Molecularly imprinted polymers for the recognition of peptides using peptide-metal interactions, *J. Am. Chem. Soc.,* 123, 2072–2073, 2001.

17. Kempe, M., Glad, M., and Mosbach, K., An approach towards surface imprinting using the enzyme ribonuclease A, *J. Mol. Recognit.,* 8, 35–39, 1995.

18. Klein, J.U., Whitcombe, M.J., Mulholland, F., and Vulfson, E.N., Template-mediated synthesis of a polymeric receptor specific to amino acid sequences, *Angew. Chem. Int. Ed.,* 38, 2057–2060, 1999.

19. Takeuchi, T., Fukuma, D., and Matsui, J., Combinatorial molecular imprinting: An approach to synthetic polymer receptors, *Anal. Chem.,* 71, 285–290, 1999.

20. Lanza, F. and Sellergren, B., Method for synthesis and screening of large groups of molecularly imprinted polymers, *Anal. Chem.,* 71, 2092–2096, 1999.

21. Panasyuk, T.L., Mirsky, V.M., Piletsky, S.A., and Wolfbeis, O.S., Electropolymerized molecularly imprinted polymers as receptor layers in capacitive chemical sensors, *Anal. Chem.,* 71, 4609–4613, 1999.

22. Piletsky, S.A., Piletska, E.V., Chen, B., Karim, K., Weston, D., Barrett, G., Lowe, P., and Turner, A.P.F., Chemical grafting of molecularly imprinted homopolymers to the surface of microplates: Application of artificial adrenergic receptor in enzyme-linked assay for β-agonists determination, *Anal. Chem.,* 72, 4381–4385, 2000.

23. Malitesta, C., Losito, I., and Zambonin, P.G., Molecularly imprinted electrosynthesized polymers: New materials for biomimetic sensors, *Anal. Chem.,* 71, 1366–1370, 1999.

24. Peng, H., Liang, C., Zhou, A., Zhang, Y., Xie, Q., and Yao, S., Development of a new atropine sulfate bulk acoustic wave sensor based on a molecularly imprinted electrosynthesized copolymer of aniline with *O*-phenylenediamine, *Anal. Chim. Acta,* 423, 221–228, 2000.

25. Dickert, F.L., Tortschanoff, M., Bulst, W.E., and Fischerauer, G., Molecularly imprinted sensor layers for the detection of polycyclic aromatic hydrocarbons in water, *Anal. Chem.,* 71, 4559–4563, 1999.

26. Deore, B., Chen, Z., and Nagaoka, T., Potential-induced enantioselective uptake of amino acid into molecularly imprinted overoxidized polypyrrole, *Anal. Chem.,* 72, 3989–3994, 2000.

27. Dai, S., Shin, Y., Barnes, C.E., and Toth, L.M., Enhancement of uranyl adsorption capacity and selectivity on silica sol-gel glasses via molecular imprinting, *Chem. Mater.,* 9, 2521–2525, 1997.

28. Glad, M., Norrlöw, O., Sellergren, B., Siegbahn, N., and Mosbach, K., Use of silane monomers for molecular imprinting and enzyme entrapment in polysiloxane-coated porous silica, *J. Chromatogr.,* 347, 11–23, 1985.

29. Makote, R. and Collinson, M.M., Dopamine recognition in templated silicate films, *Chem. Commun.,* 3, 425–426, 1998.

30. Sasaki, D.Y., Rush, D.J., Daitch, C.E., Alam, T.M., Assink, R.A., Ashley, C.S., Brinker, C.J., and Shea, K.J., Molecular imprinted receptors in sol-gel materials for aqueous phase recognition of phosphates and phosphonates, *ACS Symp. Ser.,* 703, 314–323, 1998.

31. Markowitz, M.A., Kust, P.R., Deng, G., Schoen, P.E., Dordick, J.S., Clark, D.S., and Gaber, B.P., Catalytic silica particles via template-directed molecular imprinting, *Langmuir,* 16, 1759–1765, 2000.

32. Sasaki, D.Y. and Alam, T.M., Solid-state ^{31}P-NMR study of phosphonate binding sites in guanidine-functionalized, molecular imprinted silica xerogels, *Chem. Mater.,* 12, 1400–1407, 2000.

33. Katz, A. and Davis, M.E., Molecular imprinting of bulk, microporous silica, *Nature,* 403, 286–289, 2000.

34. Lulka, M.F., Chambers, J.P., Valdes, E.R., Thompson, R.G., and Valdes, J.J., Molecular imprinting of small molecules with organic silanes: Fluorescence detection, *Anal. Lett.,* 30, 2301–2313, 1997.

35. Lulka, M.F., Iqbal, S.S., Chambers, J.P., Valdes, E.R., Thompson, R.G., Goode, M.T., and Valdes, J.J., Molecular imprinting of ricin and its A and B chains to organic silanes: Fluorescence detection, *Mater. Sci. Eng. C-Bio S*, 11, 101–105, 2000.

36. Iqbal, S.S., Lulka, M.F., Chambers, J.P., Thompson, R.G., and Valdes, J.J., Artificial receptors: Molecular imprints discern closely related toxins, *Mater. Sci. Eng. C-Bio S*, 7, 77–81, 2000.

37. Lahav, M., Kharitonov, A.B., Katz, O., Kunitake, T., and Willner, I., Tailored chemosensors for chloroaromatic acids using molecular imprinted TiO_2 thin films on ion-sensitive field-effect transistors, *Anal. Chem.*, 73, 720–723, 2001.

38. Lee, S.W., Ichinose, I., and Kunitake, T., Molecular imprinting of azobenzene carboxylic acid on a TiO_2 ultrathin film by the surface sol-gel process, *Langmuir*, 14, 2857–2863, 1998.

39. Lee, S.W., Ichinose, I., and Kunitake, T., Molecular imprinting of protected amino acids in ultrathin multilayers of TiO_2 gel, *Chem. Lett.*, 12, 1193–1194, 1998.

40. Kato, M., Nishide, H., Tsuchida, E., and Sasaki, T., Complexation of metal ion with poly (1-vinylimidazole) resin prepared by radiation-induced polymerization with template metal ion, *J. Polym. Sci. Polym. Chem.*, 19, 1803–1809, 1981.

41. Kido, H., Miyama, T., Tsukagoshi, K., Maeda, M., and Takagi, M., Metal-ion complexation behavior of resins prepared by a novel template polymerization technique, *Anal. Sci.*, 8, 749–753, 1992.

42. Chen, H., Olmstead, M.M., Albright, R.L., Devenyi, J., and Fish, R.H., Metal-ion templated polymers: Synthesis and structure of n-(4-vinylbenzyl)-1,4,7-triazcyclonanezinc(ii) complexes, their copolymerization with divinylbenzene, and metal-ion selectivity studies of the demetalated resins: Evidence for a sandwich complex in the polymer matrix, *Angew. Chem. Int. Ed.*, 36, 642–645, 1997.

43. Aherne, A., Alexander, C., Payne, M.J., Perez, N., and Vulfson, E.N., Bacteria-mediated lithography of polymer surfaces, *J. Am. Chem. Soc.*, 118, 8771–8772, 1996.

44. D'Souza, S.M., Alexander, C., Carr, S.W., Waller, A.M., Whitcombe, M.J., and Vulfson, E.N., Directed nucleation of calcite at a crystal-imprinted polymer surface, *Nature*, 398, 312–316, 1999.

45. Shi, H.Q., Tsai, W.B., Garrison, M.D., Ferrari, S., and Ratner, B.D., Template-imprinted nanostructured surfaces for protein recognition, *Nature*, 398, 593–597, 1999.

46. Ye, L., Weiss, R., and Mosbach, K., Synthesis and characterization of molecularly imprinted microspheres, *Macromolecules*, 33, 8239–8245, 2000.

47. Ye, L., Cormack, P.A.G., and Mosbach, K., Molecularly imprinted monodisperse microspheres for competitive radioassay, *Anal. Commun.*, 36, 35–38, 1999.

48. Surugiu, I., Ye, L., Yilmaz, E., Dzgoev, A., Danielsson, B., Mosbach, K., and Haupt, K., An enzyme-linked molecularly imprinted sorbent assay, *Analyst*, 125, 13–16, 2000.

49. Biffis, A., Graham, N.B., Siedlaczek, G., Stalberg, S., and Wulff, G., The synthesis, characterization and molecular recognition properties of imprinted microgels. *Macromol. Chem. Phys.*, 202, 163–171, 2001.

50. Ishi-i, T., Nakashima, K., and Shinkai, S., Regioselective introduction of two boronic acid groups into [60]fullerene using saccharides as imprinting templates, *Chem. Commun.*, 9, 1047–1048, 1998.

51. Ishi-i, T., Iguchi, R., and Shinkai, S., D/L-Selective re-binding of saccharide-imprinted [60]fullerene-bisadducts based on a saccharide-boronic acid interaction: Development of a molecular imprinting technique useful in a homogeneous system, *Tetrahedron*, 55, 3883–3892, 1999.

52. Zimmerman, S.C., Wendland, M.S., Rakow, N.A., Zharov, I., and Suslick, K.S., Synthetic hosts by monomolecular imprinting inside dendrimers, *Nature*, 418, 399–403, 2002.

53. Xia, Y. and Whitesides, G.M., Soft lithography, *Angew. Chem. Int. Ed.*, 37, 550–575, 1998.

54. Yan, M. and Kapua, A., Fabrication of molecularly imprinted polymer microstructures, *Anal. Chim. Acta*, 435, 163–167, 2001.

55. Piletsky, S.A., Matuschewski, H., Schedler, U., Wilpert, A., Piletska, E.V., Thiele, T.A., and Ulbricht, M., Surface functionalization of porous polypropylene membranes with molecularly imprinted polymers by photograft copolymerization in water, *Macromolecules,* 33, 3092–3098, 2000.

56. Sulitzky, C., Ruckert, B., Hall, A.J., Lanza, F., Unger, K., and Sellergren, B., Grafting of molecularly imprinted polymer films on silica supports containing surface-bound free radical initiators, *Macromolecules,* 35, 79–91, 2002.

57. Panasyuk-Delaney, T., Mirsky, V.M., Ulbricht, M., and Wolfbeis, O.S., Impedometric herbicide chemosensors based on molecularly imprinted polymers, *Anal. Chim. Acta,* 435, 157–162, 2001.

58. Pérez, N., Whitcombe, M.J., and Vulfson, E.N., Surface imprinting of cholesterol on submicrometer core-shell emulsion particles, *Macromolecules,* 34, 830–836, 2001.

59. Yilmaz, E., Haupt, K., and Mosbach, K., The use of immobilized templates: A new approach in molecular imprinting, *Angew. Chem. Int. Ed.,* 39, 2115–2118, 2000.

60. Matsui, J., Nicholls, I.A., and Takeuchi, T., Highly stereoselective molecularly imprinted polymer synthetic receptor for cinchona alkaloids, *Tetrahedron Asymmetry,* 7, 1357–1361, 1996.

61. Ramström, O., Nicholls, I.A., and Mosbach, K., Synthetic peptide receptor mimics: Highly stereoselective recognition in non-covalent molecularly imprinted polymers, *Tetrahedron Asymmetry,* 5, 649–656, 1994.

62. Mayes, A.G., Andersson, L.I., and Mosbach, K., Sugar binding polymers showing high anomeric and epimeric discrimination obtained by non-covalent molecular imprinting, *Anal. Biochem.,* 222, 483–488, 1994.

63. Sellergren, B. and Shea, K.J., Origin of peak asymmetry and the effect of temperature on solute retention in enantiomer separations on imprinted chiral stationary phases, *J. Chromatogr. A,* 690, 29–39, 1995.

64. Kempe, M., Antibody-mimicking polymers as chiral stationary phases in HPLC, *Anal. Chem.,* 68, 1948–1953, 1996.

65. Mayes, A.G. and Mosbach, K., Molecularly imprinted polymer beads: Suspension polymerization using a liquid perfluorocarbon as the dispersing phase, *Anal. Chem.,* 68, 3769–3774, 1996.

66. Hosoya, K., Yoshihako, K., Shirasu, Y., Kimata, K., Araki, T., Tanaka, N., and Haginaka, J., Molecularly imprinted uniform-size polymer-based stationary phase for high-performance liquid chromatography: Structural contribution of cross-linked polymer network on specific molecular recognition, *J. Chromatogr.,* 728, 139–148, 1996.

67. Nilsson, K.G.I., Lindell, J., Norrlöw, O., and Sellergren, B., Imprinted polymers as antibody mimics and new affinity gels for selective separations in capillary electrophoresis, *J. Chromatogr. A,* 680, 57–61, 1994.

68. Vallano, P.T. and Remcho, V.T., Highly selective separations by capillary electrochromatography: Molecular imprint polymer sorbents, *J. Chromatogr. A,* 887, 125–135, 2000.

69. Schweitz, L., Andersson, L.I., and Nilsson, S., Capillary electrochromatography with predetermined selectivity obtained through molecular imprinting, *Anal. Chem.,* 69, 1179–1183, 1997.

70. Lin, J.M., Nakagama, T., Uchiyama, K., and Hobo, T., Capillary electrochromatographic separation of amino acid enantiomers using on-column prepared molecularly imprinted polymer, *J. Pharmaceut. Biomed. Anal.,* 15, 1351–1358, 1997.

71. Lin, J.M., Nakagama, T., Uchiyama, K., and Hobo, T., Molecularly imprinted polymer as chiral selector for enantioseparation of amino acids by capillary gel electrophoresis, *Chromatographia,* 43, 585–591, 1996.

72. Spégel, P., Schweitz, L., and Nilsson, S., Molecularly imprinted microparticles for capillary electrochromatography: Studies on microparticle synthesis and electrolyte composition, *Electrophoresis,* 22, 3833–3841, 2001.

73. Kriz, D., Berggren-Kriz, C., Andersson, L.I., and Mosbach, K., Thin-layer chromatography based on the molecular imprinting technique, *Anal. Chem.,* 66, 2636–2639, 1994.

74. Suedee, R., Songkram, C., Petmoreekul, A., Sangkunakup, S., Sankasa, S., and Kongyarit, N., Thin-layer chromatography using synthetic polymers imprinted with quinine as chiral stationary phase, *J. Planar Chromatogr.,* 11, 272–276, 1998.

75. Suedee, R., Songkram, C., Petmoreekul, A., Sangkunakup, S., Sankasa, S., and Kongyarit, N., Direct enantioseparation of adrenergic drugs via thin-layer chromatography using molecularly imprinted polymers, *J. Pharmaceut. Biomed. Anal.,* 19, 519–527, 1999.

76. Sellergren, B., Direct drug determination by selective sample enrichment on an imprinted polymer, *Anal. Chem.,* 66, 1578–1582, 1994.

77. Muldoon, M.T. and Stanker, L.H., Molecularly imprinted solid phase extraction of atrazine from beef liver extracts, *Anal. Chem.,* 69, 803–808, 1997.

78. Zander, A., Findlay, P., Renner, T., Sellergren, B., and Swietlow, A., Analysis of nicotine and its oxidation products in nicotine chewing gum by a molecularly imprinted solid phase extraction, *Anal. Chem.,* 70, 3304–3314, 1998.

79. Andersson, L.I., Paprica, A., and Arvidsson, T., A highly selective solid phase extraction sorbent for pre-concentration of sameridine made by molecular imprinting, *Chromatographia,* 46, 57–62, 1997.

80. Martin, P., Wilson, I.D., Morgan, D.E., Jones, G.R., and Jones, K., Evaluation of a molecularly imprinted polymer for use in the solid phase extraction of propranolol from biological fluids, *Anal. Commun.,* 34, 45–47, 1997.

81. Matsui, J., Okada, M., Tsuruoka, M., and Takeuchi, T., Solid phase extraction of a triazine herbicide using a molecularly imprinted synthetic receptor, *Anal. Commun.,* 34, 85–87, 1997.

82. Mullett, W.M., Lai, E.P.C., and Sellergren, B., Determination of nicotine in tobacco by molecularly imprinted solid phase extraction with differential pulsed elution, *Anal. Commun.,* 36, 217–220, 1999.

83. Pichon, V., Bouzige, M., Miège, C., and Hennion, M.-C., Immunosorbents: Natural molecular recognition materials for sample preparation of complex environmental matrices, *Trends Anal. Chem.,* 18, 219–235, 1999.

84. Mullett, W.M. and Lai, E.P.C., Determination of theophylline in serum by molecularly imprinted solid-phase extraction with pulsed elution, *Anal. Chem.,* 70, 3636–3641, 1998.

85. Berggren, C., Bayoudh, S., Sherrington, D., and Ensing, K., Use of molecularly imprinted solid-phase extraction for the selective clean-up of clenbuterol from calf urine, *J. Chromatogr. A,* 889, 105–110, 2000.

86. Ferrer, I., Lanza, F., Tolokan, A., Horvath, V., Sellergren, B., Horvai, G., and Barceló, D., Selective trace enrichment of chlorotriazine pesticides from natural waters and sediment samples using terbuthylazine molecularly imprinted polymers, *Anal. Chem.,* 72, 3934–3941, 2000.

87. Ellwanger, A., Berggren, C., Bayoudh, S., Crecenzi, C., Karlsson, L., Owens, P.K., Ensing, K., Cormack, P., Sherrington, D., and Sellergren, B., Evaluation of methods aimed at complete removal of template from molecularly imprinted polymers, *Analyst,* 126, 784–792, 2001.

88. Matsui, J., Fujiwara, K., and Takeuchi, T., Atrazine-selective polymers prepared by molecular imprinting of trialkylmelamines as dummy template species of atrazine, *Anal. Chem.,* 72, 1810–1813, 2000.

89. Quaglia, M., Chenon, K., Hall, A.J., Lorenzi, E.D., and Sellergren, B., Target analogue imprinted polymers with affinity for folic acid and related compounds, *J. Am. Chem. Soc.,* 123, 2146–2154, 2001.

90. Mathew-Krotz, J. and Shea, K.J., Imprinted polymer membranes for the selective transport of targeted neutral molecules, *J. Am. Chem. Soc.,* 118, 8154–8155, 1996.

91. Dzgoev, A. and Haupt, K., Molecularly imprinted polymer membranes for chiral separation, *Chirality,* 11, 465–469, 1999.

92. Wang, H.Y., Kobayashi, T., and Fujii, N., Molecular imprint membranes prepared by the phase inversion precipitation technique, *Langmuir,* 12, 4850–4856, 1996.

93. Hedborg, E., Winquist, F., Lundström, I., Andersson, L.I., and Mosbach, K., Some studies of molecularly imprinted polymer membranes in combination with field-effect devices, *Sensor Actuators A,* 36–38, 796–799, 1993.

94. Sergeyeva, T.A., Piletsky, S.A., Brovko, A.A., Slinchenko, E.A., Sergeeva, L.M., Panasyuk, T.L., and Elskaya, A.V., Conductometric sensor for atrazine detection based on molecularly imprinted polymer membranes, *Analyst,* 124, 331–334, 1999.

95. Kochkodan, V., Weigel, W., and Ulbricht, M., Thin layer molecularly imprinted microfiltration membranes by photofunctionalization using a coated α-cleavage photoinitiator, *Analyst,* 126, 803–809, 2001.

96. Hong, J.-M., Anderson, P.E., Qian, J., and Martin, C.E., Selectively permeable ultrathin film composite membranes based on molecularly imprinted polymers, *Chem. Mater.,* 10, 1029–1033, 1998.

Index

A

Ab, see Antibodies
ABI/Perseptive Biosystems, supports for
    immunoaffinity chromatography, 138
Absorbance detection
  applications
    affinity capillary electrophoresis, 708, 714, 753
    analytical affinity chromatography, 685–686
    chromatographic immunoassays, 791
    competitive-binding chromatographic
      immunoassays, 154, 810
    direct analyte detection, 151–152, 519
    displacement immunoassays, 817
    measuring immobilized ligands, 43
    microanalytical systems, 779
    off-line immunoextraction, 147
    on-line immunoextraction, 148
    one-site immunometric assays, 826
  disadvantages in capillary electrophoresis, 719
  enzyme labels, 794
    alkaline phosphatase, 795
    β-galactosidase, 795
    horseradish peroxidase, 796
Absorption throughput, importance in affinity
    purifications, 290
AcA34, purification of DNA repair enzymes, 184
ACE, see Affinity capillary electrophoresis
  see also Angiotensin-converting enzyme
Acenocoumarol, binding to human serum albumin, 675
AC-ESI-MS, see Affinity capture-release
    electrospray-ionization mass spectrometry
Acetate, IMAC, 265
Acetazolamide
  binding to carbonic anhydrase II, 689–690
  kinetic analysis with a biosensor, 689–690
  structure, 690
Acetic acid
  elution in immunoaffinity chromatography, 143
  elution of atrazine from an imprinted polymer, 849
Acetonitrile
  α_1-acid glycoprotein, 587

α-chymotrypsin, 573
  elution of atrazine from an imprinted polymer, 849
N-Acetoxy-d₃-succinimide, acetylation, 507
N-Acetoxysuccinimide, acetylation, 507
Acetylcholine receptors, analysis with autoantibodies,
    136
Acetylcholinesterase
  direct-detection chromatographic immunoassays,
    474, 803
  postcolumn immunodetection, 804
  SEC and postcolumn immunodetection, 157, 463
Acetyl-CoA, elution on CoA-Sepharose, 332
Acetyldigitoxin
  binding studies, 115
  binding to human serum albumin, 475, 605
N-Acetyl-D-galactosamine, ligand on agarose, 113
N-Acetyl-D-glucosamine
  action as ligand, 112
  interactions with galactosyltransferase, 631
  ligand on agarose, 113
N-Acetylglucosaminyl residues, binding to wheat
    germ agglutinin, 318
N-Acetylglucosaminyltransferase V, binding studies,
    752–753
N-Acetyl-tryptophan, chiral separation by
    α-chymotrypsin, 576
AChE, see Acetylcholinesterase
Achiral columns, combined use with chiral stationary
    phases, 582
ACHT, see α-Chymotrypsin
Acid dissociation constants, values for compounds
    binding to cellobiohydrolase I, 579
Acid elution, immunosorbents, 292–293
α_1-Acid glycoprotein, 115
  application & elution conditions, 116
  applications in affinity chromatography, 115
  chiral stationary phase, 115, 470
  commercial columns, 116
  comparison of soluble & immobilized forms, 597
  general properties, 115
  immobilization, 116
ACL, affinity supports, 17

857

Application conditions
    factors related to the column and sample, 85–86
    solvent considerations, 84
Applied Biosystems
    array SPR, 694–695
    flow-through beads, 27
Aptamers, 204
    affinity chromatography, 204–205
    applications, 204
    definition, 250
    in vitro selection, 251
    isolation of RNA-binding protein, 204
    screening by SELEX, 250
    selection process, 204
AR, see Adsorption rate
Arabinofuranosyluracil-5′-triphosphate
    enzyme purification, 334
    immobilization to Affi-Gel 10, 334
    immobilized inhibitor for enzymes, 106
    inhibitor of DNA polymerase, 334
Arachidonate 5-lipoxygenase, purification by
    immunoaffinity chromatography, 145
Arachidonic acids, purification by immunoaffinity
    chromatography, 145
araUTP, see Arabinofuranosyluracil-5′-triphosphate
AraUTP-Affi-Gel 10, structure, 334
Arg, see Arginine
Arginine
    chiral separation of derivatives by trypsin, 577–578
    elution for L-arginine-Sepharose, 319
Arginine ethyl ester, chiral separation by trypsin, 578
Arginine methyl ester, chiral separation by trypsin,
    578
Arginyl tRNA synthetase, purification using
    blue-Sephadex G-150, 337
Arg-Oet, see Arginine ethyl ester
Arg-Ome, see Arginine methyl ester
Argon
    FAB, 744
    ICP-MS, 744
Arica, Y., microporous protein A membrane, 379
Aromatic boronates, affinity columns, 217
Aromatic molecules, imprinting, 842
Array technology, definition, 768
Arsenis, C., early use of affinity supports in
    enzymology, 7
Artocarpus integrifolia, see also Jack fruit
    source of jacalin, 381
Arylcarboxylic acids, chiral recognition by human
    serum albumin, 672
α-Arylpropionic acids, chiral separation by serum
    albumins, 581
Ascaris suum, source of carboxypeptidase inhibitors,
    338
Ascites fluid, production of monoclonal antibodies,
    350
Ascorbate peroxidase isoenzymes, purification by
    immunoaffinity chromatography, 145
Aspartame, chiral separation by α-chymotrypsin,
    576

Aspartate, elution with hadacidin-Sepharose, 333
N-α-Aspartyl-phenylalanine methyl ester, chiral
    separation by α-chymotrypsin, 576
Association equilibrium constant
    antibody-antigen binding, 130
    biospecific elution, 90–91
    determination by affinity chromatography, 596
    determination by frontal analysis, 536–537
    effects on retention in immunoaffinity
      chromatography, 130
    general scheme for solute-ligand binding, 81
    measurement for immobilized ligands, 44
    model for ACE, 703
    monomer and template, 840
    relationship to retention factor, 599
    role during elution in immunoaffinity
      chromatography, 142
    typical values for antibodies, 790
Association kinetics
    covalent versus noncovalent interactions, 840
    effect on sample retention, 86
    requirements in affinity chromatography, 82
Association phase, 687–688
Association rate constant
    antibody-antigen binding, 130
    description of a simultaneous-injection
      competitive-binding immunoassay, 809
    description of direct detection in chromatographic
      immunoassays, 805
    general scheme for solute-ligand binding, 81
    model for ACE, 703
Astec, α_1-acid glycoprotein columns, 116
Atenolol, bioanalysis using α_1-acid glycoprotein, 585
Atmospheric-pressure ESI, description, 740
Atomic force microscopy
    affinity-based static arrays, 770
    analysis of imprints, 843
ATP, see Adenosine triphosphate
ATP-agarose
    helicases, 190
    primases, 193
N6-ATP-Sepharose, purification of adenylyl cyclase,
    409
Atraton, interaction with humic acids, 539
Atrazine degradation products
    homogeneous immunoassay, 524–525
    RPLC and on-line immunoextraction, 148
Atrazine
    binding with human serum albumin, 536–537
    competitive-binding chromatographic
      immunoassay, 154
    degradation products, 530
    homogeneous immunoassay, 524–526
    molecularly imprinted polymer, 534
    on-line immunoextraction and GC, 532–534
    on-line immunoextraction with RPLC, 148,
      530–532
    optical biosensors, 538
    sequential-injection competitive-binding
      immunoassay, 522–524, 814

O